Reinforced Concrete
Mechanics and Design

REINFORCED CONCRETE
Mechanics and Design

FOURTH EDITION

JAMES G. MACGREGOR

PhD, P. Eng., Honorary Member ACI
D. Eng. (Hon.), D.Sc. (Hon.), FRSC
University Professor Emeritus
Department of Civil Engineering
University of Alberta

JAMES K. WIGHT

Department of Civil & Environmental Engineering
University of Michigan

PEARSON

Prentice
Hall

Upper Saddle River, New Jersey 07458

Library of Congress Cataloging-in-Publication Data on File

Vice President and Editorial Director, ECS: *Marcia J. Horton*
Executive Editor: *Eric Svendsen*
Acquisitions Editor: *Dorothy Marrero*
Vice President and Director of Production and Manufacturing, ESM: *David W. Riccardi*
Executive Managing Editor: *Vince O'Brien*
Managing Editor: *David A. George*
Production Editor: *Craig Little*
Director of Creative Services: *Paul Belfanti*
Creative Director: *Jayne Conte*
Cover Designer: *Bruce Kenselaar*
Art Editor: *Greg Dulles*
Manufacturing Buyer: *Lisa McDowell*
Senior Marketing Manager: *Holly Stark*

Cover photo: This is a photo of the Confederation Bridge from New Brunswick to Prince Edward Island, Canada, under construction. Courtesy of Alain & Buffie Boily.

© 2005, 1997, 1992, 1988 by Pearson Education, Inc.
Pearson Prentice Hall
Pearson Education, Inc.
Upper Saddle River, New Jersey 07458

Printed in the United States of America

10 9 8 7 6 5 4 3 2 1

ISBN 0-13-142994-9

Pearson Education Ltd., *London*
Pearson Education Australia Pty. Ltd., *Sydney*
Pearson Education Singapore, Pte. Ltd.
Pearson Education North Asia Ltd., *Hong Kong*
Pearson Education Canada, Inc., *Toronto*
Pearson Educación de Mexico, S.A. de C.V.
Pearson Education—Japan, *Tokyo*
Pearson Education Malaysia, Pte. Ltd.
Pearson Education, Inc., *Upper Saddle River, New Jersey*

Contents

Preface

Reinforced concrete design is both an art and a science. This book presents the theory of reinforced concrete as a direct application of the laws of statics and mechanics of materials. In addition, it emphasizes that a successful design not only satisfies the design rules, but also is capable of being built at a reasonable price.

Philosophy of Reinforced Concrete: Mechanics and Design

A special three-tiered design makes *Reinforced Concrete: Mechanics and Design* an outstanding textbook for university courses on reinforced concrete design. The book can be used as a textbook in successive courses of varying difficulty. In the teaching of reinforced concrete design, different parts of the subject require different tiers of education and experience. The flexural strength of beams is presented in Sections 4-2, 4-3, and 5-2. Because these sections are among the first to be taught, they are presented at a beginner's level. The closely related Sections 11-3 and 11-4 present the essential theory of concrete columns at a somewhat higher level, but still at a level suitable for a first course on reinforced concrete design. Still more advanced subjects are presented at levels suitable for advanced undergraduate or postgraduate students. These topics include, for example, unsymmetrical beams and columns, strain-compatibility solutions of beams, $P - \Delta$ analyses of frames, regions adjacent to discontinuities such as concentrated loads or abrupt changes in section, deep beams, column–beam joints, and shear walls. These second- (or third-) tier subjects make this book a valuable text for graduate courses and a first rate reference volume in design offices.

1. The various core topics—flexure, shear, columns, and so on—are presented at several levels. Each core subject starts with a basic level of presentation suitable for a first undergraduate course on reinforced concrete. More advanced and optional topics are presented at a higher level. The topics covered extend beyond the bounds of the ACI Code. This helps maintain the up-to-date coverage of emerging topics in structural design and reinforced concrete design.

2. Extensive figures are used to illustrate aspects of design or theory.

3. Emphasis is placed on the logical order and completeness of the many design examples presented in the book. Examples are done in a step-by-step order, with every step worked out completely from first principles at least once.

4. Guidance is given in the text and in the examples, to help students make the many judgement decisions required in reinforced concrete design

5. Reinforced concrete design is just one aspect of structural engineering. To properly carry out designs the engineer must have a knowledge of aspects of design and construction, including the design process, economics, loads, and serviceability. These topics are presented in Chapters 2, 3, and 9 and are used in examples of design for wind and earthquake forces.

The widespread adoption of the first three editions of this book as the textbook in both introductory and graduate level courses on reinforced concrete in American universities affirms the great success of this book in meeting these goals.

An Overview of Reinforced Concrete: Mechanics and Design—What Is New in the Fourth Edition?

New Co-author

During the fourth revision of *Reinforced Concrete: Mechanics and Design*, the original author invited Professor J. K. Wight of the University of Michigan to join him as a co-author of the fourth edition. Dr. Wight brings extensive experience as a teacher of reinforced concrete design, as a researcher, and as a specialist in the field of design of concrete buildings for seismic regions. Professor Wight has served both as a member and as Chair of the ACI Technical Activities Committee, the ultimate arbiters of the technical accuracy of ACI publications. Currently he is Chair of ACI Committee 318, the committee responsible for updating the ACI Code. He is committed to maintaining the high standards that both the Code and this book have met.

The 2002 revisions to the ACI Code were more extensive and far reaching than any code revision since 1963. As a result, there have been significant changes to this book, some of which are listed along with the following chapter summaries.

Chapters 1–3

The first three Chapters of the book are introductory.

Chapter 1 defines and illustrates various types of reinforced concrete members.

Chapter 2 sets the stage for the rest of the book and extends the introduction through a discussion of the goals of structural design and the concepts of limit-states design. Chapter 2 also presents an introduction to safety theory, a review of the properties of design loads, and a discussion of design for economy. Changes in this edition include the following:

• In the 2002 ACI Code, the load and resistance factors from the body of the 1995 code and those in an Appendix C of that code were interchanged. This change and the corresponding change in load and resistance factors to be used are explained in Chapters 2, 4, 5, and 11 and are used in a number of extended examples.

Chapter 3 presents many of the significant properties of concrete and reinforcement. These are the basis for the development of the theories of flexure, shear, and bond of concrete in subsequent chapters. Chapter 3 is also intended to be a reference for information on structural aspects of concrete technology.

• The fourth edition extends and updates the discussion of basic properties of concrete, revising the sections on statistical evaluation of concrete strengths and on stress–strain curves for concrete. New topics include the quality of mixing water needed to make durable concrete. Shotcrete and high-alumina cement are discussed briefly. Data from recently reported studies of the variability of concrete strengths, including data on the relationship between 28-day and 56-day strengths, are presented.

• A new section in Chapter 3 summarizes properties of Fiber Reinforced Polymer (FRP) reinforcement.

Chapters 4–10

The next group of chapters develops the underlying concepts that affect the design of concrete structures.

Chapters 4 and 5 present the basic theory of flexure and flexural design for beams. Each topic starts with a review of related aspects of the behavior of reinforced concrete based on statics and mechanics of materials

• The concepts of tension-controlled beam sections and compression-controlled beam sections, introduced in the third edition of this book, have been expanded in this edition. The strain, ε_t, in the layer of tension reinforcement farthest from the neutral axis (the axis of zero strains) is used as an index to the behavior of beams and columns and the selection of resistance factors, ϕ. Previously, this classification was based on the ratio of the beam steel to the amount of steel corresponding to the balanced steel ratio of beams and columns. This replaces the classification beam and column behavior based on the ratio of the amount of steel present to the amount corresponding to the balanced failure condition. All the examples encountered in an introductory concrete design course have been recalculated based on the new ϕ factors. Several long examples on advanced topics brought forward from the third edition have not been converted to the new load and resistance factors.

• A new section in Chapter 5 introduces design of beams with FRP reinforcement.

Chapter 6, Shear describes the behavior of beams loaded in shear. It has been extensively revised in this edition. Chapters 7 (torsion) and 8 (bond and anchorage) were extensively revised in the third edition.

• The explanation of the mechanics of the shear failure of concrete beams has been rearranged and clarified. A brief section on modern theories of shear failure has been added.

• The discussion of the effect of overall member size on shear strength has been amplified.

• A new example of the design of a continuous beam for shear has been added.

Previous editions of this book have presented one example of the design of beams for shear. In this edition, a new example of the design of a continuous beam for shear has been added.

Chapter 7, Torsion was extensively revised in the third edition.

• The introduction to Chapter 7 has been revised to distinguish between circulatory torsion and warping torsion.

• An example of the design of a hollow beam for torsion has been added.

Chapter 8, Bond and Anchorage has been revised somewhat in this edition.

The calculation of bar cutoff points has been rearranged to express the bar cutoff rules as functions of the mechanical reasons for each to be easier to apply. The selection of the distance that reinforcement is extended beyond flexural bar cutoff points is based on four technical reasons why bars should be extended. For a given case, the designer selects the verbal description of the applicable reason of why bars are extended, and then chooses the lengths of the corresponding extensions. It also discusses the concept of interaction diagrams and the shape of such diagrams. An interaction diagram for a concrete column is derived and interaction diagrams are used in examples.

Chapter 9, Serviceability reviews serviceability limit states, particularly those of deflection and crack control. Several new tables list aspects of the design for serviceability.

• The alternate design method, also known as working stress design, was deleted from the 2002 ACI Code. Because this method of design analysis is used in serviceability calculations in Chapter 9, it has been introduced here.

• Tables have been added in Chapter 9, listing serviceability limit states considered in the design of concrete buildings.

The next two groups of chapters use the theory from Chapters 4 through 10 and synthesize it to design continuous beams or slabs and columns in buildings.

Chapters 10–12

Chapter 10 deals with the design of continuous concrete structures, such as one-way slabs and beams. This combines the various limit states from Chapters 4, 5, 6, 8, and 9 as required in design.

Chapter 11 deals with short columns. It discusses the concept of interaction diagrams and their shape. An interaction diagram for a concrete column is derived and used in a design.

• Chapter 11 ends with a discussion and example of strain-compatibility calculations of the strength of biaxially loaded columns, using a strain-compatibility solution based on the rectangular stress block.

Chapter 12 defines three classes of slender columns: (a) hinged-end columns, (b) columns in braced frames, and (c) columns in sway frames. Chapter 12 was extensively revised in the 1997 edition of this book.

• A brief section has been added on the lateral and torsional stability of frames.

Chapters 13–15

Chapters 13, 14, and 15 deal with two-way slabs. Two-way slabs are a mode of construction unique to reinforced concrete. This type of floor represents a major fraction of the floor structures in concrete buildings in North America.

Chapter 13 starts with reviews of the moments in two-way slabs. This is followed by the derivation of the direct design method, used to compute the magnitude and distribution of the moments in a slab. Chapter 13 also discusses the shear strength of two-way slabs.

• The design of headed-stud shear reinforcement in two-way slabs, both with and without moment transfer to the columns, has been added to Chapter 13. This type of reinforcement is widely used but is not covered by the ACI code. Two examples have been extended to show the steps in design using such reinforcement.

Chapter 14 presents the Equivalent Frame Method for the design of two-way slabs. This was originally derived to give a computer method based on the same designs as the direct design method. Although the use of the direct design method in design offices has largely been superseded by that of the equivalent frame method, an understanding of the direct design method is essential to engineers using the equivalent frame method in Chapter 14.

Chapter 15   presents two inelastic analyses of two-way slabs, including the yield line method and the strip method.

•   The combination of bending moments and twisting moments from finite element analyses of slabs in order to design the steel is discussed in Chapter 15.

Chapters 16–20

The rest of the book deals with other types of structural members and loads.

Chapter 16   discusses the design of footings.

Chapter 17   introduces shear friction and composite beams.

Chapter 18   covers the concept of D-regions, which are the regions of beams or columns located near structural or geometrical discontinuities, such as concentrated loads or changes of member section; this concept was incorporated in Appendix *A* of the 2002 ACI Code. The use of D-region theory in the design of deep beams was first introduced to American engineers in Chapter 18 of the 1988 edition of this book. Chapter 18 is a detailed review of D-region design based on ACI 318-02 Appendix A. Among the other additions is an expanded discussion of nodal zones. Several new design examples are included in this chapter, including one of a continuous beam, and the others have been enlarged and re-vised to agree with ACI Appendix A.

Chapter 19   on walls is new in this edition. Walls are classified according to the forces acting on them and according to whether the walls resist lateral forces acting on the build-ing (and so brace the building) or the walls themselves are supported and braced by the frame. Design examples illustrate the design of a bearing wall and the design of a shear wall subjected to wind loads.

In this new chapter, various types of walls are discussed, the stability of axially com-pressed plates is addressed, and the structural design of walls, and the thickness and rein-forcement in walls are examined. Lateral-force resisting mechanisms for medium height buildings are discussed. A design example follows the design of a shear wall in a ten-story building. In addition to a review of the design of structural walls loaded in flexure and shear, this example illustrates the calculation of wind loads on buildings from the ASCE 7 loading standard.

Chapter 20   presents aspects of design for seismic forces. This chapter has been com-pletely revised to conform with the extensively revised Chapter 21 of the ACI Code. Ex-amples illustrating the design of a beam, of a column, and of a beam–column joint from a special-moment frame are presented.

Appendices and References

Appendix A   contains 33 design tables and 14 design charts referred to throughout the rest of the book.

Appendix B   lists the notation used in the ACI Code and this book.

References   provides an extensive set of sources to aid readers who wish to read more about reinforced concrete design.

Use of Reinforced Concrete: Mechanics and Design in Undergraduate Courses

The prerequisites for the course sequences described are a university-level course in statics and a course in mechanics of materials.

A one-semester course on reinforced concrete might cover Sections 2-1 to 2-4 and 2-6 to 2-8 on the basis for design, safety factors, loads, and design for economy; Sections 3-2, 3-3, and 3-9 on material properties (all of these could be used as reading assignments); Chapter 4 and Sections 5-1 through 5-3 on flexure; Sections 6-1 to 6-3 and 6-5 on shear; Chapter 8 on bond and anchorage; Sections 9-1 to 9-5 on serviceability; Chapter 10 on continuous slabs and beams; Sections 11-1 to 11-5 on short columns; and Sections 16-1 through 16-5 on footings.

A subsequent one-semester course might cover Chapter 7 on torsion; Sections 12-1 to 12-12 on slender columns; Chapter 13 on two-way slabs; and Section 17-2 on shear friction. Chapter 18 would be optional. The prerequisites for these courses are a course in statics and one in mechanics of materials.

The first, one-quarter undergraduate course on reinforced concrete might cover Sections 2-1 to 2-4 and 2-6 to 2-8 on the bases for design; Sections 3-2 to 3-5, 3-9, and 3-13 on material properties; Chapter 4 and Sections 5-1 to 5-3 on flexure; Sections 6-1 through 6-3 and 6-5 on shear; Sections 8-1 through 8-7 on anchorage; Sections 10-3 and 10-4 on one-way slabs; and Sections 11-1 through 11-5 on columns.

A subsequent one-quarter course might include Sections 8-1 to 9-5 on serviceability; the rest of Chapter 10; Sections 16-1 to 16-5 on footings; Chapter 7; Sections 12-1 and 12-2 on slender columns; and Sections 13-3 to 13-10 on two-way slabs.

This text makes frequent reference to the 2002 ACI Code and assumes that the reader will have a copy of this code.

Although the foot-pound-second system of units is the system of units prevalent throughout the book, eight examples are repeated completely in SI (metric) units. A number of metric design charts are given in Appendix A.

Acknowledgments and Dedication

This book is dedicated to the members of ACI Committee 318, who work as volunteers to develop and update the ACI Code. In particular, we wish to acknowledge the chairmanships of Eugene P. Holland, Chester P. Siess, W. G. Corley, John E. Breen, and, most recently, Jim Cagley chair of the extensively revised 2002 code. We also thank Basile Rabbat, the long-suffering secretary of ACI Committee 318, for his succesful task of keeping the paperwork in check.

The manuscript of the entire Fourth Edition book was reviewed by Sharon Wood and Cathy French, both of whom are active members of ACI Committee 318. It was also thoroughly reviewed by Dr. Keith Thompson, Dr. Robert W. Barnes, and Dr. Anton K. Schindler of Auburn University. This book is greatly improved by their suggestions.

Finally, we thank Barb for her support and encouragement.

JAMES G. MACGREGOR
University Professor Emeritus,
University of Alberta

JAMES K. WIGHT
University of Michigan

About the Authors

James G. MacGregor, University Professor of Civil Engineering at the University of Alberta, Canada, retired in 1993 after 33 years of teaching, research, and service, including three years as Chair of the Department of Civil Engineering. He has a B.Sc. from the University of Alberta and an M.S. and a Ph.D. from the University of Illinois. In 1998 and 1999 he received a Doctor of Engineering (Hon) from Lakehead University, and in 1999 a Doctor of Science (Hon) from the University of Alberta. Dr. MacGregor is a Fellow of the Academy of Science of the Royal Society of Canada and a Fellow of the Canadian Academy of Engineering. A Past President and Honorary Member of the American Concrete Institute, Dr. MacGregor has been an active member of ACI since 1958. He has served on ACI technical committees including the ACI Building Code Committee and its subcommittees on flexure, shear, and stability and the ACI Technical Activities Committee. This involvement and his research has been recognized by honors jointly awarded to MacGregor, his colleagues, and students. These included the ACI Wason Medal for the Most Meritorious Paper (1972, and 1999), the ACI Raymond C. Reese Medal, and the ACI Structural Research Award (1972 and 1999). His work on the developing the Strut-and-Tie model into the ACI Code was recognized by the ACI Structural Research Award (2004). In addition, he has received several ASCE Awards, including the prestigious ASCE Norman Medal with three colleagues (1983). Dr. MacGregor chaired the Canadian Committee on Reinforced Concrete Design from 1977 through 1989, moving on to chair the Standing Committee on Structural Design for the National Building Code of Canada from 1990 through 1995. He is a registered professional engineer in two Canadian provinces. From 1973 to 1976 he was a member of the Council of the Association of Professional Engineers, Geologists, and Geophysicists of Alberta. At the time of his retirement from the University of Alberta, Professor MacGregor was a principal in MKM Engineering Consultants. His last project with that firm was the derivation of site-specific load and resistance factors for the eight-mile long bridge, shown on the cover of this book MacGregor has written or co-authored over 100 technical papers. This culminated with this book, now in it's fourth American edition and first Canadian edition.

James K. Wight received his B.S. and M.S. degrees in Civil Engineering from Michigan State University in 1969 and 1970, and his Ph.D. from the University of Illinois at Urbana-Champaign in 1973. He has been a professor of structural engineering in the Civil and Environmental Engineering Department at the University of Michigan since 1973. He teaches undergraduate and graduate classes on analysis and design of reinforced concrete structures. He is well known for his work in earthquake-resistant design of concrete structures and spent a one-year sabbatical leave in Japan where he was involved in the construction and simulated earthquake testing of a full-scale reinforced concrete building. Professor Wight has been an active member of the American Concrete Institute since 1973 and was named a Fellow of the Institute in 1984. He is currently Chairman of the ACI Building Code Committee 318 and past Chair of Subcommittee 318-E. He is also past Chair of the ACI Technical Activities Committee and Committee 352 on Joints and Connections in Concrete Structures. He has received several awards from the American Concrete Institute including the Delmar Bloem Distinguished Service Award (1991), the Joe Kelly Award for outstanding efforts for the education of students in design of reinforced concrete structures (1999), the Boise Award for outstanding accomplishments in research, teaching and service in the field of structural concrete (2002), and the Structural Research Award (2003) for a paper he co-authored with a former student. Professor Wight has also received numerous awards for his teaching and service at the University of Michigan including the ASCE Student Chapter Teacher of the Year Award, the College of Engineering's Distinguished Service and Teaching Excellence Awards, and the Chi Epsilon-Great Lakes District Excellence in Teaching Award.

1

Introduction

1-1   REINFORCED CONCRETE STRUCTURES

Concrete and reinforced concrete are used as building materials in every country. In many, including the United States and Canada, reinforced concrete is a dominant structural material in engineered construction. The universal nature of reinforced concrete construction stems from the wide availability of reinforcing bars and of the constituents of concrete (gravel, sand, water and cement), from the relatively simple skills required in concrete construction, and from the economy of reinforced concrete compared with other forms of construction. Concrete and reinforced concrete are used in buildings of all sorts (Fig. 1-1), underground structures, water tanks, television towers, offshore oil exploration and production structures (Fig. 1-2), dams, bridges (Fig. 1-3), and even ships.

1-2   MECHANICS OF REINFORCED CONCRETE

Concrete is strong in compression, but weak in tension. As a result, cracks develop whenever loads, restrained shrinkage, or temperature changes give rise to tensile stresses in excess of the tensile strength of the concrete. In the plain concrete beam shown in Fig. 1-4b, the moments about O due to applied loads are resisted by an internal tension–compression couple involving tension in the concrete. Such a beam fails very suddenly and completely when the first crack forms. In a *reinforced concrete* beam (Fig. 1-4c), steel bars are embedded in the concrete in such a way that the tension forces needed for moment equilibrium after the concrete cracks can be developed in the bars.

Alternatively, the reinforcement could be placed in a longitudinal duct near the bottom of the beam, as shown in Fig. 1-5, and stretched or *prestressed*, reacting on the concrete in the beam. This would put the reinforcement into tension and the concrete into compression. This compression would delay cracking of the beam. Such a member is said to be a *prestressed concrete* beam. The reinforcement in such a beam is referred to as *prestressing tendons* and must be of high-strength steel.

The construction of a reinforced concrete member involves building a form or mould in the shape of the member being built. The form must be strong enough to support the weight and hydrostatic pressure of the wet concrete, plus any forces applied to it

1

Fig. 1-1
City Hall, Toronto, Canada.

The Toronto City Hall consists of two towers, 20 and 27 stories in height, with a circular council chamber cradled between them. These structures and the surrounding terraces, pools, and plaza illustrate the degree to which architecture and structural engineering can combine to create a living sculpture. This complex has become the trademark and social hub of the city of Toronto. The council chamber consists of a reinforced concrete bowl containing seating for the council, press, and citizens. This is covered by a concrete dome. The wind resistance of the two towers results largely from the two vertical curved walls forming the backs of the towers. The architectural concept was by Viljo Revell, of Finland, winner of an international design competition. Mr. Revell entered into an association with John B. Parkin Associates, who developed the initial concept and carried out the structural design. The structural design is described in [1-1]. (Photograph used with permission of Neish Owen Rowland & Roy, Architects Engineers, Toronto.)

Fig. 1-2
Glomar Beaufort Sea 1
(CIDS) being towed through
the Bering Strait to the
Beaufort Sea, Alaska.

This concrete island oil drilling structure consists of a steel mud base, a 230-ft-square cellular concrete segment, and a deck assembly with drilling rig, quarters for 80 workers, and supplies for 10 months. The structure is designed to operate in 35 to 60 ft of water in the Arctic Ocean. Forces from the polar sea ice are resisted by the thick walls of the concrete segment. Design was carried out by Global Marine Development Inc. Engineering and construction support to Global Marine was provided by A. A. Yee Inc. and ABAM Engineers Inc. (Photograph courtesy of Global Marine Development Inc.)

by workers, concrete buggies, wind, and so on. The reinforcement is placed in this form and held in place during the concreting operation. After the concrete has hardened, the forms are removed.

1-3   REINFORCED CONCRETE MEMBERS

Reinforced concrete structures consist of a series of "members" that interact to support the loads placed on the structure. The second floor of the building in Fig. 1-6 is built of concrete joist–slab construction. Here, a series of parallel ribs or *joists* supports the load from the top slab. The reactions supporting the joists apply loads to the beams, which in turn are supported by columns. In such a floor, the top slab has two functions: (1) it transfers load laterally to the joists, and (2) it serves as the top flange of the joists, which act as T-shaped beams that transmit the load to the beams running at right angles to the joists. The first floor of the building in Fig. 1-6 has a slab-and-beam design in which the slab spans between beams, which in turn apply loads to the columns. The column loads are applied to *spread footings*, which distribute the load over an area of soil sufficient to prevent overloading of the soil. Some soil conditions require the use of pile foundations or other deep foundations. At the perimeter of the building, the floor loads are supported either directly on the walls, as shown in Fig. 1-6, or on exterior columns, as shown in Fig. 1-7. The walls or columns, in turn, are supported by a basement wall and wall footings.

The slabs in Fig. 1-6 are assumed to carry the loads in a north–south direction (see direction arrow) to the joists or beams, which carry the loads in an east–west direction to other beams, girders, columns, or walls. This is referred to as *one-way slab* action and is analogous to a wooden floor in a house, in which the floor decking transmits loads to perpendicular floor joists, which carry the loads to supporting beams, and so on.

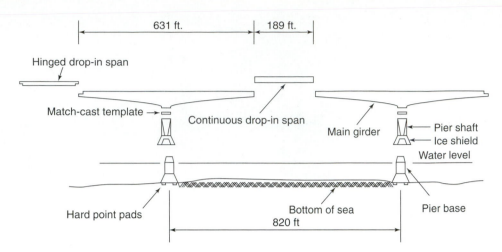

631 ft. | 189 ft.

Hinged drop-in span

Match-cast template →

Continuous drop-in span

Main girder

Pier shaft
Ice shield
Water level

Hard point pads

Bottom of sea
820 ft

Pier base

Fig. 1-3
Confederation Bridge.

(a) Confederation Bridge is an 8.1-mile, prestressed concrete bridge linking New Brunswick and Prince Edward Island, Canada. It consists of segmentally precast prestressed approach spans at both ends and forty-four 820-ft spans forming the main portion of the bridge. Each 820-ft span consists of six major parts:

1. A set of three hard-point pads, which were placed on the bedrock at the location of each pier and were brought to the desired level by pumping tremie concrete into bags between the pads and the bedrock.
2. A conical pier base and lower pier shaft, which sits on the hard-point pads.
3. A pier shaft with a conical ice shield at water level, to break the sea ice as it moves past the bridge.
4. A match-cast section between the top of the pier shaft and the bottom of the girder, to ensure a match between the top of the pier and the bottom of the girder.
5. A 631-ft-long double-cantilever section cantilevering out on each side of the pier. The cantilever sections are 46 ft deep at the piers and weigh 17,500 kips.
6. A drop-in span to complete the 820-ft span. The drop-in spans are alternately simply supported or made continuous with the double-cantilever sections. The cover photograph shows the heavy-lift vessel Svanen placing a double-cantilever section on a pier shaft.

The ability to form and construct concrete slabs makes possible the slab or plate type of structure shown in Fig. 1-7. Here, the loads applied to the roof and the floor are transmitted in two directions to the columns by plate action. Such slabs are referred to as *two-way slabs*.

The first floor in Fig. 1-7 is a *flat slab* with thickened areas called *drop panels* at the columns. In addition, the tops of the columns are enlarged in the form of *capitals* or *brackets*. The thickening provides extra depth for moment and shear resistance adjacent to the columns. It also tends to reduce the slab deflections.

The roof of the building shown in Fig. 1-7 is of uniform thickness throughout without drop panels or column capitals. Such a floor is a special type of *flat slab* referred to as a *flat plate*. Flat-plate floors are widely used in apartments because the underside of the slab is flat and hence can be used as the ceiling of the room below. Of equal importance, the forming for a flat plate is generally cheaper than that for flat slabs with drop panels or for one-way slab-and-beam floors.

1-4 FACTORS AFFECTING CHOICE OF CONCRETE FOR A STRUCTURE

The choice of whether a structure should be built of concrete, steel, masonry, or timber depends on the availability of materials and on a number of value decisions.

Fig. 1-3
(Continued)

(b) Erection of Confederation Bridge. This photograph shows the various stages in the erection of the Confederation Bridge. Starting from the left, there is a dredging barge preparing the bottom of the channel prior to the placing of the hard-point pads. The heavy-lift barge Svanen, capable of lifting 18,000 kips, is shown lowering a pier shaft into place. Next are two pier shafts. The double-cantilever sections have been erected on the next three piers. The continuous drop-in spans are in place on the final two groups of piers, and the joints have been concreted to make them continuous. The hinged drop-in spans will be erected next, filling in the gaps. The photograph shows truck-mounted cranes and other equipment on the bridge. These were placed on the double-cantilever section before it left the casting yard. The Confederation Bridge was designed by J. Muller International Stanley Joint Venture Inc. of Calgary and San Diego and was built by Strait Crossing Joint Venture, also of Calgary. (Photograph courtesy of J. G. MacGregor.)

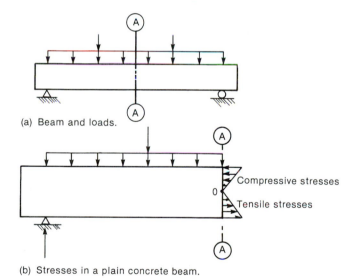

(a) Beam and loads.

(b) Stresses in a plain concrete beam.

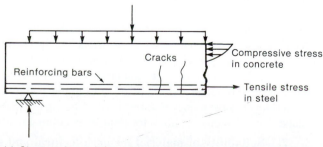

(c) Stresses in a reinforced concrete beam.

Fig. 1-4
Plain and reinforced concrete beams.

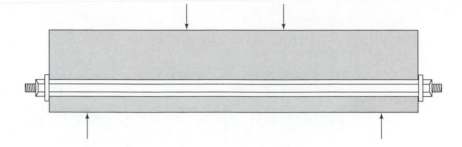

Fig. 1-5
Prestressed concrete beam.

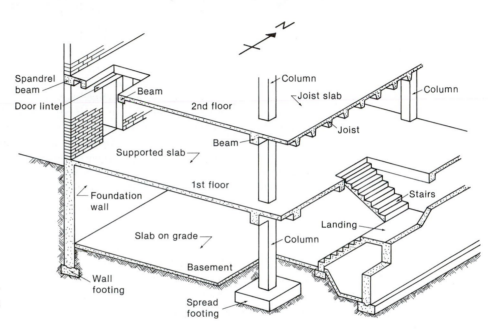

Fig. 1-6
Reinforced concrete building elements. (Adapted from [1-2].)

1. Economy. Frequently, the foremost consideration is the overall cost of the structure. This is, of course, a function of the costs of the materials and of the labor and time necessary to erect them. Concrete floor systems tend to be thinner than structural steel systems because the girders and beams or joists all fit within the same depth, as shown in the second floor in Fig. 1-6, or the floors are flat plates, as shown in Fig. 1-7. This produces an overall reduction in the height of a building compared to a steel building, which leads to (a) lower wind loads because there is less area exposed to wind, and (b) savings in cladding and mechanical and electrical risers.

Frequently, however, the overall cost is affected as much or more by the overall construction time, since the contractor and the owner must allocate money to carry out the construction and will not receive a return on their investment until the building is ready for occupancy. As a result, financial savings due to rapid construction may more than offset increased material and forming costs. The materials for reinforced concrete structures are widely available and can be produced as they are needed in the construction, whereas structural steel must be ordered and partially paid for in advance to schedule the job in a steel-fabricating yard.

Any measures the designer can take to standardize the design and forming will generally pay off in reduced overall costs. For example, column sizes may be kept the same for several floors to save money in form costs, while changing the concrete strength or the percentage of reinforcement allows for changes in column loads.

2. Suitability of material for architectural and structural function. A reinforced concrete system frequently allows the designer to combine the architectural and

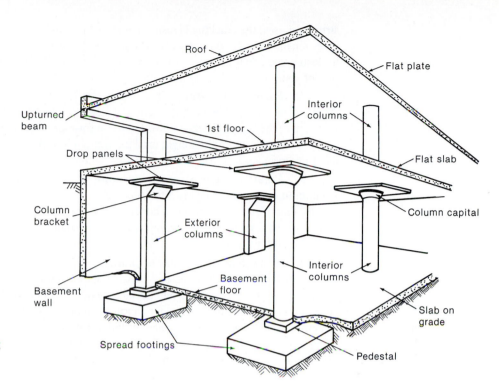

Fig. 1-7
Reinforced concrete building elements. (Adapted from [1-2].)

structural functions. Concrete has the advantage that it is placed in a plastic condition and is given the desired shape and texture by means of the forms and the finishing techniques. This allows such elements as flat plates or other types of slabs to serve as load-bearing elements while providing the finished floor and ceiling surfaces. Similarly, reinforced concrete walls can provide architecturally attractive surfaces in addition to having the ability to resist gravity, wind, or seismic loads. Finally, the choice of size or shape is governed by the designer and not by the availability of standard manufactured members.

3.   Fire resistance. The structure in a building must withstand the effects of a fire and remain standing while the building is being evacuated and the fire extinguished. A concrete building inherently has a 1- to 3-hour fire rating without special fireproofing or other details. Structural steel or timber buildings must be fireproofed to attain similar fire ratings.

4.   Rigidity. The occupants of a building may be disturbed if their building oscillates in the wind or the floors vibrate as people walk by. Due to the greater stiffness and mass of a concrete structure, vibrations are seldom a problem.

5.   Low maintenance. Concrete members inherently require less maintenance than do structural steel or timber members. This is particularly true if dense, air-entrained concrete has been used for surfaces exposed to the atmosphere and if care has been taken in the design to provide adequate drainage from the structure.

6.   Availability of materials. Sand, gravel, water, cement, and concrete mixing facilities are very widely available, and reinforcing steel can be transported to most job sites more easily than can structural steel. As a result, reinforced concrete is frequently used in remote areas.

On the other hand, there are a number of factors that may cause one to select a material other than reinforced concrete. These include:

1.   Low tensile strength. As stated earlier, the tensile strength of concrete is much lower than its compressive strength (about $\frac{1}{10}$); hence, concrete is subject to cracking. In structural uses, the cracking is restrained by using reinforcement, as shown in Fig. 1-4c, to carry tensile forces and limit crack widths to within acceptable values. Unless care is taken in design and construction, however, these cracks may be unsightly or may allow penetration of water.

2. Forms and shoring. The construction of a cast-in-place structure involves three steps not encountered in the construction of steel or timber structures. These are (a) the construction of the forms, (b) the removal of these forms, and (c) the propping or shoring of the new concrete to support its weight until its strength is adequate. Each of these steps involves labor and/or materials that are not necessary with other forms of construction.

3. Relatively low strength per unit of weight or volume. The compressive strength of concrete is roughly 5 to 10 percent that of steel, while its unit density is roughly 30% that of steel. As a result, a concrete structure requires a larger volume and a greater weight of material than does a comparable steel structure. As a result, long-span structures are often built from steel.

4. Time-dependent volume changes. Both concrete and steel undergo approximately the same amount of thermal expansion and contraction. Because there is less mass of steel to be heated or cooled, and because steel is a better conductor than concrete, a steel structure is generally affected by temperature changes to a greater extent than is a concrete structure. On the other hand, concrete undergoes drying shrinkage, which, if restrained, may cause deflections or cracking. Furthermore, deflections will tend to increase with time, possibly doubling, due to creep of the concrete under sustained loads.

1-5 HISTORICAL DEVELOPMENT OF CONCRETE AND REINFORCED CONCRETE AS STRUCTURAL MATERIALS

Cement and Concrete

Lime mortar was first used in structures in the Minoan civilization in Crete about 2000 B.C. and is still used in some areas. This type of mortar had the disadvantage of gradually dissolving when immersed in water and hence could not be used for exposed joints or underwater joints. About the third century B.C., the Romans discovered a fine sandy volcanic ash that, when mixed with lime mortar, gave a much stronger mortar, which could be used under water.

The most remarkable concrete structure built by the Romans was the dome of the Pantheon in Rome, completed in A.D. 126. This dome has a span of 144 ft, a span not exceeded until the nineteenth century. The lowest part of the dome was concrete with aggregate consisting of broken bricks. As the builders approached the top of the dome they used lighter and lighter aggregates, using pumice at the top to reduce the dead-load moments. Although the outside of the dome was, and still is, covered with decorations, the marks of the forms are still visible on the inside [1-3], [1-4].

While designing the Eddystone Lighthouse off the south coast of England just before A.D. 1800, the English engineer John Smeaton discovered that a mixture of burned limestone and clay could be used to make a cement that would set under water and be water resistant. Owing to the exposed nature of this lighthouse, however, Smeaton reverted to the tried-and-true Roman cement and mortised stonework.

In the ensuing years a number of people used Smeaton's material, but the difficulty of finding limestone and clay in the same quarry greatly restricted its use. In 1824, Joseph Aspdin mixed ground limestone and clay from different quarries and heated them in a kiln to make cement. Aspdin named his product Portland cement because concrete made from it resembled Portland stone, a high-grade limestone from the Isle of Portland in the south of England.

This cement was used by Brunel in 1828 for the mortar in the masonry liner of a tunnel under the Thames River and in 1835 for mass concrete piers for a bridge. Occasionally in the production of cement, the mixture would be overheated, forming a hard clinker which was considered to be spoiled and was discarded. In 1845, I. C. Johnson found that the best cement resulted from grinding this clinker. This is the material now known as portland cement. Portland cement was produced in Pennsylvania in 1871 by D. O. Saylor and about the same time in Indiana by T. Millen of South Bend, but it was not until the early 1880s that significant amounts were produced in the United States.

Reinforced Concrete

W. B. Wilkinson of Newcastle-upon-Tyne obtained a patent in 1854 for a reinforced concrete floor system that used hollow plaster domes as forms. The ribs between the forms were filled with concrete and were reinforced with discarded steel mine-hoist ropes in the center of the ribs. In France, Lambot built a rowboat of concrete reinforced with wire in 1848 and patented it in 1855. His patent included drawings of a reinforced concrete beam and a column reinforced with four round iron bars. In 1861, another Frenchman, Coignet, published a book illustrating uses of reinforced concrete.

The American lawyer and engineer Thaddeus Hyatt experimented with reinforced concrete beams in the 1850s. His beams had longitudinal bars in the tension zone and vertical stirrups for shear. Unfortunately, Hyatt's work was not known until he privately published a book describing his tests and building system in 1877.

Perhaps the greatest incentive to the early development of the scientific knowledge of reinforced concrete came from the work of Joseph Monier, owner of a French nursery garden. Monier began experimenting in about 1850 with concrete tubs reinforced with iron for planting trees. He patented his idea in 1867. This patent was rapidly followed by patents for reinforced pipes and tanks (1868), flat plates (1869), bridges (1873), and stairs (1875). In 1880–1881, Monier received German patents for many of the same applications. These were licensed to the construction firm Wayss and Freitag, which commissioned Professors Mörsch and Bach of the University of Stuttgart to test the strength of reinforced concrete and commissioned Mr. Koenen, chief building inspector for Prussia, to develop a method for computing the strength of reinforced concrete. Koenen's book, published in 1886, presented an analysis which assumed that the neutral axis was at the midheight of the member.

The first reinforced concrete building in the United States was a house built on Long Island in 1875 by W. E. Ward, a mechanical engineer. E. L. Ransome of California experimented with reinforced concrete in the 1870s and patented a twisted steel reinforcing bar in 1884. In the same year, Ransome independently developed his own set of design procedures. In 1888, he constructed a building having cast-iron columns and a reinforced concrete floor system consisting of beams and a slab made from flat metal arches covered with concrete. In 1890, Ransome built the Leland Stanford, Jr. Museum in San Francisco. This two-story building used discarded cable car rope as beam reinforcement. In 1903 in Pennsylvania, he built the first building in the United States completely framed with reinforced concrete.

In the period from 1875 to 1900, the science of reinforced concrete developed through a series of patents. An English textbook published in 1904 listed 43 patented systems, 15 in France, 14 in Germany or Austria–Hungary, 8 in the United States, 3 in the United Kingdom, and 3 elsewhere. Most of these differed in the shape of the bars and the manner in which the bars were bent.

From 1890 to 1920, practicing engineers gradually gained a knowledge of the mechanics of reinforced concrete, as books, technical articles, and codes presented the theories. In an 1894 paper to the French Society of Civil Engineers, Coignet (son of the earlier

Coignet) and de Tedeskko extended Koenen's theories to develop the working-stress design method for flexure, which was used universally from 1900 to 1950. During the past seven decades, extensive research has been carried out on various aspects of reinforced concrete behavior, resulting in the current design procedures.

Prestressed concrete was pioneered by E. Freyssinet, who in 1928 concluded that it was necessary to use high-strength steel wire for prestressing because the creep of concrete dissipated most of the prestress force if normal reinforcing bars were used to develop the prestressing force. Freyssinet developed anchorages for the tendons and designed and built a number of pioneering bridges and structures.

Design Specifications for Reinforced Concrete

The first set of building regulations for reinforced concrete were drafted under the leadership of Professor Mörsch of the University of Stuttgart and were issued in Prussia in 1904. Design regulations were issued in Britain, France, Austria, and Switzerland between 1907 and 1909.

The American Railway Engineering Association appointed a Committee on Masonry in 1890. In 1903 this committee presented specifications for portland cement concrete. Between 1908 and 1910, a series of committee reports led to the *Standard Building Regulations for the Use of Reinforced Concrete*, published in 1910 [1-5] by the National Association of Cement Users, which subsequently became the American Concrete Institute.

A Joint Committee on Concrete and Reinforced Concrete was established in 1904 by the American Society of Civil Engineers, the American Society for Testing and Materials, the American Railway Engineering Association, and the Association of American Portland Cement Manufacturers. This group was later joined by the American Concrete Institute. Between 1904 and 1910, the Joint Committee carried out research. A preliminary report issued in 1913 [1-6] lists the more important papers and books on reinforced concrete published between 1898 and 1911. The final report of this committee was published in 1916 [1-7]. The history of reinforced concrete building codes in the United States was reviewed in 1954 by Kerekes and Reid [1-8].

1-6 BUILDING CODES AND THE ACI CODE

The design and construction of buildings is regulated by municipal bylaws called *building codes*. These exist to protect the public's health and safety. Each city and town is free to write or adopt its own building code, and in that city or town, only that particular code has legal status. Because of the complexity of writing building codes, cities in the United States generally base their building codes on model codes. Prior to the year 2000, there were three model codes: the *Uniform Building Code* [1-9], the *Standard Building Code* [1-10], and the *Basic Building Code* [1-11]. These codes covered such things as use and occupancy requirements, fire requirements, heating and ventilating requirements, and structural design. In 2000, these three codes were replaced by the *International Building Code, IBC* [1-12], which is to be updated every three years.

The definitive design specification for reinforced concrete buildings in North America is the *Building Code Requirements for Structural Concrete (ACI-318-02)* [1-13], which is explained in a *Commentary* [1-13]. The code and the commentary are bound together in one volume.

This code, generally referred to as the *ACI Code*, has been incorporated by reference in the International Building Code and serves as the basis for comparable codes in Canada, New Zealand, Australia, and much of Latin America. The ACI Code has legal status only if adopted in a local building code.

In recent years, the ACI Code has undergone a major revision every six years. ACI 318-95 is the revision published in 1995. An interim revision is published halfway between the major revisions. The interim revision scheduled for 1998 was delayed until 1999, to mesh with the IBC 2000. Subsequent ACI Code revisions are to be issued every three years, starting with 2002. This book refers extensively to the 2002 ACI Code. It is recommended that the reader have a copy available.

The term *structural concrete* is used to refer to the entire range of concrete structures: from *plain concrete* without any reinforcement; through ordinary reinforced concrete, reinforced with normal reinforcing bars; through *partially prestressed concrete*, generally containing both reinforcing bars and prestressing tendons; to *fully prestressed concrete*, with enough prestress to prevent cracking in everyday service. In 1995, the title of the ACI Code was changed from *Building Code Requirements for Reinforced Concrete* to *Building Code Requirements for Structural Concrete* to emphasize that the code deals with the entire spectrum of structural concrete.

The rules for the design of concrete highway bridges are specified in the *Standard Specifications for Highway Bridges*, American Association of State Highway and Transportation Officials, Washington, D.C [1-14]. A limit states version of these specifications was published in 1994 and the next revision is planned for 2005 [1-15].

Each nation or group of nations in Europe has its own building code for reinforced concrete. The *CEB–FIP Model Code for Concrete Structures* [1-16], published in 1978 and revised in 1990 by the Comité Euro-International du Béton, Lausanne, is intended to serve as the basis for future attempts to unify European codes. This code and the ACI Code are similar in many ways. The European Community has published a draft *Eurocode No. 2, Design of Concrete Structures* [1-17]. Eventually, it is intended that this code will govern concrete design throughout the European Community.

Another document that will be used extensively in Chapters 2 and 19 is the ASCE standard *ASCE 7-98*, entitled *Minimum Design Loads for Buildings and Other Structures* [1-18], published in 1998.

2
The Design Process

2-1 OBJECTIVES OF DESIGN

The structural engineer is a member of a team whose members work together to design a building, bridge, or other structure. In the case of a building, an architect generally provides the overall layout, and mechanical, electrical, and structural engineers design individual systems within the building.

The structure should satisfy four major criteria:

1. **Appropriateness.** The arrangement of spaces, spans, ceiling heights, access, and traffic flow must complement the intended use. The structure should fit its environment and be aesthetically pleasing.

2. **Economy.** The overall cost of the structure should not exceed the client's budget. Frequently, teamwork in design will lead to overall economies.

3. **Structural adequacy.** Structural adequacy involves two major aspects.

    (a) A structure must be strong enough to support all anticipated loadings safely.

    (b) A structure must not deflect, tilt, vibrate, or crack in a manner that impairs its usefulness.

4. **Maintainability.** A structure should be designed so as to require a minimum of maintenance and to be able to be maintained in a simple fashion.

2-2 THE DESIGN PROCESS

The design process is a sequential and iterative decision-making process. The three major phases are the following:

1. **Definition of the client's needs and priorities.** All buildings or other structures are built to fulfill a need. It is important that the owner or user be involved in determining the attributes of the proposed building. These include functional requirements, aesthetic requirements, and budgetary requirements. The latter include first cost, rapid construction to allow early occupancy, minimum upkeep, and other factors.

2. **Development of project concept.** Based on the client's needs and priorities, a number of possible layouts are developed. Preliminary cost estimates are made, and the final choice of the system to be used is based on how well the overall design satisfies the client's needs within the budget available. Generally, systems that are conceptually simple and have standardized geometries and details that allow construction to proceed as a series of identical cycles are the most cost effective.

During this stage, the overall structural concept is selected. From approximate analyses of the moments, shears, and axial forces, preliminary member sizes are selected for each potential scheme. Once this is done, it is possible to estimate costs and select the most desirable structural system.

The overall thrust in this stage of the structural design is to satisfy the design criteria dealing with appropriateness, economy, and, to some extent, maintainability.

3. **Design of individual systems.** Once the overall layout and general structural concept have been selected, the structural system can be designed. Structural design involves three main steps. Based on the preliminary design selected in phase 2, a *structural analysis* is carried out to determine the moments, shears, and axial forces in the structure. The individual members are then *proportioned* to resist these forces. The proportioning, sometimes referred to as *member design*, must also consider overall aesthetics, the constructability of the design, and the maintainability of the final structure. The final stage in the design process is to prepare construction drawings and specifications.

2-3   LIMIT STATES AND THE DESIGN OF REINFORCED CONCRETE

Limit States

When a structure or structural element becomes unfit for its intended use, it is said to have reached a *limit state*. The limit states for reinforced concrete structures can be divided into three basic groups:

1. **Ultimate limit states.** These involve a structural collapse of part or all of the structure. Such a limit state should have a very low probability of occurrence, since it may lead to loss of life and major financial losses. The major ultimate limit states are as follows:

(a)   **Loss of equilibrium** of a part or all of the structure as a rigid body. Such a failure would generally involve tipping or sliding of the entire structure and would occur if the reactions necessary for equilibrium could not be developed.

(b)   **Rupture** of critical parts of the structure, leading to partial or complete collapse. The majority of this book deals with this limit state. Chapters 4 and 5 consider flexural failures; Chapter 6, shear failures; and so on.

(c)   **Progressive collapse.** In some structures, an overload on one member may cause that member to fail. The load acting on it is transferred to adjacent members which, in turn, may be overloaded and fail, causing them to shed their load to adjacent members, causing them to fail one after another, until a major part of the structure has collapsed. This is called a *progressive collapse* [2-1], [2-2]. Progressive collapse is prevented, or at least is limited, by one or more of the following:

(i)   Controlling accidental events by taking measures such as protection against vehicle collisions or gas explosions.

(ii)   Providing local resistance by designing key members to resist accidental events.

(iii)   Providing minimum horizontal and vertical ties to transfer forces.

(iv) Providing alternative lines of support to anchor the tie forces.

(v) Limiting the spread of damage by subdividing the building with planes of weakness, sometimes referred to as *structural fuses*.

The ACI Code requires structural detailing to provide ties that allow alternative load paths to support the load if the primary load paths are unable to carry them [2-1], [2-2]. A structure is said to have *general structural integrity* if it is resistant to progressive collapse. See also Section 8-6.

For example, a terrorist bomb or a vehicle collision may accidentally remove a column that supports an interior support of a two-span continuous beam. If properly detailed, the structural system may change from two spans to one long span. This would entail large deflections and a change in the load path from beam action to *catenary* or tension membrane action. ACI Section 7.13 requires continuous ties of tensile reinforcement around the perimeter of the building at each floor to reduce the risk of progressive collapse. The ties provide reactions to anchor the catenary forces and limit the spread of damage. Because such failures are most apt to occur during construction, the designer should be aware of the applicable construction loads and procedures. See also Section 8-6.

(d) **Formation of a plastic mechanism.** A mechanism is formed when the reinforcement yields to form plastic hinges at enough sections to make the structure unstable.

(e) **Instability** due to deformations of the structure. This type of failure involves buckling and is discussed more fully in Chapter 12.

(f) **Fatigue.** Fracture of members due to repeated stress cycles of service loads may cause collapse. Fatigue is discussed in Sections 3-13 and 9-8.

2. **Serviceability limit states.** These involve disruption of the functional use of the structure, but not collapse per se. Since there is less danger of loss of life, a higher probability of occurrence can generally be tolerated than in the case of an ultimate limit state. Design for serviceability is discussed in Chapter 9. The major serviceability limit states include the following:

(a) **Excessive deflections** for normal service. Excessive deflections may cause machinery to malfunction, may be visually unacceptable, and may lead to damage to nonstructural elements or to changes in the distribution of forces. In the case of very flexible roofs, deflections due to the weight of water on the roof may lead to increased depth of water, increased deflections, and so on, until the capacity of the roof is exceeded. This is a *ponding failure* and in essence is a collapse brought about by a lack of serviceability.

(b) **Excessive crack widths.** Although reinforced concrete must crack before the reinforcement can function effectively, it is possible to detail the reinforcement to minimize the crack widths. Excessive crack widths may be unsightly and may allow leakage through the cracks, corrosion of the reinforcement, and gradual deterioration of the concrete.

(c) **Undesirable vibrations.** Vertical vibrations of floors or bridges and lateral and torsional vibrations of tall buildings may disturb the users. Vibration has rarely been a problem in reinforced concrete buildings.

3. **Special limit states.** This class of limit states involves damage or failure due to abnormal conditions or abnormal loadings and includes

(a) damage or collapse in extreme earthquakes,

(b) structural effects of fire, explosions, or vehicular collisions,

(c) structural effects of corrosion or deterioration, and

(d) long-term physical or chemical instability (normally not a problem with concrete structures).

Limit-States Design

Limit-states design is a process that involves

1. the identification of all potential modes of failure (i.e., identification of the significant limit states),

2. the determination of acceptable levels of safety against occurrence of each limit state, and

3. a consideration of the significant limit states by the designer.

For normal structures, step 2 is carried out by the building-code authorities, who specify the load combinations and the load factors to be used. For unusual structures, the engineer may need to check whether the normal levels of safety are adequate.

For buildings, a limit-states design starts by selecting the concrete strength, cement content, water–cementitious materials ratio, air content, and cover to the reinforcement to satisfy the durability requirements of ACI Chapter 4. Next, the minimum member sizes and minimum covers are chosen to satisfy the fire-protection requirements of the local building code. Design is then carried out, starting by proportioning for the ultimate limit states followed by a check of whether the structure will exceed any of the serviceability limit states. This sequence is followed because the major function of structural members in buildings is to resist loads without endangering the occupants. For a water tank, however, the limit state of excessive crack width is of equal importance to any of the ultimate limit states if the structure is to remain watertight. In such a structure, the design for the limit state of crack width might be considered before the ultimate limit states are checked. In the design of support beams for an elevated monorail, the smoothness of the ride is extremely important, and the limit state of deflection may govern the design.

Basic Design Relationship

Figure 2-1a shows a beam that supports its own dead weight, w, plus some applied loads, P_1, P_2, and P_3. These cause bending moments, distributed as shown in Fig. 2-1b. The bending moments are obtained directly from the loads by using the laws of statics, and for a given span and combination of loads w, P_1, P_2, and P_3, the moment diagram is independent of the composition or size of the beam. The bending moment is referred to as a *load effect*. Other load effects include shear force, axial force, torque, deflection, and vibration.

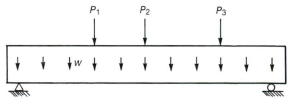

(a) Beam

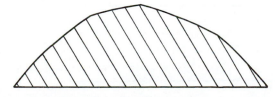

(b) Load effect — bending moment

Fig. 2-1
Beam with loads and a load effect.

Figure 2-2a shows flexural stresses acting on a beam cross section. The compressive stresses and tensile stresses in Fig. 2-2a caused by forces C and T that are separated by a distance jd, can be replaced by their resultants, C and T, as shown in Fig. 2-2b. The resulting couple is called an *internal resisting moment*. The internal resisting moment when the cross section fails is referred to as the *moment capacity* or *moment resistance*. The word "resistance" can also be used to describe shear resistance or axial load resistance.

The beam shown in Fig. 2-2 will safely support the loads if, at every section, the resistance of the member exceeds the effects of the loads:

$$\text{resistances} \geq \text{load effects} \tag{2-1}$$

To allow for the possibility that the resistances will be less than computed or the load effects larger than computed, *strength-reduction factors*, ϕ, less than 1, and *load factors*, α, greater than 1, are introduced:

$$\phi R_n \geq \alpha_1 S_1 + \alpha_2 S_2 + \cdots \tag{2-2a}$$

Here, R_n stands for nominal resistance and S stands for load effects based on the specified loads. Written in terms of moments, (2-2a) becomes

$$\phi_M M_n \geq \alpha_D M_D + \alpha_L M_L + \cdots \tag{2-2b}$$

where M_n is the *nominal moment resistance*. The word "nominal" implies that this resistance is a computed value based on the specified concrete and steel strengths and the dimensions shown on the drawings. M_D and M_L are the bending moments (load effects) due to the specified dead load and specified live load, respectively; ϕ_M is a strength-reduction factor for moment; and α_D and α_L are load factors for dead and live load, respectively. The strength-reduction factors are sometimes referred to as *resistance factors*.

Similar equations can be written for shear, V, and axial force, P:

$$\phi_V V_n \geq \alpha_D V_D + \alpha_L V_L + \cdots \tag{2-2c}$$

$$\phi_P P_n \geq \alpha_D P_D + \alpha_L P_L + \cdots \tag{2-2d}$$

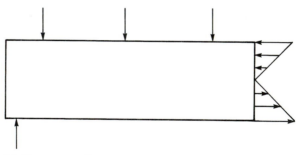

(a) Stresses acting on a cross section.

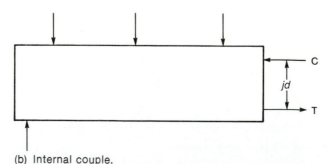

Fig. 2-2
Internal resisting moment.

(b) Internal couple.

Equation (2-1) is the basic limit-states design equation. Equations (2-2a) to (2-2d) are special forms of this basic equation. Equation (11-1) of the ACI Code, for example, is the same as (2-2c) except that, in that equation, the group of terms $(\alpha_D V_D + \alpha_L V_L + \cdots)$ is expressed as V_u, which is defined as the *factored shear force*. Throughout the ACI Code, the symbol U is used to refer to the combination $(\alpha_D D + \alpha_L L + \cdots)$. This combination is referred to as the *required strength* or the *factored loads*. The symbols M_u, V_u, T_u, and so on, refer to *factored-load effects* calculated from the factored loads, U—hence the subscript u.

2-4   STRUCTURAL SAFETY

There are three main reasons that some sorts of safety factors, such as load and resistance factors, are necessary in structural design:

1.   Variability in resistance. The actual strengths (resistances) of beams, columns, or other structural members will almost always differ from the values calculated by the designer. The main reasons for this are as follows [2-3]:

(a)   variability of the strengths of concrete and reinforcement,

(b)   differences between the as-built dimensions and those shown on the structural drawings, and

(c)   effects of simplifying assumptions made in deriving the equations for member resistance.

A histogram of the ratio of beam moment capacities observed in tests, M_{test}, to the nominal strengths computed by the designer, M_n, is plotted in Fig. 2-3. Although the mean strength is roughly 1.05 times the nominal strength in this sample, there is a definite chance that some beam cross sections will have a lower capacity than computed. The variability shown here is due largely to the simplifying assumptions made in computing the resisting moment, M_n.

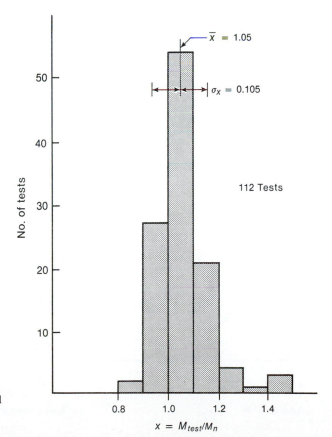

Fig. 2-3
Comparison of measured and computed failure moments, based on all data for reinforced concrete beams with $f'_c > 2000$ psi, [2-4].

2. Variability in loadings. All loadings are variable, especially live loads and environmental loads due to snow, wind, or earthquakes. Figure 2-4a compares the sustained component of live loads measured in a family of 151-ft^2 areas in offices. Although the average sustained live load was 13 psf in this sample, 1 percent of the measured loads exceeded 44 psf. For this type of occupancy and area, building codes specify live loads of 50 psf. For larger areas, the mean sustained live load remains close to 13 psf, but the variability decreases, as shown in Fig. 2-4b. A transient live load representing unusual loadings due to parties, temporary storage, and so on, must be added to get the total live load. As a result, the maximum live load on a given office will generally exceed the 13 to 44 psf quoted above.

In addition to actual variations in the loads themselves, the assumptions and approximations made in carrying out structural analyses lead to differences between the actual forces and moments and those computed by the designer [2-3]. Due to the variabilities of resistances and load effects, there is a definite chance that a weaker-than-average structure will be subjected to a higher-than-average load. In extreme cases, failure may occur. The load factors and resistance factors in (2-2a through d) are selected to reduce the probability of failure to a very small level.

The consequences of failure are a third factor that must be considered in establishing the level of safety required in a particular structure.

3. Consequences of failure. A number of subjective factors must be considered in determining an acceptable level of safety for a particular class of structure. These include such things as

    (a) the cost of clearing the debris and replacing the structure and its contents;

    (b) the potential loss of life—it may be desirable to have a higher factor of safety for an auditorium than for a storage building;

    (c) the cost to society in lost time, lost revenue, or indirect loss of life or property due to a failure—for example, the failure of a bridge may result in intangible costs due to traffic jams that could approach the replacement cost; and

    (d) the type of failure, warning of failure, and existence of alternative load paths. If the failure of a member is preceded by excessive deflections, as in the case of a flexural failure of a reinforced concrete beam, the persons endangered by the impending collapse will be warned and will have a chance to leave the building prior to failure. This may not be possible if a member fails suddenly without warning, as may be the case with a tied column. Thus, the required level of safety may not need to be

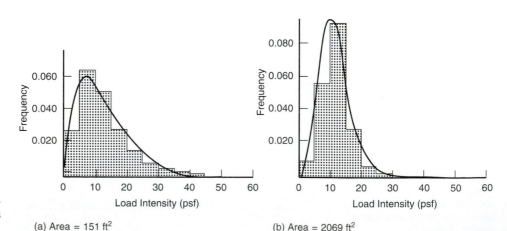

Fig. 2-4
Frequency distribution of sustained component of live loads in offices. (From [2-5].)

(a) Area = 151 ft^2

(b) Area = 2069 ft^2

as high for a beam as for a column. In some structures, the yielding or failure of one member causes a redistribution of load to adjacent members. In other structures, the failure of one member causes complete collapse. If no redistribution is possible, a higher level of safety is required.

2-5   PROBABILISTIC CALCULATION OF SAFETY FACTORS

The distribution of a population of resistances, R, of a group of similar structures is plotted on the horizontal axis in Fig. 2-5. This is compared to the distribution of the maximum load effects, S, expected to occur on those structures during their lifetimes, plotted on the vertical axis in the same figure. For consistency, both the resistances and the load effects are expressed in terms of a quantity such as bending moment. The 45° line in this figure corresponds to a load effect equal to the resistance. Combinations of S and R falling above this line correspond to $S > R$ and, hence, failure. Thus, load effect S_1 acting on a structure having strength R_1 would cause failure, whereas load effect S_2 acting on a structure having resistance R_2 represents a safe combination.

For a given distribution of load effects, the probability of failure can be reduced by increasing the resistances. This would correspond to shifting the distribution of resistances to the right in Fig. 2-5. The probability of failure could also be reduced by reducing the dispersion of the resistances.

The term $Y = R - S$ is called the *safety margin*. By definition, failure will occur if Y is negative, represented by the shaded area in Fig. 2-6. The *probability of failure*, P_f, is the chance that a particular combination of R and S will give a negative value of Y. This probability is equal to the ratio of the shaded area to the total area under the curve in Fig. 2-6. This can be expressed as

$$P_f = \text{probability that } [Y < 0] \tag{2-3}$$

The function Y has mean value $\overline{Y}$ and standard deviation σ_Y. From Fig. 2-6, it can be seen that $\overline{Y} = 0 + \beta\sigma_Y$, where $\beta = \overline{Y}/\sigma_Y$. If the distribution is shifted to the right by increasing the resistance, thereby making $\overline{Y}$ larger, β will increase, and the shaded area, P_f, will decrease. Thus, P_f is a function of β. The factor β is called the *safety index*.

If Y follows a standard statistical distribution, and if $\overline{Y}$ and σ_Y are known, the probability of failure can be calculated or obtained from statistical tables as a function of the type

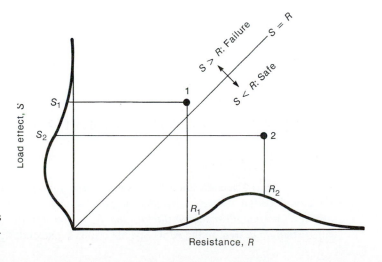

Fig. 2-5
Safe and unsafe combinations of loads and resistances. (From [2-6].)

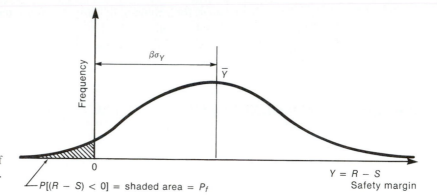

Fig. 2-6
Safety margin, probability of
failure, and safety index.
(From [2-6].)

of distribution and the value of β. Consequently, if Y follows a normal distribution and β is 3.5, then $\overline{Y} = 3.5\sigma_Y$, and, from tables of the normal distribution, P_f is 1/909, or 1.1×10^{-4}. This suggests that roughly 1 in every 10,000 structural members designed on the basis that $\beta = 3.5$ will fail due to excessive load or understrength sometime during its lifetime.

The appropriate values of P_f and hence of β are chosen by bearing in mind the consequences of failure. Based on current design practice, β is taken between 3 and 3.5 for ductile failures with average consequences of failure and between 3.5 and 4 for sudden failures or failures having serious consequences [2-6], [2-7].

Because the strengths and loads vary independently, it is desirable to have one factor, or a series of factors, to account for the variability in resistances and a second series of factors to account for the variability in load effects. These are referred to, respectively, as *strength-reduction factors* (also called *resistance factors*), ϕ, and *load factors*, α. The resulting design equations are (2-2a) through (2-2d).

The derivation of probabilistic equations for calculating values of ϕ and α is summarized and applied in [2-6], [2-7], and [2-8].

The resistance and load factors in the 1971 through 1995 ACI Codes were based on a statistical model which assumed that if there were a 1/1000 chance of an "overload" and a 1/100 chance of "understrength," the chance that an "overload" and an "understrength" would occur simultaneously is $1/1000 \times 1/100$ or 1×10^{-5}. Thus the ϕ factors were originally derived so that a strength of ϕR_n would be exceeded 99 out of 100 times. The ϕ factors for columns were then divided by 1.1, since the failure of a column has serious consequences. The ϕ factors for tied columns that fail in a brittle manner were divided by 1.1 a second time to reflect the consequences of the mode of failure. The original derivation is summarized in the appendix to [2-6]. Although this model is simplified by ignoring the overlap in the distributions of R and S in Figs. (2-5) and (2-6), it gives an intuitive estimate of the relative magnitudes of the understrengths and overloads.

2-6 DESIGN PROCEDURES SPECIFIED IN THE ACI BUILDING CODE

Strength Design

In the 2002 ACI Code, design is based on *required strengths* computed from combinations of factored loads and *design strengths* computed as ϕR_n, where ϕ is a *resistance factor*, also known as a *strength-reduction factor*, and R_n is the nominal resistance. This process is called *strength design*. In the AISC Specifications for steel design, the same design process is known as LRFD (Load and Resistance Factor Design). Strength design and LRFD are methods of limit-states design, except that primary attention is always placed on the ultimate limit states, with the serviceability limit states being checked after the original design is completed.

ACI Sections 9.1.1 and 9.1.2 present the basic limit-states design philosophy of that code.

> 9.1.1—Structures and structural members shall be designed to have design strengths at all sections at least equal to the required strengths calculated for the factored loads and forces in such combinations as are stipulated in this code.

The term *design strength* refers to ϕR_n, and the term *required strength* refers to the load effects calculated from factored loads, $\alpha_D D + \alpha_L L + \cdots$.

> 9.1.2—Members also shall meet all other requirements of this code to insure adequate performance at service load levels.

This clause refers primarily to control of deflections and excessive crack widths.

Working-Stress Design

Prior to 2002, Appendix A of the ACI Code allowed design of concrete structures either by strength design or by *working-stress design*. In 2002, this appendix was deleted. The commentary to the 2002 ACI Section 1.1 still allows the use of working-stress design, provided that the local jurisdiction adopts an exception to the ACI Code allowing the use of working-stress design. Chapter 9 on serviceability presents some concepts from working-stress design. Here, design is based on *working loads*, also referred to as *service loads* or *unfactored loads*. In flexure, the maximum elastically computed stresses cannot exceed *allowable stresses* or *working stresses* of 0.4 to 0.5 times the concrete and steel strengths. The 1999 ACI Code referred to this procedure as the *alternate design procedure*.

The working-stress design method assumes that the ultimate limit states will automatically be satisfied by basing design on allowable stresses at working load levels. Depending on the variability of the materials and loads, this is not necessarily so.

The drawbacks of working-stress design are discussed in [2-6] and [2-7]. The most serious drawbacks stem from its inability as currently formulated to

(a)   account properly for the variability of the resistances and loads,

(b)   from the lack of any knowledge of the level of safety, and

(c)   from its inability to deal with groups of loads where one load increases at a rate different from that of the others.

This last criticism is especially serious when a relatively constant load such as dead load counteracts the effects of a highly variable load such as wind, as illustrated in Fig. 2-7. Here a 20 percent increase in the wind causes a 20 percent increase in the maximum flexural stresses (from 500 psi to 600 psi) as expected, but causes a 100 percent increase in the stresses at point *A* in Fig. 2-7d.

Plastic Design

Plastic design, also referred to as *limit design* (not to be confused with limit-states design) or *capacity design*, is a design process that considers the redistribution of moments as successive cross sections yield, thereby forming *plastic hinges* which lead to a plastic mechanism. These concepts are of considerable importance in seismic design, where the ductility of the structure leads to a decrease in the forces that must be resisted by the structure.

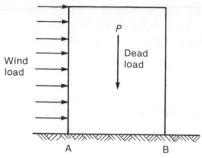

(a) Structure.

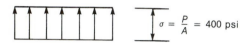

(b) Dead load stresses.

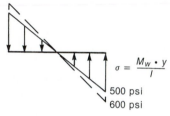

(c) Wind load stresses.

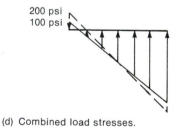

Fig. 2-7
Stresses due to counteracting
loads. (From [2-7].)

(d) Combined load stresses.

Plasticity Theorems

Several aspects of the design of statically indeterminate concrete structures are justified, in part, by using the theory of plasticity. These include the ultimate strength design of continuous frames and two-way slabs for elastically computed loads and moments and the use of strut-and-tie models for concrete design. Before the theorems of plasticity are presented, several definitions are required:

- A distribution of internal forces (moments, axial forces, and shears) or corresponding stresses is said to be *statically admissible* if it is in equilibrium with the applied loads and reactions.

- A distribution of cross-sectional strengths that equals or exceeds the statically admissible forces, moments, or stresses at every cross section in the structure is said to be a *safe* distribution of strengths.

- A structure is said to be a *collapse mechanism* if there is one more hinge, or plastic hinge, than required for stable equilibrium.

- A distribution of applied loads, forces, and moments that results in sufficient plastic hinges to produce a collapse mechanism is said to be *kinematically admissible*.

The theory of plasticity is expressed in terms of the following three theorems:

1.   Lower-bound theorem. If a structure is subjected to a statically admissible distribution of internal forces and if the member cross sections are chosen to provide a safe distribution of strength for the given structure and loading, the structure either will not collapse or will be just at the point of collapsing. The resulting distribution of internal forces and moments corresponds to a failure load that is a lower bound to the load at failure. This is called a *lower bound* because the computed failure load is less than or equal to the actual collapse load.

2.   Upper-bound theorem. A structure will collapse if there is a kinematically admissible set of plastic hinges that results in a plastic collapse mechanism. For any kinematically admissible plastic collapse mechanism, a collapse load can be calculated by equating external and internal work. The load calculated by this method will be greater than or equal to the actual collapse load. Thus, the calculated load is an *upper bound* to the failure load.

3.   Uniqueness theorem. If the lower-bound theorem involves the same forces, hinges, and displacements as the upper-bound solution, the resulting failure load is the true or *unique* collapse load.

For the upper- and lower-bound solutions to occur, the structure must have enough ductility to allow the moments and forces from the original set to redistribute to those corresponding to the bounds of plasticity solutions.

Reinforced concrete design is usually based on elastic analyses. Cross sections are proportioned to have factored nominal strengths, ϕM_n, ϕP_n, and ϕV_n, equal to or greater than the M_u, P_u, and V_u from an elastic analysis. Because the elastic moments and forces are a statically admissible distribution of forces, and because the resisting-moment diagram is chosen by the designer to be a safe distribution, the strength of the resulting structure is a lower bound.

Similarly, the strut-and-tie models presented in Chapter 18 (ACI Appendix A) give lower-bound estimates of the capacity of concrete structures if

(a)   the strut-and-tie model of the structure represents a statically admissible distribution of forces,

(b)   the strengths of the struts, ties, and nodal zones are chosen to be safe, relative to the computed forces in the strut-and-tie model, and

(c)   the members and joint regions have enough ductility to allow the internal forces, moments, and stresses to make the transition from the strut-and-tie forces and moments to the final force and moment distribution.

Thus, if adequate ductility is provided the strut-and-tie model will give a so-called safe estimate, which is a lower-bound estimate of the strength of the strut-and-tie model. Plasticity solutions are used to develop the yield-line method and the strip method of analysis for slabs, presented in Chapter 15.

2-7   REVISIONS TO LOAD FACTORS AND LOAD COMBINATIONS IN THE 2002 ACI CODE

The 2002 ACI Code presents two separate but intertwined revisions to the load factors, load combinations, and resistance factors. The new load factors and load combinations in ACI 318-02 Sections 9.2.1 through 9.2.5 are from ASCE 7-98, *Minimum Design Loads for Buildings and Other Structures* [2-2], with slight modifications. The load factors from ACI 318-02 Section 9.2 *must* be used with the strength reduction factors in ACI 318-02 Sections 9.3.1 through 9.3.5. The previous ACI load combinations and resistance factors have been moved to code Appendix C of the 2002 Code. Sections C.2.1 through C.2.7 and Sections C.3.2.1 through C.3.5.

The new load factors and strength reduction factors were derived in [2-7] for use in the design of steel, timber, masonry, and concrete structures and are used in the AISC LRFD Specification for steel structures. For concrete structures, new resistance factors compatible with the new load factors were derived by ACI Committee 318 and Nowak and Szerszen [2-10].

The primary changes are as follows:

1. The load factors and load combinations in Appendix C of the 1999 Code were interchanged with those in Sections 9.2 and 9.3 of the 1999 Code. It was the intent of ACI Committee 318 that the traditional ACI load factors and load combinations, currently in Appendix C of the 2002 Code, would be deleted in the 2008 Code leaving Sections 9.2.1 and 9.3.2 as the only source of load factors, load combinations and strength reduction factors in future editions of ACI 318.

2. The second revision changed the way that the ACI code distinguishes between ductile and brittle members when selecting strength reduction factors, ϕ. Traditionally, the 1999 and earlier codes assigned a ϕ factor of 0.9 for ductile flexural failures of beams and slabs, and ϕ factors of 0.7 to 0.75 for brittle failures of tied and spiral columns, respectively. These factors were based on arbitrary conceptions of what constituted ductile and brittle failures. The definitions of ductility used to define the strength reduction factors for beams, slabs, and columns in the 2002 code are based on the curvatures computed from the strains in the tension steel at the ultimate limit state. In this sense the curvature is used as a measure of the ductility of beams, slabs, or columns. Load factors and load combinations are discussed in this chapter. The new strength reduction factors, ϕ, will not be discussed until the necessary flexural theory has been developed in Chapters 4, 5, and 11.

Terminology and Notation

The ACI Code uses the subscript u to designate the *required strength*, which is a load effect computed from combinations of factored loads. The sum of the combination of factored loads is U as, for example, in

$$U = 1.2D + 1.6L \tag{2-4}$$

where the symbol U and subscript u are used to refer to the sum of the factored loads in terms of loads, or in terms of the effects of the factored loads, M_u, V_u, and P_u.

The member strengths computed using the specified material strengths, f'_c and f_y, and the nominal dimensions, b and d shown on the drawings, are referred to as the *nominal moment capacity*, M_n, or *nominal shear capacity*, V_n, and so on. The *usable strength* or design strength is the nominal strength multiplied by a strength-reduction factor, ϕ. The design equation is thus:

$$\phi M_n \geq M_u \tag{2-2b}$$

$$\phi V_n \geq V_u \tag{2-2c}$$

and so on.

Load Factors and Load Combinations from ACI 318-02 Sections 9.2.1 through 9.2.5. (The New Provisions)

Load Combinations

The safety provisions from ACI Chapter 9 will be presented first, followed by the provisions in ACI Appendix C.

Structural failures usually occur under combinations of several loads. In recent years these combinations have been presented in what is referred to as the *companion action format*. This is an attempt to model the expected load combinations.

The load combinations in ACI 318-02 Section 9.2.1 are examples of *companion action load combinations* chosen to represent realistic load combinations that might occur. In principle, each of these combinations includes one or more *permanent loads* (*D*, *F*, and *T*) with load factors of 1.2, plus the dominant or *principal variable load* (*L, S, W*, or others) with a load factor of 1.6, plus one or more *companion-action variable loads*. The companion action loads are computed by multiplying the specified loads (*L, S, W*, or others) by *companion action load factors* between 0.2 and 1.0. The companion-action load factors were chosen to provide results for the companion action loads that would be likely during an instance in which the principal variable load is maximized.

In the design of structural members in buildings that are not subjected to significant wind or earthquake forces, and members unaffected by wind or earthquakes, the factored loads are computed from:

$$U = 1.4(D + F) \tag{2-5}$$
(ACI Eq. 9-1)

where *D* is the specified dead load and *F* is the load due to the weight and pressures of fluids with well-defined densities and in tanks in which the maximum height of the fluid is controlled.

Because these equations are used in many chapters we will refer to them by their ACI numbers.

For combinations including dead load; live load, *L*; and/roof loads:

$$U = 1.2(D + F + T) + 1.6(L + H) + 0.5(L_r, \text{ or } S, \text{ or } R) \tag{2-6}$$
(ACI Eq. 9-2)

where:

T = the load effect produced by the combined actions of temperature, creep, shrinkage, differential settlement, and shrinkage compensating cement

L = live load that is a function of use and occupancy

M = load from soil pressure or soil weight lateral

L_r = roof live load

S = roof snow load

R = roof rain load

The terms in (2-5) through (2-12) may be expressed as *direct loads* (such as distributed loads from dead and live weight) or *load effects* (such as moments and shears caused by the given loads). The design of a roof structure, or the columns and footings supporting a roof and one or more floors, would take the roof live load equal to the largest of the three loads (*L_r*, *S*, or *R*), with the other two roof loads in the brackets taken as zero. If any of *T*, *F*, or *H* is zero, the corresponding term drops out of ACI Eq. (9-2). Thus, for the common case of a member supporting dead and live load only, ACI Eq. (9-2) is written as:

$$U = 1.2D + 1.6L \tag{2-4}$$

If the roof load exceeds the floor live loads, or if a column supports a total roof load that exceeds the total floor live load supported by the column:

$$U = 1.2D + 1.6(L_r \text{ or } S \text{ or } R) + (1.0L \text{ or } 0.8W) \tag{2-7}$$
(ACI Eq. 9-3)

The roof loads are *principal variable actions* in ACI Eq. (9-3), and they are *companion variable actions* in ACI Eq. (9-4) and (9-2).

$$U = 1.2D + 1.6W + 1.0L + 0.5(L_r \text{ or } S \text{ or } R) \tag{2-8}$$
(ACI Eq. 9-4)

Wind load is the principal variable action in ACI Eq. (9-4) and is a companion variable action in ACI Eq. (9-3).

Earthquake Loads

If earthquake loads are significant:

$$U = 1.2D + 1.0E + 1.0L + 0.2S \qquad\qquad (2\text{-}9)$$

$$(\text{ACI Eq. 9-5})$$

where the load factor of 1.0 for the earthquake loads corresponds to a *strength level earthquake* which has a much longer return period, and hence is larger than a *service load earthquake*. If the loading code used in a jurisdiction is based on the smaller service load earthquake, the load factor on E is 1.4 instead of 1.0 (ACI Section 9.2.1(c)).

Dead Loads that Stabilize Overturning and Sliding

If the effects of dead loads stabilize the structure against wind or earthquake loads, as shown in Fig 2-7:

$$U = 0.9D + 1.6W + 1.6H \qquad\qquad (2\text{-}10)$$

$$(\text{ACI Eq. 9-6})$$

or

$$U = 0.9D + 1.0E + 1.6H \qquad\qquad (2\text{-}11)$$

$$(\text{ACI Eq. 9-7})$$

Load Factor for Small Live Loads

ACI Section 9.2.1(a) allows that the load factor of 1.0 for L in ACI Eqs. (9-3), (9-4), and (9-5) may be reduced to 0.5 except for

    (a) garages,

    (b) areas occupied as places of public assembly, and

    (c) all areas where the live load is greater than 100 psf.

Directionality of Wind Load

Prior to 1998, ASCE 7 assumed that the direction of the highest wind loads coincided with the direction of the weakest structural framing. This procedure required a wind load factor of 1.3. In ASCE 7-98, the equations used to compute the wind velocity pressure, q, are multiplied by a *directionality factor*, K_d, equal to 0.85 for rectangular buildings and 0.9 or 0.95 for circular chimneys or tanks. If the computed wind load does not include this factor, the load factor 1.6 for wind in ACI Eqs. (9-4) and (9-6) can be reduced to 1.3. To give the same values for the design winds as in previous ACI Codes, the load factor for wind should be $(1.3/0.85) = 1.53$. This was rounded up to 1.6. The load factor on H shall be set equal to zero in ACI Eqs. (9-6) and (9-7) if the effect of H counteracts the effects of W or E. When lateral earth pressure provides resistance to structural effects from other forces, it is not included in H. Instead, the resistance components from earth pressure are included in the design resistance.

Strength-Reduction Factors, ϕ, ACI 318-02 Section 9.3 (The New Provisions)

Just as the 2002 ACI Code allows the use of either of two sets of load combinations in design, it also gives two sets of strength-reduction factors. In ACI 318-02 one set of load factors is given in ACI Section 9.2.1, with the corresponding strength-reduction factors, ϕ, given in ACI Section 9.3.1. Alternatively, the load factors in ACI 318-02. Section C.2.1 and the corresponding strength-reduction factors in ACI Section C.3.1 may be used. If ACI Section 9.2 is used, the resistance factors in ACI Section 9.3 *must* be used. Similarly, if ACI Section C.3.1 is used, C.3.2 must be used also. In both cases, the resistance factors are computed from the strain, ε_t, in the extreme tension layer of steel, located at a depth

d_t from the extreme compressive face of the member [2-11], [2-12]. This is a change from the selection of ϕ factors in previous codes. These are derived and discussed in Sections 4-3 and 11-4.

Flexure or Combined Flexure

9.3.2.1   Tension-controlled sections as defined in
ACI 318-02 Section 10.3.4 $\phi = 0.90$

9.3.2.2   Compression-controlled sections:

(a) Members with spiral reinforcement $\phi = 0.70$

(b) Other compression-controlled sections $\phi = 0.65$

There is a transition region between tension-controlled and compression-controlled sections.

Other actions

9.3.2.3   Shear and torsion $\phi = 0.75$

9.3.2.4   Bearing on concrete $\phi = 0.65$

The concept of tension-controlled and compression-controlled sections, and the resulting strength-reduction factors, will be presented for beams in flexure, axially loaded columns, and columns loaded in combined axial load and bending in Chapters 4, 5, and 11. The derivation of the ϕ factors will be introduced at that time.

Load Factors and Load Combinations from ACI 318-02, Appendix C—Traditional ACI Load Factors

The ACI Code presents two different sets of load factors and load combinations. The load factors and combinations introduced in the 1963 ACI Code and amended in the 1971 ACI Code are presented in ACI Sections C.2 and C.3 in Appendix C of the 2002 code. These load factors and load combinations differ from those in ACI Sections 9.2 and 9.3.

In the design of buildings that are not subjected to significant wind or earthquake forces, or for members unaffected by wind or earthquakes, the factored loads are computed from

$$U = 1.4D + 1.7L \tag{2-12}$$
(ACI Eq. C-1)

where D and L are the specified dead and live loads.

If wind loads do affect the design, ACI Section C.2.2 requires that three combinations of loads be considered and the design based on the largest values of U of either sign at each critical section:

1.   Where the load effects due to wind add to those due to dead or live loads:

$$U = 0.75(1.4D + 1.7L) + 1.6W) \tag{2-13}$$
(ACI Eq. C-2)

2.   Where the effects of dead loads stabilize the structure against wind loads, as in Fig. 2-7:

$$U = 0.9D + 1.6W \tag{2-14}$$
(ACI Eq. C-3)

But for any combination of D, L, and W, the required strength shall not be less than as given by ACI Eq. (C-6).

Similar load combinations are given in ACI Section C.2.2 for earthquake loadings, Section C.2.3 for lateral earth pressure, Section C.2.4 for fluid pressures (in tanks, etc.), Section C.2.5 for impact loads, and Section C.2.6 for differential settlement, creep, shrinkage, and temperature change.

Equation 2-10 would be used to compute the stresses at point A in Fig. 2-7. At point B the most severe of Eqs. 2-5 to 2-9 would apply.

Traditional ACI Strength-Reduction Factors, ACI 318-02 Appendix C

Prior to 2002 the traditional ACI strength-reduction factors were functions of the assumed ductility and the importance assumed for the member being designed. These were arbitrarily assumed in the 1963 code and have not been changed much since. Beams developing tension failures were assigned $\phi = 0.90$ because they were most ductile. Tied columns were assumed to have the least ductility and were assigned $\phi = 0.70$. Spiral columns were assigned $\phi = 0.75$ to reflect the somewhat greater ductility of spiral columns. The strength-reduction factors in the 2002 ACI Code were based on a different way to estimate the ductility. The strength-reduction factors were computed as a function of the ductility, which in turn was computed as a function of the steel strain in the extreme tension fiber, as will be explained in Chapters 4, 5, and 11. The final set of strength-reduction factors from Appendix C are:

C.3.2.1 Tension-controlled sections as defined in
ACI 318-02 Section 10.3.4 $\phi = 0.90$

C.3.2.2 Compression-controlled sections:
(a) Members with spiral reinforcement $\phi = 0.75$
(b) Other compression-controlled sections $\phi = 0.70$

There is a transition region between tension-controlled and compression-controlled sections.

C.3.2.3 Shear and torsion $\phi = 0.85$
C.3.2.4 Bearing on concrete $\phi = 0.70$

Comparison of New and Old Load and Resistance Factors

The new set of load and resistance factors in ACI 318-02 Sections 9.2 and 9.3 were chosen to give similar amounts of reinforcement in designs carried out by the two alternative sets of load and resistance factors [2-10]. There is one very significant departure from this goal, however. The resistance factor for flexure in the 2002 code was set at $\phi = 0.9$ [2-10] which is higher than the $\phi = 0.80$ or 0.85 needed to give the same designs as obtained previously. The new strength-reduction factor for flexure was established by a *calibration* of the ACI Code, which suggested that $\phi = 0.90$ gave safety levels for tension-controlled beams that were comparable to those for other structural actions [2-10].

For a hypothetical case of a beam loaded in flexure by nominal dead and live loads D and L, with L half as big as D, the load and resistance factors in ACI Sections 9.2.1 through 9.2.5 and 9.3.1 through 9.3.5 give

$$(\phi_{2002} \text{ times the 2002 nominal resistance}) \geq (\text{product of 2002 load factors and loads})$$

$$(\phi_{2002}R_{n,2002}) \geq [1.2 \times D + (1.6 \times L)] = 1.2D + 1.6 \times 0.5D$$

$$R_{n,2002} = 2.0D/(\phi_{2002})$$

where for tension-controlled flexural failures, ϕ_{2002} is taken equal to 0.90, giving

$$R_n \geq 2.22D$$

This compares with the factored strength ratio from Chapter 9 of ACI 318-99

$$(\phi_{1999}R_{n,1999}) \geq (1.4 \times D + 1.7 \times 0.5D)$$

or

$$R_{n,1999} \geq 2.25D/0.9 = 2.5D$$

Thus, the 2002 load and resistance factors give structures with about $2.22/2.50 = 0.89$ times as much flexural reinforcement as required by the ACI Codes from 1971 through 1999. A smaller amount of reinforcement resisting the same load means the service load stress in the steel will be higher in the 2002 design than in the 1999 design. This will result in larger crack widths and larger deflections than those expected for the 1999 code. Although fatigue is seldom a problem in reinforced concrete buildings, higher service load steel stresses may require that fatigue strength be considered if over 20,000 cycles of load are expected.

A design must use one set of load and resistance factors either all from ACI Code Chapter 9 or all from Appendix C. For other limit states the difference between the traditional ACI load factors (in ACI 318-02 Sections C.9.2.1 through C.9.2.5) and the factors based on ASCE 7 (in ACI 318-02 Sections 9.2.1 through 9.2.5) is much less than the 9 percent provided for flexure. In examples in this book, the load and resistance factors in ACI 318-02 Sections 9.2 and 9.3 will generally be used.

In the analysis of a building frame, it is frequently best to analyze the structure elastically three times—once each for $1.0D$, $1.0L$, and $1.0W$—and to combine the resulting moments, shears, and so on for each member according to (2-4) to (2-12). (Exceptions to this are analyses of cases in which linear superposition does not apply, such as second-order analyses of frames. These must be carried out at the factored-load level.) The procedure used is illustrated in Example 2-1.

Example 2-1   Computation of Factored-Load Effects

Figure 2-8 shows a beam and column from a concrete building frame. The loads per foot on the beam are dead load, $D = 1.58$ kips/ft, and live load, $L = 0.75$ kip/ft. Additionally, wind load is represented by the concentrated loads at the joints. The moments and shears in a beam and in the columns over and under the beam due to $1.0D$, $1.0L$, and $1.0W$ are shown in Fig. 2-8b to d.

Compute the required strengths, using (2-4) through (2-12). For the moment at section A, four load cases must be considered

(a) $U = 1.4(D + F)$ (2-3)
(ACI Eq. 9-1)

Since the beam does not support a tank containing fluid, $F = 0$, and $U = 1.4 \times -39 = -54.6$ ft-kips

(b) $U = 1.2(D + F + T) + 1.6(L + H) + 0.5(L_r \text{ or } S \text{ or } R)$ (2-6)
(ACI Eq. 9-2)

- Again, for the fluid forces, $F = 0$.
- Assuming that there is no differential settlement of the interior columns relative to the exterior columns and assuming there is no restrained shrinkage, the self-equilibrating actions, T, will be taken to be zero.
- The soil weight and pressure $H = 0$.
- Since the beam being considered is not a roof beam, L_r or S, or R are all equal to zero. (Note that the axial loads in the columns support axial forces from the roof load and the slab live load.)

ACI Eq. 9-2 becomes

$$U = 1.2D + 1.6L$$ (2-4)
$$= 1.2 \times -39 + 1.6 \times -19 = -77.2 \text{ ft-kips}$$

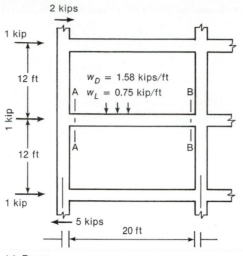

(a) Frame.

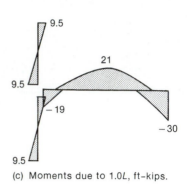

(c) Moments due to 1.0L, ft–kips.

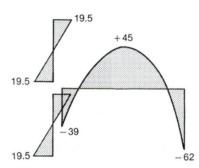

Fig. 2-8
Moment diagrams—
Example 2-1.

(b) Moments due to 1.0D, ft–kips.

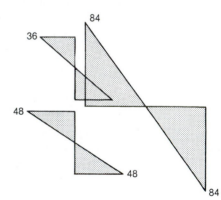

(d) Moments due to 1.0W, ft–kips.

(c) Equation (2-8) does not govern because this is not a roof beam.

(d) $U = 1.2D + 1.6W + 1.0L + 1.0(L_r \text{ or } S \text{ or } R)$ (2-9)

where ACI Section 9.2.1(a) allows 1.0L to be reduced to 0.5L Eq. 9-2, and ACI Eq. 9-2 becomes

$$U = 1.2D + 1.6W + 0.5L$$
$$= 1.2 \times -39 \pm 1.6 \times 84 + 0.5 \times -19 = -56.3 \pm 134.4$$
$$= -56.3 \pm 134.4$$
$$= -190.7 \text{ or } +78.1 \text{ kip-ft}$$

The positive and negative values of the wind-load moment are due to winds alternately blowing from the two sides of the building.

(e) The dead-load moments can counteract a portion of the wind- and live-load moments. This makes it necessary to consider (2-10):

$$U = 0.9D + 1.6W + 1.6H$$ (2-10)
$$= 0.9 \times -39 \pm 1.6 \times 84 = -35.1 \pm 134.4$$
$$= +99.3 \text{ or } -169.5 \text{ ft-kips}$$

It is not necessary to check the fourth load case because this problem does not involve uplift or overturning. Thus the required strengths, M_u, at section A–A are $+99.3$ ft-kips and -190.7 ft-kips. ■

This computation is repeated for a sufficient number of sections to make it possible to draw shearing-force and bending-moment envelopes for the beam. (Bending-moment envelopes are discussed in Section 10-3.) The solution of the four equations given above can easily be programmed for a programmable calculator with D, L, and W as input values and the seven values of U and/or the maximum positive and negative values of the factored load effect as output.

2-8   LOADINGS AND ACTIONS

Direct and Indirect Actions

An *action* is anything that gives rise to stresses in a structure. The term *load* or *direct action* refers to concentrated or distributed forces resulting from the weight of the structure and its contents, or pressures due to wind, water, or earth. An *indirect action* or *imposed deformation* is a movement or deformation that does not result from applied loads, but that causes stresses in a structure. Examples are uneven support settlements of continuous beams and shrinkage of concrete if it is not free to shorten.

Because the stresses due to imposed deformations do not resist an applied load, they are generally *self-equilibrating*. Consider, for example, a prism of concrete with a reinforcing bar along its axis. As the concrete shrinks, its shortening is resisted by the reinforcement. As a result, a compressive force develops in the steel and an equal and opposite tensile force develops in the concrete, as shown in Fig. 2-9. If the concrete cracks from this tension, the tensile force in the concrete at the crack is zero, and for equilibrium, the steel force must also disappear at the cracked section. Section 1.3.3 of ASCE 7 refers to imposed deformations as *self-straining forces*.

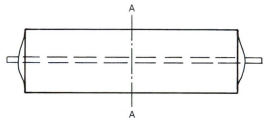

(a) Prism after shrinkage.

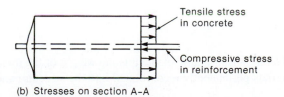

Tensile stress in concrete

Compressive stress in reinforcement

Fig. 2-9
Self-equilibrating stresses due to shrinkage.

(b) Stresses on section A–A

Classifications of Loads

Loads may be described by their variability with respect to time and location. A *permanent* load remains roughly constant once the structure is completed. Examples are the self-weight of the structure and soil pressure against foundations. *Variable* loads, such as occupancy loads and wind loads, change from time to time. Variable loads may be *sustained loads* of long duration, such as the weight of filing cabinets in an office, or loads of *short duration*, such as the weight of people in the same office. Creep deformations of concrete structures result from the permanent loads and the sustained portion of the variable loads. A third category is *accidental loads*, which include vehicular collisions and explosions.

Variable loads may be *fixed* or *free* in location. Thus, the loading in an office building is free, since it can occur at any point in the loaded area. A train load on a bridge is not fixed longitudinally, but is fixed laterally by the rails.

Loads are frequently classified as *static loads* if they do not cause any appreciable acceleration or vibration of the structure or structural elements and as *dynamic loads* if they do. Small accelerations are often taken into account by increasing the specified static loads to account for the increases in stress due to such accelerations and vibrations. Larger accelerations, such as those which might occur in highway bridges, crane rails, or elevator supports are accounted for by multiplying the effect of the live load by an *impact factor*. Alternatively, dynamic analyses may be used.

Three levels of live load or wind load may be of importance. The load used in calculations involving the ultimate limit states should represent the maximum load on the structure in its lifetime. Wherever possible, therefore, the specified live, snow, and wind loadings should represent the mean value of the corresponding maximum lifetime load. A *companion action load* is the portion of a variable load that is present on a structure when some other variable load is at its maximum. In checking the serviceability limit states, it may be desirable to use a *frequent* live load, which is some fraction of the mean maximum lifetime load (generally, 50 to 60 percent); for estimating sustained load deflections, it may be desirable to consider a *sustained* or *quasi-permanent* live load, which is generally between 20 and 30 percent of the specified live load. This differentiation is not made in the ACI Code, which assumes that the entire specified load will be the load present in service. As a result, service-load deflections and creep deflections of slender columns tend to be overestimated.

Loading Specifications

In the future, it is expected that cities in the United States will base their building codes on the *International Building Code* [2-9]. The loadings specified in this code are based in part on the loads recommended in *ASCE Minimum Design Loads for Buildings and Other Structures (ASCE 7-98)*, formerly ANSI A 58.1.

It should be emphasized that the basic structural design equation (2-2) implies that if the fractiles of the distributions of the the loads S_1, S_2, and so on were to differ from code to code, then the load factors α_1, α_2, and so on must also differ.

In the following sections, the types of loadings presented in ASCE 7-98 will be reviewed very briefly. This review is intended to describe the characteristics of the various loads. For specific values, the reader should consult the code in effect in his or her own locality.

Dead Loads

The *dead load* on a structural element is the weight of the member itself, plus the weights of all materials permanently incorporated into the structure and supported by the member in question. This includes the weights of permanent partitions or walls, the weights of plumbing stacks, electrical feeders, permanent mechanical equipment, and so on. Tables of dead loads are given in ASCE 7-98 [2-2].

In the design of a reinforced concrete member, it is necessary initially to estimate the weight of the member. Methods of making this estimate are given in Chapters 4 and 10. Once the member size has been computed, its weight is calculated by multiplying the volume by the density of concrete, taken as 145 lb/ft^3 for plain concrete and 150 lb/ft^3 for reinforced concrete. (5 lb/ft^3 is added to account for reinforcement.) For lightweight concrete members, the density of the concrete must be determined from trial batches or as specified by the producer. In heavily reinforced members, the density of the reinforced concrete may exceed 150 lb/ft^3 when the weight of stirrups and longitudinal steel are included. In extreme cases, design should be based on an estimate of the density for the members in question.

In working with SI units (metric units), the weight of a member is calculated by multiplying the volume by the mass density of concrete and the gravitational constant, 9.81 N/kg. In this calculation, it is customary to take the mass density of normal-density concrete containing an average amount of reinforcement (roughly, 2 percent by volume) as 2450 kg/m^3, made up of 2300 kg/m^3 for the concrete and 150 kg/m^3 for the reinforcement. The weight of a cubic meter of reinforced concrete is thus $(1 \text{ m}^3 \times 2450 \text{ kg/m}^3 \times 9.81 \text{ N/kg})/1000 = 24.0 \text{ kN}$, and its weight density is 24 kN/m^3.

The dead load referred to in ACI Eqs. (9-1) to (9-7) is the load computed from the dimensions shown on drawings and the assumed densities. It is therefore close to the mean value of this load. Actual dead loads will vary from the calculated values, because the actual dimensions and densities may differ from those used in the calculations. Sometimes the materials for the roof, partitions, or walls are chosen on the basis of a separate call for tenders, and their actual weights may be unknown at the time of the design. Tabulated densities of materials frequently tend to underestimate the actual dead loads of the material in place in a structure.

Some types of dead load tend to be highly uncertain. These include pavement on bridges, which may be paved several times over a period of time, or where a greater thickness of pavement may be applied to correct sag or alignment problems. Similarly, earth fill over an underground structure may be up to several feet thicker than assumed and may or may not be saturated with water. In the construction of thin curved-shell roofs or other lightweight roofs, the concrete thickness may exceed the design values and the roofing may be heavier than assumed, leading to overloads.

If dead-load moments, forces, or stresses tend to counteract those due to live loads or wind loads, the designer should carefully examine whether the counteracting dead load will always exist. Thus, dead loads due to soil or machinery may be applied late in the construction process and may not be applied evenly to all parts of the structure at the same time, leading to a critical set of moments, forces, or stresses under partial loads.

It is generally not necessary to checkerboard the self-weight of the structure by using dead-load factors of $\alpha_D = 0.9$ and 1.2 or 1.4 in successive spans, because the structural dead loads in successive spans of a beam tend to be highly correlated. On the other hand, it may be necessary to checkerboard the superimposed dead load by using load factors of $\alpha_D = 0$ or 1.2 in cases where counteracting dead load is absent at some stages of construction or use.

Live Loads Due to Use and Occupancy

Most building codes contain a table of design or specified live loads. To simplify the calculations, these are expressed as uniform loads on the floor area. In general, a building live load consists of a sustained portion due to day-to-day use (see Fig. 2-4) and a variable portion

TABLE 2-1 Typical Live Loads Specified in ASCE 7-98

	Uniform, psf	Concentration, lb
Apartment buildings		
Private rooms and corridors serving them	40	
Public rooms and corridors serving them	100	
Office buildings		
Lobbies and first-floor corridors	100	2000
Offices	50	2000
Corridors above first floor	80	2000
File and computer rooms shall be designed for heavier loads based on anticipated occupancy		
Schools		
Classrooms	40	1000
Corridors above first floor	80	1000
First-floor corridors	100	1000
Stairs and exitways	100	
Storage warehouses		
Light	125	
Heavy	250	
Stores		
Retail		
Ground floor	100	1000
Upper floors	75	1000
Wholesale, all floors	125	1000

Source: Based on *Minimum Design Loads for Buildings and Other Structures*, ASCE Standard ASCE 7-98, with the permission of the publisher, the American Society of Civil Engineers.

generated by unusual events. The sustained portion changes a number of times during the life of the building—when tenants change, when the offices are rearranged, and so on. Occasionally, high concentrations of live loading occur during periods when adjacent spaces are remodeled, when office parties are held, or when material is stored temporarily. The loading given in building codes is intended to represent the maximum sum of these loads that will occur on a small area during the life of the building. Typical specified live loads are given in Table 2-1.

In buildings where nonpermanent partitions might be erected or rearranged during the life of the building, allowance should be made for the weight of these partitions. ASCE 7-98 specifies that provision for partition weight should be made, regardless of whether partitions are shown on the plans, unless the specified live load exceeds 80 psf. It is customary to represent the partition weight with a uniform load of 20 psf or a uniform load computed from the actual or anticipated weights of the partitions placed in any probable position. ASCE 7-98 considers this a live load, because it may or may not be present in a given case.

As the loaded area increases, the average maximum lifetime load decreases because, although it is quite possible to have a heavy load on a small area, it is unlikely that this would occur in a large area (Fig. 2-4). This is taken into account by multiplying the specified live loads by a *live-load reduction factor*.

In ASCE 7-98, this factor is based on the *influence area*, A_I, for the member being designed. The concept of influence lines and influence areas is explained in Section 10-3. To figure out the influence area of a given member, one imagines that the member in question is raised by a unit amount, say, 1 in. as shown in Fig. 2-10. The portion of the loaded area that is raised when this is done is called the *influence area*, A_I, since loads acting anywhere in this area will have a significant impact on the load effects in the member in question. This concept is illustrated in Fig. 2-10 for an interior floor beam and an edge column.

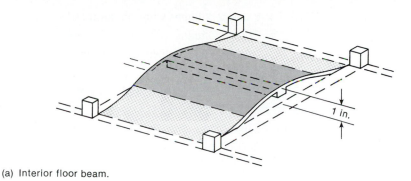

(a) Interior floor beam.

(b) Edge column.

Fig. 2-10
Influence areas.

In contrast, the *tributary area*, A_T, extends out from the beam or column to the lines of zero shear in the floor around the member under consideration. For the beam in Fig. 2-10a, the limits on A_T are given by the dashed lines halfway to the next beam on each side. The tributary areas are shown in a darker shading in Figs. 2-10a and b. An examination of Fig. 2-10a shows that A_T is half of A_I for an interior beam. For the column in Fig. 2-10b, it is four times. Since two-way slab design is based on the total moments in one slab panel, the influence area for such a slab is defined by ASCE 7-98 as the panel area.

ASCE 7-95 allowed the use of reduced live loads, L, in the design of members, based on the influence area A_I. However, the influence-area concept is not widely known compared with that of the tributary area, A_T. In ASCE 7-98, the influence area is given as $A_I = K_{LL}A_T$, where A_T is the tributary area of the member being designed and K_{LL} is the ratio A_I/A_T. The reduced live load, L, is given by

$$L = L_o \left[0.25 + \frac{15}{\sqrt{K_{LL}A_T}} \right]$$

(2-19)

(ASCE 7 Eq. 4-1)

where L_o is the unreduced live load. Values of K_{LL} are given as follows:

Interior columns and exterior columns
without cantilever slabs $K_{LL} = 4$

Exterior columns with cantilever slabs	$K_{LL} = 3$
Corner columns with cantilever slabs	$K_{LL} = 2$
Interior beams and edge beams without cantilever slabs	$K_{LL} = 2$
All other members, including panels in two-way slabs	$K_{LL} = 1$

The use of live-load reduction factors is illustrated in step 1 of Example 10-2 and in step 8 of Example 19-4.

The live-load reduction applies only to live loads due to use and occupancy (not for snow, etc.). No reduction is made for areas used as places of public assembly, for garages, or for roofs. In ASCE 7-98, the reduced live load cannot be less than 50 percent of the unreduced live load for columns supporting one floor or for flexural members, and no less than 40 percent for other members.

For live loads exceeding 100 psf, no reduction is allowed by ASCE 7-98, except that the design live load on columns supporting more than one floor can be reduced by 20 percent.

The reduced uniform live loads are then applied to those spans or parts of spans that will give the maximum shears, moments, and so on, at each critical section. This approach is illustrated in Chapter 10.

The ASCE document requires that office and garage floors and sidewalks be designed to safely support either the reduced uniform design loads or a concentrated load of 1000 to 8000 lb (depending on occupancy), spread over an area of 30 in. by 30 in., whichever causes the worst effect. The concentrated loads are intended to represent heavy items such as office safes, pianos, car wheels, and so on.

In checking the concentrated load capacity, it generally is necessary to assume an effective width of floor to carry the load to the supports. For one-way floors, this is usually the width of the concentrated load reaction plus one slab effective depth on each side of the load. For two-way slabs, Section 13-6 shows that a concentrated load applied at various points in the slab gives maximum moments (at midspan and near the support columns) that are similar in magnitude to those computed for a complete panel loaded with a uniform load. In many cases this makes it unnecessary to check the concentrated load effects on two-way slabs.

The live loads are assumed to be large enough to account for the impact effects of normal use and traffic. Special impact factors are given in the loading specifications for supports of elevator machinery, large reciprocating or rotating machines, and cranes.

Classification of Buildings for Wind, Snow, and Earthquake Loads

The ASCE 7-98 requirements for design for wind, snow, and earthquake become progressively more restrictive as the level of risk to human life in the event of a collapse increases. These are referred to as *use categories*:

I.   Buildings and other structures that represent a low hazard to human life in the event of failure, such as agricultural facilities.

II.   Buildings and other structures that do not fall into categories I, III, or IV.

III.   Buildings or other structures that represent a substantial hazard to human life in the event of failure, such as assembly occupancies, schools, colleges, jails, and buildings containing significant quantities of toxic or explosive substances.

IV.   Buildings and other structures designated as essential facilities, such as hospitals, fire and police stations, communication centers, and power-generating stations and facilities.

Snow Loads, *S*

Snow accumulation on roofs is influenced by climatic factors, roof geometry, and the exposure of the roof to the wind. Unbalanced snow loads due to drifting or sliding of snow or uneven removal of snow by workers are very common. Large accumulations of snow will often occur adjacent to parapets or other points where roof heights change. ASCE 7-98 gives detailed rules for calculating snow loads to account for the effects of snow drifts. It is necessary to design for either a uniform or an unbalanced snow load, whichever gives the worst effect.

Snow load is considered to be a live load when applying the load factors in ACI Appendix C. A live-load reduction factor is not applied to snow loads.

Roof Loads, L_r, and Rain Loads, *R*

In addition to snow loads, roofs should be designed for certain minimum loads (L_r) to account for workers or construction materials on the roof during erection or when repairs are made. Consideration must also be given to loads due to rainwater, *R*. Since roof drains are rarely inspected to remove leaves or other debris, ASCE 7-98 requires that roofs be able to support the load of all rainwater that could accumulate on a particular portion of a roof if the primary roof drains were blocked. Frequently, controlled-flow roof drains are used. These slow the flow of rainwater off a roof. This reduces plumbing and storm sewage costs but adds to the costs of the roof structure.

If the design snow load is small and the roof span is longer than about 25 ft, rainwater will tend to form ponds in the areas of maximum deflection. The weight of the water in these regions will cause an increase in the deflections, allowing more water to collect, and so on. If the roof is not sufficiently stiff, a *ponding failure* will occur when the weight of ponded water reaches the capacity of the roof members [2-13].

Construction Loads

During the construction of concrete buildings, the weight of the fresh concrete is supported by formwork, which frequently rests on floors lower down in the structure. In addition, construction materials are often piled on floors or roofs during construction. ACI Section 6.2.2 states the following:

> No construction loads exceeding the combination of superimposed dead load plus specified live load shall be supported on any unshored portion of the structure under construction, unless analysis indicates adequate strength to support such additional loads.

Wind Loads

The pressure exerted by the wind is related to the square of its velocity. Due to the roughness of the earth's surface, the wind velocity at any particular instant consists of an average velocity plus superimposed turbulence, referred to as *gusts*. As a result, a structure subjected to wind loads assumes an average deflected position due to the average velocity pressure and vibrates from this position in response to the gust pressure. In addition, there will generally be deflections transverse to the wind (due to vortex shedding) as the wind passes the building. The vibrations due to the wind gusts are a function of (1) the relationship between the natural energy of the wind gusts and the energy necessary to displace the building, (2) the relationship between the gust frequencies and the natural frequency of the building, and (3) the damping of the building [2-13].

Three procedures are specified in ASCE 7-98 for the calculation of wind pressures on buildings: the *simplified procedure*, limited in application to buildings of five stories or fewer; the *analytical procedure*, limited to regular buildings that are not subject to across-wind loading, vortex shedding, or channeling of the wind due to upwind obstructions; and the *wind tunnel procedure*, used for complex buildings. We shall consider the analytical procedure. Variations of this method apply to design of the main wind-force-resisting systems of buildings and to the design of components and cladding. The wind loads on a 10-story building are computed in Example 19-4.

In the analytical procedure, the wind pressure on the main wind-force-resisting system is

$$p = qGC_p - q_i(GC_{pi}) \qquad \text{(2-20)}$$

$$\text{(ASCE 7 Eq. 6-15)}$$

where $q = q_z$ is the velocity pressure evaluated at height z above the ground on the windward wall, $q = q_h$ is the pressure on the roof, leeward walls, and sidewalls, evaluated at the mean roof height, h, and $q_i = q_h$ is the internal pressure or suction on the interior of the walls and roof of the building, also evaluated at the mean roof height.

The total wind pressure p, on a surface is the sum of the external pressure on the windward wall which is given by the first term on the right-hand side of (2-20) plus the second term, p_i, which accounts for the internal pressure. The internal pressure, p_i, is the same on all internal surfaces at any given time. Thus, the internal pressure or suction on the inside of the windward wall is equal but opposite in direction to the internal pressure or suction on the inside of the leeward wall. As a result, the interior wind forces on opposite walls cancel out in many cases, leaving only the external pressure to be resisted by the main wind-force-resisting system. The terms in (2-20) are defined as:

1. **Design pressure, p.** The *design pressure* is an equivalent static pressure or suction in psf assumed to act perpendicular to the surface in question. On some surfaces, it varies over the height; on others, it is assumed to be constant.

2. **Wind Velocity pressure, q.** The *wind velocity pressure*, q psf, is the pressure exerted by the wind on a flat plate suspended in the wind stream. It is calculated as

$$q_z = 0.00256 K_z K_{zt} K_d V^2 I \qquad \text{(2-21)}$$
$$\text{(ASCE 7 Eq. 6-13)}$$

where

$V =$ basic 3-sec gust wind speed in miles per hour at a height of 33 ft (10 m) above the ground in open terrain

$K_z =$ velocity pressure exposure coefficient, which increases with height above the surface and reflects the roughness of the surface terrain

$K_{zt} =$ accounts for increases in wind speed as if passes over hills.

$K_d =$ directionality factor equal to 0.85 for rectangular buildings and 0.90 to 0.95 for circular tanks and the like

$I =$ importance factor, which is a function of the building *use category* discussed earlier.

The constant 0.00256 reflects the mass density of the air and accounts for the mixture of units in (2-20).

Prior to 1995, V was based on the "fastest mile wind," which had a chance of 1 in 50 of being exceeded in any one year. This was the velocity corresponding to the time it took for a 1-mile-long piece of air to pass the wind gauge. In ASCE 7-95, the definition of V was changed to the velocity of a 3-sec gust that has a 1-in-50 chance of being exceeded in any one year. The 1995 definition gives a much higher value of V than the earlier definition. However, since the gust effect has largely been accounted for by using the 3-sec gust speed, the gust factor, G, in (2-20) is close to 0.85. The overall result causes relatively little change in the design pressure, p.

Maps and tables of V are given in the ASCE standard. Special attention must be given to mountainous terrain, gorges, and promontories subject to unusual wind conditions and regions subject to tornadoes. The importance factor, I, ranges from 0.87 for building-use category I, to 1.0 for normal buildings (building-use category II), to 1.15 for building-use categories III and IV. These values correspond to mean recurrence intervals of 1 in 25 years for use category I buildings, 1 in 50 for use category II buildings, and 1 in 100 years for use category III and IV buildings.

At any location, the mean wind velocity is affected by the roughness of the terrain upwind from the structure in question. At a height of 700 to 1500 ft, the wind reaches a steady velocity, as shown by the vertical lines in the plots of K_z in Fig. 2-11. Below this height, the velocity decreases and the turbulence, or gustiness, increases as one approaches the surface. These effects are greater in urban areas than in rural areas, due to the greater surface roughness in built-up areas. The factor K_z in (2-21) relates the wind pressure at any elevation z feet to that at 33 ft (10 m) above the surface for open exposure. ASCE 7-98 gives tables and equations for K_z as a function of the type of exposure (urban, country, etc.) and the height above the surface.

Another change in ASCE 7-98 was the recognition that the direction of the fastest wind did not necessarily coincide with an axis of the building being designed. A *wind directionality factor*, K_d, equal to 0.85 was included in (2-21), provided that design was based on a wind-load factor of 1.6. If design was based on the wind-load factor of 1.3, the directionality factor remains equal to 1.0.

The velocity pressure on the windward wall varies with height, z, measured from the ground to the height in question. For side walls, leeward wall, and roof surfaces, q_h is a suction (negative pressure) evaluated by using h equal to the average height of the roof.

3.   Gust response factor, G. The gust factor, G, in (2-20) relates the dynamic properties of the wind and the structure. For flexible buildings, it is calculated. For most buildings that tend to be stiff, it is taken to be equal to 0.85.

4.   External pressure coefficient, C_p. When wind blows past a structure, it exerts a positive pressure on the windward wall and a negative pressure (suction) on the leeward wall, side walls, and roof as shown in Fig. 2-12. The overall pressures to be used in the

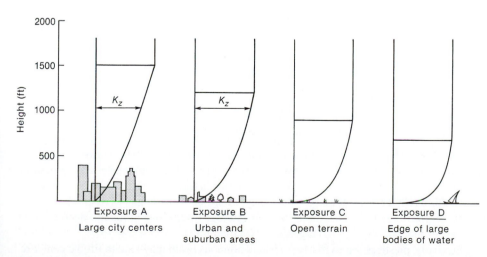

Fig. 2-11
Profiles of velocity pressure exposure coefficient, K_z, for differing terrain.

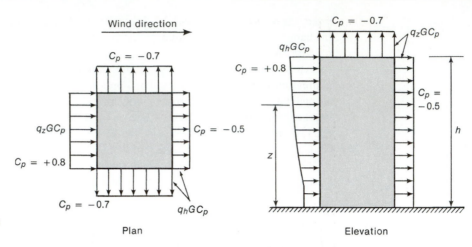

Fig. 2-12
Wind pressures and suctions
on a building.

design of a structural frame are computed via (2-20), where C_p is the sum or difference in the pressure coefficients for the windward and leeward walls. Thus, $C_p = +0.80$ (pressure) on the left-hand (windward) wall in Fig. 2-12 and $C_p = -0.50$ (suction) on the right-hand (leeward) wall add together to produce the load on the frame because they have the same direction. Values of the pressure coefficients are given in the ASCE 7. Typical values are shown in Fig. 2-12 for a building having the shape and proportions shown. For a rectangular building with the wind on the narrow side, C_p for the leeward wall varies between -0.5 and -0.2.

Earthquake Loads

Earthquake loads and design for earthquakes are discussed in Chapter 20.

Self-Equilibrating Loadings

Most loads result from things like the weight of the structure or externally applied loads such as live load or wind load. These loads cause internal forces and moments that are in equilibrium with the external loads. Many structures are subjected to *imposed or restrained deformations*, which are independent of the applied loads. Examples include differential settlements, nonlinear thermal stresses in bridge decks, restrained shrinkage, and prestressing of indeterminate structures. These deformations cause a set of internal forces or moments that are in equilibrium with themselves, as shown in Fig. 2-9. ASCE 7 refers to these as *self-straining forces*. Because these loading cases do not involve applied loads, the magnitude of the internal forces and moments results from

(a) the magnitude of the imposed deformation, and

(b) the resistance of the structure to the deformation (a function of the stiffness of the structure at the time that the deformation occurs).

Consider a two-span beam in which the central support settles relative to the line joining the end supports. The structure resists the differential settlement, setting up internal forces and moments. If the beam is uncracked when it is forced through the differential settlement, the internal forces are larger than they would be if the beam were cracked. If the beam undergoes creep, the magnitude of the internal forces and moments decreases, as shown experimentally by Ghali, Dilger, and Neville [2-14].

Similarly, prestress forces in a two-span continuous beam may tend to lift the center reaction off its support, changing the reactions. This, in turn, causes internal forces and so-called *secondary moments* that are in equilibrium with the change in the reactions. The magnitude of these forces and moments is larger in an uncracked beam than in a cracked beam. They may be partially dissipated by creep.

ACI Section 9.2.1 uses the symbol T to refer to the forces and moments resulting from imposed or restrained deformations and assigns them a load factor of 1.2. This section states,

> Estimations of differential settlement, creep, shrinkage, expansion of shrinkage-compensating concrete, or temperature change shall be based on a realistic assessment of such effects occurring in service.

The commentary for this section states,

> The term realistic assessment is used to indicate that the most probable values rather than the upper bound values of the variable should be used.

Designers often account for imposed deformations by detailing such things as relief joints or other means of dissipating the effects of the deformations.

Other Loads

ASCE 7-98 also gives soil loads on basement walls, loads due to floods, and loads due to ice accretion.

2-9   DESIGN FOR ECONOMY

A major aim of structural design is economy. The overall cost of a building project is strongly affected by both the cost of the structure and the financing charges, which are a function of the rate of construction.

In a cast-in-place building, the costs of the floor and roof systems make up roughly 90 percent of the total structural costs. The cost of a floor system is divided between the costs of building and stripping the *forms*; providing, bending, and placing the *reinforcement*; and providing, placing, and finishing the *concrete*. Table 2-2 lists typical *relative* costs per square foot for several floor-framing systems. The floor systems are listed in order of increasing complexity of form construction. Two things are noticeable: (1) the amount of material goes up (and, as a result, the cost increases) as the column spacing increases, and (2) the cost of the forms is the biggest single item in the total costs, accounting for 40 to 60 percent of the total. The major differences between the systems then come from increased amounts of materials as spans increase and from increased costs of forming as the complexity of the forms increases. In the case of the one-way joist floor, a portion of the form cost was for rental of prefabricated forms.

The cost data given in Table 2-2 suggest that floor-forming costs should be a major consideration in the layout of the structural system. Formwork costs can be reduced by reusing the forms from area to area or floor to floor. Beam, slab, and column sizes should be chosen to allow the maximum reuse of the forms. It is generally uneconomical to try to save concrete and steel by meticulously calculating the size of every beam and column to fit the loads exactly, because, although this could save cents in materials, it will cost dollars in forming costs.

Furthermore, changing section sizes often leads to increased design complexity, which in turn leads to a greater chance of design error and a greater chance of construction error. A simple design that achieves all the critical requirements saves design and construction time and generally gives an economical structure.

TABLE 2-2 Breakdown of Floor-Construction Costs[a]

	20 ft × 20 ft Panel		25 ft × 25 ft Panel	
	Relative Cost/ft^2	Percent	Relative Cost/ft^2	Percent
Flat plate				
Forms	0.43	43	0.44	38
Concrete	0.32	32	0.38	33
Reinforcement	0.25	25	0.33	29
	1.00		1.15	
One-way joists[b]				
Forms	0.62	58	0.61	54
Concrete	0.27	25	0.29	26
Reinforcement	0.19	17	0.23	20
	1.08		1.13	
One-way slab and beams				
Forms	0.58	48	0.62	45
Concrete	0.29	25	0.33	24
Reinforcement	0.32	17	0.42	31
	1.19		1.37	

[a]The costs of the various floor systems are expressed relative to the cost of a 20 ft × 20 ft panel flat plate. These cost ratios are based on a particular set of designs [2-15] for a common live load and a given set of costs [2-16].

[b]The panel sizes considered are 20 ft × 20 ft and 20 ft × 30 ft.

Wherever possible, haunched beams should be avoided. If practical, beams should be the same width as or a little wider than the columns into which they frame, to simplify the formwork for column-beam joints. Deep spandrel beams make it difficult to move forms from floor to floor and should be avoided if possible. In one-way joist floors, it is advisable to use the same depth of joist throughout rather than switching from deep joists for long spans to shallow joists for short spans. The saving in concrete due to such a change is negligible and generally is more than offset by the extra labor of materials required, plus the need to rent and schedule two different sizes of joist forms. In joist floors, the beams should be the same depth as the joists.

If possible, a few standard column sizes should be chosen, with each column size maintained for three or four stories or the entire building. The amount of reinforcement and the concrete strength used can vary as the load varies. Columns should be aligned on a regular grid, if possible, and constant story heights should be maintained.

Economies are also possible in reinforcement placing. Complex or congested reinforcement will lead to higher per-pound charges for placement of the bars. It is frequently best, therefore, to design columns for 1.5 to 2 percent reinforcement and beams for no more than one-half to two-thirds of the maximum allowable reinforcement ratios. Grade-60 reinforcement is almost universally used for column reinforcement and flexural reinforcement in beams. In slabs where reinforcement quantities are controlled by minimum reinforcement ratios, there may be a slight advantage in using Grade-40 reinforcement. The same may be true for stirrups in beams if the stirrup spacings tend to be governed by the maximum spacings. However, before specifying Grade-40 steel, the designer should check whether it is available locally in the sizes needed.

Since the flexural strength of a floor is relatively insensitive to concrete strength, there is no major advantage in using high-strength concrete in floor systems. An exception to this would be a flat-plate system, where the shear capacity may govern the thickness. On the other hand, column strengths are directly related to concrete strength, and the most economical columns tend to result from the use of high-strength concrete.

2-10   HANDBOOKS AND DESIGN AIDS

Since a great many repetitive computations are necessary to proportion reinforced concrete members, handbooks containing tables or graphs of the more common quantities are available from several sources. The American Concrete Institute publishes its *Design Handbook* in several volumes [2-17], [2-18], [2-19], and the Concrete Reinforcing Steel Institute publishes the *CRSI Handbook* [2-20].

Once a design has been completed, it is necessary for the details to be communicated to the reinforcing-bar suppliers and placers and to the construction crew. The *ACI Detailing Manual* [2-21] presents drafting standards and is an excellent guide to field practice. ACI Standard 301, *Specifications for Structural Concrete for Buildings,* [2-22] indicates the items to be included in construction specifications. Finally, the ACI publication *Formwork for Concrete* [2-23] gives guidance for form design.

The ACI *Manual of Concrete Practice* [2-24] collects together most of the ACI committee reports on concrete and structural concrete and is an invaluable reference on all aspects of concrete technology. It is published annually in hard copy and on a CD-ROM. The current edition includes more than 150 ACI committee reports.

2-11   CUSTOMARY DIMENSIONS AND CONSTRUCTION TOLERANCES

The selection of dimensions for reinforced concrete members is based on the size required for strength and for other aspects arising from construction considerations. Beam widths and depths and column sizes are generally varied in increments of 1, 2, 3, or 4 in., slab thicknesses in $\frac{1}{2}$-in. increments.

The actual as-built dimensions will differ slightly from those shown on the drawings, due to construction inaccuracies. ACI Standard 347 [2-23] on formwork gives the accepted tolerances on cross-sectional dimensions of concrete columns and beams as $\pm\frac{1}{2}$ in. and on the thickness of slabs and walls as in $\pm\frac{1}{4}$ in. For footings, they recommend tolerances of $+2$ in. and $-\frac{1}{2}$ in. on plan dimensions and -5 percent of the specified thickness.

The lengths of reinforcing bars are generally given in 2-in. increments. The tolerances for reinforcement placing concern the variation in the depth, d, of beams, the minimum reinforcement cover, and the longitudinal location of bends and ends of bars. These are specified in ACI Sections 7.5.2.1 and 7.5.2.2. ACI Committee 117 has published a comprehensive list of tolerances for concrete construction and materials [2-25].

2-12   ACCURACY OF CALCULATIONS

Structural loads, with the exception of dead loads or fluid loads in a tank, are rarely known to more than two significant figures. Thus, although calculations should include three significant figures, it is seldom necessary to record more than this. In this book, three significant figures are used, except that four are used when the number starts with one. Care should be taken in calculations where loads, forces, or stresses offset each other, because the final answer may be the difference of two similar large numbers.

Most mistakes in structural design arise from three sources: errors in looking up or writing down numbers, errors due to unit conversions, and failure to understand fully the statics or behavior of the structure being analyzed and designed. The last type of mistake is especially serious, since failure to consider a particular type of loading or the use of the wrong statical model may lead to serious maintenance problems or collapse. For this reason, designers are urged to use the limit-states design process to consider all possible modes of failure and to use free-body diagrams to study the equilibrium of parts or all of the structure.

2-13 "SHALL BE PERMITTED"

Throughout the ACI Code, the word "may" has been replaced with the phrase "shall be permitted" or something equivalent, at the request of the three model codes. The phrase "shall be permitted" implies that the designer is permitted to use the alternative material or design method mentioned in the section in question.

2-14 INSPECTION

The quality of construction depends in part on the workmanship during construction. Inspection is necessary to confirm that the construction is in accordance with the project drawings and specifications. ACI Section 1.3.1 requires that concrete construction be inspected throughout the various work stages by, or under the supervision of, a registered design professional, or by a qualified inspector. More stringent requirements are given in ACI Section 1.3.5 for inspection of moment-resisting frames in seismic regions.

The ACI and other organizations certify the qualifications of construction inspectors. Inspection reports should be distributed to the owner, the designer, the contractor, and the building official. The inspecting engineer or architect preserves these reports for at least two years after the completion of the project.

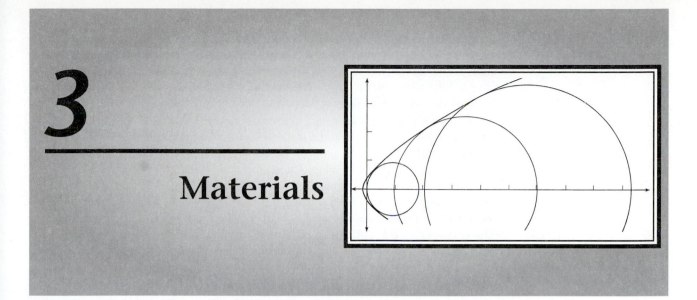

3
Materials

3-1 CONCRETE

Concrete is a composite material composed of aggregate, generally sand and gravel, chemically bound together by hydrated portland *cement*. The aggregate generally is graded in size from sand to gravel, with the maximum gravel size in structural concrete commonly being $\frac{3}{4}$ in., although $\frac{3}{8}$-in. or $1\frac{1}{2}$-in. aggregate may be used.

3-2 BEHAVIOR OF CONCRETE FAILING IN COMPRESSION

Mechanism of Failure in Concrete Loaded in Compression

Concrete is a mixture of cement paste and aggregate, each of which has an essentially linear and brittle stress–strain relationship in compression. Brittle materials tend to develop tensile fractures perpendicular to the direction of the largest tensile strain. Thus, when concrete is subjected to uniaxial compressive loading, cracks tend to develop parallel to the maximum compressive stress. In a cylinder test, the friction between the heads of the testing machine and the ends of the cylinder prevents lateral expansion of the ends of the cylinder and in doing so restrains the vertical cracking in those regions. This strengthens conical regions at each end of the cylinder. The vertical cracks that occur at midheight of the cylinder do not enter these conical regions and the failure surface appears to consist of two cones.

Although concrete is made up of essentially elastic, brittle materials, its stress–strain curve is nonlinear and appears to be somewhat ductile. This can be explained by the gradual development of *microcracking* within the concrete and the resulting redistribution of stress from element to element in the concrete [3-1]. Microcracks are internal cracks $\frac{1}{8}$ to $\frac{1}{2}$ in. in length. Microcracks that occur along the interface between paste and aggregate are called *bond cracks*; those that cross the mortar between pieces of aggregate are known as *mortar cracks*.

There are four major stages in the development of microcracking and failure in concrete subjected to uniaxial compressive loading:

1. Shrinkage of the paste occurs during hydration and this volume change of the concrete is restrained by the aggregate. The resulting tensile stresses lead to *no-load bond cracks*, before the concrete is loaded. These cracks have little effect on the concrete at low loads, and the stress–strain curve remains linear up to 30 percent of the compressive strength of the concrete, as shown by the solid line in Fig. 3-1.

2. When concrete is subjected to stresses greater than 30 to 40 percent of its compressive strength, the stresses on the inclined surfaces of the aggregate particles will exceed the tensile and shear strengths of the paste–aggregate interfaces, and new cracks, known as *bond cracks*, will develop. These cracks are stable; they propagate only if the load is increased. Once such a crack has formed, however, any additional load that would have been transferred across the cracked interface is redistributed to the remaining unbroken interfaces and to the mortar. This redistribution of load causes a gradual bending of the stress–strain curve for stresses above 40 percent of the short-time strength. The loss of bond leads to a wedging action, causing transverse tensions above and below the aggregates.

3. As the load is increased beyond 50 or 60 percent of ultimate, localized *mortar cracks* develop between bond cracks. These cracks develop parallel to the compressive loading and are due to the transverse tensile strains. During this stage, there is stable crack propagation; cracking increases with increasing load but does not increase under constant load. The onset of this stage of loading is called the *discontinuity limit* [3-2].

4. At 75 to 80 percent of the ultimate load, the number of mortar cracks begins to increase, and a continuous pattern of microcracks begins to form. As a result, there are fewer undamaged portions to carry the load, and the stress vs. longitudinal-strain curve becomes even more markedly nonlinear. The onset of this stage of cracking is called the *critical stress* [3-3].

If the lateral strains, ϵ_3, are plotted against the longitudinal compressive stress, the dashed curve in Fig. 3-1 results. The lateral strains are tensile and initially increase, as is expected from the poisson's effect. As microcracking becomes more extensive, these cracks contribute to the apparent lateral strains. As the load exceeds 75 to 80 percent of the ultimate

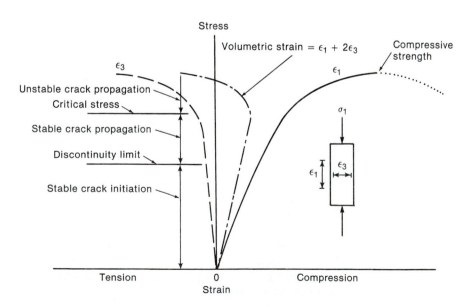

Fig. 3-1
Stress–strain curves for concrete loaded in uniaxial compression. (From [3-2].)

compressive strength, the cracks and lateral strains increase rapidly, and the volumetric strain (relative increase in volume), ϵ_v, begins to increase, as shown by the broken line in Fig. 3-1.

The critical stress is significant for several reasons. The ensuing increase in volume causes an outward pressure on ties, spirals, or other confining reinforcement, and these in turn act to restrain the lateral expansion of the concrete, thus delaying its disintegration.

Equally important is the fact that the structure of the concrete tends to become unstable at loads greater than the critical load. Under stresses greater than about 75 percent of the short-time strength, the strains increase more and more rapidly until failure occurs. Figure 3-2a shows the stress–strain–time response of concrete loaded rapidly to various fractions of its short-time strength, with this load being sustained for a long period of time or until failure occurred. As shown in Fig. 3-2b, concrete subjected to a sustained axial load greater than the critical load will eventually fail under that load. The critical stress is between 0.75 and $0.80 f_c'$.

Under cyclic compressive loads, axially loaded concrete has a *shake-down* limit approximately equal to the point of onset of significant mortar cracking at the critical stress. Cyclic axial stresses higher than the critical stress will eventually cause failure.

As mortar cracking extends through the concrete, less and less of the structure remains. Eventually, the load-carrying capacity of the uncracked portions of the concrete reaches a maximum value referred to as the *compressive strength* (Fig. 3-1). Further straining is accompanied by a drop in the stress that the concrete can resist, as shown by the dotted portion of the line for ϵ_1 in Fig. 3-1.

When concrete is subjected to compression with a strain gradient, as would occur in the compression zone of a beam, the effect of the unstable crack propagation stage shown in Fig. 3-1 is reduced because, as mortar cracking softens the highly strained concrete, the load is transferred to the stiffer, more stable concrete at points of lower strain nearer the neutral axis. In addition, continued straining and the associated mortar cracking of the highly stressed regions is prevented by the stable state of strain in the concrete closer to the neutral axis. As a result, the stable-crack-propagation stage extends almost up to the ultimate strength of the concrete.

Tests [3-5] suggest that there is no significant difference between the stress–strain curves of concrete loaded with or without a strain gradient up to the point of maximum stress. The presence of a strain gradient does appear to increase the maximum strains that can be attained in the member, however.

The dashed line in Fig. 3-2c represents the gain in short-time compressive strength with time. The dipping solid lines are the failure limit line from Fig. 3-2b plotted against a log time scale. These lines indicate that there is a permanent reduction in strength due to sustained high loads. For concrete loaded at a young age, the minimum strength is reached after a few hours. If the concrete does not fail at this time, it can sustain the load indefinitely. For concrete loaded at an advanced age, the decrease in strength due to sustained high loads may not be recovered.

The *CEB–FIP Model Code 1990* [3-6] gives equations for both the dashed curve and the solid curves in Fig. 3-2c. The dashed curve (short-time compressive strength with time) can also be represented by (3-5), presented later in this chapter.

Under uniaxial tensile loadings, small localized cracks are initiated at tensile-strain concentrations and these relieve these strain concentrations. This initial stage of loading results in an essentially linear stress–strain curve during the stage of stable crack initiation. Following a very brief interval of stable crack propagation, unstable crack propagation and fracture occur. The direction of cracking is perpendicular to the principal tensile stress and strain.

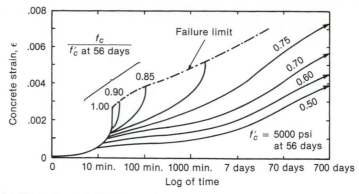

(a) Strain–time relationship.

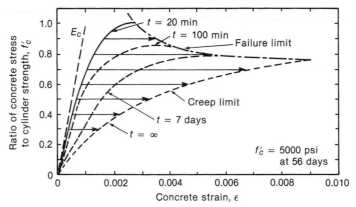

(b) Stress–strain–time relationship.

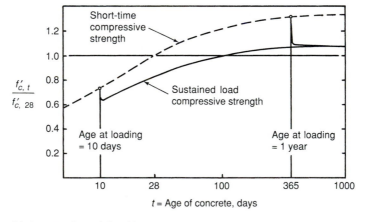

Fig. 3-2
Effect of sustained loads on
the behavior of concrete in
uniaxial compression.
(From [3-4].)

(c) Strength–time relationship.

3-3   COMPRESSIVE STRENGTH OF CONCRETE

Generally, the term *concrete strength* is taken to refer to the uniaxial compressive strength as measured by a compression test of a standard test cylinder, because this test is used to monitor the concrete strength for quality control or acceptance purposes. For convenience, other strength parameters, such as tensile or bond strength, are expressed relative to the compressive strength.

Standard Compressive-Strength Tests

The standard acceptance test for measuring the strength of concrete involves short-time compression tests on cylinders 6 in. in diameter by 12 in. high, made, cured, and tested in accordance with ASTM Standards C 31 and C 39.

The test cylinders for an acceptance test must be allowed to harden in their molds for 24 hours at the job site at 60 to 80°F, protected from loss of moisture and excessive heat, and then must be cured at 73°F in a moist room or immersed in water saturated with lime. The standard acceptance test is carried out when the concrete is 28 days old.

Field-cured test cylinders are frequently used to determine when the forms may be removed or when the structure may be used. These should be stored as near the location of that concrete in the structure as is practical and should be cured in a manner as close as possible to that used for the concrete in the structure.

The standard strength "test" is the average of the strengths of two cylinders from the same concrete batch tested at 28 days (or an earlier age, if specified). These are tested at a loading rate of about 35 psi per second, producing failure of the cylinder at $1\frac{1}{2}$ to 3 minutes. For high-strength concrete, acceptance tests are sometimes carried out at 56 or 90 days, because some high-strength concretes take longer than normal concretes to reach their design strength.

Traditionally, the compressive strength has been tested by using 6-in.-diameter by 12-in.-to cylinders. For high-strength concretes, the axial stiffness of some testing machines is close to the axial stiffness of the cylinders being tested. In such cases, the strain energy released by the machine at the onset of crushing of the test cylinder leads to a brittle failure of the cylinder. This can cause a decrease in the measured f_c'. This is alleviated by testing 4-by-8-in. cylinders, which have an axial stiffness less than a fifth of that of 6-by-12-in. cylinders. Aïtcin et al. [3-7] report tests on 8-in.-, 6-in.-, and 4-in.-diameter cylinders of concretes with nominal strengths of 5000, 13,000, and 17,500 psi; some of each strength were cured in air, or sealed, or cured in lime-water baths.

The water-cured specimens and the sealed specimens had approximately the same strengths at ages of 7, 28, and 91 days of curing. Aïtcin et al. [3-7] concluded the strengths of the 4-in.- and 6-in.-diameter cylinders were similar. This suggests that the strengths of 4-by-8-in. cylinders will be similar to the strengths of 6-by-12-in. cylinders, and that 4-in. cylinders can be used as control tests.

Other studies quoted in the 1993 report on high-strength-concrete by ACI Committee 363 [3-8] gave different conversion factors. The report concluded that 4-by-8-in. control cylinders give a higher strength and a larger coefficient of variation than 6-by-12-in. cylinders.

Statistical Variations in Concrete Strength

Concrete is a mixture of water, cement, aggregate, and air. Variations in the properties or proportions of these constituents, as well as variations in the transporting, placing, and compaction of the concrete, lead to variations in the strength of the finished concrete. In addition, discrepancies in the tests will lead to apparent differences in strength. The shaded area in Fig. 3-3 shows the distribution of the strengths in a sample of 176 concrete-strength tests.

The mean or average strength is 3940 psi, but one test has a strength as low as 2020 psi and one is as high as 6090 psi.

If more than about 30 tests are available, the strengths will generally approximate a normal distribution. The normal distribution curve, shown by the curved line in Fig. 3-3, is symmetrical about the mean value, $\bar{x}$, of the data. The dispersion of the data can be measured

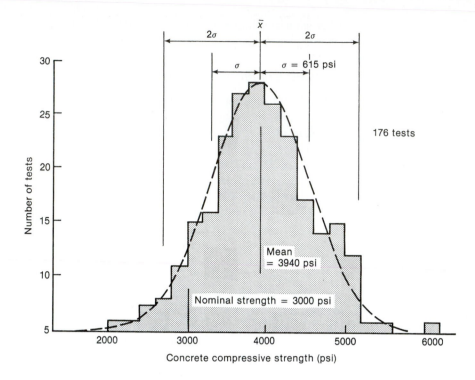

Fig. 3-3
Distribution of concrete
strengths.

by the *sample standard deviation, s,* which is the root-mean-square deviation of the strengths from their mean value:

$$s = \sqrt{\frac{(x_1 - \overline{x})^2 + (x_2 - \overline{x})^2 + (x_3 - \overline{x})^2 + \cdots + (x_n - \overline{x})^2}{n - 1}} \qquad (3\text{-}1)$$

The standard deviation divided by the mean value is called the *coefficient of variation, V:*

$$V = \frac{s}{\overline{x}} \qquad (3\text{-}2)$$

This makes it possible to express the degree of dispersion on a fractional or percentage basis rather than an absolute basis. The concrete test data in Fig. 3-3 have a standard deviation of 615 psi and a coefficient of variation of 615/3940 = 0.156, or 15.6 percent.

If the data correspond to a normal distribution, their distribution can be predicted from the properties of such a curve. Thus, 68.3 percent of the data will lie within 1 standard deviation above or below the mean. Alternatively, 15.9 percent of the data will have values less than $(\overline{x} - s)$. Similarly, for a normal distribution, 10 percent of the data, or 1 test in 10, will have values less than $\overline{x}(1 - aV)$, where $a = 1.282$. Values of a corresponding to other probabilities can be found in statistics texts.

Figure 3-4 shows the mean concrete strength, f_{cr}, required for various values of the coefficient of variation if no more than 1 test in 10 is to have a strength less than 3000 psi. As shown in this figure, as the coefficient of variation is reduced, the value of the mean strength, f_{cr}, required to satisfy this requirement can also be reduced.

Based on the experience of the U.S. Bureau of Reclamation on large projects, ACI Committee 214 [3-9] has defined various standards of control for moderate-strength concretes. A coefficient of variation of 15 percent represents *average control.* (See Fig. 3-4.) About one-tenth of the projects studied had coefficients of variation less than 10 percent, which was termed *excellent control,* and another tenth had values greater than

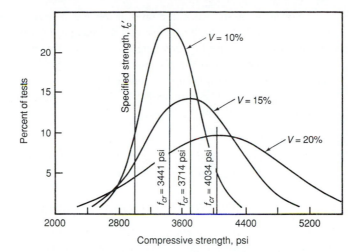

Fig. 3-4
Normal frequency curves for coefficients of variation of 10, 15, and 20 percent. (From [3-10].)

about 20 percent, which was termed *poor control*. For low-strength concrete, the coefficient of variation corresponding to average control has a value of $V = 0.15 f_c'$. Above a mean strength of about 4000 psi, the standard deviation tends to be independent of the mean strength, and for average control s is about 600 psi [3-9]. The test data plotted in Fig. 3-3 correspond to average control as defined by the Committee 214 definition of *average control*.

In 2001, Nowak and Szerszen [3-11] collected concrete control data from sources around the United States. The data are summarized in Table 3-1. The degree of concrete control was considerably better than that assumed by ACI Committee 214. In particular, the mean of the coefficients of variation reported by Nowak is much lower than the $V = 15$ percent that ACI 214 assumed to be representative of good control. In Table 3-1, the coefficients of variation range from 0.07 to 0.115, with one exception (lightweight concrete). This range of concrete variability appears to be representative of concrete produced in modern ready-mix plants, which represents the vast majority of concrete in North America. Nowak and Szerszen recommend a single value of $V = 0.10$. It would appear that this is a "property" of modern ready-mix concretes.

TABLE 3-1   Statistical Parameters for f_c' for Concrete

Type of concrete	Number of Tests	Specified Strengths	Mean Strengths	Mean/ Specified	Coefficient of Variation
Ordinary ready mix concrete	317	3000 to 6000 psi	4060 to 6700 psi	3000 psi—1.38 6000 psi—1.14	3000psi—0.111 6000psi—0.080
Ordinary plant-precast concrete	1174	5000 to 6500 psi	6910 to 7420 psi	5000 psi—1.38 6500 psi—1.14	0.10
Lightweight concrete	769	3000 to 5000 psi	4310 to 5500 psi	3000psi—1.44 5000psi—1.10	3000psi—0.185 5000psi—0.070
High-strength concrete—28 days	2052	7000 to 12,000 psi	8340 to 12,400 psi	7000 psi—1.19 12,000psi—1.04	7000psi—0.115 12,000psi—0.105
High-strength concrete—56 days	914	7000 to 12,000 psi	10,430 to 14,000 psi	7000 psi—1.49 12,000psi—1.17	7000 psi—0.080 12,000psi—0.105

Source: From data presented in [3-11].

Nowak suggests that λ, the ratio of mean test strength to specified strength, can be taken as $\lambda = 1.35$ for 3000 psi concrete, decreasing linearly to 1.14 at $f'_c = 5000$ psi and constant at $\lambda = 1.14$ for higher strengths. However, the mean strength ratio cannot be considered a property of modern concrete, since it is easy for a designer to increase or decrease this while proportioning the concrete mix.

The data in Table 3-1 suggest the following coefficients of variation for various degrees of concrete control:

Poor control	$V > 0.140$
Average control	$V = 0.105$
Excellent control	$V < 0.070$

Building-Code Definition of Compressive Strength

The *specified compressive strength*, f'_c, is measured by compression tests on 6-by-12-in. cylinders tested after 28 days of moist curing. This is the strength specified on the construction drawings and used in the calculations. As shown in Fig. 3-4, the specified strength is less than the average strength. The required mean strength of the concrete, f_{cr}, must be at least (ACI Section 5.3.2.1)

Specified compressive strength, f'_c, less than or equal to 5000 psi:

Use the larger value of

$$f'_{cr} = f'_c + 1.34s \qquad (3\text{-}3a)$$

and
$$\text{(ACI Eq. 5-1)}$$

$$f'_{cr} = f'_c + 2.33s - 500 \qquad (3\text{-}3b)$$

$$\text{(ACI Eq. 5-2)}$$

Specified compressive strength, f'_c, greater than 5000 psi:

Use the larger value of

$$f'_{cr} = f'_c + 1.34s \qquad (3\text{-}4a)$$

and
$$\text{(ACI Eq. 5-1)}$$

$$f'_{cr} = 0.90f'_c + 2.33s \qquad (3\text{-}4b)$$

$$\text{(ACI Eq. 5-3)}$$

where s is the standard deviation determined in accordance with ACI Section 5.3.1. Special rules are given if the standard deviation is not known.

Equations (3-3a) and (3-4a) give the lowest average strengths required to ensure a probability of not more than 1 in 100 that the average of any three consecutive strength tests will be below the specified strength. Alternatively, it ensures a probability of not more than 1 in 11 that any one test will fall below f'_c. Equation (3-3b) gives the lowest mean strength to ensure a probability of not more than 1 in 100 that any individual strength test will be more than 500 psi below the specified strength. Lines indicating the corresponding required average strengths, f_{cr}, are plotted in Fig. 3-4. In these definitions, a test is the average of two cylinder tests.

For any one test (average of two cylinder tests at the same age), (3-3a and b) and (3-4a and b) give a probability of 0.99 that a single test will fall more than 500 psi below the specified strength, equivalent to a 0.01 chance of understrength. This does not ensure that the number of low tests will be acceptable, however. Given a structure requiring 4000 cubic yards of concrete with 80 concrete tests during the construction period, the probability of a single test falling more than 500 psi below the specified strength is $1 - 0.99^{80}$, or about 55 percent [3-12].

This may be an excessive number of understrength test results in projects where owners refuse to pay for concretes that have lower strengths than specified. A higher target for the mean concrete strength than that currently required by (3-3) and (3-4) will frequently be specified to reduce the probability of low strength tests.

Factors Affecting Concrete Compressive Strength

Among the large number of factors affecting the compressive strength of concrete, the following are probably the most important for concretes used in structures.

   1.   **Water/cement ratio.** The strength of concrete is governed in large part by the ratio of the weight of the water to the weight of the cement for a given volume of concrete. A lower water/cement ratio reduces the porosity of the hardened concrete and thus increases the number of interlocking solids. The introduction of tiny, well-distributed air bubbles in the cement paste, referred to as *air entrainment*, tends to increase the freeze–thaw durability of the concrete. When the water in the concrete freezes, pressure is generated in the capillaries and pores in the hardened cement paste. The presence of tiny, well-distributed air bubbles provides a way to dissipate the pressures due to freezing. However, the air voids introduced by air entrainment reduce the strength of the concrete. A water/cement ratio of 0.40 corresponds to 28-day strengths in the neighborhood of 4700 psi for air-entrained concrete and 5700 psi for non-air-entrained concrete. For a water/cement ratio of 0.55, the corresponding strengths are 3500 and 4000 psi, respectively. Voids due to improper consolidation tend to reduce the strength below that corresponding to the water/cement ratio.

   2.   **Type of cement.** Traditionally, five basic types of portland cement have been produced:

   > *Normal, Type I*: used in ordinary construction, where special properties are not required;
   >
   > *Modified, Type II*: lower heat of hydration than Type I; used where moderate exposure to sulfate attack exists or where moderate heat of hydration is desirable;
   >
   > *High early strength, Type III*: used when high early strength is desired; has considerably higher heat of hydration than Type I;
   >
   > *Low heat, Type IV*: developed for use in mass concrete dams and other structures where heat of hydration is dissipated slowly. In recent years, very little Type IV cement has been produced. It has been replaced with a combination of Types I and II cement with fly ash.
   >
   > *Sulfate resisting, Type V*: used in footings, basement walls, sewers, and so on that are exposed to soils containing sulfates.

   In recent years, blended portland cements produced to satisfy ASTM C 1157-00 *Standard Performance Specification for Hydraulic Cement* have partially replaced the traditional five basic cements. This in effect allows the designer to select different blends of cement.

   Figure 3-5 illustrates the rate of strength gain with different cements. Concrete made with Type III (high early strength) cement gains strength more rapidly than does concrete made with Type I (normal) cement, reaching about the same strength at 7 days as a corresponding mix containing Type I cement would reach at 28 days. All five types tend to approach the same strength after a long period of time, however.

   3.   **Supplementary cementitious materials.** Sometimes, a portion of the cement is replaced by materials such as fly ash, ground granulated blast-furnace slag, or silica fume to achieve economy, reduction of heat of hydration, and, depending on the materials, improved workability. Fly ash and silica fume are referred to as *pozzolans*, which are defined as siliceous, or siliceous and aluminous, materials that in themselves possess little or no cementitious properties, but that will, in the presence of moisture, react with calcium hydroxide to form compounds with such properties. When supplementary cementitious materials are used in mix design, the water/cement ratio, *w/c*, is restated in terms of the *water/cementitious materials ratio*, *w/cm*, where *cm* represents the total weight of the

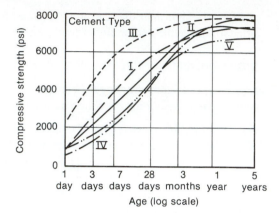

Fig. 3-5
Effect of type of cement on strength gain of concrete (moist cured, water/cement ratio = 0.49). (From [3-13] copyright ASTM; reprinted with permission.)

cement and the supplementary cementitious materials, as defined in ACI Section 4.1.1. Upper limits on the amounts of fly ash, slag, or silica fume are given in ACI Section 4.2.3 for concretes exposed to freeze–thaw conditions. The design of concrete mixes containing supplementary cementitious materials is explained in [3-14].

Fly ash, precipitated from the chimney gases from coal-fired power plants, frequently leads to improved workability of the fresh concrete. It often slows the rate of strength gain of concrete, but generally not the final strength, and, depending on composition of the fly ash, might reduce or improve the durability of the hardened concrete [3-15]. Fly ashes from different sources vary widely in composition and have different effects on concrete properties. They also affect the color of the concrete.

Ground granulated blast-furnace slag tends to reduce the early-age strength and heat of hydration of concrete. Strengths at older ages will generally exceed those for normal concretes with similar *w/cm* ratios. Slag tends to reduce the permeability of concrete and its resistance to attack by certain chemicals [3-15].

Silica fume consists of very fine spherical particles of silica produced as a by-product in the manufacture of ferrosilicon alloys. The extreme fineness and high silica content of the silica fume make it a highly effective pozzolanic material. It is used to produce low-permeability concrete with enhanced durability and/or high-strength [3-14].

4. Aggregate. The strength of concrete is affected by the strength of the aggregate, its surface texture, its grading, and, to a lesser extent, by the maximum size of the aggregate. Strong aggregates, such as felsite, traprock, or quartzite, are needed to make very-high-strength concretes. Weak aggregates include sandstone, marble, and some metamorphic rocks, while limestone and granite aggregates have intermediate strength. Normal-strength concrete made with high-strength aggregates fails due to mortar cracking, with very little aggregate failure. The stress–strain curves of such concretes tend to have an appreciable declining branch after reaching the maximum stress. On the other hand, if aggregate failure precedes mortar cracking, failure tends to occur abruptly with a very steep declining branch. This occurs in very-high-strength concretes (see Fig. 3-18) and in some lightweight concretes (see Fig. 3-26).

Concrete strength is affected by the bond between the aggregate and the cement paste. The bond tends to be better with crushed, angular pieces of aggregate.

A well-graded aggregate produces a concrete that is less porous. Such a concrete tends to be stronger. The strength of concrete tends to decrease as the maximum aggregate size increases. This appears to result from higher stresses at the paste–aggregate interface.

Some aggregates react with alkali in cement, causing a long-term expansion of the concrete that destroys the structure of the concrete. Unwashed marine aggregates also lead to a breakdown of the structure with time.

5. Mixing water. There are no standards governing the quality of water for use in mixing concrete. In most cases, water that is suitable for drinking and that has no pronounced taste or odor may be used [3-19]. It is generally thought that the pH of the water should be between 6.0 and 8.0. Salt water or brackish water must not be used as mixing water, because chlorides and other salts in such water will attack the structure of the concrete and may lead to corrosion of prestressing tendons. Strands and wires used as tendons are particularly susceptible to corrosion due to their small diameter and higher stresses compared to reinforcing bars [3-20].

6. Moisture conditions during curing. The development of the compressive strength of concrete is strongly affected by the moisture conditions during curing. Prolonged moist curing leads to the highest concrete strength, as shown in Fig. 3-6.

7. Temperature conditions during curing. The effect of curing temperature on strength gain is shown in Fig. 3-7 for specimens placed and moist-cured for 28 days under the constant temperatures shown in the figure and then moist-cured at 73°F. The 7- and 28-day strengths are reduced by cold curing temperatures, although the long-term strength tends to be enhanced. On the other hand, high temperatures during the first month increase the 1- and 3-day strengths but tend to reduce the one-year strength.

The temperature during the setting period is especially important. Concrete placed and allowed to set at temperatures greater than 80°F will never reach the 28-day strength of concrete placed at lower temperatures. Concrete that freezes soon after it has been placed will have a severe strength loss.

Occasionally, control cylinders are left in closed boxes at the job site for the first 24 hours. If the temperature is higher than ambient inside these boxes, the strength of the control cylinders may be affected.

8. Age of concrete. Concrete gains strength with age, as shown in Figs. 3-5 to 3-7. Prior to 1975, the 7-day strength of concrete made with Type I cement was generally 65 to 70 percent of the 28-day strength. Changes in cement production since then have resulted in a more rapid early strength gain and less long-term strength gain. ACI Committee 209 [3-21] has proposed the following equation to represent the rate of strength gain for concrete made from Type I cement and moist-cured at 70°F:

$$f'_{c(t)} = f'_{c(28)} \left(\frac{t}{4 + 0.85t} \right) \tag{3-5}$$

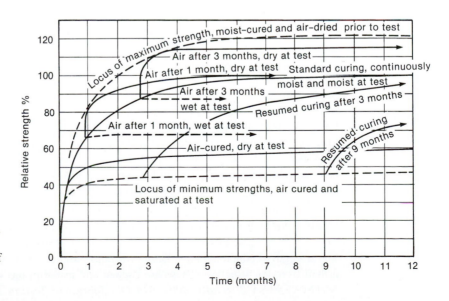

Fig. 3-6
Effect of moist-curing conditions at 70°F and moisture content of concrete at time of test on compressive strength of concrete. (From [3-17].)

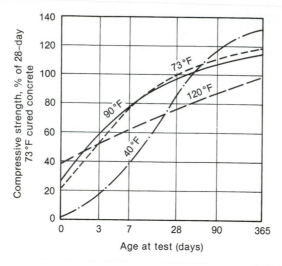

Fig. 3-7
Effect of temperature during the first 28 days on the strength of concrete (water/cement ratio = 0.41, air content = 4.5 percent, Type I cement, specimens cast and moist-cured at temperature indicated for first 28 days (all moist-cured at 73°F thereafter). (From [3-18].)

Here, $f'_{c(t)}$ is the compressive strength at age t. For Type III cement, the coefficients 4 and 0.85 become 2.3 and 0.92.

Concrete cured under temperatures other than 70°F may set faster or slower than indicated by these equations, as shown in Fig. 3-7.

9. Maturity of concrete. Young concrete gains strength as long as the concrete remains about a *threshold temperature* of −10 to −12°C or +11 to +14°F. Maturity is the summation of the product of the difference between the curing temperature and the threshold temperature, and the time the concrete has cured at that temperature, [3-22], [3-23]. The summation of the product of the curing temperature and the time the concrete has cured at that temperature is called the *maturity* [3-22], [3-23] of the concrete:

$$\text{maturity} = M = \sum_{i=1}^{n}(T_i - 10)(t_i) \tag{3-6}$$

In this equation, T_i is the temperature in Celsius of Fahrenheit during the ith interval and t_i is the number of days spent curing at that temperature. The minus sign is used for both Celsius and Fahrenheit. Figure 3-8 shows the form of the relationship between maturity and compressive strength of concrete. Although no unique relationship exists, Fig. 3-8 may be used for guidance in determining when forms can be removed. Maturity should not be used as the sole determinant of adequate strength. It will detect neither such errors in the concrete batching as inadequate cement or excess water nor excessive delays in placing the concrete after batching.

10. Rate of loading. The standard cylinder test is carried out at a loading rate of roughly 35 psi per second, and the maximum load is reached in $1\frac{1}{2}$ to 2 minutes, corresponding to a strain rate of about 10 microstrain/sec. Under very slow rates of loading, the axial compressive strength is reduced to about 75 percent of the standard test strength, as shown in Fig. 3-2. A portion of this reduction is offset by continued maturing of the concrete during the loading period [3-4]. At high rates of loading, the strength increases, reaching 115 percent of the standard test strength when tested at a rate of 30,000 psi/sec (strain rate of 20,000 microstrain/sec). This corresponds to loading a cylinder to failure in roughly 0.10 to 0.15 second and would approximate the rate of loading experienced in a severe earthquake. [3-10]

Core Tests

The strength of concrete in a structure (*in-place strength*) is frequently measured on cores drilled from the structure. These are capped and tested in the same manner as cylinders. ASTM C 42-94 *Standard Method of Obtaining and Testing Drilled Cores and Sawed*

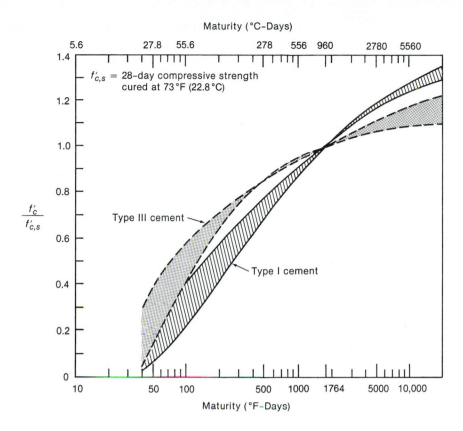

Fig. 3-8
Normalized compressive
strength versus maturity.
(From [3-23].)

Beams of Concrete specifies how such tests should be carried out. Core-test strengths show a great amount of scatter, because core strengths are affected by a wide range of variables.

Core tests have two main uses. The most frequent use of core tests is to assess whether concrete in a new structure is acceptable. ACI Section 5.6.5.2 permits the use of core tests in such cases and requires three cores for each strength test more than 500 psi below the specified value of f'_c. Cores obtained by using a water-cooled bit have a moisture gradient from the wet outside surface to the dry interior concrete. This causes stress gradients that reduce the test strength of the core. ACI Section 5.6.5.3 was changed in 2002 to require that cores be prepared for shipping to the testing lab by wiping the drilling water from the surfaces and wrapping them in watertight bags or containers immediately after drilling [3-24].

Cores should not be tested earlier than 48 hours after drilling, nor later than 7 days after drilling. Waiting 48 hours enables the moisture gradients in the cores to dissipate. This reduces the stress gradient in the core. ACI Section 5.6.4.4 states that concrete evaluated with the use of cores has adequate strength if the average strength of the cores is at least 85 percent of the specified strength and no single core shows a strength less than 75 percent of f'_c. This is just an acceptance rule. Because the 85 percent value tends to be smaller than the actual ratio of core strength to cylinder strength, the widespread practice of taking the in-place strength equal to (core strength)/0.85 overestimates the in-place strength.

The second use of core-test data is to determine the in-place strength of concrete that is equivalent to the f'_c used in the design equations given in the code. This is referred to as the *equivalent specified strength* and is used when evaluating the strength of an existing member or structure.

Neville [3-25] discusses core testing of in-place concrete and points out the advantages and drawbacks to using cores to estimate the concrete strength in a structure. Bartlett and MacGregor [3-24], [3-26] suggest the following procedure for estimating the equivalent specified strength of concrete in a structure by using core tests:

1. **Plan the scope of the investigation.** The regions that are cored must be consistent with the information sought. That is, either the member in question should be cored, or, if this is impractical, the regions that are cored should contain the same type of concrete, of about the same age, and cured in the same way as the suspect region. The number of cores taken depends, on one hand, on the cost and the hazard from taking cores out of critical parts of the structure, and on the other hand, on the desired accuracy of the strength estimate. If possible, at least six cores should be taken from a given grade of concrete in question. It is not possible to detect outliers (spurious values) in smaller samples, and the penalty for small sample sizes (given by k_1 in (3-8)) is significant. The diameter of the core should not be less than 3 times the nominal maximum size of the coarse aggregate, and the length of the core should be between 1 and 2 times the diameter. If possible, the core diameter should not be less than 4 in., because the variability of the core strengths increases significantly for smaller diameters.

2. **Obtain and test the cores.** Use standard methods to obtain and test the cores as given in ASTM C 42-90. Carefully record the location in the structure of each core, the conditions of the cores before testing, and the mode of failure. This information may be useful in explaining individual low core strengths. A load–stroke plot from the core test may be useful in this regard. It is particularly important that the moisture condition of the core correspond to one of the two standard conditions prescribed in ASTM C 42-90 and be recorded.

3. **Convert the core strengths, f_{core}, to equivalent in-place strengths, f_{cis}.** This is done by using

$$f_{cis} = f_{core}(F_{\ell/d} \times F_{dia} \times F_r)(F_{mc} \times F_d) \tag{3-7}$$

where the factors in the first set of parentheses correct the core strength to that of a standard 4-in.-diameter core, with length/diameter ratio equal to 2, not containing reinforcement:

$F_{\ell/d}$ = correction for length/diameter ratio as given in ASTM C 42-87
= 0.87, 0.93, 0.96, 0.98, and 1.00 for ℓ/d = 1.0, 1.25, 1.50, 1.75, and 2.0, respectively

F_{dia} = correction for diameter of core
= 1.06 for 2-in. cores, 1.00 for 4-in. cores, and 0.98 for 6-in. cores

F_r = correction for the presence of reinforcing bars
= 1.00 for no bars, 1.08 for one bar, and 1.13 for two

It is generally prudent to cut off parts of a core that contain reinforcing bars, provided the specimen that remains for testing has a length/diameter ratio equal to at least 1.0.

The factors in the second set of parentheses account for differences between the condition of the core and that of the concrete in the structure:

F_{mc} = accounts for the effect of the moisture condition of the core at the time of the core test
= 1.09 if the core was soaked before testing, and 0.96 if the core was air-dried at the time of the test

F_d = accounts for damage to the surface of the core due to drilling
= 1.06

4. **Check for outliers in the set of equivalent in-place strengths.** Reference [3-26] gives a technique for doing this. If an outlier is detected via a statistical test, one should try to determine a physical reason for the anomalous strength.

5. **Compute the equivalent specified strength from the in-place strengths.** The *equivalent specified strength*, f'_{ceq}, is the strength that should be used in design equations

when checking the capacity of the member in question. To calculate it, one first computes the mean, $\overline{f}_{cis}$, and sample standard deviation, s_{cis}, of the set of equivalent in-place strengths, f_{cis}, which remains after any outliers have been removed. Bartlett and MacGregor [3-26] present the following equation for f'_{ceq}, which uses the core test data to obtain a lower-bound estimate of the 10 percent fractile of the in-place strength:

$$f'_{ceq} = k_2\left[\overline{f}_{cis} - 1.282\sqrt{\frac{(k_1 S_{cis})^2}{n} + \overline{f}_{cis}^2(V_{\ell/d}^2 + V_{dia}^2 + V_r^2 + V_{mc}^2 + V_d^2)}\right] \quad (3\text{-}8)$$

Here,

k_1 = a factor dependent on the number of core tests, after removal of outliers, equal to 2.40 for 2 tests, 1.47 for 3 tests, 1.20 for 5 tests, 1.10 for 8 tests, 1.05 for 16 tests, and 1.03 for 25 tests

k_2 = a factor dependent on the number of batches of concrete in the member or structure being evaluated, equal to 0.90 and 0.85, respectively, for a cast-in-place member or structure that contains one batch or many batches, and equal to 0.90 for a precast member or structure

n = number of cores after removal of outliers

$V_{\ell/d}$ = coefficient of variation due to length/diameter correction, equal to 0.025 for $\ell/d = 1$, 0.006 for $\ell/d = 1.5$, and zero for $\ell/d = 2$

V_{dia} = coefficient of variation due to diameter correction, equal to 0.12 for 2-in.-diameter cores, zero for 4-in. cores, and 0.02 for 6-in. cores

V_r = coefficient of variation due to presence of reinforcing bars in the core, equal to zero if none of the cores contained bars, and to 0.03 if more than a third of them did

V_{mc} = coefficient of variation due to correction for moisture condition of core at time of testing, equal to 0.025

V_d = coefficient of variation due to damage to core during drilling, equal to 0.025

The individual coefficients of variation in the second term of (3-8) are taken equal to zero if the corresponding correction factor, F, is taken equal to 1.0 in (3-7).

Example 3-1   Computation of an Equivalent Specified Strength from Core Tests

As a part of an evaluation of an existing structure, it is necessary to compute the strength of a 6-in.-thick slab. To do so, it is necessary to have an equivalent specified compressive strength, f'_{ceq}, to use in place of f'_c in the design equations. Several batches of concrete were placed in the slab.

1.   **Plan the scope of the investigation.** From a site visit, it is learned that five cores can be taken. These are 4-in.-diameter cores drilled vertically through the slab, giving cores that are 6 in. long. They are taken from randomly selected locations around the entire floor in question.

2.   **Obtain and test the cores.** The cores were tested in an air-dried condition. None of them contained reinforcing bars. The individual core strengths were 5950, 5850, 5740, 5420, and 4830 psi.

3.   **Convert the core strengths to equivalent in-place strengths.** From (3-7),

$$f_{cis} = f_{core}(F_{\ell/d} \times F_{dia} \times F_r)(F_{mc} \times F_d)$$

The ℓ/d of the cores was 6 in./4 in. = 1.50. From ASTM C 42-90, $F_{\ell/d} = 0.96$, and we have

$$f_{cis} = f_{core}(0.96 \times 1.0 \times 1.0)(0.96 \times 1.06)$$
$$= f_{core} \times 0.977$$

The individual strengths, f_{cis}, are 5812, 5715, 5607, 5295, and 4720 psi.

4. Check for low outliers. Although there is quite a difference between the lowest and second-lowest values, we shall assume that all five tests are valid.

5. Compute the equivalent specified strength.

$$f'_{ceq} = k_2\left[\overline{f}_{cis} - 1.282\sqrt{\frac{(k_1 s_{cis})^2}{n} + \overline{f}_{cis}^2(V_{\ell/d}^2 + V_{dia}^2 + V_r^2 + V_{mc}^2 + V_d^2)}\right] \quad (3\text{-}8)$$

The mean and sample standard deviation of the f_{cis} values are $\overline{f}_{cis} = 5430$ psi and $s_{cis} = 422$ psi, respectively. Other terms in (3-8) are $k_1 = 1.20$ for 5 tests, $k_2 = 0.85$ for several batches, $n = 5$ tests. Because no correction was made in step 3 for the effects of core diameter or reinforcement in the core (F_{dia} and $F_r = 1.0$), V_{dia} and V_r are equal to zero. The terms under the square-root sign in (3-8) are

$$\frac{(k_1 s_{cis})^2}{n} = \frac{(1.20 \times 442)^2}{5} = 56{,}265$$

$$\overline{f}_{cis}^2(V_{\ell/d}^2 + V_{dia}^2 + V_r^2 + V_{mc}^2 + V_d^2) = 5430^2(0.006^2 + 0.0^2 + 0.0^2 + 0.025^2 + 0.025^2)$$

$$= 37{,}918$$

$$f'_{ceq} = 0.85(5430 - 1.282\sqrt{56{,}265 + 37{,}918})$$

$$= 4281 \text{ psi}$$

The concrete strength in the slab should be taken as 4280 psi when calculating the capacity of the slab. ∎

Strength of Concrete in a Structure

The strength of concrete in a structure tends to be somewhat lower than the strength of control cylinders made from the same concrete. This difference is due to the effects of different placing, compaction, and curing procedures; the effects of vertical migration of water during the placing of the concrete in deep members; the effects of difference in size and shape; and the effects of different stress regimes in the structure and the specimens.

The concrete near the top of deep members tends to be weaker than the concrete lower down, probably due to the increased water/cement ratio at the top due to upward water migration after the concrete is placed and by the greater compaction of the concrete near the bottom due to the weight of the concrete higher in the form [3-27].

3-4 STRENGTH UNDER TENSILE AND MULTIAXIAL LOADS

Tensile Strength of Concrete

The tensile strength of concrete falls between 8 and 15 percent of the compressive strength. The actual value is strongly affected by the type of test carried out to determine the tensile strength, the type of aggregate, the compressive strength of the concrete, and the presence of a compressive stress transverse to the tensile stress [3-28], [3-29], [3-30].

Standard Tension Tests

Two types of tests are widely used. The first of these is the *modulus of rupture* or flexural test (ASTM C 78 or C 293), in which a plain concrete beam, generally 6 in. × 6 in. × 30 in. long, is loaded in flexure at the third points of a 24-in. span until it fails due to cracking on the tension face. The flexural tensile strength or modulus of rupture, f_r, from a modulus-of-rupture test is calculated from the following equation, assuming that the concrete is linearly elastic:

$$f_r = \frac{6M}{bh^2} \tag{3-9}$$

In this equation,

M = moment

b = width of specimen

h = overall depth of specimen

The second common tensile test is the *split cylinder* test (ASTM C 496), in which a standard 6 in. × 12 in. compression test cylinder is placed on its side and loaded in compression along a diameter, as shown in Fig. 3-9a.

In a split-cylinder test, an element on the vertical diameter of the specimen is stressed in biaxial tension and compression, as shown in Fig. 3-9c. The stresses acting across the vertical diameter range from high transverse compressions at the top and bottom to a nearly uniform tension across the rest of the diameter, as shown in Fig. 3-9d. The splitting tensile strength, f_{ct}, from a split-cylinder test is computed as

$$f_{ct} = \frac{2P}{\pi \ell d} \tag{3-10}$$

where

P = maximum applied load in the test

ℓ = length of specimen

d = diameter of specimen

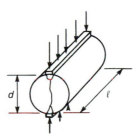

(a) Test procedure.

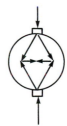

(b) Simplified force system.

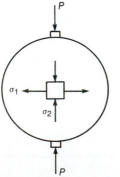

(c) Stresses on element.

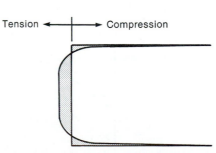

(d) Distribution of σ_1 on vertical diameter.

Fig. 3-9
Split-cylinder test.

Various types of tension tests give different strengths. In general, the strength decreases as the volume of concrete that is highly stressed in tension is increased. A third-point-loaded modulus-of-rupture test on a 6-in.-square beam gives a modulus-of-rupture strength f_r that averages 1.5 times f_{ct}, while a 6-in.-square prism tested in pure tension gives a direct tensile strength that averages about 86 percent of f_{ct} [3-30].

Relationship between Compressive and Tensile Strengths of Concrete

Although the tensile strength of concrete increases with an increase in the compressive strength, the ratio of the tensile strength to the compressive strength decreases as the compressive strength increases. Thus, the tensile strength is approximately proportional to the square root of the compressive strength. The mean split cylinder strength, $\overline{f}_{ct}$, from a large number of tests of concrete from various localities has been found to be [3-10]

$$\overline{f}_{ct} = 6.4\sqrt{f'_c} \qquad (3\text{-}11)$$

where f_{ct}, f'_c and $\sqrt{f'_c}$ are all in psi. Values from (3-11) are compared with split-cylinder test data in Fig. 3-10. It is important to note the wide scatter in the test data. The ratio of measured to computed splitting strength is essentially normally distributed.

Similarly, the mean modulus of rupture, $\overline{f}_r$, can be expressed as [3-10]

$$\overline{f}_r = 8.3\sqrt{f'_c} \qquad (3\text{-}12a)$$

Again, there is scatter in the modulus of rupture. Raphael[3-28] discusses the reasons for this, as do McNeely and Lash [3-29]. The distribution of the ratio of measured to computed modulus-of-rupture strength approaches a log-normal distribution.

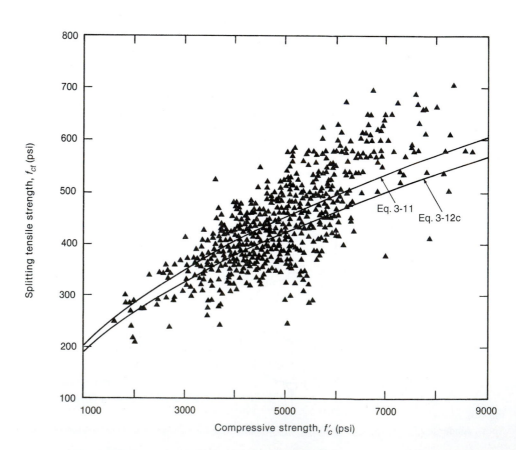

Fig. 3-10
Relationship between splitting tensile strengths and compression strengths. (From [3-10].)

ACI Section 9.5.2.3 defines the modulus of rupture for use in calculating deflections as

$$f_r = 7.5\sqrt{f_c'} \tag{3-12b}$$

A lower value is used in strength calculations (ACI Section 11.4.2.1):

$$f_r = 6\sqrt{f_c'} \tag{3-12c}$$

Factors Affecting the Tensile Strength of Concrete

The tensile strength of concrete is affected by the same factors that affect the compressive strength. In addition, the tensile strength of concrete made from crushed rock may be up to 20 percent greater than that from rounded gravels. The tensile strength of concrete made from lightweight aggregate tends to be less than that for normal sand-and-gravel concrete, although this varies widely, depending on the properties of the particular aggregate under consideration.

The tensile strength of concrete develops more quickly than the compressive strength. As a result, such things as shear strength and bond strength, which are strongly affected by the tensile strength of concrete, tend to develop more quickly than the compressive strength. At the same time, however, the tensile strength increases more slowly than would be suggested by the square root of the compressive strength at the age in question. Thus, concrete having a 28-day compressive strength of 3000 psi would have a splitting tensile strength of about $6.4\sqrt{f_c'} = 350$ psi. At 7 days this concrete would have compressive strength of about 2100 psi (0.70 times 3000 psi) and a tensile strength of about 260 psi (0.75 times 350 psi). This is less than the tensile strength of $6.4\sqrt{2100} = 293$ psi that one would compute from the 7-day compressive strength. This is of importance in choosing form-removal times for flat slab floors, which tend to be governed by the shear strength of the column–slab connections [3-31].

Strength under Biaxial and Triaxial Loadings

Biaxial Loading of Uncracked, Unreinforced Concrete   Concrete is said to be *loaded biaxially* when it is loaded in two mutually perpendicular directions with essentially no stress or restraint of deformation in the third direction, as shown in Fig. 3-11a. A common example is shown in Fig. 3-11b.

The strength and mode of failure of concrete subjected to biaxial states of stress varies as a function of the combination of stresses as shown in Fig. 3-12. The pear-shaped line in Fig. 3-12a represents the combinations of the biaxial stresses, σ_1 and σ_2, which cause cracking or compression failure of the concrete. This line passes through the uniaxial compressive strength, f_c', at A and A' and the uniaxial tensile strength, f_t', at B and B'.

Under biaxial tension (σ_1 and σ_2 both tensile stresses) the strength is close to that in uniaxial tension, as shown by the region B–D–B' (zone 1) in Fig. 3-12a. Here, failure occurs by tensile fracture perpendicular to the maximum principal tensile stress, as shown in Fig. 3-12b, which corresponds to point B in Fig. 3-12a.

When one principal stress is tensile and the other is compressive, as shown in Fig. 3-11a, the concrete cracks at lower stresses than it would if stressed uniaxially in tension or compression [3-32]. This is shown by regions A–B and A'–B' in Fig. 3-12a. In this region, zone 2 in Fig. 3-12a, failure occurs due to tensile fractures on planes perpendicular to the principal tensile stresses. The lower strengths in this region suggest that failure is governed by a limiting tensile strain rather than a limiting tensile stress.

Under uniaxial compression (points A and A' and zone 3 in Fig. 3-12a), failure is initiated by the formation of tensile cracks on planes parallel to the direction of the compressive stresses. These planes are planes of maximum principal tensile strain.

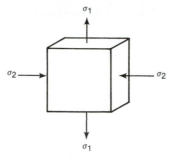

(a) Biaxial state of stress.

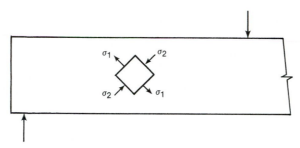

Fig. 3-11
Biaxial stresses.

(b) Biaxial state of stress in the web of a beam.

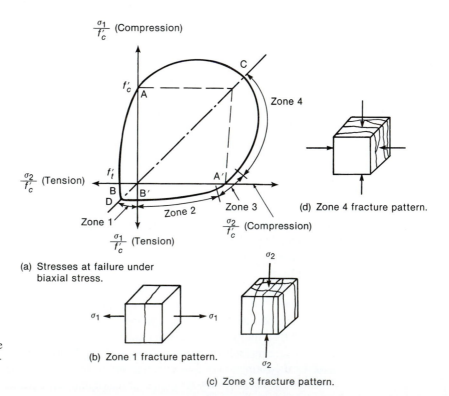

Fig. 3-12
Strength and modes of failure
of unreinforced concrete subjected to biaxial stresses.
(From [3-2].)

(a) Stresses at failure under
biaxial stress.

(b) Zone 1 fracture pattern.

(c) Zone 3 fracture pattern.

(d) Zone 4 fracture pattern.

Under biaxial compression (region $A-C-A'$ and zone 4 in Fig. 3-12a), the failure pattern changes to a series of parallel fracture surfaces on planes parallel to the unloaded sides of the member, as shown. Such planes are acted on by the maximum tensile strains. Biaxial and triaxial compression loads delay the formation of bond cracks and mortar cracks. As a result, the period of stable crack propagation is longer and the concrete is more ductile. As shown in Fig. 3-12, the strength of concrete under biaxial compression is greater than the uniaxial compressive strength. Under equal biaxial compressive stresses, the strength is about 107 percent of f'_c, as shown by point C.

In the webs of beams, the principal tensile and principal compressive stresses lead to a biaxial tension–compression state of stress, as shown in Fig. 3-11b. Under such a loading, the tensile and compressive strengths are less than they would be under uniaxial stress, as shown by the quadrant AB or $A'B'$ in Fig. 3-12a. A similar biaxial stress state exists in a split-cylinder test, as shown in Fig. 3-9c. This explains in part why the splitting tensile strength is less than the flexural tensile strength.

In zones 1 and 2 in Fig. 3-12, failure occurred when the concrete cracked, and in zones 3 and 4, failure occurred when the concrete crushed. In a reinforced concrete member with sufficient reinforcement parallel to the tensile stresses, cracking does not represent failure of the member because the reinforcement resists the tensile forces after cracking. The biaxial load strength of cracked reinforced concrete is discussed in the next subsection.

Compressive Strength of Cracked Reinforced Concrete

If cracking occurs in reinforced concrete under a biaxial tension–compression loading and there is reinforcement across the cracks, the strength and stiffness of the concrete under compression parallel to the cracks is reduced. Figure 3-13a shows a concrete element that has been cracked by horizontal tensile stresses. The natural irregularity of the shape of the cracks leads to variations in the width of a piece between two cracks, as shown. The compressive stress acting on the top of the shaded portion is equilibrated by compressive stresses and probably some bending stresses on the bottom and shearing stresses along the edges, as shown in Fig. 3-13b. When the crack widths are small, the shearing stresses transfer sufficient load across the cracks that the compressive stress on the bottom of the shaded portion is not significantly larger than that on the top, and the strength is unaffected by the cracks. As the crack widths increase, the ability to transfer shear across them decreases. For equilibrium, the compressive stress on the bottom of the shaded portion must then increase. Failure occurs when the highest stress in the element approaches the uniaxial compressive strength of the concrete.

Tests of concrete panels loaded in in-plane shear, carried out by Vecchio and Collins [3-33], have shown a relationship between the transverse tensile strain, ϵ_1, and the compressive strength parallel to the cracks, $f_{2\text{max}}$:

$$\frac{f_{2\text{max}}}{f'_c} = \frac{1}{0.8 + 170\epsilon_1} \tag{3-13}$$

where the subscripts 1 and 2 refer to the major (tensile) and minor (compressive) principal stresses and strains. The average transverse strain, ϵ_1, is the average transverse strain measured on a gauge length that includes one or more cracks. Equation 3-13 is plotted in Fig. 3-14a. An increase in the strain ϵ_1 leads to a decrease in strength. The same authors [3-34] have suggested a stress–strain relationship, $f_2-\epsilon_2$, for transversely cracked concrete:

$$f_2 = f_{2\text{max}}\left[2\left(\frac{\epsilon_2}{\varepsilon_o}\right) - \left(\frac{\epsilon_2}{\varepsilon_o}\right)^2\right] \tag{3-14}$$

where $f_{2\text{max}}$ is given by (3-13), and ε_o is the strain at the highest point in the compressive stress–strain curve, which the authors took as 0.002. The term in brackets describes a parabolic stress–strain curve with apex at ε_o and a peak stress that decreases as ϵ_1 increases.

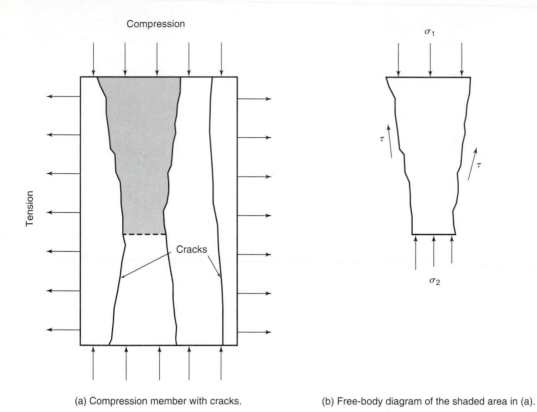

(a) Compression member with cracks.

(b) Free-body diagram of the shaded area in (a).

Fig. 3-13
Stresses in a biaxially loaded, cracked-concrete panel with cracks parallel to the direction of the
principal compression stress.

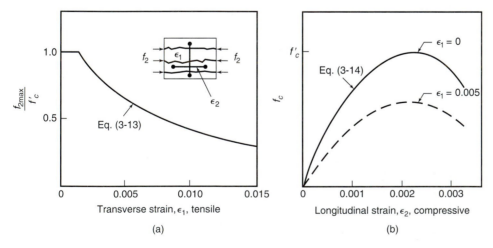

Fig. 3-14
Effect of transverse tensile strains
on the compressive strength of
cracked concrete.

If the parabolic stress–strain curve given by (3-14) is used, the strain for any given stress can be computed from

$$\epsilon_c = \epsilon_c'\left(1 - \sqrt{\frac{f_2}{f_c'}}\right) \qquad (3\text{-}15)$$

If the descending branch of the curve is also assumed to be a parabola, (3-15) can be used to compute strains on the postpeak portion of the stress–strain curve if the minus sign before the radical is changed to a plus.

The stress–strain relationships given by (3-13) and (3-14) represent stresses and strains averaged over a large area of a shear panel or beam web. The strains computed in this way include the widths of cracks in the computation of tensile strains, ϵ_1, as shown in the inset to Fig. 3-14a. These equations are said to represent *smeared* properties. Through smearing, the peaks and hollows in the strains have been attenuated by using the averaged stresses and strains. In this way, (3-13) and (3-14) are an attempt to replace the stress analysis of a cracked beam web having finite cracks with the analysis of a continuum. This substitution was a breakthrough in the analysis of concrete structures.

Triaxial Loadings

Under triaxial compressive stresses, the mode of failure involves either tensile fracture parallel to the maximum compressive stress (and thus orthogonal to the maximum tensile strain if such exists) or a shear mode of failure. The strength and ductility of concrete under triaxial compression exceed those under uniaxial compression, as shown in Fig. 3-15. This figure presents the stress–longitudinal strain curves for cylinders each subjected to a constant lateral fluid pressure $\sigma_2 = \sigma_3$ while the longitudinal stress, σ_1, was increased to failure. These tests suggested that the longitudinal stress at failure was

$$\sigma_1 = f_c' + 4.1\sigma_3 \qquad (3\text{-}16)$$

Tests of lightweight and high-strength concretes ([3-35] and [3-8]) suggest that their compressive strengths are less influenced by the confining pressure, with the result that the coefficient 4.1 in (3-16) drops to about 2.0.

The strength of concrete under combined stresses can also be expressed via a *Mohr rupture envelope*. The Mohr's circles plotted in Fig. 3-16 correspond to three of the cases

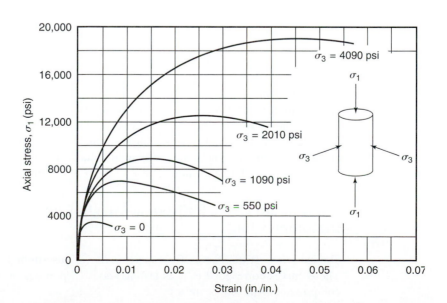

Fig. 3-15
Axial stress–strain curves from triaxial compression tests on concrete cylinders; unconfined compressive strength $f_c' = 3600$ psi. (From [3-3].)

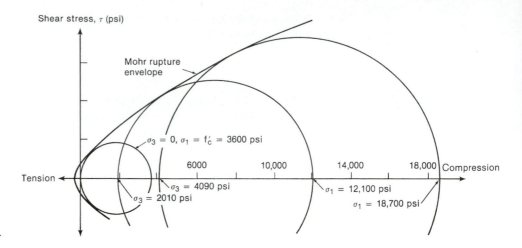

Fig. 3-16
Mohr rupture envelope for
concrete tests from Fig. 3-15.

plotted in Fig. 3-15. The Mohr's circles are tangent to the Mohr rupture envelope shown with the outer line.

In concrete columns or in beam–column joints, concrete in compression is sometimes enclosed by closely spaced hoops or spirals. When the width of the concrete element increases due to Poisson's ratio and microcracking, these hoops or spirals are stressed in tension, causing an offsetting compressive stress in the enclosed concrete. The resulting triaxial state of stress in the concrete enclosed or *confined* by the hoops or spirals increases the ductility and strength of the confined concrete. This effect is discussed in Chapters 11 and 20.

3-5 STRESS–STRAIN CURVES FOR CONCRETE

The behavior and strength of reinforced concrete members is controlled by the size and shape of the members and by the stress–strain properties of the concrete and the reinforcement. The stress–strain behavior discussed in this section will be used in subsequent chapters to develop relationships for the strength and behavior of reinforced concrete beams and columns.

Tangent and Secant Moduli of Elasticity

Three ways of defining the modulus of elasticity are illustrated in Fig. 3-17. The slope of a line that is tangent to a point on the stress–strain curve, such as A, is called the *tangent modulus of elasticity*, E_T, at the stress corresponding to point A. The slope of the stress–strain curve at the origin is the *initial tangent modulus of elasticity*. The *secant modulus of elasticity at a given stress* is the slope of a line through the origin and through the point on the curve representing that stress (for example, point B in Fig. 3-17). Frequently, the secant modulus is defined by using the point corresponding to $0.4f_c'$, representing the service-load stresses. The slopes of these lines have units of psi/strain, where strain is unitless, with the result that the units of the modulus of elasticity are psi.

Stress–Strain Curve for Normal-Weight Concrete in Compression

Typical stress–strain curves for concretes of various strengths are shown in Fig. 3-18. These curves correspond to tests lasting about 15 minutes on specimens resembling the compression zone of a beam.

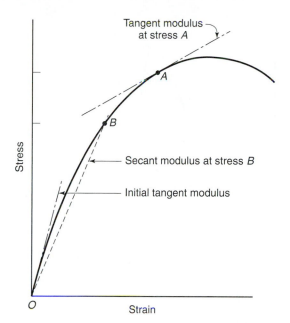

Fig. 3-17
Tangent and secant moduli of elasticity.

The stress–strain curves in Fig. 3-18 all rise to a maximum stress, reached at a strain between 0.0015 and 0.003, followed by a descending branch. The shape of this curve results from the gradual formation of microcracks within the structure of the concrete, as discussed in Section 3-2.

The length of the descending branch of the curve is strongly affected by the test conditions. Frequently, an axially loaded concrete test cylinder will fail explosively at the point of maximum stress. This will occur in axially flexible testing machines if the strain energy released by the testing machine as the load drops exceeds the energy that the specimen can absorb. If a member is loaded in compression due to bending (or bending plus axial load), the descending branch may exist because, as the stress drops in the most highly strained fibers, other less highly strained fibers can resist the load, thus delaying the failure of the highly strained fibers.

The stress–strain curves in Fig. 3-18 show five properties used in establishing mathematical models for the stress–strain curve of concrete in compression [3-37]:

1. The initial slope of the curves (initial tangent modulus of elasticity) increases with an increase in compressive strength.

The modulus of elasticity of the concrete, E_c, is affected by the modulus of elasticity of the cement paste and by that of the aggregate. An increase in the water–cement ratio increases the porosity of the paste, reducing its modulus of elasticity and strength. This is accounted for in design by expressing E_c, as a function of f'_c.

Of equal importance is the modulus of elasticity of the aggregate. Normal-weight aggregates have modulus-of-elasticity values ranging from 1.5 to 5 times that of the cement paste. Because of this, the fraction of the total mix that is aggregate also affects E_c. Lightweight aggregates have modulus-of-elasticity values comparable to that of the paste; hence, the aggregate fraction has little effect on E_c for lightweight concrete.

The modulus of elasticity of concrete is frequently taken as given in ACI Section 8.5.1, namely,

$$E_c = 33(w^{1.5})\sqrt{f'_c}\,\text{psi} \qquad\qquad (3\text{-}17)$$

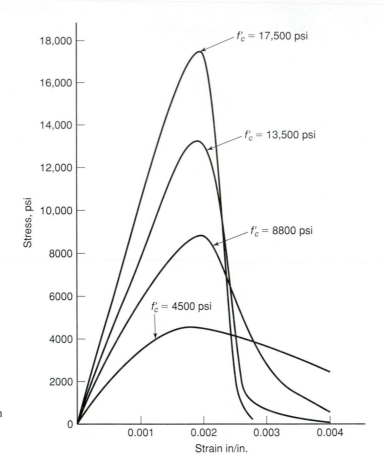

Fig. 3-18
Typical concrete stress–strain
curves in compression.
(Plotted using (3-21)
to (3-24).)

where w is the weight of the concrete in lb/ft^3. This equation was derived from short-time tests on concretes with densities ranging from 90 to 155 lb/ft^3 and corresponds to the secant modulus of elasticity at approximately $0.50f'_c$ [3-38]. The initial tangent modulus is about 10% greater. Because this equation ignores the type of aggregate, the scatter of data is very wide. Equation (3-17) systematically overestimates E_c in regions where low-modulus aggregates are prevalent. If deflections or vibration characteristics are critical in a design, E_c should be measured for the concrete to be used.

For normal-weight concrete with a density of 145 lb/ft^3, ACI Section 8.5.1 gives the modulus of elasticity as

$$E_c = 57,000\sqrt{f'_c}\ \text{psi} \tag{3-18}$$

ACI Committee 363 [3-8] proposed the following equation for high-strength concretes:

$$E_c = 40,000\sqrt{f'_c} + 1.0 \times 10^6\,\text{psi} \tag{3-19}$$

2. The rising portion of the stress–strain curve resembles a parabola with its vertex at the maximum stress.

For computational purposes the rising portion of the curves is frequently approximated by a parabola [3-34] [3-36] [3-39]. This curve tends to become straighter as the concrete strength increases [3-38].

3. The strain, ϵ_0, at maximum stress increases as the concrete strength increases.

4.   As explained in Section 3-2, the slope of the descending branch of the stress–strain curve results from the destruction of the structure of the concrete, caused by the spread of microcracking and overall cracking. For concrete strengths up to about 6000 psi, the slope of the descending branch of the stress–strain curve tends to be flatter than that of the ascending branch. The slope of the descending branch increases with an increase in the concrete strength, as shown in Fig. 3-18. For concretes with f_c' greater than about 10,000 psi, the descending branch is a nearly vertical, discontinuous "curve." This is because the structure of the concrete is destroyed by major longitudinal cracking.

5.   The maximum strain reached, ϵ_{cu}, decreases with an increase in concrete strength.

The descending portion of the stress–strain curve after the maximum stress has been reached is highly variable and is strongly dependent on the testing procedure. Similarly, the maximum or limiting strain, ϵ_{cu}, is very strongly dependent on the type of specimen, type of loading, and rate of testing. The limiting strain tends to be higher if there is a possibility of load redistribution at high loads. In flexural tests, values from 0.0025 to 0.006 have been measured. (See Section 4-1.)

Equations for Compressive Stress–Strain Diagrams

A common representation of the stress–strain curve for concretes with strengths up to about 6000 psi is the *modified Hognestad* stress–strain curve shown in Fig. 3-19a. This consists of a second-degree parabola with apex at a strain of $1.8f_c''/E_c$, where $f_c'' = 0.9f_c'$, followed by a downward-sloping line terminating at a stress of $0.85f_c''$ and a limiting strain of 0.0038 [3-36]. Equation (3-14) describes a second-order parabola with its apex at the strain ε_o. The reduced strength, $f_c'' = 0.9f_c'$, at the peak of the stress–strain curve corresponds to the 0.85 term in ACI Section 10.2.7.1 and is borne out by tests. The $0.9f_c'$ accounts for the differences between cylinder strength and member strength. These differences result from different curing and placing, which give rise to different water-gain effects due to vertical migration of bleed water, and differences between the strengths of rapidly loaded cylinders and the strength of the same concrete loaded more slowly, as shown in Fig. 3-2.

Two other expressions for the stress–strain curve will be presented. The stress–strain curve shown in Fig. 3-19b is convenient for use in analytical studies involving concrete

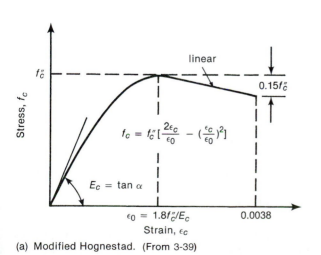

(a) Modified Hognestad. (From 3-39)   (b) Todeschini. (From 3-40)

Fig. 3-19
Analytical approximations to the compressive stress–strain curve for concrete.

strengths up to about 6000 psi because the entire stress–strain curve is given by one continuous function. The highest point in the curve, f_c'', is taken to equal $0.9f_c'$ to give stress-block properties similar to that of the rectangular stress block of Section 4-2 when $\epsilon_{ult} = 0.003$ for f_c' up to 5000 psi. The strain ε_o, corresponding to maximum stress, is taken as $1.71f_c'/E_c$. For any given strain ϵ, $x = \epsilon/\varepsilon_o$. The stress corresponding to that strain is

$$f_c = \frac{2f_c''x}{1 + x^2} \tag{3-20}$$

For a compression zone of constant width, the average stress under the stress block from $\epsilon = 0$ to ϵ is $\beta_1 f_c''$, where

$$\beta_1 = \frac{\ln(1 + x^2)}{x} \tag{3-21}$$

The center of gravity of the area of the stress–strain curve between $\epsilon = 0$ and ϵ is at $k_2\epsilon$ from the point where ϵ exists, where

$$k_2 = 1 - \frac{2(x - \tan^{-1}x)}{x^2\beta_1} \tag{3-22}$$

where x is in radians when computing $\tan^{-1}x$. The stress–strain curve is satisfactory for concretes with stress–strain curves that display a gradually descending stress–strain curve at strains greater than ε_0. Hence, it is applicable to f_c' up to about 5000 psi for normal-weight concrete and about 4000 psi for lightweight concrete.

Expressions for the compressive stress–strain curve for concrete are reviewed by Popovics [3-41]. Thorenfeldt, Tomaszewicz, and Jensen [3-42] generalized two of these expressions to derive a stress–strain curve that applies to concrete strengths from 15 to 125 MPa. The relationship between a stress, f_c, and the corresponding strain, ϵ_c, is

$$\frac{f_c}{f_c'} = \frac{n(\epsilon_c/\varepsilon_o)}{n - 1 + (\epsilon_c/\varepsilon_o)^{nk}} \tag{3-23}$$

where

f_c' = peak stress obtained from a cylinder test
ε_o = strain when f_c reaches f_c' (see (3-27))
n = a curve-fitting factor equal to $E_c/(E_c - E_c')$ (see (3-24))
E_c = initial tangent modulus (when $\epsilon_c = 0$)
$E_c' = f_c'/\varepsilon_o$

k = a factor to control the slopes of the ascending and descending branches of the stress–strain curve, taken equal to 1.0 for ϵ_c/ε_o less than 1.0 and taken greater than 1.0 for ϵ_c/ε_o greater than 1.0. (See (3-25).)

The four constants ε_o, E_c, n, and k can be derived directly from a stress–strain curve for the concrete if one is available. If not, they can be computed from (3-25) to (3-27), given by Collins and Mitchell [3-43]. Equations (3-17) and (3-18) can be used to compute E_c, although they were derived for the secant modulus through the origin and points representing 0.4 to $0.5f_c'$. For normal-density concrete,

$$n = 0.8 + \left(\frac{f_c'}{2500}\right) \tag{3-24}$$

where f_c' is in psi. For ϵ_c/ε_o less than or equal to 1.0,

$$k = 1.0 \tag{3-25}$$

and for $\epsilon_c/\varepsilon_o > 1.0$,

$$k = 0.67 + \left(\frac{f'_c}{9000}\right) \geq 1.0 \text{ (psi)} \tag{3-26}$$

If n, f'_c, and E_c are known, the strain at peak stress can be computed from

$$\varepsilon_o = \frac{f'_c}{E_c}\left(\frac{n}{n-1}\right) \tag{3-27}$$

A family of stress–strain curves calculated from (3-23) is shown in Fig. 3-18. Equation (3-23) produces a smooth continuous descending branch. Actually, the descending branch for high-strength concretes tends to drop in a series of jagged steps as the structure of the concrete is destroyed. Equation 3-23 approximates this with a smooth curve as shown in Fig. 3-18. Numerical integration can be used to solve for the stress-block parameters k_1 and k_2 for a given concrete strength. (See Section 4-2.) Values of k_1 and k_2 are given by Collins and Mitchell [3-43].

Traditionally, stress blocks based directly on stress–strain curves have the peak stress equal to f''_c, which is $0.85f'_c$ to $0.9f'_c$, to allow for differences between the in-place strength and the cylinder strength. For prediction of experimentally obtained behavior, the ordinates of the stress–strain curve should be computed for a strength f'_c and then multiplied by 0.90. For design based on stress–strain relationships, the stress–strain curve should be derived for a strength of f'_c and the ordinates multiplied by 0.90.

As shown in Fig. 3-15, a lateral confining pressure causes an increase in the compressive strength of concrete and a large increase in the strains at failure. The additional strength and ductility of confined concrete are utilized in hinging regions in structures in seismic regions. Stress–strain curves for confined concrete are described in [3-44].

When a compression specimen is loaded, unloaded, and reloaded, it has the stress–strain response shown in Fig. 3-20. The envelope to this curve is very close to the stress–strain curve for a monotonic test. This, and the large residual strains that remain after unloading, suggest that the inelastic response is due to damage to the internal structure of the concrete, as is suggested by the microcracking theory presented earlier.

Stress–Strain Curve for Normal-Weight Concrete in Tension

The stress–strain response of concrete loaded in axial tension can be divided into two phases. Prior to the maximum stress, the stress–strain relationship is slightly curved. The diagram is linear to roughly 50 percent of the tensile strength. The strain at peak stress is

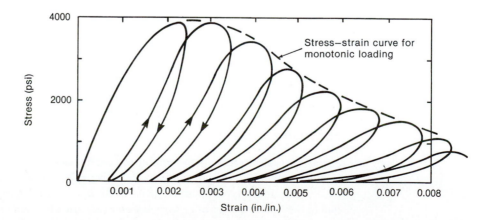

Fig. 3-20
Compressive stress–strain
curves for cyclic loads.
(From [3-45].)

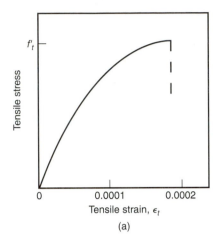
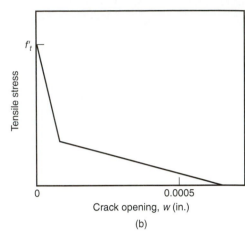

Fig. 3-21
Stress–strain curve and
stress–crack opening curves
for concrete loaded in tension.

about 0.0001 in pure tension and 0.00014 to 0.0002 in flexure. The rising part of the
stress–strain curve may be approximated either as a straight line with slope E_c and a maximum stress equal to the tensile strength f'_t or as a parabola with a maximum strain
$\epsilon'_t = 1.8 f'_t/E_c$ and a maximum stress f'_t. The latter curve is illustrated in Fig. 3-21a with f'_t
and E_c based on (3-11) and (3-18).

After the tensile strength is reached, microcracking occurs in a *fracture process zone*
adjacent to the point of highest tensile stress, and the tensile capacity of this concrete drops
very rapidly with increasing elongation. In this stage of behavior, elongations are concentrated in the fracture process zone while the rest of the concrete is unloading elastically.
The unloading response is best described by a *stress-versus-crack-opening diagram*, idealized in Fig. 3-21b as two straight lines. The crack widths shown in this figure are of the
right magnitude. The actual values depend on the situation. The tensile capacity drops to
zero when the crack is completely formed. This occurs at a very small crack width. A more
detailed discussion is given in [3-46].

Poisson's Ratio

At stresses below the critical stress (see Fig. 3-1), the Poisson's ratio for concrete varies
from about 0.11 to 0.21 and usually falls in the range from 0.15 to 0.20. On the basis of
tests of biaxially loaded concrete, Kupfer et al. [3-32] report values of 0.20 for Poisson's
ratio for concrete loaded in compression in one or two directions, 0.18 for concrete loaded
in tension in one or two directions, and 0.18 to 0.20 for concrete loaded in tension and
compression. Poisson's ratio remains approximately constant under sustained loads.

3-6   TIME-DEPENDENT VOLUME CHANGES

Concrete undergoes three main types of volume change, which may cause stresses, cracking, or deflections that affect the in-service behavior of reinforced concrete structures.
These are shrinkage, creep, and thermal expansion.

Shrinkage

Shrinkage is the decrease in the volume of concrete during hardening and drying under constant temperature. The amount of shrinkage increases with time, as shown in Fig. 3-22a.

The primary type of shrinkage is called *drying shrinkage* or simply *shrinkage* and is
due to the loss of a layer of *adsorbed water* (electrically bound water molecules) from the

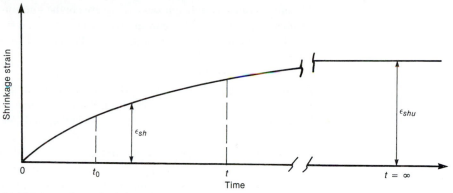

(a) Shrinkage of an unloaded specimen.

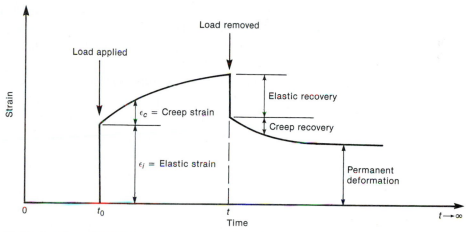

Fig. 3-22
Time-dependent strains.

(b) Elastic and creep strains due to loading at time, t_0, and unloading at time, t.

surface of the gel particles. This layer is roughly one water molecule thick, or about 1 percent of the size of the gel particles. The loss of free unadsorbed water has little effect on the magnitude of the shrinkage.

Shrinkage strains are dependent on the relative humidity and are largest for relative humidities of 40 percent or less. They are partially recoverable upon rewetting the concrete, and structures exposed to seasonal changes in humidity may expand and contract slightly due to changes in shrinkage strains.

The magnitude of shrinkage strains also depends on the composition of the concrete. The hardened cement paste shrinks, whereas the aggregate does not. Thus, the larger the fraction of the total volume of the concrete that is made up of hydrated cement paste, the greater the shrinkage. The aggregates act to restrain the shrinkage. There is less shrinkage with quartz or granite aggregates than with sandstone aggregate, because the quartz has a higher modulus of elasticity. The water/cement ratio affects the amount of shrinkage, because high water content reduces the volume of aggregate, thus reducing the restraint of the shrinkage by the aggregate. The more finely a cement is ground, the more surface area it has, and as a result, there is more adsorbed water to be lost during shrinkage and hence more shrinkage.

Drying shrinkage occurs as the moisture diffuses out of the concrete. As a result, the exterior shrinks more rapidly than the interior. This leads to tensile stresses in the outer skin of the concrete and compressive stresses in the interior as shown in Fig. 2-9. For large members, the ratio of volume to surface area increases, resulting in less shrinkage, because there is more moist concrete to restrain the shrinkage. The shrinkage develops more slowly in large members.

A secondary form of shrinkage called *carbonation shrinkage* occurs in carbon dioxide–rich atmospheres, such as those found in parking garages. At 50 percent relative humidity, the amount of carbonation shrinkage can equal the drying shrinkage, effectively doubling the total amount of shrinkage. At higher and lower humidities, the carbonation shrinkage decreases.

The ultimate drying shrinkage strain, ϵ_{shu}, for a 6-by-12-in. cylinder maintained for a very long time at a relative humidity of 40 percent ranges from 0.000400 to 0.001100 (400 to 1100 $\times$ 10^{-6} strain), with an average of about 0.000800 [3-19]. Thus, in a 25-ft bay in a building, the average shrinkage strain would cause a shortening of about $\frac{1}{4}$ in. in unreinforced concrete. In a structure, however, the shrinkage strains will tend to be less for the same concrete, for the following reasons:

1. The ratio of volume to surface area will generally be larger than for the cylinder; as a result, drying takes place much more slowly.

2. A structure is built in stages, and some of the shrinkage is dissipated before adjacent stages are completed.

3. The reinforcement restrains the development of the shrinkage.

The Euro-International Concrete Committee (CEB) [3-6] and the American Concrete Institute [3-21] have both published procedures for estimating shrinkage strains. The CEB method is more recent than the ACI procedure and accounts for member size in a better fashion. It will be presented here. The equations that follow apply only to the longitudinal shrinkage deformations of plain or lightly reinforced normal-weight concrete elements. Extensive recent unpublished data (Dilger, W. H. Personal communication, 2000) shows that the CEB method underestimates the shrinkage of North American concretes. To correct for this, a factor of 1.2 has been included in (3-30).

The axial shrinkage strains, ϵ_{cs}, occurring between times t_s at the start of shrinkage and t in plain concrete can be predicted from the formula

$$\epsilon_{cs}(t, t_s) = \epsilon_{cso}\beta_s(t, t_s) \tag{3-28}$$

where ϵ_{cso} is the basic shrinkage strain for a particular concrete and relative humidity, given by (3-29), and $\beta_s(t, t_s)$ is a coefficient given by (3-31) to describe the development of shrinkage between time t_s and t as a function of the effective thickness of the member. Note that:

$$\epsilon_{cso} = \epsilon_s(f_{cm})\beta_{RH} \tag{3-29}$$

where

$$\epsilon_s(f_{cm}) = 1.2[160 + \beta_{sc}(9 - f_{cm}/f_{cmo})] \times 10^{-6} \tag{3-30}$$

in which f_{cm} is the mean compressive strength at 28 days, psi. This can be taken equal to f_{cr} as given by (3-3) and (3-4). For concrete with a standard deviation, s, equal to $0.15f_c'$, f_{cm} would be $1.20f_c'$. We shall take f_{cm} as the smaller of $1.2f_c'$ or $(f_c' + 1200 \text{ psi})$. Shrinkage is not a function of compressive strength per se. It decreases with decreasing water/cement ratio and decreasing cement content. The strength f_{cm} in (3-29) is used as an empirical measure of these quantities:

$$f_{cmo} = 1450 \text{ psi},$$
$$\beta_{sc} = \text{coefficient that depends on the type of cement}$$
$$= 50 \text{ for Type I cement and 80 for Type III cement, and}$$

β_{RH} = coefficient that accounts for the effect of relative humidity on shrinkage.

For RH between 40 and 99 percent,

$$\beta_{RH} = -1.55\left[1 - \left(\frac{RH}{RH_0}\right)^3\right] \tag{3-31}$$

For RH is the equal to or greater than 99 percent,

$$\beta_{RH} = +0.25$$

where RH is the relative humidity of the ambient atmosphere in percent

$$RH_0 = 100 \text{ percent}$$

The effect of relative humidity on the total shrinkage is illustrated in Fig. 3-23. More shrinkage occurs in dry ambient conditions. The shrinkage reaches an lower limit at RH = 40 percent.

The development of shrinkage with time is given by

$$\beta_s(t, t_s) = \left[\frac{(t - t_s)/t_1}{350(h_e/h_0)^2 + (t - t_s)/t_1}\right]^{0.5} \tag{3-32}$$

where h_e is the effective thickness in inches to account for the volume/surface ratio and is given by:

$$h_e = 2A_c/u \tag{3-33}$$

and where,

A_c is the area of the cross section, in.^{2}

u is the perimeter of the cross section exposed to the atmosphere, in.

$h_0 = 4$ in.

t is the age of the concrete, days

t_s is the age of the concrete in days when shrinkage or swelling started, generally taken as the age at the end of moist-curing, and

$t_1 = 1$ day.

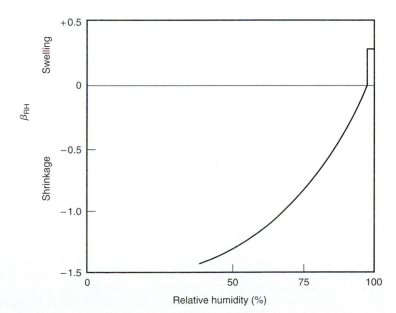

Fig. 3-23
Effect of relative humidity on shrinkage.

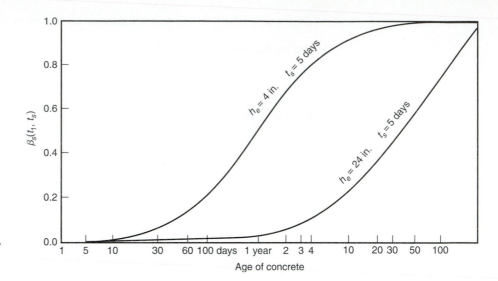

Fig. 3-24
Effect of effective thickness, h_e, on the rate of development of shrinkage.

The development of shrinkage with time predicted from (3-31) is shown in Fig. 3-24 for effective thicknesses of 4 in. and 24 in. Shrinkage develops much more rapidly in thin members, because moisture diffuses out of the concrete more rapidly.

Because β_{RH} is negative, the computed shrinkage strain is also negative, implying that the concrete shortens due to shrinkage. In atmospheres with relative humidities greater than 99 percent, β_{RH} is positive, indicating that the concrete swells in such environments.

Variability of Shrinkage

The shrinkage predicted by (3-28) has a coefficient of variation of about 35 percent, which means that 10 percent of the time the shrinkage will be less than 0.55 times the predicted shrinkage and 10 percent of the time it will exceed 1.45 times the predicted value. This is a very large spread. If shrinkage has a critical effect on a given structure, shrinkage tests should be carried out on the concrete in question.

If a lower level of accuracy is acceptable, the following values are representative of the shrinkage that would occur in 70 years in normal-weight structural concrete having strengths between 3000 and 7500 psi: [3-6]

Dry atmospheric conditions or inside, RH = 50 percent:

Effective thickness, h_e = 6 in., $\epsilon_{cs}(70y)$ = −0.000560 strain

Effective thickness = 24 in., $\epsilon_{cs}(70y)$ = −0.000470 strain

Humid atmospheric conditions, RH = 80 percent:

Effective thickness = 6 in., $\epsilon_{cs}(70y)$ = −0.000310 strain

Effective thickness = 24 in., $\epsilon_{cs}(70y)$ = −0.000260 strain

Example 3-2 Calculation of Shrinkage Strains

A lightly reinforced 6-in.-thick floor in an underground parking garage is supported around the outside edge by a 16-in.-thick basement wall. Cracks have developed in the slab perpendicular to the basement wall at roughly 6 ft on centers. The slab is 24 months old and the wall is 26 months old. The concrete is 3000 psi, made from Type I cement, and was moist-cured for 5 days in each case. The average relative humidity is 50 percent. Compute the width of these cracks, assuming that they result from the restraint of the slab shrinkage parallel to the wall by the basement wall.

FLOOR SLAB

1. **Compute the basic shrinkage strain, ϵ_{cso}.**

$$\epsilon_{cso} = \epsilon_s(f_{cm})\beta_{RH} \qquad (3\text{-}29)$$

$$\epsilon_s(f_{cm}) = 1.2[160 + \beta_{sc}(9 - f_{cm}/f_{cmo})] \times 10^{-6} \qquad (3\text{-}30)$$

where

$$\beta_{sc} = 50 \text{ for Type I cement}$$

$$f_{cm} = \text{mean concrete strength} \approx 1.20f'_c$$

$$= 3600 \text{ psi}$$

$$f_{cmo} = 1450 \text{ psi}$$

$$\epsilon_s(f_{cm}) = 1.2[160 + 50(9 - 3600/1450)] \times 10^{-6}$$

$$= 583 \times 10^{-6} = 0.000583 \text{ strain}$$

$$\beta_{RH} = -1.55\left[1 - \left(\frac{RH}{RH_0}\right)^3\right] \qquad (3\text{-}31)$$

where $RH = 50$ percent, $RH_0 = 100$ percent, and

$$\beta_{RH} = -1.55[1 - (50/100)^3]$$

$$= -1.356$$

Therefore, the basic shrinkage is

$$\epsilon_{cso} = 0.000583 \times -1.356$$

$$= -0.000791 \text{ strain}$$

2. **Compute the coefficient for the development of shrinkage with time.**

$$\beta_s(t, t_s) = \left[\frac{(t - t_s)/t_1}{350(h_e/h_0)^2 + (t - t_s)/t_1}\right]^{0.5} \qquad (3\text{-}32)$$

where h_e is the effective thickness $= 2A_c/u$. Consider a 1-ft-wide strip of slab exposed to the air on the top and bottom:

$$h_e = \frac{2 \times (6 \times 12)}{2 \times 12} = 6 \text{ in.} \qquad h_0 = 4 \text{ in.}$$

$$t = 730 \text{ days} \qquad t_s = 5 \text{ days} \qquad t_1 = 1 \text{ day}$$

$$\beta_s(t, t_s) = \left[\frac{(730 - 5)/1}{350 \times (6/4)^2 + (730 - 5)/1}\right]^{0.5}$$

$$= 0.692$$

This means that, after two years, 69 percent of the slab shrinkage will have occurred.

3. **Compute the shrinkage strain, $\epsilon_{cs}(t, t_s)$.**

$$\epsilon_{cs}(t, t_s) = \epsilon_{cso}\beta_s(t, t_s) \qquad (3\text{-}28)$$

$$= -0.000791 \times 0.692$$

$$\epsilon_{cs}(t, t_s) = -0.000547 \text{ strain}$$

BASEMENT WALL

1.   **Compute the basic shrinkage strain, ϵ_{cso}.** This will be the same as for the floor slab, because f_{cm}, β_{sc}, and RH are the same.

2.   **Compute the coefficient for the development of shrinkage with time.** Compute the effective thickness, $h_e = 2A_c/u$. Again, consider a 1-ft-wide strip of wall that is exposed to air only on the inside face. Then

$$h_e = \frac{2 \times (16 \times 12)}{12} = 32 \text{ in.} \qquad h_0 = 4 \text{ in.}$$

$$t = 791 \text{ days} \qquad t_s = 5 \text{ days} \qquad t_1 = 1 \text{ day}$$

$$\beta_s(t, t_s) = \left[\frac{(791 - 5)/1}{350 \times (32/4)^2 + (791 - 5)/1} \right]^{0.5}$$

$$= 0.184$$

3.   **Compute the shrinkage strain, $\epsilon_{cs}(t, t_s)$**

$$e_{cs}(t, t_s) = -0.000791 \times 0.184$$

$$= -0.000146 \text{ strain}$$

RELATIVE SHRINKAGE AND CRACK WIDTH

Thus, the relative shrinkage strain between the slab and the wall when the slab is 24 months old is $-0.000547 - (-0.000146) = -0.000401$. If the average crack spacing is 6 ft, the shortening between cracks will be about $6 \times 12 \times -0.000401 = -0.029$ in. Hence the cracks will be about 0.029 in. wide on average.

This calculation does not allow for the effect of the reinforcement in restraining the shrinkage strains. The actual shrinkage would be from 75 to 100 percent of the calculated values.   ■

Creep of Unrestrained Concrete

When concrete is loaded, an instantaneous elastic strain develops, as shown in Fig. 3-22b. If this load remains on the member, creep strains develop with time. These occur because the adsorbed water layers tend to become thinner between gel particles transmitting compressive stress. This change in thickness occurs rapidly at first, slowing down with time. With time, bonds form between the gel particles in their new position. If the load is eventually removed, a portion of the strain is recovered elastically and another portion by creep, but a residual strain remains (see Fig. 3-22b), due to the bonding of the gel particles in the deformed position.

The creep strains, ϵ_c, are on the order of one to three times the instantaneous elastic strains. Creep strains lead to an increase in deflections with time, may lead to a redistribution of stresses within cross sections, cause a decrease in prestressing forces, and so on.

The ratio of creep strain after a very long time to elastic strain, ϵ_c/ϵ_i, is called the *creep coefficient, ϕ*. The magnitude of the creep coefficient is affected by the ratio of the sustained stress to the strength of the concrete, the humidity of the environment, the dimensions of the element, and the composition of the concrete. Creep is greatest in concretes with a high cement–paste content. Concretes containing a large aggregate fraction creep less, because only the paste creeps and because creep is restrained by the aggregate. The rate of development of the creep strains is also affected by the temperature, reaching a plateau at about 160°F. At the high temperatures encountered in fires, very large creep strains occur. The type of cement (i.e., normal or high-early-strength cement) and the water/cement ratio are important only in that they affect the strength at the time when the concrete is loaded.

For creep, as for shrinkage, several calculation procedures exist [3-6], [3-21]. The method given here is from the *CEB–FIB Model Code 1990* [3-6]. It is applicable for concretes up to compressive strengths of about 10,000 psi subjected to a compressive loading up to about $0.40f'_c$ at an age t_0, exposed to relative humidities of 40 percent or higher and mean temperatures between 40° and 90°F. For stresses less than $0.40f'_c$, creep is assumed to be linearly related to stress. Beyond this stress, creep strains increase more rapidly and may lead to failure of the member at stresses greater than $0.75f'_c$, as shown in Fig. 3-2a. Similarly, creep increases significantly at mean temperatures in excess of 90°F.

The total strain, $\epsilon_c(t)$, at time t in a concrete member uniaxially loaded with a constant stress $\sigma_c(t_0)$ at time t_0 is

$$\epsilon_c(t) = \epsilon_{ci}(t_0) + \epsilon_{cc}(t) + \epsilon_{sc}(t) + \epsilon_{cT}(t) \tag{3-34}$$

where

$\epsilon_{ci}(t_0)$ = initial strain at loading = $\sigma_c(t_0)/E_c(t_0)$
$\epsilon_{cc}(t)$ = creep strain at time t where t is greater than t_0
$\epsilon_{cs}(t)$ = shrinkage strain at time t
$\epsilon_{cT}(t)$ = thermal strain at time t
$E_c(t_0)$ = modulus of elasticity at the age of loading

The stress-dependent strain at time t is

$$\epsilon_{cs}(t) = \epsilon_{ci}(t_0) + \epsilon_{cc}(t) \tag{3-35}$$

For a stress σ_c applied at time t_0 and remaining constant until time t, the creep strain ϵ_{cc} between time t_0 and t is

$$\epsilon_{cc}(t, t_0) = \frac{\sigma_c(t_0)}{E_c(28)}\phi(t, t_0) \tag{3-36}$$

where $E_c(28)$ is the modulus of elasticity at the age of 28 days, given by (3-17) or (3-18), and $\phi(t, t_0)$ is the creep coefficient, given by

$$\phi(t, t_0) = \phi_0\beta_c(t, t_0) \tag{3-37}$$

where ϕ_0 is the basic creep given by (3-38) and $\beta_c(t, t_0)$ is a coefficient to account for the development of creep with time, given by (3-42). We have

$$\phi_0 = \phi_{RH}\beta(f_{cm})\beta(t_0) \tag{3-38}$$

where

$$\phi_{RH} = 1 + \frac{1 - RH/RH_0}{0.46(h_e/h_0)^{1/3}} \tag{3-39}$$

$$\beta(f_{cm}) = \frac{5.3}{(f_{cm}/f_{cmo})^{0.5}} \tag{3-40}$$

$$\beta(t_0) = \frac{1}{0.1 + (t_0/t_1)^{0.2}} \tag{3-41}$$

in which h_e, h_0, RH, RH_0, f_{cm}, f_{cmo}, and t_1 are as defined in connection with (3-29) to (3-33).

The basic creep coefficient, ϕ_0, is actually a function of the relative humidity, the composition of the concrete, and the degree of hydration at the start of loading. The last two of these are expressed empirically in (3-39) and (3-40) as functions of the mean 28-day strength, f_{cm}, and the age at loading, t_0.

The development of creep with time is given by

$$\beta_c(t, t_0) = \left[\frac{(t - t_0)/t_1}{\beta_H + (t - t_0)/t_1}\right]^{0.3} \tag{3-42}$$

with

$$\beta_H = 150\left[1 + \left(1.2\frac{RH}{RH_0}\right)^{18}\right]\frac{h_e}{h_0} + 250 \leq 1500 \tag{3-43}$$

The effects of the effective thickness and age at the time of loading on the creep coefficient $\phi(t, t_0)$ are illustrated in Fig. 3-25. The creep coefficient is about half as big for concrete loaded at one year as it is for concrete loaded at 7 days. The effective thickness has less effect than it did on shrinkage (Fig. 3-24), reducing the value of $\phi(t, t_0)$ by about 20 percent for the example shown.

Variability of Creep Strains

When compared to creep-test data, the creep coefficient $\phi(t, t_0)$ computed in this way has a coefficient of variation of about 20 percent [3-6]. Ten percent of the time the actual value of $\phi(t, t_0)$ will be less than 75 percent of the computed value, and 10 percent of the time it will exceed 125 percent of the computed value. If creep deflections are anticipated to be a serious problem for a particular structure, consideration should be given to carrying out creep tests on the concrete to be used.

In cases where a lower level of accuracy is acceptable, the creep coefficient at 70 years, $\phi(70y, t_0)$, can be taken from Table 3-2.

The total shortening of a plain concrete member at time t due to elastic and creep strains resulting from a constant stress σ_c applied at time t_0 can be computed from (3-35), which becomes

$$\epsilon_{c\sigma}(t) = \sigma_c(t_0)\left[\frac{1}{E_c(t_0)} + \frac{\phi(t, t_0)}{E_c(28)}\right] \tag{3-44}$$

The term in the square brackets is the *creep compliance function*, $J(t, t_0)$, representing the total stress-dependent strain per unit stress. Since creep strains involve the entire member, the term E_c in (3-44) should be based on the average concrete strength, f_{cm}. The modulus

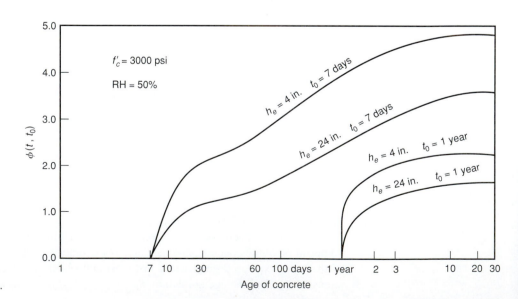

Fig. 3-25
Effect of effective thickness, h_e, and of age at loading, t_0, on creep coefficient, $\phi(t, t_0)$.

TABLE 3-2   Creep Coefficient, $\phi(70y, t_0)$ for Normal-Weight Concrete after 70 Years of Loading

Age at Loading, t_0 (days)	Dry Atmospheric Conditions (RH = 50%)		Humid Atmospheric Conditions (RH = 80%)	
	Effective Thickness, h_e (Eq. 3-32)			
	6 in.	24 in.	6 in.	24 in.
1	4.8	3.9	3.4	3.0
7	3.3	2.7	2.4	2.1
28	2.6	2.1	1.8	1.6
90	2.1	1.7	1.5	1.3
365	1.6	1.3	1.1	1.0

Source: [3-6].

of elasticity should be based on the value given by (3-17) or (3-18) with the average concrete strength, $f_{cm} \approx 1.2f_c'$.

EXAMPLE 3-3   Calculation of Unrestrained Creep Strains

A plain concrete pedestal 24 in. × 24 in. × 10 ft high is subjected to an average stress of 1000 psi. Compute the total shortening in 5 years if the load is applied 2 weeks after the concrete is cast. The properties of the concrete and the exposure are the same as in Example 3-2.

1.   **Compute the basic creep coefficient, ϕ_0.**

$$\phi_0 = \phi_{RH}\beta(f_{cm})\beta(t_0) \tag{3-38}$$

$$\phi_{RH} = 1 + \frac{1 - RH/RH_0}{0.46(h_e/h_0)^{1/3}} \tag{3-39}$$

where

$$RH = 50\% \qquad RH_0 = 100\% \qquad h_0 = 4 \text{ in.}$$

$$h_e = 2A_c/u = 2 \times 24 \times 24/(4 \times 24) = 12 \text{ in.}$$

$$\phi_{RH} = 1 + \frac{1 - 50/100}{0.46(12/4)^{1/3}} = 1.754$$

$$\beta(f_{cm}) = \frac{5.3}{(f_{cm}/f_{cmo})^{0.5}} \tag{3-40}$$

where

$$f_{cm} = 1.20f_c' = 3600 \text{ psi and } f_{cmo} = 1450 \text{ psi}$$

$$\beta(f_{cm}) = \frac{5.3}{(3600/1450)^{0.5}} = 3.364$$

$$\beta(t_0) = \frac{1}{0.1 + (t_0/t_1)^{0.2}} \tag{3-41}$$

where

$$t_0 = 14 \text{ days and } t_1 = 1 \text{ day}$$

$$\beta(t_0) = \frac{1}{0.1 + (14/1)^{0.2}} = 0.557$$

Thus,

$$\phi_0 = 1.754 \times 3.364 \times 0.557 = 3.29$$

The basic creep coefficient is 3.29.

2. **Compute the development of creep with time, $\beta_c(t, t_0)$.**

$$\beta_c(t, t_0) = \left[\frac{(t - t_0)/t_1}{\beta_H + (t - t_0)/t_1} \right]^{0.3} \tag{3-42}$$

where

$$\beta_H = 150\left[1 + \left(1.2\frac{RH}{RH_0} \right)^{18} \right]\frac{h_e}{h_0} + 250 \le 1500 \tag{3-43}$$

$$= 150\left[1 + \left(1.2\frac{50}{100} \right)^{18} \right]\frac{12}{4} + 250 = 700$$

$$t = 5 \times 365 = 1825 \text{ days} \qquad t_0 = 14 \text{ days} \qquad t_1 = 1 \text{ day}$$

$$\beta_c(t, t_0) = \left[\frac{(1825 - 14)/1}{700 + (1825 - 14)/1} \right]^{0.3} = 0.907$$

This indicates that 90.7 percent of the total creep will have occurred by the end of 5 years.

3. **Compute the creep coefficient, $\phi(t, t_0)$.**

$$\phi(t, t_0) = \phi_0\beta_c(t, t_0) = 3.29 \times 0.907 \tag{3-37}$$

$$= 2.98$$

4. **Compute the total stress-dependent strain, $\epsilon_{c\sigma}(t, t_0)$**

$$\epsilon_{c\sigma}(t) = \sigma_c(t_0)\left[\frac{1}{E_c(t_0)} + \frac{\phi(t, t_0)}{E_{ci}} \right] \tag{3-44}$$

where

$$\sigma_c(t_0) = 1000 \text{ psi}$$

$$E_c = 57,000\sqrt{f'_c} \tag{3-18}$$

$E_c(t_0)$ is the modulus of elasticity at 14 days, where the concrete strength at 14 days is given by (3-5) with $t = 14$ days:

$$f'_c(t) = f'_c(28)\left(\frac{t}{4 + 0.85t} \right)$$

$$= 3000\left(\frac{14}{4 + 0.85 \times 14} \right) = 2642 \text{ psi}$$

$$E_c(t_0) = 57,000 \times \sqrt{2642} = 2,930,000 \text{ psi}$$

E_{ci} is the modulus of elasticity at 28 days, based on the mean concrete strength, $f_{cm} = 3600$ psi.

$$E_{ci} = 57,000\sqrt{3600} = 3,420,000 \text{ psi}$$

$$\epsilon_{c\sigma}(5y) = 1000\left(\frac{1}{2,930,000} + \frac{2.98}{3,420,000} \right)$$

$$= 0.000341 + 0.000871 = 0.00121 \text{ strain}$$

The unrestrained creep strain is almost three times the instantaneous strain.

5. **Compute the total shortening.**

$$\Delta\ell = \ell \times \epsilon_{c\sigma}(5y) = 120 \times 0.00121$$

$$= 0.145 \text{ in.}$$

The pedestal would shorten by 0.145 in. in 5 years.   ∎

Restrained Creep

In an axially loaded reinforced concrete column, the creep shortening of the concrete causes compressive strains in the longitudinal reinforcement, increasing the load in the steel and reducing the load, and hence the stress, in the concrete. As a result, a portion of the elastic strain in the concrete is recovered and, in addition, the creep strains are smaller than they would be in a plain concrete column with the same initial concrete stress. A similar redistribution occurs in the compression zone of a beam with compression steel.

This effect can be modeled using an *age-adjusted effective modulus*, $E_{caa}(t, t_0)$, and an *age-adjusted transformed section* in the calculations [3-47], [3-48] and [3-49], where

$$E_{caa}(t, t_0) = \frac{E_c(t_0)}{1 + \chi(t, t_0)[E_c(t_0)/E_c(28)]\phi(t, t_0)} \tag{3-45}$$

in which $\chi(t, t_0)$ is an aging coefficient that can be approximated by Eq. (3-45) [3-50]

$$\chi(t, t_0) = \frac{t_0^{0.5}}{1 + t_0^{0.5}} \tag{3-46}$$

The axial strain at time t in a column loaded at age t_0 with a constant load P is

$$\epsilon_c(t, t_0) = \frac{P}{A_{traa} \times E_{caa}(t, t_0)} \tag{3-47}$$

where A_{traa} is the age-adjusted transformed area of the column cross section. The concept of the transformed sections is presented in Section 9-2. For more information on the use of the age-adjusted effected modulus, see [3-47] through [3-49].

EXAMPLE 3-4   Computation of the Strains and Stresses in an Axially Loaded Reinforced Concrete Column

A concrete pedestal 24 in. × 24 in. × 10 ft high has eight No. 8 longitudinal bars and is loaded with a load of 630 kips at an age of 2 weeks. Compute the elastic stresses in the concrete and steel at the time of loading and the stresses and strains at an age of 5 years. The properties of the concrete and the exposure are the same as in Examples 3-2 and 3-3.

Steps 1, 2, and **3** are the same as in Example 3-3. The following quantities are computed:

$$f'_c(14) = 2830 \text{ psi} \qquad f'_{cm}(28) = 3600 \text{ psi}$$

$$E_c(14) = 2,930,000 \text{ psi} \qquad E_c(28) = 3,420,000 \text{ psi}$$

$$\phi(t, t_0) = 2.98$$

4. **Compute the transformed area at the instant of loading, A_{tr}.** (Transformed sections are discussed in Section 9-2.)

$$\text{Elastic modular ratio} = n = \frac{E_s}{E_c(14)} = \frac{29,000,000}{2,930,000}$$

$$= 9.90$$

The steel will be "transformed" into concrete, giving the transformed area

$$A_{tr} = A_c + (n - 1)A_s = 576 \text{ in.}^2 + (9.90 - 1) \times 6.32 \text{ in}^2$$
$$= 632 \text{ in.}^2$$

The stress in the concrete is 630,000 lb/632 in.^{2} = 997 psi. The stress in the steel is n times the stress in the concrete = 9.90 × 997 psi = 9869 psi.

5. **Compute the age-adjusted effective modulus, $E_{caa}(t, t_0)$, and the age-adjusted modular ratio, n_{aa}.**

$$E_{caa}(t, t_0) = \frac{E_c(t_0)}{1 + \chi(t, t_0)[E_c(t_0)/E_{cm}(28)]\phi(t, t_0)} \tag{3-45}$$

where

$$\chi(t, t_0) = \frac{t_0^{0.5}}{1 + t_0^{0.5}} = \frac{14^{0.5}}{1 + 14^{0.5}} \tag{3-46}$$

$$= 0.789$$

$$E_{caa}(t, t_0) = \frac{2{,}930{,}000}{1 + 0.789 \times \dfrac{2{,}930{,}000}{3{,}420{,}000} \times 2.98}$$

$$= 972{,}000 \text{ psi}$$

$$\text{Age-adjusted modular ratio, } n_{aa} = \frac{E_s}{E_{caa}(t, t_0)} = \frac{29{,}000{,}000}{972{,}000}$$

$$= 29.8$$

6. **Compute the age-adjusted transformed area, A_{traa}, the stresses in the concrete and in the steel, and the shortening.** Again, the steel will be transformed to concrete.

$$A_{traa} = A_c + (n_{aa} - 1)A_s = 576 \text{ in.}^2 + (29.8 - 1) \times 6.32 \text{ in.}^2$$

$$= 758 \text{ in.}^2$$

$$\text{Stress in concrete} = f_c = \frac{P}{A_{traa}} = \frac{630{,}000 \text{ lb}}{758 \text{ in.}^2}$$

$$= 831 \text{ psi}$$

$$\text{Stress in steel} = n_{aa} \times f_c = 29.8 \times 831 \text{ psi}$$

$$= 24{,}800 \text{ psi}$$

$$\text{Strain} = \frac{f_c}{E_{caa}} = \frac{831}{972{,}000}$$

$$= 0.000855 \text{ strain}$$

$$\text{Shortening } \epsilon \times \ell = 0.000855 \times 120 \text{ in.}$$

$$= 0.103 \text{ in.}$$

The creep has reduced the stress in the concrete from 1000 psi at the time of loading to 831 psi at 5 years. During the same period, the steel stress has increased from 9560 psi to 24,800 psi. A column with less reinforcement would experience a larger increase in the reinforcement stress. To prevent yielding of the steel under sustained loads, ACI Section 10.9.1 sets a lower limit of 1 percent on the reinforcement ratio in columns.

The plain concrete pedestal in Example 3-3, which had a constant concrete stress of 1000 psi throughout the 5-year period, shortened 0.145 in. The pedestal in this example, which had an initial concrete stress of 1000 psi, shortened 70 percent as much. ■

Thermal Expansion

The coefficient of thermal expansion or contraction, α, is affected by such factors as composition of the concrete, moisture content of the concrete, and age of the concrete. Ranges from normal-weight concretes are 5 to 7×10^{-6} *strain*/°F for those made with siliceous aggregates and 3.5 to 5×10^{-6}/°F for concretes made from limestone or calcareous aggregates. Approximate values for lightweight concrete are 3.6 to 6.2×10^{-6}/°F. An all-around value of 5.5×10^{-6}/°F may be used. The coefficient of thermal expansion for reinforcing steel is 6×10^{-6}/°F. In calculations of thermal effects, it is necessary to allow for the time lag between air temperatures and concrete temperatures.

As the temperature rises, so does the coefficient of expansion and at the temperatures experienced in building fires, it may be several times the value at normal operating temperatures [3-51]. The thermal expansion of a floor slab in a fire may be large enough to exert large shear-forces on the supporting columns.

3-7   HIGH-STRENGTH CONCRETE

Concretes with 28-day strengths in excess of 6000 psi are referred to as *high-strength concretes*. Strengths of up to 18,000 psi have been used in buildings. Reference [3-8] presents the state of the art of the production and use of high-strength concrete.

Admixtures such as superplasticizers improve the dispersion of cement in the mix and produce workable concretes with much lower water/cement ratios than were previously possible. The resulting concrete has a lower void ratio and is stronger than normal concretes. Most high-strength concretes have water-to-cementitious-materials ratios (w/cm ratios) of 0.40 or less. Many have w/cm ratios in the range from 0.25 to 0.35, with 0.30 close to an optimum value. Workable concrete with these low w/cm ratios is made possible through the use of large amounts of superplasticizers to improve the workability. Only the amount of water needed to hydrate the cement in the mix is provided. This results in concrete with a dense amorphous structure without voids. Some high-strength concretes have less water than this and some of the cement does not hydrate. Coarse aggregates should consist of strong fine-grained gravel with a rough surface. Smooth river gravels give a lower paste–aggregate bond strength and a weaker concrete.

Rigid concrete control must be enforced at the job site, because all shortcomings in selection of aggregates, amounts of water used in mixes, placing, curing, and the like, lead to weaker concrete. Attention should be given to limiting and controlling the temperature rise due to hydration.

High-Performance Concrete

The term *high-performance concrete* is used to refer to concrete with special properties, such as ease of placement and consolidation, high early-age strength to allow early stripping of forms, durability, and high strength. High-strength concrete is only one type of high-performance concrete.

Mechanical Properties

Many of the mechanical properties of high-strength concretes are reviewed in [3-8], [3-52], and [3-53]. It is important to remember that high-strength concrete is not a unique material with a unique set of properties. For example, the modulus of elasticity is strongly affected by the modulus of elasticity of the coarse aggregate.

As shown in Fig. 3-18, the stress–strain curves for higher-strength concretes tend to have a more linear loading branch and a steep descending branch. High-strength concrete exhibits less internal microcracking for a given strain than does normal concrete. In normal concrete, unstable microcracking starts to develop at a compressive stress of about $0.75 f_c'$, referred to as the critical stress (See Section 3-2.) In high-strength concrete, the critical stress is about $0.85 f_c'$. Failure occurs by fracture of the aggregate on relatively smooth planes parallel to the direction of the applied stress. The lateral strains tend to be considerably smaller than in lower-strength concrete. One implication of this is that spiral and confining reinforcement may be less effective in increasing the strength and ductility of high-strength concrete column cores.

Equations (3-17) and (3-18) overestimate the modulus of elasticity of concretes with strengths in excess of about 6000 psi. Reference [3-8] proposes that

$$E_c = 40{,}000 \sqrt{f_c'} + 1.0 \times 10^6 \text{ (psi)} \qquad (3\text{-}19)$$

As noted earlier, E_c varies as a function of the modulus of the coarse aggregate.

The modulus of rupture of high-strength concretes ranges from $(7.5 \text{ to } 12)\sqrt{f_c'}$. A lower bound to the split-cylinder tensile test data is given by $6\sqrt{f_c'}$.

28-Day and 56-Day Compression Strengths

High-strength concrete frequently contains admixtures that delay the final strength gain. As a result, the concrete is still gaining strength at 56 days, rather than reaching a maximum at about 28 days. In 2001, Nowak and Szerszen [3-11] collected cylinder-test data on high-strength concretes, including tests of companion cylinders at 28 and 56 days. The data is summarized in Table 3-3.

On average, λ was 1.11 at 28 days, increasing to 1.20 at 56 days, an increase of 8.7 percent between 28 days and 56 days.

In the development of resistance factors for the design of reinforced concrete members such as columns, the strength gain after 28 days is generally ignored, giving a strength reserve of about 5 to 9 percent. If the concrete control is based on reaching the desired strength at 56 days, some or all of this strength reserve has already been used if the nominal section strength is calculated using the 56-day concrete strength. As a result, the strength reserve is partially used up and the true safety index for the structure is less than expected.

Shrinkage and Creep

Shrinkage of concrete is approximately proportional to the percentage of water by volume in the concrete. High-strength concrete has a higher paste content, but the paste has a lower

TABLE 3-3   Differences between 28-day and 56-day Concrete Strengths

Specified f_c'	Age	No. of tests	$\lambda = f_{c,test}'/f_{c,specified}'$	V = coefficient of variation
7000 psi	28 days	210	1.19	0.115
	56 days	58	1.49	0.080
8000 psi	28 days	753	1.09	0.090
	56 days	428	1.09	0.095
10,000 psi	28 days	635	1.13	0.115
	56 days	238	1.18	0.105
12,000 psi	28 days	381	1.04	0.105
	56 days	190	1.17	0.105

water/cement ratio. As a result, the shrinkage of high-strength concrete is about the same as that of normal concrete.

Test data suggest that the creep coefficient, ϕ, for high-strength concrete is considerably less than that for normal concrete [3-8].

3-8 LIGHTWEIGHT CONCRETE

Structural lightweight concrete is concrete having a density between 90 and 120 lb/ft^3 and containing naturally occurring lightweight aggregates such as pumice; artificial aggregates made from shales, slates, or clays that have been expanded by heating; or sintered blast-furnace slag or cinders. Such concrete is used when a saving in dead load is important. Lightweight concrete costs about 20 percent more than normal concrete. The terms "all-lightweight concrete" and "sand-lightweight concrete" refer to mixes having lightweight sand or natural sand, respectively.

The modulus of elasticity of lightweight concrete is less than that of normal concrete and can be computed from (3-17).

The stress–strain curve of lightweight concrete is affected by the lower modulus of elasticity and relative strength of the aggregates and the cement paste. If the aggregate is the weaker of the two, failure tends to occur suddenly in the aggregate, and the descending branch of the stress–strain curve is very short or nonexistent, as shown by the upper solid line in Fig. 3-26. On the other hand, if the aggregate does not fail, the stress–strain curve will have a well-defined descending branch, as shown by the curved lower solid line in this figure. As a result of the lower modulus of elasticity of lightweight concrete, the strain at which the maximum compressive stress is reached is higher than for normal-weight concrete.

The tensile strength of all-lightweight concrete is 70 to 100 percent of that of normal-weight concrete. Sand-lightweight concrete has tensile strengths in the range from 80 to 100 percent of those of normal-weight concrete. The fracture surface of lightweight concrete tends to be smoother than for normal concrete, because the aggregate fractures.

The shrinkage and creep of lightweight concrete are similar to or slightly greater than those for normal concrete. The creep coefficients computed from (3-37) can be used for lightweight concrete.

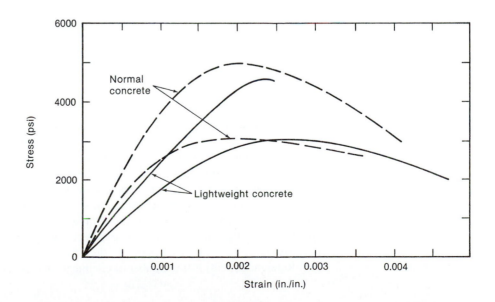

Fig. 3-26
Compressive stress–strain curves for normal-weight and lightweight concretes, $f'_c = 3000$ and 5000 psi. (From [3-54].)

3-9 DURABILITY OF CONCRETE

The durability of concrete structures is discussed in [3-55]. The three most common durability problems in concrete structures are the following:

1. **Corrosion of steel in the concrete.** Corrosion involves oxidation of the reinforcement. For corrosion to occur, there must be a source of oxygen and moisture, both of which diffuse through the concrete. Typically, the pH value of new concrete is on the order of 13. The alkaline nature of concrete tends to prevent corrosion from occurring. If there is a source of chloride ions, these also diffuse through the concrete, decreasing the pH of the part the chloride ions have penetrated. When the pH of the concrete adjacent to the bars drops below about 10 or 11, corrosion can start. The thicker and less permeable the cover concrete is, the longer it takes for moisture, oxygen, and chloride ions to reach the bars. Shrinkage or flexural cracks penetrating the cover allow these agents to reach the bars more rapidly. The rust products that are formed when reinforcement corrodes have several times the volume of the metal that has corroded. This increase in volume causes cracking and spalling of the concrete adjacent to the bars. Factors affecting corrosion are discussed in [3-56].

ACI Section 4.4 attempts to control corrosion of steel in concrete by requiring a minimum strength and a maximum water-to-cementitious-materials ratio to reduce the permeability of the concrete and by requiring at least a minimum cover to the reinforcing bars. The amount of chlorides in the mix is also restricted. Epoxy-coated bars are sometimes used to delay or prevent corrosion.

Corrosion is most serious under conditions of intermittent wetting and drying. Adequate drainage should be provided to allow water to drain off structures. Corrosion is seldom a problem for permanently submerged portions of structures.

2. **Breakdown of the structure of the concrete due to freezing and thawing.** When concrete freezes, pressures develop in the water in the pores, leading to a breakdown of the structure of the concrete. *Entrained air* provides closely spaced microscopic voids, which relieve these pressures. ACI Section 4.2.1 requires minimum air contents to reduce the effects of freezing and thawing exposures. The spacing of the air voids is also important, and some specifications specify spacing factors. ACI Section 4.2.2 sets maximum water/cementitious materials ratios of 0.40 to 0.50 and minimum concrete strengths of 4000 to 5000 psi for concretes, depending on the severity of the exposure. These can give strengths higher than would otherwise be used in structural design. Thus a water/cement ratio of 0.40 will generally correspond to a strength of 4500 to 5000 psi for air-entrained concrete. This additional strength can be utilized in computing the strength of the structure. If the degree of saturation exceeds 85 percent, air entrainment will not provide freeze–thaw resistance [3-57].

Again, drainage should be provided so that water does not collect on the surface of the concrete. Concrete should not be allowed to freeze at a very young age and should be allowed to dry out before severe freezing.

3. **Breakdown of the structure of the concrete due to chemical attack.** Sulfates cause disintegration of concrete unless special cements are used. ACI Section 4.3.1 specifies cement type, minimum water/cementitious materials ratios, and minimum compressive strengths for various sulfate exposures. Geotechnical reports will generally give sulfate levels.

Some aggregates containing silica react with the alkalies in the cement, causing a disruptive expansion of the concrete, leading to severe random cracking. This *alkali silica reaction* is counteracted by changing the source of the aggregate or by using low-alkali cements [3-58]. It is most serious if the concrete is warm in service and if there is a source of moisture.

Reference [3-59] lists a number of chemicals that attack concretes.

ACI 318 Chapter 4 presents requirements for concrete that is exposed to freezing, thawing, deicing chemicals, sulfates, and chlorides. Examples are pavements, bridge decks, parking garages, water tanks, and foundations in sulfate-rich soils.

Sulfates cause disintegration of concrete unless special cements are used. ACI 318 Table 4.3.1 gives special requirements for concrete in contact with sulfates in soils or in water. In many areas in the western United States, soils contain sulfates.

3-10   BEHAVIOR OF CONCRETE EXPOSED TO HIGH AND LOW TEMPERATURES

High Temperatures and Fire

When a concrete member is exposed to high temperatures such as occur in a building fire, for example, it will behave satisfactorily for a considerable period of time. During a fire, high thermal gradients occur, however, and as a result, the surface layers expand and eventually crack or spall off the cooler, interior part of the concrete. The spalling is aggravated if water from fire hoses suddenly cools the surface.

The modulus of elasticity and the strength of concrete decrease at high temperatures, whereas the coefficient of thermal expansion increases [3-51]. The type of aggregate affects the strength reduction, as is shown in Fig. 3-27. Most structural concretes can be classified into one of three aggregate types: carbonate, siliceous, or lightweight. Concretes made with carbonate aggregates, such as limestone and dolomite, are relatively unaffected by temperature until they reach about 1200 to 1300°F, at which time they undergo a chemical change and rapidly lose strength. The quartz in siliceous aggregates, such as quartzite, granite, sandstones, and schists, undergoes a phase change at about 800 to 1000°F, which causes an abrupt change in volume and spalling of the surface. Lightweight aggregates gradually lose their strength at temperatures above 1200°F.

The reduction in strength and the extent of spalling due to heat are most pronounced in wet concrete and, as a result, fire is most critical with young concrete. The tensile strength tends to be affected more by temperature than does the compressive strength.

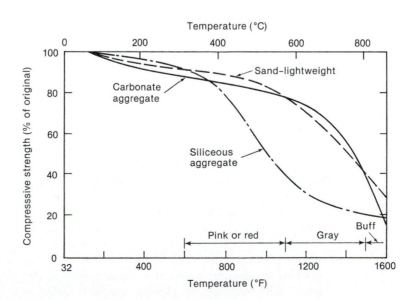

Fig. 3-27
Compressive strength of concretes at high temperatures.
(From [3-51].)

Concretes made with limestone and siliceous aggregates tend to change color when heated, as indicated in Fig. 3-27, and the color of the concrete after a fire can be used as a rough guide to the temperature reached by the concrete. As a general rule, concrete whose color has changed beyond pink is suspect. Concrete that has passed the pink stage and gone into the gray stage is probably badly damaged. Such concrete should be chipped away and replaced with a layer of new concrete or shotcrete.

Very Cold Temperatures

In low temperatures, the strength of hardened concrete tends to increase, the increase being greatest for moist concrete, as long as the water does not freeze [3-59]. Very cold temperatures are encountered in liquid-natural-gas storage facilities.

Subfreezing temperatures can significantly increase the compressive and tensile strengths and the modulus of elasticity of moist concrete. Dry concrete properties are not affected as much by low temperature. Reference [3-60] reports compression tests of moist concrete with a strength of 5000 psi at 75°F which reached a strength of 17,000 psi at −150°F. The same concrete tested oven-dry or at an interior relative humidity of 50 percent showed a 20 percent increase in compressive strength relative to the strength at 75°F.

The split-cylinder tensile strength of the same concrete increased from 600 psi at +75°F to 1350 psi at −75°F [3-60].

3-11 SHOTCRETE

Shotcrete is concrete or mortar that is pneumatically projected onto a surface at high velocity. A mixture of sand, water, and cement is sprayed through a nozzle. Shotcrete is used as new structural concrete or as repair material. It has properties similar to those of cast-in-place concrete, except that the properties depend on the skill of the nozzleperson who applies the material. Further information on shotcrete is available in [3-61].

3-12 HIGH-ALUMINA CEMENT

High-alumina cement is occasionally used in structures. Concretes made from this type of cement have an unstable crystalline structure that could lose its strength over time, especially if exposed to moderate to high humidities and temperatures [3-62]. In general, high-alumina cements should be avoided in structural applications.

3-13 REINFORCEMENT

Because concrete is weak in tension, it is reinforced with steel bars or wires that resist the tensile stresses. The most common types of reinforcement for nonprestressed members are hot-rolled deformed bars and wire fabric. In this book, only the former will be used in examples, although the design principles apply with very few exceptions to members reinforced with welded wire mesh or cold-worked deformed bars.

The ACI Code requires that reinforcement be steel bars or steel wires. Significant modifications to the design process are required if materials such as fiber-reinforced-plastic (FRP) rods are used for reinforcement because such materials are brittle and do not have the ductility assumed in the derivation of design procedures for concrete reinforced with steel bars. In addition, special attention must be given to the anchorage of FRP reinforcement. See Section 3-14.

Hot-Rolled Deformed Bars

Grades, Types, and Sizes

Steel reinforcing bars are basically round in cross section, with lugs or deformations rolled into the surface to aid in anchoring the bars in the concrete (Fig. 3-28). They are produced according to the following ASTM specifications, which specify certain dimensions and certain chemical and mechanical properties:

1.   **ASTM A 615:** *Standard Specification for Deformed and Plain Billet-Steel Bars for Concrete Reinforcement.* This specification covers the most commonly used reinforcing bars. They are available in sizes 3 to 18 in Grade 60 (yield strength of 60 ksi) plus sizes 3 to 6 in Grade 40 and sizes 6 to 18 in Grade 75. The specified mechanical properties are summarized in Table 3-4. The diameters, areas, and weights are listed in Table A-1 in Appendix A. The phosphorus content is limited to ≤0.06 percent.

2.   **ASTM A 706:** *Standard Specification for Low-Alloy Steel Deformed Bars for Concrete Reinforcement.* This specification covers bars intended for special applications where weldability, bendability, or ductility is important. As indicated in Table 3-4, the A 706 specification requires a larger elongation at failure and a more stringent bend test than A 615. ACI Section 21.2.5.1 requires the use of A 615 bars meeting special requirements or A 706 bars in seismic applications. A 706 limits the amounts of carbon, manganese, phosphorus, sulfur, and silicon and limits the carbon equivalent to ≤0.55 percent. These bars are available in sizes 3 through 18 in Grade 60. There are both a lower and an upper limit on the yield strength.

3.   **ASTM A 996:** *Standard Specification for Rail-Steel and Axle-Steel Deformed Bars for Concrete Reinforcement.* This specification covers bars rolled from discarded railroad rails or from discarded train car axles. It is less ductile and less bendable than A 615 steel. Only Type R rail-steel bars with R rolled into the bar are permitted by the ACI Code. These bars are not widely available.

Reinforcing bars are available in four grades, with yield strengths at 40, 50, 60, and 75 ksi, referred to as Grades 40, 50, 60, and 75, respectively. Grade 60 is the steel most commonly used in buildings and bridges. Other grades may not be available in some areas. Grade 75 is used in large columns. Grade 40 is the most ductile, followed by Grades 60, 75, and 50, in that order.

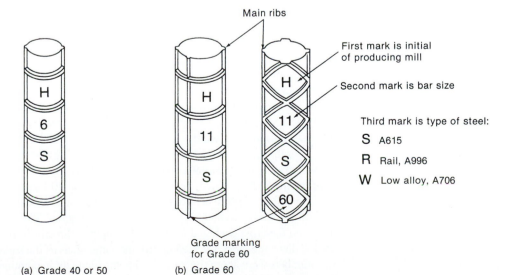

Fig. 3-28
Standard reinforcing-bar
markings. (Courtesy of
Concrete Reinforcing Steel
Institute.)

TABLE 3-4   Summary of Mechanical Properties of Reinforcing
Bars from ASTM A 615 and ASTM A 706

	Billet-Steel A 615			Low-alloy steel, A 706
	Grade 40	Grade 60	Grade 75	Grade 60
Minimum tensile strength, psi	70,000	90,000	100,000	80,000[a]
Minimum yield strength, psi	40,000	60,000	75,000	60,000
Maximum yield strength, psi	—	—	—	78,000
Minimum elongation in 8-in. gauge length, percent				
No. 3	11	9	—	14
No. 4 and 5	12	9	—	14
No. 6	12	9	7	14
No. 7 and 8	—	8	7	12
No. 9, 10, and 11	—	7	6	12
No. 14 and 18	—	7	6	10
Pin diameter for bend test,[b] where d = nominal bar diameter				
No. 3, 4, and 5	$3.5d$	$3.5d$	—	$3d$
No. 6	$5d$	$5d$	$5d$	$4d$
No. 7 and 8	—	$5d$	$5d$	$4d$
No. 9, 10, and 11	—	$7d$	$7d$	$6d$
No. 14 and 18	—	$9d$	$9d$	$8d$

[a] But not more than 1.25 times the actual yield.

[b] Bend tests are 180°, except that 90° bends are permitted for No. 14 and 18 A 615 bars.

Grade-60 deformed reinforcing bars are available in the 11 sizes listed in Table A-1. The sizes are referred to by their nominal diameter expressed in eighths of an inch. Thus, a No. 4 bar has a diameter of $\frac{4}{8}$ in. (or $\frac{1}{2}$ in.). The nominal cross-sectional area can be computed directly from the nominal diameter, except for that of the No. 10 and larger bars, which have diameters slightly larger than $\frac{10}{8}$ in., $\frac{11}{8}$ in., and so on. Size and grade marks are rolled into the bars for identification purposes, as shown in Fig. 3-28. Grade-40 bars are available only in sizes 3 through 6. Grade-75 steel is available only in sizes 6 to 18.

ASTM A 615 and A 706 also specify metric (SI) bar sizes. They are available in 11 sizes. Each is the same as an existing inch–pound bar size, but is referred to by its nominal diameter in whole millimeters. The sizes are #10, #13, #16, #19, #22, #25, #29, #32, #36, #43, and #57, corresponding to the nominal diameters 10 mm, 13 mm, 16 mm, and so on. The nominal diameters of metric reinforcement are the traditional U.S. Customary unit diameters—$\frac{3}{8}$ in. (9.5 mm), $\frac{4}{8}$ in. (12.7 mm), $\frac{5}{8}$ in. (15.9 mm) and so on—rounded to the nearest whole millimeter. During a transition period in which metric bars are being introduced, the bar size designation will include an "M" to denote a metric size bar. The diameters, areas, and weights of SI bar sizes are listed in Table A-1M in Appendix A.

ASTM A 615 defines three grades of metric reinforcing bars: Grades 300, 420, and 520, having specified yield strengths of 300, 420 and 520 MPa, respectively.

For the review of the strength of existing buildings the yield strength of the bars must be known. Prior to the late 1960s, reinforcing bars were available in *structural, intermediate,* and *hard* grades with specified yield strengths of 33 ksi, 40 ksi, and 50 ksi (228 MPa, 276 MPa, and 345 MPa), respectively. Reinforcing bars were available in inch–pound sizes 3 to 11, 14, and 18. For sizes 3 to 8, the size number was the nominal diameter of the bar in eighths of an inch, and the cross-sectional areas were computed directly from this diameter. For sizes 9 to 18, the diameters were selected to give the same areas as the square bars used earlier, and the size numbers were approximately equal to the diameter in eighths of an inch. In the 1970s, the 33-ksi and 50-ksi bars were dropped, and a new 60-ksi yield strength was introduced.

Mechanical Properties

Idealized stress–strain relationships are given in Fig. 3-29 for Grade-40, -60, and -75 reinforcing bars, and for welded-wire fabric. The initial tangent modulus of elasticity, E_s, for all reinforcing bars can be taken as 29×10^6 psi. Grade 40 bars display a pronounced yield plateau, as shown in Fig. 3-29. Although this plateau is generally present for Grade-60 bars, it is typically much shorter. High-strength bars generally do not have a well-defined yield point.

Figure 3-30 is a histogram of mill-test yield strengths of Grade-60 reinforcement having a nominal yield strength of 60 ksi. As shown in this figure, there is a considerable variation in yield strength, with about 10 percent of the tests having a yield strength equal to or greater than 80 ksi, 133 percent of the nominal yield strength. The coefficient of variation of the yield strengths plotted in Fig. 3-30 is 9.3 percent.

ASTM specifications base the yield strength on *mill tests* that are carried out at a high rate of loading. For the slow loading rates associated with dead loads or for many live loads, the *static yield strength* is applicable. This is roughly 4 ksi less than the mill-test yield strength [3-63].

Fatigue Strength

Some reinforced concrete elements, such as bridge decks, are subjected to a large number of loading cycles. In such cases, the reinforcement may fail in fatigue. Fatigue failures of the reinforcement will occur only if one or both of the extreme stresses in the stress cycle is tensile. The relationship between the range of stress, S_r, and the number of cycles is shown in Fig. 3-31. For practical purposes there is a fatigue threshold or *endurance limit* below which fatigue failures will normally not occur. For straight ASTM A 615 bars, this is about 24 ksi and is essentially the same for Grade-40 and Grade-60 bars. If there are fewer than 20,000 cycles, fatigue will not be a problem with deformed-bar reinforcement.

The fatigue strength of deformed bars decreases:

    (a)   as the stress range (the maximum tensile stress in a cycle minus the algebraic minimum stress) increases,

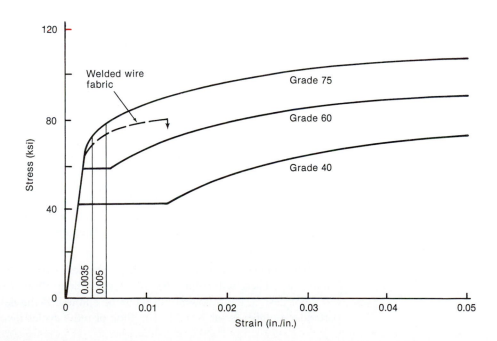

Fig. 3-29
Stress–strain curves for reinforcement.

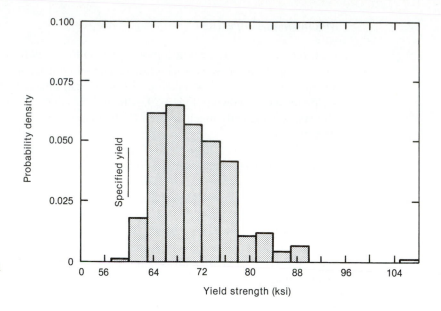

Fig. 3-30
Distribution of mill-test yield
strengths for Grade-60 steel.
(From [3-63].)

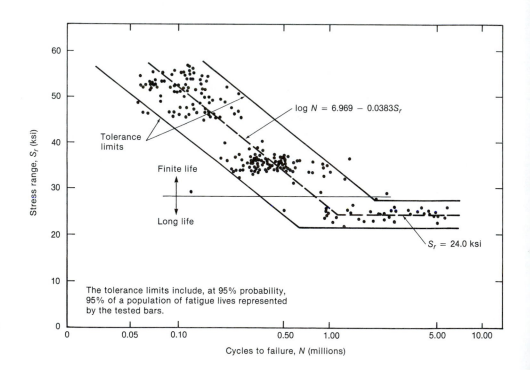

Fig. 3-31
Test data on fatigue of
deformed bars from a single
U.S. manufacturer.
(From [3-64].)

(b) as the level of the lower (less tensile) stress in the cycle is reduced, and

(c) as the ratio of the radius of the fillet at the base of the deformation lugs to the height of the lugs is decreased. The fatigue strength is essentially independent of the yield strength.

The fatigue strength is greatly reduced by bends or tack welds in the region of maximum stress. These will reduce the fatigue strength by about 50 percent.

For design, the following rules can be applied: If the deformed reinforcement in a particular member is subjected to 1 million or more cycles involving tensile stresses, or a combination of tension and compression stresses, fatigue failures may occur if the difference between the maximum and minimum stresses under the repeated loading exceeds 20 ksi. In the vicinity of bends or in locations where auxiliary reinforcement has been tack-welded to the main reinforcement,

      (d)   In the vicinity of welds or bends, fatigue failures may occur if the stress range exceeds 10 ksi. Further guidance is given in [3-65]. Design of reinforced concrete for fatigue is discussed in Section 9-8.

Strength at High Temperatures   Deformed-steel reinforcement subjected to high temperatures in fires tends to lose some of its strength, as shown in Fig. 3-32 [3-51]. When the temperature of the reinforcement exceeds about 850°F, both the yield and ultimate strengths drop significantly. One of the functions of concrete cover on reinforcement is to prevent the reinforcement from getting hot enough to lose strength.

Welded-Wire Reinforcement

Welded-wire reinforcement is a prefabricated reinforcement consisting of smooth or deformed wires welded together in square or rectangular grids. Sheets of wires are welded in electric-resistance welding machines in a production line. This type of reinforcement is used in walls or slabs where relatively regular reinforcement patterns are possible. The ability to place a large amount of reinforcement with a minimum of work frequently makes welded-wire fabric economical.

      The wire for welded-wire fabric is produced in accordance with the following specifications: ASTM A 82 *Specification for Steel Wire, Plain, for Concrete Reinforcement*, and ASTM A 496 *Specification for Steel Wire, Deformed, for Concrete Reinforcement*. The deformations are typically two or more lines of *indentations* of about 4 to 5 percent of the bar diameter, rolled into the wire surface. As a result, the deformations on wires are less pronounced than on deformed bars. Wire sizes range from about 0.125 in. diameter to 0.625 in. diameter and are referred to as W or D, for plain or deformed wires, respectively, followed by a number that corresponds to the cross-sectional area of the wire in 0.01-in.^{2} increments. Thus a W2 wire is a smooth wire with a cross-sectional area of

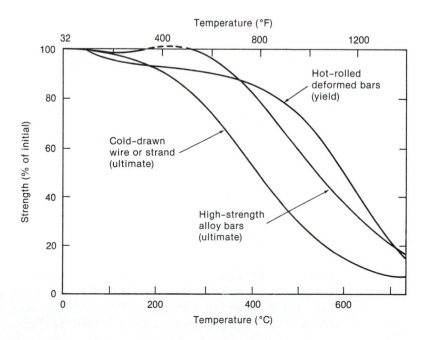

Fig. 3-32
Strength of reinforcing steels
at high temperatures.
(From [3-51].)

0.02 in.^{2} ACI Section 3.5, 3.4 does not allow wires smaller than size D4. Diameters and areas of typical wire sizes are given in Table A-2a.

Welded-wire fabric satisfies the following specifications: ASTM A 185 *Specification for Steel Wire Fabric, Plain, for Concrete Reinforcement*, and ASTM A 497 *Specification for Steel Wire Fabric, Deformed, for Concrete Reinforcement*. Deformed welded-wire fabric may contain some smooth wires in either direction. Welded-wire fabric is available in standard or custom patterns, referred to by a style designation (such as 6 × 6—W4 × W4). The numbers in the style designation refer to: (spacing of longitudinal wires × spacing of transverse wires—size of longitudinal wires × size of transverse wires). Thus a 6 × 6–W4 × W4 fabric has W4 wires at 6 in. on centers each way. Areas and weights of common welded-wire fabric patterns are given in Table A-2b.

Welded smooth-wire fabric depends on the cross-wires to provide a mechanical anchorage with the concrete, while welded deformed-wire fabric utilizes both the wire deformations and the cross-wires for bond and anchorage. In smooth wires, two cross-wires are needed to mechanically anchor the bar for its yield strength.

Smooth and deformed fabric are made from wires ranging from about 0.125 in. to 0.625 in. in diameter. The minimum yield and tensile strength of smooth wire for wire fabric is 65 ksi and 75 ksi. For deformed wires, the minimum yield and tensile strengths are 70 ksi and 80 ksi. According to ASTM A 497, these yield strengths are measured at a strain of 0.5 percent. ACI Sections 3.5.3.5 and 3.5.3.6 define the yield strength of both smooth and deformed wires as 60 ksi, except that if the yield strength at a strain of 0.35 percent has been measured, that value can be used.

The elongation at failure decreases as the wire size decreases, because the cold-working process used in drawing the small-diameter wires strain-hardens the steel. Reference [3-66] quotes tests indicating that the mean elongation at failure ranges from about 1.25 percent for W1.4 wires (0.133 in. diameter) to about 6 percent for W31 wires (0.628 in. diameter). These are smaller than the elongations at failure of reinforcing bars, given in Table 3-4, which range from 6 to 11 percent. There is no ACI Code limitation on minimum elongation at failure in tension tests. If it is assumed that 3 percent elongation is adequate for moment redistribution in structures reinforced with A 615 bars, wires of sizes W8.5 or D8.5 (0.328 in. diameter) or bigger should have adequate ductility. References [3-67] and [3-68] describe tests in which welded-wire fabric showed adequate ductility for use as stirrups or joint ties in members tested under cyclic loads.

3-14   FIBER-REINFORCED POLYMER (FRP) REINFORCEMENT

Since 1990, extensive research has been carried out on structures reinforced with *fiber-reinforced polymer reinforcement* (FRP) in the form of bars or preformed two-dimensional grids. These consist of aligned fibers encased in a hardened resin and are made by a number of processes, including pultrusion, braiding, and weaving. FRP reinforcement has been used in structures subject to corrosion and in applications that require nonmagnetic bars, such as floors supporting some medical devices (such as MRI machines). Common types are GFRP (made with glass fibers), AFRP (made with Aramid fibers), and CFRP (made with carbon fibers).

Properties of FRP Reinforcement

All types of FRP reinforcement have elastic–brittle stress–strain curves, with ultimate tensile strengths between 60,000 psi and 300,000 psi [3-69]. The strengths and the moduli of elasticity vary, depending on the type of fibers and on the ratio of the volume of fibers to the volume of the FRP bars. Typical values of the modulus of elasticity in tension, expressed as a percentage of the modulus for steel reinforcement, range from 20 to 25 percent for GFRP, from 20 to 60 percent for AFRP, and from 60 to 300 percent for CFRP. The modulus of elasticity in compression ranges from 80 to 85 percent of the modulus of GFRP in tension for CFRP and to 100 percent for AFRP. In a similar manner, compressive strengths on the order of 55 percent, 78 percent, and 20 percent of the tensile strength have been reported [3-69] for GFRP, CFRP, and AFRP, respectively. In some bars, there is a size effect due to shear lag between the surface and the center of the bars, which leads to a lower apparent tensile strength because the interior fibers are not fully stressed at the onset of rupture of the bars.

FRP is susceptible to creep rupture under high sustained tensile loads. Extrapolated strengths after 500,000 hours of sustained loads vary. They are on the order of 47 to 66 percent of the initial ultimate strength for AFRP and 79 to 91 percent for CFRP, depending on the test method.

The bond strength of FRP bars and concrete is affected by the smooth surface of the resin bars. Some bars are manufactured with windings of FRP cords or are coated with sand to improve bond. There is no standardized deformation pattern, however. FRP bars tend to be susceptible to surface damage during construction. FRP bars cannot be bent once the polymer has set. If bent bars are required, they must be bent during manufacture. The polymer resins used to make FRP bars undergo a phase change between 150° and 250°F, causing a reduction in strength. By the time the temperature of the bar reaches 480°F, the tensile strength has dropped to about 20 percent of the strength at room temperature.

The elastic–brittle stress–strain behavior of FRP bars affects the beam-design philosophy. In the ACI beam-design philosophy, the value of the strength-reduction factor, ϕ, ranges from 0.65 for members in which the strain in the extreme tensile layer of steel is zero or compression to 0.90 for beams in which the bar strain at ultimate exceeds 0.005 strain in tension. Design of FRP reinforced beams is discussed in Section 5-6.

PROBLEMS

3-1 What is the significance of the "critical stress"
  (a) with respect to the structure of the concrete?
  (b) with respect to spiral reinforcement?
  (c) with respect to strength under sustained loads?

3-2 A group of 43 tests on a given type of concrete had a mean strength of 3622 psi and a standard deviation of 421 psi. Does this concrete satisfy the requirements of ACI Section 5.3.2 for 3000-psi concrete?

3-3 The concrete containing Type I cement in a structure is cured for 3 days at 70°F, followed by 6 days at 40°F. Use the maturity concept to estimate its strength as a fraction of the 28-day strength under standard curing.

3-4 Use Fig. 3-12a to estimate the compressive strength σ_2 for biaxially loaded concrete subjected to
  (a) $\sigma_1 = 0$.
  (b) $\sigma_1 = 0.75$ times the tensile strength, in tension.
  (c) $\sigma_1 = 0.5$ times the compressive strength, in compression.

3-5 The concrete in the core of a spiral column is subjected to a uniform confining stress σ_3 of 800 psi. What will the compressive strength σ_1 be? The uniaxial compressive strength is 4000 psi.

3-6 What factors affect the shrinkage of concrete?

3-7 What factors affect the creep of concrete?

3-8 A structure is made from concrete containing Type I cement. The average ambient relative humidity is 70 percent. The concrete was moist-cured for 4 days. $f'_c = 4000$ psi.

  (a) Compute the unrestrained shrinkage strain of a rectangular beam with cross-sectional dimensions 8 in. × 20 in. at 3 years after the concrete was placed.

  (b) Compute the stress-dependent strain in the concrete in a 20 in. × 20 in. plain concrete column at age 2 years. A load of 400 kips was applied to the column at age 60 days.

4

Flexure: Basic Concepts, Rectangular Beams

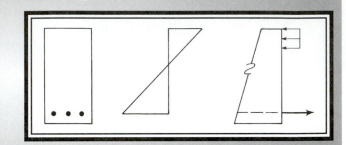

4-1   INTRODUCTION

In this chapter, the stress–strain curves for concrete and reinforcement from Chapter 3 are used to develop a theory for flexure. This theory is applied to rectangular-beam cross sections, and examples of the design of beam cross sections are given. In Chapter 5, this is extended to T beams and beams with compressive steel.

Because the complete design of a beam requires an understanding of shear, bond, and other aspects, a complete design example is deferred until Chapter 10, where all these aspects are considered. In Chapter 11, the flexure theory is extended to combined bending and axial load, to permit the design of columns.

Most reinforced concrete structures can be subdivided into beams and slabs subjected primarily to flexure (bending) and columns subjected to axial compression (accompanied, in most cases, by flexure). Typical examples of flexural members are the slab and beams shown in Fig. 4-1. The load, P, applied at point A is carried by the strip of slab shown shaded. The end reactions from this slab strip load the beams at B and C. The beams, in turn, carry the slab reactions to the columns at $D, E, F,$ and G. The beam reactions cause axial loads in the columns.

The slab in Fig. 4-1 is assumed to transfer loads in one direction and hence is called a *one-way slab*. If there were no beams, the slab would carry the load in two directions. Such a slab is referred to as a *two-way slab*. (See Chapters 13 to 15.)

B-Regions and D-Regions

Through most of the length of a beam or column, a straight-line distribution of strains will exist, and the normal flexure theory can be applied. Such regions are referred to as *B-regions*, where the B stands for beam or for Bernoulli, who first postulated the straight-line strain distribution. Adjacent to discontinuities, concentrated loads, holes, or changes in cross section, the strain distribution is not linear, and different types of analysis must be used. Such

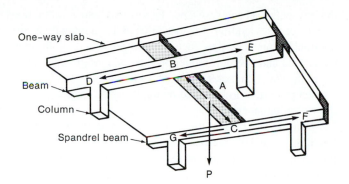

Fig. 4-1
One-way flexure.

regions are referred to as *D-regions*, where the D stands for discontinuity or for disturbed. Chapters 4 to 15 deal primarily with B-regions. D-regions are discussed in Chapter 18.

Analysis versus Design

Two different types of problems arise in the study of reinforced concrete:

1. *Analysis.* Given a cross section, concrete strength, reinforcement size, location, and yield strength, compute the resistance or capacity.

2. *Design.* Given a factored load effect such as M_u, select a suitable cross section, including dimensions, concrete strength, reinforcement, and so on.

Although both types of problem are based on the same principles, the procedure is different in each case. Analysis is easier, because all the decisions concerning reinforcement, beam size, and so on have been made and it is only necessary to apply the strength-calculation principles to determine the capacity. Design, on the other hand, involves the choice of beam sizes, material strengths, and reinforcement to produce a cross section that can resist the loads and moments. Because the analysis problem is easier, most sections in this and other chapters start with the analysis to develop the fundamental concepts and then move to consider design.

Required Strength and Design Strength

The basic safety equation for flexure is

$$\text{factored resistance} \geq \text{factored load effects} \tag{4-1a}$$

or

$$\phi M_n \geq M_u \tag{4-1b}$$

where M_u is the *moment due to the factored loads*, which the ACI Code refers to as the *required ultimate moment*. This is a load effect computed by structural analysis from the governing combination of factored loads given in ACI Section 9.2. The term M_n refers to the *nominal moment capacity* of a cross section, computed from the nominal dimensions and specified material strengths. The factor ϕ in (4-1b) is a *strength-reduction factor* (ACI Section 9.3) to account for possible variations in dimensions and material strengths and possible inaccuracies in the strength equations. A major revision in the ACI Code *load factors* and *strength-reduction factors* was started in the 1995 code with the inclusion of an Appendix C that presents load and resistance factors and load combinations from ASCE 7, the standard which specifies loadings for structures [4-1], along with a corresponding set of strength-reduction factors. The new factors were presented as an alternative to the traditional ACI Code load and strength-reduction factors. Design was either to be based on the load and strength-reduction factors from ACI 318-02, Sections 9.2 and 9.3, throughout the design of a given project, or to be based on the load and resistance factors from ACI 318-02, Appendix C, throughout.

In the 2002 code, the load factors and strength-reduction factors in the 1999 ACI Sections 9.2 and 9.3 were interchanged with those in the 1999 ACI Appendix C. As part of this change, a reevaluation of the total safety level for beams resulted in having the strength reduction factor, ϕ, for flexure remain equal to 0.90, as it had been, even though the 2002 load factors were lower than the 1999 and earlier load factors.

It is anticipated that starting with the 2008 code, the traditional load and resistance factors will no longer be sanctioned for design use.

For flexure without axial load, both Section 9.3.2.1 and Section C.3.2.1 of ACI 318-02 give $\phi = 0.90$ for what are called *tension-controlled sections*. Most practical beams will be tension-controlled sections, and ϕ will be equal to 0.90. The concept of tension-controlled sections will be discussed later in the chapter. The product, ϕM_n, is referred to as the *design moment, design strength*, or *factored moment resistance*.

Positive and Negative Moments

A moment that causes compression on the top surface of a beam and tension on the bottom surface will be called a *positive moment*. The compression zones for positive and negative moments are shown shaded in Fig. 4-2. In this textbook, bending-moment diagrams will be plotted on the compression side of the member.

Symbols and Abbreviations

Although symbols are defined as they are first used and are summarized in Appendix B, several *must be understood completely* if one is to understand the theory developed in this

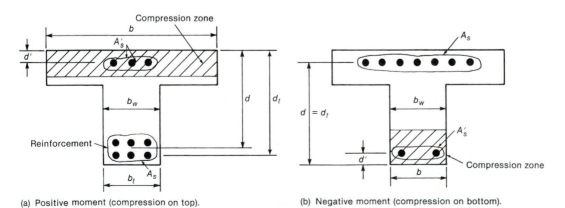

(a) Positive moment (compression on top).　　(b) Negative moment (compression on bottom).

Fig. 4-2
Cross-sectional dimensions.

book. These include the terms M_u and M_n (defined earlier) and the cross-sectional dimensions illustrated in Fig. 4-2. The following is a list of symbols used in this book:

The prime symbol ($'$) generally refers to compressions, as in A'_s and d'.

A_s is the area of reinforcement on the tension face of the beam, tension reinforcement, in.^{2}.

A'_s is the area of reinforcement on the compression side of the beam, compression reinforcement, in.^{2}.

b is the width of the compression face of the beam. This is illustrated in Fig. 4-2 for positive and negative moment regions, in.

b_w is the width of the web of the beam (and may or may not be the same as b), in.

d is the distance from the extreme fiber in compression to the centroid of the steel on the tension side of the member, in. In the positive-moment region (Fig. 4-2a), the tension steel is near the bottom of the beam, while in the negative-moment region (Fig. 4-2b) it is near the top.

d_t is the distance from the extreme compression fiber to the farthest layer of tension steel, in.

f'_c is the specified 28-day compressive strength of the concrete, psi.

f_s is the stress in the tension reinforcement, psi.

f_y is the specified yield strength of the reinforcement, psi.

h is the overall height of a beam cross section.

jd is the *lever arm*, the distance between the resultant compressive force and the resultant tensile force, in.

j is a dimensionless ratio used to define the lever arm, jd. It varies during the life of the beam.

ϵ_{cu} is the assumed concrete strain on the compression face of the beam at flexural failure.

ϵ_s is the strain in the tension reinforcement.

ϵ_t is the strain in the extreme tension reinforcement.

ρ is the longitudinal tension reinforcement ratio, $\rho = A_s/bd$.

4-2   FLEXURE THEORY

Statics of Beam Action

A *beam* is a structural member that supports applied loads and its own weight primarily by internal moments and shears. Figure 4-3a shows a simple beam that supports its own dead weight, w per unit length, plus an applied load, P. If the axial applied load, N, is equal to zero, as shown, the member is referred to as a beam. If N is a compressive force, the member is called a *beam–column*. If it were tensile, the member would be a tension tie. Chapter 4 will be restricted to the very common case where $N = 0$.

The loads w and P cause *bending moments*, distributed as shown in Fig. 4-3b. The bending moment is a *load effect* calculated from the loads by using the laws of statics. For a simply supported beam of a given span and for a given set of loads w and P, the moments are independent of the composition and size of the beam.

At any section within the beam, the *internal resisting moment*, M, shown in Fig. 4-3c is necessary to equilibrate the bending moment. An internal resisting shear, V, is also required, as shown.

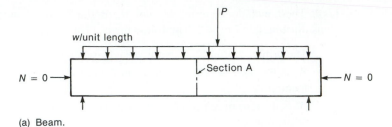

(a) Beam.

(b) Bending moment diagram.

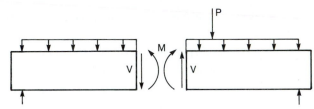

(c) Free body diagrams showing internal moment and shear force.

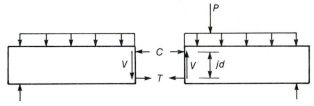

Fig. 4-3
Internal forces in a beam.

(d) Free body diagrams showing internal moment as a compression–tension force couple.

The internal resisting moment, M, results from an internal compressive force, C, and an internal tensile force, T, separated by a lever arm, jd, as shown in Fig. 4-3d. Since there are no external axial loads, N, summation of the horizontal forces gives

$$C - T = 0 \qquad \text{or} \qquad C = T \tag{4-2}$$

If moments are summed about an axis through the point of application of the compressive force, C, the moment equilibrium of the free body gives

$$M = T \times jd \tag{4-3a}$$

Similarly, if moments are summed about the point of application of the tensile force, T,

$$M = C \times jd \tag{4-3b}$$

Because $C = T$, these two equations are identical. Equations 4-2 and 4-3 come directly from statics and are equally applicable to beams made of steel, wood, or reinforced concrete.

The conventional *elastic* beam theory results in the equation $\sigma = My/I$, which, for an uncracked, homogeneous rectangular beam without reinforcement, gives the distribution of stresses shown in Fig. 4-4. The stress diagram shown in Fig. 4-4c and d may be visualized

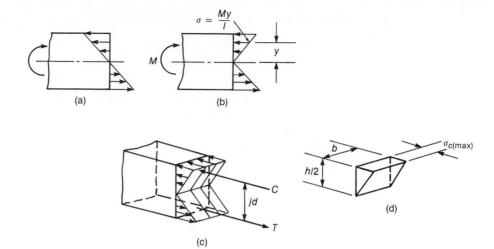

Fig. 4-4
Elastic beam stresses and
stress blocks.

as having a "volume"; hence, one frequently refers to the *compressive stress block* and the *tensile stress block*. The resultant of the compressive stresses is the force C given by

$$C = \frac{\sigma_{c(\max)}}{2}\left(b\frac{h}{2}\right)$$ (4-4)

This is equal to the volume of the compressive stress block shown in Fig. 4-4d. In a similar manner, one could compute the force T from the tensile stress block. The forces C and T act through the centroids of the volumes of the respective stress blocks. In the *elastic* case, these forces act at $h/3$ above or below the neutral axis, so that $jd = 2h/3$. From (4-3b) and (4-4) and Fig. 4-4, we can write

$$M = \sigma_{c(\max)}\frac{bh}{4}\left(\frac{2h}{3}\right)$$ (4-5a)

$$M = \sigma_{c(\max)}\frac{bh^3/12}{h/2}$$ (4-5b)

or, since

$$I = \frac{bh^3}{12}$$

and

$$y_{\max} = h/2$$

it follows that

$$M = \frac{\sigma_{c,\max}I}{y_{\max}}$$ (4-5c)

Thus, for the elastic case, identical answers are obtained from the traditional beam stress equation, (4-5d), and from (4-5a) when the stress block concept is used.

The elastic beam theory (4-5c) is not used in the design of reinforced concrete beams—first, because the compressive stress–strain curve of concrete is nonlinear, as shown in Fig. 3-15, and, what is even more important, because the concrete cracks at low tensile stresses, making it necessary to provide steel reinforcement to transfer the tensile force, T. These two factors are easily handled by the stress-block concept, combined with (4-4) and (4-5).

Flexure Theory for Reinforced Concrete

The theory of flexure for reinforced concrete is based on three basic assumptions, which are sufficient to allow one to calculate the moment resistance of a beam. These are presented first and used to illustrate the behavior of a beam cross section under increasing moment. Following this, four additional simplifying assumptions from the ACI Code are presented to simplify the analysis for practical application.

Flexural Behavior

The cracking pattern and strains measured in a laboratory test of a reinforced concrete beam are shown in Fig. 4-5. The cracks are indicated by the short vertical and inclined lines in Fig. 4-5b and c. A photograph of the beam after failure is shown in Fig. 4-6. The cracks have been marked with black ink so that they show in the photograph.

The strains plotted in Fig. 4-5 were measured on a 16-in. gauge length extending 8 in. on either side of the midspan of the beam. This region is shown shaded in Fig. 4.5a. Figures 4-5b and c, 4-6, and 4-8 all show that there were two cracks in this region at failure. The strains were computed using

$$\epsilon = \frac{\Delta_\ell}{\ell}$$

where ϵ is the strain, ℓ is the gauge length, and Δ_ℓ is the change in length of the gauge length. The widths of the cracks were included in the elongation, Δ_ℓ, used to compute in the strain, ϵ.

The measured strains were used to compute the curvature corresponding to each load level. Curvature, $1/r$, is defined as the angle change over a known length and, as shown in the inset of Fig. 4-7, is computed as

$$\frac{1}{r} = \frac{\epsilon}{y} \tag{4-6}$$

where ϵ is the strain at a distance y from the axis of zero strain at the load stage in question and r is the radius of curvature. Figure 4-7 relates the bending moment, M, at midspan of the beam to the curvatures at the same location. This is a *moment–curvature diagram*.

Initially, the beam was uncracked, as shown in Fig. 4-5a. The strains at this stage were very small, and the stress distribution was essentially linear. The moment and curvature are shown by point A in Fig. 4-7. The moment–curvature diagram for this stage (segment O–B in Fig. 4-7) was linear.

When the stresses at the bottom of the beam reached the tensile strength of the concrete, cracking occurred. After cracking, the tensile force in the concrete was transferred to the steel. As a result, less of the concrete section was effective in resisting moments, as shown by the distribution of stresses in Fig. 4-5b, and the stiffness of the beam decreased. Thus the slope of the moment–curvature diagram (shown by B–C–D in Fig. 4-7) also decreased. The crack pattern and strains in Fig. 4-5b correspond to the behavior expected under the loads applied in everyday service (stage (b) in Fig. 4-5 and point C in Fig. 4-7). The stress distribution in the concrete was computed from the measured strains and the stress-strain curve for concrete given in Fig. 3-19b and (3-20). It is still close to linear at this stage: the largest crack had a width of 0.006 in. Such a crack is entirely acceptable, as will be discussed in Chapter 9.

Eventually, the reinforcement reached the yield point shown by point D in Fig. 4-7. The compressive stresses were still close to being linear at this stage. Once yielding had occurred, the curvatures increased rapidly with very little increase in moment, as shown in Fig. 4-7. The increase in curvature can also be seen from the difference between the strain diagram in Fig. 4-5b and that in Fig. 4-5c. The beam failed as a result of the crushing of the concrete at the top of the beam.

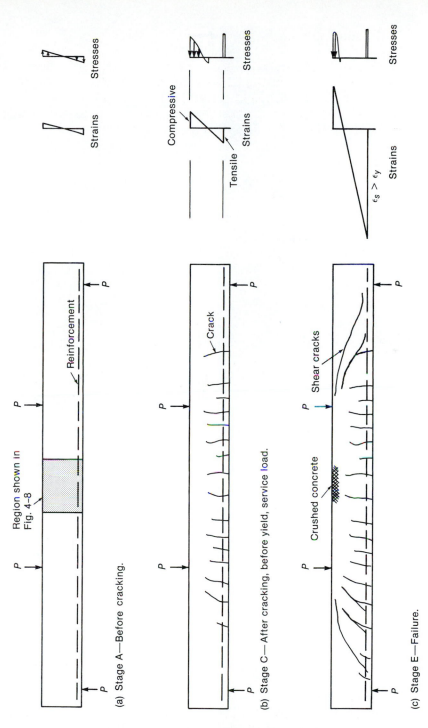

(a) Stage A—Before cracking.

(b) Stage C—After cracking, before yield, service load.

(c) Stage E—Failure.

Fig. 4-5
Cracks, strains, and stresses in test beam.

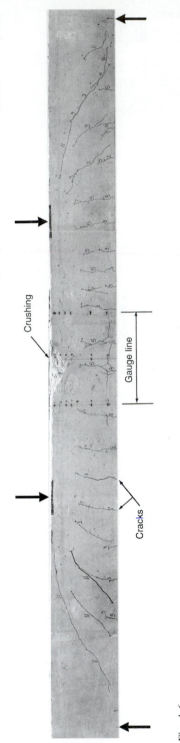

Fig. 4-6
Test beam after failure.
(Photograph courtesy of J. G. MacGregor.)

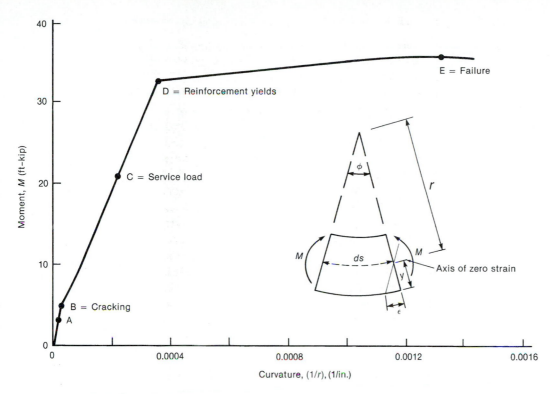

Fig. 4-7
Moment–curvature diagram for test beam.

It is worth noting that, although concrete itself is not a very ductile material (as indicated in the stress–strain curves in Fig. 3-17), reinforced concrete beams can exhibit large ductilities, as is shown by the long, almost flat postyield portion of the moment–curvature diagram. At service load (stage *C*), the midspan deflection of the beam was 0.31 in., or 1/383 of the span. At ultimate, this had increased to 2 in., or 1/60 of the span, showing the great ductility of this beam.

In practice, reinforced concrete design has been carried out in one of two ways. Until the mid-1960s, designers considered the loads expected in service and carried out calculations, assuming a linear stress distribution for concrete in compression. This was called *working-stress design* and corresponded to the load stage plotted in Fig. 4-5b and point *C* in Fig. 4-7. Since then, calculations have been carried out at the failure state (Fig. 4-5c and point *E* in Fig. 4-7) for loads larger than those expected in service (factored loads), and checks are made of the deflections and cracking at service-load levels. This is called *limit-states design* or *strength design*, as explained in Section 2-3. In this book, we concentrate on limit-states design. Working-stress design principles are used to calculate the deflections and steel stresses at service loads, however, and are explained in Chapter 9.

Basic Assumptions in Flexure Theory

Three basic assumptions are made:

1. Sections perpendicular to the axis of bending that are plane before bending remain plane after bending.

2. The strain in the reinforcement is equal to the strain in the concrete at the same level.

3. The stresses in the concrete and reinforcement can be computed from the strains by using stress–strain curves for concrete and steel.

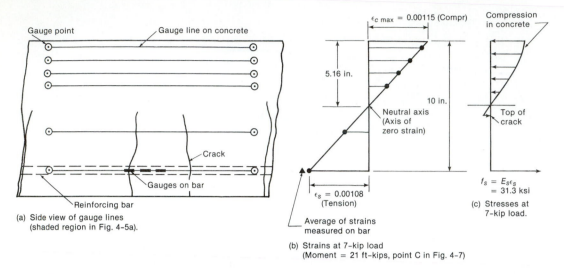

Fig. 4-8
Distribution of strains and stresses in test beam at service load.

The first of these is the traditional "plane sections remain plane" assumption made in the development of beam theory. Figure 4-7b is an enlargement of the strain distribution plotted in Fig. 4-5b. The dots represent strains measured on the 16-in. gauge lines shown in Fig. 4-8a. The measured strains are seen to be linear. Figure 4-9 shows strain distributions measured in tests of two columns subjected to combined axial load and moment. Again the distributions are linear as assumed. This assumption does not hold for so-called deep beams (with spans shorter than about four times their depths), because such members tend to act as tied arches rather than beams. (See Chapter 18.)

The second assumption is necessary because the concrete and the reinforcement must act together to carry load. This assumption implies a perfect bond between the concrete and

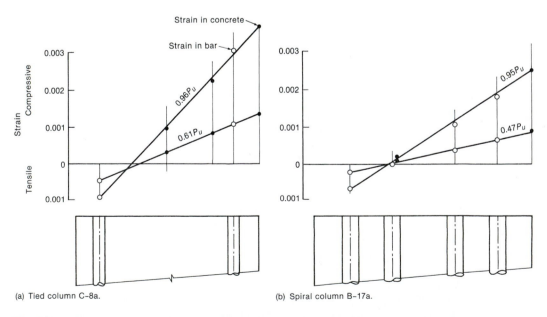

Fig. 4-9
Strains measured in tests of eccentrically loaded columns. (From [4-2].)

the steel. The agreement between the strains measured in the steel, plotted with a triangle in Fig. 4-8b, and the line through the dots representing the measured concrete strains shows that the steel and concrete do act together as postulated in this assumption. The ACI Code combines these first two assumptions as follows:

> 10.2.2—Strain in reinforcement and concrete shall be assumed directly proportional to the distance from the neutral axis. . . .

The stress diagram in Fig. 4-8c was obtained from the strain distribution by using the theoretical stress–strain curve for concrete presented in Fig. 3-19b. The moment computed from this stress distribution balances the load on the beam.

Additional Assumptions in Flexure Theory for Design

The three assumptions already made are sufficient to allow calculation of the strength and behavior of reinforced concrete elements. For design purposes, however, several additional assumptions are introduced to simplify the problem, with little loss of accuracy.

4.   The tensile strength of concrete is neglected in flexural-strength calculations (ACI Section 10.2.5).

The strength of concrete in tension is roughly one-tenth of the compressive strength, and the tensile force in the concrete below the zero strain axis is small compared with the tensile force in the steel. Hence, the contribution of the tensile stresses in the concrete to the flexural capacity of the beam is small and can be neglected. It should be noted that this assumption is made primarily to simplify flexural calculations. In some instances, particularly shear, bond, deflection, and service-load calculations for prestressed concrete, the tensile resistance of concrete is not neglected.

5.   Concrete is assumed to fail when the maximum compressive strain reaches a limiting value.

Strictly speaking, there is no such thing as a limiting compressive strain for concrete. As shown in Fig. 4-7, a simply supported reinforced concrete beam reaches its maximum capacity when the slope, $dM/d\phi$, of the moment–curvature diagram equals zero (point E). Failure occurs when $dM/d\phi$ becomes negative, corresponding to an unstable situation in which further deformations occur at decreasing loads [4-3]. Design calculations are very much simplified, however, if a limiting strain is assumed. Since the moment and curvature at the maximum moment point on the moment–curvature diagram correspond to one particular value of the extreme compressive strain, the moments corresponding to any other strain at the extreme fiber will be smaller. As a result, this assumption will always give conservative estimates of the strength.

The maximum compressive strains, ϵ_{cu}, from tests of beams and eccentrically loaded columns of normal-strength, normal-density concrete are plotted in Fig. 4-10a [4-4], [4-5]. Similar data from tests of normal-density and lightweight concrete are compared in Fig. 4-10b. ACI Section 10.2.3 specifies a limiting compressive strain, ϵ_{cu}, equal to 0.003, which approximates the smallest measured values plotted in Fig. 4-10a and b. In Europe, the CEB Model Code [4-6] uses a limiting strain of 0.002 for columns under concentric axial load. This corresponds to the highest point in a typical stress–strain curve. (See Fig. 3-18 for an example.) The CEB Model Code uses $\epsilon_{cu} = 0.0035$ for beams and eccentrically loaded columns. Higher limiting strains have been measured in members with significant moment gradient, [4-7] and in members in which the concrete is confined by spirals or closely spaced hoops [4-8]. On the other hand, the shell concrete outside the ties of high-strength concrete columns has frequently been observed to spall off at compressive strains between 0.0025 and 0.003. Throughout this book, the limiting compressive strain will be assumed constant and equal to 0.003.

6.   The compressive stress–strain relationship for concrete may be based on stress–strain curves or may be assumed to be rectangular, trapezoidal, parabolic, or any

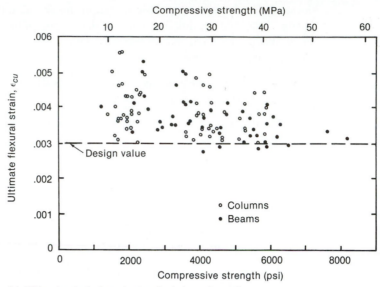

(a) Ultimate strain from tests of reinforced members.

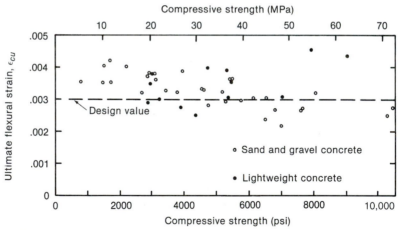

Fig. 4-10
Limiting compressive strain
(From [4-4], [4-5].)

(b) Ultimate strain from tests of plain concrete specimens.

other shape that results in prediction of strength in substantial agreement with the results of comprehensive tests (ACI Section 10.2.6).

Thus, rather than using a representative stress–strain curve, such as that given in Fig. 3-19b, other diagrams which are easier to use in computations are acceptable, provided that they adequately predict test results. As is illustrated in Fig. 4-11, the shape of the stress block in a beam at the ultimate moment can be expressed mathematically in terms of three constants:

k_3 = ratio of the maximum stress f_c'' in the compression zone of a beam to the cylinder strength, f_c'

k_1 = ratio of the average compressive stress to the maximum stress (this is equal to the ratio of the shaded area in Fig. 4-12 to the area of the rectangle ck_3f_c')

k_2 = ratio of the distance between the extreme compression fiber and the resultant of the compressive force to the depth of the neutral axis, c

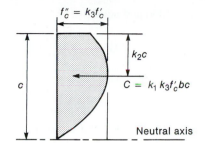

Fig. 4-11
Mathematical description of compression stress block.

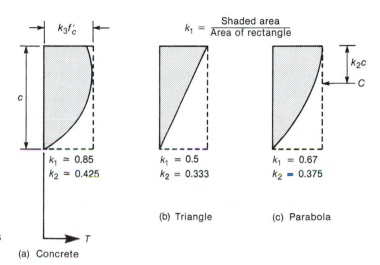

Fig. 4-12
Values of k_1 and k_2 for various stress distributions.

For a rectangular compression zone of width b and depth to the neutral axis, c, the resultant compressive force is

$$C = k_1 k_3 f_c' bc \tag{4-7a}$$

and the bending moment about the neutral axis (axis of zero strain) due to this force is

$$M = k_1 k_3 (1 - k_2) f_c' bc^2 \tag{4-8a}$$

Values of k_1 and k_2 are given in Fig. 4-12 for various assumed compressive stress–strain diagrams or "stress blocks."

As a further simplification, ACI Section 10.2.7 permits the use of the equivalent rectangular concrete stress distribution shown in Fig. 4-14 for ultimate strength calculations. The rectangular distribution is defined by the following:

1. A uniform compressive stress of $\alpha_1 f_c'$ shall be assumed distributed over an equivalent compression zone bounded by the edges of the cross section and a straight line located parallel to the neutral axis at a distance $a = \beta_1 c$ from the fiber with the maximum compressive strain.

2. The distance c from the fiber of maximum compressive strain to the neutral axis is measured perpendicular to that axis.

3. The factor β_1 shall be taken as follows [4-5]:

(a) For concrete strengths f_c up to and including 4000 psi,

$$\beta_1 = 0.85 \tag{4-9a}$$

(b) For f'_c between 4000 and 8000 psi,

$$\beta_1 = 1.05 - 0.05\frac{f'_c}{1000} \tag{4-9b}$$

(c) For f'_c greater than 8000 psi,

$$\beta_1 = 0.65 \tag{4-9c}$$

The symbols α_1 and β_1 used to describe the rectangular stress block are different from the symbols k_1, k_2, and k_3 used to describe the stress block obtained from tests because the rectangular stress block is described by two symbols, whereas three are used to describe the stress block from tests. For a rectangular compression zone of constant width b and depth to the neutral axis c, the resultant compressive force is

$$C = \alpha_1\beta_1 f'_c\, bc \tag{4-7b}$$

and the moment of this force about the neutral axis is

$$M = \alpha_1\beta_1\left(1 - \frac{\beta_1}{2}\right)f'_c\, bc^2 \tag{4-8b}$$

The term α_1 accounts for at least two factors affecting the flexural strength:

1. The compressive strength of concrete is reduced by sustained loads, as discussed in Section 3-2, and

2. the upward migration of water during concrete placement increases the w/cm ratio of the concrete near the top of the member.

The first of these is probably the dominant factor.

Studies of the effects of sustained loads on the strength of concrete [4-3] (see Section 3-2) and tests of columns [4-2] suggest that α_1 can be taken equal to 0.85 for commonly occurring concrete strengths. The dashed line in Fig. 4-13 is a lower-bound line corresponding to $k_3 = \alpha_1 = 0.85$ and β_1, given by (4-9).

When working in the SI system of units, the factor β_1 is taken as

$$\beta_1 = 1.09 - 0.008f'_c \tag{4-9M}$$

but not more than 0.85 nor less than 0.65. (The equation number includes M, which stands for metric.)

The equivalent rectangular stress block with $\alpha_1 = 0.85$ and β_1 from (4-9) has been shown [4-4] [4-5] [4-6] to give very good agreement with test data for *beams*. (See Fig. 2-3.) For columns, the agreement is good up to a concrete strength of about 6000 psi. For columns loaded with small eccentricities and having strengths greater than 6000 psi, the moment capacity tends to be overestimated by the ACI Code stress block. This is because Eq. (4-9) for β_1 was chosen as a lower bound on the test data, as suggested by the dashed line in Fig. 4-13. The internal moment arm of the compression force in the concrete about the *centroidal axis* of a rectangular column is $(h/2 - \beta_1 c/2)$, where c is the depth to the neutral axis (axis of zero strain). If β_1 is too small, the moment arm will be too large and the moment capacity will be overestimated.

Various authors have suggested equations for β_1 that correspond to a line through the data in Fig. 4-13, rather than lines representing the lower bound to the data. Alternatively, for high-strength concretes, it may be desirable to base the stress block on numerical integration using a representative stress–strain curve for the concrete in question. In analytical studies of columns and beams [4-7], the height (maximum stress) of the stress–strain curve is taken as $f''_c = 0.85$ to 0.90 times the height measured in cylinder tests. This is equivalent to multiplying the maximum stress by $\alpha_1 = 0.85$ to 0.9. Column strengths computed using $f''_c = 0.85$ to 0.90 f'_c have shown good agreement with tests [4-6], [4-7].

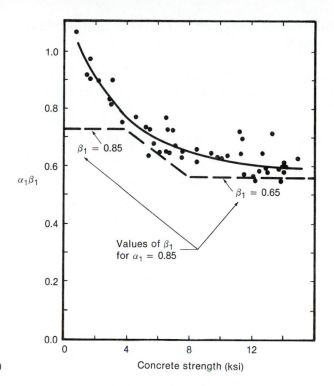

Fig. 4-13
Values of $\alpha_1\beta_1$ from tests of
concrete prisms. (From [4-5].)

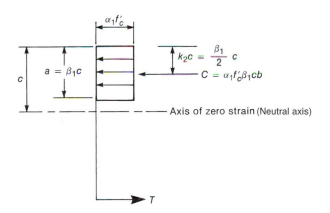

Fig. 4-14
Equivalent rectangular stress
block.

4-3   ANALYSIS OF REINFORCED CONCRETE BEAMS

Stress and Strain Compatibility and Equilibrium

Two requirements are satisfied throughout the analysis and design of reinforced concrete beams and columns:

1.   *Stress and strain compatibility.* The stress at any point in a member must correspond to the strain at that point. Except for short, deep beams, the distribution of strains over the depth of the member must be linear to satisfy assumptions 1 and 2 presented earlier in this chapter.

2.   *Equilibrium.* The internal forces must balance the external load effects, as illustrated in Fig. 4-3 and (4-2) and (4-3).

Analysis of the Flexural Capacity of a General Cross Section

The use of equilibrium and strain compatibility in the computation of the capacity of an arbitrary cross section such as the one shown in Fig. 4-15 involves four steps and will be illustrated with an example.

Example 4-1   Calculation of the Nominal Moment Capacity of a Beam

The beam shown in Fig. 4-15 is made of concrete with a compressive strength, f'_c, of 3000 psi and has three No. 8 bars with a yield strength, f_y, of 60,000 psi.

1. **Initially, assume that the stress f_s in the tension reinforcement equals the yield strength f_y, and compute the tension force $T = A_s f_y$:**

$$A_s = 3 \text{ No. 8 bars} = 3 \times 0.79 \text{ in.}^2 = 2.37 \text{ in.}^2$$

$$f_y = 60{,}000 \text{ psi for Grade 60 bars}$$

$$T = A_s f_y = 2.37 \text{ in.}^2 \times 60{,}000 \text{ psi} = 142{,}200 \text{ lb}$$

The assumption that $f_s = f_y$ will be checked in step 3. If the steel has yielded, a simple solution exists; if not, a more complex solution must be used. This assumption will generally be true, because the ACI Code requires that the steel percentage be small enough that the steel will yield.

2. **Compute the area of the compression stress block so that $C = T$.** This is done for the equivalent rectangular stress block shown in Fig. 4-15b. The stress block consists of a uniform stress of $\alpha_1 f'_c$ distributed over a depth $a = \beta_1 c$, measured from the extreme compression fiber. For $f'_c = 3000$ psi, ACI Section 10.2.7.1 gives $\alpha_1 = 0.85$, and ACI Section 10.2.7.3 gives $\beta_1 = 0.85$. (See (4-9a).) The magnitude of the compression force is obtained from equilibrium:

$$C = T = 142{,}200 \text{ lb}$$

Note that T was computed in step 1.

By the geometry of this particular triangular beam, shown in Fig. 4-15a, if the depth of the compression zone is a, the width is also a, and the area is $a^2/2$. This is, of course, true only for a beam of this particular triangular shape.

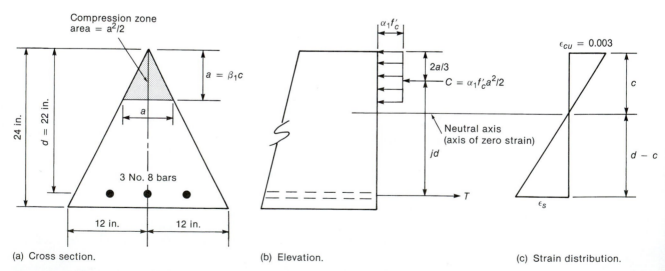

(a) Cross section.   (b) Elevation.   (c) Strain distribution.

Fig. 4-15
Analysis of arbitrary cross section—Example 4-1

Therefore, $C = (0.85f'_c)(a^2/2)$ and

$$a = \sqrt{\frac{142,200 \text{ lb} \times 2}{0.85 \times 3000}} = 10.56 \text{ in.}$$

3.   Check whether $f_s = f_y$. This is done by using strain compatibility. The strain distribution at ultimate is shown in Fig. 4-15c. To plot this, it is necessary to know ϵ_{cu} and c:

$$\epsilon_{cu} = 0.003 \qquad\qquad \text{(assumption 5, Section 4-2)}$$

$$c = \frac{a}{\beta_1}$$

For $f'_c = 3000$ psi, $\beta_1 = 0.85$ (from (4-8). Therefore, $c = 10.56/0.85 = 12.42$ in. measured perpendicular to the neutral axis.

By similar triangles, as shown in Fig. 4-15c,

$$\frac{\epsilon_s}{d - c} = \frac{0.003}{c}$$

or

$$\epsilon_s = 0.003\left(\frac{22 - 12.42}{12.42}\right) = 0.00231$$

For Grade-60 reinforcement,

$$\epsilon_y = \frac{60,000 \text{ psi}}{29,000,000 \text{ psi}} = 0.00207$$

Therefore, $\epsilon_s > \epsilon_y$ and $f_s = f_y$. Thus the assumption made in step 1 is satisfied.

4.   Compute M_n.

$$M_n = Cjd = Tjd$$

where jd is the lever arm, the distance from the resultant tensile force (at the centroid of the reinforcement) to the resultant compressive force C. Because the area on which the compression stress block acts is triangular in this example, C acts at $2a/3$ from top of the beam. Therefore,

$$jd = d - \frac{2a}{3}$$

$$M_n = A_s f_y\left(d - \frac{2a}{3}\right)$$

$$= 2.37 \text{ in.}^2 \times 60,000 \text{ psi}\left(22 - \frac{2 \times 10.56}{3}\right) \text{ in.}$$

$$= 2.127 \times 10^6 \text{ in.-lb}$$

$$M_n = \frac{2.13 \times 10^6}{12,000}\text{ft-kips} = 177 \text{ ft-kips}$$

Thus, the nominal moment capacity of this beam cross section is 177 ft-kips.   ∎

This analysis is not complete because the strength reduction factor, ϕ, has not been computed. This will be concluded after several terms have been defined.

This general analysis procedure can be used to compute the moment capacity for beams of any shape. It should be noted that the equations used to compute a, jd, and M_u in Example 4-1 were specifically derived for a beam with a *triangular section* and apply only to such a beam. Since beams are frequently rectangular in cross section, the analysis of rectangular beams is considered later.

Universal Procedure for Computing Nominal Moment Capacity

The calculation of the nominal moment capacity of this beam involved four steps:

First, it was assumed that the tension steel yields before the concrete crushes. The tensile force at yield, T, equals the yield stress times the area of the tension steel, $A_s f_y$. The line of action of the tension steel force, T, coincides with the centroid of the steel.

Second, because this is a beam, $C = T$. The force in the concrete, C, equals the tension in the steel. Knowing the strength of the concrete, the size and line of action of the compression stress block can be computed, along with the distance between C and T, referred to as the internal lever arm, jd, where j is the ratio of the distance between the resultants of C and T to the effective depth, d, of the tension reinforcement. The term j ranges from about 0.8 to 0.98.

Third, the internal resisting moment is equal to the couple consisting of the forces C and T given by:

$$M_n = T \times jd \tag{4-3a}$$

$$M_n = C \times jd \tag{4-3b}$$

Fourth, because we assumed the steel had yielded $f_s = f_y$ in step 1, it is necessary to check this assumption. This will be discussed shortly.

This four-step procedure can be applied to any reinforced concrete beam, regardless of the shape of the section or the concrete and steel.

Definitions of Effective Depth and Extreme Effective Depth

1. In a flexural failure, the effective depth, d, is measured from the extreme compressive fiber to the *centroid* of the longitudinal reinforcement. This definition is used in calculations of the nominal moment capacity.

2. For any of:

    (a) a compression-controlled failure,

    (b) a transition failure, or

    (c) a tension-controlled failure,

as defined in the next subsection, the governing strain is ϵ_t, in the extreme tensile layer of steel at the *extreme tension depth, d_t*, as defined in ACI Section R10.3.3. The extreme tensile layer is the layer of reinforcement located farthest from the extreme compression face. See Fig. 4-2.

Traditional Definitions of Tension, Compression, and Balanced Failures

Depending on the properties of a beam, flexural failures may occur in three different ways:

1. *Balanced failure.* Concrete crushes and steel yields *simultaneously*. Such a beam has *balanced reinforcement*.

2. *Tension Failure.* Reinforcement yields before concrete crushes (reaches its limiting compressive strain). Such a beam is said to be *under-reinforced.*

3. *Compression failure.* Concrete crushes before steel yields. Such a beam is said to be *over-reinforced.*

In the test specimen shown in Figs. 4-5 to 4-8, the reinforcement yielded before failure occurred, and hence the beam developed a *tension failure.* At failure (point *E* in Fig. 4-8) the curvature at the section of maximum moment was almost four times that at yielding (point *D*). As a result, the beam deflected extensively and developed wide cracks in the final loading stages. This type of behavior is said to be *ductile* since the moment–curvature or the load–deflection diagram has a long plastic region (*D–E* in Fig. 4-8). If such a beam in a building were to fail, it would do so in a ductile manner, giving the occupants of the building warning of any impending failure and giving them an opportunity to evacuate the building, thereby reducing the consequences of collapse.

Moment–curvature diagrams are presented in Fig. 4-16d for the three beams shown in Fig. 4-16a, b, and c. The beams differ only in the amount of reinforcement and the number of layers of reinforcement. At failure, the reinforcement in beam *A* has yielded, as shown by the strain diagram in Fig. 4-16a. This beam develops a tension failure and has a ductile

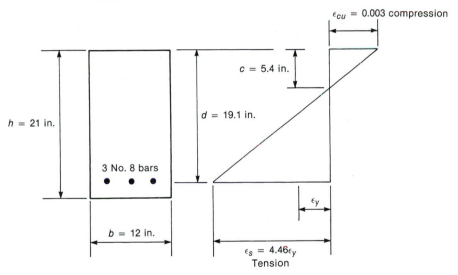

(a) Beam A—Tension failure.

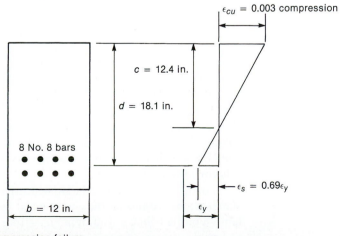

Fig. 4-16
Tension, compression, and balanced failures.

(b) Beam B—Compression failure.

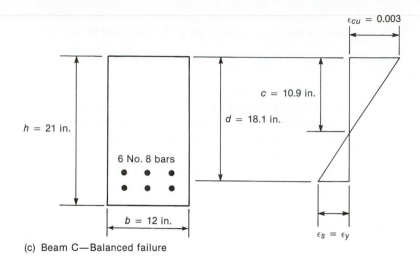

(c) Beam C—Balanced failure

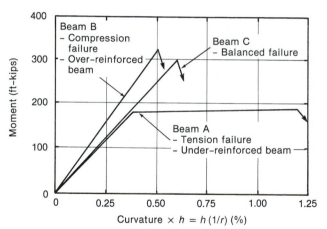

Fig. 4-16
(Continued)

(d) Moment–curvature diagram

moment–curvature response as shown in Fig. 4-16d. As will be seen later in this section, *tension failures* occur when the *mechanical reinforcement ratio* $\omega = \rho f_y / f_c'$ is small.

In the case of beam *B* in Fig. 4-16, the concrete in the extreme compression fiber reaches the assumed crushing strain of 0.003 before the steel starts to yield. This is called a *compression failure*. The moment–curvature diagram for such a beam does not have the ductile postyielding response displayed by beam *A*. If overloaded, this beam may fail suddenly in a *brittle* manner without warning to the occupants and, as a result, such a failure may have serious consequences. Compression failures occur for high values of the mechanical ratio ω. This beam is said to be *over-reinforced* since the concrete crushed before the reinforcement could yield. The opposite occurs in an *under-reinforced* beam.

In the case of beam *C*, the strain distribution at failure, shown in Fig. 4-16c, involves simultaneous crushing of the concrete and yielding of the steel. This case, which also exhibits a brittle failure as shown in Fig. 4-16d, marks the boundary between ductile tension and brittle compression failures—hence the name *balanced failure*.

New Definitions of Flexural Behavior Cases—ACI 318-02 Sections 10.3.2 through 10.3.5

The design equation for flexure is:

$$\phi M_n \geq M_u \tag{4-1b}$$

where ϕ is a *strength reduction factor*. For beams the factor ϕ is defined in ACI 318-02 Section 9.3.2, based on ACI 318-02 Sections 10.3.2 through 10.3.5, in terms of the *net strain* in the extreme layer of tension steel in the cross section of the beam or column. The *net strain* is the steel strain due to loads, not including strains due to shrinkage, temperature changes, or creep, or strains resulting from prestress. The term *extreme tension layer* of steel refers to the bar or layer of reinforcement farthest from the extreme compression fiber, measured perpendicular to the neutral axis (the axis of zero strains).

The code sections listed define four types of beams depending on the anticipated mode of failure. See [4-19].

1. *Balanced failure*—As stated in the preceeding section, beam is said to have developed a *balanced failure* when the strain at the *centroid* of the tension reinforcement reaches the yield strain in tension simultaneously with the strain in the concrete in the extreme compression fiber first reaching the limiting compressive strain, ϵ_{cu}. A balanced failure is thus characterized as the simultaneous onset of crushing of the compression face and yielding of the longitudinal tension reinforcement.

The subscript b will be used to denote the balanced condition.

2. *Tension-controlled failure*—A tension-controlled failure is said to have occurred if the reinforcement in the *extreme layer of tensile steel* (i.e., the layer or group of bars farthest from the compression face) has a net tensile strain of 0.005 or larger (more tensile) when the beam reaches its nominal flexural strength. For Grade 60 reinforcement with a yield strength $f_y = 60$ ksi, the tensile yield strain is $\epsilon_y = 60/29{,}000 = 0.00207$ strain. The tension-controlled limit strain of 0.005 was chosen to be several times the yield strain of the beam reinforcement so that the moment curvature diagram resembles Fig. 4-17. ACI Section 10.3.3 allows ϵ_y to be rounded to $\epsilon = 0.002$ for convenience. Because the tension-controlled limit strain of 0.005 exceeds the yield strain, the longitudinal reinforcement will have yielded before the nominal strength is reached, giving a moment-curvature diagram similar to the measured curve in Fig. 4-7 or the calculated curve for beam A in Fig. 4-16. The subscript TC will be used to denote tension-controlled, as in ϵ_{TC}.

3. *Compression-controlled failure*—A cross section is said to develop a *compression-controlled failure* if the net strain in the reinforcement in the extreme tension layer is less tensile than $\epsilon_t = 0.002$ tension at the nominal strength of the beam. The subscript CC will be used to denote compression-controlled, as in ϵ_{CC}.

4. *Transition failure*—A beam with a net tensile strain at nominal strength in the extreme layer of tension reinforcement that is between ϵ_y in tension, $\epsilon_y = 0.00207$ (or rounded off to 0.002 for Grade-60 steel) and 0.005.

These definitions of types of failures were proposed by Mast [4-10] and strongly criticized by Darwin [4-12] and Naaman [4-12].

5. *Compression-controlled limit*—Except for the different locations of the fibers in which the strains are evaluated, the traditional balanced failure strain distribution corresponds to the boundary between transition failures and compression-controlled failures and is referred to as the *compression-controlled limit*.

Upper Limit on Beam Reinforcement

Prior to 2002, the maximum steel ratio in beams was limited to 0.75 times the reinforcement ratio corresponding to a balanced section, that is, $0.75\rho_b$. In ACI 318-02 Section 10.3.5, the upper limit on beam reinforcement has been accomplished by limiting the longitudinal steel in a beam to the amount corresponding to a tensile steel strain of 0.004 at the nominal strength. ACI Section 10.3.5 states

> For nonprestressed flexural members and nonprestressed members with [small or zero axial loads] the net tensile strain ϵ_t at nominal strength shall not be less than 0.004.

Section 10.3.4 requires ϵ_t not less than 0.005 for ϕ to be equal to 0.90. So, for net tensile strains between 0.004 and 0.005, we shall use ϕ as calculated in accordance with 9.3.2.2.

This is discussed more fully later in this chapter.

Analysis of Rectangular Beams with Tension Reinforcement Only

Equations for M_n—Tension Steel Yielding

In the preceding section, equilibrium and strain compatibility were used to compute the moment capacity of a particular beam cross section. For the particular case of a rectangular beam, the same procedure can be used to derive equations for computing the moment capacity.

Consider the beam shown in Fig. 4-17. The compressive force in the concrete is

$$C = (0.85f_c')ba$$

The tension force in the steel is

$$T = A_s f_s$$

and for equilibrium, $C = T$. Therefore, the depth of the equivalent rectangular stress block is

$$a = \frac{A_s f_s}{0.85 f_c' b}$$

If $f_s = f_y$ as assumed in step 1 of Example 4-1, this becomes

$$a = \frac{A_s f_y}{0.85 f_c' b} \tag{4-10}$$

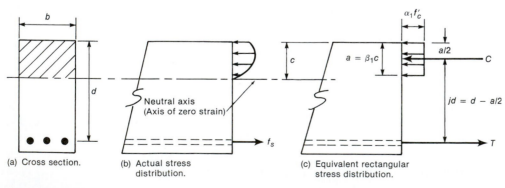

(a) Cross section. (b) Actual stress distribution. (c) Equivalent rectangular stress distribution.

Fig. 4-17
Stresses and forces in a rectangular beam.

It is possible to express the equations of M_n and ϕM_n in several ways, based on $M_n = T \times jd$ or $M_n = C \times jd$, or in a nondimensionalized fashion. These are considered in turn in the following paragraphs.

Equation for M_n Based on $M_n = T \times jd$.   Summing moments about the line of action of the compressive force C in Fig. 4-17c gives

$$M_n = T \times jd$$

Substituting $T = A_s f_s$, where f_s is equal to f_y and $jd = (d - a/2)$, yields

$$M_n = A_s f_y \left(d - \frac{a}{2} \right) \tag{4-11a}$$

and

$$\phi M_n = \phi \left[A_s f_y \left(d - \frac{a}{2} \right) \right] \tag{4-11b}$$

This is the basic equation for the flexural capacity of beams. The calculation of ϕ will follow.

Equation for M_n Based on $M_n = C \times jd$.   Alternatively, one could sum moments about the line of action of the tensile force, T. Thus,

$$M_n = C \times jd \tag{4-3b}$$

Substituting $C = (0.85 f_c') ba$ and $jd = (d - a/2)$ gives

$$M_n = (0.85 f_c') ba \left(d - \frac{a}{2} \right) \tag{4-12a}$$

and

$$\phi M_n = \phi \left[(0.85 f_c') ba \left(d - \frac{a}{2} \right) \right] \tag{4-12b}$$

Nondimensionalized Equations for M_n.   If we substitute $A_s = \rho bd$ into (4-10), we get

$$a = \frac{\rho f_y}{f_c'} \left(\frac{d}{0.85} \right) \tag{4-13a}$$

where $\rho f_y / f_c' = \omega$ and is referred to as the *mechanical reinforcement ratio*. The term ω is frequently used as a measure of the flexural behavior of a beam, since it incorporates the three major variables affecting that behavior (ρ, f_y and f_c'). Thus,

$$a = \frac{\omega d}{0.85} \tag{4-13b}$$

Substituting (4-13b) into (4-12) gives

$$\phi M_n = \phi \left[f_c' bd^2 \omega \left(1 - \frac{\omega}{2 \times 0.85} \right) \right]$$

or

$$\phi M_n = \phi [f_c' bd^2 \omega (1 - 0.59\omega)] \tag{4-14}$$

For design this is frequently expressed as

$$\phi M_n = \phi\left(\frac{bd^2}{12,000}k_n\right) \tag{4-15}$$

or (because the most economical design corresponds to $\phi M_n = M_u$) as

$$\frac{M_u}{\phi k_n} = \frac{bd^2}{12,000} \tag{4-16}$$

where M_u and M_n are in ft-kips and b and d are in inches. In SI units, these equations become

$$\phi M_n = \phi\left(\frac{bd^2}{10^6}k_n\right) \tag{4-15M}$$

and

$$\frac{M_u}{\phi k_n} = \frac{bd^2}{10^6} \tag{4-16M}$$

where M_u and M_n are in kN–m, and b and d are in mm.
    In (4-15) and (4-16), the term ϕk_n is

$$\phi k_n = \phi[f_c'\omega(1 - 0.59\omega)] \tag{4-17}$$

with f_c' in psi or MPa. Values of ϕk_n are given in Table A-3 (see Appendix A) for various concrete and steel strengths and reinforcement ratios.

Neutral Axis Depth for Balanced Steel Ratios.   From similar triangles, we can say that

$$\frac{c_b}{d} = \frac{\epsilon_{cu}}{\epsilon_{cu} + \epsilon_y} \tag{4-18}$$

Substituting $\epsilon_{cu} = 0.003$ and multiplying numerator and denominator by $E_s = 29,000,000$ psi gives

$$\frac{c_b}{d} = \frac{87,000}{87,000 + f_y} \tag{4-19}$$

where f_y is in psi. If a beam has a neutral-axis depth c less than c_b at ultimate, the steel strains will exceed ϵ_y, and vice versa. Thus, if $c \le c_b$ at failure, $f_s = f_y$.
    Equations 4-11 and 4-12 include $a = \beta_1 c$, rather than c. Substituting into (4-19) gives

$$\frac{a_b}{d} = \beta_1\left(\frac{87,000}{87,000 + f_y}\right) \tag{4-20}$$

where f_y is in psi.
    In SI units, $E_S = 200,000$ MPa, and (4-20) becomes

$$\frac{a_b}{d} = \beta_1\left(\frac{600}{600 + f_y}\right) \tag{4-20M}$$

where f_y is in MPa.
    To check whether $f_s = f_y$ in design, we shall either check whether $a/d \le a_b/d$, or we shall compute ϵ_t in the extreme tensile reinforcement and check whether it exceeds ϵ_y. Table A-4 gives values of a_b/d from (4-20) for various concrete and steel strengths.

Extreme Tension Depth and Net Tensile Strain.   The net tensile strain in the layer of steel farthest from the compression face is ϵ_t. The words *net tensile strain* refer to the steel strain at nominal strength, exclusive of strains due to effective prestress, creep, shrinkage, or temperature. This is the strain, ϵ_s, due to the factored live and dead loads on the beam. ACI Section 10.3.3 defines a section as being *compression-controlled* if the net tensile strain, ϵ_t, is less than or equal to the *yield strain in tension*, ϵ_y. For Grade 60 reinforcement

$$\epsilon_y = \frac{60,000}{29,000,000} = 0.00207$$

ACI Sec. 10.3.3 allows this to be rounded off to 0.002. The strain distribution corresponding to the *compression-controlled limit* is shown in Fig. 4-18c. Here the neutral axis depth is c_{CCL} and the strain ϵ_t is ϵ_y and occurs in the extreme tension layer, which is at a depth d_t.
    Derived in a similar fashion to (4-20), the depth of the rectangular stress block at failure at the compression-controlled limit, a_{CCL}, is

$$\frac{a_{CCL}}{d_t} = \beta_1 \left(\frac{87,000}{87,000 + f_y} \right) \tag{4-21}$$

where f_y is in psi, and

$$\frac{a_{CCL}}{d_t} = \beta_1 \left(\frac{600}{600 + f_y} \right) \tag{4-21M}$$

where f_y is in MPa.
    The subscripts CCL, TCL, and BRL refer to compression-controlled limit, tension-controlled limit, and beam reinforcement limit, respectively.
    ACI Section 10.3.3 defines a section as being *tension-controlled* if the net tensile strain in the layer of steel farthest from the compression face of the beam equals or exceeds 0.005 in tension. The strain distribution corresponding to the *tension-controlled limit* is shown in Fig. 4-18e, provided $\epsilon_t = 0.005$. Here, the neutral axis depth is c_{TCL}. From Fig. 4-18e, using similar triangles,

$$\frac{c_{TCL}}{d_t} = \frac{0.003}{0.003 + 0.005}$$

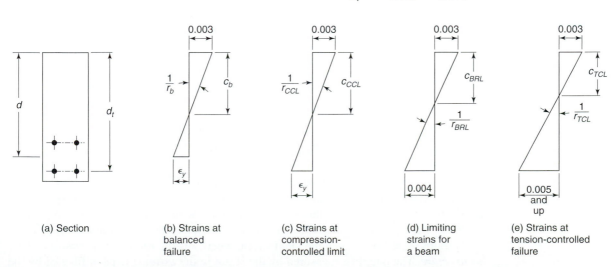

| (a) Section | (b) Strains at balanced failure | (c) Strains at compression-controlled limit | (d) Limiting strains for a beam | (e) Strains at tension-controlled failure |

Fig. 4-18
Balanced, compression-controlled, beam limit, and tension-controlled sections. Note that Fig. 4-18b is based on d while Fig. 4-18c through e are based on d_t.

or

$$\frac{c_{TCL}}{d_t} = 0.375 \tag{4-22}$$

and

$$\frac{a_{TCL}}{d_t} = 0.375\beta_1 \tag{4-23}$$

The ratios c_{TCL}/d_t and a_{TCL}/d_t are independent of the system of units. Values of a_{CCL}/d_t and a_{TCL}/d_t are given in Table A-4.

Traditional ACI Determination of Whether $f_s = f_y$ Based on ρ_b

Equation (4-19) gives a relationship for c_b/d, where c_b is the neutral axis depth at balanced failure. But $c = a/\beta_1$, and from (4-13a),

$$a_b = \frac{\rho_b f_y}{0.85 f_c'} d$$

Therefore, at balanced failure,

$$\frac{c_b}{d} = \frac{\rho_b f_y}{0.85\beta_1 f_c'} \tag{4-24}$$

where c_b and ρ_b are the neutral-axis depth and steel ratio corresponding to a balanced failure. From (4-18) and (4-24),

$$\rho_b = \frac{0.85\beta_1 f_c'}{f_y} \left(\frac{\epsilon_{cu}}{\epsilon_{cu} + \epsilon_y} \right)$$

If we substitute $\epsilon_{cu} = 0.003$ and multiply numerator and denominator by $E_s = 29,000,000$ psi, we obtain

$$\rho_b = \frac{0.85\beta_1 f_c'}{f_y} \left(\frac{87,000}{87,000 + f_y} \right) \tag{4-25}$$

where f_y and f_c' are in psi.

In SI units, $E_s = 200,000$ MPa, and (4-25) becomes

$$\rho_b = \frac{0.85\beta_1 f_c'}{f_y} \left(\frac{600}{600 + f_y} \right) \tag{4-25M}$$

where f_y and f_c' are in MPa. This is called the balanced steel ratio.

A comparison of the strain distributions in Figs. 4-18b, c, d, and e, respectively, shows that the ductility of the section increases with increasing curvatures, from $1/r_b = \epsilon/y$ at balanced failure, through increasing curvatures, $1/r_{CCL}$ and $1/r_{TCL}$, as the steel strains reach and exceed the yield strain, where r is the *radius of curvature* and $1/r$ is *curvature*. The ductility, measured by the strain in the extreme tension fiber or by the corresponding curvature $1/r = \epsilon/y$, is greater for the tension-controlled limit in Fig. 4-18e than it is for the compression-controlled limit section in Fig. 4-18c. In a member with one layer of steel, the effective depth and the extreme tension to the extreme tension steel, d and d_t, are both measured to the same layer of tension steel, and hence are equal, giving the same nominal moment capacities.

For beams or columns with several layers of steel, the effective depth d and the depth to the bars farthest from the neutral axis differ. After the steel has all yielded, however, the moment capacity is computed using the normal effective depth, d.

The equations for calculating the capacity of beams at the various strain limits were derived for a rectangular cross section. The equations for net tensile strain do not apply to beams with nonrectangular compression zones, such as Example 4-1. Because the width of the compression zone varied in the example, the amount of reinforcement required to balance the compression force is a function of the shape of the compression zone and the strain distribution, and it will be necessary to correct for the shape.

Beams should have properties corresponding to a *tension-controlled* flexural failure initiated by yield of the steel reinforcement, because such a beam has adequate ductility to allow redistribution of stresses and to give a warning of impending failure by deflecting excessively. This behavior is more desirable than is either a *compression-controlled failure*, initiated by crushing of the concrete, suddenly and without warning, or a *transition failure*, initiated by yield of the steel, but at a smaller curvature than a tension-controlled failure and hence giving less warning of failure than tension-controlled failure.

ACI 318-02 Section 10.3.5.   The 1999 ACI Code did not specify an upper limit on the amount of tension reinforcement in beams when design was based on the strain in the extreme tension steel layer. To reduce the chance of brittle beam failures, Section 10.3.5 was added in ACI 318-02 to require that beams have properties that correspond to tension failures with $f_s = f_y$ and a minimum amount of ductility. This was accomplished by limiting that the net strain at ultimate, ϵ_t, be equal to or exceed 0.0040 in tension in the extreme tensile reinforcement.

ACI 318-02 Section B.10.3.3.   Alternatively, ACI 318-02 Section B.10.3.3 limits the steel ratio ρ to a maximum of $0.75\rho_b$. This value of ρ corresponds to a net tensile strain at ultimate of 0.00376 at the centroid of the reinforcement, which is very close to the limit of 0.004 in ACI Section 10.3.5.

Strength-Reduction Factor, ϕ, for Beams—Design Based on ACI 318-02, Sections 9.3.2.1, 9.3.2.2, 10.3.3, 10.3.4, and 10.3.5

For compression-controlled beams and columns, ACI 318-02 Section 9.3.2.2, gives $\phi = 0.70$ if the compression zone is enclosed by spiral reinforcement and $\phi = 0.65$ for *other reinforced members* (i.e., for members without spiral reinforcement). Because spirals are not used in beams, we shall disregard $\phi = 0.70$ for beams. Equations (4-26a) and (4-27a) are for spiral columns and are given here for completeness.

For tension-controlled beams, in which the net tensile strain, ϵ_t, in the extreme tension layer of reinforcement is greater than or equal to 0.005 strain in tension, $\phi = 0.90$ (ACI 318-02 Section 9.3.2.2). Between $\phi = 0.65$ and $\phi = 0.90$, there is a transition region [4-9].

Equations for ϕ in the transition region are given in the commentary to ACI Section 9.3.2.2. These can be expressed in terms of the net tensile strain, ϵ_t, or the neutral-axis depth ratio, (c/d_t), at ultimate, as follows:

For members with $f_y = 60$ ksi, and spiral reinforcement,

$$\phi = 0.57 + 67\epsilon_t \tag{4-26a}$$

or

$$\phi = 0.37 + \frac{0.20}{c/d_t} \tag{4-27a}$$

For *other members* with $f_y = 60$ ksi,

$$\phi = 0.48 + 83\epsilon_t \tag{4-26b}$$

or

$$\phi = 0.23 + \frac{0.25}{c/d_t} \tag{4-27b}$$

For *other members* with $f_y = 40$ ksi, the strain at the compression failure limit is $f_y/E_s = 0.0014$. By similar triangles,

$$\phi = 0.55 + 69.4\epsilon_t \tag{4-26c}$$

or

$$\phi = 0.34 + \frac{0.21}{c/d_t} \tag{4-27c}$$

In ACI 318-02 Section 9.2 the words *other members* are meant to refer to all flexural members other than spiral columns. Equations (4-26) and (4-27) were derived by fitting straight lines to the limits on the variation of the strength reduction factors, ϕ, in the transition region.

Strength-Reduction Factor, ϕ, for Beams—Design Based on ACI 318-02, Sections C.3.2.2 to C.3.2.7 and Sections 10.3.3 to 10.3.5

The strength-reduction factors used when designing with the alternative load and resistance factors in ACI 318-02 Appendix C for flexure are as follows:

- For tension-controlled sections, ACI Section C.3.2.1 gives $\phi = 0.90$.
- For spirally reinforced compression-controlled sections, ACI Section C.3.2.2(a) gives $\phi = 0.75$.
- For other reinforced members (tied columns), ACI Section C.3.2.2(b) gives $\phi = 0.70$.

Appendix C also requires a transition between $\phi = 0.65$ to $\phi = 0.90$. This is discussed in Chapter 11.

EXAMPLE 4-2 Calculation of the Design Moment Capacity, ϕM_n, of a Beam

The design moment capacity, or factored resistance, ϕM_n of the beam shown in Fig. 4-15 and considered in Example 4-1 will now be calculated. Steps 1 and 2 will be identical to those used to compute the nominal moment capacity in Example 4-1.

1. Initially, assume that the stress f_s in the tension reinforcement equals the yield strength f_y, and compute the tension force $T = A_s f_y$ (see step 1 in Example 4-1):

$$T = 142{,}200 \text{ lb}$$

Steps 1 and 2 are the same as in Example 4-1.

$$M_n = A_s f_y \left(d - \frac{2a}{3} \right)$$

$$= 2.37 \text{ in.}^2 \times 60{,}000 \text{ psi} \left(22 - \frac{2 \times 10.56}{3} \right) \text{ in.}$$

$$= 2.127 \times 10^6 \text{ in.-lb}$$

$$M_n = \frac{2.13 \times 10^6}{12{,}000} \text{ ft-kips} = 177 \text{ ft-kips}$$

Thus, the nominal moment capacity of this beam cross section is 177 ft-kips.

   3.   **Check whether $f_s = f_y$.** From step 3 of Example 4-1, the strain in the layer of bars closest to the tension face of the beam is $\epsilon_t = 0.00231$.
   For Grade-60 reinforcement, $\epsilon_y = 0.00207$. Because ϵ_t exceeds ϵ_y in tension, but is less than 0.005, the section is a transition section.

   4.   **Compute Strength Reduction Factor, ϕ.** ACI Section 10.3.5 limits the extreme tensile strain at the nominal moment of flexural members to 0.004 tensile strain. In Example 4-1, the computed strain was $\epsilon_s = \epsilon_t = 0.00231$. As a result, this beam violates ACI Section 10.3.5 and the beam cross section would have to be changed. Since this calculation deals with a hypothetical section we shall retain the cross section and compute the corresponding strength reduction factor. From ACI Fig. R9.3.2:
   From (4-26a) for an "other" section with Grade-60 reinforcement (without spiral reinforcement),

$$\phi = 0.48 + 83\epsilon_t = 0.48 + 83 \times 0.00231$$
$$= 0.672$$

and

$$\phi M_n = 0.672 \times 177 = 119 \text{ ft-kips} \qquad\blacksquare$$

This is much lower than $M_n = 177$ ft-kips and illustrates the major penalty incurred if ϵ_t is less than 0.005. The selection of strength reduction factors is discussed later in this chapter.
   Calculation of ϵ_t, f_s, and ϕ Based on Extreme Tensile Strain at Nominal Strength.
   There are several different ways to compute ϕ. The most common method is to compute it from the extreme tensile strain at nominal strength, ϵ_t. The extreme tensile strain is a function of the extreme compressive strain (0.003), the depth of the neutral axis ($c = a/\beta_1$), and the depth of the extreme tensile steel (d_t). The calculation is based on similar triangles:

$$\frac{c}{0.003} = \frac{d_t - c}{\epsilon_t}$$

Rearranging gives

$$\epsilon_t = \frac{(d_t - c)}{c} \times 0.003 \qquad\qquad (4\text{-}28)$$

Once ϵ_t is known, f_s, ϵ_s, and ϕ can be obtained. ϕ is computed using (4-26) and (4-27).
   Calculation of ϵ_t, f_s, and ϕ based on a_{CCL} and a_{TCL}, or c_{CCL} and c_{TCL}.
   In the calculation of the moment capacity of a beam it is necessary to compute the depth of the rectangular stress block, a, or the depth to the neutral axis, c, or a/d_t and c/d_t. Values of these quantities are given in Table A-5 or A-5M.

EXAMPLE 4-3   Analysis of Singly Reinforced Beams: Tension Steel Yielding

Compute the nominal moment capacities, M_n, of three beams, each with $b = 10$ in., $d = 20$ in. and three No. 8 bars giving $A_s = 3 \times 0.79 = 2.37$ in.^{2} and $\rho = A_s/bd = 2.37/(10 \times 20) = 0.0119$. Assume the beam has $1\frac{1}{2}$ in. clear cover, No. 3 stirrups, and 3 No. 8 bars

$$d_t = 22.5 - (1.5 + 0.375 + 1.0/2)$$
$$= 20.13 \text{ in.} \quad \text{Say } d_t = 20 \text{ in.}$$

The beam cross section is shown in Fig. 4-19a.

BEAM 1: $f'_c = 3000$ psi AND $f_y = 60,000$ psi

1.   Compute a. Assume that steel stress, f_s, equals f_y (which corresponds to $\rho \leq \rho_b$). This will be checked in step 2. From (4-10), the depth of the equivalent rectangular stress block is

$$a = \frac{A_s f_y}{0.85 f'_c b} \tag{4-10}$$

$$= \frac{2.37 \text{ in.}^2 \times 60,000 \text{ psi}}{0.85 \times 3000 \text{ psi} \times 10 \text{ in.}} = 5.58 \text{ in.}$$

Therefore, $a = 5.58$ in. (Fig. 4-19a)

2a.   Compute ϵ_t, f_s, and ϕ. Two methods of computing ϕ are available; select one or the other. The first method starts with $c = \frac{a}{\beta_1} = 5.58/0.85 = 6.56$ in. and, from (4-28) $\epsilon_t = \dfrac{(d_t - c)}{c} \times 0.003 = 0.00615$. Because $\epsilon_t > \epsilon_y$, $f_s = f_y$. Because $\epsilon_t \geq 0.005$, the section is tension-controlled

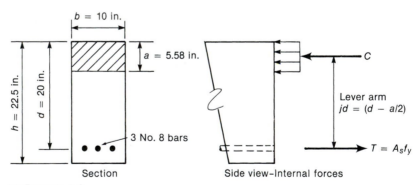

(a) Example 4-3.

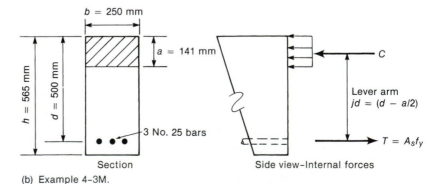

Fig. 4-19
Beams—Examples 4-3 and 4-3M.

(b) Example 4-3M.

and $\phi = 0.90$. Alternatively, base check on the extreme tensile strain, ϵ_t:

$$\epsilon_t = \frac{(d_t - c)}{c} \times 0.003 = \frac{(20 - 6.56)}{6.56} \times 0.003 \tag{4-28}$$

$$\epsilon_t = 0.0061$$

2b.   Check using a/d and c/d If $a/d \leq a_b/d$, f_s will be equal to f_y, where a_b/d is given by (4-21). We have

$$\frac{a}{d} = \frac{5.58 \text{ in.}}{20 \text{ in.}} = 0.279$$

From (4-20),

$$\frac{a_b}{d} = \beta_1 \left(\frac{87,000}{87,000 + f_y} \right)$$

From (4-9a), $\beta_1 = 0.85$ for $f'_c = 3000$ psi and

$$\frac{a_b}{d} = 0.85 \left(\frac{87,000}{87,000 + 60,000} \right) = 0.503$$

Since the actual $a/d_t = 0.279$ is less than $a_b/d = 0.503$, it follows that $f_s = f_y$.
To check whether the section is tension-controlled, check whether $a/d_t \leq a_{TCL}/d_t$. Note that this check is made in terms of a/d_t rather than a/d. For this beam, d_t and d are the same. If the steel were in several layers, this would not be true. From (4-23),

$$a_{TCL}/d_t = 0.375\beta_1 = 0.375 \times 0.85 = 0.319$$

Because $a/d_t = 0.279$ is less than 0.319, the section is tension-controlled and $\phi = 0.90$. Note that values of a_b/d and a_{TCL}/d_t are given in Table A-4.

3.   Compute the nominal moment capacity, M_n. Summing moments about the resultant compressive force, C, the moment capacity is $M_n = T\,jd$, where the tension force is $T = A_s f_y$ and the lever arm $jd = (d - a/2)$, giving (4-11a), or

$$M_n = A_s f_y \left(d - \frac{a}{2} \right) \tag{4-11a}$$

Thus,

$$M_n = 2.37 \text{ in.}^2 \times 60,000 \text{ psi} \left(20 - \frac{5.58 \text{ in.}}{2} \right)$$

$$= 2,447,000 \text{ in.-lb} = \frac{2,447,000}{12,000} \text{ ft-kips} = 204 \text{ ft-kips}$$

The nominal moment capacity of beam 1 is $M_n = 204$ ft-kips.
Since the section is tension-controlled, ACI Section 9.3.2.1 gives $\phi = 0.90$. The design or factored moment capacity of this beam is

$$\phi M_n = 0.9 \times 204 \text{ ft-kips} = 184 \text{ ft-kips}$$

BEAM 2: SAME AS BEAM 1, EXCEPT THAT f_c' = 6000 psi

1. **Compute a, c, and d_t**

$$a = \frac{2.37 \times 60{,}000}{0.85 \times 6000 \times 10} = 2.79 \text{ in.}$$

From Beam, $d_t = 20$ in. and where, from (4-9b),

$$\beta_1 = 1.05 - 0.05\left(\frac{f_c'}{1000}\right) = 0.75,$$

but not more than 0.85 nor less than 0.65

$$c = \frac{a}{\beta_1} = \frac{2.79}{0.75} = 3.72 \text{ in.}$$

2. **Check whether $f_s = f_y$ and whether the section is tension-controlled.** Again, we will base the check of whether $f_s = f_y$ on ϵ_t, where

$$\epsilon_t = \frac{(d_t - c)}{c} \times 0.003$$

and

$$\epsilon_t = \frac{(20 - 3.72)}{3.72} \times 0.003$$
$$= 0.0131 \text{ and } \phi = 0.90$$

Alternatively, the checks can be based on a_b/d, where

$$\frac{a}{d} = \frac{2.79}{20} = 0.139$$

From (4-20),

$$\frac{a_b}{d} = \beta_1\left(\frac{87{,}000}{87{,}000 + f_y}\right)$$

$$\frac{a_b}{d} = 0.75\left(\frac{87{,}000}{87{,}000 + 60{,}000}\right) = 0.444$$

Since $a/d = 0.139$ is less than $a_b/d = 0.444$, it follows that $f_s = f_y$, and we also have

$$\frac{a_{TCL}}{d_t} = 0.375\beta_1 = 0.281$$

Again, $d_t = d$. Because $a/d_t = 0.139$ is less than $a_{TCL}/d_t = 0.281$, the section is tension-controlled, and $\phi = 0.90$.

3. **Compute M_n and ϕM_n:**

$$M_n = \frac{2.37 \times 60{,}000(20 - 2.79/2)}{12{,}000} = 220 \text{ ft-kips}$$

$$\phi M_n = 0.9 \times 220 \text{ ft-kips} = 198 \text{ ft-kips}$$

Note that doubling the concrete strength increased M_n by only 8 percent.

BEAM 3: SAME AS BEAM 1, EXCEPT THAT $f_y = 40,000$ psi

1.   **Compute a and c/d**

$$a = \frac{2.37 \times 40,000}{0.85 \times 3000 \times 10} = 3.72 \text{ in.}$$

$$c = \frac{3.72}{0.85} = 4.38 \text{ in.}$$

$$\frac{c}{d_t} = \frac{4.38}{20} = 0.219$$

2a.   **Compute ϵ_t and check whether $f_s = f_y$ and compute ϕ**

$$\epsilon_t = \frac{(d_t - c)}{c} \times 0.003$$
$$= \frac{(20 - 4.38)}{4.38} \times 0.003$$
$$= 0.0107$$

Since this exceeds 0.005, $f_s = f_y$ and $\phi = 0.90$.

2b.   **Check whether $f_s = f_y$ and whether the section is tension-controlled using c/d_t**

$$\frac{a}{d_t} = \frac{3.72}{20} = 0.186$$

$$\frac{a_b}{d} = 0.85\left(\frac{87,000}{87,000 + 40,000}\right) = 0.582 > \frac{a}{d}$$

$$\frac{a_{TCL}}{d_t} = 0.375\beta_1 = 0.319 > \frac{a}{d}$$

Thus, $f_s = f_y$, and the section is tension-controlled. Hence, $\phi = 0.90$.

3.   **Compute M_n and ϕM_n:**

$$M_n = \frac{2.37 \times 40,000(20 - 3.72/2)}{12,000} = 143 \text{ ft-kips}$$

$$\phi M_n = 0.9 \times 143 \text{ ft-kips} = 129 \text{ ft-kips}$$

Note that reducing f_y by 33 percent (compared with beam 1) reduced M_n by 30 percent. ∎

EXAMPLE 4-3M   Analysis of Singly Reinforced Beams: Tension Steel Yielding—SI Units

Compute the nominal moment capacity, M_n, of a beam with $f'_c = 20$ MPa, $f_y = 420$ MPa, $b = 250$ mm, $d = 500$ mm, and three No. 25 bars giving $A_s = 3 \times 500 = 1500$ mm² and $\rho = A_s/bd = 1500/(250 \times 500) = 0.0120$. See Fig. 4-19b.

1.   **Compute a:**

$$a = \frac{A_s f_y}{0.85 f'_c b}$$

$$= \frac{1500 \text{ mm}^2 \times 420 \text{ MPa}}{0.85 \times 20 \text{ MPa} \times 250 \text{ mm}} = 148 \text{ mm}$$

Therefore, $a = 148$ mm and $c = a/\beta_1 = 148/0.85 = 174$ mm.

2a. **Check whether** $f_s = f_y$ **and whether the section is tension-controlled.**

$$\epsilon_t = \left(\frac{500 - 174}{174}\right) \times 0.003 = 0.0056.$$ Thus, $f_s = f_y = 420$ MPa and the section is tension-controlled with $\phi = 0.90$. Alternatively:

2b. **Check whether** $f_s = f_y$ **using neutral axis depth** $a/d \leq a_{TCL}/d, f_s = f_y$, where a_b/d is given by (4-20M). Hence,

$$\frac{a}{d} = \frac{148 \text{ mm}}{500 \text{ mm}} = 0.296$$

From (4-20M),

$$\frac{a_b}{d} = \beta_1\left(\frac{600}{600 + f_y}\right)$$

From (4-9M), $\beta_1 = 0.85$ for $f_c' = 20$ MPa, and

$$\frac{a_b}{d} = 0.85\left(\frac{600}{600 + 420}\right) = 0.50$$

Since the actual $a/d = 0.296$ is less than $a_b/d = 0.50$, it follows that $f_s = f_y$. To check whether the section is tension-controlled, check whether $a_{TCL}/d_t \leq a/d_t$. Note that this check is in terms of a/d_t. For this beam, d_t and d are the same. If the steel were in several layers, this would not be true. From (4-23),

$$\frac{a_{TCL}}{d_t} = 0.375\beta_1 = 0.375 \times 0.85 = 0.319$$

Because $a/d_t = 0.296$ is less than 0.319, the section is tension-controlled, and $\phi = 0.90$.

3. **Compute the nominal moment capacity,** M_n. From (4-11a), M_n is

$$M_n = A_n f_y\left(d - \frac{a}{2}\right)$$

$$= 1500 \text{ mm}^2 \times 420 \text{ N/mm}^2\left(500 - \frac{148}{2}\right) \text{mm}$$

(where 1 MPa $= 1$ N/mm^2). Therefore,

$$M_n = 268 \times 10^6 \text{ N-mm} = 268 \text{ kN-m}$$

The nominal moment capacity of the beam is $M_n = 268$ kN $\cdot$ m. The design or factored moment capacity, ϕM_n, of this beam is $0.9 \times 268 = 242$ kN-m. ∎

Equations for M_n and ϕM_n: Tension Steel Elastic at Failure

From statics, we once again find that

$$C = T$$

and

$$0.85 f_c' ba = A_s f_s$$
$$= \rho E_s \epsilon_s bd$$

From strain compatibility (see Fig. 4-15c),

$$\epsilon_s = \epsilon_{cu}\left(\frac{d - c}{c}\right)$$

Solving these together and observing that $a = \beta_1 c$ gives

$$0.85 f_c' a^2 = \rho E_s \epsilon_{cu} \beta_1 d^2 - \rho E_s \epsilon_{cu} a d$$

or

$$\left(\frac{0.85 f_c'}{\rho E_s \epsilon_{cu}} \right) a^2 + (d)a - \beta_1 d^2 = 0 \qquad (4\text{-}29)$$

This can be solved for a, and from (4-12), ϕM_n or M_n can be computed.

Beams with $\rho > \rho_b$ are not allowed by ACI Section 10.3.5.

When ρ is greater than ρ_b in an existing beam, the value of M_n is relatively insensitive to changes in ρ. This is because both f_s and jd decrease as A_s increases.

In 1937, Whitney [4-14] used a semiempirical analysis to determine that the moment capacity for compression failures was

$$M_n = 0.333 f_c' b d^2 \qquad (4\text{-}30)$$

From (4-29) and (4-12a), the constant in (4-30) is found to range from 0.29 to 0.35 for beams with $\rho = \rho_b$, increasing by roughly 10 percent for beams with $\rho = 2\rho_b$.

EXAMPLE 4-4   Analysis of Singly Reinforced Beam: Tension-Steel Elastic

Compute the nominal moment capacity, M_n, of a beam having $b = 10$ in., $d = 20$ in., $A_s = 4.74$ in.^{2} (six No. 8 bars), $f_c' = 3000$ psi, and $f_y = 60,000$ psi.

1.   Compute a. When analyzing the capacity of a beam, one does not know at the start whether the steel will yield. Because $f_s = f_y$ in most beams encountered in practice, we will make this assumption and correct it later if necessary. From (4-10),

$$\text{trial } a = \frac{A_s f_y}{0.85 f_c' b}$$

$$= \frac{4.74 \times 60,000}{0.85 \times 3000 \times 10} = 11.2''$$

Therefore, trial $a = 11.2$ in.

2.   Check whether $f_s = f_y$ and whether the section is tension-controlled. These checks will be based on (4-20). We have

$$\frac{a}{d} = \frac{11.2}{20} = 0.558 \ (\text{based on trial value of } a)$$

$$\frac{a_b}{d} = \beta_1 \left(\frac{87,000}{87,000 + f_y} \right) = 0.503$$

Since $a/d = 0.558$ is greater than $a_b/d = 0.503$, this beam will fail in compression with f_s less than f_y. As a result, the value of a computed in step 1 is incorrect, and we must start again. Because f_s is less than f_y, the section is compression-controlled.

3.   Recompute a by using (4-29). We have

$$\left(\frac{0.85 f_c'}{\rho E_s \epsilon_{cu}} \right) a^2 + (d)a - \beta_1 d^2 = 0$$

where $\rho = A_s / bd = 0.0237$.

$$\left(\frac{0.85 \times 3000 \text{ psi}}{0.0237 \times 29 \times 10^6 \text{ psi} \times 0.003} \right) a^2 + 20a - 0.85 \times 20^2 = 0$$

Therefore, $1.237a^2 + 20a - 340 = 0$, and

$$a = \frac{-20 \pm \sqrt{20^2 - (4 \times -340 \times 1.237)}}{2 \times 1.237} = 10.36 \text{ in.}$$

This computed value of a is less than the 11.2 in. computed in step 1, because the actual steel stress is less than f_y.

4.   Compute M_n by using (4-12a). Because the steel stress f_s is not known, it is necessary to use (4-12a), rather than (4-11a), to compute M_n. We obtain

$$M_n = 0.85 f_c' ab\left(d - \frac{a}{2}\right)$$

$$= \frac{0.85 \times 3000 \times 10.36 \times 10(20 - 10.36/2)}{12,000} = 326 \text{ ft-kips}$$

Thus $M_n = 326$ ft-kips. Since the section is compression-controlled and does not contain spiral reinforcement, ACI Section C.3.2.2(b) gives $\phi = 0.65$ and $\phi M_n = 212$ ft-kips.

Whitney's equation for M_n, (4-30), gives $M_n = 333$ ft-kips [4-12]. For this example, this is essentially the same as the $M_n = 326$ ft-kips computed from the exact equation.   ■

Effect of Variables on M_n for Singly Reinforced Beams

Figure 4-20 compares the moment capacities of a rectangular cross section with varying material strengths and amounts of reinforcement. The upper and lower solid lines are plotted for $f_y = 60,000$ psi and concrete strengths, $f_c' = 6000$ and 3000 psi, respectively. The dashed curve is for $f_y = 40,000$ psi and $f_c' = 3000$ psi. Each curve consists of two parts: a steep portion to the left of point B for ρ less than ρ_b (tension failures), and a flatter portion to the right of point B for compression failures when ρ is greater than ρ_b.

The major difference between the two solid curves is the bending moment at which the change from tension to compression failures occurs. This is almost directly proportional

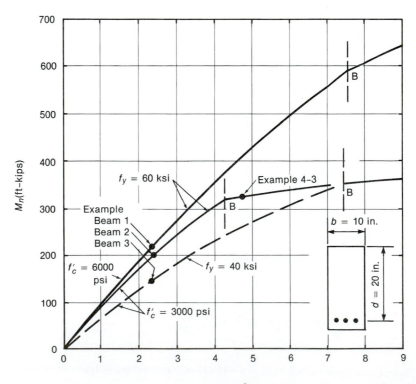

Fig. 4-20
Effect of variables on
strength of beams.

to the concrete strength. For values A_s less than about 3 in.[2], however, the two solid curves in Fig. 4-20 are very close to one another, indicating that the 100 percent difference between their concrete strengths has relatively little effect on their flexural capacities in this range. This can be seen from (4-11a), which, rearranged slightly, becomes

$$M_n = A_s f_y d \left(1 - \frac{A_s f_y}{1.7 f_c' b d}\right)$$

The three main variables in this equation are A_s, f_y, and d, and the moment capacity varies almost linearly with these three. Thus the moment capacities plotted in Fig. 4-20 increase almost linearly with $\rho = A_s/bd$ until ρ_b is reached. On the other hand, the concrete strength, f_c', has a much smaller effect on the strength of under-reinforced beams ($\rho < \rho_b$), acting only to change the depth of the compression zone, a, and hence to change the lever arm, $jd = (d - a/2)$. The points labeled beams 1 and 2 in this figure refer to the strengths computed for these beams in Example 4-3. Here, a 100 percent change in concrete strength, from 3000 psi to 6000 psi, increased the moment capacity by only 8 percent.

The effect of the yield strength, f_y, can be seen by comparing the dashed curve and the lower solid curve, both of which are plotted for the same concrete strength. The ratio of the ordinates of these two curves is roughly 40/60 for values of ρ less than ρ_b. For example, beam 3 (from Example 4-3) had a capacity 70 percent of that of beam 1.

In summary, for steel ratios up to 0.015 or so, the value of M_n is affected almost linearly by A_s, by f_y, and (although it is not shown in Fig. 4-20) by d. In this range, M_n is roughly proportional to f_y. The value of M_n at which the behavior changes from tension failures to compression failures and the value of M_n for compression failures are roughly proportional to f_c' and are essentially independent of f_y. Thus, the most effective ways to increase the strength of a beam while maintaining a tension failure mode are to increase A_s, f_y, or d. Increasing f_c' is effective only if it allows higher steel percentages to be used.

4-4   DESIGN OF RECTANGULAR BEAMS

General Factors Affecting the Design of Rectangular Beams

Location of Reinforcement

Concrete cracks due to tension, and as a result, reinforcement is required where flexure, axial loads, or shrinkage effects cause tensile stresses.

A uniformly loaded, simply supported beam deflects as shown in Fig. 4-21a and has the moment diagram shown in Fig. 4-21b. Because this beam is in positive moment

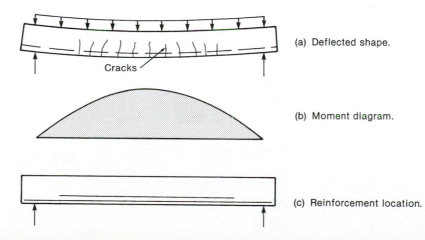

(a) Deflected shape.

(b) Moment diagram.

(c) Reinforcement location.

Fig. 4-21
Simply supported beam.

throughout, tensile flexural stresses and cracks are developed along the bottom of the beam. Longitudinal reinforcement is required to resist these tensile stresses and is placed close to the bottom side of the beam, as shown in Fig. 4-21c. Since the moments are greatest at midspan, more reinforcement is required at the midspan than at the ends, and it may not be necessary to extend all the bars into the supports. In Fig. 4-21c, some of the bars are *cut off* within the span.

A cantilever beam develops negative moment throughout and deflects as shown in Fig. 4-22, with the concave surface downward, so that flexural tensions and cracks develop on the top surface. In this case, the reinforcement is placed near the top surface, as shown in Fig. 4-22c. Since the moments are largest at the fixed end, more reinforcement is required here than at any other point. In some cases, some of the bars may be terminated before the free end of the beam. Note that the bars must be anchored into the support.

Commonly, reinforced concrete beams are continuous over several supports, and under gravity loads they develop the moment diagram and deflected shape shown in Fig. 4-23. Again, reinforcement is needed on the tensile face of the beam, which is at the top of the beam in the negative-moment regions at the supports and at the bottom in the positive-moment regions at the midspans. Two possible arrangements of this reinforcement are shown in Fig. 4-23c and d. Prior to 1965, it was common practice to bend the bottom reinforcement up to the top of the beam when it was no longer required at the bottom. In this way a *bent-up* or *truss bar* could serve as negative and positive reinforcement in the same beam. Such a system is illustrated in Fig. 4-23d. Today, the straight bar arrangement shown in Fig. 4-23c is used almost exclusively. In some cases, a portion of the positive-moment or negative-moment reinforcement is terminated or cut off when no longer needed. Note, however, that a portion of the steel is extended past the points of inflection, as shown. This is done primarily to account for shifts in the points of inflection due to shear cracking and to allow for changes in loadings and loading patterns. The calculation of bar-cutoff points is discussed in Chapter 8.

In addition to longitudinal reinforcement, additional bars, referred to as *stirrups*, are provided to resist shear forces and to hold the various layers of bars in place during construction. These are shown in the cross section in Fig. 4-23. The design of stirrups is discussed in Chapter 6.

In conclusion, it is important that designers be able to visualize the deflected shape of a structure. The reinforcing bars for flexure are placed on the tensile face of the member. This is the convex side of the deflected shape.

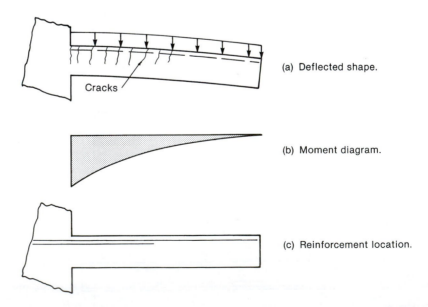

(a) Deflected shape.

Cracks

(b) Moment diagram.

(c) Reinforcement location.

Fig. 4-22
Cantilever beam.

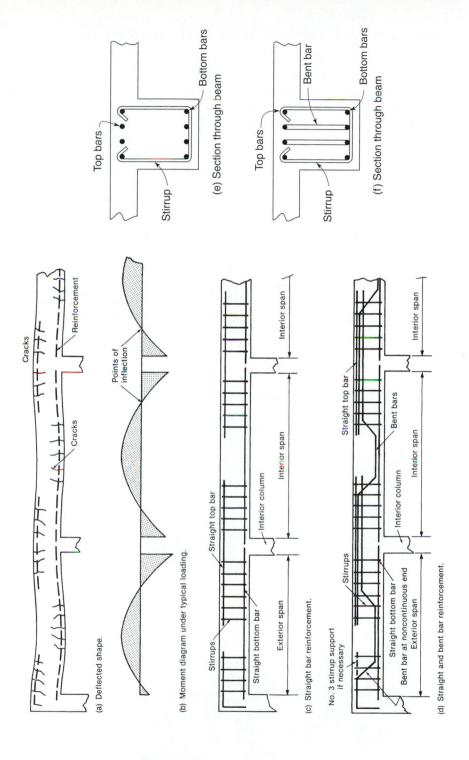

(a) Deflected shape.

(b) Moment diagram under typical loading.

(c) Straight bar reinforcement.

No. 3 stirrup support if necessary

(d) Straight and bent bar reinforcement.

(e) Section through beam

(f) Section through beam

Fig. 4-23
Continuous Beam

Construction of Reinforced Concrete Beams and Slabs

The simplest concrete flexural member is the one-way slab shown in Fig. 4-1. The form for such a slab consists of a flat surface, generally built of plywood supported on wooden or steel joists. Whenever possible, the forms are constructed in such a way that they can be reused on several floors. The forms must be strong enough to support the weight of the wet concrete plus construction loads such as workers, concrete buggies, and so on. In addition, the forms must be aligned correctly and *cambered* (arched upward), if necessary, so that the finished floor is flat after the forms are removed.

The reinforcement is supported in the form on wire supports referred to as *bolsters* or *chairs*, which holds the bars at the correct distance above the forms until the concrete has hardened. If the finished slab is expected to be exposed to moisture, wire bolsters may rust, staining the surface. In such a case, small precast concrete blocks or plastic bar chairs may be used instead. Wire bolsters can be seen in the photograph in Fig. 4-24.

Beam forms are most often built of plywood supported by scaffolding or by wooden supports. The size of beam forms is generally chosen to allow maximum reuse of the forms, since the cost of building the forms is a significant part of the total cost of a concrete floor system, as was discussed in Section 2-8. Some designers prefer to choose 12- or 16-in. beam widths, because these widths fit evenly into the width of a standard 4 ft × 8 ft sheet of plywood.

Fig. 4-24
Intersection of column and two beams.
(Photograph courtesy of J. G. MacGregor.)

Reinforcement for two beams and some slabs is shown in Fig. 4-24. Here, closed stirrups have been used and the top bars are supported by the top of the closed stirrups. The negative-moment bars in the slabs still must be placed. Frequently, the positive-moment steel, stirrups, and stirrup support bars for a beam are preassembled into a cage that is dropped into the form.

Relationship between Beam Depth and Deflections

The deflections of a beam can be calculated from equations of the form

$$\Delta_{max} = \frac{C_1 w \ell^4}{EI} \tag{4-31a}$$

Rearranging this and making assumptions concerning steel strains and neutral-axis depth eventually gives an equation of the form

$$\frac{\Delta}{\ell} = C \frac{\ell}{d} \tag{4-31b}$$

Thus, for any acceptable ratio of deflection to span lengths, Δ/ℓ, it should be possible to specify span-to-depth ratios, ℓ/d, which, if exceeded, may result in unacceptable deflections. ACI Table 9.5(a) gives minimum thicknesses computed in this way for members *not supporting partitions* or other construction that are liable to be damaged by deflection. (See Table A-14 of this book.) These minimum thicknesses are frequently used in selecting the overall depths of beams or slabs. Deflections are discussed in Chapter 9.

Concrete Cover and Bar Spacing

It is necessary to have cover (concrete between the surface of the slab or beam and the reinforcing bars), for four primary reasons:

1.   To bond the reinforcement to the concrete so that the two elements act together. The efficiency of the bond increases as the cover increases. A cover of at least one bar diameter is required for this purpose in beams and columns. (See Chapter 8.)

2.   To protect the reinforcement against corrosion. Depending on the environment and the type of member, varying amounts of cover ranging from $\frac{3}{8}$ to 3 in. are required (ACI Section 7.7). In highly corrosive environments such as slabs or bridges exposed to deicing salts or ocean spray, the cover should be increased. ACI Commentary Section R7.7 allows alternative methods of satisfying the increased cover requirements for elements exposed to the weather. An example of an alternative method might be a waterproof membrane on the exposed surface.

3.   To protect the reinforcement from strength loss due to overheating in the case of fire. The cover for fire protection is specified in the local building code. Generally speaking, $\frac{3}{4}$ in. cover to the reinforcement in a structural slab will provide a 1-hour fire rating, while a $1\frac{1}{2}$-in. cover to the stirrups or ties of beams corresponds to a 2-hour fire rating.

4.   Additional cover is sometimes provided on the top of slabs, particularly in garages and factories, so that abrasion and wear due to traffic will not reduce the cover below that required for structural and other purposes.

In this book, the amounts of clear cover will be based on ACI Section 7.7.1 unless specified otherwise.

The arrangement of bars within a beam must allow sufficient concrete on all sides of each bar to transfer forces into or out of the bars; sufficient space so that the fresh concrete can be placed or compacted around all the bars; and sufficient space to allow a vibrator to reach through to the bottom of the beam.

Pencil-type concrete immersion vibrators used in consolidation of the fresh concrete are $1\frac{1}{2}$ to $2\frac{1}{2}$ in. in diameter. Enough space should be provided between the beam bars to allow a vibrator to reach the bottom forms in at least one place in the beam width.

Figure 4-24 shows the reinforcement at an intersection of two beams and a column. The longitudinal steel in the beams is at the top of the beams because this is a negative-moment region. Although this region looks congested, there are adequate openings to place and vibrate the concrete. Reference [4-16] discusses the congestion of reinforcement in regions such as this and recommends design measures to reduce the congestion.

ACI Sections 3.3.2, 7.6.1, and 7.6.2 specify the spacings and arrangements shown in Fig. 4-25. When bars are placed in two or more layers, the bars in the top layer must be

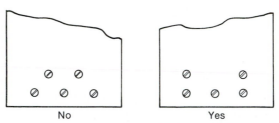

(a) Arrangement of bars in two layers (ACI Section 7.6.2).

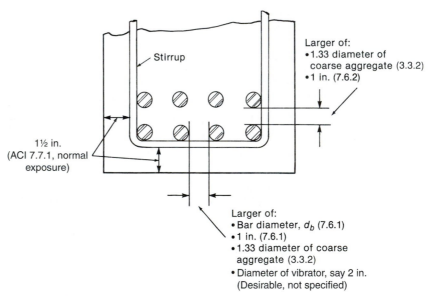

(b) Minimum bar spacing and cover limits in ACI Code.

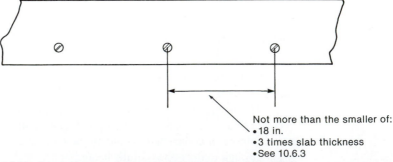

Fig. 4-25
Bar spacing limits in ACI
Code.

(c) Maximum spacing of flexural reinforcement in slabs (7.6.5).

directly over those in the other layers, to allow the concrete and vibrators to pass through the layers. Potential conflicts between the reinforcement in a beam and the bars in the columns or other beams should be considered. Figure 4-24 shows moderate congestion between the negative-moment reinforcement in both beams and the vertical column steel. Figure 4-26, based on an actual case history [4-13], shows what can happen if potential conflicts at a joint are ignored. The left-hand side shows how the design was envisioned, and the right-hand side the way the joint was built. Placement tolerances and the need to resolve the interference problems have reduced the effective depth of the negative-moment reinforcement from $9\frac{1}{2}$ in. to $7\frac{3}{4}$ in., an 18 percent reduction in depth and a corresponding reduction in moment capacity. To identify and rectify bar conflicts, it sometimes is necessary to draw the joint to scale, showing the actual width of the bars (see, for example, Fig. 18-44). Conflicts between bars in the columns and other beams should be considered (Figs. 4-24 and 4-26).

Calculation of Effective Depth and Minimum Web Width for a Given Bar Arrangement

The effective depth, d, of a beam is defined as the distance from the extreme compression fiber to the centroid of the longitudinal tensile reinforcement, as shown in Fig. 4-2.

EXAMPLE 4-5   Calculation of d and of Minimum b

Compute d and the minimum value of b for a beam having bars arranged as shown in Fig. 4-27. The maximum size of coarse aggregate is specified as $\frac{3}{4}$ in. The overall depth of the beam is 24 in.

This beam has two different bar sizes. The larger bars are in the bottom layer, to maximize the effective depth and hence the lever arm. Note also that the bars are symmetrically arranged about the centerline of the beam. The bars in the upper layer are directly above those in the lower layer. Placing them on the outside of the section allows the top layer of bars to be supported by tying them directly to the stirrups.

1.   Compute clear cover. From ACI Section 7.7.1, the clear cover to the stirrups is 1.5 in. (Fig. 4-25b). From ACI Sections 7.6.2 and 3.3.2, the minimum distance between layers of bars is the larger of 1 in. or $1\frac{1}{3}$ times the aggregate size, which in this case gives $1\frac{1}{3} \times \frac{3}{4} = 1$ in.

2.   Compute the centroid of the bars.

Layer	Area, $A(\text{in.}^2)$	Distance from Bottom, y (in.)	Ay in.^{3}
Bottom	$3 \times 1.00 = 3.00$	$1.5 + \frac{3}{8} + (\frac{1}{2} \times \frac{9}{8}) = 2.44$	7.31
Top	$2 \times 0.79 = \underline{1.58}$	$2.44 + (\frac{1}{2} \times \frac{9}{8}) + 1 + (\frac{1}{2} \times \frac{8}{8}) = 4.50$	$\underline{7.11}$
	Total $A = 4.58$		Total $Ay = 14.42$

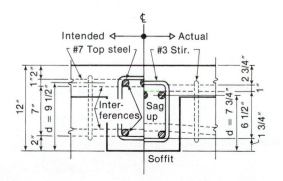

Fig. 4-26
Bar placing problems at the intersection of two beams. (From [4-13].)

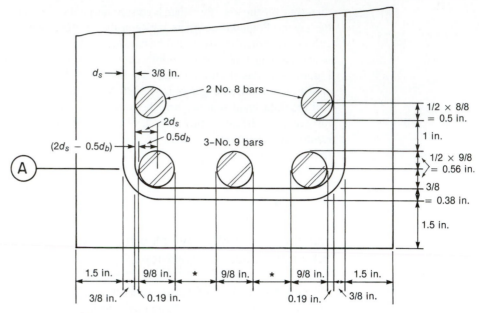

Fig. 4-27
Example 4-4.

*Not less than 1 in. nor d_b = 9/8 in.

The centroid is located at $Ay/A = \bar{y} = 14.42/4.58 = 3.15$ in. from the bottom of the beam. The effective depth $d = 24 - 3.15$ in. = 20.85 in.— say, $d = 20.8$ in.

    3. Compute the minimum web width. This is computed by summing the widths along the most congested layer. The minimum inside radius of a stirrup bend is two times the stirrup diameter, d_s, which for a No. 3 stirrup is $\frac{3}{4}$ in. (ACI Section 7.2.3). For No. 11 or smaller bars, there will be a space between the bar and the tie, as shown in Fig. 4-27:

$$\text{space} = 2d_s - 0.5d_b$$
$$= 2 \times \tfrac{3}{8} - 0.5 \times \tfrac{9}{8} = 0.19 \text{ in.}$$

The minimum horizontal distance between bars is the largest of 1 in., or $1\frac{1}{3}$ times the aggregate size, or the bar diameter. (See Fig. 4-25b.) In this case the largest bars are No. 9 bars with a nominal diameter of $\frac{9}{8}$ in. Summing the widths along a section at A–A and ignoring space for the vibrator gives

$$b_{\min} = 1.5 + \tfrac{3}{8} + 0.19 + 5\left(\tfrac{9}{8}\right) + 0.19 + \tfrac{3}{8} + 1.5$$
$$= 9.76 \text{ in.}$$

    Thus, the minimum width is 10 in., and design should be based on $d = 20.8$ in. ∎

Estimating the Effective Depth of a Beam

It is generally satisfactory to estimate the effective depth of a beam using the following approximations:

For beams with one layer of reinforcement,

$$d \simeq h - 2.5 \text{ in.} \tag{4-32a}$$

For beams with two layers of reinforcement,

$$d \simeq h - 3.5 \text{ in.} \tag{4-32b}$$

(The value 3.5 in. given by (4-32b) corresponds to the 3.15 in. computed in Example 4-5.) The error introduced by using (4-32b) to compute d is in the order

$$\frac{(24 - 3.5)}{(24 - 3.15)} = 0.983$$

Thus, (4-32b) underestimates d by 1.7 percent. This is acceptable.

For reinforced concrete slabs, the minimum clear cover is $\frac{3}{4}$ in. rather than $1\frac{1}{2}$ in., and the positive moment steel is all in one layer, with the negative moment steel in another layer. This steel will generally be No. 3, 4, or 5 bars. Stirrups are seldom, if ever, used in one-way slabs in buildings. For this case, (4-32a) and (4-32b) can be rewritten as follows:

For one-way slab spans up to 12 ft,

$$d \simeq h - 1 \text{ in.} \tag{4-32c}$$

For one-way slab spans over 12 ft,

$$d \simeq h - 1.1 \text{ in.} \tag{4-32d}$$

In SI units, (4-31) works out to be the following:

For beams with one layer of tension reinforcement,

$$d \simeq h - 65 \text{ mm} \tag{4-32aM}$$

For beams with two layers of tension reinforcement,

$$d \simeq h - 90 \text{ mm} \tag{4-32bM}$$

For one-way slabs with spans up to 3.5 m,

$$d \simeq h - 25 \text{ mm} \tag{4-32cM}$$

For one-way slabs with spans over 3.5 m,

$$d \simeq h - 30 \text{ mm} \tag{4-32dM}$$

It is important not to overestimate d, because normal construction practices lead to smaller values of d than are shown on the drawings. Studies of construction accuracy show that, on the average, the effective depth of the negative-moment reinforcement in slabs is 0.75 in. less than specified [4-15]. In a 5-in.-thick slab, an error of 0.75 in. in the steel placement will reduce the flexural capacity by 19 percent.

Generally speaking, b should not be less than 10 in. and preferably not less than 12 in. for beams, although with two bars, beam widths as low as 7 in. can be used in extreme cases. The use of a layer of closely spaced bars may lead to a splitting failure along the plane of the bars, as will be explained in Chapter 8. Since such a failure may lead to a loss of bar anchorage or to corrosion, care should be taken to have at least the minimum bar spacings. (A horizontal crack of this sort can be seen at midspan of the beam shown in Fig. 4-6.) Where there are several layers of bars, a continuous vertical opening large enough for the concrete vibrator to pass through should be provided. See Fig. 4-25b. Minimum web widths for various arrangements are given in Table A-6 (see Appendix A).

Minimum Reinforcement

If the cracking moment of a beam exceeds the strength of the beam after cracking, a sudden failure could occur with little or no warning when the beam cracks. For this reason, ACI Section 10.5 requires a minimum amount of flexural reinforcement equal to

$$A_{s,\,min} = \frac{3\sqrt{f_c'}}{f_y} b_w d, \text{ and } \geq \frac{200 b_w d}{f_y} \tag{4-33}$$

$$\text{(ACI Eq. 10-3)}$$

where f_c' and f_y are in psi. In SI units, this becomes

$$A_{s,\,min} = \frac{\sqrt{f_c'}}{4f_y} b_w d, \text{ and } \geq \frac{1.4b_w d}{f_y} \tag{4-33M}$$

$$\text{(ACI Eq. 10-3M)}$$

where f_c' and f_y are in MPa.

For statically determinate T beams with the flange in tension, the minimum reinforcement is equal to the smaller of

$$A_{s,\,min} = \frac{6\sqrt{f_c'}}{f_y} b_w d \tag{4-34}$$

or the value given by (4-33), with b_w taken equal to the width of the flange, where f_c' and f_y are in psi. In SI units, (4-34) becomes

$$A_{s,\,min} = \frac{\sqrt{f_c'}}{2f_y} b_w d \tag{4-34M}$$

$$\text{(ACI Eq. 10-4M)}$$

The requirements of (4-33) and (4-34) need not be applied if the area of reinforcement provided is at least one-third greater than that required to provide the required moment capacity. (ACI Section 10.5.2.)

Design of Rectangular Beams with Tension Reinforcement

In analysis of beam cross sections, is based on (4-1b):

$$\phi M_n \geq M_u \tag{4-1b}$$

Here, M_u represented factored moments due to loads, which, for gravity loads, is (ACI Section 9.2.1)

$$M_u = 1.2M_D + 1.6M_L$$

where M_D and M_L are the moments due to the unfactored dead and live loads, respectively. The factored resisting moment, ϕM_n, is the couple formed by the internal tensile and compressive forces and can be calculated from

$$\phi M_n = \phi[f_c'\,bd^2\omega(1 - 0.59\omega)] \tag{4-14}$$

where $\omega = \rho f_y / f_c'$.

In the design, the problem to be solved involves the selection of a beam cross section to support a given value of the live-load moment, M_L, plus its own dead-load moment and moments due to other loads it may be required to support. In this calculation, there are *six* unknowns: b, d, ρ, f_y, f_c' and the dead-load moment of the beam, but only *two* independent equations, (4-11a) or (4-14) and the relationship between the beam size and the dead-load moment of the beam. As a result, it is not possible to design a beam uniquely. The design procedure is thus an iterative process in which four assumptions must be made. Although this requires some intuition and understanding of the construction process, the resulting design freedom makes reinforced concrete the universal and valuable material it is.

The value of f_c' to be used in the design is chosen at the start of the design and used throughout the project. The choice of concrete strength and amount of air entraining are based on durability considerations if the member is exposed to freezing and thawing, deicing salts, or other aggressive environments. ACI Chapter 4, "Durability Requirements," particularly ACI Tables 4.2.1 to 4.3.1, specifies minimum concrete strengths ranging from 4000 to 5000 psi for various exposures. The strength chosen for durability reasons may be utilized in the design for strength. If durability is not a problem, reinforced concrete beams and slabs are generally constructed of 3000-, 3750-, or 4000-psi concrete, with 3000-psi being the most common. The strength of the concrete in columns may be higher, as will be discussed in Chapter 11.

The yield strength most commonly used in the United States is 60,000 psi. Steel with a yield strength of 40,000 psi is occasionally used for flexural reinforcement. Only No. 3 to No. 6 bars are available in Grade-40 steel.

Once f_c' and f_y have been chosen for the beam or slab in question, three major independent variables remain: the width, b; the depth, d (or the overall height, h); and the reinforcement ratio, ρ. If b and d are known, it is possible to go directly to the computation of ρ and $A_s = \rho b d$. If not, the equation used to arrive at b and d is

$$\frac{M_u}{\phi k_n} = \frac{b d^2}{12{,}000} \tag{4-16}$$

where $\phi k_n = \phi[f_c' \omega(1 - 0.59\omega)]$ and M_u is in ft-kips. To solve for b and d, it is necessary to assume a value of ω or ρ, compute ϕk_n, and compute $b d^2/12{,}000$ and eventually b and d. Table A-3, which lists ϕk_n, and Table A-7, which lists $b d^2/12{,}000$, may be used to aid in this calculation. A value of ρ equal to 0.01 (corresponding to $\omega \approx 0.12$ to 0.24 or $\rho = 0.30$ to $0.50\rho_b$, depending on the concrete strength) is generally used to get the first value of ϕk_n. Also, $\rho = 0.010$ will correspond to a tension-controlled section that will have $\phi = 0.90$. It is possible to round off the values of b and d to practical sizes at this stage, since the values of b and d chosen in this step are then used in a recomputation of ρ and A_s in the next stage.

For designing in metric (SI) units, reinforced concrete beams and slabs are generally constructed of 20-, 25-, or 30-MPa concrete, with 20-MPa being the most common. In locations where durability is important, ACI Section 4.2.2 requires the use of air-entrained concrete having specified maximum water/cementitious materials ratios and minimum strengths ranging from 28 to 35 MPa. The most common yield strength is 420 MPa. Grade-300 reinforcement is available only in sizes 10, 15, and 20.

If b and d are not known, the equation used to arrive at b and d when working in SI units is

$$\frac{M_u \times 10^6}{\phi k_{nm}} = b d^2 \tag{4-16M}$$

where $\phi k_{nm} = \phi[f_c' \omega(1 - 0.59\omega)]$ in metric units, M_u is in kN-m and b and d are in mm. To solve for b and d, it is necessary to assume a value of ω or ρ, compute ϕk_{nm}, and compute $b d^2$ and eventually b and d. A value of ρ equal to 0.01 is generally used to get the first value of ϕk_{nm}.

Two different procedures exist for calculating the area of reinforcement, A_s. If the values of b and d chosen for the beam are close to the calculated values, the steel area can be calculated directly from ρ by using

$$A_s = \rho b d$$

The resulting A_s should not be less than $A_{s,\,min}$ (4-33), and ρ should not exceed $0.75\rho_b$ ((4-25) and Table A-5). It is then necessary to check whether this amount of reinforcement is adequate to resist M_u. This is done by using (4-11a) as in Example 4-2.

In most cases, however, it is much better to calculate the area of reinforcement by using

$$\phi M_n = \phi\left[A_s f_y\left(d - \frac{a}{2}\right)\right] \tag{4-11b}$$

where $(d - a/2)$ is referred to as jd. The terms A_s and a are unknown in this equation. It is necessary to assume j, compute A_s, recompute a and $(d - a/2)$ using this value of A_s, and recompute A_s until convergence is obtained. For beams with Grade-60 reinforcement, j can range from about 0.95 for minimum reinforcement to about 0.84 for $\rho = 0.75\rho_b$. (See Table A-3.) For the most common steel percentages in beams, j is generally between 0.87 and 0.91. For one-way slabs, which generally have a lower reinforcement ratio than beams, j will generally vary between 0.90 and 0.95. In design problems in this book, j will initially be assumed equal to 0.875 for rectangular beams and 0.925 for slabs. Thus, for the initial trial, A_s can be computed from (4-11a) by replacing $(d - a/2)$ with jd, where $j \simeq 0.875$ for beams and $j \simeq 0.925$ for slabs, and by replacing M_n with M_u/ϕ:

$$A_s = \frac{M_u}{\phi f_y jd} \tag{4-35}$$

Since this computed value of A_s is based on an estimate of j, it is necessary to check the accuracy of the estimate by using the selected A_s to compute ϕM_n in (4-11b). If ϕM_n is not close enough, iterations may be needed. Three general types of design problems exist. These will be discussed in the next three subsections.

Design of Reinforcement When b and h Are Known

The first type of design problem is the case in which the dimensions of the concrete section have been established for nonstructural reasons, such as architectural appearance, reuse of standard forms, and fire resistance. In this case, b and d (or h) are known, and it is only necessary to compute A_s.

EXAMPLE 4-6   Design of Reinforcement When b and h Are Known

For architectural reasons, it is necessary that the beam shown in Fig. 4-28 be 24 in. wide by 24 in. deep. The strengths of the concrete and steel are 3000 psi and 60,000 psi, respectively. In addition to its own dead load, this beam carries a superimposed service (unfactored) dead load of 1.0 kip/ft and a service live load of 2.45 kips/ft.

Compute the area of reinforcement required at midspan, and select the reinforcement.

1. **Estimate the factored moment, M_u:**

$$\text{Weight/ft of beam} = \frac{(2 \times 2 \times 1)\ \text{ft}^3 \times 150\ \text{lb/ft}^3}{1\ \text{ft of length}} = 600\ \text{lb/ft} = 0.60\ \text{kip/ft}$$

The factored load is

$$U = 1.2D + 1.6L \tag{2-4}$$

$$= 1.2(0.6 + 1.0) + 1.6 \times 2.45 \tag{ACI Eq. 9-2}$$

$$= 5.84\ \text{kips/ft}.$$

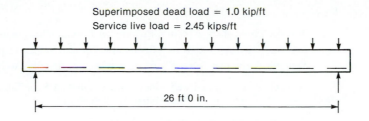

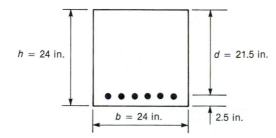

Fig. 4-28
Beam designed in
Example 4-6.

or

$$w_u = 1.2(0.6 + 1.0 \text{ kip/ft}) + 1.6(2.45 \text{ kip/ft})$$
$$= 5.84 \text{ kip/ft}$$

The factored load effect (factored ultimate moment) is

$$M_u = \frac{w_u \ell_n^2}{8}$$

$$= \frac{5.84 \text{ kip/ft} \times 26^2 \text{ ft}^2}{8} = 493 \text{ ft-kips}$$

It is necessary, therefore, to provide $\phi M_n \geq M_u$ or $\phi M_n \geq 493$ ft-kips.

2. Compute the effective depth, d. Because the beam is quite wide, assume that all the bars will be in one layer. From (4-32a), d can be estimated as

$$d \simeq h - 2.5 \text{ in.}$$
$$\simeq 24 - 2.5 \text{ in.}$$

Therefore, try $d = 21.5$ in.

3. Compute the area of reinforcement, A_s. Assume that

$$jd = d - \frac{a}{2} = 0.875d \text{ (this is equivalent to assuming that } a = 0.25d)$$

$$= 0.875 \times 21.5 \text{ in.} = 18.8 \text{ in.}$$

From (4-35), assuming $\phi = 0.9$ gives

$$A_s = \frac{M_u}{\phi f_y jd}$$

$$= \frac{493 \text{ ft-kips} \times 12 \text{ in./ft}}{0.9 \times 60 \text{ ksi} \times 18.8 \text{ in.}} = 5.83 \text{ in.}^2$$

Possible choices are (Table A-8):

Ten No. 7 bars, $A_s = 6.00$ in.^{2}, Required web width if bars in one layer, $b_w = 21.5$ in.
Six No. 9 bars, $A_s = 6.00$ in.^{2}, Required web width for one layer, $b_w = 16.5$ in.
Four No. 9 bars plus 3 No. 8 bars, $A_s = 5.95$ in.^{2}, Required web width, $b_w = 57.5$ in.

A check of the required web width as per Example 4-5 (or Table A-6) shows that all of these choices are acceptable. **Try six No. 9 bars, $A_s = 6.00$ in.^{2}.**

4. **Check whether $A_s \geq A_{s, min}$.** From (4-33),

$$A_{s, min} = \frac{3\sqrt{f_c'}}{f_y} b_w d, \text{ and } \geq \frac{200 b_w d}{f_y}$$

$$= \frac{3\sqrt{3000}}{60,000 \text{ psi}} \times 24 \times 21.5, \text{ and } \geq \frac{200 \times 24 \times 21.5}{60,000}$$

$$= 1.41 \text{ in.}^2, \text{ and } \geq 1.72 \text{ in.}^2.$$

The minimum area of steel is the larger of 1.41 in.^{2} and 1.72 in.^{2}.

Since 6.00 in.^{2} exceeds 1.72 in.^{2}, $A_s > A_{s, min}$. (If not OK, increase A_s to $A_{s, min}$ or satisfy ACI Section 10.5.3.)

5. **Compute a for $A_s = 6.00$ in.^{2}, and check if $f_s = f_y$ and whether the section is tension-controlled.** From (4-10),

$$a = \frac{A_s f_y}{0.85 f_c' b} \tag{4-10}$$

$$= \frac{6.00 \text{ in.}^2 \times 60,000 \text{ psi}}{0.85 \times 3000 \text{ psi} \times 24 \text{ in.}} = 5.88 \text{ in.}$$

$$= \frac{a}{d} = \frac{5.88 \text{ in.}}{21.5 \text{ in.}} = 0.273$$

to check whether $f_s = f_y$ we shall either check whether a/d is less than a_b/d or compute ϵ_t. From (4-20),

$$\frac{a_b}{d} = 0.85\left(\frac{87,000}{87,000 + 60,000}\right) = 0.503 \text{ (see also Table A-4)}$$

Since $a/d = 0.273$ is less than $a_b/d = 0.503$, $f_s = f_y$.

To check whether the section is tension-controlled, we shall check whether ϵ_t is more tensile than or equal to 0.005. Since all the steel is in one layer, $d = d_t$, $a/d_t = 0.273$.

$$a = 5.88 \text{ in.}$$

$$c = \frac{5.88}{\beta_1} = \frac{5.88}{0.85} = 6.92 \text{ in.}$$

$$\epsilon_t = \frac{(d_t - c)}{c} \times 0.003 \tag{4-28}$$

$$= \frac{(21.5 - 6.92)}{6.92} \times 0.003$$

$$= 0.0063$$

Thus the stress $f_s = 60,000$ psi and the section is tension-controlled. As a result $\phi = 0.90$. Alternatively,

$$\frac{a_{TCL}}{d_t} = 0.375\beta_1 = 0.375 \times 0.85$$

$$\frac{a_{TCL}}{d_t} = 0.319 \tag{4-23}$$

Since $a/d_t = 0.273$ is less than $a_{TCL}/d_t = 0.319$, the section is tension-controlled and from ACI Section 9.3.2.1, $\phi = 0.90$. If the section was compression-controlled or transitional, a lower value of ϕ would be required. Such a beam would be less ductile than a tension-controlled section. If this was not considered acceptable, the section should be enlarged or compression steel added. Compression steel is discussed in Chapter 5.

6.   **Compute M_n and ϕM_n.** Because A_s was calculated using an estimated value of jd, it is necessary to check whether the reinforcement selected provides adequate moment capacity. It may also be desirable to recompute d. From (4-11a),

$$M_n = A_s f_y \left(d - \frac{a}{2} \right)$$

$$= 6.00 \text{ in.}^2 \times 60 \text{ ksi} \left(21.5 \text{ in.} - \frac{5.88}{2} \right)$$

$$= 6680 \text{ in.-kips} = 557 \text{ ft-kips}$$

$$\phi M_n = 0.9 \times 557 \text{ ft-kips} = 501 \text{ ft-kips}$$

Since $\phi M_n = 501$ ft-kips exceeds $M_u = 493$ ft-kips this should be OK.

7.   **Check the area of steel required.** Because A_s was computed using an estimate of jd, it must be checked using the computed value of a. To do so, recompute the area of steel using the lever arm $(d - a/2)$ based on the value of a computed in Step 5.

Because $\phi M_n = $ exceeds $M_u = 493$ ft-kips M_u, A_s is OK. **Therefore, use six No. 9 bars.** ■

Design of Beams when b and h Are Not Known

The second type of design problem involves finding b, d, and A_s. Three sets of decisions not encountered in Example 4-6 must be made here. These are a preliminary estimate of the dead load of the beam, selection of a trial steel percentage, and the final selection of the beam dimensions b and h.

Although no dependable rule exists for guessing the weight of beams, the weight of a rectangular beam will be roughly 10 to 20 percent of the loads it must carry. Alternatively, one can estimate h as being roughly 8 to 10 percent of the span, corresponding to the traditional rule of thumb (1 in. deep per foot of span), estimate b as $0.5h$, and use these dimensions to compute a trial weight. These two procedures frequently give quite different values, in which case an intermediate value can be chosen. The dead load estimated at this stage is corrected when the dimensions are finally chosen.

It is then necessary

(1) to select a trial steel ratio ρ, which is used to calculate ϕk_n

(2) to use (4-16), and

(3) to get b and d.

This choice is affected by

(a) economic considerations (generally, $\rho \simeq 0.01$ is an economical choice),

(b) ductility (generally, $\rho \simeq 0.35\rho_b$ to $\rho \simeq 0.4\rho_b$ gives a desirable level of ductility), and

(c) by placing considerations (it may be hard to place the reinforcement if ρ exceeds 0.015).

For Grade-60 reinforcement, $0.4\rho_b$ is 0.0086 for 3000 psi, 0.0114 for 4000 psi, and 0.0134 for 5000 psi. For this reason, we start our design by assuming that $\rho = 0.010$ at the point of maximum moment in all cases involving rectangular beams.

Some factors need to be considered in choosing the beam dimensions b and h:

1. A deeper beam requires less reinforcement. The savings here are offset by increased forming costs and by either reduced headroom in the story below or an increase in the overall height of the building.

2. Longer development lengths are required for closely spaced bars. This is normally not a problem, except for top bars in short rectangular beams.

3. It may be possible to avoid deflection calculations if the overall height of the beam exceeds the values given in ACI Table 9.5(a). (See Table A-14 of this book.)

4. For rectangular beams, past practice has been to select sections with d/b between 1.5 and 2. Now, beam cross sections are often almost square, or even wider than their depth. A deeper beam requires less flexural reinforcement, whereas a shallow beam with the same flexural capacity is more efficient for shear and torsion, is easier to form, is stiffer and hence deflects less, and reduces the story height and overall height of the building. A wide beam may also be wider than the column and thereby ease the forming of the beam–column joint considerably.

The beam size chosen at this stage is rounded off to fit convenient form sizes as discussed earlier. A considerable amount of rounding can be carried out at this stage, because the final stage in the design is to compute the required area of steel corresponding to the b and d selected at this stage.

EXAMPLE 4-7 Design of a Beam for Which b and d Are Not Known

A beam is to carry its own dead load plus a uniform service live load of 1.75 kips/ft and a uniform superimposed service dead load of 1 kip/ft on a 33-ft span. Select b, d, and A_s if f'_c is 3500 psi and f_y is 60,000 psi.

1. **Estimate the dead load of the beam.** Estimate the weight of a rectangular beam as 10 to 20 percent of the loads it must carry. This range corresponds to 0.3 to 0.6 kip/ft. Alternatively, estimate h as being roughly 8 to 10 percent of the span, and b as 0.5h, and compute the weight. This gives $h = 2.6$ to 3.3 ft, and following the procedure in step 1 of Example 4-4, these correspond to weights equal to 0.52 to 0.82 kip/ft. Given these four values, estimate the beam weight at 0.5 kip/ft.

2. **Compute the factored moment, M_u.**

$$w_u = 1.2(1 + 0.5) + 1.6(1.75) = 4.6 \text{ kip/ft}$$

$$M_u = \frac{w_u \ell_n^2}{8} = 626 \text{ ft-kips}$$

3. **Compute b and d.** From (4-16),

$$\frac{M_u}{\phi k_n} = \frac{bd^2}{12,000}$$

where $k_n = f'_c \omega(1 - 0.59\omega)$, $\omega = \rho f_y/f'_c$, and M_u is in ft-kips. To start this calculation, assume either ρ or ω. We shall try $\rho \simeq 0.010$. The value of ω for $\rho = 0.01$ is

$$\omega = 0.01 \times \frac{60,000}{3500} = 0.171$$

Therefore,

$$k_n = 3500 \times 0.171(1 - 0.59 \times 0.171) = 538$$

From (4-16), assume $\phi = 0.9$ because this is a beam.

$$\frac{bd^2}{12,000} = \frac{M_u}{\phi k_n}$$

$$\frac{bd^2}{12,000} = \frac{626}{0.9 \times 538} = 1.293$$

or

$$bd^2 = 15{,}510 \text{ in.}^3.$$

Minimum overall depth to avoid deflection calculations if beam is not supporting brittle parti-tions [from ACI Table 9.5(a)] for a simple beam is $\ell/16$, which, in this case, is 24.75 in. All the choices listed exceed this, so deflection should not be a problem.

Try d/b between 1.5 and 2. On this basis, we choose $b = 16$ in. and $h = 36$ in. The size cho-sen has been rounded off to aid in construction of forms, and so on. Assuming two layers of rein-forcement, $d = 36 - 3.5 = 32.5$ in. **Use $b = 16$ in., $h = 36$ in., and $d = 32.5$ in.**

Possible choices (with h calculated from (4-30) assuming two layers of reinforcement for the narrower beams and one layer in the 18-in.-wide beam) are

$$b = 12 \text{ in. by } d = 36 \text{ in. and } h = 36 + 3.5 = 39.5 \text{ in.}$$
$$b = 16 \text{ in. by } d = 31.1 \text{ in. and } h = 31.1 + 3.5 = 34.6 \text{ in.}$$
$$b = 18 \text{ in. by } d = 29.3 \text{ in. and } h = 29.3 + 3.5 = 32.8 \text{ in.}$$

Try a rectangular cross section with $b = 16$ in., $h = 36$ in., and $d = 32.5$ in.

4. Check the dead load and revise M_u. For $b = 16$ in. and $h = 36$ in., the self-weight per foot is

$$(1.33 \times 3 \times 1)\text{ft}^3/\text{ft} \times 0.15 \text{ kip/ft}^3 = 0.600 \text{ kip/ft}$$

Total load/ft $= 1.2(1 + 0.6) + 1.6 \times 1.75 = 4.72$ kips/ft and the total factored moment, M_u, becomes 643 ft-kips, compared with the original estimate of 626 ft-kips. If the moment M_u increased by more than about 10 percent, it may be desirable to repeat steps 3 and 4.

5. Compute the area of reinforcement A_s. Assume that

$$jd = \left(d - \frac{a}{2}\right) = 0.875d$$
$$= 28.4 \text{ in.}$$

From (4-35),

$$A_s = \frac{M_u}{\phi f_y jd}$$

$$= \frac{643 \text{ ft-kips} \times 12 \text{ in./ft}}{0.9 \times 60 \text{ ksi} \times 28.4 \text{ in.}} = 5.03 \text{ in.}^2$$

6. Minimum Reinforcement. From (4-33),

$$A_{s,\text{min}} = \frac{3\sqrt{f_c'}}{f_y} b_w d = \frac{3\sqrt{3500}}{60{,}000} 16 \times 33 = 1.56 \text{ in.}^2$$

but not less than $A_{s,\text{min}} = 200 b_w d/f_y = 200 \times 16 \times 33/60{,}000$

$$= 1.76 \text{ in.}^2$$

7. Select steel

Because $A_{s,\text{reqd}} = 5.03$ in.^{2} exceeds $A_{s,\text{min}} = 1.76$ in.^{2} and $A_{s,\text{reqd}} = 5.03$ in.^{2} governs.

Possible choices are:

Five No. 9 bars, $A_s = 5.00$ in.^{2}

Seven No. 8 bars, $A_s = 5.53$ in.^{2}

Ten No. 7 bars, $A_s = 6.00$ in.^{2}

Four No. 9 bars plus two No. 8 bars, $A_s = 5.58$ in.^{2}

Four No. 8 bars plus four No. 7 bars, $A_s = 5.56$ in.^{2}

All of these will fit into two layers in $b = 16$ in. if the maximum size aggregate is $\frac{3}{4}$ in. (Table A-6). We shall try four No. 9 bars plus two No. 8 bars arranged as shown in Fig. 4-29. The larger bars are placed in the bottom row to give them the largest possible lever arm.

If the values of b and d chosen in step 3 are close to the computed ones, A_s could theoretically be computed directly from the value of ρ chosen in step 3 since $A_s = \rho bd$. Generally, however, the rounding of b and d makes it necessary to compute A_s as done in this example.

7. **Compute d.**

$$d_t = 36 \text{ in.} - (1.5 - 0.375 - 1.128/2) = 33.6 \text{ in.}$$

From (4-37a) and (4-37b), $d = 36 - 2.5$ to 3.5 in. Use $d = 32.5$ in. For the steel placement in Fig. 4-29, d can be computed using the technique given in Example 4-5. In most cases, this step can be omitted since $d = 32.5$ in. from (4-32b) is close enough.

8. **Compute ϵ_t and check whether $f_s = f_y$ and whether the section is tension-controlled.**

$$a = \frac{A_s f_y}{0.85 f_c'} = 8.21 \text{ in.}$$

$$c = a/\beta_1 = 8.21/0.85 = 9.65 \text{ in.}$$

From the strain distribution and similar triangles:

$$\epsilon_t = 0.003 \frac{(d_t - c)}{c} = 0.003 \frac{(33.6 - 9.65)}{9.65} = 0.00745.$$

Because $\epsilon_t = 0.00745$ exceeds 0.005, the section is tension-controlled, $f_s = f_y$, and $\phi = 0.90$.

Alternatively, we could base the check of whether the beam is tension-controlled on a/d_t where $a = 8.21$ in. and $a_{TCL}/d_t = 0.375 \beta_1 = 0.319$. The value of a/d_t the ratio is $8.21/33.6 = 0.244$. Because this is less than $a_{TCL}/d = 0.319$, the section is tension-controlled, $f_s = f_y$, and $\phi = 0.90$.

From (4-20) or Table A-4, a_b/d is 0.503. Since a/d is less than a_b/d, $f_s = f_y$, d is 33 in., $d_t = 36$ in. $- (1.50 + 0.375 + 1.125/2)$ in. $= 33.5$ in. Thus

$$\frac{a}{d_t} = \frac{8.21}{33.6} = 0.244 \qquad \frac{a_{TCL}}{d_t} = 0.319$$

since a/d_t is less than a_{TCL}/d_t, the section is tension-controlled and $\phi = 0.90$.

9. **Compute M_n and ϕM_n.** Because the area of steel calculated in step 5 was based on an estimate of jd, it is necessary to check whether the reinforcement chosen provides the required moment resistance. From (4-11),

$$M_n = A_s f_y \left(d - \frac{a}{2} \right)$$

$$= \frac{5.58 \times 60,000 \left(32.5 - \dfrac{8.21}{2} \right)}{12,000} = 792 \text{ ft-kips}$$

$$\phi M_n = 0.9 \times 792 \text{ ft-kips} = 713 \text{ ft-kips}$$

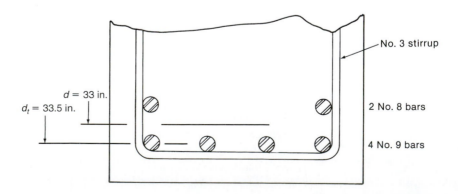

Fig. 4-29
Reinforcement location—
Example 4-7.

Since $\phi M_n \geq M_u$, the design is OK. **Therefore, use b = 16 in., h = 36 in., with f'_c = 3500 psi, f_y = 60,000 psi, and reinforcement as shown in Fig. 4-29.** If ϕM_n were less than M_u or if ϕM_n were much greater than M_u, it would be necessary to recompute the area of steel required. (See step 7 of Example 4-5). ∎

EXAMPLE 4-7M   Design of a Beam for Which b and h Are Not Known—SI Units

A beam is to carry its own dead load plus a uniform service live load of 25.5 kN/m and a uniform superimposed service dead load of 14.5 kN/m on a 10-m span. Select b, d, and A_s if f'_c is 25 MPa and f_y is 420 MPa.

1. **Estimate the dead load of the beam.** Estimate the weight of a rectangular beam as 10 to 20 percent of the loads it must carry. This range corresponds to 4 to 8 kN/m. Alternatively, estimate h as being roughly 8 to 10 percent of the span, and b as $0.5h$, and compute the weight. This gives h = 0.8 to 1.0 m, and the trial weight as

$$\text{trial weight of beam} = (0.8 \times 0.4 \times 1)\text{m}^3 \times 2450 \text{ kg/m}^3 \times (9.81/1000)\text{N/kg}$$
$$= 7.69 \text{ kN/m for } h = 0.8 \text{ m or } 12.0 \text{ kN/m for } h = 1.0 \text{ m.}$$

Given these four values (4, 7.69, 8, and 12 kN/m), estimate the beam weight at 8 kN/m.

2. **Compute the factored moment, M_u:**

$$w_u = 1.2(14.5 + 8.0) + 1.6(25.5) = 67.8 \text{ kN/m}$$

$$M_u = \frac{w_u \ell_n^2}{8} = 848 \text{ kN-m}$$

3. **Compute b and d.** From (4-16M),

$$\frac{M_u}{\phi k_{nM}} = \frac{bd^2}{10^6}$$

where $\phi k_{nM} = \phi[f'_c \omega(1 - 0.59\omega)]$, $\omega = \rho f_y/f'_c$, and M_u is in kN-m. To start this calculation, assume either ρ or ω. We shall assume that ρ = 0.01. For flexure, ϕ = 0.9, except when using the ϕ values from C9.3.2, when it is 0.9 for tension-controlled sections. As we will see later, ρ = 0.01 will always give a tension-controlled beam section. The value for ω for ρ = 0.01 is

$$\omega = 0.01 \times \frac{420}{25} = 0.168$$

Therefore,

$$\phi k_{nM} = 0.9[25 \times 0.168(1 - 0.59 \times 0.168)] = 3.41$$

From (4-16M),

$$bd^2 = \frac{848 \times 10^6}{3.41} = 248.7 \times 10^6 \text{ mm}^2$$

Possible choices (with h calculated using (4-39M) assuming two layers of reinforcement for the narrower beams and one layer in the 450-mm-wide beam) are

b = 300 mm by d = 910 mm and h = 910 + 90 = 1000 mm
b = 400 mm by d = 788 mm and h = 788 + 90 = 878 mm
b = 450 mm by d = 743 mm and h = 743 + 65 = 808 mm

Minimum overall depth to avoid deflection calculations if the beam is not supporting brittle partitions [from ACI Table 9.5(a)] for a simple beam is $\ell/16$, which in this case is 625 mm. Since all the choices listed exceed this value and, deflections should not be a problem.

Try d/b between 1.5 and 2. On this basis, we shall choose $b = 400$ mm and $h = 900$ mm. The size chosen was rounded off to aid in the construction of forms, and so on. Assuming two layers of steel, $d = 900 - 90 = 810$ mm. **Use $b = 400$ mm, $h = 900$ mm, and $d = 810$ mm.**

4. Check the dead load and revise M_u: For $b = 400$ mm and $h = 900$ mm, the self-weight per meter is

$$(0.40 \times 0.90 \times 1.0) \text{ m}^3/\text{m} \times 2450 \times (9.81/1000) \text{ kN/m}^3 = 8.65 \text{ kN/m}$$

Therefore, the total factored moment, M_u, becomes 857 kN-m compared to the original estimate of 843 kN-m. If the moment, M_u, is increased by more than about 10 percent, it may be desirable to repeat steps 3 and 4.

5. Compute the area of reinforcement, A_s. Assume that

$$jd = \left(d - \frac{a}{2}\right) = 0.875d$$
$$= 709 \text{ mm}$$

From (4-35),

$$\text{required } A_s = \frac{M_u}{\phi f_y jd}$$
$$= \frac{857 \times 10^6 \text{ N-mm}}{0.9 \times 420 \text{ MPa} \times 709 \text{ mm}} = 3200 \text{ mm}^2$$

6. Minimum reinforcement.

From (4-31M),

$$A_{s, \min} = \frac{\sqrt{f_c'}}{4f_y} b_w d, \text{ and } \geq \frac{1.4 b_w d}{f_y}$$
$$= \frac{\sqrt{25}}{4 \times 420} \times 400 \times 810 = 964 \text{ mm}^2$$

but not less than

$$\frac{1.4 \times 400 \times 810}{420} = 1080 \text{ mm}^2$$

7. Select steel.

Since $A_{s \text{ req}} = 3200$ mm^2 exceeds 1080 mm^2, use the required $A_s = 3200$ mm^2. Possible choices are

5 No. 29 bars, $A_s = 3220$ mm^2
7 No. 25 bars, $A_s = 3570$ mm^2

Both of these will fit in one layer in $b = 400$ mm if the maximum size aggregate is 20 mm. Use five No. 29 bars in one layer.

If the values of b and d chosen in step 3 are close to the computed ones, A_s could theoretically be calculated directly from the value of ρ chosen in step 3 since $A_s = \rho bd$. Generally, however, the rounding of b and d makes it necessary to compute A_s as done in this example.

8. Compute d. For steel placed in the manner chosen, d can be computed in the manner given in Example 4-4. This gives $d = 835$ mm. In most cases, this step can be omitted, because $d = 810$ mm (from (4-32bM)) is close enough.

9.   **Check whether is ϵ_t is less than 0.004 or greater than 0.005**

$$a = \frac{A_s f_y}{0.85 f'_c b}$$

$$= \frac{3220 \times 420}{0.85 \times 25 \times 400} = 159 \text{ mm}$$

$$\epsilon_t = \frac{(d_t - c)}{c} \times 0.003$$

$$\text{where } c = \frac{a}{\beta_1} = \frac{159}{0.85}$$

$$= 187 \text{ mm}$$

$$\epsilon_t = \frac{(835 - 187)}{187} \times 0.003$$

$$= 0.0104$$

Because 0.0104 exceeds 0.004, the section satisfies the definition of a beam in ACI Section 10.3.5. Because 0.0104 exceeds 0.005 it is tension-controlled. Thus, $\phi = 0.90$. Alternatively this check could be based on a/d as in the following:

$$d = 835 \text{ mm and } d_t = 835 \text{ mm}$$

$$\frac{a}{d} = \frac{159}{835} = 0.191$$

From (4-23), the section will be tension-controlled if a/d_t is less than or equal to

$$\frac{a_{TCL}}{d_t} = 0.375\beta_1 = 0.319$$

10.   **Compute M_u and ϕM_n.** Because the area of steel calculated in step 5 was based on an estimate of jd, it is necessary to check whether the reinforcement chosen provides the required moment resistance. From (4-11),

$$M_n = A_s f_y \left(d - \frac{a}{2} \right)$$

$$= \frac{3220 \times 420(835 - 159/2)}{10^6} = 1022 \text{ kN-m}$$

$$\phi M_n = 0.90 \times 1022 \text{ kN-m} = 920 \text{ kN-m}$$

Since $\phi M_n \geq M_u$, the design is OK. **Therefore, use $b = 400$ mm, $h = 900$ mm, with $f'_c = 25$ MPa and $f_y = 420$ MPa, and five No. 29 bars in one layer, similar to layers in Fig. 4-29.** If ϕM_n were less than M_u, or if ϕM_n were much greater than M_u, it would be necessary to re-compute the area of steel required (see step 7 of Example 4-6). ∎

The third type of rectangular beam design problem occurs when the overall height, h, of the member is predetermined, either to maintain the desired floor to floor clearance or to limit deflections. In this case, the design procedure is identical to Example 4-7 except that in the choice of the concrete section, the value of h, and hence d, is known and bd^2 is solved to find b.

Direct Solution of Required Area of Steel

An alternative method of solving for the required area of steel can be obtained from (4-1), (4-10), and (4-11). From (4-1), the smallest acceptable value of ϕM_n is

$$M_u = \phi M_n$$

Substituting this and (4-10) into (4-11) gives

$$M_u = \phi\left[A_sf_y\left(d - \frac{A_sf_y}{1.7f'_cb}\right)\right]$$

Rearranging yields

$$\left(\frac{\phi f_y^2}{1.7f'_cb}\right)A_s^2 - (\phi f_y d)A_s + M_u = 0 \tag{4-36}$$

This is a quadratic equation in x of the type

$$Ax^2 + Bx + C = 0$$

where $x = A_s$. Once f_y, f'_c, b, and d have been chosen, (4-38) can be solved for a value of A_s. This replaces the iterative calculation used in Examples 4-6 and 4-7 and is useful if programmable calculators are employed.

EXAMPLE 4-8 Use (4-34) to Solve for A_s

The loadings, material strengths, and dimensions are as in Example 4-6. Thus,

$f_y = 60,000$ psi $\quad f'_c = 3000$ psi
$b = 24$ in. $\quad d = 21.5$ in.
$M_u = 541$ ft-kips $\quad$ Assume $\phi = 0.9$
$\quad = 6.492 \times 10^6$ in.-kips

1. **Compute the terms in (4-36):**

$$A = \frac{\phi f_y^2}{1.7f'_cb}$$

$$= \frac{0.9 \times 60,000^2}{1.7 \times 3000 \times 24} = 26.5 \times 10^3$$

$$B = -\phi f_y d$$
$$= -0.9 \times 60,000 \times 21.5 = -1.16 \times 10^6$$
$$C = 5.92 \times 10^6$$

2. **Solve (4-36) for A_s:**

$$A_s = \frac{+1.16 \times 10^6 \pm \sqrt{(-1.16 \times 10^6)^2 - 4(26.5 \times 10^3 \times 5.92 \times 10^6)}}{2(26.5 \times 10^3)}$$

$$= 5.89 \text{ in.}^2 \text{ or } 38.0 \text{ in.}^2$$

The higher root is several times the balanced steel ratio and will be discarded. Thus the required $A_s = 5.89$ in.^{2}.

3. **Select the reinforcement, and compute ϕM_n as a check.** Choose six No. 9 bars, $A_s = 6.00$ in.^{2}:

$$a = \frac{A_sf_y}{0.85f'_cb} = 5.88 \text{ in.}$$

$a/d = 0.274$ is less than $a_{TCL}/d_t = 0.319$ (Table A-4); therefore, $\phi = 0.9$.

Finally, it is good practice to compute ϕM_n to check whether the A_s is adequate. This guards against errors in solving (4-36). Thus,

$$\phi M_n = \frac{0.9[6.00 \times 60(21.5 - 5.88/2)]}{12} = 501 \text{ ft-kips}$$

Since $\phi M_n = M_u$, A_s is OK. Therefore, use six No. 9 bars. ∎

PROBLEMS

4-1 Figure P4-1 shows a simply supported beam and the cross section at midspan. The beam supports a uniform service (unfactored) dead load consisting of its own weight plus 1.4 kips/ft and a uniform service (unfactored) live load of 1.5 kips/ft. The concrete strength is 3000 psi, and the yield strength of the reinforcement is 60,000 psi. The concrete is normal-weight concrete. Use load and strength-reduction factors from ACI Sections 9.2 and 9.3.

(a) Compute the weight/ft of the beam, the factored load per foot, w_u, and the moment due to the factored loads, M_u, and sketch the bending moment diagram.

(b) Compute ϕM_n for the cross section shown. Is the beam safe?

(c) Draw the cross section at midspan showing
(1) the location of the compression zone.
(2) the dimensions of b, d, h, a.

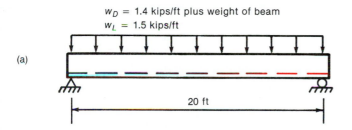

Fig. P4-1

4-2 Repeat Problem 4-1 for the cantilever beam shown in Fig. P4-2. The beam supports a uniform service (unfactored) dead load of 1 kip/ft plus its own dead load and it supports a concentrated service (unfactored) live load of 12 kips as shown. The concrete is normal-weight concrete with $f_c' = 4000$ psi and the steel is Grade 60. Draw the section at the support. Use load and strength-reduction factors from ACI Sections 9.2 and 9.3.

4-3 Assuming that the maximum concrete compressive strain is 0.003, compute the steel strain corresponding to the moment M_n for the beam shown in Fig. P4 -1. Is $f_s = f_y$ in this beam?

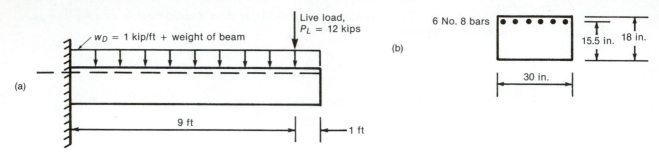

Fig. P4-2

4-4 (a) Compare ϕM_n for singly reinforced rectangular beams having the following properties. Use
load and strength reduction factors from ACI
Sections 9.2 and 9.3.

Beam No.	b (in.)	d (in.)	Bars	f'_c (psi)	f_y (psi)
1	12	22	3 No. 7	3,000	60,000
2	12	22	2 No. 9 plus 1 No. 8	3,000	60,000
3	12	22	3 No. 7	3,000	40,000
4	12	22	3 No. 7	4,500	60,000
5	12	33	3 No. 7	3,000	60,000

(b) Taking beam 1 as the reference point, discuss
the effects of changing A_s, f_y, f'_c, and d on
ϕM_n. (Note that each beam has the same properties as beam 1 except for the italicized quantity.)

(c) What is the most effective way of increasing
ϕM_n? What is the least effective way?

4-5 For each of the beams shown in Fig. P4-5 and without doing any calculations:

(a) Draw the deflected shape.

(b) Sketch

(1) the bending moment diagram due to the
weight of the beam.

(2) the bending moment diagram for the other
loads shown.

(3) the sum of the two diagrams.

(c) Show on an elevation view of the beam the location of flexural reinforcement for the final
moment diagram from part (b).

4-6 A 16-ft-span simply supported beam has a rectangular cross section with $b = 14$ in., $d = 19.5$ in., and
$h = 22$ in. The beam is made from normal-weight
3500-psi concrete and has six No. 6 Grade-40 bars.
This beam supports its own dead load plus a uniform

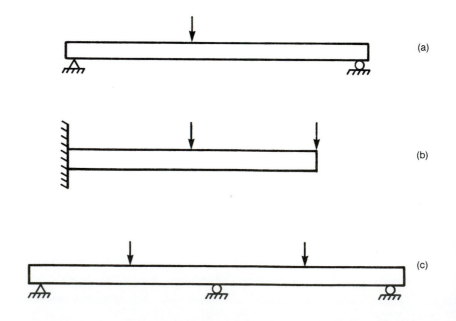

Fig. P4-5

service (unfactored) additional dead load of 1.0 kip/ft. Compute the maximum uniform service live load that the beam can support. Use load and strength reduction factors from ACI Sections 9.2 and 9.3.

4-7 A 12-ft-long cantilever supports its own dead load plus an additional uniform service (unfactored) dead load of 0.5 kip/ft. The beam is made from normal-weight 4000-psi concrete and has $b = 16$ in., $d = 15.5$ in., and $h = 18$ in. It is reinforced with four No. 7 Grade-60 bars. Compute the maximum service (unfactored) concentrated live load that can be applied at 1 ft from the free end of the cantilever. Use load and strength-reduction factors from ACI Sections 9.2 and 9.3.

4-8 Explain why a rectangular stress block with a maximum stress of $0.85f_c'$ and a depth $a = \beta_1 c$ is used in design.

4-9 Explain the meaning of "balanced failure" and "tension-controlled failure."

4-10 Explain the meaning of
(a) compression-controlled section
(b) tension-controlled section.
(c) Either explain why ρ is limited to $0.75\rho_b$ (with a limited to $0.75a_b$) or explain why a/d_t is limited to a_{TCL}/d_t.

4-11 (a) Compute the effective depth, d, and the minimum allowable web width, b_w, of the beam shown in Fig. P4-11. Use the minimum bar spacings allowed by the ACI Code for concrete with $\frac{3}{4}$-in. coarse aggregate. Select the cover for concrete not exposed to weather or in contact with the ground.

(b) Compare the computed depth to that given by the approximate (4-30).

(c) Check the minimum web width using Table A-6.

4-12 Give three reasons for the minimum cover requirements in the ACI Code. Under what circumstances are greater covers used?

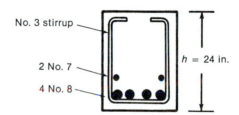

No. 3 stirrup

2 No. 7

4 No. 8

$h = 24$ in.

Fig. P4-11

4-13 Give two reasons for the minimum bar spacing requirements in the ACI Code.

4-14 A rectangular beam has $b = 12$ in., $h = 20$ in., and eight No. eight bars in two layers of four bars. $f_c' = 3750$ psi and $f_y = 60,000$ psi. Compute ϕM_n.

4-15 Select reinforcement for a 20-ft-span rectangular beam with $b = 16$ in. and $h = 21$ in. The beam supports its own weight plus a superimposed service (unfactored) uniform dead load of 0.6 kip/ft and a uniform service live load of 2 kips/ft. Use $f_c' = 3750$ psi and $f_y = 60,000$ psi.

4-16 Select b, d, h, and the reinforcement for a 24-ft-span simply supported rectangular beam that supports its own dead load plus an additional service dead load of 1.5 kips/ft plus a service live load which consists of two concentrated loads of 10 kips, each located at the third points of the span. Use $f_c' = 3000$ psi and $f_y = 60,000$ psi.

4-17 Select b, d, h, and the reinforcement for a 22-ft-span simply supported rectangular beam which supports its own dead load plus a superimposed service dead load of 1.25 kips/ft plus a uniform service load of 2 kips/ft. Use $f_c' = 3000$ psi and $f_y = 60,000$ psi.

4-18 The beam shown in Fig. P4-18 carries its own dead load plus an additional uniform service dead load of 0.5 kip/ft and a uniform service live load of 1.5 kips/ft. The dead load acts on the entire beam, of course, but the live load can act on parts of the span. Three possible loading cases are shown in Fig. P4-18. Use load and strength reduction factors from ACI Sections 9.2 and 9.3.

(a) Draw factored bending-moment diagrams for the three loading cases shown and superimpose them to draw a bending-moment envelope.

(b) Design the beam, selecting b, d, h, and the reinforcing bars. Use $f_c' = 3750$ psi and $f_y = 60,000$ psi and interior exposure.

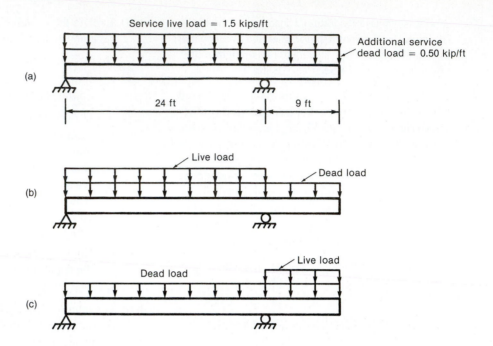

Service live load = 1.5 kips/ft

Additional service dead load = 0.50 kip/ft

(a)

24 ft

9 ft

Live load

Dead load

(b)

Live load

Dead load

(c)

Fig. P4-18

(c) Draw an elevation of the beam showing the reinforcement. Estimate the lengths of the top and bottom bars from the bending moment envelope.

(d) Draw cross sections at the points of maximum positive and negative moment.

4-19, 4-20, and 4-21 Use design tables to design the beams described in Problems 4-16, 4-17, and 4-18.

4-22 Write a computer or calculator program to solve for the area of steel, A_s, directly using (4-36).

4-23 Select the largest possible b and d and the corresponding A_s (based on ACI Section 10.5) and the smallest allowable b and d and the corresponding A_s (based on ACI Section 10.3.3) to give a resisting moment of $\phi M_n = 250$ ft-kips. In both cases select a section with $b = 0.5d$. Use $f_c' = 3000$ psi and $f_y = 60,000$ psi.

5

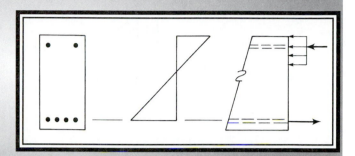

Flexure: T Beams, Beams with Compression Reinforcement, and Special Cases

5-1 INTRODUCTION

In Chapter 4, the theory of flexure for reinforced concrete was developed and was applied to rectangular beams with flexural reinforcement in the tension zone. Frequently, concrete beams are T- or I-shaped, and sometimes they have reinforcement in both the tension and compression zones. In this chapter, the theory of flexure is extended to cover these types of problems.

Beams whose cross sections are not symmetrical about the loading axis and beams bent about two axes require special treatment, because the axis of zero strain (neutral axis) generally is not parallel to the axis about which the resultant moment acts. The analysis of such beams is discussed in Section 5-4.

When beams have tension reinforcement in several layers spread over the depth of the beam, or are built of two types of concrete, or contain reinforcement that is not elastic–plastic, strain compatibility must be considered in the calculations. This is discussed in Section 5-5.

5-2 T BEAMS

Practical Applications of T Beams

In the floor system shown in Fig. 5-1, the slab is assumed to carry the loads in one direction to beams that carry them in the perpendicular direction. During construction, the concrete in the columns is placed and allowed to harden before the concrete in the floor is placed (ACI Section 6.4.5). In the next operation, concrete is placed in the slab and beams in a monolithic pour (ACI Section 6.4.6). As a result, the slab serves as the top flange of the beams, as indicated by the shading in Fig. 5-1. Such a beam is referred to as a *T beam*. The interior beam, *AB*, has a flange on both sides. The *spandrel beam, CD*, with a flange on one side only, is also referred to as a T beam.

163

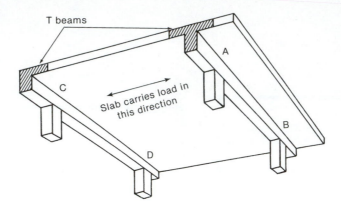

Fig. 5-1
T beams in a one-way
beam-and-slab floor.

An exaggerated deflected view of the interior beam is shown in Fig. 5-2. This beam develops positive moments at midspan (section *A–A*) and negative moments over the supports (section *B–B*). A photograph of the region between A–A and the left end of such a beam is shown in Fig. 5-3. At midspan, the compression zone is in the flange, as shown in Fig. 5-2b and d. Generally, it is rectangular, as shown in b, although, in a few cases, the neutral axis may shift down into the web, giving a T-shaped compression zone, as shown in Fig. 5-2d. At the support, the compression zone is at the bottom of the beam and is rectangular, as shown in Fig. 5-2c. For computational purposes, such a beam will be classed as a "rectangular beam" if the compression zone is rectangular (Fig. 5-2b and c) and as a "T beam" if the zone is T-shaped (Fig. 5-2d).

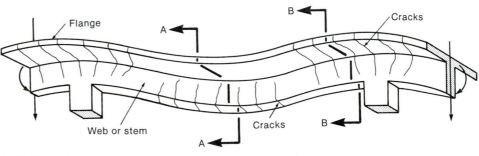

(a) Deflected beam.

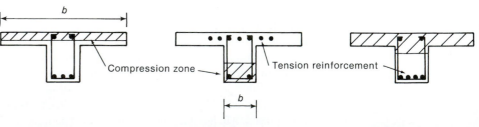

Fig. 5-2
Positive and negative moment
regions in a T beam.

(b) Section A–A
(rectangular
compression zone).

(c) Section B–B
(negative moment).

(d) Section A–A
(T-shaped
compression zone).

Fig. 5-3
Photograph of a test
specimen representing half of
the beam shown in Fig. 5-2a.
(Photograph courtesy of J. G.
MacGregor.)

Frequently, a beam-and-slab floor involves slabs supported by beams which, in turn, are supported by other beams referred to as *girders* (Fig. 5-4). Again, all the concrete above the top of the column is placed at one time. Note that the slab acts as a flange for both the beams and girders.

Effective Flange Width and Transverse Reinforcement

The forces acting on the flange of a simply supported T beam are illustrated in Fig. 5-5a. At the support, there are no compressive stresses in the flange, but at midspan, the full width is stressed in compression. The transition requires horizontal shear stresses on the web–flange interface as shown in Fig 5-5a. As a result there is a "shear-lag" effect, and the portions of the flange closest to the web are more highly stressed than those portions farther away, as shown in Figs. 5-5a and 5-6.

Figure 5-6 shows the distribution of the flexural compressive stresses in a slab that forms the flanges of a series of parallel beams at a section of maximum positive moment. The compressive stress is a maximum over each web, dropping between the webs. Toward the supports, the variation from maximum to minimum becomes more pronounced.

When proportioning the section for positive moments, an "effective width" is used (Fig. 5-6b). This is the width, b, that, when stressed uniformly to $f_{c(\max)}$, gives the compression *force* that is actually developed in the real compression zone of width b_0.

ACI Section 8.10 presents rules for estimating this width for design purposes. For an interior beam, ACI Section 8.10.2 states that

1. the width of slab effective as a T-beam flange shall not exceed one-fourth the span length of the beam, and

2. the effective overhanging slab width on each side of the web shall not exceed the smaller of either eight times the slab thickness or one-half the clear distance to the next beam web.

ACI Sections 8.10.3 and 8.10.4 give considerably more stringent rules for beams with slabs on one side only and for isolated T beams. In general, the code rules are a conservative approximation to the elastic solutions for effective width.

A number of elastic solutions have been used to estimate the effective flange width [5-2], [5-3]. These solutions suggest that this width is affected by the type of loading (uniform, concentrated), the type of supports, the spacing of the beams, and the relative stiffness

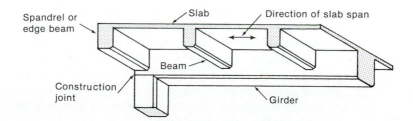

Fig. 5-4
Slab, beam, and girder floor.
(From [5-1].)

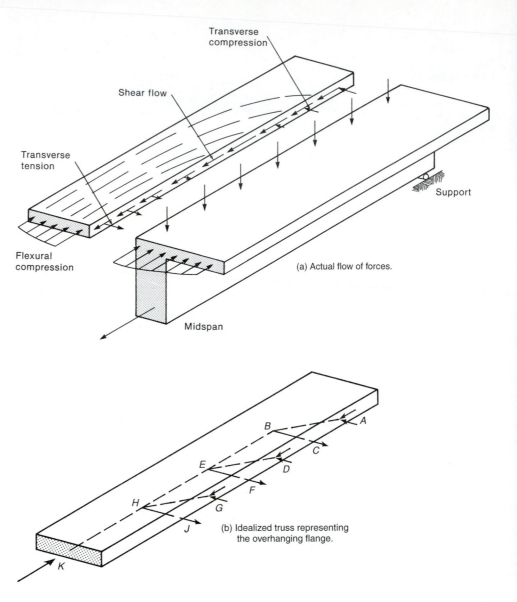

Fig. 5-5
Forces on a T-beam flange.

of the slabs and beams. However, it must be noted that all such studies have ignored the cracking of the flange observed in tests.

Tests studying the web-to-flange connection in reinforced concrete T beams are reviewed by Razaqpur and Ghali [5-4]. They present two analyses that could be used to design reinforcement for such joints. The first uses shear friction reinforcement, presented in Chapter 17, to transfer the shear and normal stresses shown in Fig. 5-5a from the web to the flange. The second analysis makes use of the strut-and-tie model illustrated in Fig. 5-5b to design the connection. As will be discussed in Section 18-9, transverse reinforcement, perpendicular to the web–flange interface, is required to resist the transverse tensile forces. If this reinforcement is not provided, the beam may crack along a longitudinal section along the web-to-flange interface. The truss shown in Fig. 5-5b is referred to as a *strut-and-tie model*. Strut-and-tie models of compression and tension flanges, respectively, are given in Figs. 18-54 and 18-55 in Chapter 18.

Section 18-9 and [5-4] show that significant amounts of transverse steel may be needed, especially in tension flanges, to transfer the forces into the flange and the longitudinal reinforcement in the flange. The arrangement of this steel may control the effectiveness of the longitudinal tension reinforcement placed in the flanges and the degree of utilization of the width of the flange.

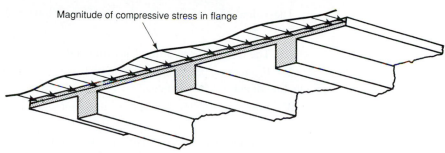

(a) Distribution of maximum flexural compressive stresses

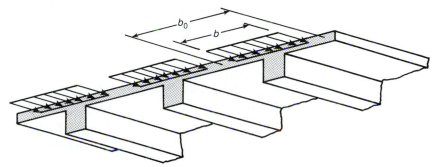

Fig. 5-6
Effective width of T beams.

(b) Flexural compressive stress distribution assumed in design

The spread of the compression force across the width of the left-hand proportion of the flange in Fig. 5-5 can be idealized by a truss mechanism within the plane of the flange. The truss consists of a series of compression struts, shown by dashed lines in Fig. 5-5b, and transverse tension ties shown by solid lines. The resultant compression force in one overhanging flange at midspan is represented by the force at K. The horizontal shear force applied to the flange at A is transferred out into the flange by the compression strut A–B. At B, the longitudinal force in strut A–B is transferred to the strut B–E. The transverse component of the force in strut A–B must be resisted by the transverse tension tie B–C, and so on. The ACI Code does not give rules for the design of this transverse steel in the flanges. Significant amounts of transverse steel may be needed in tension flanges. The arrangement of this steel may control the effectiveness of the longitudinal tension reinforcement placed in the flanges.

Loads applied to the flange will cause negative moments in the flange where it joins the web. If the floor slab is continuous and spans perpendicular to the beam, as in Fig. 5-1, or perpendicular to the "beams," as in Fig. 5-4, the slab reinforcement coincidentally will be adequate to resist these moments. If, however, the slab is not continuous (in an isolated T beam) or if the slab reinforcement is parallel to the beam, as is the case of the "girders" in Fig. 5-4, additional reinforcement is required at the top of the slab, perpendicular to the beam stem. (ACI Section 8.10.5). This reinforcement is designed by assuming that the flange acts as a cantilever loaded with the factored dead and live loads. For an isolated T beam, the full overhanging flange width is considered. For a girder in a monolithic floor system (Fig. 5-4), the overhanging part of the effective width is used in this calculation. (See Section 10-6.)

Analysis of T Beams

Generally, the compression zone of a T beam is rectangular, as shown in Fig. 5-2b or c. Beams with such flanges may be analyzed as "rectangular beams" with a width, b, as shown. In the unusual case in which the compression zone is T-shaped, as shown in Fig. 5-2d, the analysis separately considers the resistance provided by the overhanging flanges and that provided by the remaining rectangular beam.

The calculation of the moment capacity follows the four-step procedure used in Chapter 4.

1. The internal lever arm, jd, is estimated.

2. From the moment and the assumed lever arm, the compressive force resultant C and its line of action are computed.

3. Since $T = C$, the tensile force resultant T and its line of action can be computed as well.

4. The moment capacity is then computed as $M_n = C \times jd$, or $M_n = T \times jd$. This analysis applies to all types of reinforced concrete beams.

Consider the beam shown in Fig. 5-7a, with the depth of the stress block, a, greater than the flange thickness, h_f. The internal forces in this beam consist of a compressive force C at the centroid of the compression zone (centroid of shaded area in Fig. 5-7a) and a tensile force $T = A_s f_y$, assuming that the steel yields. These form a resisting moment, $M_n = C jd$ or $M_n = T jd$.

To avoid the need for locating the centroid of the shaded area (where a is not yet known), it is convenient to consider two hypothetical beams:

1. **Beam F,** where the subscript f refers to the beam F (Fig. 5-7c), with a compression zone consisting of the overhanging flanges, area A_f, stressed to $\alpha_1 f_c'$ where α_1 is

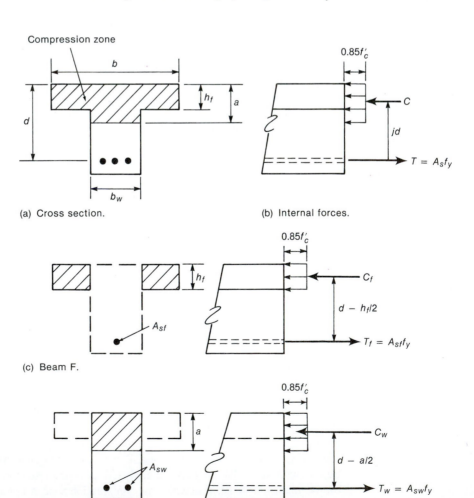

(a) Cross section.

(b) Internal forces.

(c) Beam F.

(d) Beam W.

Fig. 5-7
Subdivision of a T beam
for analysis.

equal to 0.85, giving a compressive force, C_f, equal to 0.85 f'_c times the shaded area in Fig. 5-7c. The force C_f acts at the centroid of the area of the overhanging flanges. For equilibrium, beam F has a tensile steel area A_{sf} chosen such that $C_f = T_f$ or $A_{sf}f_y = C_f$. This area of steel, A_{sf}, is a portion of the total A_s and is chosen to have the same centroid as A_s. The moment capacity of this beam, M_{nf}, is the moment of C_f about the tension steel.

2.   **Beam W** (Fig. 5-7d), where the subscript w refers to beam W, which is a rectangular beam of width b_w having a compression zone of area b_wa and utilizing the remaining tensile steel, $(A_s - A_{sf}) = A_{sw}$. The compressive force in this beam, C_w, acts through the centroid of its compression area. The moment capacity of this beam, M_{nw}, is the moment of C_w about the tensile steel.

The total moment capacity of the T beam is the sum of the moment capacities of the two individual beams, $M_n = M_{nf} + M_{nw}$, calculated as follows:

Beam F

$$\text{Area of compression zone} = (b - b_w)h_f$$

$$\text{Force in compression zone } C_f = (0.85f'_c)(b - b_w)h_f$$

To compute the area of reinforcement required in beam F, set $T_f = C_f$, assuming that $f_s = f_y$. Then

$$A_{sf}f_y = 0.85f'_c(b - b_w)h_f \tag{5-1a}$$

or

$$A_{sf} = \frac{0.85f'_c(b - b_w)h_f}{f_y} \tag{5-1b}$$

The lever arm is $(d - h_f/2)$.

Summing the moments about the centroid of the tension reinforcement gives

$$M_{nf} = 0.85f'_c(b - b_w)h_f\left(d - \frac{h_f}{2}\right) \tag{5-2a}$$

Alternatively, summing the moments about the line of action of C_f gives

$$M_{nf} = A_{sf}f_y\left(d - \frac{h_f}{2}\right) \tag{5-2b}$$

Beam W

$$\text{Area of tension steel } A_{sw} = A_s - A_{sf}$$

$$\text{Compression force } C_w = 0.85f'_cb_wa$$

or

$$a = \frac{A_{sw}f_y}{0.85f'_cb_w} \tag{5-3}$$

The lever arm is $d - a/2$. Thus,

$$M_{nw} = 0.85f'_cb_wa\left(d - \frac{a}{2}\right) \tag{5-4a}$$

or

$$M_{nw} = A_{sw}f_y\left(d - \frac{a}{2}\right) \tag{5-4b}$$

T Beam = Beam F + Beam W. The nominal moment capacity of the T beam is the sum of the nominal moment capacities of beam F and beam W. Hence,

$$M_n = M_{nf} + M_{nw}$$

giving

$$M_n = \left[0.85f_c'(b - b_w)h_f\left(d - \frac{h_f}{2}\right)\right] + \left[0.85f_c'b_w a\left(d - \frac{a}{2}\right)\right] \tag{5-5a}$$

or

$$M_n = \left[A_{sf}f_y\left(d - \frac{h_f}{2}\right)\right] + \left[A_{sw}f_y\left(d - \frac{a}{2}\right)\right] \tag{5-5b}$$

Finally, the factored moment capacity is ϕM_n.

Occasionally, a will turn out to be equal to h_f. This may be considered to be a "rectangular beam" for calculation purposes.

Determination of Whether $f_s = f_y$

In the derivation of (5-5a and b), it was assumed that $f_s = f_y$. As was discussed in Section 4-3, this can be checked by comparing the computed c/d or a/d ratios to c_b/d or a_b/d, given by

$$\frac{c_b}{d} = \frac{87,000}{87,000 + f_y} \tag{4-19}$$

and

$$\frac{a_b}{d} = \beta_1\left(\frac{87,000}{87,000 + f_y}\right) \tag{4-20}$$

where f_y is in psi. If f_y is in MPa, 87,000 becomes 600 in both equations. If the computed c/d or a/d ratios are less than those given by (4-19) and (4-20), $f_s = f_y$ at failure.

The ratio of a_b/d corresponding to ρ_b is given by (4-20). This equation is independent of section shape and hence applies to T beams. It should be noted, however, that (4-25) for ρ_b was derived for a rectangular section where the area of the compression zone is ab_w and needs to be modified to apply to a T section or other section with a width that varies over the height of the compression zone.

Tension-Controlled Sections and Compression-Controlled Sections

As was explained in Section 4-3, ACI 318-02 Section 10.3.4 defines a *tension-controlled section* as a section in which the net tensile strain in the extreme layer of tension steel equals or exceeds 0.005 tension strain at ultimate. Thus, a *tension-controlled failure* will occur if the ratio, c/d_t, of the neutral axis depth, c, to the depth from the extreme compression fiber to the farthest layer of tension reinforcement, d_t, is less than or equal to the *tension-controlled limit* given by

$$\frac{c_{TCL}}{d_t} = 0.375 \tag{4-22b}$$

This can also be expressed in terms of the ratio of the depth, a, of the rectangular stress block to d_t at the tension-controlled limit:

$$\frac{a_{TCL}}{d_t} = 0.375\beta_1 \tag{4-23}$$

A T beam with the flange in compression is almost always a tension-controlled section.

ACI 318.02 Section 10.3.3 defines a *compression-controlled section* as a section in which the *net tensile strain*, ϵ_t, in the *extreme tension steel* (the steel farthest from the extreme compression fiber) at ultimate is less tensile than the yield strain in tension, $\epsilon_y = f_y/E_s$. For Grade 60 reinforcement, $\epsilon_y = f_y/E_s = 60{,}000$ psi/$29{,}000{,}000$ psi $= 0.00207$, which the code allows, but does not require, to be rounded off to 0.002 for Grade-60 steel. For other yield strengths, the compression-controlled limit is based on the specified yield strain, $\epsilon_y = f_y/E_s$, without rounding. The term *net tensile strain* refers to the strain caused by flexure and does not include strain induced by prestressing, shrinkage, or creep.

A *compression-controlled failure* with $f_s = f_y$ will occur if the ratio a/d_t is greater than or equal to that at the *compression-controlled limit*:

$$\frac{a_{CCL}}{d_t} = \beta_1\left(\frac{87{,}000}{87{,}000 + f_y}\right) \tag{4-21}$$

Calculation of Strength-Reduction Factor ϕ

Beams having a net tensile strain in the extreme tension steel, ϵ_t, that is more compressive than, or equal to, the compression-controlled limit of $\epsilon_t = 0.002$ tension are designed for flexure using $\phi = 0.65$, provided that the compression zone does not contain spiral reinforcement. Compression-controlled sections containing spiral reinforcement are designed by using $\phi = 0.70$. Spirals are used in columns, but they are seldom, if ever, used in beams.

Beams in which the net tensile strain ϵ_t in the extreme tension steel is more tensile than the tension-controlled limit at ultimate are designed with $\phi = 0.90$. For *transition sections* with compression steel enclosed in column ties, for which ϵ_t is between 0.0020 and 0.0050, ϕ is interpolated between 0.65 (or, rarely, 0.70) and 0.90. For beams with Grade 60 steel, but without spiral reinforcement, ACI Commentary Section R.9.3.2 gives

$$\phi = 0.48 + 83\epsilon_t \tag{4-26b}$$

or

$$\phi = 0.23 + \frac{0.25}{c/d_t} \tag{4-27b}$$

As a result, there is up to a 28 percent strength penalty for using more steel than the tension-controlled limit, because of the lower ϕ factor.

Upper Limit on Longitudinal Reinforcement in T Beams— ACI 318-02 Section 10.3.5

ACI 318-02 Section 10.3.5 limits the amount of flexural reinforcement in beams so that the net tensile strain, ϵ_t, is not less than 0.004 tensile strain at nominal strength. This limit, $\epsilon_{tBL} = 0.0040$ strain, referred to as the beam reinforcement limit, *BRL*, corresponds to a transition section. Previous ACI Codes limited ρ to less than or equal to 0.75 times the balanced reinforcement ratio; thus, $\rho \leq 0.75\rho_b$. This is equivalent to limiting the net tensile strain in the extreme tension steel to a strain of 0.00376. ACI 318-02 Section 10.3.5 has changed in wording from the 1999 ACI Code, which did not limit the maximum reinforcement in beams. This was not, however, intended to be a change in philosophy. It was always intended that beams should have the minimum ductility corresponding to $\rho \leq 0.75\rho_b$.

Upper Limit on Reinforcement in T Beams—ACI 318-02 Section B.10.3.3

An alternative method of setting the maximum steel ratio is given in ACI 318-02 Section B.10.3.3. This procedure was presented in Section 4.3 for rectangular beams. It will not be derived or presented for T beams because it is tedious to use and is expected to be removed from the code in 2005 or 2008.

Minimum Reinforcement

It is necessary to have sufficient tension reinforcement so that the moment capacity after cracking exceeds the cracking moment. For a T beam with its flange in compression and for negative-moment regions of continuous T beams where the flange is in tension, this is done according to ACI Section 10.5.1, by checking whether $A_s = A_{sf} + A_{sw}$ exceeds $A_{s,min}$ given by

$$A_{s,min} = \frac{3\sqrt{f_c'}}{f_y}b_w d, \text{ and } \geq \frac{200 b_w d}{f_y} \qquad (4\text{-}32)$$

$$(\text{ACI Eq. 10-3})$$

where f_c' and f_y are in psi, or

$$A_{s,min} = \frac{\sqrt{f_c'}}{4f_y}b_w d, \text{ and } \geq \frac{1.4 b_w d}{f_y} \qquad (4\text{-}32\text{M})$$

where f_c' and f_y are in MPa.

When a statically determinate T beam, such as a cantilever with a T-shaped cross section, has its *flange in tension*, ACI Section 10.5.2 gives

$$A_{s,min} = \frac{6\sqrt{f_c'}}{f_y}b_w d \qquad (4\text{-}34)$$

where f_c' and f_y are in psi, or

$$A_{s,min} = \frac{\sqrt{f_c'}}{2f_y}b_w d \qquad (4\text{-}34\text{M})$$

where f_c' and f_y are in MPa. The amount of minimum reinforcing in a statically determinate T beam with the flange in tension need not exceed the amount given by (4-34) and (4-34M), with b_w set equal to the width of the flange.

Alternatively, ACI Section 10.5.3 allows (4-33) and (4-34) to be waived at sections where the longitudinal tension reinforcement provided is at least 1.33 times that required by analysis.

EXAMPLE 5-1 Analysis of the Positive-Moment Capacity of a T Beam: *a* Less Than h_f

An interior T beam in floor system has a clear span, from face to face of the columns, of 18 ft and the cross section shown in Fig. 5-8. The concrete and steel strengths are 3000 psi and 40,000 psi, respectively. Compute the design moment capacity of this beam in the positive-moment region.

1. **Compute the effective width of the flange, *b*.** From ACI Section 8.10.2,

    (a) *b* shall not exceed one-fourth of the span length = 18/4 = 4.5 ft = 54 in.

    (b) The effective overhanging slab width on each side of the web shall not exceed eight times the slab thickness = 8 × 5 in. = 40 in., which gives b = 40 + 12 + 40 = 92 in., or

    (c) half the clear distance to the next beam web on each side. This gives b = 108/2 + 12 + 130/2 = 131 in.

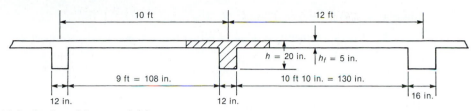

(a) Section through beams and slab.

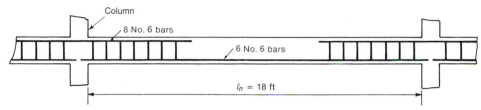

(b) Elevation of beam.

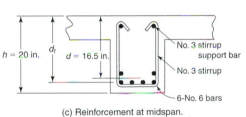

(c) Reinforcement at midspan.

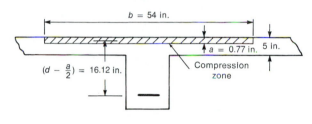

(d) Compression zone and lever arm.

Fig. 5-8
Beam—Examples 5-1 and 5-2.

The smallest of these is b. Therefore, $b = 54$ in.

2. **Compute d and d_t.** At midspan assuming two layers of steel, with $h = 20$ in. from Fig. 5-8c.

$$d \approx h - 3.5 \text{ in.} = 16.5 \text{ in}$$
$$d_t = h - (1.5 + 0.375 + 0.75/2) \quad \text{(4-32b)}$$
$$= 17.75 \text{ in.}$$

3. **Compute a.** Since almost all T beams will have the depth of compression stress block, a, less than the flange thickness, h_f, we shall assume "rectangular beam action" as shown in Fig. 5-2b. If the resulting value of a is less than or equal to the flange thickness, h_f, we will continue. If $a > h_f$, it is necessary to analyze the beam for a "T-beam action" as in Example 5-3. We have

$$a = \frac{A_s f_y}{0.85 f_c' b}$$
$$= \frac{2.64 \times 40,000}{0.85 \times 3000 \times 54} = 0.767 \text{ in.}$$

This is less than $h_f = 5$ in., as shown in Fig. 5-8d, so the beam has "rectangular beam action" and ϕM_n can be computed via the basic procedure for rectangular beams (4-11b).

4. **Check whether $A_s \geq A_{s,\min}$.** The flange is in compression; hence,

$$A_{s,\min} = \frac{3\sqrt{f_c'}}{f_y} b_w d, \text{ and } \geq \frac{200 b_w d}{f_y}$$
$$= \frac{3\sqrt{3000}}{40,000} 12 \times 16.5, \text{ and } \geq \frac{200 \times 12 \times 16.5}{40,000}$$
$$= 0.813 \geq 0.990 \text{ in.}^2 \quad \text{(4-33)}$$

Since $A_s = 2.64$ in.^{2} exceeds 0.99 in.^{2}, $A_s > A_{s,min}$.

5. Check whether $f_s = f_y$ and whether the section is tension-controlled:

$$\frac{a}{d} = \frac{0.767}{16.5} = 0.0465$$

For No. 6 negative moment bars in one layer:

$$d_t = 20 - (1.5 + 0.375 + 0.75/2) = 17.75 \text{ in.}$$

$$\frac{a}{d_t} = \frac{0.767}{17.75} = 0.0432$$

This is much less than a_{TCL}/d_t at the tension-controlled limit:

$$\frac{a_{TCL}}{dt} = 0.375\beta_1 = 0.319 \tag{4-23}$$

Thus, the section is tension-controlled. As a result, $f_s = f_y$, and $\phi = 0.90$.

Design has been carried out by using ACI Sections 10.3.3 to 10.3.5. Alternatively, design could be carried out by ACI Section 10.3.3, which limits $\rho < 0.75\rho_b$. Because $a < h_f$, the compression zone is rectangular, and this check can be made by examining whether a/d exceeds the value of a/d for $0.75\rho_b$, as was done for rectangular beams. We have

$$\frac{a}{d} = \frac{0.767}{16.5} = 0.047$$

From (4-20) (or from Table A-5),

$$\frac{a_b}{d} = \beta_1\left(\frac{87,000}{87,000 + f_y}\right) = 0.582$$

The value of a/d for $0.75\rho_b$ is approximately $0.75 \times 0.582 = 0.437$. Since $0.0470 < 0.437$, $\rho < 0.75\rho_b$—therefore, OK.

If $a > h_f$, a different series of calculations is needed to check that $\rho \le 0.75\rho_b$. For this reason, we will not use the $0.75\rho_b$ method from Appendix B of the 2002 code.

The limits in steps 4 and 5 are very seldom reached in the case of a T beam with the flange in compression. In the event that they were, however, it would be necessary to revise the cross section.

6. Compute ϕM_n:

$$\phi M_n = \phi\left[A_s f_y\left(d - \frac{a}{2}\right)\right]$$

where $\phi = 0.90$ because $a/d = 0.047$ is much smaller than $a_{TCL}/d_t = 0.319$

$$= \frac{0.9[2.64 \times 40,000(16.5 - 0.767/2)]}{12,000}$$

$$= 128 \text{ ft-kips}$$

The positive design moment capacity of the beam shown in Fig. 5-8 is 128 ft-kips. ∎

EXAMPLE 5-2 Analysis of the Negative-Moment Capacity of a T Beam

Compute the design negative-moment capacity of the T beam shown in Fig. 5-8. The arrangement of the negative-moment reinforcement is shown in Fig. 5-9. The concrete and steel strengths are 3000 psi and 40,000 psi, as before.

Because this section is subjected to a negative moment, cracking develops in the top flange (Fig. 5-2) and the compressive zone is at the *bottom* of the beam, as shown in Figs. 5-2c and 5-9. Note that ACI Section 10.6.6 requires that a portion of the tension reinforcement be in the flange, which coincidentally allows all the negative-moment reinforcement to be placed in one layer. Two bars

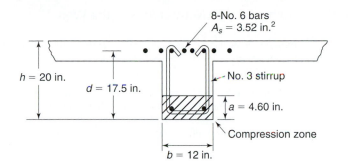

Fig. 5-9
Negative-moment section of
beam shown in Fig. 5-8—
Example 5-2.

are shown in the lower corners of the stirrups. These are provided in response to ACI Section 12.13.3, which requires that every bend in the stirrup away from the anchored ends of the stirrup shall enclose a longitudinal bar. In addition, ACI Section 12.11.1 requires that at least a third of the positive-moment reinforcement be extended at least 6 in. into the supports. Unless the extended bars are adequately anchored in the support to develop compressive stresses (which would not be true if the bars were interrupted at the supports in the manner shown in Fig. 5-8b), they should not be included in the computations.

1. **Compute b.** Since the compression zone is at the bottom of the beam, $b = 12$ in.

2. **Compute d and d_t.** Since the tension reinforcement is in one layer,

$$d \simeq h - 2.5 \text{ in.} = 17.5 \text{ in.}$$

$$d_t = h - 2.25 = 17.75 \text{ in.} \qquad (4\text{-}32a)$$

3. **Compute a:**

$$a = \frac{A_s f_y}{0.85 f'_c b}$$

$$= \frac{3.52 \times 40,000}{0.85 \times 3000 \times 12} = 4.60 \text{ in.} \qquad (4\text{-}10)$$

4. **Check whether $A_s \geq A_{s,\min}$.** ACI Section 10.5.2 applies to statically determinate T beams with the flange in tension. ACI Section 10.5.1 applies to "every section of a flexural member where tensile reinforcement is required by analysis except as provided in 10.5.2." Since the beam is continuous, ACI Section 10.5.1 applies, and

$$A_{s,\min} = \frac{3\sqrt{3000}}{40,000} 12 \times 17.5, \text{ and } \geq \frac{200 \times 12 \times 17.5}{40,000}$$

$$= 0.863 \text{ in.}^2 \geq 1.05 \text{ in.}^2$$

Because 3.52 in.^{2} exceeds 1.05 in.^{2}, $A_s \geq A_{s,\min}$—therefore, OK.

5. **Check whether $f_s = f_y$ and whether the section is tension-controlled:**

$$d_t = 17.5 \text{ in.} \quad \frac{a}{d_t} = 0.263$$

$$\frac{a_{TCL}}{d_t} = 0.375\beta_1 = 0.319 \qquad (4\text{-}23)$$

Since 0.263 is less than 0.319, the section is tension-controlled, $f_s = f_y$, and $\phi = 0.90$.

6. **Compute ϕM_n:**

$$\phi M_n = \phi \left[A_s f_y \left(d - \frac{a}{2} \right) \right]$$

$$= \frac{0.9[3.52 \times 40,000(17.5 - 4.60/2)]}{12,000} = 161 \text{ ft-kips}$$

Therefore, the design negative moment is $\phi M_n = 161$ ft-kips.   ∎

EXAMPLE 5-3   Analysis of a T Beam with the Neutral Axis in the Web

Compute the positive design moment capacity of the beam shown in Fig. 5-10. The concrete and steel strengths are 3000 psi and 60,000 psi, respectively. The beam contains No. 3 stirrups, not shown in Fig. 5-10.

1.   **Compute b.** This beam is an isolated T beam in which a T-shaped flange is used to increase the area of the compression zone. For such a beam, ACI Section 8.10.4 states that the flange thickness shall not be less than one-half the width of the web and that the effective flange width shall not exceed four times the width of the web. By observation, the flange dimensions satisfy this. Thus $b = 18$ in.

2.   **Compute d.** $d = 24.5$ in., as shown in Fig. 5-10a.

3.   **Compute a.** Assume that the compression zone will be rectangular. Accordingly,

$$a = \frac{A_s f_y}{0.85 f_c' b}$$

$$= \frac{4.74 \times 60,000}{0.85 \times 3000 \times 18} = 6.20 \text{ in.}$$

As shown in Fig. 5-10b, $a = 6.20$ in. is greater than the thickness of the flange, $h_f = 5$ in. This implies that compressive stresses exist in the shaded regions below the flanges. This cannot occur. Because $a > h_f$, our assumption that the compression zone is rectangular is wrong, and our calculated value of a is incorrect. It is therefore necessary to analyze this beam as a "T beam."

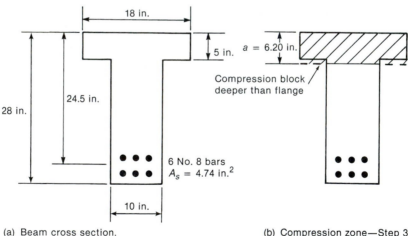

(a) Beam cross section.   (b) Compression zone—Step 3.

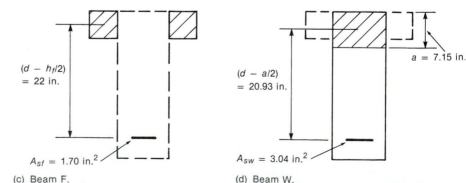

Fig. 5-10
Beam—Example 5-3.

(c) Beam F.   (d) Beam W.

4. **Divide the beam into "beam F" and "beam W."** Beam F consists of the overhanging portions of the flanges plus an area of steel, A_{sf}, such that $A_{sf}f_y$ balances the compression in the overhanging flanges (Fig. 5-10c). The subscripts f and w refer to "flange" and "web," respectively.

(a) **Beam F.** The compression force, C_f, in the overhanging flanges is

$$C_f = (0.85f_c')(b - b_w)h_f$$
$$= (0.85 \times 3000)(18 - 10) \times 5 = 102,000 \text{ lb}$$

The force in the steel in beam F is

$$A_{sf}f_y = C_f \qquad (5\text{-}1a)$$

and the area of steel in beam F is

$$A_{sf} = \frac{102,000}{60,000} = 1.70 \text{ in.}^2$$

Beam F is shown in Fig. 5-10c. Summing the moments about the centroid of the tensile steel gives us

$$M_{nf} = C_f\left(d - \frac{h_f}{2}\right) \qquad (5\text{-}2a)$$

$$= \frac{102,000(24.5 - 5/2)}{12,000} = 187 \text{ ft-kips}$$

(b) **Beam W.** Beam W consists of the concrete in the web plus the remainder of the tensile reinforcement, as shown in Fig. 5-10d. Thus,

$$A_{sw} = A_s - A_{sf}$$
$$= 4.74 - 1.70 = 3.04 \text{ in.}^2$$

For beam W, $b = b_w$, and

$$a = \frac{A_{sw}f_y}{0.85f_c'b_w} \qquad (5\text{-}3)$$

$$= \frac{3.04 \times 60,000}{0.85 \times 3000 \times 10} = 7.15 \text{ in.}$$

Thus, the depth of the compression stress block is actually 7.15 in., rather than the 6.20 in. computed in step 3, assuming "rectangular beam action." For beam W, the design moment capacity is

$$M_{nw} = A_{sw}f_y\left(d - \frac{a}{2}\right) \qquad (5\text{-}4b)$$

$$= \frac{3.04 \times 60,000(24.5 - 7.15/2)}{12,000} = 318 \text{ ft-kips}$$

The total positive nominal moment capacity is thus

$$M_n = M_{nf} + M_{nw} = 187 + 318 \text{ ft-kips}$$
$$= 505 \text{ ft-kips}$$

5. **Check whether $A_s \geq A_{s,min}$.** From (4-33)

$$A_{s,min} = \frac{3\sqrt{3000}}{60,000}10 \times 2.45, \text{ and } \geq \frac{200 \times 10 \times 24.5}{60,000}$$

$$= 0.671 \text{ in.}^2, \text{ and } \geq 0.817 \text{ in.}^2$$

Since $A_s = 4.74$ in.^{2} exceeds this, $A_s \geq A_{s,min}$—therefore, OK.

6. Check whether $f_s = f_y$ and whether the section is tension-controlled:

$$d_t = 28 - (1.5 + 0.375 + 1.0/2) = 25.62 \text{ in.} \qquad \frac{a}{d_t} = \frac{7.15 \text{ in.}}{25.62 \text{ in.}} = 0.279$$

$$\frac{a_{TCL}}{d_t} = 0.85 \times 0.375 = 0.319 \qquad\qquad\qquad\qquad\qquad (4\text{-}23)$$

Because 0.280 is less than 0.319, the section is tension-controlled, $f_s = f_y$, and $\phi = 0.90$.

7. Compute ϕM_n:

$$\phi M_n = 0.90 \times 505 = 455 \text{ ft-kips} \qquad\blacksquare$$

EXAMPLE 5-3M   Analysis of a T Beam with the Neutral Axis in the Web—SI Units

Compute the positive design moment capacity of the beam shown in Fig. 5-11. The concrete and steel strengths are 20 MPa and 420 MPa, respectively. The beam contains No. 10 stirrups, not shown in Fig. 5-11.

1. Compute b. This beam is an isolated T beam in which a T-shaped flange is used to increase the area of the compression zone. For such a beam, ACI Section 8.10.4 states that the flange thickness shall not be less than one-half the width of the web and that the effective flange width shall not exceed four times the width of the web. By observation, the flange dimensions satisfy this. Thus $b = 500$ mm.

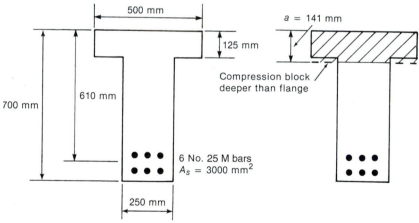

(a) Beam cross section.

(b) Compression zone—Step 3.

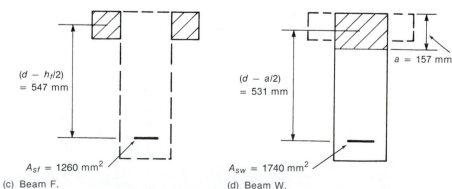

(c) Beam F.

(d) Beam W.

Fig. 5-11
Beam—Example 5-3M.

2. **Compute** d. $d = 610$ mm, as shown in Fig. 5-11a.

3. **Compute** a. Assume that the compression zone will be rectangular. Accordingly,

$$a = \frac{A_s f_y}{0.85 f_c' b}$$

$$= \frac{3000 \times 420}{0.85 \times 20 \times 500} = 148 \text{ mm}$$

As shown in Fig. 5-11b, $a = 148$ mm is greater than the thickness of the flange. This implies that compression stresses exist in the shaded regions below the flanges. This cannot occur. Because $a > h_f$, our assumption that the compression zone is rectangular is wrong, and the calculated value of a is incorrect. It is therefore necessary to analyze this beam as a "T beam."

4. **Divide the beam into "beam F" and "beam W."** Beam F is shown in Fig. 5-11c.

   (a) **Beam F.** The compression force, C_f, in the overhanging flanges is

$$C_f = (0.85 f_c')(b - b_w) h_f$$

$$= 0.85 \times 20(500 - 250)125$$

$$= 531{,}000 \text{ N}$$

The force in the steel in beam F is

$$A_{sf} f_y = C_f \tag{5-1a}$$

and the area is

$$A_{sf} = \frac{531{,}000 \text{ N}}{420 \text{ MPa}} = 1260 \text{ mm}^2$$

Thus beam F consists of the overhanging flanges plus tension reinforcement with an area of 1260 mm^2, as shown in Fig. 5-11c. To calculate the moment capacity of beam F, sum moments about the centroid of the tension steel:

$$\text{lever arm} = d - \frac{h_f}{2}$$

$$= 610 - \frac{125}{2} = 547 \text{ mm}$$

$$M_{nf} = \frac{531{,}000 \times 547}{10^6} = 290 \text{ kN-m} \tag{5-2a}$$

   (b) **Beam W.** Beam W consists of the concrete in the web plus the remainder of the tension reinforcement, A_{sw}, as shown in Fig. 5-11d. Hence,

$$A_{sw} = 3000 - 1260 = 1740 \text{ mm}^2$$

For beam W, $b = b_w$ and

$$a = \frac{A_{sw} f_y}{0.85 f_c' b_w} \tag{5-3}$$

$$= \frac{1740 \times 420}{0.85 \times 20 \times 250} = 172 \text{ mm}$$

Thus, the depth of the compression stress block is actually 172 mm, rather than the 148 mm computed in step 3, assuming "rectangular beam action." For beam W, the design moment capacity is

$$M_{nw} = A_{sw} f_y \left(d - \frac{a}{2} \right) \tag{5-4b}$$

$$= \frac{1740 \times 420(610 - 172/2)}{10^6} = 383 \text{ kN-m}$$

The total positive nominal moment capacity is thus

$$M_n = M_{nf} + M_{nw} = 290 + 373$$

$$= 663 \text{ kN-m}$$

5. **Check whether $A_s > A_{s,\min}$.** From (4-32M),

$$A_{s,\min} = \frac{\sqrt{f_c'}}{4f_y} b_w d, \text{ and } \geq \frac{1.4 b_w d}{f_y}$$

$$= \frac{\sqrt{20}}{4 \times 420} 250 \times 610, \text{ and } \geq \frac{1.4 \times 250 \times 610}{420}$$

$$= 406 \text{ mm}^2 \geq 508 \text{ mm}^2$$

Since $A_s = 3000 \text{ mm}^2$ exceeds this, $A_s > A_{s,\min}$—therefore, A_s is OK.

6. **Check whether $f_s = f_y$ and whether the section is tension-controlled:**

$$d_t = 700 - (40 + 10 + 25/2) = 637.5 \text{ mm} \qquad \frac{a}{d_t} = \frac{172 \text{ mm}}{637.5 \text{ mm}} = 0.270$$

$$\frac{a_{TCL}}{d_t} = 0.375\beta_1 = 0.319 \tag{4-23}$$

Because 0.270 is less than 0.319, the section is tension-controlled, $f_s = f_y$, and $\phi = 0.90$.

7. **Compute ϕM_n.**

$$\phi M_n = 0.90 \times 673 = 606 \text{ kN-m} \qquad\blacksquare$$

Design of T Beams

The design of a T beam involves the choice of the cross section and the reinforcement required. The flange thickness and width are usually established during the design of the floor slab. The size of the beam stem is influenced by the same factors that affect the size of a rectangular beam. In the case of a continuous T beam, the concrete compressive stresses are most critical in the negative-moment regions, where the compression zone is in the beam stem. Frequently, the stem size is chosen so that $\rho \simeq 0.5\rho_b$ at the point of maximum negative moment.

Once the size of the cross section has been determined, it is possible to compute the area of reinforcement required from (4-35). This calculation is similar to Example 4-6. In the negative-moment region, assume that $j = 0.875$ as a first trial value, as was done in Example 4-6. The large flange width in the positive-moment region results in a small value of a (see Fig. 5-8d, for example) and hence a larger value of j where $j = 1 - a/2$. As a first trial value, assume that $j = 0.95$ when designing the positive-moment reinforcement. This is equivalent to assuming that $a = 0.1d$.

EXAMPLE 5-4   Design Reinforcement in a T Beam

A T beam with the overall dimensions shown in Fig. 5-8 is subjected to a factored positive moment, M_u, of 230 ft-kips. Using $f_c' = 3000$ psi and $f_y = 60,000$ psi, design the required reinforcement.

1. **Compute the effective flange width.** Following the computations in Example 5-1, b is 60 in.

2. **Compute d.** Assuming that there will be two layers of reinforcement, use $d = h - 3.5$ in. or $d = 16.5$ in.

3. **Compute the area of reinforcement, A_s.** From (4-34),

$$A_s = \frac{M_u}{\phi f_y jd}$$

Since this is a positive-moment region in a T beam, assume that $j = 0.95$. Then

$$A_s = \frac{230 \times 12,000}{0.9 \times 60,000 \times (0.95 \times 16.5)} = 3.26 \text{ in.}^2$$

Possible choices from Table A-8:

4 No. 8 bars in one layer, $A_s = 3.16 \text{ in.}^2$
2 No. 9 bars plus 2 No. 8 bars, $A_s = 3.58 \text{ in.}^2$
6 No. 7 bars, 4 in one layer, $A_s = 3.60 \text{ in.}^2$

A check of the web width shows that these will fit into a 12-in. web. Although four No. 8 bars give a little less area than required, try this combination, because it can go in one layer rather than two. This will increase d from the 16.5 in. estimated in step 2 to 17.5 in. **Try four No. 8 bars, $A_s = 3.16 \text{ in.}^2$.**

4. **Check whether $A_s \geq A_{s,\min}$.** From (4-32),

$$A_{s,\min} = \frac{3\sqrt{f_c'}}{f_y} b_w d, \text{ and } \geq \frac{200 b_w d}{f_y}$$

$$= \frac{3\sqrt{3000}}{60,000} \times 12 \times 16.5, \text{ and } \geq \frac{200 \times 12 \times 16.5}{60,000}$$

$$= 0.542 \text{ in.}^2 \geq 0.66 \text{ in.}^2$$

Therefore, $A_s = 3.26 \text{ in.}^2 > A_{s,\min} = 0.66 \text{ in.}^2$—OK.

5. **Compute a and check whether $f_s = f_y$ and whether the section is tension-controlled.** Assuming rectangular beam action,

$$a = \frac{A_s f_y}{0.85 f_c' b}$$

$$= \frac{3.16 \times 60,000}{0.85 \times 3000 \times 60} = 1.239 \text{ in.}$$

Since this is less than the flange thickness, rectangular beam action exists. Thus,

$$d_t = 20 \text{ in.} - (1.5 + 0.375 + 1.00/2) = 17.62 \text{ in.} \qquad \frac{a}{d_t} = \frac{1.239}{17.62} = 0.0703$$

This is very much less than

$$\frac{a_{TCL}}{d_t} = 0.375\beta_1 = 0.319 \qquad\qquad (4\text{-}23)$$

Therefore, the section is tension controlled, $f_s = f_y$, and $\phi = 0.90$.

6. **Compute ϕM_n:**

$$\phi M_n = \phi\left[A_s f_y\left(d - \frac{a}{2}\right)\right]$$

$$= \frac{0.9[3.16 \times 60,000(17.5 - 1.24/2)]}{12,000} = 240 \text{ ft-kips}$$

Because $\phi M_n > M_u$, this is OK. **Use four No. 8 Grade 60 bars in one layer.** The lever arm $(d - a/2) = jd$ was $0.965d$ in this case. The value of $0.95d$ assumed in step 3 normally gives a satisfactory first trial. ∎

A complete design of a continuous T beam, including calculation of moment diagrams and proportioning for flexure, shear, and anchorage, is carried out in Chapter 10.

5-3   BEAMS WITH COMPRESSION REINFORCEMENT

Occasionally, beams are built with both tension reinforcement and compression reinforcement. The effect of compression reinforcement on the behavior of beams and the reasons it is used are discussed in this section, followed by methods of analyzing such beams.

Effect of Compression Reinforcement on Strength and Behavior

The resultant internal forces at ultimate load, in beams with and without compression reinforcement, are compared in Fig. 5-12. The beam in Fig. 5-12b has compression steel of area A'_s located at d' from the extreme compression fiber. The area of the tension reinforcement, A_s, is the same in both beams. In both beams, the total compressive force

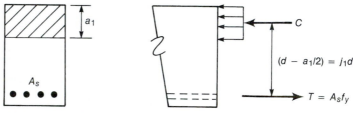

(a) Beam with tension steel only.

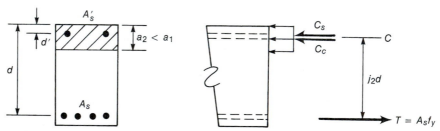

(b) Beam with tension and compression steel.

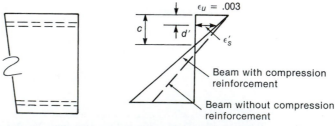

(c) Effect of compression reinforcement on strain distribution in two beams with the same area of tension reinforcement.

Fig. 5-12
Effect of compression reinforcement on moment capacity.

$C = T$, where $T = A_s f_y$. In the beam without compression reinforcement (Fig. 5-12a), this compressive force, C, is resisted entirely by concrete. In the other case, C is the sum of C_c provided by the concrete and C_s provided by the steel. Because some of the compression is resisted by compression reinforcement, C_c will be less than C, with the result that the depth of the compression zone, a_2, in Fig. 5-12b is less than a_1 in Fig. 5-12a.

Summing moments about the resultant compressive force C gives the following results:

For the beam without compression steel,

$$M_n = A_s f_y(j_1 d)$$

For the beam with compression steel,

$$M_n = A_s f_y(j_2 d)$$

where $j_2 d$ is the distance from the tensile force to the resultant of C_s and C_c.

The only difference between these two expressions is that j_2 is a little larger than j_1, because a_2 is smaller than a_1. Thus, for a given amount of tension reinforcement, the addition of compression steel has little effect on the usable ultimate moment capacity, provided the tension steel yields in the beam without compression reinforcement. This is illustrated in Fig. 5-13. For normal ratios of tension reinforcement ($\rho \leq 0.015$) the increase in moment is generally less than 5 percent.

The effectiveness of compression steel decreases as it is moved away from the compression face. As shown in Fig. 5-12c, if the distance d' from the extreme compression fiber to the compression steel is increased, the strain ϵ'_s in the compression steel is decreased. As a result, the stress in this steel may be reduced below the yield stress. The effect of increasing d'/d from 0.1 to 0.2 can be seen in Fig. 5-13. The symbol ρ' is the longitudinal compression steel ratio, $\rho' = A'_s/bd$.

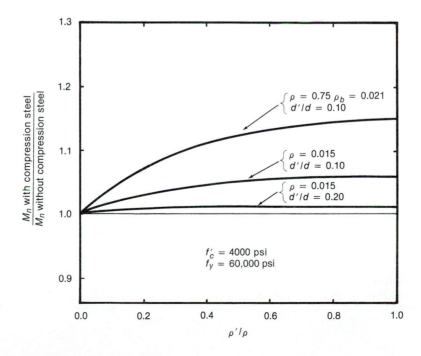

Fig. 5-13
Increase in moment capacity
due to compression
reinforcement.

Reasons for Providing Compression Reinforcement

There are four primary reasons for using compression reinforcement in beams:

1. **Reduced sustained-load deflections.** First and most important, the addition of compression reinforcement reduces the long-term deflections of a beam subjected to sustained loads. Figure 5-14 presents deflection–time diagrams for beams with and without compression reinforcement. The beams were gradually loaded, over a period of several hours, to the service-load level. This load was then maintained for 2 years. At the time of loading (time = 0 in Fig. 5-14), the three beams deflected between 1.6 and 1.9 in., or approximately the same amount. As time passed, the deflections of all three beams increased. The *additional* deflection with time is 195 percent of the initial deflection for the beam without compression steel ($\rho' = 0$), but only 99 percent of the initial deflection for the beam with compression steel equal to the tension steel ($\rho' = \rho$). The ACI Code accounts for this in the deflection-calculation procedures as outlined in Chapter 9.

Creep of the concrete in the compression zone transfers load from the concrete to the compression steel, reducing the stress in the concrete (as occurred in Example 3-4). Because of the lower compression stress in the concrete, it creeps less, leading to a reduction in sustained-load deflections.

2. **Increased ductility.** The addition of compression reinforcement causes a reduction in the depth of the compression stress block, a. Thus, as shown in Fig. 5-12, a_2 is smaller than a_1. As a decreases, the strain in the tension reinforcement at failure increases, as shown in Fig. 5-12c, resulting in more ductile behavior. Figure 5-15 compares moment-curvature diagrams for three beams with $\rho < \rho_b$ and varying amounts of compression reinforcement, ρ', where $\rho' = A_s'/bd$. The moment at first yielding of the tension reinforcement is seen to change very little when compression steel is added to these beams. The increase in moment after yielding is largely due to strain hardening of the reinforcement. Because this occurs at very high curvatures and deflections, it is ignored in design. On the other hand, the ductility increases significantly, as shown in Fig. 5-15. This is particularly important in seismic regions or if moment redistribution is desired.

3. **Change of mode of failure from compression to tension.** When $\rho > \rho_b$, a beam fails in a brittle manner, through crushing of the compression zone before the steel yields. The moment–curvature diagram for such a beam is shown in Fig. 5-16 ($\rho' = 0$). When enough compression steel is added to such a beam, the compression zone is strengthened sufficiently to allow the tension steel to yield before the concrete crushes.

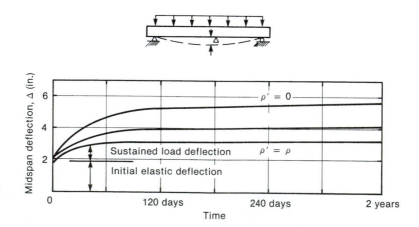

Fig. 5-14
Effect of compression reinforcement on sustained load deflections. (Adapted from [5-5].)

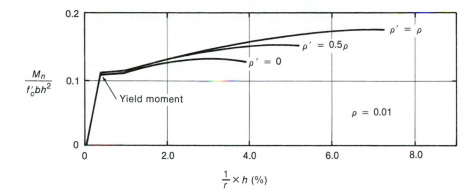

Fig. 5-15
Effect of compression reinforcement on strength and ductility of under-reinforced beams. $\rho < \rho_b$. (From [5-6])

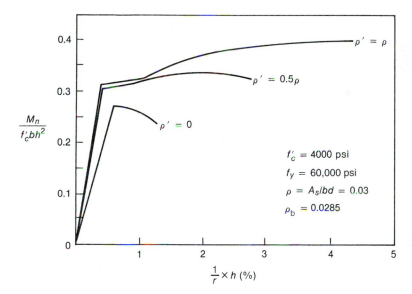

Fig. 5-16
Moment–curvature diagrams for beams with $\rho > \rho_b$, with and without compression reinforcement. (Adapted from [5-6].)

The beam then displays a ductile mode of failure, as shown in Fig. 5-16. If the compression steel yields, the strain distributions and curvatures at failure in a beam with compression reinforcement will be essentially the same as those in a singly reinforced beam (tension steel only) having a reinforcement ratio of $(\rho - \rho')$. The term $(\rho - \rho')$ is referred to as the *effective reinforcement ratio*. Frequently, designers will add compression steel so that $(\rho - \rho') \leq 0.5\rho_b$.

Two cases where compression steel are frequently used are the negative-moment region of continuous T beams and midspan regions of the inverted T beams used to support precast floor panels.

   4.   **Fabrication ease.** When assembling the reinforcing cage for a beam, it is customary to provide bars in the corners of the stirrups to hold the stirrups in place in the form and also to help anchor the stirrups. If developed properly, these bars in effect are compression reinforcement, although they are generally disregarded in design, since they have only a small effect on the moment capacity and because they must be enclosed in closely spaced closed stirrups to prevent buckling of the compression steel.

Analysis of Beams with Tension and Compression Reinforcement

In the analysis of T beams in the preceding section, the cross section was hypothetically divided into two beams. A similar procedure will be used for a beam with compression reinforcement (Fig. 5-17a). The strain distribution, stresses, and internal forces in this beam are shown in Fig. 5-17b to d. For analysis, we will imagine that this beam is divided into *beam 1*, consisting of the compression reinforcement at the top and sufficient steel at the bottom so that $T_1 = C_s$, and *beam 2*, consisting of the concrete web and the remaining tensile reinforcement, as shown in Fig. 5-17e and f.

The stress in the compression reinforcement has been shown as f_s'. Figure 5-17b shows the strain distribution for a beam with compression steel. From similar triangles,

$$\epsilon_s' = \left(\frac{c - d'}{c}\right)0.003$$

If $\epsilon_s' \geq \epsilon_y$ then $f_s' = f_y$. Replacing c with $c = a/\beta_1$ gives

$$\epsilon_s' = \left(1 - \frac{\beta_1 d'}{a}\right)0.003 \tag{5-6}$$

Setting $\epsilon_s' = \epsilon_y$ and $\epsilon_y = f_y/E_s$, where $E_s = 29 \times 10^6$ psi, we can solve for the limiting value of d'/a, for which the compression reinforcement will yield

$$\left(\frac{d'}{a}\right)_{\text{lim}} = \frac{1}{\beta_1}\left(1 - \frac{f_y}{87,000}\right) \tag{5-7}$$

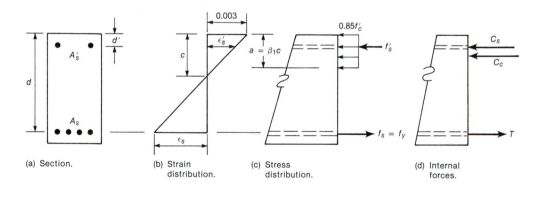

(a) Section.   (b) Strain distribution.   (c) Stress distribution.   (d) Internal forces.

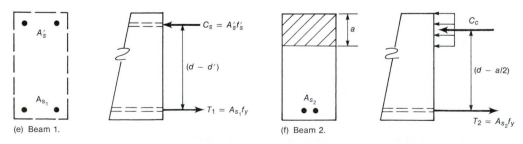

(e) Beam 1.   (f) Beam 2.

Fig. 5-17
Strains, stresses, and forces in a beam with compression reinforcement.

where f_y is in psi. In SI units, with f_y in MPa, we have

$$\left(\frac{d'}{a}\right)_{lim} = \frac{1}{\beta_1}\left(1 - \frac{f_y}{600}\right)$$

(5-7M)

If the value of d'/a is *greater* than this value, the compression steel will not yield at ultimate [i.e., $f_s' = f_y$ only if $d'/a \le (d'/a)_{lim}$]. Values of $(d'/a)_{lim}$ are given in Table A-10.

The procedure followed in computing the moment capacity of a beam with compression reinforcement varies with whether the reinforcement yields. These two cases are discussed separately.

Case 1: Compression Steel Yields

If the compression steel yields in a particular beam, the analysis is straightforward. It is assumed that the beam can be divided into two imaginary beams, each with $C = T$.

Beam 1 consists of reinforcement in tension and compression and resists moment as a steel force couple. The area of tension steel in this beam is obtained by setting $C_s = T_1$, or

$$A_s'f_y = A_{s1}f_y$$

which gives $A_{s1} = A_s'$. From Fig. 5-17e, the nominal moment capacity of this beam is

$$M_{n1} = A_s'f_y(d - d')$$

(5-8)

Beam 2 consists of the concrete plus the remaining steel:

$$A_{s2} = A_s - A_{s1}$$

If $f_s' = f_y$, then $A_{s1} = A_s'$. The compression force in the concrete in beam 2 is

$$C_c = 0.85f_c'ba$$

Since $C = T$ for beam 2, where $T = (A_s - A_s')f_y$, the depth of the compression stress block, a, is

$$a = \frac{(A_s - A_s')f_y}{0.85f_c'b}$$

(5-9a)

From Fig. 5-17f, the nominal moment capacity of beam 2 is

$$M_{n2} = (A_s - A_s')f_y\left(d - \frac{a}{2}\right)$$

(5-10)

and the total nominal moment capacity of a beam with compression steel is

$$M_n = A_s'f_y(d - d') + (A_s - A_s')f_y\left(d - \frac{a}{2}\right)$$

(5-11a)

In the derivation of (5-11a), the compression in the concrete, C_c, was computed by using the full rectangular compression zone, ab (Fig. 5-17f), including the area displaced by the compression steel, A_s'. As a result, the steel is stressed to $0.85f_c'$ in beam 2, and the additional stress needed to yield it in beam 1 is $(f_y - 0.85f_c')$. To correct for this, we have

$$A_{s1} = A_s'\left(1 - \frac{0.85f_c'}{f_y}\right)$$

and

$$M_n = A_s'\left(1 - \frac{0.85f_c'}{f_y}\right)f_y(d - d')$$ (5-11b)

$$+ \left[A_s - A_s'\left(1 - \frac{0.85f_c'}{f_y}\right)\right]f_y\left(d - \frac{a}{2}\right)$$

where

$$a = \frac{[A_s - A_s'(1 - 0.85f_c'/f_y)]f_y}{0.85f_c'b}$$ (5-9b)

The main effect of $(1 - 0.85f_c'/f_y)$ is to reduce the first term in (5-11b) and increase the second term compared with (5-11a). These largely offset each other, with the result that, for beams with Grade-60 reinforcement, (5-11a) will overestimate M_n by up to 0.3 percent (and normally by less). For Grade-40 steel, the increase will be about twice as much. Because this increase is so small, the term $(1 - 0.85f_c'/f_y)$ will be ignored for simplicity.

Determination of Whether $f_s = f_y$ in Tension Reinforcement. The derivation of (5-11a) assumed that both the compression steel and the tension steel yielded. It is necessary to check whether this is true. If d'/a is less than or equal to the limiting value given in (5-7) or (5-7M), the compression steel will yield. The tension steel will yield if a tension failure or a balanced failure occurs. Thus, the tension steel will yield if $a/d \leq a_b/d$, where a_b/d is given by (4-20).

Alternative Determination of Whether $f_s = f_y$ in Tension Reinforcement. The tension steel will yield if a tension or balanced failure occurs. A balanced failure corresponds to a strain distribution with $\epsilon_{cu} = 0.003$ at the extreme compression fiber and ϵ_y at the centroid of the tension steel. Assuming that both the compression steel and the tension steel yield, and substituting $c = a/\beta_1$ and a from (5-9a) into (4-19), the balanced condition is defined as

$$\frac{(A_s - A_s')f_y}{0.85\beta_1 f_c'bd} = \frac{87,000}{87,000 + f_y}$$ (5-12a)

where f_c' and f_y are in psi. Substituting $\rho = A_s/bd$ and $\rho' = A_s'/bd$ gives

$$(\rho - \rho')_b = \frac{0.85\beta_1 f_c'}{f_y}\left(\frac{87,000}{87,000 + f_y}\right)$$ (5-13a)

In SI units, 87,000 becomes 600, and f_c' and f_y are in MPa.

Equation (5-13a) is similar to (4-25) for ρ_b. Note that the effect of compression reinforcement is to reduce the effective value of ρ to $(\rho - \rho')$. Equation (5-13a) applies only if $f_s' = f_y$. A similar equation will be derived later for the case when $f_s' < f_y$.

Case 2: Compression Steel Does Not Yield

If the compression reinforcement does not yield, f_s' is not known, and a different solution is required. Assuming that the tensile steel yields, the internal forces in the beam (Fig. 5-17) are

$$T = A_s F_y$$

$$C_c = 0.85f_c'ba$$ (5-14a)

Ignoring the correction to the compressive stresses in the steel incorporated into (5-11b), we have

$$C_s = (E_s \epsilon'_s) A'_s \tag{5-14b}$$

where ϵ'_s is given by (5-6). From equilibrium,

$$C_c + C_s = T$$

or

$$0.85 f'_c ba + E_s A'_s \left(1 - \frac{\beta_1 d'}{a} \right) 0.003 = A_s f_y$$

This can be reduced to a quadratic equation in a:

$$(0.85 f'_c b) a^2 + (0.003 E_s A'_s - A_s f_y) a - (0.003 E_s A'_s \beta_1 d') = 0 \tag{5-15}$$

Once the depth of the stress block, a, is known, the nominal moment capacity of the section is

$$M_n = C_c \left(d - \frac{a}{2} \right) + C_s (d - d') \tag{5-16}$$

where C_c and C_s are defined by (5-14a) and (5-14b). Note that (5-15) applies only if $f'_s \leq f_y$.

Determination of Whether $f_s = f_y$ in the Tension Reinforcement.   The derivation of (5-15) assumed that the tension steel yielded. It is necessary to check whether this is true. The tension steel will yield if a tension failure or a balanced failure occurs. Thus the tension steel will yield if $a/d \leq a_b/d$, where a_b/d is given by (4-20).

Alternative Determination of Whether $f_s = f_y$ in the Tension Reinforcement.   The tension steel will yield if a tension or balanced failure occurs. Equation (5-13a) gives the value of $(\rho - \rho')$ corresponding to a balanced failure provided that $f'_s = f_y$. If this is not true, (5-12a) becomes

$$\frac{(A_s - A'_s f'_s / f_y) f_y}{0.85 \beta_1 f'_c bd} = \frac{87,000}{87,000 + f_y} \tag{5-12b}$$

and (5-13a) becomes

$$\left(\rho - \frac{\rho' f'_s}{f_y} \right)_b = \frac{0.85 \beta_1 f'_c}{f_y} \left(\frac{87,000}{87,000 + f_y} \right) \tag{5-13b}$$

In SI units, replace 87,000 with 600 in these equations.

If $(\rho - \rho' f'_s / f_y)$ is less than or equal to $(\rho - \rho' f'_s / f_y)_b$, then $f_s = f_y$ for the tension steel.

Upper Limit on Tension Reinforcement in Beams with Compression Steel—ACI Sections 10.3.5 and 10.3.5.1.   The 1999 ACI Code did not have an upper limit on the amount of tensile reinforcement in a beam, because the code committee believed that the penalty resulting from the drop in the strength-reduction factor ϕ from 0.90, for flexure in tension-controlled sections, to 0.65, for compression-controlled sections, was a sufficient penalty to deter the use of compression-controlled beams. This was misinterpreted by some designers to be a signal that ACI had abandoned its traditional ban on over-reinforced beams. ACI 318-02 Section 10.3.5 reintroduced an upper limit on the tension reinforcement in a beam as that corresponding to a net tensile strain of 0.004 in the extreme tension reinforcement. ACI Section 10.3.5.1 allows the use of compression reinforcement to increase the net tensile strain. Enough compression steel would generally be used to ensure a tension-controlled section to take advantage of the higher value of ϕ. A section will be tension-controlled if the a/d ratio at ultimate is less than or equal to

$$\frac{a_{TCL}}{d_t} = 0.375\beta_1 \tag{4-23}$$

Upper Limit on Tension Reinforcement in Beams with Compression Steel—ACI Section B.10.3.3 ACI Section B.10.3.3 limits the amount of tension steel in beams to 0.75 times that corresponding to a balanced failure, that is, to 0.75 times the amount given by (5-13a) and (5-13b). The code goes on to say that the portion of ρ_b equalized by compression steel need not be reduced by the 0.75 factor.

Minimum Tension Reinforcement

Minimum tension reinforcement should correspond to (4-33).

Ties for Compression Reinforcement

As the ultimate load is approached, the compression steel in a beam may buckle, causing the surface layer of concrete to spall off, possibly leading to failure. For this reason, ACI Section 7.11 requires compression reinforcement to be enclosed within stirrups or ties over the length the bars are in compression. The spacing and size of the ties is similar to that of column ties. (See Chapter 11.) Frequently, longitudinal reinforcement that has been detailed to satisfy the bar cutoff rules in Chapter 8 is stressed in compression near points of maximum moment. These bars are not normally enclosed in ties if the compression in them is not included in the calculation of the strength of the beam. Ties are required throughout the length that the compression steel is needed in compression. This can be interpreted to allow the ties to be cut off where ϕM_n for the section without compression steel exceeds the factored moment, M_u.

If the compression reinforcement undergoes stress reversals, the stirrups shall either be closed stirrups with 135° hooks around corner bars or two-piece closed stirrups made up of a U stirrup with 135° bends, closed with a J-shaped cap bar, as shown in Fig. 6-32d. The stirrup hooks should be 135° hooks with the tails embedded in the core of the column. This allows them to act as ties between the top and bottom beam reinforcement so that a severely damaged beam can develop tensile membrane action. The cap bars are placed with the 90° hook alternately on one side of the web or the other. The 135° bends give the cap bars and U stirrups some resistance against uplift in a situation requiring structural integrity reinforcement to satisfy ACI Section 7.13.2.

Examples of the Analysis of Beams with Compression and Tension Reinforcement

Two examples are presented, one for each of the two cases given above. In each example, the solution starts by assuming that $f'_s = f_y$ and $f_s = f_y$ since this is the easiest solution. As soon as a has been computed, the assumptions are checked. If this check shows that

$f'_s < f_y$, it is necessary to change the solution and base the calculations on (5-15). If $f_s < f_y$, more compression steel should be added.

EXAMPLE 5-5   Analysis of a Beam with Compression Reinforcement: Compression-Reinforcement Yields

The beam shown in Fig. 5-18 has $f'_c = 3000$ psi and $f_y = 60,000$ psi. For this beam, based on the tension steel only, $a = 10.14$ in., $d_t = 23.63$ in., giving $a/d_t = 0.429$, which exceeds the tension-controlled limit $a_{TCL}/d_t = 0.319$. As a result, ϕ would be less than 0.90. To allow the use of $\phi = 0.9$ and to give more ductility, 2 No. 7 bars have been added as compression steel. Compute the design moment capacity. Assume design is in accordance with ACI 318-02 Sections 9.2 and 9.3.

1.   **Assume that $f'_s = f_y$ and $f_s = f_y$, and divide the beam into two components.** The beam is divided into beam 1 and beam 2 (Fig. 5-18b and c). Since the steel is all assumed to yield, $A_{s1} = A'_s$. The area of steel in beam 2 is

$$A_{s2} = A_s - A_{s1}$$

$$= 4.74 - 1.20 = 3.54 \text{ in.}^2$$

2.   **Compute a for beam 2:**

$$a = \frac{(A_s - A'_s)f_y}{0.85f'_c b} \tag{5-9a}$$

$$= \frac{3.54 \times 60,000}{0.85 \times 3000 \times 11} = 7.57 \text{ in.}$$

3.   **Check whether compression steel yields.** In step 1, we assumed that $f'_s = f_y$. It is necessary to check this assumption:

$$d' = 2.5 \text{ in.}$$

$$\frac{d'}{a} = \frac{2.5}{7.57} = 0.330$$

From (5-7) or Table A-10,

$$\left(\frac{d'}{a}\right)_{\lim} = \frac{1}{\beta_1}\left(1 - \frac{f_y}{87,000}\right) \tag{5-7}$$

$$= \frac{1}{0.85}\left(1 - \frac{60,000}{87,000}\right) = 0.365$$

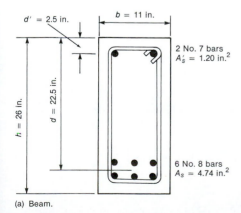

(a) Beam.

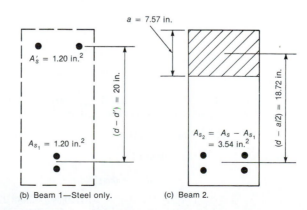

(b) Beam 1—Steel only.   (c) Beam 2.

Fig. 5-18
Example 5-5.

Since $d'/a = 0.330$ is *less* than the limiting value of 0.365, the compression steel yields and $f'_s = f_y$.

4. Check whether $f_s = f_y$ for the tension steel and whether the section is tension-controlled:

$$a = 7.57 \text{ in.}$$

and

$$\frac{a}{d} = \frac{7.57}{22.5} = 0.336$$

From Table A–4, $a_b/d = 0.503$. Since 0.336 is less than 0.503, the tension steel yields. Thus,

$$d_t = 26.0 - 1.5 - 0.375 - 1.0/2 = 23.63 \text{ in.}$$

$$\frac{a}{d_t} = \frac{7.57}{23.63} = 0.320$$

The tension-controlled limit is

$$\frac{a_{TCL}}{d_t} = 0.375\beta_1 = 0.319$$

Since 0.320 is marginally larger than 0.319, the section is a transition section, and ϕ will be less than 0.90. From (4-27b),

$$\phi = 0.23 + \frac{0.25}{a/(\beta_1 d_t)} = 0.23 + \frac{0.25}{7.57/(0.85 \times 23.63)}$$

$$= 0.893$$

Alternatively, if design was carried out by ACI Section B. 10.3.3, the upper limit on the steel would be such that $\rho - \rho' \leq 0.75\rho_b$. To check this, compute the compression forces C_c and C_s in the concrete and the compression steel for the balanced condition with $a = a_b$. Then, compute the area of steel needed to balance these compressions:

$$A_{sb} = \frac{C_c + C_s}{f_y}$$

This comes out to 6.49 in.^{2} and $0.75A_{sb} = 4.87$ in.^{2} This is a little more than $A_s = 4.74$ in.^{2}. Thus the beam satisfies the upper limit on steel in ACI Section B. 10.3.3. For design by B. 10.3.3, $\phi = 0.90$.

Which of these two checks needs to be made depends on whether design is by ACI 318-02 Sections 9.2 and 9.3, or ACI Section B. 10.3.3. We shall assume that design was being carried out by Chapter 9 and will use $\phi = 0.893$.

5. Check whether $A_s \geq A_{s,min}$. From (4-33),

$$A_{s,min} = \frac{3\sqrt{3000}}{60,000} \times 11 \times 22.5, \text{ and } \geq \frac{200 \times 11 \times 22.5}{60,000}$$

$$= 0.68 \geq 0.83 \text{ in.}^2$$

Thus, $A_s = 4.74$ in.^{2} exceeds $A_{s,min} = 0.83$ in^2. This should have been obvious, since the reinforcement exceeded the tension-controlled limit.

6. Compute ϕM_n:

(a) Beam 1:

$$\phi M_{n1} = \phi[A'_s f_y(d - d')]$$

$$= 0.893\left[\frac{1.20 \times 60,000(22.5 - 2.5)}{12,000}\right] = 107.1 \text{ ft-kips} \qquad (5\text{-}8)$$

(b)   Beam 2:

$$\phi M_{n2} = \phi\left[(A_s - A'_s)f_y\left(d - \frac{a}{2}\right)\right]$$

$$= 0.893\left[\frac{3.54 \times 60,000(22.5 - 7.57/2)}{12,000}\right] = 295.8 \text{ ft-kips} \qquad (5\text{-}10)$$

The total moment capacity is

$$\phi M_n = \phi M_{n1} + \phi M_{n2} = 403 \text{ ft-kips}$$

The design moment capacity of the beam shown in Fig. 5-18 is 403 ft-kips.

7.   **Design ties for compression reinforcement.** Compression reinforcement must be enclosed within ties satisfying the size and spacing requirements for column ties from ACI Section 7.10.5. If the longitudinal bars are subjected to stress reversals or the beam resists torsion, the stirrups should be closed. In this example, the compression reinforcement is two No. 7 bars. From ACI Section 7.10.5, No. 3 ties can be used. The tie spacing parallel to the axis of the beam is the smallest of the following:

16 longitudinal bar diameters = 16 × 0.875 in. = 14 in.

48 tie diameters = 48 × 0.375 in. = 18 in., or

least side of member = 11 in.

Since the beam is not subject to stress reversals or torque, closed stirrups are not required by ACI Section 7.11.2. Stirrups are required over the length that the bars are needed as compression reinforcement. The length of beam reinforced with these ties will extend each way from the point of maximum moment. The stirrups already included for shear can be used to tie the compression steel provided they are one- or two-piece closed ties. (ACI Sections 7.10.5, 7.13.2.1, and 7.13.2.3). **Use No. 3 U stirrups with 135° hooks at 11 in. o.c plus No. 3 cap bars.**   ∎

EXAMPLE 5-5M   Analysis of a Beam with Compression-Reinforcement: Compression-Reinforcement Yields—SI Units

The beam shown in Fig. 5-19 has $f'_c = 20$ MPa and $f_y = 420$ MPa. If this beam were reinforced in tension only with the six No. 25M bars shown, the compression stress-block depth would be $a = 275$ mm, and the extreme tensile depth would be $d_t = 538$ mm, giving $a/d_t = 0.511$. This exceeds the tension-controlled-limit of $a_{TCL}/d_t = 0.319$. As a result f'_s would be less than f_y and ϕ would be less than 0.90. To allow the use of $\phi = 0.90$ and to increase the ductility, three No. 19M bars have been added as compression steel. Compute the design moment capacity.

1.   **Assume that $f'_s = f_y$ and $f_s = f_y$, and divide the beam into two components.** The beam is divided into beam 1 and beam 2 (Fig. 5-19b and c). Since both the top and bottom reinforcement are assumed to yield, $A_{s1} = A'_s$. The area of steel in beam 2 is

$$A_{s2} = A_s - A_{s1} = 3060 - 852 \text{ mm}^2$$

$$= 2208 \text{ mm}^2$$

2.   **Compute a for beam 2 using (5-9a)**

$$a = \frac{(A_s - A'_s)f_y}{0.85f'_c b} = \frac{(3060 - 852) \times 420}{0.85 \times 20 \times 275} = 198 \text{ mm}.$$

and $c = a/\beta_1 = 233$ mm.

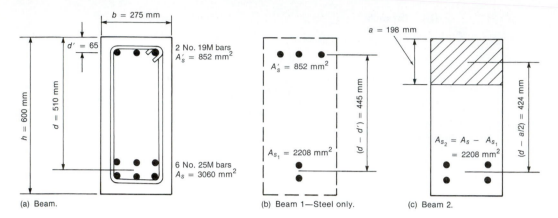

(a) Beam. (b) Beam 1—Steel only. (c) Beam 2.

Fig. 5-19
Example 5-5M.

3. Check whether compression steel yields: For usual covers, the distance d' from the top of the beam to the centroid of the layer of compression steel can be taken as

$$d' = 65 \text{ mm}.$$

The compression steel will yield if it has a strain ε_s' greater than or equal to ε_y. This can be checked using similar triangles representing the strains in the flexural compression zone. Thus, the compression reinforcement will yield if

$$\frac{\varepsilon_{cu}}{c} \le \frac{\varepsilon_y}{c - d'}.$$

Alternatively, the compression steel will yield if $d'/a = 65/198 = 0. 0.328$, based $d' = 65$ mm computed from the cover is less than the limiting value of d'/a from (5-7M)

$$\left(\frac{d'}{a}\right)\text{lim} = \frac{1}{\beta_1}\left(1 - \frac{f_y}{600}\right) = 0.353 \tag{5-7M}$$

Thus, the compression reinforcement will reach its yield strain, as assumed, if

$$d' \le 0.353\,a = 0.353 \times 198 \text{ mm} = 69.9 \text{ mm}$$

is less than the specified $d' = 65$ mm. Because the specified $d' = 65$ mm is less than the limiting $d' = 69.9$ mm, the compression steel will yield.

Alternatively, because $d'/a = 0.328$ is less than $(d'/a)_{\text{lim}} = 0.353$, the compression steel will yield giving $f_s' = f_y$. Thus, the compression steel yields as assumed.

4. Check whether $f_s = f_y$ for the tension reinforcement and whether the section is tension-controlled:

$$d_t = 600 - (40 + 10 + 19/2) = 540 \text{ mm}$$

$$a/d_t = 198/540 = 0.367$$

Compute ε_t and f_s in the extreme tension layer

$$c = a/\beta_1 = 198/0.85 = 233 \text{ mm}$$

$$\varepsilon_t = \frac{d_t - c}{c} \times 0.003 = \frac{540 - 233}{233} \times 0.003$$

$$= 0.00395$$

$f_s = E_s \times \varepsilon_t$ but $= 790$ Mpa, but not more than $f_y = 420$ Mpa.

Thus, $f_s = 420$ Mpa.

Compute ϕ

Because ε_t is between 0.0020 and 0.005, the beam is in the transition region for ϕ.

$$\phi = 0.48 + 83\,\varepsilon_t \qquad \text{(4-26b)}$$

Thus, $\phi = 0.808$.

     Alternatively, if design was carried out by ACI Section B. 10.3.3, the upper limit on the steel would be that $\rho - \rho' \leq 0.75\rho_b$. To check this, compute the compression forces C_c in the concrete and C_s in the compression steel for the balanced condition with $a = a_b$. Then, compute the area of steel needed to balance these compressions:

$$A_{sb} = \frac{C_c + C_s}{f_y}$$

This comes out to 3850 mm^2 and $0.75A_{sb} = 2890$ mm^2, which is five percent less than $A_s = 3060$ mm^2. Thus the beam satisfies the upper limit on steel in B. 10.3.3. For design by B. 10.3.3, $\phi = 0.90$.

     Which of these two checks needs to be made depends on whether design is based on ACI 318-02 Sections 9.2 and 9.3 or by ACI 318-02 Appendix C and Section B. 10.3.3. We shall assume that design was carried out by ACI 318-02 Sections 9.2 and 9.3.

    5.   **Check whether $A_s \geq A_{s,min}$.** From (4-33M),

$$A_{s,min} = (392, \text{ and } \geq 491) \text{ mm}^2$$

$$= 491 \text{ mm}^2$$

Thus $A_s = 3060$ mm^2 exceeds $A_{s,min}$. This should have been obvious, since the reinforcement exceeded the tension-controlled limit.

    6.   **Compute ϕM_n:**

      (a)   **Beam 1:**

$$\phi M_{n1} = \phi[A'_s f_y(d - d')] \qquad \text{(5-8)}$$

$$= \frac{0.808[852 \times 420(510 - 65)]}{10^6} = 122 \text{ kN-m}$$

      (b)   **Beam 2:**

$$\phi M_{n2} = \phi\left[(A_s - A'_s)f_y\left(d - \frac{a}{2}\right)\right] \qquad \text{(5-10)}$$

$$= \frac{0.808[(3060 - 852) \times 420(510 - 65)]}{10^6}$$

$$= 308 \text{ kN-m}$$

The total moment capacity is

$$\phi M_n = \phi M_{n1} + \phi M_{n2} = 430 \text{ kN-m}$$

Therefore, the design moment capacity of the beam shown in Fig. 5-19 is 430 kN-m. ■

EXAMPLE 5-6   Analysis of a Beam with Compression Reinforcement: Compression Reinforcement Does Not Yield

     The beam shown in Fig. 5-20 has $f'_c = 3000$ psi and $f_y = 60,000$ psi. It is similar to the beam considered in Example 5-5, except that there is more compression steel. Compute the design moment capacity, ϕM_n. The solution will start in the same way as Example 5-5. If the assumptions made in step 1 prove to be incorrect, a different solution must be used.

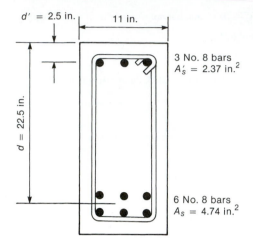

Fig. 5-20
Example 5-6.

1. **Assume that** $f'_s = f_y$ **and** $f_s = f_y$ **and divide the beam into two components.** Beam 1 has three No. 8 bars as compression reinforcement and an area of tension reinforcement, A_{s1}, equal to three No. 8 bars concentrated at d below the top of the beam:

$$A_{s2} = A_s - A_{s1}$$
$$= 4.74 - 2.37 = 2.37 \text{ in.}$$

2. **Compute** a **for beam 2: Assume the top and bottom steel yields.**

$$a = \frac{(A_s - A'_s)f_y}{0.85f'_c b}$$

$$= \frac{2.37 \times 60,000}{0.85 \times 3000 \times 11} = 5.07 \text{ in.}$$

(5-9a)

3. **Check whether compression steel yields.** From Fig. 5-20, $d' = 2.5$ in., and

$$\frac{d'}{a} = \frac{2.5}{5.07} = 0.493$$

But this exceeds the limiting value of $d'/a = 0.365$ from (5-7) or Table A-10. Therefore, the compression steel does not yield. As a result, a from step 2 is incorrect, and a must be reevaluated, by using (5-15), assuming that the compression steel is elastic.

4. **Solve (5-17) for** a_2:

$$(0.85f'_c b)a^2 + (0.003E_s A'_s - A_s f_y)a - (0.003E_s A'_s \beta_1 d') = 0$$

or $28,050a^2 - 78,210a - 438,200 = 0$ and

$$a = \frac{78,210 \pm \sqrt{78,210^2 - (4 \times 28,050 \times -438,200)}}{2 \times 28,050}$$
$$= 5.59 \text{ in.}$$

Note that this is larger than the $a = 5.07$ in. computed in step 2. This is to be expected, since the stress in the compression steel is lower than assumed in step 2, and as a result, there is a larger compression force in the concrete.

5. **Check whether** $f_s = f_y$ **for the tension steel and whether the section is tension-controlled:**

$$a = 5.59 \text{ in.} \quad \text{and} \quad \frac{a}{d} = \frac{5.59}{22.5} = 0.248$$

From Table A-4, $a_b/d = 0.503$. Because 0.248 is less than 0.503, the tension steel yields, and we have

$$d_t = 26 - 1.5 - 0.375 - \frac{1.0}{2} = 23.63 \text{ in.}$$

$$\frac{a}{d_t} = \frac{5.59}{23.63} = 0.237$$

The tension-controlled limit (based on the strain in the tension steel) is

$$\frac{a_{TC_L}}{d_t} = 0.375\beta_1 = 0.319$$

Thus, the section is tension-controlled and $\phi = 0.90$.

6. **Check whether $A_s \geq A_{s,\text{min}}$.** By inspection, A_s exceeds $A_{s,\text{min}}$.

7. **Compute ϕM_n.** We have

$$\phi M_n = \phi\left[C_c\left(d - \frac{a}{2}\right) + C_s(d - d')\right] \tag{5-16}$$

where

$$C_c = 0.85 f_c' ba \tag{5-14a}$$

$$= \frac{0.85 \times 3000 \times 11 \times 5.59}{1000} = 157 \text{ kips}$$

$$C_s = (E_s \epsilon_s') A_s' \tag{5-14b}$$

where

$$\epsilon_s' = \left(1 - \frac{\beta_1 d'}{a}\right)0.003 \tag{5-6}$$

Thus,

$$C_s = E_s A_s'\left(1 - \frac{\beta_1 d'}{a}\right)0.003$$

$$= \frac{29 \times 10^6 \times 2.37\left(1 - \frac{0.85 \times 2.5}{5.59}\right)0.003}{1000}$$

$$= 128 \text{ kips}$$

and

$$\phi M_n = \frac{0.9[157(22.5 - 5.59/2) + 128(22.5 - 2.5)]}{12}$$

$$= 423 \text{ ft-kips}$$

Thus the design moment capacity of the beam shown in Fig. 5-20 is 423 ft-kips.

A worthwhile partial check on the calculations can be obtained by comparing $T_s = A_s f_y = 4.74 \times 60 = 284.4$ kips and $C_c + C_s = 156.8 + 127.8 = 284.5$ kips. These should be the same, since $C = T$ and they are, within the accuracy of the calculations. ∎

A comparison of the strengths of the beams considered in Examples 5-5 and 5-6 shows that almost doubling the area of compression steel, resulting in an increase of 20 percent in the total area of steel in the beam, increased the moment capacity by only 4 percent. This illustrates the fact that additional compression reinforcement is generally not an effective method of increasing the moment capacity of a beam.

5-4 UNSYMMETRICAL BEAM SECTIONS OR BEAMS BENT ABOUT TWO AXES

Figure 5-21 shows one half of a simply supported beam with an unsymmetrical cross section. The loads lie in a plane referred to as the *plane of loading*, and it is assumed that this passes through the shear center of the unsymmetrical section. This beam is free to deflect vertically and laterally between its supports. The applied loads cause moments that must be resisted by an internal resisting moment about a horizontal axis, shown by the moment in Fig. 5-21a. This internal resisting moment results from compressive and tensile forces C and T, as shown in Fig. 5-21b. Because the applied loads do not cause a moment about an axis parallel to the plane of loading (such as A–A), the internal force resultants C and T cannot do so either. As a result, C and T must both lie in the plane of loading or in a plane parallel to it. The distances z in Fig. 5-21b must be equal.

Figure 5-22 shows a cross section of an inverted L-shaped beam loaded with gravity loads. Because this beam is loaded with vertical loads, leading to moments about a horizontal axis, the line joining the centroids of the compressive and tensile forces must be vertical as shown (both are a distance f from the right-hand side of the beam). As a result, the compression zone must be triangular and the neutral axis must be inclined as shown in Fig. 5-22.

Since $C = T$, and assuming that $f_s = f_y$,

$$\tfrac{1}{2}(3f \times g \times 0.85f'_c) = A_s f_y \qquad (5\text{-}16)$$

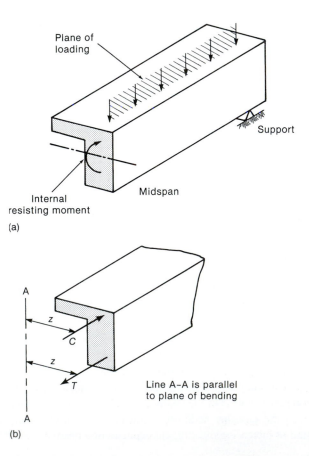

Fig. 5-21
Location of C and T forces in unsymmetrical beam.

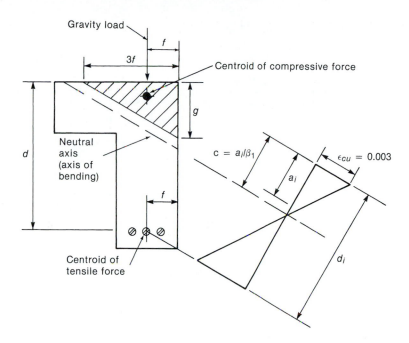

Fig. 5-22
Unsymmetrical beam.

Since the moment is about a horizontal axis, the lever arm must be vertical. Therefore, the case shown in Fig. 5-22,

$$jd = d - \frac{g}{3} \qquad (5\text{-}17)$$

and

$$M_n = A_s f_y \left(d - \frac{g}{3} \right) \qquad (5\text{-}18)$$

These equations apply only to the triangular compression zone shown in Fig. 5-22. Different equations or a trial-and-error solution will generally be necessary for other shapes.

The rectangular stress block was derived for rectangular beams with the neutral axis parallel to the compression face of the beam. This assumes the resultant compression force C can be reached without crushing of the extreme fibers. Rüsch [3-7] and Mattock and Kriz [5-8] have studied this and conclude the rectangular stress block is widely applicable.

The checks of whether $f_s = f_y$ or whether the section is tension-controlled respectively, are done by checking whether a_i/d_i and a_i/d_{ti}, measured perpendicular to the neutral axis, are less than the appropriate values from Table A-4.

The discussion to this point has dealt with isolated beams which are free to deflect both vertically and laterally. Such a beam would deflect perpendicular to the axis of bending, that is, both vertically and laterally. If the beam in Fig. 5-22 were the edge beam for a continuous slab that extended to the left to other beams, this slab would prevent lateral deflections. As a result, the neutral axis would be forced to be very close to horizontal and the beam could be designed in the normal fashion.

EXAMPLE 5-7   Analysis of an Unsymmetrical Beam

The beam shown in Fig. 5-23 has an unsymmetrical cross section and an unsymmetrical arrangement of reinforcement. This beam is subjected to vertical loads only. Compute the design moment capacity of this cross section if $f_c' = 3000$ psi and $f_y = 60,000$ psi.

1.   **Assume that $f_s = f_y$ and compute the size of the compression zone.** The centroid of the three bars is computed to lie at 6.27 in. from the right side of the web. The centroid of the compression

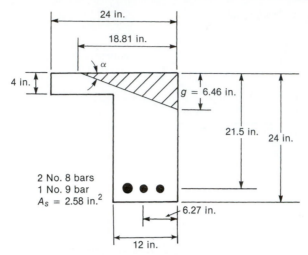

(a) Geometry and stress block.

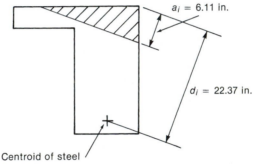

Fig. 5-23
Example 5-7.

(b) Values of a_i and d_i measured perpendicular to the neutral axis.

zone must also be located this distance from the side of the web. Thus the width of the compression zone is $3 \times 6.27 = 18.8$ in.

Since $C = T$,

$$\tfrac{1}{2}(18.8 \times g \times 0.85f_c') = A_s f_y \tag{5-16}$$

or

$$g = \frac{2.58 \times 60{,}000 \times 2}{18.8 \times 0.85 \times 3000}$$

$$= 6.46 \text{ in.}$$

The compression zone is shown shaded in Fig. 5-23. If the compression zone were deeper than shown and cut across the reentrant corner, a more complex trial-and-error solution would be required.

2. Check if $f_s = f_y$ and whether the section is tension-controlled. From Fig. 5-23b, $a_i = 6.11$ in. and $d_i = 22.37$ in., giving $a_i/d_i = 0.273$. Since a_i/d_i is less than $a_b/d = 0.503$, $f_s = f_y$.

d_{ti} is the inclined depth from the extreme compression fiber to the extreme tension steel, in this case the No. 9 bar. From the geometry of the section, $d_{ti} = 23.44$ in. and $a_i/d_{ti} = 0.261$. Since this is less than $a_{TCL}/d_t = 0.319$, the section is tension-controlled and $\phi = 0.90$.

3. Check if $A_s \geq A_{s,\min}$. Since $A_s = 2.58$ in.^{2} exceeds $A_{s,\min} = (0.71 \geq 0.86)$ in.^{2} $A_s > A_{s,\min}$.

4. **Compute ϕM_n.**

$$\phi M_n = \phi\left[A_s f_y\left(d - \frac{g}{3}\right)\right]$$

$$= \frac{0.9[2.58 \times 60{,}000(21.5 - 6.46/3)]}{12{,}000}$$

$$= 225 \text{ ft-kips}$$

(5-18)

Thus the design moment capacity of the section shown in Fig. 5-23 is 225 ft-kips. Note that the moment calculation is based on the lever arm measured *vertically* (parallel to the plane of loading). ■

5-5   ANALYSIS OF MOMENT CAPACITY BASED ON STRAIN COMPATIBILITY

The analysis procedures presented so far in this chapter and Chapter 4 have been restricted to problems involving

1.   Elastic–plastic reinforcement with a constant yield strength;

2.   Tension reinforcement and compression reinforcement in two groups of bars that can be represented by compact layers at the centroids of the respective groups;

3.   All concrete of the same strength;

4.   A rectangular, T, or other easily definable cross-sectional shape.

If any of these restrictions does not apply, a trial-and-error solution based on strain compatibility can be used. In the strain-compatibility calculations, *tensile* stresses, strains, and forces are taken to be *negative*, as subsequently assumed in Chapter 11. As noted in step 7 of the calculation, moments will be summed about the centroidal axis of the gross cross section. The following steps are required in such a solution:

1.   Assume the strain distributions is defined by a strain ϵ_{cu} of 0.003 in the extreme compressive fiber and an assumed value of the depth c to the neutral axis.

2.   Compute the depth of the rectangular stress block, $a = \beta_1 c$.

3.   Compute the strains in each layer of reinforcement from the assumed strain distribution.

4.   From the stress–strain curve for the reinforcement and the strains from step 3, determine the stress in each layer of reinforcement.

5.   Compute the force in the compression zone and in each layer of reinforcement.

6.   Compute $P = C - T$. For a beam without axial force, P equals zero. If the calculated value of P is not equal to zero, adjust the strain distribution and repeat steps 1 to 6 until P is as close to zero as desired. The imbalance should not exceed 0.1 to 0.5 percent of C.

7.   Sum the moments of the internal forces. If $P = 0$, this can be about any convenient axis. We shall sum the moments about the centroid of the cross section. This axis is normally used in columns where P is not zero, as explained in Chapter 11.

Example 5-8   Strain-Compatibility Analysis of Moment Capacity

Compute the design moment capacity, ϕM_n, of the cross section shown in Fig. 5-24a. The concrete strength is 3500 psi. The reinforcement has the stress–strain curve shown in Fig. 5-24b.

In this solution, *compressive* strains and stresses are taken as *positive*, tensile strains and stresses as negative. As a result, the stress–strain curve for the reinforcement in tension has the following equations:

Part *O–A*, $\epsilon \geq -0.002$:

$$f_s = (29 \times 10^3 \epsilon) \text{ ksi}$$

(5-19a)

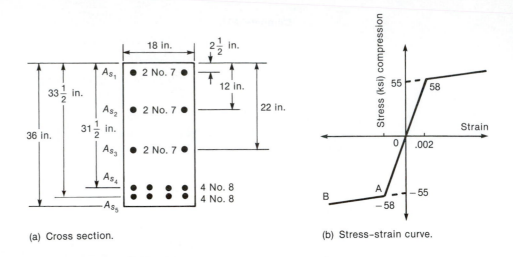

(a) Cross section.

(b) Stress–strain curve.

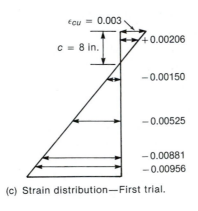

(c) Strain distribution—First trial.

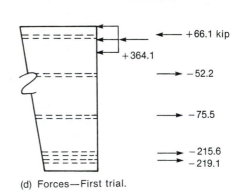

(d) Forces—First trial.

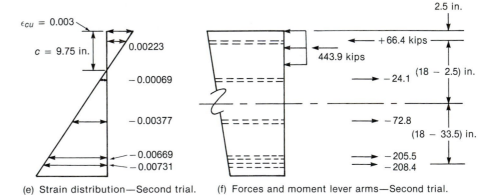

Fig. 5-24
Strain compatibility solution—Example 5-8.

(e) Strain distribution—Second trial.

(f) Forces and moment lever arms—Second trial.

Part A–B, $\epsilon < -0.002$:

$$f_s = (-55 + 1.5 \times 10^3 \epsilon) \text{ ksi} \qquad (5\text{-}19\text{b})$$

Similar equations can be derived for the compressive branch of the stress–strain curve of the steel:

1. Assume a strain distribution. The first trial strain distribution in Fig. 5-24c is defined by $\epsilon_{cu} = 0.003$ and $c = 8$ in.

2. Compute the depth of the equivalent rectangular stress block.

$$a = \beta_1 c$$
$$= 0.85 \times 8 = 6.8 \text{ inches}$$

Therefore, $a = 6.8$ in. for first trial.

3. **Compute the strains in each layer of reinforcement.**
4. **Compute stress in each layer of reinforcement.**
5. **Compute forces in the compression zone and in each layer of reinforcement.**
6. **Compute $P = C - T$.**

Steps 3 to 6 are carried out in Table 5-1. For a bar located at distance y below the top of the beam, the strain is

$$\epsilon = 0.003 - \left(0.003\frac{y}{c}\right) \tag{5-20}$$

For A_{s1}, y is less than a. As a result, this layer of steel displaces concrete assumed to be stressed in compression. If $y < a$,

$$f_s = (f_s \text{ from } (5\text{-}19)) - 0.85f'_c$$

The strains and forces in the various layers are illustrated in Fig. 5-24c and d. The sum of the forces in the bars and the concrete is 130.2 kips tension. Since there is no axial force in this member, the sum should be zero. Thus the assumed compressed zone is too small.

As a second trial, try $c = 9.75$ in., which results in $a = 8.29$ in. Steps 3 to 6 of the second trial are also given in Table 5-1. At the end of the second trial, the forces have converged to within 1 kip. Since this is less than 0.1 percent of the force in the compression zone, it will be assumed to

TABLE 5-1   Calculation of Internal Forces—Example 5-8[a]

Layer	y (in.)	ϵ	f_s (ksi)	A_s(in.²)	F_s (kips)	C_c (kips)
			First trial: Assume that $c = 8$ in., $a = \beta_1 c = 6.8$ in.			
Compression zone	—	—	—	—	—	$6.8 \times 18 \times 0.85$ $\times 3.5 = +364.1$
A_{s1}	2.5	+0.00206	$58.1 - 0.85$ $\times 3.5 = 55.1$	1.20	+66.1	—
A_{s2}	12	−0.00150	−43.5	1.20	−52.2	—
A_{s3}	22	−0.00525	−62.9	1.20	−75.5	—
A_{s4}	31.5	−0.00881	−68.2	3.16	−215.6	—
A_{s5}	33.5	−0.00956	−69.3	3.16	−219.1	—

$$\Sigma F_s = -496.3 \text{ kips} \quad \Sigma C_c = +364.1 \text{ kips}$$
$$\Sigma F_s + \Sigma C_c = -132.2 \text{ kips}$$

Layer	y (in.)	ϵ	f_s (ksi)	A_s(in.²)	F_s (kips)	C_c (kips)
			Second trial: Assume that $c = 9.75$ in., $a = 8.29$ in.			
Compression zone	—	—	—	—	—	$8.29 \times 18 \times 0.85$ $\times 3.5 = 443.9$
A_{s1}	2.5	+0.00223	$58.3 - 0.85$ $\times 3.5 = 55.4$	1.20	+66.4	—
A_{s2}	12	−0.00069	−20.1	1.20	−24.1	—
A_{s3}	22	−0.00377	−60.7	1.20	−72.8	—
A_{s4}	31.5	−0.00669	−65.0	3.16	−205.5	—
A_{s5}	33.5	−0.00731	−66.0	3.16	−208.4	—

$$\Sigma F_s = -444.4 \text{ kips} \quad \Sigma F_c = +443.9 \text{ kips}$$
$$\Sigma F_s + \Sigma C_c = -0.5 \text{ kips}$$

[a] ϵ, f_s, and F_s are positive in compression.

be close enough. It is sometimes useful to plot $P = C - T$ versus c to help in the choice of c for future trials.

7. Check whether $f_s = f_y$ and whether the section is tension-controlled. The stresses in each layer of steel have been computed in Table 5-1, and no further check of f_s is needed. $a/d_t = 9.75/36 = 0.271$. Since this is less than $a_{TCL}/d_t = 0.319$, the section is tension-controlled and $\phi = 0.90$.

If the design were carried out according to ACI Section 10.3.3, it would be necessary to check whether ϵ_t was larger than 0.004 tensile strain, or whether $\rho \leq 0.75 \rho_b$. From Fig. 5-24c the strain in the extreme tensile layer of steel at ultimate is 0.00956 tensile. Therefore the section is tension-controled and $\phi = 0.9$.

8. Compute the moments about midheight. Once P has converged to zero, the moments can be computed. The forces from the second trial are shown in Fig. 5-24f. The distances from the midheight are taken positive upward, negative downward. A counterclockwise moment is taken as positive. We have

$$M_n = 443.9\left(\frac{36}{2} - \frac{8.29}{2}\right) + 66.4(18 - 2.5)$$
$$+ [-24.1(18 - 12)] + [-72.8(18 - 22)]$$
$$+ (-205.5(18 - 31.5)] + [-208.4(18 - 33.5)]$$
$$= 13{,}330 \text{ in.-kips}$$

Thus, the nominal moment capacity of the section shown in Fig. 5-24 is $M_n = 13{,}330$ in.-kips and the design moment capacity $\phi M_n = 12{,}000$ ft-kips. ∎

This type of problem is ideally suited for solution via a spreadsheet.

5-6   DESIGN OF MEMBERS WITH FRP REINFORCEMENT-DESIGN ACCORDING TO ACI 440 REPORT

Properties of FRP Reinforcement

The three most common types of FRP are GFRP with glass fibers, AFRP with Aramid fibers, and CFRP with carbon fibers. The material properties are described briefly in Section 3-14. In tests, FRP bars have a brittle, nearly linear stress–strain relationship in compression or tension. The design procedure is controlled by the brittle nature of the FRP reinforcement.

Most types of fiber-reinforced polymer reinforcement (FRP) cannot be bent once the polymer has set. As a result, if bent or hooked bars are required, they must be bent during manufacture.

The polymer resins used to make FRP bars undergo a phase change between 150° and 250°F, causing a reduction in strength. By the time the temperature of the bar reaches 480°F, the tensile strength of a typical FRP drops to about 20 percent of the strength at room temperature.

Design of Members with FRP Reinforcement

This section is based on the recent reports by ACI Committee 440. [5-9] There is no accepted American consensus standard covering the design of members reinforced with fiber-reinforced polymer (FRP) reinforcement. Currently, design is based on ultimate-strength design using FRP material properties supplied by the manufacturer of the reinforcement. Generally the properties provided are low fractiles of the distributions of strengths and moduli. The three most important properties are the initial ultimate tension strength, f^*_{fu}, the modulus of elasticity, E_f, and the initial rupture strain, ε^*_{fu}.

The design philosophy for FRP-reinforced structures differs significantly from that used for steel-reinforced structures. For the latter, the designer ensures that the steel reinforcement will yield in tension before the concrete crushes. This provides some ductility to redistribute moments and provides warning of impending failure. This warning, in part, allowed the strength reduction factor to be increased to 0.9. On the other hand, the elastic-brittle FRP reinforcement does not yield. Instead, it fractures suddenly without warning.

In steel-reinforced beams, the balanced steel ratio, ρ_b, is defined as the steel ratio at which the steel yields simultaneously with the concrete reaching its crushing strain. Steel-reinforced beams are designed with ρ less than about 0.5 to $0.75\rho_b$ or compression reinforcement is added so that the tension steel yields before the concrete crushes. This is intended to assure some ductility.

The definition of *balanced failure* differs for steel-reinforced beams and beams with FRP. A beam reinforced with FRP is said to have a balanced failure if the FRP bars reach their design rupture strain at the same time that the concrete in the extreme compression fiber reach its assumed limiting compressive strain. The reinforcement ratio corresponding to this point is called the *FRP balanced-reinforcement ratio*, ρ_{fb}. A strength reduction factor of $\phi = 0.5$ has been proposed for beams with FRP reinforcement having ρ less than ρ_{fb} because beams with more reinforcement fail suddenly without warning [5-9]. For beams with ρ greater than $1.4\rho_{fb}$, $\phi = 0.7$ has been *proposed*, with a linear transition of ϕ for beams with ρ between ρ_{fb} and $1.4\rho_{fb}$. For beams with $\rho > \rho_{fb}$, the concrete is assumed to develop a normal rectangular compression stress block. It should be emphasized that the ACI Code has not sanctioned the use of FRP reinforcement and has not approved these values of ϕ.

Following selection of the beam and reinforcement, it is necessary to check for cracking and deflections. In addition, it is necessary to check the possibility of creep rupture of the FRP. This is done by computing the stresses in the reinforcement at service loads, assuming a straight-line distribution of concrete stresses. The resulting tension stress in the reinforcement must not exceed the appropriate creep rupture stress limits. Reference [5-7] suggests the tensile stress due to sustained loads should be limited to $0.2f_{fu}$ for GFRP, $0.3f_{fu}$ for AFRP, and $0.55f_{fu}$ for CFRP.

FRP bars loaded in compression are subject to localized buckling of the fibers; as a result, they should not be used as longitudinal bars in columns.

The shear strength of FRP beams has not been adequately studied. The newer theories of shear strength, such as the Modified Compression Field Theory method discussed in Section 6-6, suggest that the shear, V_c, carried by the concrete will be less than that for comparable beams with steel reinforcement. The shear strength is strongly affected by the longitudinal strain, ε_x, in the flexural reinforcement. Because the modulus of elasticity of FRP reinforcement is lower than for steel, ε_x will be larger than in a beam with steel reinforcement. This allows the cracks to open wider for a given amount of flexural reinforcement, reducing the shears transmitted across the cracks.

Prior to embarking on the design of an FRP-reinforced beam, the latest edition of the ACI Committee 440 report [5-9] should be reviewed to determine recent improvements in the design procedures.

PROBLEMS

5-1 and 5-2 Compute ϕM_n for the beams shown in Fig. P5-1 and P5-2. Use $f'_c = 3750$ psi for 5-1 and 3000 psi for 5-2 and $f_y = 60,000$ psi.

5-3 Compute the negative-moment capacity, ϕM_n, for the beam shown in Fig. P5-3. Use $f'_c = 3000$ psi and $f_y = 40,000$ psi.

5-4 For the beam shown in Fig. P5-4, $f'_c = 3000$ psi and $f_y = 60,000$ psi.

(a) Compute the effective flange width at midspan.

(b) Compute ϕM_n for the positive- and negative-moment regions. Check whether $f_s = f_y$ at ultimate. At the supports, the bars are in one

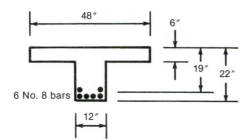

Fig. P5-1

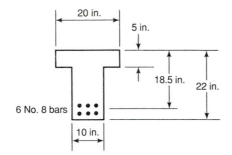

Fig. P5-2

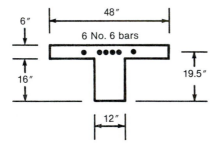

Fig. P5-3

layer; at midspan, the No. 8 bars are in the bottom layer, the No. 7 bars in a second layer. (Note that the bottom reinforcement is not adequately anchored in the support to serve as compression reinforcement at the face of the support. Therefore, it can be ignored.)

5-5 The beam shown in Fig. P5-5 carries its own dead load plus an additional service (unfactored) dead load of 1.5 kips/ft plus a service live load of 3.5 kips/ft.

(a) Draw bending moment diagrams for the three loading cases shown in Fig. P4-18 and superimpose them to get a bending-moment envelope.

(b) Select reinforcement at the negative- and positive-moment regions. Use $f_c' = 4000$ psi and $f_y = 60,000$ psi.

5-6 Compute ϕM_n for the beam shown in Fig. P5-6. Use $f_c' = 3000$ psi and $f_y = 60,000$ psi, and

(a) the reinforcement is six No. 8 bars.

(b) the reinforcement is nine No. 8 bars.

5-7 Compute ϕM_n for the beam shown in Fig. P5-7. Use $f_c' = 4000$ psi and $f_y = 60,000$ psi.

5-8 Give three reasons why compression reinforcement is used in beams.

5-9 (a) Compute ϕM_n for the three beams shown in Fig. P5-9. In each case, $f_c' = 3000$ psi, $f_y = 60,000$ psi, $b = 12$ in., $d = 32.5$ in., and $h = 36$ in.

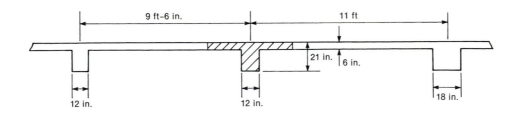

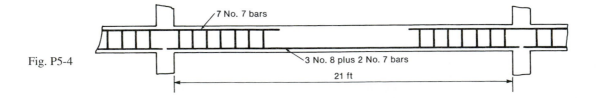

Fig. P5-4

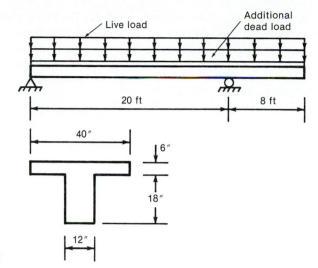

Fig. P5-5

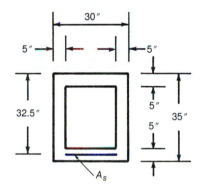

Fig. P5-6

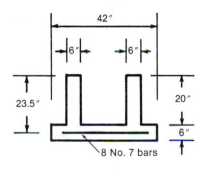

Fig. P5-7

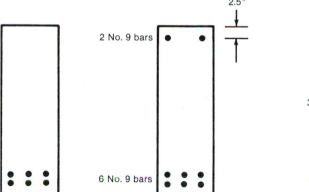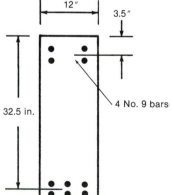

Fig. P5-9

(a) From the results of part (a), comment on whether adding compression reinforcement is a cost-effective way of increasing the strength, ϕM_n, of a beam.

5-10 Compute ϕM_n for the beam shown in Fig. P5-10. Use $f_c' = 2500$ psi and $f_y = 60,000$ psi. Does the steel yield in this beam at ultimate?

5-11 The beam shown in Fig. P5-11 has elastic–plastic reinforcement ($f_s = f_y$ when $\epsilon_s \geq \epsilon_y$) with a yield strength of 60,000 psi. The concrete has $f_c' = 4000$ psi. Compute ϕM_n, using a strain-compatibility solution.

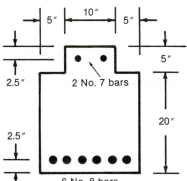

Fig. P5-10

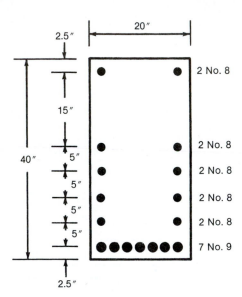

Fig. P5-11

6

Shear in Beams

A beam resists loads primarily by means of internal moments, M, and shears, V, as shown in Fig. 6-1. In the design of a reinforced concrete member, flexure is usually considered first, leading to the size of the section and the arrangement of reinforcement to provide the necessary moment resistance. Limits are placed on the amounts of flexural reinforcement which can be used, to ensure that if failure was ever to occur, it would develop gradually, giving warning to the occupants. The beam is then proportioned for shear. Because a shear failure is frequently sudden and brittle, as suggested by the damage sustained by the building in Fig. 6-2 [6-1], the design for shear must ensure that the shear strength equals or exceeds the flexural strength at all points in the beam.

The manner in which shear failures can occur varies widely with the dimensions, geometry, loading, and properties of the members. For this reason, there is no unique way to design for shear. In this chapter, we deal with the internal shear force, V, in relatively slender beams and the effect of the shear on the behavior and strength of beams. Examples of the design of such beams for shear are given in this chapter. Footings and two-way slabs supported on isolated columns develop shearing stresses on sections around the circumference of the columns, leading to failures in which the column and a conical piece of the slab punch through the slab (Chapter 13). Short, deep members such as brackets, corbels, and deep beams transfer shear to the support by in-plane compressive stresses rather than shear stresses. Such members are considered in Chapter 18.

Chapter 21 of the ACI Code gives special rules for shear reinforcement in members resisting seismic loads. These are reviewed in Chapter 20.

This chapter uses four different models of the shear strength of beams. Each highlights a different aspect of the behavior and strength of beams failing in shear:

1. The stresses in uncracked beams are presented to explain the onset of shear cracking.

2. This is followed by plastic truss models of beams with shear cracks. The truss model is used to explain the effect of shear cracks on the forces in the longitudinal tension reinforcement and in the compression flanges of the beam.

3. The ACI Code design procedure for shear in beams is presented and is illustrated by examples.

209

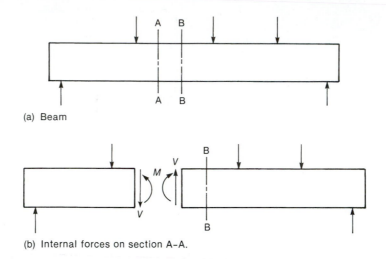

(a) Beam

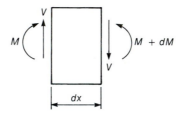

(b) Internal forces on section A–A.

Fig. 6-1
Internal forces in a beam.

(c) Internal forces on portion between sections A–A and B–B.

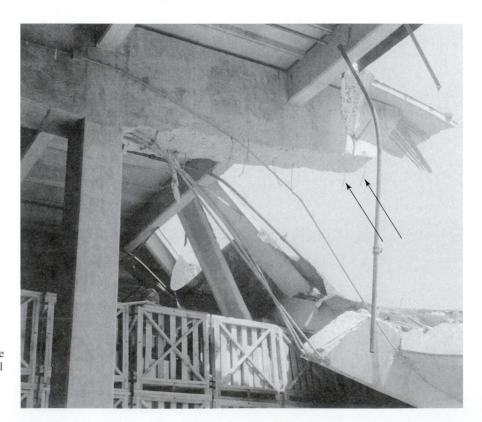

Fig. 6-2
Shear failure: U.S. Air Force
warehouse. Note the small size
and large spacing of the vertical
web reinforcement that has
fractured. (Photograph courtesy of C. P. Siess.)

4. Several comprehensive models of cracked beams loaded in shear are reviewed, on the basis of recent revisions to shear design theory. These models are mentioned because, in the author's opinion, they come close to being the final explanation of the shear strength of reinforced concrete. [6-2]

Items 1 and 2 in this list are included to provide background for the ACI Code design methods. Item 4 shows the effects of other variables that affect the shear strength of slender beams.

6-2   BASIC THEORY

Stresses in an Uncracked Elastic Beam

From the free-body diagram in Fig. 6-1c, it can be seen that $dM/dx = V$. Thus shear forces and shear stresses will exist in those parts of a beam where the moment changes from section to section. By the traditional theory for *homogeneous, elastic, uncracked* beams, we can calculate the shear stresses, v, on elements 1 and 2 cut out of a beam (Fig. 6-3a), using the equation

$$v = \frac{VQ}{Ib} \tag{6-1}$$

where

V = shear force on the cross section

I = moment of inertia of the cross section

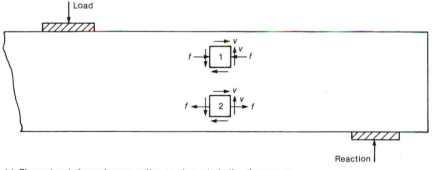

(a) Flexural and shear stresses acting on elements in the shear span.

(b) Distribution of shear stresses.

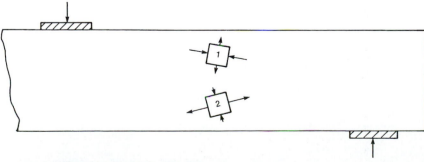

(c) Principal stresses on elements in shear span.

Fig. 6-3
Normal, shear, and principal stresses in a homogeneous uncracked beam.

Q = first moment about the centroidal axis of the part of the cross-sectional area lying farther from the centroidal axis than the point where the shear stresses are being calculated

b = width of the member at the section where the stresses are being calculated

Equal shearing stresses exist on both the horizontal and vertical planes through an element, as shown in Fig. 6-3a. The shear stresses on the top and bottom of the elements cause a clockwise couple, and those on the vertical sides of the element cause an anticlockwise couple. These two couples are equal and opposite in magnitude and hence cancel each other out. The horizontal shear stresses are important in the design of construction joints, web-to-flange joints, and regions adjacent to holes in beams. For an *uncracked rectangular* beam, (6-1) gives the distribution of shear stresses shown in Fig. 6-3b.

The elements in Fig. 6-3a are subjected to combined normal stresses due to flexure, f, and shearing stresses, v. The largest and smallest normal stresses acting on such an element are referred to as *principal stresses*. The principal stresses and the planes they act on are found by using a Mohr's circle for stress, as explained in any mechanics-of-materials textbook. The orientations of the principal stresses on the elements in Fig. 6-3a are shown in Fig. 6-3c.

The surfaces on which principal tension stresses act in the *uncracked* beam are plotted by the curved lines in Fig. 6-4a. These surfaces or *stress trajectories* are steep near the bottom of the beam and flatter near the top. This corresponds with the orientation of the elements shown in Fig. 6-3c. Since concrete cracks when the principal tensile stresses exceed the tensile strength of the concrete, the initial cracking pattern should resemble the family of lines shown in Fig. 6-4a.

The cracking pattern in a test beam with longitudinal flexural reinforcement, but no shear reinforcement, is shown in Fig. 6-4b. Two types of cracks can be seen. The vertical cracks occurred first, due to flexural stresses. These start at the bottom of the beam where the flexural stresses are the largest. The inclined cracks at the ends of the beam are due to combined shear and flexure. These are commonly referred to as *inclined cracks, shear cracks*, or *diagonal tension cracks*. Such a crack must exist before a beam can fail in shear. Several of the inclined cracks have extended along the reinforcement toward the support, weakening the anchorage of the reinforcement.

Although there is a similarity between the planes of maximum principal tensile stress and the cracking pattern, it is by no means perfect. In reinforced concrete beams, flexural cracks generally occur before the principal tensile stresses at midheight become critical. Once a flexural crack has occurred, the tensile stress perpendicular to the crack

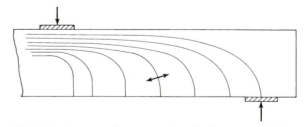

(a) Principal compressive stress trajectories in an uncracked beam.

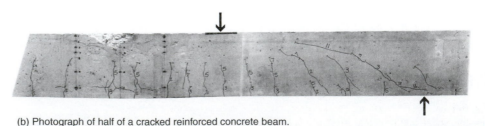

Fig. 6-4
Principal compressive stress trajectories and inclined cracks. (Photograph courtesy of J. G. MacGregor.)

(b) Photograph of half of a cracked reinforced concrete beam.

drops to zero. To maintain equilibrium, a major redistribution of stresses is necessary. As a result, the onset of inclined cracking in a beam cannot be predicted from the principal stresses unless shear cracking precedes flexural cracking. This very rarely happens in reinforced concrete, but it does occur in some prestressed beams.

Average Shear Stress between Cracks

The initial stage of cracking generally results in vertical cracks which, with increasing load, extend in a diagonal manner, as shown in Fig. 6-4b. The equilibrium of the section of beam between two such cracks (Fig. 6-5b) can be written as

$$T = \frac{M}{jd} \qquad \text{and} \qquad T + \Delta T = \frac{M + \Delta M}{jd}$$

or

$$\Delta T = \frac{\Delta M}{jd}$$

where jd is the lever arm, which is assumed to be constant. For the moment equilibrium of the element,

$$\Delta M = V \Delta x \tag{6-2}$$

and

$$\Delta T = \frac{V \Delta x}{jd} \tag{6-3}$$

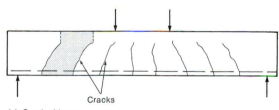

(a) Cracked beam.

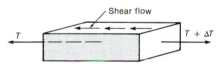

Shear flow

(c) Bottom part of beam.

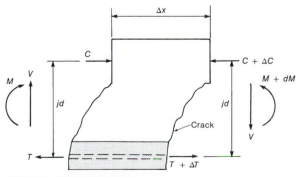

(b) Portion of beam between two cracks.

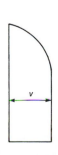

(d) Average shear stresses.

Fig. 6-5
Calculation of average shear stress between cracks.

If the shaded portion of Fig. 6-5b is isolated as shown in Fig. 6-5c, the force ΔT must be transferred by horizontal shear stresses on the top of the element. The *average* value of these stresses below the top of the crack is

$$v = \frac{\Delta T}{b_w \Delta x}$$

or

$$v = \frac{V}{b_w jd} \tag{6-4}$$

where $jd \simeq 0.875d$ and b_w is the thickness of the web. The distribution of *average* horizontal shear stresses is shown in Fig. 6-5d. Since the vertical shear stresses on an element are equal to the horizontal shear stresses on the same element, the distribution of vertical shear stresses will be as shown in Fig. 6-5d. This assumes that about 30 percent of the shear is transferred in the compression zone. The rest of the shear is transferred across the cracks. In 1970, Taylor [6-3] reported tests of beams without web reinforcement in which he found that about 25 percent of the shear was transferred by the compression zone, about 25 percent by doweling action of the flexural reinforcement, and about 50 percent by aggregate interlock along the cracks. (See Fig. 6-13, discussed later.) Modern shear failure theories assume that a significant amount of the shear is transferred in the web of the beam, most of this across inclined cracks.

The ACI design procedure arbitrarily replaces jd in (6-4) with d to simplify the calculations, giving

$$v = \frac{V}{b_w d} \tag{6-5}$$

In some of the more recent design methods, presented in Section 6-4, jd is retained but is renamed the *depth for shear calculations*, d_v. This is defined as the distance between the resultant flexural compression and tension forces acting on the cross section, as shown in Fig. 6-5b, except that d_v need not be taken smaller than $0.9d$.

Beam Action and Arch Action

In the derivation of (6-4), it was assumed that the beam was prismatic and the lever arm jd was constant. The relationship between shear and bar force (6-3) can be rewritten as [6-4]

$$V = \frac{d}{dx}(Tjd) \tag{6-6}$$

which can be expanded as

$$V = \frac{d(T)}{dx}jd + \frac{d(jd)}{dx}T \tag{6-7}$$

Two extreme cases can be identified. If the lever arm, jd, remains constant, as assumed in normal elastic beam theory, then

$$\frac{d(jd)}{dx} = 0 \quad \text{and} \quad V = \frac{d(T)}{dx}jd$$

where $d(T)/dx$ is the *shear flow* across any horizontal plane between the reinforcement and the compression zone, as shown in Fig. 6-5c. For beam action to exist, this shear flow must exist. The other extreme occurs if the shear flow, $d(T)/dx$, equals zero, giving

$$V = T\frac{d(jd)}{dx}$$

or

$$V = C\frac{d(jd)}{dx}$$

This occurs if the shear flow cannot be transmitted, because the steel is unbonded, or if the transfer of shear flow is disrupted by an inclined crack extending from the load to the reactions. In such a case, the shear is transferred by *arch action* rather than beam action, as illustrated in Fig. 6-6. In this member, the compression force C in the inclined strut and the tension force T in the reinforcement are constant over the length of the shear span.

Shear Reinforcement

In Chapter 4, we saw that horizontal reinforcement was required to restrain the opening of a vertical flexural crack, as shown in Fig. 6-7a. An inclined crack opens approximately perpendicular to itself, as shown in Fig. 6-7b, and either a combination of horizontal flexural

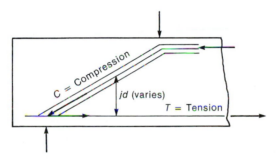

Fig. 6-6
Arch action in a beam.

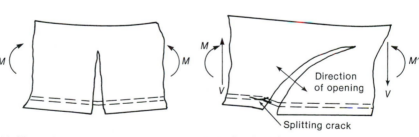

(a) Flexural crack.

(b) Inclined crack.

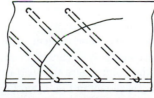

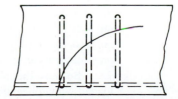

Fig. 6-7
Inclined cracks and shear reinforcement.

(c) Inclined shear reinforcement.

(d) Vertical shear reinforcement.

reinforcement and inclined reinforcement (Fig. 6-7c) or a combination of horizontal and vertical reinforcement (Fig. 6-7d) is required to restrain it from opening. The inclined or vertical reinforcement is referred to as *shear reinforcement* or *web reinforcement* and may be provided by inclined or vertical *stirrups*, as shown in Fig. 6-28 or 6-29. Most often, vertical stirrups are used in North America. The arrangement of stirrups in a beam is illustrated in Fig. 4-23. Inclined stirrups are not effective in beams resisting shear reversals, such as seismic loads, because the reversals will cause cracking parallel to the inclined reinforcement, rendering it ineffective.

6-3   BEHAVIOR OF BEAMS FAILING IN SHEAR

The behavior of beams failing in shear varies widely, depending on the relative contributions of beam action and arch action and the amount of web reinforcement.

Behavior of Beams without Web Reinforcement

The moments and shears at inclined cracking and failure of rectangular beams without web reinforcement are plotted in Fig. 6-8b and c as a function of the ratio of the shear span a to the depth d. (See Fig. 6-8a.) The beam cross section remains constant as the span is varied. The maximum moment and shear that can be developed correspond to the nominal moment capacity, M_n, of the cross section plotted as a horizontal line in Fig. 6-8b. The shaded areas in this figure show the reduction in strength due to shear. Web reinforcement is provided to ensure that the beam reaches the full flexural capacity, M_n [6-5].

Figure 6-8b suggests that the shear spans can be divided into three types: short, slender, and very slender shear spans. The term *deep beam* is also used to describe beams with short shear spans. Very short shear spans, with a/d from 0 to 1, develop inclined cracks joining the load and the support. These cracks, in effect, destroy the horizontal shear flow from the longitudinal steel to the compression zone, and the behavior changes from beam action to arch action, as shown in Fig. 6-6 or 6-9. Here, the reinforcement serves as the tension tie of a tied arch and has a uniform tensile force from support to support. The most common mode of failure in such a beam is an anchorage failure at the ends of the tension tie.

Short shear spans with a/d from 1 to 2.5 develop inclined cracks and, after a redistribution of internal forces, are able to carry additional load, in part by arch action. The final failure of such beams will be caused by a bond failure, a splitting failure, or a dowel failure along the tension reinforcement, as shown in Fig. 6-10a, or by crushing of the compression zone over the crack, as shown in Fig. 6-10b. The latter is referred to as a *shear compression failure*. Because the inclined crack generally extends higher into the beam than does a flexural crack, failure occurs at less than the flexural moment capacity.

In slender shear spans, those having a/d from about 2.5 to about 6, the inclined cracks disrupt equilibrium to such an extent that the beam fails at the inclined cracking load, as shown in Fig. 6-8b. Very slender beams, with a/d greater than about 6, will fail in flexure prior to the formation of inclined cracks.

Figures 6-9 and 6-10 come from [6-5], which presents an excellent discussion of the behavior of beams failing in shear and the factors affecting their strengths. It is important to note that, for short and very short beams, a major portion of the load capacity after inclined cracking is due to load transfer by the compression struts shown in Fig. 6-9 and 6-10. If the beam is not loaded on the top and supported on the bottom in the manner shown in Fig. 6-9, these compression struts are less effective, and failure occurs at, or close to, the inclined cracking load.

Because the moment at the point where the load is applied is $M = Va$ for a beam loaded with concentrated loads, as shown in Fig. 6-8a, Fig. 6-8b can be replotted in terms of shear capacity, as shown in Fig. 6-8c. The shear corresponding to a flexural failure is the

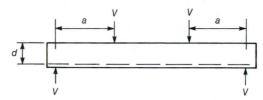

(a) Beam.

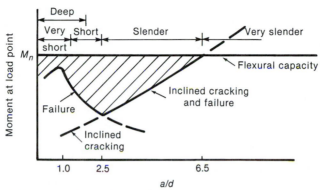

(b) Moments at cracking and failure.

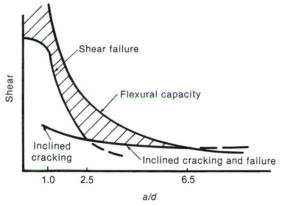

(c) Shear at cracking and failure.

Fig. 6-8
Effect of *a/d* ratio on shear strength of beams without stirrups.

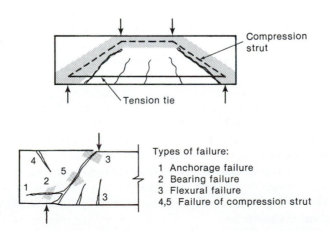

Types of failure:

1 Anchorage failure
2 Bearing failure
3 Flexural failure
4,5 Failure of compression strut

Fig. 6-9
Modes of failure of deep beams, *a/d* = 0.5 to 2.0. (Adapted from [6-5].)

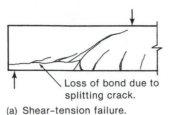

(a) Shear–tension failure.

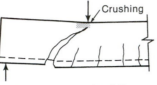

(b) Shear compression failure.

Fig. 6-10
Modes of failure of short
shear spans, *a/d* = 1.5 to 2.5.
(Adapted from [6-5].)

upper curved line. If stirrups are not provided, the beam will fail at the shear given by the "shear failure" line. This is roughly constant for *a/d* greater than about 2. Again, the shaded area indicates the loss in capacity due to shear. Note that the inclined cracking loads of the short shear spans and slender shear spans are roughly constant. This is recognized in design by ignoring *a/d* in the equations for the shear at inclined cracking. In the case of slender beams, inclined cracking causes immediate failure if no web reinforcement is provided. For very slender beams, the shear required to form an inclined crack exceeds the shear corresponding to flexural failure, and the beam will fail in flexure before inclined cracking occurs.

B-Regions and D-Regions

Figure 6-8 indicates that there is a major change in behavior at a shear span ratio, *a/d*, of about 2 to 2.5. Longer shear spans carry load by beam action and are referred to as *B-regions*, where the B stands for beam or for Bernoulli, who postulated the linear strain distribution in beams. Shorter shear spans carry load primarily by arch action involving in-plane forces. Such regions are referred to as *D-regions*, where the D stands for discontinuity or disturbed [6-6].

St. Venant's principle suggests that a local disturbance, such as a concentrated load or reaction, will dissipate within about one beam depth from the point at which the load is applied. On the basis of this principle, it is customary to assume that D-regions extend about one member depth each way from concentrated loads, reactions, or abrupt changes in section or direction, as shown in Fig. 6-11. The regions between D-regions can be treated as B-regions.

In general, arch action enhances the "shear" strength of a section. As a result, B-regions tend to be weaker than corresponding D-regions, as shown by the lower line in Fig. 6-8c for *a/d* greater than 2 to 2.5. If a shear span consists entirely of D-regions that meet or overlap, as shown by the left end of Fig. 6-11a, its behavior will be governed by arch action. This accounts for the increase in shear strength when *a/d* is less than 2.

For longer shear spans, such as the right-hand end of the beam in Fig. 6-11a, the shear strength of the right end is governed by the B-region and is relatively constant, as shown in Fig. 6-8c. This type of member is discussed in this chapter. D-regions are discussed in Chapter 18.

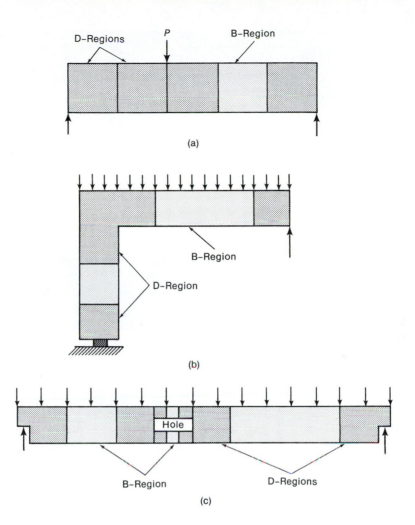

Fig. 6-11
B-regions and D-regions.

Inclined Cracking

Inclined cracks must exist before a shear failure can occur. Inclined cracks form in the two different ways shown in Fig. 6-12. In thin-walled I beams in which the a/d ratio is small, the shear stresses in the web are high, while the flexural stresses are low. In a few extreme cases and in some prestressed beams, the principal tension stresses at the neutral axis may exceed those at the bottom flange. In such a case, a *web-shear crack* occurs (Fig. 6-12a). The corresponding inclined cracking shear can be calculated as the shear necessary to cause a principal tensile stress equal to the tensile strength of the concrete at the centroid of the beam.

In most reinforced concrete beams, however, flexural cracks occur first and extend more or less vertically into the beam, as shown in Fig. 6-4b or 6-12b. These alter the state of stress in the beam, causing a stress concentration near the head of the crack. In time, either

1.   the flexural cracks extend to become *flexure-shear cracks* (Fig. 6-12b), or

2.   flexure-shear cracks develop in the uncracked region over the flexural cracks (Fig. 6-4b).

Flexure-shear cracking *cannot* be predicted by calculating the principal stresses in an uncracked beam. For this reason, empirical equations have been derived to calculate the flexure-shear cracking load.

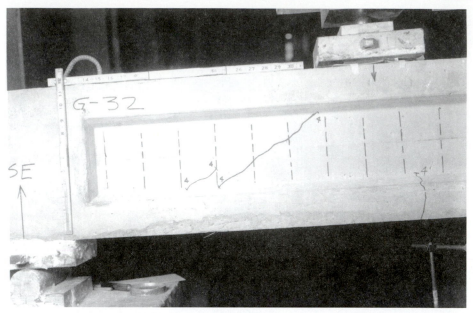

(a) Web-shear cracks.

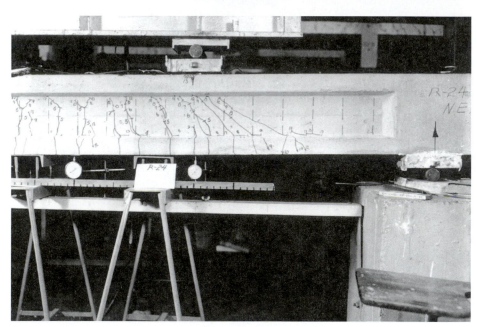

Fig. 6-12
Types of inclined cracks.
(Photographs courtesy of
J. G. MacGregor.)

(b) Flexure-shear cracks.

The inclined cracks in a T beam loaded to produce positive and negative moments are shown in Fig. 5-3. The slope of the inclined cracks in the negative-moment regions changes direction over the support because the shear force changes sign here. All of the inclined cracks in this beam are flexure-shear cracks.

Internal Forces in a Beam without Stirrups

The forces transferring shear across an inclined crack in a beam without stirrups are illustrated in Fig. 6-13. Shear is transferred across line A–B–C by V_{cz}, the shear in the compression

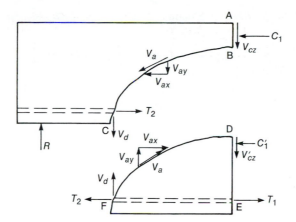

Fig. 6-13
Internal forces in a cracked
beam without stirrups.

zone, by V_{ay}, the vertical component of the shear transferred across the crack by interlock of the aggregate particles on the two faces of the crack, and by V_d, the dowel action of the longitudinal reinforcement. Immediately after inclined cracking, as much as 40 to 60 percent of the total shear is carried by V_d and V_{ay} together [6-3].

Considering the portion D–E–F below the crack and summing moments about the reinforcement at point E shows that V_d and V_a cause a moment about E that must be equilibrated by a compression force C_1'. Horizontal force equilibrium on section A–B–D–E shows that $T_1 = C_1 + C_1'$, and finally, T_1 and $C_1 + C_1'$ must equilibrate the external moment at this section.

As the crack widens, V_a decreases, increasing the fraction of the shear resisted by V_{cz} and V_d. The dowel shear, V_d, leads to a splitting crack in the concrete along the reinforcement (Fig. 6-10a). When this crack occurs, V_d drops, approaching zero. When V_a and V_d disappear, so do V_{cz}' and C_1', with the result that all the shear and compression are transmitted in the width AB above the crack. At this point in the life of the beam, the section A–B is too shallow to resist the compression forces needed for equilibrium. As a result, this region crushes or buckles.

Note also that if $C_1' = 0$, then $T_2 = T_1$, and as a result, $T_2 = C_1$. In other words, the inclined crack has made the tensile force at point C a function of the moment at section A–B–D–E. This shift in the tensile force must be considered in detailing the bar cutoff points and in anchoring the bars.

The shear failure of a slender beam without stirrups is sudden and dramatic. This is evident from Fig. 6-2. Although this beam had stirrups (which have broken and are hanging down from the upper part of the beam), they were so small as to be useless.

Factors Affecting the Shear Strength of Beams without Web Reinforcement

Beams without web reinforcement will fail when inclined cracking occurs or shortly afterwards. For this reason, the shear capacity of such members is taken equal to the inclined cracking shear. The inclined cracking load of a beam is affected by five principal variables, some included in design equations and others not.

Tensile Strength of Concrete.   The inclined cracking load is a function of the tensile strength of the concrete. The stress state in the web of the beam involves the biaxial principal tension and the compression stresses, as shown in Fig. 6-30. (See also Section 3-2.) A similar biaxial state of stress exists in a split-cylinder tension test (Fig. 3-9), and the inclined cracking load is frequently related to the strength from such a test. As discussed earlier, the flexural cracking which precedes the inclined cracking disrupts the elastic-stress field to such an extent that inclined cracking occurs at a principal tensile stress roughly half of f_{ct} for the uncracked section.

Longitudinal Reinforcement Ratio, ρ_w. Figure 6-14 presents the shear capacities of simply supported beams without stirrups as a function of the steel ratio, $\rho_w = A_s/b_w d$. The practical range of ρ_w for beams developing shear failures is about 0.0075 to 0.025. In this range, the shear strength is approximately

$$V_c = 2\sqrt{f_c'}b_w d \quad \text{lb} \tag{6-8}$$

(ACI Eq. 11-3)

or, in SI units,

$$V_c = \frac{\sqrt{f_c'}b_w d}{6} \quad \text{N} \tag{6-8M}$$

as indicated by the horizontal dashed line in Fig. 6-14. This equation tends to overestimate V_c for beams with small steel percentages [6-7].

When the steel ratio, ρ_w, is small, flexural cracks extend higher into the beam and open wider than would be the case for large values of ρ_w. An increase in crack width causes a decrease in the maximum values of the components of shear, V_d and V_{ay}, that are transferred across the inclined cracks by dowel action or by shear stresses on the crack surfaces. Eventually, the sliding resistance along the crack, $V_{ci} = V_{ay} + V_d$, drops below that required to resist the loads, and the beam fails in shear. In a beam without stirrups, the failure is sudden. As a result, inclined cracking occurs earlier.

Shear Span-to-Depth Ratio, a/d. The shear span-to-depth ratio, a/d or M/Vd, affects the inclined cracking shears and ultimate shears of shear spans with a/d less than 2, as shown in Fig. 6-8c. Such shear spans are "deep" shear spans (D-regions) and are discussed in Chapter 18. For longer shear spans, where B-region behavior dominates, a/d has little effect on the inclined cracking shear (Fig. 6-8c) and can be neglected.

Size of Beam. An increase in the overall depth of a beam with very little (or no) web reinforcement results in a decrease in the shear at failure for a given f_c', ρ_w, and a/d. The

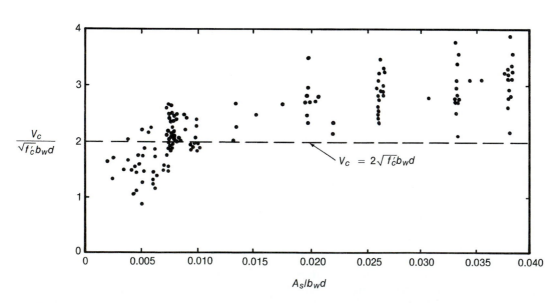

Fig. 6-14
Effect of reinforcement ratio, ρ_w, on shear capacity, V_c, of beams without stirrups. [6-7]

width of an inclined crack depends on the product of the strain in the reinforcement crossing the crack and the spacing of the cracks. With increasing beam depth, the crack spacings and the crack widths tend to increase. (See Section 9-2.) This leads to a reduction in the maximum shear stress, $v_{ci\,max}$, that can be transferred across the crack by aggregate interlock. An unstable situation develops when the shear stresses transferred across the crack exceed the shear strength, $v_{ci\,max}$. When this occurs, the faces of the crack slip, one relative to the other. Figure 6-15 is based on a figure presented by Collins and Kuchma [6-8]. It shows a significant decrease in the shear strengths of geometrically similar, uniformly loaded beams with effective depths d ranging from 4 in. to 118 in. and made with 0.1-in., 0.4-in., and 1-in. maximum size coarse aggregate.

The dashed lines in Fig. 6-15 show the variation in shear strength of beams without stirrups in tests. The beams were uniformly loaded and simply supported as shown in the inset. Each black circular dot in the figure corresponds to the strength of a beam having the section plotted directly below it. None of these beams had web reinforcement.

The open circle labelled Air Force Warehouse Beams refers to the beam in Fig. 6-2. There is very good agreement between the shears at failure and the dashed lines in Fig. 6-15. The horizontal line at $V_u/\sqrt{f'_c}b_w d = 2.0$ shows the shear V_c that ACI 318 assumes to be carried by the concrete. Figure 6-15 shows a very strong size dependency in uniformly loaded beams *without* web reinforcement [6-8].

In beams *with* at least the minimum required web reinforcement, the web reinforcement holds the crack faces together so that the shear transfer across the cracks by aggregate interlock is not lost. As a result, the reduction in shear strength due to size shown in Fig. 6-15 is not observed in beams with web reinforcement [6-8].

Using fracture mechanics, Bazant [6-9] has explained the size effect on the basis of energy release on cracking. The amount of energy released increases with an increase in size,

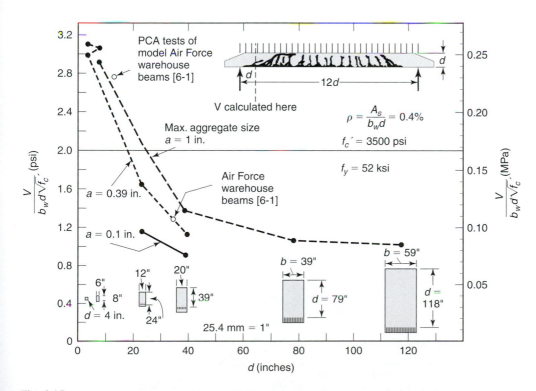

Fig. 6-15
Effect of beam depth, d, on failure shear for beams of various sizes. From [6-8]

particularly in the depth. Although a portion of the size effect results from energy release, the author believes crack-width explanation of size effects fits the test data trends more closely.

Two points plotted in Fig. 6-15 represent the shear stress at failure in the roof beams in the U.S. Air Force warehouse collapse [6-1] shown in Fig. 6-2, and the shear stress at failure of a scale model of these beams. The fact that these points fall close to the experimental data suggests that the U.S. Air Force warehouse failures were strongly affected by size [6-8].

Axial Forces. Axial tensile forces tend to decrease the inclined cracking load, while axial compressive forces tend to increase it (Fig. 6-16). As the axial compressive force is increased, the onset of flexural cracking is delayed, and the flexural cracks do not penetrate as far into the beam. Axial tension forces directly increase the stress, and hence the strain, ϵ_x, in the longitudinal reinforcement. This causes an increase in the inclined crack width, w, which, in turn, results in a decrease in the maximum shear stress, $v_{ci\,max}$, that can be transmitted across the crack. This reduces the shear failure load. The opposite is true for axial compressive forces.

A similar increase is observed in prestressed concrete beams. The compression due to prestressing reduces the londitudinal strain, ε_x, leading to a higher failure load.

Coarse Aggregate Size. As the size (diameter) of the coarse aggregate increases, the roughness of the crack surfaces increases, allowing higher shear stresses to be transferred across the cracks. As shown in Fig. 6-15, a beam with 1-in. coarse aggregate and 40-in. effective depth failed at about 150 percent of the failure load of a beam with $d = 40$ in. and 0.1-in. maximum aggregate size. In high-strength concrete beams and some lightweight concrete beams, the cracks penetrate pieces of the aggregate rather than going around them, resulting in a smoother crack surface. This decreases the shear transferred by aggregate interlock along the cracks, thereby decreasing V_c [6-8].

Behavior of Beams with Web Reinforcement

Inclined cracking causes the shear strength of beams to drop below the flexural capacity, as shown in Fig. 6-8b and c. The purpose of web reinforcement is to ensure that the full flexural capacity can be developed.

Prior to inclined cracking, the strain in the stirrups is equal to the corresponding strain of the concrete. Since concrete cracks at a very small strain, the stress in the stirrups

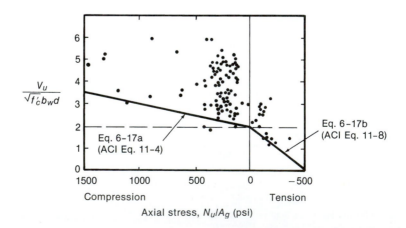

Fig. 6-16
Effect of axial loads on inclined cracking shear.
From [6-7].

prior to inclined cracking will not exceed 3 to 6 ksi. Thus, stirrups do not prevent inclined cracks from forming; they come into play only after the cracks have formed.

The forces in a beam with stirrups and an inclined crack are shown in Fig. 6-17. The terminology is the same as in Fig. 6-13. The shear transferred by tension in the stirrups, V_s, does not disappear when the crack opens, so there will always be a compression force C_1' and a shear force V_{cz}' acting on the part of the beam below the crack. As a result, T_2 will be less than T_1, the difference depending on the amount of web reinforcement. The force T_2 will, however, be larger than the flexural tension $T = M/jd$ based on the moment at C.

The loading history of such a beam is shown qualitatively in Fig. 6-18. The components of the internal shear resistance must equal the applied shear, indicated by the upper 45° line. Prior to flexural cracking, all the shear is carried by the uncracked concrete. Between

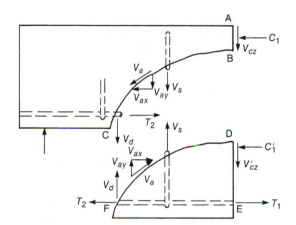

Fig. 6-17
Internal forces in a cracked beam with stirrups.

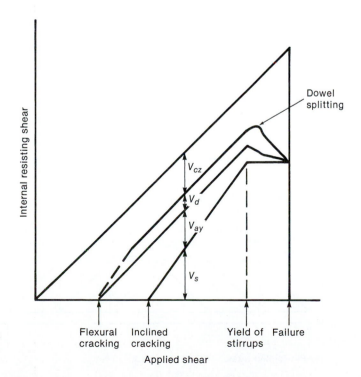

Fig. 6-18
Distribution of internal shears in a beam with web reinforcement.
From [6-5].

flexural and inclined cracking, the external shear is resisted by V_{cz}, V_{ay}, and V_d. Eventually, the stirrups crossing the crack yield, and V_s stays constant for higher applied shears. Once the stirrups yield, the inclined crack opens more rapidly. As the inclined crack widens, V_{ay} decreases further, forcing V_d and V_{cz} to increase at an accelerated rate, until a splitting (dowel) failure occurs, the compression zone crushes due to combined shear and compression, or the web crushes.

Each of the components of this process except V_s has a brittle load–deflection response. As a result, it is difficult to quantify the contributions of V_{cz}, V_d, and V_{ay}. In design, these are lumped together as V_c, referred to somewhat incorrectly as "the shear carried by the concrete." Thus, the nominal shear strength, V_n, is assumed to be

$$V_n = V_c + V_s \tag{6-9}$$

Traditionally in North American design practice, V_c is taken equal to the failure capacity of a beam without stirrups, which, in turn, is taken equal to the inclined cracking shear, as suggested by the line indicating inclined cracking and failure for a/d from 2.5 to 6.5 in Fig. 6-8c. This is discussed more fully in Section 6-5.

6-4 TRUSS MODEL OF THE BEHAVIOR OF SLENDER BEAMS FAILING IN SHEAR

The behavior of beams failing in shear must be expressed in terms of a mechanical–mathematical model before designers can make use of this knowledge in design. The best model for beams with web reinforcement is the truss model. This is applied to slender beams in this chapter and to deep beams in Chapter 18.

In 1899 and 1902, respectively, the Swiss engineer Ritter and the German engineer Mörsch, independently, published papers proposing the truss analogy for the design of reinforced concrete beams for shear. These procedures provide an excellent conceptual model to show the forces that exist in a cracked concrete beam.

As shown in Fig. 6-19a, a beam with inclined cracks develops compressive and tensile forces, C and T, in its top and bottom "flanges," vertical tensions in the stirrups, and inclined

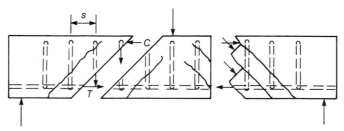

(a) Internal forces in a cracked beam.

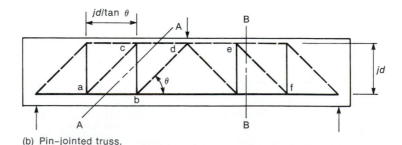

Fig. 6-19
Truss analogy.

(b) Pin–jointed truss.

compressive forces in the concrete "diagonals" between the inclined cracks. This highly indeterminate system of forces is replaced by an analogous truss. The simplest truss is shown in Fig. 6-19b; a more complicated truss is shown in Fig. 6-20b.

Several assumptions and simplifications are needed to derive the analogous truss. In Fig. 6-19b, the truss has been formed by lumping all of the stirrups cut by section A–A into one vertical member b–c and all the diagonal concrete members cut by section B–B into one diagonal member e–f. This diagonal member is stressed in compression to resist the shear on section B–B. The compression chord along the top of the truss is actually a force in the concrete but is shown as a truss member. The compressive members in the truss are shown with dashed lines to imply that they are really forces in the concrete, not separate truss members. The tensile members are shown with solid lines.

Figure 6-20a shows a beam with inclined cracks. The left end of this beam can be replaced by the truss shown in Fig. 6-20b. In design, the ideal distribution of stirrups would correspond to all stirrups reaching yield by the time the failure load is reached. It will be assumed, therefore, that all the stirrups have yielded and that each transmits a force of $A_v f_y$ across the crack, where A_v is the area of the stirrup legs. When this is done, the truss becomes statically determinate. The truss in Fig. 6-20b is referred to as the *plastic-truss model*, since we are depending on plasticity in the stirrups to make it statically determinate. The beam will be proportioned so that the stirrups yield before the concrete crushes, so that it will not depend on plastic action in the concrete.

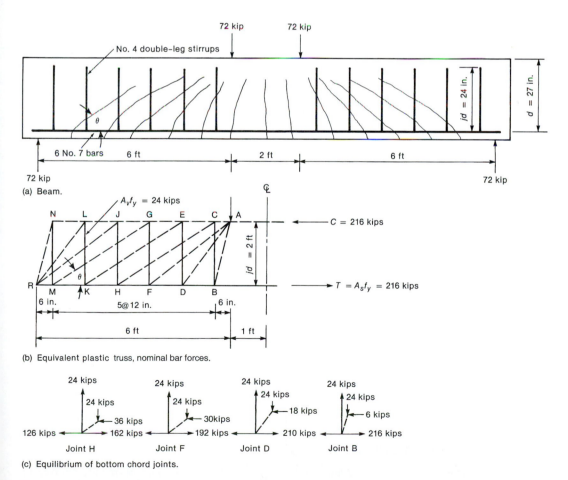

Fig. 6-20
Construction of an analogous plastic truss—Example 6-1.

This truss model ignores the shear components V_{cz}, V_{ay}, and V_d in Fig. 6-17. Thus it does not assign any shear "to the concrete." A truss analogy that includes such a term will be discussed briefly later.

EXAMPLE 6-1   Construction of a Plastic-Truss Model

A beam with cross section 30 in. high by 12 in. wide carries unfactored concentrated loads of 30 kips dead load and 11.25 kips live load acting at points A located 6 ft from each end support. Draw a plastic-truss model for the beam shown in Fig 6-20a. Compute the forces in the tension and compression chords.

Assume that:

1.   All the shear is resisted by the stirrups.

2.   The factored weight of the beam can be ignored or, as discussed later, it can be added to the concentrated loads at A and R.

3.   The beam is on the verge of a simultaneous shear and flexural failure. Thus it is assumed that all of the stirrups have yielded, each carrying equal vertical forces of $A_v f_y = 0.4$ in.^{2} × 60 ksi = 24 kips.

The factored concentrated loads are

$$P_u = 1.2 \times 30 + 1.6 \times 11.25 = 54 \text{ kips at each point } A.$$

The factored self weight of the beam is

$$w_u = 1.2(2.5 \times 1 \times 0.15) = 0.45 \text{ kips/ft}$$

$$\phi V_n \geq V_u \text{ and } V_n = V_u/\phi = 54/0.75 = 72 \text{ kips}.$$

Thus, if we ignore the self weight, the nominal shear is 72 kips.

The moment at midspan is 72 kips × 6 ft = 432 ft-kips. Assuming that $jd = (d - a/2)$ is 2 ft, the compression and tension forces C and T at midspan are $M/jd = 432/2 = 216$ kips as shown in Fig. 6-20b.

The factored vertical applied load of 72 kips at A must be transferred by diagonal compressive struts (shown by dashed inclined lines in Fig. 6-20b) to enough stirrups (shown by solid vertical lines) to equilibrate this force. Since each stirrup can resist a vertical force of $A_v f_y = 24$ kips, three stirrups are required to equilibrate 72 kips. The vertical applied load of 72 kips will be transmitted by the three diagonals AB, AD, and AF to joints B, D, and F at the bottom of the truss. The right-hand diagram in Fig. 6-20c shows the equilibrium of joint B.

The vertical force in the stirrup is 24 kips. The vertical component of the force in the diagonal AB must be 24 kips to satisfy equilibrium. From the slope of AB, we find that it has a horizontal component of 6 kips. At midspan, the tension force, T, in the reinforcement at the bottom of the beam is 216 kips. Summing horizontal forces at joint B, we find that the force in BD is 210 kips. Carrying out similar summations at each of the lower chord joints gives the distribution of force in the lower chord shown by the shaded area in Fig. 6-21a. The forces in the lower chord from beam theory are given by the dashed line labeled $T = M/jd$. Note that the computed forces in the lower chord exceed those from beam theory. This shift will be discussed later in this section and in Chapter 8.

The stirrup BC transmits the vertical force of $A_v f_y = 24$ kips to the top of the truss at joint C, where it is resisted by the vertical component of the force in diagonal CH, and so on. Repeating the process for the top chord gives the distribution of compression forces shown shaded in Fig. 6-21b. ∎

If one were to use the truss model in design, the strut and tie forces would be computed for *applied load* shears of $\phi V_n \geq P_u$ where $P_u = 54/0.75 = 72.0$ kips.

Two ways of including the *dead load* are:

1.   Apply half of the dead load at each of the load points A, or

2.   Apply $P_{Du} = (1.0 \text{ ft} \times w_D)/\phi = 0.60$ kips at the top of each stirrup. For this to work, the areas of the stirrups must increase from panel to panel so that the stirrups have a yield force equivalent to $V_u + \Sigma 0.6$ kips.

The compression diagonals originating at the load (AB, AD, and AF) are referred to as a *compression fan*. The number of diagonal struts in the fan must be such that the entire vertical load at A is resisted by the vertical force components in the diagonals meeting at A.

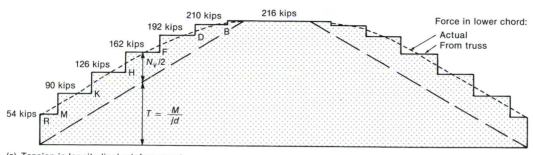

(a) Tension in longitudinal reinforcement.

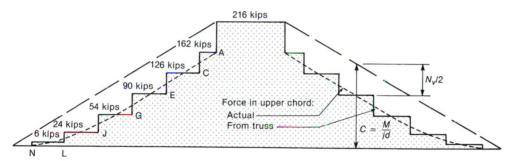

(b) Compression in upper chord.

Fig. 6-21
Forces in lower and upper chords of truss in Fig. 6-20.

A similar compression fan exists at the support R (*RN, RL, RJ*). Between the compression fans is a *compression field* consisting of the parallel diagonal struts *CH, EK,* and *GM*. The angle θ of the compression field is determined by the number of stirrups needed to equilibrate the vertical loads in the fans.

Each of the compression fans occurs in a D-region (discontinuity region). The compression field is a B-region (beam region).

Figure 6-22a shows the crack pattern in a two-span continuous beam. The corresponding truss model is shown in Fig. 6-22b. Figure 6-23 is a close-up of the compression fan over the interior support after failure. The radiating struts in the fan can be clearly seen.

Simplified Truss Analogy

A statically determinate truss analogy can be derived via the method suggested by Marti [6–11], [6–12]. Figures 6-24a and b show a uniformly loaded beam with stirrups and a truss model incorporating all the stirrups and representing the uniform load as a series of concentrated loads at the panel points. The truss in Fig. 6-24b is statically indeterminate, but can be solved if it is assumed that the forces in each stirrup cause that stirrup to just reach yield, as was done in the preceding paragraphs. For design, it is easier to represent the truss as shown in Fig. 6-24c, where the tension force in each vertical member represents the force in all the stirrups within a length $jd \cot \theta$. Similarly, each inclined compression strut represents a width of web equal to $jd \cos \theta$. The uniform load has been idealized as concentrated loads of $w(jd \cot \theta)$ acting at the panel points. The truss in Fig. 6-24c is statically determinate. To draw such a truss, it is necessary to choose θ. This will be discussed later.

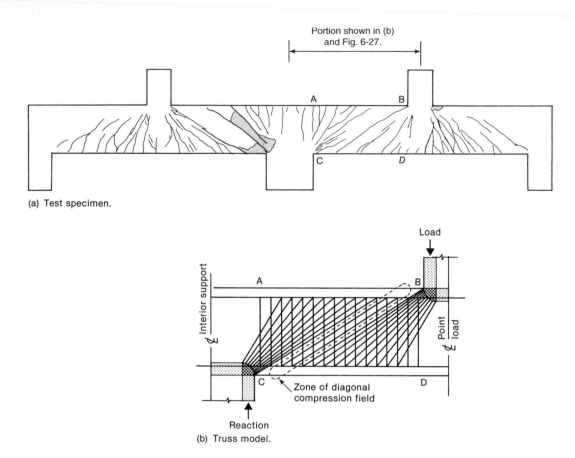

(a) Test specimen.

(b) Truss model.

Fig. 6-22
Crack pattern and truss model for a two-span beam. (From [6-10].)

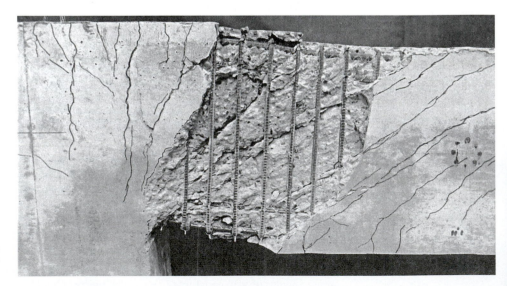

Fig. 6-23
Compression fan at interior support of the beam shown in Fig. 6-22b. (Photograph courtesy of J. G. MacGregor.)

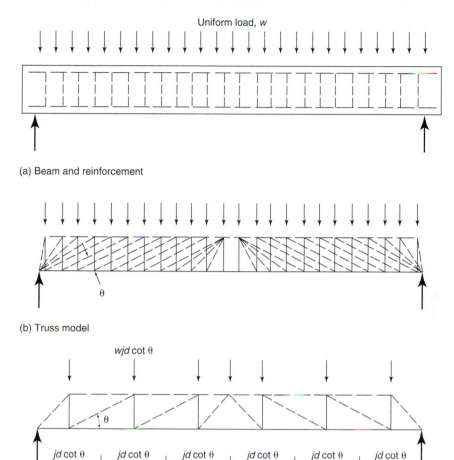

Fig. 6-24
Truss model for design.
(From Collins/Mitchell, *Pre-stressed Concrete Structures*, (c) 1990, p. 339. [6-13] Reprinted by permission of Prentice Hall, Upper Saddle River, New Jersey.)

(a) Beam and reinforcement

(b) Truss model

(c) Statically determinate truss

Internal Forces in the Plastic-Truss Model

If we consider the free-body diagram cut by section A–A parallel to the diagonals in the compression field region in Fig. 6-25a, the entire vertical component of the shear force is resisted by tension forces in the stirrups crossing this section. The horizontal projection of section A–A is $jd \cot \theta$, and the number of stirrups it cuts is $jd \cot \theta/s$. The force in one stirrup is $A_v f_y$, which can be calculated from

$$A_v f_y = \frac{V_s}{jd \cot \theta} \tag{6-10}$$

The free body shown in Fig. 6-25b is cut by a vertical section between G and J in Fig. 6-20b. Here, the vertical force, V, acting on the section must be resisted by an inclined compressive force $D = V \sin \theta$ in the diagonals (Fig. 6-25c). The width of the diagonals is $jd \cos \theta$, as shown in Fig. 6-25b, and the average compressive stress in the diagonals is

$$f_2 = \frac{V}{b_w jd \cos \theta \sin \theta} \tag{6-11a}$$

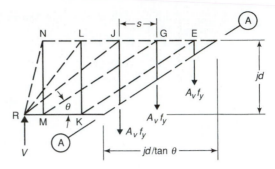

(a) Calculation of forces in stirrups.

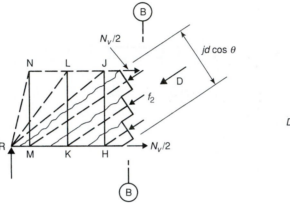

Fig. 6-25
Forces in stirrups and compression diagonals.

(b) Calculation of stress in compression diagonals.

(c) Replacement of V with internal forces of D and N.

with the use of trigonometric identities, this equation becomes

$$f_2 = \frac{V}{b_w jd}\left(\tan\theta + \frac{1}{\tan\theta}\right) \tag{6-11b}$$

where b_w is the thickness of the web. If the web is very thin, this stress may cause the web to crush, as shown in Fig. 6-26.

The shear V on section B–B has been replaced by the diagonal compression force

$$D = \frac{V}{\sin\theta} \tag{6-12}$$

and an axial tension force

$$N_v = \frac{V}{\tan\theta} \tag{6-13}$$

as shown in Fig. 6-25c.

If it is assumed that the shear stress is constant over the height of the beam, the resultants of D and N_v act at midheight. As a result, a tensile force of $N_v/2$ acts in each of the top and bottom chords, as shown in Fig. 6-27b. This reduces the force in the compression chord and increases the force in the tension chord.

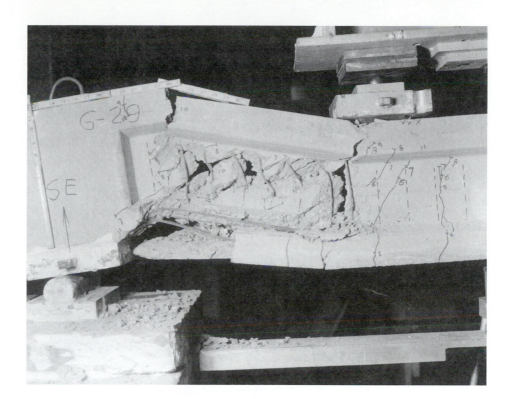

Fig. 6-26
Web crushing failure.
(Photograph courtesy of
J. G. MacGregor.)

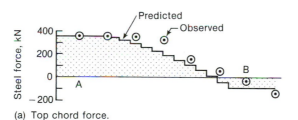

(a) Top chord force.

Fig. 6-27
Measured and computed
forces in the top and bottom
reinforcement of the portion
of the beam modeled in
Fig. 6-22b. From [6-10].

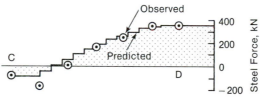

(b) Bottom chord force.

In the compression-fan regions, the angle θ varies, and hence N_v varies, approaching zero immediately under the load. The distribution of forces in the tension chord and compression chord of the truss in Fig. 6-22 are shown in Fig. 6-21. The force distribution, C or $T = M/jd$, due to flexure is shown by the dashed lines.

In the compression-field region (F to R on the lower chord and A to J on the upper chord), the force in the tension chord halfway between each panel point is larger than $T = M/jd$ by the amount $N_v/2$, and the force in the compression chord is smaller than $C = M/jd$ by the same amount, as shown by the short dashed lines in Fig. 6-21. For this truss, $\cot \theta = 1.5$, and hence $N_v/2 = 0.75V$, which, for $V = 72$ kips, is 54 kips. At the end of the beam, it is necessary to anchor the longitudinal bars for a tension force of $N_v/2$ even though the moment is zero.

In the compression-fan region under the load, the value $N_v/2$ in the tension chord gradually reduces to zero, as shown in Fig. 6-21a, so that, at the point of maximum moment, the force in the reinforcement is $T = M/jd$. The shift in the tension-force diagram is equivalent to computing T from a moment diagram that has been shifted away from the points of maximum moment by an amount $(jd \cot \theta)/2$. This is discussed more fully in Chapter 8.

Figure 6-27 compares the measured and computed forces in the top and bottom bars in the beam shown in Fig. 6-22. The diagrams correspond to portion A–B–C–D of the beam modeled in Fig. 6-22b. The truss model accurately predicted the longitudinal bar forces. The radiating struts in the compression fan are clearly visible in Fig. 6-23.

Value of θ in Compression Field Region

When a reinforced concrete beam with stirrups is loaded to failure, inclined cracks initially develop at an angle of 35 to 45° with the horizontal. With further loading, the angle of the compression stresses may cross some of the cracks [6-14]. For this to occur, aggregate interlock must exist.

The allowable range of θ is expressed as $0.5 \leq \cot \theta \leq 2.0$ ($\theta = 26$ to $64°$) in the Swiss code [6-15]. This range was selected to limit crack widths. Only a more restricted range, $\frac{3}{5} \leq \cot \theta \leq \frac{5}{3}$ ($\theta = 31$ to $59°$), is allowed in the European Concrete Committee's Model Code [6-16].

In design, the value of θ should be in the range $25° \leq \theta \leq 65°$. The choice of a small value of θ reduces the number of stirrups required (6-10), but increases the compression stresses in the web (6-11) and increases N_v and, hence, the shift in the moment diagram. The opposite is true for large angles.

In the *analysis* of a given beam, as done in Figs. 6-20, 6-21, and 6-22, the angle θ is determined by the number of stirrups needed to equilibrate the applied loads and reactions. The angle should be within the limits given, except in compression-fan regions. In the *design* of a beam, the crack angle is a free choice which leads to values of the other unknowns.

Crushing Strength of Concrete in the Web

The web of the beam will crush if the inclined compressive stress, f_2 from (6-11), exceeds the strength of the concrete. The compressive strength, $f_{2\,\text{max}}$, of the concrete in a web that has previously been cracked and that contains stirrups stressed in tension at an angle to the cracks will tend to be less than f_c', as was explained in Section 3-2. A reasonable limit is $0.25f_c'$ for $\theta = 30°$, increasing to $0.45f_c'$ for $\theta = 45°$. This problem is discussed more fully in [6-14] and in Section 18-2.

6-5 ANALYSIS AND DESIGN OF REINFORCED CONCRETE BEAMS FOR SHEAR—ACI CODE

In the ACI Code, the basic design equation for the shear capacity of slender concrete beams (beams with shear spans containing B-regions) is

$$\phi V_n \geq V_u \tag{6-14}$$

(ACI Eq. 11-1)

where V_u is the shear force due to the factored loads, ϕ is a strength-reduction factor, taken equal to 0.75 for shear, provided the load factors are from ACI 318-02 Section 9.2. It is taken equal to 0.85 when the load factors from ACI Section C.2 are used. The nominal shear resistance is

$$V_n = V_c + V_s \tag{6-9}$$

(ACI Eq. 11-2)

where V_c is the shear carried by the concrete and V_s is the shear carried by the stirrups.

A shear failure is said to occur when one of several shear limit states is reached. The following paragraphs list the principal limit states and describe how these are accounted for in the ACI Code.

Shear-Failure Limit States: Beams without Web Reinforcement

Slender beams without web reinforcement will fail when inclined cracking occurs or shortly afterward. For this reason, the shear strength of such members is taken equal to the inclined cracking shear. The factors affecting the inclined cracking load were discussed in Section 6-3.

Design Equations for the Shear Strength of Members without Web Reinforcement

In 1962, the ACI–ASCE Committee on Shear and Diagonal Tension [6-17] presented an equation for calculating the shear at inclined cracking in beams without web reinforcement:

$$V_c = \left(1.9\sqrt{f_c'} + \frac{2500\rho_w V_u d}{M_u}\right) b_w d \text{ (psi)} \tag{6-15}$$

(ACI Eq. 11-5)

The derivation of this equation followed two steps. First, a rudimentary analysis of the stresses at the head of a flexural crack in a shear span was carried out to identify the significant parameters. Then, the existing test data were statistically analyzed to establish the constants, 1.9 and 2500, and to drop other terms. The data used in the statistical analysis included "short" and "slender" beams, thereby mixing data from two different behavior types. In addition, most of the beams had high reinforcement ratios, ρ_w. More recent studies have suggested that (6-15) underestimates the effect of ρ_w for beams without web reinforcement and is not entirely correct in its treatment of the variable a/d, expressed as $V_u d/M_u$ in the equation.

For the normal range of variables, the second term in the parentheses in (6-15) will be equal to about $0.1\sqrt{f_c'}$. If this value is substituted into (6-15), then the following equation results:

$$V_c = 2\sqrt{f_c'} b_w d \tag{6-8}$$

(ACI Eq. 11-3)

In 1977, the ACI–ASCE Committee on Shear and Diagonal Tension recommended that (6-15) no longer be used for the reasons given in the paragraph after (6-15) [6-7]. For this reason, it will not be employed in this book.

On the basis of statistical studies of beam data for slender beams without web reinforcement, Zsutty [6-18] derived the following equation, which much more closely models the actual effects of f'_c, ρ_w, and a/d than does (6-15):

$$v_c = 59\left(f'_c \rho_w \frac{d}{a}\right)^{1/3} \text{ psi} \tag{6-16}$$

For design, the ACI Code presents both (6-8) and (6-15) for computing V_c (ACI Sections 11.3.1.1 and 11.3.2.1).

For axially loaded members, the ACI Code modifies (6-8) as follows:

Axial compression (ACI Section 11.3.1.2):

$$V_c = 2\left(1 + \frac{N_u}{2000A_g}\right)\sqrt{f'_c}\, b_w d \tag{6-17a}$$

$$\text{(ACI Eq. 11-4)}$$

Axial tension (ACI Section 11.3.2.3):

$$V_c = 2\left(1 + \frac{N_u}{500A_g}\right)\sqrt{f'_c}\, b_w d \tag{6-17b}$$

$$\text{(ACI Eq. 11-8)}$$

In both of these equations, N_u is positive in compression and $\sqrt{f'_c}$, N_u/A_g, 500, and 2000 all have units of psi. Axially loaded members are discussed more fully in Section 6-8.

Circular cross sections. Members with circular cross sections, such as some columns, may have to be designed for shear. When circular ties or spirals are used as web reinforcement, the calculation of V_c can be based on (6-8), (6-17a), and (6-17b), with b_w taken equal to the diameter of the circular section and d taken equal to the distance from the extreme compression fiber to the centroid of the longitudinal tension reinforcement, but need not be taken greater than $0.8h$, and A_v is taken equal to twice the area, A_b of the bar used as a circular tie or as a spiral. (See ACI Sections 11.0 and 11.3.3.) These definitions are based on tests reported in [6-19] and [6-20].

There is a tendency for ties located close to the inside surface of a hollow member to straighten and pull out through the inside surface of the tube. Means of preventing this must be considered in the design of hollow members.

In SI units, (6-8), (6-17a), and (6-17b) become, respectively,

$$V_c = \frac{\sqrt{f'_c}}{6} b_w d \tag{6-8M}$$

$$\text{(ACI Eq. 11-3M)}$$

For Axial compression

$$V_c = \left(1 + \frac{N_u}{14A_g}\right)\left(\frac{\sqrt{f'_c}}{6}\right) b_w d \tag{6-17aM}$$

$$\text{(ACI Eq. 11-4M)}$$

and for axial tension

$$V_c = \left(1 + \frac{0.3N_u}{A_g}\right)\left(\frac{\sqrt{f'_c}}{6}\right) b_w d \tag{6-17bM}$$

$$\text{(ACI Eq. 11-8M)}$$

where N_u is positive in compression and the terms $\sqrt{f'_c}$, $N_u/14A_g$, and 0.3 all have units of MPa.

Shear Failure Limit States: Beams with Web Reinforcement

1.   Failure due to yielding of the stirrups.   In Fig. 6-17, shear was transferred across the surface A–B–C by shear in the compression zone, V_{cz}, by the vertical component of the aggregate interlock, V_{ay}, by dowel action, V_d, and by stirrups, V_s. In the ACI Code V_{cz}, V_{ay}, and V_d are lumped together as V_c, which is referred to as the "shear carried by the concrete." Thus, the nominal shear strength, V_n, is assumed to be

$$V_n = V_c + V_s \qquad (6\text{-}9)$$

$$(\text{ACI Eq. 11-2})$$

The ACI Code further assumes that V_c is equal to the shear strength of a beam without stirrups, which, in turn, is taken equal to the inclined cracking load, as given by (6-8), (6-15), or (6-17). It should be emphasized that taking V_c equal to the shear at inclined cracking is an *empirical* observation from tests, which is *approximately true* if it is assumed that the horizontal projection of the inclined crack is d, as shown in Fig. 6-28. If a

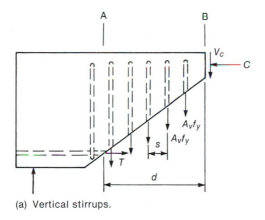

(a) Vertical stirrups.

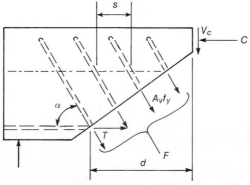

Fig. 6-28
Shear resisted by stirrups.

(b) Inclined stirrups.

flatter crack is used, so that $jd \cot \theta$ is greater than d, a smaller value of V_c must be used. For θ approaching $30°$ in the plastic truss model, V_c approaches zero, as assumed in that model.

Figure 6-28a shows a free body between the end of a beam and an inclined crack. The horizontal projection of the crack is taken as d, suggesting that the crack is slightly flatter than $45°$. If s is the stirrup spacing, the number of stirrups cut by the crack is d/s. Assuming that all the stirrups yield at failure, the shear resisted by the stirrups is

$$V_s = \frac{A_v f_y d}{s} \tag{6-18}$$

$$(\text{ACI Eq. 11-15})$$

This is equivalent to (6-10), derived from the truss model, if $\theta = 45°$ and jd is replaced by d.

If the stirrups are inclined at an angle α to the horizontal, as shown in Fig. 6-28b, the number of stirrups crossing the crack is approximately $d(1 + \cot \alpha)/s$, where s is the horizontal spacing of the stirrups. The inclined force is

$$F = A_v f_y \left[\frac{d(1 + \cot \alpha)}{s} \right] \tag{6-19}$$

The shear resisted by the stirrups, V_s, is the vertical component of F, which is $F \sin \alpha$, so that

$$V_s = A_v f_y \left(\sin \alpha + \cos \alpha \right) \frac{d}{s} \tag{6-20}$$

$$(\text{ACI Eq. 11-16})$$

Figures 6-28 and 6-17 also show that the inclined crack affects the longitudinal tension force, T, making it larger than the moment diagram would suggest. This is immediately obvious from the truss analogy as shown in Fig. 6-29, but is less obvious in the ACI design method.

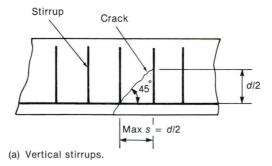

(a) Vertical stirrups.

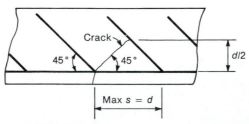

Fig. 6-29
Maximum spacing of stirrups.   (b) Inclined stirrups.

If V_u exceeds ϕV_c, stirrups must be provided so that

$$V_u \leq \phi V_n \tag{6-14}$$

(ACI Eq. 11-1)

where V_n is given by (6-9). In design, this is generally rearranged to the form

$$\phi V_s \geq V_u - \phi V_c \quad \text{or} \quad V_s \geq \frac{V_u}{\phi} - V_c$$

Introducing (6-18) gives the stirrup spacing:

$$s = \frac{A_v f_y d}{V_u/\phi - V_c} \tag{6-21}$$

This equation applies only to vertical stirrups.

Stirrups are unable to resist shear unless they are crossed by an inclined crack. For this reason, ACI Section 11.5.4.1 sets the maximum spacing of vertical stirrups as the smaller of $d/2$ or 24 in., so that each 45° crack will be intercepted by at least one stirrup (Fig. 6-29a). The maximum spacing of inclined stirrups is such that a 45° crack extending from midheight of the member to the tension reinforcement will intercept at least one stirrup, as shown in Fig. 6-29b.

If $V_u/\phi - V_c = V_s$ exceeds $4\sqrt{f_c'}b_w d$, the maximum allowable stirrup spacings are reduced to half those just described. For vertical stirrups, the maximum is the smaller of $d/4$ or 12 in. This is done for two reasons. Closer stirrup spacing leads to narrower inclined cracks and provides better anchorage for the lower ends of the compression diagonals.

In a wide beam with stirrups around the perimeter, the diagonal compression in the web tends to be supported by the bars in the corners of the stirrups, as shown in Fig. 6-30a. The situation is improved if there are more than two stirrup legs, as shown in Fig. 6-30b. ACI Commentary Section R11.5.6 suggests that the transverse spacing of stirrup legs in wide beams should be limited to a fraction of the width by placing several overlapping stirrups. The *CEB–FIB Model Code 1990* [6-16] suggests that the maximum transverse spacing of stirrup legs should be limited to $2d/3$ or 800 mm (32 in.), whichever is smaller.

2. Shear failure initiated by failure of the stirrup anchorages. Equations (6-21) and (6-18) are based on the assumption that the stirrups will yield at ultimate. This will be true only if the stirrups are well anchored. Generally, the upper end of the inclined crack approaches very close to the compression face of the beam, as shown in Figs. 6-4, 6-30, and 6-31. At ultimate, the stress in the stirrups approaches or equals the yield strength, f_y, at every

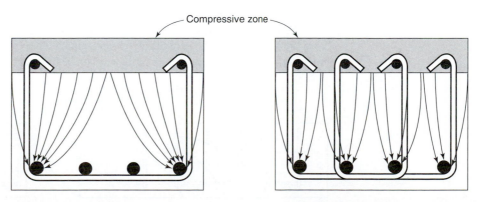

Compressive zone

Fig. 6-30
Flow of diagonal compression force in the cross sections of beams with stirrups.

(a) Widely spaced stirrup legs

(b) Closely spaced stirrup legs

point where an inclined crack intercepts a stirrup. Thus the portions of the stirrups shown shaded in Fig. 6-31 must be able to anchor f_y. For this reason, ACI Section 12.13.1 both requires that the stirrups extend as close to the compression and tension faces as cover and bar spacing requirements permit and, in addition, specifies certain types of hooks to anchor the stirrups. The ACI Code requirements for stirrup anchorage are illustrated in Fig. 6-32:

(a) ACI Section 12.13.3 requires that each bend away from the ends of a stirrup enclose a longitudinal bar, as shown in Fig. 6-32a.

(b) For No. 5 bar or D31 wire stirrups and smaller with any yield strength, ACI Section 12.13.2.1 allows the use of a standard hook around longitudinal reinforcement without any specified embedment length. The hooks may be 90°, 135°, or 180°, as shown in Fig. 6-32b. Either 135° or 180° hooks are preferred.

(c) For No. 6, 7, and 8 stirrups with f_y of 40,000 psi, ACI Section 12.13.2.1 allows the details shown in Fig. 6-32b.

(d) For No. 6, 7, and 8 stirrups with f_y greater than 40,000 psi, ACI Section 12.13.2.2 requires a standard hook around a longitudinal bar plus embedment between midheight of the member and the outside of the hook of at least $0.014 d_b f_y / \sqrt{f'_c}$.

(e) Requirements for welded-wire fabric stirrups formed of sheets bent in U shape or vertical flat sheets are illustrated in ACI Commentary Fig. 12.13.2.3.

(f) In deep members, particularly where the depth varies gradually, it is sometimes advantageous to use lap-spliced stirrups, described in ACI Section 12.13.5 and shown in Fig. 6-32c. This type of stirrup has proven unsuitable in seismic areas.

(g) ACI Section 7.11 requires closed stirrups in beams with compression reinforcement, beams subjected to stress reversals, and beams subjected to torsion.

(h) ACI Section 7.13.2.3 requires U-stirrups with 135° hooks around longitudinal bars in all perimeter beams. Closed stirrups may be constructed as shown in Fig. 6-32d. Such a stirrup would not fulfill the force transfer intended in ACI Sections 7.13.2.2 and 7.13.2.3 to provide some measure of structural integrity by resisting upward forces in severely cracked beams. This is because the cap bar in Fig. 6-32d cannot be counted on to resist upward load, see Section 2-3.

(i) Although it is not mentioned in ACI 318-02, stirrups may be anchored by mechanical anchorages capable of developing the yield strength of the bars. One possibility is the use of bars welded or forged to end plates, which serve to anchor the bars. This detail, known as a *headed bar*, is useful in regions where large amounts of shear reinforcement are required—for example, in the walls of an offshore oil production structure—or where member depth is inadequate to anchor stirrups—for example, in slabs, see Section 13-7.

Standard hooks and the development length ℓ_d are discussed in Chapter 8 of this book and in ACI Sections 7.1.3 and 12.2.

Standard stirrup hooks are bent around a smaller-diameter pin than normal bar bends. Very-high-strength steels may develop small cracks during this bending operation. These cracks may in turn lead to fracture of the bar before the yield strength can be developed. For

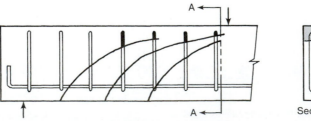

Fig. 6-31
Anchorage of stirrups.

Section A-A

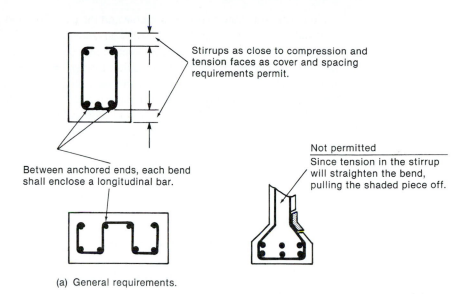

Stirrups as close to compression and tension faces as cover and spacing requirements permit.

Between anchored ends, each bend shall enclose a longitudinal bar.

Not permitted
Since tension in the stirrup will straighten the bend, pulling the shaded piece off.

(a) General requirements.

Standard stirrup hook, ACI Section 7.1.3, Must enclose a bar, ACI Section 12.13.2.1

(b) Stirrup anchorage requirements for No. 5 and smaller bars as per ACI Sections 7.1.3 and 12.13.2.1. 135° or 180° hooks are preferred.

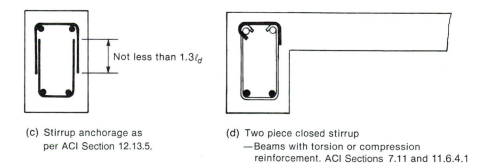

Not less than $1.3\ell_d$

(c) Stirrup anchorage as per ACI Section 12.13.5.

(d) Two piece closed stirrup —Beams with torsion or compression reinforcement. ACI Sections 7.11 and 11.6.4.1

Fig. 6-32
Stirrup detailing requirements.

this reason, ACI Section 11.5.2 limits the yield strength used in design calculations to 60,000 psi, except for welded-wire fabric stirrups, for which the limit is 80,000 psi. This difference is justified by the fact that the bend test for the wire used to make welded-wire fabric is more stringent than that for bars. In addition, welded-wire stirrups tend to be more closely spaced than stirrups made from reinforcing bars and give better control of inclined-crack widths.

Because the anchorage length available between the inclined crack and the top of the stirrup is generally very short, as shown in Fig. 6-31, the author recommends both the use of the smallest-diameter stirrups possible and the use of 40,000-psi steel in stirrups where this grade is available. This has the added advantage that the more closely spaced stirrups will help prevent excessively wide inclined cracks, and the lower-yield strength is easier to develop. In addition, it is recommended that 135° stirrup hooks be used throughout, except in very narrow beams, where 180° stirrups may be needed to avoid having the tails of the hooks cross, which would make it difficult to place the longitudinal bars inside the stirrups.

For a beam with 135° stirrup hooks, as shown in the middle cross section in Fig. 6-32b, note that ACI Section 7.1.3(a) and (c) define stirrup hooks for No. 5 and smaller bars, the usual sizes of bars used for stirrups, as either a 90° bend plus $6d_b$ extension at the free end of the bar or a 135° bend plus $6d_b$ extension. Note also that ACI Section 7.2.2 sets the minimum inside diameter of a stirrup bend as $4d_b$. The minimum beam width to avoid crossing of the 135° tails of the stirrup hooks No. 3 stirrups with 135° stirrup hooks can be computed as follows:

- cover to outside of the stirrups = 1.5 in.
- outside radius of a No. 3 stirrup bend = $(4 + 2) \times 0.375$ in./2 = 1.125 in.
- width of bend from 90° to 135° = $1.125/\sqrt{2}$ = 0.80 in.
- bar extension = 6×0.375 in./$\sqrt{2}$ = 1.60 in.

This totals 5.03 in. on each side of the beam. Leaving a clear gap of about 2 in. to lower the longitudinal steel through gives a minimum beam width of 12 in. when using 135° stirrup bends. For 90° stirrup bends, the corresponding total is 4.88 in. each side, also giving a minimum beam width of 12 in.

3. **Serviceability failure due to excessive crack widths at service loads.** Wide inclined cracks in beams are unsightly and may allow water to penetrate the beam, possibly causing corrosion of the stirrups. In tests of three similar beams [6-21], the maximum service-load crack width in a beam with the shear reinforcement provided entirely by bent-up bars was 150 percent of that in a beam with vertical stirrups. The maximum service-load crack width in a beam with inclined stirrups was only 80 percent of that in a beam with vertical stirrups. In addition, the crack widths were less with closely spaced small-diameter stirrups than with widely spaced large-diameter stirrups.

ACI Section 11.5.6.9 attempts to guard against excessive crack widths by limiting the maximum shear that can be transmitted by stirrups to $V_{s(\max)} = 8\sqrt{f_c'}b_w d$. In a beam with $V_{s(\max)}$, the stirrup stress will be 34 ksi at service loads, corresponding to a maximum crack width of about 0.014 in. [6-5]. Although this limit generally gives satisfactory crack widths, the use of closely spaced stirrups and horizontal steel near the faces of beam webs is also effective in reducing crack widths.

4. **Shear failure due to crushing of the web.** As indicated in the discussion of the truss analogy, compression stresses exist in the compression diagonals in the web of a beam. In very thin-walled beams, these may lead to crushing of the web. Since the diagonal compression stress is related to the shear stress, v, a number of codes limit the ultimate shear stress to 0.2 to 0.25 times the compression strength of the concrete. The ACI Code limit on V_s for crack control $(V_{s(\max)} = 8\sqrt{f_c'}b_w d)$ provides adequate safety against web crushing in reinforced concrete beams.

5. **Shear failure initiated by failure of the tension chord.** The truss analogy shows that the force in the longitudinal tensile reinforcement at a given point in the shear span is a function of the moment at a section located approximately d closer to the nearest section of maximum moment. Partly for this reason, ACI Section 12.10.3 requires that flexural reinforcement extend the larger of d or 12 bar diameters past the point where it is no longer needed (except at the supports of simple spans or at the ends of cantilevers).

The exemption of supports of simple beams and the free ends of cantilevers recognizes that it is difficult to extend the bars at these locations. At a simple support, however, this overlooks the force in the bars at the supports. At each support of the analogous plastic truss in Example 6-1, there was a tensile force of 54 kips $= 0.75V_u$ in the longitudinal steel. This beam had 6 No. 7 bars, all of which extended into the support. A force of 54 kips would cause a stress of 15,000 psi in this steel at the support. Assuming $f'_c = 3000$ psi, a development length of $15,000/60,000 = 0.25$ times ℓ_d would be required. Thus, there must be an extension of 0.25×47.9 in. $= 12$ in. beyond the support. Alternatively, the bars need hooked anchorages. (See Fig. 8-22 for details.)

Figure 6-33 shows a free-body diagram of a simple support. Summing moments about point 0 and ignoring any moment resulting from the shearing stresses acting along the inclined crack gives

$$\frac{V_u}{\phi} \times d = T_n \times d + V_s \times 0.5d$$

or

$$T_n = \frac{V_u}{\phi} - 0.5V_s \qquad (6\text{-}22)$$

where V_s is the shear resisted by the stirrups at the face of the support and T_n is the longitudinal bar force to be anchored. Frequently, there are more stirrups than needed, as a result of satisfying maximum spacing requirements; in such a case, V_s would be less than the value given by (6-22), because the stress in the stirrups would be less than yield. From (6-9),

$$V_s = \frac{V_u}{\phi} - V_c$$

Substituting this result into (6-22) gives

$$T_n = 0.5\left(\frac{V_u}{\phi} + V_c\right) \qquad (6\text{-}23)$$

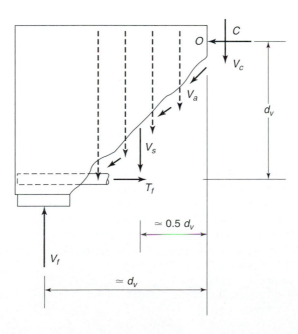

Fig. 6-33
Free-body diagram of beam reaction zone.

This requirement is generally assumed to be satisfied by extending the bars a distance d past where they are required for flexure. Although ACI specifically exempts simple beam supports from this requirement, it is good practice to anchor the flexural steel at the end of a simple beam.

Types of Web Reinforcement

ACI Sections 11.5.1.1 and 11.5.1.2 allow shear reinforcement for nonprestressed beams to consist of

(a) stirrups or ties perpendicular to the axis of the member. This is by far the most common type of web reinforcement.

(b) welded wire fabric with the wires perpendicular to the axis of the member serving as shear reinforcement. The welded wire fabric may be in the form of a sheet bent into a U shape or may be a ladderlike sheet with the rungs of the ladder transverse to the member. The latter are widely used in precast members. ACI Sections 12.13.2, 12.13.3, and 12.13.4 describe how welded wire fabric stirrups are anchored.

(c) stirrups inclined at an angle of 45° or more to the longitudinal tension reinforcement. The inclination of these stirrups must be such that they intercept potential inclined cracks. Inclined stirrups are difficult to detail near the ends of a beam and are not widely used. It is believed that stirrups flatter than 45° may slip along the longitudinal bars at high loads.

(d) a portion of the longitudinal flexural reinforcement may be bent up where no longer needed for flexure, as shown in Fig. 4-23d. Bent-up bars are no longer widely used because of the added cost of the labor needed to bend the bars. Because major inclined cracks tend to occur at the bend points, only the center 75 percent of the inclined portion of the bent bar is considered to be effective as shear reinforcement.

(e) combinations of spirals, circular ties, and hoops.

Minimum Web Reinforcement

Because a shear failure of a beam without web reinforcement is sudden and brittle, and because shear-failure loads vary widely about the values given by the design equations, ACI Section 11.5.5.1 requires a minimum amount of web reinforcement to be provided if the applied shear force, V_u, exceeds half of the factored inclined cracking shear, $\phi(0.5V_c)$, except in

1. slabs and footings;
2. concrete joist construction; and
3. beams with a total depth not greater than 10 in. (250 mm), 2.5 times the thickness of the flange, or one-half the width of the web, whichever is greatest.

The exceptions each represent a type of member in which load redistribution can occur across the width of the member or, in the case of joist floors, to adjacent members.

ACI Section 7.13.2.3 requires either one-piece closed stirrups or U stirrups with 135° hooks around top steel in all perimeter beams around a building. These stirrups, acting with the top and bottom reinforcement in the perimeter beams, are intended to provide a tension tie around the building to limit the extent of collapse arising from the failure of an interior beam.

Where required, the minimum web reinforcement shall be (ACI Section 11.5.5.3)

$$A_{v,min} = 0.75\sqrt{f_c'}\,\frac{b_w s}{f_y} \tag{6-24}$$

$$\text{(ACI Eq. 11-13)}$$

but not less than

$$A_{v,min} = \frac{50b_w s}{f_y} \qquad (6\text{-}25)$$

From (6-24) and (6-25), the minimum web reinforcement provides web reinforcement to transmit shear stresses of $0.75\sqrt{f'_c}$ psi and 50 psi, respectively. From (6-8), for $f'_c = 2500$ psi, 50 psi is half of the shear stress at inclined cracking. Equation (6-24) governs when f'_c exceeds 4440 psi.

Equation (6-24) is new in the 2002 code. It reflects the fact that the tensile stress at cracking increases as the compressive strength increases [6-22].

In SI units, (6-24) and (6-25) respectively become

$$A_{v,min} = \frac{1}{16}\sqrt{f'_c}\frac{b_w s}{f_y} \qquad (6\text{-}24\text{M})$$

$$(\text{ACI Eq. } 11\text{-}13\text{M})$$

but not less than

$$A_{v,min} = \frac{b_w s}{3f_y} \qquad (6\text{-}25\text{M})$$

In seismic regions, web reinforcement is required in most beams, since V_c is taken equal to zero if earthquake-induced shear exceeds half the total shear. (See Section 19-6 and ACI Section 21.3.4.2.)

Strength-Reduction Factor for Shear

In the 2002 ACI Code the strength-reduction factor, ϕ, for shear and torsion is 0.75 (ACI Section 9.3.2.3), provided the load factors in Section 9.2.1 are used. This value is lower than for flexure, because shear-failure loads are more variable than flexure-failure loads. Alternatively, $\phi = 0.85$ may be used along with the load factors in Appendix C of ACI 318-02. Special strength-reduction factors are required for shear in some members subjected to seismic loads. (See Section 19-5 and ACI Section 9.3.4.)

Location of Maximum Shear for the Design of Beams

In a beam loaded on the top flange and supported on the bottom as shown in Fig. 6-34a, the closest inclined cracks that can occur adjacent to the supports will extend outward from the supports at roughly 45°. Loads applied to the beam within a distance d from the support in such a beam will be transmitted directly to the support by the compression fan above the 45° cracks and will not affect the stresses in the stirrups crossing the cracks shown in Fig. 6-34. As a result, ACI Section 11.1.3.1 states,

> For nonprestressed members, sections located less than a distance d from the face of the support may be designed for the same shear, V_u, as that computed at a distance d.

This is permitted only when

1.   the support reaction, in the direction of the applied shear, introduces compression into the end regions of a member,

2.   the loads are applied at or near the top of the beam, and

3.   no concentrated load occurs within d from the face of the support.

Thus, for the beam shown in Fig. 6-34a, the values of V_u used in design are shown shaded in the shear force diagram of Fig. 6-34b.

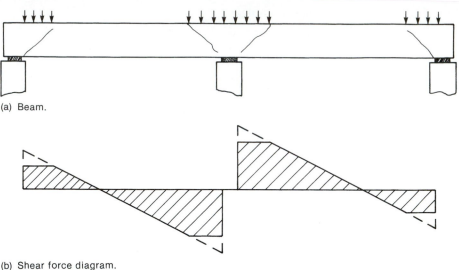

(a) Beam.

(b) Shear force diagram.

Fig. 6-34
Shear force diagram for
design.

This allowance must be applied carefully since it is not applicable in all cases. Fig. 6-35 shows five other typical cases that arise in design. If the beam shown in Fig. 6-34 were loaded on the *lower* flange, as indicated in Fig. 6-35a, the critical section for design would be at the face of the support, since loads applied within d of the support must be transferred across the inclined crack before they reach the support.

A typical beam-to-column joint is shown in Fig. 6-35b. Here the critical section for design is d away from the section as shown.

If the beam is supported by a girder of essentially the same depth, as shown in Fig. 6-35c, the compression fans that form in the supported beams will tend to push the bottom off the supporting beam. The critical sections in the supported beams can be taken at d from the end of the beam *provided* that "hanger reinforcement" is provided to support the reactions from the compression fans. The design of hanger reinforcement is discussed in Section 6-7.

Generally, if the beam is supported by a tensile force rather than a compressive force, the critical section will be at the face of the support, and the joint must be carefully detailed, since the crack will extend into the joint, as shown in Fig. 6-35d.

Occasionally, a significant part of the shear at the end of the beam will be caused by a concentrated load acting less than d from the face of the column, as shown in Fig. 6-35e. In such a case, the end portion of the beam should be considered to act as a *deep beam* with respect to shear and flexure, and the concentrated load must be considered. The design of deep beams is discussed in Chapter 18.

Shear at Midspan of Uniformly Loaded Beams

In a normal building, the dead and live loads are assumed to be uniform loads. Although the dead load is always present over the full span, the live load may act over the full span, as shown in Fig. 6-36c, or over part of the span, as shown in Fig. 6-36d. Full uniform load over the full span gives the maximum shear at the ends of the beam. Full uniform load over half the span plus dead load on the remaining half gives the maximum shear at midspan. The maximum shears at other points in the span are closely approximated by the *shear-force envelope* resulting from these cases (Fig. 6-36e).

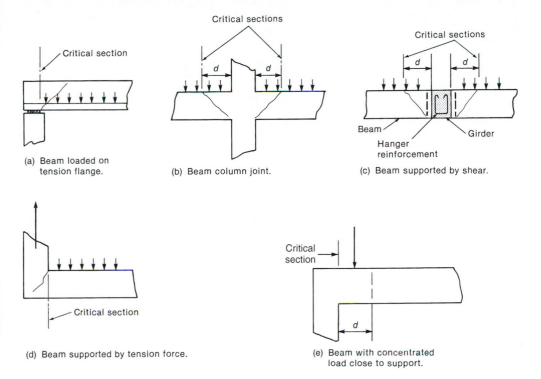

(a) Beam loaded on
    tension flange.

(b) Beam column joint.

(c) Beam supported by shear.

(d) Beam supported by tension force.

(e) Beam with concentrated
    load close to support.

Fig. 6-35
Application of ACI Section 11.1.3.

The shear at midspan due to a uniform live load on half the span is

$$V_{u,\text{midspan}} = \frac{w_{Lu}\ell}{8} \tag{6-26}$$

This can be positive or negative. Although this has been derived for a simple beam, it is acceptable to apply (6-26) to continuous beams also.

High-Strength Concrete

Tests suggest that the inclined cracking load of beams increases less rapidly than $\sqrt{f'_c}$ for f'_c greater than about 8000 psi. This was offset by an increased effectiveness of stirrups in high-strength concrete beams [6-23], [6-24]. Other tests suggest that the required amount of minimum web reinforcement increases as f'_c increases. ACI Section 11.1.2.1 was changed in the 2002 code to allow the use of $\sqrt{f'_c}$ in excess of 100 when computing the shear carried by the concrete (V_c, V_{ci}, and V_{cw}) for reinforced concrete and prestressed concrete beams and reinforced concrete joists, provided that they had minimum web reinforcement in accordance with ACI Sections 11.5.5.3, 11.5.5.4, and 11.6.5.2. For high-strength two-way concrete slabs, see ACI Commentary Section R11.5.5.1. $\sqrt{f'_c}$ is limited to 100 psi by the lack of test data on higher strength two-way slabs.

Lightweight Concrete

The inclined cracking load of beams made from lightweight concrete is generally less than that of beams made of normal sand and gravel concrete because the tensile strength, f'_t, is

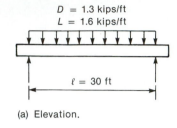

(a) Elevation.

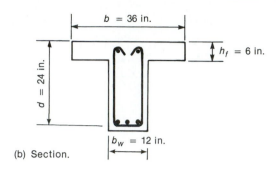

(b) Section.

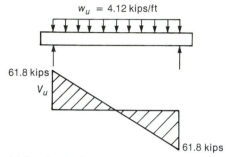

(c) Load case 1.

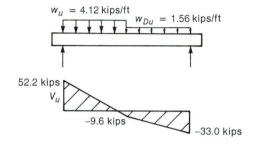

(d) Load case 2.

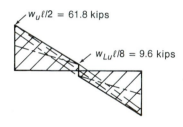

(e) Shear force envelope.

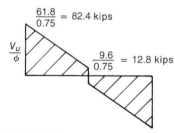

(f) V_u/ϕ diagram.

Fig. 6-36
Beam and shear force envelope—Example 6-2.

generally less for a given f_c' than it is for normal-density concrete, as discussed in Chapter 3. ACI Section 11.2 requires that the values of V_c be reduced in either one of two ways:

1. Substitute $f_{ct}/6.7$ for $\sqrt{f_c'}$ in (6-8), (6-15), and (6-17), where f_{ct} is the split-cylinder tensile strength of the concrete and where $f_{ct}/6.7$ shall not exceed $\sqrt{f_c'}$.

2. Multiply V_c from (6-8), (6-15), and (6-17) by 0.75 if the concrete is made with lightweight sand and lightweight gravel or by 0.85 if the concrete is made with normal sand and lightweight gravel.

Comparison of the Truss Model and the ACI Procedure

For beams with large amounts of web reinforcement, failing at high shear stresses, the behavior closely approaches that predicted by the plastic-truss model. (See Fig. 6-27.) On the other hand, the truss model predicts zero shear strength for beams without web reinforcement and tends to underestimate the shear capacity for beams with V_s less than about V_c.

The truss model emphasizes two aspects of shear behavior frequently overlooked in the ACI procedure: the shift in the tensile force diagram, which is of great importance in detailing the longitudinal steel (as discussed in Chapter 8), and the need for adequate anchorage of the stirrups in the top and bottom flanges.

Examples of the Design of Beams for Shear

EXAMPLE 6-2   Design of Vertical Stirrups in a Simply Supported Beam

Figure 6-36b shows the cross section of a simply supported T beam. This beam supports a uniformly distributed service (unfactored) dead load of 1.3 kips/ft, including its own weight, and a uniformly distributed service live load of 1.6 kips/ft. Design vertical stirrups for this beam. The concrete strength is 4000 psi, the yield strength of flexural reinforcement is 60,000 psi, and that of the stirrups is 40,000 psi. It is assumed that the longitudinal bars are properly detailed to prevent anchorage and flexural failures. A complete design example including these aspects is presented in Chapter 10. Use load and resistance factors from ACI 318-02 Sections 9.2.1 and 9.3.2.3.

1.   **Compute the factored shear-force envelope.** From ACI Eq. 9.2 with F, T, H, L_r, S, and R all equal to zero total factored load.

$$w_u = 1.2 \times 1.3 \text{ kips/ft} + 1.6 \times 1.6 \text{ kips/ft}$$

$$= 4.12 \text{ kips/ft}$$

Factored dead load:

$$w_{Du} = 1.2 \times 1.3 \text{ kips/ft} = 1.56 \text{ kips/ft}$$

Three loading cases should be considered: Fig. 6-36c, Fig. 6-36d, and the mirror opposite of Fig. 6-36d. The three shear-force diagrams are superimposed in Fig. 6-36e. For simplicity, we approximate the shear-force envelope with straight lines and design simply supported beams for $V_u = w_u \ell/2$ at the ends and $V_u = w_{Lu}\ell/8$ at midspan, where w_u is the total factored live and dead load and w_{Lu} is the factored live load. From (6-14),

$$V_u \le \phi V_n$$

Setting these equal, we find that the smallest value of V_n that satisfies (6-14) is

$$V_n = \frac{V_u}{\phi}$$

V_u/ϕ is plotted in Fig. 6-36f.

Since this beam is loaded on the top and supported on the bottom, the critical section for shear is located at $d = 2$ ft from the support. From Fig. 6-36f and similar triangles, the shear at d from the support is

$$\frac{V_u}{\phi} \text{ at } d = 82.4 \text{ kips} - \frac{2 \text{ ft}}{15 \text{ ft}}(82.4 - 12.8) \text{ kips}$$

$$= 73.1 \text{ kips}$$

Therefore,

$$\frac{V_u}{\phi} \text{ at } d = 73.1 \text{ kips and min. } V_n = 73.1 \text{ kips}$$

2.   **Are stirrups required by ACI Section 11.5.5.1?** No stirrups are required if $V_n \le V_c/2$, where

$$V_c = 2\sqrt{f'_c}b_w d \qquad (6\text{-}8)$$

$$= 2\sqrt{4000} \text{ psi} \times 12 \text{ in.} \times \frac{24 \text{ in.}}{1000} \qquad (\text{ACI Eq. 11-3})$$

$$= 36.4 \text{ kips}$$

Since $V_n = 73.1$ kips exceeds $V_c/2 = 18.2$ kips, stirrups are required.

3. Is the cross section large enough? ACI Section 11.5.6.9 gives the maximum shear in the stirrups as

$$V_{s,\text{max}} = 8\sqrt{f_c'}b_w d = 4 \times (2\sqrt{f_c'}b_w d) \tag{6-27a}$$
$$= 4V_c$$

Thus, the maximum V_n is

$$V_{n,\text{max}} = V_c + V_{s,\text{max}} = (1 + 4)(2\sqrt{f_c'}b_w d) = 5V_c \tag{6-27b}$$
$$= 182 \text{ kips}$$

Since V_n at $d = 73.1$ kips is less than $V_{n,\text{max}} = 182$ kips, the section is big enough.

4. Check the anchorage of stirrups and maximum spacing. Try No. 3 double-leg stirrups, $f_y = 40,000$ psi:

$$A_v = 2 \times 0.11 \text{ in.}^2 = 0.22 \text{ in.}^2$$

(a) Check the anchorage of the stirrups. Since the bar size of the stirrups is less than No. 6, ACI Section 12.13.2.1 states that such stirrups can be anchored by a 90° or 135° stirrup hook around a longitudinal bar. Provide a No. 4 bar in the upper corners of the stirrups to anchor them.

(b) Find the maximum stirrup spacing.

Based on the beam depth: ACI Section 11.5.4.1 sets the maximum spacing as the smaller of $0.5d = 12$ in. or 24 in. ACI Section 11.5.4.3 requires half this spacing in regions where V_s exceeds $4\sqrt{f_c'}\,b_w d$. Thus, the stirrup spacing must be cut in half if V_n exceeds

$$V_c + 4\sqrt{f_c'}b_w d = 6\sqrt{f_c'}b_w d = 109 \text{ kips}$$

Since the maximum V_n is less than 109 kips, the maximum spacing based on the beam depth is $s = 12$ in.

Based on minimum A_v (6-29), ACI Eq. (11-13),

$$A_{v,\text{min}} = 0.75\sqrt{f_c'}\,\frac{b_w s}{f_y} \tag{6-24}$$

(ACI Eq. 11-13)

but not less than

$$A_{v,\text{min}} = \frac{50b_w s}{f_y} \tag{6-25}$$

Rearranging (6-24) gives

$$s_{\text{max}} = \frac{A_v f_y}{0.75\sqrt{f_c'}b_w}$$

$$= \frac{0.22 \times 40,000}{0.75\sqrt{4000} \times 12} = 15.5 \text{ in.}$$

but not more than

$$s_{\text{max}} = \frac{A_v f_y}{50b_w}$$

$$= \frac{0.22 \text{ in.}^2 \times 40,000 \text{ psi}}{50 \text{ psi} \times 12 \text{ in.}} = 14.7 \text{ in.}$$

Therefore, the maximum spacing based on the beam depth governs. **Maximum $s = 12$ in.**

5.   Compute the stirrup spacing required to resist the shear forces. For vertical stirrups, from (6-21),

$$s = \frac{A_v f_y d}{V_u/\phi - V_c}$$

where $V_c = 36.4$ kips.
At d from the support, $V_u/\phi = 73.1$ kips and

$$s = \frac{0.22 \text{ in.}^2 \times 40,000 \text{ psi} \times 24 \text{ in.}}{(73.1 - 36.4) \times 1000 \text{ lb}} = 6.10 \text{ in.}$$

Use $s = 6$ in. at d from the support. Because this is a reasonable spacing, we can use No. 3 double-leg stirrups as assumed. The stirrup spacing will be changed to 8 in. at the point where this is possible and then to the maximum spacing of 12 in. The intermediate spacings selected are up to the designer. Generally, no more than three different spacings are used in a given beam, and the spacings are varied in multiples of 2 or 3 in.
Compute V_u/ϕ, at the section where s can be increased to 8 in. Rearranging (6-21) gives

$$\frac{V_u}{\phi} = \frac{A_v f_y d}{s} + V_c \tag{6-21}$$

$$= \frac{0.22 \times 40,000 \times 24}{8} + 36,400 = 62,800 \text{ lb}$$

$$= 62.8 \text{ kips}$$

From Fig. 6-38f and similar triangles, this shear occurs at

$$x = \frac{82.4 - 62.8}{82.4 - 12.8} \times 15 \text{ ft}$$

$$= 4.18 \text{ ft} = 50.1 \text{ in. from the end of the beam}$$

Compute V_u/ϕ, at the section where s can be increased to 12 in. From (6-26),

$$\frac{V_u}{\phi} = \frac{0.22 \times 40,000 \times 24}{12} + 36,400 = 54,000 \text{ lb}$$

$$= 54 \text{ kips}$$

This occurs at

$$x = \frac{82.4 - 54.0}{82.4 - 12.8} 15 \times 12$$

$$= 72.6 \text{ in. from the end of the beam}$$

Stirrups must be continued to the point where $V_u/\phi = V_c/2 = 18.2$ kips. This occurs at

$$x = \frac{82.4 - 18.2}{82.4 - 12.8} 15 \times 12$$

$$= 166 \text{ in. from the end of the beam}$$

To summarize, $s = 6$ in. to 51 in. from the support, $s = 8$ in. from that point to 74 in. from the support, and $s = 12$ in. from that point to 73 in. from the support, continuing to 166 in. from the end of the beam. In choosing the numbers of stirrups at each spacing, it is assumed that each stirrup reinforces a length of web extending $s/2$ on each side of the stirrup. For this reason, the first stirrup is placed at $s/2$ from the support. We shall select the following spacings:

1 at 3 in.
8 at 6 in. (extending to $3 + 8 \times 6 = 51$ in. from the support)

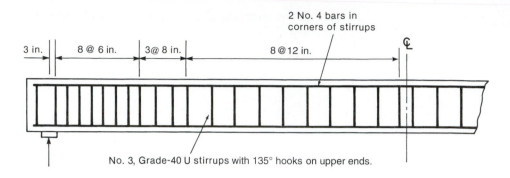

Fig. 6-37
Stirrups in beam—
Example 6-2.

3 at 8 in. (extending to 51 + 24 = 75 in. from the support)
8 at 12 in. (extending to 75 + 96 = 171 in. from the support)

This leaves an 18-in. length at the center without stirrups. Although not required by the code, we shall place an additional stirrup at each end. The final selection is: **Use No. 3 Grade-40 double-leg stirrups: one at 3 in., 8 at 6 in., three at 8 in., and eight at 12 in., each end**. The total number of stirrups in half the beam is 20. Another 20 are required in the other half at similar spacings. The beam is drawn to scale in Fig. 6-37. The cross section is shown in Fig. 6-36b. ∎

EXAMPLE 6-2M   Design of Vertical Stirrups in a Simply Supported Beam—SI Units

Figure 6-38 shows the elevation and cross section of a simply supported T beam. This beam supports a uniformly distributed service (unfactored) dead load of 20 kN/m, including its own weight, and a uniformly distributed service live load of 24 kN/m. Design vertical stirrups for this beam. The concrete strength is 25 MPa, the yield strength of the flexural reinforcement is 420 MPa, and the yield strength of the stirrups is 300 MPa. Use load and resistance factors from ACI 318-02 Sections 9.2.3 and 9.3.2.3.

1.   **Compute the design factored shear-force envelope.** For ACI (Eq. 9-2) with F, T, H, L_r, S and R all equal to zero. Total factored load:

$$w_u = 1.2 \times 20 \text{ kN/m} + 1.6 \times 24 \text{ kN/m}$$

$$= 62.4 \text{ kN/m}$$

Factored dead load:

$$w_{Du} = 1.2 \times 20 \text{ kN/m} = 24.0 \text{ kN/m}$$

Three loading cases should be considered: Fig. 6-38c, Fig. 6-38d, and the mirror opposite of Fig. 6-38d. The three shear-force diagrams are superimposed in Fig. 6-38e. For simplicity, we approximate the shear-force envelope with straight lines and design simply supported beams for $V_u = w_u \ell / 2$ at the ends and $V_u = w_{Lu} \ell / 8$ at midspan, where w_u is the total factored live and dead load and w_{Lu} is the factored live load. From (6-14),

$$V_u \le \phi V_n$$

Setting these equal, we obtain the smallest value of V_n that satisfies (6-14):

$$V_n = \frac{V_u}{\phi}$$

This is plotted in Fig. 6-38f.

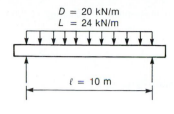

(a) Elevation.

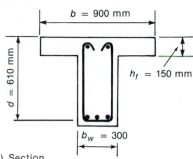

(b) Section.

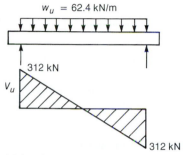

(c) Load case 1.

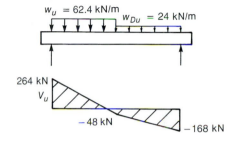

(d) Load case 2.

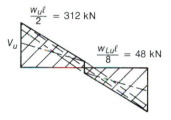

(e) Shear force envelope.

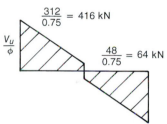

(f) V_u/ϕ diagram.

Fig. 6-38
Beam and shear force
envelope—Example 6-2M.

Since this beam is loaded on the top flange and supported on the bottom flange, the critical section is located at $d = 0.61$ m from the support. From Fig. 6-38f and similar triangles, the shear at d from the support is

$$\frac{V_u}{\phi} \text{ at } d = 416 \text{ kN} - \frac{0.61 \text{ m}}{5 \text{ m}}(416 - 64)$$

$$= 373 \text{ kN}$$

Therefore, V_u/ϕ at $d = 373$ kN and min. $V_n = 373$ kN.

2.   Are stirrups required by ACI Section 11.5.5.1? No stirrups are required if $V_n \leq V_c/2$, where

$$V_c = \frac{\sqrt{f'_c}\, b_w d}{6} \tag{6-8M}$$

$$= \frac{\sqrt{25} \text{ MPa} \times 300 \text{ mm} \times 610 \text{ mm}}{6 \times 1000}$$

$$= 153 \text{ kN}$$

Because $V_n = 373$ kN exceeds $V_c/2 = 76.3$ kN, stirrups are required.

3. Is the cross section large enough? ACI Section 11.5.6.9 gives the maximum shear in the stirrups as

$$V_{s,max} = \frac{2}{3}(\sqrt{f_c'}\, b_w d) \tag{6-27aM}$$

Thus, the maximum allowable V_n is

$$V_{n,max} = V_c + V_{s,max} = \left(\frac{2}{3} + \frac{1}{6}\right)\sqrt{f_c'}\, b_w d = 5V_c \tag{6-27bM}$$

$$= 765 \text{ kN}$$

Because V_n at $d = 373$ kN is less than $V_{n,\,max} = 765$ kN, the section is big enough.

4. Check anchorage of stirrups and maximum spacing. Try No. 10 double-leg stirrups, $f_y = 300$ MPa:

$$A_v = 2 \times 71 = 142 \text{ mm}^2$$

(a) Check the anchorage of the stirrups. ACI Section 12.13.2.1 allows No. 25M and smaller stirrups to be anchored by a standard 90° or 135° hook stirrup hook around a longitudinal bar. Provide a No. 10M or larger bar in each of the upper corners of the stirrup to anchor them,

(b) Find the maximum stirrup spacing.

Based on the beam depth: ACI Section 11.5.4.1 requires the smaller of $0.5d = 305$ mm or 600 mm. ACI Section 11.5.4.3 requires half these spacings if V_s exceeds $\left(\frac{1}{3}\right)\sqrt{f_c'}b_w d$:

$$\frac{\sqrt{f_c'}\, b_w d}{3} = 305 \text{ kN}$$

Because the maximum V_n is less than 305 kN, the maximum spacing is 305 mm.
Based on minimum allowed A_v ((6-29M), ACI Eq. (11-14)),

$$A_{v,min} = \frac{1}{16}\sqrt{f_c'}\,\frac{b_w s}{f_y} \tag{6-24M}$$
$$\text{(ACI Eq. 11-13M)}$$

but not less than

$$A_{v,min} = \frac{1}{3}\frac{b_w s}{f_y} \tag{6-25M}$$

Rearranging gives

$$s_{max} = \frac{16 \times A_v f_y}{\sqrt{f_c'}b_w} = \frac{16 \times (2 \times 71) \text{ mm}^2 \times 300 \text{ MPa}}{\sqrt{25} \times 300 \text{ mm}} = 454 \text{ mm}$$

but not less than

$$s_{max} = \frac{3A_v f_y}{b_w} = \frac{3 \times (2 \times 71) \times 300}{300} = 426 \text{ mm}$$

Therefore, the maximum spacing based on the beam depth governs. **Maximum $s = 305$ mm.**

5. Compute the stirrup spacing required to resist the shear forces. For vertical stirrups, (6-21) applies; that is,

$$s = \frac{A_v f_y d}{V_u/\phi - V_c} \tag{6-21}$$

where $V_c = 153$ kN. At d from the support, $V_u/\phi = 373$ kN and

$$s = \frac{2 \times 71 \text{ mm}^2 \times 300 \text{ MPa} \times 610 \text{ mm}}{(373 - 153) \times 1000 \text{ N}} = 118 \text{ mm}$$

Try No. 10 U stirrups at $s = 100$ mm. Change spacing to $s = 150$ mm where this is acceptable, and then to the maximum spacing of 300 mm. The intermediate spacings selected are up to the designer. Generally, no more than three different spacings are used and the spacings are varied in multiples of 50 or 75 mm. For No. 10M U-stirrups $A_v = 2 \times 71 \text{ mm}^2 = 142 \text{ mm}^2$.

At d from support the spacing required to support the factored shears is 118 mm. Use $s = 100$ mm. Change s to 150 mm.

Compute V_u/ϕ, where s can be increased to 150 mm. Rearranging (6-21) gives

$$\frac{V_u}{\phi} = \frac{A_v f_y d}{s} + V_c$$

$$= \frac{(2 \times 71) \times 300 \times 610}{150 \times 1000} + 153 = 326 \text{ kN}$$

From Fig. 6-38f and similar triangles, this shear occurs at

$$x = \frac{416 - 326}{416 - 64} \times 5000 \text{ mm}$$

$$= 1280 \text{ mm from the end of the beam}$$

Change s to 300 mm. Compute V_u/ϕ, where s can be increased to 300 mm. From (6-21),

$$\frac{V_u}{\phi} = \frac{2 \times 71 \times 300 \times 610}{300 \times 1000} + 153 = 240 \text{ kN}$$

This occurs at

$$x = \frac{416 - 240}{416 - 64} 5000$$

$$= 2500 \text{ mm from the end of the beam}$$

Stirrups must be continued to the point where $V_u/\phi = V_c/2 = 76.5$ kN. This occurs at

$$x = \frac{416 - 76.5}{416 - 64} 5000$$

$$= 4820 \text{ mm from the end of the beam}$$

In choosing the numbers of stirrups at each spacing, each stirrup is assumed to reinforce a length of web extending $s/2$ on each side of the stirrup. For this reason, the first stirrup is placed at $s/2$ from the support. We shall select the following spacings: To summarize, s = one at 50, 13 at 100 to 1350 mm from the support, eight at 150 to 2550 mm and eight at 300 to 4950 mm each end. This leaves a space of 50 mm × 2 at midspan.

The total number of stirrups in half of the beam is 30. Another 30 stirrups are required in the other half at similar spacings. The elevation and cross section would resemble those in Figs. 6-37 and 6-34b.   ∎

EXAMPLE 6-3   Design of Stirrups in a Continuous Beam

Figure 6-39 shows the elevation and cross section of an interior span of a continuous T beam. This beam supports a floor supporting an unfactored dead load of 2.45 kips/ft (including its own weight) and an unfactored live load of 2.4 kips/ft. The concrete is of normal density, with $f_c' = 3500$ psi. The yield strength of the web reinforcement is 40 ksi. Design vertical stirrups, using ACI Sections 11.1 through 11.5. Use the load and resistance factors from Sections 9.2.1 and 9.3.2 of the 2002 ACI Code.

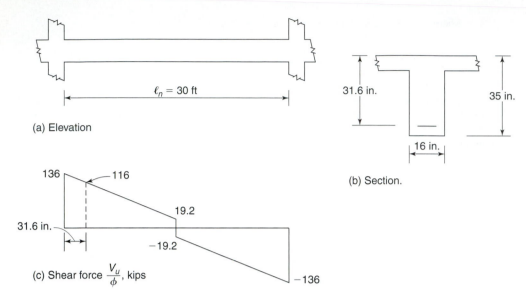

(a) Elevation

$\ell_n = 30$ ft

31.6 in.

35 in.

16 in.

(b) Section.

136

116

19.2

31.6 in.

−19.2

(c) Shear force $\dfrac{V_u}{\phi}$, kips

−136

Fig. 6-39
Beam and shear force envelope—Example 6-3.

1. **Compute the factored shear-force envelope.** ACI Section 9.2.1 gives the following load combinations for beams loaded with dead and live loads:

$$U = 1.4(D + F) \qquad \text{(ACI Eq. 9-1)}$$

$$U = 1.2(D + F + T) + 1.6(L + H) + 0.5(L_r \text{ or } S \text{ or } R) \qquad \text{(ACI Eq. 9-2)}$$

Here, F is the load due to the weight and pressures of fluids with well-defined densities and controllable maximum heights or due to related internal moments and forces. The beam does not support tanks containing fluids; thus, F is zero.

T denotes the cumulative effect of temperature, creep, shrinkage, differential settlement, and shrinkage-compensating concrete. ACI 318-02 Section 9.2.3 and ACI 318-02 Commentary Section R9.2.3 suggest that T should be based on a realistic assessment of such effects occurring in service. T loads tend to be dissipated by inelastic action in the beam at high loads and will be ignored here. Thus, T is zero.

H represents the loads due to the weight and pressure of soil, water in soil, or other materials. The beam being designed does not support soils; hence, H is zero.

L_r, S, and R are live load, snow load, and rain load on a roof. Since this beam supports a floor, these loads are not applicable.

ACI 318-02 Equations (9-1) and (9-2) respectively become

$$U = 1.4D = 1.4 \times 2.45 = 3.43 \text{ kip/ft}$$

and

$$U = 1.2D + 1.6L = 1.2 \times 2.45 + 1.6 \times 2.4 = 6.78 \text{ kip/ft}$$

The larger of these governs, so we have

$$\text{Factored total load } w_u = 6.78 \text{ kip/ft}$$

and

$$\text{Factored live load } w_{Lu} = 3.84 \text{ kip/ft.}$$

This is an interior span, and ACI Section 8.3.3.3 gives the shear at the face of the support as

$$V_u = w_u \ell_n / 2$$

where $\ell_n = 15$ ft is the clear span.

$$V_u = 6.78 \text{ kip/ft} \times 30 \text{ ft}/2$$

$$= 101.7 \text{ kips}$$

is the clear span of the beam and $V_u/\phi = 101.7/0.75 = 136$ kips at the faces of the supports. ACI 318-02 Section 9.3.2.3 gives ϕ for shear and torsion as 0.75.

For simply supported beams, we have taken the shear at midspan as

$$V_u = w_{Lu} \ell_n / 8 \tag{6-26}$$

$$= 3.84 \text{ kip/ft} \times 30 \text{ ft}/8$$

$$= 14.4 \text{ kips}$$

$$\text{so } V_u/\phi = 14.4/0.75 = 19.2 \text{ kips}$$

We shall use the same equation to compute the shear at midspan of a continuous beam. The shear-force envelope is shown in Fig. 6-39c.

$$V_c = 2\sqrt{f'_c} b_w d \tag{ACI Eq. 11-3}$$

$$= 2 \times \sqrt{3500} \times 16 \times 31.6$$

$$= 59,800 \text{ lb} = 59.8 \text{ kips}$$

and

$$V_c/2 = 29.9 \text{ kips}$$

The maximum shear at d from the face of the support is $V_u/\phi = 116$ Kips. Since 116 kips exceeds 29.9 kips, web reinforcement is required. Using similar triangles V_u/ϕ is equal to $V_c/2$ at

$$x = \frac{136 \text{ kips} - 29.9 \text{ kips}}{136 \text{ kips} - 19.2 \text{ kips}} \times 15 \times 12 \text{ in.} = 164 \text{ in.}$$

from the faces of the supports.

4.   Anchorage of stirrups. Try No. 3 Grade-40 U stirrups with 135° hooks around a longitudinal bar in each upper corner. ACI Section 12.13.2.1 considers such stirrups to be anchored.

5.   Find the maximum stirrup spacing.

Based on beam depth: ACI Section 11.5.4.1 gives a maximum spacing of $d/2 = 31.6$ in./2 $= 15.8$ in. If V_s exceeds $4\sqrt{f'_c} b_w d = 120$ kips, which corresponds to $V_u/\phi = (4 + 2)\sqrt{f'_c} b_w d$ $= 179$ kips, ACI Section 11.5.4.3 requires the maximum spacing to be reduced to $d/4$.

The maximum shear at d from the face of a support is $V_u/\phi = 116$ kips. Since this value is less than 179 kips, the maximum spacing based on beam depth is $d/2 = 15.8$ in.

Based on minimum A_v,

$$A_{v,\text{min}} = 0.75\sqrt{f'_c} \frac{b_w s}{f_y} \tag{6-24}$$

$$\tag{ACI Eq. 11-13}$$

but not more than

$$A_{v,\text{min}} = \frac{50 b_w s}{f_y} \tag{6-25}$$

Rearranging (6-24) gives

$$s_{max} = \frac{A_v f_y}{0.75\sqrt{f_c'}b_w}$$

$$= \frac{2 \times 0.11 \times 40,000}{0.75\sqrt{3500} \times 16}$$

$$= 12.4 \text{ in.}$$

and, from (6-25),

$$s_{max} = \frac{A_v f_y}{50 b_w} = \frac{2 \times 0.11 \times 40,000}{50 \times 16} = 11 \text{ in.}$$

Therefore, the maximum spacing s based on $A_{v,min}$ governs. Accordingly, we try $s = 10$ in.

6.   Compute the stirrup spacing required to resist the factored shear force.

$$s = \frac{A_v f_y d}{\dfrac{V_u}{\phi} - V_c} \tag{6-21}$$

where, from step 3, $V_c = 59.8$ kips.

At d from support,

$$\frac{V_u}{\phi} = 116 \text{ kips and}$$

$$s = \frac{2 \times 0.11 \text{ in.} \times 40,000 \text{ psi} \times 31.6 \text{ in.}}{(116 \text{ kips} - 59.8 \text{ kips}) \times 1000 \text{ lbs.}}$$

$$= 4.95 \text{ in.}$$

At d from support, use $s = 4$ in.
    Change s to 6 in.:

$$\frac{V_u}{\phi} = \frac{A_v f_y d}{s} + V_c \tag{6-21}$$

$$\frac{V_u}{\phi} \text{ for } s = 6 \text{ in.} = \frac{2 \times 0.11 \times 40,000 \times 31.6}{6 \times 1000 \text{ lbs}} + 59.8$$

$$= 106 \text{ kips}$$

This occurs at $x = \dfrac{136 - 106}{136 - 19.2} \times 15 \times 12 = 46.2$ in. from the faces of the supports.
    Change s to 10 in.:

$$\frac{V_u}{\phi} \text{ for } s = 10 \text{ in.} = \frac{2 \times 0.11 \times 40,000}{10 \times 1000} + 59.8$$

$$= 87.6 \text{ kips}$$

This occurs at $x = \dfrac{136 - 87.6}{136 - 19.2} \times 15 \times 12 = 74.6$ in. from the faces of the supports.
    Select stirrups:

one at 2 in. from face of support

twelve at 4 in. (extending to $2 + 12 \times 4 = 50$ in. from the face of the support)

five at 6 in. (extending to $50 + 30 = 80$ in. from the face of the support)

nine at 10 in. (extending to $80 + 90 = 170$ in. from the support)

This leaves a space of $2 \times (180 - 170) = 20$ in. without any stirrups near midspan. We could arbitrarily add one extra stirrup at 10 in. at mid span.

7.   Anchorage of flexural reinforcement. ACI Section 12.10.3 requires that flexural reinforcement extend d or $12d_b$ past the point where it is no longer required for flexure, except at the supports of simple spans and at the free ends of cantilevers. Extend all flexural reinforcement a distance $d = 31.6$ in., say 32 in., past the flexural points of inflection.

Summary.   **Use No. 3 Grade-40 U stirrups with 135° hooks around bars in the top corners of the section one at 2 in., 12 at 4 in., five at 6 in., and nine at 10 in., each half.** The total number of stirrups in half the beam is 27. Another 27 stirrups are required in the other half at similar spacings.   ■

6-6   MODERN SHEAR DESIGN METHODS

In most current concrete codes, shear design is based on (6-9),

$$V_n = V_c + V_s \qquad (6\text{-}9)$$

where V_c is the "shear carried by the concrete." The currently used value of V_c was empirically derived from test results. Ever since reinforced concrete began to be studied, a mechanical explanation of V_c has been the Holy Grail sought by researchers. Today, the underlying mechanisms of shear transfer are understood enough to allow shear design based at least partially on theory, instead of on equations extrapolated from test results. An excellent review of the current status of modern shear design methods is given in [6-2].

The details of six methods of designing for shear in slender beams are compared in Table 6-1. The first and second methods are the *traditional truss analogy*, discussed in Section 6-4, and the *traditional ACI method*, examined in Section 6-5. Three of the remaining four methods are successively improved versions of the compression field theory:

(a)   the compression field theory (CFT-84), from the 1984 Canadian concrete code [6-25] [6-14],

(b)   the modified compression field theory, from the 1994 Canadian Code [6-26] and the AASHTO Limit States Design Specification, and

(c)   the revision of MCFT currently being considered for inclusion in the 2004 Canadian code (MCFT-04). See [6-27], [6-28], [6-29].

The sixth procedure listed is based on an application of shear friction to slender beams failing in shear. Except for the truss analogy and the CFT-84 method, the design methods are based on (6-9), although each of the methods has a unique way of calculating V_c, as shown in Table 6-1.

Definitions and Design Methods

Design procedures for shear fall into two classifications:

1.   *Whole-member design methods*, such as the truss analogy, idealize the member as a *truss* or a *strut-and-tie model* that represents the complete load-resisting mechanism for one load case.

2.   *Sectional design methods,* in which shear-force and bending-moment envelopes are derived for all significant load cases and design is carried out cross section by cross section for the envelope values at sections along the length of the beam.

Both methods are needed for design, because the sectional design methods are preempted in D-regions by the in-plane compression forces in the struts that dominate the behavior. As a result, sectional design methods do not work well in such D-regions. The truss analogy is included in Table 6-1 to permit a comparison of a whole-member design method with the sectional design procedures.

TABLE 6-1 Characteristics of Several Modern Shear Design Methods

Features	Design Methods					
			Compression Field Theories			Shear Friction [6-31]
	Truss Analogy	ACI Design Method	CFT-84 [6-25]	MCFT-94 [6-26]	MCFT-04 [6-27, 30, 31]	
Notes		1.	2.	3.	4.	5.
1. Includes additive V_c and V_s terms giving $V_n = V_c + V_s$	No	Yes	Yes	Yes	Yes	Yes
2. Uses a Mohr's circle for strain to derive equations used to check compatibility.	No	No	Yes	Yes	Yes	No
3. Considers web crushing. Uses smeared properties of cracked web-concrete, based on Vecchio [6-33], or similar equations to check crushing of the web. Onset of crushing is a function of an *index stress ratio*, v/f'_c, and an *index strain*, ϵ_x	No	No	Yes	Yes	Yes	No
4. Considers slip on inclined crack in web. Uses shear friction based on slip equations proposed [6-30] [6-31]	No	No	Walraven [6-30]	Walraven [6-30]	Walraven [6-30]	Loov [6-31]
5. Longitudinal reinforcement proportioned for combined shear and moment	Yes	No, except for empirical extensions of d away from the section.		Yes	Yes	Yes
Size Effect	No	No	No	Yes	Yes	No

Notes on Table 6-1

1. In the traditional ACI design method, $V_c = \beta\lambda\sqrt{f'_c}b_wd$, where β is an empirical function of $\sqrt{f'_c}$.

2. In the 1984 [6-25] version of the compression field theory, $V_c = \beta\sqrt{f'_c}b_wd$, where β is a function of the angle of the compression field, θ, the shear stress ratio, v/f'_c, and the strain in the longitudinal reinforcement, ϵ_ℓ, and λ is the modifier for lightweight concrete. The angle θ could have any value between 15 and 75° as long as the same angle is used for all calculations.

3. In the 1994 [6-26] version of the modified compression field theory, $V_c = \beta\lambda\sqrt{f'_c}b_wd$, where, for beams with at least minimum stirrups, β and θ are tabulated functions of the index variable $v/\lambda fc'$, where v is the shear stress at the section and ϵ_x the index longitudinal strain. The tabulated values of β and θ were dependent on a number of variables and based on an angle θ selected from minimum reinforcement cost considerations. This set of β and θ values oscillated widely and for practical cases of beams it had to be tabulated. For members without minimum stirrups, β and θ are functions of the computed crack width which is taken as the product of the strain perpendicular to the crack spacing.

4. The 2004 version of the MCFT [6-27] allows a range of θ values and requires calculation of the corresponding β values. It is still necessary to compute the longitudinal index strain ϵ_x. Once again, $V_c = \beta\lambda\sqrt{f'_c}b_wd$.

5. In the shear friction method [6-31], the user selects a series of inclined sections through the member and computes the compressive clamping force perpendicular to the inclined section from yielding of the longitudinal and transverse reinforcement crossing one particular inclined section. The clamping force gives rise to an inclined frictional force along the inclined section under consideration. The shear friction method does not consider crushing of the web as a failure mechanism. The clamping force gives rise to an inclined frictional force along the inclined section under consideration. A number of trial inclined sections are considered to find the one having the least capacity. The shear friction method does not consider crushing of the web as a failure mechanism.

Two other concepts that need to be defined are as follows:

1. *Smeared-crack models* replace cracked reinforced concrete with a hypothetical new material that does not crack, as such. The material has properties that are *averaged* or *smeared* over a long gauge length that includes the width of one or more cracks. This approach reduces the discontinuities in the postcracking behavior.

2. A shear-sensitive region must satisfy both equilibrium and *compatibility*. Compatibility relationships for shear cracking of concrete and postcracking deformations and strengths have been derived from Mohr's circles. The Mohr's circle of strain is used to determine relationships between deformations [6-14]. The Mohr's circle of stress is used to determine relationships between stresses.

Traditional ACI Design Procedure

The traditional ACI design method uses V_c which we shall write as

$$V_c = \beta\lambda\sqrt{f'_c}b_w d \qquad\qquad (6\text{-}8a)$$

where β is an empirically derived function whose value is taken to be 2 in (6-8), ACI Eq. (11-3), λ is a modification factor used to allow for lightweight concrete, and d is the distance from the extreme compression face to the centroid of the longitudinal reinforcement. In the ACI Code, $V_s = A_v f_y\, d/s$, which implies that the cracks are at an angle close to $45°$. Other design expressions have similar equations for V_s, except for the shear friction method, which computes V_s on the basis of the number of stirrups and other bars crossed by an inclined shear plane or crack crossing the web at a given slope.

Compression Field Theory

Three of the design procedures in Table 6-1 use the compression field theory to model the given structure. This theory is the inverse of the *tension field theory* developed by Wagner [6-30] in 1929 for the design of light-gauge plates in metal airplane fuselages. If a light-gauge web is loaded in shear, it buckles due to the diagonal compression in the web. Once buckling has occurred, further increases in shear require the web shear mechanism to be replaced by a truss or a field of inclined tension forces between the buckles in the web. This diagonal *tension field* in turn requires a truss that includes vertical compression struts and longitudinal compression chords to resist the reactions from the tension diagonals in the web.

The compression field theory (CFT) is just the opposite of the tension field theory. In the CFT, the web of the beam cracks due to the principal tension stresses in the web. Cracking reduces the ability of the web to transmit diagonal tension forces across the web. After cracking, loads are carried by a trusslike mechanism with a field of hypothetical diagonal compression members between the cracks and tensions in the stirrups and the longitudinal chords.

The CFT from the 1984 Canadian Code [6-25] did not have a V_c term. Instead, stirrups were provided for the full shear. In designing a structure, it was necessary to check web crushing by using stresses and strains derived from Mohr's circles. In design, the angle θ was assumed and was used to compute web stresses and the capacity of the concrete. In the CFT-84, the angle θ could have any value between 15 and $75°$, as long as the same angle was used for all the calculations at a given section.

In the 1994 version of the modified compression field theory, [6-26], [6-28]

$$V_c = \beta\lambda\sqrt{f'_c}b_w\, d_v \qquad\qquad (6\text{-}8a)$$

where, for beams with at least the minimum shear reinforcement required by the code, the factor β is a function of the strut angle θ. Tables of values of β and θ which had been computed iteratively were use to select β and θ to minimize the total reinforcement costs. The resulting set of θ and β values oscillated widely for practical beams. The equations for β and θ were so complex that the values had to be given in tables.

For members with less than minimum web reinforcement, β and θ were functions of the widths of shear cracks, taken as the product of the longitudinal strain and the crack spacing. In MCFT-94, β was a function of

(a) an *index value* of the *shear stress ratio* v/f'_c.

(b) an *index value* of the *longitudinal strain* ε_x, computed as the strain in the longitudinal reinforcement due to flexure, shear, and normal force.

(c) a *crack spacing factor* s_{ze}, related to the distance s_z between reinforcement crossing the cracks, and

(d) the crack angle θ.

Special values of s_{ze} were given for members without stirrups or for members with less than the minimum stirrups specified in ACI Section 11-5. Items (a) and (b) are referred to as *index values* because the exact values of v/f'_c and ε_x are less important in selecting β and θ than the trends in these variables. The longitudinal strain ε_x was taken to act at the level of the longitudinal tension reinforcement. The 2004 version of the MCFT (called MCFT-04 in Table 6-1) is proposed for inclusion in the 2004 Canadian code. In it, equations are given for θ, from which β can be calculated. The objective function used to determine the most economical value was relaxed, allowing much simpler equations for β.

Shear Friction Method

The shear friction method for beam shear takes V_c to be the shear force transferred across each of a series of inclined sections cut through the entire beam, representing inclined cracks or shear slip planes. For simplicity, the inclined sections are taken to be straight. The shear force V_{ci} on the inclined plane is based on one of several equations derived by Walraven [6-31] or Loov and Patnaik [6-32] for the shear transferred across the shear plane in a composite beam at the onset of slip.

The clamping force needed to mobilize shear friction stress v_{ci} along the crack is taken as the sum of the components, perpendicular to the slip plane, of the yield forces $A_s f_y$ from the longitudinal reinforcement and the stirrups $A_v f_y$ crossed by the crack.

Equivalence of Shears on Inclined and Vertical Planes

For an element from the web, the shear force V_{ci} on the inclined slip plane can be related to the shear on a vertical plane as shown in Fig. 6-40. On the vertical plane,

$$V_c = v_c\, b_w\, d_v. \qquad (6\text{-}29)$$

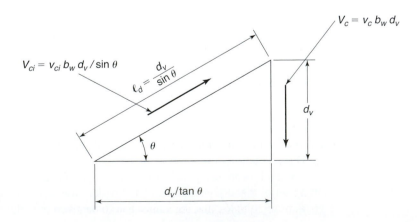

Fig. 6-40
Equivalence of shear stresses acting on an element.

The length of the inclined section is

$$\ell_d = (d_v/\sin\theta)$$

The inclined shear force on the sloping side of the element is

$$V_{ci} = v_{ci}b_w(d_v/\sin\theta) \tag{6-30}$$

where v_{ci} is the average shear stress parallel to the inclined plane, b_w is the width of the web of the beam, d_v is the effective depth for shear, ℓ_d is the length of the inclined slip plane, and θ is the angle between the slip plane and the longitudinal flexural reinforcement. The vertical component of V_{ci} is $V_c = V_{ci}\sin\theta$. Substituting this expression into (6-29) gives

$$V_c = (v_{ci}b_wd_v\sin\theta/\sin\theta) \tag{6-31}$$

The $\sin\theta$ terms cancel out, leaving

$$V_c = v_{ci}b_wd_v$$

From this equation and (6-29), it follows that $v_{ci} = v_c$. Thus, for a given shear force, the average shear stresses v_{ci} on an inclined plane are the same as the vertical shear stresses v_c on the vertical plane due to the same loading.

Two relationships for the shear stress that can be transferred across an inclined crack are, from Walraven [6-31]

$$v_{ci} = \frac{2.16\sqrt{f_c'}}{0.3 + \dfrac{24w}{a + 0.63}} \tag{6-32}$$

and, from Loov and Patnaik [6-32],

$$v_{ci} = 0.6\lambda\sqrt{\alpha f_c'} \tag{6-33}$$

For a given set of loads, concrete stresses, and crack widths, taking into account the different assumptions, we find that the magnitude of V_{ci} from the MCFT-04 and that from the shear friction method are similar. This suggests that in many cases the "shear carried by the concrete" is closely related to shear friction on the crack surfaces and V_c is due to shear friction along the crack.

6-7   HANGER REINFORCEMENT

When a beam is supported by a girder or other beam of essentially the same depth, as shown in Fig. 6-41 or Fig. 6-42, hanger reinforcement should be provided in the joint. Compression fans form in the supported beams, as shown in Fig. 6-41a. The inclined compressive forces will push the bottom off the supporting beam unless they are resisted by hanger reinforcement designed to equilibrate the downward component of the compressive forces in the members of the fan.

No rules are given in the ACI Code for the design of such reinforcement. The following proposals are based on the 1984 Canadian concrete code [6-25] and on [6-33]. In addition to the stirrups provided in the supporting beam for shear, hanger reinforcement with a tensile capacity of

$$\phi A_h f_y \geq \left(1 - \frac{h_b}{h_1}\right)V_{u2} \tag{6-34}$$

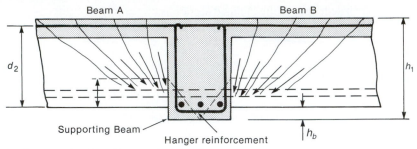

(a) Compression fan at beam–girder joint.

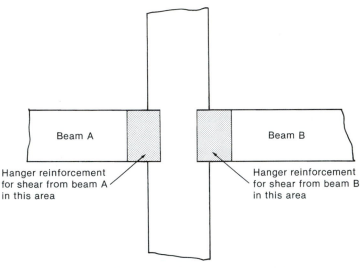

Fig. 6-41
Hanger reinforcement.

(b) Plan of joint area showing location of hanger reinforcement.

should be provided within a length of $b_{w2} + 2h_b$ in the supporting beam and $d_2/4$ in the supported beam, adjacent to each face of the supporting beam where shear is being transferred, where

A_h = the area of hanger reinforcement adjacent to one face of the supporting beam,

b_{w2} = the width of the supported beam,

d_2 = the effective depth of the supported beam,

h_b = the vertical distance from the bottom of the supporting beam to the bottom of the supported beam,

h_1 = the overall depth of the supporting beam,

h_2 = the overall depth of the supported beam, and

V_{u2} = the factored shear at the end of the supported beam.

If shears are transferred to both side faces of the supporting beam, (6-35) is evaluated separately for each face.

    The additional hanger reinforcement, A_h, is placed in the supporting beam to intercept 45° planes starting on the shear interface at one-quarter of the depth of the supported beam, h_2, above its bottom face and spreading down into the supporting beam, as shown by the 45°. dashed lines in Figs. 6-41a and 6-42a . These provisions can be waived if the shear, V_{u2}, at the end of the supported beam is less than $3\sqrt{f_c'}b_{w2}d_2$, since the inclined cracking is not fully developed at this shear.

    The hanger reinforcement should be well anchored, top and bottom. The lower layer of reinforcement in the supported beam should be above the reinforcement in the supporting beam. See also [6-33].

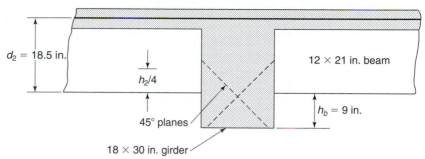

(a) Beams.

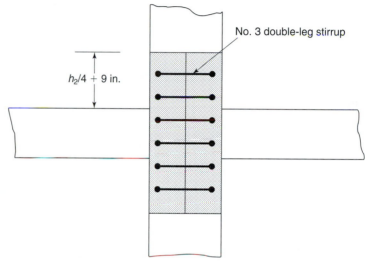

No. 3 double-leg stirrup

$h_2/4$ + 9 in.

Fig. 6-42
Example 6-4.

(b) Plan view of joint zone.

EXAMPLE 6-4   Design of Hanger Reinforcement in a Beam-Girder Junction

Figure 6-42a shows a beam-girder joint. Each beam transfers a factored end shear of 45 kips to the girder. Design hanger reinforcement, assuming the yield strength of the reinforcement, is 60 ksi. The shear reinforcement in the beam and girder is No. 3 Grade-60 double-leg stirrups.

$$h_b = 30 \text{ in.} - 21 \text{ in.} = 9.0 \text{ in.}$$

$$h_1 = 30 \text{ in.}$$

The factored tensile force $\phi A_h f_y$ to be transferred by hanger reinforcement is

$$\phi A_h f_y = 2 \times 45 \text{ kips} \left(1 - \frac{9}{30}\right) = 63.0 \text{ kips} \tag{6-35}$$

The area of hanger reinforcement required is

$$\phi A_h f_y = 63.0 \text{ kips} \tag{6-34}$$

$$A_h = \frac{T_h/\phi}{f_y}$$

$$= \frac{63.0/0.75}{60} = 1.40 \text{ in.}^2$$

We could use

four No. 4 double-leg stirrups, $A_s = 1.60$ in.^{2}
seven No. 3 double-leg stirrups, $A_s = 1.54$ in.^{2}

Since the shear reinforcement is No. 3 stirrups, we shall select No. 3 stirrups for the hanger steel. It will be placed as shown in Fig. 6-42b. This is in addition to the shear reinforcement already provided. ∎

6-8 TAPERED BEAMS

In a prismatic beam, the average shear stress between two cracks is calculated as

$$v = \frac{V}{b_w jd} \qquad (6\text{-}4)$$

which is simplified to

$$v = \frac{V}{b_w d} \qquad (6\text{-}5)$$

In the derivation of (6-4), it was assumed that jd was constant. If the depth of the beam varies, the compressive and tensile forces due to flexure will have vertical components. A segment of a tapered beam is shown in Fig. 6-43. The moment, M, at the left end of the section can be represented by two horizontal force components, C and T, separated by the lever arm jd. The tension force actually acts parallel to the centroid of the reinforcement and hence has a vertical component $T \tan \alpha_T$, where α_T is the angle between the tensile force and horizontal. Similarly, the compressive force acts along a line joining the centroids of the stress blocks at the two sections and hence has a vertical component $C \tan \alpha_c$. The shear force on the left end of the element can be represented as

$$V = V_R + C \tan \alpha_c + T \tan \alpha_T$$

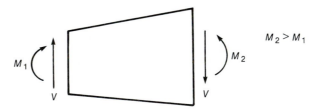

(a) Forces on segment of beam.

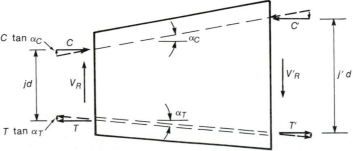

Fig. 6-43
Reduced shear force in non-prismatic beam.

(b) Internal forces and shears.

where V_R is the reduced shear force resisted by the stirrups and the concrete. Substituting $C = T = M/jd$ and letting $\alpha = \alpha_c + \alpha_T$ gives

$$V_R = V - \frac{|M|}{jd} \tan \alpha \tag{6-35}$$

where $|M|$ represents the absolute value of the moment and α is positive if the lever arm jd increases in the same direction as $|M|$ increases.

The shear stresses in a tapered beam then become

$$v = \frac{V_R}{b_w d} \tag{6-36}$$

Several examples of the use of (6-36) are shown in Fig. 6-44.

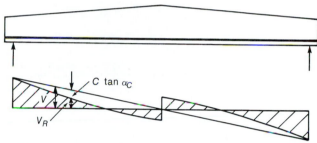

(a) Roof beam.

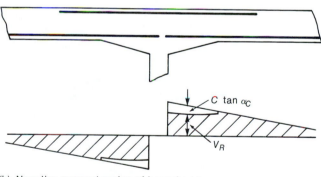

(b) Negative moment region of haunched beam.

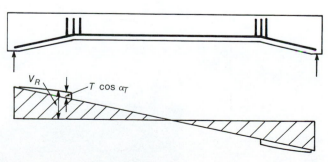

Fig. 6-44
Examples of V_R.

(c) Haunched simply supported beam.

6-9   SHEAR IN AXIALLY LOADED BEAMS OR COLUMNS

Reinforced concrete beams can be subjected to shear plus axial tensile or compressive forces due to such causes as gravity load effects in inclined members and stresses resulting from restrained shrinkage or thermal deformations. Similarly, wind or seismic forces cause shear forces in axially loaded columns. Figure 6-45 shows a tied column that failed in shear during an earthquake. The inclined crack in this column resembles Fig. 6-4b rotated through 90°.

Axial forces have three major effects on the shear strength. An axial compressive or tensile force will increase or reduce the load at which flexural and inclined cracks occur. If

Fig. 6-45
Shear failure in a tied column, 1971 San Fernando earthquake. (Photograph courtesy of U.S. National Bureau of Standards.)

V_c is assumed to be related to the inclined-cracking shear, as is done in the ACI Code, this will directly affect the design. If they have not been considered in the design, axial tensile forces may lead to premature yielding of the longitudinal reinforcement, which in turn will effectively do away with any transfer of shear by aggregate interlock.

Axial Tension

For axial tensile loadings, the nominal shear carried by the concrete is given by

$$V_c = 2\left(1 + \frac{N_u}{500A_g}\right)\sqrt{f'_c}b_w d \text{ lb} \tag{6-17b}$$
$$\text{(ACI Eq. 11-8)}$$

where N_u/A_g is expressed in psi and is negative in tension. The term inside the parentheses becomes zero when the average axial stress on the section reaches or exceeds 500 psi in tension, which is roughly the tensile strength of concrete.

In SI units, (6-17b) becomes

$$V_c = \left(1 + \frac{0.3N_u}{A_g}\right)\left(\frac{\sqrt{f'_c}}{6}\right)b_w d \tag{6-17bM}$$

This expression tends to be conservative, especially for high tensions, as shown in Fig. 6-16, and although the evidence is ambiguous, tests have shown that beams subjected to tensions large enough to crack them completely through can resist shears approaching those for beams not subjected to axial tensions [6-7]. This shear capacity results largely from aggregate interlock along the tension cracks. It should be noted, however, that if the longitudinal tension reinforcement yields under the action of shear, moment, and axial tension, the shear capacity drops very significantly. This is believed to have affected the failure of the beam shown in Fig. 6-2.

Axial Compression

Axial compression tends to increase the shear strength. The ACI Code provides two expressions to account for this increase. One of these procedures involves an extrapolation of (6-15). For reasons given when (6-15) was presented, this equation will not be used in this book. In addition, recent tests by Gupta and Collins [6-34] indicate that this procedure may be unsafe in some cases. The ACI Code also presents the following equation for calculating V_c for members subjected to combined shear, moment, and axial compression:

$$V_c = 2\left(1 + \frac{N_u}{2000A_g}\right)\sqrt{f'_c}b_w d \text{ lb} \tag{6-17a}$$
$$\text{(ACI Eq. 11-4)}$$

Here, N_u/A_g is positive in compression and has units of psi.

In SI units, (6-17a) becomes

$$V_c = \left(1 + \frac{N_u}{14A_g}\right)\left(\frac{\sqrt{f'_c}}{6}\right)b_w d \tag{6-17aM}$$

The design of a beam subjected to axial compression or tension is identical to that for a beam without such forces, except that the value of V_c is modified.

EXAMPLE 6-5 Checking the Shear Capacity of a Column Subjected to Axial Compression Plus Shear and Moments

A 12-in. × 12-in. column with $f'_c = 3000$ psi and longitudinal steel and ties having $f_y = 60,000$ psi is subjected to factored axial forces, moments, and shears, as shown in Fig. 6-46.

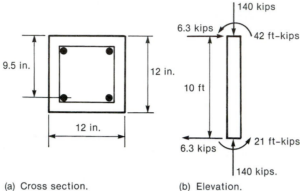

Fig. 6-46
Shear in column—
Example 6-5.

(a) Cross section.   (b) Elevation.

1. **Compute the nominal shear forces in the column.** Summing moments about the centroid at one end of the column, we find that the factored shear is

$$V_u = \frac{(42 + 21)\text{ ft-kips}}{10\text{ ft}} = 6.3\text{ kips}$$

$$V_n = \frac{V_u}{\phi} = \frac{6.3}{0.75} = 8.40\text{ kips}$$

2. **Are stirrups required by ACI Section 11.5.5.1?** No stirrups are required if $V_n < V_c/2$, where

$$V_c = 2\left(1 + \frac{N_u}{2000A_g}\right)\sqrt{f'_c}b_w d \tag{6-17a}$$

$$= 2\left(1 + \frac{140,000}{2000 \times 144}\right)\sqrt{3000} \times 12 \times \frac{9.5}{1000} = 18.6\text{ kips}$$

and $V_c/2 = 9.3$ kips. Since $V_n = 8.40$ kips is less than $V_c/2$, shear reinforcement is not necessary. If it were required, ties at a spacing of not more than $d/2$ would serve as shear reinforcement. These should be used over one-fourth of the column length from each end. ∎

6-10 SHEAR IN SEISMIC REGIONS

In seismic regions, beams and columns are particularly vulnerable to shear forces. Reversed loading cycles cause crisscrossing inclined cracks, which cause V_c to decrease and disappear. As a result, special calculation procedures and special details are required in seismic regions. (See Chapter 19.)

PROBLEMS

6-1 For the beam shown in Fig. P6-1,

    (a) Draw a shear-force diagram.

    (b) Show the direction of the principal tensile stresses at middepth at points A, B, and C.

    (c) Sketch, on a drawing of the beam, the inclined cracks that would develop at A, B, and C.

$$\phi V_n$$

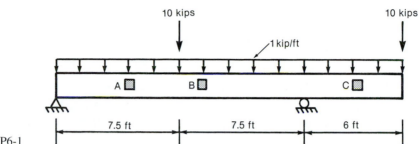

Fig. P6-1

6-2, 6-3, 6-4, and 6-5 Compute   for the cross sections shown in Figs. P6-2, P6-3, P6-4, and P6-5. In each case, use $f'_c = 3000$ psi and f_y (of the stirrups) $= 40,000$ psi.

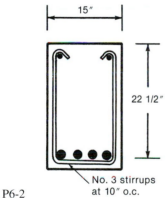

Fig. P6-2

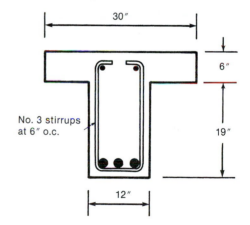

Fig. P6-3

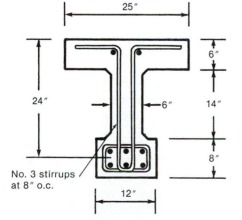

Fig. P6-4

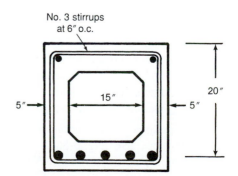

Fig. P6-5

6-6 ACI Section 12.13.1 states that "Web reinforcement shall be carried as close to the compression and tension surfaces of a member as cover requirements and proximity of other reinforcement will permit." Explain why.

6-7 ACI Section 11.5.4.1 sets the maximum spacing of vertical stirrups at $d/2$. Explain why.

6-8 Figure P6-8 shows a simply supported beam. The beam has No. 3 Grade-40 double-leg stirrups with $A_v f_y = 8.8$ kips and four No. 8 Grade-60 longitudinal bars with $A_s f_y = 190$ kips. The plastic truss model for the left end is shown in the figure. Assuming that the stirrups are all loaded to $A_v f_y$,

  (a) Use the method of joints to compute the forces in each panel of the compression and tension

chords and plot them. (See Fig. 6-21.) The force in member $L_{11}-L_{13}$ is M_{U12}/jd.

  (b) Plot $A_s f_s = M/jd$ on the diagram from part (a) and compare the bar forces from the truss model to those computed from M/jd.

  (c) Compute the compression stresses in the diagonal member L_1-U_7. (See (6-11).) The beam width, b_w, is 12 in.

6-9 The beam shown in Fig. P6-9 supports the unfactored loads shown. The dead load includes the weight of the beam.

  (a) Draw shearing-force diagrams for

    (1) factored dead and live load on the entire beam;

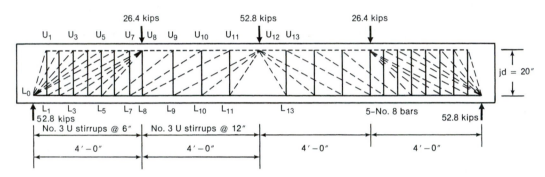

Fig. P6-8

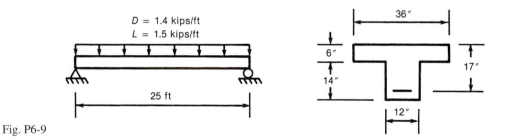

Fig. P6-9

  (2) factored dead load on the entire beam plus factored live load on the left half-span; and

  (3) factored dead load on the entire beam plus factored live load on the right half-span.

  (b) Superimpose the diagrams to get a shear-force envelope. (See Fig. 6-36e.) Compare the shear at midspan to that from (6-26).

  (c) Design stirrups. Use $f'_c = 3000$ psi and No. 3 double-leg stirrups with $f_y = 40{,}000$ psi.

6-10 The beam shown in Fig. P6-10 supports the unfactored loads shown in the figure. The dead load includes the weight of the beam.

  (a) Draw shearing-force diagrams for

    (1) factored dead and live load on the entire length of beam;

    (2) factored dead load on the entire beam plus factored live load between B and C; and

    (3) factored dead load on the entire beam plus factored live load between A and B and

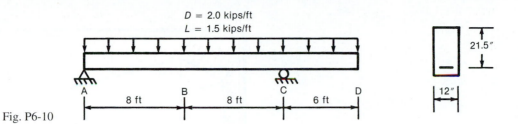

Fig. P6-10

between *C* and *D*. Loadings (2) and (3) will give the maximum positive and negative shears at *B*.

(b) Draw the shear-force envelope. The shear at B should be the dead-load shear plus or minus the shear from (6-26).

(c) Design stirrups. Use $f'_c = 3500$ psi, and let f_y of the stirrups $= 40,000$ psi.

6-11 Figure P6-11 shows an interior span of a continuous beam. The shears at the ends are $\pm w_u \ell_n/2$. The shear at midspan is from (6-26).

(a) Draw a shear-force envelope.

(b) Design stirrups, using $f'_c = 2500$ psi and $f_y = 40,000$ psi for the stirrups.

6-12 Figure P6-12 shows a rigid frame and the factored loads acting on the frame. The 7-kip horizontal load can act from the left or the right. $f'_c = 3000$ psi; $f_y = 40,000$ psi.

(a) Design stirrups in the beam.

(b) Are stirrups required in the columns? If so, design the stirrups for the columns.

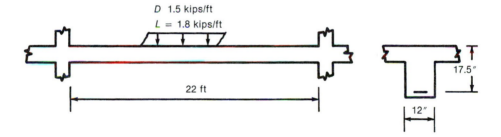

Fig. P6-11

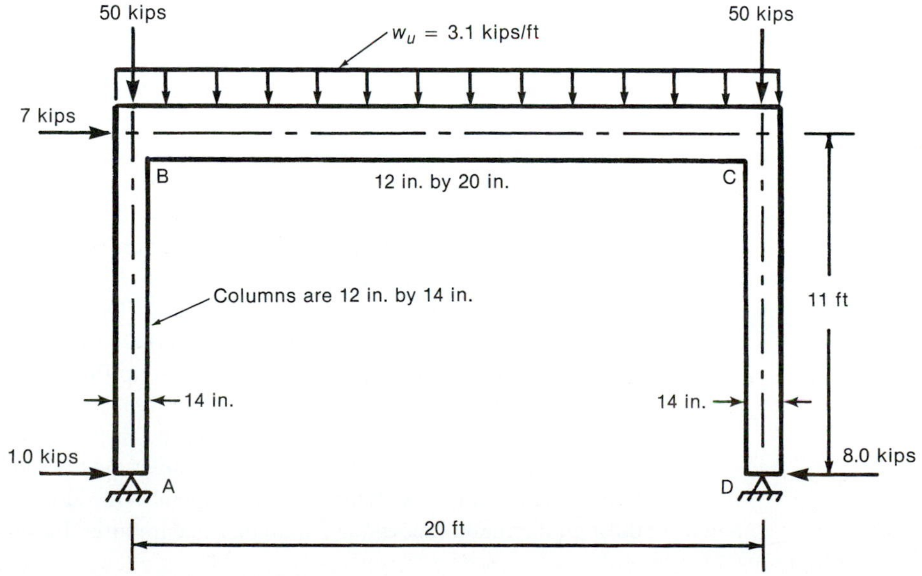

Fig. P6-12

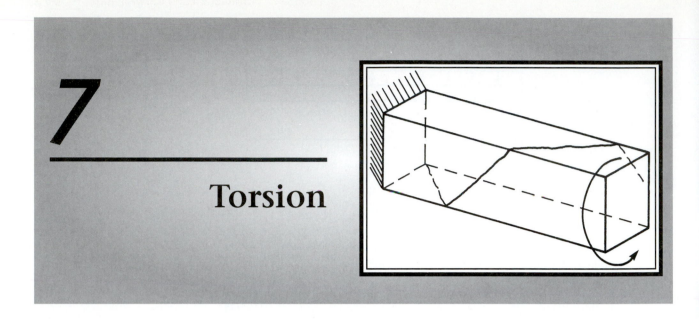

7

Torsion

A moment acting about the longitudinal axis of a member is called a *twisting moment*, a *torque*, or a *torsional moment, T*. In structures, torsion results from eccentric loading of beams, as shown in Fig. 7-20 (which will be discussed later), or from deformations resulting from the continuity of beams or similar members that join at an angle to each other, as shown in Fig. 7-21 (also discussed later).

Shearing Stresses Due to Torsion in Uncracked Members

Solid Members

In a member subjected to torsion, a torsional moment causes shearing stresses on cross-sectional planes and on radial planes extending from the axis of the member to the surface. The element shown in Fig. 7-1 is stressed in shear, τ, by the applied torque, T. In a circular member, the shearing stresses are zero at the axis of the bar and increase linearly to a maximum stress at the outside of the bar, as shown in Fig. 7-2a. In a rectangular bar, the shearing stresses vary from zero at the center to a maximum at the centers of the long sides. Around the perimeter of a square bar, the shearing stresses vary from zero at the corners to a maximum at the center of each side, as shown in Fig. 7-2b.

The distribution of shearing stresses on a cross section can be visualized by using the *soap-film analogy*. The equations for the slope of an inflated membrane are analogous to the equations for shearing stress due to torsion. Thus, the distribution of shearing stresses can be visualized by cutting an opening in a plate to the shape of the cross section that is loaded in torsion, stretching a membrane or soap film over this opening, and inflating the membrane. Figure 7-3 shows an inflated membrane over a circular opening, representing a circular shaft. The maximum slope at each point in the membrane is proportional to the shearing stress at that point. The shearing stress acts perpendicular to the direction of the line of maximum slope. Such a line is tangent to the slope at point A in Fig. 7-3. A section through the membrane along a diameter is parabolic. Its slope varies linearly from zero at

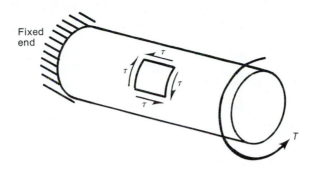

Fig. 7-1
Shear stresses due to torsion.

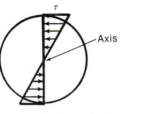

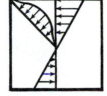

Fig. 7-2
Distribution of torsional
shear stresses in a circular bar
and in a square bar.

(a) Stress distribution
in a circular bar.

(b) Stress distribution
in a square bar.

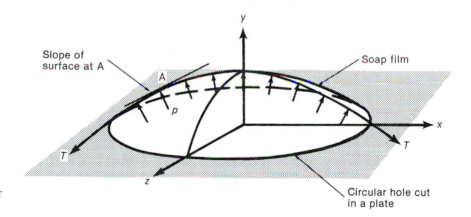

Fig. 7-3
Soap-film analogy: circular
bar.

the center to a maximum at the edge in the same way as the stress distribution plotted in
Fig. 7-2a. Figure 7-4 shows a membrane over a square opening. Here, the slopes of the radial lines correspond to the stress distribution shown in Fig. 7-2b. A similar membrane for
a U-shaped cross section made up from a series of rectangles is shown in Fig. 7-5a. The
corresponding stress distribution is shown in Fig. 7-5b. For a hollow member with continuous walls, the membrane is similar to Fig. 7-3 or 7-4, except that the region inside the hollow part is represented by a rigid plate having the shape of the hole.

The torsional moment is proportional to the volume under the membrane. A comparison of Fig. 7-4 and Fig. 7-5a shows that, for a given maximum slope corresponding to a
given maximum shearing stress, the volume under a solid figure is much greater than that
under an open figure. Thus, for a given maximum shearing stress, a solid rectangular cross
section can transmit a much higher torsional moment than can an open section. The same
is true of a hollow cross section without slits in the walls.

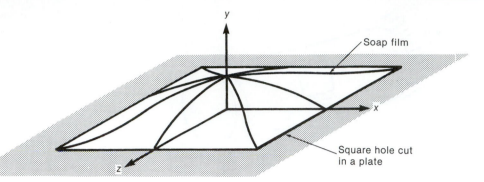

Fig. 7-4
Soap-film analogy: square bar.

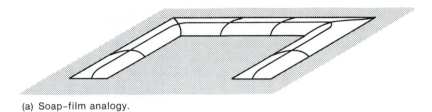

(a) Soap–film analogy.

Fig. 7-5
Soap-film analogy: channel-
shaped member.

(b) Distribution of shearing stresses.

The maximum shearing stress in an elastic circular shaft is

$$\tau_{max} = \frac{Tr}{J} \qquad (7\text{-}1)$$

where

τ_{max} = maximum shearing stress
T = torsional moment
r = radius of the bar
J = polar moment of inertia, $\pi r^4/2$

In a similar manner, the maximum shearing stress in a rectangular elastic shaft occurs at the center of the long side and can be written as

$$\tau_{max} = \frac{T}{\alpha x^2 y} \qquad (7\text{-}2)$$

where x is the shorter overall dimension of the rectangle, y is the longer overall dimension, and α varies from 0.208 for $y/x = 1.0$ (square bar) to 0.333 for $y/x = \infty$ (an infinitely wide plate [7-1]. An approximation to α is

$$\alpha = \frac{1}{3 + 1.8x/y} \tag{7-3}$$

For a cross section made up of a series of thin rectangles, such as that in Fig. 7-5,

$$\tau_{max} \simeq \frac{T}{\Sigma(x^2y/3)} \tag{7-4}$$

where the term $x^2y/3$ is evaluated for each of the rectangles.

The soap-film analogy and (7-1) to (7-4) apply to elastic bodies. For fully plastic bodies, the shearing stress will be the same at all points. Thus, the soap-film analogy must be replaced by a figure having a slope that is constant at the value corresponding to the fully plastic case, producing a cone for a circular shaft or a pyramid for a square member. Such a figure would be formed by pouring sand onto a plate having the same shape as the cross section. This is referred to as the *sand-heap analogy*. For a solid rectangular cross section, the fully plastic shearing stress is

$$\tau_p = \frac{T}{\alpha_p x^2 y} \tag{7-5}$$

where α_p varies from 0.33 for $y/x = 1.0$ to 0.5 for $y/x = \infty$.

The behavior of uncracked concrete members in torsion is neither perfectly elastic nor perfectly plastic, as assumed by the soap-film and sand-heap analogies. However, solutions based on each of these models have been used successfully to predict torsional behavior.

Hollow Members

Figure 7-6a shows a thin-walled tube with continuous walls subjected to a torque about its longitudinal axis. An element $ABCD$ cut from the wall is shown in Fig. 7-6b. The thicknesses of the walls along sides AB and CD are t_1 and t_2, respectively. The applied torque causes shearing forces V_{AB}, V_{BC}, V_{CD}, and V_{DA} on the sides of the element as shown, each equal to the shearing stresses on that side of the element times the area of the side. From $\Sigma F_x = 0$, we find that $V_{AB} = V_{CD}$; but $V_{AB} = \tau_1 t_1 dx$ and $V_{CD} = \tau_2 t_2 dx$, which together give $\tau_1 t_1 = \tau_2 t_2$, where τ_1 and τ_2 are the shearing stresses acting on sides AB and CD, respectively. The product τt is referred to as the *shear flow, q*.

For equilibrium of a smaller element at corner B of the element in Fig. 7-6b, $\tau_1 = \tau_3$; similarly, at C, $\tau_2 = \tau_4$, as shown in Fig. 7-6c and d. Thus, at points B and C on the perimeter of the tube, $\tau_3 t_1 = \tau_4 t_2$. This shows that, for a given applied torque, T, the shear flow, q, is constant around the perimeter of the tube. The shear flow has units of (stress × length) pounds force per inch (N/mm). The name *shear flow* comes from an analogy to water flowing around a circular flume. The volume of water flowing past any given point in the flume is constant at any given period of time.

Figure 7-6e shows an end view of the tube. The torsional shear force acting on the length ds of wall is $q\,ds$. The perpendicular distance from this force to the centroidal axis of the tube is r, and the moment of this force about the axis is $rq\,ds$, where r is measured from midplane of the wall because that is the line of action of the force $q\,ds$. Integrating around the perimeter gives the torque in the tube:

$$T = \int_p rq\,ds \tag{7-6}$$

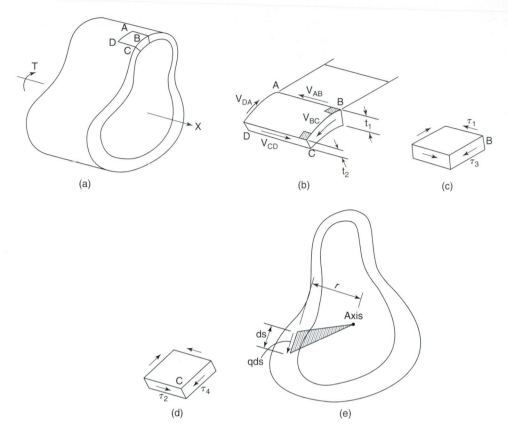

Fig. 7-6
Shear stresses in a thin-walled tube. (From [7-2] Popov, E. P., *Mechanics of Materials*, SI Version, 2/e (c) 1978, p. 80. Reprinted by permission of Prentice Hall, Upper Saddle River, New Jersey.)

where $\int_p$ denotes integration around the perimeter of the tube. However, q is constant around the perimeter of the tube, so q can be moved outside the integral giving:

$$T = q \int_p r\, ds \tag{7-7}$$

The shaded triangle in Fig. 7-6e has an area of $r\,ds/2$. Thus, $r\,ds$ in (7-7) is two times the area of the shaded triangle stretching between the elemental length of perimeter, ds, and the axis of the tube. Furthermore, $\int_p r\,ds$ is equal to two times the area enclosed by the centerline of the wall thickness. This area is referred to as the *area enclosed by the shear flow path*, A_o. For the cross section shown in Fig. 7-7, A_o is the area, including the area of the hole in the center of the tube. Equation (7-7) becomes

$$T = 2qA_o \tag{7-8}$$

where $q = \tau t$. Rearranging gives

$$\tau = \frac{T}{2A_o t} \tag{7-9}$$

where t is the wall thickness at the point where the shear stress, τ, due to torsion is being computed. The maximum torsional shear stress occurs where the wall thickness is the least. For the hollow trapezoidal bridge cross section shown in Fig. 7-7, for example, this would be in the lower flange.

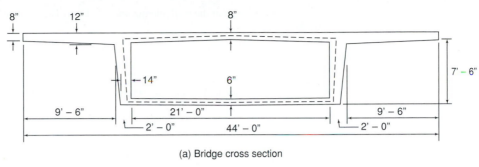

(a) Bridge cross section

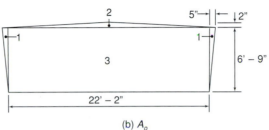

(b) A_o

Fig. 7-7
Cross section of a bridge—
Example 7-1.

This analysis applies only if the walls of the tube are continuous (no slits parallel to the axis of the tube) or if, in the case of a solid member, the member can be approximated as a tube with continuous walls. Equation (7-9) can be applied to either elastic or inelastic sections. Consider the shape of the membrane in Fig. 7-3. A tube can be called thin walled if the change in slope of the membrane is small across the thickness of the wall and can be ignored without serious loss of accuracy.

EXAMPLE 7-1   Compute Torsional Shear Stresses in a Bridge Cross Section, Using Thin-Walled Tube Theory

Figure 7-7a shows the cross section of a bridge. Compute the shear stresses, τ, at the top and bottom of the walls and in the lower flange that are due to an applied torque of 1650 ft-kips.

1.   Compute A_o. A_o is the area enclosed by the midplane of the walls of the tube. The dashed line in Fig. 7-7a is the perimeter of A_o. The protruding deck flanges are not part of the tube and are ignored in computing A_o. Divide A_o into triangles and a rectangle as shown in Fig. 7-7b. Then

$$A_O = (2 \times 6'9'' \times 5''/2) + (23' \times 2''/2) + (22'2'' \times 6'9'')$$

$$= 405 + 276 + 21{,}546$$

$$= 22{,}227 \text{ in.}^2$$

2.   Compute the shear flow, q. From (7-8),

$$q = \frac{T}{2A_O} = \frac{1650 \times 12{,}000}{2 \times 22{,}227}$$

$$= 445 \text{ lb/in.}$$

3.   Compute the shear stresses. At the top of the wall, the thickness, t, is 24 in. The torsional shear stress at the top of the walls is

$$\tau = q/t = 445/24$$

$$= 18.6 \text{ psi}$$

At the bottom of the wall, the thickness is 14 in. The torsional shear stress at the bottom of the walls is

$$\tau = q/t = 445/14$$
$$= 31.8 \text{ psi}$$

The thickness of the bottom flange is 6 in. The torsional shear stress in the bottom flange is

$$\tau = q/t = 445/6$$
$$= 74.2 \text{ psi} \qquad \blacksquare$$

This example illustrates the calculation of the torsional shear stress, τ. In design, (7-8) is written in a slightly different form, as is discussed in Section 7-4.

Principal Stresses Due to Torsion

When the beam shown in Fig. 7-8 is subjected to a torsional moment, T, shearing stresses develop on the top and front faces, as shown by the elements in Fig. 7-8a. The principal stresses on these elements are shown in Fig. 7-8b. The principal tensile stress equals the principal compressive stress, and both are equal to the shear stress if T is the only loading. The principal tensile stresses eventually cause cracking that spirals around the body, as shown by the line A–B–C–D–E in Fig. 7-8c.

In a reinforced concrete member, such a crack would cause failure unless it was crossed by reinforcement. This generally takes the form of longitudinal bars in the corners and stirrups. Since the crack spirals around the body, four-sided (closed) stirrups are required.

Principal Stresses due to Torsion and Shear

If a beam is subjected to combined shear and torsion as shown in Fig. 7-9, the two shearing-stress components add on one side face and counteract each other on the other, as shown in Fig. 7-9a and b. As a result, inclined cracking starts on the face where the stresses add (crack AB) and extends across the flexural tensile face of the beam (in this case the top since this is a cantilever beam). If the bending moments are sufficiently large, the cracks will extend almost vertically across the back face, as shown by crack CD in Fig. 7-9c. The flexural compression zone near the bottom of the beam prevents the cracks from extending the full height of the front and back faces.

Circulatory Torsion and Warping Torsion

Members subjected to torsion can be divided into two families, distinguished by how the torsion is resisted. If the cross section is solid (either square, rectangular, circular, or polygonal) or is a closed tube, torsion is resisted by torsional stresses, τ, which act in a continuous manner around the section, as shown in Figs. 7-2, 7-5, and 7-6. In the closed tube, these stresses can be represented by a shear flow, $q = \tau t$, which is constant around the circumference. This is referred to as *circulatory torsion* or *St. Venant torsion* after the French mathematician who derived the equations for torsional stresses in noncircular cross sections in 1853 and developed the soap-film analogy.

In a circular bar, the torsional stresses are constant around the circumference of the bar, as shown in Figs. 7-2a and 7-3. As a result, the shear strain is constant around the circumference, so planar cross sections perpendicular to the axis of the bar remain planar under load. In a bar with a rectangular cross section, however, the torsional stresses vary from a maximum at the middle of the long sides of the rectangle to zero at the corners, as

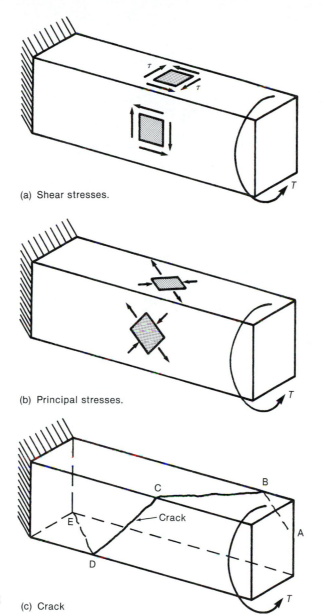

(a) Shear stresses.

(b) Principal stresses.

(c) Crack

Fig. 7-8
Principal stresses and cracking
due to pure torsion.

shown in Figs. 7-2b and 7-4. As a result, the shear strain varies around the circumference of the section, causing the section to deform in such a manner that plane sections through the bar do not remain plane. This distortion is referred to as *warping*. If the warping deformations are restrained, a part (or all) of the torsion is resisted by *warping torsion*. Warping torsion generally occurs in a cross section consisting of three or more walls connected together to form a channel section or an I-beam section.

There is no absolute demarcation between members with circulatory torsion and those with warping torsion. Frequently, both types of torsion will be present in the same member, the relative amounts changing from section to section. Two cases will be considered in the following paragraphs, each giving a different distribution of the warping torsion and circulatory torsion.

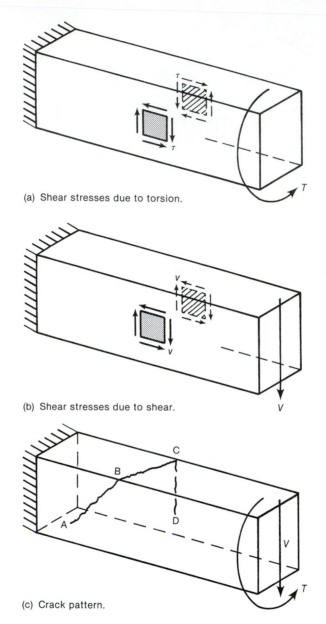

(a) Shear stresses due to torsion.

(b) Shear stresses due to shear.

(c) Crack pattern.

Fig. 7-9
Combined shear, torsion, and
moment.

Figure 7-10 shows a steel cantilever I beam loaded by a torque, T_u, at the free end. Near the free end, end $A-B$, T_u is resisted mainly by circulatory torsion. Near the support end $C-D$ the torque is resisted mainly by shear forces in the flanges:

$$V_{fu} = T_u/h$$

These forces act at the midthickness of the flanges, with h the distance between the resultant forces in the two flanges. The forces V_{fu} can be idealized as acting at a distance a from the end, causing moments of $M_u = V_{fu} \times a$ in each flange at the fixed end.

$$a = (h/2)\sqrt{\frac{EI}{JG}} \qquad (7\text{-}10)$$

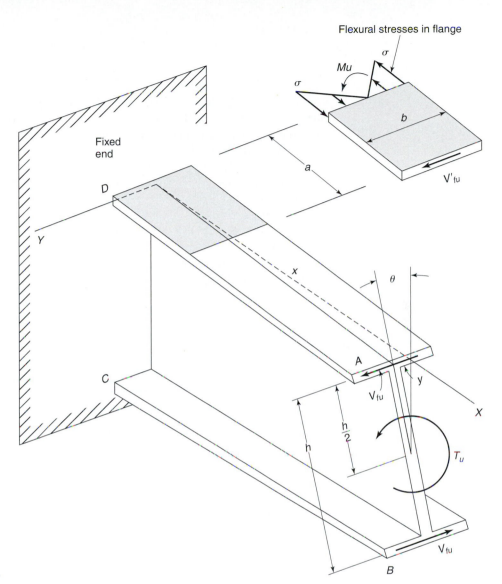

Fig. 7-10
Warping torsion of an I beam.

where

> E and G are the modulus of elasticity and the shearing modulus, respectively;
>
> I is the moment of inertia of the entire section about a plane of symmetry in the web so that the moment of inertia of one flange is approximately $I/2$; and
>
> J is the polar moment of inertia of the cross section.

The moment, M_u, causes flexural stresses, σ, at the fixed end of the flanges equal to

$$\sigma = \frac{M_u \times b/2}{I_{f\ell}} = \left(\frac{T_u a}{h}\right) \times \left(\frac{b/2}{tb^3/12}\right) \tag{7-11}$$

or

$$\sigma = \frac{6T_u a}{tb^3/12} \qquad (7\text{-}12)$$

where t and b are the thickness and width of the flange and h is the height of the I beam, center to center of flanges. Beyond a distance a from the fixed end, all of the torsion can be assumed to be resisted by circulatory torsion.

Another frequent case is a bridge consisting of two girders and a slab, as shown in Fig. 7-11. The bridge is loaded by a torque, T_u, at midspan. Half of this is resisted by each end of the bridge, as shown by the torque diagram in Fig. 7-11b. The cross sections immediately to the left of the applied torque are restrained against warping by the cross sections on the right of the loading point, which have torsional stresses of the opposite sign. The cross sections near the ends of the beam are free to warp.

The balance of this chapter will consider only circulatory torsion. Analyses of structures subjected to warping torsion are presented in books on advanced strength of materials or on bridge design [7-3].

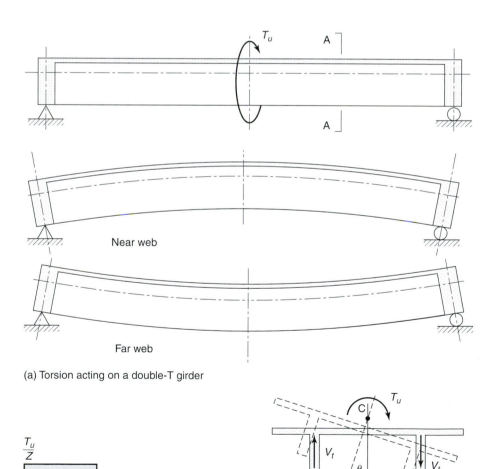

(a) Torsion acting on a double-T girder

Fig. 7-11
Warping torsion on a bridge.
(From [7-3].)

(b) Torque diagram

(c) Deflected position of Section A–A

7-2   BEHAVIOR OF REINFORCED CONCRETE MEMBERS SUBJECTED TO TORSION

Pure Torsion

When a concrete member is loaded in pure torsion, shearing stresses, and principal stresses develop as shown in Fig. 7-8a and b. One or more inclined cracks develop when the maximum principal tensile stress reaches the tensile strength of the concrete. The onset of cracking causes failure of an unreinforced member. Furthermore, the addition of longitudinal steel without stirrups has little effect on the strength of a beam loaded in pure torsion because it is effective only in resisting the longitudinal component of the diagonal tension forces.

A rectangular beam with longitudinal bars in the corners and closed stirrups can resist increased load after cracking. Figure 7-12 is a torque-twist curve for such a beam. At the cracking load, point A in Fig. 7-12, the angle of twist increases without an increase in torque as some of the forces formerly in the uncracked concrete are redistributed to the reinforcement. The cracking extends into the central core of the member, rendering the core ineffective. Figure 7-13 compares the strengths of a series of solid and hollow rectangular beams with the same exterior size and increasing amounts of both longitudinal and stirrup reinforcement [7-4]. Although the cracking torque was lower for the hollow beams, the ultimate strengths were the same for solid and hollow beams having the same reinforcement, indicating that the strength of a cracked reinforced concrete member loaded in pure torsion is governed by the outer skin or tube of concrete containing the reinforcement.

After the cracking of a reinforced beam, failure may occur in several ways. The stirrups, or longitudinal reinforcement, or both, may yield, or, for beams that are *over-reinforced* in torsion, the concrete between the inclined cracks may be crushed by the principal compression stresses prior to yield of the steel. The most ductile behavior results when both reinforcements yield.

Combined Torsion, Moment, and Shear

Torsion seldom occurs by itself. Generally, there are also bending moments and shearing forces. Test results for beams without stirrups, loaded with various ratios of torsion and

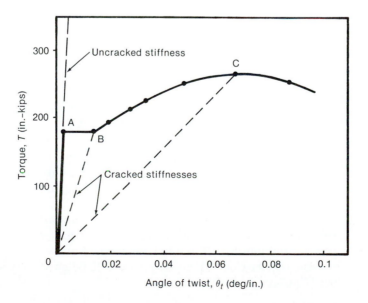

Fig. 7-12
Torque twist curve for a rectangular beam. (From [7-4].)

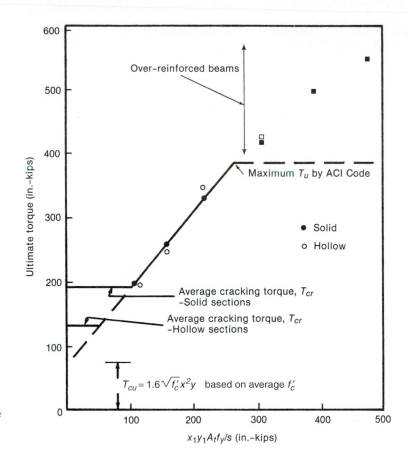

Fig. 7-13
Torsional strength of solid and hollow sections with the same outside dimensions. (From [7-4].)

shear, are plotted in Fig. 7-14. [7-5] The lower envelope to the data is given by the quarter ellipse

$$\left(\frac{T_c}{T_{cu}}\right)^2 + \left(\frac{V_c}{V_{cu}}\right)^2 = 1 \qquad (7\text{-}13a)$$

where, in this graph, $T_{cu} = 1.6\sqrt{f'_c}\,x^2 y$ and $V_{cu} = 2.68\sqrt{f'_c}\,b_w d$. These beams failed at or soon after inclined cracking.

7-3   DESIGN METHODS FOR TORSION

Two very different theories are used to explain the strength of reinforced concrete members. The first, based on a *skew bending theory* developed by Lessig [7-6] and extended by Hsu [7-4] was the basis for the torsion-design provisions in the 1971 through 1989 ACI Codes. This theory assumes that some shear and torsion is resisted by the concrete, the rest by shear and torsion reinforcement. The mode of failure is assumed to involve bending on a skew surface resulting from the crack's spiraling around three of the four sides of the member, as shown in Fig. 7-9c.

The second design theory is based on a *thin-walled tube/plastic space truss model*, similar to the plastic-truss analogy presented in Chapter 6. This theory, presented by Lampert and Thürlimann [7-7] and by Lampert and Collins [7-8], forms the basis of the torsion provisions in the Comité Euro-International du Béton Model code [7-9] and, since 1995, in the ACI Code Section 11.6.

7-4   THIN-WALLED TUBE/PLASTIC SPACE TRUSS DESIGN METHOD

This model for the torsional strength of beams combines the thin-walled tube analogy from Fig. 7-6 with the plastic-truss analogy for shear presented in Section 6-4. This gives a mechanics-based model of the behavior that is easy to visualize and leads to much simpler calculations than the skew bending theory.

Both solid and hollow members are considered as tubes. Test data for solid and hollow beams in Fig. 7-13 suggest that, once torsional cracking has occurred, the concrete in the center of the member has little effect on the torsional strength of the cross section and hence can be ignored. This, in effect, produces an equivalent tubular member.

Torsion is assumed to be resisted by shear flow, q, around the perimeter of the member as shown in Fig. 7-15a. The beam is idealized as a thin-walled tube. After cracking, the tube is idealized as a hollow truss consisting of closed stirrups, longitudinal bars in the corners, and compression diagonals approximately centered on the stirrups, as shown in Fig. 7-15b. The diagonals are idealized as being between cracks that are at an angle θ, generally taken as 45° for reinforced concrete.

The derivation of the thin-walled tube/plastic space truss method as used in the ACI Code was presented and compared to tests in [7-10]. In this book, the analogy is referred to as the thin-walled tube analogy because this is the terminology used in the derivation of (7-9) in mechanics of materials textbooks. The walls of the tube are actually quite thick, being on the order of one-sixth to one-quarter of the smaller side of a rectangular member.

Lower Limit on Consideration of Torsion

Torsional reinforcement is not required if torsional cracks do not occur. In pure torsion, the principal tensile stress, σ_1, is equal to the shear stress, τ, at a given location. Thus, from (7-9) for a thin-walled tube,

$$\sigma_1 = \tau = \frac{T}{2A_o t} \tag{7-14}$$

Solid section

To apply this to a solid section, it is necessary to define the wall thickness and enclosed area of the equivalent tube prior to cracking. ACI Section 11.6.1 is based on the assumption that, prior to any cracking, the wall thickness, t, can be taken equal to $3A_{cp}/4p_{cp}$, where p_{cp} is the perimeter of the concrete section and A_{cp} is the area enclosed by this perimeter. The area, A_o, enclosed by the centerline of the walls of the tube is taken as $2A_{cp}/3$. Substituting these expressions into (7-14) gives

$$\sigma_1 = \tau = \frac{T p_{cp}}{A_{cp}^2} \tag{7-15}$$

Torsional cracking is assumed to occur when the principal tensile stress reaches the tensile strength of the concrete in biaxial tension–compression, taken as $4\sqrt{f_c'}$. Thus, the torque at cracking is

$$T_{cr} = 4\sqrt{f_c'}\left(\frac{A_{cp}^2}{p_{cp}}\right) \tag{7-16}$$

The tensile strength was taken as $4\sqrt{f_c'}$, which is smaller than the $6\sqrt{f_c'}$ used elsewhere because, as is shown in Fig. 3-12a, the tensile strength under biaxial compression and tension is less than that in uniaxial tension.

In combined shear and torsion, the inclined cracking load follows a circular interaction diagram similar to that in Fig. 7-14:

$$\left(\frac{T}{T_{cr}}\right)^2 + \left(\frac{V}{V_{cr}}\right)^2 = 1 \qquad (7\text{-}13b)$$

where V_{cr} is the inclined cracking shear in the absence of torque and T_{cr} is the cracking torque in the absence of shear. If $T = 0.25T_{cr}$, the reduction in the inclined cracking shear is:

$$\left(\frac{V}{V_{cr}}\right) = \sqrt{1 - \left(\frac{0.25T_{cr}}{T_{cr}}\right)^2} \qquad (7\text{-}17)$$

and for this case

$$V = 0.97V_{cr}$$

Thus, the existence of a torque equal to a quarter of the inclined cracking torque will reduce the inclined cracking shear by only 3 percent. This was deemed to be negligible. The *threshold torsion* below which torsion can be ignored in a solid cross section is

$$T_{th} = \phi\sqrt{f_c'}\left(\frac{A_{cp}^2}{p_{cp}}\right) \qquad (7\text{-}18)$$

Thin-walled hollow section

For a *thin-walled hollow section* the interaction diagram approaches a straight line joining $V/V_{cr} = 1.0$ and $T/T_{cr} = 1.0$. As a result, a torsion equal to $0.25T_{cr}$ would reduce the

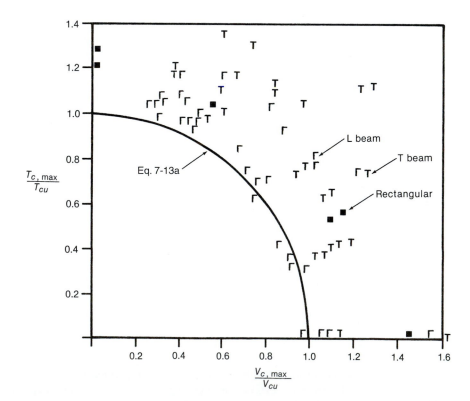

Fig. 7-14
Interaction of torsion and shear. (From [7-5].)

inclined cracking shear to $0.75V_{cr}$, a 25 percent reduction. This was believed to overestimate the reduction in the shear at cracking. ACI Section 11.6.1 replaces A_{cp} in (7-18) with A_g, the area of the concrete in the cross section, not including the area of the voids. This is intended to do two things: First, in tests in [7-4] the cracking load was reduced to A_g/A_{cp} times the cracking load of a solid section. Second, the interaction diagram is somewhere between a circular arc and a straight line depending on the wall thickness relative to the overall dimensions of the member. This is approximated by multiplying the threshold torque by A_g/A_{cp} a second time.

For an isolated beam, A_{cp} is the area enclosed by the perimeter of the section, including the area of any holes, and p_{cp} is the perimeter of the section. For a beam cast monolithically with a floor slab, ACI Section 11.6.1 states that the overhanging flange width included is that defined in ACI Section 13.2.4, which assumes that the overhanging flange extends the greater of the distances that the beam web projects above or below the flange, but not more than four times the slab thickness.

In SI units, (7-18) becomes

$$T_{thM} = \frac{\phi\sqrt{f'_c}}{12}\left(\frac{A_{cp}^2}{P_{cp}}\right) \tag{7-18M}$$

The torsion computed with (7-18) is referred to as the *threshold torque*.

The inclined cracking load of a hollow member in pure torsion is approximately A_g/A_{cp} times the cracking load of a solid member with the same outside dimensions [7-4], as shown in Fig. 7-13. The interaction diagram for cracking due to shear and torsion plots as a quarter of a circle for a solid member, as shown in Fig. 7-14.

For a hollow section, the interaction diagram approaches a straight line as A_g/A_{cp} decreases, where A_g is the area of the concrete only in a cross section and A_{cp} is the total area enclosed by the perimeter. As a result, a smaller torsion is required to cause a significant reduction in the shear cracking load of a hollow section compared to that of a solid section. The effects of the reduced cracking load and the change in the shape of the interaction diagram on the shear cracking load are estimated for design by replacing A_{cp} in (7-18) with A_g, giving

$$T_{th,H} = \phi\sqrt{f'_c}\left(\frac{A_g^2}{p_{cp}}\right) \tag{7-19}$$

Area of Stirrups for Torsion

A beam subjected to pure torsion can be modeled as shown in Fig. 7-15a and b. A rectangular beam will be considered for simplicity, but a similar derivation could be applied to any cross-sectional shape. The beam is idealized as a space truss consisting of longitudinal bars in the corners, closed stirrups, and diagonal concrete compression members that spiral around the beam between the cracks. The height and width of the truss are y_o and x_o, measured between the centers of the corner bars. The angle of the cracks is θ, which initially is close to 45°, but may become flatter at high torques.

To calculate the required area of stirrups, it is necessary to resolve the shear flow into shear forces acting on the four walls of the tube, as shown in Fig. 7-15b. From (7-8), the shear force per unit length of the perimeter of the tube or truss, referred to as the shear flow, q, is given by

$$q = \frac{T}{2A_o} \tag{7-8b}$$

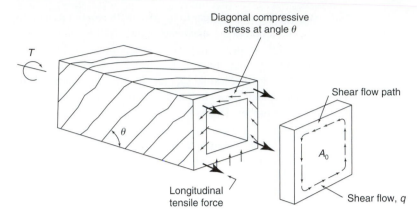

(a) Thin-walled tube analogy

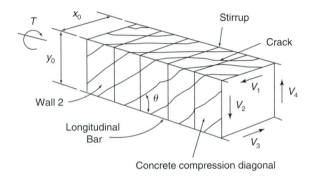

Fig. 7-15
Thin-walled tube analogy and
space truss analogy.

(b) Space truss analogy

The total shear force due to torsion along each of the top and bottom sides of the truss is

$$V_1 = V_3 = \frac{T}{2A_o} x_o \qquad (7\text{-}20a)$$

Similarly, the shear forces due to torsion along each of the two vertical sides are

$$V_2 = V_4 = \frac{T}{2A_o} y_o \qquad (7\text{-}20b)$$

Summing moments about one corner of the truss, we find that the internal torque is

$$T = V_1 y_o + V_2 x_o$$

Substituting for V_1 and V_2 in (7-20a) and (7-20b) gives

$$T = \left(\frac{T}{2A_o} x_o\right) y_o + \left(\frac{T}{2A_o} y_o\right) x_o \qquad (7\text{-}21)$$

or

$$T = \frac{2T(x_o y_o)}{2A_o} \qquad (7\text{-}22)$$

By definition, however, $x_o y_o = A_o$. Thus, we have shown that the internal forces V_1 through V_4 equilibrate the applied torque, T.

A portion of one of the vertical sides is shown in Fig. 7-16. The inclined crack cuts

$$n_2 = \frac{y_o \cot \theta}{s}$$

stirrups, where s is the spacing of the stirrups. The force in the stirrups must equilibrate V_2. Assuming that all the stirrups yield at ultimate, we have

$$V_2 = \frac{A_t f_{yv} y_o}{s} \cot \theta \qquad (7\text{-}23)$$

where f_{yv} is the yield strength of the stirrups. Replacing V_2 with (7-20b) and taking T equal to the nominal torsion capacity, T_n, gives

$$T_n = \frac{2 A_o A_t f_{yv}}{s} \cot \theta \qquad (7\text{-}24)$$
$$\text{(ACI Eq. 11-21)}$$

where θ may be taken as any angle between 30 and 60°. For nonprestressed concrete, ACI Section 11.6.3.6 suggests that θ be taken as 45°, because this corresponds to the angle assumed in the derivation of the equation for designing stirrups for shear. The factors affecting the choice of θ are discussed later in this section.

After inclined cracking, the torsional resistance is provided by the tube formed by the walls of the space truss. ACI Section 11.6.3.6 allows the area A_o to be taken as $0.85 A_{oh}$, where A_{oh} is the area enclosed by the outermost closed stirrups.

Hsu [7-11] has given the following more accurate values of t and A_o:

$$t = \frac{4 T_n}{A_c f'_c} \qquad (7\text{-}25)$$

$$A_o = A_{cp} - \frac{2 T_n p_{cp}}{A_{cp} f'_c} \qquad (7\text{-}26)$$

Here, p_{cp} and A_{cp} are the perimeter of the concrete section and the area enclosed by that perimeter, respectively, and T_n is the nominal torsion resistance. For very large members, the use of (7-25) and (7-26) may lead to smaller amounts of torsional reinforcement than that of (7-24). A_o will be taken as $0.85 A_{oh}$ in this book.

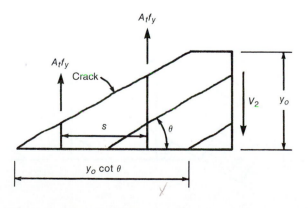

Fig. 7-16
Forces in stirrups.

Area of Longitudinal Reinforcement

The longitudinal reinforcement must be proportioned to resist the longitudinal tension forces that occur in the space truss. As shown by the force triangle in Fig. 7-17, the shear force V_2 can be replaced with a diagonal compression force, D_2, parallel to the concrete struts and an axial tension force, N_2, where D_2 and N_2 are respectively given by

$$D_2 = \frac{V_2}{\sin \theta} \tag{7-27}$$

and

$$N_2 = V_2 \cot \theta \tag{7-28}$$

Because the shear flow, q, is constant from point to point along side 2, the force N_2 acts along the centroidal axis of side 2. For a beam with longitudinal bars in the top and bottom corners of side 2, half of N_2 will be resisted by each corner bar. A similar resolution of forces occurs on each side of the truss. For a rectangular member, as shown in Fig. 7-15b, the total longitudinal force is

$$N = 2(N_1 + N_2)$$

Substituting (7-20a and b) and (7-28) and taking T equal to T_n gives

$$N = \frac{T_n}{2A_o} 2(x_o + y_o) \cot \theta \tag{7-29}$$

where $2(x_o + y_o)$ is the perimeter of the closed stirrup, p_h. Longitudinal reinforcement with a total area of A_ℓ must be provided for the longitudinal force, N. Assuming that this reinforcement yields at ultimate, with a yield strength of $f_{y\ell}$, produces

$$A_\ell f_{y\ell} = N$$

or

$$A_\ell = \frac{T_n p_h}{2A_o f_{y\ell}} \cot \theta$$

and substituting $A_o = 0.85 \, A_{oh}$ and T_u/ϕ for T_n gives

$$A_\ell = \frac{(T_u/\phi)}{1.7 A_{oh} f_{y\ell}} \cot\theta \tag{7-30}$$

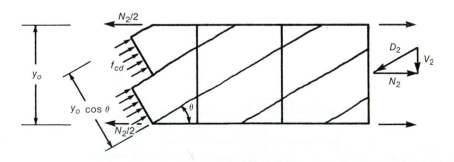

Fig. 7-17
Side of space truss—replacement of shear force V_2.

Alternatively, A_ℓ can be expressed in terms of the area of the torsional stirrups. Substituting (7-24) into (7-30) gives

$$A_\ell = \left(\frac{A_t}{s}\right) p_h \left(\frac{f_{yv}}{f_{y\ell}}\right) \cot^2 \theta \tag{7-31}$$

(ACI Eq. 11-22)

Although the ACI Code gives (7-31) to compute A_ℓ, it frequently is easier to compute A_ℓ from (7-30) instead, because the term (A_t/s) is avoided. Because the individual wall tension forces N_1, N_2, N_3, and N_4 act along the centroidal axes of the side in question, the total force, N, acts along the centroidal axis of the member. For this reason, the longitudinal torsional reinforcement must be distributed evenly around the perimeter of the cross section so that the centroid of the bar areas coincides approximately with the centroid of the member. One bar is placed in each corner of the stirrups to anchor the compression struts where the compressive forces change direction around the corner.

Combined Shear and Torsion

In ACI Codes prior to 1995, a portion T_c of the torsion was carried by concrete and a portion V_c of the shear was carried by concrete. When both shear and torsion acted, an elliptical interaction diagram was assumed between T_c and V_c, and stirrups were provided for the rest of the torsion and shear. The derivation of (7-24) and (7-31) for the space truss analogy assumed that all the torsion was carried by reinforcement, T_s, without any "torsion carried by concrete," T_c. When shear and torsion act together, the 1995 and subsequent ACI Codes assume that V_c remains constant and T_c remains equal to zero, so that

$$V_n = V_c + V_s \tag{6-9}$$

(ACI Eq. 11-2)

$$T_n = T_s \tag{7-32}$$

where V_c is given by (6-9) (ACI Eq. (11-2)). The assumption that there is no interaction between V_c and T_c greatly simplifies the calculations, compared with those required by the ACI codes prior to 1995. Design comparisons carried out by ACI Committee 318 showed that, for combinations of low V_u and high T_u, with v_u less than about $0.8(\phi 2\sqrt{f_c'})$ psi, the 1995 code method requires more stirrups than are required by previous ACI Codes. For v_u greater than this value, the thin-walled-tube method requires the same or marginally fewer stirrups than the skew-bending method.

Maximum Shear and Torsion

A member loaded by torsion or by combined shear and torsion may fail by yielding of the stirrups and longitudinal reinforcement, as assumed in the derivation of (7-24) and (7-31), or by crushing of the concrete due to the diagonal compressive forces, D_2, shown in Fig. 7-17. A serviceability failure may occur if the inclined cracks are too wide at service loads. The limit on combined shear and torsion in ACI Section 11.6.3.1 was derived to limit service-load crack widths, but as is shown later, it also gives a lower bound on the web's crushing capacity.

Crack Width Limit

As was explained in Section 6-5, ACI Section 11.5.6.9 attempts to guard against excessive crack widths by limiting the maximum shear, V_s, that can be transferred by stirrups to on upper limit of $8\sqrt{f_c'}b_w d$. In ACI Section 11.6.3.1, the same concept is used, expressed in terms of stresses. The shear stress, v due to direct shear is $V_u/b_w d$. From (7-9), with A_o after torsional cracking taken as $0.85A_{oh}$ and $t = A_{oh}/p_h$, the shear stress, τ, due to torsion is $T_u p_h/(1.7A_{oh}^2)$. In a hollow section, these two shear stresses are additive on one side, at point A in Fig. 7-18a, and the limit is given by

$$\frac{V_u}{b_w d} + \frac{T_u p_h}{1.7A_{oh}^2} \leq \phi\left(\frac{V_c}{b_w d} + 8\sqrt{f_c'}\right) \qquad (7\text{-}33a)$$
$$(\text{ACI Eq. 11-19})$$

If the wall thickness varies around the cross section, as for example in Fig. 7-7, ACI Section 11.6.3.2 states that (7-33a) is evaluated at the location where the left-hand side is the greatest.

If a hollow section has a wall thickness, t, less than A_{oh}/p_h, ACI Section 11.6.3.3 requires that the actual wall thickness be used. Thus, the second term of (7-33a) becomes $T_u/(1.7A_{oh}t)$. Alternatively, the second term on the left-hand side of (7-33a) can be taken as $T_u/(A_o t)$, where A_o and t are computed as in Example 7-1.

In a solid section, the shear stresses due to direct shear are assumed to be distributed uniformly across the width of the web, while the torsional shear stresses exist only in the walls of the thin-walled tube, as shown in Fig. 7-18b. In this case, a direct addition of the two terms tends to be conservative, so a root-square summation is used instead:

$$\sqrt{\left(\frac{V_u}{b_w d}\right)^2 + \left(\frac{T_u p_h}{1.7A_{oh}^2}\right)^2} \leq \phi\left(\frac{V_c}{b_w d} + 8\sqrt{f_c'}\right) \qquad (7\text{-}33b)$$
$$(\text{ACI Eq. 11-18})$$

The right-hand sides of (7-33a) and (7-33b) include the term $V_c/b_w d$; hence, the same equation can be used for prestressed concrete members and for members with axial tension or compression that have different values of V_c.

In SI units, the right-hand sides of (7-33a) and (7-33b) become

$$\phi\left(\frac{V_c}{b_w d} + \frac{8\sqrt{f_c'}}{12}\right) \qquad (7\text{-}33bM)$$

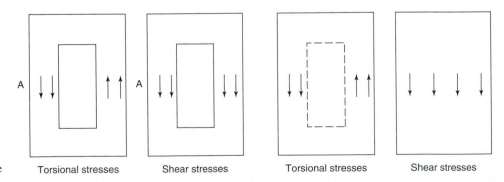

Fig. 7-18
Addition of shear stresses due to torsion and shear.
(From [7-10].)

| Torsional stresses | Shear stresses | Torsional stresses | Shear stresses |

(a) Hollow section   (b) Solid section

Web Crushing Limit

Failure can also occur due to crushing of the concrete in the walls of the tube due to the inclined compressive forces in the struts between cracks. As will now be shown, this sets a higher limit on the stresses than do (7-33a) and (7-33b).

The diagonal compressive force in a vertical side of the member shown in Fig. 7-17 is given by (7-27). This force acts on a width $y_o \cos \theta$, as shown in Fig. 7-17. The resulting compressive stress due to torsion is

$$f_{cd} = \frac{V_2}{t y_o \cos \theta \sin \theta} \tag{7-34}$$

Substituting (7-9), again taking A_o equal to $0.85 A_{oh}$ and approximating t as A_{oh}/p_h, gives

$$f_{cd} = \frac{T_u p_h}{1.7 A_{oh}^2 \cos \theta \sin \theta} \tag{7-35}$$

The compressive stresses due to shear may be calculated in a similar manner as

$$f_{cd} = \frac{V_u}{b_w d \cos \theta \sin \theta} \tag{7-36}$$

For a solid section, these will be "added" via root square, as explained in the derivation of (7-33b), giving

$$f_{cd} = \sqrt{\left(\frac{V_u}{b_w d \cos \theta \sin \theta} \right)^2 + \left(\frac{T_u p_h}{1.7 A_{oh}^2 \cos \theta \sin \theta} \right)^2} \tag{7-37}$$

The value of f_{cd} from (7-35) should not exceed the crushing strength of the cracked concrete in the tube, f_{ce}. As was discussed in Section 3-2, Collins and Mitchell [7-12] have related f_{ce} to the strains in the longitudinal and transverse reinforcement in the tube. For $\theta = 45°$ and for longitudinal and transverse strains, $\epsilon = 0.002$, equal to the yield strain of Grade-60 steel, Collins and Mitchell predict $f_{ce} = 0.549 f_c'$. Setting f_{cd} in (7-34) equal to $0.549 f_c'$ and evaluating $\cos \theta \sin \theta$ for $\theta = 45°$ gives the upper limit on the shears and torques, as determined by crushing of the concrete in the walls of the tube:

$$\sqrt{\left(\frac{V_u}{b_w d} \right)^2 + \left(\frac{T_u p_h}{1.7 A_{oh}^2} \right)^2} \leq \phi(0.275 f_c') \tag{7-38}$$

The limit in (7-33a) and (7-33b) has been set at $\phi(v_c + 8\sqrt{f_c'})$ to limit crack widths where, for reinforced concrete, v_c can be assumed to be $2\sqrt{f_c'}$, giving a limit of $\phi 10 \sqrt{f_c'}$. The limit $\phi(0.275 f_c')$ in (7-38) will always exceed $\phi 10 \sqrt{f_c'}$ for f_c' greater than 1324 psi. Since reinforced concrete members will always have f_c' greater than 1324 psi, only the crack width limits, (7-33a) and (7-33b), are included in the ACI Code. Two simplifications were made in the derivation of (7-38). First, the calculation of f_{cd} in (7-36) involved the effective depth, d, while the calculation of f_{cd} in (7-34) used the height of a wall of the space truss, y_o, which is about $0.9d$. Second, all the shear was assumed to be carried by truss action without a V_c term. These were considered to be reasonable approximations in view of the levels of accuracy of the right-hand sides of (7-33) and (7-38).

In [7-10], the code limit, (7-33b), is compared with tests of reinforced concrete beams in pure torsion that failed due to crushing of the concrete in the tube. The limit gave an acceptable lower bound on the test results.

Value of θ

ACI Section 11.6.3.6 allows the value of θ to be taken as any value between 30° and 60°, inclusive. ACI Section 11.6.3.7 requires that the value of θ used in calculating the area of longitudinal steel, A_ℓ, be the same as used to calculate A_t. This is because a reduction in θ leads to (a) a reduction in the required area of stirrups, A_t, as shown by (7-24); (b) an increase in the required area of longitudinal steel, A_ℓ, as shown by (7-30); and (c) an increase in f_{cd}, as shown by (7-35).

ACI Section 11.6.3.6 suggests a default value of $\theta = 45°$ for non-prestressed reinforced concrete members. This value will be used in the examples.

Combined Moment and Torsion

Torsion causes an axial tensile force N, given by (7-29). Half of this, $N/2$, is assumed to act in the top chord of the space truss, half in the bottom chord, as shown in Fig. 7-19a. Flexure causes a compression–tension couple, $C = T = M_u/jd$, shown in Fig. 7-19b, where $j \approx 0.9$. For combined moment and torsion, these internal forces add together, as shown in Fig. 7-19c. The reinforcement provided for the flexural tension force, T, and that provided for the tension force in the lower chord due to torsion, $N/2$, must be added together, as required by ACI Section 11.6.3.8.

In the flexural compression zone, the force C tends to cancel out some, or all, of $N/2$. ACI Section 11.6.3.9 allows the area of the longitudinal torsion reinforcement in the compression zone to be reduced by an amount equal to $M_u/(0.9df_{y\ell})$, where M_u is the moment that acts in conjunction with the torsion at the section being designed. It is necessary to compute this reduction at a number of sections, because the bending moment varies along the length of the member. If several loading cases must be considered in design, M_u and T_u must be from the same loading case. Normally, the reduction in the area of the compression steel is not significant, as will be shown in Example 7-2.

Torsional Stiffness

The torsional stiffness, K_t, of a member of length ℓ is defined as the torsional moment, T, required to cause a unit twist in the length ℓ; that is,

$$K_t = \frac{T}{\phi_t \ell} \tag{7-39}$$

or

$$K_t = \frac{T}{\theta_t} \tag{7-40}$$

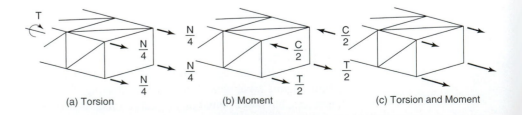

Fig. 7-19
Internal forces due to combined torsion and moment.

(a) Torsion   (b) Moment   (c) Torsion and Moment

where ϕ_t is the angle of twist per unit length and $\theta_t = \phi_t\ell$ is the total twist in the length ℓ. For a thin-walled tube of length ℓ, the total twist can be found by virtual work by equating the external work done when the torque, T, acts through a virtual angle change θ_t to the internal work done when the shearing stresses due to torsion, τ, act through a shear strain $\gamma = \tau/G$. The resulting integral equation is

$$T\theta_t = \int_V \tau\gamma \, dV$$

where $\int_V$ implies integration over the volume. Replacing γ with τ/G, τ with (7-9), and d_V with $\ell t \, ds$ gives

$$T\theta_t = \ell \int_p \frac{T}{2A_ot} \left(\frac{T}{2A_otG}\right) t \, ds$$

where $\int_p$ implies integration around the perimeter of the tube. This reduces to

$$\theta_t = \ell \frac{T}{4A_o^2 G} \int_p \frac{ds}{t} \tag{7-41}$$

If the wall thickness t is constant this becomes,

$$\theta_t = \frac{Tp_o}{4A_o^2 tG} \tag{7-42}$$

where p_o is the perimeter of A_o. Substituting (7-42) into (7-40) gives the torsional stiffness as

$$K_t = \left(\frac{4A_o^2 t}{p_o}\right) \frac{G}{\ell} = \frac{CG}{\ell} \tag{7-43}$$

where C, the torsional constant, refers to the term in parentheses. Equation (7-43) is similar to the equation for the flexural stiffness of a beam that is fixed at the far end:

$$K = \frac{4EI}{\ell}$$

In this equation, EI is the *flexural rigidity* of the section. The term CG in (7-43) is the *torsional rigidity* of the section.

Figure 7-12 shows a measured torque-twist curve for pure torsion for a member having a moderate amount of torsional reinforcement (longitudinal steel, plus stirrups at roughly $d/3$). From (7-40) and (7-43), the slope, T/ϕ_t, of a radial line through the origin and any point on the torque–twist curve gives the effective value of the term CG corresponding to that torque.

Prior to torsional cracking, the value of CG corresponds closely to the uncracked value, $\beta_t x^2 yG$. At torsional cracking, there is a sudden increase in ϕ_t and hence a sudden drop in the effective value of CG. In this test, the value of CG immediately after cracking (line 0–B in Fig. 7-12) was one-fifth of the value before cracking. At failure (line 0–C), the effective CG was roughly one-sixteenth of the uncracked value. This drastic drop in torsional stiffness allows a significant redistribution of torsion in certain indeterminate beam systems. Methods of estimating the postcracking torsional stiffness of beams are given by Collins and Lampert [7-14].

Equilibrium and Compatibility Torsion

Torsional loadings can be separated into two basic categories: *equilibrium torsion,* where the torsional moment is required for the equilibrium of the structure, and *compatibility torsion,* where the torsional moment results from the compatibility of deformations between members meeting at a joint.

Figure 7-20 shows three examples of equilibrium torsion. Figure 7-20a shows a cantilever beam supporting an eccentrically applied load P that causes torsion. Figure 7-20b shows the cross section of a beam supporting precast floor slabs. Torsion will result if the dead loads of slabs A and B differ, or if one supports a live load and the other does not. The torsion in the beams in Fig. 7-20a and b must be resisted by the structural system if the beam is to remain in equilibrium. If the applied torsion is not resisted, the beam will rotate about its axis until the structure collapses. Similarly, the canopy shown in Fig. 7-20c applies a torsional moment to the beam. For this structure to stand, the beam must resist the torsional moment, and the columns must resist the resulting bending moments.

By contrast, Fig. 7-21 shows an example of compatibility torsion. The beam A–B in Fig. 7-21a develops a slope at each end when loaded and develops the bending moment diagram shown in Fig. 7-21b. If, however, end A is built monolithically with a cross beam C–D, as shown in Fig. 7-21a, beam A–B can develop an end slope at A only if beam C–D twists about its own axis. If ends C and D are restrained against rotation, a torsional moment, T, will be applied to beam C–D at A, as shown in Fig. 7-21d. An equal and opposite moment, M_A, acts on A–B. The magnitude of these moments

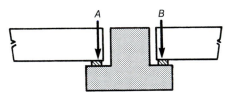

(a) Cantilever beam with eccentrically applied load.

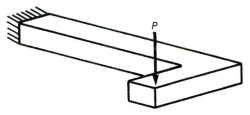

(b) Section through a beam supporting precast floor slabs.

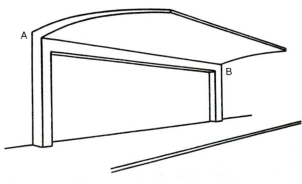

Fig. 7-20
Examples of equilibrium torsion.

(c) Canopy.

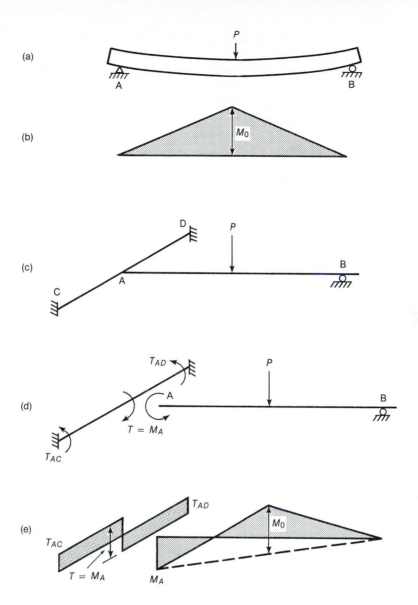

Fig. 7-21
Compatibility torsion.

depends on the relative magnitudes of the torsional stiffness of *C–D* and the flexural stiffness of *A–B*. If *C* and *D* were free to rotate about the axis *C–D*, *T* would be zero. On the other hand, if *C* and *D* could not rotate, and if the torsional stiffness of *C–D* was very much greater than the flexural stiffness of *A–B*, the moment M_A would approach a maximum equal to the moment that would be developed if *A* were a fixed end. Thus, the moment M_A and the twisting moments *T* result from the need for the end slope of beam *A–B* at *A* to be *compatible* with the angle of twist of beam *C–D* at point *A*. Note that the moment M_A causes a reduction in the moment at midspan of beam *A–B*, as shown in Fig. 7-21e.

When a beam cracks in torsion, its torsional stiffness drops significantly, as was discussed in the preceding section. As load is applied to beam *A–B*, torsional moments build up in member *C–D* until it cracks due to torsion. With the onset of cracking, the torsional stiffness of *C–D* decreases, and the torque, *T*, and the moment, M_A, drop. When this happens, the moment at the midspan of *A–B* must increase. This phenomenon is discussed by Collins and Lampert [7-14] and is the basis of ACI Sections 11.6.2 and 11.6.3.

If the torsional moment, T_u, is required to maintain equilibrium, ACI Section 11.6.2.1 requires that the members involved be designed for T_u. On the other hand, in those cases where compatibility torsion exists and a reduction of the torsional moment can occur as a result of redistribution of moments, ACI Section 11.6.2.2 permits T_u for non prestressed member to be reduced to

$$T_{cr} = \phi 4\sqrt{f_c'}\left(\frac{A_{cp}^2}{p_{cp}}\right) \tag{7-44}$$

at the sections located d away from the faces of the supports. This is approximately equal to the cracking torque of a member loaded in pure torsion. The resulting torsional reinforcement will help limit crack widths to acceptable values at service loads. If the torsional moments are reduced, it is necessary to redistribute these moments to adjoining members.

In SI units, (7-44) becomes

$$T_{cr} = \frac{\phi\sqrt{f_c'}}{3}\left(\frac{A_{cp}^2}{p_{cp}}\right) \tag{7-44M}$$

Calculation of Torsional Moments

Equilibrium Torsion: Statically Determinate Case

In torsionally statically determinate beams, such as shown in Fig. 7-20a, the torsional moment at any section can be calculated by cutting a free-body diagram at that section.

Equilibrium Torsion: Statically Indeterminate Case

In the case shown in Fig. 7-20c, the torsional moment, t, transmitted to the beam per foot of length of beam $A–B$ is the moment of the weight of a 1-ft strip of the projecting canopy taken about the line of action of the vertical reactions. The distribution of the torque along beam $A–B$ will be such that

$$\text{change in slope between the columns at } A \text{ and } B = \int_A^B \frac{t\,dx}{CG}$$

If this change in angle is zero and the distribution of CG along the member is symmetrical about the midspan, the torque at A and B will be $\pm t\ell/2$ (Fig. 7-22a). If the flexural stiffness of column B were less than column A, end B would rotate more than end A, and the torque diagram would not be symmetrical, since more torque would go to the stiffer end, end A.

On the other hand, the increased torsion at end A will lead to earlier torsional cracking at that end. When this occurs, the effective torsional rigidity, CG, at end A will decrease, reducing the stiffness at A. This will cause a redistribution of the torsional moments along the beam so that the final distribution approaches the symmetrical distribution.

The interaction between the torque, T, the column stiffness, K_c, and the torsional stiffness, K_t, along the beams makes it extremely difficult to estimate the torque diagram in statically indeterminate cases. At the same time, however, it is necessary to design the beam for the full torque necessary to equilibrate the loads on the overhand. Provided that the beam is detailed to have adequate ductility, a safe design will result for any reasonable distribution of torque, T, that is in equilibrium with the loads. This can vary from $T = t\ell$ at one end and $T = 0$ at the other (Fig. 7-22b) to the reverse. In such cases, however, wide torsional cracks would develop at the end where T was assumed to be zero, since that end

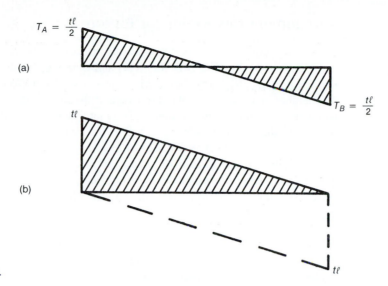

Fig. 7-22
Torques in beam $A–B$ of the
structure shown in Fig. 7-20c.

has to twist through the angle necessary to reduce T from its initial value to zero. In most cases, it is sufficiently accurate to design stirrups in this class of beams for $T = \pm t\ell/2$ at each end by using the torque diagram shown in Fig. 7-22a.

Compatibility Torsion

Based on some assumed set of torsional and flexural stiffnesses, an elastic grid or plate analysis or an approximation to such an analysis leads to torsional moments in the edge members. These can then be redistributed to account for the effects of torsional stiffness. This process was illustrated in Fig. 7-21 and will be considered in Example 7-3.

7-5   DESIGN FOR TORSION, SHEAR, AND MOMENT—ACI CODE

The design procedure for combined torsion, shear, and moment involves designing for the moment, ignoring the torsion and shear, and then providing stirrups and longitudinal reinforcement to give adequate shear and torsional strength. The basic design equations are

$$\phi V_n \geq V_u$$

$$(6\text{-}14)$$
$$\text{(ACI Eq. 11-1)}$$

$$V_n = V_c + V_s$$

$$(6\text{-}9)$$
$$\text{(ACI Eq. 11-2)}$$

and

$$\phi T_n \geq T_u$$

$$(7\text{-}45)$$
$$\text{(ACI Eq. 11-20)}$$

where ϕ is the strength reduction factor for shear and torsion, taken equal to 0.75 if design is based on the load combinations in ACI Section 9.2 and 0.85 for design based on the load combinations in ACI Section C-2. T_n is given by (7-24) and (7-30) (ACI Eq. 11-21).

Selection of Cross Section for Torsion

A torsional moment is resisted by shearing stresses in the uncracked member (Figs. 7-2 and 7-5b) and by the shear flow forces (V_1 and V_2 in Fig. 7-15b) in the member after cracking. For greatest efficiency, the shearing stresses and shear flow forces should flow around the member in the same circular direction and should be located as far from the axis of the member as possible. Thus the solid square member in Figs. 7-2b and 7-4 is more efficient than the U-shaped member in Fig. 7-5. For equal volumes of material, a closed tube will be much more efficient than a solid section. For building members, solid rectangular sections are generally used for practical reasons. For bridges, box sections like Fig. 7-7a are frequently used. Open U sections like the one shown in Fig. 7-11, made up of beams and a deck, are common, but they are weak and lack stiffness in torsion. Much of the torsion is resisted by warping torsion in such cases.

For proportioning hollow sections, some codes require that the distance from the centerline of the transverse reinforcement to the inside face of the wall not be less than $0.5A_{oh}/ph$. This ensures that the diagonal compression struts will develop their required thickness within the wall, corresponding to the assumption that $A_o = 0.85A_{oh}$. For hollow sections with very thin walls, such as the bridge girder shown in Fig. 7-7, the term $(T_u ph/2A_{oh}^2)$ in (7-33a) is replaced by $(T_u/(1.7A_{oh}t))$, where t is the distance from the centerline of the stirrup to the inside face of the wall. (See ACI Section 11.6.3.3.) Hollow sections are more difficult to form and to place reinforcement and concrete in than are solid sections. The form for the void must be built; then it must be held in position to prevent it from floating upwards in the fresh concrete; finally, it must be removed after the concrete has hardened. Sometimes, styrofoam void forms are left in place.

Although the ACI Code does not require fillets at the inside corners of a hollow section, it is good practice to provide them. Fillets reduce the stress concentrations where the inclined compressive forces flow around inside corners and also aid in the removal of the formwork. Section 11.6.1.2 of the 1989 ACI Commentary suggested that each side of a fillet should by $x/6$ in length if there are fewer than 8 longitudinal bars, where x is the smaller dimension of a rectangular cross section, and $x/12$ if the section has 8 or more longitudinal bars, but not necessarily more than 4 in.

Location of Critical Section for Torsion

In Section 6-5 and Figs. 6-34 and 6-35, the critical section for shear was found to be located at a distance d away from the face of the support. For an analogous reason, ACI Section 11.6.2.4 allows sections located at less than d from the support to be designed for the same torque, T_u, that exists at a distance d from the support. This would not apply if a large torque were applied within a distance d from the support.

Definitions of A_{cp} and p_{cp}

A_{cp} is the total area enclosed by the perimeter of the cross section and includes the area of any holes, provided that there are no gaps or slits in the perimeter. In the case of a torsional member cast monolithically with a slab, A_{cp} includes the area of the portion of the slab defined in ACI Section 13.2.4 on each side of the beam where there is a flange. The perimeter p_{cp} is the perimeter of A_{cp}, including the flange portions.

Definition of A_{oh}

ACI Section 11.6.3 states that the area enclosed by the shear flow path, A_o, shall be worked out by analysis, except that it is permissible to take A_o as $0.85A_{oh}$, where A_{oh} is the area enclosed by the centerline of the outermost closed stirrups as shown in Fig. 7-23.

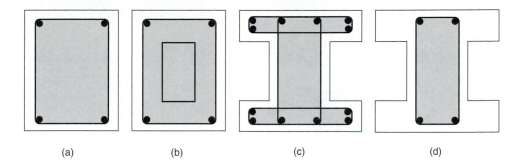

Fig. 7-23
Examples of A_{oh}.

(a)   (b)   (c)   (d)

In most cases, it is adequate to use this definition, although it may be on the conservative side for very large box girders. Equations (7-25) and (7-26), derived by Hsu [7-12], could be used. They satisfy the requirement for analysis. Figure 7-23 shows A_{oh} for several cross sections and illustrates the importance of having a square or wide rectangular section to resist torsion.

Torsional Reinforcement

Amounts and Details of Torsional Reinforcement

Torsional reinforcement consists of closed stirrups satisfying (7-24) (ACI Eq. (11-21)) and longitudinal bars satisfying (7-31) (ACI Eq. (11-22)). According to ACI Section 11.6.3.8, these are added to the longitudinal bars and stirrups provided for flexure and shear. It is possible to reduce the area of longitudinal torsional reinforcement in the flexural compression zone, as was discussed earlier in this section (ACI Section 11.6.3.9).

In designing for shear, a given size of stirrup, with the area of each leg being A_v, is chosen, and the required spacing, s, is computed. In considering combined shear and torsion, it is necessary to add the stirrups required for shear to those required for torsion. The area of stirrups required for shear and torsion will be computed in terms of A_v/s and A_t/s, both with units of in.^{2}/in. of length of beam. These can be added. Since A_v refers to both legs of a stirrup, while A_t refers to only one leg, the total A_{v+t}/s is

$$\frac{A_{v+t}}{s} = \frac{A_v}{s} + \frac{2A_t}{s} \qquad (7\text{-}46)$$

where A_{v+t} refers to the cross-sectional area of both legs of a stirrup. It is now possible to select A_{v+t} and compute a spacing s. If a stirrup in a wide beam had more than two legs for shear, only the outer legs would be included in the summation in (7-46).

Types of Torsional Reinforcement and Its Anchorage

Because the inclined cracks can spiral around the beam, as shown in Fig. 7-8c, 7-9c, or 7-15b, stirrups are required in all four faces of the beam. For this reason, ACI Section 11.6.4.1 requires the use of longitudinal bars plus either (a) closed stirrups perpendicular to the axis of the member, (b) closed cages of welded-wire fabric with wires transverse to the axis of the member, or (c) spirals. These should extend as close to the perimeter of the member as cover requirements will allow, so as to make A_{oh} as large as possible.

Tests by Mitchell and Collins [7-13] have examined the types of stirrup anchorages required. Figure 7-24a shows one corner of the space truss model shown in Fig. 7-15b. The inclined compressive stresses in the concrete, f_{cd}, have components parallel to the top and side surfaces, as shown in Fig. 7-24b. The components acting toward the corner are balanced by tensions in the stirrups. The concrete outside the reinforcing cage is anchored only

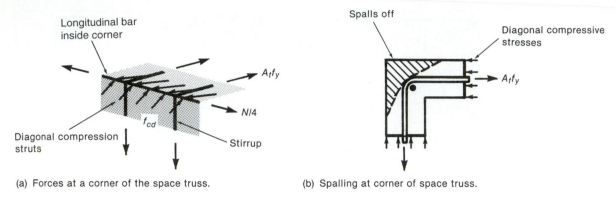

(a) Forces at a corner of the space truss.   (b) Spalling at corner of space truss.

Fig. 7-24
Compressive strut forces at a corner of a torsional member.

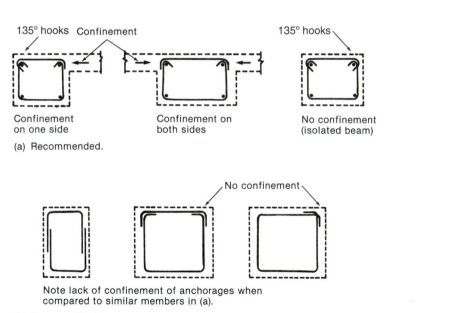

Fig. 7-25
Anchorage of closed stirrups.
(From [7-15].)

poorly, at best, however, and the shaded region will spall off if the compression in the outer shell is large. For this reason, ACI Section 11.6.4.2(a) requires that stirrups be anchored with 135° hooks around a longitudinal bar if the corner can spall. If the concrete around the stirrup anchorage is restrained against spalling by a flange or slab or similar member, ACI Section 11.6.4.2(b) allows the use of the anchorage details shown in Fig. 7-25a.

ACI Section 11.6.4.3 requires that longitudinal reinforcement for torsion be developed at both ends. Since the maximum torsions generally act at the ends of a beam, it is generally necessary to anchor the longitudinal torsional reinforcement for its yield strength at the face of the support. This may require hooks or horizontal U-shaped bars lap spliced with the longitudinal torsion reinforcement. A common error is to extend the bottom reinforcement in spandrel beams loaded in torsion 6 in. into the support, as allowed in ACI Section 12.11.1. Generally, this is not adequate to develop the longitudinal bars needed to resist torsion. Figure 7-26 shows a spandrel beam in a test slab that failed in torsion due, in part, to inadequate anchorage of the bottom reinforcement in the support.

Fig. 7-26
Torsional failure of a spandrel beam in a test of a flat
plate with edge beams. (Photograph courtesy of J. G.
MacGregor.)

Minimum Torsional Reinforcement

When the factored torsional moment exceeds

$$T_{th} = \phi\sqrt{f_c'}\left(\frac{A_{cp}^2}{p_{cp}}\right) \tag{7-18}$$

or, in SI units,

$$T_{th} = \frac{\phi\sqrt{f_c'}}{12}\left(\frac{A_{cp}^2}{p_{cp}}\right) \tag{7-18M}$$

the larger of; (a) the torsional reinforcement satisfying the strength requirements of ACI Section 11.6.3, and (b) the minimum reinforcement required by ACI Section 11.6.5 must be provided. ACI Section 11.6.5.2 specifies that the minimum area of closed stirrups shall be

$$A_v + A_t = 0.75\sqrt{f_c'}\,\frac{b_w s}{f_{yv}}, \text{ and } \geq \frac{50 b_w s}{f_{yv}} \tag{7-47}$$

$$\text{(ACI Eq. 11-23)}$$

In SI units, this becomes

$$A_v + 2A_t, \text{ and } \geq \frac{b_w s}{3 f_y} \tag{7-47M}$$

In Hsu's tests [7-4] of rectangular reinforced concrete members subjected to pure torsion, two beams failed at the torsional cracking load. In these beams, the total ratio of the volume of the stirrups and longitudinal reinforcement to the volume of the concrete was 0.802 and 0.88 percent. A third beam, with a volumetric ratio of 1.07 percent, failed at 1.08 times the torsional cracking torque. All the other beams tested by Hsu had volumetric ratios of 1.07 percent or greater and failed at torques in excess of 1.2 times the cracking torque. This suggests that beams with similar concrete and steel strengths loaded in pure torsion should have a volumetric ratio of torsional reinforcement in the order of 0.9 to 1.0 percent. Thus, the minimum volumetric ratio should be set at about 1 percent; that is,

$$\frac{A_{\ell,\min}s}{A_{cp}s} + \frac{A_t p_h}{A_{cp}s} \geq 0.01$$

or

$$A_{\ell,\min} = 0.01 A_{cp} - \frac{A_t p_h}{s}$$

If the constant 0.01 is assumed to be a function of the material strengths in the test specimens, the first term on the right-hand side of this equation can be rewritten as $7.5\sqrt{f'_c}f_{y\ell}$. In the 1971 to 1989 ACI Codes, a transition was provided between the total volume of reinforcement required by the equation for $A_{\ell,\min}$ for pure torsion and the much smaller amount of minimum reinforcement required in beams subjected to shear without torsion. This was accomplished by multiplying the same term by $\tau(\tau + v)$, giving

$$A_{\ell,\min} = \frac{7.5\sqrt{f'_c}}{f_{y\ell}}A_{cp}\left(\frac{\tau}{\tau + v}\right) - \left(\frac{A_t}{s}\right)p_h\left(\frac{f_{yv}}{f_{y\ell}}\right) \tag{7-48}$$

During the development of the 1995 torsion provisions, it was assumed that a practical limit on $\tau/(\tau + v)$ was 2/3 for beams that satisfied (7-31). When this was introduced, (7-48) became

$$A_{\ell,\min} = \frac{5\sqrt{f'_c}}{f_{y\ell}}A_{cp} - \left(\frac{A_t}{s}\right)p_h\left(\frac{f_{yv}}{f_{y\ell}}\right) \tag{7-49}$$
$$\text{(ACI Eq. 11-24)}$$

This equation was derived for the case of pure torsion. When it is applied to combined shear, moment, and torsion, it is not clear how much of the area of the stirrups should be included in A_t/s. In this book, we shall assume that A_t/s in (7-49) is the actual amount of transverse reinforcement provided for shear and torsion, where A_t is for one leg of a closed stirrup. The value of A_t/s should not be taken less than $25b_w/f_{yv}$ of the amount corresponding to (7-47).

In SI units, (7-49) becomes

$$A_{\ell,\min} = \frac{5\sqrt{f'_c}}{12f_{y\ell}}A_{cp} - \left(\frac{A_t}{s}\right)p_h\left(\frac{f_{yv}}{f_{y\ell}}\right) \tag{7-49M}$$

Spacing of Torsional Reinforcement

Figure 7-24 shows that the corner bars in a beam help to anchor the compressive forces in the struts between cracks. If the stirrups are too far apart, or if the longitudinal bars in the corners are too small in diameter, the compressive forces will tend to bend the longitudinal bars outward, weakening the beam. ACI Section 11.6.6.1 limits the stirrup spacing to the smaller of $p_h/8$ or 12 in., where p_h is the perimeter of the outermost closed stirrups.

Because the axial force due to torsion, N, acts along the axis of the beam, ACI Section 11.6.6.2 specifies that the longitudinal torsional reinforcement be distributed around the perimeter of the closed stirrups, with the centroid of the steel approximately at the centroid of the cross section. The longitudinal steel has a maximum spacing of 12 in. The longitudinal reinforcement should be inside the stirrups, with a bar inside each corner of the stirrups. The diameter of the longitudinal bars should be at least 1/24 of the stirrup spacing, but not less than 0.375 in. In tests, [7-13] corner bars with a diameter of 1/31 of the stirrup spacing bent outward at failure.

ACI Section 11.6.6.3 requires that torsional reinforcement continue a distance $(b_t + d)$ past the point where the torque is less than

$$T_{th} = \phi\sqrt{f_c'}\left(\frac{A_{cp}^2}{p_{cp}}\right) \tag{7-18}$$

where b_t is the width of that part of the cross section containing the closed stirrups. This length takes into account the fact that torsional cracks spiral around the beam. In SI units, (7-16) is divided by 12.

Maximum Yield Strength of Torsional Reinforcement

ACI Section 11.6.3.4 limits the yield strength used in design calculations to 60 ksi. This is done to limit crack widths at service loads.

High-Strength Concrete and Lightweight Concrete

In the absence of tests of high-strength concrete beams loaded in torsion, ACI Section 11.1.2 limits the value of $\sqrt{f_c'}$ to 100 psi in all torsional calculations. This affects only ACI Sections 11.6.1, 11.6.2.2, 11.6.3.1, and 11.6.5.3.

For the reasons presented in Section 6-5, the quantity $\sqrt{f_c'}$, which is related to the tensile strength, is reduced by ACI Section 11.2 for beams constructed of lightweight concrete. Unfortunately, this section was not updated when the torsion section was revised. We shall assume that it applies to the code sections mentioned in the preceding paragraph. The reduction is applied in the manner presented in Section 6-5.

7-6   APPLICATION OF ACI CODE DESIGN METHOD FOR TORSION

Review of the Steps in the Design Method

1. Calculate the factored bending-moment (M_u) diagram or envelope for the member.

2. Select b, d, and h based on the ultimate flexural moment. For problems involving torsion, square cross sections are preferable.

3. Given b and h, draw final M_u, V_u, and T_u diagrams or envelopes. Calculate the area of reinforcement required for flexure.

4. Determine whether torsion must be considered. Torsion must be considered if T_u exceeds the torque given by (7-18). Otherwise, it can be neglected, and the design carried out according to Chapter 6 of this book.

5. Determine whether the case involves equilibrium or compatibility torsion. If it is the latter, the torque may be reduced to the value given by (7-44) at the sections d from the faces of the supports. If this is done, the moments and shears in the other members must be adjusted accordingly.

6. Check whether the section is large enough for torsion. If the combination of V_u and T_u exceeds the values given by (7-33a) or (7-33b), enlarge the section.

7. Compute the area of stirrups required for shear. This is done by using (6-8), (6-9), (6-14), and (6-18) (ACI Eqs. (11-1), (11-2), (11-3), and (11-17)). To facilitate the addition of stirrups for shear and torsion, calculate

$$\frac{A_v}{s} = \frac{V_s}{f_{yv}d}$$

If V_s exceeds $8\sqrt{f'_c}b_w d$, the cross section is too small and must be enlarged. This is satisfied automatically by step 6.

8. Compute the area of stirrups required for torsion by using (7-45) and (7-24) (ACI Eqs. (11-20) and (11-21)). Again, these will be computed in terms of A_t/s.

9. Add the stirrup amounts together, using (7-46), and select the stirrups. The area of the stirrups must exceed the minimum given by (7-47) (ACI Eq. 11-23). The spacing must satisfy ACI Sections 11.6.4.4, 11.6.6.1, and 11.6.6.3. The stirrups must be closed.

10. Design the longitudinal reinforcement for torsion, using (7-30) or (7-31) (ACI Eq. (11-22)), and add it to that provided for flexure. The longitudinal reinforcement for torsion must exceed the minimum given by (7-49) (ACI Eq. (11-24)) and must satisfy ACI Sections 11.6.4.3, 11.6.6.2, and 11.6.6.3.

EXAMPLE 7-2   Design for Torsion, Shear, and Moment: Equilibrium Torsion

The cantilever beam shown in Fig. 7-27a supports its own dead load plus a concentrated load as shown. The beam is 54 in. long, and the concentrated load acts at a point 6 in. from the end of the beam and 6 in. away from the vertical axis of the member. The *unfactored* concentrated load consists of a 20-kip dead load and a 20-kip live load. Use $f'_c = 3000$ psi and f_y and $f_{yv} = 60,000$ psi. Use load combination and strength-reduction factor from ACI 318-02 Chapter 9. This specifies $\phi = 0.75$ for shear torsion.

1. Compute the bending-moment diagram. Estimate the size of the member. The minimum depth of control flexural deflections is $\ell/8 = 6.75$ in. (ACI Table 9.5a or Table A-14 of this book). This seems too small, in view of the loads involved. As a first trial, use a 16-in.-wide-by-24-in. section, with $d = 21.5$ in.:

$$w = \frac{16 \times 24}{144} \times 0.15 = 0.40 \text{ kip/ft}$$

$$\text{factored uniform load} = 0.48 \text{ kip/ft}$$

$$\text{factored concentrated load} = 1.2 \times 20 + 1.6 \times 20 = 56 \text{ kips}$$

The bending-moment diagram is as shown in Fig. 7-27b, with the maximum $M_u = 229$ ft-kips.

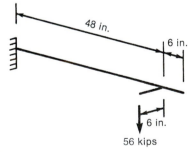

(a) Beam.

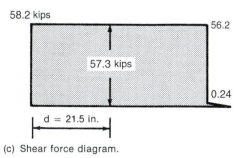

(c) Shear force diagram.

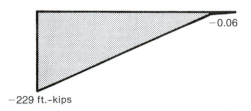

(b) Bending moment diagram.

(d) Torque diagram.

Fig. 7-27
Cantilever beam—
Example 7-2.

2.   **Select b, d, and h for flexure.** (See Example 4-6.)
From (4-17),

$$\frac{M_u}{\phi k_n} = \frac{bd^2}{12{,}000}$$

where $\phi = 0.9$ as a first trial for flexure. For shear and torsion use $\phi = 0.75$.
where $\phi k_n = \phi[f'_c \omega(1 - 0.59\omega)]$, $\omega = \rho f_y/f'_c$, and M_u is in ft-kips. Try $\rho \approx 0.01$:

$$\omega = \frac{0.01 \times 60{,}000}{3000} = 0.20$$

$$\phi k_n = 0.9 \times 3000 \times 0.20(1 - 0.59 \times 0.20)$$

$$= 476 \text{ (see also Table A-3)}$$

Therefore,

$$\frac{bd^2}{12{,}000} = \frac{229}{476} = 0.481$$

$$bd^2 = 5773 \text{ in.}^3$$

Possible choices are

$b = 12$ in., $d = 21.9$, and $h = 24.4$ in.
$b = 14$ in., $d = 20.3$, and $h = 22.8$ in.
$b = 16$ in., $d = 19.0$, and $h = 21.5$ in.

Use $b = 14$ in., $d = 21.5$ in., and $h = 24$ in. as shown in Fig. 7-28. Since this is smaller than the section originally chosen, it is necessary to recompute w and M_u:

$$w = 0.35 \text{ kip/ft} \quad \text{and} \quad w_u = 0.42 \text{ kip/ft}$$

$$M_u = 228 \text{ ft-kips at root of cantilever}$$

$$A_{s(req'd)} = \frac{M_u \times 12{,}000}{\phi f_y jd}$$

Assume that $j = 0.875$. Then

$$A_{s(req'd)} = \frac{228 \times 12{,}000}{0.9 \times 60{,}000(0.875 \times 21.5)}$$

$$= 2.69 \text{ in.}^2$$

Check M_n for $A_s = 2.69$ in.^{2}. From (4-11),

$$a = 4.53 \text{ in.}$$

Check whether the section is tension-controlled and whether $\phi = 0.90$ for tension-controlled sections in flexure. From ACI Section 9.3.2.3, $\phi = 0.75$ for shear and tension.

$$a = 4.53 \text{ in.}$$

$$c = a/\beta_1 = 5.33 \text{ in.}$$

$d_t = 21.5$ in. and from similar triangles the flexural strain at the level of the steel is

$$\varepsilon_t = \frac{(21.5 - 5.33)}{5.33} \times 0.003 = 0.0091$$

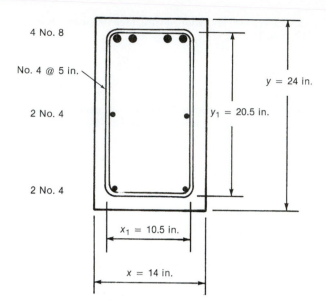

4 No. 8

No. 4 @ 5 in.

2 No. 4

2 No. 4

$x_1 = 10.5$ in.

$x = 14$ in.

$y = 24$ in.

$y_1 = 20.5$ in.

Fig. 7-28
Cross Section of a cantilever
beam—Example 7-2.

Because 0.0091 exceeds 0.005, the section is tension-controlled and $\phi = 0.90$ for flexure.

$$\phi M_n = \frac{0.9 \times 2.69 \times 60{,}000(21.5 - 4.53/2)}{12{,}000}$$
$$= 233 \text{ ft-kips}$$

Therefore, use a 14-in.-by-24-in. section with $d = 21.5$ in. $A_s = 2.69$ in.^{2} is required at the top for flexure.

3. Compute the final M_u, V_u, and T_u diagrams.
The shear force and torque diagrams are shown in Fig. 7-27. The shear and torque at d from the face of the support are shown.

4. Should torsion be considered?
For the cross section, $A_{cp} = 14 \times 24 = 336$ in.^{2} and $p_{cp} = 2(14 + 24) = 76$ in. From ACI Section 11.6.1, torsion can be neglected if T_u is less than

$$T_\theta = \phi\sqrt{f'_c}\left(\frac{A_{cp}^2}{p_{cp}}\right) = 0.75\sqrt{3000}\left(\frac{336^2}{76}\right) = 61{,}000 \text{ in.-lb} = 5.09 \text{ ft-kips} \qquad (7\text{-}18)$$

Since $T_u = 28.0$ ft-kips exceeds the threshold torque, 5.09 ft-kips, torsion must be considered.

5. Equilibrium or compatibility torsion?
The torsion is needed for equilibrium; therefore, design for $T_u = 28.0$ ft-kips.

6. Is the section big enough to resist the torsion?
For a solid cross section, ACI Section 11.6.3.1(a) requires the section to satisfy

$$\sqrt{\left(\frac{V_u}{b_w d}\right)^2 + \left(\frac{T_u p_h}{1.7 A_{oh}^2}\right)^2} \le \phi\left(\frac{V_c}{b_w d} + 8\sqrt{f'_c}\right)$$

(7-33b)

(ACI Eq. 11-18)

where ϕ is 0.75 for shear and torsion.

From ACI Section 11.3.1.1, take $V_c = 2\sqrt{f'_c}b_w d$. $A_{oh} = $ area within centerline of closed stirrups. Assume 1.5 in. of cover and No. 4 stirrups, as shown in Fig. 7-28. Then

$$A_{oh} = (14 - 2 \times 1.5 - 0.5)(24 - 2 \times 1.5 - 0.5) = 215 \text{ in.}^2$$

$$p_h = 2(10.5 + 20.5) = 62 \text{ in.}$$

$$\sqrt{\left(\frac{57,300}{14 \times 21.5}\right)^2 + \left(\frac{28.0 \times 12,000 \times 62}{1.7 \times 215^2}\right)^2} \leq 0.75(2\sqrt{3000} + 8\sqrt{3000})$$

$$\sqrt{36,200 + 70,300} = 326 \text{ psi} \leq 0.75 \times 10\sqrt{3000} = 411 \text{ psi}$$

Since 326 psi is less than 411 psi, the cross section is large enough.

7.   Compute the stirrup area required for shear.
From (6-9) and (6-14),

$$V_u \leq \phi(V_c + V_s)$$

$$V_c = 2\sqrt{f_c'}b_w d = 2 \times \sqrt{3000} \times 14 \times 21.5$$

$$= 33.0 \text{ kips}$$

$$V_s = \frac{57.3}{0.75} - 32.97 = 43.4 \text{ kips}$$

From (6-18) (ACI Eq. (11-15)),

$$V_s = \frac{A_v f_y d}{s} \quad \text{or} \quad \frac{A_v}{s} = \frac{V_s}{f_y d}$$

$$\frac{A_v}{s} = \frac{43,400}{60,000 \times 21.5} = 0.034 \text{ in.}^2/\text{in.}$$

For shear, we require stirrups with $A_v/s = 0.034$.

8.   Compute the stirrup area required for torsion.
From (7-45) (ACI Eq. (11-20)), $\phi T_n \geq T_u$. Therefore,

$$T_n = \frac{28.0 \times 12,000}{0.75} = 448,000 \text{ in.-lb}$$

From (7-24) (ACI Eq. (11-21)),

$$T_n = \frac{2A_o A_t f_{yv}}{s} \cot \theta \quad \text{or} \quad \frac{A_t}{s} = \frac{T_n}{2A_o f_{yv}} \cot \theta$$

From ACI Section 11.6.3.6,

$$A_o = 0.85 A_{oh} = 0.85 \times 215$$

$$= 183 \text{ in.}^2$$

$$\theta = 45°$$

$$\frac{A_t}{s} = \frac{448,000}{2 \times 183 \times 60,000} = 0.020 \text{ in.}^2/\text{in.}$$

For torsion, we require stirrups with $A_t/s = 0.020$.

9.   Add the stirrup areas and select stirrups.
From (7-46),

$$\frac{A_{v+t}}{s} = \frac{A_v}{s} + \frac{2A_t}{s}$$

$$= 0.034 + 2 \times 0.020 = 0.074 \text{ in.}^2/\text{in.}$$

Check minimum stirrups: From (7-47) (ACI Eq. (11-23)),

$$\frac{A_v + 2A_t}{s} \geq 0.75\sqrt{f_c'} \frac{b_w}{f_{yv}}, \text{ and } \geq \frac{50b_w}{f_{yv}}$$

$$\text{minimum}\frac{A_{v+t}}{s} = \frac{0.75\sqrt{3000} \times 14}{60{,}000} = 0.010 \text{ in.}^2/\text{in., and } \geq 0.012 \text{ in.}^2/\text{in.}$$

Since 0.074 in.²/in. exceeds 0.012 in.²/in., the minimum does not govern.

For No. 3 stirrups, A_{v+t}(two legs) = 0.22 in.², and the required s = 2.97 in. For No. 4 stirrups, A_{v+t}(two legs) = 0.40 in.², and the required s = 5.41 in. The minimum stirrup spacing (ACI Section 11.6.6.1) is the smaller of $p_h/8 = 62/8 = 7.75$ in., or 12 in.

Use No. 4 closed stirrups at 5 in. on centers.

10. Design the longitudinal reinforcement for torsion. From (7-30) or (7-31) (ACI Eq. (11-22)),

$$A_\ell = \left(\frac{A_t}{s}\right) p_h \left(\frac{f_{yv}}{f_{y\ell}}\right) \cot^2 \theta \tag{7-31}$$

where A_t/s is the amount computed in step 8. Then

$$A_\ell = (0.020) \times 62 \times \frac{60{,}000}{60{,}000} \cot^2 45°$$

$$= 1.24 \text{ in.}^2$$

From (7-49) (ACI Eq. (11-24)), the minimum A_ℓ is

$$A_{\ell\,\text{min.}} = \frac{5\sqrt{f_c'}\, A_{cp}}{f_{y\ell}} - \left(\frac{A_t}{s}\right) p_h \left(\frac{f_{yv}}{f_{y\ell}}\right)$$

where A_t/s will be based on the stirrup area required for torsion in step 8, i.e. 0.020 in.²/in. No. 4 double-leg stirrups at 5 in. on centers. Since A_t is one leg of a stirrup, A_t/s = 0.20/5 = 0.040.

$$A_\ell = \frac{5\sqrt{3000} \times 336}{60{,}000} - (0.020) \times 62 \times \frac{60{,}000}{60{,}000}$$

$$= -0.29 \text{ in.}^2$$

This value is less than the strength requirement, so the minimum A_ℓ does not govern. Provide $A_\ell = 1.24$ in.².

To satisfy the 12-in. maximum spacing specified in ACI Section 11.6.6.2, we need at least 6 bars each of area 0.207 in.². There must be a longitudinal bar in each corner of the stirrups. The minimum bar diameter is $\frac{1}{24} = 0.042$ times the stirrup spacing = 0.042 × 5 = 0.21 in.

Provide 4 No. 4 bars in the bottom half of the beam and add 1.24 − (4 × 0.20) = 0.44 in.² to the flexural steel. The total area of steel required in the top face of the beam is 2.69 + 0.44 = 3.13 in.². Thus, we have

6 No. 7 bars: A_s = 3.60 in.², will not fit in one layer.
4 No. 8 bars: A_s = 3.16 in.², will fit in one layer.
3 No. 8 and 2 No. 7 bars: A_s = 3.57 in.², will fit in one layer.
2 No. 8 and 3 No. 7 bars: A_s = 3.38 in.², will fit in one layer.

Provide 4 No. 8 bars at the top of the beam.

ACI Section 11.6.3.9 allows one to subtract $M_u/(0.9df_{y\ell})$ from the area of longitudinal steel in the flexural compression zone. We shall not do this, for three reasons: First, M_u varies along the length of the beam; second, we need two bars of minimum diameter 0.21 in. in the corners of the stirrups; third, we need bars to support the corners of the stirrups.

The final beam design is shown in Fig. 7-28. It is interesting to note that the cross section is identical to the one designed by using the 1989 ACI Code in the preceding edition of this book. ∎

EXAMPLE 7-3   Design for Torsion, Shear, and Moment: Equilibrium Torsion, Hollow Section

Redesign the cantilever beam from Example 7-2, using a hollow cross section. This is a hypothetical example to show the differences between the design for a solid section and that for a hollow section. For a beam of this size, the additional costs of forming the void, holding the void forms in place when the concrete is placed, and removing them when the concrete has hardened would more than offset any material savings due to the void, if we ignore the smaller dead load. On the other hand, the cross section of the bridge shown in Fig. 7-7 would be more economical if built as a hollow section.

1. **Estimate the size and weight of the beam.**
The most efficient cross section for torsion is a circle; slightly poorer is a rectangular section approaching a square. To estimate the weight of the beam, we shall try a 20-in.-by-24-in. cross section with 5-in.-thick walls on all four sides, leaving a 10-in.-by-14-in. void. The corners of the void have 3-in.-by-3-in. fillets. The area of concrete in the cross section is

$$A_g = (20 \times 24) - (10 \times 14) + 4(3 \times 3/2) = 358 \text{ in.}^2$$

The beam self-weight is

$$w = (358/144) \times 0.150 = 0.373 \text{ kips/ft}$$

The factored uniform self-weight is

$$1.2 \times 0.373 = 0.448 \text{ kips/ft}$$

The factored concentrated load is

$$1.2 \times 20 + 1.6 \times 20 = 56 \text{ kips}$$

The bending-moment diagram is shown in Fig. 7-27b, with the maximum $M_u = 229$ ft-kips.

2. **Select b, d, and h for flexure.** (See Example 4-2.) From (4-17),

$$\frac{M_u}{\phi k_n} = \frac{bd^2}{12,000}$$

where b and d are in in., M_u is in ft-kips, and

$$\phi k_n = \phi[f_c' \omega(1 - 0.59\omega)]$$

in which $\omega = \dfrac{\rho f_y}{f_c'}$

Try $\rho = 0.01$. For $f_c' = 3000$ psi,

$$\omega = 0.01 \times 60,000/3000 = 0.20$$

Assume beam is tension-controlled and $\phi = 0.90$ for flexure.

$$\phi k_n = 0.9 \times 3000 \times 0.20(1 - 0.59 \times 0.20)$$
$$= 476 \text{ lb/in.}^2$$

$$\frac{bd^2}{12,000} = \frac{229}{476} = 0.481 \text{ and } bd^2 = 5773 \text{ in.}^3$$

A possible choice is $b = 20$ in., $d = 17.0$ in., and $h = 19.5$ in.

For this reason, we shall **try a hollow section with** b = **20 in.,** h = **24 in., and** d = **21.5 in. and having 5-in.-thick walls**. Since this is the section chosen earlier, w and M_u are OK as computed. To estimate the flexural reinforcement required, assume that j = 0.875. Then

$$A_{s,\,reqd} = \frac{M_u x 12{,}000}{\phi f_y j d} = \frac{229 x 12{,}000}{0.9 \times 60{,}000 \times (0.875 x 21.5)}$$

$$= 2.71 \text{ in.}^2$$

Check M_n for A_s = 2.71 in.^{2}:

$$a = A_s f_y / (0.85 f'_c b)$$

$$= 3.18 \text{ in.}$$

This will fit inside the top flange of the box, which is 5 in. thick. So

$$\phi M_n = \phi A_s f_y (d - a/2)/12{,}000 = 0.9 \times 2.71 \times 60{,}000\left(21.5 - \frac{3.18}{2}\right)/12{,}000$$

$$= 243 \text{ ft-kips}$$

Since this is greater than 228 ft-kips, A_s = 2.71 in.^{2}, will be used.

Try a 20-in.-by-24-in. section with a 10-in.-by-14-in. void, with d = **21.5 in. and** A_s = **2.71 in.^{2} in the top flange for flexure.**

3. **Compute the final M_u, V_u, and T_u diagrams.**
The moment, shear force, and torque diagrams are essentially the same as in Fig. 7-27. The shear and torque at d from the face of the support are shown.

4. **Should torsion be considered?**
Torsion must be considered if T_u exceeds the threshold torque given by $\phi\sqrt{f'_c}\left(\frac{A_{cp}^2}{p_{cp}}\right)$.

Because the cracking torque, and hence the threshold torque, is lower in a hollow section than in a solid section, ACI Section 11.6.1 requires replacing A_{cp} in (7-18) with A_g, giving

$$\phi\sqrt{f'_c}\left(\frac{A_g^2}{p_{cp}}\right) \tag{7-19}$$

$$A_g = 358 \text{ in.}^2$$

$$p_{cp} = 2(20 + 24) = 88 \text{ in.}$$

$$\phi\sqrt{f'_c}\left(\frac{A_g^2}{p_{cp}}\right) = 0.75\sqrt{3000}\left(\frac{358^2}{88}\right) = 59{,}800 \text{ in.-lbs} = 4.99 \text{ ft-kips.}$$

Since T_u = 28.0 ft-kips exceeds 4.99 ft-kips, torsion must be considered.

5. **Equilibrium or compatibility torsion?**
The torsion is needed for equilibrium; therefore, design the beam for T_u = 28.0 ft-kips.

6. **Is the section big enough to resist the torsion?**
For a hollow section,

$$\left(\frac{V_u}{b_w d}\right) + \left(\frac{T_u P_h}{1.7 A_{oh}^2}\right) \le \phi\left(\frac{V_c}{b_w d} + 8\sqrt{f'_c}\right) \tag{7-33a}$$

(ACI Eq. 11-19)

Assuming that the beam is not exposed to weather and the minimum cover to the No. 4 stirrups is 1.5 in., the center-to-center widths of the stirrups are as follows:

$$\text{horizontally} = 20 - 2(1.5 + 0.25) = 16.5 \text{ in.}$$

$$\text{vertically} = 24 - 2(1.5 + 0.25) = 20.5 \text{ in.}$$

$$p_h = 2(16.5 + 20.5) = 74.0 \text{ in.}$$

$$A_{oh} = 16.5 \times 20.5 = 338 \text{ in.}^2$$

$$b_w = 20 - 10 = 10 \text{ in.}$$

ACI Section 11.6.3.3 requires that the second term on the left side of (7-33a) be replaced with $T_u/1.7A_{oh}t$ if A_{oh}/p_h exceeds the wall thickness. $A_{oh}/p_h = 338/74 = 4.57$. Therefore, (7-33a) may be used as given. Substituting into (7-33a) gives

$$\text{left-hand side: } \left(\frac{57.3 \times 1000}{10 \times 21.5} \right) + \left(\frac{28.0 \times 12{,}000 \times 74.0}{1.7 \times 338^2} \right)$$

$$= 267 + 128 = 395 \text{ psi}$$

$$\text{right-hand side: } \phi\left(\frac{V_c}{b_w d} + 8\sqrt{f_c'} \right) = 0.75(10\sqrt{3000}) = 411 \text{ psi}$$

Since 395 psi is less than 411 psi, the section is big enough.

7.   Compute the stirrup area required for shear.

$$\phi V_n \geq V_u \tag{6-14}$$
$$\text{(ACI Eq. 11-1)}$$

where

$$V_n = \frac{V_u}{\phi} = \frac{57.3}{0.75} = 76.4 \text{ kips} \tag{6-9}$$

$$V_n = V_c + V_s \tag{ACI Eq. 11-2}$$

$$V_c = 2\sqrt{f_c'}\, b_w d = 2 \times \sqrt{3000} \times 10 \times 21.5 \tag{6-8}$$

$$= 23{,}600 \text{ lbs} = 23.6 \text{ kips} \tag{ACI Eq. 11-3}$$

$$V_s \text{ required at } d \text{ from support} = 76.4 - 23.6 = 52.8 \text{ kips}$$

$$V_s = \frac{A_s f_y d}{s} \tag{6-18}$$

Rearranging gives

$$\frac{A_v}{s} = \frac{V_s}{f_y d} = \frac{52{,}800}{60{,}000 \times 21.5}$$

$$= 0.0409 \text{ in.}^2/\text{in.}$$

For shear, we need stirrups with $A_v/s = 0.0409$ in.^{2}/in.

8.   Compute the stirrup area required for torsion.

$$\phi T_n \geq T_u \tag{7-45}$$

$$T_n = \frac{T_u}{\phi} = \frac{28.0}{0.75} = 37.3 \text{ kips} \tag{ACI Eq. 11-20}$$

$$T_n = 2A_o \frac{A_t f_y}{s} \cot \theta \tag{7-24}$$
$$\text{(ACI Eq. 11-21)}$$

Rearranging terms and setting θ equal to the default value of 45° from ACI Section 11.6.3.6(a) gives

$$\frac{A_t}{s} = \frac{T_n}{2A_o f_y}$$

where A_o may be taken as $0.85\,A_{oh}$ and $T_n = T_u/\phi = 28.0/0.75 = 37.3$ ft-kips

$$\frac{A_t}{s} = \frac{37.3 \times 12{,}000}{2(0.85 \times 338) \times 60{,}000}$$

$$= 0.0130 \text{ in.}^2/\text{in.}$$

For torsion, we need stirrups with $A_t/s = 0.0130$ in.^{2}/in.

9. Add the stirrup areas and select stirrups.

Stirrups required for combined shear and torsion

$$\frac{A_{v+t}}{s} = \frac{A_v}{s} + 2\frac{A_t}{s} \qquad (7\text{-}46)$$

$$= 0.0409 + 2 \times 0.0130 = 0.0669 \text{ in.}^2/\text{in.}$$

Minimum stirrups required from ACI Section 11.6.5.2

From ACI Section 11.6.5.2, the minimum stirrups required for combined shear and torsion are the larger of

$$A_{v+t,\,\min} = \frac{50 b_w}{f_y} = 0.0083 \text{ in.}^2/\text{in.} \qquad (7\text{-}47)$$

and

$$\frac{A_{v+t,\,\min}}{s} = \frac{0.75\sqrt{3000}}{60{,}000} \times b_w = 0.00685 \text{ in.}^2/\text{in.} \qquad \begin{array}{c}(7\text{-}47)\\ \text{(ACI Eq. 11-23)}\end{array}$$

Since the required A_{v+t}/s exceeds this, use the amount required for combined shear and torsion. Try No. 3 closed stirrups, area of two legs $= 2 \times 0.11 = 0.22$ in.^{2}. From the combined strength requirement, the spacing required for the factored shear and torsion is

$$\frac{A_v}{s} + \frac{2A_t}{s} = 0.0669 \text{ in.}^2/\text{in.}$$

so

$$s = \frac{0.22 \text{ in.}^2}{0.0669} = 3.29 \text{ in.}$$

For No. 4 closed stirrups, the required spacing is $0.40/0.0669 = 5.98$ in.

From ACI Section 11.5.4.1, the maximum spacing of stirrups for shear is $d/2$ for V_s less than $4\sqrt{f_c'}\,b_w d = 4\sqrt{3000} \times 10 \times 21.5 = 47{,}100$ lbs $= 47.1$ kips. Since $V_s = 52.8$ kips, the maximum spacing for shear must be reduced to $d/4 = 21.5/4 = 5.38$ in. $= d/2 = 21.5/2 = 10.75$ in. From ACI Section 11.6.6.1, the maximum spacing of stirrups for shear and torsion is the smaller of $p_h/8 = 74.0/8 = 9.25$ in. and 12 in. Therefore, the maximum stirrup spacing is 5.38 in. but $s = 6.2$ in. is required for V_s.

Use No. 4 closed stirrups: One at 2.5 in. from face of support and ten at 5 in. on centers.

10. Design the longitudinal reinforcement for torsion.

$$A_\ell = \left(\frac{A_t}{s}\right) p_h \left(\frac{f_{yv}}{f_{y\ell}}\right) \cot^2 \theta \qquad \begin{array}{c}(7\text{-}31)\\ \text{(ACI Eq. 11-22)}\end{array}$$

where A_t/s is 0.0130 in.^{2}/in., the amount computed in step 8, and $\theta = 45°$.

$$A_\ell = 0.0130 \times 74.0 \times 1.0 = 0.962 \text{ in.}^2$$

From ACI Section 11.6.5.3, the minimum area of longitudinal steel is

$$A_\ell = \frac{5\sqrt{f_c'}}{f_{y\ell}} A_{cp} - \left(\frac{A_t}{s}\right) p_h \left(\frac{f_{yv}}{f_{y\ell}}\right)$$

$$= \frac{5\sqrt{3000}}{60,000} 480 - (0.0130) \times 74 \times \left(\frac{60,000}{60,000}\right)$$

$$= 2.19 - 0.962 = 1.23 \text{ in.}^2$$

Therefore, we need to provide 1.23 in.^{2} of longitudinal reinforcement for torsion. From ACI Section 11.6.6.2, the maximum spacing of the longitudinal bars is 12.0 in. We will put one longitudinal bar in each corner of the stirrups, one at the middle of the top and bottom sides, and one at midheight of each vertical side, a total of eight bars. The area per bar is 1.23/8 = 0.154 in.^{2}. From ACI Section 11.6.6.2, the minimum diameter of the corner bar is 0.042 × the stirrup spacing = 0.042 × 5 = 0.21 in. We could use either No. 3 or No. 4 bars. No. 4 bars are a little more robust, so we use No. 4 bars.

For the top of the beam,

for flexure, we need $A_s = 2.71$ in.^{2}

for torsion, we need 3×0.154 in.$^2 = 0.461$ in.^{2}

total steel at top of beam = $2.71 + 0.461 = 3.17$ in.^{2}

The possible choices are as follows:

six No. 7 bars, $A_s = 3.60$ in.^{2}. These will fit in the 20 in. width in one layer.

two No. 8 bars plus 3 No. 7, $A_s = 3.38$ in.^{2}. These will fit.

Use two No. 8 bars plus three No. 7 bars at the top of the beam plus five additional No. 4 longitudinal bars, two at midheight and three in the bottom flange.

ACI Section 11.6.3.9 allows the longitudinal steel in the compression zone to be reduced by the compression due to flexure. We will not do this for three reasons: first, M_u varies along the beam while T_u remains constant; second, we need No. 4 corner bars to satisfy the 0.042s rule; third, we need bars to support the corners of the stirrups. ∎

EXAMPLE 7-4 Compatibility Torsion

The one-way joist system shown in Fig. 7-29 supports a total factored dead load of 157 psf and a factored live load of 170 psf, totaling 327 psf. Design the end span, AB, of the exterior spandrel beam on grid line 1. The factored dead load of the beam and the factored loads applied directly to it total 1.1 kips/ft. The spans and loadings are such that the moments and shears can be calculated by using the moment coefficients from ACI Section 8.3.3 (see Section 10-2 of this book). Use $f_y = f_{yv} = 60,000$ psi and $f_c' = 4000$ psi.

1. Compute the bending moments for the beam. In laying out the floor, it was found that joists with an overall depth of 18.5 in. would be required. The slab thickness is 4.5 in. The spandrel beam was made the same depth, to save forming costs. The columns supporting the beam are 24 in. square. For simplicity in forming the joists, the beam overhangs the inside face of the columns by $1\frac{1}{2}$ in. Thus, the initial choice of beam size is $h = 18.5$ in., $b = 25.5$ in., and $d = 16$ in.

Although the joist loads are transferred to the beam by the joist webs, we shall assume a uniform load for simplicity. Very little error is introduced by this assumption. The joist reaction per foot of length of beam is

$$\frac{w\ell}{2} = \frac{0.327 \text{ ksf} \times 29.75 \text{ ft}}{2} = 4.86 \text{ kips/ft}$$

The total load on the beam is

$$w = 4.86 + 1.1 = 5.96 \text{ kips/ft}$$

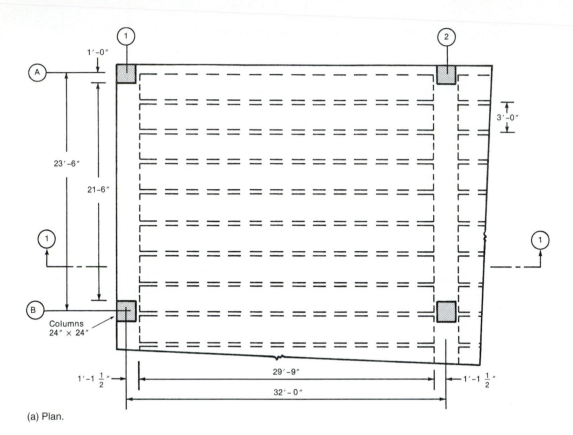

(a) Plan.

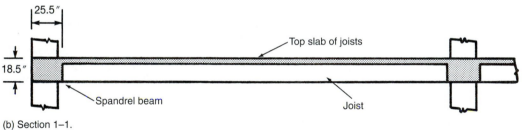

(b) Section 1–1.

Fig. 7-29
Joist floor—Example 7-4.

The moments in the edge beam are as follows:

exterior end negative: $-M_u = \dfrac{w\ell_n^2}{16} = -172$ ft-kips

midspan positive: $+M_u = \dfrac{w\ell_n^2}{14} = +197$ ft-kips

first interior negative: $-M_u = \dfrac{w\ell_n^2}{10} = -276$ ft-kips

2. **Compute b, d, and h.** Since b and h have already been selected, we shall check whether they are sufficiently large to ensure a ductile flexural behavior. Going through such a check, we find that $\rho \simeq 0.4\rho_b$ at the first interior negative moment point and that the ratio, ρ, is smaller at other points. Thus, the section has adequate size for flexure. The areas of steel required for flexure are as follows:

exterior end negative: $A_s = 2.64$ in.^{2}

midspan positive: $A_s = 3.02$ in.^{2}

first interior negative: $A_s = 4.23$ in.^{2}

The actual steel will be chosen when the longitudinal torsion reinforcement has been calculated.

    3.   Compute the final M_u, V_u, and T_u diagrams. The moment and shear diagrams for the edge beam, computed from the ACI moment coefficients (ACI Section 8.3.3; Section 10-2 of this book), are plotted in Fig. 7-30a and b.

    The joists are designed as having a clear span of 29.75 ft from the face of one beam to the face of the other beam. Because the exterior ends of the joists are "built integrally with" a "spandrel beam," ACI Section 8.3.3 gives the exterior negative moment in the joists as

$$-M_u = \frac{w\ell_n^2}{24}$$

Rather than consider the moments in each individual joist, we shall compute an average moment per foot of width of support:

$$-M_u = \frac{0.327 \text{ ksf} \times 29.75 \text{ ft}^2}{24} = -12.1 \text{ ft-kips/ft}$$

Although this is a bending moment in the joist, it acts as a twisting moment on the edge beam. As shown in Fig. 7-31a, this moment and the end shear of 4.86 kips/ft act at the face of the edge beam. Summing moments about the center of the columns (point A in Fig. 7-31a) gives the moment transferred to the column as 17.6 ft-kips/ft.

    For the design of the edge beam for torsion, we need the torque about the axis of the beam. Summing moments about the centroid of the edge beam (Fig. 7-31b) gives the torque:

$$t = 17.6 \text{ ft-kips/ft} - 5.96 \text{ kips/ft} \times \frac{0.75}{12} \text{ft}$$

$$= 17.2 \text{ ft-kips/ft}$$

The forces and torque acting on the edge beam per foot of length are shown in Fig. 7-31b.

    If the two ends of the beam A–B are fixed against rotation by the columns, the total torque at each end will be

$$T = \frac{t\ell_n}{2}$$

If this is not true, the torque diagram can vary within the range illustrated in Fig. 7-22. For the reasons given earlier, we shall assume that $T = t\ell_n/2$ at each end of member A–B. This gives the torque diagram shown in Fig. 7-30c.

    The shear forces in the spandrel beam are:

End A—$V_u = 5.96 \times 21.5/2 = 64.1$ kips

At d from end A, $V_u = 5.96 \times (21.5/2) - 1.33 = 56.1$ kips

End B—$V_u = 1.15 \times 64.1 = 73.7$ kips

At d from End B—$V_u = 5.96[1.15 \times (21.5/2) - 1.33] = 65.8$ kips

    4.   Should torsion be considered? If T_u exceeds the following, it must be considered:

$$T_{th} = \phi\sqrt{f_c'}\left(\frac{A_{cp}^2}{p_{cp}}\right)$$

    The effective cross section for torsion is shown in Fig. 7-32. ACI Section 11.6.1 states that the overhanging flange shall be as defined in ACI Section 13.2.4. The projection of the flange is the smaller of the height of the web below the flange (14 in.) and four times the thickness of the flange (18 in.):

$$A_{cp} = 18.5 \times 25.5 + 4.5 \times 14 = 535 \text{ in.}^2$$

$$p_{cp} = 18.5 + 25.5 + 14 + 14 + 4.5 + 39.5 = 116 \text{ in.}$$

$$T_{th} = 0.75\sqrt{4000}\left(\frac{535^2}{116}\right) = 117,000 \text{ in.-lb}$$

$$= 9.75 \text{ ft-kips}$$

Since the maximum torque of 185 ft-kips exceeds this value, torsion must be considered.

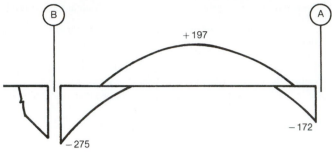

(a) Moments, M_u (ft–kips).

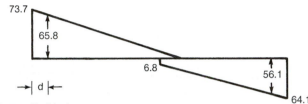

(b) Shears, V_u (kips).

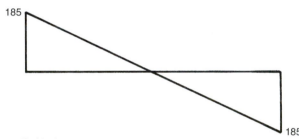

(c) Torque (ft–kips).

(d) Reduced Torque, T_u (ft–kips).

Fig. 7-30
Moments, shears, and torques
in end span of edge beam—
Example 7-4.

5. (a) Equilibrium or compatibility torsion? The torque resulting from the 0.75-in. offset of the axes of the beam and column (see Fig. 7-31a) is necessary for the equilibrium of the structure and hence is equilibrium torque. The torque at the ends of the beam due to this is

$$5.96 \times \frac{0.75}{12} \times \frac{21.5}{2} = 4.00 \text{ ft-kips}$$

On the other hand, the torque resulting from the moments at the ends of the joists exists only because the joint is monolithic and the edge beam has a torsional stiffness. If the torsional stiffness were to decrease to zero, this torque would disappear. This part of the torque is therefore compatibility torsion.

Because the loading involves compatibility torsion, we can reduce the maximum torsional moment, T_u, in the spandrel beam, at d from the faces of the columns to:

$$T_u = \phi 4 \sqrt{f_c'} \left(\frac{A_{cp}^2}{p_{cp}} \right) = 0.75 \times 4 \sqrt{4000} \times \frac{535^2}{116} = 468{,}200 \text{ in.-kips} = 39.0 \text{ ft-kips} \quad (7\text{-}44)$$

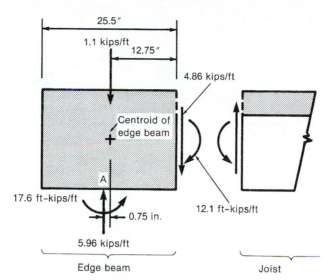

(a) Freebody diagram of edge beam.

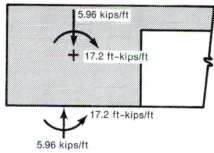

Fig. 7-31
Forces on edge beam—
Example 7-4.

(b) Forces on edge beam resolved through
centroid of edge beam.

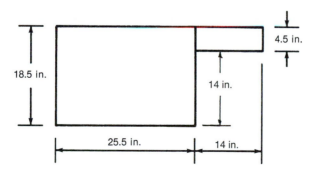

Fig. 7-32
Effective section for torsion—
Example 7-4.

but not less than the equilibrium torque of 4.0 ft-kips/ft. Assuming the remaining torque after redistribution is evenly distributed along the length of the spandrel beam. The distributed reduced torque, t, due to moments at the ends of the joists has decreased to

$$t = \frac{39.0}{21.5 - 2 \times 1.33} = 4.14 \text{ ft-kips/ft}$$

5. (b) Adjust the moments in the joists. The moment diagram for the joists, with the exterior negative moment of $-w\ell_n^2/24$ per foot of width of floor, is plotted in Fig. 7-33a. The torsional

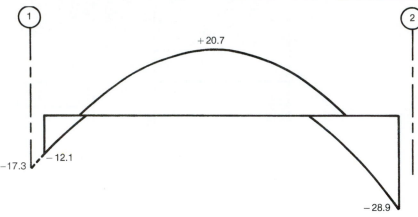

(a) Moment in joists per foot of width before adjustment (ft-kips).

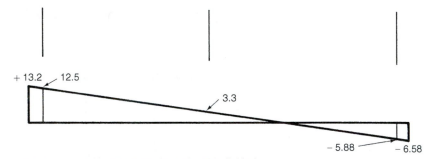

(b) Redistribution of moment per foot of width (ft-kips).

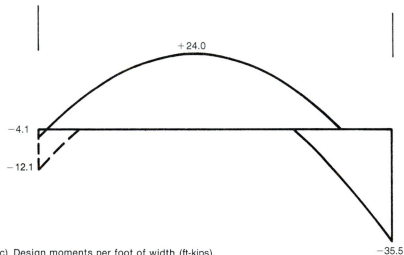

(c) Design moments per foot of width (ft-kips).

Fig. 7-33
Moments in end span of
joist—Example 7-4.

moments in the spandrel beam—mainly compatibility moments—can be dissipated by torsional cracking of the spandrel beam. ACI Section 11.6.2.2 allows the negative moment at the joint between the joists and the spandrel beam to be decreased to the value given by (7-31) decreasing from −17.3 ft-kip/ft to −4.14 ft-kip/ft, for a reduction of 12.4 ft-kip/ft in the moment in the one-ft wide strip of joists. This causes a redistribution of the end moment. The moment at the spandrel beam end of the joist, end 1, will decrease by 12.5 ft-kip/ft. Half of this, 6.3 ft-kip/ft, is carried over to the far end of the joist, as shown in Fig. 7.33b. The changes in the joist end moments at the faces of the spandrel beam and interior beams are +13.2 ft-kip/ft and −6.6 ft-kip/ft. At midspan, the change is +3.3 ft-kip/ft. The resulting moment diagram per foot of width is shown in Fig. 7-33c. Each joist supports a 3-ft-wide strip and hence supports three times these moments. The exterior negative-moment steel in the joist should be designed for a negative moment since it is necessary to develop torsional cracks in the spandrel beam before the redistribution can occur. A good rule of thumb is to design the exterior negative steel for the moment computed from $w\ell_n^2/24$, as shown by the dashed line in Fig. 7-33c.

6.   Is the section big enough for the torsion? For a solid section, the limit on shear and torsion is given by

$$\sqrt{\left(\frac{V_u}{b_wd}\right)^2 + \left(\frac{T_up_h}{1.7A_{oh}^2}\right)^2} \le \phi\left(\frac{V_c}{b_wd} + 8\sqrt{f_c'}\right) \qquad (7\text{-}33b)$$

(ACI Eq. 11-18)

From Fig. 7-32,

$$A_{oh} = (18.5 - 2 \times 1.5 - 0.5)(25.5 - 2 \times 1.5 - 0.5)$$
$$= 330 \text{ in.}^2$$
$$p_h = 2(15.0 + 22.0)$$
$$= 74 \text{ in.}$$

$$\sqrt{\left(\frac{67,700}{25.5 \times 16}\right)^2 + \left(\frac{39.0 \times 12,000 \times 74}{1.7 \times 330^2}\right)^2} = \sqrt{27,530 + 35,000}$$
$$= 250 \text{ psi}$$

From (6-8) (ACI Eq. (11-3)),

$$V_c = 2\sqrt{f_c'}\,b_wd$$
$$\phi(2\sqrt{f_c'} + 8\sqrt{f_c'}) = 0.75\left(10\sqrt{4000}\right)$$
$$= 474 \text{ psi}$$

Since 250 psi is less than 474 psi, the section is large enough.

7.   Compute the stirrup area required for shear in the edge beam. From (6-9) and (6-14) (ACI Eqs. (11-1) and (11-2)),

$$V_s = \frac{V_u}{\phi} - V_c$$

and from (6-18) (ACI Eq. (11-15)),

$$\frac{A_v}{s} = \frac{V_u/\phi - V_c}{f_{yv}d}$$

where, from (6-8) (ACI Eq. (11-3)),

$$V_c = 2\sqrt{4000} \times 25.5 \times \frac{16}{1000} = 51.6 \text{ kips}$$

Thus,

$$\frac{A_v}{s} = \frac{V_u/0.75 - 51.6}{60 \times 16}$$

where V_u is in kips.
At the left of the beam (End *B*)

$$\frac{A_v}{s} = \frac{\dfrac{73.7}{0.75} - 51.6}{60 \times 16} = 0.0486$$

At *d* from end *B*

$$\frac{A_v}{s} = \frac{\dfrac{67.7}{0.75} - 51.6}{60 \times 16} = 0.0403$$

Figure 7-35a illustrates the calculation of $V_u/\phi - V_c$. Figure 7-35b is a plot of the A_v/s required for shear along the length of the beam. The values of A_v/s for shear and A_t/s for torsion (step 8) will be superimposed in step 9.

8. Compute the stirrups required for torsion. From (7-24) (ACI Eq. (11-21)), taking $\theta = 45°$ and $A_o = 0.85A_{oh}$ gives

$$\frac{A_t}{s} = \frac{T_u/\phi}{2 \times 0.85A_{oh}f_{yv}}$$

$$= \frac{T_u/\phi \times 12,000}{2 \times 0.85 \times 330 \times 60,000}$$

$$= 0.000357\frac{T_u}{\phi}$$

$$t = \frac{39.0}{21.5 - 2 \times 1.33} = 4.14 \text{ ft-kips/ft.}$$

At end B, $T_u = 54.99$ ft-kips, $T_u/\phi = 73.3$ ft-kips, and $A_t/s = 0.0261$
At d from B, $T_u = 50.3$ ft-kips, $T_u/\phi = 67.1$ ft-kips, and $A_t/s = 0.0239$
At d from end A, $T_u = 76.0$ ft-kips, $T_u/\phi = 101.3$ ft-kips, and $A_t/s = 0.0361$

where T_u is in ft-kips. These are is plotted in Fig. 7-35c. $A_{t/s}$ is plotted in Fig. (7-35d).

9. Add the stirrup areas and select the stirrups.

$$\frac{A_{v+t}}{s} = \frac{A_v}{s} + \frac{2A_t}{s} \tag{7-46}$$

At d from end B

$$\frac{A_{v+t}}{s} = 0.0403 + 2 \times 0.0239 = 0.0881$$

For No. 4 double-leg stirrups, $s = 4.58$ in.

A_{v+t}/s is plotted in Fig. 7-35e. The maximum allowable spacings are as follows:

for shear (ACI Section 11.5.4.1), $d/2 = 8$ in.;
for torsion (ACI Section 11.6.6.1), the smaller of 12 in. and $p_h/8 = 74/8 = 9.25$ in.

The dashed horizontal lines in Fig. 7-35e are the values of A_{v+t}/s for No. 4 closed stirrups at spacings of 5 in.($= 2 \times 0.20/5.0 = 0.080$), 7 in. and 8 in. Stirrups must extend to points where $V_u/\phi = V_c/2$, or to $(d + b_t)$, where b_t is the width of the portion of the edge beam with closed stirrups, which is $16 + 25.5 = 41.5$ in. past the point where torsional reinforcement is no longer needed, that is, past the points where $T_u/\phi = $ (the torque given by (7-18))/$\phi = 9.75/0.75 = 13.0$ ft-kips. These points are indicated in Fig. 7-35c to e. Since they are closer than 41.5 in. to midspan, stirrups are required over the entire span.

Provide No. 4 closed stirrups:

End A: One @ 3 in., six @ 7 in.
End B: One @ 3 in., 12 @ 5 in., then @ 8 in. on centers throughout the rest of the span.

10. Design the longitudinal reinforcement for torsion.

(a) Longitudinal reinforcement required to resist T_n

$$A_\ell = \left(\frac{A_t}{s}\right)p_h\left(\frac{f_{yv}}{f_{y\ell}}\right)\cot^2\theta \tag{7-31}$$

where A_t/s is the amount computed in step 8. This varies along the length of the beam. For simplicity, we shall keep the longitudinal steel constant along the length of the span and shall base it on the maximum $A_t/s = 0.0185$ in.^{2}/in. Again, $\theta = 45°$. We have

$$A_\ell = 0.0185 \times 74 \times 1.0 \times 1.0 = 1.37 \text{ in.}^2$$

Alternatively, use (7-30) to compute the required amount of longitudinal reinforcement. instead of (7-31),

$$A_\ell = \frac{T_n p_h}{2 A_o f_{y\ell}} \cot \theta$$

(7-31)

where T_n = nominal resisting torque.
p_h = perimeter of closed stirrup = $2(15 + 22) = 74$ in.,
A_o = area enclosed by centerline of the shear flow path = $0.85 A_{oh}$, and
A_{oh} is the area inside the centerline of the closed stirrups = $15 \times 22 = 330$ in.^{2}
θ = inclination of cracks. The same valve of θ must be used in (7-30) and (7-31).
ACI Section 11.6.3.6 (a) suggests the use of $\theta = 45°$. Substituting in (7.30) gives

$$A_\ell = \frac{T_n \times 74}{2 \times 0.85 \times 330 \times 60} \cot 45° = 0.00220 \, T_n$$

The minimum A_ℓ is given by (7-49) (ACI Eq. (11-24)):

$$A_{\ell, \min} = \frac{5\sqrt{f'_c} A_{cp}}{f_{y\ell}} - \left(\frac{A_t}{s}\right) p_h \left(\frac{f_{yv}}{f_{y\ell}}\right)$$

where A_t/s shall not be less than $25 b_w/f_{yv} = 25 \times 25.5/60{,}000 = 0.0106$. Again, A_t/s varies along the span. The maximum A_ℓ will correspond to the minimum A_t/s. In the center region of the beam, No. 4 stirrups at 8 in. have been chosen. (See Fig. 7-35e.) Assuming half of those stirrups are for torsion, we shall take $A_t/s = 1/2 \times 0.20/8 = 0.0125$ in (7-49):

$$A_{\ell, \min} = \frac{5\sqrt{4000} \times 535}{60{,}000} - 0.0125 \times 74 \times 1.0$$

$$= 2.82 - 0.92 = 1.90 \text{ in.}^2$$

Since $A_\ell = 1.77$ in.^{2} is less than $A_{\ell, \min} = 1.90$ in.^{2}, use 1.90 in.^{2}.

From ACI Section 11.6.6.2, the longitudinal steel is distributed around the perimeter of the stirrups with a maximum spacing of 12 in. There must be a bar in each corner of the stirrups, and these bars have a minimum diameter of 1/24 of the stirrup spacing, but not less than a No. 3 bar.

The minimum bar diameter corresponds to the maximum stirrup spacing: For 8 in., $8/24 = 0.33$ in.

To satisfy the 12-in.-maximum spacing, we need **3 bars at the top and bottom and one halfway up each side.** A_s per bar = $1.90/8 = 0.24$ in.^{2}. **Use No. 5 bars for longitudinal steel A_ℓ.**

The longitudinal torsion steel required at the top of the beam is provided by increasing the area of flexural steel provided at each end and by lap-splicing 3 No. 5 bars with the negative-moment steel. The lap splices should be at least a Class B tension lap for a No. 5 top bar (see Table 8-4), since all the bars are spliced at the same point.

Exterior end negative moment: $A_s = 2.64 + 3 \times 0.24 = 3.36$ in.^{2} Use No. 6 bars because bars must be anchored in column.

Use 8 No. 6 $= 3.52$ in.^{2}. These fit in one layer.
First interior negative moment: $A_s = 4.23 + 3 \times 0.24 = 4.95$ in.^{2}.
Use 4 No. 8 + 3 No. 7 $= 4.96$ in.^{2}. These fit in one layer, minimum bars 17.5 in.

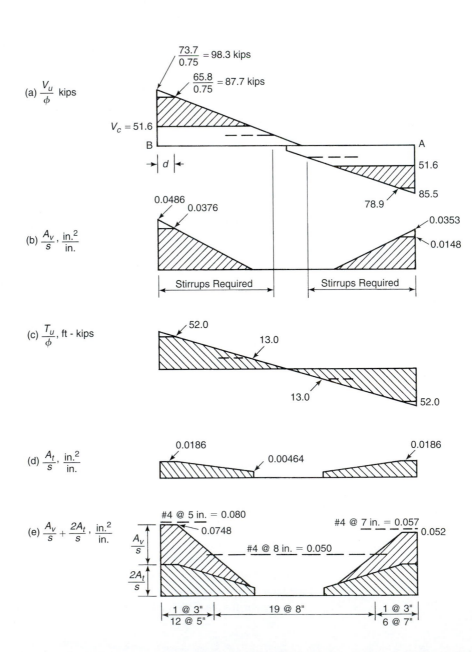

Fig. 7-34
Section *A–A* from Fig. 7-31 through edge beam at first interior negative moment section. Joist reinforcement omitted for clarity.

Fig. 7-35
Calculation of stirrups for shear and torsion—Example 7-4.

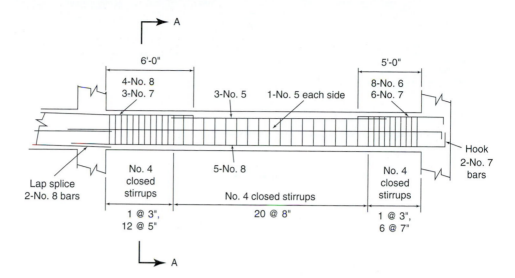

Fig. 7-36
Reinforcement in edge
beam—Example 7-4.

The longitudinal torsional steel required at the bottom is obtained by increasing the area of steel at midspan. The increased area of steel will be extended from support to support.

Midspan positive moment: $A_s = 3.02 + 3 \times 0.24 = 3.74$ in.^{2}
Use 5 No. 8 $= 3.95$ in.^{2}. These fit in one layer.

The steel finally chosen is shown in Fig. 7-36. A section through the beam at the first interior support is shown in Fig. 7-34. The cutoff points for the flexural steel were based on Fig. A-5b, except that the area of positive moment steel anchored in the supports by hooks and lap splices was taken equal to the larger of the amounts given in Fig. A-5b and the bottom layer of $A_\ell = 3 \times 0.22 = 0.66$ in.^{2}. This was rounded up arbitrarily to 2 No. 7 bars. ∎

PROBLEMS

7-1 A cantilever beam 8 ft long and 18 in. wide supports its own dead load plus a concentrated load located 6 in. from the end of the beam and 4.5 in. away from the vertical axis of the beam. The concentrated load is 15 kips dead load and 20 kips live load. Design reinforcement for flexure, shear, and torsion. Use $f_y = 60,000$ psi for all steel and $f'_c = 3750$ psi.

7-2 Explain why the torsion in the edge beam A–B in Fig. 7-22c is called "equilibrium torsion," while the torsion in the edge beam $A1$–$B1$ in Fig. 7-31 is called "compatibility torsion."

8 Development, Anchorage, and Splicing of Reinforcement

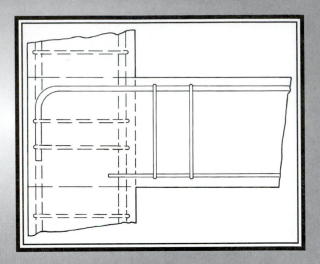

8-1   INTRODUCTION

In a reinforced concrete beam, the flexural compressive forces are resisted by concrete, while the flexural tensile forces are provided by reinforcement, as shown in Fig. 8-1. For this process to exist, there must be a force transfer, or *bond*, between the two materials. The forces acting on the bar are shown in Fig. 8-1b. For the bar to be in equilibrium, bond stresses must exist. If these disappear, the bar will pull out of the concrete and the tensile force, T, will drop to zero, causing the beam to fail.

Bond stresses must be present whenever the stress or force in a reinforcing bar changes from point to point along the length of the bar. This is illustrated by the free-body diagram in Fig. 8-2. If f_{s2} is greater than f_{s1}, bond stresses, μ, must act on the surface of the bar to maintain equilibrium. Summing forces parallel to the bar, one finds that the average bond stress, μ_{avg}, is

$$(f_{s2} - f_{s1})\frac{\pi d_b^2}{4} = \mu_{avg}(\pi d_b)\ell$$

and taking $(f_{s2} - f_{s1}) = \Delta f_s$ gives

$$\mu_{avg} = \frac{\Delta f_s d_b}{4\ell} \tag{8-1}$$

If ℓ is taken as a very short length, dx, this equation can be written as

$$\frac{df_s}{dx} = \frac{4\mu}{d_b} \tag{8-2}$$

where μ is the *true bond stress* acting in the length dx.

(a) Internal forces in beam.

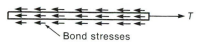

Bond stresses

(b) Forces on reinforcing bar.

Fig. 8-1
Need for bond stresses.

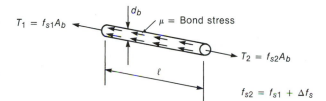

Fig. 8-2
Relationship between change
in bar stress and average
bond stress.

Average Bond Stress in a Beam

In a beam, the force in the steel at a crack can be expressed as

$$T = \frac{M}{jd} \tag{8-3}$$

where jd is the internal lever arm and M is the moment acting at the section. If we consider a length of beam between two cracks, as shown in Fig. 8-3, the moments acting at the two cracks are M_1 and M_2. If the beam is reinforced with one bar of diameter d_b, the forces on the bar are as shown in Fig. 8-3c. Summing horizontal forces gives

$$\Delta T = (\pi d_b)\mu_{avg}\,\Delta x \tag{8-4}$$

where d_b is the diameter of the bar, or

$$\frac{\Delta T}{\Delta x} = (\pi d_b)\mu_{avg}$$

But

$$\Delta T = \frac{\Delta M}{jd}$$

giving

$$\frac{\Delta M}{\Delta x} = (\pi d_b)\mu_{avg}\,jd$$

From the free-body diagram in Fig. 8-3d, we can see that $\Delta M = V\Delta x$ or $\Delta M / \Delta x = V$. Therefore,

$$\mu_{avg} = \frac{V}{(\pi d_b)jd} \tag{8-5}$$

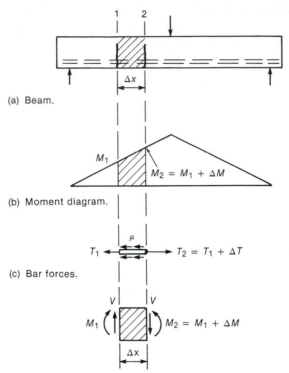

(a) Beam.

(b) Moment diagram.

(c) Bar forces.

Fig. 8-3
Average flexural bond stress.

(d) Part of beam between sections 1 and 2.

If there is more than one bar, the bar perimeter (πd_b) is replaced with the sum of the perimeters, ΣO, giving

$$\mu_{\text{avg}} = \frac{V}{\Sigma O jd} \tag{8-6}$$

Equations (8-5) and (8-6) give the *average bond stress* between two cracks in a beam. As shown later, the actual bond stresses vary from point to point between the cracks.

Bond Stresses in an Axially Loaded Prism

Figure 8-4a shows a prism of concrete containing one reinforcing bar, which is loaded in tension. At the cracks, the stress in the bar is $f_s = T/A_s$. Between the cracks, a portion of the load is transferred to the concrete by bond, and the resulting distributions of steel and concrete stresses are shown in Fig. 8-4b and c. From (8-2), we see that the bond stress at any point is proportional to the slope of the steel stress diagram at that same point. Thus, the bond-stress distribution is shown in Fig. 8-4d. Since the stress in the steel is equal at each of the cracks, the force is also equal, so that $\Delta T = 0$ at the two cracks, and from (8-4), we see that the average bond stress, μ_{avg}, is also equal to zero. Thus, for the average bond stress to equal zero, the total area under the bond-stress diagram between any two cracks in Fig. 8-4d must equal zero when $\Delta T = 0$.

The bond stresses given by (8-2) and plotted in Fig. 8-4d are referred to as *true bond stresses* or *in-and-out bond stresses* (they transfer stress into the bar and back out again) to distinguish them from the *average bond stresses* calculated from (8-1).

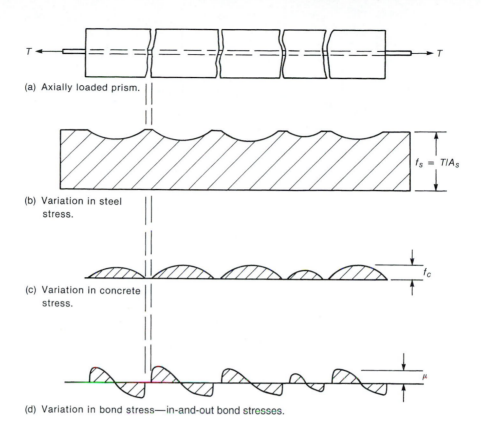

(a) Axially loaded prism.

$f_s = T/A_s$

(b) Variation in steel stress.

f_c

(c) Variation in concrete stress.

μ

Fig. 8-4
Steel, concrete, and bond stresses in a cracked prism.

(d) Variation in bond stress—in-and-out bond stresses.

True Bond Stresses in a Beam

At the cracks in a beam, the bar force can be computed from (8-3). If the concrete and the bar are bonded together, a portion of the tensile force will be resisted by the concrete at points between the cracks. As a result, the tensile stresses in the steel and the concrete at the level of the steel will vary, as shown in Fig. 8-5c and d. This gives rise to the bond-stress distribution plotted in Fig. 8-5e. In the constant-moment region between the two loads, the shear is zero, and the average bond stress from the in-and-out bond-stress diagram in Fig. 8-5(e) is zero. Between a support and the nearest load, there is shear. Once again, there are in-and-out bond stresses, but now the total area under the bond-stress diagram is not zero. The average bond stress in Fig. 8-5e must equal the value given by (8-5).

Bond Stresses in a Pull-Out Test

The easiest way to test the bond strength of bars in a laboratory is by means of the *pull-out test*. Here, a concrete cylinder containing the bar is mounted on a stiff plate and a jack is used to pull the bar out of the cylinder, as shown in Fig. 8-6a. In such a test, the concrete is compressed and hence does not crack. The stress in the bar varies as shown in Fig. 8-6b, and the bond stress (see (8-2)) varies as shown in Fig. 8-6c. This test does not give values representative of the bond strength of beams, because the concrete is not cracked and hence there is no in-and-out bond-stress distribution. Also, the bearing stresses of the concrete against the plate cause a frictional component that resists the transverse expansion that would reflect Poisson's ratio. Prior to 1950, pull-out tests were used extensively to evaluate the bond strength of bars. Since then, various types of beam tests have been used to study bond strength [8-1].

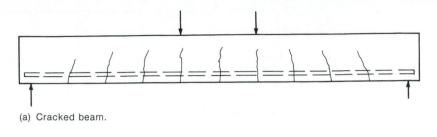

(a) Cracked beam.

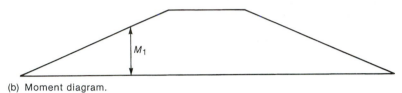

(b) Moment diagram.

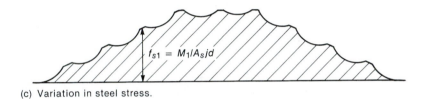

(c) Variation in steel stress.

(d) Tensile stress in concrete.

Fig. 8-5
Steel, concrete, and bond
stresses in a cracked beam.

(e) In-and-out bond stresses.

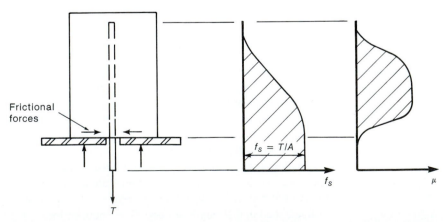

Fig. 8-6
Stress distribution in a pull-
out test.

(a) Test method.   (b) Bar stress.   (c) Bond stress.

8-2   MECHANISM OF BOND TRANSFER

A smooth bar embedded in concrete develops bond by adhesion between the concrete and the bar and by a small amount of friction. Both of these effects are quickly lost when the bar is loaded in tension, particularly because the diameter of the bar decreases slightly, due to Poisson's ratio. For this reason, smooth bars are generally not used as reinforcement. In cases where smooth bars must be embedded in concrete (anchor bolts, stirrups made of small diameter bars, etc.), mechanical anchorage in the form of hooks, nuts, and washers on the embedded end (or similar devices) are used.

Although adhesion and friction are present when a deformed bar is loaded for the first time, these bond-transfer mechanisms are quickly lost, leaving the bond to be transferred by bearing on the deformations of the bar as shown in Fig. 8-7a. Equal and opposite bearing stresses act on the concrete, as shown in Fig. 8-7b. The forces on the concrete have both a longitudinal and a radial component (Fig. 8-7c and d). The latter causes circumferential tensile stresses in the concrete around the bar. Eventually, the concrete will split parallel to the bar, and the resulting crack will propagate out to the surface of the beam. The splitting cracks follow the reinforcing bars along the bottom or side surfaces of the beam,

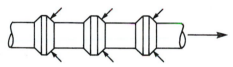

(a) Forces on bar.

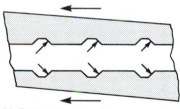

(b) Forces on concrete.

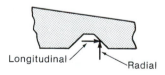

Longitudinal   Radial

(c) Components of force on concrete.

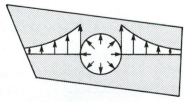

(d) Radial forces on concrete and splitting stresses shown on a section through the bar.

Fig. 8-7
Bond-transfer mechanism.

as shown in Fig. 9-3e. Once these cracks develop, the bond transfer drops rapidly unless reinforcement is provided to restrain the opening of the splitting crack.

The load at which splitting failure develops is a function of

1. the minimum distance from the bar to the surface of the concrete or to the next bar—the smaller this distance, the smaller is the splitting load;

2. the tensile strength of the concrete; and

3. the average bond stress—as this increases, the wedging forces increase, leading to a splitting failure.

These factors are discussed more fully in [8-2], [8-3], and [8-4]. Typical splitting-failure surfaces are shown in Fig. 8-8. The splitting cracks tend to develop along the shortest distance between a bar and the surface or between two bars. In Fig. 8-8 the circles touch the edges of the beam where the distances are shortest.

If the cover and bar spacings are large compared to the bar diameter, a *pull-out failure* can occur, where the bar and the annulus of concrete between successive deformations pull out along a cylindrical failure surface joining the tips of the deformations.

8-3   DEVELOPMENT LENGTH

Because the actual bond stress varies along the length of a bar anchored in a zone of tension as shown in Fig. 8-5e, the ACI Code uses the concept of *development length* rather than bond stress. The development length, ℓ_d, is the shortest length of bar in which the bar stress can increase from zero to the yield strength, f_y. If the distance from a point where the bar stress equals f_y to the end of the bar is less than the development length, the bar will pull out of the concrete. The development lengths are different in tension and compression, because a bar loaded in tension is subject to in-and-out bond stresses and hence requires a considerably longer development length.

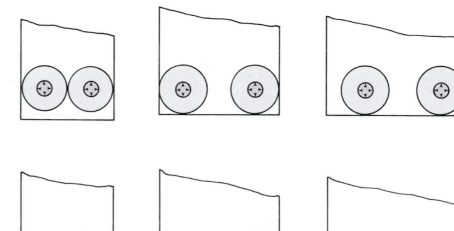

Fig. 8-8
Typical splitting-failure surfaces.

(a) Side cover and half the bar spacing both less than bottom cover.

(b) Side cover = bottom cover, both less than half the bar spacing.

(c) Bottom cover less than side cover and half the bar spacing.

The development length can be expressed in terms of the ultimate value of the average bond stress by setting $(f_{s2} - f_{s1})$ in (8-1) equal to f_y:

$$\ell_d = \frac{f_y d_b}{4\mu_{\text{avg},u}} \tag{8-7}$$

Here, $\mu_{\text{avg},u}$ is the value of μ_{avg} at bond failure in a beam test.

Tension-Development Lengths

Analysis of Bond Splitting Load

Although the code equations for bond strength were derived from statistical analyses of test results, the following analysis illustrates the factors affecting the splitting load. Consider a cylindrical concrete prism of diameter $2c$, containing a bar of diameter d_b, as shown in Fig. 8-9a. The radial components of the forces on the concrete, shown in Fig. 8-9b and c, cause a pressure p on a portion of the cross section of the prism, as shown in Fig. 8-9b. This is equilibrated by tensile stresses in the concrete on either side of the bar. In Fig. 8-9c, the distribution of these stresses has arbitrarily been assumed to be triangular. The circular prism in Fig. 8-9 represents the zones of highest radial tensile stresses, shown by the larger circles in Fig. 8-8. Splitting is assumed to occur when the maximum stress is equal to the tensile strength of the concrete, f_{ct}. For equilibrium in the vertical direction in a prism of length equal to ℓ,

$$\frac{p d_b \ell}{2} = K\left(c - \frac{d_b}{2}\right) f_{ct} \, \ell$$

where K is the ratio of the average tensile stress to the maximum tensile stress and equals 0.5 for the triangular stress distribution shown in Fig. 8-9c. Rearranging gives

$$p = \left(\frac{c}{d_b} - \frac{1}{2}\right) f_{ct}$$

If the forces shown in Fig. 8-7b and c are assumed to act at 45°, the average bond stress, $\mu_{\text{avg},u}$, at the onset of splitting is equal to p. Taking $f_{ct} = 6\sqrt{f_c'}$ gives

$$\mu_{\text{avg},u} = 6\sqrt{f_c'}\left(\frac{c}{d_b} - \frac{1}{2}\right) \tag{8-8}$$

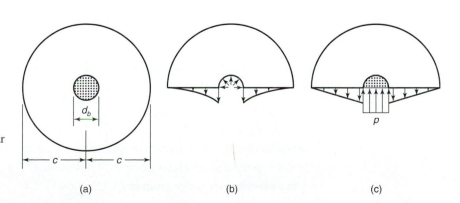

Fig. 8-9
Concrete stresses in a circular concrete prism containing a reinforcing bar that is subjected to bond stresses, shown in section.

The length of bar required to raise the stress in the bar from zero to f_y is called the development length, ℓ_d. From (8-7),

$$\ell_d = \frac{f_y}{4\mu_{\text{avg},u}} d_b$$

Substituting (8-8) for $\mu_{\text{avg},u}$ gives

$$\ell_d = \frac{f_y}{24\sqrt{f_c'}}\left(\frac{c}{d_b} - \frac{1}{2}\right)d_b$$

Arbitrarily taking $c = 1.5d_b$ (for the reasons given in the derivation of (8-13) and (8-14) later in the chapter) and rearranging yields

$$\ell_d = \frac{f_y}{24\sqrt{f_c'}} d_b \tag{8-9}$$

As we shall see in the next few pages, (8-9) is very similar to (8-11) if α, β, γ, and λ are taken equal to 1.0. Four major assumptions were made in this derivation:

(a)   the distribution of the tensile stresses in the concrete,

(b)   the angle of inclination of the forces on the deformations,

(c)   the replacement of the concentrated forces on the deformations with a force that is uniformly distributed along the length and around the circumference of the bar, and

(d)   the neglect of the effect of in-and-out bond stresses between cracks.

The similarity of the final result here to (8-11), derived later, reinforces the validity of the splitting model.

Basic Tension-Development Equation

In 1977, Orangun et al. [8-2] fitted a regression equation through the results of a large number of bond and splice tests. The resulting equation for bar development length, ℓ_d, included terms for the bar diameter, d_b; the bar stress to be developed, $\sqrt{f_c'}$; the cover or bar spacing; and the transverse steel ratio. It served as the basis of the development-length provisions in the 1989 ACI Code. These provisions proved difficult to use, however, and between 1989 and 1995, ACI Committee 318 and the ACI bond committee simplified the design expressions, in two stages. First, a basic expression was developed for the development length, ℓ_d, given in ACI Section 12.2.3 as

$$\ell_d = \frac{3}{40}\frac{f_y}{\sqrt{f_c'}}\frac{\alpha\beta\gamma\lambda}{\left(\frac{c+K_{tr}}{d_b}\right)}d_b \tag{8-10}$$

(ACI Eq. 12-1)

where the term $(c + K_{tr})/d_b$ is limited to 2.5 or smaller, to prevent pull-out bond failures, and the length ℓ_d is not taken less than 12 in. Also,

ℓ_d is the development length, in.
d_b is the bar diameter, in.
α is a bar-location factor given in ACI Section 12.2.4.
β is an epoxy-coating factor given in ACI Section 12.2.4.
γ is a bar-diameter factor given in ACI Section 12.2.4.

λ is a lightweight-concrete factor given in ACI Section 12.2.4.

c is the smaller of   (a)   the smallest distance measured from the surface of the concrete to the *center* of a bar being developed, and

(b)   one-half of the *center-to-center* spacing of the bars or wires being developed

K_{tr} is a transverse reinforcement factor given in ACI Section 12.2.4.

Values of the factors α, β, γ, λ, and K_{tr} will be presented later.

The second stage in the derivation of (8-10) was to substitute common values of c and K_{tr} into (8-10), as described next.

Simplified Tension-Development-Length Equations

In most cases, (8-10) would be difficult to use in design because c and K_{tr} vary along the length of a member. This equation was simplified by substituting lower limit values of c and K_{tr} for common design cases, to get more widely applicable equations that did not explicitly include these factors. For deformed bars or deformed wire, ACI Section 12.2.2 defines the development length, as given in Tables 8-1 and 8-1M.

The cases 1 and 2 described in the top row of Tables 8-1 and 8-1M are illustrated in Fig. 8-10a and b. The "code minimum" stirrups and ties mentioned in case 1 correspond to the minimum amounts and maximum spacings specified in ACI Sections 11.5.4.1, 11.5.5.3, and 7.10.5.

Bar-Spacing Factor, c

The factor c in (8-10) is the smaller of two quantities:

1.   In the first definition, c is the smallest distance from the surface of the concrete to the *center* of the bar being developed. (See Fig. 8-10a.)

ACI Section 7.7.1(c) gives the minimum cover to principal reinforcement as $1\frac{1}{2}$ in. For a beam stem not exposed to weather, with No. 11 bars enclosed in No. 3 stirrups or ties, c will be (1.5-in. cover to the stirrups + 0.375-in. stirrup) + (half of the bar diameter, $1.41/2 = 0.71$ in.) $= 2.58$ in.

TABLE 8-1   Equations for Development Lengths, $\ell_d{}^a$

	No. 6 and Smaller Bars and Deformed Wires	No. 7 and Larger Bars
Case 1: Clear spacing of bars being developed or spliced not less than d_b, **and** stirrups or ties throughout ℓ_d not less than the code minimum	$\ell_d = \dfrac{f_y \alpha \beta \lambda}{25\sqrt{f'_c}} d_b$ (8-11)	$\ell_d = \dfrac{f_y \alpha \beta \lambda}{20\sqrt{f'_c}} d_b$ (8-12)
or		
Case 2: Clear spacing of bars being developed or spliced not less than $2d_b$ and clear cover not less than d_b		
Other cases	$\ell_d = \dfrac{3 f_y \alpha \beta \lambda}{50\sqrt{f'_c}} d_b$ (8-13)	$\ell_d = \dfrac{3 f_y \alpha \beta \lambda}{40\sqrt{f'_c}} d_b$ (8-14)

^{a}The length ℓ_d computed using (8-11) to (8-14) shall not be taken less than 12 in.

TABLE 8-1M Equations for Development Lengths—SI Units[a]

	No. 20 and Smaller Bars and Deformed Wires	No. 25 and Larger Bars
Case 1: Clear spacing of bars being developed or spliced not less than d_b, **and** stirrups or ties throughout ℓ_d not less than the code minimum	$$\ell_d = \frac{12f_y\alpha\beta\lambda}{25\sqrt{f_c'}}d_b$$ (8-11M)	$$\ell_d = \frac{12f_y\alpha\beta\lambda}{20\sqrt{f_c'}}d_b$$ (8-12M)
or		
Case 2: Clear spacing of bars being developed or spliced not less than $2d_b$ and clear cover not less than d_b		
Other cases	$$\ell_d = \frac{18f_y\alpha\beta\lambda}{25\sqrt{f_c'}}d_b$$ (8-13M)	$$\ell_d = \frac{18f_y\alpha\beta\lambda}{20\sqrt{f_c'}}d_b$$ (8-14M)

[a]The length ℓ_d computed using (8-11M) to (8-14M) shall not be taken less than 300 mm.

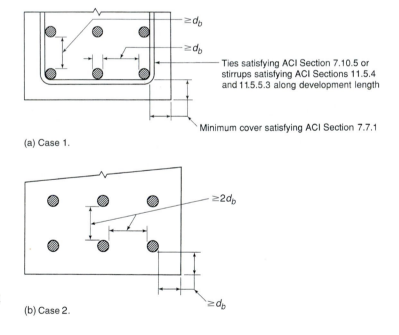

(a) Case 1.

(b) Case 2.

Fig. 8-10
Explanation of cases 1 and 2
in ACI Section 12.2.2.

2. In the second definition, c is equal to one-half of the center-to-center spacing of the bars.

ACI Section 7.6.1 gives the minimum clear spacing of bars as d_b, but not less than 1 in. For No. 11 bars, the diameter is 1.41 in., giving the center-to-center spacing of the bars as $1.41/2 + 1.41 + 1.41/2 = 2.82$ in. and $c = 1.41$ in.

The smaller of these two is $c = 1.41$ in. $= 1.0d_b$. The minimum stirrups or ties given in ACI Sections 7.10.5, 11.5.4.1, and 11.5.5.3 correspond to K_{tr} between $0.1d_b$ and $0.5d_b$, depending on a wide range of factors. Thus, for this case, $(c + K_{tr})/d_b \approx 1.5$. Substituting this and the appropriate bar size factor, γ, into (8-10) gives (8-11) and (8-12) in Table 8-1.

For case 2 in Table 8-1, consider a slab with clear cover to the outer layer of bars of d_b and a clear spacing between the bars of $2d_b$. It is assumed that splitting in the cover will be restrained by bars perpendicular to the bars being developed. As a result, c is governed by the bar spacing. If the clear spacing is $2d_b$, the center-to-center spacing is $3d_b$. Thus, $c = 3d_b/2 = 1.5d_b$. Substituting this into (8-10) and taking $K_{tr} = 0$ gives 8-11 or 8-12 in Table 8-1.

For the situation where the minimum clear cover to the bar being developed is $1.5d_b$ and the minimum clear spacing is d_b, c is the smaller of $1.5d_b$ and d_b. Substituting γ and $c = d_b$ into (8-10), assuming that $K_{tr} = 0$ gives (8-13) and (8-14), which apply to cases other than 1 and 2, as shown in Table 8-1.

Values of ℓ_d computed from (8-11) to (8-14) are given in Table A-11. Typically, ℓ_d is about $30d_b$ for No. 3 to No. 6 bottom bars and about $40d_b$ for No. 7 and larger bottom bars.

Values of ℓ_d computed from (8-11) to (8-14) are tabulated in Tables A-11 and A-11M in Appendix A. Table A-6 can be used to get the minimum web widths corresponding to a $1.0d_b$ spacing (case 1). If the actual web width exceeds the value from this table, it is not necessary to check the spacings further in determining whether the beam is of case 1.

Factors in (8-10) through (8-14)

The Greek-letter factors in (8-10) through (8-14) are defined in ACI Section 12.2.4 as follows:

α = **bar-location factor**

   Horizontal reinforcement so placed that more than 12 in. of fresh concrete is cast in the member below the development length or splice 1.3

   Other reinforcement 1.0

Horizontal reinforcement with more than 12 in. of fresh concrete below it at the time the bar is embedded in concrete is referred to as *top reinforcement*. During the placement of the concrete, water and mortar migrate vertically upward through the concrete, collecting on the underside of reinforcing bars. If the depth below the bar exceeds 12 in., sufficient mortar will collect to weaken the bond significantly. This applies to the top reinforcement in beams with depths greater than 12 in. and to horizontal steel in walls cast in lifts greater than 12 in. The factor was reduced from 1.4 to 1.3 in 1989, as a result of tests in [8-5].

β = **coating factor**

   Epoxy-coated bars or wires with cover less than $3d_b$, or clear spacing less than $6d_b$ 1.5

   All other epoxy-coated bars or wires..1.2

   Uncoated reinforcement 1.0

The product of $\alpha\beta$ need not be taken greater than 1.7.

Tests of epoxy-coated bars have indicated that there is negligible friction between concrete and the epoxy-coated bar deformations. As a result, the forces acting on the deformations and the concrete with epoxy-coated bars in Fig. 8-7a and b act in a direction perpendicular to the surface of the deformations. In a bar without an epoxy coating, friction between the deformation and the concrete allows the forces on the deformation and the concrete to act at an angle flatter than the one shown in Fig. 8-7. Because of this, the radial-force components are larger in an epoxy-coated bar than in a normal bar for a given longitudinal force component; hence, splitting occurs at a lower longitudinal force [8-6]. The 1.5 value of β corresponds to cases where splitting failures occur. For larger covers and spacings, pull-out failures tend to occur, and the effect of epoxy coating is smaller.

γ = **bar-size factor**

No. 6 and smaller bars and deformed wires 0.8

No. 7 and larger bars..1.0

Comparison of (8-10) with a large collection of bond and splice tests showed that a shorter development length was possible for smaller bars.

λ = **lightweight-aggregate-concrete factor**

When lightweight-aggregate concrete is used 1.3

However, when f_{ct} is specified, λ shall be permitted to be taken as $6.7\sqrt{f_c'}/f_{ct}$ but not less than 1.0

When normal-weight concrete is used 1.0

The tensile strength of lightweight concrete is generally less than that of normal-weight concrete; hence, the splitting load will be less. In addition, in some lightweight concretes, the wedging forces that the bar deformations exert on the concrete can cause localized crushing, which allows bar slip to occur. The factor λ does not differentiate between sand-lightweight and all-lightweight concretes, as is done in shear calculations. This decision is based on the results of tests of hooked bar anchorages, which did not show differences between the two types of lightweight concrete.

as defined earlier, c = **spacing or cover dimension, in.**

c is the smaller of

(a)   the smallest distance measured from the surface of the concrete to the *center* of the bar and

(b)   one-half of the *center-to-center* spacing of the bars.

It is important to note that c is defined relative to the center of the bars in both cases.

K_{tr} = **transverse reinforcement index**

$$= \frac{A_{tr}f_{yt}}{1500sn}$$

where the factor 1500 has units of lb/in.^2

A_{tr} = total cross-sectional area of all transverse reinforcement within the spacing s, which crosses the potential plane of splitting along the reinforcement being developed within the development length, in.^2 (illustrated in Fig. 8-11).

f_{yt} = specified yield strength of the transverse reinforcement, psi

s = maximum center-to-center spacing of transverse reinforcement within ℓ_d, in.

n = number of bars or wires being developed along the plane of splitting

ACI Section 12.2.4 allows K_{tr} to be taken equal to zero to simplify the calculations, even if there is transverse reinforcement.

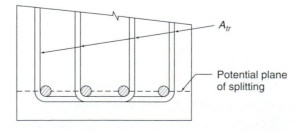

Fig. 8-11
Definition of A_{tr}.

Excess Flexural Reinforcement

If the flexural reinforcement provided exceeds the amount required to resist the factored moment, the bar stress that must be developed is less than f_y. In such a case, ACI Section 12.2.5 allows ℓ_d to be multiplied by (A_s required/A_s provided). This multiplier could also be expressed as f_s/f_y. If room is available, it is good practice to ignore this factor, thus ensuring that the steel is fully anchored even if a change in the use of the structure requires the bars to be fully stressed. In statically indeterminate structures, the increased stiffness resulting from the additional reinforcement can lead to higher moments at the section with excess reinforcement. In such a case, the steel is more highly stressed than would be expected from the ratio of areas. This multiplier is not applied in the design of members resisting seismic loads.

The development length, ℓ_d, calculated as the product of ℓ_d from ACI Section 12.2.2 or Section 12.2.3 and the factors given above, shall not be taken less than 12 in., except when computing the length of lap slices (ACI Section 12.2.1).

Compression-Development Lengths

Compression-development lengths are considerably shorter than tension-development lengths, because some force is transferred to the concrete by the bearing at the end of the bar and because there are no cracks in such an anchorage region (and hence no in-and-out bond). The basic compression-development length is (ACI Section 12.3)

$$\ell_{dc} = \frac{0.02 d_b f_y}{\sqrt{f'_c}} \text{ but not less than } 0.0003 d_b f_y \qquad (8\text{-}15)$$

where the constant 0.0003 has units of "in.^{2}/lb". Values of ℓ_{dc} are given in Tables A-12 and A-12M. The development length in compression may be reduced by multiplying ℓ_{dc} by the applicable modification factors given in ACI Section 12.3.3 for excess reinforcement and enclosure by spirals or ties. The resulting development length shall not be less than 8 in. (ACI Section 12.3.1).

Development Lengths for Bundled Bars

Where a large number of bars are required in a beam or column, the bars are sometimes placed in bundles of 2, 3, or 4 bars (ACI Section 7.6.6). The effective perimeter for bond failure of bundles is less than the total perimeter of the individual bars in the bundle. ACI Section 12.4 accounts for this by requiring an increased development length of 1.2 times the individual bar development lengths for bars in a 3-bar bundle and 1.33 times for bars in a 4-bar bundle. The value of d_b used in ACI Section 12.2 shall be taken as the diameter of a hypothetical single bar having the same area as the bundle.

Development Lengths for Coated Bars

In bridge decks and parking garages, epoxy-coated or galvanized reinforcement are frequently used to reduce corrosion problems. Epoxy-coated bars are covered by the factor β in ACI Section 12.2.4. There is no value of β for zinc-coated bars. The zinc coating on galvanized bars can affect the bond properties via a chemical reaction with the concrete. This effect can be prevented by treating the bars with a solution of chromate after galvanizing. If this is done, the bond is essentially the same as that for normal reinforcement.

The recent introduction of MMFX chromium-alloy reinforcing bars may reduce the fraction of the bar market taken by epoxy-coated bars.

Development Lengths for Welded-Wire Fabric

ACI Section 12.8 provides rules for development of welded plain-wire fabric in tension. The development of *plain-wire fabric* depends on the mechanical anchorage from at least two cross wires. The nearest of these cross wires must be located 2 in. or farther from the critical section. Wires that are closer to the critical section cannot be counted for anchorage. The second cross wire must not be located closer to the critical section than

$$\ell_{dw} = 0.27 \frac{A_w}{s_w} \frac{f_y}{\sqrt{f'_c}} \lambda \qquad (8\text{-}16)$$

where A_w and s_w are, respectively, the cross-sectional area, and spacing of the wire being developed. The development length may be reduced by multiplying by the factor in ACI Section 12.2.5 for excess reinforcement, but may not be taken as less than 6 in. except when computing the length of splices according to ACI Section 12.19.

Deformed fabric derives anchorage from bond stresses along the deformed wires and from mechanical anchorage from the cross wires. The ASTM specification for welded deformed-wire fabric does not require as strong welds for deformed fabric as for plain-wire fabric. ACI Section 12.7 gives the basic development length of deformed fabric with at least one cross wire in the development length, but not closer than 2 in. to the critical section, as ω times the development length computed from ACI Sections 12.2.2, 12.2.4, and 12.2.5, where ω is the larger of the factors given by (8-17a) and (8-17b). (ACI Section 12.7.1 refers to "the wire-fabric factor", but does not use the symbol "ω." We have introduced the symbol ω for the wire-fabric factor to clarify the presentation.) From ACI Section 12.7.2,

$$\omega = \frac{f_y - 35{,}000}{f_y} \qquad (8\text{-}17a)$$

or

$$\omega = \frac{5d_b}{f_y} \qquad (8\text{-}17b)$$

but ω need not be taken greater than 1.0.

ACI Section 12.7.3 applies to deformed fabric with no cross wires in the development length or with a single cross wire less than 2 in. from the critical section (i.e., the section where f_y must be developed). In this case,

$$\omega = 1.0 \qquad (8\text{-}17c)$$

The development length computed shall not be taken less than 8 in., except when computing the length of splices according to ACI Section 12.18 or stirrup anchorages according to ACI Section 12.13.

Tests have shown that the development length of deformed-wire fabric is not affected by epoxy coating; for this reason, the epoxy coating factor, β, is taken equal to 1.0 for epoxy-coated deformed-wire fabric.

Deformed welded-wire fabric may have some plain wires in one or both directions. For the purpose of determining the development length, such a fabric shall be considered to be a plain-wire fabric if any of the wires in the direction that development is being considered is a plain wire (ACI Section 12.7.4).

8-4   HOOKED ANCHORAGES

Behavior of Hooked Anchorages

Hooks are used to provide additional anchorage when there is insufficient length available to develop a bar. Unless otherwise specified, the so-called *standard hooks* described in ACI Section 7.1 are used. Details of 90° and 180° standard hooks and standard stirrup and tie hooks are given in Fig. 8-12. It is important to note that a standard hook on a large bar takes up a lot of room, and the actual size of the hook is frequently quite critical in detailing a structure.

A 90° hook loaded in tension develops forces in the manner shown in Fig. 8-13a. The stress in the bar is resisted by the bond on the surface of the bar and by the bearing on the concrete inside the hook [8-7]. The hook moves inward, leaving a gap between it and the concrete outside the bend. Because the compressive force inside the bend is not collinear with the applied tensile force, the bar tends to straighten out, producing compressive stresses on the outside of the tail. Failure of a hook almost always involves crushing of the concrete inside the hook. If the hook is close to a side face, the crushing will extend to

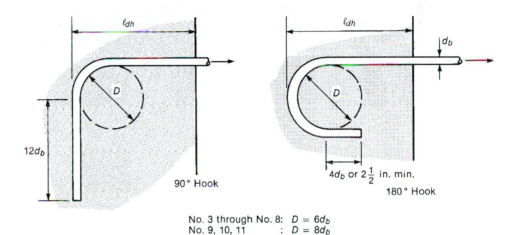

No. 3 through No. 8: $D = 6d_b$
No. 9, 10, 11   : $D = 8d_b$
No. 14 and 18   : $D = 10d_b$

(a) Standard hooks—ACI Sections 7.1 and 7.2.1.

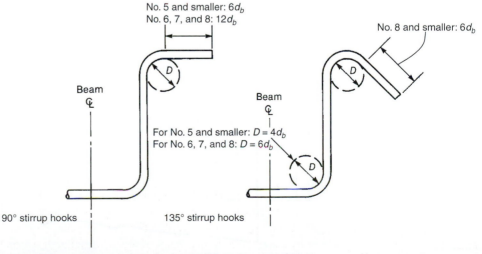

Fig. 8-12
Standard hooks.

(b) Stirrup and tie hooks—ACI Section 7.1.3.

the surface of the concrete, removing the side cover. Occasionally, the concrete outside the tail will crack, allowing the tail to straighten.

The stresses and slip measured at points along a hook at a bar stress of $1.25f_y$ (75 ksi) in tests of hooks in No. 7 bars are plotted in Fig. 8-13b and c [8-7]. The axial stresses in the bar decrease due to the bond on the lead-in length and the bond and friction on the inside of the bar. The magnitude and direction of slip at A, B, and C are shown by the arrows. For the 90° hook, the slip measured at A was 1.75 times that measured at A in the 180° hook.

The amount of slip depends on, among other things, the angle of the bend and the orientation of the hook relative to the direction of concrete placing. The slip of hooks displays a top-bar effect which is not recognized in the calculation of ℓ_{dh}. In tests, top-cast hooks, oriented so that weaker mortar was trapped inside the bend during casting, slipped 50 to 100 percent more at a given bar stress than did bottom-cast bars [8-8].

Tests on bars hooked around a corner bar show that tensile stresses can be developed at a given end slip that are 10 percent to 30 percent larger than can be developed if the hook does not contain a bar.

Design of Hooked Anchorages

The design process described in ACI Section 12.5.1 does not distinguish between 90° and 180° hooks or between top and bottom bar hooks. The development length of a hook, ℓ_{dh} (illustrated in Fig. 8-12a), is computed as by (8-18), which may be reduced by appropriate multipliers given in ACI Section 12.5.3, except as limited in ACI Section 12.5.4. The final development length shall not be less than $8d_b$ or 6 in., whichever is smaller. Accordingly,

$$\ell_{dh} = [(0.02\beta\lambda f_y/\sqrt{f_c'})]\, d_b \times \text{(applicable factor from ACI Section 12.5.3)} \quad (8\text{-}18)$$

where $\beta = 1.2$ for epoxy-coated bars or wires and 1.0 for uncoated reinforcement, and λ is the lightweight-aggregate factor given in ACI Section 12.2.4, which is equal to 1.3 for lightweight-aggregate concrete. In other cases, β and λ shall be taken to be 1.0. Values of ℓ_{dh} for uncoated bars in normal-weight concrete are given in Tables A-13 and A-13M.

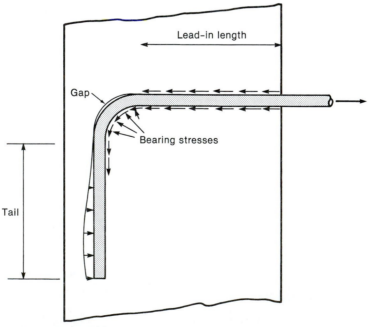

Fig. 8-13
Behavior of hooks.

(a) Forces acting on bar.

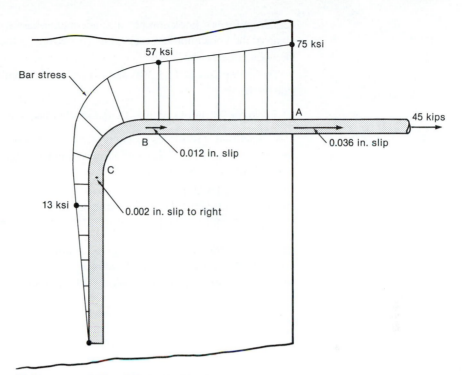

(b) Stresses and slip—90° standard hook.

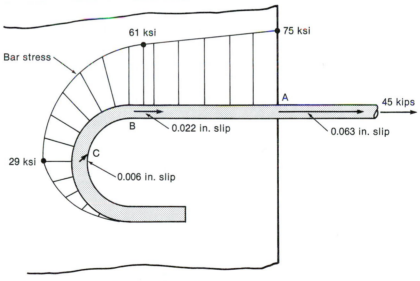

Fig. 8-13
(Continued)

(c) Stresses and slip—180° standard hook

Multipliers from ACI Section 12.5.3

The factors from ACI Section 12.5.3 account for the confinement of the hook by stirrups. Confinement reduces the chance that the concrete between the hook and the concrete surface will spall off, leading to a premature hook failure. For clarity, ACI Section 12.5.3(a) has been divided here into 12.5.3(a_1) and (a_2). The factors are as follows:

12.5.3(a_1) for 180° hooks on No. 11 and smaller bars with side cover (normal to the plane of the hook) not less than $2\frac{1}{2}$ in. .... ×0.7

12.5.3(a₂) for 90° hooks on No. 11 and smaller bars with side cover (normal to the plane of the hook) not less than $2\frac{1}{2}$ in. *and* cover on the bar extension (tail) beyond the hook not less than 2 in. . . . . ×0.7

The multipliers in 12.5.3(b) and (c) reflect the confinement of the concrete outside the bend.

12.5.3(b) for 90° hooks on No. 11 and smaller bars that are *either*

• enclosed within ties or stirrups *perpendicular* to the bar being developed, spaced not greater than $3d_b$ along the development length, ℓ_{dh}, of the hook, as shown in Fig. 8-14a [R12.5.3(a)], *or*

• enclosed within ties or stirrups *parallel* to the bar being developed, spaced not greater than $3d_b$ along the length of the tail extension of the hook plus bend, as shown in Fig. 8-14b [R12.5.3(b)] . . . ×0.8, except as given in ACI Section 12.5.4.

12.5.3(c) for 180° hooks on No. 11 or smaller bars enclosed within ties or stirrups perpendicular to the bar being developed, spaced not greater than $3d_b$ along the development length, ℓ_{dh}, of the hook . . . ×0.8, except as given in ACI Section 12.5.4.

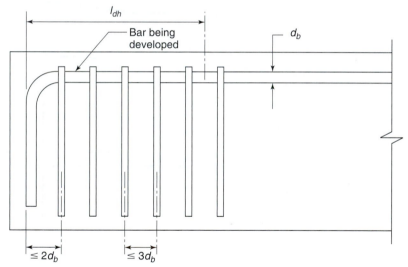

(a) Ties or stirrups placed perpendicular to the bar being developed. (See row 3 in Table 8-2).

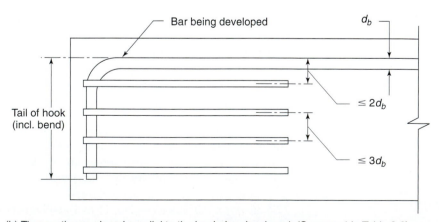

Fig. 8-14
Confinement of hooks by stirrups and ties.

(b) Ties or stirrups placed parallel to the bar being developed. (See row 4 in Table 8-2).

TABLE 8-2   Hook Lengths, ℓ_{dh} in. = ℓ_{dh} times Factors from ACI 318-02 Sections 12.5.3 and 12.5.4, but Not Less Than $8d_b$ or 6 in.

	Location	Type	d_b = Hooked Bar Size[a]	Side Cover, in.	Top or Bottom Cover, in.	Tail Cover, in.	Stirrups or Ties	Factor[c]
1.	Anywhere, 12.5.3(a)	180°	≤ No. 11	≥ 2.5 in.	Any	Any	Not required	× 0.7
2.	Anywhere, 12.5.3(a)	90°	≤ No. 11	≥ 2.5 in.	Any	≥ 2 in.	Not required	× 0.7
3.	Anywhere, 12.5.3(b)	90°	≤ No. 11	Any	Any	Any	Enclosed in stirrups or ties perpendicular to hooked bar, spaced ≤ $3d_b$ along ℓ_{dh}.[b]	× 0.8 except as in line 6.
4.	Anywhere, 12.5.3(b)	90°	≤ No. 11	Any	Any	Any	Enclosed in stirrups or ties parallel to hooked bar, spaced ≤ $3d_b$ along ℓ_{dh}.[b]	× 0.8 except as in line 6.
5.	Anywhere, 12.5.3(c)	180°	≤ No. 11	Any	Any	Any	Enclosed in stirrups or ties perpendicular to hooked bar, spaced ≤ $3d_b$ along ℓ_{dh}.[b]	× 0.8
6.	At the ends of members, 12.5.4[d]	90° or 180°	≤ No. 11	≤ 2.5 in.	≤ 2.5 in.	Any	Enclosed in stirrups or ties perpendicular to hooked bar, spaced ≤ $3d_b$[b]	× 1.0

[a] d_b is the diameter of the bar being developed by the hook.

[b] The first stirrup or tie should enclose the hook within $2d_b$ of the outside of the bend.

[c] If two or more factors apply, ℓ_{dh} is multiplied by the product of the factors.

[d] Line 6 (ACI Section 12.5.4) applies at the discontinuous ends of members.

12.5.3(d) where anchorage or development for f_y is not specifically required, reinforcement in excess of that required by analysis

$$\ldots \times (A_s \text{ required})/(A_s \text{ provided})$$

12.5.4 for bars being developed by a standard hook at discontinuous ends of members with both side cover and top (or bottom) cover over a hook of less than $2\frac{1}{2}$ in., the hooked bar shall be enclosed within ties or stirrups perpendicular to the bar being developed, spaced not greater than $3d_b$ along the the development length of the hook, ℓ_{dh}, of the hook. In this case, the factors of ACI Section 12.5.3(b) and (c) shall not apply.

Figure 8-14 shows the meaning of the words "ties or stirrups parallel to" or "perpendicular to the bar being developed" in ACI Sections 12.5.3(b) and (c), and 12.5.4. In ACI Sections 12.5.3 and 12.5.4, d_b is the diameter of the hooked bar, and the first tie or stirrup shall enclose the bent portion of the hook, within $2d_b$ of the outside of the bend. If a hook satisfies more than one of the cases in ACI Section 12.5.3, ℓ_{dh} from (8-18) is multiplied by each of the applicable factors. Thus, if a 90° hook satisfies both the covers from ACI Section 12.5.3(a_2) and the stirrups from 12.5.3(b), ℓ_{dh} is the length from the term in square brackets in (8-18) multiplied by 0.7 from 12.5.3(a_2) and 0.8 from 12.5.3(b), making the total reduction $0.7 \times 0.8 = 0.56$.

For No. 14 and 18 bars, the factors from ACI Sections 12.5.3 and 12.5.4 are taken equal to 1.0. In other words, the development lengths are not reduced. Even so, it is still desirable to provide stirrups and ample cover.

Finally, the length of the hook, ℓ_{dh}, shall not be less than 8 bar diameters or 6 in., whichever is greater, after all the reduction factors have been applied.

ACI Section 12.5.4 applies at the discontinuous ends of members where *both* the side cover to the hook and the top (or bottom) cover to the lead-in length are less than $2\frac{1}{2}$ in. This section applies if the bar stresses are such that hooks are required at such points as the ends of simply supported beams (particularly if these are deep beams), at the free ends of cantilevers, and at the ends of members that terminate in a joint. (See Table 8-2.) Hooked bars at discontinuous ends of slabs are assumed to have confinement from the slab on each side of the hook; hence, ACI Section 12.5.4 is not applied. Although ACI Section 12.5.4 requires confinement of the full length ℓ_{dh}, the confinement is effective only in the region that will spall off if crushing occurs inside the hook.

Hooks may not be used to develop bars in compression, because bearing on the outside of the hook is not efficient.

Special requirements in ACI Section 21.5.4.1 govern hooks in joints in frames resisting seismic loads. Because such hooks must be inside the column, these rules implicitly allow for confinement of the hooks.

8-5 DESIGN FOR ANCHORAGE

The basic rule governing the development and anchorage of bars is as follows:

The calculated tension or compression in reinforcement at each section of reinforced concrete members shall be developed on each side of that section by embedment length, hooks, or mechanical anchorages, or a combination thereof. (ACI Section 12.1)

This requirement is satisfied in various ways for different types of members. Three examples of bar anchorage, two for straight bars and one for hooks, are presented in this section. The anchorage and cutoff of bars in beams is discussed in Section 8-6.

EXAMPLE 8-1 Anchorage of a Straight Bar

A 16-in.-wide cantilever beam frames into the edge of a 16-in.-thick wall, as shown in Fig. 8-15. At ultimate, the No. 8 bars at the top of the cantilever are stressed to their yield strength at point A at the face of the wall. Compute the minimum embedment of the bars into the wall. The concrete is sand-lightweight concrete with a strength of 3000 psi. The yield strength of the flexural reinforcement is 60,000 psi. Construction joints are located at the bottom and top of the beam, as shown in Fig. 8-15. The beam has closed No. 3 stirrups with $f_y = 40,000$ psi at a spacing of 7.5 in. throughout its length. The cover is 1.5 in. to the stirrups. The 3 No. 8 bars are inside the No. 4 at 12 in. vertical steel in each face of the wall. The wall steel is Grade 60.

We shall do this problem twice, first using ACI Section 12.2.2 (Table 8-1) and then using ACI Section 12.2.3 (8-10).

The stirrups are not shown.

1. Find the spacing and confinement case. The clear side cover to the No. 8 bars in the wall is $1.5 + 0.5 = 2$ in. $= 2d_b$. The clear spacing of the bars is

$$\frac{16 - 2(1.5 + 0.5) - 3 \times 1.0}{2} = 4.5 \text{ in.} = 4.5d_b$$

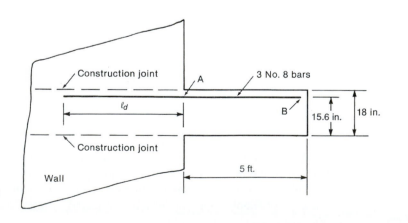

Fig. 8-15
Cantilever beam—Examples 8-1 and 8-2.

There are no stirrups or ties in the wall, but there are No. 4 at 12 in. o.c. vertical bars outside of the 3 No. 8 bars in the cantilever. Since the clear spacing between the bars is not less than $2d_b$ and the clear cover to the No. 8 bars exceeds d_b, this is case 2, and for No. 8 bars, (8-12) applies.

2.   **Compute the development length.** From (8-12),

$$\ell_d = \frac{f_y \alpha \beta \lambda}{20\sqrt{f'_c}} d_b$$

where

$\alpha = 1.3$ because there will be more than 12 in. of fresh concrete under the bar when the fresh concrete in the beam covers the bars

$\beta = 1.0$ because the bars are not epoxy coated

$\lambda = 1.3$ because the concrete is sand lightweight

$$\ell_d = \frac{60,000 \times 1.3 \times 1.0 \times 1.3}{20\sqrt{3000}} \times 1.0 = 92.6 \text{ in.}$$

Thus, the development length is $\ell_d = 92.6$ in.

The bars must extend 92.6 in. into the wall to develop the full yield strength. **Extend the bars 8 ft into the wall.**

Alternatively, this could be solved by using ACI Section 12.2.3 and (8-10). The necessary steps will be repeated.

1.   **Compute the development length.** From (8-10),

$$\ell_d = \frac{3}{40} \frac{f_y}{\sqrt{f'_c}} \frac{\alpha\beta\gamma\lambda}{\dfrac{c + K_{tr}}{d_b}} d_b$$

where α, β, and λ are as before and $\gamma = 1.0$ because the bars are No. 8.

$c =$ the smaller of

(a)   the distance from the center of the bar to the nearest concrete surface:

the side cover is $1.50 + 0.50 + 1.0/2 = 2.5$ in.;

(b)   half the center-to-center spacing of the bars, or $0.5\left(\dfrac{16 - 2 \times 2.5}{2}\right) = 2.75$ in.

Therefore, $c = 2.5$ in.

$$K_{tr} = \frac{A_{tr} f_{yt}}{1500 sn}$$

where

$s =$ spacing of the transverse reinforcement within the development length, ℓ_d
  $= 12$ in.

$A_{tr} =$ total cross-sectional area of reinforcement crossing the plane of splitting within the spacing s (one vertical No. 4 bar at 12 in.) each face
  $= 2 \times 0.20 = 0.40 \text{ in.}^2$

$f_{yt} = 60,000$ psi for the wall steel

$n =$ number of bars being anchored $= 3$

Thus,

$$K_{tr} = \frac{0.40 \text{ in.}^2 \times 60,000 \text{ psi}}{1500 \text{ lb/in.}^2 \times 12 \text{ in.} \times 3} = 0.444 \text{ in.}$$

The term

$$\frac{c + K_{tr}}{d_b} = \frac{2.5 \text{ in.} + 0.444 \text{ in.}}{1.0 \text{ in.}} = 2.94 \nless 2.5$$

is set equal to 2.5. Substituting into (8-10) gives

$$\ell_d = \frac{3}{40} \frac{60{,}000}{\sqrt{3000}} \frac{1.3 \times 1.0 \times 1.0 \times 1.3}{2.5} \times 1.0 = 55.5$$

Thus, $\ell_d = 55.5$ in. so the bars should be extended 55.5 in. into the wall, say, 5 ft. **Extend the bars 5 ft into the wall.**

In this case, there is a large difference between the ℓ_d computed from (8-12) and the ℓ_d computed from (8-10). This is because (8-14) was derived by using $(c + K_{tr})/d_b$ equal to 1.5. In this case, it is actually equal to 2.5. ∎

EXAMPLE 8-1M Anchorage of a Straight Bar—SI Units

A 400-mm-wide cantilever beam frames into the edge of a 400-mm-thick wall similar to Fig. 8-15. At ultimate, the three No. 25 bars at the top of the wall are stressed to their yield strength at point A at the face of the wall. Compute the minimum embedment of the bars into the wall. The concrete is sand/low-density concrete with a strength of 20 MPa. The yield strength of the flexural reinforcement is 420 MPa. Construction joints are located at the bottom and top of the beam, as shown in Fig. 8-15. The beam has closed No. 10 stirrups with $f_y = 420$ MPa at a spacing of 180 mm throughout its length. (The stirrups are not shown.) The cover is 40 mm to the stirrups. The three No. 25 bars are inside No. 13 vertical steel in each face of the wall.

1. **Find the spacing and confinement case.** The clear side cover to the No. 25 bars in the wall is $40 + 13 = 53$ mm $= 2.1d_b$. The clear spacing of the bars is

$$\frac{400 - 2(40 + 13) - 3 \times 25}{2} = 110 \text{ mm} = 4.4\,d_b$$

There are no stirrups or ties in the wall.

Since the clear spacing between the bars is not less than $2d_b$ and the clear cover to the No. 8 bars exceeds d_b, this is case 2, and for No. 25 bars, (8-12M) applies.

2. **Compute the development length.** From (8-12M),

$$\ell_d = \frac{12 f_y \alpha \beta \lambda}{20\sqrt{f_c'}} d_b$$

where

$\alpha = 1.3$ because there will be more than 300 mm of fresh concrete under the bar when the concrete in the beam covers the bars
$\beta = 1.0$ because the bars are not epoxy coated
$\lambda = 1.3$ because the concrete is sand lightweight

$$\ell_d = \frac{12 \times 420 \times 1.3 \times 1.0 \times 1.3}{20\sqrt{20}} \times 25 = 2380 \text{ mm}$$

Thus, the development length is $\ell_d = 2380$ mm.

The bars must extend 2380 mm into the wall to develop the full yield strength. **Extend the bars 2.4 m into the wall.** ∎

Example 8-1 was also solved by using (8-10). This will not be done in the SI example since the process is the same.

EXAMPLE 8-2   Development of a Bar in a Cantilever

The cantilever shown in Fig. 8-15 extends 60 in. from the face of the wall. The bars shown are stressed to their yield strength at the face of the wall. Is there adequate development length for No. 8 bars in the span? If not, what is the largest-size bar that can be used? The beam has No. 3 Grade-40 closed stirrups at 7.5 in. on centers along its full length.

The point of maximum bar force occurs at the face of the wall (point A in Fig. 8-15). The bar must be developed on each side of this point. To accomplish this, the bar must extend a minimum of ℓ_d into the support (Example 8-1) and a minimum of ℓ_d into the span.

1.   Find the spacing and confinement case. From Example 8-1, the clear spacing between the No. 8 bars in the beam is $4.5d_b$. For the 7.5-in. stirrup spacing, the minimum area of stirrups, by

ACI Section 11.5.5.3, is

$$A_{v,min} = 0.75\sqrt{f_c'}\frac{b_w s}{f_y} \text{ but not less than}$$

$$A_{v,min} = 50\frac{b_w s}{f_y} \tag{6-25}$$

(ACI Eq. 11-13)

$$= 50\frac{16 \times 7.5}{40,000} = 0.15 \text{ in.}^2$$

The stirrups provided are No. 3 double-leg stirrups, for which $A_v = 0.22$ in.^{2}. The spacing does not exceed the minimum of $d/2 = 7.80$ in.

Since the clear spacing between the bars is at least d_b and the stirrups exceed the minimum amount required, this is case 1, and for No. 8 bars, (8-12) applies.

2.   Compute the development length. From Example 8-1, the development length for a No. 8 top bar is 92.6 in.

Since the bars extend $60 - 1.5 = 58.5$ in. into the beam from the face of the wall, there is insufficient length to develop a No. 8 bar. We must use smaller bars, with shorter ℓ_d, or hook the No. 8 bars at B. Try six No. 6 bars ($A_s = 2.64$ in.^{2} compared with 2.37 in.^{2} for three No. 8 bars). For the new bar arrangement, we must start over.

1.   (Repeated) Find the spacing and confinement case. The clear side cover to the bars in the wall portion is $1.5 + 0.5 = 2$ in. The clear spacing between the six No. 6 bars is

$$\frac{16 - 2 \times 2 - 6 \times 0.75}{5} = 1.50 \text{ in.} = 2d_b$$

Since the stirrups exceed the ACI Code minimum and the bar spacing is not less than d_b, this is case 1, and (8-11) applies.

2.   (Repeated) Compute the development length. From (8-11),

$$\ell_d = \frac{f_y \alpha \beta \lambda}{25\sqrt{f_c'}}d_b = \frac{60,000 \times 1.3 \times 1.0 \times 1.3}{25\sqrt{3000}} \times 0.75 = 55.5 \text{ in.}$$

and $\ell_d = 55.5$ in. Since 58.5 in. > 55.5 in., No. 6 bars can be developed without hooks at the free end. **Use six No. 6 bars.** ∎

ACI Section 12.2.5 allows the development length to be reduced by multiplying ℓ_d by the ratio $(A_s \text{ required})/(A_s \text{ provided}) = 2.37$ in.$^2/2.64$ in.$^2 = 0.898$. We will not take advantage of this, because the reduced anchorage length will not allow the bars to be used for their full capacity in the event of a change in use of the structure.

EXAMPLE 8-2M   Development of a Bar in a Cantilever—SI Units

A cantilever similar to the one shown in Fig. 8-15 extends 1600 mm from the face of the wall. The three No. 25 Grade-420 bars are stressed to their yield strength at the face of the wall. Is there adequate development length for No. 25 bars in the span? If not, what is the largest-size bar that can be used? The beam has No. 10 Grade-300 closed stirrups at 180 mm on centers along its full length.

The point of maximum bar force occurs at the face of the wall (point A in Fig. 8-15). The bar must be developed on each side of this point. To accomplish this, the bar must extend a minimum of ℓ_d into the support (Example 8-1M) and a minimum of ℓ_d into the span.

1. Find the spacing and confinement case. From Example 8-1M, the clear spacing between the No. 25 bars in the beam is $4.3d_b$. For the 180-mm stirrup spacing, the minimum area of stirrups by ACI Section 11.5.5.3 is

$$A_{v,\min} = \frac{1}{16}\sqrt{f_c'}\,b_w s/fy \text{ but not less than:}$$

$$A_v = \frac{b_w s}{3 f_y} \tag{6-25M}$$

$$\tag{ACI Eq. 11-13}$$

$$= \frac{400 \times 180}{3 \times 300} = 80 \text{ mm}^2$$

The stirrups provided are No. 10 double-leg stirrups, for which $A_v = 200$ mm^2. The spacing does not exceed the minimum of $d/2 = 192$ mm.

Since the clear spacing between the bars is at least d_b and the stirrups exceed the minimum amount required, this is case 1, and for No. 8 bars, (8-12M) applies.

2. Compute the development length. From Example 8-1, the development length for a No. 25 top bar is 2380 mm. Since the bars extend $1600 - 40 = 1560$ mm into the beam from the face of the wall, there is insufficient length to develop a No. 25 bar. We must use smaller bars with shorter ℓ_d, or hook the No. 25 bars at B. Try six No. 19 bars ($A_s = 1704$ mm^2 compared to 1530 mm^2 for three No. 25 bars). For the new bar arrangement, we must start over.

1. (Repeated) Find the spacing and confinement case. The clear side cover to the bars in the wall portion is $40 + 15 = 55$ mm $= 2.75d_b$. The clear spacing between the six No. 19 bars is

$$\frac{400 - 2 \times 55 - 6 \times 19}{5} = 35.2 \text{ mm} = 1.85d_b$$

Since the stirrups exceed the code minimum and the bar spacing is not less than d_b, this is case 1, and (8-11M) applies.

2. (Repeated) Compute the development length. From (8-11M),

$$\ell_d = \frac{12 f_y \alpha \beta \lambda}{25 \sqrt{f_c'}} d_b = \frac{12 \times 420 \times 1.3 \times 1.0 \times 1.3}{25\sqrt{20}} \times 19 = 1450 \text{ mm}$$

and $\ell_d = 1450$ mm. Since 1560 mm $>$ 1450 mm, No. 19 bars can be developed without hooks at the free end. **Use six No. 19 bars.** ∎

EXAMPLE 8-3 Hooked Bar Anchorage into a Column

The exterior end of a 16-in.-wide-by-24-in.-deep continuous beam frames into a 24-in.-square column, as shown in Fig. 8-16. The column has four No. 11 longitudinal bars. The negative moment reinforcement at the exterior end of the beam consists of four No. 8 bars. The concrete is 3000-psi normal-weight concrete. The steel strength is 60,000 psi. Design the anchorage of the four No. 8 bars into the column.

1. Compute the development length for the beam bars. If the No. 8 bars extended straight into the column, they would be confined by the vertical steel, not the column ties. As a result, either this is an "other case" in Table 8-1, or it must be solved by (8-10). We shall use (8-10) because the column steel provides considerable confinement. We have

$$\ell_d = \frac{3}{40}\frac{f_y}{\sqrt{f_c'}}\frac{\alpha \beta \gamma \lambda}{\left(\dfrac{c + K_{tr}}{d_b}\right)}d_b$$

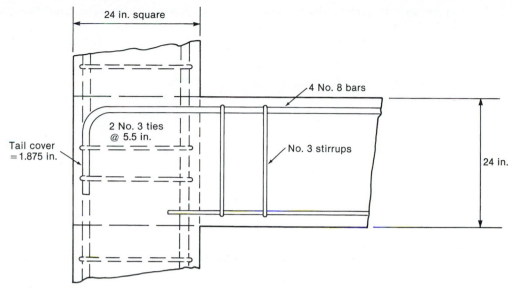

(a) Section A–A.

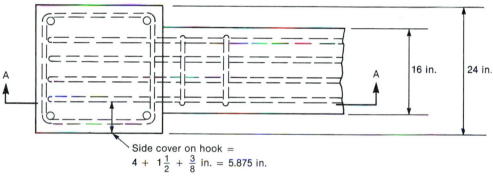

Side cover on hook =
$4 + 1\frac{1}{2} + \frac{3}{8}$ in. = 5.875 in.

(b) Plan

Fig. 8-16
Column–beam joint—Example 8-3.

where

$\alpha = 1.3$ (top bars)
$\beta = 1.0$ (no coating)
$\gamma = 1.0$ (No. 8 bars)
$\lambda = 1.0$ (normal-weight concrete)
$c =$ the smaller of

(a) the distance from the center of the bar to the nearest edge of the concrete:

$$\text{side distance} = 4.0 + 1.50 + 0.375 = 5.875 \text{ in.;}$$

(b) half the center-to-center spacing of the bars, or $\dfrac{1}{2}\left(\dfrac{16 - 2 \times 2.375}{3}\right) = 1.875$ in.

Therefore, $c = 1.875$ in.

$$K_{tr} = \frac{A_{tr}f_{yt}}{1500sn}$$

A_{tr} = the column reinforcement crossing the splitting plane in the plane of the layer of bars:
= two No. 11 bars = $2 \times 1.56 = 3.12$ in.^{2}
s = spacing of transverse reinforcement $24 - 2(1.5 + 0.375 + 1.41/2) = 18.8$ in.):

$$K_{tr} = \frac{3.12 \times 60,000}{1500 \times 18.8 \times 4} = 1.66 \text{ in.}$$

$$\frac{c + K_{tr}}{d_b} = \frac{1.875 + 1.66}{1.0} = 3.53 \text{ but not more than } 2.5$$

Substituting into (8-10) gives

$$\ell_d = \frac{3}{40} \frac{60,000}{\sqrt{3000}} \frac{1.3 \times 1.0 \times 1.0 \times 1.0}{2.5} \times 1.0 = 42.7$$

Thus, the development length is 42.7 in. Since this greatly exceeds the width of the column, it is necessary to use hooks to anchor the bars.

2. Compute the development length for hooked beam bars. The basic development length for a Grade-60 hooked bar is

$$\frac{0.02 \beta \lambda f_y}{\sqrt{f_c'}} d_b \qquad (8\text{-}18)$$

Therefore,

$$\ell_{dh} = \frac{0.02(1.0)(1.0)(60000)}{\sqrt{3000}}(1.0) = 21.9 \text{ in.}$$

Assume that the four No. 8 bars will extend into the column inside the vertical column bars, as shown in Fig. 8-16. ACI Section 11.11.2 requires minimum ties in the joint area. Assume that these are No. 3 ties. The required spacing of No. 3 closed ties by ACI Section 11.11.2 is computed via ACI Eq. (11-13):

$$A_v = \frac{0.75\sqrt{f_c'} b_w s}{f_y}, \text{ and } \geq \frac{50 b_w s}{f_y}$$

or

$$s = \frac{0.22 \text{ in.}^2 \times 60,000 \text{ psi}}{50 \text{ psi} \times 24 \text{ in.}}$$

$$= 11 \text{ in.}$$

The side cover to hooked bars is

4-in. offset in the side of the beam + 1.5-in. cover + 0.375-in. ties = 5.875 in.

This exceeds $2\frac{1}{2}$ in. and is therefore OK. The top cover to the lead-in length in the joint exceeds $2\frac{1}{2}$ in., since the joint is in the column.

The cover on the bar extension beyond the hook (the tail of the hook) is

1.5-in. cover to ties + 0.375-in. ties = 1.875 in.

$$\ell_{dh} = \ell_{hb} \times \text{multipliers in ACI Section 12.5.3}$$

12.5.3.2(a): The side cover exceeds 2.5 in., but the cover on the bar extension is less than 2 in.; therefore, the multiplier = 1.0. ACI Section 12.5.4 does not apply, since the side cover and top cover both exceed 2.5 in. Therefore, only the minimum ties required by ACI Section 11.11.2 are required: No. 3 ties at 11 in. These are spaced farther apart than $3d_b = 3 \times 1.0$ in.; therefore, ACI Section 12.5.3.(b) does not apply, and so the multiplier is 1.0. Thus,

$$\ell_{dh} = (\ell_{dh} \text{ from ACI Section 12.5.2}) \times 1.0 \times 1.0$$

$$= 21.9 \text{ in.} \geq 8d_b \text{ or 6 in.—therefore, OK.}$$

The hook-development length available is

$$24 \text{ in.} - \text{cover on bar extension-ties} = 22.1 \text{ in.}$$

Since 22.1 in. exceeds 21.9 in., the hook extension is OK.

Check the vertical height of a standard hook on a No. 8 bar. From Fig. 8-12a, the vertical height of a 90° standard hook is $4d_b + 12d_b = 16$ in. This will fit into the joint. Therefore, anchor the four No. 8 bars into the joint, as shown in Fig. 8-16.   ■

8-6   BAR CUTOFFS AND DEVELOPMENT OF BARS IN FLEXURAL MEMBERS

Why Bars Are Cut Off

In reinforced concrete, reinforcement is provided near the tensile face of beams to provide the tension component of the internal resisting couple. A continuous beam and its moment diagram are shown in Fig. 8-17. At midspan, the moments are positive, and reinforcement is required near the bottom face of the member, as shown in Fig. 8-17a. The opposite is true at the supports. For economy, some of the bars can be *terminated* or *cut off* where they are no longer needed. The location of the cutoff points is discussed in this section.

Four major factors affect the location of bar cutoffs:

1.   Bars can be cut off where they are no longer needed to resist tensile forces or where the remaining bars are adequate to do so. The location of points where bars are no longer needed is a function of the flexural tensions resulting from the bending moments *and* the effects of shear on these tensile forces.

2.   There must be sufficient extension of each bar, on each side of every section, to develop the force in that bar at that section. This is the basic rule governing the development of reinforcement, presented in Section 8-5 (ACI Section 12.1).

3.   Tension bars, cut off in a region of moderately high shear force, cause a major stress concentration, which can lead to major inclined cracks at the bar cutoff.

4.   Certain constructional requirements are specified in the code as good practice.

Generally speaking, bar cutoffs should be kept to a minimum to simplify design and construction, particularly in zones where the bars are stressed in tension.

In the following sections, the location of theoretical cutoff points for flexure, referred to as *flexural cutoff points*, is discussed. This is followed by a discussion of how

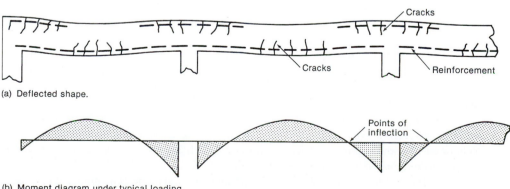

(a) Deflected shape.

(b) Moment diagram under typical loading.

Fig. 8-17
Moments and reinforcement in a continuous beam.

these flexural cutoff locations must be modified to account for shear, development, and constructional requirements to get the *actual cutoff points* used in construction.

Location of Flexural Cutoff Points

The calculation of the flexural cutoff points will be illustrated with the simply supported beam shown in Fig. 8-18a. At midspan, this beam has five No. 8 reinforcing bars, shown in section in Fig. 8-18c. At points C and C', two of these bars are cut off, leaving three No. 8 bars in the end portions of the beam, as shown in Fig. 8-17b.

The beam is loaded with a uniform load of 6.6 kips/ft, including its self-weight, which gives the diagram of ultimate moments, M_u, shown in Fig. 8-18d. This is referred to as the *required-moment diagram*, since, at each section, the beam must have a factored capacity, ϕM_n, at least equal to M_u. The maximum required moment at midspan is

$$M_u = \frac{w_u \ell_n^2}{8} = 330 \text{ ft-kips}$$

Assuming 3000-psi concrete and Grade-60 reinforcement, the moment capacity, ϕM_n, of the section with five No. 8 bars is 331 ft-kips, which is adequate at midspan. At points away from midspan, the required M_u is less than 330 ft-kips, as shown by the moment diagram in Fig. 8-18d. Since $\phi M_n = \phi A_s f_y jd$ ((4-1b) and (4-12a)), less reinforcement (less A_s) is required at points away from midspan. This is accomplished by "cutting off" some of the bars where they are no longer needed. In the example illustrated in Fig. 8-18, it has been arbitrarily decided that two No. 8 bars will be cut off where they are no longer needed. The remaining three No. 8 bars give a moment capacity $\phi M_n = 211$ ft-kips. Thus, the two bars *theoretically* can be cut off when $M_u \leq 211$ ft-kips, because the remaining three bars will be strong enough to resist M_u. From the equation for the required-moment diagram (Fig. 8-18d), we find $M_u = 211$ ft-kips at 3.99 ft from each support. Consequently, the two bars that are to be cut off are no longer needed for flexure in the outer 3.99 ft of each end of the beam and *theoretically* can be cut off at those points, as shown in Fig. 8-18e.

Figure 8-18f is a plot of the moment capacity, ϕM_n, at each point in the beam and is referred to as a *moment-capacity diagram*. At midspan (point E in Fig. 8-18e), the beam has five bars and hence has a capacity of 331 ft-kips. To the left of point C, the beam contains three bars, giving it a capacity of 211 ft-kips. The distance CD represents the development length, ℓ_d, for the two bars cut off at C. At the ends of the bars at point C, these two bars are undeveloped and thus cannot resist stresses. As a result, they do not add to the moment capacity at C. On the other hand, the bars are fully developed at D, and in the region from D to D' they could be stressed to f_y if required. In this region, the moment capacity is 331 ft-kips.

The three bars that extend into the supports are cut off at points A and A'. At A and A', these bars are undeveloped, and as a result, the moment capacity is $\phi M_n = 0$ at A and A'. At points B and B', the bars are fully developed, and the moment capacity $\phi M_n = 211$ ft-kips.

In Fig. 8-18g, the moment-capacity diagram from Fig. 8-18f and the required moment diagram from Fig. 8-18d are superimposed. Because the moment capacity is greater than or equal to the required moment at all points, the beam has adequate capacity *for flexure, neglecting the effects of shear*.

In the calculation of the moment capacity and required moment diagrams in Fig. 8-18, only flexure was considered. Shear has a significant effect on the stresses in the longitudinal tensile reinforcement and must be considered in computing the cutoff points. This effect is discussed in the next section.

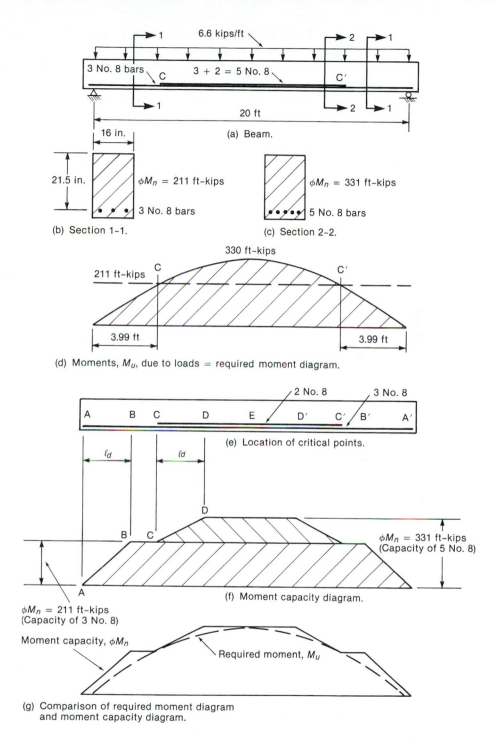

Fig. 8-18
Required moment and moment capacity.

(a) Beam.

(b) Section 1-1.

(c) Section 2-2.

(d) Moments, M_u, due to loads = required moment diagram.

(e) Location of critical points.

(f) Moment capacity diagram.

(g) Comparison of required moment diagram and moment capacity diagram.

Effect of Shear on Bar Forces and Location of Bar-Cutoff Points

In Section 6-4, the truss analogy was presented to model the shear strength of beams. It was shown in Figs. 6-19, 6-20, and 6-25a that inclined cracking increased the tension force in the flexural reinforcement. This effect will be examined more deeply later in this chapter.

Figure 8-19a shows a beam with inclined and flexural cracks. From flexure theory, the tensile force in the longitudinal reinforcement is $T = M/jd$. If jd is assumed to be constant, the distribution of T is the same as the distribution of moment, as shown in Fig. 8-19b. The maximum value of T is 216 kips between the loads.

In Fig. 6-19b, the same beam was idealized as a truss. The distribution of the tensile force in the longitudinal reinforcement from the truss model is shown by the solid stepped line in Fig. 8-19c. For comparison, the diagram of steel force due to flexure is shown by the line labeled $T = M/jd$.

The presence of inclined cracks has increased the force in the tension reinforcement at all points in the shear span except in the region of maximum moment, where the tensile force of 216 kips equals that computed from flexure. The increase in tensile force gets larger as one moves away from the point of maximum moment, as the slope of the compression diagonals decreases. For the 34° struts in the truss in Fig. 6-19b (1.5 horizontal to 1 vertical), the force in the tension reinforcement has been increased by $0.75V_u$ in the end portion of the shear span. Another way of looking at this is to assume that the force in the tension steel corresponds to a moment diagram which has been *shifted* $0.75jd$ toward the support.

For beams having struts at other angles, θ, the increase in the tensile force is $V_u/(2 \tan \theta)$, corresponding to a shift of $jd/(2 \tan \theta)$ in the moment diagram.

The ACI Code does not explicitly treat the effect of shear on the tensile force. Instead, ACI Section 12.10.3 arbitrarily requires that longitudinal tension bars be extended a

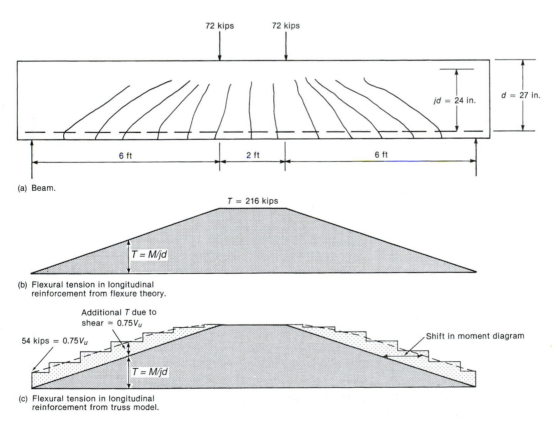

(a) Beam.

(b) Flexural tension in longitudinal reinforcement from flexure theory.

(c) Flexural tension in longitudinal reinforcement from truss model.

Fig. 8-19
Tension in longitudinal reinforcement.

minimum distance equal to the greater of d or 12 bar diameters past the theoretical cutoff point for flexure. This accounts for the shift due to shear, plus

> contingencies arising from unexpected loads, yielding of supports, shifting of points of inflection or other lack of agreement with assumed conditions governing the design of elastic structures [8-9] ....

In the beam in Fig. 8-19, there is a tensile force of $0.75V_u$ in the tensile reinforcement at the face of the support. If the shear stresses are large enough to cause significant inclined cracking (say, greater than $v_u = 4\sqrt{f_c'}$), it is good practice to anchor these bars for this force. The actual force depends on the angle θ, but $0.75V_u$ is a reasonable value. This is especially important for short, deep beams, as pointed out in ACI Section 12.10.6 and illustrated in Chapter 18 of this book.

Development of Bars at Points of Maximum Bar Force

For reinforcement and concrete to act together, each bar must have adequate embedment on both sides of each section to develop the force in the bar at that section. In beams, this is critical at

   1.   points of maximum positive and negative moment, which are points of maximum bar stress, and

   2.   points where reinforcing bars adjacent to the bar under consideration are cut off or bent (ACI Section 12.10.2).

Thus, bars must extend at least a development length, ℓ_d, each way from such points or be anchored with hooks or mechanical anchorages.

It is clear why this applies at points of maximum bar stress, such as point E in Fig. 8-18e, but the situation of bar cutoffs needs more explanation. In Fig. 8-18, the selection of bar cutoffs for flexure alone was discussed. To account for bar forces resulting from shear effects, cut-off bars are then extended away from the point of maximum moment a distance d or 12 bar diameters past the flexural cutoff point. This is equivalent to using the modified bending moment, M_u', shown in dashed lines in Fig. 8-20a, to select the cutoffs. The beam in Fig. 8-18 required five bars at midspan. If all five bars extended the full length of the beam, the bar-stress diagram would resemble the modified-moment diagram as shown in Fig. 8-20b. Instead, two bars will be cut off at points C and C', where the moment $M_u' = 211$ ft-kips, which is equal to the moment capacity, ϕM_n, of the cross section with three No. 8 bars (Fig. 8-20c). The stresses in the cut-off bars and in the full-length bars are plotted in Fig. 8-20d and e, respectively. Points D and D' are located at a development length, ℓ_d, away from the ends of the cut-off bars. By this point in the beam, the cut-off bars are fully effective. As a result, all five bars act to resist the applied moments between D and D', and the stresses in both sets of bars are the same (Fig. 8-20d and e) and furthermore are the same as if all bars extended the full length of the beam (Fig. 8-20b). Between D and C, the stress in the cut-off bars reduces to zero, while the stress in the remaining three bars increases. At point C, the stress in the remaining three bars reaches the yield strength, f_y, as assumed in selecting the cutoff points. For the bars to reach their yield strength at point C, the distance A–C must not be less than the development length, ℓ_d. If A–C is less than ℓ_d, the required anchorage can be obtained by hooking the bars at A, by using smaller bars, or by extending all five bars into the support.

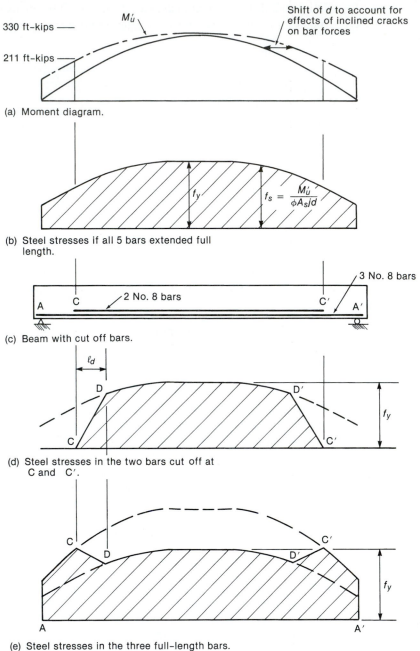

Fig. 8-20
Steel stresses in vicinity of
bar cutoff points.

Development of Bars in Positive-Moment Regions

Figure 8-21 shows a uniformly loaded, simple beam and its bending-moment diagram. As a first trial, the designer has selected two No. 14 bars as reinforcement. These run the full length of the beam and are enclosed in minimum stirrups. The development length of a No. 14 Grade-60 bar in 3000-psi concrete is 93 in. The point of maximum bar stress is at midspan, and since the bars extend 9 ft 6 in. = 114 in. each way from midspan, they are developed at midspan.

Because the bending-moment diagram for a uniformly loaded beam is a parabola, it is possible for the bar stress to be developed at midspan and not be developed at, for example,

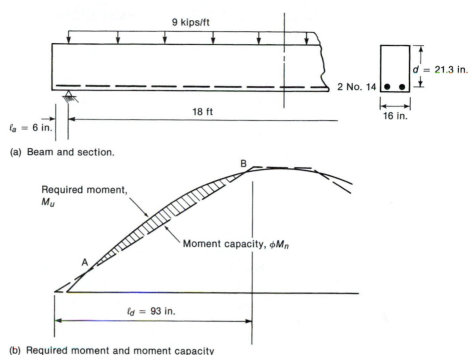

9 kips/ft

2 No. 14

$d = 21.3$ in.

16 in.

$\ell_a = 6$ in.

18 ft

(a) Beam and section.

B

Required moment, M_u

Moment capacity, ϕM_n

A

$\ell_d = 93$ in.

(b) Required moment and moment capacity diagrams.

Fig. 8-21
Anchorage of positive-moment reinforcement.

the quarter points of the span, where the moment is three-fourths of the maximum. This is illustrated in Fig. 8-21b, where the moment capacity and the required-moment diagrams are compared. The moment capacity is assumed to increase linearly, from zero at the ends of the bars to $\phi M_n = 363$ ft-kips at a distance $\ell_d = 93$ in. from the ends of the bars. Between points A and B, the required moment exceeds the moment capacity. Stated in a different way, the bar stresses required at points between A and B are larger than those which can be developed in the bar.

Ignoring the extension of the bar into the support for simplicity, it can be seen from Fig. 8-22 that the slope of the rising portion of the moment-capacity diagram cannot be less than that indicated by line O–A. If the moment-capacity diagram had the slope O–B, the bars would have insufficient development for the required stresses in the shaded region of Fig. 8-22. Thus, the slope of the moment-capacity diagram, $d(\phi M_n)/dx$, cannot be less than that of the tangent to the required-moment diagram, dM_u/dx, at $x = 0$. The slope of the moment-capacity diagram is $\phi M_n/\ell_d$. The slope of the required-moment diagram is $dM_u/dx = V_u$. Thus, the least slope the moment-capacity diagram can have is

$$\frac{\phi M_n}{\ell_d} = V_u$$

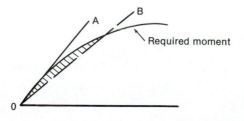

B

A

Required moment

0

Fig. 8-22
Anchorage at point of zero moment.

so the longest development length that can be tolerated is

$$\ell_d = \frac{\phi M_n}{V_u}$$

where M_n is the moment capacity based on the bars in the beam at 0 and V_u is the shear at 0.

ACI Section 12.11.3 requires that, *at simple supports* where the reaction induces compressive confining stresses in the bars (as would be the case in Fig. 8-21), the size of the positive-moment reinforcement should be small enough that the development length, ℓ_d, satisfies the relation

$$\ell_d \leq \frac{1.3 M_n}{V_u} + \ell_a \qquad (8\text{-}19)$$

where ℓ_a is the embedment length past the centerline of the support. The factor 1.3 accounts for the fact that transverse compression tends to increase the bond strength by offsetting some of the splitting stresses. When the beam is supported in such a way that there are no bearing stresses above the support, the factor 1.3 becomes 1.0 (giving (8-20), ACI (Eq. 12-3)).

When the bars are hooked with the point of tangency of the hook outside the centerline of the support, or if mechanical anchors are provided, ACI Section 12.11.3 does not require that (8-19) be satisfied. It should be noted that hooked bars at a support can lead to bearing failures unless they are carefully detailed. Figure 8-23a shows, to scale, a support of a simple beam. The potential crack illustrated does not encounter any reinforcement. In precast beams, the end of the beam is often reinforced as shown in Fig. 8-23b.

At positive-moment *points of inflection* (points where the positive-moment envelope passes through zero), a similar situation exists, except that there are no transverse bearing stresses. Here, the code requires that the diameter of the positive-moment reinforcement should be small enough that the development length, ℓ_d, satisfies

$$\ell_d \leq \frac{M_n}{V_u} + \ell_a \qquad (8\text{-}20)$$

where ℓ_a is the longer of the effective depth, d, or 12 bar diameters, but not more than the actual embedment of the bar in the negative-moment region past the point of inflection. The ACI Code does not specify this last condition, which is added here for completeness.

Equations (8-19) and (8-20) are written in terms of M_n rather than ϕM_n, since M_n leads to a slightly more conservative value. It should be noted that the derivation of (8-19) and (8-20) did not consider the shift in the bar force due to shear. As a result, these equations do not provide a sufficient check of the end anchorage of bars at simple supports of short, deep beams or beams supporting shear forces larger than about $V_u = 4\sqrt{f_c'} b_w d$. In such cases, the bars should be anchored in the support for a force of at least $V_u/2$, and preferably $0.75V_u$, as discussed in the preceding section.

Equations (8-19) and (8-20) are not applied in negative-moment regions, because the shape of the moment diagram is concave downward such that the only critical point for anchorage is the point of maximum bar stress.

EXAMPLE 8-4   Checking the Development of a Bar in a Positive-Moment Region

The beam in Fig. 8-21 has two No. 14 bars and No. 3 U stirrups at 10 in. o.c. The material strengths are $f_c' = 3000$ psi and $f_y = 60,000$ psi. The beam supports a total factored load of 9.0 kips/ft. Check whether ACI Section 12.11.3 is satisfied. The relevant equation is

$$\ell_d \leq 1.3 \frac{M_n}{V_u} + \ell_a \qquad (8\text{-}19)$$

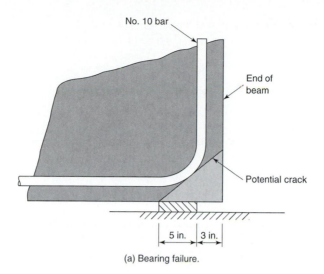

(a) Bearing failure.

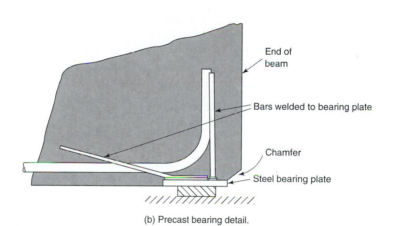

Fig. 8-23
Simple beam supports.

(b) Precast bearing detail.

    1.   Find the spacing and confinement case for No. 14 bars (diameter = 1.69 in.). The bar spacing $= 16 - 2 \times (1.5 + 0.375) - (2 \times 1.69) = 8.87$ in.

The beam has code-minimum stirrups. Therefore, the beam is case 1 in Table 8-1.
    2.   Compute development length for No. 14 bars. From (8-12),

$$\ell_d = \frac{f_y \alpha \beta \lambda}{20\sqrt{f_c'}} d_b$$

$$= \frac{60{,}000 \times 1.0 \times 1.0 \times 1.0}{20 \times \sqrt{3000}} \times 1.69 = 92.7$$

Thus, the development length is 92.7 in.
    3.   Solve (8-19) for the required ℓ_d. At the support, there are two No. 14 bars:

$$M_n = 2 \times 2.25 \times 60{,}000 \left(21.3 - \frac{2 \times 2.25 \times 60{,}000}{1.7 \times 3000 \times 16} \right)$$

$$= 4.86 \times 10^6 \text{ in.-lb} = 4860 \text{ in.-kips}$$

At the support,

$$V_u = \frac{w\ell}{2} = \frac{9.0 \times 18}{2}$$
$$= 81 \text{ kips}$$

ℓ_a = extension of bar past centerline of support = 6 in.

Thus,

$$1.3\frac{M_n}{V_u} + \ell_a = \frac{1.3 \times 4860 \text{ in.-kips}}{81 \text{ kips}} + 6 \text{ in.}$$
$$= 84 \text{ in.}$$

But $\ell_d = 92.7$ in. is greater than 84 in.; therefore, No. 14 bars cannot be used. Try six No. 8 bars. It is necessary to recompute ℓ_d.

4.   Find the spacing and confinement case for No. 8 bars. Either compute the spacing, which works out to 1.25 in. $= 1.25d_b$, or use Table A-6 to find that the minimum web width for six No. 8 bars is 15.5 in. Since 16 in. exceeds 15.5 in., the bar spacing exceeds d_b. The beam has code-minimum stirrups. Therefore, this is case 1.

5.   Compute development length for No. 8 bars. From (8-12),

$$\ell_d = \frac{f_y\alpha\beta\lambda}{20\sqrt{f_c'}}d_b = \frac{60{,}000 \times 1.0 \times 1.0 \times 1.0}{20 \times \sqrt{3000}} \times 1.0$$
$$= 54.8$$

Thus, $\ell_d = 54.8$ in. (See also Table A-9.)

6.   Solve (8-19) for the required ℓ_d. $M_n = 5067$ in.-kips; thus,

$$1.3\frac{M_n}{V_u} + \ell_a = \frac{1.3 \times 5067}{81} + 6 = 87.3 \text{ in.}$$

Since 54.8 in. $<$ 87.3 in., this is acceptable. **Use six No. 8 bars.**   ∎

Effect of Discontinuities at Bar Cutoff Points

The bar-stress diagrams in Fig. 8-20d and e suggest that a severe discontinuity in bar stresses exists in the vicinity of points where bars are cut off in a region of flexural tension. One effect of this discontinuity is a reduction in the inclined cracking shear in this vicinity [8-10]. The resulting inclined crack starts at, or near, the end of the cut-off bars. ACI Section 12.10.5 prohibits bar cutoffs in a zone of flexural tension unless *one* of the following is satisfied:

1.   The factored shear, V_u, at the cutoff point is not greater than

$$\frac{2}{3}\phi(V_c + V_s) \tag{8-21}$$

(ACI Section 12.10.5.1)

2.   Extra stirrups are provided over a length of $0.75d$, starting at the end of the cutoff bar and extending along it. The maximum spacing of the extra stirrups is $s = d/8\beta_b$, where β_b is the ratio of the area of the bars that are cut off to the area of the bars immediately before the cutoff. The area of the stirrups, A_v, is not to be less than $60b_w s/f_y$ (ACI Section 12.10.5.2).

3.   For No. 11 bars and smaller, the continuing reinforcement provides twice the area required for flexure at the cutoff point, and V_u is not greater than $0.75\phi(V_c + V_s)$ (ACI Section 12.10.5.3).

As a result of the difficulties in satisfying these sections, designers frequently will extend all bars into the supports in simple beams or past the points of inflection in continuous beams.

Requirements for Structural Integrity

A structure is said to have *structural integrity* if localized damage does not spread progressively to other parts of the structure. The 1989 ACI Code introduced Section 7.13, which provides details to improve the integrity of joist construction, beams without stirrups, perimeter beams, and precast buildings. See Section 2-3. These requirements were updated in the 2002 ACI Code.

There are two underlying concepts: *catenary action* and the provision of *ties* around buildings to anchor the catenary forces. A catenary is the shape taken by a loaded, hanging cable. If a support is damaged or removed from a continuous beam, the beam should be able to bridge the gap by catenary action, admittedly at large deflections. In continuous beams without stirrups, the top steel over the damaged support will tear out through the top of the beam, and catenary action will not develop. ACI Section 7.13.2.1, which deals with joist construction (joists do not usually have stirrups), and ACI Section 7.13.2.3, which applies to beams without closed stirrups, require a portion of the bottom reinforcement to be continuous over the support or lap spliced at the support to provide catenary action. At the outside of the building the catenary must be anchored into the supporting members with hooks.

ACI Section 7.13.2.2 requires that some of the negative- and positive-moment reinforcement in perimeter beams be continuous around the structure and be tied with closed stirrups or stirrups with 135° hooks around top bars, so that these beams will serve to anchor catenary forces from interior beams.

8-7   CALCULATION OF BAR CUTOFF POINTS

General Procedure

Bar cutoff points will be calculated in a five-stage procedure. For a perimeter beam the stages include:

1. The flexural reinforcement required at the points of maximum positive moment and maximum negative moment is computed and bars are selected.

2. From the top and bottom bars chosen in step 1, select bars to make up the made-continuous structural integrity tie. These bars extend throughout the span and will be lap spliced as required by ACI Section 7.13.2.2. These bars are separated from the rest of the bars at this stage because they will not have bar cutoffs as such.

3. The *flexural cutoff points* are computed for the remaining longitudinal bars. These are then detailed to satisfy the various rules. The resulting cutoff points are referred to as the *actual cutoff points*.

4. The stirrups needed in the structural integrity ties are selected, as are extra stirrups required near the cutoff points.

Two examples will be presented. The first is based on the calculation of flexural cutoff points from the equations of the moment diagrams. The second is based on graphic solutions of these diagrams.

The ACI Code presents guidelines for detailing flexural reinforcement in seventeen subsections spread over ACI Sections 7.13, 12.10, 12.11, and 12.12. To apply these sections when computing bar cutoffs, they have been collected in groups, covering aspects of detailing the reinforcement. We will call these "rules." These are presented in approximately the order used in design.

The steps between selecting the flexural cutoffs and choosing the actual bar cutoffs involves detailing rules for (1) structural integrity, (2) extension of bars into supports, (3) effects of shear on bar cutoff locations, and (4) anchorage or development of the bars.

1. **Design for Structural Integrity.** This group of rules, first introduced in the 1989 ACI Code, provides a modest degree of structural integrity to reduce the likelihood that localized damage can cause the collapse of a major part of the structure. The structural integrity rules are based on the concept of arranging and splicing the beam reinforcement to allow it to convert to a series of tension ties that support load as a catenary. The designer must decide which bars will be used as ties. This may preempt some of the subsequent cut-off decisions. For example, the rules require that at least two of the bottom and two of the top longitudinal bars in the perimeter beams must be spliced to form continuous steel tension ties around the perimeter of the building.

In the event of severe damage to the structure, these ties provide a means of anchoring the membrane or catenary forces resulting from the damage, so that the remaining structure can bridge over two spans or change the direction of the span, thus reducing the chance of total collapse of the structure.

The structural integrity rules do not define specific tie forces to be resisted; instead, the number of bars in a tie is given as fractions of the numbers of positive and negative bars in the maximum moment regions. As a result the tie forces increase with the size of the bars in the member. At discontinuous or simply supported ends the ties must be anchored into the supports.

Structural Integrity—Rule 1

The designer selects a number of members to provide structural integrity (see Section 2-3) and reinforces these members to enable them to undergo large deformations while supporting significant loads. ACI Section 7.13 presents four types of beam that could develop structural integrity in a given floor structure:

Structural Integrity Rule 1(a) Perimeter Beams. At least one-quarter of the total positive-moment (bottom) reinforcement at midspan, but not less than two bars, shall be spliced over, or near, the supports. Similarly, at least one-sixth of the negative-moment (top) reinforcement shall be spliced at or near midspan. In perimeter beams these two groups of spliced bars are enclosed in stirrups. The tie resulting from the combination of the top and bottom bars and the stirrups is assumed to have the toughness to act as a tensile tie that will be robust enough to anchor catenary forces from the partially collapsed building. The tie continues around the perimeter of the building at the floor level in question. ACI Section 7.13.2(b) specifies that the splices should be Class A tension lap splices or welding or mechanical splices, In perimeter beams this minimum group of four bars must be enclosed within U stirrups with 135° hooks at both upper ends of the U stirrups, or in one-piece closed stirrups. If U stirrups are used, the two bottom longitudinal bars are placed in the lower corners of the stirrups and the two top bars are placed inside the bent portion of the 135° stirrup hooks. This allows much of the tensile strength of the stirrups to be developed, thereby preventing the top reinforcement from pulling out of the top of the beam. Two-piece closed stirrups or stirrups with 90° hooks cannot develop the vertical forces that would pull the top bars out of the top of the beam. These stirrups need not extend through joints. (ACI Section 7.13.2.3.)

Structural Integrity, Rule 1(b) Continuous Interior Beams. At least one-quarter of the *positive-moment reinforcement* required at midspan, but not less than two bars, must be continuous over the length of the span, or spliced near the supports with a Class A tension lap splice or equivalent. This steel is assumed to form a tensile tie. This tie is weaker than the tie used in perimeter beams. In the event of serious damage to a beam or its supports this tie provides a possible anchor for membrane or catenary forces. At discontinuous supports the tie shall be terminated with a standard hook into the spandrel beam, column, or other support. ACI Section 7.13.2.4 does not require stirrups in this class of beam. However, if stirrups are needed for the applied shear forces, U stirrups with 135° ties on the upper ends, with the longitudinal tie bars placed in the lower corners of the stirrups, would greatly improve the ductility of the structure, especially if the 135° hooks enclose longitudinal top bars.

Structural Integrity, Rule 1(c) One-Way Joists. In one-way joist construction, at least one bottom bar should be continuous over the supports or should be spliced at or near the supports with a Class A splice. The continuous tie should terminate with a standard hook into a support. (ACI Section 7.13.2.1.)

Structural Integrity, Rule 1(d) Supports of Beams Forming Part of a Frame That Is a Primary Lateral Load-Resisting System for the Building. When a moment resisting frame is part of a primary lateral load system, the positive-moment reinforcement required at the supports should be anchored for the specified yield strength of the reinforcement at the faces of the support.

Structural Integrity, Rule 1(e) All Buildings Must Be Tied Together. A clear path must be provided for each load from the point where it acts on the structure to the supports.

2.   Extension of Bottom or Top Steel into Supports.

Extension of Bars, Rule 2(a) Simple Supports. At least one-third of the bottom flexural reinforcement must extend along the bottom face of the beam into the supports. This steel intercepts cracks that may develop in the soffit of the beam, between the face of the support and the positive-moment point of inflection. In Fig (8-26a) the bottom steel is extended, partly so that cracks in the soffit are held closed by reinforcement.

At simple supports at least one-third of the positive-moment reinforcement, but not less than two bars, must extend at least 6 in. into the support. (ACI Section 12.11.1.)

Extension of bars, Rule 2(b) Continuous Supports. At least one-quarter of the positive-moment reinforcement must extend at least 6 in. into continuous supports. (ACI Section 12.11.1.)

3.   Effects of Shear on the Bending-Moment Diagram. Inclined cracking and shear cause a shift in the envelope of maximum bar forces at a given section, as was shown in Figs. 6-21 and 8-19. This shift results in larger bar forces at a given location than those due to flexure alone. As a result, reinforcement must be extended further from the maximum moment sections, relative to the moment diagram.

Effects of Shear, Rule 3(a) Positive-Moment Regions. Positive-moment reinforcement must extend the longer of d or $12d_b$ past the flexural cutoff points, except at supports or at the ends of cantilevers. See Figs. 8-19 and 8-20.

Effects of Shear, Rule 3(b) Negative-Moment Regions. Negative-moment reinforcement must extend the longest of d, $12d_b$ or $\ell_n/16$ past the flexural cutoff points, except at supports or at the ends of cantilevers. The origin of the term $\ell_n/16$ is not known.

4.   Anchorage and Development of Reinforcement. All bars must be anchored in, or through, the supports such that the full tensile force in the bar at any section can be developed on both sides of that section. Maximum bar forces occur at points of maximum positive and negative moments in the bar being developed at or near cross sections where adjacent bars are cut off. See Fig. 8-20.

Anchorage, Rule 4(a) Positive-Moment Reinforcement. Positive-moment reinforcement must extend at least ℓ_d from the point of maximum bar stress or from the flexural cutoff points of adjacent bars. (ACI Sections 12.10.2 and 12.10.4, See Fig. 8-20.)

Anchorage, Rule 4(b) Negative-Moment Reinforcement. Negative-moment reinforcement must be anchored into or through supporting columns or members. (ACI Section 12.12.1.)

Anchorage, Rule 4(c) Negative-Moment at End Supports. Negative-moment reinforcement must be anchored for the yield force at end supports.

Anchorage, Rule 4(d) Positive-Moment Reinforcement, Development at Point of Inflection. At the positive-moment point of inflection and at simple supports, the positive-moment reinforcement must satisfy either (8-19) (8-20).

Structural Integrity, Negative-Moment Bars, Rule 1(a) —Neg: Interior beams. At least one-third of the negative-moment reinforcement must be extended by the greatest of d, $12d_b$, or $\ell_n/16$ past the negative-moment point of inflection (ACI Section 12.12.3).

Structural Integrity, Negative Moment Bars, Rule 1(b) Perimeter beams. In addition to satisfying rule 1—*Neg* (a), one-sixth of the negative moment reinforcement required at the support must be made continuous at midspan. This can be achieved by means of a Class A tension splice at midspan (ACI Section 7.13.2.2).

Extension of Bars, Rule 2—Neg: Bars must extend the longer of d or $12d_b$ past the flexural cutoff points, except at supports or at the ends of cantilevers (ACI Section 12.10.3).

Effects of Shear, Rule 3—Neg: Bars must extend at least ℓ_d from the point of maximum bar stress or from the flexural bar cutoffs of adjacent bars (ACI Sections 12.10.2 and 12.10.4).

Anchorage, Rule 4—Neg: Negative-moment reinforcement must be anchored into or through supporting columns or members (ACI Section 12.12.1).

EXAMPLE 8-5   Calculation of Bar Cutoff Points from Equations of Moment Diagrams

The beam shown in Fig. 8-24a is constructed of 3000-psi concrete and Grade-60 reinforcement. It supports a factored dead load of 0.42 kip/ft and a factored live load of 3.4 kips/ft. The cross sections at the points of maximum positive and negative moment, as given in Figs. 8-25c and 8-27c are shown in Fig. 8-24b.

1.   Locate flexural cutoffs for positive-moment reinforcement. The positive moment in span *AB* is governed by the loading case in Fig. 8-25a. From a free-body analysis of a part of span *AB* (Fig. 8-25d), the equation for M_u at a distance x from A is

$$M_u = 46.5x - \frac{3.82x^2}{2}\text{ ft-kips}$$

At midspan, the beam has two No. 9 plus two No. 8 bars. The two No. 8 bars will be cut off. The capacity of the remaining bars is

$$\phi M_n = \frac{0.9 \times (2 \times 1.0) \times 60,000\left(21.5 - \dfrac{2.0 \times 60,000}{1.7 \times 3000 \times 12}\right)}{12,000}$$

$$= 176 \text{ ft-kips}$$

Therefore, flexural cutoff points occur where $M_u = 176$ ft-kips. Setting $M_u = 176$ ft-kips, the equation for M_u can be rearranged to

$$1.91x^2 - 46.5x + 176 = 0$$

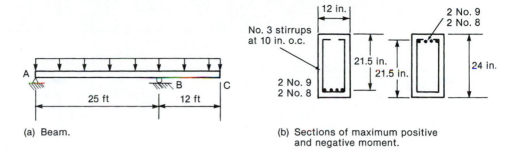

(a) Beam.

(b) Sections of maximum positive and negative moment.

Fig. 8-24
Beam—Example 8-5.

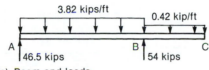

(a) Beam and loads.

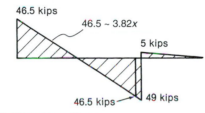

(b) Shear force diagram.

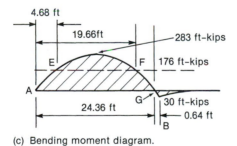

(c) Bending moment diagram.

Fig. 8-25
Calculation of flexural cutoff points for a statically determinate beam with an overhang, loaded to produce maximum positive moment—Example 8-5.

(d) Free body diagram.

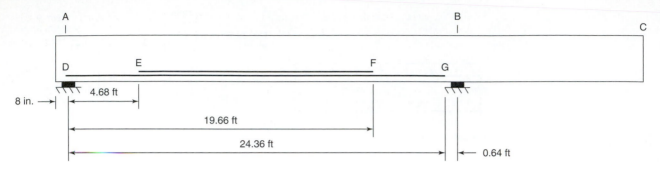

(a) Flexural cutoff points for positive moment steel.

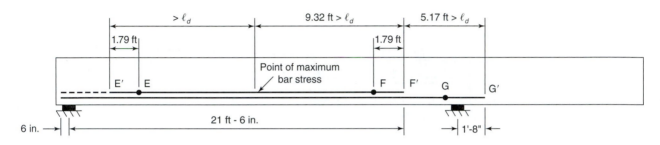

(b) Actual cutoff points for positive moment steel.

Fig. 8-26
Location of positive-moment cutoff points—Example 8-5.

This is a quadratic equation of the form

$$Ax^2 + Bx + C = 0$$

and so has the solution

$$x = \frac{-B \pm \sqrt{B^2 - 4AC}}{2A} = \frac{46.5 \pm \sqrt{(-46.5)^2 - 4(1.91)(176)}}{2(1.91)}$$

Thus $x = 4.68$ ft or 19.66 ft from A. These flexural cutoff points are shown in Fig. 8-25c and 8-26a. They will be referred to as *flexural cutoff points E and F*. In a similar fashion, flexural cutoff point G is found to be 24.4 ft from A or 0.64 ft from B.

2. **Compute the development lengths for the bottom bars.**

$$\text{Bar spacing} = \frac{12 - 2(1.5 + 0.375) - 2 \times 1.128 - 2 \times 1.0}{3} = 1.33 \text{ in.}$$

Since the bar spacing exceeds d_b for both the No. 8 and 9 bars, and since the beam has minimum stirrups, the bars satisfy case 1 in Table 8-1. From (8-12) (or from Table A-11),

$$\frac{\ell_d}{d_b} = \frac{f_y \alpha \beta \lambda}{20 \sqrt{f_c'}} = \frac{60,000 \times 1.0 \times 1.0 \times 1.0 \times 1.0}{20 \times \sqrt{3000}} = 54.8$$

Thus for the No. 8 bars, $\ell_d = 54.8 \times 1.0 = 54.8$ in., and for the No. 9 bars, $\ell_d = 54.8 \times 1.128 = 61.8$ in.

3.   Locate actual cutoff points for positive-moment reinforcement. The actual cutoff points are determined from the flexural cutoff points using the "rules" stated earlier. Since the location of cutoffs G and D are affected by the locations of cutoffs E and F, the latter are established first, starting with F. Because the beam is simply supported it is not included in the ACI Code listing of members that are susceptible to actions requiring structural integrity.

(a)   Cutoff F. Two No. 8 bars are cut off. They must satisfy rules 2 (extension of bars into the supports), 3 (effect of shear on moment diagrams), and 4 (anchorage).

Extension of bars into the supports, Rule 2. At least one-third of the positive-moment reinforcement, but not less than two bars, must extend at least 6 in. into the supports. We shall extend two No. 9 bars into each of the supports A and B.

Effect of shear, Rule 3. Extend the bars by the larger of $d = 21.5$ in. $= 1.79$ ft, or $12d_b = 1$ ft. Therefore, the first trial position of the actual cutoff is at $19.66 + 1.79 = 21.45$ ft from the center of the support at A, say, 21 ft-6 in. (see point F' in Fig. 8-26b).

Anchorage, Rule 4. Bars must extend at least ℓ_d past the points of maximum bar stress. For the bars cut off at F', the maximum bar stress occurs at midspan, at 12.18 ft from A. The distance from the point of maximum bar stress to the actual bar cutoff is $21.5 - 12.18 = 9.32$ ft. ℓ_d for the No. 8 bars is 54.8 in. The distance available is more than ℓ_d—therefore OK. **Cut off two No. 8 bars at 21 ft 6 in. from A** (shown as point F' in Fig. 8-26b).

(b)   Cutoff G. Two No. 9 bars are cut off; must satisfy rules 2 and 3 at G, and rule 4 at the positive-moment point of inflection.

Extension of bars into simple supports, Rule 2. In step (a) we extended two No. 9 bars 6 in. into support A.

Effects of shear, Rule 3. Because the cutoff is at the support, we do not need to extend the bars further.

Anchorage, Rule 4a. Bars must extend at least ℓ_d past actual cutoffs of adjacent bars. ℓ_d for No. 9 bottom bar $= 61.8$ in. $= 5.15$ ft. Distance from F' to $G' = (25$ ft 6 in.$) - (21$ ft 6 in.$) = 4$ ft. Bar does not extend ℓ_d therefore, extend the bars to 21.45 ft $+ 5.15$ ft $= 26.6$ ft, say, 26 ft 8 in.

Anchorage at point of inflection, Rule 4d. Must satisfy (8-20) at point of inflection (point where the moment is zero). Therefore, at G, $\ell_d \leq M_n/V_u + \ell_a$. The point of inflection is 0.64 ft from the support (Fig. 8-25c). At this point V_u is 46.5 kips (Fig. 8-25b) and the moment capacity M_n for the bars in the beam at the point of inflection (two No. 9 bars) is

$$M_n = 176 \text{ ft-kips} \times \frac{12}{0.9} = 2345 \text{ in.-kips}$$

$\ell_a =$ larger of d (21.5 in.) or $12d_b$ (13.5 in.) but not more than the actual extension of the bar past the point of inflection ($26.67 - 24.36 = 2.31$ ft $= 27.7$ in.). Therefore, $\ell_a = 21.5$ in., and

$$\frac{M_n}{V_u} + \ell_a = \frac{2345}{46.5} + 21.5$$
$$= 71.9 \text{ in.}$$

Since this length exceeds $\ell_d = 61.8$ in., OK. **Cut off two No. 9 bars 1 ft 8 in. from B** (shown as point G' in Fig. 8-26b).

(c)   Cutoff E. Two No. 8 bars are cut off; they must satisfy rules 3 and 4.

Rule 3a Effects of shear, positive moment. Extend the bars $d = 1.79$ ft. past the flexural cutoff point. Therefore, the actual cutoff E' is at $4.68 - 1.79 = 2.89$ ft (2 ft 10 in.) from A.

Rule 4a Anchorage, positive moment. The distance from the point of maximum moment to the actual cutoff exceeds $\ell_d = 54.8$ in.—therefore OK. **Cut off two No. 8 bars at 2 ft 10 in. from A** (point E' in Fig. 8-26b; note that this is changed later).

(d)   Cutoff D. Two No. 9 bars are cut off; must satisfy rules 2, 3, and 4.

Rule 2 Extension into simple support. This was done in step (a).

Rule 3a Effects of shear on moment diagram. Because the cutoff is at the support we do not need to extend the bars by d or $12d_b$.

Rule 2 Bars must extend ℓ_d from the actual cutoff E', where $\ell_d = 61.8$ in. (No. 9 bars). The maximum possible length available is 2 ft 10 in. + 6 in. = 40 in. Since this is less than ℓ_d, we must either extend the end of the beam, hook the ends of the bars, use smaller bars, or eliminate the cutoff E'. We shall do the latter. Therefore, **extend all four bars 6 in. past point A.**

Rule 4d, Development of bars at simple support. We must satisfy (8-19) at the support.

$$\ell d \le \frac{1.3 M_n}{V_u} + \ell_a$$

$$V_u = 46.5 \text{ kips}$$

$$M_n = \frac{3.58 \times 60,000 \left(21.5 - \dfrac{3.58 \times 60,000}{1.7 \times 3000 \times 12}\right)}{1000}$$

$$= 3860 \text{ in.-kips}$$

$$\ell_a = 6 \text{ in.}$$

$$1.3 \frac{M_n}{V_n} + \ell_a = 1.3 \frac{3860}{46.5} + 6 = 113.9 \text{ in.}$$

Since this exceeds ℓ_d, rule 4d is satisfied. The actual cutoff points are illustrated in Fig. 8-26b.

4. **Locate flexural cutoffs for negative-moment reinforcement.** The negative moment is governed by the loading case in Fig. 8-27a. The equations for the negative bending moments are as follows:

Between A and B, with x measured from A,

$$M_u = -5.8x - \frac{0.42x^2}{2} \text{ ft-kips}$$

and between C and B, with x_1 measured from C,

$$M_u = \frac{-3.82x_1^2}{2} \text{ ft-kips}$$

Over the support at B, the reinforcement is two No. 9 bars plus two No. 8 bars. The two No. 8 bars are no longer required when the moment is less than $\phi M_n = 176$ ft-kips (capacity of the beam with two No. 9 bars).

So, between A and B,

$$-176 = -5.8x - 0.21x^2$$

$$x = -45.9 \text{ ft or } 18.26 \text{ ft from } A$$

Therefore, the flexural cutoff point for two No. 8 top bars in span AB is at 18.28 ft from A. Finally, between B and C,

$$-176 = -1.91x_1^2$$

$$x_1 = 9.60 \text{ ft from } C$$

Therefore, the flexural cutoff point for two No. 8 top bars in span BC is at 9.60 ft from C. The flexural cutoff points for the negative-moment steel are shown in Fig. 8-28a and are lettered H, J, K, and L.

5. **Compute development lengths for the top bars.** Since there is more than 12 in. of concrete below the top bars, $\alpha = 1.3$. For the No. 8 bars, $\ell_d = 71.2$ in., and for the No. 9 bars, $\ell_d = 80.3$ in.

6. **Locate the actual cutoff points for the negative-moment reinforcement.** Again, the inner cutoffs will be considered first, because their location affects the design of the outer cutoffs. The choice of actual cutoff points is illustrated in Fig. 8-28b.

Effects of shear (a) Cutoff J. Two No. 8 bars are cut off; they must satisfy rules 3 and 4.

Rule 3a. Extend bars by $d = 1.79$ ft past the flexural cutoff. Cut off at $18.26 - 1.79 = 16.47$ ft from A and 8.53 ft from B, say, 8 ft 6 in.

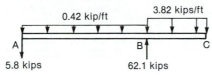

(a) Beam and loads.

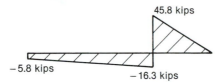

(b) Shear force diagram.

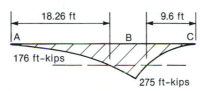

(c) Bending moment diagram.

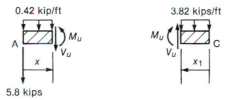

Fig. 8-27
Calculation of flexural cutoff
points for negative
moment–Example 8-5.

(d) Free-body diagrams.

Anchorage, Rule 4b negative moment steel. The bars must extend ℓ_d from the point of maximum bar stress. For the two No. 8 top bars, the maximum bar stress is at *B*. The actual bar extension is 8.5 ft = 102 in. This exceeds ℓ_d = 71.2 in.—therefore OK.

Anchorage, Rule 4b negative-moment steel. Bar must be anchored at support. This will be satisfied when cutoff *K* is selected. Therefore, **cut off two No. 8 bars at 8 ft 6 in. from *B*** (point *J'* in Fig. 8-28b).

(b) **Cutoff *H*.** Two No. 9 bars cut off; must satisfy rule 2.

Anchorage, Rule 4b. Bar must extend ℓ_d past *J'*, where ℓ_d = 80.3 in. Length available = 17 ft —therefore OK. **Extend two No. 9 bars to 2 in. from the end of the beam** (point *H'* in Fig. 8-28b).

(c) **Cutoff *K*.** Two No. 8 bars are cut off; They must satisfy rules 3 and 4. The flexural cutoff is at 9.60 ft from *C* (2.40 ft from *B*).

Effect of shear, Rule 3b. Extend bars *d* = 1.79 ft. The end of the bars is at 2.40 + 1.79 = 4.19 ft from *B*, say, 4 ft 3 in.

Anchorage, Rule 4b. Extend ℓ_d past *B*. ℓ_d for a No. 8 top bar = 71.2 in. The extension of 4 ft 3 in. is thus not enough. Try extending the No. 8 top bars 6 ft past *B* to point *K'*.

Anchorage, Rule 4b. The bars must be anchored at the support. Since the No. 8 bars extend more than ℓ_d on either side of the support, they are adequately anchored. Therefore, cut off two No. 8 bars at 6 ft from *B* (point *K'* in Fig. 8-28b). Note that this is changed in the next step.

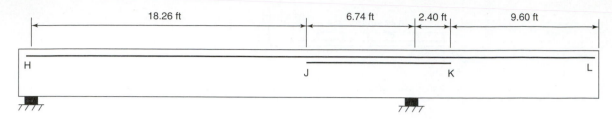

(a) Flexural cutoff points for negative moment steel.

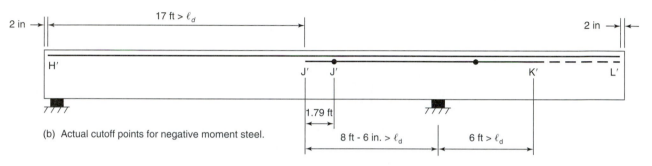

(b) Actual cutoff points for negative moment steel.

Fig. 8-28
Location of negative-moment cutoff points—Example 8-5.

(d) Cutoff L. two No. 9 bars are cut off. Rule 2 applies.

Anchorage, Rule 4b. The bars must extend ℓ_d past K'. For a No. 9 top bar $\ell_d = 80.3$ in. $= 6.69\ ft$. The actual extension is $11.83 - 6 = 5.83$ ft, which is less than ℓ_d—therefore, not OK. Two solutions are available: either extend all the bars to the end of the beam, or change the bars to six No. 7 bars in two layers. We shall do the former. The final actual cutoff points are shown in Fig. 8-29.

 7.   **Check whether extra stirrups are required at cutoffs.** ACI Section 12.10.5 prohibits bar cutoffs in a tension zone, unless

12.10.5.1: V_u at actual cutoff $\leq \frac{2}{3}\phi(V_c + V_s)$ at that point, or

12.10.5.2: Extra stirrups are provided at actual cutoff point, or

12.10.5.3: The continuing flexural reinforcement at the flexural cutoff has twice the required A_s and $V_u \leq 0.75\ \phi(V_c + V_s)$.

Since we have selected flexural cutoff points on the basis of the continuing steel having 1.0 times the required A_s, ACI Section 12.10.5.3 will not govern. Therefore, we will provide extra stirrups unless ACI Section 12.10.5.1 is satisfied.

As shown in Fig. 8-24b, the beam has No. 3 double-leg stirrups at 10 in. o.c. throughout its length. From (6-9), (6-14), and (6-18),

$$\phi V_n = \phi(V_c + V_s)$$

$$= \frac{0.75\left(2\sqrt{3000} \times 12 \times 21.5 + \dfrac{0.22 \times 60{,}000 \times 21.5}{10}\right)}{1000}$$

$$= 42.5 \text{ kips}$$

and $\frac{2}{3}\phi V_n = 28.3$ kips. Thus if V_u at the *actual* cutoff point exceeds 28.3 kips, extra stirrups will be required.

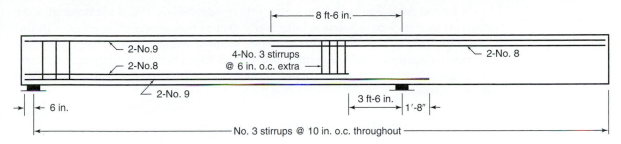

Fig. 8-29
Reinforcement details—Example 8-5.

(a)   **Cutoff F'.** This cutoff is located at 21 ft 6 in. from A or 3 ft 6 in. from B. The flexural tensions that occur at this point due to load case 1 (Fig. 8-25c). The shear at F' for load case 1 is

$$V_u = -49.0 \text{ kips} + 3.5 \text{ ft} \times 3.82 \text{ kips} = -35.6 \text{ kips}$$

where the sign simply indicates the direction of the shear force. Since $V_u = 35.6$ kips exceeds $\frac{2}{3}\phi V_n = 28.3$ kips, extra stirrups are required. Extra stirrups must extend $0.75d = 0.75 \times 21.5 = 16.1$ in. along the No. 8 bars from cutoff point F'. Thus,

$$\text{Spacing} \le \frac{d}{8\beta_b}$$

where

$$\beta_b = \frac{\text{area of bars cut off}}{\text{total area immediately prior to cutoff}}$$

$$= \frac{2 \times 0.79 \text{ in.}^2}{2 \times 1.0 \text{ in.}^2 + 2 \times 0.79 \text{ in.}^2} = 0.44$$

Therefore, the stirrup spacing is

$$s \le \frac{21.5 \text{ in.}}{8 \times 0.44} = 6.09 \text{ in.}$$

Try four extra No. 3 double-leg stirrups at 6 in. o.c. Check:

$$\text{required } A_v = \frac{60 b_w s}{f_y}$$

$$= \frac{60 \times 12 \times 6}{60,000} = 0.072 \text{ in.}^2$$

The A_v provided is 0.22 in.²—therefore, OK. **Use three extra No. 3 double-leg stirrups at 6 in. o.c., placing the first one at 2 in. from end of cutoff bars at F'.**

(b)   **Cutoff E.** All bars were extended into the support at A, so supplementary shear reinforcement is not required here.

(c)   **Cutoff J'.** The cutoff is located at 8 ft 6 in. from B. The flexural tension that occurs in these bars is due to load case 2 (Fig. 8-27). By inspection, V_u is considerably less than $\frac{2}{3}\phi V_n = 28.3$ kips. Therefore, no extra stirrups are required at this cutoff.

The final reinforcement details are shown in Fig. 8-29. For nonstandard beams such as this one, a detail of this sort should be shown in the contract drawings.   ■

The calculations just carried out are tedious, and if the underlying concepts are not understood, the detailing provisions become a meaningless mumbo jumbo. The reader is advised to reread Section 8-6 to clarify the concepts involved.

Several things can be done to simplify these calculations. One is to extend all of the bars past their respective points of inflection so that no bars are cut off in zones of flexural tension. This reduces the number of cutoffs required and eliminates the need for extra stirrups, on one hand, while requiring more flexural reinforcement, on the other. A second method is to work out the flexural cutoff points graphically. This approach is discussed in the next section.

Graphical Calculation of Flexural Cutoff Points

The flexural capacity of a beam is $\phi M_n = \phi A_s f_y jd$, where jd is the internal level arm and is relatively insensitive to the amount of reinforcement. If it is assumed that jd is constant, then ϕM_n is directly proportional to A_s. Since, in design, ϕM_n is set equal to M_u, we can then say that the amount of steel, A_s, required at any section is directly proportional to M_u at that section. If it is desired to cut off a third of the bars at a particular cutoff point, the remaining two-thirds of the bars would have a capacity of two-thirds of the maximum ϕM_n, and hence this cutoff would be located where M_u was two-thirds of the maximum M_u.

Figure A-1 in Appendix A is a schematic graph of the bending-moment envelope for a typical interior span of a multispan continuous beam designed for maximum negative moments of $-w\ell_n^2/11$ and a maximum positive moment of $w\ell_n^2/16$ (as per ACI Section 8.3.3). Similar graphs for end spans are given in Figs. A-2 to A-4.

Figure A-1 can be used to locate the flexural cutoff points and points of inflection for typical interior uniformly loaded beams, *provided they satisfy the limitations of ACI Section 8.3.3.* Thus, the extreme points of inflection for positive moments (points where the positive-moment diagram equals zero) are at $0.146\ell_n$ from the faces of the two supports, while the corresponding negative-moment points of inflection are at $0.24\ell_n$ from the supports. This means that positive-moment steel must extend from midspan to at least $0.146\ell_n$ from the supports, while negative-moment steel must extend at least $0.24\ell_n$ from the supports. The use of Figs. A-1 through A-4 is illustrated in Example 8-6. A more complete example is given in Chapter 10.

EXAMPLE 8-6   Use of Bending-Moment Diagrams to Select Bar Cutoffs

A continuous T beam having the section shown in Fig. 8-30a carries a factored load of 3.26 kips/ft and spans 22 ft from center to center of 16-in.-square columns. The design has been carried out with the moment coefficients in ACI Section 8.3.3. Locate the bar cutoff points. For simplicity, all the bars will be carried past the points of inflection before cutting them off, except in the case of the positive-moment steel in span *BC*, where the No. 7 bar will be cut off earlier to illustrate the use of the bar cutoff graph. The concrete strength is 3000 psi, and the steel yield strength is 60,000 psi. Double-leg No. 3 stirrups are provided at 7.5 in. throughout.

1.   Structural integrity provisions. The beam is a continuous interior beam. Design should be by means of ACI Section 7.13.2.4. This section requires at least one-quarter of the positive-moment flexural reinforcement, but not less than two bars be made continuous by Class A lap splices over, or near, the supports, terminating at discontinuous ends with a standard hook. For span *AB*, the two No. 7 bars will serve as the tie. In spans *BC* and span *CD* we will use the two No. 6 bottom bars as structural-integrity steel. The lap splice length can be taken as 1.0 times ℓ_d for the No. 7 bottom bars.

2.   Determine the positive-moment steel cutoff points for a typical interior span—span *BC*.

(a)   Development lengths of bottom bars. From Table A-6, the minimum web width for two No. 7 bars and one No. 6 is 9 in. Because the beam width exceeds this value, the bar spacing is at least d_b, and because the beam has at least code-minimum stirrups, this is case 1 development. From Table A-11 for $\beta = 1.0$ and $\lambda = 1.0$,

No. 6 bar, $\ell_d = 43.8 d_b = 32.9$ in. $= 2.74$ ft. Class A splice $= 1.0\ell_d = 2.74$ ft
No. 7 bar, $\ell_d = 54.8 d_b = 48$ in. $= 4$ ft. Class A splice $= 4$ ft.

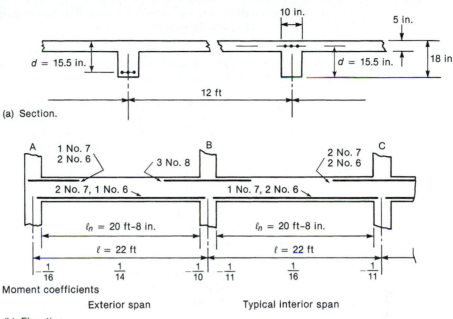

(a) Section.

(b) Elevation.

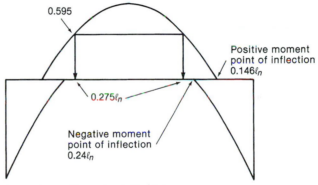

(c) Use of Fig. A–1 to locate flexural cutoff points.

Fig. 8-30
Calculation of flexural cutoff
points—Example 8-6.

(b)   Cutoff point for one No. 7 bottom bar—span *AB* or *BC*. Cut off one No. 7 bar when it is no longer needed on each side of column B, and extend the remaining two No. 6 bars into the supports.

After the No. 7 bar is cut off, the remaining A_s is 0.88 in.^{2} or $0.88/1.48 = 0.595$ times A_s at midspan. Therefore, the No. 7 bar can be cut off where M_u is 0.595 times the maximum moment. From Fig. A-1, this occurs at $0.275\ell_n$ from each end for the positive moment in a typical interior span. This is illustrated in Fig. 8-30c.

Therefore, the flexural cutoff point for the No. 7 bar is at

$$0.275\ell_n = 0.275 \times 20.67 \text{ ft}$$

$$= 5.68 \text{ ft from the faces of the columns}$$

To compute the actual cutoff points, we must satisfy detailing rules 3 and 4.

Effects of shear, Rule 3a. The bar must extend by the longer of $d = 15.5$ in. or $12d_b = 12 \times 0.875$ in. $= 10.5$ in. past the flexural cutoff. Therefore, extend the No. 7 bar 15.5 in. $= 1.29$ ft. The end of the bar will be at $5.68 - 1.29$ ft $= 4.39$ ft, say, 4 ft 4 in. from face of column as shown in Fig. 8.31b.

Anchorage, Rule 4a Positive movement. Bars must extend ℓ_d from the point of maximum bar stress. For No. 7 bottom bar $\ell_d = 48$ in. Clearly, the distance from the point of maximum bar stress at midspan to the end of bar exceeds this. Therefore, **cut off the No. 7 bar at 4 ft 4 in. from the column faces in the interior span.**

(c) *Cutoff point for remaining positive-moment steel.* The flexural cutoff points for the two No. 6 bars are at the positive-moment points of inflection. From Fig. A-1, these are at $0.146\ell_n = 3.02$ ft from the face of the supports. At these points we must satisfy detailing rules 1, 2, 3, and 4.

Structural integrity, Rule 1c. The application of rule 1 depends on whether the beam is an interior beam or perimeter beam and whether it has U stirrups with 135° stirrup hooks, on one hand, or no stirrups or open stirrups, on the other. This is an interior beam.

Structural integrity, Rule 1b. Does not require stirrups. Since there is little penalty to providing U stirrups with 135 ° hooks on the upper ends of the stirrup legs, we shall provide such stirrups. At least one-fourth of the positive moment steel is lap spliced at or near the supports with a Class A splice lapped $1.0\ell_d$ (see Section 8-8). We will extend the two No. 6 bars into the support, so as to have a bar in each lower corner of the stirrups, and for simplicity, we will lap splice both of them $\ell_d = 32.9$ in.—say, 36 in.

Effects of shear, Rule 3a. The bars must extend a distance d past the flexural cutoff. This is satisfied by running the bars into the support.

Anchorage, Rule 4a. The bars must extend by ℓ_d past the cutoff point of the No. 7 bar. For a No. 6 bottom bar ℓ_d is 29.3 in. More than ℓ_d is available—therefore, OK.

Anchorage, Rule 4d at positive moment point of inflection. At the positive-moment point of inflection (8-20) must be satisfied: $\ell_d \leq M_n/V_u + \ell_a$.

At the point of inflection, there are two No. 6 bars. For these bars M_n is

$$M_n = 2 \times 0.44 \times \dfrac{60{,}000\left(15.5 - \dfrac{2 \times 0.44 \times 60{,}000}{1.7 \times 3000 \times 62}\right)}{1000}$$

$$= 810 \text{ in.-kips}$$

The points of inflection are at $0.146\ell_n = 3.02$ ft from the faces of the columns:

$$V_u \text{ at point of inflection} = w\left(\dfrac{\ell_n}{2} - 0.146\ell_n\right)$$

$$V_u = 3.26\left(\dfrac{20.67}{2} - 3.02\right)$$

$$= 23.8 \text{ kips}$$

ℓ_a is the greater of $d = 15.5$ in. or $12d_b = 12 \times 0.75 = 9$ in., but not more than the actual extension of the bar past the point of inflection, which is 3.02 ft plus half of the lap-splice length (3.02 ft + 0.5 × 36 in. = 54 in.) Therefore, $\ell_a = 15.5$ in.

$$\dfrac{M_n}{V_u} + \ell_a = \dfrac{810 \text{ in.-kips}}{23.8 \text{ kips}} + 15.5 \text{ in.}$$

$$= 49.5 \text{ in.}$$

Since $\ell_d = 32.9$ in. is less than 49.5 in., the anchorage at the point of inflection is OK. Therefore, **extend two No. 6 bars 16.5 in.—say, 18 in.—past the center of the support.**
Check the shear at the points where No. 7 bars are cut off (ACI Section 12.10.5):

$$\phi(V_c + V_s) = 0.75\left(2 \times \sqrt{3000} \times 10 \times 15.5 + \dfrac{0.22 \times 60{,}000 \times 15.5}{7.5}\right)/1000$$

$$= 33.2 \text{ kips}$$

The shear at the cutoff point for the 7 bar is

$$V_u = 3.26\left(\dfrac{20.67}{2} - 4.33\right)$$

$$= 19.6 \text{ kips}$$

Since $V_u = 19.6$ kips is less than $\frac{2}{3} \times 33.2 = 22.1$ kips, no extra stirrups are required at the points where the No. 7 bar is cut off.

3.   Determine the negative-moment steel cutoffs for end B, span BC. To simplify the calculations, detailing, and construction, all the bars will be extended past the negative-moment point of inflection and cutoff. From Fig. A-1, the negative-moment point of inflection is at $0.24\ell_n = 0.24 \times 20.67$ ft $= 4.96$ ft from the face of the column. Therefore, the flexural cutoff point is at 4.96 ft from face of column.

Development lengths of top bars.   Again, this is case 1. From Table A-11 for $\beta = 1.0$ and $\lambda = 1.0$,

ℓ_d for No. 6 $= 57.0d_b = 42.8$ in.

ℓ_d for No. 7 $= 71.2d_b = 62.3$ in.

ℓ_d for No. 8 $= 71.2d_b = 71.2$ in.

The actual cutoff point is determined by applying rules 1, 2, 3, and 4.

Structural integrity, Rule 1c. This beam will form a tie.

Effects of shear, Rule 3b. At least one-third of the negative moment reinforcement must extend $d = 15.5$ in. $= 1.29$ ft, $12d_b = 12$ in. (at B, less at C), or $\ell_n/16 = 20.67/16 = 1.29$ ft past the point of inflection. Therefore, the bars must be extended by $4.96 + 1.29 = 6.25$ ft.

Therefore, try a cutoff point at 6 ft 3 in. from the face of the columns.

Structural Integrity, Rule 1. Already chosen as rule 1c.

Anchorage, Rule 4a. Bars must extend ℓ_d past the point of maximum bar stress. For the No. 8 bars at support B, $\ell_d = 71.2$ in. The actual bar extension of 6 ft. 3 in. is adequate.

Anchorage, Rule 4b. Negative-moment bars must be anchored into the support. This will be checked by checking the anchorage of the No. 8 bars on the opposite side of columns B and C.

Therefore, **cut off top bars at 6 ft 3 in. from the face of column B** (see Fig. 8-31b).

Repeat the calculations at end C of span BC; the development length of the No. 7 top bars is 5 ft 3 in. Since this is less than the 6 ft 3 in. chosen earlier by using rule 1c, we shall **cut off top bars at 6 ft 3 in. from the face of column C.**

4.   Determine the positive-moment cutoffs for the end span AB. Since the beam AB frames into a column at the exterior end, it must be designed using moment coefficients of $\frac{1}{16}$, $\frac{1}{14}$, and $\frac{1}{10}$ (ACI Section 8.3.3). Therefore, use Fig. A-2 to select the cutoff points. For simplicity, extend all the positive-moment bars 6 in. into the support. This will satisfy rules 1 and 2.

Structural Integrity, Rule 1c. To satisfy rule 1c we shall hook the two No. 7 bars into the support at A and will lap splice these bars $1.0\ell_d$ for a No. 7 bar $= 48$ in., with the two No. 6 bars from span BC.

Anchorage, Rule 4d. At the positive-moment point of inflection, $\ell_d \leq M_n/V_u + \ell_a$. For two No. 7 $M_n = 1100$ in.-kips. From Fig. A-2, the positive-moment point of inflection is at $0.1\ell_n = 2.07$ ft from the face of the exterior column. We thus have

$$V_u = w\left(\frac{\ell_n}{2} - 2.07\right) = 27.0 \text{ kips}$$

$$\ell_a = d = 15.5$$

Therefore,

$$\frac{M_n}{V_u} + \ell_a = \frac{1100}{27.0} + 15.5$$
$$= 56.2 \text{ in.}$$

This exceeds ℓ_d, which is 48 in. for No. 7 bottom bar—therefore OK.

Therefore, **hook the two No. 7 bars at support A, and lap-splice them 48 in. at support B.**

5.   Determine the negative-moment cutoffs for the interior end of the end span. At the interior end of the end span (at end B of span AB), the negative-moment point of inflection is at $0.24\ell_n$ from the face of the interior column (Fig. A-2). Following the calculations for the negative-moment bars in the interior span, we can **cut off the top bars at end B or span AB at 6 ft 3 in. from the face of the column** (see Fig. 8-31b).

6. Determine the negative-moment cutoffs for the exterior end of the end span. It is frequently difficult to anchor the top bars at the exterior end of a beam. These bars develop a stress of f_y at the inside face of the exterior column and hence must be anchored into the column for this stress (rule 4b).

The exterior column is 16 in. square. Since the development length for a No. 7 top bar is 62.3 in., we clearly cannot extend the bars straight into the column. Try hooks:

$$\text{basic development length } \ell_{dh} \text{ (8-18b)} = \frac{0.02\beta\lambda f_y}{\sqrt{f'_c}} \, d_b$$

$$= \frac{0.02 \times 1.0 \times 1.0 \times 60{,}000 \times 0.875}{\sqrt{3000}} = 19.2 \text{ in.}$$

This is multiplied by 0.7 if the hooks have adequate cover (ACI Section 12.5.3). Figure 8-31a shows a plan view of the exterior column-beam joint. The side cover to the hooked bars exceeds 2.5 in. and the bars can be placed with 2 in. of cover to the tail of the hook. Therefore,

$$\ell_{dh} = 19.2 \times 0.7 = 13.4 \text{ in.}$$

There is a space of 14 in. available in the column; therefore, we can use the top steel selected earlier with standard hooks in the column. Note that in some cases ACI Section 12.5.4 will require that the

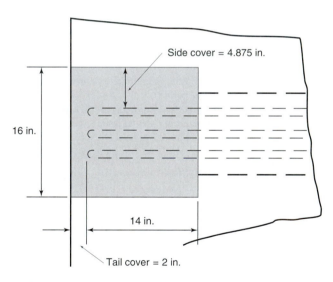

Side cover = 4.875 in.

16 in.

14 in.

Tail cover = 2 in.

(a) Plan of exterior column-beam joint.

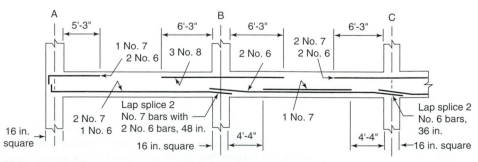

(b) Final bar details.

Fig. 8-31
Reinforcement details—
Example 8-6.

hook be enclosed within ties at $3d_b$, where d_d is the diameter of the hooked bar. Note also that the need for 2 in. of cover on the tail of the hook should be shown on the contract drawings.

Since the bars can be anchored into the column, we continue. (If the bars could not be anchored, we would repeat step 5, using the next-smaller-size bars.)

From Fig. A-2, the negative moment point of inflection is at $0.164\ell_n = 3.39$ ft from the face of the exterior column. Here anchorage rules 3 and 4 must be satisfied.

Effects of shear, Rule 3b. Extend the bars by $d = 1.29$ ft, $12d_b = 0.875$ ft, or $\ell_n/16 = 1.29$ ft beyond the point of inflection to $3.39 + 1.29 = 4.68$ ft or 4 ft 9 in. from the support.

Anchorage, Rule 4b. Bars must extend ℓ_d from the face of the column. ℓ_d for a No. 7 top bar is 62.3 in. Therefore, extend the top bars 5 ft 3 in. into the span.

Anchorage, Rule 4c. Anchorage of the negative moment steel into the supports is already satisfied by hooking the bars into the column.

Therefore, extend the top bars at the exterior end 5 ft 3 in. into the span and provide standard hooks into the column. The final reinforcement layout is shown in Fig. 8-31b. ∎

Frequently, bending moments cannot be calculated with the coefficients in ACI Section 8.3.3. This occurs, for example, in beams supporting concentrated loads, such as reactions from other beams; for beams of widely varying span lengths, such as a case that may occur in schools or similar buildings, where two wide rooms are separated by a corridor, or for beams with a change in the uniform loads along the span. For such cases, the bending-moment graphs in Figs. A-1 to A-4 cannot be used. Here, the designer must plot the relevant moment envelope from structural analyses and ensure that the moment-capacity diagram falls outside the required-moment envelope.

Standard Cutoff Points

For beams and one-way slabs that satisfy the limitations on span lengths and loadings in ACI Section 8.3.3 (two or more spans, uniform loads, roughly equal spans, factored live load not greater than 3 times factored dead load), the bar cutoff points shown in Fig. A-5 can be used. These are based on Figs. A-1 to A-4, assuming that the span-to-depth ratios are not less than 10 for beams and 18 for slabs.

These cutoff points do not apply to beams forming part of the primary lateral load-resisting system of a building. It is necessary to check whether the top bar extensions equal or exceed ℓ_d for the bar spacing actually used. Figure A-5 can be used as a guide for beams and slabs reinforced with epoxy-coated bars, but it is necessary to check the cutoff points selected.

For the beam in Example 8-6, Fig. A-5 would give negative-moment reinforcement cutoffs at 5 ft 2 in. from the face of column *A* and at 6 ft 11 in. from the face of column *B*. The extension from column B is a little longer than the computed extension.

Reinforcement Details for Beams Resisting Seismic Loads

Special cutoff requirements are given in ACI Sections 21.3.2 and 21.8.2 for longitudinal reinforcement in beams in frames resisting seismic loads. (See Section 20-7.)

8-8   SPLICES

Frequently, reinforcement in beams and columns must be spliced. There are four types of splices: lapped splices, mechanical splices, welded splices, and end-bearing splices. All four types of splices are permitted, as limited in ACI Sections 12.14, 12.15, and 12.16.

Tension Lap Splices

In a lapped splice, the force in one bar is transferred to the concrete, which transfers it to the adjacent bar. The force-transfer mechanism shown in Fig. 8-32a is clearly visible from the crack pattern sketched in Fig. 8-32b [8-11]. The transfer of forces out of the bar into the concrete causes radially outward pressures on the concrete, as shown in Fig. 8-32c; these pressures, in turn, cause splitting cracks along the bars similar to those shown in Fig. 8-8a. Once such cracks occur, the splice fails as shown in Fig. 8-33. The splitting cracks generally initiate at the ends of the splice, where the splitting pressures tend to be larger than at the middle. As shown in Fig. 8-32b, large transverse cracks occur at the discontinuities at the ends of the spliced bars. Transverse reinforcement in the splice region delays the opening of the splitting cracks and hence improves the splice capacity.

ACI Section 12.15 distinguishes between two types of tension lap splices, depending on the fraction of the bars spliced in a given length and on the reinforcement stress at the splice. Table R12.15.2 of the ACI Commentary is reproduced as Table 8-3. The equivalence between (A_s provided/A_s required) and f_s/f_y given in Table 8-3 does not appear in

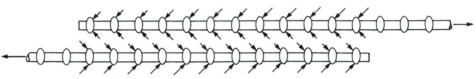

(a) Forces on bars at splice.

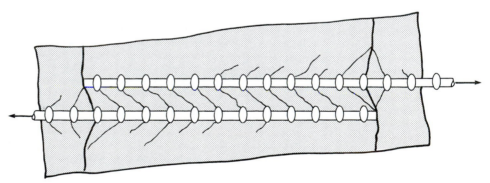

(b) Internal cracks at splice.

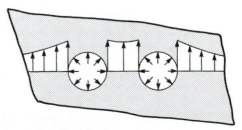

Fig. 8-32
Tension lap splice.

(c) Radial forces on concrete and splitting stresses shown on a section through the splice.

Fig. 8-33
Failure of a tension lap splice
without stirrups enclosing the
splice. (Photograph courtesy
of J. G. MacGregor.)

TABLE 8-3   Type of Tension Lap Splices Required

$\dfrac{A_s \text{ Provided}}{A_s \text{ Required}}$	or	$\dfrac{f_s}{f_y}$	Maximum Percentage of A_s Spliced within Required Lap Length	
			50%	100%
2 or more		0.5 or less	Class A	Class B
Less than 2		More than 0.5	Class B	Class B

Source: ACI Section 12.15.2.

the 2002 ACI Code, although it was expressed in terms of f_s/f_y in the 1971 code, and the intent has not changed. The splice lengths for each class of splice are as follows:

Class A splice: $1.0\ell_d$

Class B splice: $1.3\ell_d$

Because the stress level in the bar is accounted for in Table 8-3, the reduction in the development length for excess reinforcement allowed in ACI Section 12.2.5 is *not* applied in computing ℓ_d for this purpose.

The center-to-center distance between two bars in a lap splice cannot be greater than one-fifth of the splice length, with a maximum of 6 in. (ACI Section 12.14.2.3). Bars larger than No. 11 cannot be lap spliced, except at footing-to-column joints (ACI Section 15.8.2.3). Lap splices should always be enclosed within stirrups, ties, or spirals, to delay or prevent the complete loss of capacity indicated in Fig. 8-33. As indicated in ACI Sections 12.2.2 and 12.2.3, the presence of transverse steel may lead to shorter ℓ_d and hence shorter splices. ACI Section 21.3.2.3 requires that tension lap splices of flexural reinforcement in beams resisting seismic loads be enclosed by hoops or spirals.

Compression Lap Splices

In a compression lap splice, a portion of the force transfer is through the bearing of the end of the bar on the concrete [8-12], [8-13]. This transfer and the fact that no transverse tension cracks exist in the splice length allow compression lap splices to be much shorter than tension lap splices (ACI Section 12.16). Frequently, a compression lap splice will fail by spalling of the concrete under the ends of the bars. The design of column splices is discussed in Chapter 11.

Welded, Mechanical, and Butt Splices

In addition to lap splices, bars stressed in tension or compression may be spliced by welding or by various mechanical devices, such as sleeves filled with molten cadmium metal (Fig. 11-25) or threaded sleeves. The use of such splices is governed by ACI Sections 12.14.3 and 12.16.3. Descriptions of some commercially available splices are given in [8-14].

Mechanical Splices

Mechanical splices are proprietary steel devices used to provide a positive connection between bars. A common type of mechanical splice consists of a steel sleeve fitted over the joint and filled with molten metallic filler. Other common types are in the form of sleeves that are crimped onto the two bars being spliced. Still others involve tapered threads cut into the end regions of the bars to be spliced. The tapered threads are an attempt to engage the entire area of the bars in the splice, to satisfy the requirement that tests on splices develop 125 percent of the specified yield strength. Common types of mechanical splices are described in [8-14].

ACI Section 12.14.3.2 requires a *full mechanical splice* or a *full welded splice* to develop a tension or compression force, as applicable, of at least 125 percent of the specified yield strength, f_y, of the bar. Splices developing less than full tension are permitted on No. 5 and smaller bars if the splices of adjacent bars are staggered by at least 24 in. ACI Section 12.15.4 requires that such welds be able to develop twice the force required by analysis, but not less than 20,000 psi times the total area of reinforcement provided.

ACI Section 12.15.5 requires that splices in tension-tie members be made with full mechanical splices or full welded splices and that splices of adjacent bars be staggered by at least 30 in. Special requirements for splices in columns are presented in ACI Section 12.17.

PROBLEMS

8-1 Figure P8-1 shows a cantilever beam with $b = 12$ in. containing three No. 7 bars that are anchored in the column by standard 90° hooks. $f_c' = 5000$ psi and $f_y = 60,000$ psi. If the steel is stressed to f_y at the face of the column, can these bars

(a) be anchored by hooks into the column? The clear cover to the side of the hook is $2\frac{3}{4}$ in. The clear cover to the bar extension beyond the bend is $1\frac{7}{8}$ in. The joint is enclosed by ties at 6 in. o.c.

(b) be developed in the beam? The bar ends 2 in. from the end of the beam. The beam has No. 3 double-leg stirrups at 7.5 in.

8-2 Give two reasons why the tension development length is longer than the compression development length.

8-3 Why do bar spacing and cover to the surface of the bar affect bond strength?

8-4 A simply supported rectangular beam with $b = 14$ in. and $d = 17.5$ in. and with No. 3 minimum stirrups spans 14 ft and supports a total factored

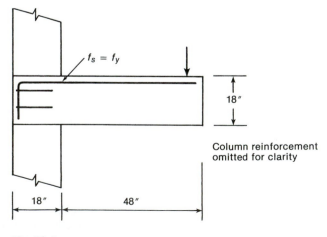

Fig. P8-1

uniform load of 6.5 kips/ft, including its own dead load. It is built of 3000-psi concrete and contains two No. 10 Grade-60 bars which extend 5 in. past the centers of the supports at each end and No. 3 minimum stirrups. Does this beam satisfy ACI Section 12.11.3? If not, what is the largest size bars which can be used?

8-5 Why do ACI Sections 12.10.3 and 12.12.3 require that bars extend d past their flexural cutoff points?

8-6 Why does ACI Section 12.10.2 define "points within the span where adjacent reinforcement terminates" as critical sections for development of reinforcement in flexural members?

8-7 A rectangular beam with cross section $b = 14$ $in.$, $h = 24$ in., and $d = 21.7$ in. supports a total factored load of 3.9 kips/ft, including its own dead load. The beam is simply supported with a 22-ft span. It is reinforced with six No. 6 Grade-60 bars, two of which are cut off between midspan and the support and four of which extend 12 in. past the centers of the supports. $f'_c = 4000$ psi. The beam has No. 3 stirrups satisfying ACI Section 11.5.4 and 11.5.5.3.

   (a) Plot to scale the factored moment diagram. $M = w\ell x/2 - wx^2/2$, where x is the distance from the support and ℓ is the span.

   (b) Redraw this diagram shifted a distance d toward the supports as shown in Fig. 8-19a.

   (c) Plot a resisting moment diagram and locate the cutoff points for the two cutoff bars.

8-8 Why does ACI Section 12.10.5 require extra stirrups at bar cutoff points in some cases?

The beam shown in Fig. P8-9 is built of 3000-psi concrete and Grade-60 steel. The effective depth $d = 18.63$ in. The beam supports a total factored uniform load of 5.25 kips/ft, including its own dead load. The frame is not part of the lateral load-resisting system for the building. Use Figs. A-1 to A-4 to select cutoff points in Problems 8-9 to 8-11.

8-9 Select cutoff points for span AB based on the following requirements:

   (a) Extend all positive moment bars into the columns before cutting them off.

   (b) Extend all negative moment bars past the negative moment point of inflection before cutting them off.

   (c) Check the anchorage of the negative moment bars at A and modify the bar size if necessary.

   (d) Compare the answer to Fig. A-5b.

8-10 Repeat Problem 8-9(a) and (b) for span BC.

8-11 Select cutoff points for span AB based on the following requirements:

   (a) Extend all negative moment bars at A past the negative moment point of inflection.

   (b) Cut off the two No. 6 positive moment bars when no longer needed at each end. Extend the remaining bars into the columns.

   (c) Cut off two of the negative moment bars at B when no longer needed. Extend the remaining bars past the point of inflection.

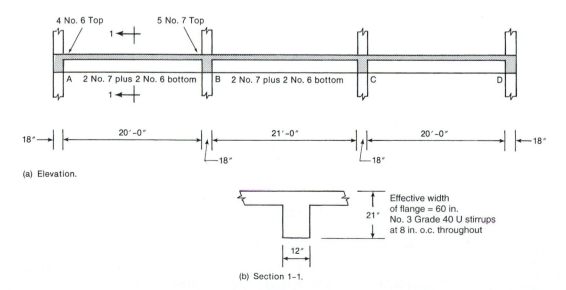

(a) Elevation.

(b) Section 1–1.

Fig. P8-9

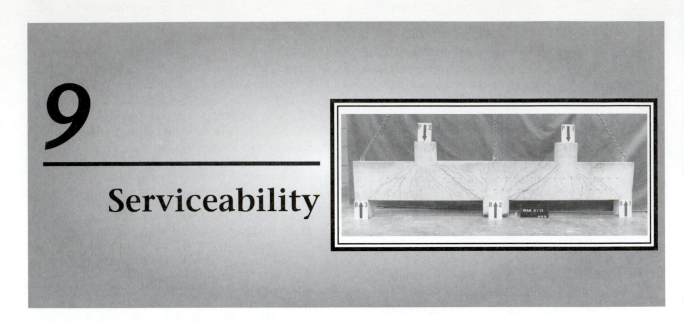

9

Serviceability

In Chapter 2, limit-states design was discussed. The limit states (states at which the structure becomes unfit for its intended function) were divided into two groups: those leading to collapse and those which disrupt the use of structures, but do not cause collapse. These were referred to as *ultimate limit states* and *serviceability limit states*, respectively. The major serviceability limit states for reinforced concrete structures are caused by excessive crack widths, excessive deflections, and undesirable vibrations. All three are discussed in this chapter. Although fatigue is an ultimate limit state, it occurs at service loads and so is considered here. The terms *service loads* and *working loads* refer to loads encountered in the everyday use of the structure. Service loads are generally taken to be the specified loads without load factors.

Historically, deflections and crack widths have not been a problem for reinforced concrete building structures. With the advent of strength design and Grade-60 reinforcement, however, the reinforcement stresses at service loads have increased by about 50 percent. Since crack widths, deflections, and fatigue are all related to steel stress, each of these has become more critical.

In the 2002 ACI Code, the net effect of the load factors and strength-reduction factor, ϕ, for flexure in ACI Section 9.3.2.1 were raised by 11 percent compared with the values in previous codes. As a result, about 11 percent less flexural steel is now needed for a given flexural strength than was required by ACI Codes published between 1971 and 1999. The steel stresses at service loads will thus be up to 11 percent higher than in beams with the reinforcement required by the 1999 and earlier codes. Crack widths and deflections both increase when the stress in the flexural steel increases. When ACI Committee 318 adopted the new strength-reduction factor for flexure, it did so knowing that limits on crack widths and deflections may need revision as future business [9-1]. The commentary for Section 2.4.1 of [9-2] explains how ASCE 7 deals with serviceability.

Serviceability Limit States

In design, a serviceability limit state consists of four parts:

1. A statement of the *limit state* causing the problem. Typical serviceability limit states (SLS) include deflection, cracking, vibrations, and so on.

2. One or more *load combinations* to be used in checking the limit state. Traditionally, the SLS load combinations used a load factor of 1.0 on all service loads. This assumes that all the specified variable loads have the values that they would have at an arbitrary point in time. Generally the variable loads are lower than this, especially for wind loads or snow loads. The selection of the load factors must consider the probability of occurrence of the loads involved, either by themselves or in conjunction with other loads. Generally, service load design codes do not adequately handle offsetting loads. (See Fig. 2-7.) In some structures, the number of loading cycles exceeding the limit state during the service life is important.

3. A *calculation procedure* to be used to check the limit state. Traditionally, serviceability-limit-state calculations for concrete structures are carried out using elastic analysis and the straight-line theory of flexure, based on average material properties.

4. A *criterion* to be used to judge whether the limit state is exceeded. Typically, this is in one of the following forms:

(Calculated deflection or other limit state) ≤ (Limiting value) [9-2]

or, for a prestressed concrete bridge pier,

"Cracks in the concrete should be open less than 100 times in the life of the structure." [9-3]

Major serviceability limit states affecting the design of reinforced concrete buildings are listed in Tables 9-1 through 9-4. These list the four parts of each serviceability limit state and suggest limits. Table 9-1 lists serviceability limit states due to cracking. Table 9-2 deals with flexural deflections. Tables 9-3 and 9-4 deal with other serviceability limit states.

9-2   ELASTIC ANALYSIS OF STRESSES IN BEAM SECTIONS

At service loads, the distribution of stresses in the compression zone of a cracked beam is close to being linear, as shown in Fig. 4-8, and the steel is elastic. As a result, an elastic calculation gives a good estimate of the concrete and steel stresses at service loads. This type of analysis is also referred to as a *straight-line theory* analysis, because a linear stress distribution is assumed. The straight-line theory is used in calculating both the stiffness, *EI*, at service loads, for deflection calculations, and steel stresses, for use in crack-width or fatigue calculations.

Calculation of *EI*

Modulus of Elasticity and Modular Ratio

Section 8.5.1 of the ACI Code assumes that concrete has a modulus of elasticity of

$$E_c = w_c^{1.5}\left(33\sqrt{f_c'}\right) \text{ psi} \tag{9-1}$$

where w_c is the unit weight of concrete, lb/ft^3. For normal-weight concrete, this reduces to $57{,}000\sqrt{f_c'}$ psi, the modulus of elasticity of concrete stressed to working-stress levels. As was discussed in Section 3-3, E_c is also affected strongly by the modulus of elasticity of the coarse aggregate. In critical situations, it may be desirable to allow for this. For American concretes E_c will generally lie between 80 percent and 120 percent of the value given by (9-1). Reinforcing steel has a modulus of elasticity of $E_s = 29 \times 10^6$ psi.

The ratio E_s/E_c is referred to as the *modular ratio, n,* and has values ranging from 9.3 for 3000-psi concrete to 6.6 for 6000-psi concrete. This means that, for a given strain less than the yield strain, the stress in steel will be six to nine times that in concrete subjected to the same strain.

Transformed Section

At service loads, the beam is assumed to act elastically. The basic assumptions in elastic bending are (a) that strains are linearly distributed over the depth of the member and (b) that the stresses can be calculated from the strains by the relationship $\sigma = E \times \epsilon$. This leads to the elastic bending equation, $\sigma = My/I$. When a beam made of two materials is loaded, the different values of E for the two materials lead to a different stress distribution, since one material is stiffer and accepts more stress for a given strain than the other. However, the elastic-beam theory can be used if the steel–concrete beam is hypothetically *transformed* to either an all-steel beam or an all-concrete beam, customarily the latter. This is done by replacing the area of steel with an area of concrete having the same axial stiffness AE. Because $E_s/E_c = n$, the resulting area of concrete will be nA_s. This transformed area of the steel is assumed to be concentrated at the same point in the cross section as the real steel area.

When the steel is in a compression zone or in an uncracked tension zone, its transformed area is nA_s, but it displaces an area of concrete equal to A_s. As a result, compression steel is transformed to an equivalent concrete area of $(n - 1)A'_s$. (In the days of working-stress design, compression steel was transformed to an equivalent concrete area of $(2n - 1)A'_s$, to reflect the effect of creep on the stresses.)

Prior to flexural cracking, the beam shown in Fig. 9-1a has the transformed section in Fig. 9-1b. Since the steel is displacing concrete, which could take stress, the transformed area is $(n - 1)A_s$ for both layers of steel. The cracked transformed section is shown in Fig. 9-1c. Here, the steel in the compression zone displaces stressed concrete and has a transformed area of $(n - 1)A'_s$, while that in the tension zone does not and hence has an area of nA_s.

The neutral axis of the cracked section occurs at a distance $c = kd$ below the top of the section. For an elastic section, the neutral axis occurs at the centroid of the area, which is defined as that point at which

$$\Sigma A_i \bar{y}_i = 0 \tag{9-2}$$

where $\bar{y}_i$ is the distance from the centroidal axis to the centroid of the ith area. The solution of (9-2) is illustrated in Example 9-1.

EXAMPLE 9-1 Calculation of the Transformed Section Properties

The beam shown in Fig. 9-1 is built of 4000-psi concrete. Compute the location of the centroid and the moment of inertia for both the uncracked section and the cracked section.

Uncracked Transformed Section

The modulus of elasticity of the concrete is

$$E_c = 57,000\sqrt{f'_c} = 3.605 \times 10^6 \text{ psi}$$

The modular ratio is

$$n = \frac{E_s}{E_c}$$

$$= \frac{29 \times 10^6}{3.605 \times 10^6} = 8.04$$

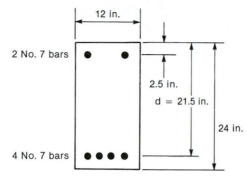

(a) Cross section.

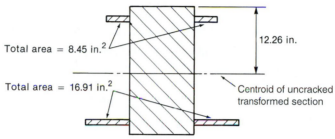

(b) Uncracked transformed section.

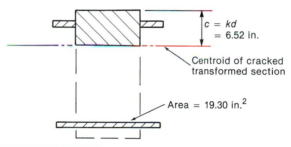

(c) Cracked transformed section.

Fig. 9-1
Transformed sections—
Example 9-1.

Since all the steel is in uncracked parts of the beam, the transformed areas of the two layers of steel are computed as follows:

$$\text{Top steel:} \quad (8.04 - 1) \times 1.20 = 8.45 \text{ in.}^2$$

$$\text{Bottom steel:} \quad (8.04 - 1) \times 2.40 = 16.91 \text{ in.}^2$$

See Fig. 9-1b.

The centroid of the transformed section is located as follows:

Part	Area (in.^{2})	y_{top} (in.)	Ay_{top} (in.^{3})
Concrete	$12 \times 24 = 288$	12	3456
Top steel	8.45	2.5	21.1
Bottom steel	16.91	21.5	363.6
	313.4		$\Sigma Ay_{top} = 3840.7$

$$\bar{y}_{top} = \frac{3840.7}{313.4} = 12.26 \text{ in.}$$

Therefore, the centroid is 12.26 in. below the top of the section. The moment of inertia is calculated as follows:

Part	Area (in.^{2})	$\bar{y}$ (in.)	$I_{own\ axis}$ (in.^{4})	$A\bar{y}^2$ (in.^{4})
Concrete	288	−0.26	13,824	19.5
Top steel	8.45	9.76	—	805
Bottom steel	16.91	−9.24	—	1440
			$I_{gt} = $	16,090 in.^{4}

The uncracked transformed moment of inertia $I_{gt} = 16{,}090$ in.^{4}. This is 16 percent larger than the gross moment of inertia of the concrete alone, referred to as I_g. The moments of inertia of the steel areas about their own centroidal axes are small and were neglected.

Cracked Transformed Section

Assume that the neutral axis is lower than the top steel. The transformed areas are computed as follows:

$$\text{Top steel:} \quad (8.04 - 1) \times 1.20 = 8.45 \text{ in.}^2$$
$$\text{Bottom steel:} \quad 8.04 \times 2.40 = 19.30 \text{ in.}^2$$

Let the depth to the neutral axis be c, as shown in Fig. 9-1c, and sum the moments of the areas about the neutral axis to compute c:

Part	Area (in.^{2})	$\bar{y}$ (in.)	$A\bar{y}$ (in.^{3})
Compression zone	$12c$	$c/2$	$12c^2/2 = 6c^2$
Top steel	8.45	$c - 2.5$	$8.45c - 21.13$
Bottom steel	19.30	$c - 21.5$	$19.30c - 414.95$

But, by definition, c is the distance to the centroid when $\Sigma A\bar{y} = 0$. Therefore, $6c^2 + 27.75c - 436.08 = 0$ and it follows that

$$c = \frac{-27.75 \pm \sqrt{27.75^2 + 4 \times 6 \times 436.08}}{2 \times 6}$$

$$= 6.52 \text{ in. or} -11.1 \text{ in.}$$

The positive value lies within the section. Since the top steel is in the compression zone, the initial assumption is OK. Therefore, **the centroidal axis (axis of zero strain) is at 6.52 in. below the top of the section.**
    Compute the moment of inertia:

Part	Area (in.^{2})	y (in.)	$I_{own\ axis}$ (in.^{4})	Ay^2 (in.^{4})
Compression zone	$12 \times 6.52 = 78.25$	$6.52/2$	277	832
Top steel	8.45	$(6.52 - 2.5) = 4.02$	—	137
Bottom steel	19.30	$(6.52 - 21.5) = -14.98$	—	4330
			$I = $	5580 in.^{4}

Thus, the cracked transformed moment of inertia is $I_{cr} = 5580$ in.^{4} ∎

In this example, I_{cr} is only 35 percent of the moment of inertia of the uncracked transformed section and 40 percent of that of the concrete section alone. This illustrates the great reduction in stiffness due to cracking.

The procedure followed in Example 9-1 can be used to locate the neutral axis for any shape of cross section bent in uniaxial bending. For the special case of rectangular beams *without* compression reinforcement, (9-2) gives

$$\frac{bc^2}{2} - nA_s(d - c) = 0$$

Substituting $c = kd$ and $\rho = A_s/bd$ results in

$$\frac{b(kd)^2}{2} - \rho nbd(d - kd) = 0$$

Dividing by bd^2 and solving for k gives

$$k = \sqrt{2\rho n + (\rho n)^2} - \rho n \tag{9-3}$$

where $\rho = A_s/bd$ and $n = E_s/E_c$. Equation (9-3) can be used to locate the neutral-axis position directly, thus simplifying the calculation of the moment of inertia. This equation applies only for rectangular beams without compressive reinforcement.

Service-Load Stresses in a Cracked Beam

The compression stresses in the beam shown in Fig. 9-2 vary linearly from zero at the neutral axis to a maximum stress of f_c at the extreme-compression fiber. The total compressive force C is

$$C = \frac{f_c bkd}{2}$$

This force acts at the centroid of the triangular stress block, $kd/3$ from the top. For the straight-line concrete stress distribution, the lever arm jd is

$$jd = d - \frac{kd}{3} = d\left(1 - \frac{k}{3}\right)$$

If the moment at service loads is M_s, we can write

$$M_s = Cjd = \frac{f_c bkd}{2}jd$$

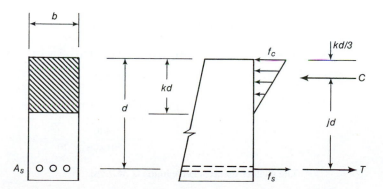

Fig. 9-2
Stress distribution in straight-line theory.

and

$$f_c = \frac{2M_s}{jkbd^2} \qquad (9\text{-}4)$$

Similarly, taking moments about C yields

$$M_s = Tjd = f_s A_s jd$$

and

$$f_s = \frac{M_s}{A_s jd} \qquad (9\text{-}5)$$

This analysis ignores the effects of creep, which will tend to increase the stress in the tension steel by a small amount.

EXAMPLE 9-2   Calculation of the Service-Load Steel Stress in a Rectangular Beam

A rectangular beam similar to the one shown in Fig. 9-2 has $b = 10$ in., $d = 20$ in., three No. 8 Grade-60 bars, and $f'_c = 3000$ psi. Compute f_s at service loads if the service live-load moment is 50 ft-kips and the service dead-load moment is 70 ft-kips.

1. **Compute k and j.**

$$E_c = 57,000 \sqrt{3000} = 3.12 \times 10^6 \text{ psi}$$

$$n = \frac{E_s}{E_c}$$

$$= \frac{29 \times 10^6}{3.12 \times 10^6} = 9.29$$

$$\rho = \frac{A_s}{bd}$$

$$= \frac{2.37}{10 \times 20} = 0.0019$$

$$\rho n = 0.110$$

$$k = \sqrt{2\rho n + (\rho n)^2} - \rho n \qquad (9\text{-}3)$$

$$= \sqrt{2 \times 0.110 + 0.110^2} - 0.110$$

$$= 0.372$$

$$j = 1 - \frac{k}{3} = 0.876$$

2. **Compute f_s at M_s.**

$$M_s = 50 + 70 = 120 \text{ ft-kips}$$
$$= 1440 \text{ in.-kips}$$

$$f_s = \frac{M_s}{A_s jd} \qquad (9\text{-}5)$$

$$= \frac{1440}{2.37 \times 0.876 \times 20} = 34.7 \text{ ksi}$$

The steel stress at service loads is 34.7 ksi. ∎

Age-Adjusted Transformed Section

Creep causes an increase in the compressive strains in a beam, a resulting drop in the neutral axis, and an increase in the strains in the tension steel. The concept of age-adjusted transformed sections, introduced in Section 3-4 for axially loaded members, can be used to estimate stresses and deflections in beams subjected to sustained loads [9-4]–[9-6].

9-3   CRACKING

Types of Cracks

Tensile stresses induced by loads, moments, and shears cause distinctive crack patterns, as shown in Fig. 9-3. Members loaded in direct tension crack right through the entire cross

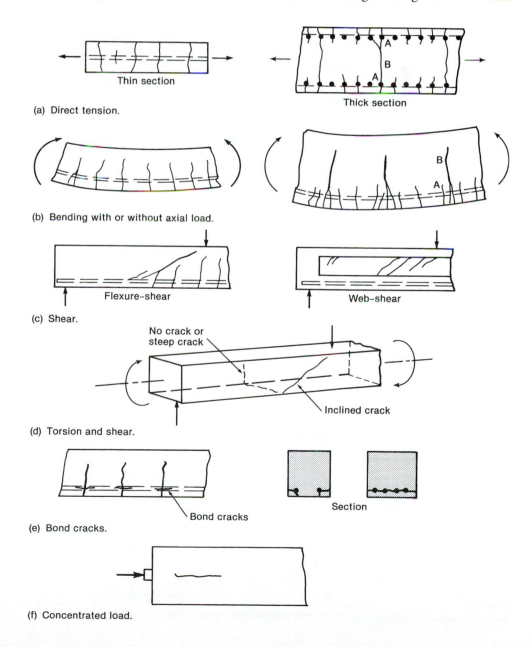

Fig. 9-3
Load-induced cracks.
(From [9-7].)

section, with a crack spacing ranging from 0.75 to 2 times the minimum thickness of the member. In the case of a very thick tension member with reinforcement in each face, small surface cracks develop in the layer containing the reinforcement (Fig. 9-3a). These join in the center of the member. As a result, for a given total change in length, the crack width at *B* is greater than at *A*.

Various mechanisms of cracking are discussed in detail in [9-8] through [9-11]. The relationship between cracking and corrosion of reinforcement is discussed in [9-12], [9-13], and [9-14]. Crack widths are discussed in [9-15], [9-16], [9-17] and [9-18]. Members subjected to bending moments develop flexural cracks, as shown in Fig. 9-3b. These vertical cracks extend almost to the zero-strain axis (neutral axis) of the member. In a beam with a web that is more than 3 to 4 ft high, the cracking is relatively closely spaced at the level of the reinforcement, with several cracks joining or disappearing above the reinforcement, as shown in Fig. 9-3b. Again, the crack width at *B* will frequently exceed that at *A*. The cracks in the middle third of the beam shown in Figs. 4-5 and 4-6 are flexural cracks.

Cracks due to shear have a characteristic inclined shape, as shown in Fig. 9-3c and Figs. 6-4 and 6-12. Such cracks extend upward as high as the neutral axis and sometimes into the compression zone. Torsion cracks are similar. In pure torsion, they spiral around the beam. In a normal beam, where shear and moment also act, they tend to be pronounced on the face where the direct shear stresses and the shear stresses due to torsion add, and less pronounced (or even absent) on the opposite face, where the stresses counteract (Fig. 9-3d).

Bond stresses lead to splitting along the reinforcement, as shown in Fig. 9-3e. Concentrated loads will sometimes cause splitting cracks or "bursting cracks" of the type shown in Fig. 9-3f. This type of cracking is discussed in Section 18-5. It occurs in bearing areas and in struts in strut-and-tie models. Such struts are called *bottle-shaped struts*.

At service loads, the final cracking pattern has generally not developed completely, with the result that there are normally only a few cracks at points of maximum stress at this load level.

Cracks also develop in response to imposed deformations, such as differential settlements, shrinkage, and temperature differentials. If shrinkage is restrained, as in the case of a thin floor slab attached at each end to a massive stiff structure, shrinkage cracks may occur. Generally, however, shrinkage simply increases the width of load-induced cracks.

A frequent cause of cracking in structures is restrained contraction resulting from the cooling down to ambient temperatures of very young members that expanded under the *heat of hydration* which developed as the concrete was setting. This most typically occurs where a length of wall is cast on a foundation cast some time before. As the wall cools, its contraction is restrained by the foundation. A typical *heat-of-hydration cracking* pattern is shown in Fig. 9-4. Such cracking can be controlled by controlling the heat rise due to the heat of hydration and the rate of cooling, or both; by placing the wall in short lengths; or by reinforcement considerably in excess of normal shrinkage reinforcement [9-18], [9-19], [9-20].

Plastic shrinkage and slumping of the concrete, which occur as newly placed concrete bleeds and the surface dries, result in settlement cracks along the reinforcement, Fig. 9-5a,

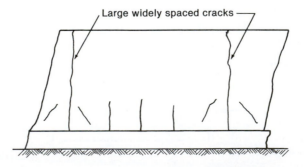
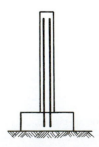

Large widely spaced cracks

Fig. 9-4
Heat-of-hydration cracking.

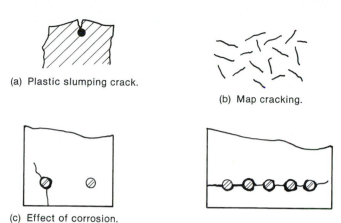

(a) Plastic slumping crack.

(b) Map cracking.

Fig. 9-5
Other types of cracks.

(c) Effect of corrosion.

or a random cracking pattern, referred to as *map cracking*, Fig. 9-5b. These types of cracks can be avoided by proper mix design and by preventing rapid drying of the surface during the first hour or so after placing. Map cracking can also be due to alkali–aggregate reaction.

Rust occupies two to three times the volume of the metal from which it is formed. As a result, if rusting occurs, a bursting force is generated at the bar location, which leads to splitting cracks and an eventual loss of cover (Fig. 9-5c). Such cracking looks similar to bond cracking (Fig. 9-3e) and may accompany bond cracking.

Development of Cracks Due to Loads

Figure 9-6 shows an axially loaded prism. Cracking starts when the tensile stress in the concrete (shown by the shaded area in Fig. 9-6b) reaches the tensile strength of the concrete (shown by the outer envelope) at some point in the bar. When this occurs, the prism cracks. At the crack, the entire force in the prism is carried by the reinforcement. Bond gradually builds up the stress in the concrete on either side of the crack until, with further loading, the stress reaches the tensile strength at some other section, which then cracks (Fig. 9-6c). With increasing load, this process continues until the distance between the cracks is not big enough for the tensile stress in the concrete to increase enough to cause cracking. Once this stage is reached, the crack pattern has stabilized, and further loading merely widens the existing cracks. The distance between stabilized cracks is a function of the overall member thickness, the cover, the efficiency of the bond, and several other things. Roughly, however, it is two to three times the bar cover. Cracks that extend completely through the member generally occur at roughly one member thickness apart.

Figure 9-7b and c show the variation in the steel and concrete stresses along an axially loaded prism with a stabilized crack pattern. At the cracks, the steel stress and strain are at a maximum and can be computed from a cracked-section analysis. Between the cracks, there is stress in the concrete. This reaches a maximum midway between two cracks. The total width, w, of a given crack is the difference in the elongation of the steel and the concrete over a length $A–B$ equal to the crack spacing:

$$w = \int_A^B (\epsilon_s - \epsilon_c)\, dx \tag{9-6}$$

where ϵ_s and ϵ_c are the strains in the steel and concrete at a given location between A and B and x is measured along the axis of the prism.

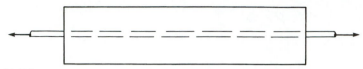

(a) Prism.

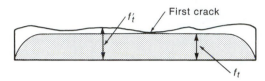

(b) Variation of tensile strength and stress along prism.

Second crack

(c) Tensile stresses after first crack.

(d) Tensile stresses after three cracks.

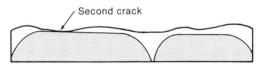

Second crack   First crack   Third crack

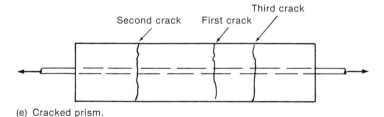

Fig. 9-6
Cracking of an axially loaded prism.

(e) Cracked prism.

The crack spacing, s, and the variation in ϵ_s and ϵ_c are difficult to calculate in practice, and empirical equations are generally used to compute the crack width. The best known of these are the Euro-International Concrete Committee (CEB) procedure [9-10], based on (9-6), and the Gergely–Lutz equation [9-11], derived statistically from a number of test series.

As discussed in Chapter 8, bond stresses are transferred from steel to concrete by means of forces acting on the deformation lugs on the surface of the bar. These lead to cracks in the concrete adjacent to the ribs, as shown in Fig. 9-8. In addition, the tensile stress in the concrete decreases as one moves away from the bar, leading to less elongation of the surface of the concrete than of the concrete at the bar. As a result, the crack width at the surface of the concrete exceeds that at the bar.

In a deep flexural member, the distribution of crack widths over the depth shows a similar effect, particularly if several cracks combine, as shown in Fig. 9-3b. The crack width at B frequently exceeds that at the level of the reinforcement.

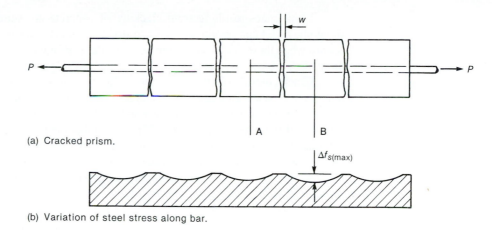

(a) Cracked prism.

(b) Variation of steel stress along bar.

Fig. 9-7
Stresses in concrete and steel
in a cracked prism.

(c) Variation of concrete stress along prism (not to scale).

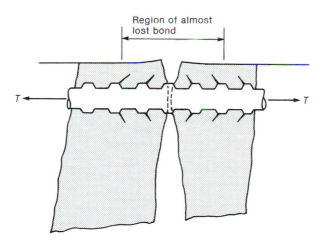

Fig. 9-8
Cracking at bar deformations.
(From [9-4].)

Reasons for Controlling Crack Widths

Crack widths are of concern for three main reasons: appearance, leakage, and corrosion. Wide cracks are unsightly and sometimes lead to concern by owners and occupants. Crack width surveys [9-9] suggest that, on easily observed, clean, smooth surfaces, cracks wider than 0.01 to 0.013 in. can lead to public concern. Wider cracks are tolerable if the surfaces are less easy to observe or are not smooth. On the other hand, cracks in exposed surfaces may be accentuated by streaks of dirt or leached materials. Sandblasting the surface will accentuate the apparent width of cracks, making small cracks much more noticeable.

Limit states resulting from cracking of concrete are summarized in Table 9-1. It should be noted that reinforced concrete structures must crack to carry the service loads. It is only when these cracks affect the usage or the appearance of the structure that cracking limit states are considered.

Crack control is important in the design of liquid-retaining structures. Leakage is a function of crack width, provided the cracks extend through the concrete member.

Corrosion of reinforcement has traditionally been related to crack width. More recent studies [9-12], [9-13], [9-14] and [9-15] suggest that the factors governing the eventual development of corrosion are independent of the crack width, although the period of time required for corrosion to start is a function of crack width. Reinforcement surrounded by concrete will not corrode until an electrolytic cell can be established. This will occur when carbonization of the concrete reaches the steel or when chlorides penetrate through the concrete to the bar surface. The time taken for this to occur will depend on a number of factors: whether the concrete is cracked, the environment, the thickness of the cover, and

TABLE 9-1 Loads and Load Combinations for Serviceability—Cracking

(1) Case	(2) Serviceability Limit State and Elements Affected	(3) Serviceability Parameter and Load Combination	(4) Calculation Method and Acceptance Criteria[9-8]	(5) References
A	**Cracking**			
A-1	**Leakage of Fluids or Gases**	Number and width of through cracks. Number of times cracks are open.	The width of through cracks must be less than a chosen maximum.	[9-8], [9-10], [9-13], [9-15], [9-20]
A-2	**Corrosion** Water and corrosive solutions reach the reinforcement.	Width of cracks that extend to the reinforcement. Number of times crack width exceeds a given limit per year.	ACI Code and ASCE 7 do not give guidance for how to compute crack widths.	ACI Section 7.7.5 and 10.6.7. Book Section 9-3 Refs. [9-4]. [9-12] to [9-14]
A-3	**Visually obvious cracks**	Cracks which can be seen. This is aggravated by sand-blasting the surface of the concrete.	ACI Code and ASCE 7 do not give guidance for how to compute crack widths. ACI 10.6.4 controls steel spacing.	ACI Section 10.6.4 and 10.6.7. Book Section 9-3 [9-7] [9-15], [9-16], [9-17], [9-18], [9-24]
A-4	**Crushing and cracking of prestressed members**	At transfer: 1.0D + 1.0 Prestress In service: 1.0D + 1.0 Prestress + 1.0L	Straight-line flexural stress calculations. Maximum compressive stress vs. long term strength, or tensile stress limit related to concrete tensile strength.	ACI Section 18-3 to 18-5
A-5	**Shrinkage and creep**	Cracks due to restrained shrinkage.	Limits based on resulting damage. 2 to 3 times ACI temperature and shrinkage steel needed to control this cracking.	ACI Section 9.2.3. Book Section 3-6 Refs. [9-19], [9-20]

the permeability of the concrete. If the concrete is cracked, the time required for a corrosion cell to be established is a function of the crack width. After a period of 5 to 10 years, however, the amount of corrosion is essentially independent of crack width.

Generally speaking, corrosion is most apt to occur if one or more of the following circumstances occurs:

1. Chlorides or other corrosive substances are present.

2. The relative humidity exceeds 60 percent. Below this value, corrosion rarely occurs except in the presence of chlorides.

3. Ambient temperatures are high, accelerating the chemical reaction.

4. Ponded water and wetting and drying cycles occur that cause the concrete at the level of the steel to be alternately wet and dry. (Corrosion is minimal in permanently saturated concrete, because the water reduces or prevents oxygen flow to the steel.)

5. Stray electrical currents occur in the bars.

Corrosion of steel embedded in concrete is discussed in [9-12], [9-13], and [9-14], and in Section 3-9 of this book. ACI Sections 4.4 or 7.7 are intended to limit or delay corrosion by limiting the permeability of the concrete. This is achieved by setting limits on the composition of the concrete and on the cover to the reinforcement.

Limits on Crack Width

There are no universally accepted rules for maximum crack widths. Prior to 1999, the ACI Code crack-control limits were based on a maximum crack width of 0.016 in. for interior exposure and 0.013 in. for exterior exposure [9-11]. What constitutes interior and exterior exposure was not defined. In addition to crack-control provisions, there are special requirements in ACI Chapter 4 for the composition of concrete subjected to special exposure conditions.

The Euro-International Concrete Committee (CEB) [9-10], [9-17] limits the mean crack width (about 60 percent of the maximum crack width) as a function of exposure condition, sensitivity of reinforcement to corrosion, and duration of the loading condition.

ACI Sections 10.6.3 to 10.6.7 handle crack widths indirectly by limiting the maximum bar spacings and bar covers for beams and one-way slabs. Prior to 1999, these limits were based on the Gergely–Lutz [9-11] equation which related the maximum crack width w at the tensile surface of a beam or slab and the cover to

(a) the stress f_s in the steel at service loads,

(b) the distance d_c from the extreme concrete fiber to the centroid of the bar closest to the tension fiber, and

(c) the area A of the prism of concrete concentric with the bar.

Two limiting crack widths were considered: $w = 0.016$ in. for interior exposure and $w = 0.013$ in. for exterior exposure. The resulting equation tended to give unacceptably small bar spacings for bar covers greater than 2.5 in. For this reason, the Gergely–Lutz equation was replaced in the 1999 ACI Code by (9-7). This equation was obtained by fitting a straight line to the Gergely–Lutz equation for a flexural crack width of 0.016 in. This resulted in the relationship [9-16]

$$s = 540/f_s - 2.5c_c \text{ but not greater than } 12(36/f_s) \text{ (psi)} \qquad (9\text{-}7)$$
$$\text{(ACI Eq. 10-4)}$$

$$s = \frac{95{,}000}{f_s} - 2.5c_c \quad \text{MPa} \qquad\qquad (9\text{-}7\text{M})$$
$$\text{but not greater than } 300 \, (2.5 \, 2/f_s) \qquad \text{(ACI Eq. 10-4M)}$$

where s is the bar spacing in in., f_s is the service-load bar stress in *ksi* or MPa in (9-7M), and c_c is the clear cover from the nearest surface of the concrete in the tension zone to the surface of the flexural tension reinforcement, in inches or mm in (9-7M).

Equation (9-7) was based on the limiting crack width for interior exposure only. This is because the eventual amount of reinforcement corrosion has been shown to be independent of surface crack width [9-12], [9-13], [9-14].

The steel stress f_s can be computed via (9-5). Alternatively, ACI 318-02 Section 10.6.4 allows the value of f_s at service loads to be taken as $f_y/(1.4/0.9) \simeq 0.64f_y$ where, (1.4/0.9) represents the average load factor divided by the strength reduction factor, ϕ, for flexure. For the beam considered in Example 9-2, this is close to the computed value.

EXAMPLE 9-3   Checking the Distribution of Reinforcement in a Beam

At the point of maximum positive moment, a beam contains the reinforcement shown in Fig. 9-9. The reinforcement has a yield strength of 60,000 psi. Is this distribution satisfactory?

Assume that $f_s = 0.6f_y = 36$ ksi. (Note that f_s is defined as the bar stress in ksi.) Let c_c be the clear cover from the nearest surface of the concrete in tension to the surface of the *flexural tension* reinforcement in inches. Then

$$c_c = 1.5 \text{ in.} + 0.375 \text{ in.} = 1.875 \text{ in.}$$

$$s = \frac{540}{36} - 2.5 \times 1.875 = 10.3 \text{ in.} < 12 \times 36/36 = 12 \text{ in.}$$

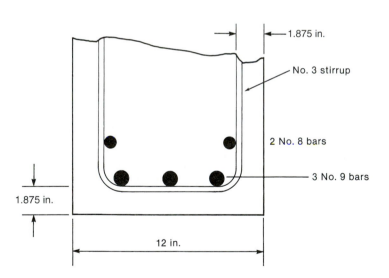

Fig. 9-9
Beam—Example 9-3.

Since the bar spacing in Fig. 9-9 is clearly less than 10.3 in., **the steel distribution is acceptable.** ∎

In theory, this should be checked at every section of maximum positive and negative moment. In practice, however, it is normally necessary to check the bar spacing only at the sections of maximum positive and negative moment having the smallest numbers of bars.

In one-way slabs, the flexural-reinforcement distribution is found by checking whether the bar spacing is less than that given by (9-7). The maximum bar spacing is calculated in Example 9-4. It is not necessary to compute z for two-way slabs (ACI Sections 10.6.1 and 10.6.2).

EXAMPLE 9-4   Calculation of the Maximum Bar Spacing
for Bars in a One-Way Slab

An 8- in.-thick slab has No. 4 bars at a spacing s. The bars have $f_y = 60,000$ psi and a minimum clear cover of $\frac{3}{4}$ in. Compute the maximum value of s.

We have

$$s = \frac{540}{f_s} - 2.5 \, c_c \tag{9-7}$$

where

$$f_s = 0.6 \, f_y = 36 \text{ ksi}$$

$$c_c = 0.75 \text{ in.}$$

Thus,

$$s = \frac{540}{36} - 2.5 \times 0.75 = 13.1 \text{ in.} < 12 \times 36/36 = 12 \text{ in.}$$

A bar spacing less than or equal to 12 in. satisfies 10.6.4. The bar spacing must also satisfy ACI Section 7.6.5, which limits s to three times the slab thickness, or 18 in. ■

Concrete covers in excess of 2 in. may be required for durability, for fire resistance, or in members such as footings.

In the negative-moment regions of T beams, the flanges will be stressed in tension and will crack as shown in Fig. 5-2. To restrict the width of the cracks in the flanges, ACI Section 10.6.6 requires that "part" of the flexural-tension reinforcement be distributed over a width equal to the smaller of the effective flange width and $\ell_n/10$. The same ACI section also requires that "some" longitudinal reinforcement be provided in the outer portions of the flange. The terms "part" and "some" are not defined. This can be accomplished by placing roughly one-fourth to one-half the reinforcement in the overhanging portions of the flange at, or near, the maximum spacing for crack control and by placing the balance over the web of the beam.

Shrinkage and Temperature Reinforcement

ACI Section 7.12 requires reinforcement perpendicular to the span in one-way slabs for tensile stresses resulting from restrained shrinkage and temperature changes. If the reinforcement is intended to replace the tensile stresses in the concrete at the time of cracking, the following simplified analysis suggests that an amount equal to

$$A_s f_y = f_t' A_g$$

is needed to replace the tensile stress lost when the concrete cracks.

$$\rho_{s,t} = \frac{A_s}{A_g} = \frac{f_t'}{f_y} \tag{9-8}$$

and

$$\rho_{s,t} \geq \frac{f_t'}{f_y}$$

For Grade-60 steel and 4000-psi concrete, $\rho_{s,t}$ is between 0.004 and 0.005. This is about three times the amount of shrinkage and temperature reinforcement specified in ACI Section 7.12.2.1(b). More complete analyses of the required amounts of shrinkage and temperature steel are presented by Gilbert [9-19] and Beeby [9-20].

In a one-way slab, as shown in section in Fig. 10-13, the area of shrinkage and temperature steel in a given direction is the sum of the areas of the top and bottom shrinkage and temperature reinforcement in that direction.

Web Face Reinforcement

In beams deeper than about 3 ft, the widths of flexural cracks may be as large, or larger, at points above the reinforcement than they are at the level of the steel, as shown in Fig. 9-3b.

To control the width of these cracks, ACI Section 10.6.7 requires skin reinforcement uniformly distributed along both faces of the web for a height of $d/2$ measured from the centroid of the longitudinal tension reinforcement toward the neutral axis. The maximum spacing, s_k, of the skin reinforcement shall not exceed the smallest of $d/6$ in., 12 in., and $1000A_b/(d - 30)$ in., where A_b in.^{2} is the area of one bar. Where other code provisions require more steel, the other provisions shall govern. These provisions grew out of research reported in [9-18].

For large beams, this can be a significant amount of reinforcement. If strain compatibility is used to calculate the stresses in the web face steel, the effect of this steel on ϕM_n can be included in design (see Section 5-5).

9-4   DEFLECTIONS: RESPONSE OF CONCRETE BEAMS

In this section, we deal with the flexural deflections of beams or slabs. Deflections of frames are considered in Section 9-6.

Load–Deflection Behavior of a Concrete Beam

Figure 9-10a traces the load–deflection history of the fixed-ended, reinforced concrete beam shown in Fig. 9-10b. Initially, the beam is uncracked and is stiff (0–A). With further load, cracking occurs when the moment at the ends exceeds the cracking moment. When a section cracks, its moment of inertia decreases, leading to a decrease in the stiffness of the beam. This causes a reduction in stiffness (A–B) in the load–deflection diagram in Fig. 9-10a. Cracking in the midspan region causes a further reduction of stiffness (point B). Eventually, the reinforcement would yield at the ends or at midspan, an effect leading to large increases in deflection with little change in load (points D and E). The service-load level is represented by point C. The beam is essentially elastic at point C, the nonlinear load deflection being caused by a progressive reduction of flexural stiffness due to increased cracking as the loads are increased.

With time, the service-load deflection would increase from C to C', due to creep of the concrete. The short-time or *instantaneous deflection* under service loads (point C) and the *long-time deflection* under service loads (point C') are both of interest in design.

Flexural Stiffness and Moment of Inertia

The deflection of a beam is calculated by integrating the curvatures along the length of the beam [9-21]. For an elastic beam, the curvature, $1/r$, is calculated as $1/r = M/EI$, where EI is the flexural stiffness of the cross section. If EI is constant, this is a relatively routine process. For reinforced concrete, however, three different EI values must be considered. These can be illustrated by the moment–curvature diagram for a length of beam, including

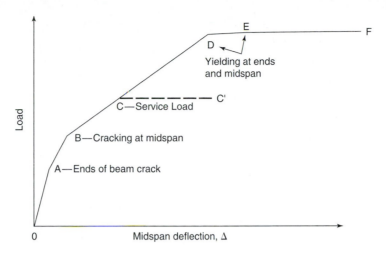

(a) Load deflection diagram.

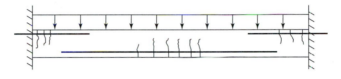

Fig. 9-10
Load–deflection behavior of
a concrete beam.

(b) Beam and loading.

several cracks, shown in Fig. 9-11d. The slope of any radial line through the origin in such a diagram is $M/\phi = EI$.

Before cracking, the entire cross section shown in Fig. 9-11b is stressed by loads. The moment of inertia of this section is called the *uncracked moment of inertia*, and the corresponding EI can be represented by the radial line O–A in Fig. 9-11d. Sometimes the uncracked transformed moment of inertia is applicable. The cross section at a crack is shown in Fig. 9-11c. The *cracked-section EI* is less than the uncracked EI and corresponds relatively well to the curvatures at loads approaching yield, as shown by the radial line O–B in Fig. 9-11d. At service loads, points C_1 and C_2 in Fig. 9-11d, the EI values are between these two extremes. The actual EI at service load levels varies considerably, as shown by the difference in the slope of the lines O–C_1 and O–C_2, depending on the relative magnitudes of the cracking moment M_{cr}, the service load moment M_a, and the yield moment M_y. The variation in EI with moment is shown in Fig. 9-11e, obtained from Fig. 9-11d.

The transition from uncracked to cracked moment of inertia reflects two different phenomena. Figure 9-7b and c show the tensile stresses in the reinforcement and the concrete in a prism. At loads only slightly above the cracking load, a significant fraction of tensile force between cracks is in the concrete, and hence the member behaves more like an uncracked section than a cracked section. As the loads are increased, internal cracking of the type shown in Fig. 9-8 occurs, with the result that the steel strains increase with no significant change in the tensile force in the concrete. At very high loads, the tensile force in the concrete is insignificant compared with that in the steel, and the member approximates a completely *cracked section*. The effect of the tensile forces in the concrete on EI is referred to as *tension stiffening*.

Figure 9-12 shows the distribution of EI along the beam shown in Fig. 9-10b. The EI varies from the uncracked value at points where the moment is less than the cracking moment to a partially cracked value at points of high moment. Since the use of such a distribution of EI values would make the deflection calculations tedious, an overall average or

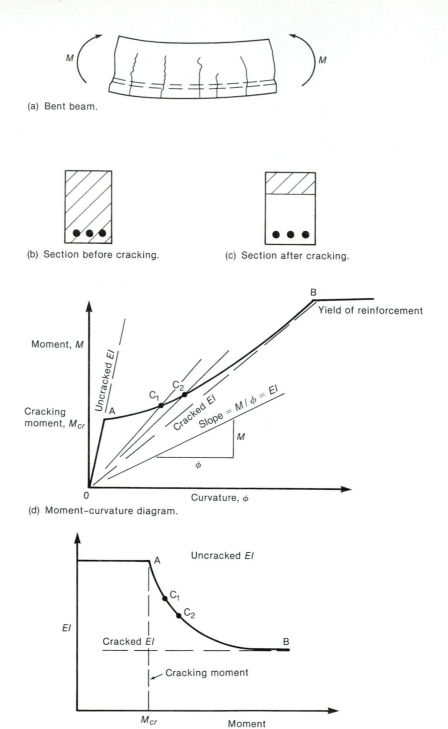

(a) Bent beam.

(b) Section before cracking.

(c) Section after cracking.

(d) Moment–curvature diagram.

(e) Variation in EI with moment.

Fig. 9-11
Moment–curvature diagram
and variation in *EI*.

effective EI value is used. The effective moment of inertia must account for both the tension stiffening and the variation of *EI* along the member.

Effective Moment of Inertia

The slope of line *OA* in Fig. 9-11d is approximately EI_{gt}, referred to as the *gross-transformed moment of inertia,* while that of line *OB* is approximately EI_{cr}, referred to as the *cracked*

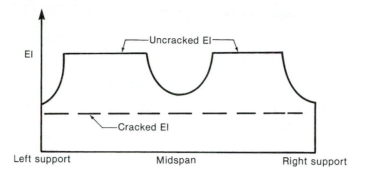

Fig. 9-12
Variation of *EI* along the
length of the beam shown in
Fig. 9-10b.

moment of inertia. At points between cracking (point *A*) and yielding of the steel (point *B*), intermediate values of *EI* exist. Branson [9-22], [9-23] derived the following equation to express the transition from I_{gt} to I_{cr} that is observed in experimental data:

$$I_e = \left(\frac{M_{cr}}{M_a}\right)^a I_{gt} + \left[1 - \left(\frac{M_{cr}}{M_a}\right)^a\right] I_{cr} \tag{9-9}$$

(ACI Eq. 9-8)

In this equation,

$$M_{cr} = \text{cracking moment} = f_r I_g / y_t$$
$$I_g t = \text{moment of inertia of the reinforced concrete section}$$
$$f_r = \text{modulus of rupture} = 7.5\sqrt{f'_c} \qquad \text{(ACI Eq. 9-9)}$$
$$y_t = \text{distance from centroid to extreme tension fiber}$$

The ACI Code defines M_a as the maximum moment in the member at the stage at which deflection is being computed. A better definition would be

M_a = maximum moment in the member at the loading stage for which the moment of inertia is being computed or at any previous loading stage

In some structures, such as two-way slabs, construction loads may exceed service loads. If the construction loads cause cracking, the effective moment of inertia will be reduced.

For a region of constant moment, Branson found the exponent *a* in (9-9) to be 4. This accounted for the tension-stiffening action. For a simply supported beam, Branson suggested that both the tension stiffening and the variation in *EI* along the length of the member could be accounted for by using *a* = 3. For simplicity, the ACI Code equation is written in terms of the moment of inertia of the gross concrete section, I_g, ignoring the small increase in the moment of inertia due to the reinforcement:

$$I_e = \left(\frac{M_{cr}}{M_a}\right)^3 I_g + \left[1 - \left(\frac{M_{cr}}{M_a}\right)^3\right] I_{cr} \tag{9-10a}$$

(ACI Eq. 9-7)

This can be rearranged to

$$I_e = I_{cr} + (I_g - I_{cr})\left(\frac{M_{cr}}{M_a}\right)^3 \tag{9-10b}$$

For a continuous beam, the I_e values may be quite different in the negative- and positive-moment regions. In such a case, the positive-moment value may be assumed to apply between the points of contraflexure and the negative-moment values in the end regions. ACI Section 9.5.2.4 suggests the use of the average I_e value. A better suggestion is given in [9-22]:

Beams with two ends continuous:

$$\text{average } I_e = 0.70 I_{em} + 0.15(I_{e1} + I_{e2}) \tag{9-11a}$$

Beams with one end continuous:

$$\text{average } I_e = 0.85 I_{em} + 0.15(I_{e \text{ continuous end}}) \tag{9-11b}$$

Here, I_{em}, I_{e1}, and I_{e2} are the values of I_e at midspan and the two ends of the beam, respectively. Moment envelopes or moment coefficients should be used in computing M_a and I_e at the positive- and negative-moment sections.

Instantaneous and Additional Sustained Load Deflections

When a concrete beam is loaded, it undergoes a deflection referred to as an *instantaneous deflection*, Δ_i. If the load remains on the beam, additional *sustained-load deflections* occur due to creep. These two types of deflections are discussed separately.

Instantaneous Deflections

Equations for calculating the instantaneous deflections of beams for common cases are summarized in Fig. 9-13. In this figure, M_{pos} and M_{neg} refer to the maximum positive and negative moments, respectively. The deflections listed are the maximum deflections along the beam, except for cases 2, 6, 7, and 8 in Fig. 9-13.

   The midspan deflection for a continuous beam with uniform loads and unequal end moments can be computed with the use of superposition, as shown in Fig. 9-14. Thus,

$$\Delta = \Delta_0 + \Delta_1 + \Delta_2$$

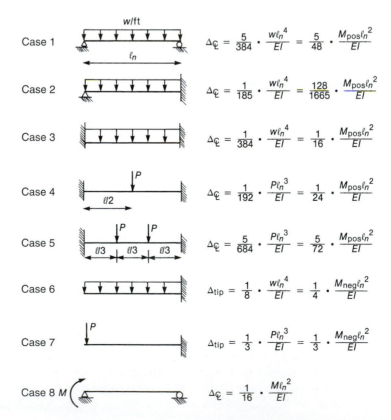

Case 1
$$\Delta_{C\!L} = \frac{5}{384} \cdot \frac{w\ell_n^4}{EI} = \frac{5}{48} \cdot \frac{M_{pos}\ell_n^2}{EI}$$

Case 2
$$\Delta_{C\!L} = \frac{1}{185} \cdot \frac{w\ell_n^4}{EI} = \frac{128}{1665} \cdot \frac{M_{pos}\ell_n^2}{EI}$$

Case 3
$$\Delta_{C\!L} = \frac{1}{384} \cdot \frac{w\ell_n^4}{EI} = \frac{1}{16} \cdot \frac{M_{pos}\ell_n^2}{EI}$$

Case 4
$$\Delta_{C\!L} = \frac{1}{192} \cdot \frac{P\ell_n^3}{EI} = \frac{1}{24} \cdot \frac{M_{pos}\ell_n^2}{EI}$$

Case 5
$$\Delta_{C\!L} = \frac{5}{684} \cdot \frac{P\ell_n^3}{EI} = \frac{5}{72} \cdot \frac{M_{pos}\ell_n^2}{EI}$$

Case 6
$$\Delta_{tip} = \frac{1}{8} \cdot \frac{w\ell_n^4}{EI} = \frac{1}{4} \cdot \frac{M_{neg}\ell_n^2}{EI}$$

Case 7
$$\Delta_{tip} = \frac{1}{3} \cdot \frac{P\ell_n^3}{EI} = \frac{1}{3} \cdot \frac{M_{neg}\ell_n^2}{EI}$$

Case 8
$$\Delta_{C\!L} = \frac{1}{16} \cdot \frac{M\ell_n^2}{EI}$$

Fig. 9-13
Deflection equations.

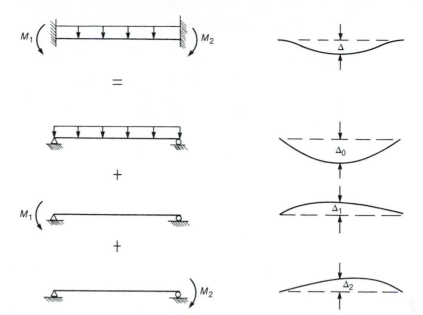

Fig. 9-14
Calculation of deflection for
a beam with unequal end
moments.

From the right-hand column in Fig. 9-14, assuming that M_0 (the moment at midspan due to uniform loads on a simple beam), M_1, and M_2 are all positive, so that the signs will be compatible, we have

$$\Delta = \frac{5}{48}\frac{M_0\ell_n^2}{EI} + \frac{3}{48}\frac{M_1\ell_n^2}{EI} + \frac{3}{48}\frac{M_2\ell_n^2}{EI}$$

The midspan moment M_m is

$$M_m = M_0 + \frac{M_1}{2} + \frac{M_2}{2}$$

Therefore, expressing the deflection in terms of M_m, M_1, and M_2, we obtain

$$\Delta = \frac{\ell_n^2}{48\,EI}\left[5M_m - \frac{5}{2}(M_1 + M_2) + 3(M_1 + M_2)\right]$$

$$= \frac{5}{48}\frac{\ell_n^2}{EI}[M_m + 0.1(M_1 + M_2)] \qquad (9\text{-}12a)$$

where Δ is the midspan deflection, M_m is the midspan moment, and M_1 and M_2 are the moments at the two ends, which must be given the correct algebraic sign. Generally, M_1 and M_2 will be negative. Alternatively, (9-12a) can be written as

$$\Delta = \frac{5\ell_n^2}{48\,EI}(1.2M_m - 0.2M_0) \qquad (9\text{-}12b)$$

The moments M_m, M_1, and M_2 in (9-12) should all result from the same loading. Equation (9-12a) suggests that the loading pattern which produces maximum positive moment will give the largest deflections. In beam design, however, moment envelopes or moment coefficients are often used to compute the critical moments at each section. (See Section 10-3.) These moments represent the worst effects of several different loading cases and, when used in (9-12a) and (9-12b), can give excessively high computed deflections. A prismatic fixed-end beam loaded with a uniform load has moments of $-\frac{1}{12}w\ell^2$ and $\frac{1}{24}w\ell^2$ at the ends and midspan, respectively. For a typical interior span, ACI Section 8.3.3 gives moments of $-\frac{1}{11}$ and $\frac{1}{16}w\ell^2$. Using these in (9-12a) gives 1.77 times the deflection computed with $-\frac{1}{12}$ and $\frac{1}{24}$. Using $\frac{1}{16}$ in (9-12b) gives two times the deflection computed with $-\frac{1}{12}$ and $\frac{1}{24}$. Frequently, the critical deflections are sustained-load deflections due to the dead load. Since dead load is not a pattern load, the appropriate moment coefficients are closer to $-\frac{1}{12}$ and $\frac{1}{24}$ than are the ACI coefficients. In conclusion, the use of the ACI moment coefficients in deflection calculations will generally overestimate the deflections, particularly when (9-12b) is used. If the ACI moment coefficients are used in calculating deflections and the resulting computed deflections are excessive, it may be desirable to recompute the deflections with more realistic moments.

Sustained-Load Deflections

Under sustained loads, concrete undergoes creep strains and the curvature of a cross section increases, as shown in Fig. 9-15. Because the lever arm is reduced, there is a small increase in the steel force, but for normally reinforced sections, this will be minimal. At the same time, the compressive stress in the concrete decreases slightly, because the compression zone is larger.

If compression steel is present, the increased compressive strains will cause an increase in stress in the compression reinforcement, thereby shifting some of the compressive force from the concrete to the compression steel. As a result, the compressive stress in the concrete decreases, resulting in reduced creep strains. The larger the ratio of compression reinforcement, $\rho' = A'_s/bd$, the greater the reduction in creep, as shown in Fig. 5-14. From test data, Branson [9-23] derived (9-13), which gives the ratio, λ, of the *additional* sustained load deflection to the instantaneous deflection (Δ_i). The total instantaneous and sustained load deflection is $(1 + \lambda)\, \Delta_i$, where

$$\lambda = \frac{\xi}{1 + 50\rho'} \tag{9-13}$$

$$\text{(ACI Eq. 9-11)}$$

where $\rho' = A'_s/bd$ at midspan for simple and continuous beams and at the support for cantilever beams and ξ is a factor between 0 and 2, depending on the time period over which sustained load deflections are of interest. Values of ξ are given in ACI Section 9.5.2.5 and Fig. 9-16.

In Section 3-6 of this book, Examples 3-3 and 3-4 illustrate the internal load transfer from compressed concrete to compression reinforcement in columns carrying sustained axial loads. These calculations were carried out by using transformed sections based on an age-adjusted effective modulus for the concrete [9-4], [9-5], [9-6]. Similar load transfer occurs in the compression zones of beams.

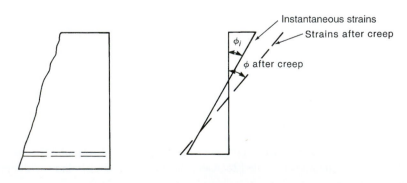

Fig. 9-15
Effect of creep on strains and curvature.

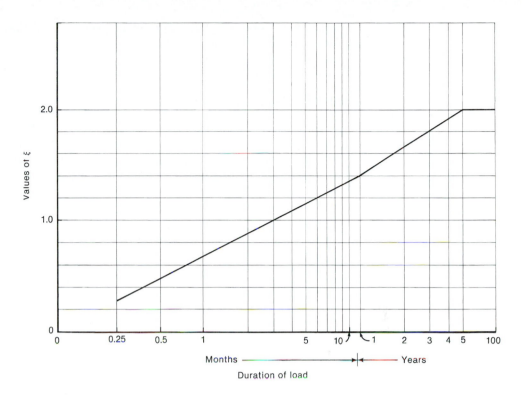

Fig. 9-16
Value of ξ.

Deflection Calculation by Integration of Curvatures

It is frequently more accurate to compute deflections by integrating the curvatures along the beam. This is particularly true for lightly reinforced sections with service-load moments close to the cracking moment. If the curvatures, ϕ, due to loads, creep, and shrinkage are known at the ends and midspan, and if they are assumed to be parabolically distributed along the length of the span, the midspan deflection, Δ, of a simple or continuous beam can be computed from

$$\Delta = \frac{\ell^2}{96}(\phi_1 + 10\phi_2 + \phi_3)$$
(9-14)

where ℓ is the span length, and ϕ_1, ϕ_2, and ϕ_3 are the curvatures at one end, at the midspan, and at the other end, respectively. Guidance in the use of (9-14) is given in [9-21] and [9-6].

9-5   CONSIDERATION OF DEFLECTIONS IN DESIGN

Serviceability Limit States Resulting from Flexural Deflections

Serviceability limit states arising from excessive flexural deflections are discussed here and are summarized in Table 9-2.

Visual Appearance

Deflections greater than about 1/250 of the span ($\ell/250$) are generally visible for simple or continuous beams, depending on a number of factors, such as floor finish [9-24]. Deflections will be most noticeable if it is possible to compare the deflected member with an unde-flected member or a horizontal line. Uneven deflections of the tips of cantilevers can be particularly noticeable. The critical deflection is the total immediate and sustained deflec-tion minus any camber. (See Case B-1 in Table 9-2.)

Damage to Nonstructural Elements

Excessive deflections can cause cracking of partitions or malfunctioning of doors and win-dows. Such problems can be handled by limiting the deflections that occur after installa-tion of the nonstructural elements by delaying their installation, or by designing the nonstructural elements to accommodate the required amount of movement. (See Cases B-2 and B-3 in Table 9-2.)

Surveys of partition damage have shown that damage to brittle partitions can occur with deflections as small as $\ell/1000$. A frequent limit on deflections that cause damage is a deflec-tion of $\ell/480$ occurring after attachment of the nonstructural element (ACI Table 9.5b). This

TABLE 9-2 Loads and Load Combinations for Serviceability Limit States—Flexural Deflections

(1) Case	(2) Serviceability Limit State and Elements Affected	(3) Serviceability Parameter and Load Combination	(4) Calculation Method and Acceptance Criteria	(5) References
B	**Excessive flexural deflections**			
B-1	Unsightly sag of floors, droopy cantilevers	Total sustained load deflections. Deflection visible when building is in service. Camber may hide sag, but large deflections may give nonstructural problems. 1.0D + 1.0H + 1.0 Prestress plus possibly 0.4L.	Deflections calculated with ACI Section 9.5. Deflection less than limit of visually obvious deflections. $\Delta/\ell \approx 1/250$ and less for cantilevers.	ACI Sections 8.2.2, 9.5, 13.5.1. Book Sections 9-4, 9-5. Refs. [9-10] [9-17], [9-22], [9-23], [9-24], [9-26], [9-30]
B-2	Damage to nonstructural components, drywall partitions, block or brick walls, jamming of doors	Deflection after nonstructural components are installed **taken equal to** sustained load deflection after installation of nonstructural components. 1.0D + 1.0H + 1.0 Prestress **plus** short-time deflection after installation of nonstructural components. 0.4L or 0.75T or 0.6S.	Deflections calculated with ACI Section 9.5. Limits depend on materials. $\Delta/\ell \leq 1/480$ ACI Table 9.5(b) Plaster, drywall, and masonry partitions $\Delta/\ell \leq 1/500$ to $1/1000$. Moveable partitions $\Delta/\ell \leq 1/240$.	ACI Sections 9.5, 13.5.1. Book Sections 9-4, 9-5. Refs. [9-10], [9-17], [9-21], [9-24], [9-26], [9-29]
B-3	Malfunction of equipment	Increase in deflection or increase in curvature of main shaft after machinery is aligned.	Deflections calculated with ACI Section 9.5. Limits to be provided by equipment manufacturer.	ACI Sections, 8.2.2, 9.5. Book Sections 9-4, 9-5. Refs. [9-22], [9-27], [9-28]
B-4	Ponding failure	Second-order increase in deflection of a roof or surface holding rain.	Deflections calculated with ACI Section 9.5. Short-time deflection under own weight and weight of collected rain.	ACI Sections R.8.2, R.9.2.5. Book Section 9-5. Refs. [9-2], [9-25], [9-27]

is the sum of the instantaneous deflection due to live loads plus the sustained portion of the deflections due to dead load and any sustained live load; that is,

$$\Delta = (\lambda_{to,\infty} - 1)\, \Delta_{iD} + (\Delta_{iL} - \Delta_{iLS}) + \lambda_\infty \Delta_{iLS} \qquad (9\text{-}15)$$

where Δ_{iD}, Δ_{iL}, and Δ_{iLS} are the instantaneous deflections due to the dead load, the live load, and the sustained portion of the live load, respectively; $\lambda_{to,\infty}$ is the value of λ (9-13) based on the value of ξ for five years or more, minus the value of ξ at the time t_0 when the partitions and so on are installed. (See Fig. 9-16.) Thus, if the nonstructural elements were installed three months after the shores were removed, the value of ξ used in calculating $\lambda_{(t_0, \infty)}$ would be $2.0 - 1.0 = 1.0$. Finally, λ_∞ is the value of λ based on $\xi = 2.0$, since it is assumed here that all the sustained *live-load* deflection affecting nonstructural damage will occur after the partitions are installed.

Racking or shear deflections of a plastered or brick wall may damage walls or cause doors to jam at very small Δ/ℓ ratios. The maximum slope, Δ/ℓ, will occur near the quarter-points of the spans.

As is evident from (9-9), the effective moment of inertia, I_e, decreases as the loads increase, once cracking has occurred. As a result, the value of I_e is larger when only dead load acts than when dead and live loads act together. There are two schools of thought about the value of I_e to be used in the calculation of the Δ_{iD} in (9-15). Some designers use the value of I_e, which is effective when only dead load is supported. Others reason that the portion of the sustained dead-load deflection which is significant is that occurring after the partitions are in place (i.e., that occurring during the live-loading period). This second line of reasoning implies that Δ_{iD} in (9-15) should be based on the value of I_e corresponding to dead plus live loads. This book supports the second procedure, particularly in view of the fact that construction loads or shoring loads may lead to premature cracking of the structure. The solution of (9-15) is illustrated in Example 9-6.

Disruption of Function

Excessive deflections may interfere with the use of the structure. This is particularly true if the members support machinery that must be carefully aligned. The corresponding deflection limits will be prescribed by the manufacturer of the machinery or by the owner of the building.

Excessive deflections may also cause problems with drainage of floors or roofs. The deflection due to the weight of water on a roof will cause additional deflections, allowing it to hold more water. In extreme cases, this may lead to progressively deeper and deeper water, resulting in a *ponding failure* of the roof [9-25]. Ponding failures are most apt to occur with long-span roofs in regions where roof design loads are small. (See Case B-4 in Table 9-2.)

Damage to Structural Elements

The deflections corresponding to structural damage are several times larger than those causing serviceability problems. If a member deflects so much that it comes in contact with other members, the load paths may change, causing cracking. Creep may redistribute the reactions of a continuous beam or frame thereby reducing the effect of differential settlements [9-26].

Allowable Deflections

In principle, the designer should select the deflection limit that applies for a given structure, basing it on the characteristics and function of the structure. In this respect, allowable deflections given in codes are only guides.

The ACI Code gives values of maximum permissible computed deflections in ACI Table 9.5(b). The ACI Deflection Committee has also proposed a more comprehensive set of allowable deflections [9-27].

TABLE 9-3   Loads and Load Combinations for Serviceability Limit States—Racking and Relative Vertical Deflections

(1) Case	(2) Serviceability Limit State and Elements Affected	(3) Serviceability Parameter Load Combination	(4) Calculation Method and Acceptance Criteria	(5) References
C	**Racking deflections**			
C-1	Cracking of partitions, jamming of doors, structural damage	Sustained and short time shear deflections after partitions, door frames, nonstructural elements are installed. 1.0D + 1.0H + 1.0 Prestress **plus** short-time deflection due to 0.4L or 0.75T or 0.6S	Deflections calculated using ACI Sec. 9.5 Limits on racking deflections.	ACI Section 9.5. Book Sections 9-5, 9-6, 13.5.1. Refs. [9-10], [9-24], [9-27], [9-29], [9-32], [9-33]
D	**Relative Vertical Displacements**			
D-1	Crushing and cracking of cladding	Relative vertical displacement of brick or other cladding, and the frame. $1.0T_t + 1.0T_{cs} + 1.0$ Prestress where T_t is thermal expansion, T_{cs} is creep and shrinkage. Crushing strain of cladding.		No ACI Sections. Book Section 9-6 Refs. [9-4], [9-14] [9-32], [9-33]
D-2	Sloping floors	Relative vertical deflection of columns and shear walls due to creep. $1.0D + 1.0T_{cs} + 1.0$ Prestress T_{cs} is creep and shrinkage Limit of visually obvious slope.		No ACI Section. Book Section 9-6. Refs. [9-24], [9-29], [9-32], [9-33]
D-3	Sloping floors	Relative vertical deflection of interior columns and exterior columns exposed to temperature changes. About $\frac{3}{4}$ in. in 25 ft $1.0D + 1.0T_t$ where T_t is thermal expansion.	Limit of visually obvious slope.	No ACI Sections. Book Section 9-6. Refs. [9-24], [9-32], [9-33]
D-4	Damage to structure	Differential settlement About $\frac{3}{4}$ in. in 25 ft	Beam and column structures can tolerate differential settlements of about $\frac{3}{4}$ in. in 25 ft	No ACI Sections. Refs. [9-10], [9-32], [9-33]

Accuracy of Deflection Calculations

The ACI Code deflection-calculation procedure gives results that are within ±20 percent of the true values, provided that M_{cr}/M_a is less than 0.7 or greater than 1.25 [9-28]. Since a small amount of cracking can make a major difference in the value of I_e, as indicated by the big decrease in slope from O–A to O–C in Fig. 9-11d, the accuracy is considerably poorer when the service-load moment is close to the cracking moment. This is because random variations in the tensile strength of the concrete will change M_{cr} and hence I_e.

For these reasons it is frequently satisfactory to use a simplified deflection calculation as a first trial and then refine the calculations if deflections appear to be a problem.

Deflection Control by Span-to-Depth Limits

The relationship between beam depth and deflections is given by

$$\frac{\Delta}{\ell} = C\frac{\ell}{d}$$

(4-31b)

where the term C allows for the support conditions, and the grade, and amount of steel. For a given desired Δ/ℓ, a table of ℓ/d values can be derived. ACI Table 9.5(a) is such a table and gives minimum thicknesses of beams and one-way slabs unless deflections are computed. This table is frequently used to select member depths. It should be noted, however, that this table applies only to *members not supporting, or attached to, partitions or other construction likely to be damaged by large deflections.* If the member supports such partitions, and so on, the ACI Code requires calculation of deflections.

Table A-14 in Appendix A summarizes the ACI deflection limits and presents other more stringent limits for the case of beams supporting brittle partitions [9-22], [9-28], [9-29], [9-30].

EXAMPLE 9-5   Calculation of Immediate Deflection

The T beam shown in Fig. 9-17 supports unfactored dead and live loads of 0.87 kip/ft and 1.2 kips/ft, respectively. It is built of 3000-psi concrete and Grade-60 reinforcement. The moments used in the

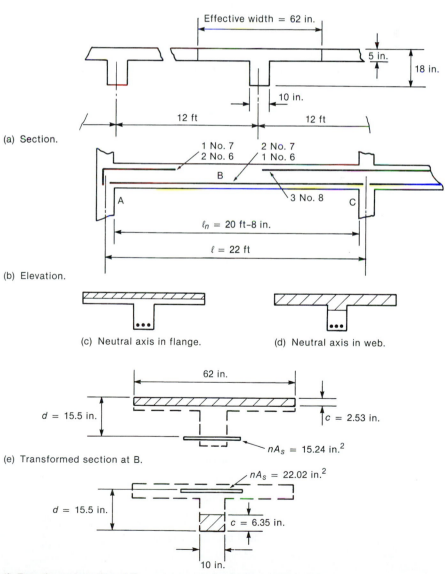

(a) Section.

(b) Elevation.

(c) Neutral axis in flange.   (d) Neutral axis in web.

(e) Transformed section at B.

(f) Transformed section at C.

Fig. 9-17
End span of beam—
Examples 9-5 and 9-6.

design of the beam were calculated from the moment coefficients in ACI Section 8.3.3. Calculate the immediate midspan deflection. Assume that the construction loads did not exceed the dead load.

1. **Is the beam cracked at service loads?** The cracking moment is

$$M_{cr} = \frac{f_r I_g}{y_t}$$

Compute I_g for the uncracked T section (ignore the effect of the reinforcement for simplicity):

flange width = effective flange width from ACI Section 8.10.2 = 62 in.

(a) Compute the centroid of the cross section.

Part	Area (in.^{2})	y_{top} (in.)	Ay_{top} (in.^{3})
Flange	$62 \times 5 = 310$	2.5	775
Web	$13 \times 10 = 130$	11.5	1495
	Area = 440 in.^{2}		$\Sigma Ay_t = 2270$ in.^{3}

The centroid is located at $2270/440 = 5.16$ in. below the top of the beam:

$$\bar{y}_{top} = 5.16 \text{ in.} \qquad \bar{y}_{bottom} = 12.84 \text{ in.}$$

(b) Compute the moment of inertia, I_g.

Part	Area (in.^{2})	$\bar{y}$ (in.)	$I_{own axis}$ (in.^{4})	$A\bar{y}^2$ (in.^{4})
Flange	310	$5.16 - 2.5 = 2.66$	646	2192
Web	130	$5.16 - 11.5 = -6.34$	1831	5227
				$I_y = 9896$ in.^{4}

(c) Find the flexural cracking moment.

$$M_{cr} = \frac{f_r I_g}{y_t}$$

where $f_r = 7.5\sqrt{f_c'} = 411$ psi.

In negative-moment regions at A and C, we have the following:

$$M_{cr} = \frac{411 \text{ psi} \times 9896 \text{ in.}^4}{5.16 \text{ in.}} = 788,200 \text{ in.-lb}$$

$$= -65.7 \text{ ft-kips}$$

$$\text{negative moment at } A = -\frac{w\ell_n^2}{16}$$

$$\text{dead-load moment} = -\frac{0.87 \times 20.67^2}{16} = -23.2 \text{ ft-kips}$$

$$\text{dead-plus-live-load moment} = \frac{(0.87 + 1.20) \times 20.67^2}{16} = -55.3 \text{ ft-kips}$$

Since both of these are less than M_{cr}, section A will not be cracked at service loads.

$$\text{negative moment at } C = \frac{w\ell_n^2}{10}$$

$$\text{dead-load moment} = -\frac{0.86 \times 20.67^2}{10} = -36.7 \text{ ft-kips (not cracked)}$$

$$\text{dead-plus-live-load moment} = -\frac{2.07 \times 20.67^2}{10} = -88.4 \text{ ft-kips (cracked)}$$

In the positive-moment region,

$$M_{cr} = \frac{411 \times 9896}{12.84} = 26.4 \text{ ft-kips}$$

$$\text{positive moment at } B = \frac{w\ell_n^2}{14}$$

$$\text{dead-load moment} = \frac{0.87 \times 20.67^2}{14}$$

$$= 26.6 \text{ ft-kips (just cracked)}$$

$$\text{dead-plus-live-load moment} = \frac{2.07 \times 20.67^2}{14} = 63.2 \text{ ft-kips (cracked)}$$

Therefore, it will be necessary to compute I_{cr} and I_e at midspan and at support C.

2.   Compute I_{cr} at midspan. It is not known whether this section will have a rectangular compression zone (Fig. 9-17c) or a T-shaped compression zone (Fig. 9-17d). We shall assume that the compression zone is rectangular for the first trial.

To locate the centroid, either follow the calculation in Example 9-1, or, because the compression zone is assumed to be rectangular and there is no compression steel, use (9-3) to compute k, where $c = kd$. Either way, we obtain

$$k = \sqrt{2\rho n + (\rho n)^2} - \rho n$$

where

$$n = \frac{E_s}{E_c}$$

$$= \frac{29 \times 10^6}{57,000\sqrt{3000}} = 9.29$$

$$\rho \text{ (based on } b = 62 \text{ in.)} = \frac{1.64}{62 \times 15.5} = 0.00171$$

$$\rho n = 0.0159$$

$$k = \sqrt{2 \times 0.0159 + 0.0159^2} - 0.0159$$

$$= 0.163$$

Therefore,

$$c = kd = 0.163 \times 15.5$$

$$= 2.53 \text{ in.}$$

Since this is less than the thickness of the flange, the compression zone is rectangular, as assumed. If it is not, use the general method, as in Example 9-1.

Compute the cracked moment of inertia at B.   Transform the reinforcement as shown in Fig. 9-17e:

$$nA_s = 9.29 \times 1.64 = 15.24 \text{ in.}^2$$

Part	Area (in.^{2})	$\bar{y}$ (in.)	$I_{own axis}$ (in.^{4})	$A\bar{y}^2$ (in.^{4})
Compression zone	$2.53 \times 62 = 156.86$	2.53/2	84	251
Reinforcement	15.24	$2.53 - 15.5 = -12.97$	—	2564
				$I_{cr} = 2899$ in.^{4}

Therefore, I_{cr} at $B = 2899$ in.^{4}

    3. **Compute I_{cr} at support C.** The cross section at support C is shown in Fig. 9-17f. The transformed area of the reinforcement is

$$nA_s = 9.29 \times 2.37 = 22.02 \text{ in.}^2$$

$$\rho = \frac{2.37}{10 \times 15.5} = 0.0153$$

From (9-3), $k = 0.410$ and

$$c = kd = 6.35 \text{ in.}$$

The positive-moment reinforcement is not developed for compression at the support and will therefore be neglected.

Part	Area (in.^{2})	$\bar{y}$ (in.)	$I_{own axis}$ (in.^{4})	$A\bar{y}^2$ (in.^{4})
Compression zone	$6.35 \times 10 = 63.5$	6.35/2	213	640
Reinforcement	22.02	$6.35 - 15.5$	—	1844
				$I_{cr} = 2697$ in.^{4}

In sum, I_g is 9896 in.^{4}, I_{cr} at B is 2899 in.^{4}, and I_{cr} at C is 2697 in.^{4}

    4. **Compute immediate dead-load deflection.**

    (a) **Compute I_e at A.** Since $M_a = -23.2$ ft-kips is less than $M_{cr} = -65.7$ ft-kips, $I_e = 9896$ in.^{4} at A for dead loads.

    (b) **Compute I_e at B.** $M_a = 26.6$ ft-kips, and $M_{cr} = 26.4$ ft-kips. Therefore, $M_{cr}/M_a = 0.992$, and $(M_{cr}/M_a)^3 = 0.978$, and we have

$$I_e = \left(\frac{M_{cr}}{M_a}\right)^3 I_g + \left[1 - \left(\frac{M_{cr}}{M_a}\right)^3\right] I_{cr} \tag{9-10a}$$

$$= 0.978 \times 9896 + (1 - 0.978) \times 2899$$

$$= 9739 \text{ in.}^4 \text{ at } B \text{ for dead loads}$$

    (c) **Compute I_e at C.** $M_a = -36.7$ ft-kips, and $M_{cr} = -65.7$ ft-kips. Therefore, the section is not cracked, and $I_e = 9896$ in.^{4}

    (d) **Compute the weighted average, I_e.**

$$I_e = 0.7I_m + 0.15(I_{e1} + I_{e2}) \tag{9-11a}$$

$$= 0.7 \times 9739 + 0.15(9896 + 9896)$$

$$= 9790 \text{ in.}^4$$

(e)   **Immediate dead-load deflection from (9-12).**

$$\Delta = \frac{5}{48}\frac{\ell_n^2}{EI}[M_m + 0.1(M_1 + M_2)]$$

where

$$\ell_n = 20.67 \text{ ft}$$
$$E = 57,000\sqrt{3000} = 3.122 \times 10^6 \text{ psi}$$
$$M_m = \text{midspan moment due to dead loads} = 26.6 \text{ ft-kips}$$
$$M_1 = \text{moment at } A \text{ due to dead loads} = -23.2 \text{ ft-kips}$$
$$M_2 = \text{moment at } C \text{ due to dead loads} = -36.7 \text{ ft-kips}$$

$$\Delta = \frac{5}{48} \times \frac{(20.67 \times 12)^2[26.6 + 0.1(-23.2 - 36.2)] \times 12,000}{3.122 \times 10^6 \times 9790}$$

$$= 0.052 \text{ in.}$$

Therefore, the immediate dead-load deflection is 0.052 in.

5.   **Compute the immediate live-load deflection.**

When the live load is applied to the beam, the moments will increase, leading to more cracking. As a result, I_e decreases. The deflection that occurs when the live load is applied is

$$\Delta_{iL} = \Delta_{i,L+D} - \Delta_{iD}$$

(a)   **Compute I_e at A.** Since $M_a = -55.3$ ft-kips is less than $M_{cr} = -65.7$ ft-kips, the section is uncracked, and $I_e = 9896$ in.[4] at A for dead plus live loads.

(b)   **Compute I_e at B.** $M_a = 63.2$ ft-kips, and $M_{cr} = 26.4$ ft-kips; therefore, $M_{cr}/M_a = 0.418$, and $(M_{cr}/M_a)^3 = 0.0729$, and we have

$$I_e = 0.0729 \times 9896 + (1 - 0.0729)\,2899 = 3409 \text{ in.}^4$$

(c)   **Compute I_e at C.** $M_a = -88.4$ ft-kips, and $M_{cr} = -65.7$ ft-kips; therefore, $M_{cr}/M_a = 0.743$, and $(M_{cr}/M_a)^3 = 0.411$, so

$$I_e = 0.411 \times 9896 + (1 - 0.411)\,2697 = 5652 \text{ in.}^4$$

(d)   **Compute the weighted average, I_e.**

$$I_e = 0.7 \times 3409 + 0.15\,(9896 + 5652) = 4720 \text{ in.}^4$$

(e)   **Compute the immediate dead-plus-live-load deflection.**

$$\Delta_{i,D+L} = \frac{5}{48}\frac{\ell_n^2}{EI}[M_m + 0.1(M_1 + M_2)]$$

$$= \frac{5(20.67 \times 12)^2[63.2 + 0.1(-55.3 - 88.4) \times 12,000]}{48 \times 3.122 \times 10^6 \times 4720}$$

$$= 0.255 \text{ in.}$$

Therefore, the immediate dead-plus-live-load deflection is 0.255 in. The deflection increased by $0.255 - 0.052 = 0.203$ in. when the live load was applied to the beam.

When the live load is removed, the beam will not return to its dead-load deflection of 0.052 in. because it is now cracked, with $I_e = 4720$ in.[4], rather than the $I_e = 9790$ in.[4] that governed the dead-load deflections. The new dead-load deflection will be $(M_D/M_{D+L}) \times 0.255$ in. $= 0.107$ in. When the live load is reapplied, the deflection will increase from 0.107 in. to 0.255 in., a gain of 0.148 in.

In the deflection calculations, M_m, M_1, and M_2 were based on the moment coefficients from ACI Section 8.3.3. These give a total moment of $M = w\ell^2/6.6$ rather than $w\ell^2/8$. As a result, the deflections tend to be overestimated. The exact amount depends on which loading case governed. ∎

ACI Table 9.5(b) limits the computed immediate live-load deflections to $\ell/180$ and $\ell/360$, respectively, for flat roofs and floors that either do not support or are not attached to elements likely to be damaged by large deflections. Conservatively, this should be taken as the initial value of 0.203 in. For this beam, $\ell/360$ is $(22 \times 12 \text{ in})/360 = 0.733$ in., where ℓ is the span from center to center of supports.

If this beam does not support partitions that can be damaged by deflections, ACI Table 9.5(a) would have indicated that it was not necessary to compute deflections, since the overall depth of the beam ($h = 18$ in.) exceeds the minimum value of $\ell/21 = (22 \times 12)/21 = 12.6$ in. given in the table.

If the structure just barely passes the allowable deflection calculations, consideration should be given to means of reducing the deflections, because deflection calculations are of dubious accuracy. The easiest ways to do this are to add compression steel or increase the depth of the section.

EXAMPLE 9-6 Calculation of Deflections Occurring after Attachment of Nonstructural Elements

The beam considered in Example 9-5 will be assumed to support partitions that would be damaged by excessive deflections. Again, it supports a dead load of 0.87 kip/ft and a live load of 1.2 kips/ft, 25 percent of which is sustained. The partitions are installed at least 3 months after the shoring is removed. Will the computed deflections exceed the allowable in the end span?

The controlling deflection is (9-15)

$$\Delta = (\lambda_{to,\infty} - 1)\, \Delta_{iD} + (\Delta_{iL} - \Delta_{iLS}) + \lambda_\infty\, \Delta_{iLS} \qquad (9\text{-}15)$$

In Example 9-5, the following quantities were calculated:

$$\Delta_{iD+L} = 0.255 \text{ in.}$$

$$M_D \text{ (at midspan)} = 26.6 \text{ ft-kips}$$

$$M_{D+L} \text{ (at midspan)} = 63.2 \text{ ft-kips}$$

Therefore, $M_L = 63.2 - 26.6$ ft-kips $= 36.6$ ft-kips.

The immediate live-load deflection is

$$\Delta_{iL} = \Delta_{iD+L} \times \frac{M_L}{M_{D+L}}$$

$$= 0.255 \times \frac{36.6}{63.2} = 0.148 \text{ in.}$$

Twenty-five percent of this results from the sustained live loads, $\Delta_{iLS} = 0.037$ in.

The immediate dead-load deflection based on I_e, which is effective when both dead and live loads are acting, is

$$\Delta_{iD} = \Delta_{iD+L} \times \frac{M_D}{M_{D+L}}$$

$$= 0.255 \times \frac{26.6}{63.2} = 0.107 \text{ in.}$$

The creep multipliers (9-13) are given by

$$\lambda = \frac{\xi}{1 + 50\rho'}$$

The beam being checked does not have compression reinforcement; hence, $\rho' = 0$. If it had such reinforcement, the value of ρ' in (9-13) would be the value at midspan. Note that the bottom steel entering the supports at A and B is not developed for compression in the support and hence does not qualify as compression steel in this calculation, which proceeds as follows:

$$\lambda_\infty = \frac{2.0}{1 + 0} = 2.0$$

$$\lambda_{(t_0, \infty)} = \frac{2.0 - 1.0}{1 + 0} = 1.0$$

The deflection occurring after the partitions are in place is

$$\Delta = 0.148 + 1.0 \times 0.107 + 2.0 \times 0.037 = 0.329 \text{ in.}$$

The maximum permissible computed deflection is $\ell/480$ (ACI Table 9.5(b)), where ℓ is the span from center to center of the supports:

$$\frac{22 \times 12 \text{ in.}}{480} = 0.55 \text{ in.}$$

Therefore, the deflections of this floor are satisfactory.   ■

9-6   FRAME DEFLECTIONS

Lateral Deflections

Section 9-4 dealt with vertical flexural deflections of floors and roofs. Several other types of deflection must be considered in the design of concrete-frame structures. The most obvious of these are lateral deflections of the frame itself. These must be controlled to prevent discomfort to occupants, damage to partitions, and so on in high frames, and because lateral deflections lead to second-order, or $P-\Delta$, effects inversely proportional to the lateral stiffness. Guidance in the selection of acceptable lateral-sway indices, Δ/H, is given in [9-31].

Member Stiffnesses

Lateral deflections are important at (a) service loads, and (b) at factoral load levels. Figures 9-11e and 9-12 show a significant reduction in EI as the members of a frame crack. ACI Section 10.11.1 specifies values of E and I for use in computing first- and second-order lateral deflections of the frame at factored load levels and PA analyses. These correspond to cases C_1 or C_2 in Figs. 9-11d and e. At service load levels the cracking will be less pronounced, between A and C_1 in Fig 9-11. ACI Section R10.11.1 suggests using $1/0.70 = 1.43$ times the EI values from ACI Section 10.11.1 for service load calculations.

Vertical Deflections

If the exterior frames of a building are exposed to atmospheric temperature changes and the interior is maintained at a constant temperature, the thermal contraction and expansion of the exterior gives rise to relative vertical deflections of interior and exterior columns. This problem is discussed in [9-32].

In tall concrete buildings with columns and shear walls, the columns will generally be more highly stressed than the walls and will shorten more, due to elastic shortening and creep. The relative shortening may exceed 1 in. after several years. As a result, the floors may slope from the walls down to the columns. Procedures for estimating and dealing with these deflections are discussed in [9-33].

Concrete buildings with brick exteriors occasionally encounter difficulties due to the incompatible long-term deformations of concrete and brick [9-34]. With time, the concrete frame shortens from creep, while the bricks expand with time due to an increase in moisture content above their kiln-dried condition. This, plus thermal deformations of the brick, leads to crushing and cracking of the brick if the joints in the brickwork close so that the bricks begin to carry load. The expansion of bricks due to moisture gain is on the order of 0.0002 in./in., while the shortening of a concrete column due to creep strains may be on the order of 0.0005 to 0.0010 in./in., giving a total relative strain between the brick and the concrete of as much as 0.0012 in./in. In a 12-ft-high story, this amounts to a shortening of the frame by 0.17 in. relative to the brick cladding. This and temperature effects must be considered in designing cladding.

9-7 VIBRATIONS

Reinforced concrete buildings are not, as a rule, subject to vibration problems. This robustness is due to their mass and stiffness. Occasionally, long-span floors or assembly structures will be susceptible to vibrations induced by persons walking, dancing, or exercising. These activities induce vibrations of approximately 2 to 4 cycles per second (hertz). If a floor system or supporting structure has a natural frequency of less than 5 cycles per second, its vibrational properties should be examined more closely, to ensure that floor vibrations will not be a problem. Guidance on this problem may be found in [9-34], [9-35].

Fortunately, it is relatively easy to estimate the natural frequency, f, of a floor if its deflection can be calculated. We obtain

$$f = 0.18\sqrt{\frac{g}{\Delta_{is}}}$$

$$= 3.5\sqrt{\frac{1}{\Delta_{is}(\text{in.})}} \qquad \text{cycles per second} \qquad (9\text{-}16)$$

TABLE 9-4   Loads and Load Combinations for Serviceability—Lateral Deflection, Vibration and Fatigue-Limit States

(1) Case	(2) Serviceability Limit State and Elements Affected	(3) Serviceability Parameter Load Combination	(4) Calculation Method and Acceptance Criteria	(5) References
D	**Lateral deflection of frames**			
D-1	Wind	Mean downwind deflection plus cyclic deflections relative to mean.	Human perception of vibration	No ACI Sections. Book Sections 9-6, 9-7. Refs. [9-10] [9-31], [9-34], [9-35]
F	**Floor vibrations** Discomfort		Human perception of vibration	No ACI Sections. Book Section 9-7. Refs. [9-24], [9-34], [9-35]
G	**Sway vibrations**	Wind-induced vibrations 1.0D + 1.0 (1-in- 2-year to 1-in-10-year wind)	Human perception of vibration	No ACI Sections, Book Section 9-7. Refs. [9-21], [9-25], [9-31], [9-34], [9-35]
H	**Fatigue**	Fracture due to large numbers of cycles and stress range. Number of cycles >20,000 Stress range: Compression or zero stress to tension stress		No ACI Sections. Book Sections 3-9, 9-8. Refs. [9-36], [9-37].

where g is the gravitational constant, and Δ_{is} is the immediate static deflection at the center of the floor due to the loads expected to be on the floor when it is vibrating [9-34]. These would include the dead load and that portion of the live load on the floor when vibrating. The factor 3.5 in (9-15) actually varies from 3.1 to 4.0 for different types of floors (simple beams, fixed slabs, etc.), but the resulting error in f is small compared to inaccuracies in the deflection calculations themselves.

For continuous beams, the first vibration mode involves alternate spans deflecting up and down at the same time. The resulting modal deflections will be larger than those due to downward loads on each span. To use (9-15) in such a case, the deflections from a special frame analysis with the dead load plus some fraction of the live load applied upward and downward on alternate spans are needed.

9-8   FATIGUE

Structures subjected to cyclic loads may fail in fatigue. This type of failure requires (1) generally in excess of 1 million load cycles, and (2) a change of reinforcement stress in each cycle in excess of about 20 ksi. Since dead-load stresses account for a significant portion of the service-load stresses in most concrete structures, the latter case is infrequent. As a result, fatigue failures of reinforced concrete structures are rare. An overview of the fatigue strength of reinforced concrete structures is given in [9-36]. (See also Section 3-13.)

Fatigue failure of the concrete itself occurs via a progressive growth of microcracking. The fatigue strength of plain concrete in compression or tension for a life of 10 million cycles is roughly 55 percent of the static strength. It is affected by the minimum stress during the cycle and by the stress gradient. Concrete loaded in flexural compression has a 15 to 20 percent higher fatigue strength than axially loaded concrete, because the development of microcracking in highly strained areas is inhibited by the restraint provided by less strained areas. The fatigue strength of concrete is not sensitive to stress concentrations.

The ACI fatigue committee [9-36] recommends that the compressive stress range, f_{cr}, should not exceed

$$f_{cr} = 0.4f'_c + 0.47f_{min} \tag{9-17}$$

where f_{min} is the minimum compressive stress in the cycle (positive in compression). The ACI bridge committee [9-37] limits the compressive stress at service load to $0.5f'_c$.

The fatigue strength of deformed reinforcing bars is affected by the stress range and the minimum stress in the stress cycle, but is almost independent of the yield strength of the bar.

In tests, bends with the minimum radius of bend permitted by the ACI Code reduced the fatigue strength in the region of the bend by a half, while tack welding of stirrups to longitudinal bars reduced the strength of the latter by more than a third [9-37]. The fatigue strength of bars is also affected by the radius at the bottom of the deformation lugs.

For bars of normal lug geometry, the ACI bridge committee [9-37] recommends that the stress range not exceed

$$f_{sr} = 23.4 - 0.33f_{min} \qquad \text{ksi} \tag{9-18}$$

where f_{sr} is the algebraic difference between the maximum and minimum stresses at service loads and f_{min} is the minimum stress at service loads. In both cases, tension is positive. At bends or near tack welds, the allowable f_{sr} should be taken as half the value given by (9-18).

The fatigue strength of stirrups is less than that of longitudinal bars [9-37], due in part to a slight kinking of the stirrup where it crosses an inclined crack and, possibly, in part to an increase in the stirrup stress resulting from a reduction in the shear carried by the concrete in cyclically loaded beams. No firm design recommendations are currently available.

A reasonable design procedure would be to limit the stirrup stress range to 0.75 times f_{sr} from (9-18), where the stirrup stress range would be calculated on the basis of the ACI shear design procedures, but using $V_c/2$.

The effect of cyclic loading on bond strength is approximately the same as that on the compressive strength of concrete [9-38]. Hence, a relationship similar to (9-17) would adequately estimate the allowable bond-stress range. During the initial stages of the cyclic load, the loaded end of the bar will slip relative to the concrete, as a result of internal cracking around the bar. The slip ceases once the cracking has stabilized.

PROBLEMS

9-1 Explain the differences in appearance of flexural cracks, shear cracks, and torsional cracks.

9-2 Why is it necessary to limit the widths of cracks?

9-3 Does the beam shown in Fig. P9-3 satisfy the ACI Code crack control provisions for interior exposure? $f_y = 60$ ksi.

9-4 Compute the maximum spacing of No. 5 bars in a one-way slab with 1 in. of clear cover that will satisfy the ACI Code crack-control provisions for exterior exposure. $f_y = 60$ ksi.

9-5 and 9-6 For the cross sections shown in Figs. P9-5 and P9-6, compute

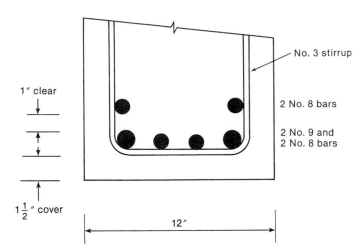

Fig. P9-3

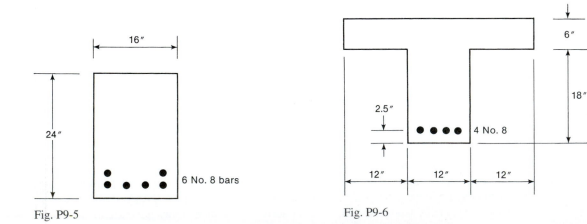

Fig. P9-5

Fig. P9-6

(a) the gross moment of inertia, I_g;

(b) the location of the neutral axis of the cracked section and I_{cr}; and

(c) I_{eff} for $M_a = 0.6\phi M_n$.

The beams have 1.5 in. of clear cover and No. 3 stirrups. The concrete strength is 3000 psi.

9-7   Why are deflections limited in design?

9-8   A simply supported beam with the cross section shown in Fig. P9-5 has a span of 25 ft and supports an unfactored dead load of 1.5 kips/ft, including its own self-weight plus an unfactored live load of 1.5 kips/ft. The concrete strength is 3000 psi. Compute

(a) the immediate dead load deflection.

(b) the immediate dead-plus-live-load deflection.

(c) the deflection occurring after partitions are installed. Assume that the partitions are installed

2 months after the shoring for the beam is removed and assume that 20 percent of the live load is sustained.

9-9   Repeat Problem 9-8 for a beam having the same dimensions and tension reinforcement, but with two No. 8 bars as compression reinforcement.

9-10 The beam shown in Fig. P9-10 is made of 4000-psi concrete and supports unfactored dead and live loads of 1 kip/ft and 1.1 kips/ft. Compute

(a) the immediate dead load deflection.

(b) the immediate dead-plus-live-load deflection.

(c) the deflection occurring after partitions are installed. Assume that the partitions are installed 4 months after the shoring is removed and assume that 10 percent of the live load is sustained.

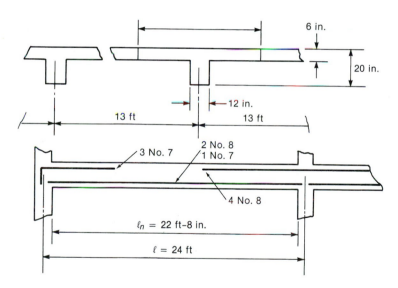

Fig. P9-10

10

Continuous Beams and One-Way Slabs

10-1   INTRODUCTION

In the design of a continuous beam or slab, it is necessary to consider several ultimate limit states—failure by flexure, shear, bond, and possibly torsion—and several serviceability limit states—excessive deflections, crack widths, and vibrations. Each of these topics has been covered in a preceding chapter. In this chapter, the distribution of moments in continuous beams is discussed, followed by a series of design examples showing the overall design of continuous one-way slabs and beams.

10-2   CONTINUITY IN REINFORCED CONCRETE STRUCTURES

The construction of two cast-in-place reinforced concrete structures is illustrated in Figs. 10-1 to 10-3. In Fig. 10-1, formwork is under construction for the beams and slabs supporting a floor. The raised platform-like areas are the bottom forms for the slabs. The rectangular openings between them are forms for the beam concrete. The beams meet at right angles at the columns.

Once the formwork is complete and the reinforcement has been placed in the forms (Fig. 10-2), the concrete for the slabs and beams will be placed in one monolithic pour. (See ACI Section 6.4.6.) Following this, the columns for the next story are erected, as shown in Fig. 10-3. ACI Section 6.4.5 requires that the concrete in the columns or walls have set before the concrete in the floor supported by those columns is placed. This sequence is required because the column concrete will tend to settle in the forms while in the plastic state. If the floor concrete had been placed, this would leave a gap between the column concrete and the beam. By placing the floor concrete after the column concrete is no longer plastic, any gap that formed as the concrete settled will be filled. The resulting construction joints can be clearly seen at the bottom and top of each column in Fig. 10-3.

As a result of this placing sequence, each floor acts as a continuous unit. Because the column reinforcement extends through the floor, the columns act with the floors to form a continuous frame.

Fig. 10-1
Formwork for a beam-and-slab floor. (Photograph courtesy of J. G. MacGregor.)

Fig. 10-2
Beam and slab reinforcement in the forms. (Photograph courtesy of J. G. MacGregor.)

Fig. 10-3
Construction of columns in a tall building. (Photograph courtesy of J. G. MacGregor.)

Braced and Unbraced Frames

A frame is said to be a *sway* frame (or an *unbraced* frame) if it relies on frame action to resist lateral loads or lateral deformations. Thus, the frame in Fig. 10-4a is unbraced, while that in Fig. 10-4b is braced by the shear wall. If the lateral stiffness of the bracing element in a story exceeds 6 to 10 times the sum of the lateral stiffnesses of all the columns in that story, a story can be considered to be braced. ACI Section 10.11.4 gives a procedure for classifying stories in a frame as sway or non-sway stories. Most concrete buildings are braced by walls, elevator shafts, or stairwells. The floor members in an unbraced frame must resist moments induced by lateral loads as well as gravity loads; in most cases, by contrast, the lateral-load moments can be ignored in the design of beams in a braced frame. The design examples in this chapter are limited to beams in braced frames.

One-Way Slab and Beam Floors

Figure 10-5 illustrates the simplest type of one-way slab and beam floor. Here the load P applied at A is carried by the shaded strip of slab to the beams at B and C. The beams, in turn, carry the load to the columns at D, E, F, and G. Frequently, such a floor will have beams in two directions, as shown by the plan layouts in Fig. 10-6. Here, the load from the slab is transmitted to the beams, which transfer the load to *girders,* which in turn are supported by the columns. (A girder is a main supporting beam, generally one that supports other beams.) In the design of such a floor, the slabs are designed first, then the beams, and finally the girders. In this way, the dead load of the slab is known when the beam is designed, and so on.

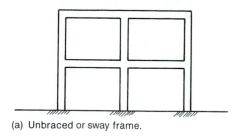

(a) Unbraced or sway frame.

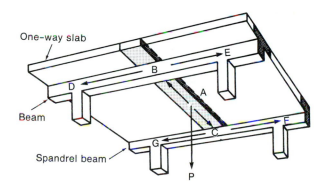

Shear wall, stair well
or elevator shaft

Fig. 10-4
Sway frames and braced
frames.

(b) Braced or non-sway frame.

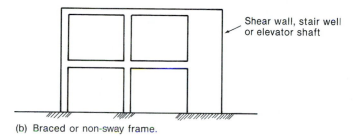

One-way slab

Beam

Fig. 10-5
One-way slab-and-beam
floor.

Spandrel beam

Slabs without beams are referred to as *flat slabs* or *flat plates*. Slabs with beams on an approximately square grid with columns at each corner of each square are referred to as *two-way beam and slab system*. Flat slabs and two-way slabs are discussed more fully in Chapters 13 to 15.

Occasionally, the beams are made wide and shallow. Such a structure is referred to as a *slab-band* floor. These are designed as one-way or two-way slab systems, depending on the geometry of the slab panels.

10-3   MOMENTS AND SHEARS IN CONTINUOUS BEAMS

Continuous slabs, beams, and frames are statically indeterminate structures. Three families of procedures are available for computing moments and shears in such members:

1. elastic analyses, such as slope-deflection, moment-distribution, and matrix methods;

2. plastic analyses; and

3. approximate analyses, such as the use of moment coefficients, or such procedures as the portal and cantilever methods of frame analysis.

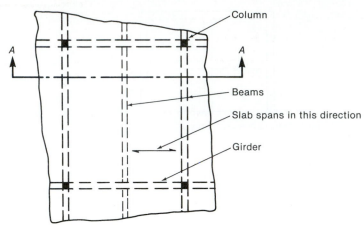

(a) Plan—beams at middle
    of panel.

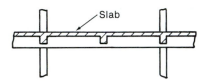

(b) Section A–A.

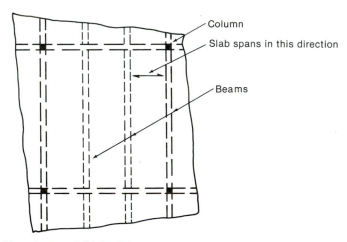

Fig. 10-6
One-way slab-and-beam
floors.

(c) Plan—beams at third points
    of panel.

The elastic analysis of reinforced concrete structures is discussed in textbooks on structural analysis. The design examples carried out in this chapter are based on the approximate moment coefficients presented in the ACI Code. The validity of these coefficients is discussed in the following paragraphs.

Pattern Loadings

Influence Lines

The largest moments in a continuous beam or a frame occur when some spans are loaded and others are not. Diagrams, referred to as *influence lines*, are used to determine which spans should and should not be loaded. An influence line is a graph of the variation in the moment, shear, or other effect, at *one particular point* in a structure, due to a unit load that moves across the beam.

Figure 10-7a is an influence line for moment at point *C* in the two-span beam shown in Fig. 10-7b. The horizontal axis refers to the *position* of a unit (1 kip) load on the beam, and the vertical ordinates are the *moment at C* due to a 1-kip load at the point in question. The derivation of the ordinates at *B, C,* and *E* is illustrated in Fig. 10-7c to e. When a unit load acts at *B*, it causes a moment of 1.93 ft-kips at *C* (Fig. 10-7c). Thus, the ordinate at *B* in Fig. 10-7a is 1.93 ft-kips. Figure 10-7d and e show that the moments at *C* due to loads at *C* and *E* are 4.06 and −0.90 ft-kip, respectively. These are the ordinates at *C* and *E* in Fig. 10-7a and are referred to as *influence ordinates.* If a concentrated load of *P* kips acted at point *E*, the moment at *C* would be *P* times the influence ordinate at *E*, or $M = -0.90P$. If a uniform load of *w* acted on the span *A–D*, the moment at *C* would be *w* times the area of the influence diagram from *A* to *D*.

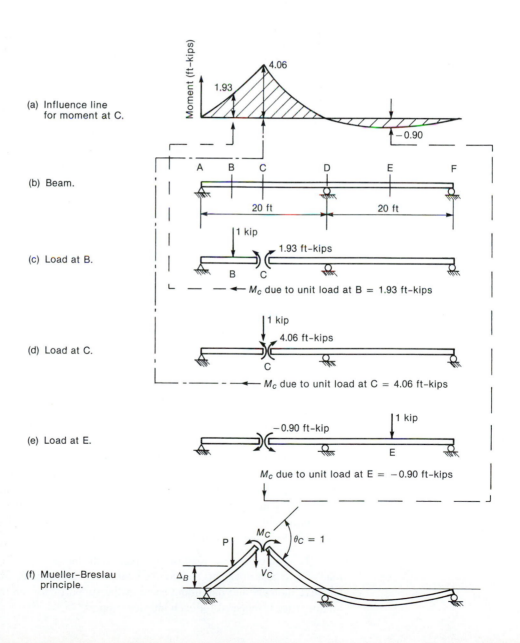

(a) Influence line for moment at C.

(b) Beam.

(c) Load at B.

(d) Load at C.

(e) Load at E.

(f) Mueller–Breslau principle.

Fig. 10-7
Concept of influence lines.

Figure 10-7a shows that a load placed anywhere between A and D will cause positive moment at point C, whereas a load placed anywhere between D and F will cause a negative moment at C. Thus, to get the maximum positive moment at C, we must load span AD only.

Two principal methods are used to calculate influence lines. In the first, a 1-kip load is placed successively at evenly spaced points across the span and the moment, or shear, at the point for which the influence line is being drawn is calculated, as was done in Fig. 10-7c to e. The second procedure, known as the Mueller–Breslau principle, is based on the principle of virtual work, which states that the total work done during a virtual displacement of a structure is zero if the structure is in equilibrium. The use of the Mueller–Breslau principle to compute an influence line for moment at C is illustrated in Fig. 10-7f. The beam is broken at point C and displaced, so that M_c does work by acting through an angle change θ_c. Note that there was no shearing displacement at C, so V_c did not do work. The load, P, at B was displaced upward by an amount Δ_B and hence did negative work. The total work done during this imaginary displacement was

$$M_c\theta_c - P\Delta_B = 0$$

so

$$M_c = P\left(\frac{\Delta_B}{\theta_c}\right) \tag{10-1}$$

where Δ_B/θ_c is the influence ordinate at B. Thus, the deflected shape of the structure for such a displacement has the same shape as the influence line for moment at C. (See Fig. 10-7a and f.)

The Mueller–Breslau principle is presented here as a qualitative guide to the shape of the influence line to determine where to load a structure to cause maximum moments or shears at various points. The ability to determine the critical loading patterns rapidly by using sketches of influence lines expedites the structural analysis considerably, even for structures that will be analyzed via computer software packages.

Pattern Loads from Influence Lines

Figure 10-8 illustrates influence lines drawn in accordance with the Mueller–Breslau principle. Figure 10-8a shows the influence line for moment at B. The loading pattern that will give the largest positive moment at B consists of loads on all spans having positive influence ordinates. Such a loading is shown in Fig. 10-8b and is referred to as an *alternate span* loading or a *checkerboard* loading.

The influence line for moment at the support C is found by breaking the structure at C and allowing a positive moment, M_c, to act through an angle change θ_c. The resulting deflected shape, as shown in Fig. 10-8c, is the influence line for M_c. The maximum *negative* moment at C will result from loading all spans having negative influence ordinates as shown in Fig. 10-8d. This is referred to as an *adjacent span* loading.

Influence lines for shear can be drawn by breaking the structure at the point in question and allowing the shear at that point to act through a unit shearing displacement, Δ, as shown in Fig. 10-9. During this displacement, the parts of the beam on the two sides of the break must remain parallel so that the moment at the section does not do work. The loadings required to cause maximum positive shear at sections A and B in Fig. 10-9 are shown in Fig. 10-9b and d.

Using this sort of reasoning, ACI Section 8.9.2 requires continuous beams and one-way slabs to be designed for the following loading patterns:

1. Factored dead load on all spans with factored partition load and factored live load on two adjacent spans and no live load on any other spans. This will give the maximum negative moment and maximum shear at the support between the two loaded spans. This loading case is repeated for each interior support.

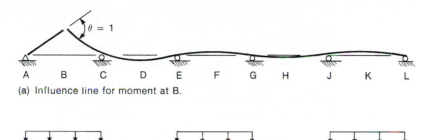

(a) Influence line for moment at B.

(b) Loading for maximum positive moment at B.

(c) Influence line for moment at C.

Fig. 10-8
Influence lines for moment
and loading patterns.

(d) Loading for maximum negative moment at C.

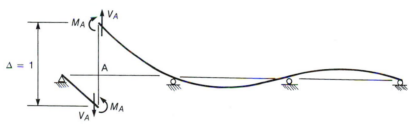

(a) Influence line for shear at A.

(b) Loading for maximum positive shear at A.

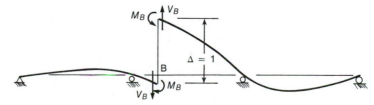

(c) Influence line for shear at B.

Fig. 10-9
Influence lines for shear.

(d) Loading for maximum positive shear at B.

2. Factored dead load on all spans with factored partition load and factored live load on alternate spans. This will give the maximum positive moments at the middle of the loaded spans, minimum positive moments (which could even be negative) at the middle of the unloaded spans, and maximum negative moment at the exterior support.

3. Factored dead and live load on all spans. Although this condition represents the maximum vertical loading possible, it is unlikely to cause the maximum reactions, shear forces, or bending moments for continuous beams.

Following this procedure for the three-bay frame shown in Fig. 10-10 gives the three loading cases and three moment diagrams shown. A fourth loading case is obtained by loading the middle and right spans. If the factored live load is larger than the factored dead load, the negative live-load moments at the middle of the unloaded spans may be larger

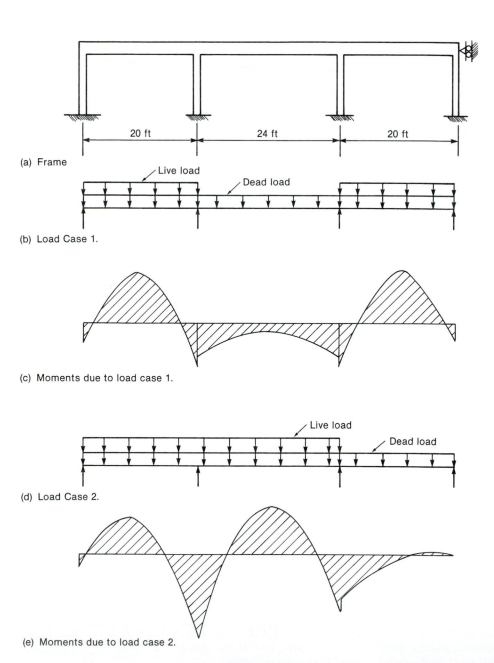

(a) Frame

(b) Load Case 1.

(c) Moments due to load case 1.

(d) Load Case 2.

Fig. 10-10
Moments in a three-span
beam.

(e) Moments due to load case 2.

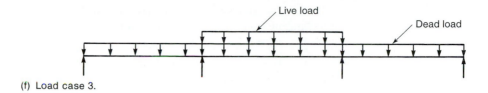

(f) Load case 3.

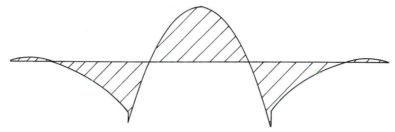

(g) Moments due to load case 3.

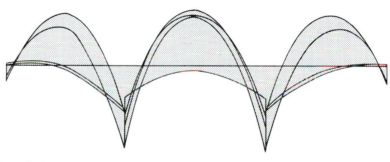

Fig. 10-10
(Continued)

(h) Moment envelope.

than the positive dead-load moments at these points, giving a total moment that is negative, as shown for the center span in Fig. 10-10c.

When the four moment diagrams are superimposed, the *moment envelope* shown in Fig. 10-10h results. At every section, the beam should be designed to have positive and negative moment capacity, ϕM_n, at least equal to the envelope values. A shear-force envelope was used in Example 6-1.

ACI Moment and Shear Coefficients

Because the calculations necessary to derive the moment envelope in Fig. 10-10h are tedious, ACI Section 8.3.3 gives approximate moment coefficients that can be used to calculate the moment and shear envelopes for nonprestressed continuous beams or one-way slabs that meet the following criteria:

1. There are two or more spans.

2. The spans are approximately equal, with the longer of two adjacent spans not more than 1.2 times the length of the shorter one.

3. The loads are uniformly distributed.

4.   The unit live load does not exceed three times the unit dead load. The word "unit" means the unfactored load per foot.

5.   The beams must be prismatic.

6.   The beam must be in a braced frame without significant moments due to lateral loads. (This requirement is not stated in the ACI Code, but it is necessary for the coefficients to apply.)

The maximum positive and negative moments and shears are computed from the expressions

$$M_u = C_m(w_u \ell_n^2) \tag{10-2}$$

$$V_u = C_v \left(\frac{w_u \ell_n}{2} \right) \tag{10-3}$$

where w_u is the total factored dead and live load per unit length, C_m and C_v are moment and shear coefficients from Fig. 10-11 (ACI Section 8.3.3), and ℓ_n is defined as follows:

1.   For negative moment at the interior face of the exterior support, for positive moment, and for shear, ℓ_n = clear span of the span in question.

2.   For negative moment at interior supports, ℓ_n = average of the clear spans of the adjacent spans.

The terminology used to identify the critical sections is illustrated in Fig. 10-11a.

For the interior span of the structure shown in Fig. 10-10, the moment envelope plotted in Fig. 10-10h has a maximum negative moment of $(1/11.1)w_u \ell_n^2$ and a maximum positive moment of $(1/17.1)w_u \ell_n^2$. The corresponding ACI moment coefficients are $\frac{1}{11}$ and $\frac{1}{16}$. These total 123 percent of $w\ell^2/8$, the extra 23 percent being the result of the different loading cases considered.

The shears at the ends of the beams are taken as the simple beam shear $w_u \ell_n/2$, except at the exterior face of the first interior support, where the shear is 1.15 times the simple beam shear. Because there is more fixity at this end of the exterior span than at the exterior end, the moments are greater here, as shown in Figs. 10-10c and e. This leads to a shear greater than $w_u \ell_n/2$ at the interior end and a shear less than $w_u \ell_n/2$ at the exterior end. The decrease of the exterior end is not recognized by the ACI shear coefficients.

Although ACI Section 8.3.3 says nothing about shear at midspan, the critical shear at this point results from the factored live load on half of the span and is approximately equal to $w_{Lu}\ell_n/8$. Thus, $C_v = 0.25 w_{Lu}/w_u$, and

$$V_u = \frac{0.25 w_{Lu}}{w_u} \left(\frac{w_u \ell_n}{2} \right) \tag{10-4a}$$

This equation is explained in Section 6-5 and illustrated in Example 6-1.

At the middle of the exterior span, the ACI shear coefficients give a shear of $1.15(w_u \ell_n/2)$ minus $w_u \ell_n/2$, or $0.15(w_u \ell_n/2)$. In design calculations, we shall use the larger of the shear given by (10-4a) and

$$V_u = 0.15 \left(\frac{w_u \ell_n}{2} \right) \tag{10-4b}$$

Frequently, the girders in a one-way slab, beam, and girder system are loaded by concentrated loads from beams framing in at midpoint or at the third points, as shown in Fig. 10-6. In such cases, the moment coefficients presented in Fig. 10-11b to d do not

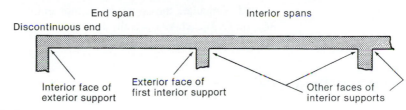

Discontinuous end

(a) Terminology

$C_m = -1/9$ if only two spans, $-1/10$ if three or more spans

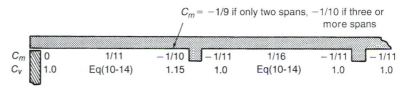

(b) Moment and shear coefficients—Discontinuous end unrestrained

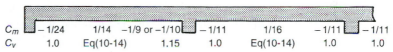

(c) Moment and shear coefficients—Discontinuous end integral with support where support is a spandrel girder

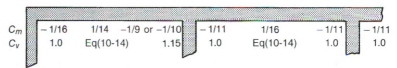

(d) Moment and shear coefficients—Discontinuous end integral with support where support is a column

Fig. 10-11
ACI moment and shear coefficients.

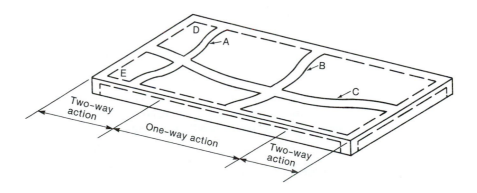

Fig. 10-12
One-way and two-way slab action.

apply, and a structural analysis is required. It is commonplace, however, to approximate the girder moments in the following manner:

1. Compute the total factored simple beam moments, M_0, due to the uniform and concentrated loads. If the beam were loaded only with uniform loads, for example, these would be $M_0 = w_u \ell_n^2 / 8$.

2. Calculate the approximate moment envelope values at the end and midspan as $8C_m$ times M_0, where C_m is the appropriate moment coefficient from Fig. 10-11.

10-4 ONE-WAY SLABS

One-way slab-and-beam systems having plans similar to those shown in Fig. 10-6 are commonly used, especially for spans of greater than 20 ft and for heavy live loads. Generally, the ratio of the long side to the short side of the slab panels exceeds 1.5.

Structural Action

For design purposes, a one-way slab is assumed to act as a series of parallel, independent 1-ft-wide strips of slab, continuous over the supporting beams. The slab strips span the short direction of the panel, as strips A and B in Fig. 10-12 do. Near the ends of the panel adjacent to the girders, some load is resisted by a bending of the longitudinal strips (strip C) and less by the transverse strips (strip A). Thus, near the girders, the load is supported by two-way slab action, which is discussed more fully in Chapter 13. In one-way slab design, this is ignored during the design of the one-way slab strips, but is accounted for by providing top reinforcement extending into the top of the slabs on each side of the girders across the ends of the panel. If this reinforcement is omitted, wide cracks may develop in the top of the slab along D–E.

Thickness of One-Way Slabs

Except for very heavily loaded slabs, such as slabs supporting several feet of earth, the thickness is chosen so that deflections will not be a problem. Occasionally, the thickness will be governed by shear or flexure, and these are checked in each design. Table 9.5(a) of the ACI Code gives minimum thicknesses of slabs *not supporting or attached to* partitions or other construction liable to be damaged by large deflections. No guidance is given for other cases. Table A-14 gives minimum thicknesses for one-way slabs that do and do not support such partitions.

Sometimes, slab thicknesses are governed by the danger of heat transmission during a fire. Thus, by this criterion, the fire rating of a floor is the number of hours necessary for the temperature of the unexposed surface to rise by a given amount, generally 250°F. For a 250°F temperature rise, a $3\frac{1}{2}$-in.-thick slab will give a 1-hour fire rating, a 5-in. slab will give a 2-hour fire rating, a $6\frac{1}{4}$-in. slab will give a 3-hour fire rating [10-1]. Generally, slab thicknesses are selected in $\frac{1}{4}$-in. increments up to 6 in., in $\frac{1}{2}$-in. increments for thicker slabs.

The slab reinforcement is supported at the right height above the forms on bent wire supports called "high chairs." The height of available chairs may control the slab thickness.

Cover

Concrete cover to the reinforcement provides corrosion resistance, fire resistance, and a wearing surface and is necessary to develop bond between steel and concrete. ACI Section 7.7.1 gives these minimum covers for corrosion protection in slabs:

1. For concrete *not exposed* to weather or in contact with the ground, No. 11 bars and smaller, $\frac{3}{4}$ in.

2. For concrete *exposed* to weather or in contact with the ground, No. 5 bars and smaller, $1\frac{1}{2}$ in.; No. 6 bars and larger, 2 in.

The words "exposed to weather" imply direct exposure to moisture changes. The undersides of exterior slabs are not considered exposed to weather unless subject to alternate wetting and drying, including that due to condensation, leakage from the exposed top surface, or runoff. ACI Commentary Section R7.7.5 recommends a minimum cover of 2 in. for slabs exposed to chlorides, such as deicing salts (as in parking garages). Alternatively, epoxy-coated bars can be used in such cases. ACI Sections 4.2 and 4.4 require special concretes in structures exposed to chlorides.

The structural endurance of a slab exposed to fire depends, among other things, on the cover to the reinforcement. Building codes give differing fire ratings for various covers. Reference [10-1] states that, for normal ratios of service-load moment to ultimate moment, $\frac{3}{4}$-in. cover will give a $1\frac{1}{4}$-hour fire rating, 1-in. cover about $1\frac{1}{2}$ hours, and $1\frac{1}{2}$-in. cover about 3 hours.

Reinforcement

Reinforcement details for one-way slabs are shown in Fig. 10-13. The straight-bar arrangement in Fig. 10-13a is almost always used in buildings. Prior to the 1960s, slab reinforcement was arranged by using the bent-bar–straight-bar arrangement shown in Fig. 10-13b. The cutoff points shown in Fig. A-5 can be used if the slab satisfies the requirements for use of the moment coefficients in ACI Section 8.3.3. Cutoff points in slabs not satisfying this clause are obtained via the procedure given in Examples 8-5 and 8-6. One-way slabs are designed by assuming a 1-ft-wide strip. The area of reinforcement is then computed as A_s/ft of width. The area of steel is the product of the area of a bar times the number of bars per foot, or

$$A_s/\text{ft} = A_b\left(\frac{12 \text{ in.}}{\text{bar spacing in inches}}\right) \tag{10-5}$$

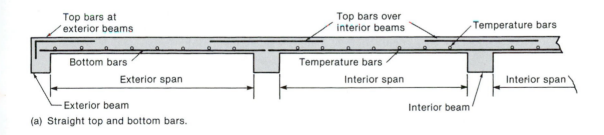

(a) Straight top and bottom bars.

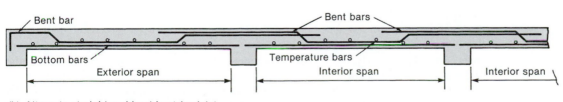

(b) Alternate straight and bent bar (obsolete).

Fig. 10-13
Sections through one-way slabs, showing reinforcement.

where A_b is the area of one bar. (See Table A-9 in Appendix A.) In SI units, (10-5) becomes (see Table A-9M)

$$A_s/\text{m} = A_b\left(\frac{1000 \text{ mm}}{\text{bar spacing in mm}}\right) \qquad (10\text{-}5M)$$

The maximum spacing of bars in one-way slabs is three times the slab thickness or 18 in., whichever is smaller (ACI Section 7.6.5). The maximum bar spacing is also governed by crack-control provisions, as was discussed in Section 9-3 (ACI Section 10.6.4).

Because a slab is thinner than the beams supporting it, the concrete in the slab shrinks more rapidly than the concrete in the beams. This may lead to shrinkage cracks in the slab. Shrinkage cracks perpendicular to the span will be crossed by flexural reinforcement, which will limit the width of these cracks. To limit the width of shrinkage cracks parallel to the span, additional *shrinkage and temperature reinforcement* is placed perpendicular to the flexural reinforcement. The amount required is specified in ACI Section 7.12, which requires the following ratios of reinforcement area to gross concrete area:

1.   Slabs with Grade-40 or -50 deformed bars: 0.0020

2.   Slabs with Grade-60 deformed bars or welded-wire fabric (smooth or deformed): 0.0018

3.   Slabs with reinforcement with a yield strength, f_y, in excess of 60,000 psi at a yield strain of 0.35 percent: $\dfrac{0.0018 \times 60,000}{f_y}$, but not less than 0.0014

Shrinkage and temperature reinforcement is spaced not farther apart than the smaller of five times the slab thickness and 18 in. (ACI Section 7.12.2.2). Splices of such reinforcement must be designed to develop the full yield strength of the bars in tension (ACI Section 7.12.2.3).

It should be noted that shrinkage cracks could be wide even when this amount of shrinkage reinforcement is provided [10-2]. In buildings, this may occur when shear walls, large columns, or other stiff elements restrain the shrinkage and temperature movements. ACI Section 7.12.1.2 states that if shrinkage and temperature movements are restrained significantly, the requirements of ACI Sections 8.2.4 and 9.2.3 shall be considered. These sections ask the designer to make a realistic assessment of the shrinkage deformations and to estimate the stresses resulting from these movements. If the shrinkage movements are restrained completely, the shrinkage and temperature steel will yield at the cracks, resulting in a few wide cracks. About three times the minimum shrinkage and temperature reinforcement specified in ACI Section 7.12 is required to limit the shrinkage cracks to reasonable widths. Alternatively, unconcreted control strips may be left during construction, to be filled in with concrete after the initial shrinkage has occurred. Methods of limiting shrinkage and temperature cracking in concrete structures are reviewed in [10-3]; see also Section 9-3.

ACI Section 10.5.4 specifies that the minimum flexural reinforcement shall be the same as the amount required in ACI Section 7.12 for shrinkage and temperature, except that, as stated previously, the maximum spacing of flexural reinforcement is three times the slab thickness (ACI Section 7.6.5).

Generally, No. 4 and larger bars are used for flexural reinforcement in slabs, since smaller bars or wires tend to be bent out of position by workers walking on the reinforcement during construction. This is more critical for top reinforcement than for bottom reinforcement, because the effective depth, d, of the top steel is reduced if it is pushed down, whereas that of the bottom steel is increased.

Examples

The examples of slabs, beams, and girders designed in this chapter are based on the floor design for a building similar to the building shown in Fig. 10-3, designed by B. James Wensley Architect, Ltd., and Read Jones Christoffersen, Ltd., Engineers. In the examples, the floor plan has been modified to simplify the presentation, and the design loads have been changed to correspond to the ASCE 7 [10-4] design-load standard.

A typical floor plan is shown in Fig. 10-14. The exterior column spacing was selected to give a uniform window width around the building. Because the exterior column spacing was greater than the width of the elevator shaft, the interior girders G3 were placed at an angle. This required more complicated formwork than a rectangular grid, but the formwork was reused floor after floor; hence, these costs were minimized. Example 10-1 considers the design of the eight-span, one-way slabs along section A–A. A partial section of these slabs is shown in Fig. 10-15. Example 10-2 considers the design of beams B3–B4–B3. The underside of a typical floor is shown in Fig. 10-16. The photograph is taken looking along beams B3–B4–B3.

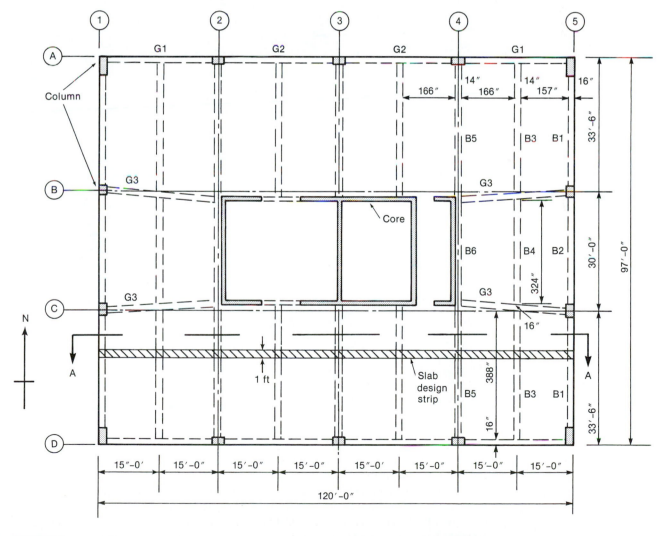

Fig. 10-14
Typical floor plan.

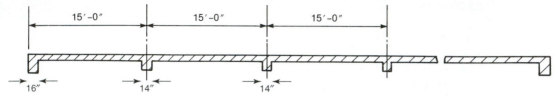

Fig. 10-15
Section *A–A*.

Fig. 10-16
Underside of beams
B3–B4–B3 in Fig. 10-14.
Photograph taken looking
north; east-wall columns are
on the right side of the pic-
ture. (Photograph courtesy of
J. G. MacGregor.)

The design loads are selected in accordance with ASCE 7, *Minimum Design Loads for Buildings and Other Structures* [10-4]. The dead loads include the weight of the floor, 0.5 psf for asphalt tile or carpet, 4 psf for mechanical equipment and lighting fixtures hung below the floor (see Fig. 10-16), and 2 psf for a suspended ceiling (not yet installed in Fig. 10-16).

ASCE 7 specifies the minimum live loads for offices as 50 psf and for lobbies as 100 psf. (See Table 2-1.) The live loads for file storage rooms and computer rooms are based on the actual usage of the room. An allowance must be made for the weight of partitions (considered to be live load in ASCE 7) if the floor loading is 80 psf or less. The design live loads can be reduced as a function of the floor area, as discussed in Section 2-7. No live-load reduction is allowed for one-way slabs. In consultation with the architect and owner, it has been decided that all floors will be designed for a live load of 100 psf, including partitions. This has been done to allow flexibility in office layouts.

A second loading case required in ASCE 7 for office buildings is a concentrated load of 2000 lb acting on a 30-in.-by-30-in. area placed on the floor to produce the maximum moments or shears in the structural members. The uniform live load is not on the structure when this load acts. For the slab spans in Fig. 10-14, the uniform live loads will always give greater moments and shears in the beams than the concentrated load. If it is assumed that the concentrated load is supported by a strip of slab extending one slab thickness on each side of the 30-in. area, it can be shown that, for this example, the uniform load governs the slab design also. For this reason, the concentrated loads are not specifically considered in Examples 10-1 and 10-2.

Transfer of Column Loads through the Floor System

In the building shown in Fig. 10-3 and 10-14 through 10-16, the columns were built of 5000-psi concrete and the floors were built of 3750-psi concrete. These strengths have been chosen on the basis of ACI Section 10.15, which allows the column concrete to be stronger than the floor concrete.

The floor concrete in the vicinity of the column–beam joint must transfer axial loads from the column above the joint to the column below. It is usual practice for the column concrete placement to be stopped at the elevation of the underside of each floor, as shown in Figs. 10-1 and 10-2, to allow the concrete to settle in the forms before the concrete in the floor over the column is placed. If the column and floor concretes are placed in the same pour, there is a chance of a gap developing between the column and slab concretes as the column concrete settles in the forms. Also there is a danger of mixing up the column and slab concretes or a danger of a cold joint between the two concretes.

ACI Section 10.15 allows the concrete in interior, edge, and corner columns to have a strength up to 1.4 times the strength of the concrete in the floor system (slab and beams) without any strengthening of the floor. On the basis of tests reported by Biachini et al. [10-5], the effective strength of the concrete in the floor system is enhanced by the lateral confinement from the floor surrounding the column. ACI Section 10.15.3 allows a column concrete strength of up to 1.4 times the floor strength for interior, edge, and corner columns without special attention. For larger strength ratios,

(a) ACI Section 10.15.1 allows the higher strength column concrete to be placed in the floor under and around the column, or

(b) ACI Section 10.15.1 allows the design of the joint based on the lower strength concrete with dowels provided to make up for the lower strength concrete, or

(c) ACI Section 10.15.3 allows the joint region to be designed by using an *effective compressive strength* of the concrete in the floor, f_{ce}, taken as

$$f_{ce} = 0.75f_{cc}' + 0.35f_{cs}' \qquad (10\text{-}6)$$

In the third case, the column concrete strength shall not be taken greater than 2.5 times the floor concrete strength; the maximum ratio is $f_{cc}'/f_{cs}' = 2.5$, where f_{cc}' is the specified strength of the column concrete and f_{cs}' is the specified strength of the slab concrete.

More realistic tests, reported by Ospina and Alexander [10-6], showed that the lateral confinement of the floor concrete is reduced when slab loads stress the slab reinforcement in tension. These tests supported the use of the 1.4 strength-ratio threshold for interior columns and edge columns, but suggested that the threshold strength ratio be lowered to 1.2

for corner columns. The joint strengths were also sensitive to the ratio of floor thickness to column width, h/c. The effective concrete strength decreased when h/c increased.

If the columns were designed using 5000-psi concrete, the strength of the floor concrete should not be less than 3570 psi for the 1.4 strength ratio to apply.

EXAMPLE 10-1 Design of a One-Way Slab

Design the eight-span floor slab spanning east–west in Fig. 10-14. A typical 1-ft-wide design strip is shown shaded. A partial section through this strip is shown in Fig. 10-15. The interior beams are assumed to be 14 in. wide. The concrete strength is 3750 psi, and the reinforcement strength is 60,000 psi, except that the yield strength of the stirrups is taken as 40,000 psi. The live load is 100 psf. The structure is in a region of low seismicity.

1. Estimate the thickness of the floor. The initial selection of the floor thickness will be based on ACI Table 9.5(a) (Table A-14), which gives the minimum thicknesses, unless deflections are computed, for members not supporting partitions likely to be damaged by large deflections. In consultation with the architect and owner, it has been decided that the partitions will be movable metal partitions that can accommodate floor deflections.

End bay:

$$\text{min. } h = \frac{\ell}{24}$$

$$= \frac{172 \text{ in.}}{24} = 7.17 \text{ in.}$$

Interior bays:

$$\text{min. } h = \frac{\ell}{28}$$

$$= \frac{180}{28} = 6.43 \text{ in.}$$

Therefore, try a 7.25-in. slab. Assuming $\frac{3}{4}$-in. clear cover and No. 4 bars,

$$d = 7.25 - \left(0.75 + \frac{0.5}{2}\right) = 6.25 \text{ in.}$$

Before the thickness is finalized, it will be necessary to check whether it is adequate for moment and shear.

2. Compute the trial unfactored loads. Given the thickness selected in step 1, it is now possible to compute a first estimate of the unfactored uniform loads. The dead load is as follows:

Slab:

$$w_D = \frac{7.25 \text{ in.}}{12 \text{ in./ft.}} \times 150 \text{ lb/ft}^3$$

$$= 90.6 \text{ lb/ft}^2 \text{ of floor surface}$$

Other dead loads:

Floor cover	0.5 psf
Mechanical equipment	4 psf
Ceiling	2 psf

Total dead load = 97.1 psf. The live load is

$$w_L = 100 \text{ psf}$$

3.   **Select load and strength-reduction factors.** Use the load factors and load combinations from ACI 318-02 Section 9.2.1 and the strength-reduction factors from ACI 318-02 Section 9.3. These factors should result in about the same amount of flexural reinforcement as the traditional ACI load combinations.

The ACI Code load combinations include several loads that do not occur often in buildings or on typical floors in a building.

These can be omitted. ACI Section 9.2.1 presents seven load combinations for concrete design. In the load combinations, U stands for the sum of the factored loads or load effects corresponding to a given load combination.

Load combination 9–1 $U = 1.4(D + F)$ $= 1.4 \times 97.1 \text{ psf} = 136 \text{ psf}$

Load combination 9–2 $U = 1.2(D + F + T) + 1.6(L + H) + 0.5(L_r \text{ or } S \text{ or } R)$

In load combination 9-2, D is the permanent load, and either L or H is the principal variable load. The roof loads would be companion action loads.

Canceling out the loads that are not present leaves:

$$U = 1.2 \times 97.1 \text{ psf} + 1.6 \times 100 \text{ psf} = 277 \text{ psf.}$$

Load combination 9–3 $U = 1.2D + 1.6 (L_r \text{ or } S \text{ or } R) + (1.0L \text{ or } 0.8W)$

Here, the roof load is the principal variable load and live load or wind load are the companion action loads. This is not a significant load combination in this example.

Load combination 9–4 $U = 1.2D + 1.6W + 1.0L + 0.5 (L_r \text{ or } S \text{ or } R)$

The principal variable load in load combination 9-4 is wind. This is not a significant load case for a floor slab.

Load combinations 9–5 and 9–7 include earthquake loads and will not be considered. Load combination 9–6 $U = 0.9D + 1.6W + 1.6H$ accounts for cases where dead load stabilizes a variable load covering overturning.

The maximum factored load on the beam is 277 psf. The factored load per foot on the strip shown shaded in Fig. 10-14 is 1 ft $\times$ 277 psf = 277 lb/ft.

4.   **Select strength-reduction factors.** Assume the slab is tension-controlled and meets the definition of a "beam." ACI Section 10.3.4 defines a tension-controlled member as one that has an extreme tensile strain greater than or equal to 0.005 at ultimate. We have chosen to use the factors in Chapter 9, which uses $\phi = 0.90$ for flexure in ACI Section 9.3.2.1, provided the extreme tensile strain exceeds 0.005.

5.   **Check whether the slab thickness is adequate for the moment.** Since $w_L < 3w_D$ and the other requirements of ACI Section 8.3.3 are met, use the ACI moment coefficients to calculate the moments.

In Chapter 4, the calculation of the required d was discussed. Equation (4-16) gave

$$\frac{bd^2}{12,000} = \frac{M_u}{\phi k_n} \text{ft-kips}$$

where $\phi k_n = \phi[f'_c \omega(1 - 0.59\omega)]$ and $\omega = \rho f_y / f'_c$. The maximum reinforcement which would normally be used would equal or exceed the tension-controlled limit for the neutral-axis depth given by (4-23). However, slabs seldom have more than $\rho = 0.01$. Check whether the slab thickness chosen is adequate if $\rho \leq 0.01$. For $f'_c = 3750$ psi and $f_y = 60,000$ psi,

$$\omega = \frac{0.01 \times 60,000}{3750} = 0.160$$

$$\phi k_n = 0.9[3750 \times 0.160(1 - 0.59 \times 0.160)] = 489$$

(See also Table A-3.) This can be used to compute bd^2 for a given M_u. The maximum moment M_u will be the negative moment at the first or second interior support (ACI Section 8.3.3). We have

Exterior face of first interior support:

$$M_u = \frac{w_u \ell_n^2}{10}$$

where

ℓ_n for computation of negative moment at *interior supports* is the average of the clear spans of the adjacent spans,

$$\ell_n = \frac{(157 \text{ in.} + 166 \text{ in.})}{2} / 12 \text{ in./ft} = 13.46 \text{ ft}$$

$$M_u = \frac{277 \text{ lb/ft} \times (13.46 \text{ ft})^2}{10}$$

$$= 5018 \text{ ft-lb/ft of width} = 5.02 \text{ ft-kips/ft.}$$

Second interior support:

$$M_u = \frac{w_u \ell_n^2}{11}$$

where

$\ell_n = 13.83 \text{ ft}$
$M_u = 4.82 \text{ ft-kips/ft.}$

Therefore, maximum $M_u = 5.02$ ft-kips/ft.
From (4-16),

$$bd^2 = \frac{M_u \times 12000}{\phi k_n}$$

$$bd^2 = \frac{5.02 \times 12{,}000}{489} = 123.2 \text{ in.}^3$$

But we are designing a 1-ft-wide strip, with $b = 12$ in. and

$$d = \sqrt{\frac{123}{12}} = 3.20 \text{ in.}$$

A minimum $d = 3.20$ in. keeps $\rho \le 0.01$. Since d (6.25 in., as computed in step 1) exceeds the minimum, the slab will be OK for flexure.

    6.   **Check whether thickness is adequate for shear.** In *beams*, ACI Section 11.5.5.1(a) requires shear reinforcement if $V_u > \phi V_c/2$. This is checked at d from the face of the supports. ACI Section 11.5.5.1 raises the threshold from $\phi V_c/2$ to ϕV_c for *slabs*. Because it is difficult to provide shear reinforcement in a slab, the upper limit on V_u is taken equal to ϕV_c for slabs. For simplicity we shall check this at the face of the support.

    If all spans were the same, the maximum shear would occur at the exterior face of the first interior support, where $V_u = 1.15 w_u \ell_n/2$. Since the clear spans are unequal, we will check that location plus a typical interior support.

Exterior face of first interior support:

$$V_u = \frac{1.15 \times 277 \text{ lb/ft} \times (157/12) \text{ ft}}{2}$$

$$= 2080 \text{ lb/ft of width}$$

Typical interior support:

$$V_u = \frac{277 \text{ lb/ft} \times (166/12)\text{ft}}{2} = 1916 \text{ lb/ft of width}$$

$$\phi V_c = 0.75(2\sqrt{f'_c}\, b_w d)$$

$$= 0.75(2\sqrt{3750} \times 12 \times 6.25) = 6890 \text{ lb/ft}$$

where ACI Section 9.3.2.3 gives $\phi = 0.75$ for shear and torsion.

Since $\phi V_c > V_u$, the slab chosen is adequate for shear. Therefore, **use $h = 7.25$ in., $d = 6.25$ in., and $w_u = 277$ lb/ft^2.** When slab thicknesses are selected on the basis of deflection control (ACI Table 9.5(a)), flexure and shear seldom govern.

7. Design of reinforcement. The calculations are done in Table 10-1 based on a 1-ft strip of slab. First, several constants used in that table must be calculated.

Table 10-1, Line 1 – The clear spans, ℓ_n, were computed as shown in Fig. 10-14.

- End bay, $\ell_n = 157$ in.
- Interior bay, $\ell_n = 166$ in.
- For interior supports, ℓ_n is the average of the two adjacent spans, so $\ell_n = 158.5$ in.

Table 10-1, Lines 2, 3, and 4 – The moments are computed as $M_u = w_n \ell_n^2 \times$ moment coefficient from line 3.

The maximum factored moment has been calculated in line 4, as 5.02 ft-kip/ft. We will calculate the reinforcement required at this location, i.e., at the exterior face of the first interior column, and at the corresponding lever arm, jd, and then use that value of jd at all other critical sections. Since the moment is smaller at all other sections, this value of jd will be on the safe (too small) side and will lead to values of A_s that are slightly too large, again on the safe side. We thus have

Table 10-1, line 5 Use (4-34) and $M_u = 5.02$ ft-kips/ft from line 4.

$$A_s = \frac{M_u}{\phi f_y jd} \tag{4-34}$$

For a first trial, assume that $jd = 0.925d$ for a slab. Also assume $\phi = 0.90$ for a tension-controlled section. Therefore,

$$A_s = \frac{5.02 \times 12,000}{0.9 \times 60,000\,(0.925 \times 6.25)}$$

$$= 0.193 \text{ in.}^2/\text{ft}$$

If this is provided exactly,

$$a = \frac{0.193 \times 60,000}{0.85 \times 3750 \times 12} = 0.303 \text{ in.}$$

TABLE 10-1 Calculations for One-Way Slab—Example 10-1

Line					
1. ℓ_n, ft	13.08	13.08	13.46	13.83	13.83
2. $(w_u \ell_n^2)$	47.40	47.40	50.18	53.0	53.0
3. M coefficients	1/24	1/14	1/10 1/11	1/16	1/11
4. M_u, ft-kip/ft	1.97	3.39	5.02 4.56	3.31	4.82
5. $A_{s(\text{req'd})}$, in.^{2}/ft	0.072	0.123	0.183	0.120	0.175
6. $A_{s(\text{min})}$, in.^{2}/ft	0.157	0.157	0.157	0.157	0.157
7. Choose steel	No. 4 @ 12"	No. 4 @ 12"	No. 4 @ 12	No. 4 @ 12"	No. 4 @ 12
8. A_s provided	0.20	0.20	0.20	0.20	0.20

and

$$jd = d - \frac{a}{2}$$

$$= 6.25 - \frac{0.303}{2} = 6.10 \text{ in.}$$

(Note that $j = \dfrac{6.10}{6.25} = 0.976$ in this case compared with the assumed value of 0.925.)

Becuase the first estimate of A_s was based on a guess of jd, we will recompute the required A_s:

$$A_s(\text{in.}^2/\text{ft}) = \frac{M_u \times 12,000}{0.9 \times 60,000 \times 6.10}$$

$$= 0.0364 \times M_u \text{ (ft-kips/ft)}$$

The area of steel required at any section will be computed as $A_s(\text{in.}^2/\text{ft}) = 0.0364 \times M_u$ (ft-kips/ft). At the first interior support, this gives $A_s = 0.183 \text{ in.}^2/\text{ft}$.

Table 10-1, Line 6 Compute the minimum flexural reinforcement, using ACI Section. 10.5.4, which refers to ACI Section 7.12 for amount and spacing:

$$A_{s(\min)} = 0.0018bh$$

$$= 0.0018 \times 12 \text{ in.} \times 7.25 \text{ in.}$$

$$= 0.157 \text{ in.}^2/\text{ft}$$

For maximum spacing, ACI Section 7.6.5 gives $3h = 21.75$ in., but not more than 18 in. Therefore, maximum spacing $= 18$ in.

8. Check reinforcement spacing for crack control. To control the width of cracks on the tension face of the slab, ACI Section 10.6.4 limits the maximum spacing of the flexural reinforcement closest to the tension face of the slab to

$$s = \frac{540}{f_s} - 2.5 \, c_c, \qquad \text{but not more than } 12\left(\frac{36}{f_s}\right) \text{ in.} \qquad (10\text{-}7)$$

$$\text{(ACI Eq. 10-5)}$$

where f_s is the stress in the tension steel in ksi, which can be taken as $0.6f_y = 36$ ksi, and c_c is the clear cover from the tension face of the slab to the surface of the reinforcement nearest to it, taken as 0.75 in. from ACI Section 7.7.1(c), together giving

$$s = \frac{540}{0.6 \times 60} - 2.5 \times 0.75 \text{ in.} = 13.1 \text{ in.}, \text{ but not more than } 12\left(\frac{36}{36}\right) \text{ in.} = 12 \text{ in.}$$

This overrides the 18-in. maximum spacing selected in the previous step in the design.

9. Select the top and bottom flexural steel. The remaining flexural calculations are given in Table 10-1. The choice of the reinforcement in line 7 of this table is made by using (10-5) or Table A-9. The resulting steel arrangement is shown in Fig. 10-17. The cutoff points have been computed by using Fig. A-5c, since the slab geometry allowed the use of the ACI moment coefficients.

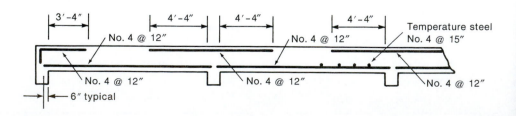

Fig. 10-17
Reinforcement—
Example 10-1.

10.   Determine the shrinkage and temperature reinforcement. ACI Section 7.12 requires shrinkage and temperature reinforcement perpendicular to the span of the slab:

$$A_s = 0.0018bh = 0.157 \text{ in.}^2/\text{ft}$$

$$\text{maximum spacing} = 18 \text{ in.}$$

Therefore, provide **No. 4 bars at 15 in. o.c., as shrinkage and temperature reinforcement.** These could be placed in the top or bottom of the slab, provided that the sum of the areas totals 0.157 in^2/ft or more. We shall place it on top of the lower layer of steel, as shown in Fig. 10-17. Here it will be supported by the bottom flexural steel during concreting.

11.   Design the transverse top steel at girders. Due to localized two-way action adjacent to the girders (G1, G2, G3, etc., in Fig. 10-14), top reinforcement is required in the slab perpendicular to the girders. This will be designed when the girders are designed.   ■

This completes the design of this slab. The slab thickness was selected in step 1 by using ACI Table 9.5(a) to limit deflections. A 6.5-in. thickness was acceptable in the six interior spans, but a 7.25-in. thickness was required in the end spans. If the entire floor was made 6 in. thick instead of 7.25 in. thick, about 45 cubic yards of concrete could be saved per floor, with a resultant saving of 180 kips of dead load per floor. In a 20-story building, that amount represents a considerable saving. With this in mind, the floor was redesigned at 6 in. thick, and the computed deflections were found to be acceptable. The calculations will not be repeated here, but the slab thickness used in the design of the beams in Example 10-2 will be taken as 6 in.

EXAMPLE 10-1M   Design of a One-Way Slab—SI Units

Design an eight-span floor slab similar to the strip shown shaded in Fig. 10-14. A 1-m-wide strip will be considered. A partial section through this strip is shown in Fig. 10-18. The beams are assumed to be 350 mm wide. The concrete strength is 25 MPa, and the reinforcement strength is 400 MPa. The live load is 4.8 kPa.

1.   Estimate the thickness of the floor. The initial selection of the floor thickness will be based on ACI Table 9.5(a) (Table A-14), which gives the minimum thickness, unless deflections are computed, for members not supporting partitions likely to be damaged by large deflections. In consultation with the architect and owner, it has been decided that the partitions will be movable metal partitions that can accommodate floor deflections. Accordingly, we have

End bay:

$$\text{min. } h = \frac{\ell}{24}$$

$$= \frac{4600 \text{ mm} - \dfrac{350 \text{ mm}}{2}}{24} = 184 \text{ mm}$$

Interior bays:

$$\text{min. } h = \frac{\ell}{28}$$

$$= \frac{4600 \text{ mm}}{28} = 164 \text{ mm.}$$

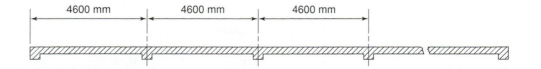

Fig. 10-18
Section A–A—Example 10-1M.

Therefore, try a 190-mm slab. Assuming 20 mm of clear cover and No. 15 bars, we obtain

$$d = 190 - \left(20 + \frac{15}{2}\right) = 162.5 \text{ mm}$$

Before the thickness is finalized, it will be necessary to check whether it is adequate for moment and shear.

2. **Compute the trial factored loads.** Given the thickness selected in step 1, it is now possible to compute a first estimate of the uniform loads. The unfactored dead load per square meter of slab is as follows:

Slab:

$$w_{Ds} = (0.190 \times 1.0 \times 1.0)\text{m}^3 \times 2450 \text{ kg/m}^3 \times (9.81/1000)\text{kN/kg}$$
$$= 4.57 \text{ kN per m}^2 \text{ of floor surface} = 4.57 \text{ kPa.}$$

The 2002 ACI Code has two sets of load and resistance factors, one set given in ACI Sections 9.2 and 9.3, the other in Appendix C. Either set can be used, provided that the load and resistance factors used throughout the design are from the same set. To illustrate their use we have chosen to use the factors in Appendix C. After removing the loads not present in the building, the basic load combination 9–1 reduces to

Load combination C.1 $U = 1.4D + 1.7L$

The corresponding strength reduction factors are 0.90 for flexure, and 0.85 for shear.

Other dead loads:

Floor cover	0.025 kPa
Mechanical equipment	0.20 kPa
Ceiling	0.10 kPa

The total dead load is $w_D = 4.90$ kPa. The live load is 4.8 kPa. The factored load is

$$w_u = 1.4 \times 4.9 + 1.7 \times 4.8 = 15.02 \text{ kPa (kN/m}^2)$$

The load per meter on the strip of slab being designed is 1 m $\times$ 15.02 kN/m^2 = 15.02 kN/m.
Since $w_L < 3w_D$ and the other requirements of ACI Section 8.3.3 are met, use the ACI moment coefficients to calculate the moments.

3. **Check the thickness required for moment.** In Chapter 4, calculation of the required d was discussed. Equation (4-16M) gives

$$\frac{bd^2}{10^6} = \frac{M_u}{\phi k_n} \text{ kN-m}$$

where $\phi k_n = \phi[f_c'\omega(1 - 0.59\omega)]$ and $\omega = \rho f_y/f_c'$. The maximum reinforcement that would normally be used is that corresponding to the tension-controlled limit for the neutral-axis depth, given by (4-23). However, slabs seldom have more than $\rho = 0.01$. Check whether the slab thickness chosen is adequate if $\rho \leq 0.01$. For $f_c' = 25$ MPa and $f_y = 400$ MPa,

$$\omega = \frac{0.01 \times 400}{25} = 0.160$$

$$\phi k_n = 0.9[25 \times 0.160(1 - 0.59 \times 0.160)] = 3.26$$

This can be used to compute bd^2 for a given M_u. The maximum moment M_u will occur at the first or second interior support (ACI Section 8.3.3).

First interior support:

$$M_u = \frac{w_u \ell_n^2}{10}$$

where

$$\ell_n = \frac{(4600 - 350 - 350/2) + (4600 - 350)}{2} = 4162 \text{ mm}$$

$$M_u = \frac{15.02 \text{ kN/m} \times 4.162^2 \text{ m}^2}{10} = 26.0 \text{ kN-m per meter width of slab}$$

Second interior support:

$$M_u = \frac{w_u \ell_n^2}{11}$$

where

$$\ell_n = 4600 - 350 = 4250 \text{ mm and}$$

$$M_u = \frac{15.02 \times 4.250^2}{11} = 24.7 \text{ kN-m/m}$$

Therefore, the maximum $M_u = 26.0$ kN-m/m, and

$$bd^2 = \frac{M_u \times 10^6}{\phi k_n}$$

$$bd^2 = \frac{26.0 \times 10^6}{3.26} = 7,975,000 \text{ mm}^3$$

But we are designing a 1-m-wide strip; hence, $b = 1000$ mm, and

$$d = \sqrt{\frac{7,975,000}{1000}} = 89.3 \text{ mm}$$

The minimum d is 89.3 mm to keep $\rho \leq 0.01$. Since d (162.5 mm, as computed in step 1) exceeds this minimum, the slab will be OK for flexure.

   4.   **Check whether thickness is adequate for shear.** Shear reinforcement is required in slabs if $V_u > \phi V_c$ (ACI Sections 11.5.5.1a and 11.5.6.1). Since it is difficult to place shear reinforcement in a slab, the upper limit on V_u will be taken as ϕV_c.

   If all the spans were the same, the maximum shear would occur at the exterior face of the first interior support, where $V_u = 1.15 w_u \ell_n/2$. Since the spans are unequal, we will check that location plus a typical interior support:

First interior support:

$$\ell_n \text{ for shear} = 4600 - 350 - \frac{350}{2} = 4075 \text{ mm}$$

$$V_u = \frac{1.15 \times 15.02 \times 4.075}{2} = 35.2 \text{ kN/m of width}$$

Typical interior support:

$$\ell_n = 4600 - 350 = 4250 \text{ mm}$$

$$V_u = \frac{15.02 \times 4.250}{2} = 31.9 \text{ kN/m}$$

$$\phi V_c = 0.85 \frac{\sqrt{f'_c} b_w d}{6}$$

$$= 0.85 \left(\frac{\sqrt{25}}{6} \times 1000 \times 162.5 \right) = 115,100 \text{ N}$$

$$= 115.1 \text{ kN}$$

Since $\phi V_c > V_u$, the slab chosen is adequate for shear. Therefore, **use $h = 190$ mm, $d = 162.5$ mm, and $w_u = 15.02$ kN/m^2**. When slab thicknesses are selected on the basis of deflection control (ACI Table 9.5(a)), flexure and shear seldom govern.

    5. Design the reinforcement. The calculations are done in Table 10-1M. First, several constants used in that table must be calculated. The maximum moment was calculated in line 4 as 26.0 kN-m/m. We will calculate the reinforcement required at that point and the corresponding lever arm, jd, and use that value of jd at all other critical sections. Since the moment is smaller at all other sections, this value of jd will be on the safe side (too small) and will lead to values of A_s that are slightly too large, again on the safe side. We thus have

$$A_s = \frac{M_u}{\phi f_y jd} \tag{4-34}$$

For a first trial, assume that $jd = 0.925d$ for a slab. Therefore,

$$A_s = \frac{26.0 \times 10^6}{0.90 \times 400 \times (0.925 \times 162.5)}$$

$$= 480 \text{ mm}^2/\text{m}$$

If this is provided exactly,

$$a = \frac{480 \times 400}{0.85 \times 25 \times 1000} = 9.04 \text{ mm}$$

and

$$jd = d - \frac{a}{2}$$

$$= 162.5 - \frac{9.04}{2} = 158.0 \text{ mm}$$

(Note that $j = 0.972$ in this case.)

    Since the first estimate of A_s was based on a guess of jd, we will recompute the required A_s:

$$A_s(\text{mm}^2/\text{m}) = \frac{M_u \times 10^6}{0.90 \times 400 \times 158}$$

$$= 17.58 \times M_u(\text{kN-m/m})$$

The area of steel required at each critical section will be computed as $A_s(\text{mm}^2/\text{m}) = 17.58 \times M_u$ (kN-m/m). At the first interior support, this gives $A_s = 457$ mm^2/m.

TABLE 10-1M Calculations for One-Way Slab—Example 10-1M

Line						
1. ℓ_n, m	4.075	4.075	4.163	4.250	4.250	4.250
2. $w_u \ell_n^2$, kN-m/m	249.4	249.4	260.3	271.3	271.3	271.3
3. M coefficients	1/24	1/14	1/10	1/11	1/16	1/11	1/16
4. M_u, kN-m/m	10.39	17.81	26.03	~~23.66~~	16.96	24.66	16.96
5. $A_{s(\text{req'd})}$, mm^2/m	~~182.7~~	~~313.1~~	457.6	~~298.2~~	433.5	~~298.2~~
6. $A_{s,\text{min}}$, mm^2/m	342	342	~~342~~	342	~~342~~	342
7. Choose steel	No. 10 @ 280	No. 10 @ 280	No. 10 @ 220	No. 10 @ 280	No. 10 @ 220	No. 10 @ 280
8. A_s provided	357	357	455	357	455	357

Compute the minimum flexural reinforcement by using ACI Section 10.5.4, which refers to ACI Section 7.12 for the amount and gives the maximum spacing as the lesser of three times the slab thickness and 500 mm:

$$A_{s,min} = 0.0018bh$$

$$= 0.0018 \times 1000 \text{ mm} \times 190 \text{ mm}$$

$$= 342 \text{ mm}^2/\text{m}$$

From ACI Section 7.6.5, the maximum spacing is $3h = 3 \times 190 = 570$ mm, but not more than 500 mm. Therefore, the maximum spacing is 500 mm.

6.   **Check reinforcement spacing for crack control.** ACI Section 10.6.4 limits the maximum spacing of reinforcement adjacent to the tension face of the slab to

$$s = \frac{95{,}000}{f_s} - 2.5\,c_c, \text{ but not more than } 300\left(\frac{252}{f_s}\right) \text{ mm} \qquad (10\text{-}7\text{M})$$

where f_s is the stress in the tension steel in MPa, which can be taken as $0.6f_y = 240$ MPa, and c_c is the clear cover from the tension face of the slab to the surface of the reinforcement nearest to it, taken as 20 mm, together giving

$$s = \left(\frac{95{,}000}{0.6 \times 400 \text{ MPa}}\right) - 2.5 \times 20 = 345 \text{ mm, but not more than}$$

$$300\left(\frac{252}{0.6 \times 400 \text{ MPa}}\right) = 315 \text{ mm}$$

This overrides the 500-mm maximum spacing selected in the previous step in the design.

7.   **Select the top and bottom flexural steel.** The remaining flexural calculations are given in Table 10-1M. The choice of the reinforcement in line 7 of this table is made by using (10-5M) or Table A-9M. The resulting steel arrangement is shown in Fig. 10-19. The cutoff points have been computed by using Fig. A-5c, since the slab geometry allowed the use of the ACI moment coefficients.

8.   **Select the shrinkage and temperature reinforcement.** ACI Section 7.12 requires shrinkage and temperature reinforcement perpendicular to the span of the slab:

$$A_s = 0.0018bh = 342 \text{ mm}^2/\text{m}$$

$$\text{maximum spacing} = 500 \text{ mm}$$

Therefore, **provide No. 10 bars at 280 mm on centers as shrinkage and temperature reinforcement.** These could be placed near the top or bottom of the slab, provided that the sum of the areas totals 342 mm^2/m or more. We shall place it on top of the lower layer of steel, as shown in Fig. 10-19.

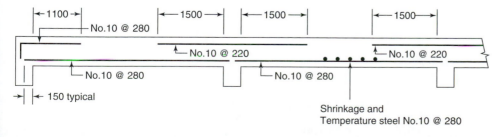

Fig. 10-19
One-way bending reinforcement—Example 10-1M.

9. Design the transverse top steel at girders. Due to localized two-way action adjacent to the girders (G1, G2, G3, etc., in Fig. 10-14), top reinforcement is required in the slab perpendicular to the girders. This will be designed when the girders are designed. ∎

10-5 CONTINUOUS BEAMS

The three major stages in the design of a continuous beam are design for flexure, design for shear, and design of bar details. In addition, it is necessary to consider deflections and crack control and, in some cases, torsion.

When the area supported by a beam exceeds 400 ft², it is frequently possible to use a reduced live load in calculating the moments and shears in the beam. The concept of live-load reduction factors was discussed in Section 2-7. Depending on the applicable building code, the live-load reduction factor will be a function of the tributary area of the beam or its influence area. The next example uses the ASCE 7-98 loading code, which is based on tributary areas. (See Section 2-7.)

EXAMPLE 10-2   Design of a Continuous T Beam

Design the beam B3–B4–B3 in Fig. 10-14. This beam supports its own dead load plus the load from a 6-in. slab. This is thinner than the slab designed in Example 10-1, as explained at the end of that example. The beam is supported on girders at lines A, B, C, and D, as shown in Fig. 10-14, and is symmetrical about the centerline of the building. Figure 10-16 is a photograph looking along the beam. Use $f'_c = 3750$ psi, $f_y = 60,000$ psi for flexural reinforcement, and $f_y = 40,000$ psi for stirrups. Base the loadings on the ASCE 7-98 design loading code. Use load factors from ACI Section 9.2.

1. Compute the trial factored loads on the beam. Because the size of the beam stem is unknown at this stage, it is not possible to compute the final loads for use in the design. For preliminary purposes, estimate the size of the beam stem. Once the size has been established, the factored load will be corrected and then used in subsequent calculations.

The ASCE 7-98 design-loading code allows live-load reductions based on tributary areas multiplied by a live-load element factor, $K_{LL} = 2$, to convert the tributary area to an influence area. (See Section 2-7.) For a beam, this is the area between the beam in question and the faces of the beams on either side of it, over the loaded length of the beam. Three tributary areas will be considered in the design of beam B3–B4–B3, as shown in Fig. 10-20.

(a) To compute the positive moment at midspan of beam B3 and negative moment at the exterior end of beam B3, it is necessary to load spans AB and CD. Since the majority of the moment in the left span results from loads on that span, we shall take the tributary area A_T to be the area over beam B3, extending halfway to the faces of the adjacent beams:

$$A_T = \frac{388(157/2 + 14 + 166/2)}{12 \times 12} = 473 \text{ ft}^2$$

From Section (2-8), the reduced live load is

$$L = L_o\left(0.25 + \frac{15}{\sqrt{K_{LL}A_T}}\right) \tag{2-4}$$

where L_o, the unreduced live load, is 100 psf and (for an interior beam) $K_{LL} = 2.0$.

$$L = L_o\left(0.25 + \frac{15}{\sqrt{2 \times 473}}\right) = 100 \times 0.738$$
$$= 73.8 \text{ psf}$$

L shall not be less than $0.50L_o$ for members supporting one floor. Thus, $L = 73.8$ psf.

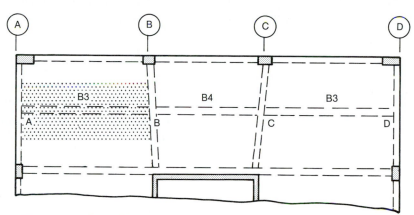

(a) Tributary area for positive moment–B3.

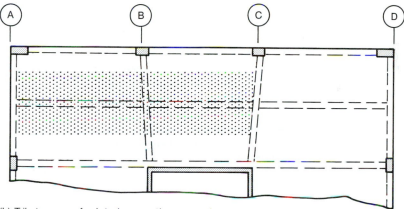

(b) Tributary area for interior negative moment.

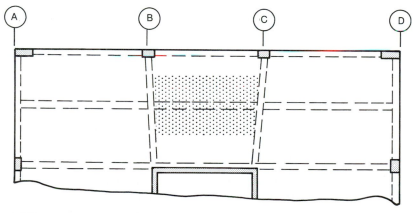

Fig. 10-20
Tributary areas—
Example 10-2.

(c) Tributary area for positive moment–B4.

(b) To compute the maximum negative moment at the interior support, B, spans AB and BC should be loaded. This corresponds to loading spans AC and CE in Fig. 10-8c and d. The corresponding tributary area, A_T, is the area between lines of zero shear parallel to the beam and on either side of it, as shown in Fig. 10-20b.

$$A_T = \frac{(157/2 + 14 + 166/2) \times (388 + 16 + 324)}{12 \times 12} = 887 \text{ ft}^2$$

and

$$K_{LL} = 2$$

$$L = 100\left(0.25 + \frac{15}{\sqrt{2 \times 887}}\right) = 60.6 \text{ psf}$$

where L shall not be less than $0.50L_o$ for members supporting one floor. Thus, $L = 60.6$ psf.

(c) To compute the positive moment in the center span (beam $B4$), span BC should be loaded. The tributary area is shown in Fig. 10-20c. Here,

$$A_T = \frac{(157/2 + 14 + 166/2) \times 324}{12 \times 12} = 395 \text{ ft}^2$$

$$L = 100\left(0.25 + \frac{15}{\sqrt{2 \times 395}}\right) = 78.4 \text{ psf}$$

In this example, three reduced live loads were calculated, in part, to illustrate the calculation. If this beam were being analyzed by computer, one might use a single reduced live load throughout, equal to the largest of the three reduced live loads.

Summary of unfactored reduced live loads

For negative moment at exterior end, positive moment at midspan, and shear in beam $B3$ = 73.8 psf.
For negative moment at support B = 60.6 psf.
For positive moment at midspan of $B4$ = 78.4 psf.

2. Select Method of Analysis of Beam and Values of the Strength-Reduction Factor

If the beam fits the requirements in ACI Section 8.3.3, we will use the moment coefficients from that section.

In this example,

(a) there are two or more spans,

(b) the ratio of the longer clear span to the shorter clear span is $388/324 = 1.198$, which is just less than 1.20,

(c) the loads are uniformly distributed,

(d) the "unit" (unfactored) live load does not exceed three times the "unit" (unfactored) dead load, and

(e) the members are prismatic.

So far, the beam fits all of these except item (d), which we will check later.

Use ACI Section 8.3.3 to compute moments and shears.

3. Select the strength-reduction factors. Because the design is based on load combinations from ACI Section 9.2, the strength reduction factors must be from ACI Section 9.3:

For flexure in tension-controlled beam sections, $\phi = 0.90$. (This may change once the reinforcement is chosen and it is possible to check the assumption of tension-controlled sections.)

Shear $\phi = 0.75$.

The size of the stem beam $B3$–$B4$–$B3$ will be chosen on the basis of negative moment at the first interior support. For this location, the factored load on the beam is as follows:

(a) Dead load (not including the portion of the beam stem below the slab):

Slab	75 psf
Floor, ceiling, mechanical	6.5 psf
	81.5 psf

Live load (reduced): 60.6 psf
Total factored load from slab:

$$w_u = 1.2 \times 81.5 + 1.6 \times 60.6 = 194.8 \text{ psf}$$

Assume that the tributary width which loads beam $B3$–$B4$–$B3$ extends halfway from this beam to the adjacent beams on each side of $B3$–$B4$–$B3$, or

$$\left(\frac{157}{2} + 14 + \frac{166}{2}\right) \text{in.} = 175.5 \text{ in.} = 14.63 \text{ ft}$$

Therefore, the factored load per foot from the slab is

$$14.63 \text{ ft} \times 194.8 \text{ psf} = 2.85 \text{ kips per foot}$$

It is now necessary to estimate the weight of the beam stem. Two very approximate methods of doing this were given in Example 4-6:

(1) The factored dead load of the stem is taken as 10 to 20 percent of the other factored loads on the beam. This gives 0.29 to 0.57 kip/ft.

(2) The overall depth of beam h is taken to be 8 to 10 percent of ℓ_n, and b_w is taken to be $0.5h$. This gives the overall h as 31 to 39 in., with the stem extending 25 to 33 in. below the slab, and gives b_w as 15 to 20 in. The factored load from stems of such sizes ranges from 0.39 to 0.69 kip/ft.

As a first trial, assume the factored weight of the stem to be 0.60 kip/ft. Then

$$\text{Total trial load per foot} = 2.85 + 0.60 \text{ kip/ft.}$$
$$= 3.45 \text{ kips/ft}$$

say, 3.4 kips/ft.

4. Choose the actual size of the beam stem. The size of the stem is governed by three factors: (a) deflections, (b) moment capacity at the point of maximum negative moment, and (c) shear capacity. In addition, the overall depth of the floor should be minimized to reduce the overall height of the building.

(a) Calculate the minimum depth based on deflections. ACI Table 9.5(a) (Table A-14) gives minimum depths, unless deflections are checked. Since the anticipated partitions will be flexible enough to undergo some deflections, this table can be used.

Beam $B3$ is one-end continuous, $f_y = 60,000$ psi, and normal-density concrete. Therefore,

$$\text{min. } h = \frac{\ell}{18.5}$$

where ℓ shall be taken as the span center to center of supports. Consequently,

$$\text{min. } h \text{ based on deflections} = \frac{388 + 16 \text{ in.}}{18.5} = 21.8 \text{ in.}$$

(b) Determine the minimum depth based on the negative moment at the exterior face of the first interior support.

$$\text{Moment} = \frac{w_u \ell_n^2}{10}$$

ℓ_n for negative moment is defined as the average of the two adjacent spans. Thus,

$$\ell_n = \frac{(388 + 324) \text{ in.}}{2} = 356 \text{ in.} = 29.67 \text{ ft}$$

Therefore,

$$M_u \simeq \frac{3.4 \times 29.67^2}{10} = 299 \text{ ft-kips}$$

In Chapter 4, beam sizes were chosen by starting with the assumption that a desirable steel ratio was $\rho = 0.01$. In a continuous beam, a higher steel ratio is often used at the point of maximum negative moment. Assume that ρ is limited to 0.013 and that there is no compression reinforcement. (From Table A-5, the ρ corresponding to the tension-controlled limit is 0.0169.) From Table A-3, $\phi k_n = 616$. Thus,

$$\frac{bd^2}{12,000} = \frac{M_u}{\phi k_n}$$

$$= \frac{299}{616} = 0.486$$

and $bd^2 = 5830 \text{ in}^3$. Possible choices are

$$b = 10 \text{ in.}, \ d = 24.1 \text{ in.}$$
$$b = 12 \text{ in.}, \ d = 22.0 \text{ in.}$$
$$b = 14 \text{ in.}, \ d = 20.4 \text{ in.}$$

Select the last of these. With one layer of reinforcement at the supports, $h \simeq 20.4 + 2.5 = 22.9$ in. This exceeds the minimum h based on deflections. **Try a 14-in. wide-by-24-in. deep extending 18 in. below the slab stem, with $d = 21.5$ in.**

(c) Check the shear capacity of the T-beam.

$$V_u = \phi(V_c + V_s)$$

Maximum shear from beam loads at interior end of B3:

$$V_u = 1.15 \, w_u \ell_n/2 = 1.15 \times 3.40 \times 16.17 = 63.2 \text{ kips}$$

From ACI Section 11.3.1.1, the nominal V_c is:

$$V_c = 2\sqrt{f_c'}b_w d = 36.9 \text{ kips}$$

ACI Section 11.5.6.8 sets the maximum nominal V_s is:

$$V_s = 8\sqrt{f_c'}b_w d = 147.6 \text{ kips}$$

Thus, the absolute maximum $\phi V_n = 0.75 \, (36.9 + 147.6 \text{ kips}) = 138.3 \text{ kips}$.

The maximum factored V_u due to the applied loads and dead loads is 63.2 kips. **Therefore, the beam stem selected has ample capacity for shear.**

(d) **Summary. Use**

$$b = \textbf{14 in.}$$

$$h = \textbf{24 in. (18 in. below slab)}$$

$$d = \textbf{21.5 in.,} \text{ assuming one layer of steel at all sections}$$

This section is shown in Fig. 10-21.

5. **Compute the dead load of the stem, and recompute the total load per foot.**

$$\text{Weight per foot of the stem below the slab} = \frac{14 \times 18 \times 12}{1728} \times 0.15 = 0.263 \text{ kip/ft}$$

Total dead load for B3–B4–B3 is:

$$w_D = 0.0815 \text{ ksf} \times 14.63 \text{ ft} + 0.263 \text{ kip/ft} = 1.46 \text{ kip/ft}$$

Live load for B3–B4–B3 is:

Beam B3,	$w_L = 0.0738 \text{ ksf} \times 14.63 \text{ ft} = 1.08 \text{ kip/ft}$	
Negative moment at B,	$w_L = 0.0606 \text{ ksf} \times 14.63 \text{ ft} = 0.887 \text{ kip/ft}$	
Positive moment for B4,	$w_L = 0.0784 \text{ ksf} \times 14.63 \text{ ft} = 1.15 \text{ kip/ft}$	

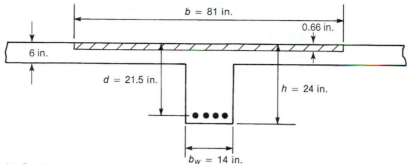

(a) Positive-moment region.

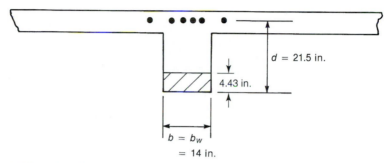

Fig. 10-21
Beam sections.

(b) Negative-moment region.

Summary of factored loads on B3–B4–B3

Beam B3, $w_u = 1.2 \times 1.46 + 1.6 \times 1.08 = 3.48$ kip/ft

Negative moment at B, $w_u = 1.2 \times 1.46 + 1.6 \times 0.887 = 3.17$ kip/ft

Positive moment for B4, $w_u = 1.2 \times 1.46 + 1.6 \times 1.15 = 3.59$ kip/ft

Note that although the weight of the beam stem is only $0.263/0.5 = 0.53$ times the original guess, the error in the original estimate of the total factored load at B is only 7 percent. If the final factored load per foot changed by more than about 10 percent, it would be desirable to recalculate the beam size and load per foot.

6.   **Calculate the flange width for the positive-moment regions.** The beam acts as a T beam in the positive-moment regions. The effective width of the overhanging flange on each side of the web is the smallest of (ACI Section 8.10.2) the following:

$0.25\ell_n$ (based on the shorter span for simplicity) $= 0.25 \times 324$ in. $= 81$ in.

$$b_w + 2(8 \times 6) = 14 \text{ in.} + 2 \times 48 \text{ in.} = 110 \text{ in.}$$

$$b_w + \frac{157 \text{ in.}}{2} + \frac{166 \text{ in.}}{2} = 175.5 \text{ in.}$$

Therefore, the effective flange width is 81 in., as shown in Fig. 10-21.

7.   **Compute the beam moments.** The moments can be computed by using frame analysis, or, if the structure fits the limitations of ACI Section 8.3.3, the moment coefficients can be used. The structure is a nonprestressed continuous beam (ACI Section 8.3.2) with more than two spans. The ratio of clear spans is 388 in./324 in. = 1.198. This is just less than the upper limit of 1.20 on the use of the moment coefficients. The loads are uniformly distributed and from step 5, the live load is less than three times the dead load. Thus, the ACI moment coefficients can be used in this design. The moment calculations are presented in Table 10-2. Since the building is symmetrical about its center, only a part of the beam is shown in this table.

Table 10-2, Line 1 The values of ℓ_n are the clear-span lengths, except that in computing the negative moment at supports B and C, ℓ_n is the average of the two adjacent spans.

TABLE 10-2 Calculation of Moments for Beam B3–B4–B3—Example 10-2

Line	B3			B4	
1. ℓ_n, ft	32.33	32.33	29.67	27.0	29.67
2. w_u, kip/ft	3.48	3.48	3.17	3.59	3.17
3. $w_u \ell_n^2$	3640	3640	2790	2620	2790
4. C_m	1/24	1/14	1/10 1/11	1/16	1/11 1/10
5. $M_u = C_m w_u \ell_n^2$, ft-kip	−152	260	−279	164	−279

Table 10-2, Line 2 The factored uniform loads differ for each span, due to the different live-load reduction factors for the various sections.

Table 10-2, Line 4 Because the exterior end of span *AB* is supported on a spandrel girder, the moment coefficient at *A* is 1/24. The resulting moment diagram is shown in Table 10-3.

8. Design the flexural reinforcement. The reinforcement will be designed in Table 10-3. This is a continuation of Table 10-2. Prior to entering Table 10-3, however, it is necessary to compute some constants for use in subsequent calculations of area of reinforcement.

(a) Compute the area of steel required at the point of maximum negative moment (first interior support).

$$A_s = \frac{M_u \text{ ft-kips}}{\phi f_y jd} \times 12{,}000 \tag{4-34}$$

where f_y is in psi.

Because there is negative moment at the support, the beam acts as a rectangular beam with compression in the web, as shown in Fig. 10-21b. Assume that $j = 0.875$ and $\phi = 0.90$

$$A_s = \frac{279 \times 12{,}000}{0.9 \times 60{,}000(0.875 \times 21.5)} = 3.30 \text{ in.}^2$$

If exactly this much is used, then

$$a = \frac{3.30 \times 60{,}000}{0.85 \times 3750 \times 14} = 4.43 \text{ in.}$$

$$\frac{a}{d} = \frac{a}{d_t} = \frac{4.43}{21.5} = 0.206$$

Since 0.206 is less than $a_b/d = 0.503$ from Table A-4, $f_s = f_y$ at ultimate. Because $a/d_t = 0.206$ is less than $a_{tcl}/d_t = 0.319$, the section is tension-controlled, and it follows that $\phi = 0.90$.

Because the original calculation of A_s was based on a guess at jd, we recompute and obtain

$$A_s = \frac{M_u \text{ (ft-kips)} \times 12{,}000}{0.9 \times 60{,}000(21.5 - 4.43/2)} = 0.00115 M_u$$

Lines 6 and 7 of Table 10-3 Calculate the area of steel required in negative moment regions as $A_s = 0.0115 M_u$. The "constant" 0.0115 was evaluated at the point of maximum negative moment and will be conservative at all other points of negative moment.

TABLE 10-3 Calculation of Reinforcement Required—Example 10-2

Line				
5. M_u, ft-kips	−152	260	−279	164
6. A_s coefficients	0.0115	0.0105	0.0115	0.0105
7. $A_{s(req'd)}$, in.^{2}	1.75	2.73	3.21	1.72
8. $A_s > A_{s(min)}$	Yes	Yes	Yes	Yes
9. Bars selected	3 No. 7	4 No. 7	6 No. 7	3 No. 7
		1 No. 6		
10. A_s provided, in.^{2}	1.80	2.84	3.60	1.80
11. b_w OK	—	Yes	—	Yes

(b)   Compute the area of steel required at the point of maximum positive moment (point B near the middle of the exterior span). In the positive-moment regions, the beam acts as a T-shaped beam with compression in the top flange. Assume that the compression zone is rectangular, as shown in Fig. 10-21a. Take $j = 0.95$ for the first calculation of A_s:

$$A_s = \frac{260 \times 12{,}000}{0.9 \times 60{,}000(0.95 \times 21.5)} = 2.83 \text{ in.}^2$$

Assume that a is less than h_f; then

$$a = \frac{A_s f_y}{0.85 f'_c b}$$

where b is the effective flange width of 81 in.

$$a = \frac{2.83 \times 60{,}000}{0.85 \times 3750 \times 81} = 0.66 \text{ in.} < h_f = 6.00 \text{ in.}$$

$$\frac{a}{d} = \frac{a}{d_t} = \frac{0.66}{21.5} = 0.0307$$

From Table A-4, it is seen that this is much less than the $a/d = 0.503$ for balanced failure; therefore, $f_s = f_y$. Similarly, $a/d_t = 0.0307$ is much less than $a/d_t = 0.319$ at the tension-controlled limit; therefore, $\phi = 0.90$.
 To compute A_s, in positive moment regions use

$$A_s = \frac{M_u \text{ (ft-kips)} \times 12{,}000}{0.9 \times 60{,}000(21.5 - 0.66/2)} = 0.0105 M_u$$

The area of steel required at positive moment regions is calculated in lines 6 and 7 of Table 10-3 by using $A_s = 0.0105 M_u$.

(c) **Calculate the minimum reinforcement.** The minimum reinforcement required is, by ACI Section 10.5.1,

$$A_{s,min} = \frac{3\sqrt{f'_c}}{f_y} b_w d, \text{ and } \geq \frac{200 b_w d}{f_y} \tag{4-32}$$

$$= \frac{3\sqrt{3750}}{60,000} \times 14 \times 21.5, \text{ and } \geq \frac{200 \times 14 \times 21.5}{60,000} \qquad \text{(ACI Eq. 10-3)}$$

$$= 0.922, \text{ and } \geq 1.00 \text{ in.}^2$$

Table 10-3, Line 10 A_s in line 10 of Table 10-3 exceeds $A_{s,min}$ in both the positive- and negative-moment regions.

(d) **Calculate the area of steel and select the bars.** The remaining calculations are done in Table 10-3. The areas computed in line 7 exceed $A_{s(min)}$ in all cases. If they did not, $A_{s(min)}$ should be used.

Lines 9 and 10 of Table 10-3 give the bars selected at each location, together with their areas. The bars were selected by using Table A-10. Small bars were selected at the exterior support, since they have to be hooked into the support and there may not be enough room for a standard hook on larger bars.

Line 11 of Table 10-3. Check whether the bars will fit into the beam web using Table A-6. In the negative-moment regions, some of the bars can be placed in the slab beside the beams; hence, it is not necessary to check whether they will fit into the web width.

9. **Check the distribution of the reinforcement.**

(a) **Positive-moment region.** Since f_y exceeds 40,000 psi, it is necessary to satisfy ACI Section 10.6.4. This will be done at the section with the smallest number of positive-moment bars: the middle of span B–C. Here, there are three No. 7 bars, as shown in Fig. 10-22a.
The clear cover to the surface of the flexural steel is

$$c_c = 1.5 \text{ in. cover} + 0.375 \text{ in. stirrups} = 1.875 \text{ in.}$$

The maximum bar spacing is

$$s = \frac{540}{f_s} - 2.5 c_c \leq 12 \frac{36}{f_y} \tag{9-7}$$

(ACI Eq. 10-4)

where $f_s = 0.6 f_y$ (in which $f_y = 60$ ksi). Thus,

$$s = \frac{540}{0.6 \times 60 \text{ ksi}} - 2.5 \times 1.875 = 10.3 \text{ in.} \leq 12 \text{ in.}$$

Since $b_w = 14$ in. and there are three bars, s is clearly smaller than 10.3 in.

(b) **Negative-moment region.** ACI Section 10.6.6 says "part" of the negative-moment steel shall be distributed over a width equal to the smaller of the effective flange width (81 in.) and $\ell_n/10 = 324/10 = 32.4$ in. At each of the interior negative-moment regions, there are six No. 7 top bars. Two of these will be placed in the corners of the stirrups, as shown in Fig. 10-22b, two over the beam web, and the other two in the slab. (Note that, for bars placed in the slab to be completely effective, there should be reinforcement perpendicular to the beam in the slab. In this case the slab reinforcement will serve this purpose.)

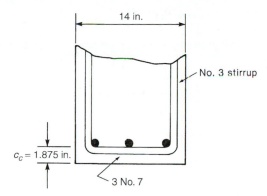

(a) Calculation of c_c in positive moment region.

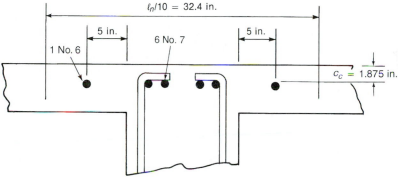

Fig. 10-22
Distribution of reinforcement.

(b) Bar spacing at interior support.

The two bars placed in the slab will be placed to give spacing, s, approaching the maximum allowed. The maximum spacing allowed is

$$s = \frac{540}{f_s} - 2.5\, c_c$$

$$= \frac{540}{0.6 \times 60 \text{ kips}} - 2.5 \times 1.875$$

$$= 10.3 \text{ in.}$$

where

$$c_c = 1.5 + 0.375 = 1.875 \text{ in.}$$

Within a width of 32.4 in. we must place six bars. These cannot be farther apart than 10.3 in. We shall arbitrarily place two bars in the slab at 5 in. outside the web of the beam.
ACI Section 10.6.6 requires "some" longitudinal reinforcement in the slab outside this band. We shall assume that the shrinkage and temperature steel already in the slab will satisfy this requirement. The final distribution of negative-moment steel is as shown in Fig. 10-22b.

10. Design the shear reinforcement. The shear-force diagrams are calculated in Table 10-4 and shown at the bottom of that table. The shear coefficients at midspan of the beams (line 5) are based on (10-4).

TABLE 10-4 Calculation of Shear Forces—Example 10-2

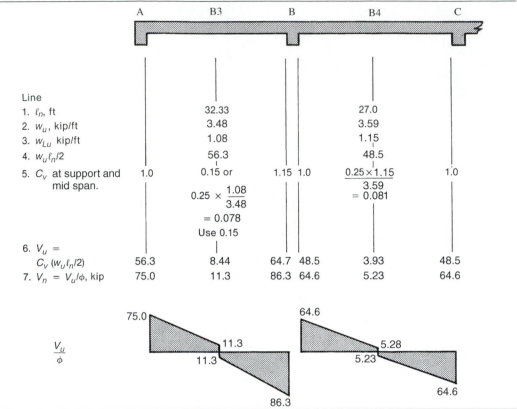

Line	A	B3	B	B4	C
1. ℓ_n, ft		32.33		27.0	
2. w_u, kip/ft		3.48		3.59	
3. w_{Lu} kip/ft		1.08		1.15	
4. $w_u \ell_n/2$		56.3		48.5	
5. C_V at support and mid span.	1.0	0.15 or $0.25 \times \dfrac{1.08}{3.48}$ $= 0.078$ Use 0.15	1.15 1.0	$\dfrac{0.25 \times 1.15}{3.59}$ $= 0.081$	1.0
6. $V_u = C_V(w_u \ell_n/2)$	56.3	8.44	64.7 48.5	3.93	48.5
7. $V_n = V_u/\phi$, kip	75.0	11.3	86.3 64.6	5.23	64.6

$$\frac{V_u}{\phi}$$

75.0

11.3

11.3

86.3

64.6

5.28

5.23

64.6

(a) **Exterior end of B3.** The critical section for shear is located at $d = 21.5$ in. from the support. From Table 10-4 and similar triangles, the shear at d from the support is

$$\frac{V_u}{\phi} \text{ at } d = 75.0 - \frac{21.5}{194}(75.0 - 11.3)$$

$$= 67.9 \text{ kips}$$

ACI Section 11.5.5.1 requires stirrups if $V_u \geq \phi V_c/2$, where

$$V_c = 2\sqrt{f'_c}\, b_w d$$

$$= 2\sqrt{3750} \times 14 \times \frac{21.5}{1000} = 36.9 \text{ kips}$$

$$\frac{V_c}{2} = 18.4 \text{ kips}$$

Since $V_u/\phi = 67.9$ kips exceeds $V_c/2 = 18.4$ kips, stirrups are required.

Try No. 3 Grade-40 double-leg stirrups with a 90° hook enclosing a No. 4 stirrup-support bar. The maximum stirrup spacing is the smaller of

$$\frac{d}{2} = 10.75 \text{ in.} \qquad\qquad \text{(ACI Section 11.5.4.1)}$$

and

$$s = \frac{A_v f_y}{50 b_w} = 12.6 \text{ in.} \qquad \text{(ACI Section 11.5.5.3)}$$

or

$$s = \frac{A_v f_y}{0.75\sqrt{f_c'} \cdot b_w} = 13.7 \text{ in.}$$

Use 10 in. as the maximum spacing.
The spacing required to support the shear forces is

$$s = \frac{A_v f_y d}{V_u/\phi - V_c} \qquad (6\text{-}21)$$

At d from end A, $V_u/\phi = 67.9$ kips, and

$$s = \frac{0.22 \times 40{,}000 \times 21.5}{(67.9 - 36.9) \times 1000}$$

$$= 6.10 \text{ in.} \text{—say, 6 in. on centers}$$

The stirrup spacing will be changed to 8 in. and then to 10 in. Compute V_u/ϕ, where $s = 8$ in.,

$$\frac{V_u}{\phi} = \frac{A_v f_y d}{s} + V_c$$

$$= \frac{0.22 \times 40 \times 21.5}{8} + 36.9$$

$$= 60.6 \text{ kips}$$

$s = 8$ in. occurs at

$$x = \frac{75.0 - 60.6}{75.0 - 11.3} \times 194 \text{ in.} = 43.9 \text{ in. from end } A$$

Compute V_u/ϕ and x where $s = 10$ in.:

$$\frac{V_u}{\phi} = 55.8 \text{ kips} \quad s = 10 \text{ at } x = 58.5 \text{ in.}$$

At the exterior end of B3, use No. 3 Grade-40 double-leg stirrups: one at 3 in., seven at 6 in., two at 8 in., and thirteen at 10 in. on centers.

(b) **Interior end of B3.** The shear at d from the support at B is

$$\frac{V_u}{\phi} = 86.3 - \frac{21.5}{194}(86.3 - 11.3)$$

$$= 78.0 \text{ kips}$$

The spacing required at this point is

$$s = \frac{0.22 \times 40{,}000 \times 21.5}{(78.0 - 36.9)1000}$$

$$= 4.6 \text{ in.}, \quad \text{so use } s = 4.5 \text{ in.}$$

Change the stirrup spacing to 6 in. and then to 10 in. V_u/ϕ, where $s = 6$ in.:

$$\frac{V_u}{\phi} = \frac{0.22 \times 40 \times 21.5}{6} + 36.9$$

$$= 68.4 \text{ kips}$$

$s = 10$ in. occurs at

$$x = \frac{86.3 - 68.4}{86.3 - 11.3} \times 194 = 46.3 \text{ in. from end } B$$

$V_u/\phi = 55.8$ kips at the point where spacing can change from 6 in. to 10 in. This occurs at

$$x = 78.8 \text{ in. from end } B$$

At the interior end of *B*3, use No. 3 Grade-40 double-leg stirrups: one at 2 in., ten at 4.5 in., five at 6 in., and eleven at 10 in. on centers.

(c)   **Ends of beam *B*4.** The calculations for beam *B*4 are carried out in the same manner as for beam *B*3. **At each end of beam *B*3, use No. 3 Grade-40 double-leg stirrups: one at 4 in., three at 8 in., and eleven at 10 in.**

11.   Check the development lengths and design-bar cutoffs.

(a)   **Perform the preliminary calculations.**

Detailing requirements:

$$d = 21.5 \text{ in.}$$

$12d_b = 10.5$ in. for No. 7 and 9 in. for No. 6.

$$\frac{\ell_n}{16} = \frac{388}{16} = 24.25 \text{ in. for } B3$$

$$= \frac{324}{16} = 20.25 \text{ in. for } B4$$

Therefore, d always exceeds $12d_b$, but $\ell_n/16$ exceeds d in beam *B*3.

(b)   **Select cutoffs for positive-moment steel in *B*3.** The reinforcement at midspan is four No. 7 plus one No. 6 bars (Table 10-3). Extend two No. 7 plus 1 No. 6 bar into the supports and cut off two No. 7 where they are no longer required. The total area of steel before the cut-offs is 2.84 in.^{2}. The cut-off bars have $A_s = 1.20$ in.^{2}. The remaining bars have 58 percent of the original area; therefore, the flexural cutoff is at the point where $M_u = 58$ percent of the maximum M_u at midspan. From Fig. A-3, this occurs at $0.21\ell_n = 0.21 \times 388 = 81.5$ in. from the exterior end, and $0.28\ell_n = 109$ in. from the interior end. These are points *B* and *C* in Fig. 10-23. To compute the actual cutoff points we must satisfy the detailing rules in ACI Sections 12.1, 12.10, 12.11, and 12.12. These are summarized in the form of four families of "rules" in Section 8-7 of this book. The following discussion will refer to that section.

Before doing so we shall calculate ℓ_d for the bottom bars.

Spacing and confinement case. The bottom bars have clear spacing and cover of at least d_b and are enclosed by at least minimum stirrups. Therefore, this is case 1 in Table 8-1 (ACI Section 12.2.2). The bars are No. 7. Thus, we have

$$\ell_d = \frac{f_y \alpha \beta \lambda}{20\sqrt{f_c'}} d_b \tag{8-10}$$

$$= \frac{60{,}000 \times 1.0 \times 1.0 \times 1.0 \times d_b}{20 \times \sqrt{3750}} = 49d_b$$

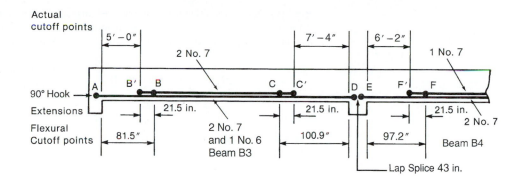

Fig. 10-23
Calculation of bar cutoff points for positive-moment steel.

For the No. 7 bars, $d_b = 0.875$ in. and thus $\ell_d = 42.9$ in. For the No. 6 bars, $d_b = 0.75$ in. and thus $\ell_d = 36.8$ in.

Application of detailing rules to beam B3

Some or all of each family of detailing rules apply to each region of the beam. The detailing process starts by selecting the rules that apply in each region. It is then easier to select the sub-rules applied in each region. *Beam B3* must satisfy rules: 1. Structural integrity; 2. Extension of bars into supports; 3. Effect of shear on bending moment diagrams; and 4. Anchorage, positive moment.

Structural integrity, Rule 1. This beam is a continuous interior beam. ACI Section 7.13.2.4 applies (see rule 1b). In beam line *B3–B4–B3*, a tension tie is created by lap splicing at least one-quarter of the positive-moment reinforcement, but not less than two bars, from span *B3* with a similar amount of steel from beam *B4*. The splices will be Class A splices, at or near the supports, with a lap of $1.0\ell_d$.

From Table A-11 the development length of a No. 7 bottom bar in 3750 psi concrete is $49d_b = 43$ in.

Lap splice two No. 7 bottom bars from span *B3*, a distance of 43 in. with two No. 7 bottom bars from span *B4*.

The two No. 7 bottom bars which will be part of the longitudinal structural integrity tie must be placed inside the lower corners of U stirrups. The longitudinal tie bars must be hooked into the supports at the two outer ends of the beam using standard 90°. hooks at the discontinuous ends.

Extension of bars into the supports, Rule 2. At least one-quarter of the positive moment reinforcement, but not less than two bars from span *B3* must extend a minimum of 6 in. into the interior supports. This extension is not long enough for straight anchorage at exterior supports. These bars shall be hooked into the supports at the discontinuous ends and shall have lap splices at or near the interior supports. This bar assignment has already been satisfied by *Rule 1.*

Effect of shear, Rule 3. The bars must extend d or $12d_b$ past the positive moment flexural cutoff points. **Extend bars d = 21.5 in. = 1.79 ft to points B' and C'** (Fig. 20-23).

Anchorage, Rule 4. The bars must extend ℓ_d past the point of maximum bar stress. The maximum bar stress occurs at midspan. By inspection, the distances from midspan to B' and C' exceed $\ell_d = 43$ in. Therefore, cut off two No. 7 bars at 81.5 in. − 21.5 in. = 5ft, from the interior face of the exterior column, and 87.5 in., say 7 ft 4 in., from the exterior face of the interior column, as determined using rule 3a.

Anchorage, Rule 4a, positive moment. Bars must extend ℓ_d past the actual cutoff points of adjacent bars. The actual extensions past B' and C' exceed 60 in. and 88 in., respectively, both of which are greater than $\ell_d = 43$ in.

Anchorage, Rule 4(e), point of inflection. At positive-moment points of inflection, the bars must satisfy

$$\ell_d \leq \frac{M_n}{V_u} + \ell_a \qquad (8\text{-}20)$$

From Fig. A-3, the positive-moment points of inflection are at $0.098\ell_n = 38$ in. from the exterior end and $0.146\ell_n = 56.6$ in. from the interior end. At each of these points, the remaining steel is 2 No. 7 plus 1 No. 6 bars, $A_s = 1.64$ in.^{2}. Accordingly we have

$$M_n = 1.64 \times 60{,}000\left(21.5 - \frac{1.64 \times 60{,}000}{1.7 \times 3750 \times 81}\right)$$

$$= 2100 \text{ in.-kips}$$

Check at the exterior end: from the shear-force diagram in Table 10-4, the V_u at 38 in. from the exterior end is

$$\frac{V_u}{\phi} = 75.0 - \frac{38}{194}(75.0 - 11.3) = 62.5 \text{ kips}$$

$$V_u = 46.9 \text{ kips}$$

and ℓ_a is the smaller of the actual extension (which exceeds 38 in.) and d, or $12d_b$, so

$$\ell_a = 21.5 \text{ in.}$$

Thus,

$$\frac{M_n}{V_u} + \ell_a = \frac{2100}{46.9} + 21.5 = 66.3 \text{ in.}$$

This exceeds $\ell_d = 43$ in.—therefore, OK.
    Check at interior end:

$$V_u \text{ at 56.6 in. from interior end} = 48.3 \text{ kips}$$

$$\frac{M_n}{V_u} + \ell_a = \frac{2100}{48.3} + 21.5$$

$$= 65.0 \text{ in.—therefore, OK}$$

Summary of the bar cutoff points for the positive moment steel in beam B3. This is the end span of a continuous interior beam. Hook two No. 7 bars in the exterior support and lap-splice two No. 7 bars in the interior support. Cut off two No. 7 bars at 5 ft from the face of the exterior column and at 7 ft 4 in. from the face of the interior column. These locations are shown as points A, B', C', and D in Fig. 10-23.

   (c)   **Select cutoffs for the positive-moment steel in B4.** This is an interior a continuous interior beam. At midspan, we have three No. 7 bars. Run two No. 7 bars into the supports and cut off the remaining No. 7 bar. The A_s remaining is $1.20/1.80 = 0.67$ times the original amount. Therefore, the flexural cutoff is located at the point where the moment is 0.67 times the midspan moment. From Fig. A-1, this occurs at $0.30\ell_n = 0.30 \times 324 = 97.2$ in. from the face of the support (Fig. 10-23). Applying rules 3a and 4e we that find the bar must extend 21.5 in. past this point, to 75.7 in. from the support. Therefore, cut off one No. 7 bar at 6 ft 2 in. from the support.
    The bars have a spacing and cover of at least d_b and are enclosed by at least minimum stirrups. Therefore, for the No. 7 bars, $\ell_d = 49 \times 0.875 = 42.9$ in.
    The remaining bars must satisfy rules 1c and 4e. Inspection shows that rule 4a is satisfied.

   Structural integrity, Rule 1. Lap splice the two No. 7 bars with the bars from beam $B3$. The length of a Class A lap splice for a No. 7 bar is $1.0\ell_d = 42.9$ in. Therefore, lap splice the bars 43 in.

   Anchorage, Rule 4e. The positive-moment points of inflection are $0.146\ell_n = 47.3$ in. from supports. At these sections the reinforcement is two No. 7 bars, and we have

$$M_n = 2 \times 0.60 \times 60{,}000\left(21.5 - \frac{2 \times 0.60 \times 60{,}000}{1.7 \times 3750 \times 81}\right)$$

$$= 1540 \text{ in.-kips}$$

$$V_u = 35.4 \text{ kips}$$

$$\frac{M_n}{V_u} + \ell_a = \frac{1540}{35.4} + 21.5 = 64.9 \text{ in.}$$

This exceeds ℓ_d—therefore, OK. The cutoff points for the positive-moment steel are as indicated in Fig. 10-23.

(d)   Select cutoffs for the negative-moment steel at the exterior end of B3. At the exterior end, there are three No. 7 bars. These must be anchored in the spandrel beam, which is 16 in. wide. The basic development lengths of standard hooks are (ACI Section 12.5) and book Section 8-4.

$$\ell_{dh} = \left(\frac{0.02 \beta \lambda f_y}{\sqrt{f'_c}} \right) d_b$$

For

$$\text{No. 7 bar: } \ell_{hb} = 17.2 \text{ in.}$$

This can be multiplied by 0.7 if the side cover (perpendicular to the plane of the hook) exceeds 2.5 in. and for 90° hooks, if the tail cover is not less than 2 in. The hooks in question enter the spandrel beam perpendicular to the axis of the beam and hence satisfy the side cover requirement. The tail cover can be set at 2 in. Therefore,

$$\ell_{dh} \text{ for No. 7 bar} = 17.2 \times 0.7 = 12.0 \text{ in.}$$

This can be accommodated in the spandrel beam. **Therefore, anchor the top bars into the spandrel beam with 90° standard hooks.**

Extend all the top bars past the negative-moment point of inflection before cutting them off. From Fig. A-3, the negative moment point of inflection is at $0.108\ell_n = 41.9$ in. from the face of the exterior support (point G in Fig. 10-24). At this cutoff point, we must satisfy rules 1, 2, and 6.

Effects of shear, Rule 3b requires at least one-third of the bars to extend the longest of d, $12d_b$, and $\ell_n/16$ past the point of inflection. $\ell_n/16 = 24.3$ in. governs here. Therefore, cut off all bars at $41.9 + 24.3 = 66.2$ in., say 5 ft 6 in. from the face of the support.

Rule 3b automatically satisfies rule 1c. Rule 4b requires that the bars extend at least ℓ_d from the face of the support.

$$\ell_d \text{ for No. 7 top bar} = 63.7d_b = 55.7 \text{ in.} - \text{say, 56 in.}$$

Since 5 ft 6 in. exceeds 56 in., rule 4b is satisfied. **Therefore, cut off all top bars at the exterior support at 5 ft 6 in. from the face of the support.**

(e)   Select cutoffs for the negative-moment steel at the interior end of B3. The negative-moment steel at the first interior support consists of six No. 7 bars. We will cut off two of these bars when they are no longer needed for flexure and extend the remaining bars $(0.67A_s)$ past the negative-moment point of inflection. From Fig. A-3, $M_u = 0.67M_{u(\max)}$ at $0.065\ell_n = 25.2$ in. from the face of the support (point H in Fig. 10-24). Here, rules 3b and 4b apply.

Effects of shear, Rule 3b. Bars must extend $d = 21.5$ in. past H to 46.7 in., say 4 ft from the face of the support.

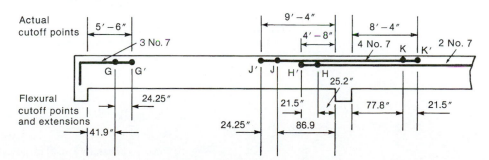

Fig. 10-24
Calculation of bar cutoff points for negative-moment steel.

Anchorage, Rule 4b, negative moment. Bars must extend a distance ℓ_d from the point of maximum bar stress (at the face of the support). 4 ft does not exceed $\ell_d = 56$ in. Therefore, extend bars so that cutoff H' is 56 in. from the face of the support.

Cut off two No. 7 top bars at 4 ft 8 in. from the exterior face of the first interior support. The remaining four bars will be extended past the negative moment point of inflection. From Fig. A-3, this is $0.224\ell_n = 86.9$ in. from the face of the support (point J in Fig. 10-24). Here, detailing rules 3b and 4b apply.

Effects of shear, Rule 3b. At least one-third of bars must extend $\ell_n/16 = 24.3$ in. past the point of inflection to 111.2 in. from the face of the support, say 9 ft 4 in. from the face of the support.

Anchorage, Rule 4b. Bars must extend $\ell_d = 56$ in. past the actual cutoff H'. Therefore, the bars must extend to 112 in. from the face of the support. This is exactly satisfied.

Cut off four No. 7 top bars at 9 ft 4 in. from the exterior face of the first interior support.

(f)   Select cutoffs for the negative moment steel in beam B4. When the interior span of a three-span continuous beam is less than about 90 percent as long as the end spans, it can have negative moment at midspan under some loads, even though this is not apparent from the ACI moment coefficients. For this reason, we will extend two No. 7 top bars the full length of the beam. The remaining four bars will be extended past the negative-moment point of inflection and will be cut off. From Fig. A-1 the negative-moment point of inflection is $0.24\ell_n = 77.8$ in. from the face of the column (point K in Fig. 10-24).

Effects of shear, Rule 3a says that these bars must extend by the larger of d, $12d_b$, or $\ell_n/16$ past this point to $77.8 + 21.5 = 99.3$ in. from the face of the support.

Anchorage, Rule 4b says that these bars must extend by $\ell_d = 56$ in. past the face of the support—therefore OK.

Cut off four No. 7 top bars at 8 ft 4 in. from the interior face of the interior support, and extend the remaining two No. 7 top bars the full length of the interior span (see Fig. 10-24).

(g)   Check shear at points where bars are cut off in a zone of flexural tension. Cutoffs B', C', F', and H' occur in zones of flexural tension. ACI Section 12.10.5 requires special consideration of these regions if V_u exceeds two-thirds of $\phi(V_c + V_s)$.

Cutoff B': 5 ft from support, $V_u = 41.5$ kips. At this point there are No. 3 double-leg stirrups at 8 in. on centers. $V_c = 36.8$ kips.

$$V_s = \frac{0.22 \times 40 \times 21.5}{8} = 23.6 \text{ kips}$$

$$\phi(V_c + V_s) = 0.75(36.9 + 23.6) = 45.4 \text{ kips}$$

Thus, V_u is 0.92 of the shear permitted and ACI Sections 12.10.5.1 and 12.10.5.3 do not apply. Therefore, we must add extra stirrups. From ACI Sections 12.10.5.2,

$$\text{maximum spacing} = \frac{d}{8\beta_b}$$

where

$$\beta_b = \frac{\text{area of bars cut off}}{\text{area immediately before cutoff}} = 0.42$$

Therefore, the maximum spacing is $21.5/(8 \times 0.42) = 6.40$ in.

The added stirrups must provide $A_v = 60b_w s/f_y$. This corresponds to a maximum spacing of

$$s = \frac{A_v f_y}{60 b_w}$$

$$= \frac{0.22 \times 40{,}000}{60 \times 14}$$

$$= 10.5 \text{ in.} \qquad (\text{Therefore, 6.40 in. governs.})$$

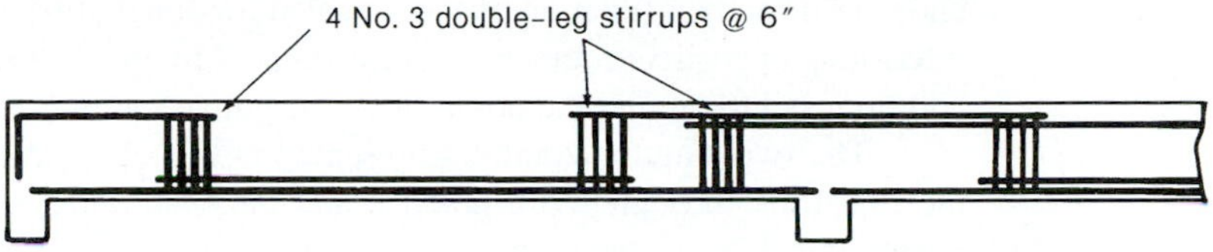

4 No. 3 double–leg stirrups @ 6″

Fig. 10-25
Additional stirrups at bar cutoff points.

Extra stirrups are required for $0.75d = 16.1$ in. **Provide four No. 3 double-leg Grade-40 stirrups at 6 in. o.c., the first stirrup at 3 in. from the end of the bar at B'.** (See Fig. 10-25.) These extend along the bar which is cut off at B'.

Cutoff C': Similar calculations show that extra stirrups are required at cutoff C'. These will be the same as at B'.

Cutoff F': At 6 ft 2 in. from the support, $V_u = 28.1$ kips. At this point, there are No. 3 double-leg stirrups at 10 in. o.c., and $\phi(V_c + V_s) = 41.8$ kips. Thus, V_u is 0.67 of the shear permitted, and again, extra stirrups are required. We shall use the same detail as before.

Cutoff H': At 4 ft 8 in. from the support, $V_u = 48.5$ kips. At this point, there are No. 3 stirrups at 6 in. o.c., and $\phi(V_c + V_s) = 51.3$ kips. Again, stirrups are required, and the same detail will be used as at B' to avoid confusion. The extra stirrups are shown in Fig. 10-25.

12.   **Design web-face steel.** Since d does not exceed 3 ft, no side-face steel is required (ACI Section 10.6.7).

This concludes the design. The bar detailing in step 11 of this example is quite lengthy. The process could be simplified considerably if all bars were extended past the points of inflection before being cut off or if the standard bar details in Fig. A-5 were used.   ■

In Example 10-2, the moments were computed by using the ACI moment coefficients. Figure 10-26 shows the elastically computed moment envelope for this beam, calculated by using the reduced live load of 78.4 psf computed in step 1c for positive moment in beam *B*4 (the center span). The beam was modeled for analysis with springs at the two discontinuous ends to represent torsional stiffness of the spandrel beams. Their spring stiffness was derived from (7-43).

The moments plotted in Fig. 10-26 can be compared to the moments from the ACI moment coefficients, shown in the moment diagram in the heading of Table 10-3 and given (in brackets) in Fig. 10-26. Moments in the end spans and at the interior supports are relatively close between the two procedures. The negative moments at the exterior

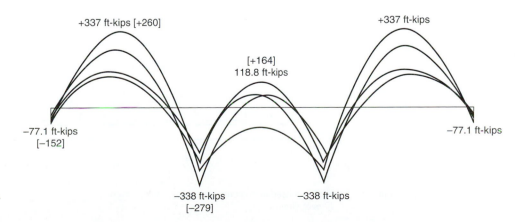

+337 ft-kips [+260]   +337 ft-kips

[+164]
118.8 ft-kips

−77.1 ft-kips   −77.1 ft-kips
[−152]

−338 ft-kips   −338 ft-kips
[−279]

Fig. 10-26
Elastic-moment envelopes—
Example 10-2.

ends are directly a function of the assumed torsional stiffness of the spandrel beams. The interior-span positive moment is overestimated by the ACI moment coefficients, which, in addition, fail to predict the possibility of negative moment at midspan of the interior span.

The two lengthy examples presented in this chapter appear somewhat cumbersome the first time through. With practice and understanding of the principles, the design of beams and slabs becomes straightforward. The beams and slabs designed for one floor can frequently be used many times throughout the building.

10-6   DESIGN OF GIRDERS

Girders support their own weight plus the concentrated loads from the beams they support. It is customary to compute the moments and shears in the girder by assuming that the girder supports concentrated loads equal to the beam reactions, plus a uniform load equal to the self-weight of the girder, plus the live load applied directly over the girder. This approach neglects two-way action in the slab adjacent to the girder. The use of moment coefficients in such a case was discussed in the last paragraph or Section 10-3.

When the flexural reinforcement in the slab runs parallel to the girder, the two-way action shown in Fig. 10-12 causes negative moments in the slab along the slab–girder connection, as indicated by the curvature of slab strip *B* in Fig. 10-12. To reinforce for these moments, ACI Section 8.10.5 requires slab reinforcement transverse to the girder, as shown in Fig. 10-27. The reinforcement is designed by assuming that the slab acts as a cantilever, projecting a distance equal to the effective overhanging slab width and carrying the factored dead load and live load supported by this portion of the slab.

Frequently, edge girders are loaded in torsion by beams framing into them from the sides between the ends of the girder. This is *compatibility torsion*, since it exists only because the free-end rotation of the beam is restrained by the torsional stiffness of the girder. In such a case, ACI Section 11.6.2.2 allows a reduction in the design torque, as discussed in Chapter 7. If torque is present, the stirrups must be closed stirrups (ACI Sections 7.11.2 and 11.6.4.1).

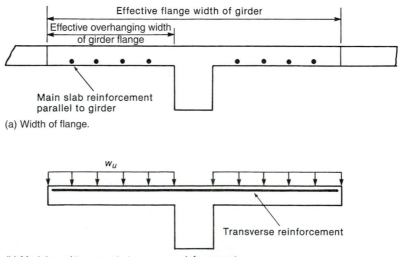

Fig. 10-27
Transverse reinforcement over girders.

10-7   JOIST FLOORS

Long-span floors for relatively light live loads can be constructed as a series of closely spaced, cast-in-place T beams (or *joists*) with a cross section as shown in Fig. 10-28. The joists span one way between beams. Most often, removable metal forms referred to as *fillers* or *pans* are used to form the joists. Occasionally, joist floors are built by using clay-tile fillers, which serve as forms for the concrete in the ribs and which are left in place to serve as the ceiling (ACI Section 8.11.5).

When the dimensions of the joists conform to ACI Sections 8.11.1 to 8.11.3, they are eligible for less cover to the reinforcement than for beams (ACI Section 7.7.1(c)) and for a 10 percent increase in the shear, V_c, carried by the concrete (ACI Section 8.11.8). The principal

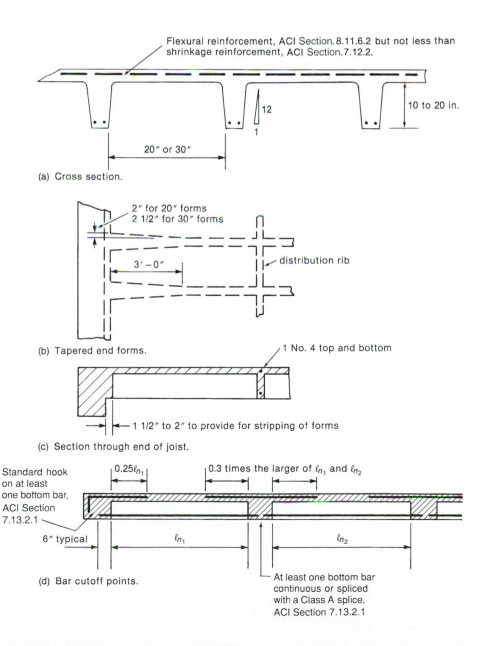

Fig. 10-28
One-way-joist construction.

requirements are that the floor be a monolithic combination of regularly spaced ribs and a top slab with

1. ribs not less than 4 in. in width,
2. depth of ribs not more than $3\frac{1}{2}$ times the minimum web width, and
3. clear spacing between ribs not greater than 30 in.

Ribbed slabs not meeting these requirements are designed as slabs and beams.

The slab thickness is governed by ACI Section 8.11.6.1, which requires a thickness of not less than 2 in. for joists formed with 20-in.-wide pans and $2\frac{1}{2}$ in. for 30-in. pans. The slab thickness and cover to the reinforcement are also a function of the fire-resistance rating required by the local building code. For a 1-hour fire rating, a $\frac{3}{4}$ in. cover and a 3- to $3\frac{1}{2}$ in. slab are required. For a 2-hour rating, a 1-in. cover and a $4\frac{1}{2}$ in. slab are required. These vary from code to code and as a function of the type of ceiling used. Joist floor systems are also used for parking garages. Here, the top cover of the slab reinforcement and the quality of concrete used must be chosen to reduce the possibility of corrosion (ACI Chapter 4). Epoxy-coated or galvanized bars are often used in such applications. Wheel loads on the slab and abrasion of the concrete in traffic paths should also be considered.

Standard Dimensions and Layout of One-Way Joists

The dimensions of standard removable pan forms are shown in Fig. 10-28. The forms are available in widths of 20 and 30 in., measured at the bottom of the ribs. Standard depths are 6, 8, 10, 12, 14, 16, and 20 in. End forms have one end filled in to serve as the side form for the supporting beams. Tapered end forms allow the width of the ends of the joists to increase, as shown in Fig. 10-28b, giving additional shear capacity at the end of the joists. The 20-in. tapered end pans reduce in width from 20 in. to 16 in. over a 3-ft length. The 30-in. pans reduce to 25 in., also over a 3-ft length. Pan forms wider than 30 in. are also available, but the resulting floors or roofs must be designed as conventional slabs and beams (ACI Section 8.11.4).

When such a floor is laid out, the rib and slab thicknesses are governed by strength, fire rating, and the available space. The overall depth and rib thickness are governed by deflections and shear. No stirrups are used in most cases. Generally, it is most economical to use a constant depth of form for all joists in a given floor. The most economical forming results if the joists and supporting beams have the same depth. Frequently, this will involve beams considerably wider than the columns. Such a system is referred to as a *joist-band* system. Frequently, it is not possible to have beams and joists of the same depth, however, and the beams will be deeper than the joists. Generally, it is most economical to use untapered end forms whenever possible, and to use tapered forms only at supports having higher-than-average shears.

Although not required by the ACI Code, load-distributing ribs perpendicular to the joists are provided at the midspan or at the third points of long spans. These have at least one continuous No. 4 bar at the top and the bottom. The CRSI Handbook suggests no load-distributing ribs in spans of up to 20 ft, one at midspan for spans of 20 to 30 ft, and two at the third points for spans over 30 ft.

For joist floors meeting the requirements of ACI Section 8.3.3, the moment and shear coefficients can be used in design, taking ℓ_n as the clear span of the joists themselves. For uneven spans, it is necessary to analyze the floor. If the joists are supported on wide beams or wide columns, the width of the supports should be modeled in the analysis. The negative moments in the ends of the joists will be underestimated if this is not done.

Reinforcing Details

The bar cutoffs given in Fig. 10-28d apply if the joists fit the requirements of ACI Section 8.3.3. The CRSI Handbook recommends cutting off half of the top and bottom bars at shorter lengths. If this is done, the shear at the points where bars are cut off in a zone of flexural tension should satisfy ACI Section 12.10.5. ACI Section 7.13.2.1 requires at least one bottom bar to be continuous or spliced over supports with a Class A tension splice. At edge beams, at least one of the bottom bars in each joist must be anchored into the beam with a standard hook.

10-8   MOMENT REDISTRIBUTION

As a continuous beam is loaded beyond its service loads, the tension reinforcement will eventually yield at some section. With further loading, this section will deform as a plastic hinge, at a constant moment taken equal to the nominal moment. Increased loads cause an increase in the sum of the positive and negative moments in the beam. Since the moment cannot increase at the plastic hinge, the increase in moments all occurs at the positive-moment hinge location. Thus, the shape of the moment diagram must shift, with more moment going to those sections which are still elastic. Eventually, a mechanism forms, and the plastic capacity is reached. Procedures for carrying out inelastic analyses of concrete structures are given in [10-8], [10-9], and [10-10].

For a uniformly loaded fixed-end beam, the maximum elastic positive moments $(w\ell^2/24)$ are half the maximum negative moment $(w\ell^2/12)$. As a result, approximately twice as much reinforcement is required at the supports compared with what is required at midspan. Sometimes, this leads to congestion of the steel at the supports. The plastic-moment distribution results in a much more variable reinforcement layout depending on the actual plastic-moment capacities in the positive- and negative-moment regions, and moments can be redistributed from the elastic distribution, allowing reductions in the peak moments, with corresponding increases in the lower moments. The amount of redistribution that can be tolerated is governed by two aspects. First, the hinging section must be able to undergo the necessary inelastic deformations. Since the inelastic rotational capacity is a function of the reinforcement ratio, as shown in Fig. 4-16, this implies an upper limit on the reinforcement ratio. Second, hinges should not have occurred at service loads, since wide cracks develop at hinge locations.

ACI Section 8.4 allows the negative moments at the supports to be increased or decreased by not more than $1000\epsilon_t$ percent, with a maximum of 20 percent, provided that $\epsilon_t \geq 0.0075$ at the section where moments are being reduced, where ϵ_t is the tensile strain in the layer of steel closest to the tension face of the member.

These moments have been computed via an elastic analysis. The modified negative moments must be used in calculating the moments at all critical sections, at the shears, and at the bar cutoff points. (See ACI Sections 8.4.1 to 8.4.3 or Sections B.8.4.1 to B.8.4.3.)

EXAMPLE 10-3   Moment Redistribution

Compute the design moments for the three-span beam from Example 10-2, using an elastic analysis (Fig. 10-26) and moment redistribution. Use ACI Sections 8.4.1 through 8.4.3. Step 7 of the example will be repeated. Using $c = a/\beta_1 = 4.84/0.85 = 5.69$ in.,

$$\epsilon_t = \frac{21.5 - 5.69}{5.69} \times 0.003$$

$$= 0.0083$$

Thus, the maximum redistribution that is allowed is $(1000\epsilon_t = 8.3)$ percent. The reduced moment is accordingly $(1 - 0.083) \times 338 = 310$ ft-kips. Check whether $A_s = 3.60$ in.^{2} is adequate:

$$\phi M_n = \frac{0.90 \times 3.60 \times 60,000\left(21.5 - \dfrac{4.84}{2}\right)}{12,000}$$

$$= 309 \text{ ft-kips}$$

We shall assume that this is close enough.

It is now necessary to compute the positive moments corresponding to this value of the negative moment. These moments correspond to a factored load of 3.56 kips/ft on two adjacent spans and an elastically computed moment of −61.8 ft-kips at the exterior end. For this loading, the maximum positive moment is 289 ft-kips. The maximum positive moment in the end spans comes from a load case with uniform loads of 3.90 kips/ft on the two end spans. For this case, the interior negative moments were −278 ft-kips, less than the negative moment after redistribution (−309 ft-kips). Hence, this moment diagram cannot be redistributed.

The final moments after redistribution are as follows:

Exterior end negative moment	−77.1 ft-kips
Exterior span positive moment	337 ft-kips
Interior support negative moment	309 ft-kips
Midspan interior span: positive	118.8 ft-kips
negative	−92.1 ft-kips

In this case, very little was accomplished by redistributing the moments. ∎

PROBLEMS

10-1 A five-span one-way slab is supported on 12-in.-wide beams with center-to-center spacing of 16 ft. The slab carries a superimposed dead load of 10 psf and a live load of 100 psf. Using $f'_c = 3000$ psi and $f_y = 60,000$ psi, design the slab. Draw a cross section showing the reinforcement. Use Fig. A-5 to locate bar -cutoff points.

10-2 A four-span one-way slab is supported on 12-in.-wide beams with center-to-center spacing of 14, 16, 16, and 14 ft. The slab carries a superimposed dead load of 20 psf and a live load of 150 psf. Design the slab, using $f'_c = 3000$ psi and $f_y = 60,000$ psi. Select bar -cutoff points, using Fig. A-5, and draw a cross section showing the reinforcement.

10-3 A three-span continuous beam supports 6-in.-thick one-way slabs that span 15 ft center to center of beams. The beams have clear spans, face to face of 16-in.-square columns, of 27, 30, and 27 ft. The

floor supports ceiling, ductwork, and lighting fixtures weighing a total of 8 psf, ceramic floor tile weighing 16 psf, partitions equivalent to a uniform dead load of 20 psf, and a live load of 100 psf. Design the beam, using $f'_c = 3000$ psi. Use $f_y = 60,000$ psi for flexural reinforcement and $f_y = 40,000$ psi for shear reinforcement. Calculate cutoff points, extending all reinforcement past points of inflection. Draw an elevation view of the beam and enough cross sections to summarize the design.

10-4 Repeat Problem 10-3, but use standard bar cutoff points from Fig. A-5.

10-5 Repeat Problem 10-3, but cut off up to 50 percent of the negative- and positive-moment bars in each span where they are no longer needed.

10-6 Explain the reason for the two live-load patterns specified in ACI Section 8.9.2.

11

Columns: Combined Axial Load and Bending

11-1   INTRODUCTION

A column is a vertical structural member supporting axial compressive loads, with or without moments. The cross-sectional dimensions of a column are generally considerably less than its height. Columns support vertical loads from the floors and roof and transmit these loads to the foundations.

In construction, the reinforcement and concrete for the beams and slabs in a floor are placed. Once this concrete has hardened, the reinforcement and concrete for the columns over that floor are placed, followed by the next-higher floor. This process is illustrated in Figs. 10-3, 11-1, and 11-2. Figure 11-1 shows a completed column prior to construction of the formwork for the next floor. This is a *tied column*, so called because the longitudinal bars are tied together with smaller bars at intervals up the column. One set of ties is visible just above the concrete. The longitudinal (vertical) bars protruding from the column will extend through the floor into the next-higher column and will be lap spliced with the bars in that column. The longitudinal bars are bent inward to fit inside the cage of bars for the next-higher column. (Other splice details are sometimes used; see Fig. 11-25.) A reinforcement cage in place ready for the column forms is shown in Fig. 11-2. The lap splice at the bottom of the column and the ties can be seen in this photograph. Typically, a column cage is assembled on sawhorses prior to erection and is then lifted into place by a crane.

The more general terms *compression members* and *members subjected to combined axial load and bending* are sometimes used to refer to columns, walls, and members in concrete trusses or frames. These may be vertical, inclined, or horizontal. A column is a special case of a compression member that is vertical.

Stability effects must be considered in the design of compression members. If the moments induced by slenderness effects weaken a column appreciably, it is referred to as a *slender column* or a *long column*. The great majority of concrete columns are sufficiently stocky that slenderness can be ignored. Such columns are referred to as *short columns*. Slenderness effects are discussed in Chapter 12.

Fig. 11-1
Tied column under construc-
tion. (Photograph courtesy of
J. G. MacGregor.)

Fig. 11-2
Reinforcement cage for a tied
column. (Photograph cour-
tesy of J. G. MacGregor.)

Although the theory developed in this chapter applies to columns in seismic regions, such columns require special detailing to resist the shear forces and repeated cyclic loads from earthquakes. This is discussed in Chapter 20.

11-2   TIED AND SPIRAL COLUMNS

Over 95 percent of all columns in buildings in nonseismic regions are tied columns similar to those shown in Figs. 11-1 and 11-2. Tied columns may be square, rectangular, L-shaped, circular, or any other required shape. Occasionally, when high strength and/or high ductility are required, the bars are placed in a circle, and the ties are replaced by a bar bent into a helix or spiral, with a pitch (distance between successive turns of the spiral—distance s in Fig. 11-4a) from $1\frac{3}{8}$ to $3\frac{3}{8}$ in. Such a column, called a *spiral column*, is illustrated in Fig. 11-3. Spiral columns are generally circular, although square or polygonal shapes are sometimes used. The spiral acts to restrain the lateral expansion of the column core under axial loads causing crushing and, in doing so, delays the failure of the core, making the column more ductile, as discussed in the next section.

In seismic regions, the ties are heavier and much more closely spaced than those shown in Figs. 11-1 and 11-2. Spiral columns are used more extensively in such regions.

Behavior of Tied and Spiral Columns

Figure 11-4a shows a portion of the core of a spiral column enclosed by one and a half turns of the spiral. Under a compressive load, the concrete in this column shortens longitudinally

Fig. 11-3
Spiral column. (Photograph courtesy of J. G. MacGregor.)

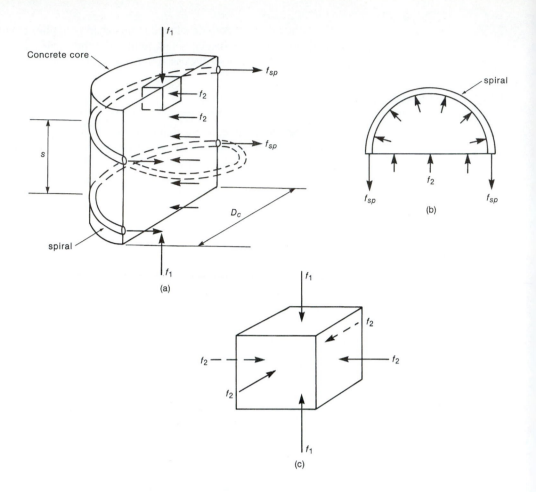

Fig. 11-4
Triaxial stresses in core of
spiral column.

under the stress f_1 and so, to satisfy Poisson's ratio, it expands laterally. This lateral expansion is especially pronounced at stresses in excess of 70 percent of the cylinder strength, as was discussed in Chapter 3. In a spiral column, the lateral expansion of the concrete inside the spiral (referred to as the *core*) is restrained by the spiral. This stresses the spiral in tension, as is shown in Fig. 11-4b. For equilibrium, the concrete is subjected to lateral compressive stresses f_2. An element taken out of the core (Fig. 11-4c) is subjected to triaxial compression. In Chapter 3, triaxial compression was shown to increase the strength of concrete:

$$f_1 = f'_c + 4.1f_2 \qquad (3\text{-}16)$$

Later in this chapter, (3-16) is used to derive an equation for the amount of spiral reinforcement needed in a column.

In a tied column in a nonseismic region, the ties are spaced roughly the width of the column apart and, as a result, provide relatively little lateral restraint to the core. Outward pressure on the sides of the ties due to lateral expansion of the core merely bends them outward, developing an insignificant hoop-stress effect. Hence, normal ties have little effect on the strength of the core in a tied column. They do, however, act to reduce the unsupported length of the longitudinal bars, thus reducing the danger of buckling of those bars as the bar stress approaches yield. The arrangement of ties is discussed in Section 11-5.

Figure 11-5 presents load-deflection diagrams for a tied column and a spiral column subjected to axial loads. The initial parts of these diagrams are similar. As the maximum load is reached, vertical cracks and crushing develop in the concrete shell outside the ties or spiral, and this concrete spalls off. When this occurs in a tied column, the capacity of the

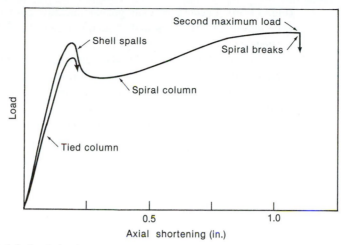

(a) Axially loaded columns.

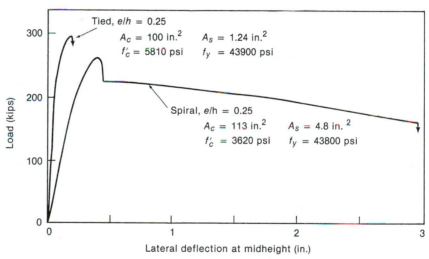

Fig. 11-5
Load–deflection behavior of
tied and spiral columns.
(Adapted from [11-1].)

(b) Eccentrically loaded columns.

core that remains is less than the load on the column. The concrete core is crushed, and the reinforcement buckles outward between ties. This occurs suddenly, without warning, in a brittle manner.

   When the shell spalls off a spiral column, the column does not fail immediately because the strength of the core has been enhanced by the triaxial stresses resulting from the effect of the spiral reinforcement. As a result, the column can undergo large deformations, eventually reaching a *second maximum load,* when the spirals yield and the column finally collapses. Such a failure is much more ductile than that of a tied column and gives warning of the impending failure, along with possible load redistribution to other members. It should be noted, however, that this is accomplished only at very high strains. For example, the strains necessary to reach the second maximum load correspond to a shortening of about 1 in. in an 8-ft-high column, as shown in Fig. 11-5a.

   When spiral columns are eccentrically loaded, the second maximum load may be less than the initial maximum, but the deformations at failure are large, allowing load redistribution

(Fig. 11-5b). Because of their greater ductility, compression-controlled failures of spiral columns are assigned a strength-reduction factor, ϕ, of 0.70, rather than the value 0.65 used for tied columns.

Spiral columns are used when ductility is important or where high loads make it economical to utilize the extra strength resulting from the higher ϕ factor. Figures 11-6 and 11-7 show a tied and a spiral column, respectively, after an earthquake. Both columns are in the same building and have undergone the same deformations. The tied column has failed completely, while the spiral column, although badly damaged, is still supporting a load. The very minimal ties in Fig. 11-6 were inadequate to confine the core concrete. Had the column ties been detailed according to ACI Section 21.4, the column would have performed much better.

Strength of Axially Loaded Columns

When a symmetrical column is subjected to a concentric axial load, P, longitudinal strains, ϵ, develop uniformly across the section, as shown in Figs. 4-9 and 11-8a. Because the steel and concrete are bonded together, the strains in the concrete and steel are equal. For any given strain, it is possible to compute the stresses in the concrete and steel using the stress–strain curves for the two materials. The forces P_c and P_s in the concrete and steel are

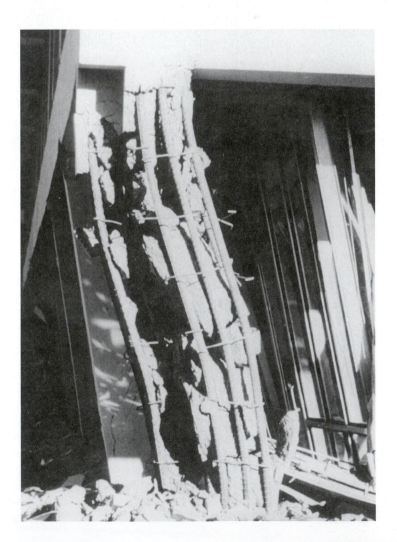

Fig. 11-6
Tied column destroyed in 1971 San Fernando earthquake. (Photograph courtesy of National Bureau of Standards.)

Fig. 11-7
Spiral column damaged by
1971 San Fernando earth-
quake. Although this column
has been deflected sideways
20 in., it is still carrying load.
(Photograph courtesy of National Bureau of Standards.)

equal to the stresses multiplied by the corresponding areas. The total load on the column,
P_0, is the sum of these two quantities. Failure occurs when P_0 reaches a maximum. For a
steel with a well-defined yield strength (Fig. 11-8c), this occurs when $P_c = f_c'' A_c$ and
$P_s = f_y A_{st}$, where $f_c'' = k_3 f_c'$ is the strength of the concrete loaded as a column. On the
basis of tests of 564 columns carried out at the University of Illinois and Lehigh University
from 1927 to 1933 [11-2], the ACI Code takes k_3 equal to 0.85. Thus, for a column with a
well-defined yield strength, the axial load capacity is

$$P_0 = 0.85 f_c'(A_g - A_{st}) + f_y A_{st} \qquad (11\text{-}1)$$

where A_g is the gross area and A_{st} is the total area of the longitudinal reinforcement. This
equation represents the summation of the fully plastic strength of the steel and the concrete. If the reinforcement is not elastic–perfectly plastic, failure occurs when P_0 reaches a

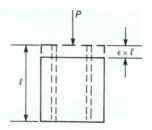

(a) Strains in column.

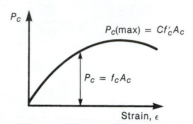

(b) Load resisted by concrete.

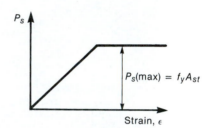

(c) Load resisted by steel.

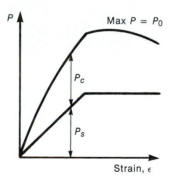

(d) Total load resisted by column.

Fig. 11-8
Resistance of an axially
loaded column.

maximum, but this may, or may not, coincide with the strain at which the maximum P_c occurs. As discussed in Section 3-5, the factor α_1 accounts, in part, for the reduction in concrete compressive strength due to (a) the slow loading or (b) the weakening of the concrete near the top of the column due to upward migration of water in the fresh concrete.

11-3 INTERACTION DIAGRAMS

Almost all compression members in concrete structures are subjected to moments in addition to axial loads. These may be due to misalignment of the load on the column, as shown in Fig. 11-9b, or may result from the column resisting a portion of the unbalanced moments at the ends of the beams supported by the columns (Fig. 11-9c). The distance e is referred to as the *eccentricity* of the load. These two cases are the same, because the eccentric load P in Fig. 11-9b can be replaced by a load P acting along the centroidal axis,

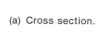

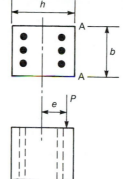

(a) Cross section.

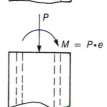

(b) Eccentric load.

(c) Axial load and moment.

Fig. 11-9
Load and moment on column.

plus a moment $M = Pe$ about the centroid. The load P and the moment M are calculated with respect to the geometric centroidal axis because the moments and forces obtained from structural analysis are referred to this axis.

To illustrate conceptually the interaction between moment and axial load in a column, an idealized homogeneous and elastic column with a compressive strength, f_{cu}, equal to its tensile strength, f_{tu}, will be considered. For such a column, failure would occur in compression when the maximum stresses reached f_{cu}, as given by

$$\frac{P}{A} + \frac{My}{I} = f_{cu} \tag{11-2}$$

where

$$A, I = \text{area and moment of inertia of the cross section, respectively}$$
$$y = \text{distance from the centroidal axis to the most highly compressed surface (surface } A\text{–}A \text{ in Fig. 11-9a), positive to the right}$$
$$P = \text{axial load, positive in compression}$$
$$M = \text{moment, positive as shown in Fig. 11-9c}$$

Dividing both sides of (11-2) by f_{cu} gives

$$\frac{P}{f_{cu}A} + \frac{My}{f_{cu}I} = 1$$

The maximum axial load the column can support occurs when $M = 0$ and is $P_{max} = f_{cu}A$. Similarly, the maximum moment that can be supported occurs when $P = 0$ and M is $M_{max} = (f_{cu}I/y)$. Substituting P_{max} and M_{max} gives

$$\frac{P}{P_{max}} + \frac{M}{M_{max}} = 1 \tag{11-3}$$

This equation is known as an *interaction equation*, because it shows the interaction of, or relationship between, P and M at failure. It is plotted as the line AB in Fig. 11-10. A similar

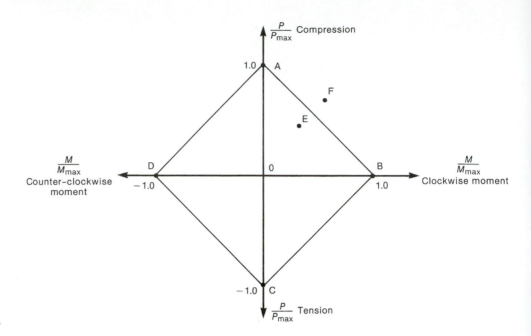

Fig. 11-10
Interaction diagram for an
elastic column, $|f_{cu}| = |f_{tu}|$.

equation for a tensile load, P, governed by f_{tu}, gives the line BC in this figure, and the lines AD and DC result if the moments have the opposite sign.

Figure 11-10 is referred to as an *interaction diagram*. Points on the lines plotted in this figure represent combinations of P and M corresponding to the resistance of the section. A point inside the diagram, such as E, represents a combination of P and M that will not cause failure. Combinations of P and M falling on the line or outside the line, such as point F, will equal or exceed the resistance of the section and hence will cause failure.

Figure 11-10 is plotted for an elastic material with $f_{tu} = -f_{cu}$. Figure 11-11a shows an interaction diagram for an elastic material with a compressive strength f_{cu}, but with the tensile strength, f_{tu}, equal to zero, and Fig. 11-11b shows a diagram for a material with $|-f_{tu}| = 0.5|f_{cu}|$. Lines AB and AD indicate load combinations corresponding to failure initiated by compression (governed by f_{cu}), while lines BC and DC indicate failures initiated by tension. In each case, the points B and D in Figs. 11-10 and 11-11 represent *balanced failures*, in which the tensile and compressive resistances of the material are reached simultaneously.

Reinforced concrete is not elastic and has a tensile strength that is lower than its compressive strength. An effective tensile strength is developed, however, by reinforcing bars on the tension face of the member. For these reasons, the calculation of an interaction diagram for reinforced concrete is more complex than that for an elastic material. The general shape of the diagram resembles Fig. 11-11b, however.

11-4 INTERACTION DIAGRAMS FOR CONCRETE COLUMNS

Strain Compatibility Solution

Concept and Assumptions

Although it is possible to derive a family of equations to evaluate the strength of columns subjected to combined bending and axial loads (see [11-3]), these equations are tedious to use. For this reason, interaction diagrams for columns are generally computed by assuming

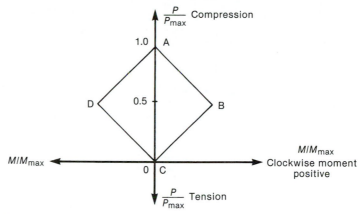

(a) Material with $f_{tu} = 0$.

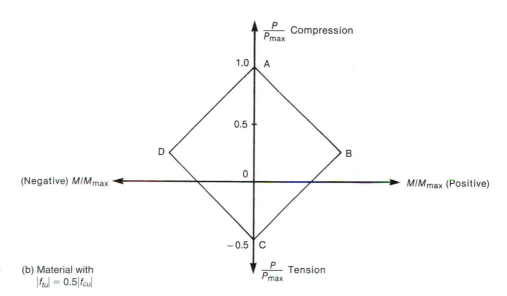

Fig. 11-11
Interaction diagrams for elas-
tic columns, $|f_{cu}|$ not equal to
$|f_{tu}|$.

(b) Material with
$|f_{tu}| = 0.5|f_{cu}|$

a series of strain distributions, each corresponding to a particular point on the interaction diagram, and computing the corresponding values of P and M. Once enough such points have been computed, the results are summarized in an interaction diagram.

The calculation process is illustrated in Fig. 11-12 for one particular strain distribution. The cross section is illustrated in Fig. 11-12a, and one assumed strain distribution is shown in Fig. 11-12b. The maximum compressive strain is set at 0.003, corresponding to compressive failure of the section. The location of the neutral axis and the strain in each level of reinforcement are computed from the strain distribution. This information is then used to compute the size of the compression stress block and the stress in each layer of reinforcement, as shown in Fig. 11-12c. The forces in the concrete and the steel layers, shown in Fig. 11-12d, are computed by multiplying the stresses by the areas on which they act. Finally, the axial force P_n is computed by summing the individual forces in the concrete and steel, and the moment M_n is computed by summing the moments of these forces about the geometric centroid of the cross section. These values of P_n and M_n represent one point on the interaction diagram.

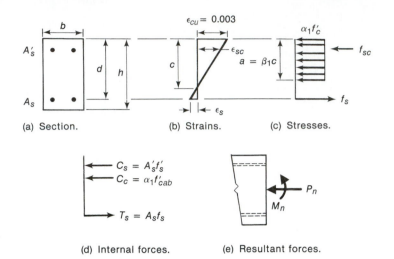

(a) Section.   (b) Strains.   (c) Stresses.

(d) Internal forces.   (e) Resultant forces.

Fig. 11-12
Calculation of P_n and M_n for a given strain distribution.

Significant Points on the Column Interaction Diagram

Figure 11-13 and Table 11-1 illustrate a series of strain distributions and the corresponding points on an interaction diagram for a typical tied column. As usual for interaction diagrams, axial load is plotted vertically and moment horizontally. Several points on the interaction diagram govern the selection of strength reduction factors, ϕ factors, for column and beam design in ACI Section 9.3.2. The method of computing the strength-reduction factor for columns was changed in the 2002 ACI code. The new and old factors are given by Table 11-1. The presentation of the new strength reduction factors was proposed by Mast [11-4]. This proposed code change received extensive discussion when it was first included in the code [11-5].

    1. Point *A*—Pure Axial Load. Point *A* in Fig. 11-13 and the corresponding strain distribution represent uniform axial compression without moment, sometimes referred to as *pure axial load*. This is the largest axial load the column can support. Later in this section the maximum usable axial load will be limited to 0.80 to 0.85 times the pure axial load capacity.

    2. Point *B*—Zero Tension, Onset of Cracking. The strain distribution at *B* in Fig. 11-13 corresponds to the axial load and moment at the onset of crushing of the concrete just as the strains in the concrete on the opposite face of the column reach zero. Case *B* represents the onset of cracking of the least compressed side of the column. Because tensile stresses in the concrete are ignored in the strength calculations, failure loads below point *B* in the interaction diagram represent cases where the section is partially cracked.

    3. Region *A–C*—Compression-Controlled Failures. Columns with axial loads P_n and moments M_n that fall on the upper branch of the interaction diagram between points *A* and *C* initially fail due to crushing of the compression face before the extreme tensile layer of reinforcement yields. Hence, they are called *compression-controlled* columns.

    4. Point *C*—Balanced Failure, Compression-Controlled Limit Strain. Point *C* in Fig. 11-13 corresponds to a strain distribution with a maximum compressive strain of 0.003 on one face of the section, and a tensile strain equal to the yield strain, ϵ_y, in the layer of reinforcement *farthest* from the compression face of the column. The *extreme tensile strain* ϵ_t occurs in the *extreme tensile layer* of steel located at d_t below the extreme compression fiber. ACI Section 10.3.2 defines this as a *balanced failure* in which both crushing of the concrete on the compressive face and yielding of the reinforcement nearest to the opposite face of the

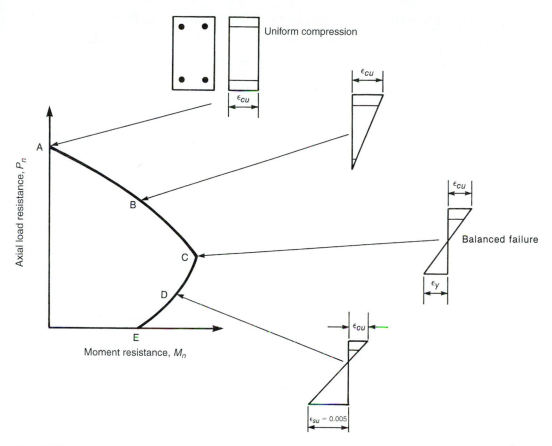

Fig. 11-13
Strain distributions corresponding to points on the interaction diagram.

column (tensile face) develop simultaneously. Traditionally, the ACI Code defined a balanced failure as one in which the steel strain at the *centroid* of the tensile reinforcement reached yield in tension when the concrete reached its crushing strain. In the 2002 ACI Code the definition of balanced failure was changed to correspond to the yield of the *extreme tensile layer* of reinforcement rather than the yield at the *centroid* of the tension reinforcement. This change provoked extensive discussion of the code revision [11-4],[11-5]. The two definitions of d_t are the same if the tensile reinforcement is all in one layer.

   5.   **Compression-Controlled Strain Limit.** As the steel strain, ϵ_s, increases beyond tensile yield, ϵ_y, the section passes through a *transition region* between *compression-controlled failures* at axial loads larger than point C, and *tension-controlled failures* due to loads that are lower than point D. As shown in Fig. 11-13, point C is the point farthest to the right in the interaction diagram or is close to that point. Thus, the maximum moment capacity can be said to occur for the strain distribution at point C in Fig. 11-13. The top and bottom strains, $\epsilon_{cu} = 0.003$ compression and $\epsilon_s = \epsilon_y$ tension, define the lower limit of strain distributions for which a crushing failure occurs before the reinforcement reaches the tensile yield strain. For this reason, ACI Section 10.3.3 labels this the *compression-controlled strain limit*. Ignoring the difference in the depth of the extreme tensile layer and the that of the centroid of the tensile reinforcement, the traditional and the new ACI 318-02 Section 10.3.2 definitions of balanced strains are effectively the same.

   6.   **Point D—Tensile-Controlled Limit.** Point D in Fig. 11-13 corresponds to a strain distribution with 0.003 compressive strain on the top face and a tensile strain of 0.005 in the *extreme layer* of tension steel (the layer closest to the tensile face of the section.) The

TABLE 11-1 Strain Regimes and Strength Reduction Factors, ϕ, for Columns and Beams

Maximum Strain	Compression-Controlled	Transition Region	Tension-Controlled
Max Compressive Strain	$\varepsilon_{cu} = 0.003$ compression	$\varepsilon_{cu} = 0.003$ compression	$\varepsilon_{cu} = 0.003$ compression
Maximum Tensile strain at ultimate	ε_t between 0.003 compression strain and 0.002 tension strain	ε_t between 0.002 tension strain and 0.005 tension strain	ε_t equal to or greater than 0.005 tension
ASCE 7 Load Factors $U = 1.2D + 1.6L$	**Values of Strength-Reduction Factor, ϕ**	**Values of Strength-Reduction Factor, ϕ**	**Values of Strength-Reduction Factor, ϕ**
ACI Section 9.2	*Tied columns* $\phi = 0.65$	*Tied columns* $\phi = 0.48 + 83\varepsilon_t$ (11-4a) $\phi = 0.23 + 0.25/(c/d_t)$ (11-4b)	*Tied columns* $\phi = 0.90$
	Spiral columns $\phi = 0.70$	*Spiral columns* $\phi = 0.57 + 67\varepsilon_t$ (11-5a) $\phi = 0.37 + 0.20/(c/d_t)$ (11-5b)	*Spiral columns* $\phi = 0.90$
	From ACI Sections 9.3.2.1 and 9.3.2.2	From ACI Sections 9.3.2.1 and 9.3.2.2	From ACI Section 9.3.2.1
Traditional ACI Load Factors $U = 1.4D + 1.7L$	**Values of Strength-Reduction Factor, ϕ** *Tied columns* $\phi = 0.70$	**Values of Strength-Reduction Factor, ϕ** *Tied columns* $\phi = 0.57 + 67\varepsilon_t$ (11-6a) $\phi = 0.37 + 0.20/(c/d_t)$ (11-6b)	**Values of Strength-Reduction Factor, ϕ** *Tied columns* $\phi = 0.90$
	Spiral columns $\phi = 0.75$	*Spiral columns* $\phi = 0.65 + 50\varepsilon_t$ (11-7a) $\phi = 0.50 + 0.15/(c/d_t)$ (11-7b)	*Spiral columns* $\phi = 0.90$
	From ACI Sections C.3.2.1 and C.3.2.2	From ACI Sections C.3.2.1 and C.3.2.2(b)	From ACI Section C.3.2.1

failure of such a column will be ductile, with steel strains at failure that are about two and a half times the yield strain (for Grade-60 steel). ACI Section 10.3.4 calls this the *tension-controlled strain limit*. The strain of 0.005 was chosen to be significantly higher than the yield strain to ensure ductile behavior. [11-4]

7. **Region *C–D*—Transition Region.** Flexural members and columns with loads and moments which would plot between points *C* and *D* in Fig. 11-13 are called *transition failures* because the magnitude of the curvatures at the critical section are in a transition between the ultimate curvatures corresponding to steel strains of 0.002 and 0.005. This is reflected in the transition from ϕ at point *C* where $\phi = 0.65$ or 0.70, and ϕ at *point D*, where the extreme tensile strain equals 0.005 tension and $\phi = 0.90$.

8. **Strength-Reduction Factor, ϕ, for Columns.** The values of the strength reduction factor, ϕ, and equations for calculating it are given in Table 11-1. The 2002 ACI Code has two sets of strength-reduction factors—one for use with the traditional ACI load factors and the other for use with the load factors from ASCE 7. The two sets are summarized and compared in Table 11-1. For compression-controlled *tied* columns, referred to in ACI Section 9.3.2.2(b) as *other* reinforced members, ACI Section 9.3.2.2(b) sets the strength-reduction factor ϕ equal to 0.65. For compression-controlled *spiral* columns, ACI Section 9.3.2.2(b) sets ϕ equal to 0.70. For *tension-controlled* columns, ACI Section 9.3.2.1 sets ϕ equal to 0.90, for both tied and spiral columns. ACI Section 9.3.2.1 specifies a linear transition from 0.65 or 0.70 to 0.90. Equations for the transition are given in Table 11-1. In these equations, a tensile strain, ε_t, is taken as positive when it is substituted in (11-4a) through (11-7b). Thus, for example, for $\varepsilon_t = 0.003$ tensile strain, $\phi = 0.48 + 83(+0.003) = 0.729$.

The 1995 code had two different transition expressions for ϕ, one for design based on the traditional ACI Code load factors and ϕ factors, and a newer one for tension-controlled columns. The traditional ACI transition equations were dropped from the 2002 code.

   9.   Strain Limit for Beams—ACI Section 10.3.5 limits the maximum amount of reinforcement in a beam by placing a lower limit on the extreme steel strain in beams, ϵ_t, to not less than 0.004 in tension. This is smaller than the tension-controlled limit strain of 0.005. The corresponding strength-reduction factor for $\varepsilon_t = 0.005$ is given in ACI Section 9.3.2.1 as 0.90. For $\varepsilon_t = 0.004$, the equations in Table 11-1 give $\phi = 0.812$. The value of ϕ for beams with tensile strains between 0.004 and 0.005 is not addressed in the code, but some users take $\phi = 0.90$ for beams failing in tension, regardless of ϵ_t. Because the extreme strain of $\varepsilon_t = 0.004$ has no significance in a column, we shall compute ϕ using (11-4).

Strength-Reduction Factor for Columns

In the design of columns, the axial load and moment capacities must satisfy

$$\phi P_n \geq P_u \quad \phi M_n \geq M_u \qquad \text{(11-8a and b)}$$

where

P_u and M_u = factored load and moment applied to the column, computed from a frame analysis

P_n and M_n = nominal strengths of the column cross section

ϕ = strength-reduction factor; the value of ϕ is the same in both relationships in (11-8)

Maximum Axial Load

As seen earlier, the strength of a column under truly concentric axial loading can be written as

$$P_{n0} = (k_3 f'_c)(A_g - A_{st}) + f_y(A_{st}) \qquad \text{(11-9)}$$

where

$k_3 f'_c$ = maximum concrete stress permitted in column design. See ACI Section 10.2.7

A_g = gross area of the section (concrete and steel)

f_y = yield strength of the reinforcement

A_{st} = total area of reinforcement in the cross section

The value of $k_3 f'_c$ was derived from tests [11-1], [11-2], [11-3].

   The strength given by (11-9) cannot normally be attained in a structure because almost always there will be moments present, and, as shown by Figs. 11-10, 11-11, and 11-13, any moment leads to a reduction in the axial load capacity. Such moments or eccentricities arise from unbalanced moments in the beams, misalignments of columns from floor to floor, uneven compaction of the concrete across the width of the section, or misalignment of the reinforcement. An examination of Fig. 11-1 will show that the reinforcement has been displaced to the left in this column. Hence in this case, the centroid of the theoretical column resistance does not coincide with the axis of the column, as built. The misalignment of the reinforcement in Fig. 11-1 is considerably greater than the allowable tolerances for reinforcement location (ACI Section 7.5.2.1), and such a column would not be acceptable.

To account for the effect of accidental moments, ACI Sections 10.3.6.1 and 10.3.6.2 specify that the maximum load on a column must not exceed 0.85 times the load from (11-1) for spiral columns and 0.8 times (11-1) for tied columns:

Spiral columns:

$$\phi P_{n(\max)} = 0.85[\text{strength from (11-1)}] = 0.85\phi[0.85f'_c(A_g - A_{st}) + f_y(A_{st})]$$

$$\text{(11-10a)}$$

$$\text{(ACI Eq. 10-1)}$$

Tied columns:

$$\phi P_{n(\max)} = 0.80\phi[0.85f'_c(A_g - A_{st}) + f_y(A_{st})] \qquad \text{(11-10b)}$$

$$\text{(ACI Eq. 10-2)}$$

These limits will be included in the interaction diagram. The difference between the allowable values for spiral and tied columns reflects the more ductile behavior of spiral columns. (See Fig. (11.5).)

Derivation of Computation Method for Interaction Diagrams

In this section, the relationships needed to compute the various points on an interaction diagram are derived by using strain compatibility and mechanics. The calculation of an interaction diagram involves the basic assumptions and simplifying assumptions stated in Section 4-2 of this book and ACI Section 10.2. For simplicity, the derivation and the computational example in the next section are limited to rectangular tied columns, as shown in Fig. 11-14a. The extension of this procedure to other cross sections is discussed later in this section.

Throughout the computations, it is necessary to rigorously observe a sign convention for stresses, strains, forces, and directions. Compression has been taken as *positive* in all cases.

Concentric Compressive Axial Load Capacity and Maximum Axial Load Capacity

The theoretical top point on the interaction diagram is calculated from (11-1). The maximum factored axial-load resistances are computed from (11-8) and (11-10). For a symmetrical section, the corresponding moment will be zero. Unsymmetrical sections are discussed briefly later in this chapter. For a symmetrical section, (11-17a) is used to compute the moment resistance.

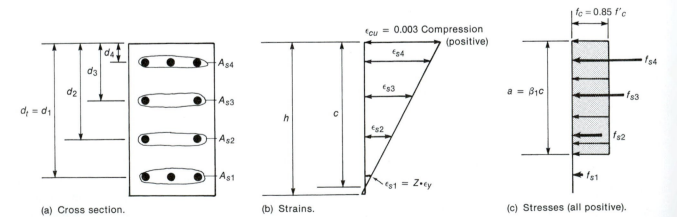

(a) Cross section.   (b) Strains.   (c) Stresses (all positive).

Fig. 11-14
Notation and sign convention for interaction diagram.

General Case

The general case involves the calculation of P_n acting at the centroid and M_n acting about the centroid of the gross cross section, for an assumed strain distribution with $\epsilon_{cu} = 0.003$. The column cross section and the assumed strain distribution are shown in Fig. 11-14a and b. Four layers of reinforcement are shown, layer 1 having strain ϵ_{s1} and area A_{s1}, and so on. Layer 1 is closest to the "least compressed" surface and is at a distance d_1 from the "most compressed" surface. Layer I is called the *extreme tension layer*. It has a depth d_t and a strain ϵ_t.

The strain distribution will be defined by setting $\epsilon_{cu} = 0.003$ and assuming a value for ϵ_{s1}. Because an iterative calculation will be necessary to consider a series of cases, as shown in Fig. 11-13, the iteration will be controlled by setting $\epsilon_{s1} = Z\epsilon_y$, where Z is an arbitrarily chosen value. Positive values of Z correspond to positive (compressive) strains (as shown in Fig. 11-14b). For example, $Z = -1$ corresponds to $\epsilon_{s1} = -1\epsilon_y$, the yield strain in tension. Such a strain distribution corresponds to the balanced-failure condition.

From Fig. 11-14b, by similar triangles,

$$c = \left(\frac{0.003}{0.003 - Z\epsilon_y} \right) d_1 \tag{11-11}$$

and

$$\epsilon_{si} = \left(\frac{c - d_i}{c} \right) 0.003 \tag{11-12}$$

where ϵ_{si} and d_i are the strain in the ith layer of steel and the depth to that layer.

Once the values of c and ϵ_{s1}, ϵ_{s2}, and so on, are known, the stresses in the concrete and in each layer of steel can be computed. For elastic–plastic reinforcement with the stress–strain curve illustrated in Fig. 11-15,

$$f_{si} = \epsilon_{si} E_s \quad \text{but} \quad -f_y \leq f_{si} \leq f_y \tag{11-13}$$

The stresses in the concrete are represented by the equivalent rectangular stress block described in assumption 6 in Section 4-2 of this book (ACI Section 10.2.7). The depth of this stress block is $a = \beta_1 c$, where a, shown in Fig. 11-14c, obviously cannot exceed the overall height of the section, h. The factor β_1 is given by

$$\beta_1 = 1.05 - 0.05 \left(\frac{f_c'}{1000} \right) \tag{4-9b}$$

but not more than 0.85 nor less than 0.65.

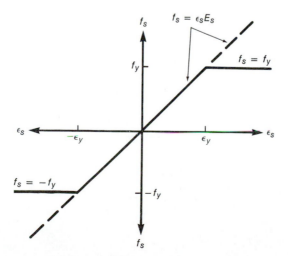

Fig. 11-15
Calculation of stress in steel
(11-13).

The next step is to compute the compressive force in the concrete, C_c, and the forces in each layer of reinforcement, F_{s1}, F_{s2}, and so on. This is done by multiplying the stresses by corresponding areas. Thus,

$$C_c = (0.85f_c')(ab) \qquad (11\text{-}14)$$

For a nonrectangular section, the area ab would be replaced by the area of the compression zone having a depth, a, measured perpendicular to the neutral axis.

If a is less than d_i,

$$F_{si} = f_{si}A_{si} \text{ (positive in compression)} \qquad (11\text{-}15a)$$

If a is greater than d_i for a particular layer of steel, the area of the reinforcement in that layer has been included in the area (ab) used to compute C_c. As a result, it is necessary to subtract $0.85f_c'$ from f_{si} before computing F_{si}:

$$F_{si} = (f_{si} - 0.85f_c')A_{si} \qquad (11\text{-}15b)$$

The resulting forces C_c and F_{s1} to F_{s4} are shown in Fig. 11-16b.

The nominal axial load capacity, P_n, for the assumed strain distribution is the summation of the axial forces:

$$P_n = C_c + \sum_{i=1}^{n} F_{si} \qquad (11\text{-}16)$$

The nominal moment capacity M_n for the assumed strain distribution is found by summing the moments of all the internal forces about the *centroid* of the column. The moments are summed about the centroid of the section, because this is the axis about which moments are computed in a conventional structural analysis. In the 1950s and 1960s, moments were sometimes calculated about the *plastic centroid*, the location of the resultant force in a column strained uniformly in compression (case A in Fig. 11-13). The centroid and plastic centroid are the same point in a symmetrical column with symmetrical reinforcement.

All the forces are shown positive (compressive) in Figs. 11-14 and 11-16. A positive internal moment corresponds to a compression at the top face, and

$$M_n = C_c\left(\frac{h}{2} - \frac{a}{2}\right) + \sum_{i=1}^{n} F_{si}\left(\frac{h}{2} - d_i\right) \qquad (11\text{-}17a)$$

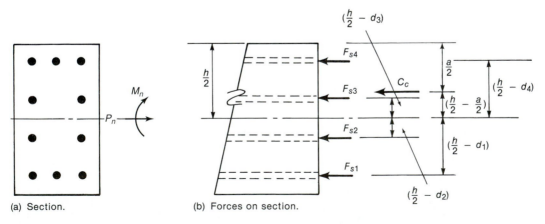

(a) Section.

(b) Forces on section.

Fig. 11-16
Internal forces and moment arms.

Equation (11-17a) is derived by assuming that the centroid of the gross (concrete) section is located at $h/2$ from the extreme-compression fiber. If the gross cross section is not symmetrical, the moments would be computed about the centroid of the gross section, and the factored moment resistance would be

$$M_n = C_c\left(\overline{y}_t - \frac{a}{2}\right) + \sum_{i=1}^{n} F_{si}\left(\overline{y}_t - d_i\right) \qquad (11\text{-}17b)$$

where $\overline{y}_t$ is the distance from the extreme-compression fiber to the centroid of the gross section.

Pure-Axial-Tension Case

The strength under pure axial tension is computed by assuming that the section is completely cracked through and subjected to a uniform strain equal to, or less (more tensile) than, $-\epsilon_y$. The stress in all the layers of reinforcement is therefore $-f_y$ (yielding in tension), and

$$P_{nt} = \sum_{i=1}^{n} -f_y A_{si} \qquad (11\text{-}18)$$

The axial tensile capacity of the concrete is, of course, ignored. For a symmetrical section, the corresponding moment will be zero. For an unsymmetrical section, (11-17b) is used to compute the moment.

EXAMPLE 11-1 Calculation of an Interaction Diagram

Compute four points on the interaction diagram for the column shown in Fig. 11-17a. Use $f'_c = 5000$ psi and $f_y = 60,000$ psi. $A_g = bh$ is 256 in.^{2}, $A_{s1} = 4$ in.^{2}, $A_{s2} = 4$ in.^{2}, $A_{st} = \Sigma A_{si} = 8$ in.^{2}, and $\rho_t = A_{st}/A_g = 0.031$. The yield strain, $\epsilon_y = f_y/E_s$, is 60,000 psi/29,000,000 psi $= 0.00207$. Use load factors and strength-reduction factors from ACI Sections 9.2.1 and 9.3.2.

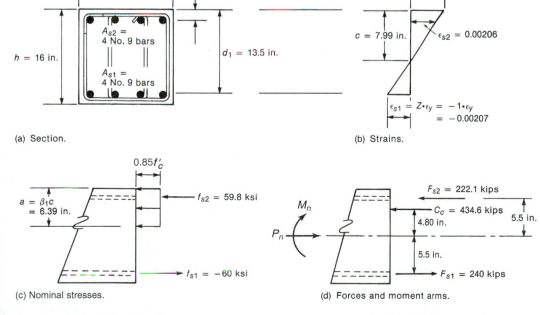

(a) Section.

(b) Strains.

(c) Nominal stresses.

(d) Forces and moment arms.

Fig. 11-17
Calculations—Example 11-1, $\epsilon_{s1} = -1 \cdot \epsilon_y$, so $Z = -1$.

1. **Select the load and resistance factors.** Use the factors from ACI Sections 9.2.1 and 9.3.2.

2. **Compute the concentric axial-load capacity and maximum axial-load capacity.** From (11-1),

$$P_0 = (0.85f'_c)(A_g - A_{st}) + f_y(A_{st})$$

$$= (0.85 \times 5 \text{ ksi})(256 - 8) \text{ in.}^2 + 60 \text{ ksi} \times 8 \text{ in.}^2$$

$$= 1054 \text{ kips} + 480 \text{ kips} = 1534 \text{ kips}$$

This is the nominal concentric axial-load capacity. The value used in drawing a design interaction diagram would be ϕP_0, where $\phi = 0.65$, because (i) the strain in the extreme tension steel, layer 1 in Fig. 11-14a or 11-17a, has a strain that is compressive, and (ii) this is a tied column. Thus,

$$\phi P_0 = 997 \text{ kips}$$

P_0 and ϕP_0 are plotted as points A and A' in Fig. 11-18 and point A in Fig. 11-13.

For this column with $\rho_t = 0.031$ (or 3.1 percent), the 480 kips carried by the reinforcement is roughly 30 percent of the 1534-kip nominal capacity of the column. For axially loaded columns, reinforcement will generally carry between 10 and 35 percent of the total capacity of the column.

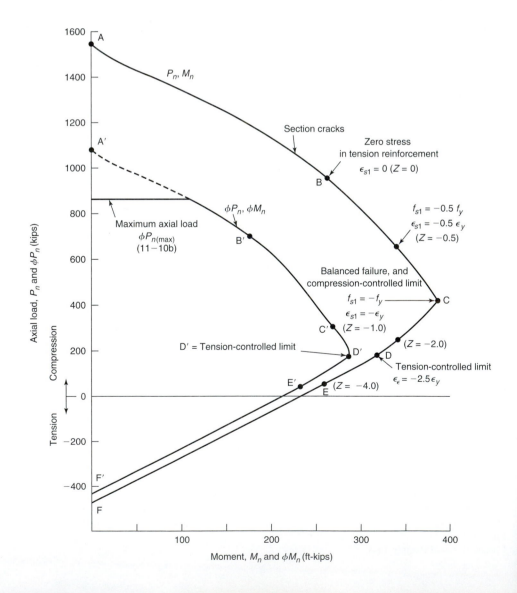

Fig. 11-18
Interaction diagram—
Example 11-1.

The maximum load allowed on this column (ACI Section 10.3.6.2) is given by (11-10b) (ACI Eq. (10-2)):

$$\phi P_{n(\max)} = 0.80\phi[(0.85f'_c)(A_g - A_{st}) + f_y(A_{st})]$$
$$= 0.80\phi P_0 = 798 \text{ kips}$$

This load is plotted as a horizontal solid line in Fig. 11-18. The portion of the ϕP_n, ϕM_n interaction diagram above this line is shown with a dashed line because this capacity cannot be used in design.

3.   Compute ϕP_n and ϕM_n for the general case. To get a more or less complete interaction diagram, a number of strain distributions must be considered and the corresponding values of P_n, M_n, ϕP_n, and ϕM_n calculated. These intervals are successively larger because the points get closer and closer together as Z gets larger. Ideally the values of Z should be chosen to agree with the *compression-controlled strain limits*, the *tensile-controlled strain limit*, and the limit on beam reinforcement as follows:

(a) $Z = +0.5$—Compression strain of $\epsilon_t = +0.5 \epsilon_y$ in the extreme tension layer of steel, not plotted in Fig. 11-13 or 11-18.

(b) $Z = +0.25$—Compression strain of $\epsilon_t = +0.25 \epsilon_y$ in the extreme tension layer of steel. This case is not plotted in Fig. 11-13 and 11-18.

(c) $Z = 0.0$—Strain ϵ_t is zero in extreme layer of tension steel. This case is considered when calculating an interaction diagram because it marks the change from compression lap splices being allowed on all longitudinal bars, to the more severe requirement of tensile lap splices. This point is plotted as points B and B' in Figs. 11-13 and 11-18.

(d) $Z = -0.5$—tension strain of $\epsilon_t = -0.5 \epsilon_y$. This strain distribution affects the length of the tension lap splices in a column and is customarily plotted on an interaction diagram. It is plotted in Fig. 11-18.

(e) $Z = -1$—Tensile strain $= - -\epsilon_y$ where ϵ_y is taken equal to 0.002, or tensile strain $\epsilon_t = Z = -0.967 \epsilon_y$ (tension) where ϵ_t is taken equal to 0.00207. For Grade 60 reinforcement $\epsilon_y = 60{,}000/29{,}000{,}000 = 0.0207$, strain. For convenience in calculations, ACI Section 10.3.3 allows ϵ_y for Grade 60 steel to be rounded to 0.002 strain. The strength corresponding to this strain is plotted in Figs. 11-13 and 11-18, points C and C'. This strain distribution is called the *balanced failure* case and *the compression-controlled strain limit*. It marks the change from compression failures originating by crushing of the compression surface of the section, to tension failures initiated by yield of the longitudinal reinforcement. It also marks the start of the *transition zone* for ϕ for columns in which ϕ increases from 0.65 or 0.70 up to 0.90.

(f) $Z = -2.5$ (or –2.417 if ϵ_y is taken as 0.00207). This corresponds to the tension-controlled strain limit of 0.005. In Fig. 11-18 this corresponds to points D and D'. It is the strain at the more tensile limit of the transition zone for ϕ, used to define a tension-controlled section.

(g) $Z = -4$ Tension strain of $-\epsilon_t = -4 e_y$. This load combination is helpful in plotting the tension–controlled branch of he interaction diagram. It plots as points E and E' in Fig. 11-18.

Finally, two or three more values of Z, say -4 and -6, are desirable to complete plotting of the tension-failure branch of the interaction diagram.

In this example, we compute the values of ϕP_n and ϕM_n for $Z = -1, -2$, and -4.

4.   Compute ϕ and ϕM_n for balanced failure ($\epsilon_{s1} = -\epsilon_y$).

(a)   Determine c and the strains in the reinforcement. The column cross section and the strain distribution corresponding to $\epsilon_{s1} = -1\epsilon_y(Z = -1)$ are shown in Fig. 11-17a and b.

The strain in the bottom layer of steel is $-1\epsilon_y = -0.00207$. From similar triangles, the depth to the neutral axis is

$$c = \frac{0.003}{0.003 - (-1 \times 0.00207)}d_1 \qquad (11\text{-}11)$$

$$= \frac{0.003}{0.003 + 0.00207} \times 13.5 \text{ in.} = 7.99 \text{ in.}$$

Again by similar triangles, the strain in the compression steel is

$$\epsilon_{s2} = \left(\frac{c - d_2}{c}\right)0.003 \qquad (11\text{-}12)$$

$$= \left(\frac{7.99 - 2.5}{7.99}\right)0.003 = 0.00206$$

(b) **Compute the stresses in the reinforcement layers.** The stress in reinforcement layer 2 is (11-13)

$$f_{s2} = \epsilon_{s2}E_s \qquad \text{but} \qquad -f_y \le f_{s2} \le f_y$$

$$\epsilon_{s2}E_s = 0.00206 \times 29{,}000 \text{ ksi} = 59.8 \text{ ksi}$$

However, $-60 \le f_{s2} \le +60$ ksi. Therefore, $f_{s2} = 59.8$ ksi. Because this is positive, it is compressive. Rounding slightly as allowed by ACI Section 10.3.3, we can take $\epsilon_y = -0.0020$. The stress in layer 1 is found to be -60 ksi.

(c) **Compute a.** The depth of the equivalent rectangular stress block is $a = \beta_1 c$, where a cannot exceed h. For $f'_c = 5000$ psi,

$$\beta_1 = 1.05 - 0.05\left(\frac{f'_c}{1000 \text{ psi}}\right) = 0.80$$

and

$$a = \beta_1 c$$

$$= 0.80 \times 7.99 \text{ in.} = 6.39 \text{ in.}$$

This is less than h; therefore, this value can be used. If a exceeded h, $a = h$ would be used. The stresses computed in steps 2 and 3 are shown in Fig. 11-17c.

(d) **Compute the forces in the concrete and steel.** The force in the concrete, C_c, is equal to the average stress, $0.85f'_c$, times the area of the rectangular stress block, ab:

$$C_c = (0.85f'_c)(ab) \tag{11-14}$$

$$= 0.85 \times 5 \text{ ksi} \times 6.39 \text{ in.} \times 16 \text{ in.} = 434.6 \text{ kips}$$

The distance $d_1 = 13.5$ in. to reinforcement layer 1 exceeds $a = 6.39$ in. Hence, this layer of steel lies outside the compression stress block and does not displace concrete included in the area (ab) when computing C_c. Thus,

$$F_{s1} = f_{s1}A_{s1}$$

$$= -60 \text{ ksi} \times 4 \text{ in.}^2 = -240 \text{ kips}$$

(Negative denotes tension.)

Reinforcement layer 2 lies in the compression zone, since $a = 6.39$ in. exceeds $d_2 = 2.5$ in. Hence, we must allow for the stress in the concrete displaced by the steel when we compute F_{s2}. From (11-15b),

$$F_{s2} = (f_{cs2} - 0.85f'_c)A_{s2}$$

$$= (59.8 - 0.85 \times 5) \text{ ksi} \times 4 \text{ in.}^2 = 222.1 \text{ kips}$$

The forces in the concrete and steel are shown in Fig. 11-17d.

(e) **Compute P_n.** The nominal axial-load capacity, P_n, is found by summing the axial-force components (11-16):

$$P_n = C_c + \Sigma F_{si}$$

$$= 434.6 - 240 + 222.1 = 417 \text{ kips}$$

Since $\epsilon_{s1} = -\epsilon_y$ (yield in tension), this is the balanced-failure condition, and $P_n = P_b$.

(f) **Compute M_n.** From Fig. 11-17d, the moment of C_c, F_{s1}, and F_{s2} about the centroid of the section is (11-17a)

$$M_n = C_c\left(\frac{h}{2} - \frac{a}{2}\right) + F_{s1}\left(\frac{h}{2} - d_1\right) + F_{s2}\left(\frac{h}{2} - d_2\right)$$

$$= 434.6 \text{ kips}\left(\frac{16}{2} - \frac{6.39}{2}\right) \text{ in.} + [-240(8 - 13.5)] + 222.1(8 - 2.5)$$

$$= 2088 + 1320 + 1222 \text{ in.-kips} = 4630 \text{ in.-kips}$$

Therefore, $M_n = M_b = 386$ ft-kips.

(g) Compute ϕ, ϕP_n, and ϕM_n. ϕ will be computed according to ACI Section 9.3.2.2. The strain ϵ_t, in the layer of reinforcement farthest from the compression face is $\epsilon_{s1} = -0.00207 = -\epsilon_y$. Thus, (11-4a) applies using $\epsilon_t = 0.00207$:

$$\phi = 0.48 + 83\epsilon_t = 0.652$$
$$\phi P_n = 0.652 \times 417 = 272$$
$$\phi M_n = 0.652 \times 386 = 252 \text{ ft-kips}$$

This completes the calculations for one value of $\epsilon_{s1} = Z(\epsilon_y)$ and gives the points B and B' in Fig. 11-18. Other values of Z are now assumed, and the calculations are repeated until one has enough points to complete the diagram.

5. Compute ϕP_n and ϕM_n for $Z = -2$. To illustrate the calculation of ϕ for cases falling between the compression-controlled limit and the tension-controlled limit, the computations will be repeated for the strain distribution corresponding to $Z = -2$.

(a) Determine c and the strains in the reinforcement. From similar triangles, $\epsilon_1 = -2\epsilon_y$ (substituting $Z = -2$ into (11-11)) gives

$$c = 5.67 \text{ in.}$$

From (11-12), the strain in the compression steel is

$$\epsilon_{s2} = 0.00168$$

The strain in the tension reinforcement is

$$\epsilon_{s1} = -2 \times 0.00207 = -0.00414$$

(b) Compute the stress in the reinforcement layers.

$$f_{s2} = 0.00168 \times 29{,}000 \text{ ksi} = 48.7 \text{ ksi}$$

This is within the range $\pm f_y$—therefore, OK.

$$f_{s1} = -60 \text{ ksi}, \text{ since } |\epsilon_{s1}| > |\epsilon_y|$$

(c) Compute a.

$$a = 0.80 \times 5.67 = 4.54 \text{ in.}$$

(d) Compute the forces in the concrete and steel. The compression force in the concrete is

$$C_c = 0.85 \times 5 \text{ ksi} \times 4.54 \text{ in.} \times 16 \text{ in.} = 308.7 \text{ kips}$$

The force in the tension reinforcement is

$$F_{s1} = -60 \text{ ksi} \times 4 \text{ in.}^2 = -240 \text{ kips}$$

Since $a = 4.54$ in. is greater than $d_2 = 2.5$ in., (11-15b) is used to compute F_{s2}.

$$F_{s2} = (48.7 - 0.85 \times 5) \times 4 = 177.8 \text{ kips}$$

(e) Compute P_n. Summing forces perpendicular to the section (11-16) gives

$$P_n = 308.7 - 240 + 177.8 = 246.5 \text{ kips}$$

(f) Compute M_n. Summing moments about the centroid of the section (11-17a) gives

$$M_n = 308.7\left(8 - \frac{4.54}{2}\right) + (-240 \times -5.50) + (177.8 \times 5.50)$$

$$= 4067 \text{ in.-kips} = 338.9 \text{ ft-kips}$$

(g) Compute ϕ, ϕP_n, and ϕM_n. The strain $\epsilon_t = \epsilon_{s1} = -0.00414$ is between $-\epsilon_y$ and -0.005. Therefore, (11-4a) applies, and using $\epsilon_t = 0.00414$:

$$\phi = 0.48 + 83\epsilon_t$$

where ϵ_t is the strain in the extreme tensile layer of steel, ϵ_{s1}. Thus,

$$\phi = 0.48 + (83 \times 0.00414) = 0.824$$
$$\phi P_n = 0.824 \times 246.5 = 203 \text{ kips}$$
$$\phi M_n = 0.824 \times 338.9 \text{ ft-kips} = 279 \text{ ft-kips}$$

P_n, M_n, ϕP_n, and ϕM_n calculated for this strain distribution are plotted as points C and C' in Fig. 11-18. The peculiar shape of this portion of the ϕP_n, ϕM_n interaction diagram is due to the transition from $\phi = 0.65$ to $\phi = 0.90$.

6.   Compute ϕP_n and ϕM_n for $Z = -4$. Repeating the calculations for $Z = -4$ (four times the yield strain) gives

$$P_n = 44.2 \text{ kips} \qquad M_n = 257.6 \text{ ft-kips}$$

Because $\epsilon_t = \epsilon_{s1} = -4 \times 0.00207 = -0.00828$, $\phi = 0.90$ is more than -0.005, $\phi = 0.90$ and

$$\phi P_n = 39.8 \text{ kips} \qquad \phi M_n = 231.8 \text{ ft-kips}$$

These are plotted as points E and E' in Fig. 11-18.

7.   Compute the capacity in axial tension. The final loading case to be considered in this example is concentric axial tension. The strength under such a loading is equal to the yield strength of the reinforcement in tension, as given by (11-18):

$$P_{nt} = \sum_{i=1}^{n} (-f_y A_{si})$$
$$= -60 \text{ ksi} (4 + 4) \text{ in.}^2 = -480 \text{ kips}$$

This is an axial tension. Because the section is symmetrical, $M = 0$.

The design capacity in pure tension is ϕP_{nt}, where $\phi = 0.9$. Thus,

$$\phi P_{nt} = 0.9 \times -480 \text{ kips} = -432 \text{ kips}$$

Points F and F' in Fig. 11-18 represent the pure-tension case. ∎

Interaction Diagrams for Circular Columns

The strain-compatibility solution described in the preceding section can also be used to calculate the points on an interaction diagram for a circular column. As shown in Fig. 11-19b, the depth to the neutral axis, c, is calculated from the assumed strain diagram by using similar triangles (or from (11-11)). The depth of the equivalent rectangular stress block, a, is again $\beta_1 c$.

The resulting compression zone is a segment of a circle having depth a, as shown in Fig. 11-19d. To compute the compressive force and its moment about the centroid of the

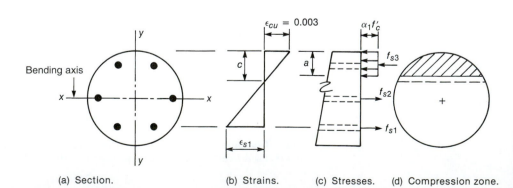

Fig. 11-19
Circular column.

(a) Section.   (b) Strains.   (c) Stresses.   (d) Compression zone.

column, it is necessary to be able to compute the area and centroid of the segment. These terms can be expressed as a function of the angle θ shown in Fig. 11-20. The area of the segment is

$$A = h^2\left(\frac{\theta - \sin\theta\cos\theta}{4}\right) \tag{11-19}$$

where θ is expressed in radians (1 radian $= 180°/\pi$). The moment of this area about the center of the column is

$$A\bar{y} = h^3\left(\frac{\sin^3\theta}{12}\right) \tag{11-20}$$

The shape of the interaction diagram of a circular column is affected by the number of bars and their orientation relative to the direction of the neutral axis. Thus, the moment capacity about axis x–x in Fig. 11-19a is less than that about axis y–y. Since the designer has little control over the arrangement of the bars in a circular column, the interaction diagram should be computed for the least favorable bar orientation. For circular columns with more than eight bars, this problem vanishes, because the bar placement approaches a continuous ring.

The interaction diagrams for the factored cross-sectional strength of a compression-controlled, circular tied column and a compression-controlled circular spiral column comparable, except that

(a)   The compression stress block acts on a portion of the column section that is a circular segment rather than a rectangular area.

(b)   the horizontal cutoff at the top of the interaction diagram has a value of $0.80P_{no}$ for tied columns and $0.85P_{no}$ for spiral columns. Again, this takes care of itself.

(c)   The value of the strength-reduction factor for compression-controlled spiral columns is $\phi = 0.70$ for design using the load factors from ACI Section 9.2.1, compared with 0.65 for compression-controlled tied columns. This requires that the ϕ factors in the transition zone be evaluated using different equations for ϕ. If the column is a circular spiral column, this change is taken into account by substituting (11-5a) and (11-5b) for (11-4a) and (11-4b).

For extreme tension strains more tensile than -0.005, $\phi = 0.9$ for spiral and tied columns.

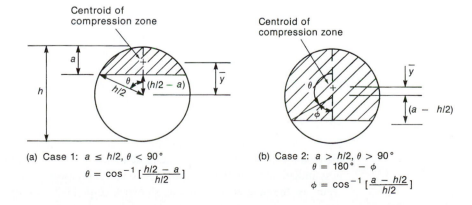

(a) Case 1: $a \le h/2$, $\theta < 90°$

$$\theta = \cos^{-1}\left[\frac{h/2 - a}{h/2}\right]$$

(b) Case 2: $a > h/2$, $\theta > 90°$

$$\theta = 180° - \phi$$

$$\phi = \cos^{-1}\left[\frac{a - h/2}{h/2}\right]$$

Fig. 11-20
Circular segments.

It should be noted that the nondimensional interaction diagrams given in Figs. A-12 to A-14 include the ϕ factors for spiral columns. They cannot be used to design circular tied columns unless an adjustment is made to ϕ.

Properties of Interaction Diagrams for Reinforced Concrete Columns

Nondimensional Interaction Diagrams

Frequently, it is useful to express interaction diagrams independently of column dimensions. This can be done by dividing the factored axial-load resistances, P_n or ϕP_n, by the column area, A_g, or by $f'_c A_g$ (equivalent to $1/0.85$ times the axial-load capacity of the concrete alone) and dividing the moment values, M_n or ϕM_n, by $A_g h$ or by $f'_c A_g h$ (which has the units of moments). A family of such curves is plotted in Figs. A-6 to A-14 in Appendix A. Each diagram is presented for a given ratio, γ, of the distance between the centers of the outermost layer of bars in the faces parallel to the axis of bending to the overall depth of the section, as shown in the inset to the diagrams. The axis of bending is shown in the inset to each graph. The use of these diagrams is illustrated in Examples 11-2 and 11-3. The steel ratio is given as $\rho_g = $ (total area of steel)/(gross area of section) in Figs. A-6 to A-17. Sometimes this is expressed as ρ_t.

Four dashed lines crossing the interaction diagrams show the loads and moments at which the layer closest to the tensile face of the column is stressed to $f_s = 0$, $f_s = 0.5 f_y$ (in tension), the compression-controlled limit $f_s = 1.0 f_y$ (in tension), and the tension-controlled limit. These are used in designing column splices, as will be discussed later in this chapter.

The interaction diagrams in Appendix A are based on the strength-reduction factors in ACI Section C.3.2 and assume that the load factors are from ACI Section C.2. Those load factors must be used if these interaction diagrams are used. Design aids, including sets of similar curves, are published by the American Concrete Institute [11-7]. Use of these diagrams is illustrated in the examples given later in this chapter.

The interaction diagrams in Appendix A are intended to be an aid to learning the use of such diagrams. They are not intended for design as such. It is more convenient to use commercially published diagrams because less interpolation is required [11-7].

Eccentricity of Load

In Fig. 11-9, it was shown that a load P applied to a column at an eccentricity e was equivalent to a load P acting through the centroid, plus a moment $M = Pe$ about the centroid. A radial line through the origin in an interaction diagram has the slope P/M, or $P/P \times e = 1/e$. For example, the balanced load and moment computed in Example 11-1 correspond to an eccentricity of 386 ft-kips/417 kips = 0.93 ft, and a radial line through this point (point B in Fig. 11-18) would have a slope of $1/0.93$. The pure-moment case may be considered to have an eccentricity $M/P = \infty$, since $P = 0$.

In a nondimensional interaction diagram such as Fig. A-6, a radial line through the origin has a slope equal to $(P/A_g)/(M/A_g h)$. Substituting $M = Pe$ shows that the line has the slope h/e or $1/(e/h)$, where e/h represents the ratio of the eccentricity to the column thickness. Radial lines corresponding to several eccentricity ratios are plotted in Fig. A-6.

Unsymmetrical Columns

Up to this point, interaction diagrams have been shown only for symmetrical columns. If the column cross section is symmetrical about the axis of bending, the interaction diagram is symmetrical about the vertical $M = 0$ axis, as shown in Fig. 11-21a. For unsymmetrical columns, the diagram is tilted as shown in Fig. 11-21b, provided that the moments are

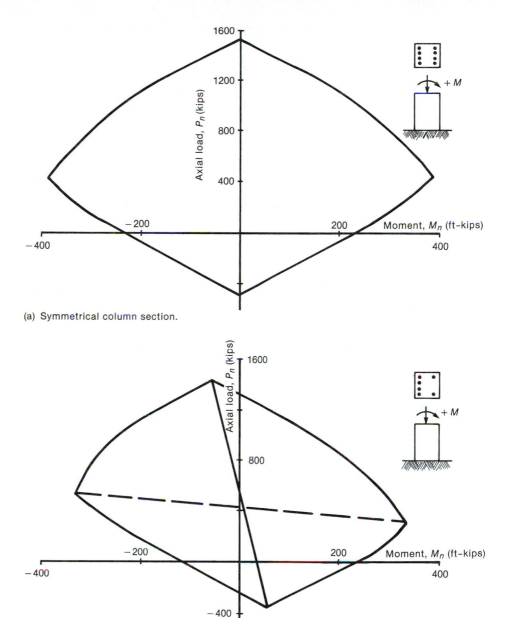

(a) Symmetrical column section.

(b) Unsymmetrical column section.

Fig. 11-21
Interaction diagrams for symmetrical and unsymmetrical
columns.

taken about the geometric centroid. The calculation of an interaction diagram for such a
member follows the same procedure as Example 11-1, except that for the cases of uniform
compressive or tensile strains (axial compression, P_0, and axial tension, P_{nt}), the unsymmetrical bar placement gives rise to a moment of the steel forces about the centroid. Figure 11-21a
is drawn for the column shown in Fig. 11-17a and is the same as the interaction diagram in
Fig. 11-18. Figure 11-21b is drawn for a similar cross section with four No. 9 bars in one face
and two No. 9 bars in the other. For positive moment, the face with four bars is in tension. As
a result, the balanced load for positive moment is less than that for negative moment.

   In a similar manner, a uniform compressive strain of 0.003 across the section, corresponding to the maximum axial-load capacity, leads to a moment, since the forces in the
two layers of steel are unequal.

Simplified Interaction Diagrams for Columns

Generally, designers have access to published interaction diagrams or computer programs to compute interaction diagrams for use in design. Occasionally, this is not true, as, for example, in the design of hollow bridge piers, elevator shafts, or unusually shaped members. Interaction diagrams for such members can be calculated by using the strain-compatibility solution presented earlier. In most cases, it is adequate to represent the interaction diagram by a series of straight lines joining the load and moment values corresponding to the following five strain distributions:

    1.   a uniform compressive strain of 0.003, giving point 1 in Fig. 11-22;

    2.   a strain diagram corresponding to incipient cracking, passing through a compressive strain of 0.003 on one face and zero strain on the other (point 2 in Fig. 11-22);

    3.   the balanced strain distribution and limiting compression-controlled strain distributions, both of which have a compressive strain of 0.003 on one face and a tensile strain of $-\epsilon_y$ at the centroid of the tension steel, or in the layer of steel farthest from the neutral axis, respectively, in the reinforcement layer nearest to the tensile face (point 3);

    4.   the limiting tension-controlled strain distribution having a compressive strain of 0.003 on one face and a tensile strain of -0.005 in the reinforcement layer nearest to the tensile face (point 4);

    5.   a uniform tensile strain of $-\epsilon_y$ in the steel with the concrete cracked (point 5).

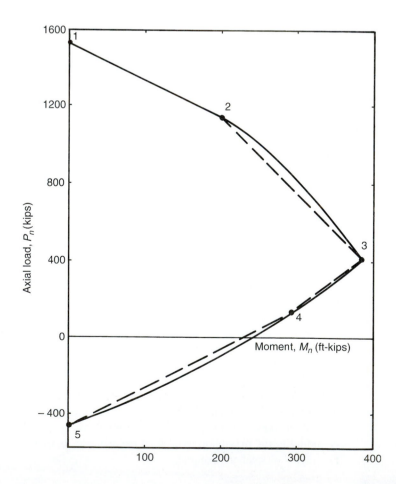

Fig. 11-22
Simplified interaction diagram.

Figure 11-22 compares the interaction diagram for Example 11-1 with an interaction diagram drawn by joining the five points just described. The five-point diagram is sufficiently accurate for design provided strength-reduction factors have been included.

For design, the top of the interaction diagram is cut off at $0.80P_{no}$, because this is a tied column. The strength-reduction factors, ϕ, have not been included in Fig. 11-22. When ϕ is computed by ACI Section 9.3.2.2 for a tied column, ϕ is 0.65 for points 1, 2, and 3 and is given by (11-4a) or (11-4b) for points 4 and 5 (for which ϕ usually is equal to 0.9).

It is difficult to calculate the pure-moment case directly. If this value is required for a symmetrical section, it can be estimated as the larger of (1) the flexural capacity ignoring the reinforcement in the compressive zone, or (2) the moment computed by ignoring the concrete and assuming a strain of $5\epsilon_y$ in the reinforcement adjacent to each face. For the column in Example 11-1, the pure-moment capacity, M_{n0}, from a strain-compatibility solution was 236 ft-kips, compared with 227 from the dashed line in Fig. 11-22, 235 computed for it as a beam (ignoring the compression reinforcement), and 220 for it as a steel couple (ignoring the concrete). The column in question has a high steel percentage. The accuracy of these approximations decreases as ρ approaches the minimum allowed.

11-5   DESIGN OF SHORT COLUMNS

Types of Calculations—Analysis and Design

In Chapter 4, two types of computations were discussed. If the cross-sectional dimensions and reinforcement are known and it is necessary to compute the capacity, an *analysis* is carried out. On the other hand, if the factored loads and moments are known and it is necessary to select a cross section to resist them, the procedure is referred to as *design* or *proportioning*. A design problem is solved by guessing a section, analyzing whether it will be satisfactory, revising the section, and reanalyzing it. In each case, the analysis portion of the problem is most easily carried out via interaction diagrams.

Factors Affecting the Choice of Column

Choice of Column Type

Figure 11-23 compares the interaction diagrams for three columns, each with the same f'_c and f_y, the same total area of longitudinal steel, A_{st}, and the same gross area, A_g. The columns differ in the arrangement of the reinforcement, as shown in the figure. To obtain the same gross area, the spiral column had a diameter of 18 in. while the tied columns were 16 in. square.

For eccentricity ratios, e/h, less than about 0.1, a spiral column is more efficient in terms of load capacity, as shown in Fig. 11-23. This is due to ϕ being 0.70 for spiral columns compared to 0.65 for tied columns and is also due to the higher axial load allowed on spiral columns by ACI Eq. (10-1). This economy tends to be offset by more expensive forming and by the cost of the spiral, which may exceed that of the ties.

For eccentricity ratios, e/h, greater than 0.2, a tied column with bars in the faces farthest from the axis of bending is most efficient. Even more efficiency can be obtained by using a rectangular column to increase the depth perpendicular to the axis of bending.

Tied columns with bars in four faces are used for e/h ratios of less than about 0.2 and also when moments exist about both axes. Many designers prefer this arrangement because there is less possibility of construction error in the field if there are equal numbers of bars in each face of the column. Spiral columns are relatively infrequent in nonseismic areas. In seismic areas or in other situations where ductility is important, spiral columns are used more extensively.

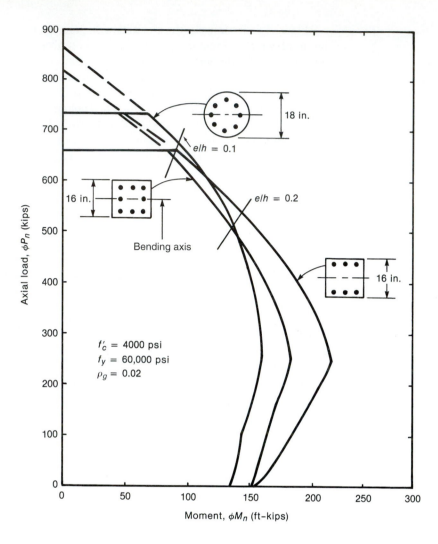

Fig. 11-23
Effect of column type on
shape of interaction diagram.

Choice of Material Properties and Reinforcement Ratios

In small buildings, the concrete strength in the columns is selected to be equal to that in the floors, so that one grade of concrete can be used throughout. Frequently, this will be 3000 or 3500 psi.

In tall buildings, the concrete strength in the columns is often higher than that in the floors, to reduce the column size. When this occurs, the designer must consider the transfer of the column loads through the weaker floor concrete. Tests of interior column–floor junctions subjected to axial column loads have shown that the lateral restraint provided by the surrounding floor members enables a floor slab to transmit the column loads, provided that the strength of the column concrete is not much higher than that of the slab [11-8], [11-9].

The code wording came from tests in which the floor slabs were not loaded. Moments resulting from dead and live loads cause the slab to crack around the column. This reduces the strengthening effect of the restraint from the slab [11-9]. Similarly, the strengthening effect is less at edge columns and corner columns that are unrestrained on one or two sides. ACI Section 10.15 allows credit for the higher strength column concrete in the design of column-to-floor joints:

(a)   ACI Section 10.15—If the specified strength of the column concrete does not exceed 1.4 times the strength of the concrete in the floor system, the code assumes that the joint can be designed using an *effective concrete strength* of 1.4 times the floor-concrete strength without special precautions; or

(b)   ACI Section 10.15.1—The joint region can be constructed with column-strength concrete in the floor in the joint, provided that the top surface of the higher strength concrete in the floor extends 24 in. into the floor system around the column; or

(c)   ACI Section 10.15.2—The computed strength of the joint can be based on the lower strength concrete, added vertical dowels and spirals to produce the needed column resistance.

(d)   ACI Section 10.15.3—For interior columns laterally restrained on four sides by slabs or beams of approximately the same depth, with the strength of the column concrete taken not larger than 2.5 times that of the floor concrete in the design calculations, the joint strength may be based on an *effective concrete strength* equal to 75 percent of the column compressive strength plus 35 percent of the floor strength.

For a 3000-psi concrete floor, ACI Section 10.15 allows up to 4200-psi floor concrete without special precautions. For higher column-concrete strengths, the maximum column strength allowed by (d) would be the smaller of $(2.5 \times 3000) = 7500$ psi and $(0.75 \times 7500 + 0.35 \times 3000) = 6675$ psi, but not more than the stronger of the two specified strengths—in this case, 7500 psi. Thus, this joint could be designed by using ACI Section 10.15.3, assuming an effective concrete strength of 6675 psi.

My own belief is that this is too liberal, and I recommend that interior joints be designed to satisfy ACI Section 10.15.3; edge joints for an effective concrete strength of 1.4 times the floor-concrete strength (but not more than the column-concrete strength); and corner columns for an effective column-concrete strength of 1.2 times the lower concrete strength. The values for edge and corner columns suggested in this paragraph are from tests reported in [11-8] [11-9], buildings of moderate height, it is not uncommon to use 3750-psi concrete in the floors and 5000-psi ($\simeq 1.4 \times 3750$) concrete in the columns.

In the vast majority of columns built, Grade-60 reinforcement is used. Unless there is a good reason for using Grade-40 reinforcement, the yield strength should be taken as 60,000 psi, except for stirrups and ties, which frequently are of Grade-40 steel.

Tests of axially loaded tied columns reported by Pfister [11-9] in 1964 included columns with high-strength-alloy-steel longitudinal reinforcement having a specified yield strength of 75 ksi and a strength of 92 ksi at a strain of 0.006. Two columns with ties spaced at the least dimension of the cross section failed more gradually than companion columns with one tie at midheight or no ties. The columns with ties were able to develop a steel stress of 65 ksi. Pfister also found that one single tie around the perimeter of the 10-by-12-in. columns was enough support for the bars.

Hudson found that the influence of ties on the ultimate capacity of axially and eccentrically loaded columns was negligible [11-10]. His columns failed by spalling of the cover on the compression side, immediately followed by buckling of the longitudinal bars.

Bresler and Gilbert [11-11] also studied tie spacing.

ACI Section 10.9.1 limits the area, A_{st}, of longitudinal reinforcement in tied and spiral columns to not less than 0.01 times the gross area, A_g (i.e., $\rho_t = A_{st}/A_g$ not less than 0.01, or 1 percent, except as allowed by ACI Section 10.8.4) and not more than $0.08A_g$ ($0.06A_g$ in the seismic regions). Under sustained loads, creep of the concrete gradually transfers load from the concrete to the reinforcement. In tests of axially loaded columns, column reinforcement yielded under sustained service loads if the ratio of longitudinal steel was less than roughly 0.01 [11-2]. (See Example 3-4.)

A recently reported reexamination [11-13] of the effect of the creep and shrinkage of modern concretes on the transfer of vertical compression stresses from the concrete to the longitudinal (vertical) bars upheld the lower limit of $\rho_g \geq 0.01$ determined in the 1928 to 1931 tests [11-2].

Although the code allows a maximum steel ratio of 0.08, it is generally very difficult to place this amount of steel in a column, particularly if lapped splices are used. Tables A-16 and A-17 give maximum steel percentages for various column sizes. These range from roughly 3 to 5 or 6 percent. In addition, the most economical tied-column section generally involves ρ_t of 1 to 2 percent. As a result, tied columns seldom have ρ_t greater than 3 percent. Exceptions to this are the lower columns in a tall building, where the column size must be limited for architectural reasons. In such a case, the bars may be tied in bundles of two to four bars. Design requirements for bundled bars are given in ACI Sections 7.6.6 and 12.14.2.2. Because they are used to resist high axial loads, spiral columns generally have steel ratios between 2.5 and 5 percent.

The minimum number of bars in a rectangular column is four, and that in a circular column or spiral column is six (ACI Section 10.9.2). Almost universally, an even number of bars is used in a rectangular column, so that the column is symmetrical about the axis of bending, and also almost universally, all the bars are the same size. Table A-18 gives the areas of various combinations of bars.

Estimating the Column Size

The initial stage in column design involves estimating the required size of the column. There is no simple rule for doing this, since the axial-load capacity of a given cross section varies with the moment acting on the section. For very small values of M, the column size is governed by the maximum axial-load capacity given by (11-10). Rearranging (11-10), simplifying, and rounding down the coefficients gives the approximate relationship

Tied columns

$$A_{g(\text{trial})} \geq \frac{P_u}{0.40(f_c' + f_y\rho_t)} \tag{11-21a}$$

where $\rho_t = A_{st}/A_g$.

Spiral Columns

Equation (11-21a) was derived for tied columns. For spiral columns, the coefficient 0.8 in (11-10b) becomes 0.85, and ϕ is 0.70. Hence, for such columns,

$$A_{g(\text{trial})} \geq \frac{P_u}{0.50(f_c' + f_y\rho_t)} \tag{11-21b}$$

Both of these equations will tend to underestimate the column size if there are moments present, since they correspond roughly to the horizontal line portion of the ϕP_n, ϕM_n interaction diagram in Fig. 11-18.

The local fire code usually specifies minimum sizes and minimum cover to the reinforcement. A conservative approximation to these values is 9-in. minimum column thickness for a 1-hour fire rating and 12-in. minimum for 2- or 3-hour ratings. The minimum clear cover to the vertical reinforcement is 1 in. for a 1-hour fire rating and 2 in. for 2-hour and 3-hour ratings. The tables in Appendix A are based on a 1.5-in. clear cover to the ties and a 2-in. clear cover to the vertical reinforcement. Column widths and depths are generally varied in increments of 2 in. Although the ACI Code does not specify a minimum column size, the minimum dimension of a cast-in-place tied column should not be less than 8 in. and preferably not less than 10 in. The diameter of a spiral column should not be less than about 12 in.

Slender Columns

A slender column deflects laterally under load. This increases the moments in the column and hence weakens the column. (See Section 12-1.) Slenderness effects are discussed in Chapter 12. Chapter 11 deals only with short columns in braced frames, the most commonly occurring case. ACI Section 10.12.2 states that it is permissible to neglect slenderness effects in braced frames if

$$\frac{k\ell_u}{r} \le 34 - 12\left(\frac{M_1}{M_2}\right) \tag{11-22}$$

where

$$k = \text{effective length factor, which, for a braced frame, will be less than or equal to 1.0}$$

$$\ell_u = \text{unsupported height of column from top of floor to the bottom of the beams or slab in the floor above}$$

$$r = \text{radius of gyration, equal to 0.3 and 0.25 times the overall depth of rectangular and circular columns, respectively}$$

$$M_1/M_2 = \text{ratio of the moments at the two ends of the column, which, for a braced frame, will generally be between } +0.5 \text{ and } -0.5$$

This limit is discussed more fully in Section 12-2. In this chapter, we shall assume that $k = 1.0$ and $M_1/M_2 = +0.5$. This will almost always be conservative. For this combination, columns are short if $k\ell_u/r \le 28$. For a square column, this corresponds to $\ell_u/h \le 8.4$.

Bar-Spacing Requirements

ACI Section 7.7.1 requires a clear concrete cover of not less than $1\frac{1}{2}$ in. to the ties or spirals in columns. More cover may be required for fire protection in some cases. The concrete for a column is placed in the core inside the bars and must be able to flow out between the bars and the form. To facilitate this, the ACI Code requires that the minimum clear distance between longitudinal bars shall not be less than the larger of 1.5 times the longitudinal bar diameter, or 1.5 in. (ACI Section 7.6.3), or $1\frac{1}{3}$ times the maximum size of the coarse aggregate (ACI Section 3.3.2). These clear-distance limitations also apply to the clear distance between lap-spliced bars and adjacent bars or lap splices (ACI Section 7.6.4). Because the maximum number of bars occurs at the splices, the spacing of bars at this location generally governs. The spacing limitations at splices in tied and spiral columns are illustrated in Fig. 11-24. Tables A-16 and A-17 give the maximum number of bars that can be used in rectangular tied columns and circular spiral columns, respectively, assuming that the bars are lapped as shown in Fig. 11-24a or b. These tables also give the area, A_{st}, of these combinations of bars and the ratio $\rho_t = A_{st}/A_g$.

Reinforcement Splices

In most buildings in nonseismic zones, the longitudinal bars in the columns are spliced just above each floor. Lap splices as shown in Fig. 11-25 are the most widely used, although, in large columns with large bars, mechanical splices or butt splices are sometimes used. (See Figs. 11-26 and 11-27.) Lap splices, welded or mechanical splices, and end-bearing splices are covered in ACI Sections 12.17.2, 12.17.3, and 12.17.4, respectively.

The requirements for lap splices vary to suit the state of stress in the bar at the ultimate load. In columns subjected to combined axial load and bending, tensile stresses may occur on one face of the column, as seen in Example 11-1. (See Fig. 11-17c, for example.)

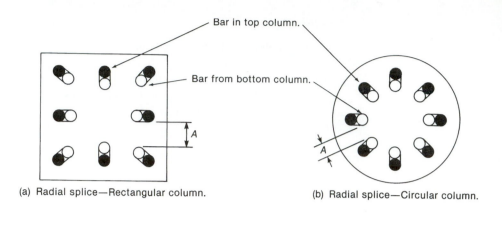

Bar in top column.

Bar from bottom column.

(a) Radial splice—Rectangular column.

(b) Radial splice—Circular column.

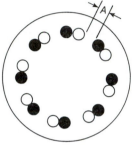

(c) Side-by-side lap splice.

A shall not be less than $1\frac{1}{2}$ in., $1\frac{1}{2}$ bar diameters or $1\frac{1}{3}$ times aggregate size.

Fig. 11-24
Arrangement of bars at lap splices in columns.

Design-interaction charts, such as Fig. A-6, frequently include lines indicating the eccentricities for which various tensile stresses occur in the reinforcement closest to the tensile face of the column. In Fig. A-6, these are labelled $f_s = 0$, $f_s = 0.5f_y$, and $f_s = f_y$ (in tension). The range of eccentricities for which various types of splices are required is shown schematically in Fig. 11-28.

Column-splice details are important to the designer for two major reasons. First, a compression lap splice will automatically be provided by the reinforcement detailer unless a different lap length is specified by the designer. Hence, if the bar stress at ultimate is tensile, compression lap splices may be inadequate, and the designer should compute and show the laps required on the drawings. Second, if Class B lap splices are required (see Fig. 11-28), the required splice lengths may be excessive. For closely spaced bars larger than No. 8 or 9, the length of such a splice may exceed 5 ft and thus may be half or more of the height of the average story. The splice lengths will be minimized by choosing the smallest practical bar sizes and the highest practical concrete strength. Alternatively, welded or mechanical splices should be used. To reduce the chance of field errors, all the bars in a column will be spliced with splices having the same length of splice, regardless of whether they are on the tension or compression face. The required splice lengths are computed via the following steps:

1.   Establish whether compression or tension lap splices are required, using the sloping dashed lines in the interaction diagrams. (See Figs. A-6 to A-14.) This is illustrated in Fig. 11-28.

2.   If compression lap splices are required, compute the basic compression lap-splice length from ACI Section 12.16.1 or 12.16.2. Where applicable, the compression lap-splice length can be multiplied by factors from ACI Section 12.17.2.4 or 12.17.5 for bars enclosed within ties or spirals. These factors apply only to compression lap splices.

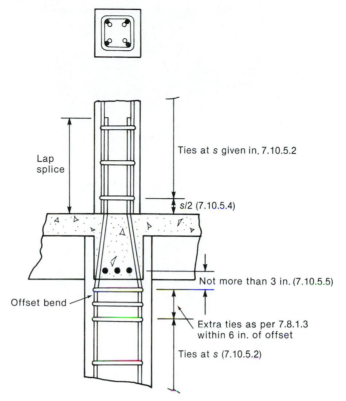

Lap splice

Ties at s given in. 7.10.5.2

$s/2$ (7.10.5.4)

Not more than 3 in. (7.10.5.5)

Offset bend

Extra ties as per 7.8.1.3
within 6 in. of offset

Ties at s (7.10.5.2)

(a) Tie spacing at interior column–beam joint.

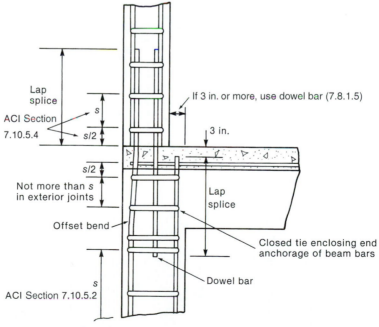

Lap splice
ACI Section
7.10.5.4

s

$s/2$

If 3 in. or more, use dowel bar (7.8.1.5)

3 in.

$s/2$

Not more than s
in exterior joints

Lap splice

Offset bend

Closed tie enclosing end
anchorage of beam bars

s
ACI Section 7.10.5.2

Dowel bar

(b) Ties at exterior column–beam joint.

Fig. 11-25
Lap Details at column-beam
joints.

Fig. 11-26
Metal-filled bar splice:
tension or compression. (Photograph courtesy of Erico
Products Inc.)

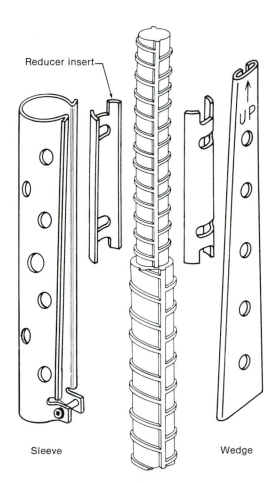

Fig. 11-27
Wedged sleeve bar splice:
compression only. (Drawing
courtesy of Gateway Building
Products.)

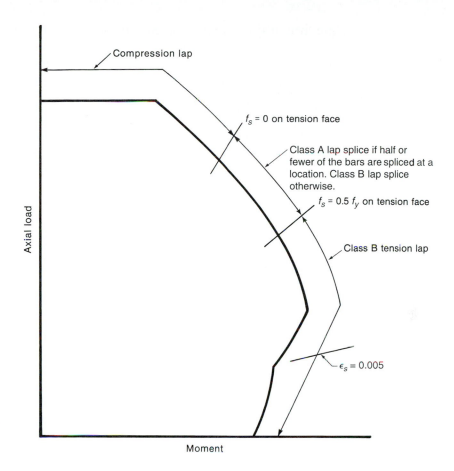

Compression lap

$f_s = 0$ on tension face

Class A lap splice if half or
fewer of the bars are spliced at a
location. Class B lap splice
otherwise.

$f_s = 0.5 f_y$ on tension face

Class B tension lap

$\epsilon_s = 0.005$

Axial load

Moment

Fig. 11-28
Types of lap splices required
if all bars are lap spliced at
every floor.

3. If *tension* lap splices are required, ℓ_d is computed from ACI Section 12.2.2 or 12.2.3. Since ACI Section 7.6.3 requires that longitudinal column bars have a minimum clear spacing of $1.5d_b$, and since they will always be enclosed by ties or spirals, ℓ_d is computed for case 1 in Table 8-1. ℓ_d is then multiplied by 1.0 or 1.3 for a Class A or a Class B splice, respectively.

It is the usual practice to lap splice all the bars in one region along the length of the column, usually just above the floor level at the bottom of the column, as shown in Figs. 11-1, 11-2, and 11-25. If bars on one side of a column need compression lap splices and those on the other side need tension lap splices, it is customary to splice all of them with tension lap splices, which are longer than compression splices.

Bars loaded only in compression can be spliced by end-bearing splices (ACI Sections 12.16.4 and 12.17.4), provided that the splices are staggered or extra bars are provided at the splice locations so that the continuing bars in each face of the column at the splice location have a tensile strength at least 25 percent of that of all the bars in that face, either by staggering the splices or by adding additional steel through the splice location. In an end-bearing splice, the ends of the two bars to be spliced are cut as squarely as possible and held in contact with a wedged sleeve device (Fig. 11-27). Butt splices are used only on vertical, or almost vertical, bars enclosed by stirrups or ties. Tests show that the force transfer in end-bearing splices is superior to that in lapped splices, even if the ends of the bars are slightly off square (up to 3° total angle between the ends of the bars).

Spacing and Construction Requirements for Ties

Ties are provided in reinforced concrete columns for four reasons [11-10], [11-11], [11-13]:

1. Ties restrain the longitudinal bars from buckling out through the surface of the column.

ACI Sections 7.10.5.1, 7.10.5.2, and 7.10.5.3 give limits on the size, the spacing, and the arrangement of the ties, so that they are adequate to restrain the bars. The minimum tie size is a No. 3 bar for longitudinal bars up to No. 10 and a No. 4 bar for larger longitudinal bars or for bundled bars. The vertical spacing of ties shall not exceed 16 longitudinal bar diameters, to limit the unsupported length of these bars, and shall not exceed 48 tie diameters, to ensure that the cross-sectional area of the ties is adequate to develop the forces needed to restrain buckling of the longitudinal bars. The maximum spacing is also limited to the least dimension of the column. In seismic regions, much closer spacings are required (ACI Section 21.4.4).

ACI Section 7.10.5.3 outlines the arrangement of ties in a cross section. These are illustrated in Fig. 11-29. A bar is adequately supported against lateral movement if it is located at a corner of a tie where the ties are spaced in accordance with 7.10.5.2 and if the dimension x in Fig. 11-29 is 6 in. or less. Diamond-shaped and octagonal-shaped ties are not uncommon and keep the center of the column open, so that the placing and vibrating of the concrete is not impeded by the cross-ties. The ends of the ties are anchored by a *standard stirrup or tie hook* around a longitudinal bar, plus an extension of at least six tie-bar diameters but not less than $2\frac{1}{2}$ in. In seismic areas, a 135° bend plus a six-tie-diameter extension is required.

2. Ties hold the reinforcement cage together during the construction process, as shown in Figs. 11-2 and 11-3.

3. Properly detailed ties confine the concrete core, providing increased ductility.

4. Ties serve as shear reinforcement for columns.

If the shear V_u exceeds $\phi V_c/2$, shear reinforcement is required (ACI Section 11.5.5.1). Ties can serve as shear reinforcement, but must satisfy both the maximum tie spacings given in ACI Section 7.10.5.2 and the maximum stirrup spacing for shear from ACI Section 11.5.4.1. The area of the legs parallel to the direction of the shear force must satisfy ACI Section 11.5.5.3.

Specially fabricated welded-wire reinforcement cages incorporating the longitudinal reinforcement and ties are sometimes an economical solution to constructing column cages. Each cage consists of several interlocking sheets bent to form one to three sides of the cage. The bars or wires forming the ties are hooked around longitudinal bars to make the ties continuous. The layout of the cages must be planned carefully, so that bars from one part of the cage do not interfere with those from another when they are assembled in the field.

Ties may be formed from continuously wound wires with a pitch and area conforming to the tie requirements.

ACI Sections 7.10.5.4 and 7.10.5.5 require that the bottom and top ties be placed as shown in Fig. 11-25. ACI Section 7.9.1 requires that bar anchorages in connections of beams and columns be enclosed by ties, spirals, or stirrups. Generally, ties are most suitable for this purpose and should be arranged as shown in Fig. 11-25b.

Finally, extra ties are required at the outside (lower) end of offset bends at column splices (see Fig. 11-25) to resist the horizontal force component in the sloping portion of the bar. The design of these ties is described in ACI Section 7.8.1. If we assume that the vertical bars in Fig. 11-25a are No. 8 Grade-60 bars and that the offset bend is at the maximum allowed slope of 1:6 (ACI Section 7.8.1.3), then the horizontal force from two bars is $(1/6) \times f_y \times 2 \times 0.79$ in.^{2} = 15.8 kips, where 0.79 in.^{2} is the area of one No. 8 bar. ACI Section 7.8.1.3 requires that ties be provided for 1.5 times this force, or 23.7 kips. The number of No. 3 Grade-60 tie legs required to resist this force is 23.7 kips/ $(60 \times 0.11$ in.$^2) = 3.6$ legs. Thus, two ties would be required within 6 in. of the bend.

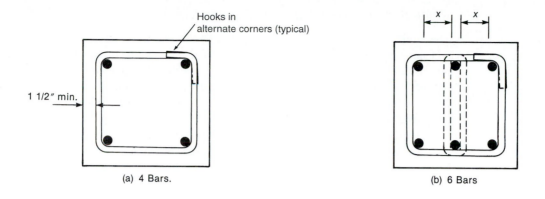

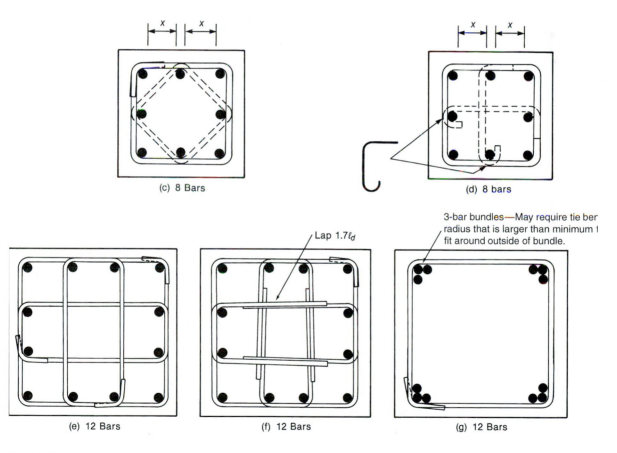

Fig. 11-29
Typical tie arrangements.

Amount of Spirals and Spacing Requirements

The minimum spiral reinforcement required by the ACI Code was chosen so that the second maximum load of the core and longitudinal reinforcement would roughly equal the initial maximum load of the entire column before the shell spalled off. (See Fig. 11-5a.) Figure 11-4 shows that the core of a spiral column is stressed in triaxial compression, and (3-16) indicates that the strength of concrete is increased by such a loading.

The amount of spiral reinforcement is defined by using a spiral reinforcement ratio, ρ_s, equal to the ratio of the volume of the spiral reinforcement to the volume of the core measured out-to-out of the spirals, enclosed by the spiral. For one turn of the spiral shown in Fig. 11-4a,

$$\rho_s = \frac{A_{sp}\ell_{sp}}{A_c\ell_c}$$

where

$\quad A_{sp}$ = area of the spiral bar = $\pi d_{sp}^2/4$
$\quad d_{sp}$ = diameter of the spiral bar
$\quad \ell_{sp}$ = length of one turn of the spiral = πD_c
$\quad D_c$ = diameter of the core, out-to-out of the spirals
$\quad A_c$ = area of the core = $\pi D_c^2/4$
$\quad \ell_c$ = spiral pitch = s

Thus,

$$\rho_s = \frac{(A_{sp})(\pi D_c)}{(\pi D_c^2/4)s}$$

or

$$\rho_s = \frac{4A_{sp}}{sD_c} \tag{11-23}$$

From the horizontal force equilibrium of the free body in Fig. 11-4a,

$$2f_{sp}A_{sp} = f_2 D_c s \tag{11-24}$$

From (11-23) and (11-24),

$$f_2 = \frac{f_{sp}\rho_s}{2} \tag{11-25}$$

From (11-1), the strength of a column at the first maximum load before the shell spalls off is

$$P_0 = 0.85f_c'(A_g - A_{st}) + f_y A_{st} \tag{11-1}$$

and the strength after the shell spalls is

$$P_2 = 0.85f_1(A_c - A_{st}) + f_y A_{st}$$

Thus, if P_2 is to equal P_0, $0.85f_1(A_c - A_{st})$ must equal $0.85f'_c(A_g - A_{st})$. Since A_{st} is small compared with A_g or A_c, we can disregard it, giving

$$f_1 = \frac{A_g f'_c}{A_c} \tag{11-26}$$

Substituting (3-16) and (11-25) into (11-26), taking f_{sp} equal to the yield strength of the spiral bar, f_y, rearranging, and rounding down the coefficient gives

$$\rho_s = 0.45\left(\frac{A_g}{A_c} - 1\right)\frac{f'_c}{f_y} \tag{11-27}$$

(ACI Eq. 10-5)

There is experimental evidence that more spiral reinforcement may be needed in high-strength concrete spiral columns than is given by (11-27) [11-14], to ensure that ductile behavior precedes any failure. To date, this evidence has not been considered by the ACI Code committee.

Design requirements for column spirals are presented in ACI Sections 10.9.3 and 7.10.4.1 through 7.10.4.9. ACI Section 7.10.4.2 requires that in cast-in-place construction, spirals be at least No. 3 in diameter. The spacing is determined by three rules:

1. Equation (11-27) (ACI Eq. (10-6)) gives the maximum spacing that will result in a second maximum load that equals or exceeds the initial maximum load. Combining (11-23) and (11-27) and rearranging gives the maximum *center-to-center* spacing of the spirals:

$$s \leq \frac{\pi d^2_{sp} f_y}{0.45 D_c f'_c[(A_g/A_c) - 1]} \tag{11-28}$$

2. For the spirals to confine the core effectively, successive turns of the spiral must be spaced relatively close together. ACI Section 7.10.4.3 limits the clear spacing of the spirals to a maximum of 3 in.

3. To avoid problems in placing the concrete in the outer shell of concrete, the spiral spacing must be as large as possible. ACI Section 7.10.4.3 limits the minimum clear spacing between successive turns to 1 in., but not less than $1\frac{1}{3}$ times the nominal size of the coarse aggregate, whichever is greater.

Table A-15 in Appendix A gives spiral spacings that satisfy all three rules just given above. The termination of spirals at the tops and bottoms of columns is governed by ACI Sections 7.10.4.6 to 10.4.8. Again, joint reinforcement satisfying ACI Section 7.9.1 is also required, and generally, ties are provided for this purpose.

Design Examples

Three examples will be considered to illustrate the design of column cross sections. In the second case, the strength will be governed by the maximum axial-load capacity, and in the third, the axial load will fall in the range where ϕ varies from the value for a column to that for a beam. Although no examples of analysis per se are presented, the analysis of a given cross section is identical to that in the three examples, except that it is not necessary to select the material properties and column size (step 1 in the examples).

EXAMPLE 11-2 Design of a Tied Column for a Given P_u and M_u

Design a tied-column cross section to support $P_u = 350$ kips, $M_u = 110$ ft-kips, and $V_u = 14$ kips. The column is in a braced frame and has an unsupported length of 10 ft 6 in.

1. **Select the material properties, trial size, and trial reinforcement ratio.** Select $f_y = 60$ ksi and $f'_c = 3$ ksi. The most economical range for ρ_t is from 1 to 2 percent. Assume that $\rho_t = 0.015$ for the first trial value. From (11-21a),

$$A_{g(\text{trial})} \geq \frac{P_u}{0.40(f'_c + f_y\rho_t)} \tag{11-21a}$$

$$\geq \frac{350}{0.40(3 + 60 \times 0.015)}$$

$$\geq 224 \text{ in.}^2 \text{ (or 15 in. square)}$$

Since moments act on this column, (11-21a) will underestimate the column size. Choose a 16-in.-square column.

To determine the preferable bar arrangement, compute the ratio e/h:

$$e = \frac{M_u}{P_u}$$

$$= \frac{110 \text{ ft-kips}}{350 \text{ kips}} = 0.314 \text{ ft} = 3.77 \text{ in.}$$

$$\frac{e}{h} = 0.236$$

For e/h in this range, Fig. 11-23 indicates that a column with bars in two faces will be most efficient. Use a tied column with bars in two faces.

Slenderness can be neglected if

$$\frac{k\ell_u}{r} \leq 34 - 12\left(\frac{M_1}{M_2}\right) \tag{11-22}$$

Since this is a braced frame, $k \leq 1.0$, and M_1/M_2 will normally be between $+0.5$ and -0.5. We shall assume that $k = 1.0$ and $M_1/M_2 = +0.5$. The left-hand side of (11-22) is

$$\frac{k\ell_u}{r} = \frac{1.0 \times 126 \text{ in.}}{0.3 \times 16 \text{ in.}} = 26.3$$

The right-hand side is

$$34 - 12\left(\frac{M_1}{M_2}\right) = 34 - 12(+0.5) = 28$$

Since 26.3 is less than 28, slenderness can be neglected.

Summary for the trial column. A 16 in. × 16 in. tied column with bars in two faces, $f'_c = 3,000$ psi, and $f_y = 60,000$ psi.

2. **Compute γ.** The interaction diagrams in Appendix A are each drawn for a particular value of the ratio, γ, of the distance between the centers of the outside layers of bars to the overall depth of the column. To estimate γ, assume that the longitudinal bars are No. 8 and the ties are No. 3 bars. For $1\frac{1}{2}$ in. of clear cover to the ties (see Fig. 11-30),

$$\gamma = \frac{16 - 2(1.5 + 0.375 + 0.5)}{16} = 0.703$$

Since the interaction diagrams in Appendix A are given for $\gamma = 0.60$ and $\gamma = 0.75$, it will be necessary to interpolate. If the bars chosen are not No. 8, it will be necessary to recompute γ.

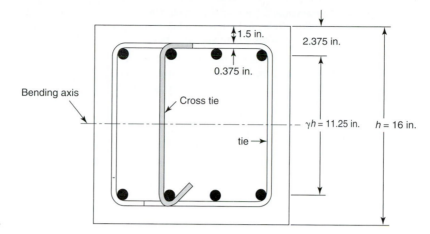

Fig. 11-30
Computation of γ and arrangement of ties—Example 11-2.

3. **Use interaction diagrams to determine ρ_t.** The interaction diagrams are entered with

$$\frac{\phi P_n}{A_g} = \frac{P_u}{A_g}$$

$$= \frac{350}{16 \times 16} = 1.367 \text{ ksi}$$

$$\frac{\phi M_n}{A_g h} = \frac{M_u}{A_g h}$$

$$= \frac{110 \times 12}{16 \times 16 \times 16} = 0.322 \text{ ksi}$$

From Fig. A-6 (interaction diagram for $\gamma = 0.6$),

$$\rho_t = 0.016$$

From Fig. A-7 (for $\gamma = 0.75$),

$$\rho_t = 0.013$$

Use linear interpolation to compute the value for $\gamma = 0.703$:

$$\rho_t = 0.016 - 0.003 \times \frac{0.103}{0.15} = 0.0139$$

If the value of ρ_t computed here exceeds 0.03 to 0.04, a larger section should be chosen. If ρ_t is less than 0.01, either use 0.01 (the minimum allowed by ACI Section 10.9.1) or recompute, using a smaller cross section.

4. **Select the reinforcement.**

$$A_{st} = \rho_t A_g$$

$$= 0.0139 \times 16 \times 16 \text{ in.}^2 = 3.56 \text{ in.}^2$$

Possible combinations are (Table A-22)

four No. 9 bars, $A_{st} = 4.00 \text{ in.}^2$, two in each face
six No. 7 bars, $A_{st} = 3.60 \text{ in.}^2$, three in each face

An even number of bars will be chosen, so that the reinforcement is symmetrical about the bending axis. From Table A-18, it is seen that none of these violates the minimum bar spacing rules. **Try a 16-in.-square column with six No. 7 bars.**

5. **Check the maximum load capacity.** P_u should not exceed $\phi P_{n(\max)}$, given by (11-10). The upper horizontal lines in the interaction diagrams represent $\phi P_{n(\max)}$, and the section chosen falls below the upper limit. Actually, this check is necessary only if one is using interaction diagrams that do not show this cutoff.

6. **Design the lap splices.** From Figs. A-6 and A-7, the stress in the bars adjacent to the tensile face for $P_u/bh = 1.367$ and $M_u/bh^2 = 0.322$ is about $0.2f_y$ in tension. From ACI Sections 12.17.2.2 and 12.17.2.3, the splice must be a Class B splice if more than half of the bars are spliced at any section or a Class A splice if half or fewer are spliced at one location. Normally, all the bars would be spliced at the same location. We shall assume that this is done. The splice length is $1.3\ell_d$. From ACI Section 12.2.2, for No. 7 bars,

$$\ell_d = \left(\frac{f_y\alpha\beta\lambda}{20\sqrt{f'_c}}\right)d_b$$

$$= \left(\frac{60,000 \times 1.0 \times 1.0 \times 1.0}{20 \times \sqrt{3000}}\right) \times 0.875$$

$$= 47.9 \text{ in.}$$

The splice length is

$$1.3\ell_d = 1.3 \times 47.9$$
$$= 62.3 \text{ in.}$$

This is a very long splice. It would approach half of the story height.

7. **Select the ties.** From ACI Section 7.10.5.1, No. 3 ties are the smallest allowed. The required spacing (ACI Section 7.10.5.2) is the smallest of the following quantities:

$$16 \text{ longitudinal bar diameters} = 16 \times \frac{7}{8} = 14 \text{ in.}$$

$$48 \text{ tie diameters} = 48 \times \frac{3}{8} = 18 \text{ in.}$$

$$\text{least dimension of column} = 16 \text{ in.}$$

If $V_u > 0.5\phi V_c$, the ties must satisfy both ACI Chapter 11 and ACI Section 7.10.5. That is,

$$V_c = 2\left(\frac{1 + N_u}{2000A_g}\right)\sqrt{f'_c}b_w d$$

$$= 2\left(1 + \frac{350,000}{2000 \times 16 \times 16}\right)\sqrt{3000} \times 16 \times 13.6$$

$$= 40.1 \text{ kips}$$

$V_u = 14$ kips is less than $0.5\phi V_c = 0.5 \times 0.75$ (for shear) $\times 40.1 = 15.0$ kips. Therefore, ACI Section 7.10.5 governs. (If $0.5\phi V_c < V_u \leq \phi V_c$, it would be necessary to satisfy ACI Sections 7.10.5, 11.5.4.1, and 11.5.5.3.) **Use No. 3 ties at 14 in. o.c.** The tie arrangement is shown in Fig. 11-30. ACI Section 7.10.5.3 requires that all corner bars and alternate other bars be at the corner of a tie. To accomplish this, we shall add a cross-tie.

Summary of the design. Use a 16 in. × 16 in. tied column with eight No. 6 bars, $f_y = 60,000$ psi, and $f'_c = 3000$ psi. Use No. 3 closed ties and cross-ties at 14 in. on centers, as shown in Fig. 11-30. Alternate the end of the cross-tie with the 90° hook. ■

EXAMPLE 11-3   Design of a Circular Spiral Column for a Large Axial Load and Small Moment

Design a spiral-column cross section to support factored forces and end moments of $P_u = 1600$ kips and $M_u = 150$ ft-kips.

1. **Select the material properties, trial size, and trial reinforcement ratio.** Use $f'_c = 4$ ksi and $f_y = 60$ ksi. Try $\rho_t = 0.04$. From (11-21b),

$$A_{g(\text{trial})} = \frac{P_u}{0.50(f'_c + f_y\rho_t)} = 500 \text{ in.}^2$$

This corresponds to a diameter of 25.2 in. We shall try 26 in.

2. **Compute γ.** Assume that the spiral is made from a No. 3 bar and the longitudinal bars are No. 10:

$$\gamma = \frac{26 \text{ in.} - 2(1.5 + 0.375 + 1.27/2)}{26 \text{ in.}} = 0.807$$

3. **Use interaction diagrams to determine ρ_t.**

$$\frac{P_u}{A_g} = \frac{1600}{\pi \times 13 \text{ in.}^2} = 3.01 \text{ ksi}$$

$$\frac{M_u}{A_g h} = \frac{150 \times 12}{\pi \times 13 \text{ in.}^2 \times 26 \text{ in.}} = 0.130 \text{ ksi}$$

Interpolating between Figs. A-13 and A-14 (interaction diagrams for spiral columns with $\gamma = 0.75$ and 0.90 and the correct material strengths) gives $\rho_t = 0.028$. Because the values of P_u/A_g and $M_u/A_g h$ fall in the upper horizontal part of the diagram, (11-9) could be used to solve for A_{st} directly.

4. **Select the reinforcement.**

$$A_{st} = \rho_t A_g$$

$$= 0.028(13 \text{ in.}^2 \times \pi) = 14.9 \text{ in.}^2$$

From Table A-18, 12 No. 10 bars give 15.2 in.^{2}, and Table A-21 shows that these will fit into a 26-in.-diameter column. Try 12 No. 10 bars.

5. **Check the maximum load capacity.** From (11-10a),

$$\phi P_{n,\,\text{max}} = 0.85 \times 0.7 \text{ (for spiral) } [0.85 \times 4(\pi \times 13 \text{ in.}^2 - 15.2) + 60 \times 15.2]$$

$$= 1600 \text{ kips—therefore, OK.}$$

6. **Select the spiral.** The minimum-size spiral is No. 3 (ACI Section 7.10.4.2). The center-to-center pitch, s, required to ensure that the second maximum load is equal to the initial maximum load, is given by (11-28), that is,

$$s < \frac{\pi d_{sp}^2 f_y}{0.45 D_c f'_c[(A_g/A_c) - 1]}$$

where

d_{sp} = diameter of spiral = 0.375 in.

f_y = yield strength of spiral bar = 60,000 psi

D_c = diameter of core, outside to outside of spirals = 26 in. − 2 × 1.5 in. = 23 in.

A_g = gross area = 13 in.$^2 \pi$ = 531 in.^{2}

A_c = area of core = 11.5 in.$^2 \pi$ = 415 in.^{2}

Thus,

$$s \leq \frac{\pi \times 0.375^2 \times 60{,}000}{0.45 \times 23 \text{ in.} \times 4000[(531/415) - 1]}$$

$$\leq 2.29 \text{ in.}$$

Hence, the center-to-center pitch cannot exceed 2.29 in., but must also satisfy detailing requirements. The maximum clear spacing (ACI Section 7.10.4.3) is 3 in. (the maximum pitch is $3.0 + 0.375 = 3.375$ in.) The minimum clear spacing (ACI Section 7.10.4.3) is 1 in. or $1\frac{1}{3}$ times the size of the coarse aggregate (ACI Section 3.3.3). For $\frac{3}{4}$-in. aggregate, the minimum pitch is 1.375 in.

Use a No. 3 spiral with $2\frac{1}{4}$-in. pitch. Table A-15 could be used to select the spiral directly.

7. Design the lap splices. From the interaction diagrams, f_s is compression. Therefore, use compression lap splices (ACI Section 12.16.1). From Table A-19, these must be 31.3 in. long. Since the bars are enclosed in a spiral, this can be multiplied by 0.75, giving a splice length of 23.5 in. (ACI Section 12.17.2.5).

Summary of the design. Use a 26-in.-diameter column with 12 No. 10 bars, $f_y = 60{,}000$ psi, and $f'_c = 4000$ psi. Use a No. 3 spiral with a pitch of $2\frac{1}{4}$ in. and $f_y = 60{,}000$ psi. Lap-splice all bars 24 in. just above every floor. ∎

EXAMPLE 11-4 Design of a Tied Column for a Small Axial Load

If the interaction diagrams used in designing the column are plotted for ϕP_n and ϕM_n and have the transition range of ϕ included, as do all the interaction diagrams in Appendix A, the solution proceeds exactly as in Example 11-2. Occasionally, however, one may obtain design tables or interaction diagrams that do not include ϕ or do not show the correct transition for ϕ. In the tension-failure region, such diagrams look like the line B–C–D in Fig. 11-18. Diagrams that correctly include ϕ have a discontinuity in the tension-failure region and resemble line B'–C'–D' in this figure. When working with diagrams that do not include ϕ, follow the procedure given in Section 11-4 and Example 11-1. ∎

11-6 CONTRIBUTIONS OF STEEL AND CONCRETE TO COLUMN STRENGTH

The interaction diagram for a column with reinforcement in two faces can be considered to be the sum of three components: (1) the load and moment resistance of the concrete, and (2) and (3) the load and moment resistance provided by each of the two layers of reinforcement.

In Fig. 11-31, the P_n, M_n interaction diagram from Example 11-1 and Fig. 11-18 is replotted to show the load and moment resistance contributed by the concrete and the two layers of steel. The center curved portion, O–J–A–G–D–L, represents the compressive force in the concrete, C_c, and the moment of this force about the centroid of the column. The shaded band represents the force in the "tensile" reinforcement adjacent to the less compressed face, force F_{s1} in Example 11-1, and the moment of this force about the centroid of the column. The outer band is the force F_{s2} in the "compressive" reinforcement adjacent to the most compressed face and its moment about the centroid. A reexamination of Example 11-1 shows that, at the balanced-failure condition, the force C_c was 434.6 kips and the moment of C_c about the centroid was 174 ft-kips. These are plotted in Fig. 11-31 as the vector OA. The force F_{s1} was -240 kips, and its moment about the centroid was 110 ft-kips, plotted as vector AB. Finally, the force F_{s2} was 222.1 kips, and its moment was 101.8 ft-kips, plotted as vector BC. An examination of line $OABC$ shows that the majority of the axial-load capacity at the balanced load comes from the concrete, with the forces in the tension and compression steel layers essentially canceling each other out. All three constituents contribute to the moment capacity.

A different case is represented by line $ODEF$ in Fig. 11-31. Here, the force F_{s1} is in compression and hence adds to the axial-load capacity and subtracts from the moment

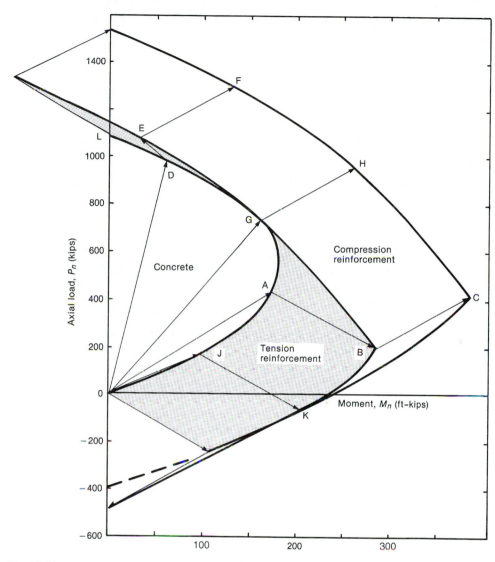

Fig. 11-31
Components of interaction diagram—Example 11-1.

capacity, as shown by the vector DE. An intermediate case, in which $F_{s1} = 0$, is shown by line OGH. A similar case, in which $F_{s2} = 0$, is shown by line OJK.

The portion of the interaction diagram due to the concrete is a continuous curve. The discontinuity in the overall interaction diagram at the balanced point is due to the reinforcement. The two branches of the interaction diagram in Figs. 11-18 and 11-31 are relatively straight. This is characteristic of a column with a high steel ratio. In the case of low steel ratios, the curve is much more curved, because the concrete portion dominates.

The effects of the individual reinforcement layers can be seen from Fig. 11-31. From the top of the diagram down to about F, both layers of steel are effective in increasing the axial capacity. At H, only the steel on the compression face is effective. At C, both layers are effective in increasing the moment capacity. At K, only the reinforcement on the tension face is effective, and finally, at M, both layers add to the tensile capacity. To optimize the design of a member subjected to axial load and bending, one may wish to provide different amounts of reinforcement in the two faces. Figure 11-31 can be used as a guide to where the reinforcement should go. Generally speaking, however, columns are built with the same reinforcement in the two faces parallel to the axis of bending, (a) to minimize the chance of putting the bars in the wrong face and (b) because end moments due to wind change sign as a result of winds blowing from alternate sides of the building. In the case of culverts, arch sections, or rigid frame legs, however, it may be desirable to have unsymmetrical reinforcement.

11-7   BIAXIALLY LOADED COLUMNS

Up to this point in the chapter we have dealt with columns subjected to axial loads accompanied by bending about one axis. It is not unusual for columns to support axial forces and bending about two perpendicular axes. One common example is a corner column in a frame.

For a given cross section and reinforcing pattern, one can draw an interaction diagram for axial load and bending about either axis. As shown in Fig. 11-32, these interaction diagrams form two edges of a three-dimensional interaction surface for axial load and bending about two axes. The calculation of *each* point on such a surface involves a double iteration: (1) the strain gradient across the section is varied, and (2) the angle of the neutral axis is varied. For the same reasons discussed in Section 5-3, the neutral axis will generally not be parallel to the resultant moment vector [11-14]. The calculation of interaction diagrams for biaxially loaded columns is discussed in [11-16].

A horizontal section through such a diagram resembles a quadrant of a circle or an ellipse at high axial loads, and depending on the arrangement of bars, it becomes considerably less circular near the balanced load, as shown in Figs. 11-32 and 11-33.

Four procedures commonly used to design rectangular columns subjected to biaxial loads will be illustrated. Notation is defined in Figs. 11-33 and 11-34.

Positive-moment vectors are shown in Fig. 11-33. By this definition of moments, a positive moment M_x causes compression at points with positive y coordinates, and a positive moment M_y causes compression at points with positive x coordinates.

In addition to the notation presented earlier, other terms include:

P_u = factored axial load.

e_x = eccentricity of applied load measured parallel to the x-axis, positive to right in the cross section in Fig. 11-33.

e_y = eccentricity of applied load measured parallel to the y-axis, positive upward on cross section in Fig. 11-33.

M_{ux} = factoral moment about x axis, equal to $P_u e_y$, positive when it causes compression in fibers in the positive y direction.

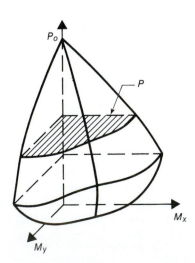

Fig. 11-32
Interaction surface for axial
load and biaxial bending.

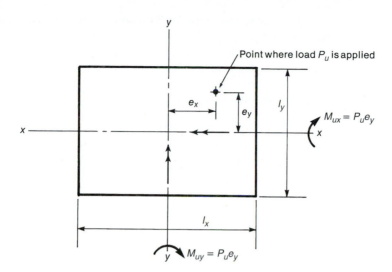

Fig. 11-33
Definition of terms: biaxially
loaded columns.

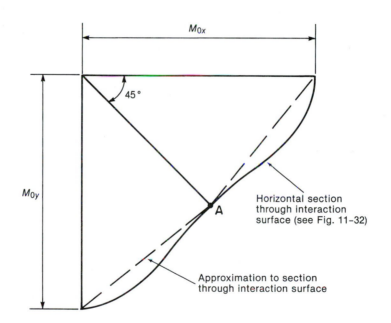

Fig. 11-34
Approximation of section
through intersection surface.

M_{uy} = factoral moment about y-axis, equal to $P_u e_x$, positive when it causes compression in fibers in positive direction from origin.

P_u = factored axial load, positive in compression.

ϕP_{nx} = nominal axial-load capacity for the moment about the x-axis corresponding to the eccentricity, e_y, and the steel and concrete section provided, with $e_x = 0$.

ϕP_{ny} = nominal axial-load capacity, for the moment about the y-axis corresponding to the eccentricity, e_x, and the steel and concrete section provided, with $e_x = 0$.

ϕP_{no} = nominal axial-load capacity, corresponding to e_x and the steel and concrete section provided, with both $e_y = 0$ and $e_x = 0$.

ℓ_x = length of side of column parallel to the x-axis.

ℓ_y = length of side of column parallel to the y axis.

1. **The strain-compatibility method.**   A strain-compatibility solution was applied to a beam section in Section 5-5. The use of an iterative strain-compatibility analysis for a biaxially loaded column is illustrated in Example 11-5. This is the most nearly theoretically correct method of solving biaxially-loaded-column problems presented in this book. Alternatively, design can be based on one of the three widely used approximate design procedures presented after Example 11-5.

2. **The equivalent eccentricity method.**   The biaxial eccentricities, e_x and e_y, can be replaced by an equivalent uniaxial eccentricity, e_{0x}, and the column designed for uniaxial bending and axial load [11-17], [11-18]. We shall define e_x as the component of the eccentricity parallel to the side l_x and the x axis, as shown in Fig. 11-33, such that the moments, M_{uy}, and M_{ux}, are

$$M_{uy} = P_u e_x \qquad M_{ux} = P_u e_y \qquad \text{(11-29a,b)}$$

If

$$\frac{e_x}{\ell_x} \geq \frac{e_y}{\ell_y} \qquad \text{(11-30)}$$

then the column can be designed for P_u and a factored moment $M_{0y} = P_u e_{0x}$, where

$$e_{0x} = e_x + \frac{\alpha e_y \ell_x}{\ell_y} \qquad \text{(11-31)}$$

where for $P_u/f_c' A_g \leq 0.4$,

$$\alpha = \left(0.5 + \frac{P_u}{f_c' A_g}\right) \frac{f_y + 40{,}000}{100{,}000} \geq 0.6 \qquad \text{(11-32a)}$$

and for $P_u/f_c' A_g > 0.4$,

$$\alpha = \left(1.3 - \frac{P_u}{f_c' A_g}\right) \frac{f_y + 40{,}000}{100{,}000} \geq 0.5 \qquad \text{(11-32b)}$$

In (11-32), f_y is in psi. If the inequality in (11-30) is not satisfied, the e_x's x's and y's are interchanged in (11-31).

This procedure is limited in application to columns that are symmetrical about two axes with a ratio of side lengths, ℓ_x/ℓ_y, between 0.5 and 2.0. The reinforcement should be in all four faces of the column. The use of (11-29) to (11-32) is illustrated in Example 11-6.

3. **Method based on 45° slice through interaction surface.**   Charts [11-6] or relationships [11-19] are available for the 45° section through the interaction surface (M_x and M_y at point A in Fig. 11-34). The design is then based on straight-line approximations to horizontal slices through the interaction surface as shown by the dashed lines in Fig. 11-34.

4. **Bresler reciprocal load method.**   ACI Commentary Sections 10.3.6 and 10.3.7 give the following equation, originally presented by Bresler [11-18], for calculating the capacity under biaxial bending:

$$\frac{1}{P_u} = \frac{1}{\phi P_{nx}} + \frac{1}{\phi P_{ny}} - \frac{1}{\phi P_{n0}} \qquad \text{(11-33)}$$

This procedure, widely used, is illustrated in Example 11-7.

EXAMPLE 11-5   Calculate the Capacity of a Biaxially Loaded Column for a Given Neutral-Axis Location

Compute the nominal axial-load capacity, P_n, and the nominal M_{nx} and M_{ny} moments corresponding to a prescribed strain distribution in the column. Use a 16-in.-by-16-in. tied column with 3000-psi concrete and eight No. 8 Grade-60 bars, with three bars in each face of the column. The neutral-axis position assumed in this example crosses the vertical axis of symmetry of the section (the y-axis) at 10 in. below the top of the section, at an angle of 30° counterclockwise from the x-axis of the cross section. Use the ACI rectangular stress block.

1.   **Select a sign convention.** The cross section of the column is a z-plane. (A z-plane is a plane that is perpendicular to the z-axis.) Positive x and y are to the right and upward, respectively. Positive moments cause compression at positive x locations and positive y locations. Compression is positive.

2.   **Locate the neutral axis.** Figure 11-35a shows the cross section and the specified neutral axis (axis of zero strain). The origin is 8 in. below the top and 8 in. from the left side. The neutral axis crosses the y-axis at 2 in. below the origin and intersects the left and right sides of the section at 14.62 in. and 5.38 in., respectively, below the top of the section. The extreme-compression fiber, A, is located in the upper-left corner of the section. The perpendicular distance from this fiber to the neutral axis is

$$c_{incl} = 14.62 \cos 30° = 12.66 \text{ in.}$$

3.   **Locate the stress block.** The stress block extends from corner A to a line at $(\beta_1 \times c_{incl}) = a_{incl} = 10.76$ in. from corner A measured perpendicular to the neutral axis. The

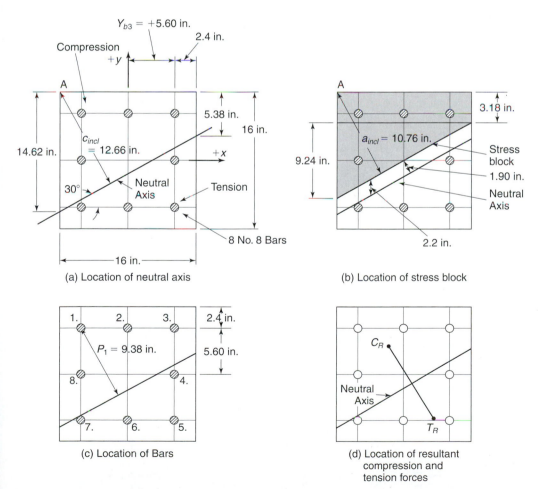

(a) Location of neutral axis

(b) Location of stress block

(c) Location of Bars

(d) Location of resultant compression and tension forces

Fig. 11-35
Location of neutral axis and stress block—Example 11-5.

column capacity is assumed to be reached when the strain at A reaches a compressive strain of 0.0030. A uniform compression stress of $0.85f'_c$ acts on the sum of the two shaded areas in Fig. 11-35b. To simplify the calculation of the forces in the concrete, the area loaded by the stress-block will be divided into a rectangle and a triangle, with dimensions as shown in Fig. 11-35b as follows:

a) **Rectangular portion**, height = 3.18 in., width = 16 in., area = 50.9 in.^{2} with centroid located at

$x = 0$, and
$y = 8.0 - 3.18/2$ in. = 6.41 in.

b) **Triangular portion**, height = 9.24 in., width = 16 in., area = 73.9 in.^{2}, with centroid located at

$x = 8 - 16/3 = -2.67$ in. to the left of the origin, and
$y = 8 - 3.19 - 9.24/3 = 1.73$ in. above origin.

4.   Compute the distance from the bars to the neutral axis. The individual reinforcing bars are numbered as shown in Fig. 11-35c. The perpendicular distance, p_i, from the neutral axis to the ith bar is

$$p_i = (y_{si} - y_{NA}) \cos \theta - x_{si} \sin \theta \qquad (11\text{-}34)$$

where

y_{si} is the vertical distance from the centroid of the section to the center of the bar, positive upward;

x_{si} is the horizontal distance from the centroid of the section to the center of the bar, positive to the right;

y_{NA} is the vertical distance from the centroid of the column to the place where the neutral axis crosses the y-axis of the section, specified in this example to be –2 in.; and

θ is the angle between the neutral axis and the x-axis, specified to be 30° counterclockwise from the x-axis.

The calculations of the bar forces are presented in Table 11-2. The calculations for bar 1 in the table are explained in the following paragraphs:

Columns 2 and 3 in Table 11-2 give the x and y distances from the column centroid to the centers of the bars. These come from Fig. 11-35c.

Column 4 gives the perpendicular distance from the neutral axis to the center of the ith bar, computed by using (11-34);

For bar 1, $x_{s1} = -5.6$ in. and $y_{s1} = +5.6$ in. Using (11-34) then gives

$$p_1 = [5.6 - (-2)] \cos 30° - (-5.6) \sin 30°$$
$$= 6.58 - (-2.8) = 9.38 \text{ in.}$$

TABLE 11-2   Computation of Bar Forces in Biaxially Loaded Column

1	2	3	4	5	6	7
Bar(i)	x_{si} in.	y_{si} in.	Perpendicular Distance = p_i in.	Steel Strain, ϵ_{si}	Steel Force, kips	Corrected Steel Force $P_{n,si}$, kips
1	−5.6	+5.6	9.38	0.0022	47.4	45.4
2	0	+5.6	6.58	0.0016	36.7	34.7
3	+5.6	+5.6	3.78	0.0009	20.6	18.6
4	+5.6	0	−1.07	−0.00025	−6.9	−6.9
5	+5.6	−5.6	−5.92	−0.00140	−32.1	−32.1
6	0	−5.6	−3.12	−0.00074	−16.0	−16.0
7	−5.6	−5.6	−0.32	−0.000076	−2.3	−2.3
8	−5.6	0	4.53	0.0011	25.2	23.2

5.   Compute the forces in the column reinforcement. Column 5 of Table 11-2, from similar triangles, gives the strain in the reinforcing bar 1 as

$$\varepsilon_{si} = \frac{p_i}{c_{incl}} \times 0.0030 = \frac{9.38}{12.66} \times 0.0030 = 0.00222 \qquad (11\text{-}35)$$

Columns 6 and 7 give the stress in reinforcing bar 1 as

$$f_{si} = \varepsilon_{si} \times E_s = 0.00222 \times 29{,}000$$
$$= 64.5 \text{ ksi}$$

but not more than $f_y = 60$ ksi not less than -60 ksi. Therefore, $f_{s1} = 60$ ksi.

If p_i exceeds the distance from the neutral axis to the lower edge of the rectangular stress block (2.2 in. vertically; see Fig. 11-35b), the bar displaces concrete in the compression stress block, an effect that will be corrected for by subtracting the force in the displaced concrete:

$$F_{si} = (f_{si} - 0.85f'_c)\, A_{si}$$

For bar number 1,

$$F_{s1} = (60 - 0.85 \times 3 \text{ ksi}) \times 0.79 \text{ in.}$$
$$= 45.4 \text{ kips}$$

6.   Compute concrete forces, bar forces, and moments.

Concrete forces:

Area 1—Rectangle: $C_{c1} = 0.85 \times f'_c \times$ area $= 0.85 \times 3 \times 50.9 = 130$ kips.
Area 2—Triangle: $C_{c2} = 188$ kips.

The calculation of moments is done in Table 11-3. The center of compression is in the upper-left quadrant of the section. From the sign convention,

M_x is positive, because it causes compression stress at positive y-coordinates (in the upper half of the cross section), and

M_y is negative, because it causes compression in the left half of the cross section.

7.   Compute ϕ factor, ϕP_n, ϕM_{nx}, and ϕM_{ny}. From step 5, calculate the extreme tension steel strain in bar 5, $\varepsilon_t = -0.00140$. Where $-$ is tension. Because this is less than the yield strain in tension, this is a compression-controlled point and $\phi = 0.65$.
The sums of columns 2, 4, and 6 of Table 11-3 are multiplied by $\phi = 0.65$, giving $\phi P_n = 249$ kips, $\phi M_{nx} = -643$ in-kips, and $\phi M_{ny} = 1300$ in-kips.

8.   Summary. The nominal load capacity and factored nominal load capacity are

$$P_n = \mathbf{383 \text{ kips}}$$
$$\phi P_n = \mathbf{249 \text{ kips}}$$
$$M_{nx} = \mathbf{-989 \text{ in-kips}}$$
$$\phi M_{nx} = \mathbf{-643 \text{ in-kips}}$$

M_{nx} **causes compression at $+y$ elements.**

$$M_{ny} = \mathbf{1995 \text{ in-kips}}$$
$$\phi M_{ny} = \mathbf{1300 \text{ in-kips}}$$ ■

The nominal moments for Example 11-5 act around the x- and y-axes. These moments can be combined into a total moment vector (Fig. 11-35). This vector has a magnitude of $\phi M_n = 1450$ k.in and is at an angle of $26.4°$ with the x-axis. In a rectangular column with uniaxial bending about the x-axis, the moment vector, M_{nx}, is parallel to the

TABLE 11-3 Computation of Moments of Bar Forces and Concrete Forces about x- and y-Axes of the Column Cross Section

1	2	3	4	5	6
Bar (i) and Concrete	Forces, $P_{n,si}$ and $P_{n,ci}$, kips	x_{si} in.	$M_{ny,si} = P_{n,si} \times x_{si}$ and $M_{ny,ci} = P_{n,ci} \times x_{ci}$ in-kips	y_{si} in.	$M_{nx,si} = P_{sn,si} \times y_{si}$ and $M_{nx,ci} = P_{cn,si} \times y_{n,ci}$ in-kips
1	45.4	−5.6	−254.2	+5.6	254.2
2	34.7	0	0	+5.6	194.3
3	18.6	+5.6	104.2	+5.6	104.2
4	−6.9	+5.6	−38.6	0	0
5	−32.1	+5.6	−179.8	−5.6	179.8
6	−16.0	0	0	−5.6	89.8
7	−2.3	−5.6	12.8	−5.6	12.9
8	23.2	−5.6	−129.9	0	0
Concrete Rectangle	130	0	0	6.41	833.9
Concrete Triangle	188	−2.67	−503.1	1.73	325.9
	Sum $= P_n$ **383** **kips**		**Sum $= M_{nx}$** **− 989** **in-kips**		**Sum $= M_{ny}$** **1995** **in-kips**

neutral axis. In the general case of biaxial bending of a noncircular, biaxially loaded section, the moment vector is at an angle with the neutral axis. In this example, the difference between the direction of the moment vector and the direction of the assumed neutral axis was small (26.4° compared with 30°). This closeness occurred because the column was square with bars in all four faces.

The internal load components can be combined into a resultant compression force, C_R, by adding all the axial forces, P_n, and the moments M_{nx} and M_{ny} of the individual bars and areas that are in compression. The same is done for the elements stressed in tension forces to get the resultant tension force, T_R. Together, these allow the eccentricities, e_x and e_y, to be calculated.

From column 2 of Table 11-3, Σ(compression-resisting forces) $= C_R = 440$ kips
From column 4,

$$\Sigma M_{ny} \text{ for compression-resisting forces} = -783 \text{ in-kips}$$
$$e_x = -783/440 = -1.78 \text{ in.}$$

By inspection of the location of those loads and compressed areas of the column which are concentrated in the upper left-hand quadrant, e_x must be in the negative x direction.

From column 6,

$$\Sigma M_{nx} \text{ for compression-resisting forces} = +1712 \text{ in-kips}$$
$$e_y = +1712/440 = +3.89 \text{ in.}$$

By inspection of the location of those loads and tensile-bar areas of the column which are located in the lower right-hand quadrant, e_y must be in the positive y direction. Thus, the resultant compression-resisting force, $C_R = 440$ kips, acts at $e_y = 3.89$ in. above and $e_x = -1.78$ in. to the left of the centroid.

From column 2,

$$\Sigma(\text{tension-resisting forces}) = T_R = -57.3 \text{ kips}$$

From column 4,

$$\Sigma M_{ny} \text{ for tension-resisting forces} = -205.6 \text{ in-kips}$$
$$e_x = -205.6/-57.3 = 3.59 \text{ in. to the right of the } y\text{-axis}$$

From column 6,

$$\Sigma M_{nx} \text{ for tension-resisting forces} = 282.5 \text{ in-kips}$$
$$e_y = 282.5/-57.3 = -4.93 \text{ in. below the } x \text{ axis.}$$

Thus, the resultant tension-resisting force, $T_R = -57.3$ kips, acts at $e_x = 3.59$ in. to the right of the y-axis, and $e_y = -4.93$ in. below the x-axis.

The line joining the points where C_R and T_R act has a horizontal projection of $1.78 + 3.59$ in. $= 5.37$ in. and a vertical projection of 3.89 in. $+ 4.93$ in. $= 8.82$ in.

The angle of the line joining these points is arctan $(8.82/5.37) = 58.7°$. For a square column with bars in all four sides, the line joining the points where the resultants C_R and T_R act should be 60° (perpendicular to the specified 30° neutral axis). The discrepancy $(60° - 58.7° = 1.3°)$ is due to round-off errors.

The resultant compressive and tensile forces, C_R, and T_R, must lie in a plane with the applied load, P_n.

In our solution for the moments, the moment vector is at an angle of 30° with the neutral axis. If one were using an iterative solution to get the position of the neutral axis for a given set of load and moments, it would be necessary to iterate two variables: (i) the distance, d_n, along the y-axis from the intersection of the neutral axis and the top of the beam and (ii) the angle, θ, between the neutral axis and the x-axis. Warner et al. [11-14] observe that P_n is affected more strongly by the height of the intercept, d_n, than by the angle, θ, and vice-versa for the angle θ and the resultant of the applied moments. They suggest following these steps in solving for the forces in a biaxial-bending problem:

1. Select the cross section.
2. Choose the depth, d_n, to the intersection of the neutral axis with the y-axis of the section and the angle, θ, between the neutral axis and the x-axis of the section.
3. Follow the computations in this example to get an initial set of P_n, M_{nx}, and M_{ny} values.
4. Keep θ constant, and iterate d_n until the computed P_n is close to the target value.
5. Iterate the angle θ until M_{nx} and M_{ny} approach the target values.

EXAMPLE 11-6   Design of a Biaxially Loaded Column: Equivalent Eccentricity Method

Select a tied column cross section to resist factored loads and moments of $P_u = 250$ kips, $M_{ux} = 55$ ft-kips, and $M_{uy} = 110$ ft-kips. Use $f_y = 60$ ksi and $f'_c = 3$ ksi. The first approximate procedure, based on (11-29) to (11-32a), will be used.

1. **Select a trial section.** Assume that $\rho_t = 0.015$. Use a section with bars in the four faces, since the column is loaded biaxially. Then

$$A_{g(\text{trial})} \geq \frac{P_u}{0.40(f_c' + f_y\rho_t)}$$

$$\geq \frac{250}{0.40(3 + 60 \times 0.015)}$$

$$\geq 160 \text{ in.}^2 \text{ or } 12.7 \text{ in. square}$$

Try a 16-in.-square column with $f_c' = 3$ ksi, $f_y = 60$ ksi, and eight No. 8 bars, three in each face.

2. **Compute γ.**

$$\gamma = \frac{(16 - 2 \times 2.4)}{16} = 0.700$$

3. **Compute e_x, e_y, and e_{0x} or e_{0y}.** From the definition of the moments and eccentricities in Fig. 11-33 and (11-29),

$$e_x = \frac{M_{uy}}{P_u}$$

$$= \frac{110 \times 12}{250} = 5.28 \text{ in.}$$

$$e_y = \frac{M_{ux}}{P_u}$$

$$= \frac{55 \times 12}{250} = 2.64 \text{ in.}$$

By inspection, $e_x/\ell_x \geq e_y/\ell_y$; therefore, use (11-31) as given. If this were not true, you would transpose the x and y terms and subscripts in (11-31) before using it. In any event, we have

$$\frac{P_u}{f_c' A_g} = \frac{250}{3 \times 256} = 0.326 < 0.4$$

Therefore, use (11-32) to compute α:

$$\alpha = \left(0.5 + \frac{P_u}{f_c' A_g}\right)\left(\frac{f_y + 40{,}000}{100{,}000}\right) \qquad (11\text{-}32\text{a})$$

$$= (0.5 + 0.326)\left(\frac{60{,}000 + 40{,}000}{100{,}000}\right)$$

$$= 0.826$$

From (11-31),

$$e_{0x} = e_x + \frac{\alpha e_y \ell x}{\ell_y}$$

$$= \left(5.28 + 0.826 \times 2.64 \times \frac{16}{16}\right) = 7.46 \text{ in.}$$

Thus, the equivalent uniaxial moment is

$$M_{0y} = P_u e_{0x}$$

$$= 250 \times 7.46 = 1865 \text{ in.-kips}$$

The column is designed for uniaxial bending for $P_u = 250$ kips and the equivalent moment, $M_{0y} = 1865$ in.-kips.

4.   Use interaction diagrams to determine ρ_t. Since the column has biaxial bending, we will select a section with bars in four faces. The interaction diagrams are entered with

$$\frac{P_u}{A_g} = \frac{250}{256} = 0.977 \text{ ksi}$$

and

$$\frac{M_{0y}}{A_g h} = \frac{1865}{16^3} = 0.455 \text{ ksi}$$

From Figs. A-9 and A-10,

$$\text{for } \gamma = 0.60, \rho_t = 0.033$$
$$\text{for } \gamma = 0.75, \rho_t = 0.024$$

By linear interpolation, $\rho_t = 0.027$ for $\gamma = 0.700$.

5.   Compute A_{st} and select the reinforcement.

$$A_{st} = \rho_t A_g = 0.027 \times 256 = 6.91 \text{ in.}^2$$

From Table A-22, select twelve No. 7 bars, four in each face, $A_{st} = 7.20$ in.^{2}. Design ties and lap splices as in earlier examples.   ∎

EXAMPLE 11-7   Design of a Biaxially Loaded Column: Bresler Reciprocal Load Method

Repeat Example 11-6, but use (11-33).

1.   Select a trial section. Select A_g as in Example 11-6. To use (11-33), it is necessary also to estimate the reinforcement required. Try a 16-in.-square column, $f'_c = 3$ ksi, $f_y = 60$ ksi, eight No. 8 bars (three in each face), and No. 3 ties.

2.   Compute γ.

$$\gamma = 0.700$$

3.   Compute ϕP_{nx}. ϕP_{nx} is the factored axial load capacity corresponding to e_x and ρ_t. We have

$$\rho_t = \frac{8 \times 0.79}{16 \times 16} = 0.0247$$

$$\frac{e_x}{\ell_x} = \frac{M_{uy}}{P_u \ell_x} = \frac{110 \text{ ft-kips} \times 12}{250 \text{ kips} \times 16 \text{ in.}}$$
$$= 0.330$$

From the interaction diagram, Fig. A-9, $(e_x/\ell_x) = 0.330$ and $\rho_t = 0.0247$ give $(\phi P_{nx}/bh) = 1.14$ for $\gamma = 0.60$. From Fig. A-10, $(\phi P_{nx}/bh) = 1.25$ for $\gamma = 0.75$. Interpolating gives $(\phi P_{nx}/bh) = 1.21$ and $\phi P_{nx} = 311$ kips.

4.   Compute ϕP_{ny}.

$$\frac{e_y}{\ell_y} = \frac{M_{ux}}{P_u \ell_y} = \frac{55 \times 12}{250 \times 16} = 0.165$$

From Fig. A-9, $(\phi P_{ny}/bh) = 2.00$ for $\gamma = 0.60$. From Fig. A-10, $(\phi P_{ny}/bh) = 1.84$ for $\gamma = 0.75$. Interpolating gives $(\phi P_{ny}/bh) = 1.947$ and $\phi P_{ny} = 498$ kips.

5.   Compute ϕP_{n0}. From Figs. A-9 and A-10, ϕP_{n0} is the point where the interaction curve for $\rho_t = 0.0247$ would intersect the vertical axis, $\phi P_{n0}/bh = 2.75$, and $\phi P_{n0} = 704$ kips.

6.   Solve for P_u.

$$\frac{1}{P_u} = \frac{1}{\phi P_{nx}} + \frac{1}{\phi P_{ny}} - \frac{1}{\phi P_{n0}}$$

$$= \frac{1}{311} + \frac{1}{498} - \frac{1}{704}$$

$$= 263 \text{ kips}$$

The required capacity is 250 kips; therefore, the column design is adequate. **Use eight No. 8 bars, three in each face, $A_{st} = 6.32$ in.^{2}.** ∎

Examples 11-5, 11-6, and 11-7 all dealt with the same column and essentially the same loads. The strain-compatibility solution in Example 11-5 computed the axial load and bending strengths about two axes for a specified 16-in.-square column with eight No. 8 bars, three per side, for one assumed neutral-axis position. This gave

$$P_n = 383 \text{ kips}$$

$$\phi P_n = 249 \text{ kips}$$

$$M_{nx} = -989 \text{ in.-kips}$$

$$\phi M_{nx} = -643 \text{ in.-kips}$$

$$M_{ny} = -1995 \text{ in.-kips}$$

$$\phi M_{ny} = -1300 \text{ in.-kips}$$

The equivalent eccentricity method was used in Example 11-6, to check whether the same cross section could support

$$P_u = 250 \text{ kips}$$

$$M_{ux} = 55 \text{ kip-ft} = 660 \text{ kip-in.}$$

$$M_{ny} = 110 \text{ kip-ft} = 1320 \text{ kip-in.}$$

After the two moments were converted to an equivalent uniaxial moment about the y-axis, interaction diagrams for ϕP_n and ϕM_n were used to compute the strength. This solution indicated that a 16-in.-square column with 12 No. 7 bars was needed to resist these loads and moments, an increase of 14 percent in the steel area. The increase was a little more than was absolutely necessary, because a bar arrangement with an equal number of bars on all four faces was chosen.

PROBLEMS

11-1 The column shown in Fig. P11-1 is made of 4000-psi concrete and Grade-60 steel.

    (a) Compute the theoretical capacity of the column for pure axial load.

    (b) Compute the maximum permissible ϕP_n for the column.

11-2 Why does a spiral improve the behavior of a column?

11-3 Compute the balanced axial load and moment capacity of the column shown in Fig. P11-1. Use $f'_c = 4000$ psi and $f_y = 60,000$ psi.

11-4 For the column shown in Fig. P11-4, use a strain-compatibility solution to compute five points on the interaction diagram corresponding to points 1 to 5 in Fig. 11-22. Plot the interaction diagram. Use $f'_c = 3000$ psi and $f_y = 60,000$ psi.

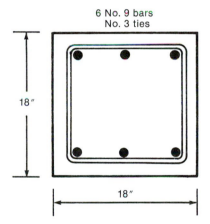

6 No. 9 bars
No. 3 ties

18″

18″

Fig. P11-1

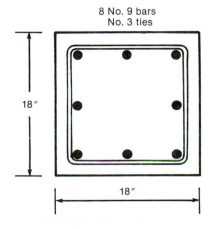

8 No. 9 bars
No. 3 ties

18″

18″

Fig. P11-4

11-5 Write a program for a calculator or computer to solve for points on an interaction diagram for rectangular columns with up to 10 layers of steel. (On a Hewlett-Packard calculator, such a program involves about 250 steps.)

11-6 Use the interaction diagrams in Appendix A to compute the maximum moment, M_u, that can be supported by the column shown in Fig. P11-1 if

    (a) $P_u = 583$ kips.

    (b) $P_u = 130$ kips.

    (c) $e = 4$ in.

    Use psi $f'_c = 3000$ and $f_y = 60,000$ psi.

11-7 Use the interaction diagrams in Appendix A to select tied-column cross sections to support the loads given in the accompanying list. In each case, use $f'_c = 3000$ psi and $f_y = 60,000$ psi. Design ties. Calculate the required splice lengths, assuming that the bars extending up from the column below are the same diameter as in the column you have designed, and draw a typical cross section of the column, showing the bars and ties.

    (a) $P_u = 390$ kips, $M_u = 220$ ft-kips, square column with bars in two faces.

    (b) $P_u = 710$ kips, $M_u = 50$ ft-kips, square column with bars in four faces.

    (c) $P_u = 130$ kips, $M_u = 240$ ft-kips, square column with bars in four faces.

11-8 Use the interaction diagrams in Appendix A to select spiral-column cross sections to support the loads given in the accompanying list. In each case, use $f'_c = 4000$ psi and $f_y = 60,000$ psi, Design spirals. Calculate the required splice lengths, and draw a typical cross section of the column, showing the bars and spiral.

    (a) $P_u = 600$ kips, $M_u = 65$ ft-kips.

    (b) $P_u = 500$ kips, $M_u = 150$ ft-kips.

11-9 Why are tension splices required in some columns?

11-10 Select a cross section and reinforcement to support $P_u = 450$ kips, $M_{ux} = 100$ ft-kips, and $M_{uy} = 130$ ft-kips. Use $f'_c = 3000$ psi and $f_y = 60,000$ psi.

12 Slender Columns

12-1 INTRODUCTION

Definition of a Slender Column

An eccentrically loaded, pin-ended column is shown in Fig. 12-1a. The moments at the ends of the column are

$$M_e = Pe \qquad (12\text{-}1)$$

When the loads P are applied, the column deflects laterally by an amount δ, as shown. For equilibrium, the internal moment at midheight must be (Fig. 12-1b)

$$M_c = P(e + \delta) \qquad (12\text{-}2)$$

The deflection increases the moments for which the column must be designed. In the symmetrical column shown here, the maximum moment occurs at midheight, where the maximum deflection occurs.

Figure 12-2 shows an interaction diagram for a reinforced concrete column. This diagram gives the combinations of axial load and moment required to cause failure of a column cross section or a very short length of column. The dashed radial line O–A is a plot of the end moment on the column in Fig. 12-1. Since this load is applied at a constant eccentricity, e, the end moment, M_e, is a linear function of P, given by (12-1). The curved, solid line O–B is the moment M_c at midheight of the column, given by (12-2). At any given load P, the moment at midheight is the sum of the end moment, Pe, and the moment due to the deflections, $P\delta$. The line O–A is referred to as a *load–moment curve* for the end moment, while the line O–B is the load–moment curve for the maximum column moment.

Failure occurs when the load–moment curve O–B for the point of maximum moment intersects the interaction diagram for the cross section. Thus the load and moment at failure are denoted by point B in Fig. 12-2. Because of the increase in maximum moment due

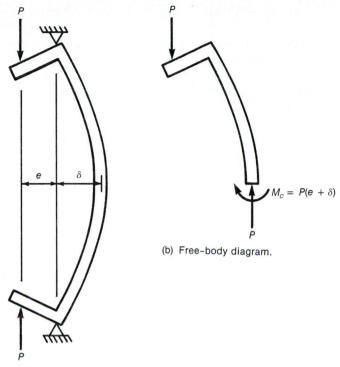

(b) Free-body diagram.

$M_c = P(e + \delta)$

Fig. 12-1
Forces in a deflected column.   (a) Column.

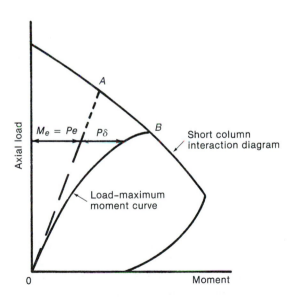

Fig. 12-2
Load and moment in a
column.

$M_e = Pe$ $P\delta$

Short column
interaction diagram

Load–maximum
moment curve

Axial load

Moment

to deflections, the axial-load capacity is reduced from *A* to *B*. This reduction in axial-load capacity results from what are referred to as *slenderness effects*.

A *slender column* is defined as a column that has a significant reduction in its axial-load capacity due to moments resulting from lateral deflections of the column. In the derivation of the ACI Code, "a significant reduction" was arbitrarily taken as anything greater than about 5 percent [12-1].

Buckling of Axially Loaded Elastic Columns

Figure 12-3 illustrates three states of equilibrium. If the ball in Fig. 12-3a is displaced laterally and released, it will return to its original position. This is *stable equilibrium*. If the ball in Fig. 12-3c is displaced laterally and released, it will roll off the hill. This is *unstable equilibrium*. The transition between stable and unstable equilibrium is *neutral equilibrium*, illustrated in Fig. 12-3b. Here, the ball will remain in the displaced position. Similar states of equilibrium exist for the axially loaded column in Fig. 12-4a. If the column returns to its original position when it is pushed laterally at midheight and released, it is in stable equilibrium; and so on.

Figure 12-4b shows a portion of a column that is in a state of neutral equilibrium. The differential equation for this column is

$$EI\frac{d^2y}{dx^2} = -Py \qquad (12\text{-}3)$$

In 1744, Leonhard Euler derived (12-3) and its solution,

$$P_c = \frac{n^2\pi^2 EI}{\ell^2} \qquad (12\text{-}4)$$

where

$$EI = \text{flexural stiffness}$$
$$\ell = \text{length of the column}$$
$$n = \text{number of half-sine waves in the length of the column}$$

Cases with $n = 1, 2$, and 3 are illustrated in Fig. 12-4c. The lowest value of P_c will occur with $n = 1.0$. This gives what is referred to as the *Euler buckling load*:

$$P_E = \frac{\pi^2 EI}{\ell^2} \qquad (12\text{-}5)$$

Such a column is shown in Fig. 12-5a. If this column were unable to move sideways at midheight, as shown in Fig. 12-5b, it would buckle with $n = 2$, and the buckling load would be

$$P_c = \frac{2^2\pi^2 EI}{\ell^2}$$

which is four times the critical load of the same column without the midheight brace.

Another way of looking at this involves the concept of the *effective length* of the column. The effective length is the length of a pin-ended column having the same buckling load. Thus the column in Fig. 12-5c has the same buckling load as that in Fig. 12-5b. The effective length of the column is $\ell/2$ in this case, where $\ell/2$ is the length of each of the half-sine waves in the deflected shape of the column in Fig. 12-5b. The effective length, $k\ell$, is equal to ℓ/n. The *effective length factor* is $k = 1/n$. Equation (12-4) is generally written as

$$P_c = \frac{\pi^2 EI}{(k\ell)^2} \qquad (12\text{-}6)$$

Four idealized cases are shown in Fig. 12-6, together with the corresponding values of the effective length, $k\ell$. Frames *a* and *b* are prevented against deflecting laterally. They are

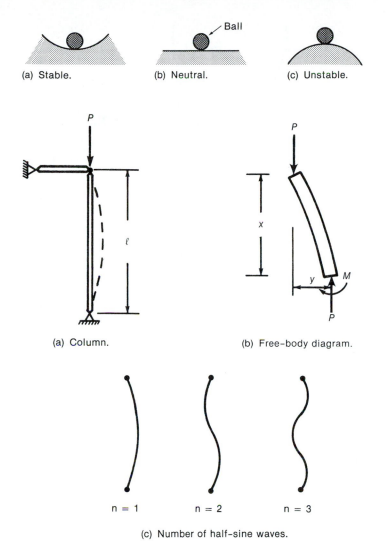

Fig. 12-3
States of equilibrium.

(a) Stable.   (b) Neutral.   (c) Unstable.

(a) Column.   (b) Free–body diagram.

n = 1   n = 2   n = 3

Fig. 12-4
Buckling of a pin-ended
column.

(c) Number of half–sine waves.

said to be *braced against sidesway*. Frames *c* and *d* are free to sway laterally when they buckle. They are called *unbraced* or *sway* frames. The critical loads of the columns shown in Fig. 12-6 are in the ratio $1:4:1:\frac{1}{4}$.

Thus it is seen that the restraints against end rotation and lateral translation have a major effect on the buckling load of axially loaded elastic columns. In actual structures, fully fixed ends, such as those shown in Fig. 12-6b to d, rarely, if ever, occur. This is discussed later in the chapter.

In the balance of this chapter we consider, in order, the behavior and design of pin-ended columns, as in Fig. 12-6a; restrained columns in frames that are braced against lateral displacement (*braced* or *nonsway frames*), Fig. 12-6b; and restrained columns in frames free to translate sideways (*unbraced frames* or *sway frames*), Fig. 12-6c and d.

Slender Columns in Structures

Pin-ended columns are rare in cast-in-place concrete construction, but do occur in precast construction. Occasionally, these will be slender, as, for example, the columns supporting the back of a precast grandstand.

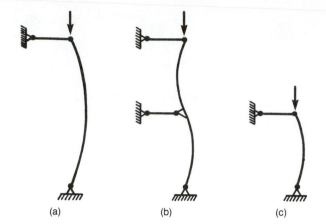

Fig. 12-5
Effective lengths of columns.

(a)   (b)   (c)

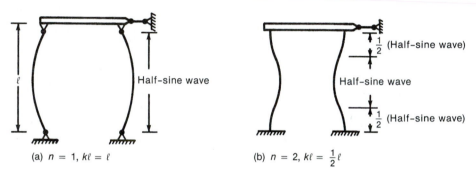

(a) $n = 1$, $k\ell = \ell$ Half–sine wave

(b) $n = 2$, $k\ell = \frac{1}{2}\ell$ $\frac{1}{2}$ (Half–sine wave)   Half–sine wave $\frac{1}{2}$ (Half–sine wave)

Frames braced against sidesway.

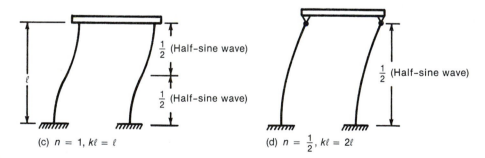

(c) $n = 1$, $k\ell = \ell$ $\frac{1}{2}$ (Half–sine wave) $\frac{1}{2}$ (Half–sine wave)

(d) $n = \frac{1}{2}$, $k\ell = 2\ell$ $\frac{1}{2}$ (Half–sine wave)

Fig. 12-6
Effective lengths of idealized
columns.

Frames free to sway laterally.

Most concrete building structures are braced frames, with the bracing provided by shear walls, stairwells, or elevator shafts that are considerably stiffer than the columns themselves (Figs. 10-4 and 10-15). Occasionally, unbraced frames are encountered near the tops of tall buildings, where the stiff elevator core may be discontinued before the top of the building, or in industrial buildings where an open bay exists to accommodate a travelling crane.

Most building columns fall in the short-column category [12-1]. Exceptions occur in industrial buildings and in buildings that have a high main-floor story for architectural or

Fig. 12-7
Bank of Brazil building,
Porto Alegre, Brazil. Each
floor extends out over the
floor below it. (Photograph
courtesy of J. G. MacGregor.)

functional reasons. An extreme example is shown in Fig. 12-7. The left corner column has a height of 50 times its least thickness. Some bridge piers and the decks of cable-stayed bridges fall into the slender-column category.

Organization of Chapter 12

The presentation of slender columns is divided into three progressively more complex parts. Slender pin-ended columns are discussed in Section 12-2. Restrained columns in nonsway frames are discussed in Sections 12-3 and 12-4. These sections deal with $P\delta$ effects and build on the material in Section 12-2. Finally, columns in sway frames are discussed in Sections 12-5 to 12-7.

12-2   BEHAVIOR AND ANALYSIS OF PIN-ENDED COLUMNS

Lateral deflections of a slender column cause an increase in the column moments, as illustrated in Figs. 12-1 and 12-2. These increased moments cause an increase in the deflections, which in turn lead to an increase in the moments. As a result, the load–moment line $O–B$ in Fig. 12-2 is nonlinear. If the axial load is below the critical load, the process will converge to a stable situation. If the load is greater than the critical load, it will not. This is referred to as a *second-order* process, since it is described by a second-order differential (12-3).

In a *first-order analysis*, the equations of equilibrium are derived by assuming that the deflections have a negligible effect on the internal forces in the members. In a *second-order analysis*, the equations of equilibrium consider the deformed shape of the structure. Instability can be investigated only via a second-order analysis, because it is the loss of equilibrium of the deformed structure that causes instability. However, because many engineering calculations and computer programs are based on first-order analyses, methods have been derived to modify the results of a first-order analysis to approximate the second-order effects.

Notation for Deflections

The notation in this chapter uses δ to refer to deflections of the column centerline away from the chord joining the ends of the column, as shown in Fig. 12-1a. The moment due to the column deflection in this figure is $P \times \delta$. In a sway frame, the joint at the top of a column deflects sideways relative to the joint at the bottom, as shown in Fig. 12-30a, which will be discussed later. This deflection is referred to as Δ, and the moment due to displacing the joint is $P \times \Delta$. We shall differentiate between δ, which is due to deflections of the column relative to the line joining the ends of the column, and Δ, which is the lateral deflection of the nodes of the frame relative to their original position.

Material Failures and Stability Failures

Load–moment curves are plotted in Fig. 12-8 for columns of three different lengths, all loaded (as shown in Fig. 12-1) with the same end eccentricity, e. The load–moment curve O–A for a relatively short column is practically the same as the line Pe. For a column of moderate length, line O–B, the deflections become significant, reducing the failure load. This column fails when the load–moment curve intersects the interaction diagram at point B. This is called a *material failure* and is the type of failure expected in most practical columns in braced frames. If a very slender column is loaded with increasing axial load, P, applied at a constant end eccentricity, e, it may reach a deflection Δ at which the value of

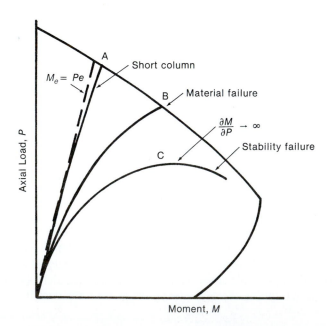

Fig. 12-8
Material and stability
failures.

the $\partial M / \partial P$ approaches infinity or becomes negative. When this occurs, the column becomes unstable, since, with further deflections, the moment capacity will drop. This type of failure is known as a *stability failure* and occurs only with very slender braced columns or with slender columns in sway frames [12-2].

Slender-Column Interaction Curves

In discussing the effects of variables on column strength, it is sometimes convenient to use slender-column interaction curves. Line $O-B_1$ in Fig. 12-9a shows the load–maximum moment curve for a column with slenderness $\ell / h = 30$ and a given end eccentricity, e_1. This column fails when the load–moment curve intersects the interaction diagram at point B_1. At the time of failure, the load and moment at the *end* of the column are given by point A_1. If this process is repeated a number of times, we get the *slender column interaction curve* shown by the broken line passing through A_1 and A_2, and so on. Such curves show the loads and maximum *end* moments causing failure of a given slender column. A family of slender-column interaction diagrams is given in Fig. 12-9b for columns with the same cross section but different slenderness ratios.

Moment Magnifier for Symmetrically Loaded Pin-Ended Beam Column

The column from Fig. 12-1 is shown in Fig. 12-10a. Under the action of the end moment, M_0, it deflects an amount δ_0. This will be referred to as the *first-order* deflection. When the axial loads P are applied, the deflection increases by the amount δ_a. The final deflection at midspan is $\delta = \delta_0 + \delta_a$. This *total* deflection will be referred to as the *second-order deflection*. It will be assumed that the final deflected shape approaches a half-sine wave. The primary moment diagram, M_0, is shown in Fig. 12-10b, and the secondary moments, $P\delta$, are shown in Fig. 12-10c. Since the deflected shape is assumed to be a sine wave, the $P - \delta$ moment diagram is also a sine wave. Using the moment area method and observing that the deflected shape is symmetrical, the deflection δ_a is computed as the moment about the support of the portion of the M/EI diagram between the support and midspan, shown shaded in Fig. 12-10c. The area of this portion is

$$\text{area} = \left[\frac{P}{EI} (\delta_0 + \delta_a) \right] \frac{\ell}{2} \times \frac{2}{\pi}$$

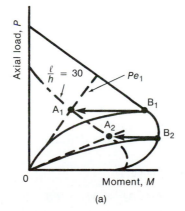

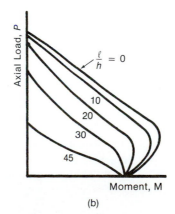

Fig. 12-9
Slender column interaction
curves. (From [12-1].)

(a)   (b)

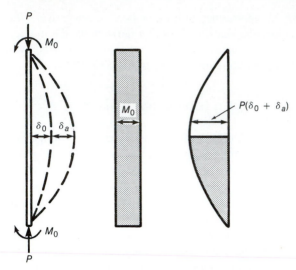

Fig. 12-10
Moments in a deflected
column.

(a) Deflected
   column

(b) M_0

(c) P-δ Moments

and the centroid of the second-order moment diagram is located at ℓ/π from the support. This equation may be simplified by replacing $\pi^2 EI/\ell^2$ with P_E, the Euler buckling load of a pin-ended column. Thus,

$$\delta_0 = \left[\frac{P}{EI}(\delta_0 + \delta_a)\frac{\ell}{2} \times \frac{2}{\pi} \right]\left(\frac{\ell}{\pi} \right)$$

$$= \frac{P\ell^2}{\pi^2 EI}(\delta_0 + \delta_a)$$

This equation can be simplified by replacing $\pi^2 EI/\ell^2$ by P_E, the Euler buckling load of a pin-ended column. Hence,

$$\delta_a = (\delta_0 + \delta_a)P/P_E$$

Rearranging gives

$$\delta_a = \delta_0\left(\frac{P/P_E}{1 - P/P_E} \right) \tag{12-7}$$

Since the final deflection δ is the sum of δ_0 and δ_a, it follows that

$$\delta = \delta_0 + \delta_0\left(\frac{P/P_E}{1 - P/P_E} \right)$$

or

$$\delta = \frac{\delta_0}{1 - P/P_E} \tag{12-8}$$

This equation shows that the second-order deflection, δ, increases as P/P_E increases, reaching infinity when $P = P_E$.

The maximum bending moment is

$$M_c = M_0 + P\delta$$

Here M_c is referred to as the *second-order moment*, and M_0 is referred to as the *first-order moment*. Substituting (12-8) gives

$$M_c = M_0 + \frac{P\delta_0}{1 - P/P_E} \tag{12-9}$$

For the moment diagram shown in Fig. 12-10b,

$$\delta_0 = \frac{M_0\ell^2}{8EI} \tag{12-10}$$

Substituting this and $P = (P/P_E)\pi^2 EI/\ell^2$ into (12-9) gives

$$M_c = \frac{M_0(1 + 0.23P/P_E)}{1 - P/P_E} \tag{12-11}$$

The coefficient 0.23 is a function of the shape of the M_0 diagram [12-3]. For example, it becomes -0.38, for a triangular moment diagram with M_0 at one end of the column and zero moment at the other and -0.18 for columns with equal and opposite end moments.

In the ACI Code, the $(1 + 0.23P/P_E)$ term is omitted because the factor 0.23 varies as a function of the moment diagram, for $P/P_E = 0.25$ to -0.18, the term $(1 + C P/P_E)$ varies from 1.06 to 0.96. and (12-11) is given essentially as

$$M_c = \delta_{ns}M_0 \tag{12-12}$$

where δ_{ns} is called the *nonsway-moment magnifier* and is given by

$$\delta_{ns} = \frac{1}{1 - P/P_c} \tag{12-13}$$

in which P_c is given by (12-6) and is equal to P_E for a pin-ended column. Equation (12-13) underestimates the moment magnifier for the column loaded with equal end moments, but approaches the truth when the end moments are not equal.

$P-\delta$ Moments and $P-\Delta$ Moments

Two different types of second-order moments act on the columns in a frame:

1. $P - \delta$ moments. These result from deflections, δ, of the axis of the bent column away from the chord joining the ends of the column, as shown in Figs. 12-1, 12-11, and 12-13. The slenderness effects in pin-ended columns and in nonsway frames result from $P-\delta$ effects. $P-\delta$ stability effects are presented in Sections 12-2 through 12-4.

2. $P - \Delta$ moments. These result from lateral deflections, Δ, of the beam–column joints from their original undeflected locations, as shown in Figs. 12-30 and 12-31. The slenderness effects in sway frames result from $P-\Delta$ moments, as discussed in Sections 12-5 through 12-7.

Effect of Unequal End Moments on the strength of a Slender Column

Up to now, we have considered only pin-ended columns subjected to equal moments at the two ends. This is a very special case, for which the maximum deflection moment, $P\delta$, occurs at a section where the applied load moment, Pe, is also a maximum. As a result, these quantities can be added directly, as done in Figs. 12-1 and 12-2.

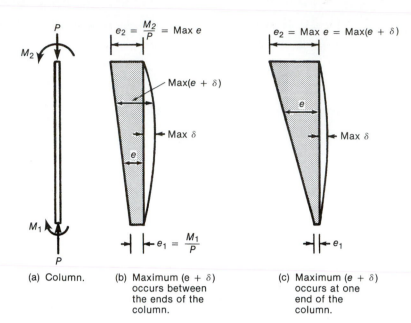

Fig. 12-11
Moments in columns with
unequal end moments.

(a) Column.

(b) Maximum $(e + \delta)$
occurs between
the ends of the
column.

(c) Maximum $(e + \delta)$
occurs at one
end of the
column.

In the usual case, the end eccentricities, $e_1 = M_1/P$ and $e_2 = M_2/P$, are not equal and so give applied moment diagrams as shown shaded in Fig. 12-11b and c for the column shown in Fig. 12-11a. The maximum value of δ occurs between the ends of the column while the maximum e occurs at one end of the column. As a result, $e_{\max}$ and $\delta_{\max}$ cannot be added directly. Two different cases exist. For a slender column with small end eccentricities, the maximum sum of $e + \delta$ may occur between the ends of the column, as shown in Fig. 12-11b. For a shorter column, or a column with large end eccentricities, the maximum sum of $e + \delta$ will occur at one end of the column, as shown in Fig. 12-11c.

These two types of behavior can be identified in the slender-column interaction diagrams shown in Figs. 12-9b and 12-12. For $e_1 = e_2$ (Fig. 12-9b), the interaction diagram for $\ell/h = 20$, for example, shows a reduction in strength throughout the range of eccentricities. For moment applied at one end only ($e_1/e_2 = 0$, Fig. 12-12a), the maximum $e + \delta$ occurs between the ends of the column for small eccentricities and at one end for large eccentricities. In the latter case, there is no slenderness effect, and the column can be considered a "short column" under the definition given in Section 12-1.

In the case of reversed curvature with $e_1/e_2 = -1$, the slender-column range is even smaller, so that a column with $\ell/h = 20$ subjected to reversed curvature has no slenderness effects for most eccentricities, as shown in Fig. 12-12b. At low loads, the deflected shape of such a column is an antisymmetrical S shape. As failure approaches, the column tends to *unwrap*, moving from the initial antisymmetrical deflected shape toward a single-curvature shape. This results from the inevitable lack of uniformity along the length of the column.

In the moment-magnifier design procedure, the column subjected to unequal end moments shown in Fig. 12-13a is replaced with a similar column subjected to equal moments of $C_m M_2$ at both ends, as shown in Fig. 12-13b. The moments $C_m M_2$ are chosen so that the maximum magnified moment is the same in both columns. The expression for the *equivalent moment factor* C_m was originally derived for use in the design of steel beam-columns [12-4] and was adopted without change for concrete design [12-1] (ACI Section 10.12.3.1):

$$C_m = 0.6 + 0.4\frac{M_1}{M_2} \geq 0.4 \qquad (12\text{-}14)$$

(ACI Eq. 10-13)

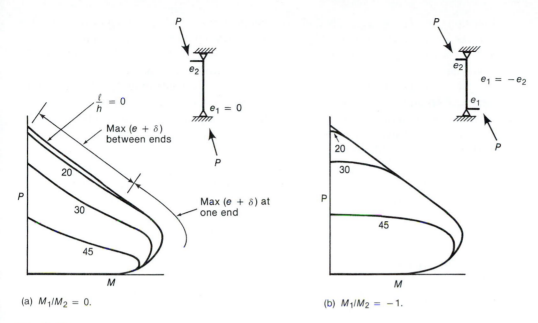

(a) $M_1/M_2 = 0$.

(b) $M_1/M_2 = -1$.

Fig. 12-12
Effect of M_1/M_2 ratio on slender column interaction curves for hinged columns. (From [12-1].)

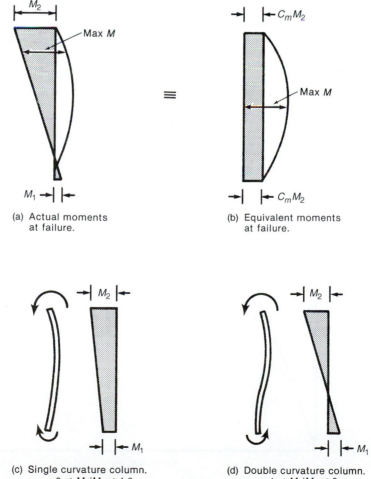

(a) Actual moments
at failure.

(b) Equivalent moments
at failure.

(c) Single curvature column.
$0 \le M_1/M_2 \le 1.0$

(d) Double curvature column.
$-1 \le M_1/M_2 \le 0$

Fig. 12-13
Equivalent moment factor, C_m.

Equation (12-14) was originally derived using the ratio of end eccentricities (e_1/e_2) rather than the ratio of end moments (M_1/M_2). When it was changed to end moments, the original eccentricity sign convention was retained. Thus, in. (12-14), M_1 and M_2 are the smaller and larger end moments, respectively, calculated from a conventional first-order elastic analysis. The sign convention for the ratio M_1/M_2 is illustrated in Fig. 12-13c and d. If the moments M_1 and M_2 cause single curvature bending without a point of contraflexure between the ends, as shown in Fig. 12-13c, M_1/M_2 is positive. If the moments M_1 and M_2 bend the column in double curvature with a point of zero moment between the two ends, as shown in Fig. 12-13d, that M_1/M_2 is negative.

Equation (12-14) applies only to hinged columns or columns in braced frames, loaded with axial loads and end moments. In all other cases, including columns subjected to transverse loads between their ends and concentrically loaded columns (no end moment), C_m is taken equal to 1.0 (ACI Section 10.12.3.1). The term C_m is not included in the equation for the moment magnifier for unbraced frames.

Column Stiffness, *EI*

The calculation of the critical load, P_c, via (12-6) involves the use of the flexural stiffness, *EI*, of the column. The value of *EI* chosen for a given column section, axial-load level, and slenderness must approximate the *EI* of the column *at the time of failure*, taking into account the type of failure (material failure or stability failure) and the effects of cracking, creep, and nonlinearity of the stress–strain curves at the time of failure.

Figure 12-14 shows moment–curvature diagrams for three different load levels for a typical column cross section. (P_b is the balanced-failure load.) A radial line in such a diagram has slope $M/\phi = EI$. The value of *EI* depends on the particular radial line selected. In a material failure, failure occurs when the most highly stressed section fails (point *B* in Fig. 12-8). For such a case, the appropriate radial line should intercept the end of the moment–curvature diagram, as shown for the $P = P_b$ (balanced-load) case in Fig. 12-14. On the other hand, a stability failure occurs before the cross section fails (point *C* in Fig. 12-8). This corresponds to a steeper line in Fig. 12-14 and thus a higher value of *EI*. The multitude of radial lines that can be drawn in Fig. 12-14 suggests that there is no all-encompassing value of *EI* for slender concrete columns.

References [12-1] and [12-5] describe empirical attempts to derive values for *EI*. The following two different sets of stiffness values, *EI*, are given in the slenderness provisions in ACI Sections 10.11.1 and 10.12.1 and will be discussed separately here:

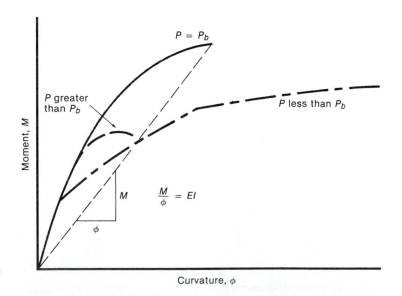

Fig. 12-14
Moment–curvature diagrams
for a column cross section.

Design Values of EI for the Computation of the Critical Loads of Individual Columns

ACI Section 10.12.1 gives the following equations for the computation of EI in calculating the critical load of an individual column:

$$EI = \frac{0.2E_cI_g + E_sI_{se}}{1 + \beta_d} \tag{12-15}$$

$$\text{(ACI Eq. 10-12)}$$

$$EI = \frac{0.40E_cI_g}{1 + \beta_d} \tag{12-16}$$

$$\text{(ACI Eq. 10-13)}$$

In both equations,

E_c, E_s = moduli of elasticity of the concrete (ACI Section 8.5.1) and the steel, respectively

I_g = gross moment of inertia of the concrete section about its centroidal axis, ignoring the reinforcement

I_{se} = moment of inertia of the reinforcement about the centroidal axis of the concrete section

The term $(1 + \beta_d)$ reflects the effect of creep on the column deflections and is discussed later.

Equation (12-15) is more "accurate" than (12-16) but is more difficult to use, since I_{se} is not known until the steel is chosen. It can be rewritten in a more usable form, however. The term I_{se} in (12-15) can be rewritten as

$$I_{se} = C\rho_t\gamma^2I_g \tag{12-17}$$

where

C = constant depending on the steel arrangement

ρ_t = total longitudinal-reinforcement ratio

γ = ratio of the distance between the centers of the outermost bars to the column thickness (illustrated in Table 12-1)

Values of C are given in Table 12-1. Substituting (12-17) into (12-15) and rearranging gives

$$EI = \frac{E_cI_g}{1 + \beta_d}\left(0.2 + \frac{C\rho_t\gamma^2E_s}{E_c}\right) \tag{12-18}$$

It is then possible to estimate EI without knowing the exact steel arrangement by choosing ρ_t, estimating γ from the column dimensions, and using the appropriate value of C from Table 12-1. For the common case of bars in four faces and $\gamma \simeq 0.75$, this reduces to

$$EI = \frac{E_cI_g}{1 + \beta_d}\left(0.2 + \frac{1.2\rho_tE_s}{E_c}\right) \tag{12-19}$$

This equation can be used for the preliminary design of columns.

EI for the Computation of Frame Deflections and for Second-Order Analyses

Two different sets of EI values are given in the slender-column sections of the ACI Code. ACI Section 10.12.3 gives (12-15) and (12-16) for use in ACI Eq. (10-10) to compute P_c

TABLE 12-1 Calculation of I_{se}[a]

Type of Column	Number of Bars	I_{se}	$\dfrac{I_{se}}{I_g} = C\rho_t\gamma^2$
(square section, γh, h, b, Bending axis[b])	—	$0.25A_{st}\gamma^2h^2$	$3\rho_t\gamma^2$
(section, γh)	3 bars per face	$0.167A_{st}\gamma^2h^2$	$2\rho_t\gamma^2$
	6 bars per face	$0.117A_{st}\gamma^2h^2$	$1.4\rho_t\gamma^2$
(section, γh)	8 bars (3 per face)	$0.187A_{st}\gamma^2h^2$	$2.2\rho_t\gamma^2$
	12 bars (4 per face)	$0.176A_{st}\gamma^2h^2$	$2.10\rho_t\gamma^2$
	16 bars (5 per face)	$0.172A_{st}\gamma^2h^2$	$2.06\rho_t\gamma^2$
(section, γh, h)	$h = 2b$ 16 bars as shown About strong axis	$0.128A_{st}\gamma^2h^2$	$1.54\rho_t\gamma^2$
(section, γh)	$b = 2h$ About weak axis	$0.219A_{st}\gamma^2h^2$	$2.63\rho_t\gamma^2$
(circular section, γh)	—	$0.125A_{st}\gamma^2h^2$	$2\rho_t\gamma^2$
(square section with circular bars, $+$)	—	$0.125A_{st}\gamma^2h^2$	$1.5\rho_t\gamma^2$

[a]Total area of steel $= A_{st} = \rho_t A_c$
[b]All sections are bent about the horizontal axis of the section which is shown by the dashed line for the top column section.
Source: [12-5]

when one is using the moment-magnifier method. These represent the behavior of a single, highly loaded column.

ACI Section 10.11.1 gives a different set of values of the moment of inertia, I, for use

(a) in elastic frame analyses, to compute the moments in columns and beams and the lateral deflections of frames, and

(b) to compute the Ψ used in computing the effective length factor, k.

The lateral deflection of a frame is affected by the stiffnesses of all the beams and columns in the frame. For this reason, the moments of inertia in ACI Section 10.11.1 are intended to represent an overall average of the moment of inertia values of EI for each type of member in a frame. In a similar manner, the effective length of a column in a frame is affected by the flexural stiffnesses of a number of beams and columns. It is incorrect to use the I values from ACI Section 10.11.1 when computing the critical load by using ACI Eq. (10-11).

Effect of Sustained Loads on Pin-Ended Columns

Up to this point, the discussion has been limited to columns failing under short-time loadings. Columns in structures, on the other hand, are subjected to sustained dead loads and sometimes to sustained live loads. The creep of the concrete under sustained loads increases the column deflections, increasing the moment $M = P(e + \delta)$ and thus weakening the column. The load–moment curve of Fig. 12-2 can be replotted, as shown in Fig. 12-15, for columns subjected to sustained loads.

Effect of Loading History

In Fig. 12-15a, the column is loaded rapidly to the service load (line $O–A$). The service load acts for a number of years, during which time the creep deflections and resulting second-order effects increase the moment, as shown by line $A–B$. Finally, the column is rapidly loaded to failure, as shown by line BC. The failure load corresponds to point C. Had the column been rapidly loaded to failure without the period of sustained service load, the load–moment curve would resemble line $O–A–D$, with failure corresponding to point D. The effect of the sustained loads has been to increase the midheight deflections and moments, causing a reduction in the failure load from D to C. On the reloading (line BC), the column deflections are governed by the EI corresponding to rapidly applied loads.

Creep Buckling

The second type of column behavior under sustained loads is referred to as *creep buckling*. Here, as shown in Fig. 12-15b, the column deflections continue to increase under the

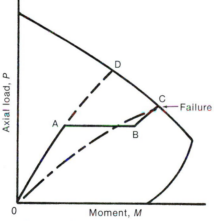

(a) "Rapid–sustained–rapid" loading history.

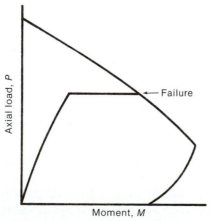

Fig. 12-15
Load-moment behavior for hinged columns subjected to sustained loads.

(b) Creep buckling.

sustained load, causing failure under the sustained load itself [12-6], [12-7]. This occurs only under high sustained loads greater than about 70 percent of the short-time capacity represented by point D. Since the sustained load will rarely exceed the strength-reduction factor, ϕ, divided by the dead-load factor, $0.65/1.2 = 0.54$ (for tied columns), times the column capacity, this type of failure is not specifically addressed in the ACI Code design procedures.

Two different design procedures are used to account for creep effects. In the *reduced-modulus procedure* [12-1], [12-7], [12-8], the value of E used to compute P_c is reduced to give the correct failure load. This procedure is illustrated by the broken line $O–C$ in Fig. 12-15a.

The second procedure replaces the column loaded at eccentricity e with one loaded at an eccentricity $e + \delta_{0,cr}$ where $\delta_{0,cr}$ is the creep deflection that would remain in the column in Fig. 12-15a if it were unloaded after reaching point B [12-9].

The ACI Code moment-magnifier procedure uses the reduced-modulus procedure. The value of EI is reduced by dividing by $(1 + \beta_d)$, as shown in (12-15) and (12-16), where, for hinged columns and columns in restrained frames, β_d is defined as the ratio of the factored axial load due to dead load to the total factored axial load. In a lightly reinforced column, creep of the concrete frequently causes a significant increase in the steel stress, so that the compression reinforcement yields at a load lower than that at which it would yield under a rapidly applied load. This effect is empirically accounted for in (12-15) by dividing both $E_c I_g$ and $E_s I_{se}$ by $(1 + \beta_d)$.

ACI Section 10.0 gives three definitions of β_d, depending primarily on whether the frame is nonsway or sway. To be stable, a pin-ended column must be in a structure that restricts sway of the ends of the column. In addition, it does not develop end moments if the structure sways sideways. In effect, a pin-ended column is always a nonsway column. For columns in a nonsway frame, ACI Section 10.0 defines β_d as the ratio of the maximum factored axial dead load to the total factored axial load. This definition applies to pin-ended columns.

Limiting Slenderness Ratios for Slender Columns

Most columns in structures are sufficiently short and stocky to be unaffected by slenderness effects. To avoid checking slenderness effects for all columns, ACI Section 10.12.2 allows slenderness effects to be ignored in the case of hinged columns and of columns in nonsway frames if

$$\frac{k\ell_u}{r} < 34 - 12\frac{M_1}{M_2} \tag{12-20}$$

$$\text{(ACI Eq. 10-7)}$$

In (12-20), k refers to the effective length factor, which is 1.0 for a pin-ended column (ACI Section 10.12.1), ℓ_u is the unsupported height (ACI Section 10.11.3.1), and r is the radius of gyration, taken as $0.3h$ for rectangular sections and $0.25h$ for circular sections (ACI Section 10.11.2). For other shapes, the value of r can be calculated from the area and moment of inertia of the cross section. By definition, $r = \sqrt{I/A}$. The sign convention for M_1/M_2 is given in Fig. 12-13.

Pin-ended columns having slenderness ratios less than the right-hand side of (12-20) should have a slender-column strength that is 95 percent or more of the short-column strength.

Arrangement of ACI Code Sections 10.10 to 10.13

ACI Section 10.11, "Magnified Moments—General," gives general requirements for the design of slender columns in both nonsway and sway frames. ACI Section 10.11.4 gives methods of distinguishing between nonsway and sway columns. If a column is in a nonsway frame, design involves ACI Sections 10.11 and 10.12, "Magnified Moments—Non-sway Frames." If a column is in a sway frame, design involves ACI Sections 10.11 and 10.13, "Magnified Moments—Sway Frames." In some relatively unusual cases involving very slender columns in sway frames, ACI Section 10.13.5 requires a further moment magnification, using ACI Section 10.12.3.

Summary of ACI Moment-Magnifier Design Procedure for Slender Pin-Ended Columns

A pin-ended column must be braced by a frame or other structure to remain stable. Hence, it will be designed under ACI Section 10.11, "Magnified Moments—General," and ACI Section 10.12, "Magnified Moments—Non-sway Frames," which give the following conditions:

1. Length of column. The unsupported length ℓ_u is the clear distance between members capable of giving lateral support (ACI Section 10.11.3.1). In the case of a pin-ended column, ℓ_u is the distance between the hinges.

2. Effective length. For a pin-ended column, the effective length k is 1.0 (ACI Section 10.11.2).

3. Radius of gyration. For a rectangular section $r = 0.3h$, and for a circular section, $r = 0.25h$. (See ACI Section 10.11.2.) For other sections, r can be calculated from the area and moment of inertia of the gross concrete section as $r = \sqrt{I_g/A_g}$.

4. Consideration of slenderness effects. For a pin-ended column, ACI Section 10.12.2 allows slenderness to be neglected if $k\ell_u/r$ satisfies (12-20). The sign convention for M_1/M_2 is given in Fig. 12-13. ACI Section 10.11.5 gives an upper limit of $k\ell_u/r = 100$ for columns designed according to ACI Sections 10.11 to 10.13.

5. Minimum moment. ACI Section 10.12.3.2 requires that the maximum end moment on the column, M_2, not be taken less than

$$M_{2,\,min} = P_u(0.6 + 0.03h) \tag{12-21}$$

$$\text{(ACI Eq. 10-14)}$$

where the 0.6 and h are in inches. When M_2 is less than $M_{2,min}$, C_m shall either be taken equal to 1.0 or evaluated from the actual end moments.

6. Moment-magnifier equation. ACI Section 10.12.3 states that the columns shall be designed for the factored axial load, P_u, and the magnified moment, M_c, defined by

$$M_c = \delta_{ns} M_2 \tag{12-22}$$

$$\text{(ACI Eq. 10-8)}$$

The subscript ns refers to nonsway. The moment M_2 is defined as the larger end moment acting on the column. ACI Section 10.12.3 goes on to define δ_{ns} as

$$\delta_{ns} = \frac{C_m}{1 - P_u/(0.75P_c)} \geq 1.0 \tag{12-23}$$

$$\text{(ACI Eq. 10-9)}$$

where

$$C_m = 0.6 + 0.4\frac{M_1}{M_2} \geq 0.4 \tag{12-14}$$

$$\text{(ACI Eq. 10-13)}$$

$$P_c = \frac{\pi^2 EI}{(k\ell_u)^2} \tag{12-24}$$

$$\text{(ACI Eq. 10-10)}$$

and

$$EI = \frac{0.2E_cI_g + E_sI_{se}}{1 + \beta_d} \qquad (12\text{-}15)$$

$$\text{(ACI Eq. 10-11)}$$

or

$$EI = \frac{0.40E_cI_g}{1 + \beta_d} \qquad (12\text{-}16)$$

$$\text{(ACI Eq. 10-12)}$$

Equation (12-23) is (12-13) rewritten to include the equivalent moment factor, C_m, and to include a *member-strength-reduction factor*, ϕ_m, taken equal to 0.75 for all slender columns [12-10]. The number 0.75 is used in (12-23) rather than the symbol ϕ, to avoid confusion with the ϕ factor used in design of the column cross section, which is 0.70 for tied columns and 0.75 for spiral columns.

If the computed value of δ_{ns} is less than 1.0, the maximum moment occurs at one end of the column. In this case, δ_{ns} is set equal to 1.0.

EXAMPLE 12-1   Design of a Slender Pin-Ended Column

Design a 20-ft-tall column to support an unfactored dead load of 90 kips and an unfactored live load of 75 kips. The loads act at an eccentricity of 3 in. at the top and 2 in. at the bottom, as shown in Fig. 12-16. Use $f'_c = 3000$ psi and $f_y = 60,000$ psi. Use the load combinations and strength-reduction factors from ACI 318-02 Sections 9.2 and 9.3.

1.   Compute the factored loads and moments and M_1/M_2.

$$P_u = 1.2D + 1.6L$$
$$= 1.2 \times 90 \text{ kips} + 1.6 \times 75 \text{ kips} = 228 \text{ kips}$$

The moment at the top is

$$M = P_u \times e$$
$$= 228 \text{ kips} \times \frac{3 \text{ in.}}{12 \text{ in.}} = 57.0 \text{ ft-kips}$$

The moment at the bottom is

$$M = 228 \text{ kips} \times \frac{2 \text{ in.}}{12 \text{ in.}}$$
$$= 38.0 \text{ ft-kips}$$

By definition, M_2 is the larger end moment in the column. Therefore, $M_2 = 57.0$ ft-kips and $M_1 = 38.0$ ft-kips. The ratio M_1/M_2 is taken to be positive, since the column is bent in single curvature (see Figs. 12-16c and 12-13c). Thus $M_1/M_2 = 0.667$.

2.   Estimate the column size. From (11-18a), assuming that $\rho_t = 0.015$,

$$A_{g(\text{trial})} \geq \frac{P_u}{0.45(f'_c + f_y\rho_t)}$$
$$A_{g(\text{trial})} \geq \frac{228 \times 1000}{0.45(3000 + 60,000 \times 0.015)}$$
$$= 130 \text{ in.}^2$$

This suggests that a 12 in. × 12 in. column would be satisfactory. It should be noted that (11-18a) was derived for short columns and will underestimate the required sizes of slender columns.

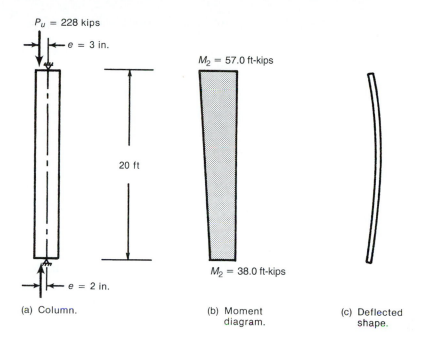

$P_u = 228$ kips

$e = 3$ in.

$M_2 = 57.0$ ft-kips

20 ft

$M_2 = 38.0$ ft-kips

$e = 2$ in.

(a) Column.

(b) Moment diagram.

(c) Deflected shape.

Fig. 12-16
Column—Example 12-1.

3.   **Is the column slender?** From (12-20), a column is short if

$$\frac{k\ell_u}{r} < 34 - 12\frac{M_1}{M_2} \qquad (12\text{-}20)$$

For the 12 in. × 12 in. section selected in step 2, $k = 1.0$ because the column is pin-ended, and where $r = 0.3h = 0.3 \times 12$ in. $= 3.6$ in.,

$$\frac{k\ell_u}{r} = \frac{1.0 \times 240 \text{ in.}}{3.6 \text{ in.}} = 66.7$$

For $M_1/M_2 = 0.667$,

$$34 - 12\frac{M_1}{M_2} = 34 - 12 \times 0.667 = 26$$

Because $k\ell_u/r = 66.7$ exceeds 26, the column is quite slender. This suggests that the 12 in. × 12 in. section may be inadequate.

We shall select a 15-in.-by-15 in.-section for the first trial.
ACI Section 10.11.5 requires a special analysis if $k\ell_u/r$ exceeds 100. Such an analysis is not required here.

4.   **Check whether the moments are less than the minimum.** ACI Section 10.12.3.2 requires that a braced column be designed for a minimum eccentricity of $0.6 + 0.03h = 1.05$ in. Since the maximum end eccentricity exceeds this, design for the moments from step 1.

5.   **Compute EI.** At this stage, the area of reinforcement is not known. Additional calculations are needed before it is possible to use (12-15) to compute EI, but either (12-16) or (12-19) can be used. From (12-16),

$$EI = \frac{0.40E_cI_g}{1 + \beta_d} \qquad (12\text{-}15)$$

where

$$E_c = 57,000\sqrt{f'_c} = 3.122 \times 10^6 \text{ psi} \quad \text{(ACI Section 8.5.1)}$$

$$I_g = bh^3/12 = 4220 \text{ in.}^4$$

The term β_d is the ratio of the factored dead load to the total factored axial load:

$$\beta_d = \frac{1.2 \times 90}{228} = 0.474$$

Thus,

$$EI = \frac{3.122 \times 10^6 \text{ psi} \times 4220 \text{ in.}^4}{2.5(1 + 0.474)}$$

$$= 3.58 \times 10^9 \text{ in.}^2\text{-lb}$$

Alternatively (12-19), assuming that $\rho_t = 0.015$ gives

$$EI = \frac{3.122 \times 10^6 \times 4220}{1 + 0.474}\left(0.2 + 1.2 \times 0.015 \times \frac{29 \times 10^6}{3.122 \times 10^6}\right)$$

$$= 3.28 \times 10^9 \text{ in.}^2\text{-lb}$$

We shall try $EI = 3.58 \times 10^9$ in.²-lb. Generally, one would use (12-15) or (12-19) if ρ_t exceeded about 0.02, because they give higher values of EI.

6.   **Compute the magnified moment.** From (12-22),

$$M_c = \delta_{ns}M_2$$

where

$$\delta_{ns} = \frac{C_m}{1 - P_u/0.75P_c} \geq 1.0 \qquad\qquad (12\text{-}23)$$

$$(\text{ACI Eq. }10\text{-}9)$$

$$C_m = 0.6 + 0.4\frac{M_1}{M_2} \geq 0.4 \qquad\qquad (12\text{-}14)$$

$$= 0.6 + 0.4 \times 0.667 = 0.867$$

$$P_c = \frac{\pi^2 EI}{(k\ell_u)^2} \qquad\qquad (12\text{-}24)$$

where $k = 1.0$ since the column is pin-ended.

$$P_c = \frac{\pi^2 \times 3.58 \times 10^9 \text{ in.}^2\text{-lb}}{(1.0 \times 240 \text{ in.})^2} = 613{,}000 \text{ lb}$$

$$= 613 \text{ kips}$$

and

$$\delta_{ns} = \frac{0.867}{1 - 228/(0.75 \times 613)} \geq 1.0$$

$$= 1.72$$

Normally, if δ_{ns} exceeds 1.75 to 2.0, a larger cross section should be selected. Continuing without selecting a larger column, the magnified moment is

$$M_c = 1.72 \times 57.0 \text{ ft-kips} = 98.0 \text{ ft-kips}$$

A 15 in. square section is probably too small. **Try a 16-in.-by-16-in. square section.**

7.   **Select the column reinforcement.** Interaction diagrams for a 16 in. × 16 in. column with four No. 7 bars, four No. 8 bars, and four No. 9 bars are given in Fig. 12-17. The column reinforcement must be designed to resist $P_u = 228$ kips and $M_c = 98.0$ ft-kips. The interaction diagrams show that four No. 7 bars would be adequate. This gives a steel ratio of 0.0094, which is less than the minimum steel ratio of 0.01 required in ACI Section 10.9.1. For this reason, we will use a **16 in. × 16 in. column with four No. 8 bars**. A 15 in. × 15 in. column will not be used, since δ_{ns} was found to be 1.72, which the author considers to be too high. ∎

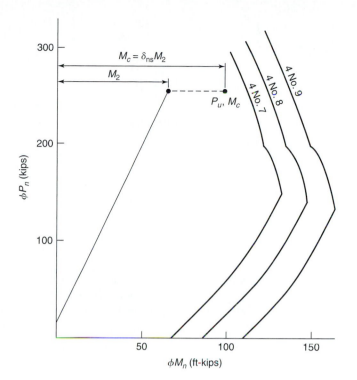

Fig. 12-17
Selection of reinforcement—
Example 12-1.

12-3   BEHAVIOR OF RESTRAINED COLUMNS IN NONSWAY FRAMES

Effect of End Restraints on Braced Frames

A simple indeterminate frame is shown in Fig. 12-18a. A load P and an unbalanced moment M_{ext} are applied to the joint at each end of the column. The moment M_{ext} is equilibrated by the moment M_c in the column and the moment M_r in the beam, as shown in Fig. 12-18b. By moment distribution,

$$M_c = \left(\frac{K_c}{K_c + K_b} \right) M_{ext} \qquad (12\text{-}25)$$

where K_c and K_b are the flexural stiffnesses of the column and the beam, respectively, at the upper joint. Thus K_c represents the moment required to bend the end of the column through a unit angle. The term in parentheses in (12-25) is the distribution factor for the column.

The total moment, M_{max}, in the column at midheight is

$$M_{max} = M_c + P\delta \qquad (12\text{-}26)$$

As was discussed earlier, the combination of the $P\delta$ moments and M_e gives rise to a larger total deflection and hence a larger rotation at the ends of the column than would be the case if just M_c acted. As a result, one effect of the axial force is to reduce the column stiffness, K_c. When this occurs, (12-25) shows that the fraction of M_{ext} assigned to the column drops, thus reducing M_e. Inelastic action in the column tends to hasten this reduction in column stiffness, again reducing the moment developed at the ends of the column. On the other hand, a reduction in the beam stiffness, K_b, due to cracking or inelastic action in the beam will redistribute moment back to the column.

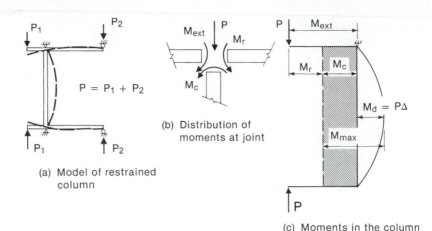

(a) Model of restrained column

(b) Distribution of moments at joint

(c) Moments in the column

Fig. 12-18
Moments in a restrained column. (From [12-1].)

This is illustrated in Fig. 12-19, which shows frame F2, tested by Furlong and Ferguson [12-11]. The columns in this frame had $\ell/h = 20$ $(k\ell_u/r = 57)$ and an initial eccentricity ratio $e/h = 0.106$. The loads bent the column in symmetrical single curvature. Failure occurred at section A at midheight of one of the columns. In Fig. 12-19b, load–moment curves are presented for section A and for section B, located at one end of the column. The moment at section B corresponds to the moment M_c in (12-26) and Fig. 12-18. Although the loads P and βP were proportionally applied, the variation in moment at B is not linear, because K_c decreases as the axial load is increased. As the moments at the ends of the columns decreased, the moments at the midspans of the beams had to increase to maintain equilibrium. The moment at A, the failure section, is equal to the sum of the moment M_c at section B and the moment due to the column deflection, $P\delta$.

Figures 12-20 to 12-22 trace the deflections and moments in slender columns in braced frames. These are based on inelastic analyses by Cranston [12-12] of reinforced concrete columns with elastic end restraints.

Figure 12-20 illustrates the behavior of a tied column having a slenderness ratio $\ell/h = 15$ $(k\ell/r = 33)$, with equal end restraints and loaded with an axial load P and an external moment of $1.5hP$ applied to the joint. A first-order analysis indicates that the end

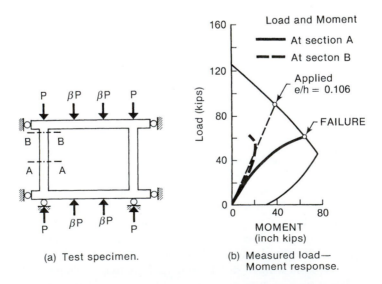

Fig. 12-19
Load-moment behavior of a column in a braced frame. (From [12-11].)

(a) Test specimen.

(b) Measured load— Moment response.

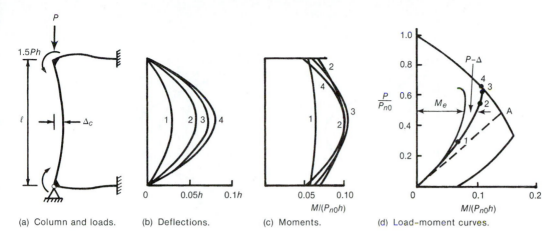

(a) Column and loads.   (b) Deflections.   (c) Moments.   (d) Load–moment curves.

Fig. 12-20
Behavior of a short restrained column bent in symmetrical single curvature. $\ell/h = 15$, $k\ell_u/r = 33$, $\rho_t = 0.01$, $K_b = 2.5K_c$ at both ends. (From [12-12].)

moments on the column itself are $0.25hP$, as shown by the dashed line $O\text{–}A$ in Fig. 12-20d. As the loads are increased, the column is deflected as shown in Fig. 12-20b. The moment diagrams in the column at the same four stages are shown in part c. The maximum midheight moment occurred at load stage 3. The increase in deflections from 3 to 4 (Fig. 12-20b) was more than offset by the decrease in end moments (Fig. 12-20c). Figure 12-20d traces the load–moment curves at midheight (centerline) and at the ends (line labeled M_e). The total moment at midheight is the sum of M_e and $P\delta$. Because the end moments decreased more rapidly than the $P\delta$ moments increased, the load–moment line for the midheight section curls upward. The failure load, point 4 in Fig. 12-20d, is higher than the failure load ignoring slenderness effects, point A. This column was *strengthened* by slenderness effects, because the beams had sufficient capacity to resist the extra end moments and allow the moments to change signs that were redistributed from the column to beams.

Figure 12-21 is a similar plot for a tied column with a slenderness ratio of $\ell/h = 40$ $(k\ell/r = 88)$. Such a column would resemble the columns in Fig. 12-7. At failure, the column deflection approached *60 percent of the overall depth* of the column, as shown in Fig. 12-21b. The moments at the ends of the columns decreased, reaching zero at load stage 2 and then becoming negative. This reduction in end moments was more than offset by the $P\delta$ moments due to the deflections. The load–moment curves for the ends of the column and for the midheight are shown in Fig. 12-21d.

The behavior shown in Figs. 12-20 and 12-21 is typical for reinforced concrete columns bent in single curvature $(M_1/M_2 \le 0)$. In such columns, both end moments decrease as P increases, possibly changing sign. The maximum moments in the column might decrease or increase, depending on the relative magnitudes of the decrease in end moments compared with the $P\delta$ moments. In either case, the beams must resist moments that may be considerably different from those produced by a first-order analysis method, such as the moment-distribution method.

For columns loaded in double curvature $(M_1/M_2 < 0)$, the behavior is different, as illustrated in Fig. 12-22. Assuming that the larger end moment at the top of the column, M_2, is positive and the smaller, M_1, is negative, it can be seen that both end moments become more negative, just as they did in Fig. 12-21. The difference, however, is that M_2 decreases and eventually becomes negative, while M_1 becomes larger (more negative). At failure of the column in Fig. 12-22, the negative moment at the bottom of the column is almost as big as the maximum positive moment.

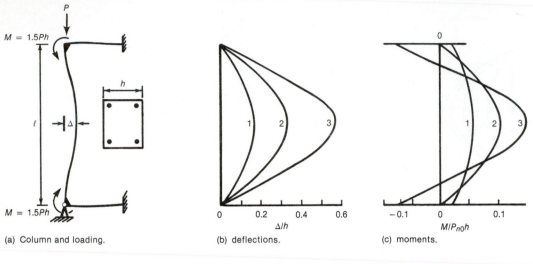

(a) Column and loading. (b) deflections. (c) moments.

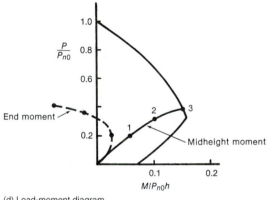

(d) Load-moment diagram.

Fig. 12-21
Behavior of a very slender restrained column bent in symmetrical single curvature. $\ell/h = 40$,
$k\ell_u/r = 88$, $\rho_t = 0.01$, $K_b = 2.5K_c$ at both ends. (From [12-12].)

Effect of Sustained Loads on Columns in Braced Frames

Consider a frame, similar to frame $F2$ shown in Fig. 12-19, that is loaded rapidly to service
load level, is held at this load level for several years, and then is loaded rapidly to failure.
If the columns are slender, the behavior plotted in Fig. 12-23a would be expected [12-13].
During the sustained-load period, the creep deflections cause a reduction in the column
stiffness K_c, which in turn leads to a reduction in the column end moments (at section B),
as shown by the horizontal line C–D in Fig. 12-23a, and a corresponding increase in the
midspan moments in the beams. At the same time, however, the $P\delta$ moment increases in
response to the increase in deflections. At the end of the sustained-load period, the end
moment is indicated by the distance E–D in Fig. 12-23a, while the total $P\delta$ moment at
midheight is shown by D–G. Failure of such a column occurs when the load–moment line
intersects the interaction diagram at H. Failure may also result from the reversal of sign of
the end moments, shown by J in Fig. 12-23a, if the end restraints are unable to resist the reversed moment. The dashed lines indicate the load–moment curve for the end and midheight sections in a column loaded to failure in a short time. The decrease in load from K
to H is due to the creep effect.

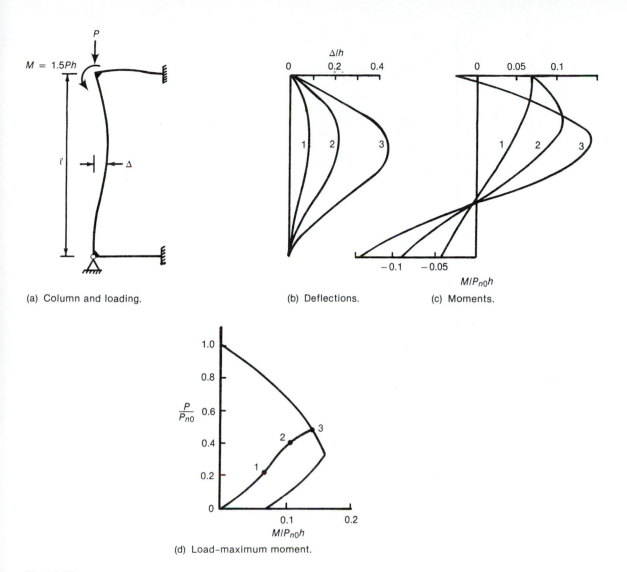

(a) Column and loading.

(b) Deflections.

(c) Moments.

(d) Load–maximum moment.

Fig. 12-22
Behavior of a very slender restrained column bent in double curvature. $\ell/h = 40$, $k\ell/r = 83$, $\rho_t = 0.01$, $K_b = 2.5K_c$ at top and $6K_c$ at bottom. (From [12-12].)

For a short column in a similar frame, the reduction in end moment due to creep may be larger than the increase in the $P\delta$ moment, resulting in a strengthening of the column [12-13], as illustrated in Fig. 12-23b. Although the axial-load capacity of the column is increased, the moments at the midspan of the beams are also increased, and they can cause failure of the frame.

The reduction of column end moments due to creep greatly reduces the risk of creep buckling (Fig. 12-15b) of columns of braced frames.

12-4   DESIGN OF COLUMNS IN NONSWAY FRAMES

This section deals with columns in continuous frames with deformations restrained in two ways. First, the frames are "nonsway" or "braced," so *the horizontal deflection of one end of a column relative to the other end* is prevented, or at least restrained, by walls or other bracing elements. Second, the columns are attached to beams that *restrain the rotations of*

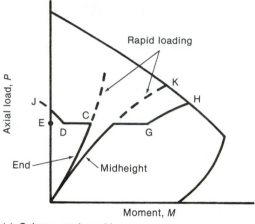

(a) Column weakened by creep.

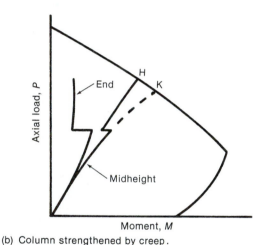

Fig. 12-23
Effect of sustained loads on
moments in columns in
braced frames.

(b) Column strengthened by creep.

the ends of the column. This section deals primarily with how the rotational restraints provided by the beams are accounted for in design.

Design Approximation for the Effect of End Restraints in Nonsway Frames

Figure 12-24a shows a restrained column in a frame. The curved solid line in Fig. 12-24b is the moment diagram (including slenderness effects) for this column at failure (similar to Fig. 12-20c or 12-21c). Superimposed on this is the corresponding first-order moment diagram for the same load level. In design, it is convenient to replace the restrained column with an equivalent hinged end column of length ℓ_i, the distance between the points on the second-order moment diagram where the moments are equal to the end moments in the first-order diagram (Fig. 12-24c). This equivalent hinged column is then designed for the axial load, P, and the end moments, M_2, from the first-order analysis.

Unfortunately, the length ℓ_i is difficult to compute. In all modern concrete and steel design codes, the *empirical assumption* is made that ℓ_i can be taken equal to the effective length for elastic buckling, $k\ell$. The accuracy of this assumption is discussed in [12-14], which concludes that $k\ell$ slightly underestimates ℓ_i for an elastically restrained, elastic column.

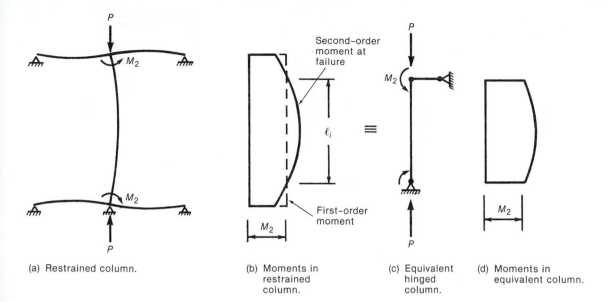

(a) Restrained column.   (b) Moments in restrained column.   (c) Equivalent hinged column.   (d) Moments in equivalent column.

Fig. 12-24
Replacement of restrained column with an equivalent hinged column for design.

The concept of effective lengths was discussed earlier in this chapter for the four idealized cases shown in Fig. 12-6. The effective length of a column, $k\ell_u$, is defined as the length of an equivalent pin-ended column having the same buckling load. When a pin-ended column buckles, its deflected shape is a half-sine wave, as shown in Fig. 12-6a. The effective length of a restrained column is taken equal to the length of a complete half-sine wave in the deflected shape.

Figures 12-6b to 12-6d are drawn by assuming truly fixed ends. This condition seldom, if ever, actually exists. In buildings, columns are restrained by beams or footings which always allow some rotation of the ends of the column. Thus the three cases considered in Fig. 12-6 will actually deflect as shown in Fig. 12-25, and the effective lengths will be greater than the values for completely fixed ends. The actual value of k for an elastic column is a function of the relative stiffnesses, ψ, of the beams and columns at each end of the column, where ψ is

$$\psi = \frac{\Sigma\,(E_cI_c/\ell_c)}{\Sigma\,(E_bI_b/\ell_b)} \tag{12-27a}$$

where the subscripts b and c refer to beams and columns, respectively, and the lengths ℓ_b and ℓ_c are measured center to center of the joints. The summation sign in the numerator refers to all the compression members meeting at a joint; the summation sign in the denominator refers to all the beams or other restraining members at the joint.

If $\psi = 0$ at one end of the column, the column is fully fixed at that end. Similarly, $\psi = \infty$ denotes a perfect hinge. Thus, as ψ approaches zero at the two ends of a column in a braced frame, k approaches 0.5, the value for a fixed-ended column. Similarly, when ψ approaches infinity at the two ends of a braced column, k approaches 1.0, the value for a pin-ended column. This is illustrated in Table 12-2. The value for columns that are fully fixed at both ends is 0.5, found in the lower-left corner of the table. The value for columns that are pinned at both ends is 1.0, found in the upper-right corner.

In practical structures, there is no such thing as a truly fixed end or a truly hinged end. Reasonable upper and lower limits on ψ are 20 and 0.2. For columns in nonsway

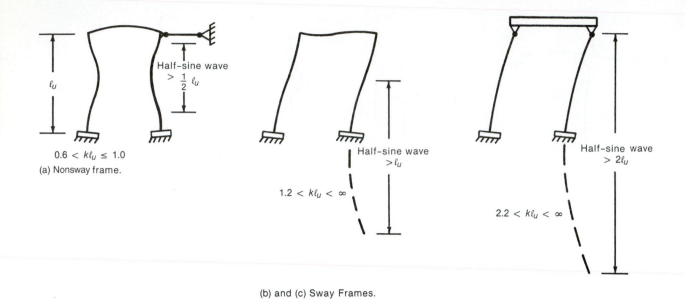

$0.6 < k\ell_u \le 1.0$

(a) Nonsway frame.

Half-sine wave $> \frac{1}{2}\ell_u$

$1.2 < k\ell_u < \infty$

Half-sine wave $> \ell_u$

$2.2 < k\ell_u < \infty$

Half-sine wave $> 2\ell_u$

(b) and (c) Sway Frames.

Fig. 12-25
Effective lengths of columns in frames with foundation rotations.

TABLE 12-2   Effective-Length Factors for Braced Frames

Top		k				
Hinged		0.70	0.81	0.91	0.95	1.00
Elastic $\psi = 3.1$		0.67	0.77	0.86	0.90	0.95
Elastic, Flexible $\psi = 1.6$		0.65	0.74	0.83	0.86	0.91
Stiff $\psi = 0.4$		0.58	0.67	0.74	0.77	0.81
Fixed		0.50	0.58	0.65	0.67	0.70
		Fixed	Stiff	Elastic, Flexible	Elastic	Hinged
		Bottom				

frames, k should never be taken less than 0.6. In sway frames, k should never be taken less than 1.2 for columns restrained at both ends.

Calculation of k from Tables

Table 12-2 can be used to select values of k for the design of braced frames. The shaded areas correspond to one or both ends truly fixed. Since such a case rarely, if ever, occurs in practice, this part of the table should not be used. The column and row labeled "Hinged", "elastic" through to "fixed" represent conservative practical degrees of end fixity. Because k values for sway frames vary widely, no similar table is given for such frames.

Calculation of k via Nomographs

The nomographs given in Fig. 12-26 are also used to compute k. To use these nomographs, ψ is calculated at both ends of the column, from (12-27), and the appropriate value of k is found as the intersection of the line labeled k and a line joining the values of ψ at the two ends of the column. The calculation of ψ is discussed in a later section.

The nomographs in Fig. 12-26 were derived [12-15], [12-16] by considering a typical interior column in an infinitely high and infinitely wide frame, in which all of the columns have the same cross section and length, as do all beams. Equal loads are applied at the tops of each of the columns, while the beams remain unloaded. All columns are assumed to buckle at the same moment. As a result of these very idealized and quite unrealistic assumptions, the nomographs tend to underestimate the value of the effective length

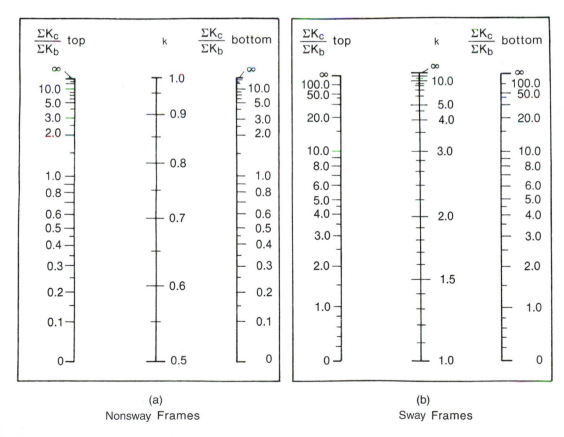

(a)
Nonsway Frames

(b)
Sway Frames

Fig. 12-26
Nomographs for effective length factors.

factor k for elastic frames of practical dimensions by up to 15 percent [12-14]. This then leads to an underestimate of the magnified moments, M_c.

The lowest practical value for k in a sway frame is about 1.2 due to friction in the hinges. When smaller values are obtained from the nomographs, it is good practice to use 1.2.

Equations for k

Approximate equations for k are presented in Section R10.12.1 of the ACI Commentary. These have been derived to give a conservative approximation of the effective-length factor.

Calculation of ψ, Column-Beam Frames

The stiffness ratio, ψ, is calculated from (12-27). The values of E_cI_c and E_bI_b should be realistic for the state of loading immediately prior to failure of the columns. Generally, at this stage of loading, the beams are extensively cracked and the columns are uncracked or slightly cracked. Ideally, the values of EI should reflect the degree of cracking and the actual reinforcement present. This is not practical, however, since this information is not known at this stage of design. ACI Sections 10.12.1 and 10.13.1 state that the calculation of k shall be based on a ψ based on the E and I values given in ACI Section 10.11.1. In computing ψ, β_d can be taken as zero.

In calculating I_b for a T beam, the flange width can be taken as defined in ACI Section 8.10.2 or 8.10.3. For common ratios of flange thickness to overall depth, h, and flange width to web width, b_w, the gross moment of inertia, I_g, is approximately twice the moment of inertia of a rectangular section with dimensions b_w and h.

In structural-steel design, the term ψ varies to account for the effects of fixity of the far ends of the beams, as expressed in the equation

$$\psi = \frac{\Sigma E_cI_c/\ell_c}{\Sigma C_bE_bI_b/\ell_b} \tag{12-27b}$$

where C_b accounts for the degree of fixity of the far ends of the beams that are attached to the ends of the columns. The nomograph for nonsway columns was derived for a frame in which the beams were deflected in single curvature with equal but opposite moments at the two ends of the beams. In such a case, $C_b = 1.0$. If the far end of a beam is hinged or fixed, the beam is stiffer than is assumed in the calculation of ψ from (12-27b). When the conditions at the far end of a beam are known definitely or when a conservative estimate can be made, C_b is evaluated theoretically. Otherwise it is taken as 1.5 if the far end of the beam is hinged and as 2.0 if the far end of the beam is fixed. These values of C_b reduce ψ and hence reduce k. This, in turn, increases the critical load of the column.

The derivation of the nomograph for columns in sway frames assumed that the beams were deflected with equal slopes at each end. Beams hinged at the far ends in sway frames have a lower stiffness than assumed. For this case, $C_b = 0.5$.

However, because of the major approximations implicit in the derivation of the nomographs, we shall take $C_b = 1.0$ in all cases.

Since hinges are never completely frictionless, a value of $\psi = 10$ is frequently used for hinged ends, rather than $\psi = \infty$.

Calculation of ψ, Column Footing Joints

The value of ψ at the lower end of a column supported on a footing can be calculated from relationships presented in the *PCI Design Handbook* [12-17]. Equation (12-27) can be rewritten as

$$\psi = \frac{\Sigma K_c}{\Sigma K_b} \tag{12-28}$$

where ΣK_c and ΣK_b are the sums of the flexural stiffnesses of the columns and the restraining members (beams) at a joint, respectively. At a column-to-footing joint, $\Sigma K_c = 4E_cI_c/\ell_c$ for a braced column restrained at its upper end, and ΣK_b is replaced by the rotational stiffness of the footing and soil, taken equal to

$$K_f = \frac{M}{\theta_f} \tag{12-29}$$

where M is the moment applied to the footing and θ_f is the rotation of the footing. The stress under the footing is the sum of $\sigma = P/A$, which causes a uniform downward settlement, and $\sigma = My/I$, which causes a rotation. The rotation θ_f is

$$\theta_f = \frac{\Delta}{y} \tag{12-30}$$

where y is the distance from the centroid of the footing area and Δ is the displacement of that point relative to the displacement of the centroid of the footing area. If k_s is the coefficient of subgrade reaction, defined as the stress required to compress the soil by a unit amount ($k_s = \sigma/\Delta$), then θ_f is

$$\theta_f = \frac{\sigma}{k_s y} = \frac{My}{I_f} \times \frac{1}{k_s y}$$

Substituting this into (12-29) gives

$$K_f = I_f k_s \tag{12-31}$$

where I_f is the moment of inertia of the contact area between the bottom of the footing and the soil and k_s is the coefficient of subgrade reaction, which can be taken from Fig. 12-27. Thus, the value of ψ at a footing-to-column joint for a column restrained at its upper end is

$$\psi = \frac{4E_cI_c/\ell_c}{I_f k_s} \tag{12-32}$$

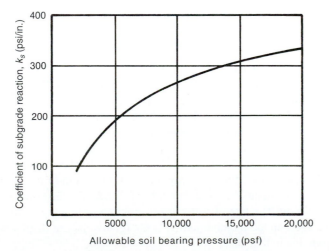

Fig. 12-27
Approximate relationship between allowable soil bearing pressure and the coefficient of subgrade reaction k_s.
(From [12-17].)

The axis about which the footing rotates is in the plane of the footing–soil interface. As a result, the length ℓ_c used to compute ψ should include the depth of the footing.

Definition of Nonsway and Sway Frames

The preceding discussions of column behavior and effective-length factors were based on the assumption that frames could be separated into "completely braced" frames and "completely unbraced" frames. A column may be considered to be "braced" or "nonsway" in a given direction if the lateral stability of the structure as a whole is provided by walls, bracing, or buttresses designed to resist all lateral forces in that direction. A column is completely "unbraced" or "free to sway" in a given plane if all resistance to lateral loads comes from bending of the columns.

In actuality, there is no such thing as a "completely braced" frame, and no clear-cut boundary exists between braced and unbraced frames. Some frames are clearly unbraced, as, for example, the frames shown in Fig. 12-25b and c. Other frames are connected to shear walls, elevator shafts, and so on, which clearly restrict the lateral movements of the frame, as shown in Fig. 12-25a. Since no such wall is completely rigid, however, there will always be some lateral movement of a braced frame, and hence some $P\Delta$ moments will result from the lateral deflections.

For the purposes of design, a story or a frame can be considered "nonsway" if horizontal displacements do not significantly reduce the vertical load capacity of the structure. Since the ACI Code moment magnifier design procedure accounts for slenderness by magnifying moments, this criterion could be restated as follows: a frame can be considered "nonsway" if the $P\Delta$ moments due to lateral deflections are small compared with the first-order moments due to lateral loads. ACI Section 10.11.4.1 allows designers to assume that a frame is nonsway if the increase in column-end moments due to second-order effects does not exceed 5 percent of the first-order moments. This test should be carried out at the end of the column where the magnified end moment is the largest.

Alternatively, ACI Section 10.11.4.2 allows designers to assume that a story in a frame is nonsway if

$$Q = \frac{\Sigma P_u \Delta_0}{V_u \ell_c} \tag{12-33}$$

(ACI Eq. 10-7)

is less than or equal to 0.05, where Q is the *stability index*, ΣP_u is the total vertical load in all the columns and walls in the story in question, V_u is the shear in the story due lateral loads, Δ_0 is the first-order relative deflection between the top and bottom of that story *due to V_u*, and ℓ_c is the height of the story measured from center to center of the joints above and below the story. This concept, which is explained and developed more fully in Section 12-6, results in a limit similar to that from ACI Section 10.11.4.1. It was originally presented in [12-18] and [12-19] and is discussed in [12-20].

ACI Commentary Section R10.11.4 suggests that, frequently, the test of whether a story is sway or nonsway can be done by inspection, by comparing the total lateral stiffness of all the columns in a story to that of the bracing elements in the story, such as walls or shear trusses. The Commentaries to the 1971 to 1989 ACI Codes [12-22] suggested that a story would be nonsway (braced) if the sum of the lateral stiffnesses, ΣK_ℓ, for the bracing elements exceeded six times the ΣK_ℓ for the columns in the direction under consideration. The lateral stiffness of a column or bracing element is $K_\ell = V/\Delta$, where V is the shear in the member and Δ is the relative lateral displacement of the ends of the column due to that shear.

Summary of Moment-Magnifier Design Procedure for Slender Columns in Braced Frames

If a column is in a nonsway frame, then design involves ACI Sections 10.11, "Magnified Moments—General," and 10.12, "Magnified Moments—Nonsway Frames," which give the following conditions:

   1.   **Length of column.** The unsupported length, ℓ_u, is defined in ACI Section 10.11.3.1 as the clear height between slabs or beams capable of giving lateral support to the column.

   2.   **Effective length.** ACI Section 10.12.1 states that the effective-length factors, k, of columns in nonsway frames shall be 1.0 or less. The effective-length factors can be estimated from Table 12-2 or from Fig. 12-26. These procedures require that the ratio, ψ, of EI/ℓ of the columns and beams be known. This factor is given by (12-27). ACI Section 10.12.1 says that ψ should be based on the E and I values in ACI Section 10.11.1.

   3.   **Evaluation of whether the frame is braced.** Frequently, this can be done by inspection, by seeing whether the bracing elements, such as walls, are considerably stiffer than the columns. Alternatively, the frame can be assumed to be nonsway if Q from (12-33) is not greater than 0.05.

   4.   **Radius of gyration.** For a rectangular cross section, $r = 0.3h$, and for a circular cross section, $r = 0.25h$. For other sections, r can be calculated from the area and moment of inertia of the gross concrete section as $r = \sqrt{I_g/A_g}$ (ACI Section 10.11.2).

   5.   **Consideration of slenderness effects.** For columns in nonsway frames, ACI Section 10.12.2 allows slenderness to be neglected if

$$\frac{k\ell_u}{r} < 34 - 12\frac{M_1}{M_2} \tag{12-20}$$

For columns in unbraced frames, ACI Section 10.12.2 allows slenderness to be neglected if $k\ell_u/r$ is less than 22. If $k\ell_u/r$ exceeds 100, design shall be based on a second-order analysis. The sign convention for M_1/M_2 is illustrated in Fig. 12-13c and d.

   6.   **Minimum moment.** For columns in braced frames, the larger end moment, M_2, shall not be taken less than

$$M_{2,\text{min}} = P_u(0.6 + 0.03h) \tag{12-21}$$

$$\text{(ACI Eq. 10-14)}$$

about each axis separately, where 0.6 and h are in inches.

   7.   **Moment-magnifier equation.** ACI Section 10.12.3 states that columns in nonsway frames shall be designed for the factored axial load, P_u, and a magnified factored moment, M_c, given by

$$M_c = \delta_{ns}M_2 \tag{12-22}$$

$$\text{(ACI Eq. 10-8)}$$

where M_2 is the larger end moment and

$$\delta_{ns} = \frac{C_m}{1 - P_u/0.75P_c} \geq 1.0 \tag{12-23}$$

$$\text{(ACI Eq. 10-9)}$$

with

$$C_m = 0.6 + 0.4\left(\frac{M_1}{M_2}\right) \geq 0.4 \tag{12-14}$$

$$\text{(ACI Eq. 10-13)}$$

where M_1/M_2 in (12-14) is positive for single-curvature bending and is negative for double-curvature bending, as illustrated in Fig. 12-13c and d.

$$P_c = \frac{\pi^2 EI}{(k\ell_u)^2} \qquad \text{(12-24)}$$
$$\text{(ACI Eq. 10-10)}$$

and

$$EI = \frac{0.2E_c I_g + E_s I_{se}}{1 + \beta_d} \qquad \text{(12-15)}$$
$$\text{(ACI Eq. 10-12)}$$

or

$$EI = \frac{0.40E_c I_g}{1 + \beta_d} \qquad \text{(12-16)}$$
$$\text{(ACI Eq. 10-13)}$$

The term β_d has three definitions, only one of which applies to columns in nonsway frames. For non-sway columns,

$$\beta_d = \frac{\text{maximum factored axial dead load in the column}}{\text{total factored axial load in the column}} \qquad \text{(12-34a)}$$

Equations (12-18) and (12-19) could also be used to compute EI for use in (12-24). The EI values given in ACI Section 10.11.1 cannot be used to compute EI for use in (12-24). The EI values in ACI Section 10.11.1 approach the average values for an entire story in a frame and are intended for use in first- and second-order frame analyses.

If P_u exceeds $0.75P_c$ in (12-23), δ_{ns} will be negative. If the stiffness were lower than expected, such a column would be unstable. Hence, if P_u exceeds $0.75P_c$, the column cross section should be enlarged. Indeed, if δ_{ns} exceeds 2.0, strong consideration should be given to enlarging the column cross section, because beyond that the calculations become very sensitive to the assumptions made.

EXAMPLE 12-2 Design of the Columns in a Braced Frame

Figure 12-28 shows part of a typical frame in an industrial building. The frames are spaced 20 ft apart. The columns rest on 4-ft-square footings. The soil bearing capacity is 4000 psf. Design columns C–D and D–E. Use $f'_c = 3000$ psi and $f_y = 60,000$ psi for beams and columns. Use lower combinations and strength-reduction factors from ACI 318.02 Sections 9.2 and 9.3.

1. Calculate the column loads from a frame analysis.
A first-order elastic analysis of the frame shown in Fig. 12-28 gave the forces and moments in the table.

	Column *CD*	Column *DE*
Service loads, P	Dead = 80 kips	Dead = 50 kips
	Live = 24 kips	Live = 14 kips
Service moments at	Dead = −60 ft-kips	Dead = 42.4 ft-kips
tops of columns	Live = −14 ft-kips	Live = 11.0 ft-kips
Service moments at	Dead = −21 ft-kips	Dead = −32.0 ft-kips
bottoms of columns	Live = −8 ft-kips	Live = −8 ft-kips

Clockwise moments on the ends of members are positive. All wind forces are assumed to be resisted by the end walls of the building.

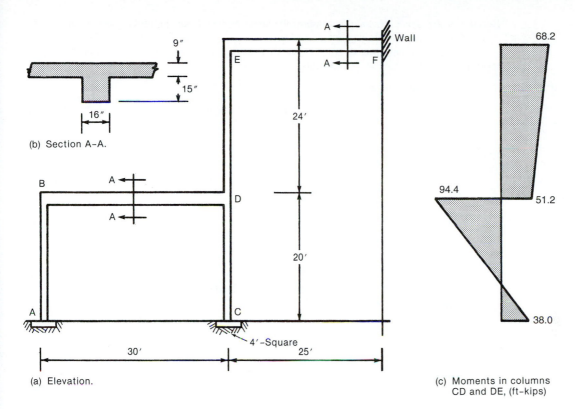

(b) Section A–A.

(a) Elevation.

(c) Moments in columns
CD and DE, (ft–kips)

Fig. 12-28
Braced frame—Example 12-2.

2.   **Determine the factored loads.**

    (a)   **Column *CD*:**

$$P_u = 1.2 \times 80 + 1.6 \times 24 = 134.4 \text{ kips}$$
$$\text{moment at top} = 1.2 \times -60 + 1.6 \times -14 = -94.4 \text{ ft-kips}$$
$$\text{moment at bottom} = 1.2 \times -21 + 1.6 \times -8 = -38.0 \text{ ft-kips}$$

The factored-moment diagram is shown in Fig. 12-28c. By definition (ACI Section 10.0), M_2 is always positive, and M_1 is positive if the column is bent in single curvature (Fig. 12-13c and d). Because column *CD* is bent in double curvature (Fig. 12-28c), M_{1b} is negative. Thus, for slender column design, $M_2 = +94.4$ ft-kips and $M_1 = -38.0$ ft-kips.

    (b)   **Column *DE*:**

$$P_u = 82.4 \text{ kips}$$
$$\text{moment at top} = +68.5 \text{ ft-kips}$$
$$\text{moment at bottom} = -51.2 \text{ ft-kips}$$

Thus $M_2 = +68.5$ ft-kips and $M_1 = +51.2$ ft-kips. M_1 is positive, because the column is in single curvature.

3.   **Make a preliminary selection of the column size.** From (11-21a) for $\rho_t = 0.015$,

$$A_{g(\text{trial})} \geq \frac{P_u}{0.40(f'_c + f_y \rho_t)}$$

$$= \frac{134.4}{0.40(3 + 0.015 \times 60)}$$

$$= 86.2 \text{ in.}^2 \text{ or } 9.3 \text{ in. square}$$

Because of the slenderness and because of the large moments, we shall take a larger column. Try 14 in. × 14 in. columns throughout.

4.   Are the columns slender? From ACI Section 10.12.2, a column in a braced frame is short if $k\ell_u/r$ is less than $34 - 12M_1/M_2$.

(a)   Column CD:

$$\ell_u = 20 \text{ ft} - 2 \text{ ft} = 18 \text{ ft} \qquad \text{(ACI Section 10.11.3.1)}$$
$$= 216 \text{ in.}$$

From Table 12-2, $k = 0.77$. Thus,

$$r = 0.3 \times 14 \text{ in.} = 4.2 \text{ in.} \qquad \text{(ACI Section 10.11.2)}$$

$$\frac{k\ell_u}{r} = \frac{0.77 \times 216}{4.2} = 39.6$$

$$34 - 12\left(\frac{M_1}{M_2}\right) = 34 - 12\left(-\frac{38.0}{94.4}\right) = 38.8$$

Since $39.6 > 38.8$, column CD is just slender.

(b)   Column DE:

$$\ell_u = 24 \text{ ft} - 2 \text{ ft} = 264 \text{ in.}$$
$$k = 0.86$$
$$\frac{k\ell_u}{r} = \frac{0.86 \times 264}{4.2} = 54.1$$
$$34 - 12\left(\frac{M_1}{M_2}\right) = 34 - 12\left(\frac{51.2}{68.5}\right) = 25.0$$

Thus, column DE is also slender. Neither column exceeds the $k\ell_u/r = 100$ limit in ACI Section 10.11.5.

5.   Check whether the moments are less than the minimum. ACI Section 10.12.3.2 requires that braced slender columns be designed for a minimum eccentricity of $(0.6 + 0.03h)$ in. For 14-in. columns, this is 1.02 in. Thus, column CD must be designed for a moment M_2 of at least

$$P_u e_{\min} = 134.4 \text{ kips} \times 1.02 \text{ in.} = 11.42 \text{ ft-kips}$$

and column DE for a moment of at least 7.0 ft-kips. Since the actual moments exceed these values, the columns shall be designed for the actual moments.

6.   Compute EI. Since the reinforcement is not known at this stage of the design, we can use either (12-16) or (12-19) to compute EI. From (12-16),

$$EI = \frac{0.40 \, E_c I_g}{1 + \beta_d}$$

where

$$E_c = 57{,}000\sqrt{f'_c} = 3.12 \times 10^6 \text{ psi} \qquad \text{(ACI Section 8.5.1)}$$
$$I_g = 14^4/12 = 3201 \text{ in.}^4$$
$$0.40 \, E_c I_g = 4.00 \times 10^9 \text{ in.}^2\text{-lb}$$

(a)   Column CD:

$$\beta_d = \frac{1.2 \times 80}{134.4} = 0.714 \qquad \text{(12-34a)}$$

$$EI = \frac{4.00 \times 10^9}{1 + 0.714} = 2.33 \times 10^9 \text{ lb-in.}^2$$

(b)   Column *DE*:

$$\beta_d = \frac{1.2 \times 50}{82.4} = 0.728$$

$$EI = \frac{4.00 \times 10^9}{1.728} = 2.31 \times 10^9 \text{ lb-in.}^2$$

7.   Compute the effective-length factors. Two methods of estimating the effective-length factors, k, have been presented. In this example, we calculate k by both methods, to illustrate their use. In practice, only one of these procedures would be used in a given set of calcultions. We begin with

$$\psi = \frac{\Sigma E_c I_c / \ell_c}{\Sigma E_b I_b / \ell_b} \tag{12-27}$$

where ACI Section 10.12.1 says that E and I shall be as in ACI Section 10.11.1. Thus, $I_c = 0.70 I_g$ and $I_b = 0.35 I_g$, where I_g is the gross moment of inertia of the cross section. For the beam section shown in Fig. 12-28b, ACI Section 8.10.2 gives the effective flange width as 90 in. Using this width gives $I_g = 36,600$ in.^{4}, so $I_b = 0.35 \times 36,600 = 12,810$ in.^{4}. Similarly, $I_c = 0.70 \times 14^4/12 = 2240$ in.^{4}. In (12-27), ℓ_c and ℓ_b are the spans of the column and beam, respectively, measured center to center of the joints in the frame.

(a)   Column *DE*: The value of ψ at E is

$$\psi_E = \frac{E_c \times 2240/(24 \times 12)}{E_b \times 12,810/300} = 0.182$$

where $E_c = E_b$. Thus, $\psi_E = 0.182$. The value of ψ at D is

$$\psi_D = \frac{E_c \times 2240/288 + E_c \times 2240/240}{E_b \times 12,810/360} = 0.481$$

The value of k from Fig. 12-26 is 0.63. The value of k from Table 12-2 is 0.86.
    As was pointed out in the discussion of Fig. 12-26, the effective-length nomographs tend to underestimate the values of k for beam columns in practical frames [12-14]. Because Table 12-2 gives reasonable values without the need to calculate ψ, it has been used to compute k in this example. Thus, we shall use $k = 0.86$ for column *DE*.

(b)   Column *CD*: The value of ψ at D is

$$\psi_D = 0.481$$

The column is restrained at C by the rotational resistance of the soil under the footing and is continuous at D. From (12-32),

$$\psi = \frac{4 E_c I_c / \ell_c}{I_f k_s}$$

where I_f is the moment of inertia of the contact area between the footing and the soil and k_s is the coefficient of subgrade reaction obtained from Fig. 12-27. Thus,

$$I_f = \frac{48^4}{12} = 442,400 \text{ in.}^4$$

$$\psi_c = \frac{4 \times 3.122 \times 10^6 \text{ lb/in.}^2 \times 3201 \text{ in.}^4/240 \text{ in.}}{48^4/12 \text{ in.}^4 \times 160 \text{ lb/in.}^3}$$

$$= 2.35$$

From Fig. 12-26: $k = 0.76$
From Table 12-2: $k = 0.77$

Use $k = 0.77$ for column *CD*.

8. **Compute the magnified moments.** From (12-22),

$$M_c = \delta_{ns} M_2 \tag{12-22}$$

where

$$\delta_{ns} = \frac{C_m}{1 - (P_u/0.75P_c)} \geq 1.0 \tag{12-23}$$
$$\text{(ACI Eq. 10-9)}$$

(a) **Column CD:**

$$C_m = 0.6 + 0.4\left(-\frac{38.0}{94.4}\right) \geq 0.4 \tag{12-14}$$
$$\text{(ACI Eq. 10-13)}$$
$$= 0.438$$

$$P_c = \frac{\pi^2 EI}{(k\ell_u)^2} \tag{12-24}$$

$$\ell_u = 18 \text{ ft} = 216 \text{ in} \tag{ACI Eq. 10-10}$$

From Table 12-2 assuming the top end is the intersection of several beams and columns, and the bottom end is "stiff," $k = 0.77$.

$$P_c = \frac{\pi^2 \times 2.33 \times 10^9 \text{ lb-in.}^2}{(0.77 \times 216 \text{ in.})^2}$$

$$= 831,000 = 831 \text{ kips}$$

$$\delta_{ns} = \frac{0.438}{1 - 134.4/(0.75 \times 831)}$$

$$= 0.558 \geq 1.0$$

Therefore, $\delta_{ns} = 1.0$. This means that the section of maximum moment remains at the end of the column, so that

$$M_c = 1.0 \times 94.4 = 94.4 \text{ ft-kips}$$

Column *CD* is designed for $P_u = 134.4$ kips and $M_u = M_c = 94.4$ ft-kips.

(b) **Column DE:**

$$C_m = 0.6 + 0.4\left(\frac{51.2}{68.5}\right) = 0.900$$

$$P_c = \frac{\pi^2 \times 2.29 \times 10^9}{(0.86 \times 264)^2} = 439 \text{ kips}$$

$$\delta_{ns} = \frac{0.900}{1 - \dfrac{82.4}{0.75 \times 439}} \geq 1.0$$

$$= 1.200$$

This column is affected by slenderness, so

$$M_c = 1.200 \times 68.5 = 82.2 \text{ ft-kips}$$

Column *DE* is designed for $P_u = 82.4$ kips and $M_u = M_c = 82.2$ ft-kips.

9. **Select the reinforcement.** Figure 12-29 gives interaction diagrams for 14 in. $\times$ 14 in. columns with four No. 8 bars, four No. 9 bars, and four No. 10 bars. We select the following for reinforcement:

Column *CD*: Use 14 in. $\times$ 14 in. column with four No. 8 bars.
Column *DE*: Use 14 in. $\times$ 14 in. column with four No. 8 bars. ∎

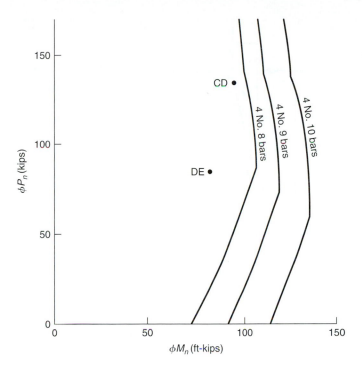

Fig. 12-29
Interaction diagrams—
Example 12-2.

12-5   BEHAVIOR OF RESTRAINED COLUMNS IN SWAY FRAMES

Statics of Sway Frames

An unbraced frame is one that depends on moments in the columns to resist lateral loads and lateral deflections. Such a frame is shown in Fig. 12-30a. The sum of the moments at the tops and bottoms of all the columns must equilibrate the applied lateral-load moment, $V\ell$, plus the moment due to the vertical loads, $\Sigma P\Delta$. Thus,

$$\Sigma\,(M_{\text{top}} + M_{btm}) = V\ell + \Sigma P\Delta \qquad (12\text{-}35)$$

It should be noted that both columns have deflected laterally by the same amount Δ. For this reason, it is not possible to consider columns independently in an unbraced frame.

    If a sway frame includes some pin-ended columns, as might be the case in a precast concrete building, the vertical loads in the pin-ended columns are included in ΣP in (12-33) and (12-35). Such columns are referred to as *leaning columns*, because they depend on the frame for stability.

    The $V - \ell$ moment diagram due to the lateral loads is shown in Fig. 12-30b and that due to the $P - \delta$ moments in Fig. 12-30c. It can be seen that these are directly additive, because the maximum lateral load moments and the maximum $P - \Delta$ moments occur at the ends of the column. Because the maximum lateral-load moments and the maximum $P - \Delta$ moments both occur at the ends of the columns and hence can be added directly, the equivalent moment factor, C_m, given by (12-14) does not apply to sway columns. On the other hand, (12-11) becomes

$$M_c = \frac{M_0(1 - 0.18P/P_E)}{1 - P/P_E} \qquad (12\text{-}36)$$

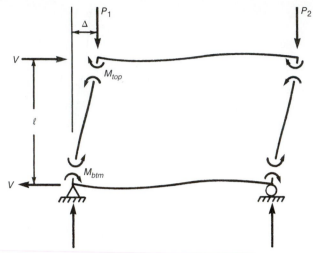

(a) Column moments in a sway frame.

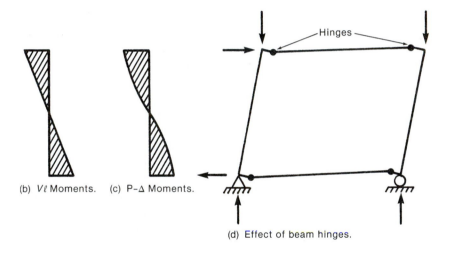

(b) $V\ell$ Moments.   (c) P-Δ Moments.

(d) Effect of beam hinges.

Fig. 12-30
Column moments in a sway
frame.

The term in brackets in (12-11) has been left out of the Moment-Magnifier equation in the ACI Code because the resulting change in the magnification does not vary significantly.

It is also important to note that if hinges were to form at the ends of the beams in the frame as shown in Fig. 12-30d, the frame would be unstable. Thus the beams must resist the *full magnified end moment* from the columns for the frame to remain stable (ACI Section 10.13.7).

Loads causing sway are seldom sustained, although such cases can be visualized, such as a frame that resists the horizontal reaction from an arch roof or a frame resisting lateral earth loads. If a sustained load acts on an unbraced frame, the deflections increase with time, leading directly to an increase in the $P-\Delta$ moment. This process is very sensitive to small variations in material properties and loadings. As a result, structures subjected to sustained lateral loads should always be braced. Indeed, braced frames should be used wherever possible, regardless of whether the lateral loads are short time or sustained.

M_{ns} and M_s Moments

Two different types of moments occur in frames:

1. moments due to loads not causing appreciable sway, M_{ns}
2. moments due to loads causing appreciable sway, M_s

The slenderness effects of these two kinds of moments are considered separately in the ACI Code design process because each is magnified differently as the individual columns deflect and as the entire frame deflects [12-21]. Column moments that cause no appreciable sway are magnified when the column deflects by an amount δ relative to its original straight axis such that the moments at points along the length of the column exceed those at the ends. In Section 12-2, this is referred to as the *member stability* effect or $P-\delta$ effect, where the lower case δ refers to deflections relative to the chord joining the ends of the column.

On the other hand, the column moments due to lateral loads can cause appreciable sway. They are magnified by the $P-\Delta$ moments resulting from the sway deflections, Δ, at joints in the frame, as indicated by (12-35). This is referred to as the $P-\Delta$ effect or *lateral drift* effect.

ACI Section 10.0 defines the *nonsway moment*, M_{ns}, as the factored end moment on a column due to loads that cause *no appreciable sway*, as computed by a first-order elastic frame analysis. These moments result from gravity loads. The 1977 through 1989 ACI Commentaries defined "no appreciable sway" as being a lateral deflection of $\Delta/\ell \le 1/1500$ of the story height at factored load levels. Gravity loads will cause small lateral deflections, except in the case of symmetrically loaded, symmetrical frames. The *sway moment*, M_s, was defined in the 1977 through 1989 Commentaries [12-22] as the factored end moment on a column due to loads which cause appreciable sway, calculated by a first-order elastic frame analysis. These moments result from lateral loads or in some cases from large unsymmetrical gravity loads or from gravity loads on highly unsymmetrical frames.

Treating the $P-\delta$ and $P-\Delta$ moments separately simplifies design. The nonsway moments frequently result from a series of pattern loads, as was discussed in Chapter 10. The pattern loads can lead to a moment envelope for the nonsway moments. The maximum end moments from the moment envelope are then combined with the magnified sway moments from a second-order analysis or from a sway moment-magnifier analysis.

12-6   CALCULATION OF MOMENTS IN SWAY FRAMES BY USING SECOND-ORDER ANALYSES

First-Order and Second-Order Analysis

A *first-order frame analysis* is one in which the effect of lateral deflections on bending moments, axial forces, and lateral deflections is ignored. The resulting moments and deflections are linearly related to the loads. In a *second-order frame analysis*, the effect of deflections on moments, and so on, is considered. The resulting moments and deflections include the effects of slenderness and hence are nonlinear with respect to the load. Because the moments are directly affected by the lateral deflections, as shown in (12-35), it is important that the stiffnesses, EI, used in the analysis be representative of the stage immediately prior to ultimate.

Second-Order Analysis

Load Level for the Analysis

In a second-order analysis, column moments and lateral frame deflections increase more rapidly than do loads. Thus, it is necessary to calculate the second-order effects at the factored load level. Either the load combinations, load factors, and strength-reduction factors from ACI Sections 9.2 and 9.3 or those from ACI 318-02 Appendix C can be used. We have chosen the first option for Examples (12-1), (12-2) and the second option for Examples (12-3) to (12-6), 318-02 option. Using the load factors from *ACI Appendix C*, and assuming that the sway deflections result from wind, the analysis would be carried out for the following two load combinations:

$$U = 0.75(1.4D + 1.7L) + (1.6W \text{ or } 1.0E) \qquad \text{(ACI Eq. C-2)}$$

If the directionality of the wind is not included, then the wind load factor changes to 1.3 and,

$$U = 1.05D + 1.275L + 1.3W$$

$$U = 0.9D + 1.3W \qquad \text{(ACI Eq. C-3)}$$

As will be discussed later, it is also necessary to check for sidesway buckling under the load combination that is limited to gravity loads:

$$U = 1.4D + 1.7L \qquad \text{(ACI Eq. C-1)}$$

Alternatively, the ASCE 7 load combinations given in ACI Section 9.2.1 could be used.

Lateral Stiffness-Reduction Factor

The term 0.75 in the denominator of (12-23) is the lateral stiffness-reduction factor, ϕ_K, which accounts for variability in the critical load, P_c, and for variability introduced by the assumptions in the moment-magnifier calculation. This factor leads to an increase in the magnified moments. A similar but higher stiffness-reduction factor appears in a second-order analysis. Two things combine to allow the use of a value of ϕ_K larger than 0.75 when a second-order analysis is carried out. First, the modulus of elasticity, E_c, used in the frame analysis is based on the specified strength, f'_c, even though the deflections are a function of the E_c that applies to the mean concrete strength, that is, a concrete strength which is 600 to 1400 psi higher than f'_c. Second, the second-order analysis is a better representation of the behavior of the frame than the sway magnifier given by (12-41). The moments of inertia given in ACI Section 10.11.1 have been multiplied by 0.875, which, when combined with the underestimate of E_c, lead to an overestimation of the second-order deflections on the order of from 20 to 25 percent, corresponding to an implicit value for ϕ_K of 0.80 to 0.85 in design based on second-order analyses.

Stiffnesses of the Members

Ultimate Limit State. The stiffnesses appropriate for strength calculations must estimate the lateral deflections accurately at the factored load level. They must be simple to apply, because a frame consists of many cross sections, with differing reinforcement ratios and differing degrees of cracking. Furthermore, the reinforcement amounts and distributions are not known at the time the analysis is carried out. Using studies of the flexural stiffness of beams with cracked and uncracked regions, MacGregor and Hage [12-20] recommend that the beam stiffnesses be taken as $0.4E_cI_g$ when carrying out a second-order analysis. In ACI Section 10.11.1, this value has been multiplied by a stiffness-reduction factor of 0.875, giving $I = 0.35I_g$.

Two levels of behavior must be distinguished in selecting the *EI* of columns. The lateral deflections of the frame are influenced by the stiffness of all the members in the frame and by the variable degree of cracking of these members. Thus, the *EI* used in the frame analysis should be an average value. On the other hand, in designing an individual column in a frame in accordance with (12-23), the *EI* used in calculating δ_{ns} must be for that column. This *EI* must reflect the greater chance that a particular column will be more cracked, or weaker, than the overall average; hence, this EI will tend to be smaller than the average EI for all the columns acting together. Reference [12-20] recommends the use of $EI = 0.8E_cI_g$ in carrying out second-order analyses of frames. ACI Section 10.11.1 gives this value multiplied by 0.875, or $EI = 0.70E_cI_g$ for this purpose. On the other hand, in calculating the moment magnifiers δ_{ns} and δ_s from (12-23) and (12-41) (ACI Eqs. 10-9 and 10-19), EI must be taken as given by (12-15) or (12-16).

The value of EI for shear walls may be taken equal to the value for beams in those parts of the structure where the wall is cracked by flexure or shear and equal to the value for columns where the wall is uncracked. If the factored moments and shears from an analysis based on $EI = 0.70E_cI_g$ for the walls indicate that a portion of the wall will crack due to stresses reaching the modulus of rupture of the wall concrete, the analysis should be repeated with $EI = 0.35E_cI_g$ for the cracked parts of the wall.

Serviceability Limit State.   The moments of inertia given in ACI Section 10.11.1 are for the ultimate limit state. At service loads, the members are cracked less than they are at ultimate. In computing deflections or vibrations at service loads, the values of I should be representative of the degree of cracking at service loads. The Commentary R10.11.1 suggests that I at service loads may be taken as $1/0.70 = 1.43$ times the values given in ACI Section 10.11.1. This reflects the low degree of cracking of columns and walls at service loads.

Foundation Rotations

The rotations of foundations subjected to column end moments reduce the fixity at the foundations and lead to larger sway deflections. These are particularly significant in the case of shear walls or large columns, which resist a major portion of the lateral loads. The effects of foundation rotations can be included in the analysis by modeling each foundation as an equivalent beam having the stiffness given by (12-31) and (12-32).

Effect of Sustained Loads

Loads causing appreciable sidesway are generally short-duration loads, such as wind or earthquake, and, as a result, do not cause creep deflections. In the unlikely event that sustained lateral loads act on an unbraced structure, the EI values used in the frame analysis should be reduced. ACI Section 10.11.1 states that in such a case, I shall be divided by $(1 + \beta_d)$, where, for this case, β_d is defined in ACI Section 10.0, definition (b) as

$$\beta_d = \frac{\text{maximum factored sustained shear within a story}}{\text{total factored shear in that story}} \qquad (12\text{-}34b)$$

Methods of Second-Order Analysis

Computer programs that carry out second-order analyses are widely available. The principles of such an analysis are presented in the following paragraphs. Methods of second-order analysis are reviewed in [12-20] and [12-23].

Iterative $P-\Delta$ Analysis

When a frame is displaced sideways under the action of lateral and vertical loads, as shown in Figs. 12-30 and 12-31, the column end moments must equilibrate the lateral loads and a moment equal to $(\Sigma P)\Delta$; that is,

$$\Sigma (M_{\text{top}} + M_{btm}) = V\ell_c + \Sigma P\Delta \qquad (12\text{-}35)$$

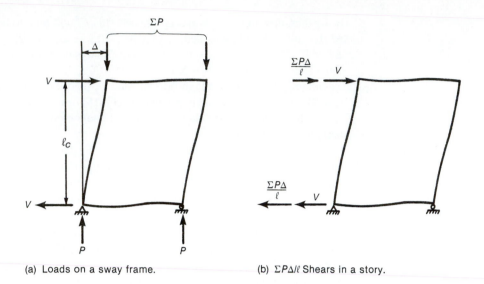

(a) Loads on a sway frame.

(b) $\Sigma P\Delta/\ell$ Shears in a story.

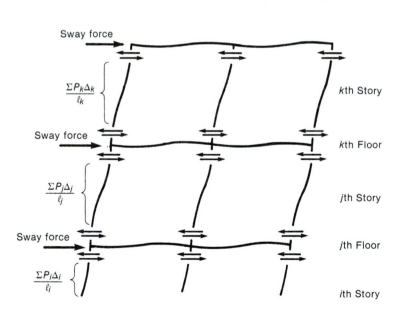

Fig. 12-31
Iterative $P-\Delta$ analyses.

(c) Calculation of sway forces.

where Δ is the lateral deflection of the top of the story relative to the bottom of the story. The moment $\Sigma P\Delta$ in a given story can be represented by shear forces, $(\Sigma P)\,\Delta/\ell_c$, where ℓ_c is the story height, as shown in Fig. 12-31b. These shears give an overturning moment of $(\Sigma P\delta/\ell_c)/(\ell_c) = (\Sigma P)\Delta$. Figure 12-31c shows the story shears in two different stories. The algebraic sum of the story shears from the columns above and below a given floor gives rise to a *sway force* acting on that floor. At the jth floor, the sway force is

$$\text{sway force}_j = \frac{(\Sigma P_i)\Delta_i}{\ell_i} - \frac{(\Sigma P_j)\Delta_j}{\ell_j} \tag{12-37}$$

where a positive $P\Delta/\ell$ moment and a positive sway force both correspond to forces that would overturn the structure in the same direction as the wind load would. The sway forces are added to the applied lateral loads at each floor level, and the structure is reanalyzed, giving new lateral deflections and larger column moments. If the deflections increase by more than about 2.5 percent, new $\Sigma P\Delta/\ell$ forces and sway forces are computed and the structure is reanalyzed for the sum of the applied lateral loads and the new sway forces. This process is continued until convergence is obtained.

At discontinuities in the stiffness of the building or discontinuities in the applied loads, the sway force may be negative. In such a case, it acts in the direction opposite that shown in Fig. 12-31c.

Ideally, one correction should be made to this process. Although the $P-\Delta$ moment diagram for a given column is the same shape as the deflected column, as shown in Fig. 12-30c, the moment diagram due to the $P\Delta/\ell$ shears is a straight-line diagram, similar to the $H\ell$ moments shown in Fig. 12-30b, compared to the curved-moment diagram. As a result, the area of the real $P-\Delta$ moment diagram is larger than that of the straight-line moment diagram. It can be shown by the moment–area theorems that the deflections due to the real diagram will be larger than those due to the $P\Delta/\ell$ shears. The increase in deflection varies from zero for a very stiff column with very flexible restraining beams, to 22 percent for a column that is fully fixed against rotation at each end. A reasonable average value is about 15 percent. The increased deflection can be accounted for by taking the story shears as $\gamma\Sigma P\Delta/\ell$, where γ is a *flexibility factor* [12-23] that ranges from 1.0 to 1.22 and can be taken equal to 1.15 for practical frames. Unfortunately, most commercially available second-order analysis programs do not include this correction. For this reason, we shall omit γ also.

Direct $P-\Delta$ Analysis for Sway Frames

The iterative calculation procedure described in the preceding section can be described mathematically as an infinite series. The sum of the terms in this series gives the second-order deflection

$$\Delta = \frac{\Delta_0}{1 - \gamma(\Sigma P_u)\Delta_0/(V_u\ell_c)} \tag{12-38}$$

where

V_u = shear in the story due to wind loads acting on the frame above the story in question

ℓ_c = story height

ΣP_u = the total axial load in all the columns in the story

$\gamma \simeq 1.15$

Δ_0 = first-order deflection of the top of a story relative to the bottom of the story. *due to the story shear,* V_u

Δ = second-order deflection

Both Δ_0 and Δ refer to the lateral deflection of the top of the story relative to the bottom of the story.

Since the moments in the frame are directly proportional to the deflections, the second-order moments are

$$M = \delta_s M_s = \frac{M_0}{1 - \gamma(\Sigma P_u)\Delta_0/(V_u\ell_c)} \tag{12-39}$$

where M_0 and M are the first- and second-order moments, respectively.

ACI Section 10.11.4.2 defines the stability index for a story as

$$Q = \frac{\Sigma P_u\Delta_0}{V_u\ell_c} \tag{12-33}$$

(ACI Eq. 10-7)

Substituting this into (12-45) and omitting the flexibility factor γ gives

$$\delta_s M_s = \frac{M_s}{1 - Q} \geq M_s \qquad \text{(12-40)}$$

$$\text{(ACI Eq. 10-18)}$$

Reference [12-23] recommends that the use of (12-40) be limited to cases where δ_s is less than or equal to 1.5, because it becomes less accurate for higher values. This corresponds to $Q \leq 1/3$. For this reason, ACI Section 10.13.4.2 limits the use of (12-40) to $Q \leq 1/3$. Example 12-6 illustrates the impact of (12-40) on magnified moments in a frame.

12-7   DESIGN OF COLUMNS IN SWAY FRAMES

Overview

In the 1995 ACI Code, the sections dealing with the design of slender columns were completely rewritten and rearranged. Subsequent codes have made few changes in this. ACI Section 10.10 is the overall umbrella section. It sets requirements for an accurate slenderness analysis (ACI Section 10.10.1) and allows the use of the more approximate moment-magnifier analysis described in ACI Sections 10.11, 10.12, and 10.13.

ACI Section 10.11, "Magnified Moments—General," gives general requirements for the design of slender columns in both nonsway and sway frames. ACI Section 10.11.4 gives methods of distinguishing between nonsway and sway columns. If a column is in a sway frame, design involves ACI Sections 10.11 and 10.13, "Magnified Moments—Sway Frames." The ACI Code design procedure for slender columns in sway frames consists of five steps:

1. **The unmagnified moments, M_{ns}, due to loads not causing appreciable sway are computed.** This is done via a regular first-order elastic-frame analysis, using the member stiffnesses given in ACI Section 10.11.1. For the load combination

$$U = 0.75(1.4D + 1.7L) + (1.6W \text{ or } 1.0E)$$

which reduces to

$$= 1.05D + 1.275L + 1.6W \qquad \text{(ACI Eq. C-2)}$$

the M_{ns} moments would result from $1.05D + 1.275L$. Alternatively, the load factors and load combinations in Sections 9.2 and 9.3 of the 2002 ACI Code can be used. The load factor on wind load in ACI Eq. C-2 is 1.6 except that it drops to 1.3 if the wind load calculation did not allow for the directionality of the wind.

2. **The magnified sway moments, $\delta_s M_s$, are computed.** Three alternative methods of carrying out this computation are given in ACI Section 10.13.4. For the load combination quoted in step 1, the M_s moments result from 1.3 or $1.6W$.

3. **The magnified sway moments, $\delta_s M_s$, are added to the unmagnified nonsway moments, M_{ns}.** This is done at each end of each column (ACI Section 10.13.3).

4. **A check is made whether the maximum moment occurs between the ends of the column.** Normally, the maximum moment in the column will be the $P - \Delta$ moment at one end, and the column is designed for this moment. However, if the axial loads on the column are high and the slenderness exceeds the limit given in ACI Section 11.13.5, it is necessary to check whether the $P - \delta$ moment at some section between the ends of the column exceeds the maximum end moment. This is done by using (12-23), the braced-frame magnifier from ACI Section 10.12.3, because this equation (ACI Eq. (10-9)) is a check on whether the $P - \delta$ moment at any point between the ends of a column exceeds the larger end moment of the column. This rarely is a problem.

5.   **A check is made whether sidesway buckling can occur under gravity loads alone** by using ACI Section 11.13.6. Again, this rarely is a problem.

Each of these steps will be discussed in the following sections, followed by examples. The design process is developed and discussed in [12-24].

Computation of $\delta_s M_s$

ACI Section 10.13.4 allows designers to compute the magnified sway moments, $\delta_s M_s$, in frames in one of three ways. In order of decreasing accuracy, these are (i) second-order analyses, (ii) a direct solution of the iterative $P-\Delta$ analyses, and (iii) the sway-moment magnifier, δ_s, used in ACI Codes since 1971.

1. Computation of $\delta_s M_s$ by Using Second-Order Analyses

ACI Section 10.13.4.1 allows the use of second-order analyses to compute $\delta_s M_s$. Second-order analyses were discussed in Section 12-6. If torsional displacements of the frame are significant, a three-dimensional second-order analysis should be used.

2. Computation of $\delta_s M_s$ by Using Direct $P-\Delta$ Analysis

ACI Section 10.13.4.2 permits the use of a direct calculation of $P-\Delta$ moments via an equation similar to (12-44). In ACI Section 10.13.4.2, this is written as

$$\delta_s M_s = \frac{M_s}{1 - Q} \geq M_s \qquad\qquad (12\text{-}40)$$

$$(\text{ACI Eq. 10-17})$$

where

$$Q = \frac{\Sigma P_u \Delta_0}{V_u \ell_c} \qquad\qquad (12\text{-}33)$$

$$(\text{ACI Eq. 10-6})$$

Although there is no easy way to incorporate torsional effects into the calculation of $\delta_s M_s$ by ACI Section 10.13.4.2 or Section 10.13.4.3, Section 12-9 discusses the extension of the Q method to second-order torsion.

3. Computation of $\delta_s M_s$ by Using Sway-Frame Moment Magnifier

ACI Section 10.13.4.3 allows the use of the traditional sway-frame moment magnifier for computing the magnified sway moments

$$\delta_s M_s = \frac{M_s}{1 - \Sigma P_u / (0.75\,\Sigma P_c)} \geq M_s \qquad\qquad (12\text{-}41)$$

$$(\text{ACI Eq. 10-18})$$

Here, ΣP_u and ΣP_c refer to the sums of the axial loads and critical loads, respectively, for all the columns in the story being analyzed. In this case, the values of P_c are calculated by using the effective lengths, $k\ell_u$, evaluated for columns in a sway frame, with β_d defined as

$$\beta_d = \frac{\text{maximum factored sustained shear in the story}}{\text{total factored shear in the story}} \qquad\qquad (12\text{-}34\text{b})$$

In most sway frames, the story shear is due to wind or seismic loads and hence is not sustained, resulting in $\beta_d = 0$. The use of the summation terms in (12-41) accounts for the fact that sway instability involves all the columns and bracing members in the story. (See Fig. 12-30.) The format including the summations was first presented in [12-1].

If $(1 - \Sigma P_u / \phi_m \Sigma P_c)$ is negative, the load on the frame, ΣP_u, exceeds the buckling load for the story, ΣP_c, indicating that the frame is unstable. A stiffer frame is required.

Moments at the Ends of the Columns

The unmagnified nonsway moments, M_{ns}, are added to the magnified sway moments, $\delta_s M_s$, at each end of each column:

$$M_1 = M_{1ns} + \delta_s M_{1s} \tag{12-42a}$$

$$\text{(ACI Eq. 10-15)}$$

$$M_2 = M_{2ns} + \delta_s M_{2s} \tag{12-42b}$$

$$\text{(ACI Eq. 10-16)}$$

The addition is carried out for the moments at the top and bottom of each column. The larger absolute sum of the resulting end moments for a given column is called M_2, and the smaller is called M_1. By definition, M_2 is always taken as positive, and M_1 is taken as negative if the column is bent in double curvature.

Maximum Moment between the Ends of the Column

In most columns in sway frames, the maximum moment will occur at one end of the column and will have the value given by (12-48a) and (12-48b). *Occasionally, for very slender, highly loaded columns*, the deflections of the column can cause the maximum column $P-\delta$ moment to exceed the $P-\Delta$ moment at one or both ends of the column, in a fashion analogous to the moments in the braced frame shown in Fig. 12-22c. At load stage 1 in this figure, the maximum moment is at the top of the column. At load stage 2, the moment at the upper third point of the column exceeds that at the top.

Because this is a rare occurrence, ACI Section 10.13.5 uses (12-43) to screen the columns, discarding those that will not have $P-\delta$ moments between the ends of the column which exceed the $P-\Delta$ moments at the ends. If

$$\frac{\ell_u}{r} > \frac{35}{\sqrt{\dfrac{P_u}{f'_c A_g}}} \tag{12-43}$$

$$\text{(ACI Eq. 10-19)}$$

then the moment at some point between the ends of the column may exceed the larger end moment [12-24].

If ℓ_u/r exceeds the value given by (12-43), there is a chance that the maximum moment on the column will exceed the larger end moment, M_2. This would occur if M_c, computed from (12-23), was larger than the end moments M_1 and M_2 from (12-42a) and (12-42b) (ACI Eqs. (10-15) and (10-16)). It is necessary to check whether the moment M_c exceeds the larger end moment, M_2. Because M_c is a $P-\delta$ moment, it is computed from ACI Eq. (10-22). If $M_c < M_2$, the maximum moment is at the end of the column and is equal to M_2. If $M_c \geq M_2$, the maximum moment occurs between the ends of the column and is equal to M_c.

Sidesway Buckling Under Gravity Loads

Another very rare event is the classical case of sidesway buckling under gravity loads alone. ACI Section 10.13.6 requires a check for this possibility, using the load combination that gives the largest gravity load, $U = 1.4D + 1.7L$. Since there are three methods to calculate $\delta_s M_s$, three corresponding methods are given to check sidesway buckling.

(a) If $\delta_s M_s$ has been computed from second-order analyses, ACI Section 10.13.6(a) limits the ratio of second-order deflections to first-order deflections to 2.5. This requires a special second-order analysis based on the EI values from ACI Section 10.11.1 divided by $(1 + \beta_d)$, with β_d from (12-34a) [ACI Section 10.0 definition (a)], with the gravity load from $1.4D + 1.7L$ or from $1.2D + 1.6L$, and with any arbitrary lateral load, such as a hypothetical lateral load of 0.005 times the factored gravity load in that story. Since we are checking effect of the gravity loads on the *ratio* of lateral deflections, it is not necessary to define a special load for this purpose. Thus, one could use a set of lateral loads already considered in the design, such as the wind loads plus two levels of gravity loads.

(b) If $\delta_s M_s$ has been computed by using the direct $P-\Delta$ analysis, it is not necessary to check for sidesway buckling under gravity loads. The upper limit of $\delta_s = 1.5$ in ACI Section 10.13.4.2 corresponds to $Q \leq 1/3$. If this is imposed, the upper limit of $Q = 0.60$ in ACI Section 10.13.6 cannot be reached, because it exceeds the $Q \leq 0.33$ limit from ACI Section 10.13.4.2.

(c) If $\delta_s M_s$ has been computed by using the sway-moment magnifier, ACI Section 10.13.6c limits δ_s from (12-41) (ACI Eq. (10-18)) to not more than 2.5, where δ_s is based on ΣP_u from $1.4D + 1.7L$ and ΣP_c is based on effective lengths from ACI Section 10.13.1 and on the EI values from (12-15) and (12-16) (ACI Eqs. (10-11) and (10-12)), with β_d from (12-34a) [ACI Section 10.0 definition (c)].

Minimum Moment

The ACI Code specifies a minimum moment $M_{2,\min}$ to be considered in the design of columns in nonsway frames, but not for columns in sway frames. This will be a problem only for the pre 2002 load combination $U = 1.4D + 1.7L$ acting on a sway frame, since this load combination does not involve $\delta_s M_s$. For this load combination, we shall design for the larger of M_2 and $M_{2,\min}$.

EXAMPLE 12-3   Design of the Columns in a Sway Frame

Figure 12-32 shows the plan of the main floor and a section through a five-story building. The building is clad with nonstructural precast panels. There are no structural walls or other bracing. The beams in the North–South direction are all 18 in. wide, with an overall depth of 30 in. The floor slabs are 7 in. thick. Design an interior and an exterior column in the ground-floor level of the frame along column line 3 for dead load, live load, and North–South wind forces. Use $f'_c = 4000$ psi and $f_y = 60,000$ psi.

1. **Make a preliminary selection of the column size.** A preliminary calculation of the gravity loads in the first-story columns, based on the respective tributary areas on the roof and each floor above the main floor, gives the following unfactored dead and unreduced live loads in the columns between the ground floor and the second floors:

Exterior column: Dead load = 176 kips
                   Live load = 10.5 kips from the roof (snow) and 104 kips of floor load

Interior column: Dead load = 381 kips

                   Live load = 22.1 kips from the roof (snow) and 223 kips floor load

Prior to 1998, ASCE 7 [12-25] allowed the floor live loads to be reduced as a function of the influence area, A_I, of floor supported by the columns. Snow loads are not reduced. In ASCE 7-98 the influence area A_I was replaced by a multiple of the tributary area. For a column, the influence area is four times the tributary area A_T of the column. (See Section 2-8, Fig. 2-10.) Thus, for an exterior code

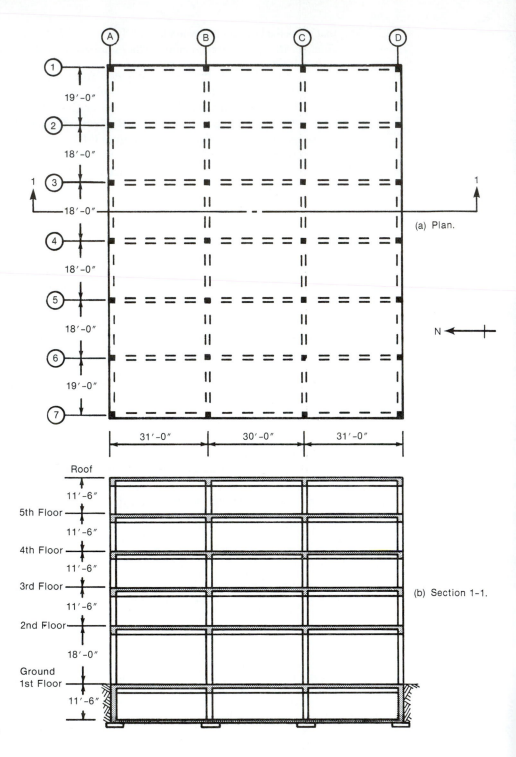

(a) Plan.

N ←——+——

31'-0"   30'-0"   31'-0"

Roof
11'-6"
5th Floor
11'-6"
4th Floor
11'-6"
3rd Floor
11'-6"
2nd Floor
18'-0"
Ground
1st Floor
11'-6"

(b) Section 1-1.

19'-0"
18'-0"
18'-0"
18'-0"
18'-0"
19'-0"

Fig. 12-32
Sway frame—Example 12-3.

$A_I = 4A_T$ is calculated as $A_I = 4(18 \text{ ft} \times 30.25/2 \text{ ft}) = 1089 \text{ ft}^2$ per floor $\times$ 4 floors $= 4356 \text{ ft}^2$. The live load due to use and occupancy can be multiplied by

$$\frac{L}{L_0} = 0.25 + \frac{15}{\sqrt{A_I}} \tag{12-9}$$
$$= 0.477$$

but not by less than 0.50 for columns supporting one floor, nor by less than 0.40 for columns supporting two or more floors. For the interior column, the influence area for four floors is $A_I = 4(18 \text{ ft} \times 30.25 \text{ ft}) \times 4 \text{ floors} = 8712 \text{ ft}^2$, and $L/L_0 = 0.411$. Thus, the reduced live loads and factored axial loads in the first-story columns are as follows:

Exterior column: Reduced live load $= 10.5 \text{ kips} + 0.477 \times 104 \text{ kips} = 60.1 \text{ kips}$
                    Factored load $= 1.4 \times 176 \text{ kips} + 1.7 \times 60.1 \text{ kips} = 349 \text{ kips}$

Interior column: Reduced live load $= 22.1 \text{ kips} + 0.411 \times 223 \text{ kips} = 113.8 \text{ kips}$
                    Factored load $= 1.4 \times 381 + 1.7 \times 113.8 = 727 \text{ kips}$

For the exterior column from (11-21a) for $\rho_t = 0.015$,

$$A_{g(\text{trial})} \geq \frac{P_u}{0.40(f_c' + f_y\rho_t)}$$
$$= \frac{349 \text{ kips}}{0.40(4 \text{ ksi} + 60 \text{ ksi} \times 0.015)}$$
$$= 178 \text{ in.}^2 \text{ or } 13.3 \text{ in. by } 13.3 \text{ in.}$$

We shall try an 18 in. $\times$ 18 in. column. This is larger than $A_{g(\text{trial})}$, to allow both for slenderness and for the moments in the column in load cases involving wind.

For the interior column, $A_{g(\text{trial})} = 371 \text{ in.}^2$, or 19.2 in. $\times$ 19.2 in. We will try 18 in. $\times$ 18 in. interior columns, so that all columns will be the same size. This may necessitate an increase in the steel ratio, ρ_t. Assume that the columns are 18 in. $\times$ 18 in. in the basement and second floor also.

FACTORED LOAD COMBINATIONS

The load factors, load combinations, and strength-reduction factors from ACI 318-02 Appendix C will be used. Three different load cases will be considered:

Case 1: gravity and wind loads, $U = 0.75(1.4D + 1.7L) + (1.6W \text{ or } 1.0E)$
                                    where $1.6W$ drops to $1.3W$ for wind loads that did not include a directionality factor. We shall assume it did not. This gives:
                      $U = 1.05D + 1.275L \pm 1.3W$,
                                      where the wind can blow from the North or the South
Case 2: dead and wind loads, $U = 0.9D \pm 1.3W$
Case 3: gravity loads only, $U = 1.4D + 1.7L$

Since this is a symmetrical frame, the gravity loads will not cause appreciable sidesway. Thus, the dead and live loads give rise to the M_{ns} moments, while the wind loads cause the M_s moments.

For clarity, all the calculations for case 1 will be presented before starting case 2, and so on.

LOAD CASE 1: $1.05D + 1.275L \pm 1.3W$

2. Is the frame sway or nonsway? ACI Section 10.11.4 defines a story in a frame as being a nonsway story if

$$Q = \frac{\Sigma P_u \Delta_0}{V_u \ell_c} \leq 0.05 \tag{12-33}$$
$$\text{(ACI Eq. 10-6)}$$

where, for the first story,

ΣP_u = total reduced factored axial load in all 28 columns in the floor = 10,565 kips

Δ_0 = 0.429 in. (from a first-order elastic analysis)

V_u = total factored shear in the first story in all seven frames = 188 kips

ℓ_c = story height center to center of joints = 18 ft × 12 = 216 in.

$$Q = \frac{10,565 \text{ kips} \times 0.429 \text{ in.}}{188 \text{ kips} \times 216 \text{ in.}} = 0.112$$

Thus, the first story of the frame is clearly a sway story. We shall treat the entire frame as a sway frame.

3.   Are the columns slender? From ACI Section 10.11.4.2, a column in an unbraced frame is slender if $k\ell_u/r \geq 22$. However, k is not known at this stage. Since k will not normally be less than 1.2, we shall use this value in our check. For the main floor,

$$\ell_u = 18.0 \text{ ft} - 2.5 \text{ ft}$$
$$= 15.5 \text{ ft} \times 12 = 186 \text{ in.}$$

From ACI Section 10.11.3, $r = 0.3h$. For 18-in.-square columns,

$$\frac{k\ell_u}{r} = \frac{1.2 \times 186}{0.3 \times 18} = 41.3$$

Thus, the columns are slender. If this calculation gives $k\ell_u/r$ values close to 22, it should be repeated with a better estimate of k.

4.   Compute the factored axial loads P_u and the M_{ns} moments from a first-order frame analysis. For the main floor columns, the values are as follows:

	Exterior Columns	Interior Columns
Factored axial force, P_u (kips)		
1.05 Dead load	184.8	399.6
1.275 Reduced live load	88.5	144.8
Factored M_{ns} moments (ft-kips)		
At top	−103.1	12.61
At bottom	−108.5	16.49

The M_s moments will be computed in the second-order analysis in step 5. The axial loads due to live load have been reduced as a function of influence area, but the live-load M_{ns} moments have not. These moments result primarily from loading on the beams and slabs in one floor, and L/L_0 for the beams is much closer to unity than for the columns.

5.   Compute $\delta_s M_s$ and M_c, using a second-order analysis. The $\delta_s M_s$ moments come directly from a second-order analysis for the loads, causing appreciable sway. A commercial second-order analysis program would normally be used. To illustrate the process, we will use an iterative $P-\Delta$ analysis.

5.1 Iteration No. 1

(a)   Compute ΣP_u, $\Sigma P_u \Delta_0/\ell_c$, and the sway forces. The total load in the columns supporting the second floor is 10,565 kips. For the columns supporting the third floor, ΣP_u = 8,589 kips.

The horizontal deflections of the ground, second, and third floors, respectively, are 0 in., 0.429 in., and 0.556 in.

The $P_u \Delta_0/\ell_c$ shears are as follows:

$$\text{First story:} \quad \frac{\Sigma P_u \Delta_0}{\ell_c} = \frac{10,565 \text{ kips} \times (0.429 - 0) \text{ in.}}{(18 \times 12) \text{ in.}}$$
$$= 20.98 \text{ kips}$$

$$\text{Second story:} \quad \frac{\Sigma P_u \Delta_0}{\ell_c} = \frac{8589 \text{ kips} \times (0.556 - 0.429) \text{ in.}}{11.5 \times 12 \text{ in.}}$$
$$= 7.90 \text{ kips}$$

Sway force at the second floor level = (20.98 kips − 7.90 kips) = 13.08 kips.

The calculation of the sway forces for the entire building can be tabulated as follows:

TABLE 12-3   Calculation of Sway Forces—Iteration No. 1

Floor	Story	ΣP_u (kips)	Floor Δ (in.)	Story Δ (in.)	$\dfrac{\Sigma P_u \Delta_0}{\ell_c}$ (kips)	Sway Force (kips)
Roof			0.721			0.356
	Fifth	2,047		0.024	0.356	
Fifth			0.697			1.422
	Fourth	4,461		0.055	1.778	
Fourth			0.642			2.300
	Third	6,544		0.086	4.078	
Third			0.556			3.826
	Second	8,589		0.127	7.904	
Second			0.429			13.076
	First	10,565		0.429	20.98	
Ground						

This frame is now reanalyzed for $1.3W$ plus the sway forces. The unfactored wind loads at each floor are as follows: roof, 25.6 kips; third, fourth, and fifth floors, 37.9 kips each; second floor, 48.7 kips. Following are the total factored wind forces plus sway forces at each floor:

Roof: $1.3 \times 25.6 + 0.356 = 33.6$ kips
Fifth floor: $1.3 \times 37.9 + 1.422 = 50.7$ kips
Fourth floor: $1.3 \times 37.9 + 2.300 = 51.6$ kips
Third floor: $1.3 \times 37.9 + 3.826 = 53.1$ kips
Second floor: $1.3 \times 48.7 + 13.08 = 76.4$ kips

These loads are resisted by the seven frames in the North–South direction.

(b)   **Check convergence.** The deflections from two cycles of iteration are as follows:

	Δ—First Cycle (in.)	Δ—Second Cycle (in.)	Second/First
Roof	0.721	0.769	1.07
Second floor	0.429	0.466	1.09

This clearly has not converged.

5.2 Iteration No. 2

(c)   **Calculate sway forces and loads based on the deflections from the second cycle.**

TABLE 12-4   Calculation of Sway Forces—Load case, Iteration No. 2

Floor	Story	ΣP_u (kips)	Floor Δ (in.)	Story Δ (in.)	$\dfrac{\Sigma P_u \Delta_0}{\ell_c}$ (kips)	Sway Force (kips)
Roof			0.769			0.371
	Fifth	2,047		0.025	0.371	
Fifth			0.744			1.407
	Fourth	4,461		0.055	1.778	
Fourth			0.689			2.442
	Third	6,544		0.089	4.220	
Third			0.600			4.120
	Second	8,589		0.134	8.340	
Second			0.466			14.45
	First	10,565		0.466	22.79	
Ground						

Reanalyze the structure for the following horizontal loads:

Roof:	$1.3 \times 25.6 + 0.371 = 33.7$ kips
Fifth floor:	$1.3 \times 37.9 + 1.407 = 50.7$ kips
Fourth floor:	$1.3 \times 37.9 + 2.442 = 51.7$ kips
Third floor:	$1.3 \times 37.9 + 4.120 = 53.4$ kips
Second floor:	$1.3 \times 48.7 + 14.45 = 77.8$ kips

(d) **Check convergence for the third cycle.**

	Δ—Second Cycle (in.)	Δ—Third Cycle (in.)	Third/Second
Roof	0.769	0.773	1.005
Second floor	0.466	0.469	1.006

This has converged adequately.

6. Compute column end moments. (a) Compute M_1 and M_2. The column end moments from the third cycle of iteration are as follows:

Exterior columns—first story: From the third cycle of the second-order analysis, we have

Top:	$\delta_s M_s = 73.9$ ft-kips
Bottom:	$= 79.5$ ft-kips

From (12-43a) and (12-43b) (ACI Eqs. (10-15) and (10-16)), with M_{ns} from step 4,

$$M_{\text{end}} = M_{ns} + \delta_s M_s$$

Because the wind can blow from either the North or the South, we shall take this as

$$M_{\text{end}} = M_{ns} \pm \delta_s M_s$$

Top: $M_{\text{top}} = -103.1 - 73.9 = -177$ ft-kips. This is the smaller end moment; therefore, this is M_1.

Bottom: $M_{\text{bottom}} = -108.5 - 79.5 = -188$ ft-kips. This is M_2.

Interior columns—first story:

Top: $\delta_s M_s = 91.1$ ft-kips

$M_{\text{top}} = 12.6 + 91.1 = 103.7$ ft-kips. This is the smaller end moment; therefore, this is M_1.

Bottom: $\delta_s M_s = 94.1$ ft-kips

$M_{\text{bottom}} = 16.5 + 94.1 = 110.6$ ft-kips. This is M_2.

7. Check whether the maximum moment is between the ends of the column. ACI Section 10.13.5 requires a check of whether the maximum moment occurs away from the ends of the column if

$$\frac{\ell_u}{r} > \frac{35}{\sqrt{\dfrac{P_u}{f'_c A_g}}} \qquad (12\text{-}43)$$

$$\text{(ACI Eq. 10-19)}$$

where $\ell_u = 18 \times 12$ in. $- 30$ in. $= 186$ in. and $r = 0.30 \times 18$ in. $= 5.4$ in.

$$\frac{\ell_u}{r} = \frac{186 \text{ in.}}{5.4 \text{ in.}} = 34.4$$

$$f'_c = 4 \text{ ksi}$$

$$A_g = 324 \text{ in.}^2$$

For the exterior column, $P_u = 273.3$ kips, and

$$\frac{35}{\sqrt{\dfrac{P_u}{f'_c A_g}}} = \frac{35}{\sqrt{\dfrac{273.3 \text{ kips}}{4 \text{ ksi} \times 324 \text{ in.}^2}}}$$

$$= 76.2$$

Since $\ell_u/r = 34.4$ is less than 76.2, the maximum moment in the exterior column is at one end of the column and is $M_2 = 188$ ft-kips.

For the interior column, $P_u = 544.4$ kips and

$$\frac{35}{\sqrt{\dfrac{544.4}{4 \times 324}}} = 54.0$$

Because $\ell_u/r = 34.4$ is less than 54.0, the maximum moment in the interior column is $M_2 = 110.6$ ft-kips.

If ℓ_u/r were greater than either of these limits, it would be necessary to compute the maximum moment along the length of that column. According to ACI Section 10.3.5, this is done by using (12-23) (ACI Eq. 10-9), with M_1 and M_2 as just computed, β_d as defined for the load combination under consideration, and k as defined in ACI Section 10.12.1. Although both columns passed the test, we shall go through this calculation for the interior column to illustrate the procedure. These calculations are italicized because they are not necessary when ℓ_u/r is less than the limit from (12-48).

The maximum moment in the column is given by

$$M_c = \delta_{ns} M_2 \tag{12-22}$$
$$\text{(ACI Eq. 10-8)}$$

where

$$\delta_{ns} = \frac{C_m}{1 - P_u/0.75P_c} \geq 1.0 \tag{12-23}$$
$$\text{(ACI Eq. 10-9)}$$

$$C_m = 0.6 + 0.4\left(\frac{M_1}{M_2}\right) \geq 0.4 \tag{12-14}$$

$$= 0.6 + 0.4\left(\frac{-103.7}{110.6}\right) \geq 0.4 \qquad \text{(ACI Eq. 10-13)}$$

$$= 0.225 \geq 0.4$$

$$= 0.4$$

$$P_c = \frac{\pi^2 EI}{(k\ell_u)^2} \tag{12-24}$$
$$\text{(ACI Eq. 10-10)}$$

$$EI = \frac{0.40 E_c I_g}{1 + \beta_d} \tag{12-16}$$
$$\text{(ACI Eq. 10-12)}$$

This is a sway frame, and definition (b) for β_d in ACI Section 10.0 applies. We thus have

$$\beta_d = \frac{maximum\ factored\ sustained\ shear\ within\ a\ story}{total\ factored\ shear\ in\ the\ story} \tag{12-34b}$$

where the shear is due to rapidly applied wind loads and the sustained shear is thus zero, giving

$$\beta_d = 0.0$$

and

$$EI = \frac{0.40 \times \left(57,000\sqrt{4000}\right) \times 18^4/12}{1 + 0.0}$$

$$= 12.61 \times 10^9 \ lb\text{-}in^2$$

ACI Section 10.13.5 says that k shall be from ACI Section 10.12.1, which gives $k \leq 1.0$. We shall assume $k = 1.0$ as a first trial. If $\delta_{ns} > 1.0$, we shall compute k and repeat the calculation. We have

$$P_c = \frac{\pi^2 \times 12.61 \times 10^6 \; lb\text{-}in.^2}{(1.0 \times 186 \; in.)^2}$$

$$= 3597 \; kips$$

$$\delta_{ns} = \frac{0.40}{1 - 544.4/0.75 \times 3597} \geq 1.0$$

$$= 0.501 \geq 1.0$$

$$= 1.0$$

Since $\delta_{ns} = 1.0$, the maximum moment in the column is $M_2 = 110.6 \; ft\text{-}kips$.

This bears out the results of the test carried out in step 7, which showed that the maximum moment would be at one end of the column. As stated earlier, all calculations in italicized print need not be done if the columns satisfy (12-44) (ACI Eq. (10-19)).

Summary for Load Case 1

Exterior columns: $P_u = 273$ kips, $M_c = 188$ ft-kips
Interior columns: $P_u = 544$ kips, $M_c = 110.6$ ft-kips

The column cross section will be chosen when all load combinations have been completed.

LOAD CASE 2: $U = 0.9D \pm 1.3W$

Steps 4, 5, and 6 will be repeated.

4. Compute the factored axial loads P_u and the M_{ns} moments from a first-order frame analysis. For the main floor columns, the values are as follows:

	Exterior Columns	Interior Columns
Factored axial force, P_u (kips)		
0.9 Dead load	158.4	342.5
Factored M_{ns} moments (ft-kips)		
At top	−47.0	5.57
At bottom	−49.4	7.37

5. Compute $\delta_s M_s$ and M_c, using a second-order analysis.

5.1 Iteration No. 1—Load case 2

(a) Compute ΣP_u, $\Sigma P_u \Delta_0/\ell_c$, and the sway forces.

TABLE 12-5 Calculation of Sway Forces—Load Case 2, Iterattion No. 1

Floor	Story	ΣP_u (kips)	Floor Δ (in.)	Story Δ (in.)	$\dfrac{\Sigma P_u \Delta_0}{\ell_c}$ (kips)	Sway Force (kips)
Roof			0.735			0.251
	Fifth	1384		0.025	0.251	
Fifth			0.710			0.780
	Fourth	2588		0.055	1.031	
Fourth			0.655			1.446
	Third	3884		0.088	2.477	
Third			0.567			2.363
	Second	5178		0.129	4.840	
Second			0.438			8.280
	First	6472		0.438	13.12	
Ground			0.00			

The total factored wind forces plus sway forces at each floor are as follows:

Roof: $1.3 \times 25.6 + 0.251 = 33.5$ kips
Fifth floor: $1.3 \times 37.9 + 0.780 = 50.1$ kips
Fourth floor: $1.3 \times 37.9 + 1.446 = 50.7$ kips
Third floor: $1.3 \times 37.9 + 2.363 = 51.6$ kips
Second floor: $1.3 \times 48.7 + 8.280 = 71.6$ kips

These loads are resisted by the seven frames in the North–South direction. The frame is now reanalyzed for these forces.

(b)   Check convergence for the second cycle. The deflections from the two cycles of iteration are as follows:

	Δ—First Cycle (in.)	Δ—Second Cycle (in.)	Second/First
Roof	0.735	0.765	1.04
Second floor	0.438	0.461	1.05

This has not converged.

(c)   Carry out a third cycle of iteration. The deflections converged after the third cycle.

(d)   Compute M_1 and M_2. The column end moments from the third cycle of iteration are as follows:

Exterior columns—first story:

Top: $\delta_s M_s = 72.6$ ft-kips
Bottom: $= 78.2$ ft-kips

$$M_{end} = M_{ns} \pm \delta_s M_s$$

Top: $M_{top} = -47.0 - 72.6 = -120$ ft-kips. This is M_1.
Bottom: $M_{bottom} = -49.4 - 78.2 = -128$ ft kips. This is M_2.

Interior columns—first story:

Top: $\delta_s M_s = 89.7$ ft-kips
            $M_{top} = 5.57 + 89.7 = 95.3$ ft-kips. This is M_1.

Bottom: $\delta_s M_s = 92.6$ ft-kips
            $M_{bottom} = 7.37 + 92.6 = 100$ ft-kips. This is M_2.

6.   Check whether the maximum moment is between the ends of the column. Since this check was satisfied in load case 1, which has a larger P_u, the maximum moment will be at one end of the column.

Summary for Load Case 2

Exterior columns: $P_u = 158.4$ kips, $M_c = 128$ ft-kips
Interior columns: $P_u = 342.5$ kips, $M_c = 100$ ft-kips

LOAD CASE 3: $U = 1.4D + 1.7L$

Although this load case does not include lateral loads, the columns must be designed for P_u and M_2 from this loading. In addition, this load case will be used to check whether the frame is subject to sidesway buckling under gravity loads, since this case involves the largest gravity loads considered in this design. ACI Section 10.13.6 presents three ways to check the tendency for sidesway buckling, depending on the method used to compute $\delta_s M_s$. When $\delta_s M_s$ has been computed by using a second-order analysis, ACI Section 10.13.6(a) requires that

$$\frac{\text{second} - \text{order lateral deflections}}{\text{first} - \text{order lateral deflections}} \leq 2.5$$

when computed for a gravity load of $1.4D + 1.7L$ plus any arbitrarily chosen lateral load. In this calculation, we shall take the lateral load as $1.275W$, since we already have deflections calculated for this loading. ACI Section 10.13.6 further requires the member stiffnesses be taken as $EI/(1 + \beta_d)$, where EI is from ACI Section 10.11.1 and where, for the gravity-load sidesway-buckling check,

$$\beta_d = \frac{\text{maximum factored sustained axial load}}{\text{total factored axial load}} \tag{12-34a}$$

(See ACI Section 10.13.6 or definition (c) in ACI Section 10.0.) Because the live-load reduction factor changes from story to story, β_d also changes from story to story. The most slender columns are in the first story, and we will use the value of β_d from this story throughout. Thus,

Total factored sustained axial load in all the columns in the first story = 10,400 kips

Total factored axial load in all the columns in the first story = 14,420 kips

$$\beta_d = \frac{10{,}400}{14{,}420} = 0.721$$

Therefore, all member stiffnesses will be divided by $(1 + \beta_d) = 1 + 0.721$. This will give lateral deflections 1.721 times those already computed for the load of $1.3W$. As a result, the second-order effects will be larger.

4. **Compute the factored axial loads P_u and the M_{ns} moments from a first-order frame analysis.** For the main floor columns, the values are as follows:

	Exterior Columns	Interior Columns
Factored axial force, P_u (kips)		
1.4D	246.4	533
1.7 reduced live load	118.0	193
Factored M_{ns} moments (ft-kips)		
At top	−137.5	16.81
At bottom	−144.6	21.99

5. **Compute $\delta_s M_s$ and M_c, using a second-order analysis.**

5.1 Iteration No. 1—Load Case 3

(a) **Compute ΣP_u, $\Sigma P_u \Delta_0/\ell_c$, and the sway forces.**

Floor	Story	ΣP_u (kips)	Floor Δ (in.)	Story Δ (in.)	$\dfrac{\Sigma P_u \Delta_0}{\ell_c}$ (kips)	Sway Force (kips)
Roof			1.241			0.81
	Fifth	2729		0.041	0.811	
Fifth			1.200			3.28
	Fourth	5948		0.095	4.095	
Fourth			1.105			5.26
	Third	8725		0.148	9.357	
Third			0.957			8.78
	Second	11,426		0.219	18.133	
Second			0.738			31.14
	First	14,420		0.738	49.300	
Ground			0.00			

The total factored wind forces plus sway forces at each floor are as follows:

Roof: $1.3 \times 25.6 + 0.81 = 34.1$ kips

Fifth floor: $1.3 \times 37.9 + 3.28 = 52.6$ kips

Fourth floor: $1.3 \times 37.9 + 5.26 = 54.5$ kips

Third floor: $1.3 \times 37.9 + 8.78 = 58.1$ kips

Second floor: $1.3 \times 48.7 + 31.14 = 94.5$ kips

These loads are resisted by the seven frames in the North–South direction. The frame is analyzed for these loads with the reduced EI values.

(b) Check convergence for the second cycle. The deflections from the two cycles of iteration are as follows:

	Δ—First Cycle (in.)	Δ—Second Cycle (in.)	Second/First
Roof	1.241	1.435	1.16
Second floor	0.738	0.887	1.20

This clearly has not converged.

(c) Carry out a third iteration. The deflections from the second and third iterations are as follows:

	Δ—Second Cycle (in.)	Δ—Third Cycle (in.)	Third/Second
Roof	1.435	1.469	1.02
Second floor	0.887	0.917	1.03

We shall assume that this is adequate convergence.

(d) Check whether sidesway buckling will occur under gravity loads. The ratio of the first-order lateral deflections to the second-order lateral deflections is $0.917/0.738 = 1.24$ at the second floor. Since this is less than 2.5, the frame is not in danger of sidesway buckling under gravity loads.

The columns must be designed for the axial loads and for first-order M_{ns} moments due to $U = 1.4D + 1.7L$ given in step 4.

6. Check whether the maximum moment is between the ends of the column. From ACI Section 10.13.5, the maximum moments will exceed those at the ends of the column if

$$\frac{\ell_u}{r} > \frac{35}{\sqrt{\dfrac{P_u}{f'_c A_g}}} \tag{12-43}$$

$$\text{(ACI Eq. 10-19)}$$

$$\text{First story: } \frac{\ell_u}{r} = \frac{216 \text{ in.} - 30 \text{ in.}}{0.3 \times 18 \text{ in.}} = 34.4$$

$$\text{Interior column: } \frac{35}{\sqrt{\dfrac{P_u}{f'_c A_g}}} = \frac{35}{\sqrt{\dfrac{726}{4 \times 324}}} = 46.8$$

Since 34.4 is less than 46.8, the maximum moment in the column is at one end.

7. Check minimum moment. The code does not require columns in sway frames to be designed for a minimum moment. However, we shall conservatively design the column for the larger of the computed moments and $M_{2,\min}$ given by (12-21) (ACI Eq. 10-14).

Exterior columns:

$$M_{2,\min} = P_u(0.6 + 0.03h) = 364 \text{ kips } (0.6 + 0.03 \times 18) \text{ in.}$$
$$= 34.6 \text{ ft-kips—does not govern}$$

Interior columns:

$$M_{2,\min} = 726 \text{ kips } (0.6 + 0.03 \times 18) \text{ in.}$$
$$= 69.0 \text{ ft-kips—governs}$$

Summary for Load Case 3

Exterior columns: $P_u = 364$ kips, $M_c = 144.6$ ft-kips
Interior columns: $P_u = 726$ kips, $M_c = 69.0$ ft-kips

SUMMARY OF P_u AND M_c FOR LOAD CASES 1 TO 3

The columns must be designed for the following combinations of axial load and moment:

Exterior columns:

Load case 1: $P_u = 273$ kips, $M_c = 188$ ft-kips
Load case 2: $P_u = 158.4$ kips, $M_c = 128$ ft-kips
Load case 3: $P_u = 364$ kips, $M_c = 144.6$ ft-kips

Interior columns:

Load case 1: $P_u = 544$ kips, $M_c = 110.6$ ft-kips
Load case 2: $P_u = 342$ kips, $M_c = 100$ ft-kips
Load case 3: $P_u = 726$ kips, $M_c = 69.0$ ft-kips

8. **Select the reinforcement.** Figure 12-33 displays interaction diagrams for an 18-in.-square column with four No. 8 and four No. 9 bars. These columns satisfy the minimum reinforcement requirements of ACI Section 10.9.1. It can be seen that an 18-in.-square column with four No. 8 bars satisfies all three load cases for the exterior columns. Four No. 9 bars are needed in an interior column.

The sections selected must also satisfy ACI Section 10.3.5.2, which gives the horizontal top part of the interaction diagrams. They are therefore adequate, and we have the following:

Exterior columns: Use 18 in. × 18 in. columns with four No. 8 bars.
Interior columns: Use 18 in. × 18 in. columns with four No. 9 bars.

9. **Check the beam capacity.** ACI Section 10.11.6 requires that beams in sway frames have adequate flexural capacity to resist the total magnified moments at each joint. In a complete design, such a check would have to be made for load cases 2 and 3. ∎

EXAMPLE 12-4 Sway Frame: Using Direct P–Δ Analysis— ACI Section 10.13.4.2

Consider the frame from Example 12-3. For brevity, we shall consider only load cases 1 and 3. The structural analysis has been carried out with the E and I values given in ACI Section 10.11.1.
Steps 1, 2, and 3 are as in Example 12-3.

LOAD CASE 1: $U = 0.75(1.4D + 1.7L) + 1.3W = 1.05D + 1.275L + 1.3W$

4. **Compute the factored axial loads P_u, the M_{ns} moments, the M_s moments, and the deflections, Δ_0, from a first-order frame analysis.** For the main floor columns, the values are as follows:

	Exterior Columns	Interior Columns
Factored axial force, P_u (kips)		
1.05 Dead load	184.8	399.6
1.3 Reduced live load	88.5	144.8
Factored M_{ns} moments (ft-kips)		
At top	−103.1	12.61
At bottom	−108.5	16.49
Factored M_s moments (ft-kips)		
At top	67.3	83.2
At bottom	72.6	86.0

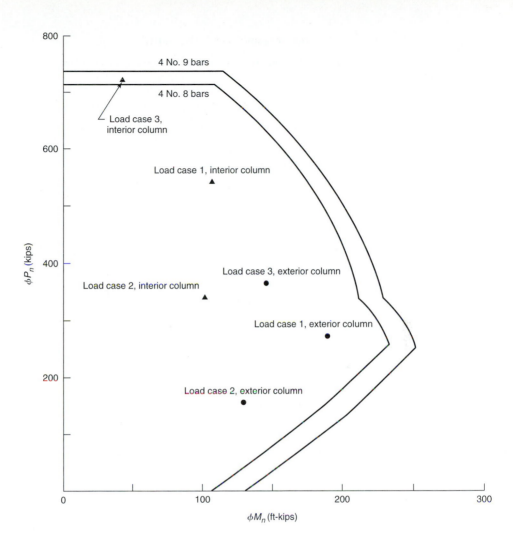

Fig. 12-33
Interaction diagrams—
Example 12-3.

The first-order lateral deflections are as follows: roof, 0.721 in.; fifth floor, 0.697 in.; fourth floor, 0.642 in.; third floor, 0.556 in.; second floor, 0.429 in.

5.   **Compute $\delta_s M_s$ and M_c according to ACI Sections 10.13.4.2 and 10.13.3.**

(a)   **Compute the story magnifiers.**
From ACI Section 10.13.4.2,

$$\delta_s M_s = \frac{M_s}{1 - Q} \ge M_s \qquad (12\text{-}40)$$
$$(\text{ACI Eq. 10-17})$$

where

$$Q = \frac{\Sigma P_u \Delta_0}{V_u \ell_c} \qquad (12\text{-}33)$$
$$(\text{ACI Eq. 10-6})$$

Story	ΣP_u (kips)	V_u (kips)	Story Δ_0 (in.)	$\dfrac{\Sigma P_u \Delta_0}{V_u \ell_c}$	δ_s
Fifth	2,047	25.6	0.024	0.014	1.014
Fourth	4,461	63.5	0.055	0.028	1.029
Third	6,544	101.4	0.086	0.040	1.042
Second	8,589	139.3	0.127	0.057	1.060
First	10,565	188	0.429	0.112	1.126

(b) Compute $\delta_s M_s$, M_1, and M_2.

Exterior columns: For the exterior columns in the first story, the column end moments are as follows:

$$M_{s,\text{top}} = \pm 67.3 \text{ ft-kips}; \delta_s M_{s,\text{top}} = (1.126 \times \pm 67.3 \text{ ft-kips}) = \pm 75.8 \text{ ft-kips}$$

$$M_{s,\text{bottom}} = \pm 72.6 \text{ ft-kips}; \delta_s M_{s,\text{bottom}} = 1.126 \times \pm 72.6 \text{ ft-kips} = \pm 81.7 \text{ ft-kips}$$

Total moment at the ends: $M_{\text{top}} = -103.1 - 75.8 = -178.9$ ft-kips—this is M_1.

$M_{\text{bottom}} = -108.5 - 81.7 = 190.2$ ft-kips—this is M_2.

Interior columns:

$$M_{s,\text{top}} = \pm 83.2 \text{ ft-kips}; \delta_s M_{s,\text{top}} = 1.126 \times \pm 83.2 \text{ ft-kips} = \pm 93.7 \text{ ft-kips}$$

$$M_{s,\text{bottom}} = \pm 86.0 \text{ ft-kips}; \delta_s M_{s,\text{bottom}} = 1.126 \times \pm 86.0 = \pm 96.8 \text{ ft-kips}$$

Total moment at the ends: $M_{\text{top}} = 12.61 + 93.7 = 106.3$ ft-kips—this is M_1.

$M_{\text{bottom}} = 16.49 + 96.8 = 113.3$ ft-kips—this is M_2.

Step 6 is the same as in Example 12-3.

Summary for Load Case 1

Exterior columns: $P_u = 273$ kips, $M_c = 190.2$ ft-kips
Interior columns: $P_u = 544$ kips, $M_c = 113.3$ ft-kips

LOAD CASE 3: $U = 1.4D + 1.7L$

Although this load case does not include lateral loads, it must be considered because the columns must be designed for the axial loads and moments from this case. In addition, it is necessary to check for whether the frame is subject to sidesway buckling under high gravity loads.

4.   Compute the factored axial loads P_u and the M_{ns} moments from a first-order frame analysis. For the main floor columns, the values are as follows:

	Exterior Columns	Interior Columns
Factored axial force, P_u (kips)		
1.4D	246	533
1.7 Reduced L	118	193
Factored M_{ns} moments (ft-kips)		
At top	−137.5	16.81
At bottom	−144.6	21.99

5.   Compute $\delta_s M_s$ and M_c according to ACI Sections 10.13.4.2 and 10.13.3. Since there are no lateral loads, $M_s = 0$ in all columns. Therefore, $\delta_s M_s = 0$.

Exterior columns:

$$M_{\text{top}} = M_{ns} + \delta_s M_s = -137.5 \text{ ft-kips} + 0 = -137.5 \text{ ft-kips. This is } M_1.$$

$M_{\text{bottom}} = -144.6$ ft-kips. This is M_2.

Interior columns:

$M_{\text{top}} = 16.81$ ft-kips. This is M_1.

$M_{\text{bottom}} = 21.99$ ft-kips. This is M_2.

6.   Check whether the maximum moment is at the end of the column. This is the same as step 6 in load case 3 in Example 12-3. The maximum moment is at one end of the column.

7.   Check whether sidesway buckling can occur under gravity loads. When $\delta_s M_s$ is computed from ACI Section 10.13.4.2, this check is done by checking whether the Q computed by using ΣP_u for $1.4D + 1.7L$ exceeds 0.60. In the calculations, EI is taken as the value corresponding to the gravity load condition [i.e., $EI/(1 + \beta_d)$, with β_d equal to the ratio of the sustained axial load to the total axial load]. Since the most slender columns are in the first story, we shall compute β_d for this story. Thus,

sustained axial load in all columns in the first story = 10,400 kips
total axial load in all the columns in the first story = 14,420 kips

$$\beta_d = \frac{10,400}{14,420} = 0.721$$

Accordingly, all member stiffnesses will be divided by $(1 + \beta_d) = 1 + 0.721$.

To compute Q, we need V_u and Δ_0 for the story, where Δ_0 is defined as the lateral story deflection due to V_u. In load case 3, $V_u = 0$, and hence $\Delta_0 = 0$. To compute Q, we need some shear and the corresponding deflections. Since we have already computed the story deflections for $1.3W$, we will use V_u and Δ_0 for this case. In that load case, the deflections were computed by using EI computed with $\beta_d = 0$. For the check of sidesway buckling, we must use $\beta_d = 0.721$. The reduced stiffness will give lateral deflections that are $1 + \beta_d = 1.721$ times those already computed for the load of $1.3W$. We thus have

V_u due to $1.3W = 188$ kips
Δ_0 in first story from step 4 of load case 1 = 0.429 in.
Δ_0 based on reduced stiffness = $0.429 \times 1.721 = 0.738$ in.

$$Q = \frac{\Sigma P_u \Delta_0}{V_u \ell_c} = \frac{14,420 \text{ kips} \times 0.738 \text{ in.}}{188 \text{ kips} \times 216 \text{ in.}}$$
$$= 0.262$$

Since $Q = 0.262$ is less than 0.60, sidesway buckling will not be a problem.

    8.   **Check minimum moment.** This is the same as step 7, load case 3, in Example 12-3.

Summary for Load Case 3

Exterior columns: $P_u = 364$ kips, $M_c = 144.6$ ft-kips
Interior columns: $P_u = 726$ kips, $M_c = 69.0$ ft-kips

This concludes Example 12-4.

EXAMPLE 12-5   Sway Frame: Using Sway-Moment Magnifier— ACI Section 10.13.4.3

Repeat Example 12-3, but using ACI Section 10.13.4.3 to compute the magnified moments. For brevity, we will consider load cases 1 and 3 only.

Steps 1, 2, and 3 are as in Example 12-3. The frame is a sway frame. The column size will be taken as 18 in. square for all columns. The same three load cases will be considered.

LOAD CASE 1: $U = 0.75(1.4D + 1.7L) \pm 1.3W) = 1.05D + 1.275L + 1.3W$

    4.   **Compute the factored axial loads P_u, the M_{ns} moments, and the M_s moments from a first-order frame analysis.** For the main floor columns, the values are as follows:

	Exterior Columns	Interior Columns
Factored axial force, P_u (kips)		
1.05 Dead load	184.8	399.6
1.275 Reduced live load	88.5	144.8
Factored M_{ns} moments (ft-kips)		
At top	−103.1	12.61
At bottom	−108.5	16.49
Factored M_s moments (ft-kips)		
At top	67.3	83.2
At bottom	72.6	86.0

5. Compute the effective-length factors. Effective-length factors can be computed by using either the nomographs in Fig. 12-26 or Table 12-2. We shall use the nomographs, with ψ based on the EI values given in ACI Section 10.11.1 for $\beta_d = 0$. The span lengths, ℓ_c and ℓ_b, are center to center of joints.

(a) Compute $E_c I_c / \ell_c$ for the columns.

$$E_c = 57{,}000\sqrt{f'_c} = 3.60 \times 10^6 \text{ psi} \qquad \text{(ACI Section 8.5.1)}$$

$$I_c = 0.70 I_g = 0.70\left(\frac{18^4}{12}\right) = 6120 \text{ in.}^4$$

Columns over the second floor and below the first floor:

$$\ell_c = 11 \text{ ft } 6 \text{ in.} = 138 \text{ in.}$$

$$\frac{E_c I_c}{\ell_c} = 160 \times 10^6 \text{ in.-lb}$$

Columns between the first and second floors:

$$\ell_c = 18 \text{ ft} = 216 \text{ in.}$$

$$\frac{E_c I_c}{\ell_c} = 102 \times 10^6 \text{ in.-lb}$$

(b) Compute $E_b I_b / \ell_b$ for the beams. Assume that the flanges of the beams are $0.25\ell = 7.5$ ft, as given in ACI Sections 8.10.2 and 8.10.3. The gross moment of inertia of the T beam is $I_g = 80{,}100 \text{ in.}^4$ (Normally, for an interior T beam, it is sufficiently accurate to take $I_g = 2(b_w h^3/12) = 81{,}000 \text{ in.}^4$). We thus have

$$I_b = 0.35 I_g = 0.35 \times 80{,}100 = 28{,}000 \text{ in.}^4$$

Beam between lines A and B, line 3:

$$\ell_b = 31 \text{ ft minus } 9 \text{ in.} = 363 \text{ in.}$$

$$\frac{E_b I_b}{\ell_b} = 278 \times 10^6 \text{ in.-lb}$$

Beam between lines B and C, line 3:

$$\ell_b = 360 \text{ in.}$$

$$\frac{E_b I_b}{\ell_b} = 280 \times 10^6 \text{ in. lb}$$

(c) Compute ψ and k.

Exterior columns, North and South walls:

$$\Psi = \frac{\Sigma(E_c I_c / \ell_c)}{\Sigma(E_b I_b / \ell_b)}$$

Two columns and one beam meet at the joint at the top of the exterior column in the first story, so

$$\Psi_{\text{top}} = \frac{160 \times 10^6 + 102 \times 10^6}{278 \times 10^6} = 0.942$$

The bottom of the exterior columns at A–2 to A–6 and D–2 to D–6 are supported on a pilaster built into the basement wall. When the frame is being deflected in the North–South direction, the stiffness of the pilaster and wall for bending about an axis parallel to the wall will be taken to be the same as that for a column of the same size. As a result,

$$\Psi_{\text{bottom}} = 0.942$$

From Fig. 12-26,

$$k = 1.28 \text{ for sway case}$$
$$k = 0.77 \text{ for braced case}$$

Interior column:

$$\Psi_{top} = \Psi_{bottom} = \frac{160 \times 10^6 + 102 \times 10^6}{278 \times 10^6 + 280 \times 10^6}$$
$$= 0.470$$

k is 1.15 for the sway case. Use $k = 1.20$ as a minimum practical value for a sway column. k is 0.68 for the braced case.

Corner columns and sidewall columns:

These columns (A–1, B–1, C–1, D–1, and A–7 through D–7) are nearly fixed at their bottom ends by the basement wall, which is perpendicular to the axis of bending being considered. Use $k = 1.2$ for the sway case, $k = 0.65$ for the braced case.

6. Compute the magnified moments for load case 1.

(a) Compute EI. Since the reinforcement is not known at this stage, use either (12-16) or (12-19) to calculate *EI*. From (12-16),

$$EI = \frac{E_c I_g / 2.5}{1 + \beta_d}$$
$$E_c = 3.60 \times 10^6 \text{ psi } I_g = 8750 \text{ in.}^4$$

$$\frac{E_c I_g}{2.5} = 12.61 \times 10^9 \text{ in.}^2/\text{lb}$$

Because this is a sway frame, definition (b) for β_d in ACI Section 10.0 applies. Since there is no sustained load shear in the story, $\beta_d = 0$. Therefore,

$$EI = \frac{12.61 \times 10^9}{1 + 0} = 12.61 \times 10^9 \text{ in.}^2\text{-lb}$$

From ACI Section 10.13.4.3,

$$\delta_s M_s = \frac{M_s}{1 - \Sigma P_u / (0.75 \Sigma P_c)} \geq M_s \qquad (12\text{-}41)$$
$$(\text{ACI Eq. 10-18})$$

where

$$\Sigma P_u = \text{the sum of the factored axial loads in all the columns in the story for load case 1}$$
$$= 10{,}565 \text{ kips}$$
$$P_c = \frac{\pi^2 EI}{(k\ell_u)^2} \qquad (12\text{-}24)$$
$$(\text{ACI Eq. 10-10})$$

For the columns in the North and South walls,

$$P_c = \frac{\pi^2 \times 12.61 \times 10^9 \text{ in.}^2\text{-lb}}{(1.28 \times 186 \text{ in.})^2} = 2196 \text{ kips}$$

For interior columns,

$$P_c = \frac{\pi^2 \times 12.61 \times 10^9}{(1.20 \times 186)^2} = 2498 \text{ kips}$$

P_c for corner and sidewall columns is also 2498 kips.

$$\Sigma P_c = 10 \times 2196 + 10 \times 2498 + 8 \times 2498 = 66,924 \text{ kips}$$

and

$$\delta_s M_s = \frac{M_s}{1 - 10,565 \text{ kips}/(0.75 \times 66,924)} \geq M_s$$
$$= 1.27 M_s$$

Exterior columns in first story:

Top of column: $M_{ns} = -103.1$ ft-kips

$M_s = \pm 67.3$ ft-kips; $\delta_s M_s = 1.27 \times \pm 67.3 = \pm 85.2$ ft-kips.

From ACI Section 10.13.3: $M_{\text{top}} = -103.1 - 85.2 = 188.3$ ft-kips. This is M_1.

Bottom of column: $M_{ns} = -108.5$ ft-kips

$M_s = \pm 72.6$ ft-kips; $\delta_s M_s = 1.27 \times \pm 72.6 = \pm 92.2$ ft-kips

$M_{\text{bottom}} = -108.5 - 92.2 = -200.7$ ft-kips. This is M_2.

Interior columns in first story:

Top of column: $M_{ns} = 12.61$ ft-kips

$M_s = \pm 83.2$ ft-kips; $\delta_s M_s = 1.27 \times \pm 83.2 = \pm 105.7$ ft-kips

$M_{\text{top}} = 12.61 + 105.7 = 118.3$ ft-kips. This is M_1.

Bottom of column: $M_{ns} = 16.49$ ft-kips

$M_s = \pm 86.0$ ft-kips; $\delta_s M_s = 1.27 \times \pm 86.0 = \pm 109.2$ ft-kips

$M_{\text{bottom}} = 16.49 + 109.2 = 125.7$ ft-kips. This is M_2.

7. Check whether the maximum column moment occurs between the ends of the column. If

$$\frac{\ell_u}{r} > \frac{35}{\sqrt{\dfrac{P_u}{f'_c A_g}}} \tag{12-43}$$

(ACI Eq. 10-19)

then it is necessary to check whether the moment at some point between the ends of the column exceeds that at the end of the column. We will check this first for interior columns, since they have the largest axial loads, P_u.

$$\frac{\ell_u}{r} = \frac{186 \text{ in.}}{0.30 \times 18 \text{ in.}} = 34.4$$

$$\frac{35}{\sqrt{\dfrac{P_u}{f'_c A_g}}} = \frac{35}{\sqrt{\dfrac{544 \text{ kips}}{4 \text{ ksi} \times 324 \text{ in.}^2}}} = 54.0$$

Because $34.4 < 54.0$, the maximum moment is at the end of the column. The same is true for the exterior column.

Summary for Load Case 1

Exterior column: $P_u = 273$ kips, $M_c = 200.7$ ft-kips

Interior column: $P_u = 544$ kips, $M_c = 125.7$ ft-kips

LOAD CASE 3: $U = 1.4D + 1.7L$

4. Compute the factored axial loads P_u and the M_{ns} moments from a first-order frame analysis. For the main floor columns, the values are as follows:

	Exterior Columns	Interior Columns
Factored axial force, P_u (kips)		
1.4D	246	533
1.7 Reduced L	118	193
Factored M_{ns} moments (ft-kips)		
At top	−137.5	16.81
At bottom	−144.6	21.99

5. Compute the effective-length factors. These will be the same as in load case 1.

6. Compute $\delta_s M_s$ and M_c according to ACI Sections 10.13.4.2 and 10.13.3. Since there are no lateral loads, $M_s = 0$ in all columns. Therefore, $\delta_s M_s = 0$ also, and we have the following:

Exterior columns:

$$M_{\text{top}} = M_{ns} + \delta_s M_s = -137.5 \text{ ft-kips} + 0 = -137.5 \text{ ft-kips. This is } M_1.$$

$$M_{\text{bottom}} = -144.6 \text{ ft-kips. This is } M_2.$$

Interior columns:

$$M_{\text{top}} = 16.81 \text{ ft-kips. This is } M_1.$$

$$M_{\text{bottom}} = 21.99 \text{ ft-kips. This is } M_2.$$

7. Check whether the maximum moment occurs between the ends of the column. This check is the same as step 7 of load case 3 in Example 12-4.

8. Check whether the frame can undergo sidesway buckling under gravity loads. When $\delta_s M_s$ is calculated from ACI Section 10.13.4.3, ACI Section 10.13.6(c) says that sidesway buckling will not be a problem if δ_s, computed by using ΣP_u for $1.4D + 1.7L$, and ΣP_c, based on $0.40\, EI/(1 + \beta_d)$, with β_d computed as the ratio of the total sustained axial loads to the total axial loads, is positive and does not exceed 2.5.

From (12-41) (ACI Eq. 10-18),

$$\delta_s = \frac{1}{1 - \Sigma P_u/(0.75 \Sigma P_c)} \geq M_s$$

where, for the first story, total sustained loads = 10,400 kips, total axial loads $\Sigma P_u = 14{,}420$ kips, and

$$\beta_d = \frac{10{,}400}{14{,}420} = 0.721$$

$$P_c = \frac{\pi^2 EI}{(k\ell_u)^2}$$

$$EI = \frac{0.40 E_c I_g}{1 + \beta_d} = \frac{12.61 \times 10^9}{1 + 0.721}$$

$$= 7.327 \times 10^9 \text{ in.}^2\text{-lb}$$

For the columns in the North and South walls,

$$P_c = \frac{\pi^2 \times 7.327 \times 10^9}{(1.28 \times 186)^2}$$

$$= 1276 \text{ kips}$$

For the interior columns and the columns in the end walls,

$$P_c = \frac{\pi^2 \times 7.327 \times 10^9}{(1.20 \times 186)^2}$$

$$= 1452 \text{ kips}$$

$$\Sigma P_c = 10 \times 1276 + 10 \times 1452 + 8 \times 1452 = 38{,}896 \text{ kips}$$

$$\delta_s = \frac{1}{1 - 14{,}420/(0.75 \times 38{,}896)} = 1.98$$

Since $\delta_s = 1.98$ is less than 2.5, sidesway buckling will not occur.

9. **Check minimum moment.** This is the same as step 7, load case 3, Example 12-3.

Summary for Load Case 3:

Exterior columns: $P_u = 364$ kips, $M_c = 144.6$ ft-kips
Interior columns: $P_u = 726$ kips, $M_c = 69.0$ ft-kips

This concludes Example 12-5. ∎

Comparison of Magnified Moments by the Three Methods for Load Case 1

For the exterior columns,

ACI Section 10.13.4.1: second-order analysis, $M_c = 188$ ft-kips
ACI Section 10.13.4.2: Q-factor analysis, $M_c = 190$ ft-kips
ACI Section 10.13.4.3: sway-moment magnifier, $M_c = 188$ ft-kips

For the interior columns,

ACI Section 10.13.4.1: second-order analysis, $M_c = 103.7$ ft-kips
ACI Section 10.13.4.2: Q-factor analysis, $M_c = 113.3$ ft-kips
ACI Section 10.13.4.3: sway-moment magnifier, $M_c = 118$ ft-kips

The three methods gave very similar design moments. If a second-order analysis is available, the easiest solution is using such an analysis. If not, the direct $P-\Delta$ (Q-magnifier) method is easiest and is recommended. The sway-magnifier method involves a lot of extra calculations to compute effective-length factors and critical loads for all the columns.

12-8 GENERAL ANALYSIS OF SLENDERNESS EFFECTS

ACI Section 10.10.1 presents requirements for a general analysis of column and frame stability. It lists a number of things the analysis must include. Such an analysis must be compared with a wide range of test results and must predict failure loads to within 15 percent of the measured failure loads. After the analysis, the structure must be reanalyzed if the final member sizes are more than 10 percent different from those assumed in the analysis.

The ACI moment-magnifier method for designing columns is limited by its derivation to prismatic columns with the same reinforcement from end to end. Tall bridge piers or other tall compression members often vary in cross section along their lengths. In such a case, the design can proceed as follows:

1. Derive moment–curvature diagrams for the various sections along the column for the factored axial loads at each section.

2. Use the Vianello method combined with Newmark's numerical method [12-26] to solve for the deflected shape and the magnified moments at each section, taking into account the loadings and end restraints. A computer program to carry out such a solution is described in [12-27].

These procedures tend to underestimate the magnified moments slightly, because the solution is discretized and shear deformations are ignored.

12-9   FRAME STABILITY

Critical Load for a Story in a Sway Frame

Figure 12-30 shows a simple frame loaded with gravity loads, ΣP, and a horizontal shear, V. We have

$$M = \Sigma(M_{top} + M_{btm}) = V\ell_c + \Sigma P\Delta \tag{12-35}$$

Assume that the moment diagram due to P has the same shape as that due to V. As shown in Fig. 12-30b and c, this is a good, but not perfect, assumption. Replace $\Sigma P\Delta$ with equivalent shear forces equal to $\gamma\Sigma P\Delta/\ell_c$, where γ is a *flexibility factor* [12-23] used to account for the difference in the moment diagrams for $\Sigma P-\Delta$ and for the shears $\Sigma P/\ell_c$. Earlier in this chapter, $Q = (\Sigma P\Delta_0)/(V\ell_c)$ was derived for a story by using the infinite-series solution of the iterative second-order analysis. The result was

$$\Delta = \frac{\Delta_0}{1 - \gamma\dfrac{\Sigma P_u\Delta_0}{V_u\ell_c}}$$

where Q was defined as $Q = \gamma\dfrac{\Sigma P_u\Delta_0}{V_u\ell_c}$

ACI Section 10.11.4.2 takes $\gamma = 1.0$ when defining Q:

$$Q = \frac{\Sigma P_u\Delta_0}{V_u\ell_c} \tag{12-33}$$

$$\text{(ACI Eq. 10-6)}$$

The sway-moment magnifier can be taken as

$$\delta_s M_s = M_s\left(\frac{1}{1 - Q}\right) \tag{12-40}$$

$$\text{(ACI Eq. 10-17)}$$

where $Q \approx \Sigma P/\Sigma P_{cr}$ for the story.

Implications of (12-33) and (12-40)

Equations (12-33) and (12-40) show that the critical load for a story in a building is a function of the drift limit that the story has been designed to meet:

$$\text{Critical load is a function of}\left[\frac{1}{\left(\Delta/\ell_c\right)}\right] \tag{12-44}$$

Frequently, buildings will be designed to satisfy a drift limit of, for example $(\Delta/h_n) = 1{:}500$, where h_n is the height of the frame. For a rectangular-plan building, this design for equal drifts in the two directions may lead to different stability indices, Q, in the two directions.

EXAMPLE 12-6 Compute the Moment Magnifiers for the Bottom Story of a Building

Consider an eight-story building with plan dimensions of 210 ft by 70 ft and a story height of 12 ft (for a total height of 96 ft), designed for a simplified uniform wind pressure of 30 psf on each windward face. For a concrete office building, an average "density" of the structure and contents would be about 15 lb/ft³. Assume that the building was designed for a story drift (deflection/ℓ_c) = 1/500 of 12 ft = 0.024 ft. in both directions. Then the total factored gravity load in the columns in the bottom story is 1.55 $\times$ (15 lb/ft³ $\times$ 96 ft $\times$ 210 ft $\times$ 70 ft) = 32,800 kips. The wind-load shears, V_u, in the bottom story would be as follows:

For wind loads blowing on the 210-ft sides,

$$V_u = 1.3(0.030 \times 210 \times 96) = 771 \text{ kips}$$

and

$$Q \approx \frac{\Sigma P_u \Delta_o}{V_u \ell_c} = \frac{32{,}800 \text{ kips} \times 0.024 \text{ ft}}{771 \text{ kips} \times 12 \text{ ft}} = 0.0851$$

$$\delta_s = \frac{1}{1 - 0.0851} = 1.093$$

For wind loads blowing on the 70-ft sides,

$$V_u = 1.275(0.030 \times 70 \times 96) = 257 \text{ kips}$$

$$Q \approx \frac{32{,}800 \times 0.024}{257 \times 12} = 0.255$$

$$\delta_s = \frac{1}{1 - 0.255} = 1.342$$

Because Q is greater than 0.05 in both directions, the structure is a sway frame in both directions. It has a high but acceptable magnification factor of $\delta_s = 1.093$ in one direction, and an uncomfortably high factor of $\delta_s = 1.342$ in the perpendicular direction. This suggests that the building should be stiffened for winds blowing parallel to the 210-ft side of the building. ∎

This example was based on an actual building that required major retrofitting repairs to stiffen the building for sway parallel to the 210 ft direction. The excessive deflections perpendicular to the short sides caused an excessive number of windows to break from wind-induced sway.

Torsional Critical Load

Figure 12-34 shows a building subjected to a torque, T, that causes the floors and roof to undergo a rotation, ϕ, per unit of height, about the center of rotation, C_R. The torsional stiffness of the building is

$$K_\theta = T/\theta$$

At height h_i, the angle of rotation is

$$\theta_i = \phi h_i$$

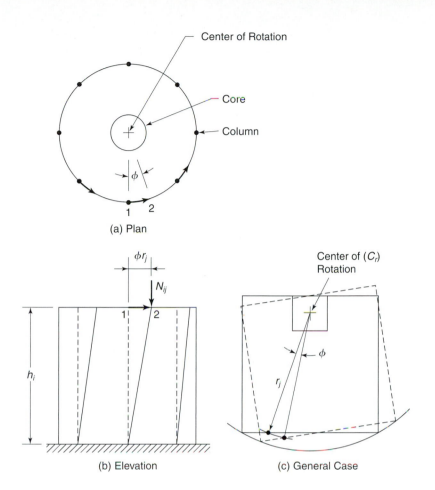

Fig. 12-34
Torsional critical load.

(a) Plan

(b) Elevation

(c) General Case

The horizontal deflection of a column-to-floor joint at height h_1 is

$$\Delta = \theta \times r_j$$

where r_j is the radial distance from the center of rotation to the point in question. A gravity load P_j at this point exerts an overturning moment $P_j\Delta = P_j(\theta \times r_j)$ on the building. The moment $P_j\Delta$ can be replaced with an equivalent horizontal force of

$$P_j\Delta/h_i = P_j(\theta \times r_j)/h_j$$

This force has a torque about C_R of

$$\frac{P_j(\theta \times r_j)}{h_i} \times r_j$$

Summing over all the columns in the story, the second-order torque is

$$T_2 = \frac{\Sigma(P_j r_j^2)\theta}{h} \qquad (12\text{-}45)$$

The applied torque, T, and the vertical loads, P_j, cause a total second-order rotation of

$$\theta_2 = \frac{(T + T_2)}{K_\theta} \qquad (12\text{-}46)$$

but

$$T_2 = \Sigma(P_j r_j^2)\theta_2/h$$

Therefore,

$$\theta_2 = \left(T - \frac{\Sigma(P_j r_j^2)}{l_c}\theta_2 \right)\frac{1}{K_\theta}$$

and

$$\frac{T}{K_\theta} = \theta_2\left(1 - \frac{\Sigma(P_j r_j^2)}{K_\theta l_c} \right)$$

Substituting $K_\theta = T/\theta_1$ gives

$$\theta_2 = \frac{\theta_1}{1 - \dfrac{\Sigma(P_j r_j^2)\theta_1}{T_1 \ell_c}} \tag{12-47}$$

At the critical load, the building is in a state of neutral equilibrium, and the rotation can have any value, including zero. Setting θ_2 equal to zero in (12-47) gives

$$\left[1 - \frac{\Sigma(P_j r_j)\theta_1}{T_1 \ell_c} \right] = 0$$

Instability occurs when the factor inside the square brackets in (12-47) equals zero. Since the factor of one cannot equal zero, then the fraction in the square brackets must be one (for unity) at the point of instability. The critical load is

$$\Sigma(P_j r_j^2) = \frac{T_1 \ell_c}{\theta_1} \tag{12-48}$$

where the *stability index for torsional buckling* is

$$Q_T = \frac{\Sigma(P_j r_j^2)}{T_1} \times \frac{\theta_1}{\ell_c} \tag{12-49}$$

Implications of (12-49)

The stability index for torsional buckling, Q_T, is similar to Q_s (12-33) for sway buckling, except that the torsional stability of a story is affected both by the total vertical load, ΣP_j, and by the square of the distance, r_j^2, from the center of rotation, C_R, to the tops of the columns on which the vertical loads act. The larger $\Sigma P r^2$ is, the lower the torsional critical load is. In design, gravity loads should be located such that $\Sigma P r^2$ is as small as possible. On the other hand, the elements resisting torsion should be placed as far away from C_R as possible, to increase the torsional stiffness.

To estimate the torsional critical load of a building, the stiffnesses of the frames, walls, and cores in the x- and y- directions are computed. The torsional stiffness, K_θ, is computed as

$$K_\theta = \frac{T}{\theta_1} \tag{12-50}$$

PROBLEMS

12-1 A hinged end column 18 ft tall supports unfactored loads of 100 kips dead load and 60 kips live load. These loads are applied at an eccentricity of 2 in. at the bottom and 4 in. at the top. Both eccentricities are on the same side of the centerline of the column. Use a tied column with f'_c = 3000 psi and f_y = 60,000 psi.

12-2 Repeat Problem 12-1, but with the top eccentricity to the right of the centerline and the bottom eccentricity to the left.

12-3 Figure P12-3 shows an exterior column in a multi-story frame. The dimensions are center to center of joints. The beams are 12 in. wide by 18 in. in overall depth. The floor slab is 6 in. thick. The building includes a service core which resists the majority of the lateral loads. Use f'_c = 3000 psi and f_y = 60,000 psi. The loads and moments on column AB are:

Factored dead load:

$$\text{axial force} = 176 \text{ kips}$$
$$\text{moment at top} = 55 \text{ ft-kips}$$
$$\text{moment at bottom} = 55 \text{ ft-kips}$$

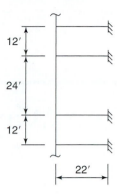

Fig. P12-3

Factored live load:

$$\text{axial force} = 210 \text{ kips}$$
$$\text{moment at top} = 72 \text{ ft-kips}$$
$$\text{moment at bottom} = 50 \text{ ft-kips}$$

Select a cross section and reinforcement.

12-4 Redesign the columns in the main floor of Example 12-3 assuming that the floor-to-floor height of the first story is 16 ft 0 in. rather than 18 ft 0 in.

13

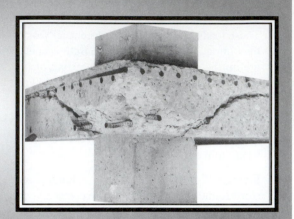

Two-Way Slabs: Behavior, Analysis, and Direct Design Method

13-1 INTRODUCTION

The structural systems designed in Chapter 10 involved one-way slabs that carried loads to beams, which, in turn, transmitted the loads to columns, as shown in Fig. 4-1. If the beams are incorporated within the depth of the slab itself, the system shown in Fig. 13-1 results. Here, the slab carries load in two directions. The load at *A* may be thought of as being carried from *A* to *B* and *C* by one strip of slab, and from *B* to *D* and *E*, and so on, by other slab strips. Since the slab must transmit loads in two directions, it is referred to as a *two-way slab*.

Two-way slabs are a form of construction unique to reinforced concrete, among the major structural materials. It is an efficient, economical, and widely used structural system. In practice, two-way slabs take various forms. For relatively light loads, as experienced in apartments or similar buildings, *flat plates* are used. As shown in Fig. 13-2a, such a plate is simply a slab of uniform thickness supported on columns. In an apartment building, the top of the slab would be carpeted, and the bottom of the slab would be finished as the ceiling for the story below. Flat plates are most economical for spans from 15 to 20 ft.

For larger spans, the thickness required to transmit the vertical loads to the columns exceeds that required for bending. As a result, the concrete at the middle of the panel is not used efficiently. To lighten the slab, reduce the slab moments, and save material, the slab at midspan can be replaced by intersecting ribs, as shown in Fig. 13-2b. Note that, near the columns, the full depth is retained to transmit loads from the slab to the columns. This type of slab is known as a *waffle slab* (or a *two-way joist* system) and is formed with fiberglass or metal "dome" forms. Waffle slabs are used for spans from 25 to 40 ft.

For heavy industrial loads, the *flat slab* system shown in Fig. 13-2c may be used. Here, the load transfer to the column is accomplished by thickening the slab near the column with *drop panels* or by flaring the top of the column to form a *column capital*. The drop panel commonly extends about one-sixth of the span each way from each column, giving extra strength in the column region while minimizing the amount of concrete at midspan. *Flat slabs* are used for loads in excess of 100 psf and for spans of 20 to 30 ft.

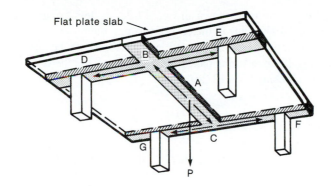

Fig. 13-1
Two-way flexure.

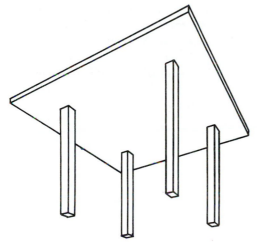

(a) Flat plate.

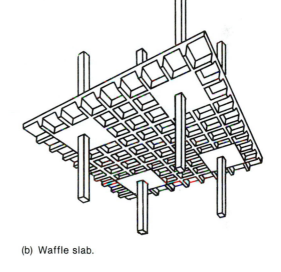

(b) Waffle slab.

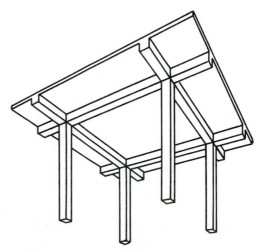

(c) Flat slab.

(d) Two–way slab with beams.

Fig. 13-2
Types of two-way slabs.

Capitals of the type shown in Fig. 13-2c are less common today than they were in the 1920s and 1930s, due to the cost of forming them.

Sometimes, a slab system will incorporate beams between some or all of the columns. If the resulting panels are roughly square, the structure is referred to as a *two-way slab with beams* (Fig. 13-2d).

13-2 HISTORY OF TWO-WAY SLABS

One of the most interesting chapters in the development of reinforced concrete structures concerns the two-way slab. Because the mechanics of slab action were not understood when the first slabs were built, a number of patented systems developed alongside a number of semiempirical design methods. The early American papers on slabs attracted copious and very colorful discussion, each patent holder attempting to prove that his theories were right and that all others were wrong.

It is not clear who built the first flat slabs. In their excellent review of the history of slabs, Sozen and Siess claim that the first American true flat slab was built by C. A. P. Turner in 1906 in Minneapolis [13-1]. In the same year, Maillart built a flat slab in Switzerland. Turner's slabs, known as mushroom slabs because the columns flared out to join the slab, had steel running in bands in four directions: the two orthogonal directions and the diagonals. These bands draped down from the top of the slab over the columns to the bottom at midspan. Some of the steel was bent down into the columns, and bars bent in a circle were placed around the columns (Fig. 13-3).

The early slab buildings were built at the risk of the designer, who frequently had to put up a bond for several years and often had to load-test the slabs before the owners would accept them. Turner based his designs on analyses carried out by H. T. Eddy, which were based on an incomplete plate-analysis theory. During this period, the use of the crossing-beam analogy in design led to a mistaken feeling that only part of the load had to be carried in each direction, so that statics somehow did not apply to slab construction.

In 1914, J. R. Nichols [13-2] used statics to compute the total moment in a slab panel. This analysis forms the basis of slab design in the current ACI Code and is presented later in this chapter. The first sentence of his paper stated "Although statics will not suffice to determine the stresses in a flat slab floor of reinforced concrete, it does impose certain lower limits on these stresses." Eddy [13-3] attacked this concept, saying "The fundamental erroneous assumption of this paper appears in the first sentence ...." Turner [13-3] thought the paper "to involve the most unique combination of multifarious absurdities imaginable from either a logical, practical or theoretical standpoint." A. W. Buel [13-3] stated that he was "unable to find a single fact in the paper nor even an explanation of facts." Rather, he felt that it was "contradicted by facts." Nichols' analysis suggested that the then current slab designs underestimated the moments by 30 to 50 percent. The emotions expressed by the discussers appear to be inversely proportional to the amount of underdesign in their favorite slab-design system.

Although Nichols' analysis is correct and was generally accepted as being correct by the mid-1920s, it was not until 1971 that the ACI Code fully recognized it and required flat slabs to be designed for 100 percent of the moments predicted from statics.

13-3 BEHAVIOR OF A SLAB LOADED TO FAILURE IN FLEXURE

There are four or more stages in the behavior of a slab loaded to failure:

1. Before cracking, the slab acts as an elastic plate and, for short-time loads, the deformations, stresses, and strains can be predicted from an elastic analysis.

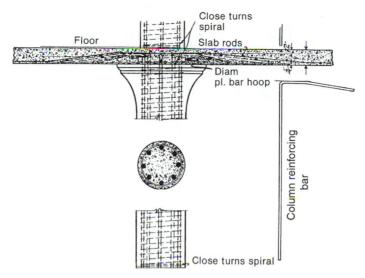

(a) Section through slab and mushroom head

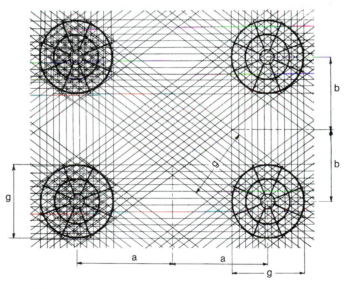

(b) Plan of reinforcement

Fig. 13-3
Reinforcement in a C. A. P.
Turner mushroom slab.

2.   After cracking and before yielding of the reinforcement, the slab is no longer of constant stiffness, since the cracked regions have a lower flexural stiffness, EI, than the uncracked regions; and the slab is no longer isotropic, since the crack pattern may differ in the two directions. Although this violates the assumptions in the elastic theory, tests indicate that the elastic theory still predicts the moments adequately. Normal building slabs are generally partially cracked at service loads.

3.   Yielding of the reinforcement eventually starts in one or more regions of high moment and spreads through the slab as moments are redistributed from yielded regions to areas that are still elastic. The progression of yielding through a slab fixed on four edges is illustrated in Fig. 13-4. In this case, the initial yielding occurs in response to negative moments that form localized plastic hinges at the centers of the long sides (Fig. 13-4b). These hinges spread along the long sides, and eventually, new hinges form at the ends of the slab (Fig. 13-4c). Meanwhile, the positive moments increase in strips across the center of the

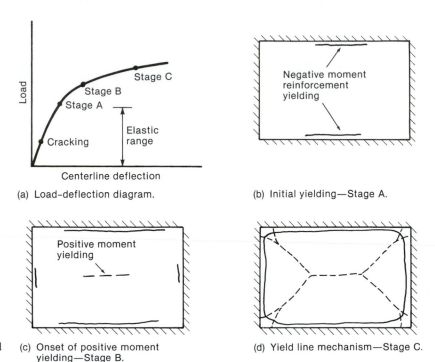

(a) Load–deflection diagram.

(b) Initial yielding—Stage A.

(c) Onset of positive moment yielding—Stage B.

(d) Yield line mechanism—Stage C.

Fig. 13-4
Inelastic action in a slab fixed on four sides.

slab in the short direction, because of the moment redistribution caused by the plastic hinges at the ends of these strips. Eventually, the reinforcement yields due to positive moments in these strips, as shown in Fig. 13-4c. With further load, the regions of yielding, known as *yield lines*, divide the slab into a series of trapezoidal or triangular elastic plates, as shown in Fig. 13-4d. The loads corresponding to this stage of behavior can be estimated by using a *yield-line analysis*.

4. Although the yield lines divide the plate to form a plastic mechanism, the hinges jam with increased deflection, and the slab forms a very flat compression arch, as shown, in Fig. 13-5. This assumes that the surrounding structure is stiff enough to provide reactions for the arch. This stage of behavior is not counted on in design at present.

This review of behavior has been presented to point out, first, that elastic analyses of slabs begin to lose their accuracy as the loads exceed the service loads, and, second, that a great deal of redistribution of moments occurs after yielding first starts. A slab supported on and continuous with stiff beams or walls has been considered here. In the case of a slab supported on isolated columns, as shown in Fig. 13-2a, similar behavior would be observed, except that the first cracking would be on the top of the slab around the column, followed by a cracking of the bottom of the slab midway between the columns.

Slabs that fail in flexure are extremely ductile. Slabs—particularly flat-plate slabs—may also fail in shear. Shear failures tend to be brittle. Shear in slabs is discussed in Sections 13-7 and 13-8.

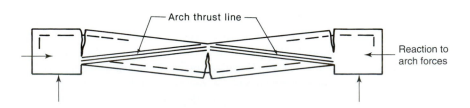

Fig. 13-5
Arch action in slab.

13-4   STATICAL EQUILIBRIUM OF TWO-WAY SLABS

Figure 13-6 shows a floor made up of simply supported planks supported by simply supported beams. The floor carries a load of w lb/ft^2. The moment per foot of width in the planks at section A–A is

$$m = \frac{w\ell_1^2}{8} \text{ ft-kips/ft}$$

The total moment in the entire width of the floor is

$$M = (w\ell_2)\frac{\ell_1^2}{8} \text{ ft-kips} \tag{13-1}$$

This is the familiar equation for the maximum moment in a simply supported floor of width ℓ_2 and span ℓ_1.

The planks apply a uniform load of $w\ell_1/2$/ft on each beam. The moment at section B–B in one beam is thus

$$M_{1b} = \left(\frac{w\ell_1}{2}\right)\frac{\ell_2^2}{8} \text{ ft-kips}$$

The total moment in both beams is

$$M = (w\ell_1)\frac{\ell_2^2}{8} \text{ ft-kips} \tag{13-2}$$

It is important to note that the full load was transferred East and West by the planks, causing a moment equivalent to $w\ell^2/8$ in the planks, and then was transferred North and South by the beams, causing a similar moment in the beams. Exactly the same thing happens in

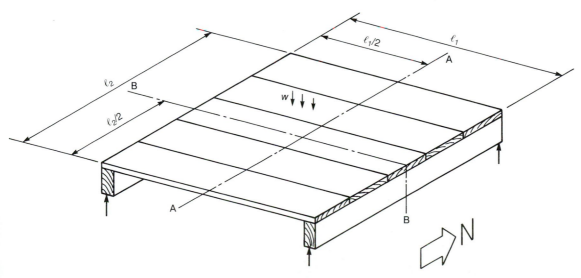

Fig. 13-6
Moments in a plank-and-beam floor.

the two-way slab shown in Fig. 13-7. The total moments required along sections *A–A* and *B–B*, respectively, are

$$M = (w\ell_2)\frac{\ell_1^2}{8} \text{ and } M = (w\ell_1)\frac{\ell_2^2}{8} \qquad (13\text{-}1, 13\text{-}2)$$

Again, the full load was transferred East and West, and then the full load was transferred North and South, this time by the slab in both cases. This, of course, must always be true regardless of whether the structure has one-way slabs and beams, two-way slabs, or some other system.

Nichols' Analysis of Moments in Slabs

The analysis used to derive (13-1) and (13-2) was first published in 1914 by Nichols [13-2]. Nichols' original analysis was presented for a slab on round columns, rather than on the point supports assumed in deriving (13-1) and (13-2). Because rectangular columns are more common today, the derivation that follows considers that case. Assume

1. the typical rectangular, interior panel in a large structure and

2. that all the panels in the structure are uniformly loaded with the same load.

These two assumptions are made to ensure that the lines of maximum moment, and hence the lines on which the shears and twisting moments are equal to zero, will be lines of symmetry in the structure. This allows one to isolate the portion of the slab shown shaded in Fig. 13-8a. This portion is bounded by lines of symmetry.

The reactions to the vertical loads are transmitted to the slab by shear around the face of the columns. It is necessary to know, or assume, the distribution of this shear to compute the moments in this slab panel. The maximum shear transfer occurs at the corners of the column, with lesser amounts transferred in the middle of the sides of the column. For this reason we shall assume that

3. the column reactions are concentrated at the four corners of each column.

Figure 13-8b shows a side view of the slab element with the forces and moments acting on it. The applied load is $(w\ell_1\ell_2/2)$ at the center of the shaded panel, minus the load on the area occupied by the column $(wc_1c_2/2)$. This is equilibrated by the upward reaction at the corners of the column.

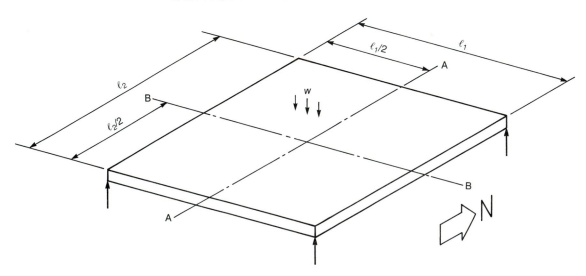

Fig. 13-7
Moments in a two-way slab.

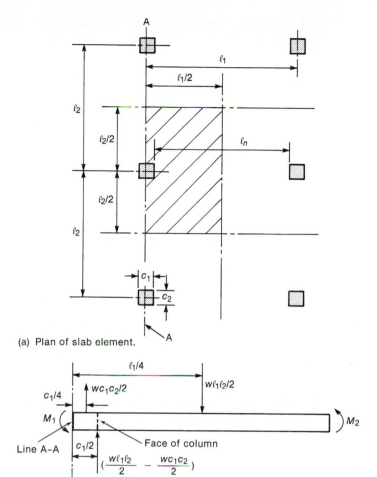

(a) Plan of slab element.

(b) Side view of slab element.

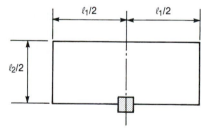

Fig. 13-8
Slab considered in Nichols'
analysis.

(c) Plan of second slab element.

The total *statical moment*, M_o, is the sum of the negative moment, M_1, and the positive moment, M_2, as computed by summing moments about line A–A:

$$M_0 = M_1 + M_2 = \left(\frac{w\ell_1\ell_2}{2}\right)\frac{\ell_1}{4} - \left(\frac{wc_1c_2}{2}\right)\frac{c_1}{4} - \left(\frac{w\ell_1\ell_2}{2} - \frac{wc_1c_2}{2}\right)\frac{c_1}{2}$$

and

$$M_o = \frac{w\ell_2}{8}\left[\ell_1^2\left(1 - \frac{2c_1}{\ell_1} + \frac{c_2c_1^2}{\ell_2\ell_1^2}\right)\right]$$ (13-3)

The ACI Code has simplified this expression slightly by replacing the term in the square brackets with ℓ_n^2, where ℓ_n is the clear span between the faces of the columns, given by

$$\ell_n = \ell_1 - c_1$$

and where

$$\ell_n^2 = \ell_1^2\left(1 - \frac{2c_1}{\ell_1} + \frac{c_1^2}{\ell_1^2}\right) \tag{13-4}$$

A comparison of (13-3) with (13-4) shows that ℓ_n^2 differs only slightly from the term in brackets in (13-3) and that the equation for the statical moment can be written as

$$M_o = \frac{w\ell_2\ell_n^2}{8} \tag{13-5}$$
$$\text{(ACI Eq. 13-3)}$$

For circular columns, Nichols assumed the shear to be uniformly distributed around the face of the column, leading to

$$M_o = \frac{w\ell_2\ell_1^2}{8}\left[1 - \frac{4d_c}{\pi\ell_1} + \frac{1}{3}\left(\frac{d_c}{\ell_1}\right)^3\right] \tag{13-6}$$

where d_c is the diameter of the column or the column capital. Nichols approximated this as

$$M_o = \frac{w\ell_2\ell_1^2}{8}\left(1 - \frac{2}{3}\frac{d_c}{\ell_1}\right)^2 \tag{13-7}$$

ACI Section 13.6.2.2 expresses this by (13-5), where ℓ_n is based on the span between equivalent square columns having the same area as the circular columns. In this case, $c_1 = d_c\sqrt{\pi}/2 = 0.886d_c$.

For square columns, the practical range of c_1/ℓ_1 is roughly from 0.05 to 0.15. For $c_1/\ell_1 = 0.05$ and $c_1 = c_2$, (13-3) and (13-5) give $M_o = Kw\ell_2\ell_1^2/8$, where $K = 0.900$ and 0.903, respectively. For $c_1/\ell_1 = 0.15$ the respective values of K are 0.703 and 0.723. Thus, (13-5) closely represents the moments in a slab supported on square columns, becoming more conservative as c_1/ℓ_1 increases.

For circular columns, the practical range of d_c/ℓ_1 is roughly from 0.05 to 0.20. For $d_c/\ell_1 = 0.05$, (13-6) gives $K = 0.936$, while (13-5), with ℓ_n defined by using $c_1 = d_c\sqrt{\pi}/2$, gives $K = 0.913$. For $d_c/\ell_1 = 0.2$, the corresponding values of K from (13-6) and (13-5) are 0.748 and 0.677, respectively. Thus, for circular columns, (13-5) tends to underestimate M_o by up to 10 percent, compared with (13-6).

If the equilibrium of the element shown in Fig. 13-8c were studied, a similar equation for M_o would result, but one having ℓ_1 and ℓ_2 interchanged and c_1 and c_2 interchanged. This indicates once again that the total load must satisfy moment equilibrium in both the ℓ_1 and ℓ_2 directions.

13-5   DISTRIBUTION OF MOMENTS IN SLABS

Relationship between Slab Curvatures and Moments

The principles of elastic analysis of two-way slabs are presented briefly in Section 15-1. The basic equation for moments is (15-6). Frequently, in studies of concrete plates, Poisson's ratio, v, is taken equal to zero. When this is done, (15-6) reduces to

$$m_x = -\frac{Et^3}{12}\left(\frac{\partial^2 z}{\partial x^2}\right)$$

$$m_y = -\frac{Et^3}{12}\left(\frac{\partial^2 z}{\partial y^2}\right)$$

$$m_{xy} = -\frac{Et^3}{12}\left(\frac{\partial^2 z}{\partial x\, \partial y}\right) \tag{13-8}$$

In these equations, $\partial^2 z/\partial x^2$ represents the curvature in a slab strip in the x direction, and $\partial^2 z/\partial y^2$ represents the curvature in a strip in the y direction where the *curvature* of a slab strip is

$$\frac{1}{r} = \frac{-\epsilon}{y}$$

where r is the *radius of curvature* (the distance from the midplane axis of the slab to the center of the curve formed by the bent slab), $1/r$ is the curvature, and y is the distance from the centoidal axis to the fiber where the strain ϵ occurs. In an elastic flexural member the curvature is

$$\frac{1}{r} = \frac{-M}{EI}$$

This shows the direct relationship between curvature and moments. Thus, by visualizing the deflected shape of a slab, one can qualitatively estimate the distribution of moments.

Figure 13-9a shows a rectangular slab that is fixed on all sides on stiff beams. One longitudinal and two transverse strips are shown. The deflected shape of these strips and the corresponding moment diagrams are shown in Fig. 13-9b to d. Where the deflected shape is concave downward, the moment causes compression on the bottom; that is, the moment is negative. This may be seen also from (13-8). Since z was taken as positive downward, a positive curvature $\partial^2 z/\partial x^2$ corresponds to a curve that is concave downward. From (13-8), a positive curvature corresponds to a negative moment. The magnitude of the moment is proportional to the curvature.

The largest deflection, Δ_2, occurs at the center of the panel. As a result, the curvatures, and hence the moments in strip B, are larger than those in strip A. The center portion of strip C is essentially straight, indicating that most of the loads in this region are being transmitted by one-way action across the short direction of the slab.

The existence of the twisting moments, m_{xy}, in a slab can be illustrated by the crossing-strip analogy. Figure 13-10 shows section B–B through the slab shown in Fig. 13-9. Here, the slab is represented by a series of crossing beams, some parallel to B–B, and the others, shown in cross section, parallel to C–C. The slab strips perpendicular to the section (shown in cross section) must twist as shown. This is due to the m_{xy} twisting moments.

Moments in Slabs Supported on Stiff Beams or Walls

The distributions of moments in a series of square and rectangular slabs will be presented in one of two graphical treatments. The distribution of the negative moments, M_A, or of the positive moments, M_B, along lines across the slab will be depicted as shown in Fig. 13-11b. These distributions may be shown as continuous curves, as shown by the solid lines and shaded areas, or as a series of steps, as shown by the dashed line. The height of the curve at any point indicates the magnitude of the moment at that point. Occasionally, the distribution of bending moments in a strip A–B–C across the slab will be plotted as shown in Fig. 13-11c or 13-9.

The moments will be expressed in terms of Cwb^2, where b is the short dimension of the panel. The value of C would be 0.125 in a square, simply supported one-way slab. The units will be ft-lb/ft of width, or kN · m/m. In all cases, the moment diagrams are for uniformly loaded slabs.

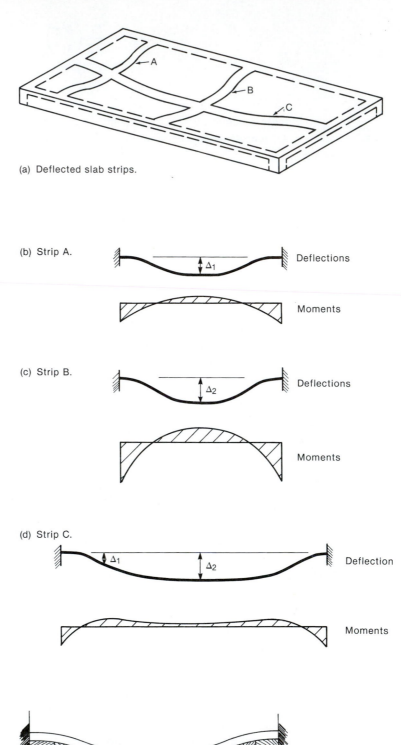

(a) Deflected slab strips.

(b) Strip A. — Deflections — Moments

(c) Strip B. — Deflections — Moments

(d) Strip C. — Deflection — Moments

Fig. 13-9
Relationship between slab
curvatures and moments.

Fig. 13-10
Deflection of strip B of
Fig. 13-9. Note twisting in
lower layer.

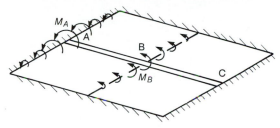

(a) Moments at edge and middle of slab.

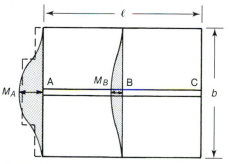

(b) Distribution of moments at edge and middle.

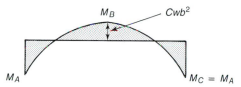

(c) Moments in strip ABC.

Fig. 13-11
Types of moment diagrams:
slab fixed on four edges.

Figure 13-12 shows moment diagrams for a simply supported square slab. The moments act *about* the lines shown (similar to Fig. 13-11a). The largest moments in the slab are about an axis along the diagonal, as shown in Fig. 13-12b.

It is important to remember that the total load must be transferred from support to support by moments in the slab or beams. Figure 13-13a shows a slab that is simply supported at two ends and is supported by stiff beams along the other two sides. The moments in the slab (which are the same as in Fig. 13-12a) account for only 19 percent of the total moments. The balance of the moments is divided between the two beams as shown.

Figure 13-13b shows the effect of reducing the stiffness of the two edge beams. Here, the stiffness of the edge beams per unit width has been reduced to be equal to that of

Fig. 13-12
Moments in a square slab
hinged on four edges. (From
[13-4].)

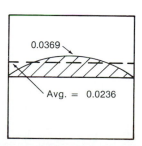

(a) Moments across center of slab.

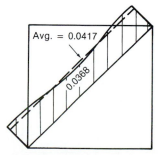

(b) Moments across diagonal.

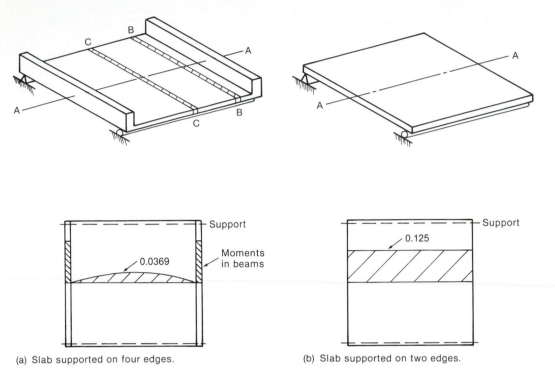

(a) Slab supported on four edges.

(b) Slab supported on two edges.

Fig. 13-13
Effect of edge-beam stiffness on moments in slab.

the slab. Now the entire moment is resisted by one-way slab action. The fact that the moments about line A–A are constant indicates that the midspan curvatures of all the strips spanning from support to support are equal. On the other hand, the curvatures in a strip along B–B in Fig. 13-13a are much smaller than those in strip C–C, since B–B is next to the stiff beam. This explains why the slab moments in Fig. 13-13a decrease towards the edge beams.

The corners of a simply supported slab tend to curl up off their supports, as is discussed later in this section. The moment diagrams presented here assume that the corners are held down by downward point loads at the corners.

The distribution of moments in several other slab configurations are plotted in Figs. 13-14 to 13-16. The moments in the square fixed-edge slab shown in Fig. 13-15a account for 36 percent of the statical moments; the balance is in the beams. As the ratio of length to width increases, the center portion of the slab approaches one-way action.

Fig. 13-14
Bending moments per unit width in rectangular slabs with simply supported edges. (From [13-4].)

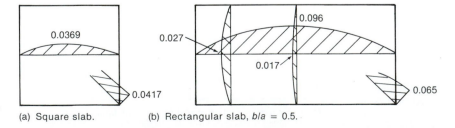

(a) Square slab.

(b) Rectangular slab, $b/a = 0.5$.

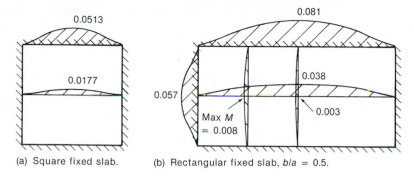

Fig. 13-15
Bending moments per unit
width in rectangular slabs with
fixed edges. (From [13-4].)

(a) Square fixed slab.   (b) Rectangular fixed slab, $b/a = 0.5$.

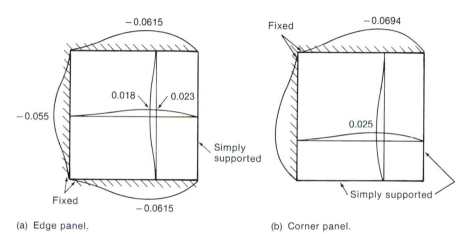

Fig. 13-16
Bending moments in edge
and corner slabs supported on
stiff beams or walls.

(a) Edge panel.   (b) Corner panel.

Moments in Slabs Supported on Isolated Columns

In a flat plate or flat slab, the slab is supported directly on the columns without any beams. Here, the stiffest portions of the slab are those running from column to column along the four sides of a panel. As a result, the moments are largest in these parts of the slab.

Figure 13-17a illustrates the moments in a typical interior panel of a very large slab in which all panels are uniformly loaded with equal loads. The slab is supported on circular columns with a diameter $c = 0.1\ell$. The largest negative and positive moments occur in the strips spanning from column to column. In Fig. 13-17b and c, the curvatures and moment diagrams are shown for strips along lines A–A and B–B. Both strips have negative moment adjacent to the columns and positive moment at midspan. In Fig. 13-17d, the moment diagram from Fig. 13-17a is replotted to show the average moments over a *column strip* of width $\ell_2/2$ and a *middle strip* between the two column strips. The ACI Code design procedures consider the average moments over the width of the middle and column strips. A comparison of Fig. 13-17a and d shows that, immediately adjacent to the columns, the theoretical elastic moments may be considerably larger than is indicated by the average values.

The total moment accounted for here is

$$w\ell_n^2[(0.122 \times 0.5\ell_2) + (0.041 \times 0.5\ell_2) + (0.053 \times 0.5\ell_2)$$

$$+ (0.034 \times 0.5\ell_2)] = 0.125w\ell_2\ell_n^2$$

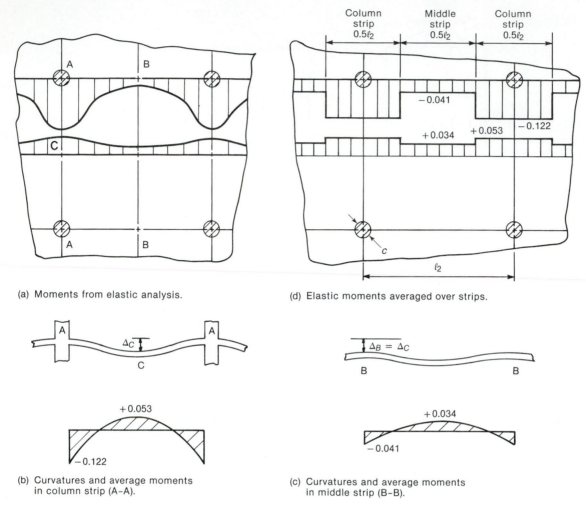

(a) Moments from elastic analysis.

(d) Elastic moments averaged over strips.

(b) Curvatures and average moments in column strip (A–A).

(c) Curvatures and average moments in middle strip (B–B).

Fig. 13-17
Moments in a flat-plate slab supported on isolated columns, $\ell_2/\ell_1 = 1.0$, $c/\ell = 0.1$.

The distribution of moments, given in Fig. 13-15, for a square slab fixed on four edges and supported on rigid beams is replotted in Fig. 13-18a with the moments averaged over column strip and middle strip bands in the same way as the flat-plate moments were in Fig. 13-17. In addition, the sum of the beam moments and the column-strip slab moments has been divided by the width of the column strip and plotted as the *total column-strip moment*. The distribution of moments in Fig. 13-17d closely resembles the distribution of middle strip and total column-strip moments in Fig. 13-18a.

An intermediate case in which the beam stiffness, I_b, equals the stiffness, I_s, of a slab of width ℓ_2 is shown in Fig. 13-18b. Although the division of moment between slab and beams differs, the distribution of total moments is again similar to that shown in Fig. 13-17d or 13-18a.

The slab-design procedures in the ACI Code take advantage of this similarity in the distributions of the total moments by presenting a unified design procedure for the whole spectrum, from slabs supported on isolated columns to slabs supported on beams in two directions.

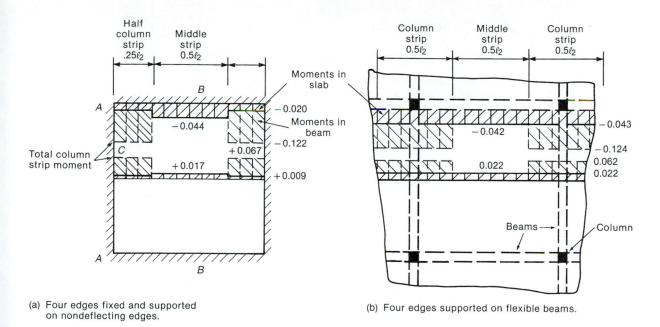

Half column strip .25ℓ₂ Middle strip 0.5ℓ₂

Column strip 0.5ℓ₂   Middle strip 0.5ℓ₂   Column strip 0.5ℓ₂

Moments in slab

B

A

−0.020

Moments in beam

−0.044

−0.042

−0.043

−0.122

−0.124

Total column strip moment

C

+0.067

0.062

+0.017

0.022

0.022

+0.009

Beams

Column

A

B

(a) Four edges fixed and supported on nondeflecting edges.

(b) Four edges supported on flexible beams.

Fig. 13-18
Moments in slabs, $\ell_2/\ell_1 = 1.0$.

13-6   DIRECT DESIGN METHOD: INTRODUCTION

ACI Section 13.5.1 allows slabs to be designed by any procedure that satisfies both equilibrium and geometric compatibility, provided that every section has a strength at least equal to the required strength and that serviceability conditions are satisfied. Two slab-design procedures are presented in detail in the ACI Code. These are the direct design method, considered in this chapter, and the equivalent frame design method, presented in Chapter 14. These two methods differ primarily in the way in which the slab moments are computed. The calculation of moments in the direct design method is based on the *statical moment*, M_o, presented earlier in this chapter. In this method, the slab is considered panel by panel, and (13-5) is used to compute the total moment in each panel. The statical moment is then divided up between positive and negative moments, and these are divided between middle strips and column strips.

The second method is called the equivalent frame method. Here, the slab is divided into a series of two-dimensional frames, and the positive and negative moments are computed via an elastic-frame analysis. Once the positive and negative moments are known, they are divided up between middle strips and column strips in exactly the same way as in the direct design method. Other methods, such as the yield-line method and the strip method, both presented in Chapter 15, are allowed under ACI Section 13.5.1. The design examples in Chapters 13 and 14 are essentially the same, to allow comparison of the two sets of design moments and other aspects.

The direct design method is emphasized in this book because an understanding of the method is essential for understanding the concepts of two-way slab design. In addition, it is an excellent method of checking slab-design calculations. In design practice, computer programs based on the equivalent frame method are usually used.

Steps in Slab Design

The steps in the design of a two-way slab are the following:

1. Choose the layout and type of slab to be used. The various types of two-way slabs and their uses have been discussed briefly in Section 13-1. The choice of type of slab is strongly affected by architectural and construction considerations.

2. Choose the slab thickness. Generally, the slab thickness is chosen to prevent excessive deflection in service. Equally important, the slab thickness chosen must be adequate for shear at both interior and exterior columns.

3. Choose the method for computing the design moments. The *equivalent frame method* uses an elastic frame analysis to compute the positive and negative moments in the various panels in the slab. The *direct design method* uses coefficients to compute these moments.

4. Compute the positive and negative moments in the slab.

5. Calculate the distribution of the moments across the width of the slab. The lateral distribution of moments within a panel depends on the geometry of the slab and the stiffness of the beams (if any). This procedure is the same whether the negative and positive moments are calculated from the direct design method or from the equivalent frame method.

6. If there are beams, a portion of the moment must be assigned to the beams.

7. Reinforcement is designed for the moments from steps 5 and 6.

8. The shear strengths at the columns are checked.

Several of the parameters used in this process will be defined prior to carrying out slab designs.

Limitations on the Use of the Direct Design Method

The direct design method is easier to use than the equivalent frame method, but can be applied only to fairly regular multipanel slabs. The limitations, given in ACI Section 13.6.1, include the following:

1. There must be a minimum of three continuous spans in each direction. Thus, a nine-panel structure (3 by 3) is the smallest that can be considered. If there are fewer than three panels, the interior negative moments from the direct design method tend to be too small.

2. Rectangular panels must have a long-span/short-span ratio not greater than 2. One-way action predominates as the span ratio reaches and exceeds 2.

3. Successive span lengths in each direction shall not differ by more than one-third of the longer span. This limit is imposed so that certain standard reinforcement cutoff details can be used.

4. Columns may be offset from the basic rectangular grid of the building by up to 0.1 times the span parallel to the offset. In a building laid out in this way, the actual column locations are used in determining the spans of the slab to be used in calculating the design moments.

5. All loads must be due to gravity only. The direct design method cannot be used for unbraced laterally loaded frames, foundation mats, or prestressed slabs.

6. The service (unfactored) live load shall not exceed two times the service dead load. Strip or checkerboard loadings with large ratios of live load to dead load may lead to moments larger than those assumed in this method of analysis.

7. For a panel with beams between supports on all sides, the relative stiffness of the beams in the two perpendicular directions given by $(\alpha_1\ell_2^2)/(\alpha_2\ell_1^2)$ shall not be less than 0.2 or greater than 5. The term α is defined in the next section, and ℓ_1 and ℓ_2 are the spans in the two directions.

Limitations 2 and 7 do not allow use of the direct design method for slab panels that transmit load as one-way slabs. Limitation 6 was changed from "three times" to "two times" in the 1995 code [13-5] to reduce the effect of pattern loads and hence to eliminate the need to check for the effect of such loads.

Beam-to-Slab Stiffness Ratio, α

Slabs are frequently built with beams spanning from column to column around the perimeter of the building. These beams act to stiffen the edge of the slab and help to reduce the deflections of the exterior panels of the slab. Very heavily loaded slabs and long-span waffle slabs sometimes have beams joining all columns in the structure.

In the ACI Code, the effects of beam stiffness on deflections and the distribution of moments are expressed as a function of α, defined as the flexural stiffness, $4EI/\ell$, of the beam divided by the flexural stiffness of a width of slab bounded laterally by the centerlines of the adjacent panels on each side of the beam:

$$\alpha = \frac{4E_{cb}I_b/\ell}{4E_{cs}I_s/\ell}$$

Since the lengths, ℓ, of the beam and slab are equal, this quantity is simplified and expressed in the code as

$$\alpha = \frac{E_{cb}I_b}{E_{cs}I_s} \tag{13-9}$$

where E_{cb} and E_{cs} are the moduli of elasticity of the beam concrete and slab concrete, respectively, and I_b and I_s are the moments of inertia of the uncracked beams and slabs. The sections considered in computing I_b and I_s are shown shaded in Fig. 13-19. The span perpendicular to the direction being designed is ℓ_2. In Fig. 13-19c, the panels adjacent to the beam under consideration have different transverse spans. The calculation of ℓ_2 in such a case is illustrated in Example 13-4. If there is no beam, $\alpha = 0$.

ACI Section 13.2.4 defines a beam in monolithic or fully composite construction as the beam stem plus a portion of the slab on each side of the beam extending a distance equal to the projection of the beam above or below the slab, whichever is greater, but not greater than four times the slab thickness. This is illustrated in Fig. 13-20.

Once the size of the slab and beam have been chosen, values of α can be computed from first principles or from Fig. 13-21. The first kinks in these curves are due to a change in scale; the second represent the point at which the flange-width limits are reached.

EXAMPLE 13–1   Calculation of α for an Edge Beam

An 8-in.-thick slab is provided with an edge beam that has a total depth of 16 in. and a width of 12 in., as shown in Fig. 13-22a. The slab and beam were cast monolithically and have the same concrete strength and the same E_c. Compute α. Since $f'_{cs} = f'_{cb}$, $E_{cb} = E_{cs}$, and (13-9) reduces to $\alpha = I_b/I_s$.

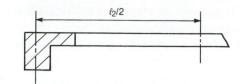

(a) Section for I_b—Edge beam.

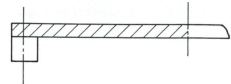

(b) Section for I_s—Edge beam.

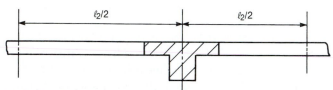

(c) Section for I_b—Interior beam.

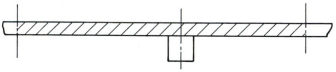

(d) Section for I_s—Interior beam.

Fig. 13-19
Beam and slab sections for
calculation of α.

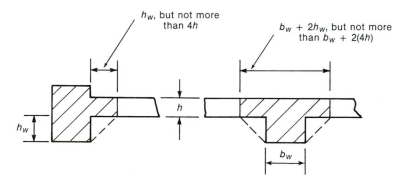

Fig. 13-20
Cross sections of beams as
defined in ACI Section 13.2.4

1. **Compute I_b.** The cross section of the beam is as shown in Fig. 13-22b. The centroid of this beam is located 7.00 in. from the top of the slab. The moment of inertia of the beam is

$$I_b = \left(12 \times \frac{16^3}{12} \right) + (12 \times 16) \times 1^2 + \left(8 \times \frac{8^3}{12} \right) + (8 \times 8) \times 3^2$$

$$= 5205 \text{ in.}^4$$

2. **Compute I_s.** I_s is computed for the shaded portion of the slab in Fig. 13-22c:

$$I_s = 126 \times \frac{8^3}{12} = 5376 \text{ in.}^4$$

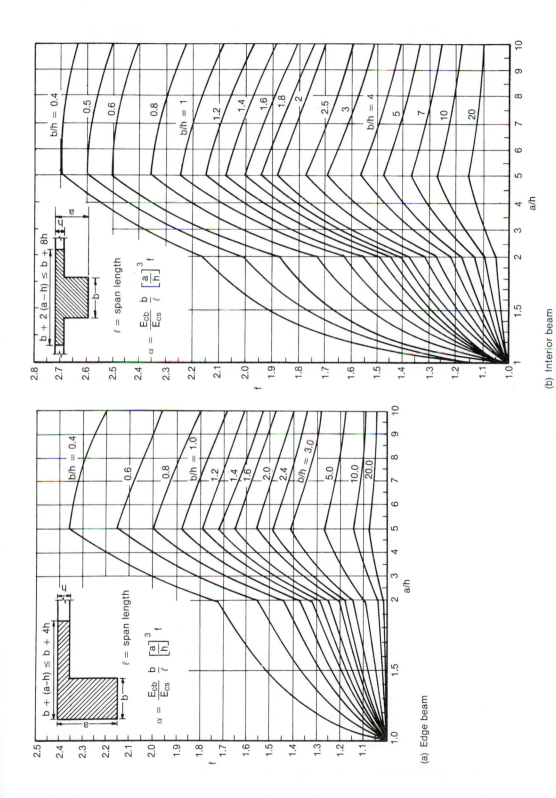

(a) Edge beam

(b) Interior beam

Fig. 13-21
Charts for computing α. (From [13-6], courtesy of the Portland Cement Association.)

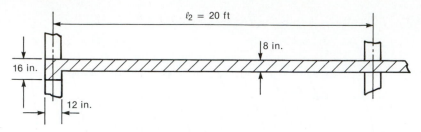

(a) Section through edge of slab.

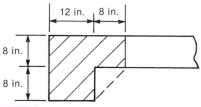

(b) Edge beam.

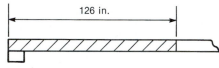

(c) Section of slab.

Fig. 13-22
Slab—Example 13-1.

3. **Compute** α.

$$\alpha = \frac{I_b}{I_s} = \frac{5205}{5376}$$

$$= 0.968$$

Alternatively, from Fig. 13-21a,

$$\alpha = \frac{E_{cb}}{E_{cs}} \frac{b}{\ell} \left(\frac{a}{h}\right)^3 f$$

where $E_{cb} = E_{cs}$. From the graph for $a/h = 2$ and $b/h = 12/8 = 1.5$, we see that $f = 1.27$ and

$$\alpha = \frac{12}{126}\left(\frac{16}{8}\right)^3 \times 1.27 = 0.968 \qquad \blacksquare$$

Minimum Thickness of Two-Way Slabs

ACI Section 9.5.3 defines minimum thicknesses that generally are sufficient to limit slab deflections to acceptable values. Thinner slabs can be used if it can be shown that the computed slab deflections will not be excessive. The computation of slab deflections is discussed in Section 13-12.

Slabs without Beams between Interior Columns

For a slab without beams between interior columns and having a ratio of long to short spans of 2 or less, the minimum thickness is as given in Table 13-1 (ACI Table 9.5(c)), but

TABLE 13-1 Minimum Thickness of Slabs without Interior Beams

| Yield Strength, $f_y{}^b$ (psi) | Without Drop Panels[a] | | | With Drop Panels[a] | | |
| | Exterior Panels | | Interior Panels | Exterior Panels | | Interior Panels |
	Without Edge Beams	With Edge Beams[c]		Without Edge Beams[c]	With Edge Beams	
40,000	$\ell_n/33$	$\ell_n/36$	$\ell_n/36$	$\ell_n/36$	$\ell_n/40$	$\ell_n/40$
60,000	$\ell_n/30$	$\ell_n/33$	$\ell_n/33$	$\ell_n/33$	$\ell_n/36$	$\ell_n/36$
75,000	$\ell_n/28$	$\ell_n/31$	$\ell_n/31$	$\ell_n/31$	$\ell_n/34$	$\ell_n/34$

Source: [13-5].

[a]The required geometry of a drop panel is defined in ACI Sections 13.3.7.1 and 13.3.7.2.

[b]For yield strengths between the values given, use linear interpolation.

[c]Slabs with beams between columns along exterior edges. The value of α for the edge beam shall not be less than 0.8.

not less than 5 in. in slabs without drop panels or 4 in. in slabs with drop panels having the dimensions defined in ACI Sections 13.3.7.1 and 13.3.7.2.

ACI Section 9.5.3.4 allows thinner slabs to be used if calculated deflections satisfy limits given in ACI Table 9.5(b). (See Section 13-12.)

As noted in the footnote to Table 13-1, a beam must have a stiffness ratio α of 0.8 or more to be called an edge beam. It can be shown that an edge beam which has a depth, a, at least twice the slab thickness, h, and an area, ab_w, at least $4h^2$, will always have α greater than 0.8. This rule of thumb can be used to simplify the selection of slab thicknesses.

Excessive slab deflections are a serious problem in many parts of North America. The thickness calculated from Table 13-1 should be rounded up to the next $\frac{1}{4}$ in. or even $\frac{1}{2}$ in. Rounding the thickness up will give a stiffer slab and hence smaller deflections. Studies of slab deflections presented in [13-7] suggest that slabs without interior beams should be about 10 percent thicker than the ACI minimum values to avoid excessive deflections.

Slabs with Beams between the Interior Supports

For slabs with beams between interior supports, ACI Section 9.5.3.3 gives the following minimum thicknesses:

(a) For $\alpha_m \le 0.2$, the minimum thicknesses in Table 13-1 shall apply.

(b) For $0.2 < \alpha_m < 2.0$, the thickness shall not be less than

$$h = \frac{\ell_n[0.8 + (f_y/200{,}000)]}{36 + 5\beta(\alpha_m - 0.2)} \qquad \text{but not less than 5 in.} \qquad (13\text{-}10)$$
$$(\text{ACI Eq. 9-12})$$

(c) For $\alpha_m > 2.0$, the thickness shall not be less than

$$h = \frac{\ell_n[0.8 + (f_y/200{,}000)]}{36 + 9\beta} \qquad \text{but not less than 3.5 in.} \qquad (13\text{-}11)$$
$$(\text{ACI Eq. 9-13})$$

(d) At discontinuous edges, either an edge beam with a stiffness ratio α not less than 0.8 shall be provided or the slab thickness shall be increased by at least 10 percent in the edge panel.

In (b) and (c),

h = overall thickness

ℓ_n = longer clear span of the slab panel under consideration

α_m = the average of the values of α for the four sides of the panel

β = longer clear span/shorter clear span of the panel

Minimum thicknesses are computed as the first step of Examples 13-7 and 13-9.

The thickness of a slab may also be governed by shear. This is particularly likely if moments are transferred to the columns at edge columns and may be serious at interior columns between two spans that are greatly different in length. The selection of slab thicknesses to satisfy shear requirements is discussed in Section 13-8. Briefly, that section suggests that the trial slab thickness be chosen such that $V_u \simeq 0.5$ to $0.55(\phi V_c)$ at edge columns and $V_u \simeq 0.85$ to $1.0(\phi V_c)$ at interior columns.

Distribution of Moments within Panels—Slabs without Beams between All Supports

Statical Moment, M_o

For design, the slab is considered to be a series of frames in the two directions, as shown in Fig. 13-23. These frames extend to the middle of the panels on each side of the column lines. In each span of each of the frames, it is necessary to compute the total statical moment, M_o. We thus have

$$M_o = \frac{w_u \ell_2 \ell_n^2}{8}$$

(13-5)

(ACI Eq. 13-3)

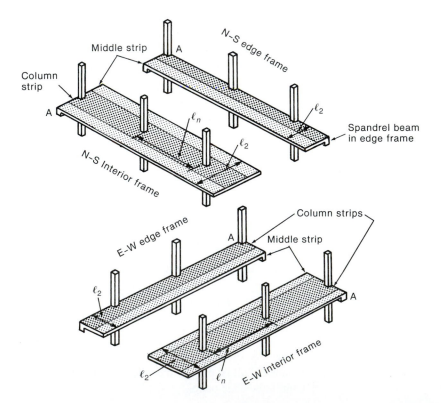

Fig. 13-23
Division of slab into frames
for design.

where

$$w_u = \text{factored load per unit area}$$
$$\ell_2 = \text{transverse width of the strip}$$
$$\ell_n = \text{clear span between columns}$$

In computing ℓ_n, circular columns or column capitals of diameter d_c are replaced by equivalent square columns with side lengths $0.886d_c$. Values of ℓ_2 and ℓ_n are shown in Fig. 13-23 for panels in each direction. Example 13-2 illustrates the calculation of M_o in a typical slab panel.

EXAMPLE 13-2   Computation of Statical Moment, M_o

Compute the statical moment, M_o, in the slab panels shown in Fig. 13-24. The slab is 8 in. thick and supports a live load of 100 psf.

1. Compute the factored uniform loads.

$$w_u = 1.2\left(\frac{8}{12} \times 0.15\right) \text{ksf} + 1.6(0.100) \text{ksf}$$
$$= 0.280 \text{ ksf}$$

Note that if the local building code allows a live-load reduction, the live load should be multiplied by the appropriate live-load-reduction factor before computing w_u.

2. Consider panel A spanning from column 1 to column 2. Slab panel A is shown shaded in Fig. 13-24a. The moments computed in this part of the example would be used to design the reinforcement running parallel to line 1–2 in this panel. From (13-5) (ACI Eq. (13-3)),

$$M_o = \frac{w\ell_2\ell_n^2}{8}$$

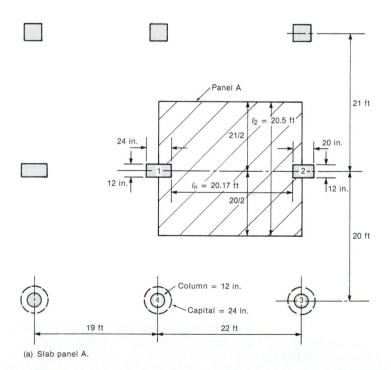

Fig. 13-24
Slab—Example 13-2.

(a) Slab panel A.

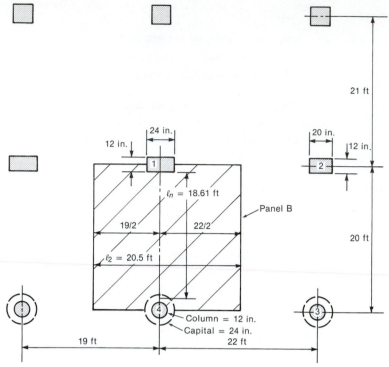

(b) Slab panel B.

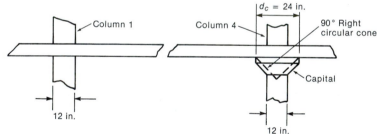

Fig. 13-24
(Continued)

(c) Section through slab panel B.

where

ℓ_n = clear span of slab panel

$$= 22 \text{ ft} - \frac{1}{2}\left(\frac{20}{12}\right) \text{ ft} - \frac{1}{2}\left(\frac{24}{12}\right) \text{ ft} = 20.17 \text{ ft}$$

ℓ_2 = width of slab panel

$$= (21/2) \text{ ft} + (20/2) \text{ ft} = 20.5 \text{ ft}$$

Therefore,

$$M_o = \frac{0.280 \text{ ksf} \times 20.5 \text{ ft} \times 20.17^2 \text{ ft}^2}{8}$$

$$= 292 \text{ ft-kips}$$

3.   Consider panel B, spanning from column 1 to column 4. Slab panel B is shown shaded in Fig. 13-24b. The moments computed here would be used to design the reinforcement running parallel to line 1–4 in this panel.

A section through the slab showing columns 1 and 4 is shown in Fig. 13-24c. Column 4 has a *column capital*. ACI Section 13.1.2 defines the effective diameter of this capital as the diameter, measured at the bottom of the slab or drop panel, of the largest right circular cone with a 90° vertex that can be included within the column and capital. The outline of such a cone is shown with dashed lines in Fig. 13-24c, and the diameter, d_c, is 24 in. For the purpose of computing ℓ_n, the circular supports are replaced by equivalent square columns having a side length $c_1 = d_c \sqrt{\pi}/2$, or $0.886 d_c$. Thus,

$$\ell_n = 20 \text{ ft} - \frac{1}{2}\left(\frac{12}{12}\right) \text{ ft} - \frac{1}{2}\left(0.886 \times \frac{24}{12}\right) \text{ ft} = 18.61 \text{ ft}$$

$$\ell_2 = \frac{19}{2} \text{ ft} + \frac{22}{2} \text{ ft} = 20.5 \text{ ft}$$

$$M_o = \frac{0.280 \text{ ksf} \times 20.5 \text{ ft} \times 18.61^2 \text{ ft}^2}{8} = 248 \text{ ft-kips} \qquad \blacksquare$$

Definition of Column Strips and Middle Strips

As seen in Section 13-5 and Fig. 13-17a, the moments vary continuously across the width of the slab panel. To aid in steel placement, the design moments are averaged over the width of column strips over the columns and middle strips between the column strips as shown in Fig. 13-17d. The widths of these strips are defined in ACI Sections 13.2.1 and 13.2.2 and are illustrated in Fig. 13-25. The column strips in both directions extend one-fourth of the smaller span, ℓ_{min}, each way from the column line.

Positive and Negative Moments in Panels

In the direct design method, the total factored statical moment M_o is divided into positive and negative factored moments according to rules given in ACI Section 13.6.3. These are illustrated in Fig. 13-26. In interior spans, 65 percent of M_o is assigned to the negative-moment region and 35 percent to the positive-moment regions. This is approximately the same as for a uniformly loaded, fixed-ended beam, where the negative moment is 67 percent of M_o and the positive moment is 33 percent.

The exterior end of an exterior span has considerably less fixity than the end at the interior support. The division of M_o in an end span into positive- and negative-moment regions is given in Table 13-2. In this table, "exterior edge unrestrained" refers to a slab whose exterior edge rested on, but was not attached to, for example, a masonry wall, while "exterior edge fully restrained" refers to a slab whose exterior edge was supported by, and was continuous with, a concrete wall with a flexural stiffness as large or larger than that of the slab.

If the computed negative moments on two sides of a support are different, the negative-moment section of the slab is designed for the larger of the two, unless a moment distribution is carried out to divide the moment between the members meeting at the joint. The use of such a moment distribution is illustrated in Example 13-9.

Provision for Pattern Loadings

In the design of continuous reinforced concrete beams, analyses are carried out for several distributions of live load to get the largest values of positive and negative moment in each span. In the case of a slab designed by the direct design method, no such analysis is done. The direct design method can be used only when the live load does not exceed two times the dead load. The effects of pattern loadings are not large in such a case [13-8].

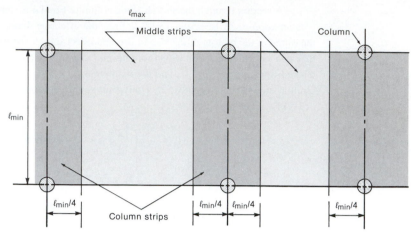

(a) Short direction of panel.

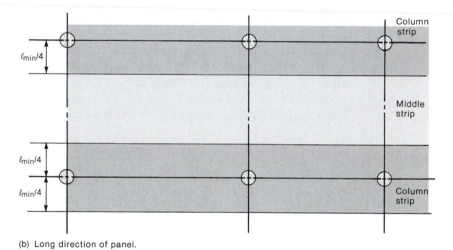

(b) Long direction of panel.

Fig. 13-25
Definition of column and
middle strips.

Provision for Concentrated Loads

Most building codes require that office floors be designed for the largest effects of either a uniform load on a full panel or a concentrated load placed anywhere on the floor, distributed over a 30-in.-by-30-in. area. The ACI slab-design provisions apply only to the uniformly loaded case. Woodring and Siess [13-9] studied the moments due to concentrated loads acting on square interior slab panels. For 20-ft-, 25-ft-, and 30-ft-square flat plates, the largest moments due to a 2000-lb concentrated load on a 30-in.-square area were equivalent to uniform loads of 39 or −23 psf, 27 or −16 psf, and 20 or −13 psf, respectively, for the three sizes of slab. The positive (downward) equivalent uniform loads are smaller than the 40- or 100-psf loads used in apartment or office floor design and hence do not govern. The negative equivalent uniform loads indicate that, at some point in the slab, the worst effect of the concentrated load is equivalent to that of an upward uniform load. Since these are all less than the dead load of a concrete floor slab, the concentrated load case does not govern here either. In sum, therefore, design for the uniform-load case will satisfy the 2000-lb concentrated-load case for 20-ft-square and larger slabs. For larger concentrated loads, the strip method can be used. (See Section 15-4.)

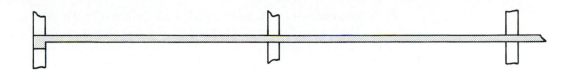

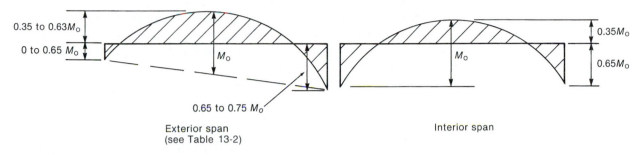

Fig. 13-26
Assignment of M_o to positive- and negative-moment regions.

TABLE 13-2 Distribution of Total Factored Static Moment, M_o, in an Exterior Span

	(1) Exterior Edge Unrestrained	(2) Slab *with* Beams between *All* Supports	(3) Slab *without* Beams between *Interior* Supports — Without Edge Beam	(4) Slab *without* Beams between *Interior* Supports — With Edge Beam	(5) Exterior Edge Fully Restrained
Interior negative factored moment	0.75	0.70	0.70	0.70	0.65
Positive factored moment	0.63	0.57	0.52	0.50	0.35
Exterior negative factored moment	0	0.16	0.26	0.30	0.65

Source: ACI Section 13.6.3.3.

Distribution of Moments between Middle Strips and Column Strips—Slabs without Beams between Interior Supports

ACI Section 13.6.4 defines the fraction of the negative and positive moments assigned to the column strips. The remaining amount of negative and positive moment is assigned to the adjacent half-middle strips. The division is a function of $\alpha_1\ell_2/\ell_1$, which depends on the aspect ratio of the panel, ℓ_2/ℓ_1, and the relative stiffness, α_1, of the beams (if any) spanning in the direction of the panel. The calculation of α was discussed earlier in this section.

For a flat plate, $\alpha_1 \ell_2 / \ell_1$ is taken equal to zero, since $\alpha = 0$ if there are no beams. In this case, 75 percent of the negative moment is in the column strip, and the remaining 25 percent is divided equally between the two adjacent half-middle strips (ACI Section 13.6.4.1). Similarly, 60 percent of the positive moment is assigned to the column strip, and the remaining 40 percent is divided, with 20 percent going to each adjacent half-middle strip (ACI Section 13.6.4.4).

At an exterior edge, the division of the exterior-end negative moment in the strip spanning perpendicular to the edge also depends on the relative torsional stiffness of the edge beam, β_t, where β_t is calculated as the shear modulus, G, times the torsional constant of the edge beam, C, divided by the EI of the slab spanning perpendicular to the edge beam (i.e., EI for a slab having a width equal to the length of the edge beam from the center of one span to the center of the other span, as shown in Fig. 13-19d). Assuming that $v = 0$ gives $G = E/2$ and that, by definition,

$$\beta_t = \frac{E_{cb}C}{2E_{cs}I_s} \tag{13-12}$$

where the cross section of the edge beam is as defined in ACI Section 13.7.5.1 and Fig. 13-27. Note that the cross section defined to compute β_t differs in some cases from that used to compute the flexural stiffness of the beam (Fig. 13-19). Conditions a, b, and c in Fig. 13-27 refer

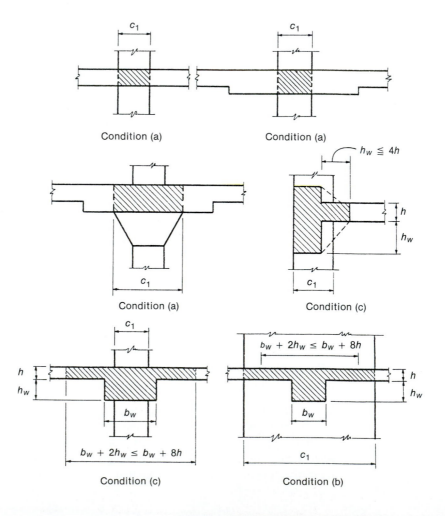

Fig. 13-27
Torsional members. (From [13-6], courtesy of the Portland Cement Association.)

to ACI Section 13.7.5.1 (a), (b), and (c), respectively. If there are no edge beams, β_t can be taken equal to zero.

The term C in (13-12) refers to the torsional constant of the edge beam. This is roughly equivalent to a polar moment of inertia. It is calculated by subdividing the cross section into rectangles and carrying out the summation

$$C = \sum \left[\left(1 - 0.63 \frac{x}{y} \right) \frac{x^3 y}{3} \right] \tag{13-13}$$

where x is the shorter side of a rectangle and y is the longer side. The subdivision of the cross section of the torsional members is illustrated in Fig. 13-28. Several possible combinations of rectangles may have to be tried to get the *maximum* value of C. To do so, the wide rectangles should be made as large as possible. Thus, the rectangles chosen in Fig. 13-28b will give a larger value of C than will those in Fig. 13-28a.

ACI Section 13.6.4.2 assigns from 45 to 100 percent of the negative moment to the column strip at the exterior end of a panel (at points A in Fig. 13-23). The exact distribution depends on ℓ_1/ℓ_2, $\alpha_1 \ell_2/\ell_1$, and β_t.

If a beam is present in the column strip (parallel to it), a portion of the column-strip moment is assigned to the beam, as specified in ACI Section 13.6.5. If the beam has $\alpha_1 \ell_2/\ell_1$ greater than 1.0, 85 percent of the column-strip moment is assigned to the beam and 15 percent to the slab. This is discussed more fully in Section 13-13 and Example 13-9.

EXAMPLE 13-3   Calculation of Moments in an Interior Panel of a Flat Plate

Figure 13-29 shows an interior panel of a flat-plate floor in an apartment building. The slab thickness is 5.5 in. The slab supports a design live load of 50 psf and a superimposed dead load of 25 psf for partitions. The columns and slab have the same strength of concrete. The story height is 9 ft. Compute the column-strip and middle-strip moments in the short direction of the panel.

1.   Compute the factored loads.

$$w_u = 1.2 \left(\frac{5.5}{12} \times 150 + 25 \right) + 1.6(50) = 193 \text{ psf}$$

Note that, if the local building code allows a live-load reduction factor, the 50-psf live load should be multiplied by the appropriate factor. The reduction should be based on the area $\ell_1 \times \ell_2$.

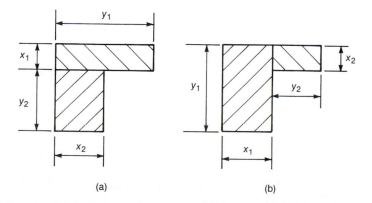

Fig. 13-28
Division of edge members for
calculation of C.

(a)   (b)

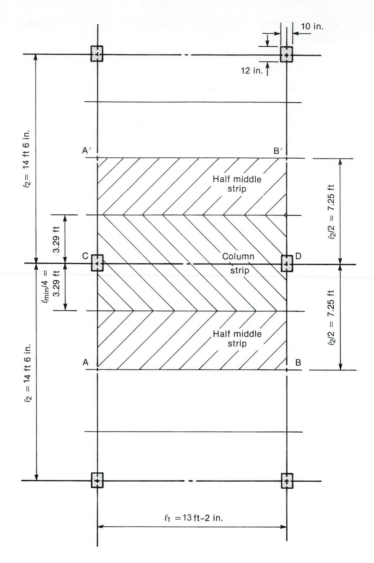

Fig. 13-29
Subdivision of two interior
panels of a flat plate—
Example 13-3.

2. **Compute the moments in the short span of the slab.**

   (a) **Compute ℓ_n and ℓ_2 and divide the slab into column and middle strips.**

$$\ell_n = 13.17 - \frac{10}{12} = 12.33 \text{ ft}$$

$$\ell_2 = 14.5 \text{ ft}$$

total width of column strips $= 6.58$ ft

total width of middle strip $= 7.92$ ft

The column strip extends the smaller of $\ell_2/4$ or $\ell_1/4$ on each side of the column centerline, as shown in Fig. 13-25 (ACI Section 13.2.1). Thus, the column strip extends $13.17/4 = 3.29$ ft on each side of column centerline. The total width of the column strip is 6.58 ft. Each half middle strip extends from the edge of the column strip to the centerline of the panel. The total width of two half middle strips is $14.5 - 6.58 = 7.92$ ft.

   (b) **Compute M_o.**

$$M_o = \frac{w_u \ell_2 \ell_n^2}{8} = \frac{0.193 \times 14.5 \times 12.33^2}{8} \qquad \text{(13-5)}$$

$$= 53.2 \text{ ft-kips} \qquad \text{(ACI Eq. 13-3)}$$

(c)   **Divide M_o into negative and positive moments.** From Code Section 13.6.3.2,

$$\text{negative moment} = 0.65M_o = -34.6 \text{ ft-kips}$$
$$\text{positive moment} = 0.35M_o = 18.6 \text{ ft-kips}$$

This process is illustrated in Fig. 13-30a, and the resulting distribution of total moments is shown in Fig. 13-30b.

(d)   **Divide the moments between the column and middle strips.**
Negative moments: From ACI Section 13.6.4.1 for $\alpha_1 \ell_2/\ell_1 = 0$ (taken equal to zero, since $\alpha_1 = 0$, because there are no beams between columns A and B in this panel),

$$\text{column-strip negative moment} = 0.75 \times -34.6 \text{ ft-kips}$$
$$= -26.0 \text{ ft-kips}$$
$$\text{middle-strip negative moment} = 0.25 \times -34.6 \text{ ft-kips}$$
$$= -8.7 \text{ ft-kips}$$

Half of this, -4.33 ft-kips, goes to each adjacent half middle strip. Since the adjacent bays have the same width, ℓ_2, a similar moment is assigned to the other half of each middle strip so that the total middle strip negative moment is 8.7 ft-kips.
Positive moments: From ACI Section 13.6.4.4, where $a_1\ell_2/\ell_2 = 0$ because there is no edge beam,

$$\text{column-strip positive moment} = 0.6 \times 18.6 = 11.2 \text{ ft-kips}$$
$$\text{middle-strip positive moment} = 0.4 \times 18.6 = 7.44 \text{ ft-kips}$$

These calculations are illustrated in Fig. 13-30a. The resulting distributions of moments in the column strip and middle strip are summarized in Fig. 13-30c. In Fig. 13-31, the moments in each strip have been divided by the width of that strip. This diagram is very similar to the theoretical elastic distribution of moments given in Fig. 13-17d.

3.   **Compute the moments in the long span of the slab.** Although it was not asked for in this example, in a slab design, it would now be necessary to repeat steps 2(a) to 2(d) for the long span.

∎

EXAMPLE 13-4   Calculation of Moments in an Exterior Panel of a Flat Plate

Compute the positive and negative moments in the column and middle strips of the exterior panel of the slab between columns B and E in Fig. 13-32. The slab is 8 in. thick and supports a superimposed service dead load of 25 psf and a service live load of 60 psf. The beam is 12 in. wide by 16 in. in overall depth and is cast monolithically with the slab.

1.   **Compute the factored loads.**

$$w_u = 1.2\left(\frac{8}{12} \times 150 + 25\right) + 1.6(60) = 246 \text{ psf}$$

2.   **Compute the moments in span *BE*.**

(a)   **Compute ℓ_n and ℓ_2 and divide the slab into middle and column strips.**

$$\ell_n = 21.0 - \frac{1}{2}\left(\frac{14}{12}\right) - \frac{1}{2}\left(\frac{16}{12}\right) = 19.75 \text{ ft}$$
$$\ell_2 = 19 \text{ ft}$$

The column strip extends the smaller of $\ell_2/4$ or $\ell_1/4$ on each side of the column centerline. Since ℓ_1 is greater than either value of ℓ_2, base this on ℓ_2. The column strip extends

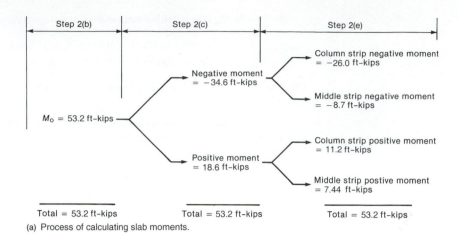

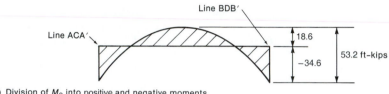

Total = 53.2 ft–kips   Total = 53.2 ft–kips   Total = 53.2 ft–kips

(a) Process of calculating slab moments.

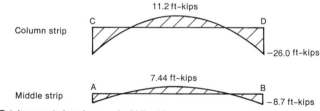

(b) Division of M_O into positive and negative moments.
— step 2(c)

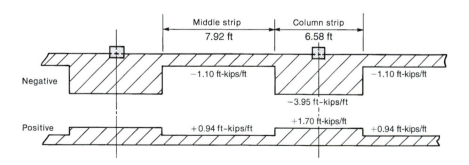

(c) Total moments in column and middle strips.

Fig. 13-30
Process of calculation of
moments—Example 13-3.

Fig. 13-31
Distribution of moments in
an interior panel of a flat
plate—Example 13-3.

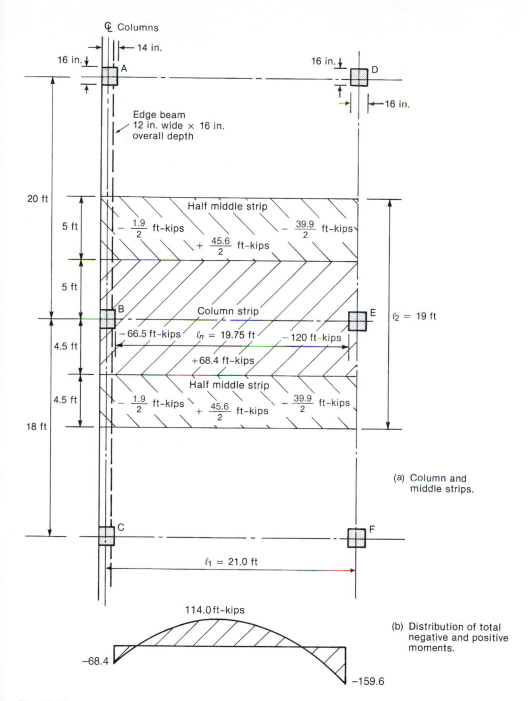

Fig. 13-32
Calculation of moments in end span—Example 13–4.

$20/4 = 5$ ft toward AD and $18/4 = 4.5$ ft toward CF from line BE, as shown in Fig. 13-32a. The total width of the column strip is 9.5 ft. The half middle strip between BE and CF has a width of 4.5 ft, and the other one is 5 ft, as shown.

(b) Compute M_o.

$$M_o = \frac{w_u \ell_2 \ell_n^2}{8} = \frac{0.246 \times 19 \times 19.75^2}{8} \tag{13-5}$$

$$= 228 \text{ ft-kips} \tag{ACI Eq. 13-3}$$

(c) Divide M_0 into positive and negative moments. The distribution of the moment to the negative and positive regions is as given in Table 13-2 (ACI Section 13.6.3.3). In the terminology of Table 13-2 this is a "slab without beams between interior supports with edge beam." From Table 13-2, the total moment is divided as follows:

$$\text{interior negative } M_u = 0.70M_o = -159.6 \text{ ft-kips}$$

$$\text{positive } M_u = 0.50M_o = +114.0 \text{ ft-kips}$$

$$\text{exterior negative } M_o = 0.30M_o = -68.4 \text{ ft-kips}$$

This calculation is illustrated in Fig. 13-33. The resulting negative and positive moments are shown in Fig. 13-32b.

(d) Divide the moments between the column and middle strips (see Fig. 13-33).

Interior negative moments: This division is a function of $\alpha_1 \ell_2 / \ell_1$, which again equals zero, since there are no beams parallel to BE. From ACI Section 13.6.4.1,

$$\text{interior column-strip negative moment} = 0.75 \times -159.6 = -120 \text{ ft-kips}$$

$$= -12.6 \text{ ft-kips/ft of width of column strip}$$

$$\text{interior middle-strip negative moment} = -39.9 \text{ ft-kips}$$

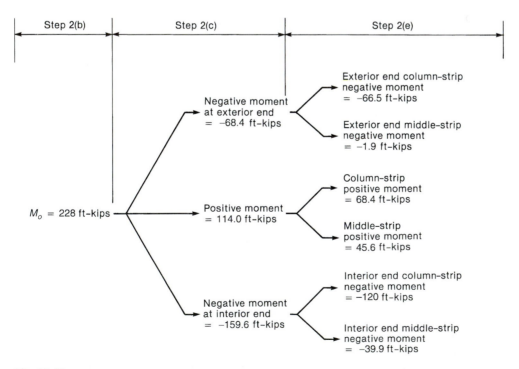

Fig. 13-33
Calculation of moments in end span—Example 13-4.

Half of this goes to each of the half middle strips beside column strip *BE*.

Positive moments: From ACI Section 13.6.4.4

$$\text{column-strip positive moment} = 0.60 \times 114.0 = 68.4 \text{ ft-kips}$$
$$= 7.20 \text{ ft-kips/ft}$$

$$\text{middle-strip positive moment} = 45.6 \text{ ft-kips}$$

Half of this goes to each half middle strip.

Exterior negative moment: From ACI Section 13.6.4.2, the exterior negative moment is divided as a function of $\alpha_1 \ell_2/\ell_1$ (again equal to zero, since there is no beam parallel to ℓ_1) and

$$\beta_t = \frac{E_{cb}C}{2E_{cs}I_s} \tag{13-12}$$

where E_{cb} and C refer to the attached torsional member shown in Fig. 13-34 and E_{cs} and I_s refer to the strip of slab being designed (the column strip and the two half middle strips shown shaded in Fig. 13-32a). To compute C, divide the edge beam into rectangles. The two possibilities shown in Fig. 13-34 will be considered. For Fig. 13-34a, (13-13) (ACI Eq. (13-7)) gives

$$C = \frac{(1 - 0.63 \times 12/16)12^3 \times 16}{3} + \frac{(1 - 0.63 \times 8/8)8^3 \times 8}{3}$$
$$= 5367 \text{ in.}^4$$

For Fig. 13-34b, $C = 3741$ in.[4]. The larger of these two values is used; therefore, $C = 5367$ in.[4].

I_s is the moment of inertia of the strip of slab being designed, which has $b = 19$ ft and $h = 8$ in. Then

$$I_s = \frac{(19 \times 12) \times 8^3}{12} = 9728 \text{ in.}^4$$

Since f'_c is the same in the slab and beam, $E_{cb} = E_{cs}$ and

$$\beta_t = \frac{5367}{2(9728)} = 0.276$$

Interpolating in the table given in ACI Section 13.6.4.2, we have

for $\beta_t = 0$: 100 percent to column strip
for $\beta_t = 2.5$: 75 percent to column strip

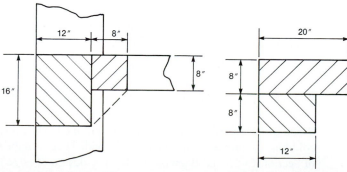

Fig. 13-34
Slab, column, and edge
beam—Example 13-4.

(a) Attached torsional member.   (b) Attached torsional member.

Therefore, for $\beta_t = 0.276$, 97.2 percent to column strip, and we have

$$\text{exterior column-strip negative moment} = 0.972(-68.4) = -66.5 \text{ ft-kips}$$

$$= -7.00 \text{ ft-kips/ft}$$

$$\text{exterior middle-strip negative moment} = -1.9 \text{ ft-kips} \qquad \blacksquare$$

Moments in Columns and Transfer of Moments to Columns

Exterior Columns

When design is carried out by the direct design method, ACI Section 13.6.3.6 specifies that the moment that is transferred from a slab without interior beams to an edge column is 0.26 to 0.30 M_o, as given in Table 13-2. This moment is used to compute the shear stresses due to moment transfer to the edge column, as discussed in Section 13-8. Although not specifically stated in the ACI Code, this moment can be assumed to be about the centroid of the shear perimeter. The exterior negative moment from the direct design method calculation is divided between the columns above and below the slab in proportion to the column stiffnesses, $4EI/\ell$. The resulting column moments are used in the design of the columns.

Interior Columns

At interior columns, the moment-transfer calculations and the total moment used in the design of the columns above and below the floor are based on an unbalanced moment resulting from an uneven distribution of live load. The unbalanced moment is computed by assuming that the longer span adjacent to the column is loaded with the factored dead load and half the factored live load, while the shorter span carries only the factored dead load. The total unbalanced negative moment at the joint is thus

$$M = 0.65\left[\frac{(w_d + 0.5w_\ell)\,\ell_2\ell_n^2}{8} - \frac{w_d'\ell_2'\,(\ell_n')^2}{8}\right]$$

where w_d and w_ℓ refer to the factored dead and live loads on the longer span and w_d', ℓ_2', and ℓ_n' refer to the shorter span adjacent to the column. The factor 0.65 is the fraction of the static moment assigned to the negative moment at an interior support. The factors 0.65 and $\frac{1}{8}$ combine to give 0.081. A portion of the unbalanced moment is distributed to the slabs, and the rest goes to the columns. Since slab stiffnesses have not been calculated, it is assumed that most of the moment is transferred to the columns, giving

$$M_{col} = 0.07[(w_d + 0.5w_\ell)\,\ell_2\ell_n^2 - w_d'\ell_2'\,(\ell_2')^2] \qquad (13\text{-}14)$$
$$(\text{ACI Eq. 13-4})$$

The moment, M_{col}, is used to design the slab-to-column joint. It is distributed between the columns above and below the joint in the ratio of their stiffnesses to determine the moments used to design the columns.

Design of Edge Beams for Shear and Moment

When a slab panel contains a beam, either an edge beam or an interior beam between the columns, the moments in the panel are divided between the slab and the beam, as discussed in Section 13-13 and Example 13-10. The design of edge beams for flat-plate floors proceeds exactly as in that example.

13-7   SHEAR STRENGTH OF TWO-WAY SLABS

A shear failure in a beam results from an inclined crack caused by flexural and shearing stresses. This crack starts at the tensile face of the beam and extends diagonally to the compression zone near a concentrated load, as explained in Chapter 6. In the case of a two-way slab or footing, the two shear-failure mechanisms shown in Fig. 13-35 are possible. *One-way shear* or *beam-action shear* (Fig. 13-35a) involves an inclined crack extending across the entire width of the structure. *Two-way shear* or *punching shear* involves a truncated cone or pyramid-shaped surface around the column as shown schematically in Fig. 13-35b. Generally, the punching-shear capacity of a slab or footing will be con-siderably less than the one-way shear capacity. In design, it is necessary to consider both failure mechanisms, however.

This section is limited to footings and slabs without beams. The shear strength of slabs with beams is discussed in Section 13-13.

Behavior of Slabs Failing in Two-Way Shear—Interior Columns

As shown in Section 13-5, the maximum moments in a uniformly loaded flat plate occur around the columns and lead to a circular crack around each column. After additional load-ing, the cracks necessary to form a fan yield-line mechanism develop (see Section 15-3), and at about the same time, inclined or shear cracks form on the truncated conical surface, shown in Fig. 13-35b. These cracks can be seen in Fig. 13-36, which shows a slab that has been sawn through along two sides of the column after the slab had failed in two-way shear.

In Chapter 6, truss models are used to explain the behavior of beams failing in shear. Alexander and Simmonds [13-10] have analyzed punching-shear failures, using a truss model similar to Fig. 13-37. Prior to the formation of the inclined cracks shown in Fig. 13-36, the shear is transferred by shear stresses in the concrete. Once the cracks have formed, only relatively small shear stresses can be transferred across them. Now the majority of the shear is transferred by inclined struts A–B and C–D extending from the bottom of the slab

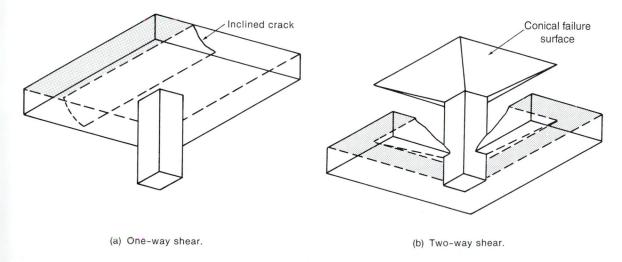

(a) One–way shear.   (b) Two–way shear.

Fig. 13-35
Shear failures in a slab.

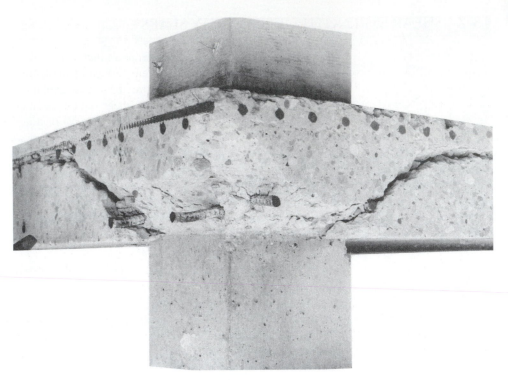

Fig. 13-36
Inclined cracks in a slab after
a shear failure. (Photograph
courtesy of J. G. MacGregor.)

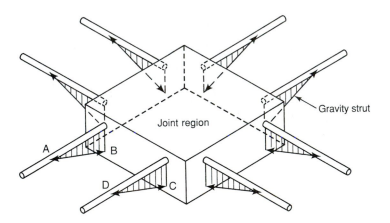

Fig. 13-37
Truss model for shear transfer at an interior column.

at the column to the reinforcement at the top of the slab at *A* and *D*. Similar struts exist on
all four sides of the column. The horizontal component of the force in the struts causes a
change in the force in the reinforcement at *A* and *D*, and the vertical component pushes up
on the bar and is resisted by tensile stresses in the concrete between the bars. Eventually,
this concrete cracks in the plane of the bars, and a punching failure results. Such a failure
occurs suddenly, with little, if any, warning.

When Alexander and Simmonds [13-10] used the truss model in Fig. 13-37 to analyze test specimens, the bond transfer from the bars to the inclined struts *A–B* and *C–D* was
too concentrated to adequately develop the forces in the bars at *A* and *D*. To overcome this,
Alexander and Simmonds assumed the gravity struts (members *A–B* and *C–D*) in Fig. 13-37
were curved, concave downwards, with slopes varying from about twice the inclination of
the struts *A–B* and *C–D* at points *B* and *C* in Fig. 13-37, to half of the strut inclination at

points *A* and *D*. For the strut to be bent in this way, downward forces must act on it. These vertical forces act downward on the strut and upward on the cover above the top steel.

Once a punching-shear failure has occurred at a slab-column joint, the shear capacity of that particular joint is almost completely lost. In the case of a two-way slab, as the slab slides down the column, load is transferred to adjacent column-slab connections, thereby overloading them and possibly causing them to fail. Thus, although a two-way slab possesses great ductility if it fails in flexure, it has very little ductility if it fails in shear. Excellent reviews of the factors affecting the shear strength of slabs are presented in [13-11] and [13-12].

Design for Two-Way Shear

General.

Initially, this discussion will consider the case of shear transfer without appreciable moment transfer. The case when both shear and moment are transferred from the slab to the column is discussed in Section 13-8.

From extensive tests, Moe [13-13] concluded that the critical section for shear was located at the face of the column. ACI–ASCE Committee 326 (now 445) [13-14] accepted Moe's conclusions, but showed that a much simpler design equation could be derived by considering a critical section at $d/2$ from the face of the column. This was referred to as the *pseudo-critical section for shear*. This simplification has been incorporated in the ACI Code. See ACI Section 11.12.1.2.

Location of the Critical Perimeter

Two-way shear is assumed to be critical on a vertical section through the slab or footing extending around the column. According to ACI Section 11.12.1.2, this section is chosen so that it is never less than $d/2$ from the face of the column and so that its length, b_o, is a minimum. Although this would imply that the corners of the shear perimeter should be rounded, the original derivation of the allowable stresses in punching shear was based on a rectangular perimeter. The intent of the code is that the critical shear section be taken as rectangular if the columns are rectangular. ACI Section 11.12.1.3 was added in a recent code edition to so state. Several examples are shown in Fig. 13-38.

Critical Sections for Slabs with Drop Panels

ACI Sections 13.3.7 through 13.3.7.3 limit the distance from the center of the column to the edge of the drop panel to one-sixth or more of the span length between the centers of the columns in each direction. This size of drop panel was originally specified to reduce the deflections of the slab panel without consideration of shear.

In slabs with drop panels, two critical sections should be considered, as shown in Fig. 13-39. Sometimes *shear capitals* are used rather than drop panels. A shear capital is a thickened square concrete "drop panel" that extends less than the one-sixth of the span required for a drop panel. It is cast monolithically with the slab.

Critical Sections near Holes and at Edges

When openings are located at less than 10 times the slab thickness from a column, ACI Section 11.12.5 requires that the critical perimeter be reduced, as shown in Fig. 13-40a.

The critical perimeter for edge or corner columns is not clearly defined in the ACI Code. In particular, in the case of a three-sided shear section at an edge column, as shown in Fig. 13-40b, the code does not state how far the sides of the critical section extend past the outer face of the column at *S* and *V*. There are at least four published limits on this extension, none of which has been endorsed by the ACI Code:

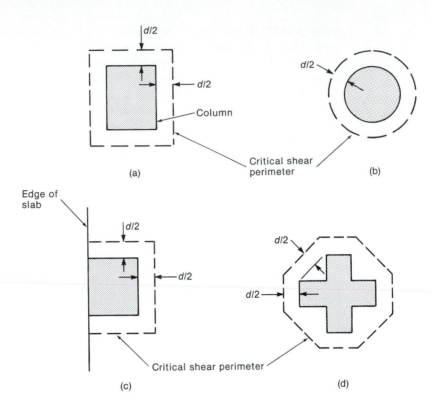

Fig. 13-38
Locations of critical shear
perimeters.

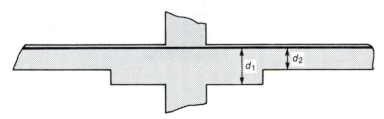

(a) Section through drop panel.

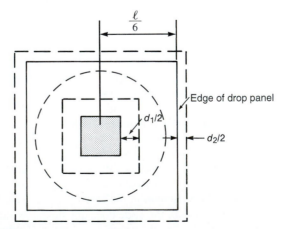

Fig. 13-39
Critical sections in a slab
with drop panels.

(b) Critical sections.

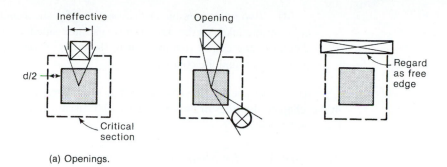

(a) Openings.

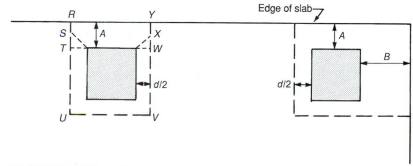

(b) Perimeters if A and B do not exceed the greater of $4h$ or $2\ell_d$.

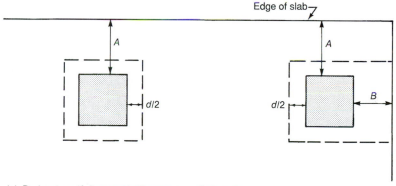

Fig. 13-40
Effect of openings and edges on critical perimeter for shear.

(c) Perimeters if A exceeds the greater of $4h$ or $2\ell_d$, but B does not.

(a)   Commonly, the sides of the critical perimeter around the column are not considered effective beyond the exterior face of the column. Thus, the critical section is T–U–V–W in Fig. 13-40b.

(b)   In a second method, the sides perpendicular to the edge are assumed to extend beyond the column to the points where they intersect 45° lines radiating out from the corner of the column. For a rectangular column this extends the sides perpendicular to the edge of the slab by a distance of $d/2_n$ giving the perimeter S–U–V–X.

(c)   In 1978, ACI Committee 426 [13-15] suggested that the side faces of the perimeter could be considered effective outside the corner of the column for the smaller of (i) four slab thicknesses, $4h$, or (ii) twice the development length, $2\ell_d$, of the flexural reinforcement perpendicular to the edge, shown by the distances labeled A and B in Fig. 13-40b and c.

(d)   Two other possible critical sections for shear and moment transfer at edge columns are taken as the shorter of (iii) a three-sided section, with the sides perpendicular to the edge extending out to the edge of the slab, and (iv) the four-sided perimeter around the column.

The authors believe the best choice lies somewhere between (b) and (d). In examples in this chapter, the sides of the perimeter will be assumed to extend the shorter of $1.5d$ and the actual extension.

Tributary Areas for Shear in Two-Way Slabs

For uniformly loaded two-way slabs, the tributary areas used to calculate V_u are bounded by lines of zero shear. For interior panels, these lines can be assumed to pass through the center of the panel. For edge panels, the moment coefficients in ACI Section 13.6.3.3 correspond to lines of zero shear at the following points:

0.44ℓ from the center of the exterior column for flat plates without edge beams, where ℓ is the span center to center of columns (see Fig. 13-41);

0.45ℓ from the center of the exterior column for flat plates with edge beams;

0.50ℓ from the center of the columns for all other panels.

Design Equations: Two-Way Shear with Negligible Moment Transfer

Unbalanced floor loads, or lateral loads, on a flat-plate building require that both moments and shears be transferred from the slab to the columns. In the case of interior columns in a braced flat-plate building, the worst loading case for shear generally corresponds to a negligible moment transfer from the slab to the column. Similarly, columns generally transfer little or no moment to footings.

Design for two-way shear without moment transfer is carried out by using (6-9), (6-14), and (13-15) to (13-17) (ACI Eqs. (11-1), (11-2), and (11-33) to (11-35)). The basic equation for shear design states that

$$V_u \le \phi V_n \tag{6-14}$$
$$\text{(ACI Eq. 11-1)}$$

where V_u is the factored shear force due to the loads and V_n is the nominal shear resistance of the slab or footing. For shear, the strength-reduction factor, ϕ is equal to 0.85 if the load factors are from ACI 318-02, Appendix C, and is equal to 0.75 if the load factors are from ACI 318-02 Section 9.2.

Punching Shear V_c, Carried by Concrete in Two-Way Slabs

ACI Eq. (11-2) defines V_n as follows:

$$V_n = V_c + V_s \tag{6-9}$$
$$\text{(ACI Eq. 11-2)}$$

where V_c and V_s are the shear resistances attributed to the concrete and the shear reinforcement, respectively. In most slabs, V_s is zero. For two-way shear, V_c is taken as the smallest of (a), (b), and (c), where

$$\text{(a)} \quad V_c = \left(2 + \frac{4}{\beta_c}\right)\sqrt{f'_c}b_o d \tag{13-15}$$
$$\text{(ACI Eq. 11-33)}$$

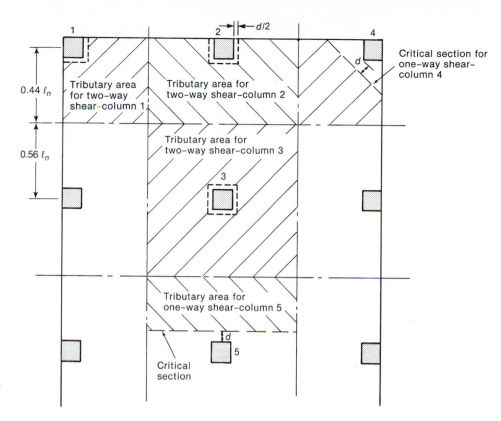

Fig. 13-41
Critical sections and tributary areas for shear in a flat plate.

where β_c is the ratio of long side to short side of the column, concentrated load, or reaction area. For nonrectangular columns, this is defined as shown in Fig. 13-42.

$$(b) \quad V_c = \left(\frac{\alpha_s d}{b_o} + 2\right)\sqrt{f'_c}b_o d \qquad\qquad (13\text{-}16)$$

$$(ACI\ Eq.\ 11\text{-}34)$$

where α_s is 40 for interior columns, 30 for edge columns, and 20 for corner columns; and

$$(c) \quad V_c = 4\sqrt{f'_c}b_o d \qquad\qquad (13\text{-}17)$$

$$(ACI\ Eq.\ 11\text{-}35)$$

The distribution of shear stresses around the column is approximately as shown in Fig. 13-43 with higher shear stresses transferred in the vicinity of the corners [13-12]. For very large columns or rectangular columns with two long sides the shear stress between the corners decreases, approaching the design value for one-way shear, $2\sqrt{f'_c}$.

Equation 13-15 (ACI Eq. (11-33)) applies to large (usually rectangular) columns having a ratio of side lengths of the column, β_c, greater than two, where the shear transfer approaches one-way action on the long sides of the column. It is compared to test data in Fig. 13-44a [13-16]. Equation (13-16) applies to very large columns having a ratio of perimeter to effective depth, b_o/d, greater than 20. It is compared to test data in Fig. 13-44b [13-17]. Equation (13-17) applies if β_c is less than 2 or in the case of an interior column if b_o/d is less than 20. In most cases, (13-17) applies.

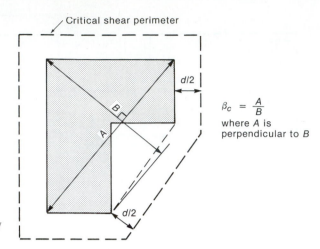

Critical shear perimeter

$$\beta_c = \frac{A}{B}$$

where A is perpendicular to B

Fig. 13-42
Definition of β_c for irregularly shaped columns.

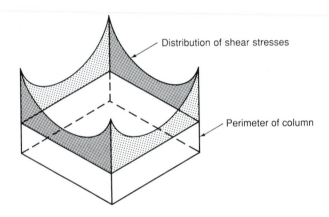

Distribution of shear stresses

Perimeter of column

Fig. 13-43
Distribution of shear stress around the perimeter of a square column.

One-Way Shear in Slabs

In the case of a uniformly loaded slab, the critical section for one-way shear is located at d from the face of the support or at d from the face of a drop panel or other change in thickness (Fig. 13-39b, ACI Section 11.12.1.1). The entire width of the slab panel is assumed to be effective in resisting one-way shear. The tributary areas for one-way shear in a slab are illustrated in Fig. 13-41 (for columns 4 and 5). The shear strength on the critical section is computed as for beams (see Chapter 6) by using (6-14), (6-9), and (6-8) (ACI Eqs. (11-1), (11-2), and (11-3)), where

$$V_c = 2\sqrt{f'_c}b_w d \quad \text{lb} \qquad (6\text{-}8)$$
$$(\text{ACI Eq. 11-3})$$

One-way shear is seldom critical in flat plates or flat slabs, as will be seen from Example 13-5.

EXAMPLE 13-5 Checking One-Way and Two-Way Shear at an Interior Column in a Flat Plate

Figure 13-45 shows an interior column in a large uniform flat-plate slab. The slab is 6 in. thick and has $d = 5$ in. for reinforcement perpendicular to the long side of the column and $d = 4.5$ in. in the other direction. The slab supports a uniform superimposed dead load of 15 psf and a uniform superimposed

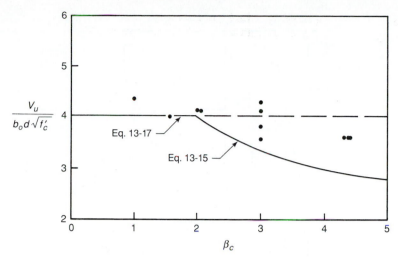

(a) Comparison of Eqs. (13-15) and (13-17) to tests of rectangular columns [13-16].

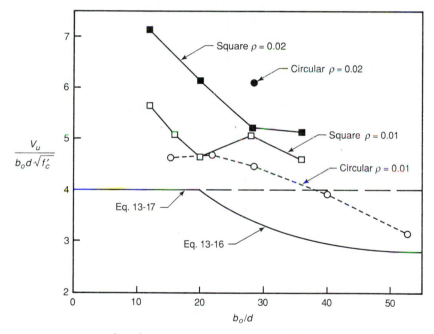

Fig. 13-44
Comparison of equations for
V_c with test data.

(b) Comparison of Eq. (13-16) to tests with large b_o/d ratios [13-17].

live load of 60 psf. The concrete strength is 3000 psi. The moments transferred from the slab to the column (or vice versa) are negligible. Check whether the shear capacity is adequate.

1.   **Determine the factored uniform load.**

$$w_u = 1.2\left(\frac{6}{12} \times 150 + 15\right) + 1.6 \times 60 = 204 \text{ psf}$$

$$= 0.204 \text{ ksf}$$

2.   **Check the one-way shear.** One-way shear is critical at a distance d from the face of the column. Thus, the critical sections for one-way shear are A–A and B–B in Fig. 13-45. The loaded areas causing shear on these sections are cross hatched. Their outer boundaries are lines of symmetry on which $V_u = 0$. Since the tributary area for section A–A is larger, this section will be more critical.

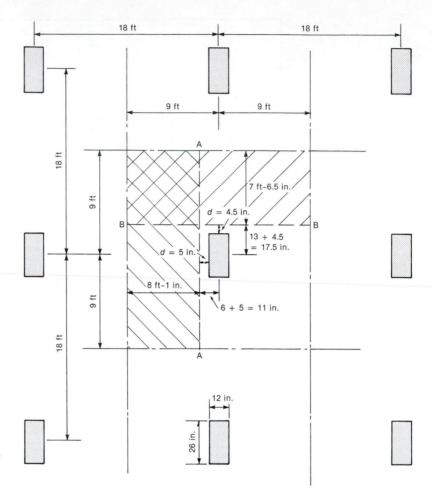

Fig. 13-45
Critical sections for one-way
shear at interior column—
Example 13-5.

(a) Compute V_u at section $A–A$.

$$V_u = 0.204 \text{ ksf} \times 8.08 \text{ ft} \times 18 \text{ ft} = 29.7 \text{ kips}$$

(b) Compute ϕV_n for one-way shear. Since there are no stirrups or other shear rein-forcement (from (6-9) and (6-14)), we have

$$\phi V_n = \phi V_c$$

where V_c for one-way shear is given by (6-8):

$$\phi V_c = 0.75(2\sqrt{f_c'}bd)$$
$$= 0.75\left(2\sqrt{3000} \times (18 \text{ ft} \times 12 \text{ in.}) \times \frac{5 \text{ in.}}{1000}\right)$$
$$= 88.7 \text{ kips}$$

Since $\phi V_c > V_u$, the slab is OK in one-way shear.

3.   Check the two-way shear. Two-way shear is critical on a rectangular section located at $d/2$ away from the face of the column, as shown in Fig. 13-46. The load on the cross-hatched area causes shear on the critical perimeter. Once again, the outer boundaries of this area are lines of symmetry, where V_u is assumed to be zero. The average depth, $d = 4.75$ in., will be used in these computations.

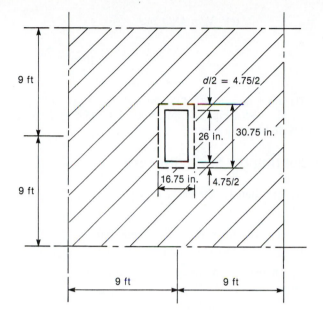

Fig. 13-46
Loaded area and critical sec-
tion for two-way shear—
Example 13-5.

(a) **Compute V_u on the critical perimeter for two-way shear.**

$$V_u = 0.204 \text{ ksf} \left[(18 \text{ ft} \times 18 \text{ ft}) - \left(\frac{16.75 \text{ in.}}{12} \times \frac{30.75 \text{ in.}}{12} \right) \text{ft}^2 \right]$$

$$= 65.4 \text{ kips}$$

(b) **Compute ϕV_c for the critical section.** V_c is the smallest of the following:

(a) $V_c = \left(2 + \dfrac{4}{\beta_c} \right) \sqrt{f_c'} b_o d$ (13-15)

                                                                                                      (ACI Eq. 11-33)

where

$$\beta_c = \frac{26}{12} = 2.167$$

$$b_o = 2(16.75 + 30.75) = 95 \text{ in.}$$

Therefore,

$$V_c = \left(2 + \frac{4}{2.167} \right) \sqrt{3000} \times 95 \times \frac{4.75}{1000}$$

$$= 95.1 \text{ kips}$$

(b) $V_c = \left(\dfrac{\alpha_s d}{b_o} + 2 \right) \sqrt{f_c'} b_o d$ (13-16)

                                                                                                      (ACI Eq. 11-34)

where $\alpha_s = 40$ for an interior column. Therefore,

$$V_c = \left(\frac{40 \times 4.75}{95} + 2 \right) \sqrt{3000} \times 95 \times \frac{4.75}{1000}$$

$$= 98.9 \text{ kips}$$

(c) $V_c = 4\sqrt{f_c'}b_o d$

$$= 4\sqrt{3000} \times 95 \times \frac{4.75}{1000} = 98.9 \text{ kips} \qquad (13\text{-}17)$$

(ACI Eq. 11-35)

Therefore, the smallest of (a), (b), and (c) is $V_c = 95.1$ kips. The loads cause $V_u = 65.4$ kips $V_u \geq \phi V_c$. $V_c = 95.1$ kips and $\phi V_c = 0.75 \times 95.1 = 71.3$ kips. **Since ϕV_c exceeds V_c, the slab is OK in two-way shear.** ∎

Augmenting the Shear Strength

If ϕV_c is less than V_u, the shear capacity can be increased by any of the following four methods:

1. Thicken the slab over the entire panel.
2. Use a drop panel to thicken the slab adjacent to the column.
3. Increase b_o by increasing the column size, or by adding a fillet or shear capital around the column.
4. Add shear reinforcement.

There are a number of schemes for increasing the two-way shear strength. Figure 13-47 shows load-deflection diagrams for five column-to-slab joints tested by Ghali and Megaly [13-18] to compare the strength and behavior of five methods of providing additional shear strength. In order of increasing ductility, specimen 1 was a flat plate without any shear strengthening, specimen 2 had a shear capital, specimen 3 had a drop panel, specimen 4 had stirrups, and specimen 5 had headed shear studs. Specimen 5, with shear studs, deflected three times as much as either of specimens 2 and 4, which failed in a brittle manner. In the test of the specimen with the capital, the outer corners of the capital spalled off, leaving an essentially round deepend region with a perimeter that was shorter than assumed for the original square capitals. The round approximate edge of this spalling is shown by the dashed circle in Fig. 13-39.

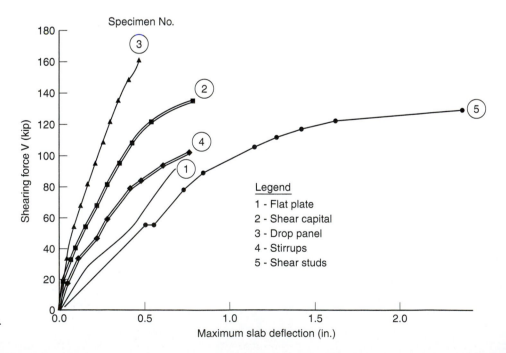

Fig. 13-47
Load-deflection curves from
tests of slabs strengthened
against two-way shear failure.
From [13-18].

Shear Reinforcement for Two-Way Slabs

Stirrups

ACI Section 11.12.3 allows the use of closed stirrups, provided there are longitudinal bars in all four corners of the stirrups as shown in Fig. 13-48a and b. Stirrups are allowed in slabs with effective depths, d, that exceed the larger of 6 in. or 16 stirrup diameters. A figure in ACI Commentary Section R.11.12.3 also suggests the use of bars bent in a "square wave" form, again with longitudinal bars enclosed within each corner of the stirrup. The precision required to bend and place either the small closed stirrups or the shallow square-wave bars makes either of these types of shear reinforcement labor-intensive and expensive. As a result, shear reinforcement consisting of stirrups or bent reinforcement is not widely used. The author does not encourage the use of shear reinforcement composed of stirrups or bent bars in two-way slabs of customary dimensions.

Based on two test series with a total of 15 slabs, the 1962 report of ACI–ASCE Committee 426, *Shear and Diagonal Tension* [13-14], stated that the measured shear strength of slabs with stirrups as shear reinforcement exceeded the shear strength of the concrete section, V_c, and also exceeded V_s from the stirrups alone. However, the report goes on to say that in most cases the measured shear strength was less than the sum of V_c and V_s. The authors of the 1962 report were very wary about the use of stirrups in slabs with thickness less than 10 in.

Structural Steel Shearheads.

Structural steel *shearheads*, shown in Fig. 13-48c, are designed according to ACI Section 11.12.4. Shearheads at exterior columns require special provisions to transfer the unbalanced moment resulting from loads on the arms perpendicular to the edge of the slab. The design of shearheads is discussed in [13-19].

Headed Shear Studs

The *headed shear studs* shown in Fig. 13-48d are not specifically recognized or permitted by the ACI Code although they act in the same machanical manner as well-anchored stirrup legs, satisfying the intent of Section 11.12.3, which allows the use of shear reinforcement consisting of bars or wires and allows single- or multiple-leg stirrups.

Headed shear-stud reinforcement at a column-slab connection consists of rows of vertical rods, each with a circular head or plate welded or forged on the top end, as shown in Figs. 13-48d and 13-49. These rows are placed extending out from the corners of the column. To aid in the handling and placement of the shear studs and to anchor the lower ends of the studs, they generally are shop-welded to flat steel bars at the desired spacing. The vertical rods are referred to as *headed shear reinforcement* or *headed shear studs*. The assembly of studs plus the bar is called a *stud rail*. Tests suggest the area of the bearing surface of the head should be at least 10 times the area of the vertical rod to enable the yield strength of the rod to be developed prior to crushing of the concrete under the head [13-20], [13-21], [13-22].

Design Using Headed Shear Studs

The following design requirements for headed shear studs are based on the ACI Committee 421 report on shear reinforcement for slabs [13-22]. They apply to stud rails placed in rows extending out into the slab from the corners of the column, as shown in Fig. (13-48d) and (13-49) [13-21], [13-22]. Headed shear stud reinforcement is designed using (6-9), (6-14), and (13-18) (ACI Eqs. (11-1), (11-2), and (11-41)). The critical sections for design are

(a)   the normal critical section for two-way shear is located at $d/2$ away from the face and corners of the column. We shall refer to this as the *inner critical section*, and

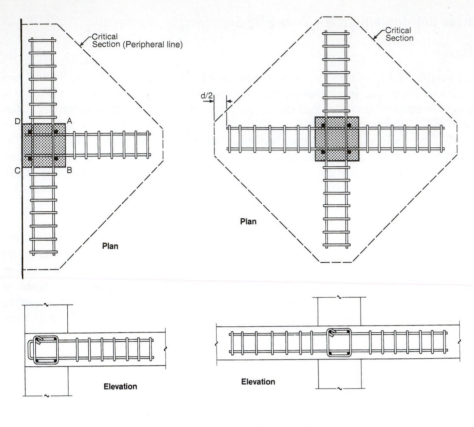

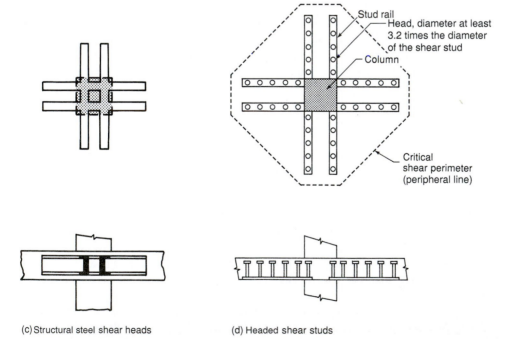

(a) Exterior column

(b) Interior column

(c) Structural steel shear heads

(d) Headed shear studs

Fig. 13-48
Shear reinforcement in slabs.

(b)   a second peripheral line is located at $d/2$ outside the outermost set of headed shear studs as shown by the dashed line in Fig. 13-48d. This is the *outer critical section*.

The bearing streeses under the heads anchoring the headed shear reinforcemnt confine the slab around the column much more effectively than normal closed stirrups do. The ACI Committee 421 report on the design of headed stud reinforcement for slabs [13-22] allows a shear of $V_c = 2\sqrt{f'_c}$ at the inner critical shear section at $d/2$ from the column in a slab with shear reinforcement. This is less than the $V_c = 4\sqrt{f'_c}$ allowed by ACI Section 11.12.2 on the same inner critical shear section in a slab *without shear reinforcement*. Refer to ACI 11.12.3. The reduction from $4\sqrt{f'_c}$ to $2\sqrt{f'_c}$ on the inner critical shear section is based, in part, on tests of slab–column connections with stirrups, quoted by ACI Committee 326 (now 445) [13-14], in which the measured shear strength, V_n, was less than the sum of V_c and V_s.

Hawkins [13-23] quotes tests of slabs with stirrups as shear reinforcement, in which the punching strength of a slab reinforced for shear was about

$$\phi V_n = V_c/2 + V_s$$

ACI Section 11.12.3.3 requires that shear reinforcement extend from the inner critical shear section located at $d/2$ from the face of the column to the outer critical shear section, which follows the peripheral line at $d/2$ beyond the location where v_c first drops below $2\sqrt{f'_c}$. For slab–column joints with stirrups, the shear stress carried by the concrete on the inner critical section is $2\sqrt{f'_c}$.

Design of headed shear reinforcement in slabs with no appreciable moment transfer

1.   Design shall be based on

$$V_u \leq \phi V_n \qquad\qquad (6\text{-}14)$$

$$(\text{ACI Eq. 11-1.})$$

$$V_n = V_c + V_s \qquad\qquad (6\text{-}9)$$

$$(\text{ACI Eq. 11-2})$$

and from ACI Section 11.12.6.2:

$$\phi v_n = \phi\,(V_c + V_s)/(b_o d) \qquad\qquad (13\text{-}18)$$

$$(\text{ACI Eq. 11-41})$$

2.   Shear reinforcement is required if the shear stress at the *inner critical section* at $d/2$ from the face of the column exceeds the smallest of the values of V_c given by (13-15), (13-16), or (13-17), provided that the shear stress, v_u, at $d/2$ from the face of the column does not exceed $\phi 8\sqrt{f'_c}$. The increase from $\phi 6\sqrt{f'_c}$ to $\phi 8\sqrt{f'_c}$ is allowed because headed shear reinforcement confines and strengthens the slab at the face of the column. In addition, v_c can be increased from $\phi 2\sqrt{f'_c}$ to $\phi 3\sqrt{f'_c}$ in the area confined by headed shear studs close to the column. Shear stud rails extend out into the slab until the maximum shear stress on a peripheral line located at $d/2$ beyond the outermost ring of shear studs drops below $\phi 2\sqrt{f'_c}$.

3.   The most common type of headed shear reinforcement consists of rows of vertical rods with flat circular heads welded or forged on their upper ends. Although shear studs have not been standardized, experience suggests the bottom surfaces of the heads should have a net bearing area of at least 10 times the cross-sectional area of the rod. Such a head allows the rod to yield before the concrete under the head crushes. This requires the diameter of circular heads to be at least 3.2 times the diameter of the vertical rods making up the studs. The lower ends of the rods are generally welded to a flat steel bar to form a *stud rail* that is nailed or otherwise fastened to the slab forms.

4. The upper limit on the distance, s_o, from the column to the first headed shear stud, and the upper limit on the spacing between subsequent headed shear studs, s, are given as:

If the total shear stress v_u is less than or equal to the index value of $V_{n,\,limit} = 6\phi\sqrt{f'_c}$ on the inner critical section at $d/2$ from the face of the column, s_o and s are limited to

$$s_o \leq 0.5d, \text{ and } s \text{ shall not exceed } 0.75d \qquad (13\text{-}19\text{a})$$

If v_u exceeds $6\phi\sqrt{f'_c}$,

$$s_o \leq 0.35d, \text{ and } s \leq 0.5d \qquad (13\text{-}19\text{b})$$

5. Headed shear stud rails are placed extending into the slab until the shear stress has decreased to $v_u \leq \phi2\sqrt{f'_c}$ on the *outer critical section* (located at $d/2$ outside of the outermost headed shear studs in a row). The outer critical section is shown by the dashed line in Fig. 13-48d. For an edge column the perimeter would be similar to the one in Fig. 13-48a.

6. ACI Section 11.12.3.1 states that the shear strength V_s carried by shear reinforcement should be in accordance with ACI Eq. (11-15) in ACI Section 11.5.6.2.

$$V_s = \frac{A_v f_y d}{s} \qquad (13\text{-}20)$$
$$(\text{ACI Eq. 11-15})$$

For shear reinforcement in a slab, A_v is equal to the sum of the cross-sectional areas of all the shear studs that cross an assumed 45° crack similar to the crack shown in Fig. 13-35b (equivalent to $A_v d/s$ in (13-20)). The crack is assumed to follow a peripheral line similar in plan to the perimeter of the column. The additional shear stress capacity provided by the headed shear reinforcement elements is

$$v_s = nA_{sh}f_y/b_o d \qquad (13\text{-}21)$$

where A_{sh} is $A_b d/s$, s is the spacing of studs along a line of shear studs, A_b is the area of one headed shear stud, n is the number of lines of shear studs intersected by the peripheral line under consideration, f_y is the yield strength of the shear studs, d represents the width of the 45° crack projected onto the slab surface, and b_o is the length of the shear perimeter. In tests, the yield strength of the studs has been utilized up to a maximum of 72,000 psi. Substituting for A_{sh}, Eq. (13-21) becomes,

$$v_s = \frac{nA_b f_y}{b_o s} \qquad (13\text{-}22)$$

ACI Section 11.12.6.2(b) states that the design shall take into account the variation of shear stress around the column. It is not clear how the array of shear studs around the column affects the shear stress resistance due to the studs.

EXAMPLE 13-6   Design an Interior Slab–Column Connection with Headed Shear Reinforcement—No Appreciable Moment Transfer

A 6-in.-thick flat-plate slab with No. 4 flexural reinforcement is supported by 12-in.-square columns spaced at 17 ft on centers *N–S* and 18 ft on centers *E–W*. The service loads on the slab are dead load = 95 psf, including self weight, and live load = 100 psf. The concrete strength is 3000 psi. Check the capacity of an interior slab–column connection. If necessary, design shear reinforcement using headed shear studs. Use the load and resistance factors in ACI 318-02 Sections 9.2 and 9.3.

1.   Select the critical section for two-way shear around the column. At this stage in the calculations the designer does not know whether shear reinforcement will be required. We will assume it is not and will redo the calculations if we are wrong. Assuming No. 4 flexural reinforcement, the average effective depth of the slab is

$$d = (6 \text{ in.} - 0.75 \text{ in. cover} - 0.5 \text{ in.}) = 4.75 \text{ in.}$$

The *critical shear section* for two-way shear in a flat plae extends around the column at $d/2 = 2.375$ in. from the face of the column as shown in Fig. 13-49.

The length of one side of the critical shear section around the column is

$$[12 + (2 \times 2.375 \text{ in.})] = 16.75 \text{ in,}$$

giving a shear perimeter $b_o = 4 \times 16.75 = 67.0$ in.

The area enclosed within the critical shear section is $(16.75 \times 16.75)/144 = 1.95 \text{ ft}^2$.

2.   Compute the shear acting on the critical shear section. The critical shear section is located at $d/2$ from the face of the column. The load combinations from ACI Section 9.2 will be used to compute the total factored dead and live loads. The basic combination for gravity load is

Load combination 9-2 $U = 1.2(D + F + T) + 1.6(L + H) + 0.5(L_r \text{ or } S \text{ or } R)$

A review of the structure indicated that F = fluid loads, T = imposed deformations, and H = soil loads are not expected on this slab. Furthermore, the slab is not a roof slab and hence the roof loads in the last brackets in this load combination do not apply, leaving

$$U = 1.2D + 1.6L$$

$$U = 1.2 \times 95 \text{ psf} + 1.6 \times 100 \text{ psf} = 274 \text{ psf}$$

The factored shear force on the critical shear section is

$$V_u = 274 \text{ psf} [17 \text{ ft} \times 18 \text{ ft} - 1.95 \text{ ft}^2] = 83.3 \text{ kips and } V_u/\phi = 111 \text{ kips}$$

Load and strength reduction factors from ACI Sections 9.2 and 9.3 were specified. Use the design equation

$$V_u \leq \phi(V_c + V_s)$$

where $\phi = 0.75$ for shear and torsion,

For this slab connection without shear reinforcement, Eq. (13-17) governs:

$$V_c = 4\sqrt{3000} \times b_o d = 4 \times 54.8 \times 67 \times 4.75 = 69.8 \text{ kips} < 111 \text{ kips}$$

Thus, shear reinforcemnt is required at the critical shear section. However, using shear reinforcement reduces V_c to 34.9 kips if stirrups are used, and 52.3 kips if shear studs are used:

$$V_c = 3\sqrt{3000} \times b_o d = 3 \times 54.8 \times 67 \times 4.75 = 52.3 \text{ kips}$$

This gives

$$V_s = V_u/\phi - V_c = 83.3/0.75 - 52.3 = 58.8 \text{ kips}$$

The maximum v_u allowed with headed shear studs is

$$v_{u,max} = \phi \times 8\sqrt{f_c'} = 0.75 \times 8 \times \sqrt{3000} = 329 \text{ psi}$$

corresponding to the maximum shear force, V_u allowed which is $329 \times 67 \times 4.75/1000 = 104.6$ kips.

Because the factored shear of $V_u = 83.3$ kips on this connection is less than 104.6 kips, the chosen slab depth and column size can be used. Web reinforcement will be needed. Because of the small depth of the slab, we shall use shear studs. Shear reinforcement is required for:

$$V_s = V_u/\phi - V_c$$
$$= 111.0 - 52.3 = 41.4 \text{ kips.}$$

3. Lay out the punching shear reinforcement—Using trial and error, we shall try eight stud rails, each with four 3/8 in diameter, headed shear studs, extending out in two orthogonal directions from each corner of the column as shown in Fig. 13-49. Rows of shear studs welded to bars will be placed parallel and perpendicular to the main slab reinforcement to cause the least disruption in the placement of the main slab steel.

Try eight stud rails, each with four, 3/8 in. diameter studs with 1.2 in. diameter heads, and $f_y = 60,000$ psi.

The spacing of the shear studs in a strip perpendicular to the column face depends on the shear stress at $d/2$ from the face of the column. A smaller spacing is required if this shear stress exceeds:

$$v_{u \text{ limit}} = \phi 6 \sqrt{f_c'} = 279 \text{ psi}$$

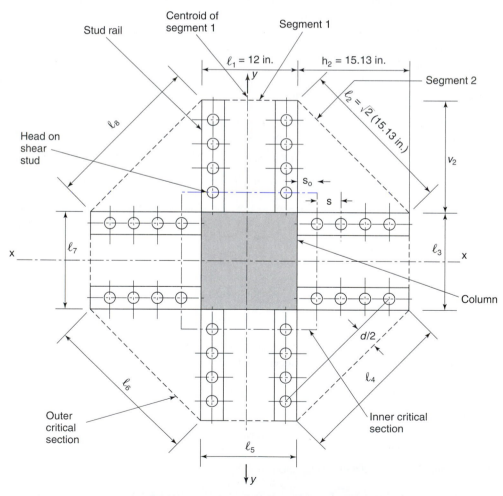

Fig. 13-49
Eight-sided critical section in a slab with headed shear studs.

The shear at $d/2$ from the face of the column is V_u = 83.3 kips, giving

$$v_{u1} = 83,300/(67.0 \times 4.75) = 262 \text{ psi.}$$

Because v_{u1} = 262 psi is less than $v_{u,\text{limit}}$ = 279 psi, (13-19a) governs and the first shear stud is located at s_o which is not taken farther than $d/2$ = 2.38 in from the face of the column., The maximum spacing of subsequent studs must not exceed $s \leq 0.75d$ = 3.56 in. These will be arranged in a fashion similar to Fig. 13-49.

Try eight stud rails with the first stud located at s_o = 2.25 in. from the column face. Subsequent studs are at a space of s = 3.5 in. with four, 3/8 in. diameter headed shear studs per rail.

The outermost studs are at 2.25 in. + 3 × 3.5 in. = 12.75 in. from the face of the column and the outer critical section is (12.75 + 2.38) = 15.13 in. from the face of the column.

The headed shear studs are arranged as shown in Fig. 13-49. The outer critical shear section is a series of straight line segments passing through points located $d/2$ outside the outer shear studs, as shown in Fig. 13-49. The perimeter of this *peripheral line* is

$$b_o = 4[(1.414 \times 15.13) + 12] = 133.5 \text{ in.}$$

The area inside this line $= [4(15.13 \times 15.13)/2 + 4 \times (12 \times 15.13) + (12 \times 12)] \text{ in.}^2$

$$= 1327 \text{ in.}^2 = 9.22 \text{ ft}^2$$

4.　Check the shear stresses on the outer critical section The factored shear force on the concrete at the outer critical section is

$$V_u = 274 \text{ psf} \times (17 \text{ ft} \times 18 \text{ ft} - 9.22 \text{ ft}^2) = 81,300 \text{ lb}$$

and,

$$v_u = \frac{81,300}{133.5 \times 4.75} = 128.2 \text{ psi}$$

The shear stress on the outer critical section is limited to $v_u = 2\phi\sqrt{3000}$ = 82.2 psi. **Since this is less than v_u = 128.2 psi for the same section, more shear studs are needed.**

Try eight stud rails each with eight shear studs. The distance from the face of the column to the outer critical section is (2.25 in. + 7 × 3.5 + 2.38) = 29.1 in.

$$b_o = 4(1.414 \times 29.1) + 4 \times 12 = 213 \text{ in.}$$

The area inside this line $= [4(12 \times 29.1) + 4 \times (29.1^2/2) + (12 \times 12)] = 3230 \text{ in.}^2 = 22.5 \text{ ft}^2.$

$$V_u = 274(17 \text{ ft} \times 18 \text{ ft} - 22.5 \text{ ft}^2) = 77.7 \text{ kips.}$$

$$v_u = \frac{77,700}{213 \times 4.75} = 76.8 \text{ psi.}$$

$$\phi v_c = 82.2 \text{ psi}$$

Because v_u = 76.8 psi is less than ϕv_c = 82.2 psi, this arrangement of headed shear studs is OK.

5.　Check shear stresses at the inner critical section.

$$\phi(v_c + v_s) \geq v_u$$

From step 1, the inner critical perimeter has a length, b_o = 67 in.

From step 2, the shear on the inner critical section is V_u = 83.6 kips. and $v_u = V_u/b_o d$ = 262 psi. The shear stress, v_s, carried by shear studs at this section is calculated with Eq. 13-22.

$$v_s = \frac{8 \times 0.11 \text{ in}^2 \times 60,000 \text{ psi}}{67.0 \text{ in.} \times 3.5 \text{ in.}} = 225 \text{ psi} \qquad (13\text{-}22)$$

The maximum shear stress allowed to be carried by the concrete in the area confined by shear studs $= 3\sqrt{f_c'} = 164$ psi. Thus, the maximum strength is

$$\phi(v_c + v_s) = 0.75(164 + 225) = 292 \text{ psi}$$

This exceeds $v_u = 262$ psi. Therefore,
Use 8 rows of eight 3/8-in. headed shear studs, 1 at 2.25 in. and 6 at 3.5 in. starting from face of column.   ∎

13-8 COMBINED SHEAR AND MOMENT TRANSFER IN TWO-WAY SLABS

Behavior of Slab–Column Connections Loaded with Shear and Moment

When lateral loads or unbalanced gravity loads cause a transfer of moment between the slab and column, or vice versa, the behavior is complex, involving flexure, shear, and torsion in the portion of the slab attached to the column as shown in Figs. 13-50 and 13-51. These will be discussed more fully later. Depending on the relative strengths in these three modes, failures can take various forms. Research on moment and shear transfer in slabs is reviewed in [13-11], [13-12], and [13-23]. The truss model discussed in Section 13-7 and Fig. 13-37 can be extended to the case of moment and shear transfer [13-24]. Figure 13-52 shows an interior slab–column connection subjected to a large moment and a small shear. As a result, the struts on the front side correspond to those in Fig. 13-37 and resist downward forces, while those on the back resist uplift.

A truss model for an edge column is shown in Fig. 13-53. It is similar to the model for the interior column except that the back face cannot transfer shear or moment. If the top reinforcement outside the column (bar A) is used to transfer moment to the column, the bar forces must be transferred to the column by horizontal anchoring struts. The components of the anchoring strut force acting perpendicular to the side face of the column must be resisted by bars such as bar B in Fig. 13-53 [13-24]. This action is relatively ineffective, with the result that bar A is not fully stressed. Figure 13-54 shows the steel strains measured at ultimate in the top bars perpendicular to the edge at the inner face of the edge column in a test specimen without an edge beam [13-25]. Only those bars that enter the column have yielded.

Moment and Shear Transfer at Slab–Column Connections—Design Methods

Four methods for designing slab–column connections transferring shear and moment are reviewed briefly, followed by an example using the design method presented by the ACI Commentary.

1. Beam method. The most fundamental method of designing for moment and shear transfer to columns in the literature considers the connection to consist of the joint region plus short portions of the slab strips and beams projecting out from each face of the joint where there is an adjacent slab [13-23]. These strips are loaded by moments, shears, and torsions, and they fail when the combined stresses correspond to one or more of the individual failure modes of the members entering the joint. For an edge beam the model gives

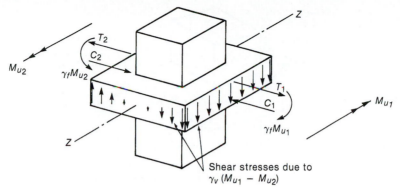

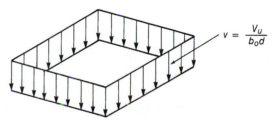

(a) Transfer of unbalanced moments to column.

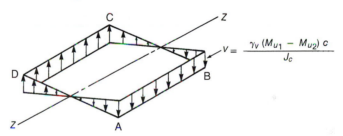

(b) Shear stresses due to V_u.

(c) Shear due to unbalanced moment.

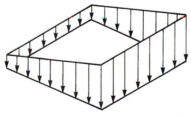

(d) Total shear stresses.

Fig. 13-50
Shear stresses due to shear
and moment transfer at an
interior column.

an estimate of the final distribution of the moments to the column. The beam model is conceptual only and has not been developed sufficiently for use.

2.   Traditional ACI Commentary Design Method (J_c Method).   The traditional ACI method for calculating the strengths of slab column joints transferring both shear and moment is by using

$$v_u = \frac{V_u}{b_o d} \pm \frac{\gamma_v M_u c}{J_c} \qquad (13\text{-}23)$$

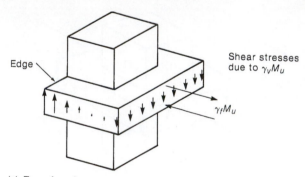

(a) Transfer of moment at edge column.

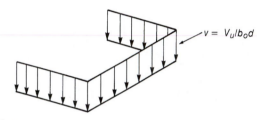

(b) Shear stresses due to V_u.

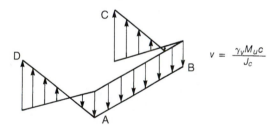

(c) Shear stresses due to M_u.

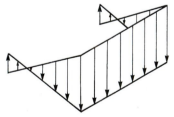

Fig. 13-51
Shear stresses due to shear and moment transfer at an edge column.

(d) Total shear stresses.

where

V_u is the shear being transferred. It acts through the centroid of the critical section for shear; γ_f is the fraction of the moment that is transferred by flexural stresses on the critical section and;

γ_v is the fraction of the moment that is transferred by shear stresses on the critical section, where

$$\gamma_v = 1 - \gamma_f$$

(13-24)
(ACI Eq. 11–39)

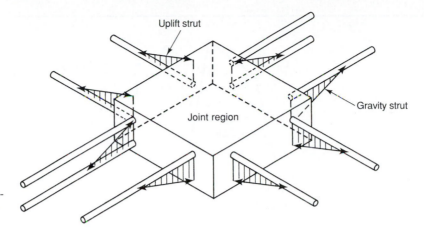

Fig. 13-52
Truss model for shear and moment shear transfer at an interior column. (From [13-10].)

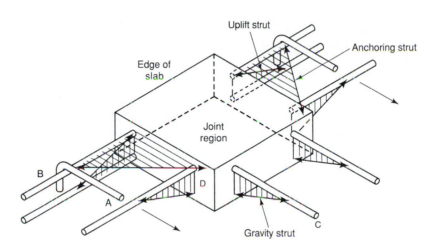

Fig. 13-53
Truss model for an edge column. (From [13-24].)

For slabs without shear reinforcement, the maximum value of v_u from (13-23) must satisfy stress limits from an empirical method [13-14], [13-26], [13-29] for designing for shear and moment transfer which assumes that the shear stresses on a critical section located at $d/2$ away from the face of the column due to the direct shear, V_u, shown in Figs. 13-50b and 13-51b, can be added to the shear stresses on the same section due to moment transfer, shown in Figs. 13-50c and 13-51c. Failure is assumed to occur when the maximum sum of the shear stresses reaches a limiting value. This analysis is highly idealized.

The term J_c in (13-23) is a section property similar to the *polar moment of inertia* of the critical shear section about the centroidal axes of the critical shear section. The calculation of γ_v and J_c is discussed in detail in the following subsections. If there are moments about two axes, a third term is added to (13-23) as is discussed later in this chapter.

A portion, $\gamma_f M_u$, of the total moment at a connection is transferred from the slab to the edge beam or column by flexure. The moment transfer to the edge beam and column is inefficient, as is shown by Fig. 13-54. In the flat-plate test illustrated by this plot, only the bars that framed into the column yielded at ultimate. In the design of a edge column slab connection it is customary to provide flexural reinforcement in a width of slab of width $3h + c_2$ centered on the edge column. Figure 13-54 shows that this steel does not transfer all the moment it is expected to. The remainder of the unbalanced moment, $\gamma_v M_u$, is assumed to be tranferred from slab to column by shear stresses on the critical shear section for shear and moment transfer.

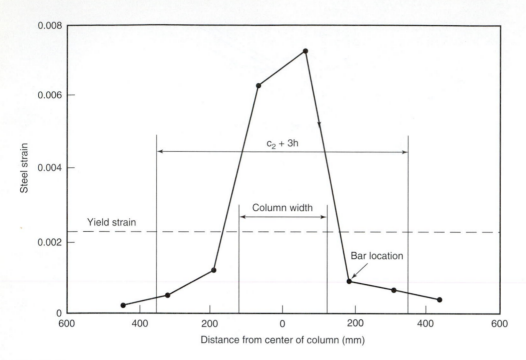

Fig. 13-54
Measured strains in top bars perpendicular to the edge of the slab within a width of $c_2 + 3h$ at the edge column at failure. From [13-25].

3. General Stress Equation Method. Prof. Ghali and his students have proposed that the rigorous *general stress equation* (13-40) should be used to solve for the shear stresses acting on the critical perimeter [13-21], [13-29]. The advantages of this approach come from the perceived strong tie between this equation and mechanics, which allows the user to extrapolate to allow new design applications. The general stress equation method is similar to the ACI Commentary method except the general stress equation (13-40) rather than (13-23) is used to solve for stresses at points along the critical shear perimeter. This requires replacement of J_c with the properties I_x, I_y, and I_{xy} about the principal axes of the connection area. The necessary equations are presented later in this section. It must be applied with care because it is very sensitive to the definitions of positive and negative directions, moments and stresses.

In connections containing shear reinforcement, the shape of the critical shear perimeter is distorted to fit around the location of the shear studs. This can be seen in the change in the critical section in Fig. 13-49, from a rectangle or square at $d/2$ from the faces of the column in Fig. 13-49 to the eight-sided critical shear perimeter at $d/2$ outside the outermost shear studs. This makes new definitions of γf and γ_v necessary. Unfortunately, these have not been accepted by ACI Committee 318.

4. Moment M and Shear V Interaction Diagram Method—at Edge Columns. ACI Committee 352 [13-30] showed that the interaction diagram for combined moment, M, and shear, V, can be represented by the two dashed lines in Fig. 13-58. The area inside those lines does not include any failures. This relationship can be used to modify the value of γ_f so that the fraction of the moment assumed to be transferred by flexure is increased while the fraction transferred by shear is reduced.

ACI Section 13.5.3.3 allows the designer to increase γ_f from 0.6 to 1.0 if the shear on the connection is less than 0.75 times the design shear, where

$$\gamma_v = 1 - \gamma_f$$

(13-24)
(ACI Eq. 11-41)

ACI Section 13.5.3.2 requires that $\gamma_f M_u$ be transferred by providing sufficient flexural reinforcement within a width extending 1.5 times the slab (or drop panel) thickness on each side of the column. The reinforcement already designed for flexure in this region can be used for this purpose.

The shearing stresses resulting from the shear, V_u, and the unbalanced moment, $\gamma_v(M_{u1} - M_{u2})$, are shown in Fig. 13-50b and c. The combined stress shown in Fig. 13-50d is given by M_u is the unbalanced moment $M_{u1} - M_{u2}$, c is the perpendicular distance from the centroidal axis (Z–Z) of the shear perimeter in Fig. 13-50c to the point where the shear stresses are being computed, and J_c is analogous to the polar moment of inertia of the shear perimeter about the axis Z–Z.

$$v_u \leq \phi v_n$$

(13-25)

$$\phi v_n = \frac{\phi V_c}{b_o d}$$

(13-26)
(ACI Eq. 11-40)

For slabs with shear reinforcement:

$$\phi v_n = \frac{\phi(V_c + V_s)}{b_o d}$$

(13-18)
(ACI Eq. 11-41)

In either case, V_c is from (13-15) through (13-17).

Fraction of Unbalanced Moment Transferred by Flexure, γ_f

In their 1962 report, ACI Committee 326 [13-14], recommended $\gamma_v = 0.20$. This implied that 80 percent of the moment was transferred by moment reinforcement and 20 percent by shear stress on the critical shear section located $d/2$ away from the face of the column.

Based on their tests on interior slab-column joints with square columns. Hanson and Hanson [13-28] set γ_f at 0.6. Thus they assumed that 60 percent of the moment was transferred by steel. This doubled γ_v for the slab, thereby doubling the shear stresses due to moment transfer.

ACI Sections 13.5.3.2 and 13.5.3.3 define the fraction of the moment transferred by flexure, γ_f, using (13-27) (ACI Eq. (13-1)):

$$\gamma_f = \frac{1}{1 + \left(\dfrac{2}{3}\right)\sqrt{b_1/b_2}}$$

(13-27)

where b_1 is the total width of the critical section measured perpendicular to the axis about which the moment acts, and b_2 is the total width parallel to the axis. In Fig. 13-50a, b_1 is perpendicular to the axis Z–Z.

For a square critical section with $b_1 = b_2$, γ_f is 0.60 and $\gamma_v = 1 - \gamma_f$ is 0.40, meaning that 60 percent of the unbalanced moment is transferred to the column by flexure and 40 percent by eccentric shear stresses. Equation 13-27 was derived to give $\gamma_f = 0.6$ for $b_1 = b_2$, as proposed by Hanson and Hanson [13-28] and to provide a transition to $\gamma_f = 1.0$ for a slab attached to the side of a wall, and to γ_f approaching zero for a slab attached to the end of a long wall. Equation (13-27) is based on a curve-fitting exercise and not on mechanics.

Fraction of Unbalanced Moment Transferred by Shear, γ_v

The moment transferred to the column by eccentric shear stresses is $\gamma_v M_u$, where

$$\gamma_v = 1 - \gamma_f \qquad (13\text{-}24)$$

Properties of the Shear Perimeter

Properties of the Critical Shear Perimeters of Column–Slab Connections—ACI Commentary Design Method

The current ACI design procedure for shear and moment transfer in slab–column connections applies to the interior-, edge-, and corner-column-to-slab connections shown in Fig. 13-56 and uses (13-23) to calculate the stresses on the critical shear section due to shear and moment. When calculating the section properties of the shear perimeters, the perimeters are generally subdivided into a number of individual sides similar to the one shown in Fig. 13-55.

The terms in (13-23) are all familiar except the following:

b_o, the length of the perimeter of the critical shear section,

J_c, defined by the ACI Commentary as a property analogous to a *polar moment of inertia* of the shear perimeter at the connection, and

γ_f and γ_v, which are the fractions of the moment, M_u, transferred by flexure and shear.

The polar moment of inertia term J_c is used to account for torsions on the faces of the shear perimeter. To compute J_c, the critical shear section is divided into two, three, four, or more individual sides as shown in Fig. 13-56. The term J_c is the sum of the individual values computed for each of the sides. Finally, the terms γ_f and γ_v divide the moment acting on the joint into a portion transferred by flexure and another transferred by shear stresses on the shear perimeter.

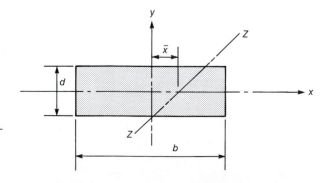

Fig. 13-55
Rectangle considered in computing the polar moment of inertia of a side of a critical section.

Polar Moment of Inertia, J_c, for an Isolated Side of a Critical Shear Perimeter

Textbooks on statics or mechanics of material discuss the calculation of the moments of inertia and polar moments of inertia of rectangular areas similar to the individual sides of the joints. Normally, these discussions of moments of inertia are limited to cases where the surface of the area under consideration can be plotted on a flat sheet of paper. The concept becomes ill-defined when it is applied to the open, there-dimensional critical shear sections shown in Fig. 13-56.

To compute the moments of inertia of a critical shear perimeter, the critical section is broken down into two to four, or more, individual plates. The rectangle in Fig. 13-55 represents side A–D of the three-sided shear perimeter in Fig. 13-56b. The polar moment of inertia of this rectangle about axis Z–Z, perpendicular to the plane of the rectangle and displaced a distance $\bar{x}$ from the centroid of the rectangle, is given by:

$$J_c = \overbrace{(I_x + I_y)} + \underbrace{A\bar{x}^2}$$

$$= \frac{bd^3}{12} + \frac{db^3}{12} + \overbrace{(bd)}\bar{x}^2 \tag{13-28a}$$

Section Properties of a Closed Rectangular Critical Shear Section

Centroid   The centroid of the four-sided critical shear section shown in Fig. 13-56a passes through the centroid of sides DA and CB.

Polar Moment of Inertia of a Rectangular Critical Shear Section

This calculation is similar to the derivation of J_c for an isolated side pierced by the axis, plus terms for the two sides parallel to the axis. in (13-28)

$$J_c = \frac{2(b_1 d^3)}{12} + \frac{2(d b_1^3)}{12} + 2(b_2 d)\left(\frac{b_1}{2}\right)^2 \tag{13-28b}$$

$$= \underbrace{I_x + I_y}_{J—\text{faces } BC \text{ and } DA} \quad \underbrace{A\bar{x}^2-\text{faces}}_{AB \text{ and } CD}$$

where

$b_1 = c_1 + 2(d/2)$

   $=$ length of the sides of the shear perimeter perpendicular to the axis of bending

$b_2 = c_2 + 2(d/2)$

   $=$ length of the sides of the shear perimeter parallel to the axis of bending

$c_1 =$ width of column perpendicular to the axis of bending

$c_2 =$ width of column parallel to the axis of bending

If large openings are present adjacent to the column, the shear perimeter will be discontinuous, as shown in Fig. 13-40. If this occurs, the calculation of the location of the centroid and J_c should include the effect of the holes.

Edge Columns

In the case of an edge column, the centroid of the critical perimeter lies closer to the inside face of the column than to the outside face. As a result, the shear stresses due to the moment shown in Fig. 13-51c are largest at the outside corners of the shear perimeter. If M_u is large and V_u is small, a negative shear stress may occur at these points. If M_u due to the combination of lateral loads and gravity loads is positive on this joint, rather than negative as shown in Fig. 13-51, the largest shear stress will occur at the outside corners.

Moments about an axis parallel to the edge (axis Z–Z in Fig. 13-56b). For the three-sided perimeter shown in Fig. 13-56b and taking b_1 as the length of the side perpendicular to the edge, as shown, the location of the centroidal axis Z–Z is calculated as follows:

Centroid

$$c_{AB} = \frac{\text{moment of area of the sides about } AB}{\text{area of the sides}}$$

$$= \frac{2(b_1 d)b_1/2}{2(b_1 d) + b_2 d} \tag{13-29}$$

The polar moment J_c of the shear perimeter is given by

$$J_c = \underbrace{2\left(\frac{b_1 d^3}{12}\right)}_{I_x} + \underbrace{2\left(\frac{d b_1^3}{12}\right)}_{I_y} + \underbrace{2(b_1 d)\left(\frac{b_1}{2} - c_{AB}\right)^2}_{A\bar{x}^2} + \underbrace{(b_2 d)c^2_{AB}}_{\substack{A\bar{x}^2 - \text{ face} \\ AB}}$$

$$\underbrace{\phantom{2\left(\frac{b_1 d^3}{12}\right) + 2\left(\frac{d b_1^3}{12}\right) + 2(b_1 d)\left(\frac{b_1}{2} - c_{AB}\right)^2}}_{J \text{ about } Z-Z \text{ of faces } DA \text{ and } BC} \tag{13-30}$$

Moments about an axis perpendicular to the edge (axis W–W in Fig. 13-56b.).

Centroid

$$c_{CB} = c_{AD} = \frac{b_2}{2} \tag{13-31}$$

Frequently, moments about an axis perpendicular to the edge are transferred from the slab to the column. In this case (13-23) becomes

$$v_u = \frac{V_u}{b_o d} \pm \frac{\gamma_v M_{u1} c}{J_{c1}} \pm \frac{\gamma_v M_{u2} c}{J_{c2}} \tag{13-23a}$$

where M_{u1} and J_{c1} refer to the moments from the span perpendicular to the edge and M_{u2} and J_{c2} refer to the moments from the span parallel to the edge. J_{c1} is given by (13-30).

A new equation is required for J_{c2} about axis $W\text{–}W$ in Fig. 13-56b:

$$c_{CB} = c_{AD} = \frac{b_2}{2} \tag{13-31}$$

$$J_{c2} = \underbrace{2(b_1 d)c_{CB}{}^2}_{\substack{A\bar{x}^2\text{—faces} \\ CB \text{ and } AD}} + \underbrace{\frac{b_2 d^3}{12}}_{I_x} + \underbrace{\frac{b_2^3 d}{12}}_{I_y} \tag{13-32}$$

$$\underbrace{\hspace{6cm}}_{J_c \text{ about } W-W \text{ of face } AB}$$

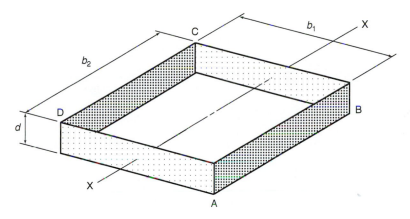

(a) Critical perimeter of an interior column

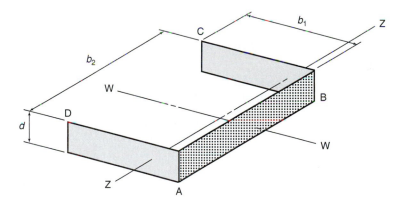

(b) Critical perimeter of edge column

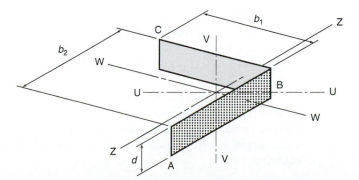

Fig. 13-56
Critical shear perimeters.

(c) Critical shear perimeter of corner column

Corner Columns

For the two-sided perimeter shown in Fig. 13-56c with sides b_1 and b_2, the location of the centroidal axis X–X is

$$c_{AB} = \frac{(b_1 d)b_1/2}{b_1 d + b_2 d} \tag{13-33}$$

The polar moment of inertia of the shear perimeter is

$$J_c = \underbrace{\frac{b_1 d^3}{12}}_{I_x} + \underbrace{\frac{b_1^3 d}{12}}_{I_y} + \underbrace{b_1 d\left(\frac{b_1}{2} - c_{AB}\right)^2}_{A\bar{x}^2} + \underbrace{(b_2 d)c_{AB}^2}_{A\bar{x}^2 \text{—face } AB}$$

$$\underbrace{\hspace{7cm}}_{J \text{ about } Z-Z \text{ of face } BC} \tag{13-34}$$

Circular Columns

For combined shear and moment calculations, ACI–ASCE Committee 426 [13-15] recommends that the shear perimeter of circular columns be based on that of a square column with the same centroid and the same length of perimeter. In this case, the equivalent square column would have sides of length $c = \sqrt{\pi} d_c/2 = 0.886 d_c$, where d_c is the diameter of the column. Two cases are illustrated in Fig. 13-57.

Fractions of Unbalanced Moment Transferred by Flexure, γ_f, and by Shear, γ_v

Design Method based on Principal Axes and Principal Moments of Inertia

Equation (13-34) for J_c of a corner column is written relative to the orthogonal x–x and y–y axes. For a square corner column, the (largest or) *major principal moment of inertia* of the shear perimeter acts about an axis on the 45° diagonal of the column through the corner, and the (smallest or) *minor principal moment of inertia* is about a perpendicular axis. In a discussion of the 1995 ACI Code revisions, Ghali [13-29] proposed two changes:

    1. Because J_c was not a standard term in mechanics it might be difficult to evaluate in some cases. He suggested that J_c be replaced by I_x and I_y, where I_x and I_y are d times (the moments of inertia about the x and y axes, respectively, of the line formed by the intersection of the critical shear perimeter and the top of the slab). It can be shown that I_x and I_y are about 3 to 6 percent smaller than J_{cx} and J_{cy}, because the term labeled I_x in (13-28) is omitted.

    2. Ghali also pointed out that the stresses computed by using moments of inertia relative to the orthogonal x and y axes do not equilibrate the applied moments for critical sections where the x and y axes are not principal axes. He suggested that the shear stresses arising from moment transfer be computed from the combined stress equation.

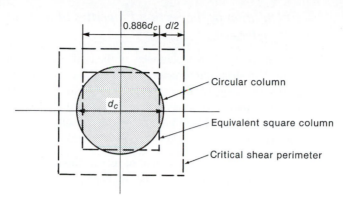

(a) Interior column.

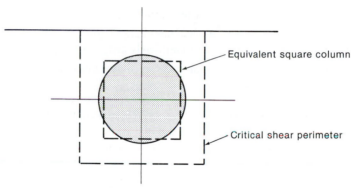

Fig. 13-57
Critical shear perimeters for
moment and shear transfer at
circular columns.

(b) Exterior column.

General Stress Equation, I_x Method

The I_x, I_y formulation is better than the J_c formulation in some cases.

For an interior column,   The moments of inertia I_x and I_y for an unbroken rectangular shear perimeter are equal to J_c minus the term $(bd^3/12)$. The moment of inertia, I, is generally about 95 percent of J_c.

For slabs with shear reinforcement,   The I_x formulation is frequently better than the J_c presentation, because the critical sections around the ends of the rows of shear reinforcement may include inclined sides for which J_c is ill defined. (See Fig. 13-49).

For corner columns,   The I_x formulation is sometimes used because the roughly 45° inclination of the principal axes of the critical section requires the use of the combined stress equation. Unfortunately, a special value of γ_f is needed and is not given by the code.

If there are large openings adjacent to the column,   The shear perimeter will be discontinuous, as shown in Fig. 13-40. In such cases, J_c should be replaced by I_x or I_y, and depending on the orientation of the inclined principal axes, either the J_c method or the combined stress equation should be used to compute the shear stresses. Again, a special value of γ_f may be needed.

Thick and thin slabs: For thin flat-plate slabs, the I_x formulation gives a moment of inertia about 95 percent of J_c, because the $(bd^3/12)$ term makes up about 5 percent of J_c. For thick slabs, such as those that occur in foundations, this term is considerably larger, and the J_c formulation may be desirable.

Calculation of the Section Properties of a Shear Perimeter—b_o, I_x, I_y, I_{xy}

Figure 13-49 shows a plan view of the critical shear perimeter for an interior slab–column joint with shear reinforcement. The eight-sided critical shear section is typical for interior joints. (See Fig. 13-48b and d and 13-49). Similarly, the critical sections for edge or corner columns may have five sides, as shown in Fig. 13-48a. To use the generalized combined stress equation, (13-40), I_x, I_y, and I_{xy} about the x-axis and y-axis through the centroid of the critical section for shear and moment transfer are needed.

To compute the moment of inertia about the x-axis, the perimeter of the critical section is broken into straight-line segments, as shown in Fig. 13-49, where

ℓ_1, ℓ_2 are the lengths of the sides, assumed straight consecutively around the perimeter

b_o is the length of the perimeter, equal to the sum of the lengths of all the segments

y_1, y_2, and x_1, x_2, and so on are the distances from the centroids of each of the segments to the x-axis and y-axis, respectively.

v_1, v_2, and so on are the projections onto the y-axis of lengths of the sides.

All these lengths are derived from the geometry of the critical shear section. The quantities we shall call I_x or I_y are not exactly moments of inertia, because all parts do not all lie in a single plane. Instead, the portions in the slab have both a length, ℓ_i, and a depth, d, extending through the thickness of the slab. We shall take this into account by rewriting the equation for the moment of inertia of a rectangular area about its own axis as

$$I_{x,\text{own}} = (d \times \ell_i)\, \ell_i^2/12$$

where $(d \times \ell_i)$ represents the area, A, of the vertical segment in the slab. This equation can be rewritten as

$$I_x = A_i \times \ell_i^2/12 \tag{13-35}$$

Equations for moments of inertia. When the centroid of a segment is located a distance y_c from the overall centroid of the critical section,

$$I_x = I_{x,\text{own}} + Ay_c^2 \tag{13-36}$$

where $I_{x,\text{own}}$ is the moment of inertia of the segment about the centroidal x-axis of the segment and y_c is the distance from the centroid of the segment to the centroid of the plan of the entire critical section.

Top and bottom segments of the critical section, Sides ℓ_1 and ℓ_5 in Fig. 13-49, are parallel to the axis about which the moment transfer occurs—in this case, the horizontal x-axis of the critical shear section, positive x to the right and positive y down. The moment of inertia of segment 1 about the x-axis is

$$I_x = I_{x,\text{own}} + Ay_c^2$$
$$= 0 + Ay_c^2 \tag{13-37}$$

where

$I_{x,\text{own}} = 0$ because the segment is parallel to the axis about which the moment of inertia is being taken

$A = \ell_i d$ is the area of the side of the segment

ℓ_i is the length of the side of the segment

d is the average effective depth of the slab at the critical section

y_c is the horizontal distance from the x-axis to the centroid of the segment

Horizontal segments, ℓ_3 and ℓ_7 in Fig. 13-49, have

$$I_x = I_{x,\text{own}} + Ay_c^2 \qquad (13\text{-}36)$$

where

$$I_{x,\text{own}} = A_3 v_3^2/12$$
$$v_3 = \ell_3$$

and the second term drops out if $y_3 = 0$. (See Fig. 13-55.)

Inclined segments: The angles between the inclined segments, 2, 4, 6, and 8, and the x-axis depend on the relative lengths of the shear reinforcement in the two directions. For segment 2,

$$I_x = I_{x,\text{own}} + Ay_c^2 \qquad (13\text{-}36)$$

where

$$I_{x,\text{own}} = (\ell_2 d)v_2^2/12 \qquad (13\text{-}38)$$

$I_{x,\text{own}}$ is the moment of inertia of segment 2 about its own centroidal axis parallel to the x-axis. The second term on the right-hand side of (13-36) transfers the I_x about the axis of the segment to the centroidal x-axis of the entire section.

Product of inertia:

$$I_{xy} = \Sigma[(\ell_i d)x_i y_i] \qquad (13\text{-}39)$$

summed over all the segments. The sign of I_{xy} depends on the signs of x_i and y_i in the quadrants around the column. For any one segment,

x positive x is to the right

y positive y is down

I_{xy} is the summation, over the cross-sectional area, of the product of the area of the segment $A = (\ell_i d)$ times the x_{ci} and y_{ci} distances from the centroid of the section to the centroid of the elemental area. For a given segment, I_{xy} is negative if the sign of either x or y is negative; that is, I_{xy} is positive if more of the area lies in the quadrants for which the product of the x and y distances is positive. $I_{xy} = 0$ if one or both axes are axes of symmetry.

General stress equation: Calculations of shear stresses using I_x, I_y, I_{xy} and principal axes must use the general stress equation (13-40).

$$v_u = \frac{V_u}{b_o d} + \left(\frac{\gamma_{vx}M_{ux}I_y - \gamma_{vy}M_{uy}I_{xy}}{I_x I_y - I_{xy}^2} \right) y + \left(\frac{\gamma_{vy}M_{uy}I_x - \gamma_{vx}M_{ux}I_{xy}}{I_x I_y - I_{xy}^2} \right) x \qquad (13\text{-}40)$$

where

γ_{vx} and γ_{vy} are the values of γ_v relative to the x- and y-axes, respectively I_x, I_y, and I_{xy} are as defined previously

Substituting this into (13-40) gives (13-23a). If calculations are carried out via the general combined stress equation, (13-40), for edge or corner columns, revised values of γ_v and γ_f should be used. See Ghali [13-21].

ACI Committee 318 did not accept the proposal from Ghali because a comparison of the two calculation procedures to test data for corner columns without shear reinforcement showed that the orthogonal axes method was adequately safe, while the principal axes procedure was quite conservative.

If either of the orthogonal x- and y-axes is an axis of symmetry of the critical section and the moments, M_{ux} and M_{uy}, under consideration are about the x- or y-axis, the product of inertia, I_{xy}, will vanish, and (13-23a) can be written as

$$v_u = \frac{V_u}{b_o d} + \frac{\gamma_{vx} M_{ux} y}{I_x} + \frac{\gamma_{vy} M_{uy} x}{I_y} \qquad (13\text{-}23b)$$

In the *general combined stress equation*, M_{ux} is the factored moment about the x-axis, where $M_{ux} = M_{u1} - M_{u2}$, shown in Fig. 13-50, and M_{uy} has a similar relationship about the y-axis. This is true only if one or both of the x- and y-axes is an axis of symmetry. By definition an axis of symmetry is a principal axis of the critical section. The original derivation of (13-23a) assumed that the x- and y-axes were principal axes, but erred in not specifically requiring them to be such.

The sign convention for directions, forces, and moments must be strictly followed to get the correct shear stresses:

x positive x is to the right, when the critical section is viewed in plan

y positive y is down

V positive shear, V, is assumed to act upwards along the axis of the column. (Positive V causes downward stresses on the slab outside the critical section and downwards on the slab between the critical section and the face of the column.)

M_x positive M_x about the x-axis of the critical section causes downward shear stresses on the portion of slab located below the x-axis of the critical section between the column face and the critical section, as shown in Figs. 13-50 and 13-51, and so causes tension at the bottom of the slab at positive y, by the right-hand rule.

M_y positive M_y causes tension at positive x, by the right-hand rule.

I_{xy} is positive if more of the area is in the $+x$, $+y$ quadrant.

Design of Headed Shear Reinforcement in Slabs Subjected to Combined Moment and Shear Transfer from Slab to Column

Design follows the rules presented in the last subsection of Section 13-7. Headed shear studs are placed in rows that extend out from the column until the shear stresses, ϕv_n, due to V and M are less than $2\sqrt{f_c'}$ at the critical section located $d/2$ outside the outermost line of shear studs that surrounds the column. Once again it should be noted that shear studs are not specifically sanctioned by the ACI Code.

Interaction Between Shear and Moment at Edge-Column to Slab Connections

Figure 13-58 is a plot of the moments and shears transferred at failure in a number of test specimens of edge columns in slabs without edge beams [13-26]. The vertical axis represents the ratio of the ultimate moment in a test, M_s, taken about the centroid of the critical shear perimeter, to the nominal moment capacity, M_n, of the flexural reinforcement within a

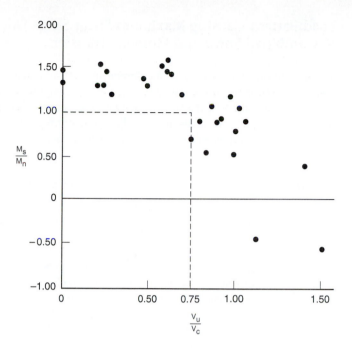

Fig. 13-58
Interaction between shear and
moment at edge columns.

width $c_2 + 3h$, centered on the column, where c_2 is the width of the column and h is the
slab thickness. (It should be noted that M_s was erroneously called M_u in [13-29] as pointed
out in the closure to the discussion of [13-30].) The horizontal axis is the ratio of the ultimate shear transferred in a test, V_u, to the shear capacity in the absence of moment
transfer, $V_c = 4\sqrt{f'_c}b_o d$. A safe lower envelope to the strengths is given by the bilinear
interaction diagram shown by the dashed lines in Fig. 13-58. This suggests that if V_u is less
than or equal to $0.75V_c$, the connection can transfer a moment at least equal to the nominal
moment capacity of the slab steel within the width $c_2 + 3h$. This does not imply that this
moment is all transferred by flexure; some will be transferred by the eccentricity of the
shear force on the front face of the critical section and some by torsion on the side faces.
The strains plotted in Fig. 13-54 indicate that the actual moment transferred by flexure in
the width $c_2 + 3h$ will be considerably less than M_n in some tests.

Adjusting γ_f

ACI Section 13.5.3.3 allows the designer to modify the value of γ_f in certain circumstances, provided that the value of the steel ratio ρ required to resist the moment $\gamma_f M_u$
within the width of the column plus $1.5h$ on each side of the column, $(c_2 + 3h)$, does not
exceed $0.375\rho_b$, based on the adjusted γ_f. For interior, edge and corner columns, ACI Section 13.5.3.3 can be paraphrased as follows:

   1.   **Interior column.** The value of γ_f from (13-24) may be increased by up to
25 percent, provided that V_u does not exceed $0.4\phi V_c$.

   2.   **Edge column, moments about an axis parallel to the edge.** The value of γ_f
from (13-24) may be increased up to 1.0, provided that V_u does not exceed $0.75\phi V_c$.

   3.   **Edge column, moments about an axis perpendicular to the edge.** The value
of γ_f from (13-24) may be increased by up to 25 percent, provided that V_u does not exceed
$0.4\phi V_c$.

   4.   **Corner column.** The value of γ_f from (13-24) may be increased up to 1.0, provided that V_u does not exceed $0.5\phi V_c$.

Load Pattern Causing Maximum Shear Stress Due to Combined Shear and Moment Transfer

The maximum shear stress at the columns supporting a slab should be computed from a consistent load case that produces a sum of shear stresses due to gravity load and shear stresses due to moment that is likely to be a maximum. The words *consistent load case* imply a combination of loads that are likely to occur together. The maximum shear generally occurs when all the panels surrounding a column are loaded with the factored dead and live loads.

For interior columns,   The maximum punching shear stresses will be approximated by full live load on all panels on the floor or roof that cause shear forces in the slab-column joint under consideration.

For edge columns,   (a) The critical loading for shear and moments about an axis parallel to the edge of the slab occurs with full factored dead and live loads simultaneously on both edge panels adjoining the column. (b) The maximum moment about axes perpendicular to the edge results from adjacent loaded and unloaded panels, or panels having significantly different lengths along the perimeter of the building. This last case produces significant moments, but the shear is less than the maximum shear that would occur with fully loaded floors.

For corner columns,   the critical loading is factored dead and live load on all panels on all floors, particularly all corner panels.

Calculation of Moment about Centroid of Shear Perimeter

The distribution of stresses calculated from (13-23) and illustrated in Fig. 13-51 assumes that V_u acts through the centroid of the shear perimeter and that M_u acts about the centroidal axis of this perimeter. In structural analyses, the calculated V_u and M_u may correspond to a different set of axes taken through the centroid of the column supporting the slab, and if so, it is necessary to compute the value of M_u corresponding to V_u acting through the centroid of the shear perimeter.

Interior Column

If the shear perimeter is symmetrical about two axes and has no gaps in it of the type shown in Fig. 13-40, the axis used to compute moments and shears will coincide with that of the shear perimeter, and no corrections are needed.

Edge Column: Moments Calculated by Direct Design Method

ACI Section 13.6.3.6 sets the gravity-load moment to be transferred between slab and column equal to $0.3M_0$. Although the axis for this moment is not defined in ACI Section 13.6.3.6, it is assumed to act about the centroid of the shear perimeter and is used without adjustment in calculating shear stresses around a column.

Edge Column: Moments Calculated by Equivalent Frame Analysis

In the equivalent frame method given in Chapter 14, moments are calculated at the point where the members meet at the center of the slab–column joint. The method used to transfer these moments to the centroidal axis of the shear perimeter is described in Section 14-2.

Shear Reinforcement for Slab–Column Connections Transferring Shear and Moment

ACI Section 11.12.6.2 states that the shear resulting from moment transfer shall be assumed to vary linearly about the centroid of the critical section defined in ACI Section 11.12.1.2. ACI Section 11.12.6.2 limits the maximum shear stress from moment and shear transfer to a column:

(a)   In slabs without shear reinforcement, the factored shear stress shall not exceed

$$\phi v_n = \phi V_c / b_o d \qquad\qquad (13\text{-}26)$$
$$(\text{ACI Eq. 11-40})$$

(b)   In slabs having shear reinforcement, the factored shear stress shall not exceed

$$\phi v_n = \phi (V_c + V_s)/(b_o d) \qquad\qquad (13\text{-}18)$$
$$(\text{ACI Eq. 11-41})$$

where V_c is computed as the smallest of the shear strengths from (13-15), (13-16), and (13-17) (ACI Eqs. (11-33), (11-34), and (11-35)). The shear V_s is defined by ACI Section 11.12.3.1 by reference to ACI Section 11.5. In the calculation of V_s, A_v is taken as the sum of the areas of the vertical rod portions of the headed studs crossed by one peripheral ring of studs, which has a shape similar to the shape of the column perimeter.

ACI Section 11.12.6.2 goes on to say that the design shall take into account the variation in shear stress around the column, and it limits the maximum shear stress to $2\sqrt{f_c'}$ at a critical section located $d/2$ outside the outermost line of stirrup legs or shear studs in the stud rails extending out from the column.

The design of a slab-to-column joint transferring shear and moment starts with a trial layout of the shear reinforcement, the calculation of the moment of inertia of the shear perimeter, and the computing of the shear stresses on the critical section. An example is given in step 11 of Example 13–8.

EXAMPLE 13–7   Checking Combined Shear and Moment Transfer at an Edge Column

A 12-in.-by-16-in. column is located 4 in. from the edge of a flat slab without edge beams, as shown in Fig. 13-59. The slab is 6.5 in. thick, with an average effective depth of 5.5 in. The concrete and reinforcement strengths are 3500 psi and 60,000 psi, respectively. The direct design method gives the statical moment, M_0, in the edge panel as 152 ft-kips. The shear from the edge panel is 31.3 kips. The portion of the slab outside the centerline of the column produces a factored shear force of 4.0 kips acting at 6 in. outside the center of the column. The loading that causes moment about the W–W axis (axis perpendicular to the edge) is less critical than the moments about the Z–Z axis (axis parallel to the edge). There is no moment about the axis perpendicular to the edge. Use load factors from ACI 318-02 Section 9.2.

1.   Locate the critical shear perimeter. The critical shear perimeter is located at $d/2$ from the sides of the column. Because the shear perimeter that intercepts the edge of the slab at D and C is shorter than one located at $d/2$ away from the outside face of the column, the perimeter shown in Fig. 13-40b is assumed to be critical.

2.   Compute the centroid of the shear perimeter.

$$c_{AB} = \frac{\Sigma Ay}{A} \qquad \text{where } y \text{ is measured from } AB$$

$$= \frac{2(18.75 \times 5.5)(18.75/2)}{2(18.75 \times 5.5) + 21.5 \times 5.5} = 5.96 \text{ in.}$$

Therefore, $c_{AB} = 5.96$ in. and $c_{CD} = 12.79$ in.

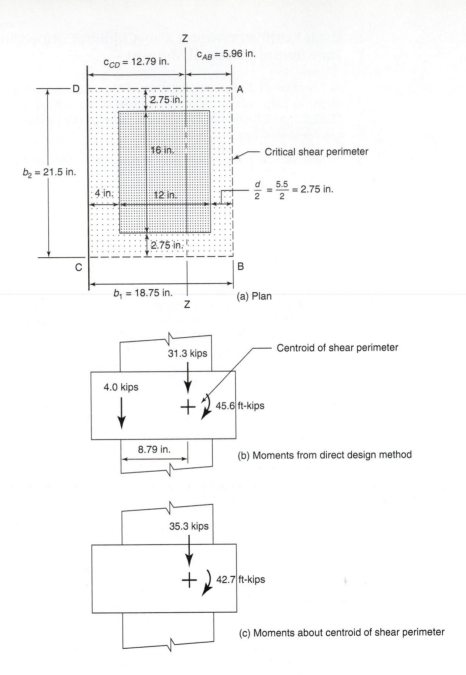

Fig. 13-59
Slab–column joint—
Example 13–7.

3. **Compute the moment about the centroid of the shear perimeter.** For the portion of the slab between the centerline of the edge columns and the centerline of the first interior columns, $M_0 = 152$ ft-kips and $V_u = 31.3$ kips. ACI Section 13.6.3.6 defines the moment to be transferred as $0.3M_0 = 45.6$ ft-kips. We shall assume that this is about the centroid of the shear perimeter and that V_u from the edge panel acts through this point. The portion of the slab outside the column centerline has a shear of $V_{uc} = 4$ kips acting at 6 in. $+ 2.79$ in. from the centroid of the shear perimeter. The total moment about the centroid of the shear perimeter is

$$M_u = 45.6 \text{ ft-kips} - 4 \text{ kips} \times 8.79/12 \text{ ft} = 42.7 \text{ ft-kips}$$

The total shear to be transferred is

$$V_u = 31.3 \text{ kips} + 4 \text{ kips} = 35.3 \text{ kips}$$

   4.   Compute ϕV_c and $V_u/\phi V_c$. V_c is the smallest value given by (13-15)–(13-17) (ACI Eqs. 11–33 to 11–35). Because the load factors from ACI Section 9.2 are used, we will use the strength reduction factor $\phi = 0.75$.

   (a) $V_c = \left(2 + \dfrac{4}{\beta_c}\right)\sqrt{f_c} b_o d$ (13-15)

where

$$\beta_c = \frac{\text{long side of the column}}{\text{short side of the column}} = \frac{16}{12}$$

$$= 1.33 \text{ and } (2 + 4/1.33) = 5.01$$

$$b_0 = 2 \times 18.75 \text{ in.} + 21.5 \text{ in.} = 59 \text{ in.}$$

$$\phi V_c = 0.75\left(2 + \frac{4}{1.33}\right)\sqrt{3500} \times 59 \times 5.5 = 72.1 \text{ kips}$$

   (b) $V_c = \left(\dfrac{\alpha_s d}{b_o} + 2\right)\sqrt{f_c'} b_o d$ (13-16)

where $\alpha_s = 30$ for an edge column and $(30 \times 5.5/59 + 2) = 4.80$

$$\phi V_c = 0.75\left(\frac{30 \times 5.5}{59} + 2\right)\sqrt{3500} \times 59 \times 5.5 = 69.1 \text{ kips}$$

   (c) $V_c = 4\sqrt{f_c'} b_o d$ (13-17)

$$\phi V_c = 0.75 \times 4\sqrt{3500} \times 59 \times 5.5 = 57.6 \text{ kips}$$

   Therefore $\phi V_c = 57.6$ kips, and $V_u/\phi V_c = 35.3$ kips/57.6 kips $= 0.613$.

   5.   Determine the fraction of the moment transferred by flexure, γ_f.

$$\gamma_f = \frac{1}{1 + \frac{2}{3}\sqrt{\dfrac{b_1}{b_2}}}$$ (13-27)

$$= \frac{1}{1 + \frac{2}{3}\sqrt{\dfrac{18.75}{21.5}}}$$ (ACI Eq. 13-1)

$$= 0.616$$

ACI Section 13.5.3.3 allows γ_f to be increased up to 1.0 if $V_u/\phi V_c$ does not exceed 0.75 and the resulting $\rho \leq 0.375\rho_b$ within a width $c_2 + 3h$ centered on the column. From step 4, $V_u/\phi V_c = 0.613$. Therefore, try γ_f equal to 1.0 and check the reinforcement needed.

   6.   Design the reinforcement for moment transfer by flexure.

$$\text{Width effective for flexure: } c_2 + 3h = 16 \text{ in.} + 3 \times 6.5 \text{ in.}$$
$$= 35.5 \text{ in.}$$
$$\text{Moment: } 1.0 \times 42.7 \text{ ft-kips} = 42.7 \text{ ft-kips}$$

Assume that $jd = 0.925d$. Then

$$A_s = \frac{M_u}{\phi f_y jd}$$

$$= \frac{42.7 \times 12,000}{0.9 \times 60,000 \times 0.925 \times 5.5}$$

$$= 1.87 \text{ in.}^2$$

Try 10 No. 4 bars $= 2.00$ in.^{2}. Since this calculation is based on a guess of jd, we should compute a for $A_S = 2.00$ in.^{2}; then recompute A_s, using that value of a:

$$a = \frac{A_s f_y}{0.85 f'_c b}$$

$$= \frac{2.00 \times 60,000}{0.85 \times 3500 \times 35.5}$$

$$= 1.136 \text{ in.}$$

$$A_s = \frac{M_u}{\phi f_y(d - a/2)} = \frac{42.7 \times 12,000}{0.9 \times 60,000(5.5 - 1.136/2)}$$

$$= 1.92 \text{ in.}^2$$

Use 10 No. 4 bars $= 2.00$ in.^{2}. For this A_s,

$$\frac{a}{d} = \frac{1.14}{5.5} = 0.207$$

From (4-21),

$$\frac{a_b}{d} = \beta_1\left(\frac{87,000}{87,000 + f_y}\right)$$

$$= 0.85\left(\frac{87,000}{87,000 + 60,000}\right)$$

$$= 0.503$$

and

$$0.375 \frac{a_b}{d} = 0.189$$

Since a/d exceeds $0.375a_b/d$, ρ exceeds $0.375\rho_b$, and γ_f cannot be increased to 1.0. We shall assume that it can be increased to some value between 0.616 and 1.0, provided the value of ρ for the resisting moment does not exceed $0.375\rho_b$. We shall arbitrarily provide eight No. 4 bars with $A_s = 1.60$ in.^{2}. These will give $\rho < 0.375\rho_b$ and will transfer $\phi M_n = 36.3$ ft-kips $= \gamma_f M_n$. The moment transferred by shear is $\gamma_v M_n = 42.7$ ft-kips $- 36.3$ ft-kips $= 6.4$ ft-kips.

7. **Compute the torsional moment of inertia, J_c.** Equation 13-28 applies to moments about the Z–Z axis. From (13-30),

$$J_c = 2\left(\frac{b_1 d^3}{12}\right) + 2\left(\frac{d b_1^3}{12}\right) + 2(b_1 d)\left(\frac{b_1}{2} - c_{AB}\right)^2 + (b_2 d)c^2_{AB}$$

$$= \frac{2 \times 18.75 \times 5.5^3}{12} + \frac{2 \times 5.5 \times 18.75^3}{12}$$

$$+ 2(18.75 \times 5.5)\left(\frac{18.75}{2} - 5.96\right)^2 + (21.5 \times 5.5)5.96^2$$

$$= 13,170 \text{ in.}^4$$

8. **Compute the shear stresses.**

$$v_u = \frac{V_u}{b_o d} \pm \frac{\gamma_v M_u c}{J_c} \tag{13-23}$$

$$v_u = \frac{35,300}{59 \times 5.5} \pm \left(\frac{6.4 \times 12,000}{13,170}\right)c$$

$$= 108.7 \pm 5.83c$$

The shear stress at AB is

$$v_{u,AB} = 108.7 + 5.83 \times 5.96$$
$$= 143.4 \text{ psi}$$

The shear stress at CD is

$$v_{u,CD} = 108.7 - 5.83 \times 12.79$$
$$= 34.1 \text{ psi}$$

From step 4, $\phi V_c = 57.6$ kips,

$$\phi v_c = \frac{\phi V_c}{b_o d} = \frac{57,600}{59 \times 5.5}$$
$$= 177 \text{ psi}$$

Since ϕv_c exceeds v_u, the shear is OK. **Use a 12-in.-by-16-in. column, as shown in Fig. 13-59, with eight No. 4 bars at 5 in. on centers, centered on the column.** ■

Shear and Moment Transfer to a Spandrel Beam

The negative moments at the exterior end of the panel cause torsional moments in the spandrel beam or in the strip of slab between the exterior columns. These are compatibility torsional moments, since they exist because the spandrel beam restrains the rotation of the edge of the panel. In such a case, ACI Section 11.6.2.2 allows the torsional moment in the spandrel beam at a section d away from the face of the columns to be reduced to

$$T_{\text{compat}} = \phi 4 \sqrt{f'_c \left(\frac{A_{cp}^2}{p_{cp}} \right)} \tag{13-41}$$

Consider the slab in Example 13-4. A_{cp} and p_{cp} are the area and perimeter of the beam cross section defined in ACI Section 13.2.4. (See Fig. 13-20.) For Example 13-4, the spandrel beam has the section shown in Fig. 13-34, and

$$A_{cp} = 12 \times 16 + 8 \times 8 = 256 \text{ in.}^2$$
$$p_{cp} = 16 + 12 + 8 + 8 + 8 + 20 = 72 \text{ in.}$$

Assuming that $f'_c = 3000$ psi, it follows that

$$\phi 4 \sqrt{f'_c} \left(\frac{A_{cp}^2}{p_{cp}} \right) = 0.75 \times 4\sqrt{3000} \left(\frac{256^2}{72} \right)$$
$$= 12.5 \text{ ft-kips}$$

Thus, the torsional moment at $d = 13.5$ in. on each side of the column can be reduced to 12.5 ft-kips. The remaining torque, (68.4 ft-kips—from step 2(c) of Example 13-4) minus $(2 \times 12.5 \text{ ft-kips}) = 43.5$ ft-kips, either is redistributed, as was done in Example 7-2, or is transferred to the column and the spandrel beam between the column and the sections d away from the column. We shall assume that it is transferred to the column and edge beam and will take the effective width for this transfer as the smaller of (a) the distance between the critical sections for torsion from ACI Section 13.5.3.2 is $16 + 3 \times 8 = 40$ in., (b) the effective width for flexure in a slab without edge beams, or (c) the width between the effective sections for torsion, $16 + 2 \times 13.5 = 43$ in., suggested by ACI–ASCE Committee

352 [13-30], as shown in Fig. 13-63. The moment transferred would require 1.68 in.^{2}, or nine No. 4 bars in this width.

Normally, the torsional moment in the spandrel beam would not be redistributed in the design of a two-way slab, especially if the slab serves as the spandrel beam. These calculations have been carried out to illustrate the process. The spandrel beam would need No. 4 closed stirrups at 6 in. on centers if designed for the shear and the full torsion and No. 3 closed stirrups at 6.25 in. on centers if designed for the shear and the reduced torsion. This reduction in the stirrups required in the spandrel beam would probably be accompanied by an increase in the widths of the torsional cracks in the spandrel beam.

Shear in Slabs: Design

In designing two-way slabs or footings with little or no transfer of moments from the slab to the column, it is customary to select the slab thickness on the basis of $V_u \simeq (0.85$ to $1.0)\phi V_c$, unless there are holes adjacent to the column. The presence of holes adjacent to the column reduces the shear perimeter, and unless the holes are placed symmetrically around the column, they introduce an eccentricity between the line of action of the shear and the centroid of the shear perimeter.

At an edge column in a two-way slab subjected to gravity loads, the shear stresses resulting from moment transfer may be of the same order of magnitude as those from direct shear transfer: hence, the initial trial slab thickness should be chosen on the basis of $V_u \simeq 0.5$ to $0.55(\phi V_c)$. The unbalanced moment to be transferred can be reduced by cantilevering the slab past the column centerline. The moment from the cantilever will offset some of the unbalanced moment to be transferred.

In unbraced, laterally loaded flat-plate frames, the shear induced by the lateral-load moments will increase the shears at either the inside or outside face of an edge column. Either case could cause serious problems.

13-9   DETAILS AND REINFORCEMENT REQUIREMENTS

Drop Panels

Drop panels are thicker portions of the slab adjacent to the columns, as shown in Fig. 13-60 or Fig. 13-2c. They are provided for three main reasons:

1.   The minimum thickness of slab required to limit deflections (see Section 13-6) may be reduced by 10 percent if the slab has drop panels conforming to ACI Section 13.3.7. The drop panel stiffens the slab in the region of highest moments and hence reduces the deflection.

2.   A drop panel with dimensions conforming to ACI Section 13.3.7 can be used to reduce the amount of negative moment reinforcement required over a column in a flat slab. By increasing the overall depth of the slab, the lever arm, jd, used in computing the area of steel is increased, resulting in less required reinforcement in this region.

3.   A drop panel gives additional depth at the column, thereby increasing the area of the critical-shear perimeter.

ACI Sections 9.5.3.2 and 13.3.7 give essentially the same requirements for the minimum size of a drop panel. These are illustrated in Fig. 13-60. The only difference between the two sections is that ACI Section 13.3.7.3 states that, in computing the negative-moment flexural reinforcement, the thickness of the drop panel below the slab used in the calculations shall not be taken greater than one-fourth of the distance from the edge of the drop panel to the edge of the column or column capital. If the drop panel were deeper than this, it is assumed that the maximum compression stresses would

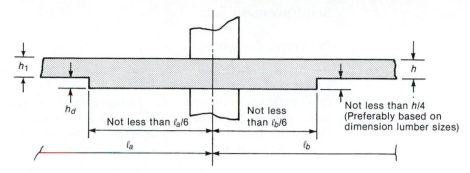

Fig. 13-60
Drop panel.

Minimum size of drop panels—ACI Sections 9.5.3.2 and 13.3.7.

occur at some point above the bottom of the drop panel so that the full depth would not be effective.

For economy in form construction, the thickness of the drop, shown as h_d in Fig. 13-60 should be related to actual lumber dimensions, such as $\frac{3}{4}$ in., $1\frac{1}{2}$ in., $3\frac{1}{2}$ in., $5\frac{1}{2}$ in. (nominal 1-in., 2-in., 4-in., or 6-in. lumber sizes) or some combination of these, plus the thickness of plywood used for forms. The drop should always be underneath the slab (in a slab loaded with gravity loads), so that the negative-moment steel will be straight over its entire length.

Column Capitals

Occasionally, the top of a column will be flared outward, as shown in Fig. 13-61 or 13-2c. This is done to provide a larger shear perimeter at the column and to reduce the clear span, ℓ_n, used in computing moments.

ACI Section 6.4.6 requires that the capital concrete be placed at the same time as the slab concrete. As a result, the floor forming becomes considerably more complicated and expensive. For this reason, other alternatives, such as drop panels or shear reinforcement in the slab, should be considered before capitals are selected. If capitals must be used, it is desirable to use the same size throughout the project.

The diameter or effective dimension of the capital is defined in ACI Section 13.1.2 as that part of the capital lying within the largest right circular cone or pyramid with a 90° vertex that can be included within the outlines of the supporting column. The diameter is measured at the bottom of the slab or drop panel, as illustrated in Fig. 13-61. This diameter c is used to define the effective width for moment transfer, $c_2 + 3h$, and to define the clear span, ℓ_n. Concrete in the capital outside the 45° lines (see Fig. 13-61) can be used to increase the shear strength.

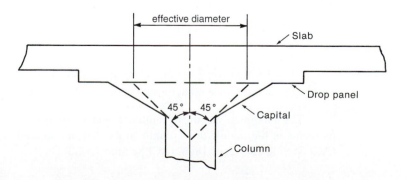

Fig. 13-61
Effective diameter of capital.

Reinforcement

Placement Sequence

In a flat plate or flat slab, the moments are largest in the slab strips spanning the long direction of the panels. As a result, the reinforcement for the long span is generally placed closer to the top and bottom of the slab than is the short-span reinforcement. This gives the largest effective depth for the largest moment. For slabs supported on beams having α greater than about 1.0, the opposite is true, and the reinforcing pattern should be reversed. If a particular placing sequence has been assumed in the reinforcement design, it should be shown or noted on the drawings. It is important, however, that the same arrangements of layers should be maintained throughout the entire floor, to avoid confusion in the field. Thus, if the East–West reinforcement is nearest the surface in one area, this arrangement should be maintained over the entire slab if at all possible.

Cover and Effective Depth

ACI Section 7.7.1 specifies the minimum clear cover to the surface of the reinforcement as $\frac{3}{4}$ in. for No. 11 and smaller bars, provided that the slab is not exposed to earth or to weather. For concrete exposed to weather, the minimum clear cover is $1\frac{1}{2}$ in. for No. 5 and smaller bars and 2 in. for larger bars. Concrete parking decks exposed to deicing salts should have greater cover and epoxy-coated bars. ACI Commentary Section 7.7.5 suggests 2 in. of cover in such a case. It may be necessary to increase the cover for fire resistance. This will be specified in the local building code.

The reinforcement in a two-way slab with spans of up to 20 ft will generally be No. 4 bars; up to 25 ft, No. 5 bars; and over 25 ft, No. 5 or No. 6 bars. For the long span of a flat plate or flat slab, $d = h - 3/4 - 0.5d_b$, and for the short span, $d = h - 3/4 - 1.5d_b$. For preliminary design these can be taken as to be the following:

for flat-plate or flat-slab spans up to 25 ft (7 m),

$$\text{long span } d \simeq h - 1.1 \text{ in. } (30 \text{ mm}) \tag{13-42a}$$

$$\text{short span } d \simeq h - 1.7 \text{ in. } (45 \text{ mm}) \tag{13-42b}$$

for flat-plate or flat-slab spans over 25 ft (7 m),

$$\text{long span } d \simeq h - 1.15 \text{ in. } (30 \text{ mm}) \tag{13-42c}$$

$$\text{short span } d \simeq h - 1.9 \text{ in. } (50 \text{ mm}) \tag{13-42d}$$

It is important not to overestimate d in slabs, because normal construction inaccuracies tend to result in smaller values of d than are shown on the drawings.

Spacing Requirements, Minimum Reinforcement, and Minimum Bar Size

ACI Section 13.4.1 requires that the minimum area of reinforcement provided for flexure should not be less than

0.0020bh if Grade 40 or 50 deformed bars are used

0.0018bh if Grade 60 deformed bars or welded-wire fabric is used

The maximum spacing of reinforcement at points of maximum positive and negative moments in the middle and column strips shall not exceed two times the slab thickness (ACI Section 13.3.2) or 18 in. (ACI Section 7.12.3).

Although there is no code limit on bar size, the Concrete Reinforcing Steel Institute recommends that top steel in slab should not be less than No. 4 bars at 12 in. on centers, to give adequate rigidity to prevent displacement of the bars under ordinary foot traffic before the concrete is placed.

Calculation of the Required Area of Steel

The calculation of the steel required is based on (4-34) and (13-43), as illustrated in Examples 4-4 and 10-1, namely,

$$A_s = \frac{M_u}{\phi f_y jd} \tag{13-43}$$

where $jd = 0.90$ to $0.95d$ for slabs of normal proportions. As recommended in Section 4-2 and Example 10-1, j can be assumed to be 0.925 for the first trial. Once a trial value of A_s has been computed for the section of maximum moment, the depth of the compression zone, a, will be computed and used to compute a more nearly correct value of $jd = d - a/2$. This will be used to compute A_s at all sections in the slab. It is also necessary to check whether A_s exceeds $A_{s,min}$ at all sections and the section is tension-controlled.

Bar Cutoffs and Anchorages

For slabs without beams, ACI Section 13.3.8.1 allows the bars to be cut off as shown in Fig. 13-62). Where adjacent spans have unequal lengths, the extension of the negative-moment bars past the face of the support is based on the length of the longer span.

Prior to 1989, ACI Fig. 13.3.8 also showed details for slabs with alternate straight and bent bars. While such a bar arrangement is still allowed, the details have been deleted from ACI Fig. 13.3.8 in the code because the bent–straight bar arrangement is rarely used now. If such a slab were designed or checked, the bar details should be based on Fig. 13.4.8 of the 1983 code, modified in accordance with 2002 ACI Section 13.3.8.5.

ACI Section 13.3.4 requires that all negative-moment steel perpendicular to an edge be bent, hooked, or otherwise anchored in spandrel beams, columns, and walls along the edge to develop f_y in tension. If there is no edge beam, this steel should still be hooked to act as torsional reinforcement and should extend to the minimum cover thickness from the edge of the slab.

Detailing Slab Reinforcement at an Edge Column

The shear and moment transfer from the slab to an exterior or corner column assumes that the edge of the slab will act as a torsional member (as with the truss in Fig. 13-53). ACI–ASCE Committee 352 [13-30] has recommended details for edge-column-to-slab connections. Working in part from these recommendations, the authors recommend the following reinforcing details:

(a) The top steel required to transfer the moment $\gamma_f M_u$ according to ACI Section 13.5.3.2 should be placed in a width equal to the smaller of $2(1.5h) + c_2$ or $2c_e + c_2$, centered on the column (Fig. 13-63), where c_e is the distance from the inner face of the column to the edge of the slab but not more than the depth of the column, c_1, and c_2 is the width of the column.

(b) Within the width defined in (a), edge reinforcement should be provided at a spacing of $1.5d$ or less. This can be provided by hooking the top steel perpendicular

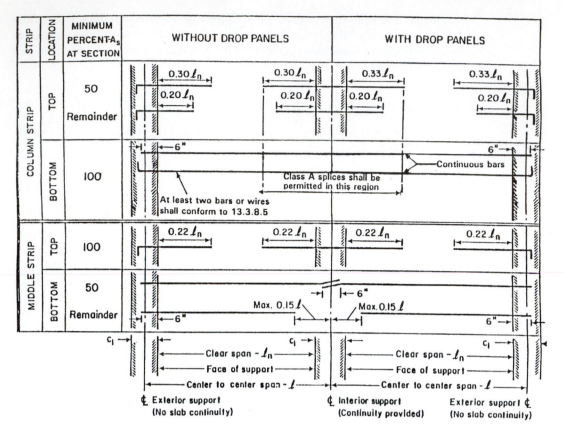

Fig. 13-62
Minimum cutoff points for slabs without beams (ACI Section 13.3.8.1).

to the edge with 180° bends or with hairpin bars at least No. 3 in size. Each leg of a hairpin should have an extension of at least ℓ_d of the top bars.

Structural Integrity Reinforcement

When a punching-shear failure occurs, it completely removes the shear capacity at a column, and the slab drops, pulling the top reinforcement out through the top of the slab, as shown in Fig. 13-36. If the slab falls on the slab below, that slab will probably fail also, causing a progressive type of failure. Research reported by Mitchell and Cook [13-31] suggests that this can be prevented by providing reinforcement through the column at the bottom of the slab. ACI Section 13.3.8.5 requires that all bottom steel in the column strip in each direction either be continuous or be lap-spliced or mechanically spliced with a Class A splice. (See Fig. 13-62.) At least two of the bottom bars in each direction must pass through the column core. In this context the words "column core" mean that these bars should be between the corner bars of the column. At exterior and corner columns, at least two bars perpendicular to the edge should be bent, hooked, or otherwise anchored within the column core. The bars passing through the column are referred to as *integrity steel*.

Corner Reinforcement

At discontinuous corners where the slab is supported on relatively stiff beams, as shown in Fig. 13-64a, there is a tendency for the corner to lift off its support unless a downward reaction is provided at the corner. In Fig. 13-64a, point B has deflected downward relative to

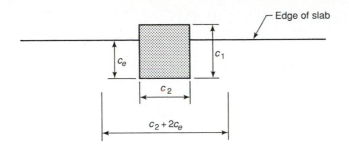

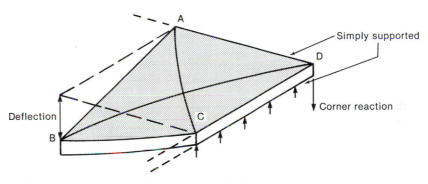

Fig. 13-63
Effective width for moment
transfer at exterior column.

(a) Deformations of corner of simply supported slab.

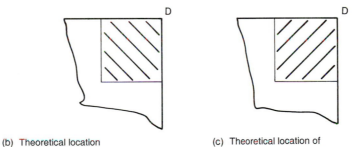

Fig. 13-64
Moments in corner of slab
supported on stiff beams.

(b) Theoretical location
of corner reinforcement
at bottom of slab.

(c) Theoretical location of
corner reinforcement at top
of slab.

the simply supported edges $A–D$ and $C–D$. If corner D is held down, strips $A–C$ and $B–D$
develop the curvatures shown. Strip $A–C$ develops positive moments and, ideally, should
be reinforced with bars parallel to $A–C$ at the bottom of the slab, as shown in Fig. 13-64b.
Strip $B–D$ develops negative moments and should be reinforced with reinforcement parallel to $B–D$ at the top of the slab, as shown in Fig. 13-64c. The diagonal bars are very awkward to place and hence are replaced by two layers of steel in two orthogonal directions,
one at the top and one at the bottom of the slab, parallel to the two sides of the slab. Two
such mats are required, one at the top of the slab and one at the bottom.

ACI Section 13.3.6 requires special corner reinforcement in the exterior corners of all
slabs with edge beams having α greater than 1.0. The reinforcement in each direction in each
face is designed for a moment per unit width equal to the largest positive moment per unit width
in the panel in question and extends one-fifth of the longer span each way from the corner.

13-10 DESIGN OF SLABS WITHOUT BEAMS

EXAMPLE 13-8 Design of a Flat-Plate Floor without Spandrel Beams: Direct Design Method

Figure 13-65 shows a plan of a part of a flat-plate floor. There are no spandrel beams. The floor supports its own dead load, plus 25 psf for partitions and finishes, plus a live load of 40 psf. The slab extends 4 in. past the exterior face of the column to support an exterior wall that weighs 300 lb/ft of length of wall. The story-to-story height is 8 ft, 10 in. Use 4000-psi normal-weight concrete and Grade-40 reinforcement. The columns are 12 in. × 12 in. or 12 in. × 20 in. and are oriented as shown. Select the thickness, compute the design moments, and select the reinforcement in the slab. Use the load combinations from ACI 318-02, Appendix C.

To fully design the portion of slab shown, it is necessary to consider two East–West strips (Fig. 13-66a) and two North–South strips (Fig. 13-66b). In steps 3 and 4, the computations are carried out for the East–West strip along column line 2 and the edge strip running East–West along line 1. Steps 5 and 6 deal similarly with the North–South strips.

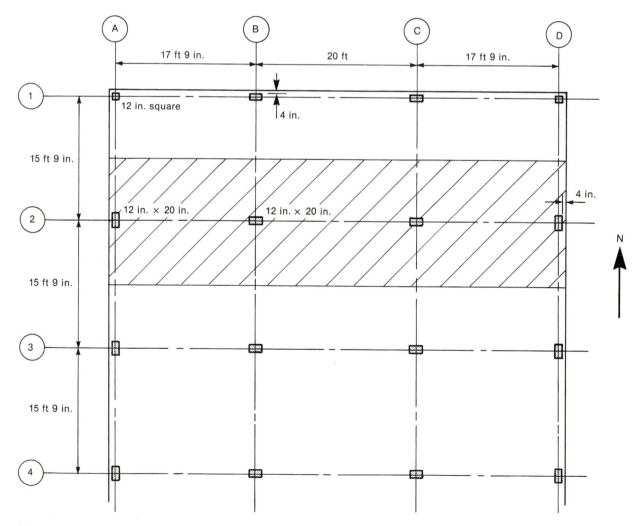

Fig. 13-65
Plan of flat-plate floor—Example 13-8.

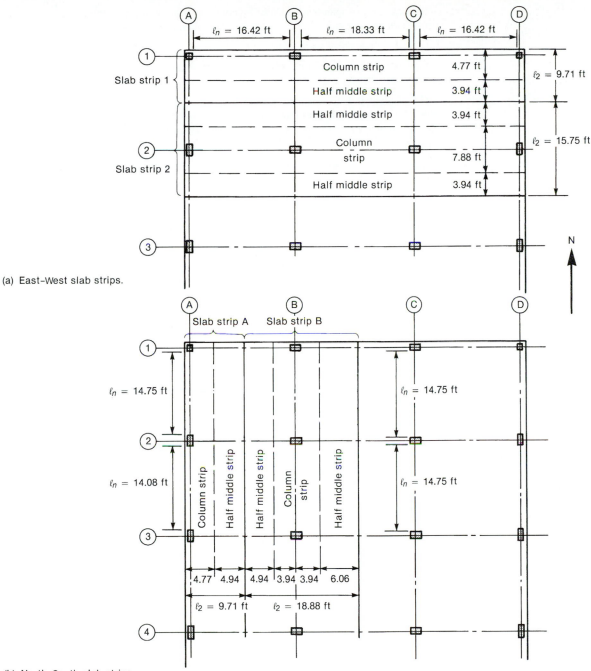

(a) East–West slab strips.

(b) North–South slab strips.

Fig. 13-66
Division of slab into strips for design—Example 13-8.

1.   Select the design method, load combinations, and strength-reduction factor. ACI Section 13.6.1 allows the use of the direct design method if the following conditions are satisfied:

(a)   Minimum of three consecutive spans each way—therefore, OK.

(b)   Longer span/shorter span shall be between 1 and 2. In this slab the radio is 1.13—therefore, OK.

(c) Successive span lengths differ by not more than one-third of the longer span. Thus, short span/long span ≥ 0.667: 17.75/20 = 0.89—therefore, OK.

(d) Columns offset up to 10 percent—OK.

(e) All loads are uniformly distributed gravity loads. Strictly speaking, the wall load is not uniformly distributed, but use DDM anyway.

(f) Unfactored live load not greater than two times the unfactored dead load. Using Table 13-1, estimate the slab thickness as $\ell/36 = 20 \times 12/36 \simeq 6.5$ in. The approximate dead load is $6.5/12 \times 150 + 25 = 106$ psf. This exceeds one-half of the live load—therefore, OK.

(g) No beams—therefore, ACI Section 13.6.1.6 does not apply, and it is not necessary to make this check.

Therefore, use the direct design method.

The problem statement requires use of load combinations from ACI Appendix C. Thus, the gravity load combination reduces to

$$U = 1.4D + 1.7L$$

and the strength-reduction factors are: Flexure, $\phi = 0.90$ shear, $\phi = 0.85$

2. **Select the thickness.**
   (a) **Determine the thickness to limit deflections.** From Table 13-1, the minimum thicknesses of panels 1 to 4 are as follows:

   Panel 1–2–A–B (corner):

   $$\text{max. } \ell_n = (17 \text{ ft } 9 \text{ in.}) - (6 + 10) \text{ in.} = 197 \text{ in.}$$

   $$\text{min. } h = \frac{\ell_n}{33} = 5.97 \text{ in.}$$

   Panel 1–2–B–C (edge):

   $$\text{max. } \ell_n = 220 \text{ in.}$$

   $$\text{min. } h = \frac{\ell_n}{33} = 6.67 \text{ in.}$$

   Panel 2–3–A–B (edge):

   $$\text{max. } \ell_n = 197 \text{ in.}$$
   $$\text{min. } h = 5.97 \text{ in.}$$

   Panel 2–3–B–C (interior):

   $$\text{max. } \ell_n = 200$$

   $$\text{min. } h = \frac{\ell_n}{36} = 6.11 \text{ in.}$$

Try $h = 6.75$ in. Because slab deflections are frequently a problem, it is best to round the thickness up rather than down.

   (b) **Check the thickness for shear.** Check at columns B2 and B1.

   Column B1: The maximum moment about the Z–Z axis occurs when the area E–F–G–H–E in Fig. 13-67 is all loaded with live load. The maximum moment about the axis perpendicular to the edge occurs with live loads on one side of the column line B. This loading will not be used, because it corresponds to a live load shear that is half that for loading of E–F–G–H–E. Assume lines of zero shear in the edge or corner panels at 0.44ℓ from the centers of the edge or corner columns.

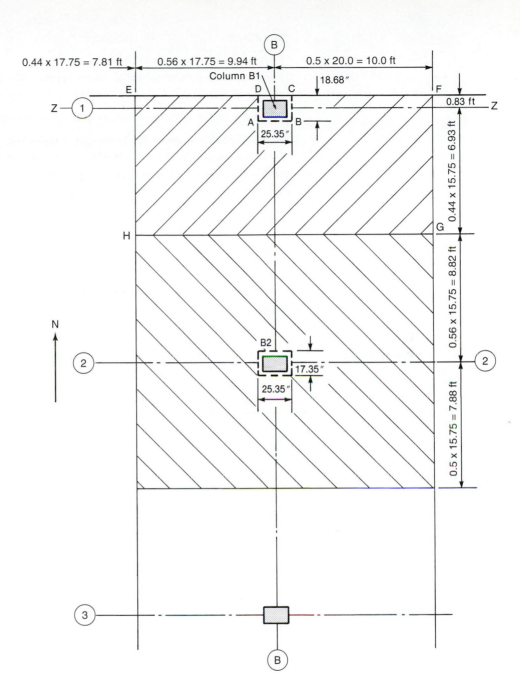

Fig. 13-67
Initial critical shear perimeters and tributary areas for
columns *B*1 and *B*2.

Tributary areas slab strip A (Fig. 13-66 and 66b): **Span *A*1–*A*2**

Width is 17.75 ft. from center line of columns in line *A* to center line of columns line *B*.
Line of zero shear is at 0.44 × 17.75 = 7.81 ft from line *A* and 9.94 ft. from line *B*.
Edge distance is 4 + 6 in. = 0.833 ft.

Length is 15.75 ft from center line of columns in line 1 to center line of columns in line 2.
Line of zero shear is at 0.44 × 15.75 = 6.93 ft from line 1 and 8.82 ft from line 2.
End distance is 4 + 6 in. = 0.833 ft.

Span A2–A3

Width is same as for span A1–A2.

Length is 15.75 ft from center line of columns in line 2 to center line of columns in line 3.

Line of zero shear is at 0.5 × 15.75 = 7.88 ft from lines 2 and 3.

Tributary areas for slab strip B (Fig. 13-67): Span B1–B2

Width is 17.75 ft from center line of columns in line A to center line of columns in line B.

Line of zero shear is at 0.56 × 17.75 = 9.94 ft from line B.

Width is 20.0 ft from center line of columns in line B to center line of columns in line C.

Line of zero shear is at 0.5 × 20.0 ft = 10.0 ft from lines B and C.

Length is same as for span A1–A2.

End distance is 0.833 ft.

Span B2–B3

Width is same as for span B1–B2.

Line of zero shear is at 9.94 ft toward line A and 10.0 ft toward line C.

Length is same as for span A1–A2.

$$w_u = 1.4\left(\frac{6.75}{12} \times 150 + 25\right) + 1.7(40) = 221 \text{ psf}$$

$$\text{Average } d \simeq 6.75 \text{ in.} - 1.4 \text{ in.} \simeq 5.35 \text{ in.}$$

Column B2: Assume lines of zero shear at $0.44\,\ell$ from the centers of the edge or corner columns. Then

$$b_o = 2(12 + 2 \cdot \frac{d}{2} + 20 + 2 \cdot \frac{d}{2}) = 85.4 \text{ in.}$$

$$V_u = 221\left[(9.94 + 10.0)(8.82 + 7.88) - \frac{17.35 \times 25.35}{144}\right]$$

$$= 72.9 \text{ kips}$$

From (13-15),

$$\phi V_c = 0.85\left(2 + \frac{4}{\beta_c}\right)\sqrt{f_c'}b_o d$$

$$= 0.85\left(2 + \frac{4}{20/12}\right)\sqrt{4000} \times 85.4 \times 5.35$$

$$= 108,000 \text{ lb}$$

From (13-16),

$$\phi V_c = 0.85\left(\frac{\alpha_s d}{b_o} + 2\right)\sqrt{f_c'}b_o d$$

$$= 0.85\left(\frac{40 \times 5.35}{85.4} + 2\right)\sqrt{4000} \times 85.4 \times 5.35$$

$$= 110,700 \text{ lb}$$

From (13-17),

$$\phi V_c = \phi 4\sqrt{f_c'}b_o d = 0.85 \times 4\sqrt{4000} \times 85.4 \times 5.35$$

$$= 98,250 \text{ lb}$$

Therefore, $\phi V_c = 98,250$ lb. Thickness is OK, since $\phi V_c > V_u$.

Because this calculation ignores the shear stresses due to moment, it is wise to consider thickening the slab if ϕV_c is less than about $1.2 V_u$ at an interior column.

Column B1:

$$b_o = 18.68 + 25.35 + 18.68 = 62.7 \text{ in.}$$

$$V_u = 221 \text{ kips}\left[(9.94 + 10.0)(6.93 + 0.833) - \left(\frac{18.68 \times 25.35}{144} \right) \right] + (1.4 \times 0.30 \times 19.94)$$

$$= 33.48 \text{ kips}$$

(The last term is the factored wall load.)

$$V_u = 33.48 \text{ kips} + 8.4 \text{ kips} = 41.9 \text{ kips}$$

ϕV_c is the smallest of the following:

$$\phi V_c = 0.85\left(2 + \frac{4}{20/12} \right)\sqrt{4000} \times 62.7 \times 5.35$$

$$= 79,350 \text{ lb}$$

$$\phi V_c = 0.85\left(\frac{30 \times 5.35}{62.7} + 2 \right)\sqrt{4000} \times 62.7 \times 5.35$$

$$= 82,200 \text{ lb}$$

$$\phi V_c = 0.85(4\sqrt{4000} \times 62.7 \times 5.35) = 72,100 \text{ lb}$$

$$\phi V_c = 1.72 \text{ times } V_u$$

Although this seems OK, this calculation ignores the portion of the moment that is transferred by shear stresses. Generally, if ϕV_c is less than 1.8 to 2 times V_u at an exterior column, the slab may have inadequate shear capacity for combined shear and moment. This will be checked in step 9 of this example. We shall increase the thickness to 7 in. **Use a 7-in. slab throughout.**

(c)   **Compute the final value of w_u.**

$$w_u = 1.4\left(\frac{7}{12} \times 150 + 25 \right) + 1.7(40)$$

$$= 225.5 \text{ psf, say 226 psf}$$

If the area of any of the panels had exceeded 400 ft^2, it would be possible to reduce the live load before factoring it in this calculation. In such a case, w_u may differ from panel to panel. Live-load reduction factors are used in step 3 of Example 13-10.

(d)   **Compute α for the exterior beams.** Since there are no edge beams, $\alpha = 0$.

3.   Compute the moments in a slab strip along column line 2 (Fig. 13-66a). This piece of slab acts as a rigid frame spanning between columns A2, B2, C2, and D2. In this slab strip, slab panels A2–B2 and C2–D2 are "end panels," and B2–C2 is an "interior panel." Columns A2 and D2 are "exterior columns," and B2 and C2 are "interior columns."

These calculations are generally tabulated as shown in Tables 13-3 to 13-6. The individual calculations will be followed in detail for slab strips 2 and 1 shown in Fig. 13-66a.

Line 1 (Table 13-3). ℓ_1 is the center-to-center span in the direction of the strip being designed.

Line 2. ℓ_n is the clear span in the direction of the strip. (See also Example 13-3.)

Line 3. ℓ_2 is the width perpendicular to ℓ_1. The lengths ℓ_2 and ℓ_n are shown in Fig. 13-66a.

Line 4. w_u may differ from panel to panel, especially if panel areas are large enough to be affected by live-load reductions.

Line 5. See also Example 13-3.

Line 6. The moment in the exterior panels is divided by using Table 13-2 (ACI Section 13.6.3.3) as in Example 13–4:

negative moment at exterior end of end span $= 0.26M_0$

positive moment in end span $= 0.52M_0$

negative moment at interior end of end span $= 0.70M_0$

TABLE 13-3 Calculation of Negative and Positive Moments for East–West Slab Strip 2—Example 13-8

	A2		B2			C2		D2
1. ℓ_1 (ft)		17.75		20			17.75	
2. ℓ_n (ft)		16.42		18.33			16.42	
3. ℓ_2 (ft)		15.75		15.75			15.75	
4. w_u (ksf) from step 2(c)		0.226		0.226			0.226	
5. $M_o = w_u \ell_2 \ell_n^2/8$ (ft-kips)		120.0		149.5			120.0	
6. Moment coefficients	−0.26	0.52	−0.70	−0.65	0.35	−0.65	−0.70	0.52	−0.26
7. Negative and positive moments (ft-kips)	−31.2	62.4	−84.0	−97.2	52.3	−97.2	−84.0	62.4	−31.2
8. Sum of column moments	31.2		24.1			24.1		31.2

TABLE 13-4 Calculation of Negative and Positive Moments for East–West Slab Strip 1—Example 13-8

	A1		B1			C1		D1
1. ℓ_1 (ft)		17.75		20			17.75	
2. ℓ_n (ft)		16.42		18.33			16.42	
3. ℓ_{2edge} (ft)		9.71		9.71			9.71	
4. w_u (ksf)		0.226		0.226			0.226	
5. $M_o = w_u \ell_{2edge} \ell_n^2/8$ (ft-kips)		74.0		92.2			74.0	
6. Moment coefficients	−0.26	0.52	0.70	−0.65	0.35	−0.65	−0.70	0.52	−0.26
7. Negative and positive moments	−19.2	38.5	−51.8	−59.9	32.3	−59.9	−51.8	38.5	−19.2
8. Column moments from slab load (ft-kips)	19.2		14.9			14.9		19.2
9. Wall load/ft (kip/ft)		0.42		0.42			0.42	
10. Wall M_o (ft/kips)		14.2		17.6			14.2	
11. Negative and positive moments from wall load (ft-kips)	−3.7	7.4	−9.9	−11.5	6.2	−11.5	−9.9	7.4	−3.7
12. Column moments from wall load (ft-kips)	3.7		1.6			1.6		3.7
13. Total column moments (ft-kips)	22.9		16.5			16.5		22.9

For the interior panel, from ACI Section 13.6.3.2,

$$\text{negative} = -0.65M_0, \quad \text{positive} = 0.35M_0$$

Line 7. This is the product of lines 5 and 6. If the negative moments on the two sides of an interior column differ by less than about 15 to 20 percent of the larger, design the slab at the joint for the larger negative moment. If the difference exceeds this, carry out a moment-distribution analysis at the joint. (See step 3 of Example 13-10.)

TABLE 13-5 Calculation of Negative and Positive Moments for North–South Slab Strip *B*—
Example 13-8

	B1			*B2*		*B3*
1. ℓ_1 (ft)		15.75			15.75	
2. ℓ_n (ft)		14.75			14.75	
3. ℓ_2 (ft)		18.88			18.88	
4. w_u (ksf)		0.226			0.226	
5. $M_o = w_u \ell_2 \ell_n^2 / 8$ (ft-kips)		116.0			116.0	
6. Moment coefficients	−0.26	0.52	−0.70	−0.65	0.35	−0.65
7. Negative and positive moments (ft-kips)	−30.2	60.3	−81.2	−75.4	40.6	−75.4
8. Σ Column moments (ft-kips)	30.2		9.8			9.8

TABLE 13-6 Calculation of Negative and Positive Moments for North–South Slab Strip *A*—
Example 13-8

	A1			*A2*		*A3*
1. ℓ_1 (ft)		15.75			15.75	
2. ℓ_n (ft)		14.42			14.08	
3. ℓ_2 (ft)		9.71			9.71	
4. w_u (ksf)		0.226			0.226	
5. $M_o = w_u \ell_2 \ell_n^2 / 8$ (ft-kips)		57.0			54.4	
6. Moment coefficients	−0.26	0.52	−0.70	−0.65	0.35	−0.65
7. Negative and positive moments (ft-kips)	−14.8	29.7	−39.9	−35.4	19.0	−35.4
8. Column moments from slab load (ft-kips)	14.8		5.8			5.8
9. Wall load/ft (kip/ft)		0.42			0.42	
10. Wall M_o (ft-kips)		10.9			10.4	
11. Negative and positive moments from wall load (ft-kips)	−2.8	5.7	−7.6	−6.7	3.6	−6.7
12. Column moments from wall load (ft-kips)	2.8		0.9			0.0
13. Total column moments (ft-kips)	17.6		6.7			5.8

Line 8. ACI Section 13.6.9.2 requires that interior columns be designed for the moment given by (13-17):

$$M_{col} = 0.07[(1.4 \times 112.5 + 0.5 \times 1.7 \times 40) \times 15.75 \times 18.33^2$$
$$- (1.4 \times 112.5)15.75 \times 16.42^2] = 24.1 \text{ ft-kips}$$

Therefore, design the interior columns for a total moment of 24.1 ft-kips divided between the columns above and below the joint in the ratio of their stiffnesses.

The unbalanced moment at the exterior edge (−31.2 ft-kips) is divided between the columns above and below the joint in the ratio of their stiffnesses.

4.   **Compute the moments in the slab strip along column line 1 (Fig. 13-66a).** This strip of slab acts as a rigid frame spanning between columns *A1, B1, C1,* and *D1*. Spans *A1* to *B1* and *C1* to

$D1$ are "end panels," and span $B1$ to $C1$ is an "interior panel" in this strip. Similarly, columns $A1$ and $D1$ are "exterior columns" and $B1$ and $C1$ are "interior columns" in this slab strip.

The calculations are carried out in Table 13-4. These are similar to those for slab strip 2, except that the panel is not as wide and there is a uniformly distributed wall load that causes additional moments in the slab. Since the wall load occurs along the edge of the slab, these moments will be assumed to be resisted entirely by the exterior column strip. The wall-load moments will be calculated separately (lines 9–12 in Table 13-4) and added in at a later step (line 4 of Table 13-7).

Line 3. ℓ_2 is the transverse span measured center to center of supports. ACI Section 13.6.2.4 states that, for edge panels, ℓ_2 in the equation for M_0 should be replaced by ℓ_{2edge}, which is the width from the edge of the slab to a line halfway between column lines 1 and 2.

TABLE 13-7 Division of Moment to Column and Middle Strips: East–West Strips— Example 13-8

	Column Strip	Middle Strip	Column Strip	Middle Strip	Edge Column Strip
	7.87	7.88	7.87	7.88	4.77
Exterior Negative Moments	A3		A2		A1
1. Slab Moment (ft–kip)	−31.2		−31.2		−19.2
2. Moment Coefficients	0.0 1.00	0.0 0.0	1.00	0.0 0.0	1.00
3. Moment to Column and Middle Strips (ft–kip)	−31.2	0	−31.2	0	−19.2
4. Wall Moment (ft–kip)					−3.7
5. Total Moment in Strip (ft–kip)	−31.2	0	−31.2	0	−22.9
6. A_s required (in.²)	1.84	0	1.84	0	1.35
7. Min A_s (in.²)	1.32	1.32	1.32	1.32	0.80
8. Choose Steel	10 #4 bars	7 #4 bars	10 #4 bars	7 #4 bars	7 #4 bars
9. A_s provided (in.²)	2.00	1.40	2.00	1.40	1.40
End Span Positive Moments					
1. Slab Moment (ft–kip)	62.4		62.4		38.5
2. Moment Coefficients	0.2 0.6	0.2 0.2	0.6	0.2 0.4	0.6
3. Moment to Column and Middle Strips (ft–kip)	12.5 37.4	12.5 12.5	37.4	12.5 15.4	23.1
4. Wall Moment (ft–kip)					7.4
5. Total Moment in Strip (ft–kip)	37.4	25.0	37.4	27.9	30.5
6. A_s required (in.²)	2.20	1.47	2.20	1.64	1.79
7. Min A_s (in.²)	1.32	1.32	1.32	1.32	0.80
8. Choose Steel	11 #4 bars	8 #4 bars	11 #4 bars	9 #4 bars	9 #4 bars
9. A_s provided (in.²)	2.20	1.60	2.20	1.80	1.80
First Interior Negative Moments	B3		B2		B1
1. Slab Moment (ft–kip)	−97.2		−97.2		−59.9
2. Moment Coefficients	0.75	0.125 0.125	0.75	0.125 0.25	0.75
3. Moment to Column and Middle Strips (ft–kip)	−72.9	−12.15 −12.15	−72.9	−12.15 15.0	−44.9
4. Wall Moment (ft–kip)					−11.5
5. Total Moment in Strip (ft–kip)	−72.9	−24.3	−72.9	−27.2	−56.4
6. A_s required (in.²)	4.29	1.43	4.29	1.60	3.32
7. Min A_s (in.²)	1.32	1.32	1.32	1.32	0.80
8. Choose Steel	14 #5 bars	8 #4 bars	14 #5 bars	8 #4 bars	11 #5 bars
9. A_s provided (in.²)	4.34	1.60	4.34	1.60	3.41
Interior Positive Moments					
1. Slab Moment (ft–kip)	52.3		52.3		32.3
2. Moment Coefficients	0.60	0.20 0.20	0.60	0.20 0.40	0.60
3. Moment to Column and Middle Strips (ft–kip)	31.4	10.45 10.45	31.4	10.45 12.9	19.4
4. Wall Moment (ft–kip)					6.1
5. Total Moment in Strip (ft–kip)	31.4	20.9	31.4	23.4	25.5
6. A_s required (in.²)	1.85	1.23	1.85	1.38	1.50
7. Min A_s (in.²)	1.32	1.32	1.32	1.32	0.8
8. Choose Steel	10 #4 bars	7 #4 bars	10 #4 bars	7 #4 bars	8 #4 bars
9. A_s provided (in.²)	2.00	1.40	2.00	1.40	1.60

Line 8. The interior columns ($B1$ and $C1$) are designed for the moment from the slab given by (13-14):

$$M_{col} = 0.07[(1.4 \times 112.5 + 0.5 \times 1.7 \times 40)9.71 \times 18.33^2$$

$$- (1.4 \times 112.5)9.71 \times 16.42^2] = 14.9 \text{ ft-kips}$$

Line 9. The factored dead load of the wall is $1.4 \times 0.3 = 0.42$ kip/ft.

Line 10. The statical moment of the wall load is $w\ell_n^2/8$. Therefore, in span $A1$–$B1$,

$$M_{0\,wall} = 0.42 \times \frac{16.42^2}{8} = 14.2 \text{ ft-kips}$$

Line 11. The negative and positive moments due to the wall load are found by multiplying the moments in line 10 by the coefficients in line 6.

Line 12. The column moments at the interior columns are assumed to be equal to the unbalanced moments at the joint. Equation (13-14) is not applied here, since there is no unbalanced live load from the wall.

Line 13. The total column moments are the sum of lines 8 and 12.

   5.   **Compute the moments in the slab strip along column line B (Fig. 13-66b).** This strip of slab acts as a rigid frame spanning between columns $B1$, $B2$, $B3$, and so on. The calculations are carried out in Table 13–5 and follow step 3 of this example.

   6.   **Compute the moments in the slab strip along column line A (Fig. 13-66b).** This strip of slab acts as a rigid frame spanning between columns $A1$, $A2$, $A3$, and so on. In this strip, span $A1$–$A2$ is an "end span," and $A2$–$A3$ and so on are "interior spans." Column $A1$ is an "exterior column," and columns $A2$ and $A3$ are "interior columns," in the slab strip. The calculations are carried out in Table 13–6 and follow step 4 of this example.

   7.   **Distribute the negative and positive moments to the column and middle strips and design the reinforcement: strips spanning East and West (strips 1 and 2).**

   In steps 3 and 4, the moments in the strips along column lines 1 and 2 were computed. These moments must be distributed to the various middle and column strips to enable the East–West reinforcement to be designed.

   (a)   **Divide the slab strips into middle and column strips.** In each panel, the column strip extends 0.25 times the smaller of ℓ_1 and ℓ_2 from the line joining the columns, as shown in Fig. 13-25 and, for this example, in Fig. 13-66a. Thus, the column strips extend 15 ft 9 in./4 = 47.25 in. on each side of the column lines. The total width of the column strip is 2×47.25 in. = 7.87 ft. The width of the middle strip is 7.88 ft. The edge strip has a width of $47.25 + 10$ in. = 4.77 ft.

   (b)   **Divide the moments between the column and middle strip and design the reinforcement.** The calculations are carried out in Table 13-7 for the East–West strips. This table is laid out to resemble a plan of the part of the slab shown shaded in Fig. 13-66a. The columns are shown by square dots and are numbered. The heavy line represents the edge of the slab. The computations leading up to the entries in each line of this table are summarized below. The calculations are repeated for each of the negative- and positive-moment regions. Since this slab has only three spans, it is not necessary to consider the typical interior negative moment.

Line 1. These moments are taken from line 7 in Tables 13-3 and 13-4. At the first interior support, the larger of the two adjacent negative moments is used (ACI Section 13.6.3.4).

Line 2. These are the moment coefficients.

(a) *Exterior negative moments.* ACI Section 13.6.4.2 gives the fraction of the exterior negative moment resisted by the column strip. This section is entered by using β_t, $\alpha_1\ell_2/\ell_1$, and ℓ_2/ℓ_1. For a slab without edge beams, $\beta_t = 0$; for a slab without beams parallel to ℓ_1 for the span being designed, $\alpha_1 = 0$. Thus, from the table in ACI Section 13.6.4.2, 100 percent of the exterior negative moment is assigned to the column strips.

(b) *Positive-moment regions.* ACI Section 13.6.4.4 gives the fraction of positive moments assigned to the column strips. Since $\alpha_1 = 0$, $\alpha_1\ell_2/\ell_1 = 0$, and 60 percent is assigned to the column strip. The remaining 40 percent is assigned to the adjacent half middle strips. At the edge, there is only one adjacent half middle strip, so 40 percent of the panel moment goes to it. At an interior column strip, there are two adjacent half middle strips, and $0.5 \times 40 = 20$ percent goes to each.

(c) *Interior negative-moment regions.* ACI Section 13.6.4.1 specifies the fraction of interior negative moments assigned to the column strips. Since $\alpha_1 = 0$, 75 percent goes to the column strip and the remaining 25 percent to the middle strips.

Line 3. This is the product of lines 1 and 2. The arrows in Table 13-7 illustrate the distribution of the moments between the various parts of the slab.

Line 4. The weight of the North–South wall along column line 1 causes moments in the edge column strip. The moments given here come from line 10 of Table 13-3.

Line 5. The calculation of this term is indicated by the arrows in Table 13-7. Line 5 is the sum of lines 3 and 4.

Line 6. The area of steel will be calculated from (4-34), as was done in the one-way slab in Example 10-1:

$$A_s = \frac{M_u}{\phi f_y j d} \tag{4-34}$$

(a) Compute d. Since the largest moments in the panel occur at support $B2$ in slab strip 2 (see Table 13–3), place reinforcement as shown in Fig. 13-68. Thus,

$$d = 7 \text{ in.} - \tfrac{3}{4} \text{ in.} - \tfrac{1}{2} \text{ bar diameter}$$
$$= 5.94 \text{ in.} \text{(assuming No. 5 bars)} \text{(or use (13--42a))}$$

(b) Compute trial A_s required at the section of maximum moment (first interior negative column strip). The largest M_u is -72.9 ft-kips. Assume that $j = 0.925$. Then

$$A_{s(\text{req'd})} = \frac{72.9 \times 12,000}{0.9 \times 40,000 \times 0.925 \times 5.94} \tag{4-33}$$
$$= 4.42 \text{ in.}^2$$

(c) Compute a and a/d and check whether the section is tension controlled:

$$a = \frac{A_s f_y}{0.85 f'_c b} = \frac{4.42 \times 40,000}{0.85 \times 4000(7.87 \times 12)}$$
$$= 0.55 \text{ in.}$$

$$\frac{a}{d} = \frac{0.55}{5.94} = 0.093$$

This is much less than the a/d in Table A-4 for the tension-controlled limit—therefore, $\phi = 0.9$.

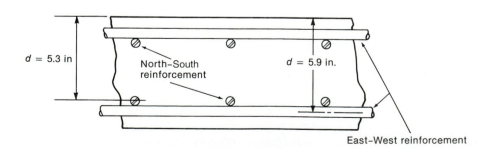

Fig. 13-68
Arrangement of bars in
slab—Example 13-8.

Alternatively, this check could be based on the strain ϵ_t in the extreme tension layer of steel, where

$$\epsilon_t = \frac{(d_t - c)}{c} \times 0.003$$

where $d_t = d = 5.94$ in.

$$c = a/\beta_1 = \frac{0.55}{0.85} = 0.647 \text{ in.}$$

and $\epsilon_t = \dfrac{5.94 - 0.647}{0.647} \times 0.003$

$$= 0.0245$$

Since $\epsilon_t = 0.0245$ exceeds 0.005, the slab is tension-controlled and $\phi = 0.90$.

(d) Compute jd and the constant for computing A_s given in Eq. (A):

$$jd = d - \frac{a}{2} = 5.94 - \frac{0.55}{2}$$

$$= 5.66 \text{ in.}$$

$$A_s(\text{in.}^2) = \frac{M_u(\text{ft-kips}) \times 12,000}{0.9 \times 40,000 \times 5.66}$$

Thus,

$$A_s(\text{in.}^2) = 0.0588 M_u(\text{ft-kips}) \tag{A}$$

The values of A_s in lines 6 of Table 13-7 are computed from (A).

Line 7. The minimum A_s is specified in ACI Section 13.3.1. (See Section 13-9.) $A_{s(\text{min})}$ = 0.002bh for Grade-40 reinforcement. Maximum bar spacing is 2h (ACI Section 13.3.2), but not more than 18 in. (ACI Section 7.12.2.2). Therefore, the maximum spacing is 14 in.

Edge column strip:

$$A_{s(\text{min})} = 0.002bh = 0.002(4.77 \times 12) \times 7$$

$$= 0.80 \text{ in.}^2$$

$$\text{minimum number of bar spaces} = \frac{4.77 \times 12}{14} = 4.09$$

Therefore, the minimum number of bars is 5.

Other strips:

$$A_{s(\text{min})} = 0.0020(7.88 \times 12) \times 7 = 1.32 \text{ in.}^2$$

The minimum number of bars is 7.

Line 8. The final bar choices are given in line 8.

Line 9. The actual areas of reinforcement provided are given in line 9. The reinforcement chosen for each East–West strip is shown in the section in Fig. 13-69. In the calculations for line 6, it was assumed that No. 5 bars would be used, giving $d = 5.94$ in. In some cases, the bars chosen were No. 4 bars. As a result, $d = 6.0$ in. could be used in some parts of the slab. This will not make enough difference to repeat the calculation.

8.   Distribute the negative and positive moments to the column and middle strips and design the reinforcement: strips spanning North and South (strips A and B). In steps 5 and 6, the moments were computed for the strips along column lines A and B. These moments must be assigned to the middle and column strips so that the North–South reinforcement can be designed. This proceeds

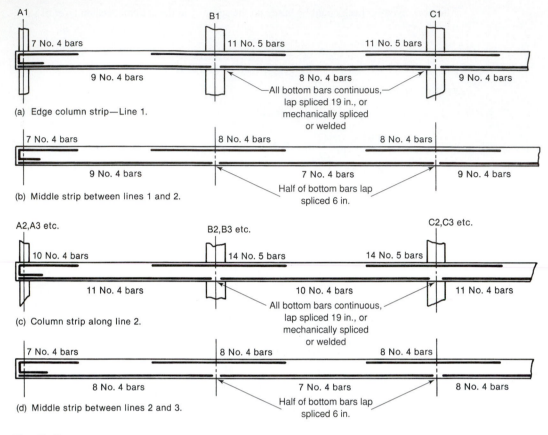

Fig. 13-69
Schematic diagram of reinforcement in East–West strips—Example 13-8.

in the same manner as did step 7, using the moments from lines 7 and 10 of Tables 13-5 and 13-6.

(a) **Divide the slab strips into middle and column strips.** See Fig. 13-66b.

(b) **Divide the moments between the column and middle strips, and design the reinforcement.** For the North–South strips, the calculations are carried out in Table 13-8. This table is laid out to represent a plan view of the shaded part of Fig. 13-66b.

Line 1. These moments are from line 7 of Tables 13-5 and 13-6.

Line 6. The area of steel is computed as in Table 13-7, except that d is smaller, as shown in Fig. 13-68. Assuming No. 5 bars, $d \simeq 7 - 0.75 - 1.5(0.625) = 5.31$ in.

(a) Compute a trial A_s:

$$\text{largest } M_u = -60.9 \text{ ft-kips}$$

$$A_{s(\text{req'd})} = \frac{60.9 \times 12,000}{0.9 \times 40,000 \times 0.925 \times 5.31} = 4.13 \text{ in.}^2$$

(b) Compute a and a/d, and check whether the section is tension-controlled:

$$a = \frac{A_s f_y}{0.85 f'_c b} = \frac{4.13 \times 40,000}{0.85 \times 4000(7.88 \times 12)} = 0.51 \text{ in.}$$

$$\frac{a}{d} = 0.097$$

TABLE 13-8 Division of Moment to Column and Middle Strips: North–South Strips— Example 13-8

	Edge Column Strip 4.77	Middle Strip 9.88	Column Strip 7.88	Middle Strip 12.12	Column Strip 7.88
	A1		B1		C1
Exterior Negative Moments					
1. Slab Moment (ft–kip)	−14.8		−30.2		−30.2
2. Moment Coefficients	1.00	0	1.00	0	1.00 0
3. Moment to Column and Middle Strips (ft–kip)	−14.8	0	−30.2	0	−30.2
4. Wall Moment (ft–kip)	−2.8				
5. Total Moment in Strip (ft–kip)	−17.6	0	−30.2	0	−30.2
6. A_s required (in.^{2})	1.15	0	2.00	0	2.00
7. Min A_s (in.^{2})	0.80	1.66	1.32	2.04	1.32
8. Choose Steel	6 #4 bars	9 #4 bars	10 #4 bars	11 #4 bars	10 #4 bars
9. A_s provided (in.^{2})	1.20	1.80	2.00	2.20	2.00
End Span Positive Moments					
1. Slab Moment (ft–kip)	29.7		60.3		60.3
2. Moment Coefficients	0.6	0.4 0.2	0.6	0.2 0.2	0.6 0.2
3. Moment to Column and Middle Strips (ft–kip)	17.8	11.9 12.1	36.2	12.1 12.1	36.2 12.1
4. Wall Moment (ft–kip)	5.7				
5. Total Moment in Strip (ft–kip)	23.5	23.9	36.2	24.1	36.2
6. A_s required (in.^{2})	1.55	1.58	2.39	1.59	2.39
7. Min A_s (in.^{2})	0.80	1.66	1.32	2.04	1.32
8. Choose Steel	8 #4 bars	9 #4 bars	12 #4 bars	11 #4 bars	12 #4 bars
9. A_s provided (in.^{2})	1.60	1.80	2.40	2.20	
	A2		B2		C2
First Interior Negative Moments					
1. Slab Moment (ft–kip)	−39.9		−81.2		−81.2
2. Moment Coefficients	0.75	0.25 0.125	0.75	0.125 0.125	0.75 0.125
3. Moment to Column and Middle Strips (ft–kip)	−29.9	−10.0 −10.2	−60.9	−10.2 −10.2	−60.9 −10.2
4. Wall Moment (ft–kip)	−7.6				
5. Total Moment in Strip (ft–kip)	−37.5	−20.2	−60.9	−20.4	−60.9
6. A_s required (in.^{2})	2.48	1.33	4.03	1.34	4.03
7. Min A_s (in.^{2})	0.80	1.66	1.32	2.04	1.32
8. Choose Steel	8 #5 bars	9 #4 bars	13 #5 bars	11 #4 bars	13 #5 bars
9. A_s provided (in.^{2})	2.48	1.80	4.03	2.20	4.03
Interior Positive Moments					
1. Slab Moment (ft–kip)	19.0		40.6		40.6
2. Moment Coefficients	0.6	0.4 0.2	0.6	0.2 0.2	0.6 0.2
3. Moment to Column and Middle Strips (ft–kip)	11.4	7.6 8.1	24.4	8.1 8.1	24.4 8.1
4. Wall Moment (ft–kip)	3.6				
5. Total Moment in Strip (ft–kip)	15.1	15.7	24.4	16.2	24.4
6. A_s required (in.^{2})	1.00	1.04	1.61	1.07	1.61
7. Min A_s (in.^{2})	0.80	1.66	1.32	2.04	1.32
8. Choose Steel	6 #4 bars	9 #4 bars	8 #4 bars	11 #4 bars	8 #4 bars
9. A_s provided (in.^{2})	1.20	1.80	1.60	2.20	1.60
	A3		B3		C3
Typical Interior Negative Moment					
1. Slab Moment (ft–kip)	−35.4		−75.4		−75.4
2. Moment Coefficients	0.75	0.25 0.125	0.75	0.125 0.125	0.75 0.125
3. Moment to Column and Middle Strips (ft–kip)	−26.6	−8.9 −9.4	−56.6	−9.4 −9.4	−56.6 −9.4
4. Wall Moment (ft–kip)	−6.7				
5. Total Moment in Strip (ft–kip)	−33.3	−18.3	−56.6	−18.8	−56.6
6. A_s required (in.^{2})	2.20	1.20	3.74	1.24	3.74
7. Min A_s (in.^{2})	0.80	1.66	1.32	2.04	1.32
8. Choose Steel	8 #5 bars	9 #4 bars	12 #5 bars	11 #4 bars	12 #5 bars
9. A_s provided (in.^{2})	2.48	1.80	3.72	2.00	3.72

This is much less than the a/d in Table A-4 for the tension-controlled limit—therefore, $\phi = 0.9$.

(c) Compute jd and the constant for computing A_s:

$$jd = d - \frac{a}{2} = 5.31 - \frac{0.51}{2} = 5.05 \text{ in.}$$

$$A_s(\text{in.}^2) = \frac{M_u(\text{ft-kips}) \times 12{,}000}{0.9 \times 40{,}000 \times 5.05}$$

Therefore,

$$A_s(\text{in.}^2) = 0.0660 M_u(\text{ft-kips}) \tag{B}$$

The values of A_s in lines 6 of Table 13-8 are computed from (B).

Line 7. For the column strips, the minimum A_s and minimum number of bars are as given in Table 13-7, because the column-strip width is the same in both directions. In the middle strips, different amounts of steel are required. We have

Middle strip between lines *A* and *B:*

$$A_{s(\text{min})} = 0.0020(9.88 \times 12) \times 7 = 1.66 \text{ in.}^2$$

$$\text{minimum number of bar spaces} = 8.5$$

$$\text{minimum number of bars} = 9$$

Middle strip between lines *B* and *C:*

$$A_{s(\text{min})} = 0.0020(12.12 \times 12) \times 7 = 2.04 \text{ in.}^2$$

$$\text{minimum number of bars} = 11$$

Line 8. The final bar choices are given in line 8.
Line 9. The reinforcement chosen for each North–South strip is given in line 9 and shown in the sections in Fig. 13-70.

9. Check the shear at the exterior columns for combined shear and moment transfer. Either column *A*2 or column *B*1 will be the most critical. Since the tributary area is largest for *B*1, we shall limit our check to that column, although both joints should be checked.

Since the reinforcement is No. 4 bars at all exterior ends, the effective depths will be a little bigger than are shown in Fig. 13-68. Thus, $d_{\text{EW}} = 6.0$ in., $d_{\text{NS}} = 5.5$ in., and $d_{\text{avg}} = 5.75$ in. Following the calculations in Example 13-7 gives these results:

(a) **Locate the critical-shear perimeter.** The critical-shear perimeter is at $d/2$ from the face of the column, where d is the average depth. Since the shortest perimeter results from the section shown in Fig. 13-71a, this perimeter will be used:

$$d_{\text{avg}} = 5.75 \text{ in.} \qquad \frac{d}{2} = 2.88 \text{ in.}$$

Thus, $b_1 = 18.88$ in. and $b_2 = 25.75$ in.

(b) **Locate the centroid of the shear perimeter.** For moments about the *Z–Z* axis,

$$y_{AB} = \frac{2 \times (18.88 \times 5.75) \times 18.88/2}{2(18.88 \times 5.75) + (25.75 \times 5.75)}$$

$$= 5.61 \text{ in.}$$

Therefore, $c_{AB} = 5.61$ in. and $c_{CD} = 13.27$ in. For moments about the *W–W* axis,

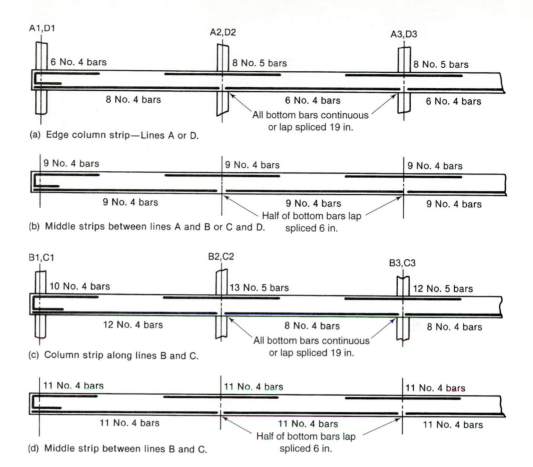

Fig. 13-70
Schematic diagram of reinforcement in North–South
strips—Example 13-8.

$$c_{CB} = c_{AD} = \frac{25.75}{2} = 12.88 \text{ in.}$$

(c)   Compute the shear and the moment about the centroid of the shear perimeter.
The tributary area of column $B1$ is shown in Fig. 13-67 and the critical-shear perimeter is
shown in Fig. 13-71a. We get

$$V_u = 0.226 \text{ ksf} \times \left[(9.94 + 10.00) \text{ ft} \times (6.93 + 0.83) \text{ ft} - \frac{18.88 \times 25.75}{144} \text{ft}^2 \right]$$

$$= 34.2 \text{ kips}$$

The shear due to the wall load outside the critical perimeter (between E and D and between C
and F in Fig. 13-67) is

$$V_{uw} = \left(18.88 \text{ ft} - \frac{25.75}{12} \text{ft} \right) \times 1.4 \times 0.3 \text{ kips/ft} = 7.0 \text{ kips}$$

total $V_u = 41.2$ kips

For slabs designed by the direct design method, the moment transferred from the slab to
the column about axis Z–Z in Fig. 13-71a is $0.3M_0$, where, from line 5 in Table 13–5,
$M_0 = 116.0$ ft-kips. Therefore, the moment to be transferred to the column is
$0.3 \times 116.0 = 34.8$ ft-kips. We shall ignore the moment from the cantilever portion of the
slab. If the slab cantilevered several feet outside the column, or if a heavy wall load existed on
the cantilever and it could be shown that this heavy wall load would be in place *before* significant construction or live loads were placed on the slab itself, the moment about Z–Z could be
reduced.

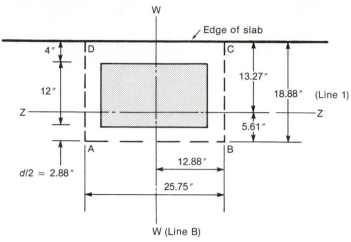

(a) Critical section—Column B1.

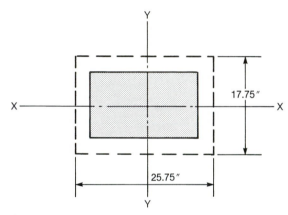

Fig. 13-71
Final critical shear perime-
ters: columns $B1$ and $B2$—
Example 13-8.

(b) Critical Section—Column B2

Moments about axis W–W in Fig. 13-71a come from the slab strips parallel to the edge. Line 7 of Table 13-4 gives the moments for all panels loaded. Line 8 gives the moments transferred to the columns when one adjacent panel is loaded and the other panel is not. Since the shear check involves the shear due to all panels loaded, we shall use the moments from line 7. From line 7 in Table 13-4, we find that the moment of the edge panel loads about axis W–W of column $B1$ is $59.9 - 51.8 = 8.1$ ft-kips. From line 13, we find that the unbalanced moment due to the wall load is 1.6 ft-kips. The total moment to be transferred is 9.7 ft-kips.

In sum, the connection at $B1$ must be designed for $V_u = 41.2$ kips, $M_{ZZ} = 34.8$ ft-kips, and $M_{WW} = 9.7$ ft-kips. These moments act about the centroids of the shear perimeter in the two directions.

(d)   Compute ϕV_c and $V_u/\phi V_c$. In step 2(b) it was shown that V_c is governed by (13-17). Use $b_0 = 2 \times 18.88$ in. $+ 25.75$ in. $= 63.51$ in. Hence,

$$\phi V_c = \phi 4 \sqrt{f'_c} b_o d \qquad (13\text{-}17)$$

$$= 0.85 \times 4 \times \sqrt{4000} \times 63.51 \times 5.75$$
$$= 78.5 \text{ kips}$$

Therefore, $\phi V_c = 78.5$ kips, and $V_u/\phi V_c = 41.2/78.5 = 0.525$.

(e)   **Determine the fraction of the moment transferred by flexure, γ_f.**

Moments about the Z–Z axis:

$$\gamma_f = \frac{1}{1 + \frac{2}{3}\sqrt{b_1/b_2}} \qquad\qquad \text{(13-27)}$$
$$\text{(ACI Eq. 13-1)}$$

where, for moment about the Z–Z axis, $b_1 = 18.88$ in. and $b_2 = 25.75$ in. Thus,

$$\gamma_f = \frac{1}{1 + \frac{2}{3}\sqrt{\frac{18.88}{25.75}}}$$
$$= 0.637$$

ACI Section 13.5.3.3 allows γ_f to be increased up to 1.0 if $V_u/\phi V_c$ does not exceed 0.75 and the resulting ρ is less than $0.375\rho_b$ within a width of $c_2 + 3h$ centered on the column. From step (d), $V_u/\phi V_c = 0.531$. Therefore, take $\gamma_f = 1.0$, and check the reinforcement required.

Moments about the W–W axis: Exchanging b_1 and b_2 in (13-27) gives $\gamma_{f2} = 0.562$. ACI Section 13.5.3.3 allows γ_f to be increased by up to 25 percent if $V_u/\phi V_c \le 0.4$. Since $V_u/\phi V_c = 0.531$, no increase can be made in γ_{f2}. The moment transferred by flexure is $\gamma_{f2}M_u = 0.562 \times 9.7$ ft-kips $= 5.5$ ft-kips.

(f)   **Design the reinforcement required for moment transfer by flexure.**

Moments about the Z–Z axis:

Width effective for flexure: $c_2 + 3h = 20$ in. $+ 3 \times 7$ in. $= 41$ in.

Moment: 1.0×34.8 ft-kips $= 34.8$ ft-kips

Assume that $jd = 0.925d$. The problem statement specified Grade 40 reinforcement.

$$A_s = \frac{M_u}{\phi f_y jd}$$
$$= \frac{34.8 \times 12,000}{0.9 \times 40,000 \times 0.925 \times 5.75}$$
$$= 2.18 \text{ in.}^2$$

The steel provided in step 8 is ten No. 4 bars (see Fig. 13-70c) in a column-strip width of 7.88 ft $= 94.5$ in., or roughly 9.5 in. on centers. The bars within the 41-in. effective width can be used for the moment transfer. Place six column strip bars into this region and add five additional bars in this region, giving $A_s = 2.20$ in.^{2} in the effective width.

Since the computed A_s was based on a guess for jd, we shall compute a for the A_s chosen and recompute A_s:

$$a = \frac{A_s f_y}{0.85 f_c' b} = \frac{2.20 \times 40,000}{0.85 \times 4000 \times 41}$$
$$= 0.631 \text{ in.}$$

$$A_s = \frac{M_u}{f_y(d - a/2)} = \frac{34.8 \times 12,000}{0.9 \times 40,000(5.75 - 0.631/2)}$$
$$= 2.13 \text{ in.}^2 \qquad \text{—therefore, the steel chosen has adequate capacity.}$$

$$\frac{a}{d} = \frac{0.631}{5.75} = 0.110$$

From (4-21),

$$\frac{a_b}{d} = \beta_1\left(\frac{87,000}{87,000 + f_y}\right) = 0.85\left(\frac{87,000}{87,000 + 40,000}\right)$$
$$= 0.582$$

and $0.375a_b/d = 0.375 \times 0.582 = 0.218$. Since $a/d = 0.110$ is less than $0.375a_b/d = 0.218$, ρ is less than $0.375\rho_b$, and we can use $\gamma_f = 1.0$. As a result, it is not necessary to transfer any of the moment about axis Z–Z by shear.

Moments about the W–W axis: Effective width for moment transfer is 12 in. + 4 in. + 1.5 × 7 in. = 26.5 in. For this moment, (5.5 ft-kips) two No. 5 bars are required in this width. This will be accomplished if the bars are uniformly distributed in this region.

(g)   Compute the shear stresses.

$$v_u = \frac{V_u}{b_o d} \pm \frac{\gamma_{y1} M_{u1} c}{J_{c1}} \pm \frac{\gamma_{v2} M_{u2} c}{J_{c2}} \tag{13-23a}$$

where M_{u1}, and so on, refer to axis Z–Z, and M_{u2}, and so on, refer to axis W–W. Here,

$$b_o = 63.51 \text{ in.}$$
$$\gamma_{v1} = 1 - \gamma_{f1} = 0.0$$
$$\gamma_{v2} = 1 - \gamma_{f2} = 0.438$$

Since $\gamma_{v1} = 0.0$, the second term in the equation for v_u drops out. As a result, it is not necessary to compute J_{c1}, and we have

$$J_{c2} = 2(b_1 d)c_{CB}^2 + \frac{b_2 d^3}{12} + \frac{b_2^3 d}{12}$$

$$= 2(18.88 \times 5.75)(12.88)^2 + \frac{25.75 \times 5.75^3}{12} + \frac{25.75^3 \times 5.75}{12}$$

$$= 44{,}600 \text{ in.}^4$$

The maximum stresses along sides CB and AD are

$$v_u = \frac{41.2 \times 1000}{63.51 \times 5.75} \pm 0.0 \pm \frac{0.438 \times 9.7 \times 12{,}000 \times 12.88}{44{,}600}$$

$$= 113 + 0.0 \pm 15$$

$$= 128 \text{ psi at } B \text{ and } 98 \text{ psi at } A$$

Since there is no shear reinforcement in the slab, $\phi v_n = \phi v_c = \phi V_c/b_o d$, where, from step (d), $\phi V_c = 78.5$ kips. Thus,

$$\phi v_n = \frac{78.5 \times 1000}{63.51 \times 5.75} = 215 \text{ psi}$$

Because $v_u < \phi v_n$, the shear is OK in this column–slab connection.

If $v_u > \phi v_c$, it would be necessary to modify the connection. Solutions would be to thicken the slab, use stronger concrete, enlarge the column, or use shear reinforcement. The choice of solution should be based on a study of the extra costs involved.

10.   Check the shear at an interior column for combined shear and moment transfer.
Tables 13-3 and 13-5 indicate that moments of 24.1 ft-kips and 9.8 ft-kips are transferred from the slab to column B_2 from strip 2 and strip B, respectively. Check the shear at this column.

(a)   Locate the critical shear perimeter. See Fig. 13-71b.

$$b_o = 2(17.75 + 25.75) = 87.0 \text{ in.}$$

(b)   Locate the centroid of the shear perimeter. Since the perimeter is continuous, the centroids pass through the centers of the sides.

(c) Compute the forces to be transferred. The tributary area for column B2 is shown in Fig. 13-67.

$$V_u = 0.226\left[(9.94 + 10.0)(8.82 + 7.88) - \frac{25.25 \times 17.25}{144}\right]$$

$$= 74.6 \text{ kips}$$

The moments to be transferred come from Tables 13-3 and 13-5. Line 7 of these tables gives the moments for all panels loaded. Line 8 of these tables gives moments transferred to the columns when one adjacent panel is loaded and the other panel is not. Since the shear check involves the shear due to all panels loaded, we shall use the moments from line 7. In line 7 of Table 13-3, the difference between the negative moments on the two sides of column B2 is $97.2 - 84.0 = 13.2$ ft-kips. From Table 13-5, it is $81.2 - 75.4 = 5.8$ ft-kips.

The forces transferred to the column are $V_u = 74.6$ kips, $M_{u1} = 5.8$ ft-kips, and $M_{u2} = 13.2$ ft-kips.

(d) Compute the fraction of the moment transferred by flexure.

$$\gamma_{f1} = 0.644 \qquad \gamma_{f2} = 0.554$$

ACI Section 13.5.3.3 allows γ_f to be increased by up to 25 percent if $V_u/\phi V_c \le 0.4$. Again, (13-17) governs, and

$$\phi V_c = \phi 4\sqrt{f'_c}\, b_o d = 0.85 \times 4\sqrt{4000} \times 87 \times 5.75$$

$$= 107.6 \text{ kips}$$

$$\frac{V_u}{\phi V_c} = \frac{74.6}{107.6} = 0.693$$

Since $V_u/\phi V_c$ exceeds 0.4, γ_f cannot be increased about either axis.

(e) Compute the torsional moment of inertia, J_c. Bending about axis Z–Z (from (13-28b)) is

$$J_{c1} = 2\left(\frac{b_1 d^3}{12}\right) + 2\left(\frac{d b_1^3}{12}\right) + 2(b_2 d)\left(\frac{b_1}{2}\right)^2$$

where $b_1 = 17.75$ in. and $b_2 = 25.75$ in. (See Fig. 13-71b.) Thus,

$$J_{c1} = 29{,}200 \text{ in.}^4$$

For bending about axis 2–2 of column B2, $b_1 = 25.75$ in., $b_2 = 17.75$ in., and $J_{c2} = 51{,}000$ in.^{4}

(f) Design the reinforcement for moment transfer. By inspection, the reinforcement already in the slab is adequate.

(g) Compute the shear stresses. Maximum stresses occur at the corner, where all three terms are additive. We have

$$b_o = 2(17.75 + 25.75) = 87.0 \text{ in.}$$

$$v_u = \frac{74.6}{87.0 \times 5.75} + \frac{0.356 \times 5.8 \times 12 \times 8.88}{29{,}200} + \frac{0.445 \times 13.2 \times 12 \times 12.88}{51{,}000}$$

$$= 0.149 + 0.008 + 0.018 = 0.175 \text{ ksi}$$

$v_u = 175$ psi and recall $\phi v_c = 215$ psi

Since $\phi v_c > v_u$, the shear is OK at this column. The moment transfer increased the shear stress by 17 percent, from 149 psi to 175 psi.

11. Check the shear at the corner column.

Tables 13–4 and 13–6 indicate that moments of 22.9 ft-kips and 17.6 ft-kips are transferred from the slab to the corner column, $A1$, from strip 1 and strip A, respectively. Two shear perimeters will be considered. Two-way shear may be critical on the perimeter shown in Fig. 13-72a, while one-way shear may be critical on the section shown in Fig. 13-72b.

Two-Way Shear

(a) Locate the critical perimeter. The critical perimeter is as shown in Fig. 13-72a. For two-way shear, we shall base the calculations of the stresses on the orthogonal x- and y- axes.

(b) Locate the centroid of the perimeter.

$$\bar{x} = \frac{(18.88 \times 5.75)(18.88/2)}{2 \times 18.88 \times 5.75}$$

$$= 4.72 \text{ in. from inside corner}$$

(c) Compute the forces to be transferred. Factored shear on the area bounded by lines of zero shear at 0.44 times the span length from the exterior end of the panel plus the distance from the center of the column to the edge:

$$V_u = 0.226 \text{ ksf} \times \left[(0.44 \times 17.75 + 0.83) \times (0.44 \times 15.75 + 0.83) - \frac{18.88 \times 18.88}{144} \right]$$

$$= 14.6 \text{ kips}$$

$$\text{Shear due to wall load} = \left[(7.81 + 0.83) + (6.93 + 0.83) - \frac{2 \times 18.88}{12} \right] \times 1.4 \times 0.30 \text{ kips/ft}$$

$$= 5.57 \text{ kips}$$

$$\text{Total shear} = 14.6 + 5.57 = 20.2 \text{ kips}$$

ACI Section 13.6.3.6 states that the amount of moment to be transferred from a slab to an edge column is $0.3M_0$. Although no similar statement is made for a corner column, we shall assume that the same moment should be transferred.

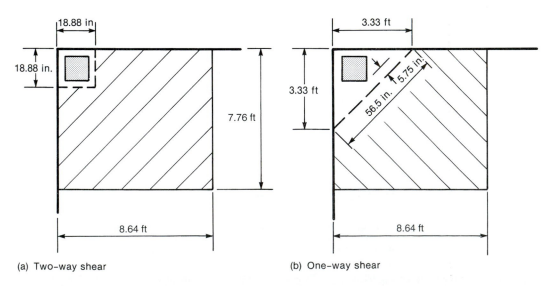

(a) Two-way shear   (b) One-way shear

Fig. 13-72
Critical shear perimeters and tributary areas: Column $A1$—Example 13-8.

For strip 1: $M_0 = 74.0$ ft-kips, $0.3M_0 = 22.2$ ft-kips
For strip A: $M_0 = 57.0$ ft-kips, $0.3M_0 = 17.1$ ft-kips

Again, it will be assumed that these moments and the shears all act through the centroid of the shear perimeter.

(d)   **Determine the fraction of the moment transferred by flexure.** Because the column is square, we shall take $\gamma_v = 0.40$ which is the value for a square interior column. The ACI Code does not give any values of γ_f for corner columns. We shall assume the same equation can be used as used for interior columns. Since the critical section is square, (13-22) gives $\gamma_f = 0.6$ about each axis. ACI Section 13.5.3.3 allows γ_f to be increased to 1.0, provided that $V_u/\phi V_c \leq 0.5$ and $\rho \leq 0.375\rho_b$ in the strip of slab which is effective for moment transfer, width $= 4 + 12 + 1.5 \times 7 = 26.5$ in.

$$b_o = 2 \times 18.88 = 37.76 \text{ in.}$$

$$\phi V_c = \phi 4\sqrt{f_c'}b_o d = 46.7 \text{ kips}$$

$$\frac{V_u}{\phi V_c} = \frac{20.2}{46.7} = 0.433$$

Because this is less than 0.5; we can use $\gamma_f = 1.0$. However, doing this would lead to a heavy concentration of slab reinforcement at the corner column. Thus, we will use $\gamma_f = 0.6$.

(e)   **Design the reinforcement for moment transfer by flexure.** Since $\gamma_f = 0.6$, 0.6 times 7 = 4.2 No. 4 bars (i.e. 5 bars) must be placed within 26.5 in. of the corner in strip 1 and 0.6 times 6 = 3.6 No. 4 bars (i.e. 4 bars) within 26.5 in. of the corner in strip A.

(f)   **Compute the torsional moment of inertia, J_c.** Bending about the Z–Z axis

$$J_c = \frac{18.88 \times 5.75^3}{12} + \frac{5.75 \times 18.88^3}{12}$$

$$+(18.88 \times 5.75)\left(\frac{18.88}{2} - 4.72\right)^2 + (5.75 \times 18.88)(4.72)^2$$

$$= 8360 \text{ in.}^4$$

(g)   **Compute the shear stresses.**

$$v_u = \frac{V_u}{b_o d} \pm \frac{\gamma_{v1}M_{u1}c}{J_{c1}} \pm \frac{\gamma_{v2}M_{u2}c}{J_{c2}}$$

All of these terms add at the inside corner to give the maximum v_u:

$$v_c = \frac{20.2}{37.75 \times 5.75} + \frac{0.4 \times 22.2 \times 12 \times 4.72}{8360} + \frac{0.4 \times 17.1 \times 12 \times 4.72}{8360}$$

$$= 0.093 + 0.060 + 0.046 = 0.199 \text{ ksi}$$

$$= 199 \text{ psi}$$

$$\phi v_c = \phi 4\sqrt{f_c'} = 215 \text{ psi}$$

Thus, $\phi v_c = 215$ psi is greater than $v_u = 199$ psi.

Therefore, the corner column *is adequate in two-way shear*.

One-Way Shear

(a)   **Locate the critical section.** This check will be made using the original column size. The critical section is as shown in Fig. 13-72b. This section is located $d_{avg} = 5.75$ from the corner of the column.

(b)   Compute the shear on the critical section.

$$V_u = 0.226[(7.81 + 0.83) \times (6.93 + 0.83) - 3.33^2/2] = 13.9 \text{ kips}$$

Shear due to wall load: $[(8.64 - 3.33) + (7.76 - 3.33)] \times 1.4 \times 0.30 = 4.09$ kips

Total shear: 18.0 kips

(c) ϕV_c **for the critical section is**

$$\phi V_c = \phi 2 \sqrt{f_c'} bd = (0.85)2\sqrt{4000} \times 56.5 \times \frac{5.75}{1000}$$

$$= 34.9 \text{ kips}$$

Therefore, the slab is OK in one-way shear.

This completes the design of the slab. The bar cutoffs are calculated by using Fig. 13-62. ∎

13-11 CONSTRUCTION LOADS ON SLABS

Most two-way slab buildings are built by using *flying forms,* which can be removed sideways out of the building and are then lifted or *flown* up to form a higher floor. When the flying form is removed from under a slab, the weight of the slab is taken by posts or *shores* which are wedged into place to take the load. Sets of flying forms and shores can be seen in Fig. 10-3. To save on the number of shores needed, it is customary to have only three to six floors of shoring below a slab at the time the concrete is placed. As a result, the weight of the fresh concrete is supported by the three to six floors below it. Because these floors are of different ages, they each take a different fraction of the load of the new slab. The calculation of the construction loads on slabs is presented in [13-32]. Depending on the number of floors that are shored and the sequence of casting and form removal, the maximum construction load on a given slab may reach 1.8 to 2.2 times the dead load of the slab. This can approach the capacity of the slab, particularly if, as is the usual case, the slab has not reached its full strength when the construction loads occur. These high loads cause cracking of the green slabs and lead to larger short- and long-time deflections than would otherwise be expected [13-33].

13-12 SLAB DEFLECTIONS

Two-way slabs frequently deflect excessively, causing sagging floors and damage to partitions, doors, and windows. ACI Section 9.5.3.2 gives minimum thicknesses of two-way slabs to avoid excessive deflections. (See Table 13-1). ACI Section 9.5.3.4 allows thinner slabs to be used if calculated deflections are less than the allowable deflections given in ACI Table 9.5(b). Generally, it is good practice to round up the thicknesses given in ACI Section 9.5.3.2 to the next-larger half-inch.

The calculation of deflections of two-way slabs is generally based on a crossing-beam analogy in which the average deflection of the midspans of the column strips in one direction are added to the midspan deflection of the perpendicular middle strip, as is shown in Fig. 13-73. The calculation follows the procedures given in Chapter 9.

The effective moment of inertia is calculated from (9-10) and (9-11) (ACI Eq. (9-7)). To use this equation, one needs M_a, defined in the code as the maximum moment in the member at the stage at which deflection is calculated. A better definition would be the maximum moment acting on the member at the stage in question or at any previous stage. In Section 13-11, it was pointed out that construction loads frequently reach two times the dead load of the slab. In computing slab deflections, M_a should be based on a factored load equal to the larger of 1.0 (dead plus live load), and 2.0 (dead load). Furthermore, M_{cr} should properly be representative of the age at which M_a acted.

ACI Table 9.5(b) gives values of maximum permissible computed deflections as a function of the span length ℓ. Unfortunately, the code does not define what ℓ is in the case

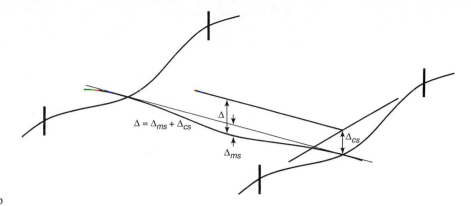

Fig. 13-73
Superposition of column-strip
and middle-strip deflections.

of a two-way slab. The author believes that the column-strip and middle-strip deflections should be limited by using the length of the appropriate strip and that the midpanel deflection should be limited by using the length of the diagonal of the panel.

EXAMPLE 13-9   Calculation of Deflections of an Interior Panel of a Flat-Plate Floor

Compute the deflections of the panel B–C–2–3 of the slab shown in Fig. 13-65.

   1.   Compute the deflections of the column strips between $B2$ and $C2$ and between $B3$ and $C3$.

   (a)   Compute M_a. Factored load on slab = 226 psf [from Example 13-8, step 2(c)]. Loads for deflection calculation:

Dead load: 1.0 (dead) = 87.5 + 25 = 112.5 psf

Service load: 1.0 (dead + live) = 87.5 + 25 + 40 = 152.5 psf

Construction load: 2.0 (dead load of slab) = 2 × 87.5 = 175 psf

Therefore, cracking would be governed by construction loads. Take M_a as 175/226 = 0.774 times the column-strip moment from Table 13-7. Then

$$\text{negative at } B2 \text{ and } C2: 0.774 \times -72.9 = 56.4 \text{ ft-kips}$$

$$\text{positive at midspan: } 0.774 \times 31.4 = 24.3 \text{ ft-kips}$$

   (b)   Compute M_{cr}. Assume that the maximum construction load on a given slab occurs when it is 14 days old. From (3-5), a 14-day-old slab has a strength of about $0.88f'_c$. From ACI Eq. (9-8),

$$M_{cr} = \frac{f_r I_g}{y_t}$$

where

$$f_r = 7.5\sqrt{f'_c} = 7.5\sqrt{0.88 \times 4000}$$
$$= 445 \text{ psi}$$

$$I_g = \frac{(7.87 \times 12) \times 7^3}{12} = 2699 \text{ in.}^4$$

$$y_t = 3.5 \text{ in.}$$

$$M_{cr} = \frac{445 \times 2699}{3.5 \times 12,000} = 28.6 \text{ ft-kips}$$

(c)   **Compute I_{cr} and I_e.**

Negative-moment region: From Fig. 13-69, A_s = 14 No. 5 bars, so

$$\rho = \frac{14 \times 0.31}{7.87 \times 12 \times 5.66} = 0.0081$$

$$n \text{ at time of cracking} = \frac{29,000,000}{57,000\sqrt{0.88 \times 4000}} = 8.58$$

Thus, $n\rho$ = 0.0696, and from (9-3), k = 0.31. It then follows that

$$I_{cr} = \frac{(7.87 \times 12) \times (0.31 \times 5.66)^3}{3}$$

$$+\ 14 \times 0.31 \times 8.58(5.66 - 0.31 \times 5.66)^2$$

$$= 170 + 568 = 738 \text{ in.}^4$$

From (9-10b),

$$I_e = I_{cr} + (I_g - I_{cr})\left(\frac{M_{cr}}{M_a}\right)^3$$

$$= 738 + (2699 - 738) \times \left(\frac{28.6}{56.4}\right)^3 = 994 \text{ in.}^4$$

Positive-moment region: Since $M_a < M_{cr}$, $I_e = I_g = 2699$ in.^{4}
Weighted average value of I_e: From (9-11a),

$$\text{average } I_e = 0.70I_{em} + 0.15(I_{e1} + I_{e2})$$

$$= 0.7 \times 2699 + 0.15(994 + 994)$$

$$= 2187 \text{ in.}^4$$

(d)   **Compute the dead-load deflections.** From (9-12), the dead-load moments are as follows:

negative: $\dfrac{112.5}{226} \times -72.9 = -36.3$ ft-kips

positive: $\dfrac{112.5}{226} \times 31.4 = 15.6$ ft-kips

From Section 9-5, Eq. (9-12a)

$$\Delta = \frac{5\,\ell_n^2}{48EI}[M_m + 0.1(M_1 + M_2)]$$

$$= \frac{5}{48}\frac{(18.33 \times 12)^2}{57,000\sqrt{4000} \times 2187}[15.6 + 0.1(-36.3 - 36.3)] \times 12,000$$

$$= 0.175 \text{ in.}$$

The midspan deflection of the column strips is 0.175 in. under unfactored dead load.

2.   **Compute the deflection of the middle strip between lines 2 and 3.**

(a)   **Compute M_a and M_{cr}.** Again, M_a is due to construction loads:

negative at line 2: $0.774 \times (-20.4) = -15.8$ ft-kips

positive at midspan: $0.774 \times 16.2 = 12.5$ ft-kips

negative at line 3: $0.774 \times (-18.8) = -14.6$ ft-kips

$$I_g = \frac{12.12 \times 12 \times 7^3}{12} = 4157 \text{ in.}^4$$

$$M_{cr} = 44.0 \text{ ft-kips}$$

(b)   **Compute I_c.** Since $M_a < M_{cr}$ at all points, $I_e = I_g = 4157$ in.^{4}

(c)   **Compute the dead-load deflections.**

Negative moment at line 2 is $0.498 \times (-20.4) = -10.2$ ft-kips.

Positive moment is $0.498 \times 16.2 = 8.1$ ft-kips.

Negative moment at line 3 is $0.498 \times (-18.8) = -9.4$ ft-kips.

$$\Delta = \frac{5}{48} \frac{(14.75 \times 12)^2}{57,000\sqrt{4000} \times 4157}[8.1 + 0.1(-10.2 - 9.4)] \times 12,000 \text{ in.}$$

$$= 0.026 \text{ in.}$$

The midspan deflection of the middle strip is 0.026 under unfactored dead load.

3.   **Compute the dead-load deflection at the middle of the panel.**

Δ is the average column-strip Δ plus the middle-strip Δ.

$\Delta = 0.175 + 0.026 = 0.201$ in.

To study the deflections of the panel completely, it would be necessary to consider several load cases. See Examples 9-5 and 9-6.   ■

13-13 DESIGN OF SLABS WITH BEAMS IN TWO DIRECTIONS

Because of its additional depth, a beam is stiffer than the adjacent slab, and as a result, moments are attracted to the beam. This was discussed in Section 13-5 and illustrated in Figs. 13-13 and 13-18. The average moments in the column strip are almost the same in a flat plate (Fig. 13-17d) and in a slab with beams between all columns (Fig. 13-18b). In the latter case, the column-strip moment is divided between the slab and the beam. This reduces the reinforcement required in the slab in the column strip, although the beam must now be reinforced.

The greater stiffness of the beams reduces the overall deflections, allowing a thinner slab to be used than in the case of a flat plate. The advantage of slabs with beams in two directions lies in their reduced weight. Also, two-way shear does not govern for most two-way slabs with beams, again allowing thinner slabs. This is offset by the increased overall depth of the floor system and increased forming and reinforcement-placing costs.

The direct design method for computing moments in the slab and beams is the same as the procedure used in slabs without beams, with one additional step. Thus, the designer will, as usual,

1.   compute M_0;
2.   divide M_0 between the positive- and negative-moment regions; and
3.   divide the positive and negative moments between the column and middle strips.

The additional step needed is

4.   divide the column-strip moments between the beam and the slab.

The amount of moment assigned to the column and middle strips in step 3 and the division of moments between the beam and slab in step 4 are a function of $\alpha_1 \ell_2/\ell_1$, where α_1

is the beam-to-slab stiffness ratio in the direction in which the reinforcement is being designed. (See Section 13-16 and Example 13-1.)

When slabs are supported on beams having $\alpha_1\ell_2/\ell_1 \geq 1.0$, the beams must be designed for shear forces computed by assuming tributary areas bounded by 45° lines at the corners of the panels and the centerlines of the panels, as is shown in Fig. 13-74. If the beams have $\alpha_1\ell_2/\ell_1$ between 0 and 1.0, the shear forces computed from these tributary areas are multiplied by $\alpha_1\ell_2/\ell_1$. In such a case, the remainder of the shear must be transmitted to the column by shear in the slab. The ACI Code is silent on how this is to be done. The most common interpretation involves using two-way shear in the slab between the beams and one-way shear in the beams, as shown in Fig. 13-75. Frequently, problems are encountered when $\alpha_1\ell_2/\ell_1$ is less than 1.0, because the two-way shear perimeter is inadequate to transfer the portion of the shear not transferred by the beams.

The size of the beams is also governed by their shear strength and flexural strength. The cross section should be large enough so that $V_u \leq \phi(V_c + V_s)$, where an upper practical limit on $V_c + V_s$ would be about $(6\sqrt{f'_c}\,b_w d)$. The critical location for flexure is the point of maximum negative moment, where the reinforcement ratio, ρ, should not exceed about $0.5\rho_b$.

The ACI Code allows any value of $\alpha_1\ell_2/\ell_1$ from zero (for no beams) to very large values (for very stiff beams). As mentioned earlier, practical difficulties arise in design, particularly with respect to shear, if $\alpha_1\ell_2/\ell_1$ is between 0 and 1.0, and it is recommended that this range of beam stiffnesses be avoided.

EXAMPLE 13-10 Design of a Two-Way Slab with Beams in Both Directions: Direct Design Method

Figure 13-76 is a plan of a part of a floor with beams between all columns. For simplicity in forming, the beams have been made the same width as the columns. The floor supports its own weight, plus

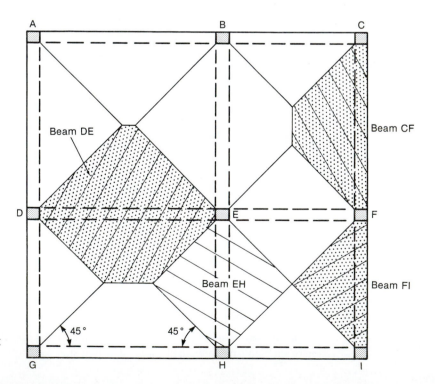

Fig. 13-74
Tributary areas for computing shear in beams supporting two-way slabs.

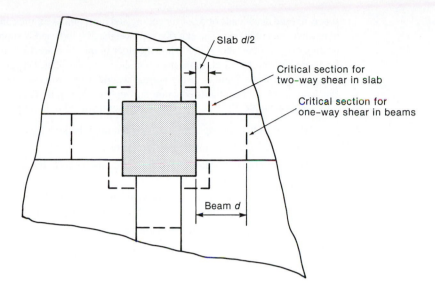

Fig. 13-75
Shear perimeters in slabs
with beams.

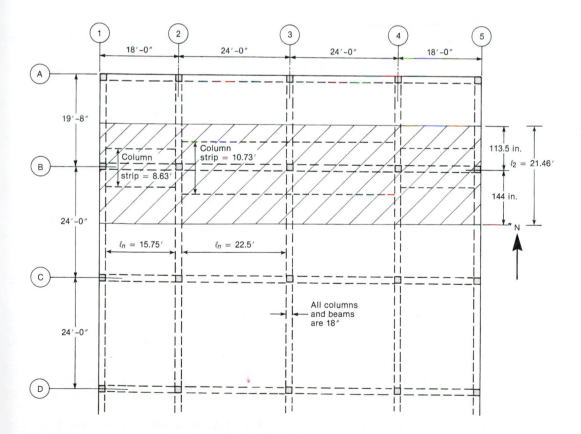

Fig. 13-76
Plan of two-way slab with beams—Example 13-10.

superimposed dead loads of 5 psf for ceiling and mechanical fixtures and 25 psf for future partitions, plus a live load of 80 psf. The exterior wall weighs 300 lb/ft and is supported by the edge beam. The heights of the stories above and below the floor in question are 12 ft and 14 ft, respectively. Lateral loads are resisted by an elevator shaft not shown in the plan. Design the East–West strips of slab along column lines A and B, using normal-weight 3000-psi concrete and Grade-60 reinforcement. Use load factors and strength-reduction factors from Appendix C of ACI 318-02.

This example illustrates several things not included in Example 13-8, including

(a) the effect of edge beams on the required thickness,

(b) the use of live-load reduction factors,

(c) the effect of beams on the division of moments between the slab and beam in the column strip,

(d) the distribution of negative moments where the direct design method gives large differences between the moments on each side of a support, and

(e) the calculation of shear in the beam and slab.

For live loads of 100 psf or less, on panels having an area greater than 400 ft^2, ASCE 7-95 allows a live-load reduction based on (2-17):

$$w = w_0\left(0.25 + \frac{15}{\sqrt{k_{LL}A_T}}\right) \tag{2-14}$$

where w and w_0 are the reduced and specified live loads, A_T is the tributary area and k_{LL} is a factor to convert the tributary area to influence area. In the case of a two-way slab, the influence area, A_I, is the panel area measured from center to center of the columns, that is, $k_{LL} = 1.0$.

1. Select the design method and load and resistance factors. ACI Code Section 13.6.1 sets limits on the use of the direct design method. These are checked as in step 1 of Example 13-8. One additional check is necessary (ACI Section 13.6.1.6): The beams will be selected so that all are the same width and depth. As a result, the ratio $\alpha_1\ell_2^2/\alpha_2\ell_1^2$ should fall within the bounds given. Therefore, use the direct design method. Use load and resistance factors from Appendix C of ACI 318-02.

2. Select the slab thickness and beam size. The slab thickness is chosen to satisfy deflection requirements once the beam size is known. If $\alpha_1\ell_2/\ell_1$ exceeds 1.0 for all beams, all the shear is transferred to the columns by the beams, making it unnecessary to check shear while selecting the slab thickness. If there were only edge beams, the minimum slab thickness for deflection would be governed by Table 13–1 (ACI Table 9.5c) and would be 8.18 in., based on $\ell_n = 22$ ft $- 6$ in. (See Table 13–1.) To select a thickness for a slab with beams between interior columns, the thickness will be arbitrarily reduced by 15 percent to account for the stiffening effect of the beams, giving a trial thickness of 7 in. Assume a beam with an overall depth of about 2.5 times that of the slab, to give a value of α a little greater than 1.0.

For the first trial, select a slab thickness of 7 in. and a beam 18 in. wide by 18 in. deep. Check the thickness, using (13-10) and (13-11). The cross sections of the beams are shown in Fig. 13-77. First compute α:

$$\alpha = \frac{E_{cb}I_b}{E_{cs}I_s} \tag{13-9}$$

For the edge beam, the centroid is 7.94 in. below the top of the section, giving $I_b = 10,944$ in.^{4}. The width of slab working with the beam along line A is 122.5 in., giving $I_s = 3501$ and $\alpha = 3.13$.

For the beam along line 1, the width of slab is 112.5 in., giving $\alpha = 3.40$. For the interior beams, $I_b = 12,534$ in.^{4}; along line B, the slab width is 275.5 in., giving $\alpha = 1.70$; along lines C and 3, the slab width is 288 in., giving $\alpha = 1.52$; along line 2, the slab width is 247.5 in., giving $\alpha = 1.77$.

The thickness computations are given in Table 13-9. A 6.92-in. slab is required in the largest panel, so the 7-in. thickness chosen is satisfactory.

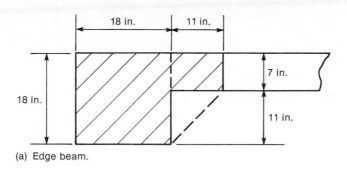

(a) Edge beam.

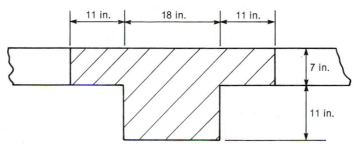

Fig. 13-77
Cross sections of beams—
Example 13-10.

(b) Interior beam.

TABLE 13-9   Computation of Minimum Thickness—Example 13-9

	Panel			
	A–B–1–2 (Corner)	A–B–2–3	B–C–1–2	B–C–2–3 (Interior)
Maximum ℓ_n	(19 ft 8 in.) − 18 in. − 9 in.	24 ft − 18 in.		
	209 in.	270 in.	270 in.	270 in.
Minimum ℓ_n	(18 ft 0 in.) − 18 in. − 9 in.			
	189 in.	209 in.	189 in.	270 in.
β	$\dfrac{209}{189} = 1.106$	1.292	1.429	1.00
α_m	$\dfrac{3.13 + 1.77 + 1.70 + 3.40}{4}$ $= 2.50$	2.03	2.10	1.63
Applicable equation	13-11, ACI 9-13	13-11, 9-13	13-11, 9-13	13-10, 9-12
h from equation	5.0	6.24	6.08	6.88
Minimum h	3.5	3.5	3.5	5.0

Before continuing, check whether the shears at the column with the largest tributary area exceed the shear capacity of the beams at that column:

$$w = 1.4\left(\frac{7}{12} \times 0.15 + 0.005 + 0.025\right) + 1.7(0.080)$$

$$= 0.301 \text{ ksf}$$

Assume that the value of d is $(18 - 2.5) = 15.5$ in. Then

$$V_u = 0.301\left[24 \text{ ft} \times 24 \text{ ft} - \left(\frac{18 + 2 \text{ beam } d}{12}\right)^2\right]$$

$$= 168.4 \text{ kips}$$

ϕV_c for the four beams is

$$\Sigma \, \phi V_c = 4\left[0.85 \times 2\sqrt{3000} \times 18 \times \frac{15.5}{1000}\right]$$

$$= 104 \text{ kips}$$

Thus, $V_u \simeq 1.6\Sigma\phi V_c$, which will not be excessive if stirrups are used. Therefore, use the beams selected earlier.

3. Compute the moments in the slab strip along column line B (Fig. 13-76). This strip of slab acts as a rigid frame spanning between columns $B1$, $B2$, $B3$, and so on. In this slab strip, the ℓ_1 direction is parallel to line B, the ℓ_2 direction is perpendicular. Slab panels $B1$–$B2$ and $B4$–$B5$ are "end panels"; the other two are "interior panels."

Values of ℓ_n, ℓ_2, and so on are shown in Fig. 13-76. Since the structure is symmetrical about line 3, only half the structure needs to be considered.

The calculations are carried out in Table 13-10. The calculations are the same as in Table 13-3, except as follows:

Line 4(a). The influence area of the panel is $\ell_n\ell_2$.

Line 4(b). The reduced live loads are based on (2-9).

Line 4(c). In computing the dead load, the weight of the beam stem has been omitted. It will be considered in line 8. We thus compute

$$w_u = 1.4\left(\frac{7}{12} \times 0.150 + 0.005 + 0.025\right) + 1.7w_\ell$$

Line 7(b). ACI Section 13.6.3.4 requires that either the negative-moment sections be designed for the larger negative moment at that support or the unbalanced negative moment be distributed in accordance with the stiffnesses of the adjoining elements. The two moments at $B2$ differ by more than 15 percent; hence, a distribution will be carried out. Because of the two-dimensional nature of slabs, the carryover of moments is not strictly the same as in beams. In the direct design method, it is sufficiently accurate to distribute the moments without any carryovers. Figure 13-78 shows the joint at $B2$.

The unbalanced moment at joint $B2$ is $257.2 - 140.2 = 117$ ft-kips.

The distribution factor to span $B1$–$B2$ is

$$\text{DF}_{B1-B2} = \frac{(K_s + K_b)_{B1-B2}}{(K_s + K_b)_{B1-B2} + (K_s + K_b)_{B2-B3} + \Sigma \, K_c}$$

where K_s, K_b, and K_c refer to the stiffnesses of the slabs, beams, and columns. In computing the stiffnesses for this purpose, the variation in the cross sections along the lengths of slabs and columns will be ignored, giving $K = 4EI/\ell$. For slab $B1$–$B2$,

$$I_s = \frac{21.46 \times 12 \times 7^3}{12}$$

$$= 7361 \text{ in.}^4$$

$$\ell \text{ (center to center of supports)} = 17.25 \text{ ft}$$

$$K_{s,B1-B2} = \frac{4 \times 7361E_c}{17.25 \times 12}$$

$$= 142 \, E_c$$

TABLE 13-10 Calculation of Negative and Positive Moments for East-West Slab Strip *B*— Example 13-10

	B1			B2			B3
1. ℓ_1 (ft)		17.25			24		
2. ℓ_n (ft)		15.75			22.5		
3. ℓ_2 (ft)		21.46			21.46		
4. (a) Area A_I (ft^2)	15.75 $\times$	21.46 =	338		483		
(b) Reduced live load (ksf)		0.080			0.0746		
(c) w_u (ksf)		0.301			0.291		
5. Slab moment, M_o (ft-kips)		200			396		
6. Moment coefficients	-0.16	0.57	-0.70	-0.65	0.35	-0.65	
7. Negative and positive moments (ft-kips)							
(a) Line 5 $\times$ line 6	-32.0	114.2	-140.2	-257.2	138.5	-257.2	
(b) From distribution of negative moments at *B*2	0	-19.2	-38.5	$+27.7$	$+13.9$	0	
(c) Sum (ft-kips)	-32.0	95.0	-178.7	-229.5	152.4	-257.2	
8. Factored load on beam (kips/ft)		0.263			0.263		
9. Beam M_o (ft-kips)		9.0			18.3		
10. Negative and positive moments from beam load							
(a) Line 6 $\times$ line 9	-1.4	5.1	-6.3	-11.9	6.4	-11.9	
(b) Distribution of negative moments at *B*2	0	-0.9	-1.8	$+1.3$	0.7		
(c) Sum (ft-kips)	-1.4	4.2	-8.1	-10.6	7.1	-11.9	
11. Column moments							
(a) From (13-14)				112.0			48.5
(b) From slab and beam	33.4			53			

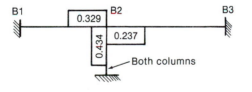

B1 —— B2 —— B3

0.329

0.434   0.237

Both columns

(a) Slab and equivalent column.

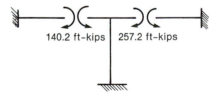

140.2 ft–kips   257.2 ft–kips

(b) Unbalanced moment at joint.

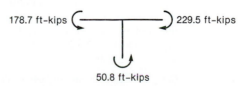

178.7 ft–kips   229.5 ft–kips

50.8 ft–kips

(c) Moments at joint after distribution.

Fig. 13-78
Moment distribution at joint *B*2.

For beam $B1$–$B2$, $I_b = 10{,}340$ in.^{4} (step 2) and $K_b = 200E_c$. Similar calculations for $B2$–$B3$ give $K_s = 102E_c$ and $K_b = 144E_c$. For column B_2 above the slab, $L = 12$ ft, $I_c = 8748$ in.^{4}, and $K_c = 243E_c$, while for column $B2$ below the slab, $K_c = 208E_c$. We then have

$$\mathrm{DF}_{B1-B2} = \frac{142E_c + 200E_c}{(142E_c + 200E_c) + (102E_c + 144E_c) + (243E_c + 208E_c)}$$

$$= 0.329$$

$$\mathrm{DF}_{B2-B3} = 0.237$$

$$\mathrm{DF}_{\text{columns}} = 0.434 \ (\text{see Fig. 13-78a})$$

The unbalanced moment to slab $B1$–$B2$ is $0.329 \times 117 = 38.5$ ft-kips.

The change in midspan moment is $38.5/2 = 19.2$ ft-kips.

Line 8. This is the weight of the beam stem below the slab plus any loads applied directly to the beam. (The live loads on the beam have been included in M_0 for the slab.) We get

$$w = 1.4\left(\frac{11 \times 18}{144} \times 0.15\right) = 0.289 \text{ kip/ft}$$

Line 9. $M_0 = w\ell_n^2/8$.

Line 10. This is calculated with the distribution factors used for line 7(b).

Line 11(b). These moments are the unbalanced moments assigned to the column in the moment distribution.

4. Compute the moments in the slab strip along column line A (Fig. 13-76). This strip of slab acts as a rigid frame spanning between columns $A1$, $A2$, and $A3$. The slab strip includes an edge beam ($A1$–$A2$, $A2$–$A3$, etc.) parallel to the slab spans. In this slab strip, panels $A1$–$A2$ and $A4$–$A5$ are "end panels"; the other two are "interior panels." The calculations are carried out in Table 13-11 and proceed in the same way as the calculations in Table 13–10, except as noted below:

Line 5. For the purpose of calculating M_0 in line 5 of this table, ACI Section 13.6.2.4 requires that, for edge panels, ℓ_2 be taken as the distance from the edge of the slab to the panel centerline.

Line 7. The unbalanced moment at $A2$ is 59.7 ft-kips. The distribution factors for spans $A1$–$A2$ and $A2$–$A3$ and for the columns are 0.371, 0.259, and 0.37, respectively. Thus, the moment distributed to span $A1$–$A2$ is $0.371 \times 59.7 = 22.1$ ft-kips.

Line 8. This is the weight of the beam stem and the wall supported by the beam:

$$w = 1.4\left(\frac{11 \times 18}{144} \times 0.15 + 0.300\right) = 0.709 \text{ kip/ft}$$

5 and 6. Compute the moments in the North–South slab strips. Although a complete solution would include computation of moments in North–South strips as well as in East–West strips, these will be omitted here.

7. Distribute the negative and positive moments to the column strips and middle strips and to the beams: strips spanning East–West. In steps 3 and 4, the moments along column lines A and B were computed. These moments must now be distributed to column and middle strips, and the column-strip moment must be divided between the slab and the beam.

(a) Divide the slabs into column and middle strips. See Fig. 13-76. The widths of the column strips as defined in ACI Section 13.2.1 vary along their lengths, as shown by the dashed lines in Fig. 13-76. Thus, along line B, the column strip width is 8.63 ft in span $B1$–$B2$ and 10.73 ft in span $B2$–$B3$. Assume a constant-width column strip of width 9 ft for simplicity in detailing the slab.

(b) Divide the moments between the column strips and middle strips. These calculations are carried out in Table 13–12, which is laid out to resemble a plan of the slab. Arrows illustrate the flow of moments to the various parts of the slab. The division of moments is a function of both the beam-stiffness ratio for the beam parallel to the strip being designed and the aspect ratio of the panel. For the East–West strips, these terms are summarized subsequently.

TABLE 13-11   Calculation of Negative and Positive Moments for East-West Slab Strip A— Example 13-10

	A1			A2			A3
1. ℓ_1 (ft)		17.25			24.0		
2. ℓ_n (ft)		15.75			22.5		
3. ℓ_2 (ft)		10.21			10.21		
4. (a) Area A_I (ft^2)		176			245		
(b) Reduced live load (ksf)		0.080			0.080		
(c) w_u (ksf)		0.301			0.301		
5. Slab moment, M_o (ft-kips)		95.3			194.5		
6. Moment coefficients	−0.16	0.57	−0.70	−0.65	0.35	−0.65	
7. Negative and positive moments from slab M_o (ft-kips) (a) Line 5 × line 6	−15.2	54.3	−66.7	−126.4	68.1	−126.4	
(b) From distribution of moments at A2	0	−11.0	−22.1	+15.5	+7.7	0	
(c) Sum (ft-kips)	−15.2	43.3	−88.8	−110.9	−75.8	−129.7	
8. Factored load on beam (kips/ft)		0.709			0.709		
9. Beam M_o (ft-kips)		22.0			44.9		
10. Negative and positive moments from beam M_o (a) Line 6 × line 9	−3.5	12.5	−15.4	−29.2	15.7	−29.2	
(b) Distribution of moments at A2	0	−2.5	−5.1	+3.5	1.7	0	
(c) Sum (ft-kips)	−3.5	10.0	−20.5	−25.7	17.4	−29.2	
11. Column moments (a) From (13-14)	—				55.0		23.1
(b) From slab and beam	19.1				27.3		

Values of α were calculated in step 2, and values of ℓ_1 and ℓ_2 are given in Tables 13-10 and 13-11. For the interior strips, ℓ_2 is taken equal to the value used in calculating the moments in the strips. We have the following calculations:

Panel B1–B2: $\ell_1 = 17.25$ ft, $\ell_2 = 21.46$ ft, $\alpha_1 = 1.70$, $\dfrac{\ell_2}{\ell_1} = 1.24$, $\dfrac{\alpha_1\ell_2}{\ell_1} = 2.11$

Panel B2–B3: $\ell_1 = 24.0$ ft, $\ell_2 = 21.46$ ft, $\alpha_1 = 1.70$, $\dfrac{\ell_2}{\ell_1} = 0.89$, $\dfrac{\alpha_1\ell_2}{\ell_1} = 1.51$

Panel C1–C2: $\ell_1 = 17.25$ ft, $\ell_2 = 24.0$ ft, $\alpha_1 = 1.52$, $\dfrac{\ell_2}{\ell_1} = 1.39$, $\dfrac{\alpha_1\ell_2}{\ell_1} = 2.11$

Panel C2–C3: $\ell_1 = 24.0$ ft, $\ell_2 = 24.0$ ft, $\alpha_1 = 1.52$, $\dfrac{\ell_2}{\ell_1} = 1.0$, $\dfrac{\alpha_1\ell_2}{\ell_1} = 1.52$

TABLE 13-12 Division of Moment to Column and Middle Strips: East–West Strips—
Example 13-10

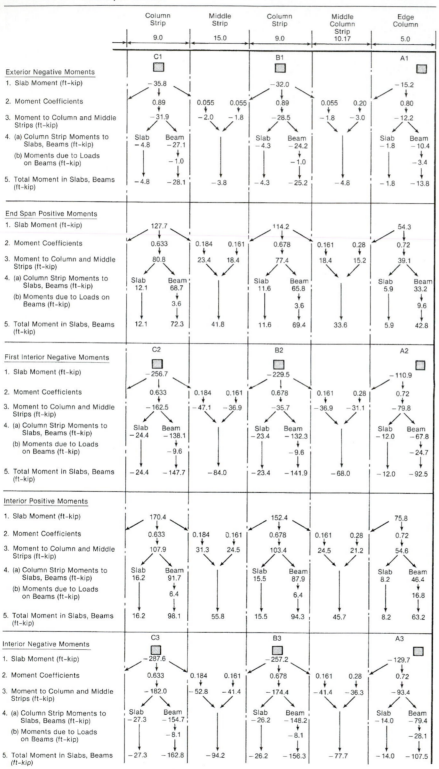

The ACI Code is not clear about which ℓ_2 should be used when considering an edge panel. Current practice involves taking ℓ_2 equal to the total width of the edge panel. Then

Panel $A1$–$A2$: $\ell_1 = 17.25$ ft, $\ell_2 = 18.92$ ft, $\alpha_1 = 3.13$, $\dfrac{\ell_2}{\ell_1} = 1.10$, $\dfrac{\alpha_1 \ell_2}{\ell_1} = 3.44$

Panel $A2$–$A3$: $\ell_1 = 24.0$ ft, $\ell_2 = 18.92$ ft, $\alpha_1 = 3.13$, $\dfrac{\ell_2}{\ell_1} = 0.79$, $\dfrac{\alpha_1 \ell_2}{\ell_1} = 2.47$

Line 1. These moments come from lines 7(d) of Tables 13-10 and 13-11. The moments in strip *C* are from a similar set of calculations. Note that the larger of the two slab moments is used at the first interior support.

Line 2. Exterior negative moments: ACI Section 13.6.4.2 gives the fraction of the exterior negative moment resisted by the column strip. This is computed from ℓ_2/ℓ_1 and $\alpha_1 \ell_2/\ell_1$, just calculated, plus

$$\beta_t = \frac{E_{cb}C}{2E_{cs}I_s} \tag{13-12}$$

where C is the torsional constant for the beam, computed from (13-13). The edge-beam cross section effective for torsion is defined in ACI Section 13.7.5. The effective cross section is shown in Fig. 13-77. To compute C, the beam is divided into rectangles. The maximum value of C corresponds to the rectangles shown in Fig. 13-77a. $C = 13{,}700$ in.[4]

The I_s in (13-12) is the moment of inertia of the slab span framing into the edge beam. Thus, $I_s = \ell_2 h^3/12$. Since the slab and beam are cast at the same time, $E_{cb} = E_{cs}$. We then have the following results:

(a) *Slab strip A:* $I_s = (10.21 \times 12)\dfrac{7^3}{12} = 3502$ in.[4], $\beta_t = \dfrac{13{,}700}{2 \times 3502} = 1.96$

Interpolating in the table in ACI Section 13.6.4.2 for $\ell_2/\ell_1 = 1.10$, $\alpha_1 \ell_2/\ell_1 = 3.44$, and $\beta_t = 1.96$ gives 0.80 of the exterior negative moment to the column strip (at column $A1$).

(b) *Slab strip B:* $I_s = 7361$ in.[4], $\beta_t = 0.93$, $\ell_2/\ell_1 = 1.24$, and $\alpha_1 \ell_2/\ell_1 = 2.11$. Interpolating in ACI Section 13.6.4.2 gives 0.89 times the exterior negative moment to the column strip (at column $B1$).

(c) *Slab strip C:* $I_s = 8232$ in.[4], $\beta_t = 0.83$, $\ell_2/\ell_1 = 1.39$, and $\alpha_1 \ell_2/\ell_1 = 2.11$ gives 0.89 times the exterior moment to the column strip.

Positive moments:

(a) *Slab strip A:* $\ell_2/\ell_1 = 1.10$, $\alpha_1 \ell_2/\ell_1 > 1.0$. From ACI Section 13.6.4.4, 72 percent of the positive moments go to the column strips.

(b) *Slab strip B:* $\ell_2/\ell_1 = 1.24$, $\alpha_1 \ell_2/\ell_1 > 1.0$. From ACI Section 13.6.4.4, 67.8 percent of the positive moments go to the column strips.

(c) *Slab strip C:* 63.3 percent of the positive moments go to the column strip.

Interior negative moments: Because $\alpha_1 \ell_2/\ell_1 > 1.0$, ACI Section 13.6.4.1 gives the same division of interior negative moment between column and middle strips as for positive moments. Thus, interpolating in ACI Section 13.6.4.1, 72, 67.8, and 63.3 percent of the negative moments in slab strips *A, B,* and *C* go to the column strip.

Line 4(a). The column-strip moments are divided between the slab and the beam, following the rules given in ACI Sections 13.6.5.1 and 13.6.5.2. If $\alpha_1 \ell_2/\ell_1 \geq 1.0$, 85 percent of the column-strip moment is assigned to the beam. For this slab, this is true in all cases.

Line 4(b). These moments come from line 10(c) of Tables 13–10 and 13–11. They are assumed to be carried entirely by the beams.

Line 5. The final moments in the middle strip between lines *A* and *B*, in the slab in the column strip along line *B*, and in the beam along line *B* are plotted in Fig.13-79.

8.   Design the slab reinforcement. The division of moments is similar in the other direction. The design of slab reinforcement is carried out in the same way as in previous examples and will not

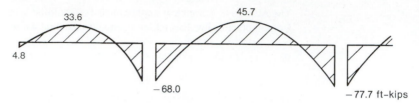

(a) Moments in middle strip between lines A and B.

(b) Moments in slab in column strip along line B.

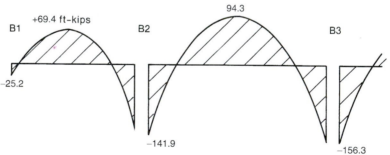

(c) Moments in beam along line B.

Fig. 13-79
Moments in slab strips—
Example 13-10.

be repeated here. ACI Section 13.3.6 requires special corner reinforcement at exterior corners. At corners $A1$ and $A5$, provide one mat each way parallel to the sides of the slab at the bottom and top of the slab. These are designed for a moment equal to the maximum positive moment per foot of width occurring in the corner panels. From Table 13-12, this is 33.6 ft-kips/10.17 ft = 3.30 ft-kips/ft. The steel should extend $\ell_n/5 = (209/5) = 42$ in. each way from the corner.

   9.   **Design the beams.** The beams must be designed for moment, shear, and bar anchorage, as in the examples in Chapter 10. The edge beams are subjected to a torque, which must be considered in the design.

      (a)   **Flexure.** The flexural design is similar to that for any continuous beam, except that the moments used are those given in line 5 of Table 13-12. Figure 13-79c shows the moments used in the design of the beam along line B. When selecting the reinforcement, the cover to the reinforcement was taken as that for a beam rather than that for a slab.

      (b)   **Torsion.** The distribution of negative moments along the edge $A1–B1–C1$ was calculated in Table 13–12. A similar analysis of the perpendicular slab strips gave the distribution of exterior negative moments in the slab along $A1–A2–A3$, shown in Fig. 13-80a. The distribution of these moments per foot of length of the edge members is shown in Fig. 13-80b. A torque diagram has been computed in the same way that a shear-force diagram would be for simply supported beams loaded with distributed loads equivalent to the distributed torques. This, in effect, assumes that the edge beams are rigidly fixed against rotation at the columns, so that the total angle change $\theta = TL/GJ$ equals zero in each span.

According to ACI Sections 11.6.1 and 11.6.2.4, torsional reinforcement is required if

$$T_u \geq \phi\sqrt{f_c'}\left(\frac{A_{cp}^{\ 2}}{p_{cp}}\right)$$

at sections located d away from the face of the columns. ACI Section 11.6.1 defines the cross section of the edge beam as given in ACI Section 13.2.4 and as illustrated in Fig. 13-77a. For this section,

$$A_{cp} = 18 \times 18 + 11 \times 7 = 401 \text{ in.}^2$$

$$p_{cp} = 3 \times 18 + 3 \times 11 + 7 = 94 \text{ in.}$$

$$\phi\sqrt{f_c'}\left(\frac{A_{cp}^{\ 2}}{p_{cp}}\right) = 0.85\sqrt{3000}\left(\frac{401^2}{94}\right)$$

$$= 79,600 \text{ in.-lb} = 6.64 \text{ ft-kips}$$

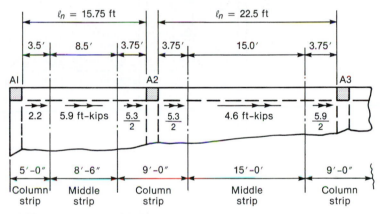

(a) Moments at edge of slab (ft–kip).

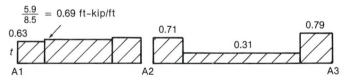

(b) Distributed torques, t, on edge beams (ft–kips/ft).

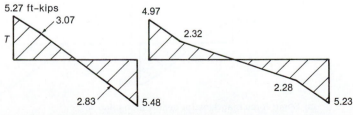

Fig. 13-80
Torque on edge beam—
Example 13-10.

(c) Torque diagram, T, for edge beams (ft–kips).

Figure 13-80c shows that T_u does not exceed this value at any point and hence that torsion can be ignored, except that closed stirrups will be used in the spandrel beams. ACI Section 11.6.2.3 allows the designer to assume a uniform distribution of t along the beam rather than the more elaborate analysis used here.

(c) **Shear.** Since $\alpha_1 \ell_2/\ell_1 \geq 1.0$ for all beams, ACI Section 13.6.8.1 requires that the beams be designed for the shear caused by loads on the tributary areas shown in Fig. 13-81a. For the beams along line A, the corresponding beam loads and shear-force diagrams are shown in Figs. 13-81b and c. Design of stirrups is in accordance with ACI Sections 11.1 to 11.5. ∎

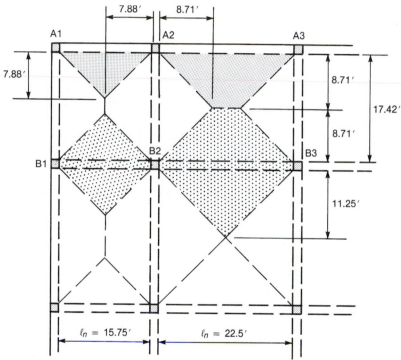

(a) Tributary areas for beams along lines A and B.

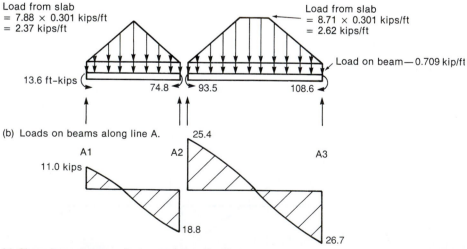

(b) Loads on beams along line A.

(c) Shear force diagrams for beams along line A.

Fig. 13-81
Shear on edge beam—
Example 13-10.

PROBLEMS

13-1 Compute α for the beam shown in Fig. P13-1. The concrete for the slab and beam was placed in one pour.

13-2 Compute the column-strip and middle-strip moments in the long-span direction for an interior panel of the flat slab floor shown in Fig. 13-29. The slab is 6 in. thick, the design live load is 40 psf, and the superimposed dead load is 5 psf for ceiling, flooring, and so on, plus 25 psf for partitions. The columns are 10 in. $\times$ 12 in., as shown in Fig. 13-29.

13-3 Compute the column-strip and middle-strip moments in an exterior bay of the flat-plate slab shown in Fig. P13-3. The slab is $7\frac{1}{2}$ in. thick and supports a superimposed service dead load of 25 psf and a service live load of 50 psf. There is no edge beam. The columns are all 18 in. square.

13-4 A 7-in.-thick flat-plate slab with spans of 20 ft in each direction is supported on 16 in. $\times$ 16 in. columns. The effective depths are 5.9 and 5.3 in. in the two directions. The slab supports its own dead load, plus 20 psf unfactored superimposed dead load and 40 psf unfactored live load. The concrete strength is 4000 psi. Check one-way and two-way shear at a typical interior support.

13-5 The slab described in Problem 13-4 is supported on 10 in. $\times$ 24 in. columns. Check one-way and two-way shear at a typical interior support.

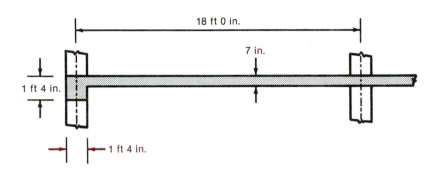

Fig. P13-1

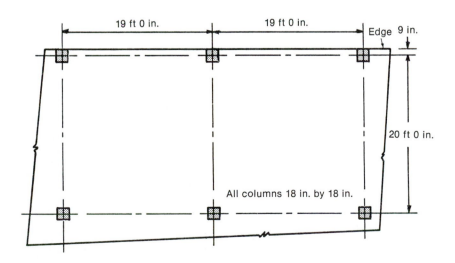

Fig. P13-3

13-6 The slab shown in Fig. P13-6 supports a superimposed dead load of 25 psf and a live load of 60 psf. The slab extends 4 in. past the exterior face of the column to support an exterior wall that weighs 400 lb/ft of length of wall. The story-to-story height is 9 ft. Use 4500-psi concrete and Grade-60 reinforcement.

(a) Select thickness.

(b) Compute moments, and design the reinforcement for slab strips running North and South.

(c) Check shear and moment transfer at columns A2 and B2. Neglect unbalanced moments about column line 2.

13-7 For the slab shown in Fig. P13-6 and the loadings and material strengths given in Problem 13-6,

(a) select the thickness;

(b) compute moments, and design the reinforcement for slab strips running East and West;

(c) check the shear and moment transfer at columns C1 and C2. Neglect unbalanced moments about column line C.

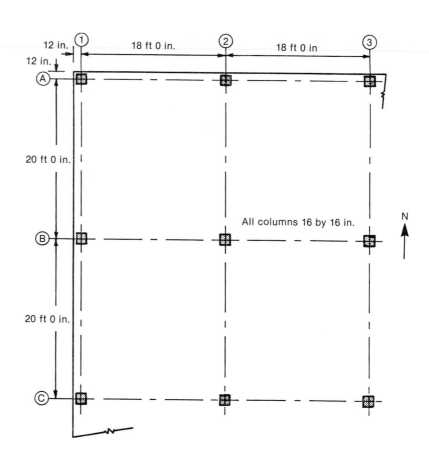

Fig. P13-6

14

Equivalent Frame Method

14-1 INTRODUCTION

The ACI Code presents two parallel methods for calculating moments in two-way slab systems. These are the direct design method, presented in Chapter 13, and the equivalent frame method, presented in this chapter. In addition, ACI Section 13.5.1 allows other methods if they meet certain requirements. The relationship between the direct design method and the equivalent frame method is explained in Section 13-6 of this book. In the direct design method, the statical moment, M_0, is calculated for each panel. This moment is then divided between positive- and negative-moment regions by using arbitrary moment coefficients, and the positive moments are adjusted to reflect pattern loadings. In the equivalent frame method, these steps are accomplished by frame analyses.

The use of frame analyses to analyze slabs was first proposed by Peabody [14-1] in 1948, and a method of slab analysis referred to as "design by elastic analysis" was incorporated in the 1956 and 1963 editions of the ACI Code. In the late 1940s, Siess and Newmark [14-2], [14-3] studied the application of moment-distribution analyses to two-way slabs on stiff beams. Following extensive research on two-way slabs carried out at the University of Illinois, Corley and Jirsa [14-4] presented a more refined method of frame analysis for slabs. This has been incorporated in the 1971 and subsequent ACI Codes. Corley and Jirsa considered only gravity loads. Studies of the use of frame analyses for laterally loaded column–slab structures [14-5] led to treatment of this problem in the 1983 and subsequent ACI Codes.

The equivalent frame method is intended for use in analyzing moments in any practical building frame. Its scope is thus wider than the direct design method, which is subject to the limitations presented in Section 13-6 (ACI Section 13.6.1). It works best for frames that can also be designed by the direct design method, however.

This chapter builds on the basic knowledge of the behavior and design of slabs in flexure and shear presented in Chapter 13.

733

14-2 EQUIVALENT FRAME ANALYSIS OF SLAB SYSTEMS FOR VERTICAL LOADS

The slab is divided into a series of equivalent frames running in the two directions of the building, as shown in Fig. 13-23 (ACI Section 13.7.2). These frames consist of the slab, any beams that are present, and the columns above and below the slab. For gravity-load analysis, the code allows analysis of an entire equivalent frame extending over the height of the building, or each floor can be considered separately, with the far ends of the columns fixed. These frames are, in turn, divided up into column and middle strips, as described in Section 13-6 and Fig. 13-25.

The original derivation of the equivalent frame method assumed that moment distribution would be the procedure used to analyze the slabs, and some of the concepts in the method are awkward to adapt to other methods of analysis. In this chapter, the equivalent frame method is presented for use with a moment-distribution analysis. This is followed by a brief discussion of how to use direct-stiffness computer programs in carrying out an equivalent frame analysis.

Calculation of Stiffness, Carryover, and Fixed-End Moments

In the moment-distribution method, it is necessary to compute *flexural stiffnesses, K; carryover factors, COF; distribution factors, DF*; and *fixed-end moments, FEM*, for each of the members in the structure. For a prismatic member, fixed at the far end, with negligible axial loads, the flexural stiffness is

$$K = \frac{kEI}{L} \tag{14-1}$$

where $k = 4$ and the carryover factor is ±0.5, the sign depending on the sign convention used for moments. For a prismatic, uniformly loaded beam, the fixed-end moments are $w\ell^2/12$.

In the equivalent frame method, the increased stiffness of members within the column–slab joint region is accounted for, as is the variation in cross section at drop panels. As a result, all members have a stiffer section at each end, as shown in Fig. 14-1b. If the EI used in (14-1) is that at the midspan of the slab strip, k will be greater than 4; similarly, the carryover factor will be greater than 0.5 and the fixed-end moments for a uniform load w will be greater than $w\ell^2/12$.

Several methods are available for computing values of k, *COF*, and fixed-end moments. Originally, these were computed by using the *column analogy* developed by Hardy Cross. Cross observed an analogy between the equations used to compute stresses in an unsymmetrical column loaded with axial loads and moments and the equations used to compute moments in a fixed-end beam. (For more details, see [14-6].)

Tables and charts for computing k, *COF*, and fixed-end moments are given in Appendix A as Tables A-20 and A-23.

Properties of Slab-Beams

The horizontal members in the equivalent frame are referred to as *slab-beams*. These consist of a slab, a slab and a drop panel, or a slab with a beam running parallel to the equivalent frame. ACI Section 13.7.3 explains how these nonprismatic beams are to be modeled for analysis:

1. At points outside of joints or column capitals, the moment of inertia may be based on the gross area of the concrete. Variations in the moment of inertia along the length shall be taken into account.

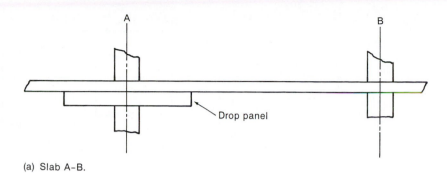

(a) Slab A–B.

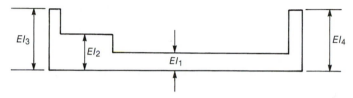

Fig. 14-1
Variation in stiffness along
span.

(b) Distribution of EI along slab.

Thus, for the slab with a drop panel shown in Fig. 14-2a, the moment of inertia at section
A–A is that for a slab of width ℓ_2 (Fig. 14-2c). At section B–B through the drop panel, the
moment of inertia is for a slab having the cross section shown in Fig. 14-2d. Similarly, for
a slab system with a beam parallel to ℓ_1, as shown in Fig. 14-3a, the moment of inertia for
section C–C is that for a slab-and-beam section, as shown in Fig. 14-3c. Section D–D is cut
through a beam running perpendicular to the page.

   2.   The moment of inertia of the slab-beams from the center of the column to the
face of the column, bracket, or capital (as defined in ACI Section 13.1.2 and Fig. 13-61)
shall be taken as the moment of inertia of the slab-beam at the face of the column, bracket,
or capital divided by the quantity $(1 - c_2/\ell_2)^2$, where ℓ_2 is the transverse width of the
equivalent frame (Fig. 13-23) and c_2 is the width of the support parallel to ℓ_2.

The application of this approach is illustrated in Figs. 14-2 and 14-3. Tables A-20 and A-22
[14-7], [14-8] present moment-distribution constants for flat plates and for slabs with drop
panels. For most practical cases, these eliminate the need to use the column-analogy solution to compute the moment-distribution constants.

EXAMPLE 14-1   Calculation of the Moment-Distribution
Constants for a Flat-Plate Floor

Figure 14-4 shows a plan of a flat-plate floor without spandrel beams. The floor is 7 in. thick. Compute the moment-distribution constants for the slab-beams in the equivalent frame along line 2,
shown shaded in Fig. 14-4.

Span A2–B2

At end A, $c_1 = 12$ in., $\ell_1 = 213$ in., $c_2 = 20$ in., and $\ell_2 = 189$ in. Thus, $c_1/\ell_1 = 0.056$ and c_2/ℓ_2
$= 0.106$. Interpolating in Table A-20, we find that the fixed-end moment is

$$M = 0.084 w\ell_2\ell_1^2$$

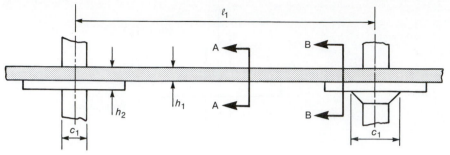

(a) Slab with drop panels.

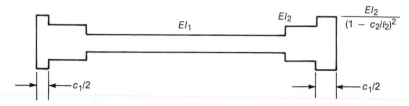

(b) Variation in EI along slab–beam.

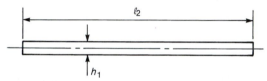

(c) Cross section used in compute I_1—Section A-A.

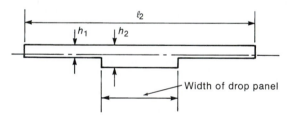

Width of drop panel

Fig. 14-2
EI values for slab with drop
panel.

(d) Cross section used to compute I_2—Section B-B.

The stiffness is

$$K = \frac{4.11EI_1}{\ell_1}$$

and the carryover factor is 0.508.

Span B2–C2

For span *B2–C2*, $c_1 = 20$ in., $\ell_1 = 240$ in., $c_2 = 12$ in., and $\ell_2 = 189$ in. Thus, $c_1/\ell_1 = 0.083$, $c_2\ell_2 = 0.064$. From Table A-20, the fixed-end moment is

$$M = 0.084w\ell_2\ell_1{}^2$$

The stiffness is

$$K = \frac{4.10EI_1}{\ell_1}$$

and the carryover factor is 0.507.

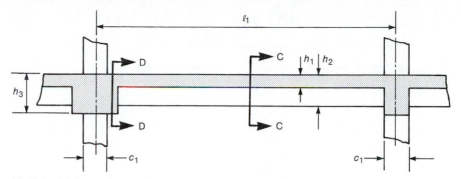

(a) Slab with beams in two directions.

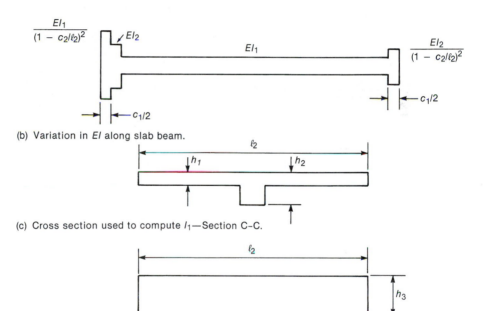

(b) Variation in *EI* along slab beam.

(c) Cross section used to compute I_1—Section C–C.

(d) Cross section used to compute I_2—Section D–D.

Fig. 14-3
EI values for slab and beam.

EXAMPLE 14-2 Calculation of the Moment-Distribution Constants for a Two-Way Slab with Beams

Figure 14-5 shows a two-way slab with beams between all columns. The slab is 7 in. thick, and all the beams are 18 in. wide by 18 in. in overall depth. Compute the moment-distribution constants for the slab-beams in the equivalent frame along line *B* (shown shaded in Fig. 14-5).

Span B1–B2

Span *B1–B2* is shown in Fig. 14-6. A cross section at midspan is shown in Fig. 14-6b. The centroid of this section lies 4.39 in. below the top of the slab, and its moment of inertia is

$$I_1 = 23,800 \text{ in.}^4$$

The columns at both ends are 18 in. square, giving $c_1/\ell_1 = 18/207 = 0.087$ and $c_2/\ell_2 = 0.070$. Since this member has uniform stiffness between the joint regions, we can use Table A-20 to get the moment-distribution constants:

$$\text{Fixed-end moment: } M = 0.084w\ell_1^2$$

$$\text{Stiffness: } K = \frac{4.10EI_1}{\ell_1}$$

$$\text{Carryover factor} = 0.507$$

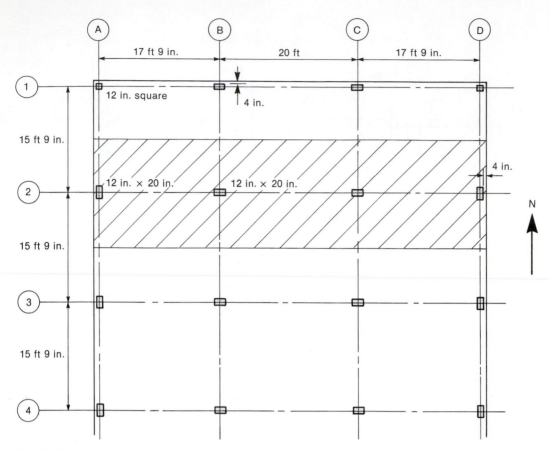

Fig. 14-4
Plan of flat plate floor—Examples 14-1, 14-3, and 14-5.

Span B2–B3

Here, $c_1/\ell_1 = 18/288 = 0.0625$, and $c_2/\ell_2 = 0.070$. From Table A-20,

$$\text{Fixed-end moment: } M = 0.084w\ell_1^2$$

$$\text{Stiffness: } K = \frac{4.07EI_1}{\ell_1}$$

$$\text{Carryover factor} = 0.505$$

■

Properties of Columns

In computing the stiffnesses and carryover factors for columns, ACI Section 13.7.4 states the following:

1. The moment of inertia of columns at any cross section outside of the joints or column capitals may be based on the gross area of the concrete, allowing for variations in the actual moment of inertia due to changes in the column cross section along the length of the column.

2. The moment of inertia of columns shall be assumed to be infinite within the depth of the slab-beam at a joint.

Figure 14-7 illustrates these points for four common cases. Again, the column analogy can be used to solve for the moment-distribution constants, or the values given in Table A-23 [14-7] can be used.

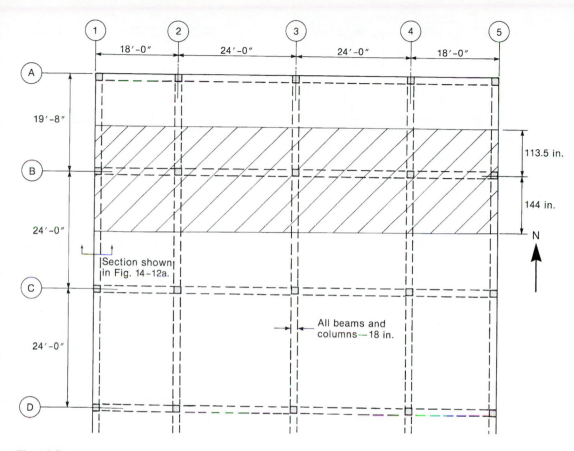

Fig. 14-5
Two-way slabs with beams—Examples 14-2 and 14-4.

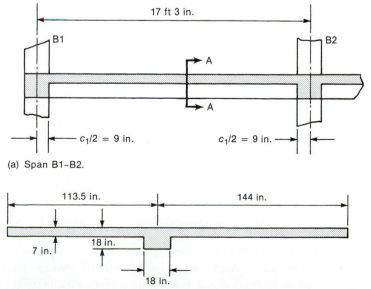

Fig. 14-6
Span $B1$-$B2$—Example 14-2.

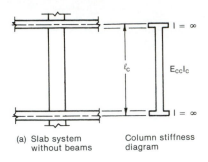

(a) Slab system without beams — Column stiffness diagram

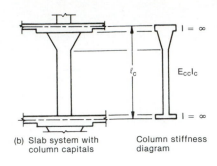

(b) Slab system with column capitals — Column stiffness diagram

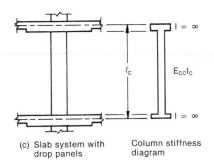

(c) Slab system with drop panels — Column stiffness diagram

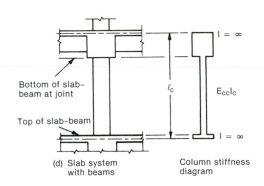

(d) Slab system with beams — Column stiffness diagram

Fig. 14-7
Sections for the calculation of column stiffness, K_c. (From [14-7].)

Torsional Members and Equivalent Columns

When the beam-and-column frame shown in Fig. 14-8a is loaded, the ends of the column and beam undergo equal rotations where they meet at the joint. If the flexural stiffness, $K = M/\theta$, is known for the two members, it is possible to calculate the joint rotations and the end moments in the members. Similarly, in the case shown in Fig. 14-8b, the ends of the slab and the wall both undergo equal end rotations when the slab is loaded. When a flat plate is connected to a column, as shown in Fig. 14-8c, the end rotation of the column is equal to the end rotation of the strip of slab C–D, which is attached to the column. The rotation at A of strip A–B is greater than the rotation at point C, however, because there is less restraint to the rotation of the slab at this point. In effect, the edge of the slab has twisted, as shown in Fig. 14-8d. As a result, the *average* rotation of the edge of the slab is greater than the rotation of the end of the column.

To account for this effect in slab analysis, the column is assumed to be attached to the slab-beam by the transverse torsional members A–C and C–A'. One way of including these members in the analysis is by the use of the concept of an *equivalent column*, which is a single element consisting of the *columns* above and below the floor and *attached torsional members*, as shown in Fig. 14-8d. The stiffness of the equivalent column, K_{ec}, represents the combined stiffnesses of the columns and attached torsional members:

$$K_{ec} = \frac{M}{\text{average rotation of the edge beam}} \qquad (14\text{-}2)$$

The inverse of a stiffness, $1/K$, is called the *flexibility*. The flexibility of the equivalent column, $1/K_{ec}$, is equal to the average rotation of the joint between the "edge beam" and the rest of the slab when a unit moment is transferred from the slab to the equivalent column. This average rotation is the rotation of the end of the columns, θ_c, plus the average twist of the beam, $\theta_{t,\text{avg}}$, both computed for a unit moment.

$$\theta_{ec} = \theta_c + \theta_{t,\text{avg}} \qquad (14\text{-}3)$$

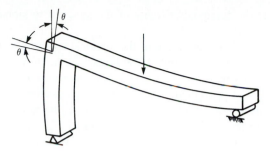

(a) Beam and column frame.

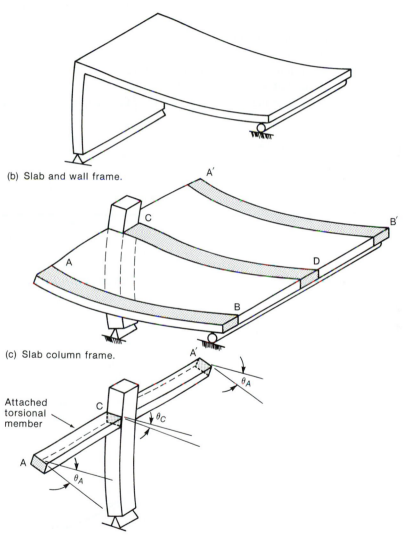

(b) Slab and wall frame.

(c) Slab column frame.

Attached
torsional
member

(d) Column and edge of slab.

Fig. 14-8
Frame action and twisting of
edge member.

The value of θ_c for a unit moment is $1/\Sigma K_c$, where ΣK_c refers to the sum of the flexural stiffnesses of the columns above and below the slab. Similarly, the value of $\theta_{t,\text{avg}}$ for a unit moment is $1/K_t$, where K_t is the torsional stiffness of the attached torsional members. Substituting into (14-3) gives

$$\frac{1}{K_{ec}} = \frac{1}{\Sigma K_c} + \frac{1}{K_t} \qquad\qquad (14\text{-}4)$$

If the torsional stiffness of the attached torsional members is small, K_{ec} will be much smaller than ΣK_c.

The derivation of the torsional stiffness of the torsional members or "edge beams" is illustrated in Fig. 14-9. Figure 14-9a shows an equivalent column with attached torsional members that extend halfway to the next column in each direction. A unit torque $T = 1$ is applied to the equivalent column, with half going to each arm. Since the assembly is stiffest adjacent to the column, the moment, t, per unit length of the edge beam is arbitrarily assumed to be as shown in Fig. 14-9b. The height of this diagram at the middle of the column has been chosen to give a total area equal to 1.0, the value of the applied moment, T.

The applied torques give rise to the twisting-moment diagram shown in Fig. 14-9c. Since half of the torque is applied to each arm, the maximum twisting moment is $\frac{1}{2}$. The twist angle per unit length of torsional member is shown in Fig. 14-9d. This is calculated by dividing the twisting moment at any point by CG, the product of the torsional constant, C (similar to a polar moment of inertia) and the modulus of rigidity, G. The total twist of the end of an arm relative to the column is the summation of the twists per unit length and is equal to the area of the diagram of twist angle per unit length in Fig. 14-9d. Since this is

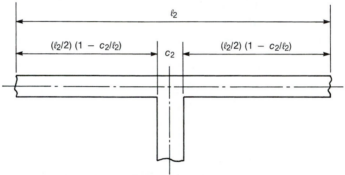

(a) Column and attached torsional member.

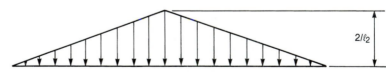

(b) Distribution of torque per unit length along column center line.

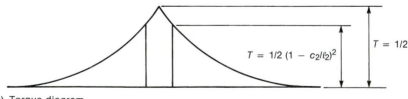

(c) Torque diagram.

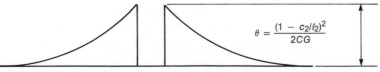

(d) Angle change per unit length.

Fig. 14-9
Calculation of K_t. (From [14-10].)

a parabolic diagram, the angle of twist at the outer end of the arm is one-third of the height times the length of the diagram:

$$\theta_{t,\text{end}} = \frac{1}{3}\frac{(1 - c_2/\ell_2)^2}{2CG}\left[\frac{\ell_2}{2}\left(1 - \frac{c_2}{\ell_2}\right)\right]$$

Replacing G with $E/2$ gives

$$\theta_{t,\text{end}} = \frac{\ell_2(1 - c_2/\ell_2)^3}{6CE}$$

This is the rotation of the end of the arm. The rotation required for use in (14-3) is the average rotation of the arm, which is assumed to be a third of the end rotation:

$$\theta_{t,\text{avg}} = \frac{\ell_2(1 - c_2/\ell_2)^3}{18CE} \tag{14-5}$$

Finally, the torsional stiffness of one arm is calculated as $K_t = M/\theta_{t,\text{avg}}$, where the moment resisted by one arm is taken as $\frac{1}{2}$, giving

$$K_t(\text{one-arm}) = \frac{9EC}{\ell_2(1 - c_2/\ell_2)^3}$$

ACI Commentary Section R13.7.5 expresses the torsional stiffness of the two arms as

$$K_t = \sum \frac{9E_{cs}C}{\ell_2(1 - c_2/\ell_2)^3} \tag{14-6}$$

where ℓ_2 refers to the transverse spans on each side of the column. For a corner column, there is only one term in the summation.

If a beam parallel to the ℓ_1 direction (a beam along C–D in Fig. 14-8) frames into the column, a major fraction of the exterior negative moment is transferred directly to the column without involving the attached torsional member. In such a case, K_{ec} underestimates the stiffness of the column. This is empirically allowed for by multiplying K_t by the ratio I_{sb}/I_s, where I_{sb} is the moment of inertia of the slab and beam together and I_s is the moment of inertia of the slab neglecting the beam stem (ACI Section 13.7.5.4).

The cross section of the torsional members is defined in ACI Section 13.7.5.1(a) to (c) and is illustrated in Fig. 13-27. Note that this cross section may be larger than that used to compute the flexural stiffness of the beam (ACI Section 13.2.4 and Fig. 13-20). It also differs from the definition of the beam section for design for torsion in ACI Section 11.6.1.

The constant C in (14-6) is calculated by subdividing the cross section into rectangles and carrying out the summation

$$C = \sum\left[\left(1 - 0.63\frac{x}{y}\right)\frac{x^3 y}{3}\right] \tag{14-7}$$

where x is the shorter side of a rectangle and y is the longer side. The subdivision of the cross section of the torsional members is illustrated in Fig. 13-28 and explained in Section 13-6.

In a moment-distribution analysis, the frame analysis is carried out for a frame with slabs having stiffnesses K_s, with equivalent columns having stiffnesses K_{ec}, and possibly with beams parallel to the slabs having stiffnesses K_b.

EXAMPLE 14-3 Calculation of K_t, ΣK_c, and K_{ec} for an Edge Column and an Interior Column

The 7-in.-thick flat plate shown in Fig. 14-4 is attached to 12-in. × 20-in. columns oriented with the 12-in. dimension perpendicular to the edge, as shown in Fig. 14-4. The story-to-story height is 8 ft 10 in. The slab and columns are of 4000-psi concrete. Compute K_t, ΣK_c, and K_{ec} for the connections between the slab strip along line 2 and columns A2 and B2.

1. **Compute the values for the exterior column, A2.**

   (a) **Define the cross section of the torsional members.** According to ACI Section 13.7.5.1, the attached torsional member at the exterior column corresponds to condition (a) in Fig. 13-27, as shown in Fig. 14-10a. Here, $x = 7$ in. and $y = 12$ in.

   (b) **Compute C.**

$$C = \Sigma\left(1 - 0.63\frac{x}{y}\right)\frac{x^3 y}{3} = \left(1 - 0.63\frac{7}{12}\right)\frac{7^3 \times 12}{3} \qquad (14\text{-}7)$$

$$= 868 \text{ in.}^4$$

   (c) **Compute K_t.**

$$K_t = \Sigma\frac{9E_{cs}C}{\ell_2(1 - c_2/\ell_2)^3} \qquad (14\text{-}6)$$

where the summation refers to the beams on either side of line 2 and ℓ_2 refers to the length ℓ_2 of the beams on each side of line 2. Since the two beams are similar,

$$K_t = 2\left[\frac{9E_{cs} \times 868}{15.75 \times 12\left(1 - \dfrac{20}{15.75 \times 12}\right)^3}\right] = 116E_{cs}$$

   (d) **Compute ΣK_c for the edge columns.** The height center to center of the floor slabs is 8 ft 10 in. = 106 in. The distribution of stiffnesses along the column is similar to that in Fig. 14-7a. The edge columns are bent about an axis parallel to the edge of the slab. Thus,

$$I_c = 20 \times \frac{12^3}{12}$$

$$= 2880 \text{ in.}^4$$

For this column, the overall height $\ell = 106$ in., the unsupported or clear height $\ell_u = 99$ in., and $\ell/\ell_u = 1.071$. The distance from the centerline of the slab to the top of the column proper,

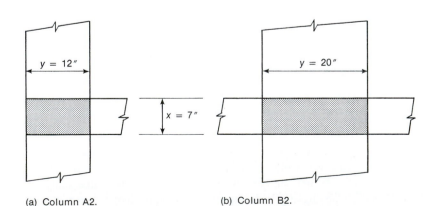

Fig. 14-10
Attached torsional members—Example 14-3.

(a) Column A2.

(b) Column B2.

(In figure: y = 12", x = 7" for Column A2; y = 20" for Column B2.)

t_a, is 3.5 in., as is the corresponding distance, t_b, at the bottom of the column. Interpolating in Table A-23 for $\ell/\ell_u = 1.071$ and $t_a/t_b = 1.0$ gives

$$K_c = \frac{4.76EI_c}{\ell_c}$$

and the carryover factor is 0.55. Since there are two columns (one above the floor and one under), each with the same stiffness, it follows that

$$\sum K_c = 2\left(\frac{4.76E_{cc} \times 2880}{106}\right) = 259E_{cc}$$

(e) **Compute the equivalent column stiffness, K_{ec}, for the edge-column connection.**

$$\frac{1}{K_{ec}} = \frac{1}{\sum K_c} + \frac{1}{K_t} = \frac{1}{259E_{cc}} + \frac{1}{116E_{cs}}$$

Because the slab and the columns have the same strength concrete, $E_{cc} = E_{cs} = E_c$. Therefore, $K_{ec} = 79.9E_c$.

Note that K_{ec} is only 30 percent of $\sum K_c$. This illustrates the large reduction in effective stiffness due to the lack of a stiff torsional member at the edge.

2. **Compute the values at the interior column, B2.** The torsional member at column B2 also has a section corresponding to condition (a) in Fig. 13-27a, with $x = 7$ in. and $y = 20$ in., as shown in Fig. 14-10b. Thus, $C = 1782$ in.^{4} and $K_t = 206.7E_c$. In the slab-strip along line 2, these columns are bent around their strong axes and have $I_c = 8000$ in.^{4}. Again,

$$\sum K_c = 2\left(\frac{4.76E_cI}{\ell}\right) = 718E_c$$

$$K_{ec} = 161E_c$$

It is important to note that unless K_t is very large, K_{ec} will be much smaller than $\sum K_c$. ■

EXAMPLE 14-4 Calculation of K_{ec} for an Edge Column in a Two-Way Slab with Beams

The two-way slab with beams between all columns in Fig. 14-5 is supported on 18-in.-square columns. The floor-to-floor height is 12 ft 0 in. above the floor in question and 14 ft 0 in. below the floor in question. All beams are 18 in. wide by 18 in. in overall depth. The concrete strength is 3000 psi in the slab and beams and 4000 psi in the columns. Compute K_{ec} for the connections between the slab strip along line B and columns B1 and B2. A vertical cross section through the structure is shown in Fig. 14-11a.

1. **Compute the values at the exterior column, B1.**

(a) **Define the cross section of the torsional member.** The torsional member at the exterior edge has the cross section shown in Fig. 14-12a (ACI Sections 13.7.5.1 and 13.2.4).

(b) **Compute C.** To compute C, divide the torsional member into rectangles to maximize C as shown in Fig. 14-12a:

$$C = \left(1 - 0.63\frac{18}{18}\right)\frac{18^3 \times 18}{3} + \left(1 - 0.63\frac{7}{11}\right)\frac{7^3 \times 11}{3}$$

$$= 13{,}700 \text{ in.}^4$$

(c) **Compute K_t.**

$$K_t = \sum \frac{9E_{cs}C}{\ell_2(1 - c_2/\ell_2)^3} \qquad (14\text{-}6)$$

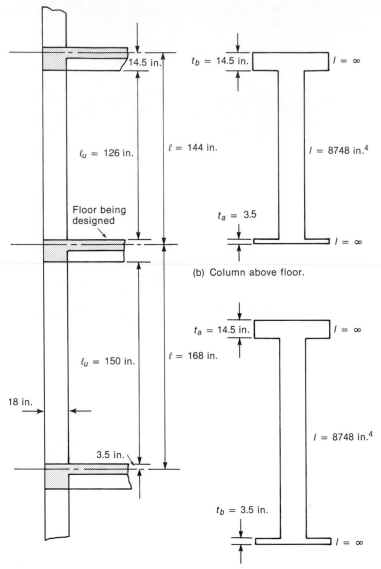

Fig. 14-11
Exterior column-to-slab joint
at $B1$—Example 14-4.

(a) Columns, beams, and slabs.

(b) Column above floor.

(c) Column below floor.

For span $A1–B1$, $\ell_2 = 18$ ft 11 in. $= 227$ in., while for span $B1–C1$, $\ell_2 = 24$ ft 0 in. $= 288$ in. Thus,

$$K_t = \frac{9E_{cs} \times 13,700}{227(1 - 18/227)^3} + \frac{9E_{cs} \times 13,700}{288(1 - 18/288)^3}$$

$$= 1216E_{cs}$$

Because the span for which moments are being determined contains a beam parallel to the span, K_t is multiplied by the ratio of the moment of inertia, I_{sb}, of a cross section including the beam stem (Fig. 14-6b) to the moment of inertia, I_s, of the slab alone. From Example 14-2, $I_{sb} = 23,800$ in.[4], so

$$I_s = \frac{257.5 \times 7^3}{12} = 7360 \text{ in.}^4$$

Therefore, $I_{sb}/I_s = 3.23$, and $K_t = 3.23 \times 1216E_{cs} = 3930E_{cs}$.

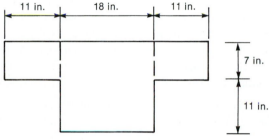

(a) Torsional member at B1.

(b) Torsional member at B2.

Fig. 14-12
Torsional members at $B1$ and
$B2$—Example 14-4.

(d)   **Compute ΣK_c for the edge columns at B1.** The columns and slabs at $B1$ are shown in Fig. 14-11a. The distributions of moments of inertia along the columns are shown in Fig. 14-11b and c. We have

$$I_c = \frac{18^4}{12} = 8748 \text{ in.}^4$$

For the bottom column, $\ell = 168$ in., $\ell_u = 150$ in., and $\ell/\ell_u = 1.11$. The ratio $t_a/t_b = 14.5$ in./3.5 in. $= 4.14$. From Table A-23,

$$K_{c,\text{lower}} = \frac{5.73EI}{\ell} = \frac{5.73E_{cc} \times 8748}{168}$$
$$= 298E_{cc}$$

For the column over the floor, $\ell = 144$ in., $\ell_u = 126$ in., and $\ell/\ell_u = 1.14$. Since moment distribution is being carried out at the lower end, $t_a = 3.5$ in., $t_b = 14.5$ in., and $t_a/t_b = 0.24$. From Table A-23,

$$K_{c,\text{upper}} = \frac{4.93E_{cc} \times 8748}{144} = 300E_c$$
$$\Sigma K_c = 598E_{cc}$$

(e)   **Compute K_{ec}.**

$$\frac{1}{K_{ec}} = \frac{1}{\Sigma K_c} + \frac{1}{K_t}$$
$$\frac{1}{K_{ec}} = \frac{1}{598E_{cc}} + \frac{1}{3930E_{cs}}$$

where

$$E_{cc} = 57,000\sqrt{4000} = 3.60 \times 10^6 \text{ psi (ACI Section 8.5.1)}$$
$$E_{cs} = 57,000\sqrt{3000} = 3.12 \times 10^6 \text{ psi}$$

Thus $E_{cc} = 1.15E_{cs}$ and $K_{ec} = 587E_{cs}$ for the edge column-to-slab connection.

2. Compute the values at the interior column, *B2*. The cross section of the torsional member is shown in Fig. 14-12b. For this section, $C = 14,450$ in.[4], and K_t of the torsional members is $K_t = 1279E_{cs}$. This is multiplied by the ratio $I_{sb}/I_s = 3.23$, giving $K_t = 4130E_{cs}$. Again, $\Sigma K_c = 598E_{cc}$ and $E_{cc} = 1.15E_{cs}$. Thus, $K_{ec} = 591E_{cs}$ at the interior column *B2*. ∎

Arrangement of Live Loads for Structural Analysis

The placement of live loads to produce maximum moments in a continuous beam or one-way slab was discussed in Section 10-2 and illustrated in Fig. 10-8. Similar loading patterns are specified in ACI Section 13.7.6 for the analysis of two-way slabs. If the unfactored live load does not exceed 0.75 times the unfactored dead load, it is not necessary to consider pattern loadings, and only the case of full factored live and dead load on all spans need be analyzed (ACI Section 13.7.6.2). This is based on the assumption that the increase in live-load moments due to pattern loadings compared to uniform live loads will be small compared with the magnitude of the dead-load moments. It also recognizes the fact that a slab is sufficiently ductile in flexure to allow moment redistribution.

If the unfactored live load exceeds 0.75 times the unfactored dead load, the pattern loadings described in ACI Section 13.7.6.3 need to be considered:

1. For maximum positive moment, factored dead load on all spans, plus 0.75 times the full factored live load on the panel in question and on alternate panels.

2. For maximum negative moment at an interior support, factored dead load on all panels, plus 0.75 times the full factored live load on the two adjacent panels.

The final design moments shall not be less than for the case of full factored dead and live load on all panels (ACI Section 13.7.6.4).

ACI Section 13.7.6.3 is an empirical attempt to acknowledge the small probability of full live load in a pattern-loading situation. Again, the possibility of moment redistribution is recognized.

Moments at the Face of Supports

The equivalent frame analysis gives moments at the point where ends of the members meet in the center of the joint. ACI Section 13.7.7 permits the moments at the face of rectilinear supports to be used in the design of the slab reinforcement. The critical sections for negative moment are illustrated in Fig. 14-13. For columns extending more than $0.175\ell_1$ from the face of the support, the moments can be reduced to the values existing at $0.175\ell_1$ from the center of the joint. This limit is necessary, since the representation of the slab-beam stiffness in the joint is not strictly applicable for very long narrow columns [14-9].

If the slab meets the requirements for slabs designed by the direct design method (see Section 13-6 and ACI Section 13.6.1) ACI Section 13.7.7.4 allows the total design moments in a panel to be reduced so far that the absolute sum of the positive moment and the average negative moment does not exceed the statical moment, M_0, for that panel, where M_0 is as given by (13-5) (ACI Eq. (13-3)). Thus, for the case illustrated in Fig. 14-14, the computed moments M_1, M_2, and M_3 would be multiplied by the ratio

$$\frac{w_u\ell_2\ell_n^2}{8} \left/ \left(\frac{M_1 + M_2}{2} + M_3\right)\right.$$

This adjustment is carried out only if this ratio is less than 1.

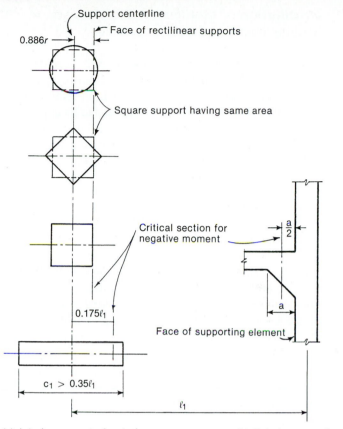

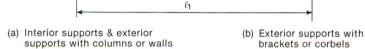

Fig. 14-13
Critical sections for negative moment. (From [14-7]; courtesy of the Portland Cement Association.)

(a) Interior supports & exterior supports with columns or walls

(b) Exterior supports with brackets or corbels

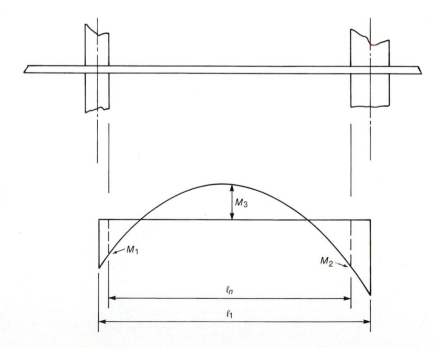

Fig. 14-14
Negative and positive moment in slab-beam.

Distribution of Moments to Column Strips, Middle Strips, and Beams

Once the negative and positive moments have been determined for each equivalent frame, these are distributed to column and middle strips in accordance with ACI Sections 13.6.4 and 13.6.6 in exactly the same way as in the direct design method. This process is discussed more fully in Section 13-6 and is illustrated in Examples 13-3, 13-4, 13-7, and 13-9.

For panels with beams between the columns on all sides, the distribution of moments to the column and middle strips according to ACI Sections 13.6.4 and 13.6.6 is valid only if $\alpha_1 \ell_2^2 / \alpha_2 \ell_1^2$ falls between 0.2 and 5.0. Cases falling outside this range tend to approach one-way action, and other methods of slab analysis are required.

Selection of Slab Thickness

Slab thicknesses are generally chosen to satisfy ACI Section 9.5.3. (See Section 13-6 and Table 13-1). Thinner slabs may be used if calculations show that deflections will not be excessive. Deflection estimates should include an allowance for long-time deflections and should allow for cracking of the beams and slabs. Frequently, construction loads will exceed the service loads. The moment M_a used in evaluating the equivalent moment of inertia of the slab should be based on the largest unfactored moments encountered prior to the time at which deflections are checked.

Calculation of Moment about Centroid of Shear Perimeter

An equivalent-frame analysis gives moments and shears at the ends of the members where they meet at the center of the column–slab joint. In Fig. 14-15b, M_{sj} is the moment in the slab at the center of the joint computed in the frame analysis, V_{slab} is the shear from the slab to the right of the column centerline, and V_{cant} is the shear from the cantilever, if one exists. The moment from the cantilever is

$$M_c = V_{cant} e_c$$

For the joint to be in equilibrium, a moment of $M_{sj} - M_c$ and a vertical force of $V_{slab} + V_{cant}$ must be transferred to the column. Figure 14-15c shows a new free body of the joint region, with the vertical reaction, $V_{cant} + V_{slab}$, acting through the centroid of the shear perimeter, which is located a distance e_{sc} from the center of the joint. Summing moments about the centroid of the shear perimeter in Fig. 14-15c gives the moment to be transferred:

$$M_{sj} - V_{cant}(e_c + e_{sc}) - V_{slab} e_{sc} - M_u = 0$$

Replacing $V_{cant} e_c$ by M_c gives

$$M_u = M_{sj} - M_c - (V_{slab} + V_{cant}) e_{sc} \qquad (14\text{-}8)$$

The connection is then designed for $(V_{slab} + V_{cant})$ and M_u.

EXAMPLE 14-5   Design of a Flat-Plate Floor Using the Equivalent Frame Method

Figure 14-4 shows a plan of a flat-plate floor without spandrel beams. Design this floor, using the equivalent frame method. Use 4000-psi concrete for the columns and slab and Grade-40 reinforcement. The story-to-story height is 8 ft 10 in. The floor supports its own dead load plus 25 psf for partitions and finishes and a live load of 40 psf.

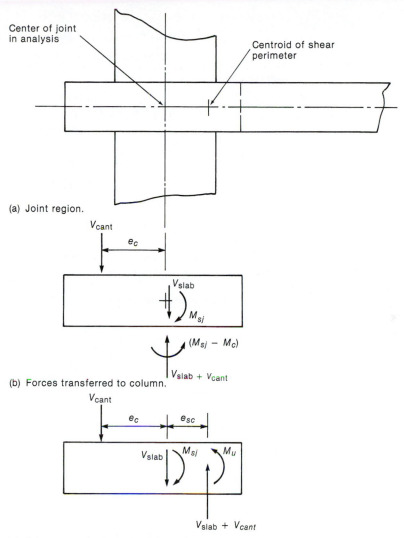

Fig. 14-15
Moment about centroid of
shear perimeter.

(a) Joint region.

(b) Forces transferred to column.

(c) Column reactions resolved through
centroid of shear

This is the same slab that was designed in Example 13-8. Only those parts of the design that differ from Example 13-8 are discussed here. The steps in the design process are numbered to match those in Example 13-8.

1.   Select the design method. Although the slab satisfies the requirements for the use of the direct design method, it has been decided to base the design on moments calculated from the equivalent frame method.

2.   Select the thickness. The selection of thickness is based on Table 13-1 and also on providing adequate shear strength. By the calculations in Example 13-8, a 7-in. slab will be used.

3.   Compute the moments in the equivalent frame along column line 2. (Examples 13-8 and 14-5 differ in this step.) The strip of slab along column line 2 acts as a rigid frame spanning between columns *A*2, *B*2, *C*2, and *D*2. For the purposes of analysis, the columns above and below the slabs will be assumed fixed at their far ends.

(a)   Calculate the moment-distribution coefficients for the slab-beams. From Example 14-1, we have *Span A2–B2:*

$$K_{A2-B2} = \frac{4.11EI_1}{\ell_1}$$

$$= \frac{4.11E_c \times 5402}{213} = 104E_c$$

$$COF_{A2-B2} = 0.508$$

$$K_{B2-A2} = 104E_c \qquad COF_{B2-A2} = 0.508$$

$$\text{fixed-end moments} = 0.084w\ell_2\ell_1^2$$

Span B2–C2:

$$K_{B2-C2} = \frac{4.10EI_1}{\ell_1}$$

$$= \frac{4.10E_c \times 5402}{288} = 76.9E_c$$

$$COF = 0.507$$

$$\text{fixed-end moments} = 0.084w\ell_2\ell_1^2$$

Span C2–D2: Same as A2–B2.

(b)   Calculate the moment-distribution coefficients for the equivalent columns. From Example 14-3,

Column A2: $K_{ec} = 79.9E_c$, $COF = 0.55$
Column B2: $K_{ec} = 161E_c$, $COF = 0.55$

(c)   Compute the distribution factors. The distribution factors are computed in the usual manner:

$$DF_{A2-B2} = \frac{K_{A2-B2}}{K_{A2-B2} + K_{ecA2}}$$

$$= \frac{104E_c}{104E_c + 79.9E_c} = 0.566$$

$$DF_{\text{column A2}} = 0.434$$

The distribution factors and carryover factors are shown in Fig. 14-16. The cantilever members projecting outward at joints A2 and D2 refer to the slab that extends outside the column to support the wall.

(d)   Select the loading cases and compute the fixed-end moments. Since $w_L = 40$ psf is less than three-fourths of $w_D = 112.5$ psf, only the case of uniform live load on each panel need be considered (ACI Section 13.7.6.2). Because the influence area of each of the panels is less than 400 ft^2, live-load reductions do not apply. Accordingly, we have

$$w_u = 1.4\left(\frac{7}{12} \times 0.15 + 0.025\right) + 1.7(0.040) = 0.226 \text{ ksf}$$

For the slab-beam in question, $\ell_2 = 15.75$ ft.

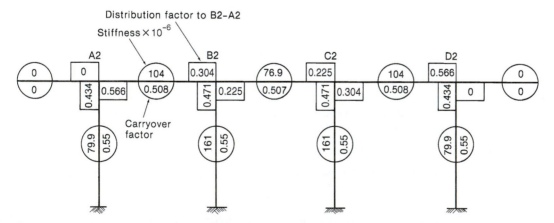

Fig. 14-16
Stiffness, carryover, and distribution factors—Example 14-5.

Span A2–B2:

$$M = 0.084 w_u \ell_2 \ell_1^2$$

$$= 0.084 \times 0.226 \times 15.75 \times 17.75^2 = 94.0 \text{ ft-kips}$$

(Note that the moment, M, is based on the center-to-center span, ℓ_1, rather than on the clear span, ℓ_n, used in the direct design method.)

Span B2–C2:

$$M = 0.084 \times 0.226 \times 15.75 \times 20^2 = 119.6 \text{ ft-kips}$$

The weight of the slab outside line A and the wall load cause a small cantilever moment at joint $A2$. It is assumed that the wall load acts 2 in. outside the exterior face of the column, or 8 in. from the center of the column. Then

$$M = (1.4 \times 15.75 \times 0.300) \times \frac{8}{12} + \left(1.4 \times 15.75 \times \frac{7}{12} \times 0.150 \right) \frac{(10/12)^2}{2}$$

$$= 5.08 \text{ ft-kips}$$

Span A2–B2:

(e)   Carry out a moment-distribution analysis. The moment-distribution analysis is carried out in Table 14-1. The sign convention used takes clockwise moments *on the joints* as positive. Thus, anticlockwise moments *on the ends of members* are positive. When this sign convention is used, the carryover factors are positive rather than negative. The resulting moment and shear diagrams are plotted in Fig. 14-17a and b. The moment diagram is plotted on

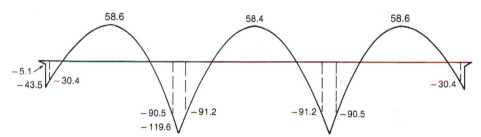

(a) Moment diagram from equivalent frame analysis (ft–kips).

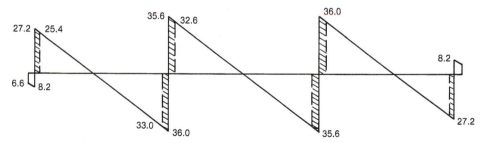

(b) Shear force diagram (kips).

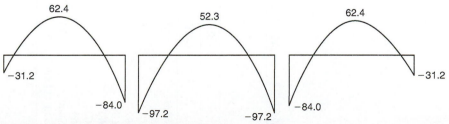

Fig. 14-17
Moments and shears in frame along line 2—Example 14-5.

(c) Moment diagram from direct design method (ft–kips)

TABLE 14-1 Moment Distribution—Example 14-5

	A2			B2			C2			D2		
	COF = 0.508			COF = 0.507			COF = 0.508					
	0	0.434	0.566	0.304	0.471	0.225	0.225	0.471	0.304	0.566	0.434	0
	Cant.	Coln.	Slab	Slab	Coln.	Slab	Slab	Coln.	Slab	Slab	Coln.	Cant.
FEM	−5.1	0	+94.0	−94.0	0	+119.6	−119.6	0	+94.0	−94.0	0	+5.1
B1		−38.6	−50.3	−7.8	−12.1	−5.8	+5.8	+12.1	+7.8	+50.3	+38.6	
C1			−3.9	−25.6		−2.9	−2.9		+25.6	+3.9		
B2		+1.7	+2.2	+6.9	+10.7	+5.1	−5.1	−10.7	−6.9	−2.2	−1.7	
C2			+3.5	+1.1		−2.6	+2.6		−1.1	−3.5		
B3		−1.5	−2.0	+0.5	+0.7	+0.3	−0.3	−0.7	−0.5	+2.0	+1.5	
C3			+0.2	−1.0		−0.2	+0.2		+1.0	−0.2		
B4		−0.1	−0.1	+0.3	+0.6	+0.3	−0.3	−0.6	−0.3	+0.1	−0.1	
Sum	−5.1	−38.5	43.5	−119.6	−0.20	+119.6	−119.6	+0.20	+119.6	−43.5	+38.5	+5.1
Sum at joint		0			0			0			0	

the compression face of the member, and positive and negative moments in the slab-beam correspond to the normal sign convention for continuous beams. The moment values calculated in Example 13-8 are plotted in Fig. 14-17c for comparison.

(f)   **Calculate the moments at the faces of the supports.** The moments at the face of the supports are calculated by subtracting the areas of the shaded parts of Fig. 14-17b from the moments at the ends of the members.

(g)   **Reduce the moments to M_0.** For slabs that fall within the limitation for the use of the direct design method, the total moments between faces of columns can be reduced to M_0. This slab satisfies ACI Section 13.6.1, so this reduction can be used.

Span A2–B2:

$$\text{Total moment} \simeq \frac{30.4 + 90.5}{2} + 58.6 = 119.1 \text{ ft-kips}$$

$$M_0 = \frac{w\ell_2\ell_n^2}{8} = 120 \text{ ft-kips}$$

Thus, the computed moments in span A2–B2 do not need to be adjusted.

Span B2–C2:

$$\text{Total moment} = \frac{91.2 + 91.2}{2} + 58.4 = 149.6 \text{ ft-kips}$$

$$M_0 = 149.5 \text{ ft-kips}$$

Because the total moment equals M_0, no reduction is made. There were no reductions in this example because we have only one load case to consider, with factored dead and live load on all spans. If the live load had exceeded three-fourths of the dead load, it would have been necessary to consider several load cases (ACI Section 13.7.6.3), and reductions of the envelope moment values would be likely.

(h)   **Calculate the column moments.** The total moment transferred to the columns at A2 is 38.5 ft-kips. This moment is distributed between the two columns in the ratio of the stiffnesses, K_c. Since the columns above and below have equal stiffnesses, half goes to each column, giving the bending-moment diagrams shown in Fig. 14-18. The columns are designed for the moments at the edge of the floor slabs. At B2, only a small moment is transferred to the columns.

4, 5, and 6. Compute the moments in the equivalent frames along column lines 1, B, and A. To complete the moment analysis, it is necessary to compute the moments in equivalent frames along the edge and interior column lines in both directions, as was done in Example 13-7.

7 and 8. Distribute the negative and positive moments to the column and middle strips and detail the reinforcement. These steps are carried out in exactly the same way as in Example 13-7.

9.   Check the shear at the exterior columns. The check of moment and shear transfer at the exterior columns differs from that carried out in the direct design method calculation because the moment and shear have been calculated at the center of the joint. These must be corrected by using (14-8) to get the moment about the centroid of the shear perimeter when the shear acts through this point.

Since we have analyzed the equivalent frame along line 2, we shall check the shear and moment transfer at column A2. The connections with columns B2 and C2 may be more critical, however.

(a)   **Locate the critical-shear perimeter.** The critical perimeter is located at $d/2$ from the face of the column, where d is the average depth, 5.75 in. (See Fig. 14-19a.) The centroid of this section is at 5.61 in. from line A–B, and the torsional moment of inertia is $J_c = 14,900 \text{ in.}^4$ about the Z–Z axis and 44,600 in.[4] about the W–W axis (Example 13-8).

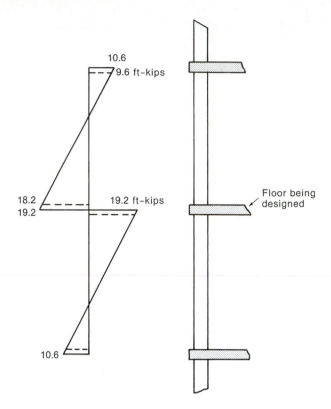

Fig. 14-18
Moments in column *A*2—
Example 14-5.

(b) Compute the shear on the section and the moment about the centroid of the section. A free-body diagram of the joint is shown in Fig. 14-19b. When the shear forces act through the centroid of the shear perimeter, the corresponding moment is (14-8)

$$M_u = 43.5 - 5.1 - (27.2 + 8.2)\frac{3.27}{12}$$

$$= 28.8 \text{ ft-kips}$$

The rest of the combined-moment and shear-transfer check proceeds as in Example 13-7.

10, 11, and 12. Check the shear at the interior and corner columns, and design the torsional reinforcement. These steps are similar to the corresponding steps in Example 13-8. ■

14-3   USE OF COMPUTERS IN THE EQUIVALENT FRAME METHOD

The equivalent frame method was derived by assuming that the structural analysis would be carried out by using the moment-distribution method, and fixed-end moments, stiffnesses, and equivalent-column stiffnesses are computed for use in such an analysis. If a standard frame-analysis program based on the stiffness method is to be used, the frame must be specially modeled to get answers that agree with those obtained from the equivalent frame method.

The Portland Cement Association has written and maintains the computer program ADOSS—"Analysis and Design of Slab Systems"—to analyze and design two-way, reinforced concrete slabs. The analysis is based on the ACI Code equivalent frame method.

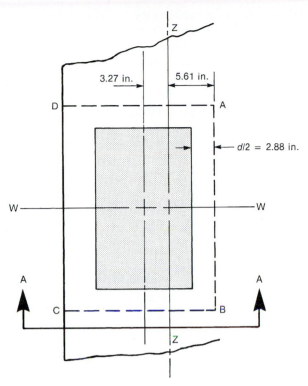

(a) Critical shear perimeter—Column A-2.

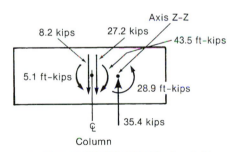

Fig. 14-19
Slab–column joint *A*2—
Example 14-5.

(b) Freebody of critical shear perimeter (Section A–A).

14-4   EQUIVALENT FRAME ANALYSIS OF LATERALLY LOADED UNBRACED FRAMES

A frame consisting of columns and either flat plates or slabs with drop panels, but lacking shear walls or other bracing elements, is inefficient in resisting lateral loads and is subject to significant lateral drift deflections. These deflections are amplified by the $P-\Delta$ moments resulting from gravity loads. As a result, flat-plate structures generally are braced by shear walls.

Unbraced slab–column frames are sometimes used for low buildings or for the top few floors in a tall building, where the story-to-story drift in the upper stories may be reduced by terminating the shear wall before the top of the building. For such cases, it is necessary to analyze equivalent-frame structures for both gravity and lateral loads and to add the results. For the gravity-load analysis, the equivalent frame analysis described in Section 14-1 can be used without modification. For lateral-load analysis, however, this equivalent frame underestimates the lateral deflections, and hence the $P-\Delta$ effects, because it is based on uncracked EI values.

For both gravity and lateral load analyses, ACI Section 13.7.2.3 requires that the slab-beam strips be attached to columns by torsional members. For laterally loaded frames, ACI Section 13.5.1.2 requires that the effects of cracking and reinforcement be taken into account in computing the stiffnesses of the frame members. The ACI Commentary Section R13.5.1.2 suggests that this can be done by using a reduced ℓ_2 of 0.25 to $0.50\ell_2$ when computing the moment of inertia of the slab-beam. Reference [14-5] suggests using $0.33I_{sb}$ instead. The procedure for analyzing an unbraced slab-and-column structure according to the ACI Code is as follows:

1. *For gravity loads*,

    (a)   Compute K_{sb} for the slab-beams, using the full width ℓ_2 and assuming that the slab is uncracked.

    (b)   Compute K_c for each of the columns at a joint, assuming that the columns are uncracked.

    (c)   Compute K_t for uncracked torsional members.

    (d)   Compute K_{ec}, using ΣK_c and K_t.

    (e)   Analyze the frame for gravity-load effects, using K_{sb} and K_{ec}, as in Section 14-1 and Example 14-5.

2. *For lateral loads*,

    (a)   Compute the cracked slab-beam stiffness, $K_{sb,cr} = 0.33K_{sb}$, where K_{sb} is from step 1(a).

    (b)   Compute an effective stiffness of each column at a joint as $(K_{ec}/\Sigma K_c)$ at that joint times the stiffness of the column in question, where K_{ec} and K_c are the values computed in steps 1(b) and 1(e).

    (c)   Analyze the frame for lateral effects, using $K_{sb,cr}$ and K_c.

3. *For combined gravity and lateral loads*, superimpose steps 1 and 2.

This is a complex series of steps. It is hoped that by the next code revision, the equivalent frame method will be restated in a clearer and a more computer-compatible form.

PROBLEMS

14-1 Design the North–South strips in the slab shown in Fig. P13-4, using the equivalent frame method. Loadings, dimensions, and material strengths are as given in Problem 13-4.

14-2 Repeat Problem 14-1 for the East–West strips in Fig. P13-4.

14-3 Compute the moments in East–West strips along lines A and B in the slab shown in Fig. 14-5. The floor supports its own weight, superimposed dead loads of 5 psf for ceiling and mechanical fixtures and 25 psf for future partitions, and a live load of 80 psf. The exterior wall weighs 300 lb/ft and is supported by the edge beam. The stories above and below the floor in question are 12 ft and 14 ft high, respectively. Lateral loads are resisted by an elevator shaft. Use 3000-psi concrete and Grade-60 reinforcement.

15

Two-Way Slabs: Elastic, Yield-Line, and Strip Method Analyses

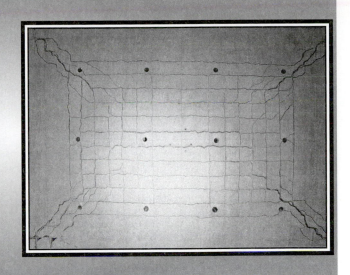

15-1 ELASTIC ANALYSIS OF SLABS

The concepts involved in the elastic analysis of slabs are reviewed very briefly here, to show the relationship between the loads and the internal moments in the slab. In addition, and even more important, this will show the relationship between moments and slab curvatures.

Slabs may be subdivided into *thick* slabs (with a thickness greater than about one-tenth of the span), *thin* slabs (with a thickness less than about one-fortieth of the span), and *medium-thick* slabs. Thick slabs transmit a portion of the loads as a flat arch and have significant in-plane compressive forces, with the result that the internal resisting compressive force C is larger than the internal tensile force T. Thin slabs transmit a portion of the loads by acting as a tension membrane; hence, T is larger than C. A medium-thick slab does not exhibit either arch action or membrane action and thus has $T = C$.

Figure 15-1 shows an element cut from a medium-thick, two-way slab. This element is acted on by the moments shown in Fig. 15-1a and by the shears and loads shown in Fig. 15-1b. (The figures are separated for clarity.)

Two types of moment exist on each edge: bending moments m_x and m_y about axes parallel to the edges, and twisting moments m_{xy} and m_{yx} about axes perpendicular to the edges. The moments are shown by moment vectors represented by double-headed arrows. The moment in question acts about the arrow according to the right-hand-screw rule. The length of vector represents the magnitude of the moment. Vectors can be added graphically or numerically. The moments m_x, and so on, are defined for a unit width of the edge they act on, as are the shears V_x, and so on. The m_x and m_y moments are positive, corresponding to compression on the top surface. The twisting moments on adjacent edges both act to cause compression on the same surface of the slab, at the corner between the two edges, as shown in Fig. 15-1a.

Summing vertical forces gives

$$\frac{\partial V_x}{\partial x} + \frac{\partial V_y}{\partial y} = -w \tag{15-1}$$

759

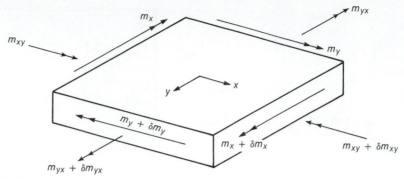

(a) Bending and twisting moments on a slab element.

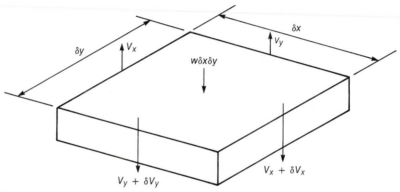

Fig. 15-1
Moments and forces in a
medium-thick plate.

(b) Shears and loads on a slab element.

Summing moments about lines parallel to the x and y axes and neglecting higher order terms gives, respectively,

$$\frac{\partial m_y}{\partial y} + \frac{\partial m_{xy}}{\partial x} = V_y \quad \text{and} \quad \frac{\partial m_x}{\partial x} + \frac{\partial m_{yx}}{\partial y} = V_x \tag{15-2}$$

It can be shown that $m_{xy} = m_{yx}$. Differentiating (15-2) and substituting into (15-1) gives the basic equilibrium equation for medium-thick slabs:

$$\frac{\partial^2 m_x}{\partial x^2} + \frac{2\partial^2 m_{xy}}{\partial x \, \partial y} + \frac{\partial^2 m_y}{\partial y^2} = -w \tag{15-3}$$

This is purely an equation of statics and applies regardless of the behavior of the plate material. For an elastic plate, the deflection, z, can be related to the applied load by means of

$$\frac{\partial^4 z}{\partial x^4} + 2\frac{\partial^4 z}{\partial x^2 \, \partial y^2} + \frac{\partial^4 z}{\partial y^4} = -\frac{w}{D} \tag{15-4}$$

where the *plate rigidity* is

$$D = \frac{Et^3}{12(1 - \nu^2)} \tag{15-5}$$

in which ν is Poisson's ratio. The term D is comparable to the EI value of a unit width of slab.

In an elastic plate analysis, (15-4) is solved to determine the deflections, z, and the moments are calculated from

$$m_x = -D\left(\frac{\partial^2 z}{\partial x^2} + \frac{\nu \partial^2 z}{\partial y^2}\right)$$

$$m_y = -D\left(\frac{\partial^2 z}{\partial y^2} + \frac{\nu \partial^2 z}{\partial x^2}\right)$$

$$m_{xy} = -D(1 - \nu)\frac{\partial^2 z}{\partial x \, \partial y} \tag{15-6}$$

where z is positive downward.

Solutions of (15-6) can be found in books on elastic plate theory.

15-2   DESIGN OF REINFORCEMENT FOR MOMENTS FROM A FINITE-ELEMENT ANALYSIS

The most common way of solving (15-6) is to use a finite-element analysis. Such an analysis gives values of m_x, m_y, and m_{xy}, in each element, where m_x, m_y, and m_{xy} are moments per unit width. A portion of an element bounded by a diagonal crack is shown in Fig. 15-2. The moments on the x and y faces, from the finite-element analysis, are shown in Fig. 15-2b. The moment about an axis parallel to the crack is m_c, given by

$$m_c \, ds = (m_x \, dy + m_{xy} k \, dy) \cos \theta + (m_y k \, dy + m_{xy} \, dy) \sin \theta$$

or

$$m_c = \left(\frac{dy}{dx}\right)^2 (m_x + k^2 m_y + 2k m_{xy}) \tag{15-7}$$

This slab is to be reinforced by bars in the x and y direction having positive moment capacities m_{rx} and m_{ry} per unit width. The corresponding moment capacity at the assumed crack is

$$m_{rc} = \left(\frac{dy}{dx}\right)^2 (m_{rx} + k^2 m_{ry}) \tag{15-8}$$

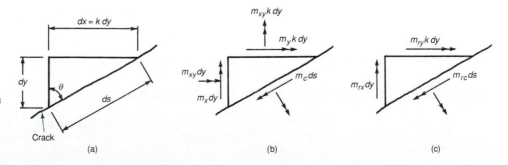

Fig. 15-2
Resolution of moments. (From [15-1]: Hillerborg, "Strip Method of Design," published by E&F, N Spon, London, 1975.)

where m_{rc} must equal or exceed m_c to provide adequate strength. Equating these and solving for the minimum, we get

$$m_{ry} = m_y + \frac{1}{k}m_{xy}$$

Since m_{ry} must equal or exceed m_y to account for the effects of m_{xy}, $(1/k)m_{xy} \geq 0$, which gives

$$m_{ry} = m_y + \frac{1}{k}|m_{xy}|$$

$$m_{rx} = m_x + k|m_{xy}| \tag{15-9}$$

where k is a positive number and $|m_{xy}|$ implies the absolute value of m_{xy}. This must be true for all crack orientations (i.e., for all values of k). As k is increased, m_{ry} goes down and m_{rx} goes up. The smallest of the sums of the two (i.e., the smallest total reinforcement) depends on the slab in question, but $k = 1$ is the best choice for a wide range of moment values [15-1], [15-2].

The reinforcement at the bottom of the slab in each direction is designed to provide positive-moment resistances of

$$m_{ry} = m_y + |m_{xy}| \tag{15-10a}$$

and

$$m_{rx} = m_x + |m_{xy}| \tag{15-10b}$$

negative-moment resistances of

$$m_{ry} = m_y - |m_{xy}| \tag{15-10c}$$

and

$$m_{rx} = m_x - |m_{xy}| \tag{15-10d}$$

Calculation of m_{rx} and m_{ry}—positive moment. When the values of m_{rx} and m_{ry} are used to calculate the areas of positive moment reinforcement (bottom steel) in a slab, the calculations are based on (15-10a) or (15-10b). The + sign and absolute number in these equations combine to make the term " $+|m_{xy}|$ " positive. Thus, (15-10a) and (15-10b) become

$$m_{ry} = m_y + m_{xy} \tag{15-10e}$$

$$m_{rx} = m_x + m_{xy} \tag{15-10f}$$

If m_{xy} is negative in either of these two equations, it is set equal to zero in that equation.

Calculation of m_{rx} and m_{ry}—negative moment. For negative moment reinforcement (top steel) the calculations are based on (15-10c) and (15-10d). The − sign and absolute number combine to make the term "$-|m_{xy}|$" negative. Thus, (15-10c) and (15-10d) become

$$m_{ry} = m_y - m_{xy} \tag{15-10g}$$

$$m_{rx} = m_x - m_{xy} \tag{15-10h}$$

If m_{xy} is positive in either of these two equations, it is set to zero in that equation.

EXAMPLE 15-1 Computation of Design Moments at a Point

A finite-element analysis gives the moments in an element as $m_x = 5$ ft-kips/ft, $m_y = -1$ ft-kips/ft, and $m_{xy} = -2$ ft-kips/ft. Compute the moments to be used to design reinforcement.

(a)   Steel at bottom of slab:

$$m_{ry} = -1 + |-2| = +1 \text{ ft-kips/ft}$$

$$m_{rx} = 5 + |-2| = +7 \text{ ft-kips/ft}$$

(b)   Steel at top of slab:

$$m_{ry} = -1 - |-2| = -3 \text{ ft-kips/ft}$$

$$m_{rx} = 5 - |-2| = +3 \text{ ft-kips/ft}$$

Since m_{rx} is positive, it is set equal to zero in designing steel at the top of the slab.   ■

Generally, the reinforcement is uniformly spaced in orthogonal bands such that

(a)   the total reinforcement provided in a band is sufficient to resist the total factored moment computed for that band, and

(b)   the moment resistance per unit width in the band is at least two-thirds of the maximum moment per unit width in the band, as computed in the finite-element analysis.

In the direct design method the widths of the bands of reinforcement were taken equal to the widths of the column strip and middle strips defined in ACI Section 13.2 This width is generally acceptable.

15-3   YIELD-LINE ANALYSIS OF SLABS

Under overload conditions in a slab failing in flexure, the reinforcement will yield first in a region of high moment. (See Section 13-3.) When that occurs, this portion of the slab acts as a plastic hinge, able to resist its hinging moment, but no more. When the load is increased further, the hinging region rotates plastically, and the moments due to additional loads are redistributed to adjacent sections, causing them to yield, as shown in Fig. 13-4. The bands in which yielding has occurred are referred to as *yield lines* and divide the slab into a series of elastic plates. Eventually, enough yield lines exist to form a plastic mechanism in which the slab can deform plastically without an increase in the applied load.

A *yield-line analysis* uses rigid plastic theory to compute the failure loads corresponding to given plastic moment resistances in various parts of the slab. It does not give any information about deflections or about the loads at which yielding first starts. Although the concept of yield-line analysis was first presented by A. Ingerslev in 1921–1923, K. W. Johansen developed modern yield-line theory [15-3]. This type of analysis is widely used for slab design in the Scandinavian countries. The yield-line concept is presented here to aid in the understanding of slab behavior between service loads and failure. Further details are given in [15-4] to [15-6].

Yield Criterion

It is assumed that the moment–curvature relationship for a slab section is elastic–plastic, with a plastic moment capacity equal to ϕM_n, the design flexural capacity of the section. To limit deflections, floor slabs are generally considerably thicker than is required for flexure, and as a result, they seldom have reinforcement ratios exceeding 0.3 to $0.4\rho_b$. In this range of reinforcement ratios, the moment–curvature response is essentially elastic–plastic.

If yielding occurs along a line at an angle α to the reinforcement, as shown in Fig. 15-3, the bending and twisting moments are assumed to be distributed uniformly along the yield line and are the maximum values provided by the flexural capacities of the reinforcement layers

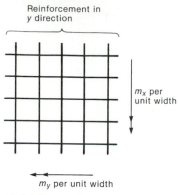

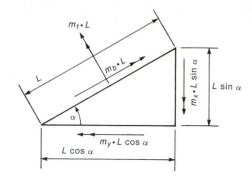

Fig. 15-3
Yield criterion.

(a) Reinforcement pattern.

(b) Moments on an element.

crossed by the yield line. It is further assumed that there is no kinking of the reinforcement as it crosses the yield line. In Fig. 15-3, the reinforcements in the x and y directions provide moment capacities of m_x and m_y per unit width (ft-lb/ft). The bending moment, m_b, and the twisting moment, m_t, per unit length of the yield line can be calculated from the moment equilibrium of the element. In this calculation, α will be measured counterclockwise from the x axis; the bending moments m_x, m_y, and m_b will be positive if they cause tension in the bottom of the slab; and the twisting moment, m_t, will be positive if the moment vector points away from the section, as shown.

Consider the equilibrium of the element in Fig. 15-3b:

$$m_b L = m_x(L \sin \alpha) \sin \alpha + m_y(L \cos \alpha) \cos \alpha$$

This equation gives the bending moment m_b as

$$m_b = m_x \sin^2 \alpha + m_y \cos^2 \alpha \tag{15-11}$$

The twisting moment m_t is

$$m_t = \left(\frac{m_x - m_y}{2}\right) \sin 2\alpha \tag{15-12}$$

These equations apply only for orthogonal reinforcement. If $m_x = m_y$, these two equations reduce to $m_b = m_x = m_y$ and $m_t = 0$, regardless of the angle of the yield line.

Locations of Axes and Yield Lines

Yield lines form in regions of maximum moment and divide the plate into a series of elastic plate segments. When the yield lines have formed, all further deformations are concentrated at the yield lines, and the slab deflects as a series of stiff plates joined together by long hinges, as shown in Fig. 15-4. The pattern of deformation is controlled by *axes* that pass along line supports and over columns, as shown in Fig. 15-5, and by the yield lines. Since the individual plates rotate about the axes and/or yield lines, these axes and lines must be straight. To satisfy compatibility of deformations at points such as A and B in Fig. 15-4, the yield line dividing two plates must intersect the intersection of the axes about which those plates are rotating. Figure 15-5 shows the locations of axes and yield lines in a number of slabs subjected to uniform loads.

Figure 15-6 shows the underside of a reinforced concrete slab similar to the slab shown in Fig. 15-4. The cracks have followed the orthogonal reinforcement pattern. The

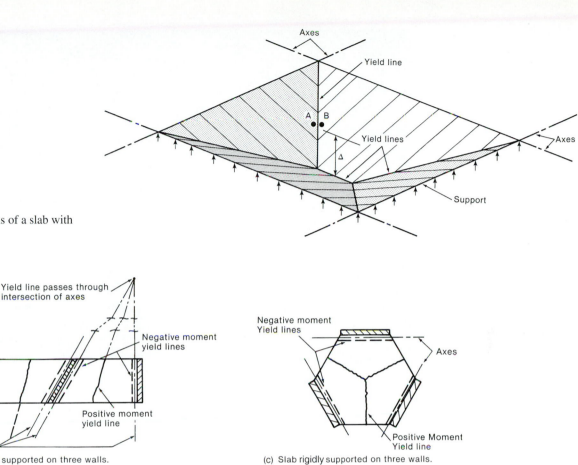

Fig. 15-4
Deformations of a slab with
yield lines.

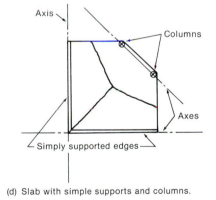

(a) Slab rigidly supported on three walls.

(c) Slab rigidly supported on three walls.

(b) Trapezoidal simply supported slab.

(d) Slab with simple supports and columns.

Fig. 15-5
Examples of yield-line patterns.

wide cracks extending in from the corners and the band of cracks along the middle of the plate mark the locations of the yield lines.

Methods of Solution

Once the yield lines have been chosen, it is possible to compute the values of m necessary to support a given set of loads, or vice versa. The solution can be carried out by the

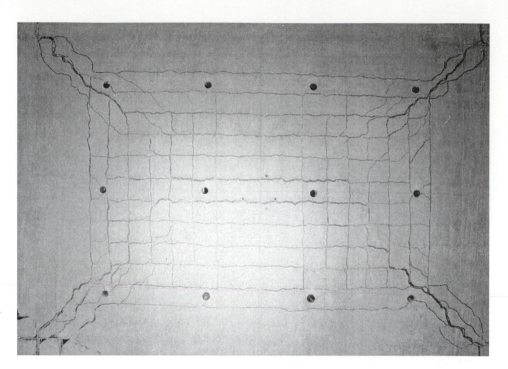

Fig. 15-6
Yield lines in a slab that is sim-
ply supported on four edges.
(Photograph courtesy of
J. G. MacGregor.)

equilibrium method, in which equilibrium equations are written for each plate segment, or by the *virtual-work method*, in which some part of the slab is given a virtual displacement and the resulting work is considered.

When the equilibrium method is used, considerable care must be taken to show all of the forces acting on each element, including the twisting moments, especially when several yield lines intersect or when yield lines intersect free edges. At such locations, off-setting vertical nodal forces are required at the intersections of yield lines. Because of the possibility that the nodal forces will be given the wrong sign or location, some building codes require that yield-line calculations be done by the virtual-work method. The introductory presentation here will be limited to the virtual-work method of solution. Further information on the equilibrium method can be obtained in [15-3] to [15-6].

Virtual-Work Method

Once the yield lines have been chosen, some point on the slab is given a virtual displacement, δ, as shown as in Fig. 15-7c. The external work done by the loads when displaced this amount is

$$\text{external work} = \int\int w\delta\, dx\, dy$$

$$= \Sigma(W\Delta_c) \tag{15-13}$$

where w = load on an element of area

δ = deflection of that element

W = total load on a plate segment

Δ_c = deflection of the centroid of that segment

The total external work is the sum of the separate work done by each plate.

The internal work done by rotating the yield lines is

$$\text{internal work} = \sum (m_b \ell \theta) \tag{15-14}$$

where m_b = bending moment per unit length of yield line

ℓ = length of the yield line

θ = angle change at that yield line

The total internal work done during the virtual displacement is the sum of the internal work done on each separate yield line. Since the yield lines are assumed to have formed before the virtual displacement is imposed, no elastic deformations occur during the virtual displacement.

The principle of virtual work states that, for equilibrium,

$$\text{external work} = \text{internal work}$$

or

$$\sum (W\Delta_c) = \sum (m_b \ell \theta) \tag{15-15}$$

The virtual work solution is an *upper-bound solution*; that is, the load W is equal to or higher than the true failure load. If an incorrect set of yield lines is chosen, W will be too large for a given m, or conversely, the value of m for a given W will be too small.

EXAMPLE 15-2   Yield-Line Analysis of One-Way Slab by Virtual Work

The one-way slab shown in Fig. 15-7 has reinforcement at the top at the ends with moment capacity m_{x1} per unit width and reinforcement at the bottom at midspan, with capacity m_{x2} per unit width. The slab is loaded with a uniform load of w per unit area. Compute the moment capacity required to support w.

1.   **Select the axes and yield lines.** Axes and negative-moment yield lines will form along the faces of the supports, and a positive-moment yield line will form at midspan, as shown by the dashed and wavy lines, respectively, in Fig. 15-7b.

2.   **Give the slab a virtual displacement.** Some point in the slab is displaced downward by an amount δ. In Fig. 15-7c, line C–D has been displaced δ.

3.   **Compute the external work.** The total load on plate segment A–B–C–D is $W = w(bL/2)$. The displacement of the resultant load W is $\Delta_c = \delta/2$. Therefore, the external work done on plate A–B–C–D is

$$W\Delta_c = w\left(\frac{bL}{2}\right)\frac{\delta}{2}$$

The total external work for the two segments is

$$\text{external work} = 2\left(\frac{wbL\delta}{4}\right)$$

4.   **Compute the internal work.** The negative-moment yield line at A–B rotates through an angle θ, where $\theta = \delta/(L/2)$. The internal work done in rotating through this angle is $m_{x1}b(2\delta/L)$. The total internal work done at all three yield lines is

$$\text{internal work} = 2\left[m_{x1}b\left(\frac{2\delta}{L}\right)\right] + m_{x2}b\left(\frac{4\delta}{L}\right)$$

5.   **Equate the external and internal work.**

$$2\left(\frac{wbL\delta}{4}\right) = 4(m_{x1} + m_{x2})\,b\frac{\delta}{L}$$

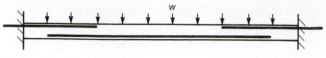

(a) Cross section.

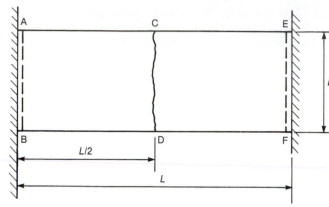

(b) Plan.

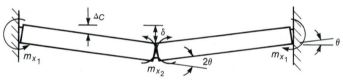

Fig. 15-7
Slab—Example 15-2.

(c) Deformed shape.

This simplifies to

$$m_{x1} + m_{x2} = \frac{wL^2}{8}$$

Thus, any combination of m_{x1} and m_{x2} that equals $wL^2/8$ will satisfy equilibrium.

If the distance A–C were taken as $0.4L$ rather than $L/2$, the values of $(m_{x1} + m_{x2})$ for the two plates would differ, indicating that an incorrect central yield line had been chosen. ■

EXAMPLE 15-3 Yield-Line Analysis of a Square Slab by Virtual Work

The simply supported square slab shown in Fig. 15-8 has a positive moment capacity, $m_x = m_y$. Compute the value of m required to support a uniform load w.

   1.   **Select the axes and yield lines.** Yield lines will form along the diagonals which, as shown in Fig. 13-12, are lines of maximum moment. The segments will rotate about axes along the four supports.

   2.   **Give the slab a virtual displacement.** Point E is displaced downward by an amount δ.

   3.   **Compute the external work.**

Load on plate segment A–B–E: $W = w\left(\dfrac{L^2}{4}\right)$

Deflection of centroid of plate: $\Delta_c = \dfrac{\delta}{3}$

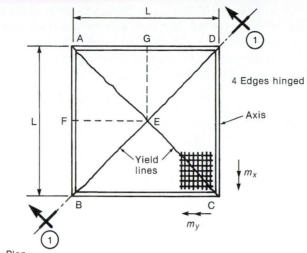

(a) Plan.

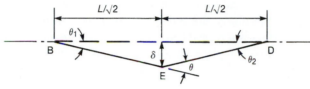

(b) Section 1–1.

Fig. 15-8
Slab—Example 15-3.

External work done on plate A–B–E: $W\Delta_c = \left(\dfrac{wL^2}{4}\right)\dfrac{\delta}{3}$

Total external work: $\displaystyle\sum W\Delta_c = 4\left(\dfrac{wL^2}{4}\right)\dfrac{\delta}{3}$

4. **Compute the internal work.** Consider the yield line A–E. Its length is $\ell = L/\sqrt{2}$. As is shown in Fig. 15-8b, the rotation of yield line A–E is

$$\theta = \theta_1 + \theta_2$$

where

$$\theta_1 = \sqrt{2}\delta/L = \theta_2$$

and thus

$$\theta = 2\sqrt{2}\delta/L$$

The internal work on line A–$E = m_b\ell\theta$. From (15-11), m_b is equal to m_x and m_y. Thus, $m_b\ell\theta = m\left(L/\sqrt{2}\right)\left(2\sqrt{2}\delta/L\right) = 2m\delta$. The total work on all four yield lines is

$$\sum (m_b\ell\theta) = 4(2m\delta)$$

5. **Equate the external and internal work.**

$$\frac{wL^2\delta}{3} = 8m\delta \quad \text{and} \quad m = \frac{wL^2}{24}$$

Consequently, the reinforcement in both directions should be designed for

$$\phi m_n = \frac{w_u L^2}{24}$$

where w_u is the factored uniform load.   ∎

In Example 15-3, it was necessary to compute the value of m_b along the yield lines and the rotation of the yield lines. Generally, it is easier to consider the rotations about the axes about which the slab segments rotate, especially if these are parallel or perpendicular to the reinforcement. Figure 15-9 shows a portion of the slab considered in Example 15-3. The rotation of yield line A–E is made up of

$$\theta_1 = \theta_y \cos \alpha_1$$

caused by the rotation θ_y of plate A–F–E about axis A–F, and

$$\theta_2 = \theta_x \cos \alpha_2$$

caused by the rotation θ_x of plate A–G–E about axis A–G. The internal work done on yield line A–E is

$$m_b \ell \theta = m_b \ell (\theta_y \cos \alpha_1 + \theta_x \cos \alpha_2)$$
$$= m_b [\theta_y (\ell \cos \alpha_1) + \theta_x (\ell \cos \alpha_2)]$$

where $\ell \cos \alpha_1$ and $\ell \cos \alpha_2$ are the projections of ℓ on the y and x axes and are equal to L_y and L_x, respectively. Since only the component of m_b about an axis parallel to A–F, m_x, can do work when rotated about the y-axis, and similarly only m_y can do work when rotated about the x-axis, replace m_b with these components, giving

$$m_b \ell \theta = m_x L_y \theta_y + m_y L_x \theta_x$$

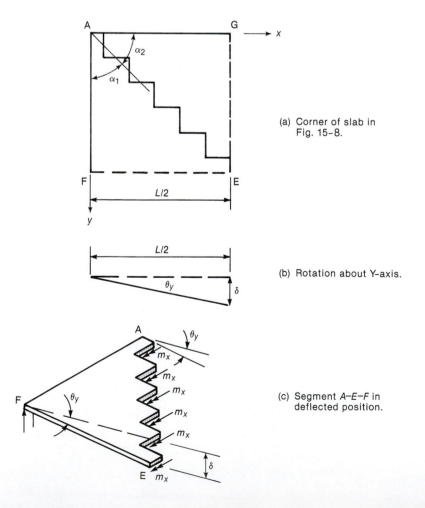

(a) Corner of slab in Fig. 15–8.

(b) Rotation about Y-axis.

(c) Segment A–E–F in deflected position.

Fig. 15-9
Rotation of segment A–E–F about Y-axis.

and hence

$$\text{total internal work} = \sum (m_x L_y \theta_y + m_y L_x \theta_x) \tag{15-16}$$

This is equivalent to considering the yield line as a series of infinitesimal steps parallel to the x and y axes with moment capacities m_x and m_y, respectively. As rotation of $A-E-F$ occurs about the x-axis ($A-F$), only the component m_x on the steps does work. Similarly, when $A-G-E$ rotates about $A-G$, only the component m_y does work.

EXAMPLE 15-4   Yield-Line Analysis of a Square Slab by Virtual Work

Redo Example 15-3, but using (15-16) to compute the internal work. Steps 1, 2, and 3 are unchanged.

4.   **Compute the internal work.** For plate $A-B-E$ (see Fig. 15-8),

$$\theta_x = 0 \qquad \theta_y = \frac{\delta}{L/2}$$

$$\text{internal work} = m_x L_y \theta_y + m_y L_x \theta_x$$

$$= m_x L \left(\frac{2\delta}{L} \right) + 0$$

$$= 2m_x \delta$$

Similarly for plate $A-G-E$,

$$\theta_x = \frac{\delta}{L/2} \qquad \theta_y = 0$$

$$\text{internal work} = 0 + m_y L \left(\frac{2\delta}{L} \right)$$

Thus,

$$\text{total internal work} = 2(2m_x \delta) + 2(2m_y \delta)$$

Since $m_x = m_y = m$, the total internal work is $8m\delta$, which is the same value calculated in step 4 of Example 15-3. Thus, $m = wL^2/24$.   ∎

EXAMPLE 15-5   Calculation of the Load Capacity of a Reinforced Concrete Slab

Figure 15-10 shows a rectangular slab that is fixed on four sides and has negative-moment (top) reinforcement with the nominal capacities shown in Fig. 15-10a and positive-moment (bottom) reinforcement, as shown in Fig. 15-10b. Compute the load w corresponding to failure of this slab.

1.   **Select the axes and yield lines.** Axes will form along the four sides of the slab. Negative-moment yield lines will form along the axes, and positive-moment yield lines are assumed as shown in Fig. 15-10c. The distance x to the intersection of the yield lines will be assumed to be 5 ft for the first trial.

2.   **Give the slab a virtual displacement.** Line $E-F$ is given a virtual displacement of δ.

3.   **Compute the external work.**

For panel $A-C-E$,

$$W\Delta_c = w \left(10 \times \frac{5}{2} \right) \frac{\delta}{3} = 8.33 w\delta \text{ ft-kips}$$

For panel $A-B-F-E$,

$$W\Delta_c = w \left(10 \times \frac{5}{2} \right) \frac{\delta}{3} + (w \times 5 \times 5) \frac{\delta}{2} = 20.83 w\delta \text{ ft-kips}$$

$$\text{Total external work} = (2 \times 8.33 + 2 \times 20.83)w$$

$$= 58.3 w\delta \text{ ft-kips}$$

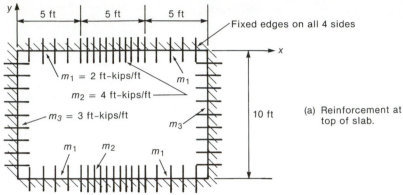

(a) Reinforcement at top of slab.

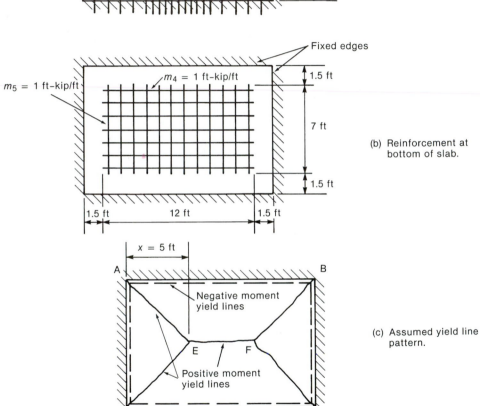

(b) Reinforcement at bottom of slab.

(c) Assumed yield line pattern.

Fig. 15-10
Slab—Example 15-5.

4. Compute the internal work.

For panel A–C–E,

$$m_x L_y \theta_y = \left(3 \text{ ft-kips/ft} \times 10 \text{ ft} \times \frac{\delta}{5} \right) + \left(1 \text{ ft-kip/ft} \times 7 \text{ ft} \times \frac{\delta}{5} \right)$$

$$= 7.4\delta \text{ ft-kips}$$

For panel A–B–F–E,

$$m_y L_x \theta_x = \left(2 \times 10 \times \frac{\delta}{5} \right) + \left(4 \times 5 \times \frac{\delta}{5} \right) + \left(1 \times 12 \times \frac{\delta}{5} \right)$$

$$= 10.4\delta \text{ ft-kips}$$

The total internal work is

$$\sum m\ell\theta = 2 \times 7.4\delta + 2 \times 10.4\delta$$

$$= 35.6\delta \text{ ft-kips}$$

5.   Equate the external and internal work. To calculate $w = w_u$, use $\phi = 0.9$ because slabs normally have very low reinforcement percentages.

$$58.3w\delta = 0.9 \times 35.6\delta \quad \text{and} \quad w = w_u = 0.550 \text{ ksf}$$

For this assumed yield-line pattern, the factored load this slab can support is 550 psf.

After trying several values of x, a value of $x = 6.1$ ft is found to give the lowest load, $w = 542$ psf. This represents the total factored dead- and live-load capacity.   ∎

Yield-Line Patterns at Discontinuous Corners

In Examples 15-3 to 15-5, the yield lines were assumed to extend along the diagonals into the corners of the slabs. In Section 13-9 and Fig. 13-64, the localized bending in the corners of a simply supported slab was discussed. As a result of this bending, the yield-line patterns in the corners of such a slab fork out to the sides of the slab, as shown in Fig. 15-11a. If the corner of the slab is free to lift, it will do so. If the corner is held down, the slab will form a crack or form a yield line across the corner as shown in Fig. 15-11b. References [15-3] and [15-4] show that inclusion of the corner segments or *corner levers* will reduce the computed uniform load capacity of a simply supported slab by up to 9 percent from the value in an analysis that ignores them. In actual slabs, this reduction is offset by increases in capacity due to strain-hardening of the reinforcement and membrane action in the slab.

Yield-Line Patterns at Columns or at Concentrated Loads

Figure 15-12 shows a slab panel supported on circular columns. The element at A is subjected to bending and twisting moments. If this element is rotated, however, it is possible to orient it so that the element is acted on by bending moments only, as shown in Fig. 15-12c. The moments calculated in this way are known as the principal moments and represent the maximum and minimum moments on any element at this point. On the element shown near

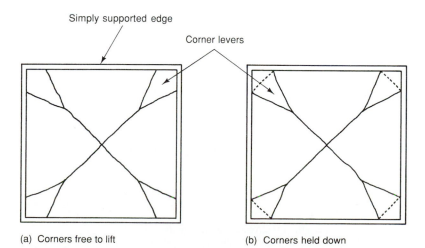

Simply supported edge

Corner levers

Fig. 15-11
Corner levers in simply supported slabs.

(a) Corners free to lift

(b) Corners held down

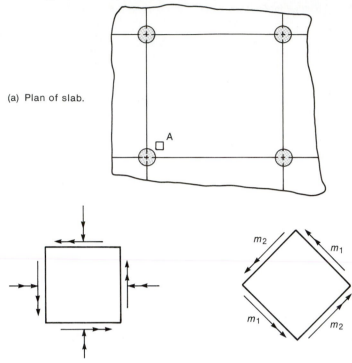

(a) Plan of slab.

Fig. 15-12
Principal moments adjacent
to a column in a flat plate.

(b) Bending and twisting
    moments on element at A.

(c) Principal moments on
    element at A.

the face of the column, the m_1 moment is the largest moment. As a result, a circular crack
develops around the column, and eventually the reinforcement crossing the circumference
of the column yields. The moments about lines radiating out from the center of the column
lead to radial cracks, and as a result, a circular *fan-shaped* yield pattern develops, as shown
in Fig. 15-13a. Under a concentrated load, a similar fan pattern also develops, as shown in
Fig. 15-13b. Note that these fans involve both positive- and negative-moment yield lines,
shown by solid lines and dashed lines, respectively.

EXAMPLE 15-6   Calculation of the Capacity of a Fan Yield Pattern

Compute the moments required to resist a concentrated load P if a fan mechanism develops. Assume
that the negative-moment capacity is m_1 and the positive moment capacity is m_2 per unit width in all
directions. Figure 15-13c shows a triangular segment from the fan in Fig. 15-13b. This segment sub-
tends an angle α and has an outside radius r.
Consider segment A–B–C.

1. **Give point A a virtual displacement of δ.**
2. **Compute the external work done.**

$$\text{External work} = P'\delta$$

where P' is the fraction of P carried by segment A–B–C.

3. **Compute the internal work done.** Consider one typical triangle. From (15-16),

$$\text{internal work} = \sum (m_x L_x \theta_x + m_y L_y \theta_y)$$

where θ_x is the rotation about line B–C:

$$\theta_x = \frac{\delta}{r} \qquad \theta_y = 0$$

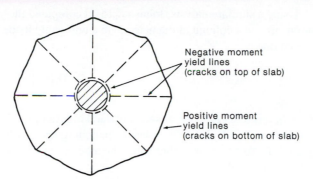

(a) Fan yield line at column in a flat plate.

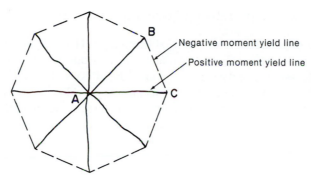

(b) Fan yield line around a downward
concentrated load at A.

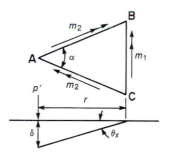

(c) Segment A–B–C.

Fig. 15-13
Fan yield lines.

For yield line A–B, the internal work is $m_2(\alpha r/2)(\delta/r)$; for yield line B–C, the internal work is $m_1(\alpha r)(\delta/r)$. The total internal work on this triangle is

$$m_1\alpha\delta + m_2\alpha\delta$$

The total number of triangles is $2\pi/\alpha$, so the total internal work is

$$2\pi\delta(m_1 + m_2)$$

Setting this expression equal to the external work gives

$$m_1 + m_2 = \frac{P}{2\pi}$$

∎

Using a similar analysis, Johnson [15-5] computed the yield-line moments for a circular fan around a column of diameter d_c in a uniformly loaded square plate with spans ℓ_1. He found that

$$m_1 + m_2 = w\ell_1^2\left[\frac{1}{2\pi} - 0.192\left(\frac{d_c}{\ell_1}\right)^{2/3}\right] \qquad (15\text{-}17)$$

If $d_c = 0$, this reduces to the answer obtained in Example 15-6.

If the slab fails due to two-way or punching shear (Chapter 13) the fan mechanism may not fully form. On the other hand, the attainment of a fan mechanism may bring on a punching-shear failure.

15-4 STRIP METHOD

Simple Strip Method: Edge-Supported Slabs

The strip method developed by Hillerborg [15-1], [15-7] is a slab-design method that gives a lower-bound equilibrium solution to the moments in a slab. As such, it gives a lower-bound (safe) estimate of the *capacity* of the slab in flexure. The equilibrium condition for an elastic slab is given by (15-3). If reinforcement is provided in the x and y directions, one possible solution that satisfies equilibrium is obtained by assuming that $m_{xy} = 0$ and dividing the total load w between the load carried in the x direction and that carried in the y direction:

$$\frac{\partial^2 m_x}{\partial x^2} = -w_x$$

$$\frac{\partial^2 m_y}{\partial y^2} = -w_y \qquad (15\text{-}18)$$

The physical significance of this is that the load w is divided into two parts, one carried by strips in the x direction, the other by strips in the y direction. The twisting of the strips due to m_{xy} (see Fig. 13-10) is ignored, resulting in a tendency to overestimate m_x and m_y.

Choice of Strips and w_x and w_y: Edge-Supported Slabs

The choice of strips and the division of the load w between those strips is entirely arbitrary. Three possible choices for a simply supported square slab are shown in Figs. 15-14 to 15-16. In Fig. 15-14, the slab is assumed to carry half the load in the x direction and half in the y direction. This gives the maximum $m_x = m_y = w\ell^2/16 = 0.0625w\ell^2$. By comparison Fig. 13-12 shows that the average maximum moment (on the diagonal) is $0.0417w\ell^2$.

A second possible choice is given in Fig. 15-15a. Here, the load is carried directly to the nearest edge, with w_x and w_y either w or 0. Thus, a narrow strip along A–A carries the load shown in Fig. 15-15b and has the moment diagram shown in Fig. 15-15c. This gives the moments about the middle of the plate as shown in Fig. 15-15d. The average moment m_x about the center of the plate is $0.0417w\ell^2$. Since each narrow strip would require different reinforcement, this choice of strips would not be a practical solution.

A third possible set of strips and distribution of w is shown in Fig. 15-16. Here, two-way action is assumed in the corners and in the middle of the slab, and one-way action is assumed along the edges. A full arrowhead indicates that all the load w is carried in the

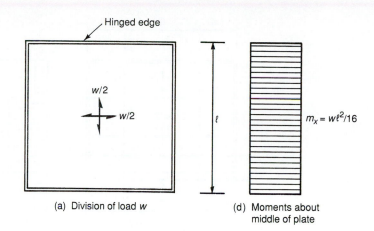

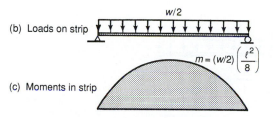

(a) Division of load w

(d) Moments about middle of plate

$$m_x = w\ell^2/16$$

(b) Loads on strip

$w/2$

(c) Moments in strip

$$m = (w/2)\left(\frac{\ell^2}{8}\right)$$

Fig. 15-14
Choice of strips and division of load w—first trial. (From [15-1]: Hillerborg, "Strip Method of Design," published by E&F, N Spon, London, 1975.)

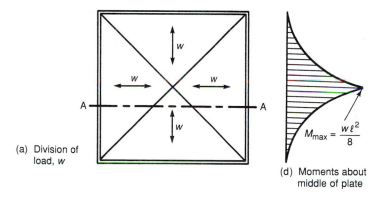

(a) Division of load, w

(d) Moments about middle of plate

$$M_{max} = \frac{w\ell^2}{8}$$

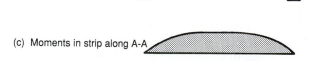

(b) Loads on strip along A-A

(c) Moments in strip along A-A

Fig. 15-15
Choice of strips and division of load w—second trial. (From [15-1]: Hillerborg, "Strip Method of Design," published by E&F, N Spon, London, 1975.)

direction of the arrow. Half arrowheads indicate the load is divided between the two directions. The average moment m_x about the center of the plate is $0.0469w\ell^2$.

All three of these patterns will give values for m_x and m_y that are on the safe side (too high), but when the combination of m_x, m_y, and m_{xy} is considered (see Section 15-2), a safe amount of reinforcement results. The pattern shown in Fig. 15-16 has the advantage that only two types of strips need to be used in each direction. As a result, it is used in practice in Europe.

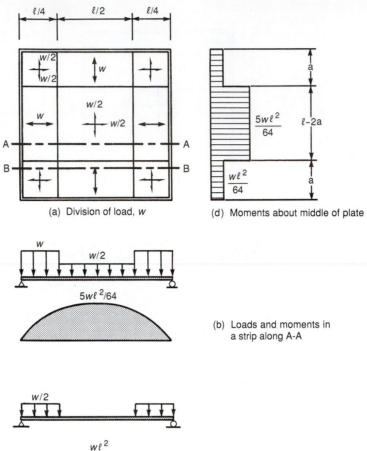

(a) Division of load, w

(d) Moments about middle of plate

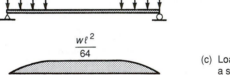

(b) Loads and moments in a strip along A-A

Fig. 15-16
Choice of strips and division of load w—third trial. (From [15-1]: Hillerborg, "Strip Method of Design," published by E&F, N Spon, London, 1975.)

(c) Loads and moments in a strip along B-B

EXAMPLE 15-7 Calculation of Moment Field in an Edge-Supported Slab

Compute the distribution of moments in the slab shown in Fig. 15-17. The total load on the slab is 150 psf.

1. Divide the slab into strips. If all four edges were hinged or all were fixed, a good choice of edge-strip width would be one-fourth of the narrow width of the panel. This will be used for the North–South edge strips. Since the ends of the panel are fixed and hence will attract more load than the hinged ends, the East–West edge strips will arbitrarily be taken 50 percent wider. The load will be assumed to be distributed in the directions shown by the arrows.

2. Compute the moments in the strips.

Strip 1–2: The loads on a 1-ft wide strip in strip 1–2 are shown in Fig. 15-17b. This gives rise to the moments in Fig. 15-17c. The maximum moment is

$$m = 375 \times 5 - \frac{75 \times 5^2}{2} = 937 \text{ ft-lb/ft}$$

Strip 3–4: The loads on strip 3–4 are shown in Fig. 15-17d. The maximum moment is

$$m = 1125 \times 10 - 150 \times 5 \times 7.5 - 75 \times 5 \times 2.5 = 4690 \text{ ft-lb/ft}$$

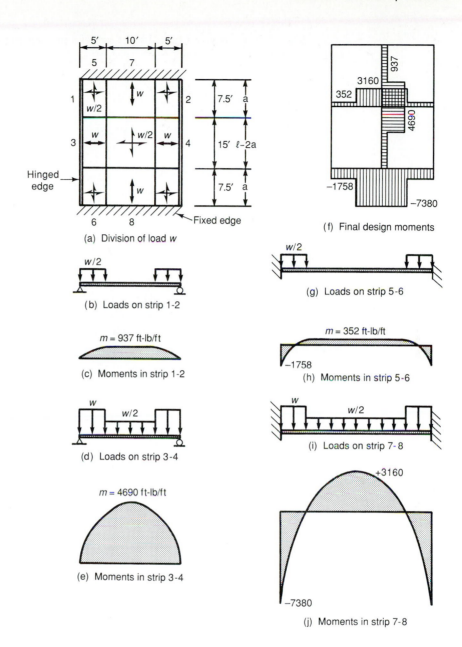

(a) Division of load w

(b) Loads on strip 1-2

$m = 937$ ft-lb/ft

(c) Moments in strip 1-2

(d) Loads on strip 3-4

$m = 4690$ ft-lb/ft

(e) Moments in strip 3-4

(f) Final design moments

(g) Loads on strip 5-6

$m = 352$ ft-lb/ft

(h) Moments in strip 5-6

(i) Loads on strip 7-8

(j) Moments in strip 7-8

Fig. 15-17
Slab—Example 15-7.

Strip 5–6: The loads are shown in Fig. 15-17g. The maximum positive and negative moments for this case are

$$\text{fixed-end moments} = \frac{wa^2}{6}\left(3 - \frac{2a}{\ell}\right)$$

$$= \frac{75 \times 7.5^2}{6}\left(3 - \frac{2 \times 7.5}{30}\right) = -1758 \text{ ft-lb/ft}$$

and

$$\text{midspan moment} = \frac{wa^2}{2} - \text{fixed-end moment}$$

$$= \frac{75 \times 7.5^2}{2} - 1758 = 352 \text{ ft-lb/ft}$$

Strip 7–8: This can be considered as the sum of a strip uniformly loaded with 75 lb/ft and a strip loaded the same way as strip 5–6. We then have

$$\text{fixed-end moments} = \frac{75 \times 30^2}{12} + 1758 = -7380 \text{ ft-lb/ft}$$

and

$$\text{midspan moment} = \frac{75 \times 30^2}{24} + 352 = 3160 \text{ ft-lb/ft}$$

The final moments are shown in Fig. 15-17f. ∎

Advanced Strip Method: Corner-Supported Elements

The simple strip method presented in the preceding section relies on the slab and the rigid supports to transmit the load in a given direction. (See Fig. 13-13.) As a result, it is possible to divide the load *w* between the two directions. For a slab on point supports, the slab must transmit the full load in both directions, as shown in Fig. 13-7. For such a case, Hillerborg has presented the *advanced strip method* [15-1]. In its simplest form [15-8], this involves dividing the slab up into (a) *simple strips* spanning one way to beam or wall supports and (b) *corner-supported elements* spanning two directions. The latter, shown shaded in Fig. 15-18, are denoted by crossed arrows with full arrowheads, which indicate that all the load is transmitted in each direction.

The corner-supported element proposed by Hillerborg satisfies the following requirements:

1. The edges are parallel to the reinforcement directions.
2. It carries a uniform load *w* per unit area.
3. It is supported on only one corner.
4. There are no shear forces on the sides of the element.
5. There are no twisting moments on the sides of the element.
6. All bending moments along one edge have the same sign or are zero.

Requirements 4 and 5 locate the edges of the corner-supported element at the point of maximum positive moment in the panel. It will further be assumed that

7. the element is square or rectangular, and
8. the bending moment is constant along each half of each edge of the corner-supported element.

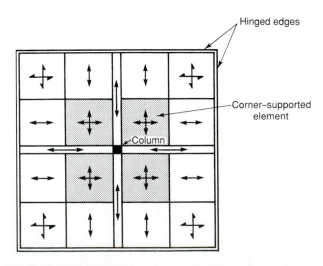

Fig. 15-18
Slab with simple strips and corner-supported elements.

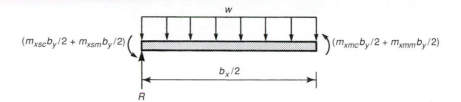

Fig. 15-19
Edge view of a corner-supported element.

In effect, the last assumption divides the elements into column strips and middle strips and allows for a constant moment over the width of each strip, as assumed in the ACI Code.

Figure 15-19 shows the forces and moments acting on a corner-supported element. The widths of the element are b_x and b_y. The vertical reaction at the corner is

$$R = wb_xb_y \tag{15-19}$$

Moment equilibrium about the y-axis gives

$$m_{xsc}\frac{b_y}{2} + m_{xsm}\frac{b_y}{2} + m_{xmc}\frac{b_y}{2} + m_{xmm}\frac{b_y}{2} = \frac{wb_yb_x^2}{2} \tag{15-20}$$

where m_{xsc} is the moment about the x-axis through the support (s) in the column strip (c), m_{xmm} is the moment about midspan of the panel (first m) in the middle strip (second m), and so on. A similar equation can be written for taking moments about the x-axis. These are essentially the same as Nichols' equation, (13-3).

The division of moment between support and midspan (between m_{xs} and m_{xm}) can be made (a) on the basis of the moment coefficients in ACI Section 13.6.3.3, (b) on the basis of the elastically computed moments from a strip as in Fig. 15-20b and loaded as shown there, or (c) arbitrarily, provided that equilibrium is satisfied. The division of moments between column and middle strips along the sides of the element could be based on ACI Section 13.6.4 as done in Example 15-8. Hillerborg assumes that the midspan moments m_{xmc} and m_{xmm} are equal, that the support moment m_{xsm} vanishes, and that the support moment m_{xsc} is twice the average m_{xs} value.

EXAMPLE 15-8   Calculation of Moment Field by Advanced Strip Method

The slab shown in Fig. 15-20a is supported on three edges by walls and by two columns as shown. Compute the distribution of moments for $w = 150$ psf.

1.   **Divide the slab into strips and corner-supported elements.** The arrangement chosen is shown in Fig. 15-20a. There are bands of simple strip elements along each wall and simple strip elements between each column. There are six corner supported elements. The boundaries of the various elements are chosen from the locations of the maximum positive moments in the strips. Assume that strip 2B–3B–3G–2G is loaded as shown in Fig. 15-20b and has the moment diagram shown. Lines C and F are located at the points of maximum positive moment. Similarly, lines 2 and 5 are at the points of maximum positive moment in strip 1C–6C–6D–1D.

2.   **Compute the moments in strip 1C–6C–6D–1D, and distribute these to the column and the middle strip.** This strip is loaded as shown in Fig. 15-20c. The width of the strip is 8.8 ft. The statical moment, considering the 8.8-ft width, is

$$M_0 = \frac{150 \times 8.8 \times 15^2}{8 \times 1000} = 37.1 \text{ ft-kips}$$

(a)   At lines 1 and 6: $M = -0.65 \times 37.1 = -24.1$ ft-kips. From ACI Section 13.6.4.3, this is uniformly distributed across the width of the strip.

(b)   At lines 2 and 5: $M = 0.35 \times 37.1 = 13.0$ ft-kips. From ACI Section 13.6.4.4, 60 percent (7.80 ft-kips) is assigned to the column strip, 5.2 ft-kips to the middle strip.

(c)   At lines 3 and 4: $M = -0.65 \times 37.1 = -24.1$ ft-kips. From ACI Section 13.6.4.1, 75 percent $= -18.1$ ft-kips is assigned to the column strip, -6.0 ft-kips to the middle strip.

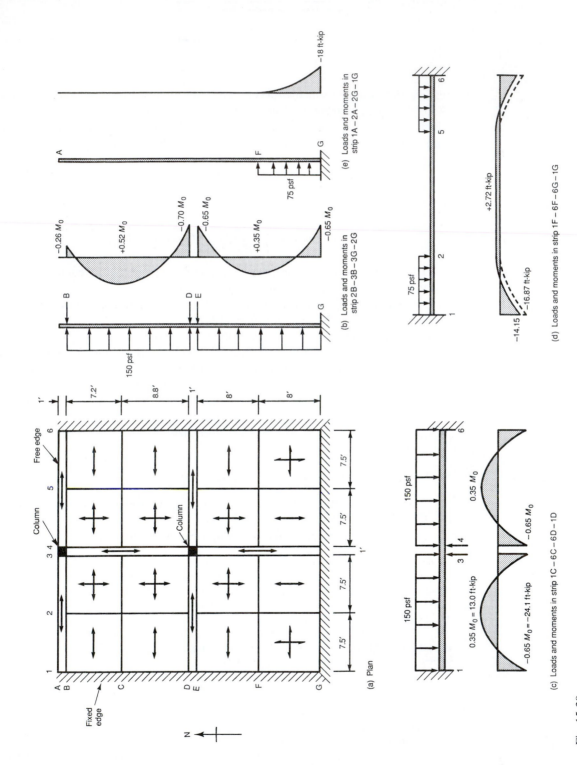

(a) Plan

(b) Loads and moments in strip 2B – 3B – 3G – 2G

(c) Loads and moments in strip 1C – 6C – 6D – 1D

(d) Loads and moments in strip 1F – 6F – 6G – 1G

(e) Loads and moments in strip 1A – 2A – 2G – 1G

Fig. 15-20
Slab—Example 15-8.

782

3.   Compute the moments in other East–West strips. For the strips between lines *A* and *B*, *B* and *C*, *D* and *E*, and *E* and *F*, these are computed in the same way, allowing for the differing widths of the strips.

The strip between lines *F* and *G* is a simple strip loaded as shown in Fig. 15-20d and having the moment diagram shown.

4.   Combine the moments from adjacent East–West middle and column strips. The column strip along edge *A* extends 1 ft plus half of 7.2 ft, totaling 4.6 ft from the edge of the slab. At the column, this strip has a total negative moment of $-24.1/8.8 = -2.74$ ft-kips in strip *A–B* and $-18.1 \times 7.2/8.8 = -14.81$ ft-kips from strip *B–C*, for a total negative moment of -17.55 ft-kips in the 4.6-ft width, or -3.82 ft-kips/ft. (See Fig. 15-21a.)

In the middle strip along *C*, the negative moment at lines 3 and 4 is -6 ft-kips from strip *B–C* for a total of -10.91 ft-kips in the 8-ft width or -1.36 ft-kips/ft.

The middle strip along line *F* consists partly of strip *E–F* and partly of the simple strip *F–G*. As shown in Fig. 15-20d, the simple strip has positive moment at lines 3 and 4. These will be disregarded. However, to satisfy equilibrium in strip *E–F*, the moment diagram will be shifted downward by 2.72 ft-kips, giving the dashed moment diagram in Fig. 15-20d. The moments in the middle strip along *F* will be taken as twice those in the middle-strip portion of strip *E–F*.

The positive moments about line 2 are computed in the same manner. From step 2(a), the negative moments along the wall (about line 1) between *C* and *D* are -24.1 ft-kips in the 8.8-ft strip. From ACI Section 13.6.4.3, the moment is uniformly distributed across the width of the strip, giving -2.74 ft-kips/ft. This moment exists from *A* to *F*. From Fig. 15-20d, the moment per unit width between *F* and *G* is $-16.87/8$ ft $= -2.11$ ft-kips/ft.

The final moments about lines 1, 2, and 3 are shown in Fig. 15-21a.

5.   Compute the moments in the North–South strips.

The calculations proceed in the same manner as for the East–West strips. The moments in the strips between 2 and 3, 3 and 4, and 4 and 5 are computed from Fig. 15-20b. The strips along the walls (between 1 and 2 and between 5 and 6) are assumed to be loaded as shown in Fig. 15-20e.

The moments are distributed across the width of the slab in the same way as in steps 2 and 4. The final design moments about lines *B, C, D, F,* and *G* are shown in Fig. 15-21b.   ■

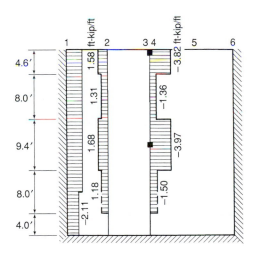

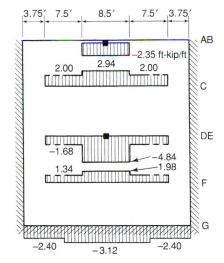

Fig. 15-21
Design moments—Example 15-8.

(a) Design moments about lines 1, 2, 3

(b) Design moments about lines AB, C, DE, F, G

PROBLEMS

Using a yield-line analysis, compute the moment capacities required in the slabs shown in the figures that follow.

15-1 Fig. P15-1: $m_x = m_y$ and uniform load $= w$.

15-2 Fig. P15-2: positive- and negative-moment capacities in both directions are m, and load $= w$. Try several yield-line patterns to get the maximum m.

15-3 Using a yield-line analysis, compute the maximum uniform ultimate load, w, that the slab in Fig. P15-3 can support. The slab is 6 in. thick. The reinforcement in the North–South direction has $\frac{3}{4}$ in. cover to the top and bottom surfaces. The East–West reinforcement is inside the North–South steel. Use $f'_c = 3000$ psi and $f_y = 40,000$ psi.

15-4 Compute the total negative and positive moment required to support an ultimate load w of 250 psf. Assume a circular-fan yield line at a column supporting an interior panel of a slab with spans $\ell_1 = 20$ ft, $d_c = 14$ in. If the overall thickness of the slab is 7 in., $f'_c = 3000$ psi, $f_y = 60,000$ psi, and the bottom steel is No. 4 at 15 in. o.c., what top reinforcement is required? Use the average d in your calculations.

15-5 Compute the moments in the slab shown in Fig. P15-5, using the strip method.

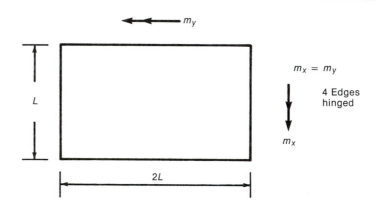

Fig. P15-1

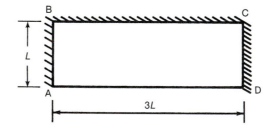

Fig. P15-2

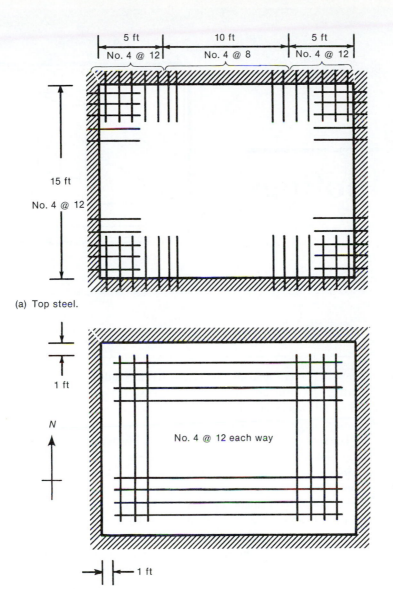

(a) Top steel.

(b) Bottom steel.

Fig. P15-3

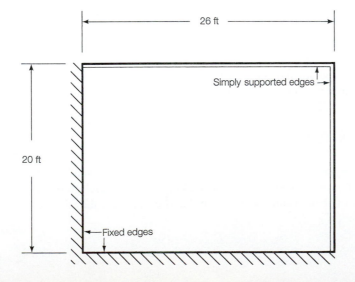

Fig. P15-5

16

Footings

16-1   INTRODUCTION

Footings and other foundation units transfer the loads from the structure to the soil or rock supporting the structure. Because the soil is generally much weaker than the concrete columns and walls that must be supported, the contact area between the soil and the footing is much larger than that between the supported member and the footing.

The more common types of footings are illustrated in Fig. 16-1. *Strip footings* or *wall footings* display essentially one-dimensional action, cantilevering out on each side of the wall. *Spread footings* are pads that distribute the column load to an area of soil around the column. These distribute the load in two directions. Sometimes spread footings have pedestals, are stepped, or are tapered to save materials. A *pile cap* transmits the column load to a series of *piles*, which, in turn, transmit the load to a strong soil layer at some depth below the surface. *Combined footings* transmit the loads from two or more columns to the soil. Such a footing is often used when one column is close to a property line. A *mat or raft foundation* transfers the loads from all the columns in a building to the underlying soil. Mat foundations are used when very weak soils are encountered. *Caissons* 2 to 5 ft in diameter are sometimes used instead of piles to transmit heavy column loads to deep foundation layers. Frequently, these are enlarged at the bottom (*belled*) to apply load to a larger area.

The choice of foundation type is selected in consultation with the geotechnical engineer. Factors to be considered are the soil strength, the soil type, the variability of the soil type over the area and with increasing depth, and the susceptibility of the soil and the building to deflections.

Strip, spread, and combined footings are considered in this chapter since these are the most basic and most common types.

16-2   SOIL PRESSURE UNDER FOOTINGS

The distribution of soil pressure under a footing is a function of the type of soil and the relative rigidity of the soil and the foundation pad. A concrete footing on sand will have a pressure distribution similar to Fig. 16-2a. The sand near the edges of the footing tends to

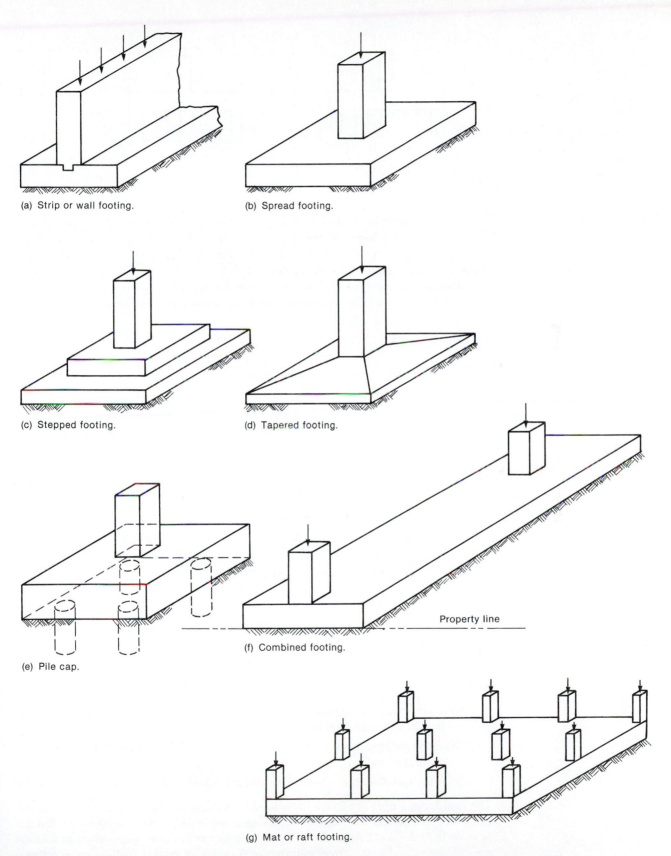

(a) Strip or wall footing.

(b) Spread footing.

(c) Stepped footing.

(d) Tapered footing.

(e) Pile cap.

(f) Combined footing.

Property line

(g) Mat or raft footing.

Fig. 16-1
Types of footings.

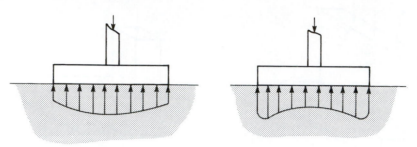

Fig. 16-2
Pressure distribution under
footings.

(a) Footing on sand.   (b) Footing on clay.

displace laterally when the footing is loaded, causing a decrease in soil pressure near the edges. On the other hand, the pressure distribution under a footing on clay is similar to Fig. 16-2b. As the footing is loaded, the soil under the footing deflects in a bowl-shaped depression, relieving the pressure under the middle of the footing. For design purposes, it is customary to assume that the soil pressures are linearly distributed in such a way that the resultant vertical soil force is collinear with the resultant downward force.

Design Methods

Allowable Stress Design

There are two different philosophies for the design of footings [16-1], [16-2]. The first is *allowable stress design*. Almost exclusively, footing design is based on the allowable stresses acting on the soil at unfactored or service loads. For a concentrically loaded spread footing,

$$\Sigma P_s \leq q_a A \tag{16-1}$$

where

P_s is the specified (unfactored) load acting on the footing. Section 2.4.1 of ASCE 7 gives an updated set of load combinations for allowable stress design [16-3]. ACI Committee 318 has not considered these for footing design as yet.

q_a is the allowable stress for the soil given by (16-3), presented in the next subsection.

A is the area of the footing in contact with the soil.

Limit-States Design

The second design philosophy is a *limit-states design* based on factored loads and factored resistances [16-1], [16-2], given by

$$\phi R_n \geq \Sigma \alpha P_s \tag{16-2}$$

where

ϕ is a resistance factor to account for the variability of the load-resisting mechanism of the soil under the footing.

R_n is the engineer's best estimate of the resistance of the soil under the footing

α is a load factor.

P_s is the specified load acting on the soil at the base of the footing.

The load factors, α, in (16-2) are those used in building design. Two sets of load factors and load combinations are available for design, those in ACI 318-02, Sections 9.2 and 9.3, or those in ACI 318-02, Appendix C. Resistance factors for limit-states design of footings are still being developed. Current estimates of ϕ values for shallow footings are as follows:

for vertical resistance, $\phi = 0.5$

for sliding resistance dependent on friction, with cohesion equal to zero, $\phi = 0.8$

for sliding resistance dependent on cohesion, with friction equal to zero, $\phi = 0.6$

Serviceability limit states should also be checked [16-1], [16-2], [16-3].

At the time of writing, virtually all building footings in North America are designed by using allowable-stress design applied to failures of the concrete foundation element or the soil itself (16-1). The rest of this chapter will apply allowable-stress design to the soil and then use ultimate-load design for the reinforced concrete foundation structure.

Limit States for the Design of Foundations

Limit States Governed by the Soil

Three primary ultimate limit states of the soil supporting an isolated foundation are [16-1], [16-3].

1. a bearing failure of the soil under the footing (Fig. 16-3),

2. a serviceability failure in which excessive differential settlement between adjacent footings causes architectural or structural damage to the structure, or

3. excessive total settlement.

Settlement occurs in two stages: *immediate settlement* as the loads are applied, and a long-term settlement known as *consolidation*.

Procedures for minimizing differential settlements involve a degree of geotechnical engineering theory outside the realm of this book.

Bearing failures are controlled by limiting the service-load stress under the footing to less than an allowable stress, q_a, as in (16-1).

Limit States Governed by the Structure

Similarly, there are three primary structural limit states for the foundations themselves [16-1], [16-2]:

4. flexural failure of the portions of the footing that project from the column or wall,

5. shear failure of the footing, and

6. inadequate anchorage of the flexural reinforcement in the footing.

As stated earlier, bearing failures of the soil supporting the foundation are prevented by limiting the service-load stress under the footing to less than an allowable stress

$$q_a = \frac{q_{\text{ult}}}{\text{FS}} \tag{16-3}$$

where q_{ult} is the stress corresponding to the failure of the soil under the footing and FS is a factor of safety in the range of 2.5 to 3. Values of q_a are obtained from the principles of

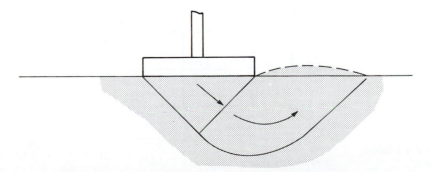

Fig. 16-3
Bearing failure of a footing.

geotechnical engineering and depend on the shape of the footing, the depth of the footing, the overburden or surcharge on top of the footing, the position of the water table, and the type of soil. When using a value of q_a provided by a geotechnical engineer, it is necessary to know what strengths were measured and in what kind of tests, and what assumptions have been made in arriving at this allowable soil pressure, particularly with respect to over-burden and depth to the base of the footing.

It should be noted that q_a in (16-3) is a service-load stress, whereas the rest of the structure has been designed by using factored loads corresponding to the ultimate limit states. The method of accounting for this difference in philosophy is explained later.

Elastic Distribution of Soil Pressure under a Footing

The soil pressure under a footing is calculated by assuming linearly elastic action in compression, but no tensile strength across the contact between the footing and the soil. If the column load is applied at, or near, the middle of the footing, as shown in Fig. 16-4, the stress, q, under the footing is

$$q = \frac{P}{A} \pm \frac{My}{I} \tag{16-4}$$

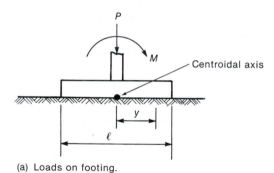

(a) Loads on footing.

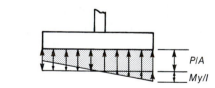

(b) Soil pressure distribution.

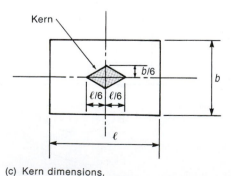

(c) Kern dimensions.

Fig. 16-4
Soil pressure under a footing: loads within kern.

where

P = vertical load, positive in compression

A = area of the contact surface between the soil and the footing

I = moment of inertia of this area

M = moment about the centroidal axis of the area

y = distance from the centroidal axis to the point where the stresses are being calculated

The moment, M, can be expressed as Pe, where e is the eccentricity of the load relative to the centroidal axis of the area A. The maximum eccentricity e for which (16-4) applies is that which first causes $q = 0$ at some point. Larger eccentricities will cause a portion of the footing to lift off the soil, since the soil–footing interface cannot resist tension. For a rectangular footing, this occurs when the eccentricity exceeds

$$e_k = \frac{\ell}{6} \qquad (16\text{-}5)$$

This is referred to as the *kern distance*. Loads applied within the *kern*, the shaded area in Fig. 16-4c, will cause compression over the entire area of the footing, and (16-4) can be used to compute q.

Various pressure distributions for rectangular footings are shown in Fig. 16-5. If the load is applied concentrically, the soil pressure q is $q_{avg} = P/A$. If the load acts through the kern point (Fig. 16-5c), $q = 0$ at one side and $q = 2q_{avg}$ at the other. If the load falls outside the kern point, the resultant upward load is equal and opposite to the resultant downward load, as shown in Fig. 16-5d. Generally, such a pressure distribution would not be acceptable, since it makes inefficient use of the footing concrete, tends to overload the soil, and may cause tilt.

Elastic and Plastic Soil-Pressure Distributions

The soil-pressure diagrams in Fig. 16-5 are based on the assumption that the soil pressure is linearly distributed under a footing. This is a satisfactory assumption at service-load levels and for footings on rock or dense glacial till. For yielding soils, the soil-pressure distribution will approach a uniform (plastic) distribution over part of the base of the footing in such a way that the resultant load on the footing and the resultant of the soil pressure coincide as is required for equilibrium.

For the design of concentrically loaded footings, the distribution of soil pressures is taken to be uniform over the entire contact area, as shown in Fig. 16-5a. For the structural design of eccentrically loaded footings, such as those for retaining walls or bridge abutments, two pressure distributions may be considered [16-2]. The first of these is a uniform, fully plastic pressure, q_p, over a portion, A_p, of the area of the contact surface. The area A_p is chosen such that the centroid of A_p and the resultant of the applied loads coincide. The second pressure distribution is a linearly varying distribution like the one in Fig. 16-5d, again with the resultant of the soil pressure coincident with the resultant of the applied loads.

The examples in this chapter are limited to the predominant case of concentric loads and a uniform or linear soil-pressure distribution over the entire contact area.

Load and Resistance Factors for Footing Design

In 1995, the ACI Code added a second set of load and resistance factors for reinforced concrete design to the traditional ACI load factors and resistance factors. In the 2002 ACI Code, the two sets of factors were interchanged: the new factors were given in ACI

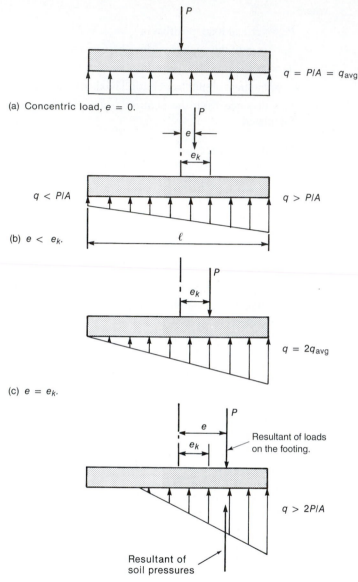

Fig. 16-5
Pressures under an eccentrically loaded footing.

Sections 9.2 and 9.3 and the traditional ACI load and resistance factors in Appendix C. In the 2002 code, both sets of load and resistance factors are acceptable for design under ACI 318.

The first three examples in this chapter are based on the load and resistance factors in ACI 2002 Sections 9.2 and 9.3. Example 16-4 uses the factors in ACI Appendix C.

Gross and Net Soil Pressures

Figure 16-6a shows a 2-ft-thick spread footing with a column at its center and with its top surface located 2 ft below the ground surface. There is no column load at this stage. The total downward load from the weights of the soil and the footing is 540 psf. This is balanced

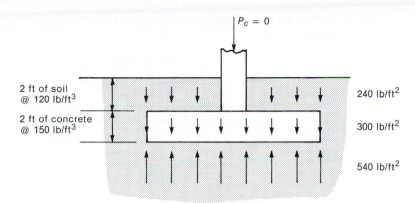

2 ft of soil
@ 120 lb/ft^3

2 ft of concrete
@ 150 lb/ft^3

$P_c = 0$

240 lb/ft^2

300 lb/ft^2

540 lb/ft^2

(a) Self weight and soil surcharge.

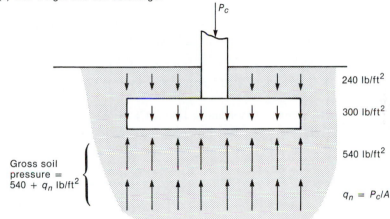

P_c

240 lb/ft^2

300 lb/ft^2

540 lb/ft^2

$q_n = P_c/A$

Gross soil
pressure =
540 + q_n lb/ft^2

(b) Gross soil pressure.

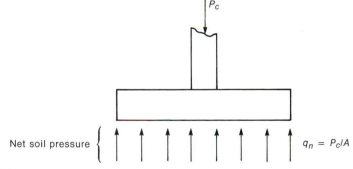

P_c

Net soil pressure

$q_n = P_c/A$

Fig. 16-6
Gross and net soil pressures.

(c) Net soil pressure.

by an equal, but opposite, upward pressure. As a result, the net effect on the concrete foot-ing is zero. There are no moments or shears in the footing due to this loading.

When the column load P_c is added, the pressure under the footing increases by $q_n = P_c/A$, as shown in Fig. 16-6b. The total soil pressure is $q = 540 + q_n$. This is re-ferred to as the *gross soil pressure* and must not exceed the allowable soil pressure, q_a. When moments and shears in the concrete footing are calculated, the upward and down-ward pressures of 540 psf cancel out, leaving only the *net soil pressure*, q_n, to cause inter-nal forces in the footing, as shown in Fig. 16-6c.

In design, the area of the footing is selected so that the *gross soil pressure* does not exceed the allowable soil pressure. The flexural reinforcement and the shear strength of the footing are then calculated by using the *net soil pressure*. Thus, the area of the footing is selected to be

$$A = \frac{D(\text{structure, footing, surcharge}) + L}{q_a} \tag{16-6}$$

where D and L refer to the unfactored service dead and live loads.

For service-load combinations including wind, W, most codes allow a 33 percent increase in q_a. For such a load combination, the required area would be

$$A = \frac{D(\text{structure, footing, surcharge}) + L + W}{1.33q_a} \tag{16-7}$$

but not less than the value given by (16-6). In (16-6) and (16-7), the loads are the unfactored service loads.

Once the area of the footing is known, the rest of the design of the footing is based on soil stresses due to the factored loads.

Load Combinations for Footing Design

The load combinations from ACI 318-02 Section 9.2 have been used in Example 16-1 through 16-3. The more significant combinations for the design of the footing itself are:

Load combinations 9-1 $\quad U = 1.4(D + F)$ (ACI Eq 9-1)

This reduces to $\quad U = 1.4D$

Load combination 9-2 $\quad U = 1.2(D + F + T) + 1.6(L + H) + 0.5(L_r \text{ or } S \text{ or } R)$
(ACI Eq 9-2)

In these combinations, D, F, and T are permanent or quasi-permanent loads, L and H are the principal variable loads, and (L_r or S or R) are companion action roof loads. In most cases, F, T, H, L_r, R and possibly S can be dropped out. Then, Eq. 9-2 reduces to:

$$U = 1.2D + 1.6L$$

Load Combination 9-3 $\quad U = 1.2D + 1.6(L_r \text{ or } S \text{ or } R) + (1.0L \text{ or } 0.8W)$
(ACI Eq 9-3)

Here the roof loads are the principal variable loads and L and W are companion action loads. The load factor on companion action L comes from ACI Section 9.2.1(a)

Load combination 9-4
$$U = 1.2D + 1.6W + 1.0L + 0.5(L_r \text{ or } S \text{ or } R) \qquad \text{(ACI Eq 9-4)}$$

In this combination, W is the principal variable load and L and the roof loads are companion action loads.

Load combination 9-6 is: $\quad U = 0.9D + 1.6W \text{ or } 1.6H$ (ACI Eq 9-6)

This combination applies when dead load stabilizes the overturning of the structure due to wind or earth pressure. (See Fig. 2-7.) Load combinations 9-5 and 9-7 deal with earthquakes and will be considered in Chapter 20.

Factored Net Soil Pressure, q_{nu}.

The factored net soil pressures used to design the footing are:

$$q_{nu} = (\text{Factored load})/A = \frac{U}{A} \tag{16-8}$$

where the factored loads come from ACI Equations 9-1, 9-2, 9-3, 9-4, and 9-6. The factored net soil pressure, q_{nu}, is based on the factored loads and will exceed q_a in most cases. This is acceptable, because the factored loads are roughly 1.5 times the service loads, whereas the factor of safety implicit in q_a is 2.5 to 3. Hence, the factored net soil pressure will be less than the pressure that would cause failure of the soil.

If both load and moment are transmitted to the footing, it is necessary to use (16-4) (if the load is within the kern) or other relationships (as illustrated in Fig. 16-5d) to compute q_{nu}. In such calculations, the factored loads would be used.

16-3   STRUCTURAL ACTION OF STRIP AND SPREAD FOOTINGS

The behavior of footings has been studied experimentally at various times. Our current design procedures have been strongly affected by the tests reported in [16-4], [16-5], and [16-6].

The design of a footing must consider bending, development of reinforcement, shear, and the transfer of load from the column or wall to the footing. Each of these is considered separately here, followed by a series of examples in the ensuing sections. In this section, only axially loaded footings with uniformly distributed soil pressures, q_{nu}, are considered.

Flexure

A spread footing is shown in Fig. 16-7. Soil pressures acting under the crosshatched portion of the footing in Fig. 16-7b cause the moments about axis A–A at the face of the column. From Fig. 16-7c, we see that these moments are

$$M_u = (q_{nu}b)\frac{f}{2} \qquad (16\text{-}9)$$

where $q_{nu}bf$ is the resultant of the soil pressure on the crosshatched area and $f/2$ is the distance from the resultant to section A–A. This moment must be resisted by reinforcement placed as shown in Fig. 16-7c. The maximum moment will occur adjacent to the face of the column on section A–A or on a similar section on the other side of the column.

In a similar manner, the soil pressures under the portion outside of section B–B in Fig. 16-7a will cause a moment about section B–B. Again, this must be resisted by flexural reinforcement perpendicular to B–B at the bottom of the footing; the result is two layers of steel, one each way, shown in section A–A in Fig. 16-7c.

The critical sections for moment in the footing are taken as follows: (ACI Sections 15.3 and 15.4.2)

1. for footings supporting square or rectangular concrete columns or walls, at the face of the column or wall;

2. for footings supporting circular or regular polygonal columns, at the face of an imaginary square column with the same area;

3. for footings supporting masonry walls, halfway between the middle and the edge of the wall;

4. for footings supporting a column with steel base plates, halfway between the face of the column and the edge of the base plate.

The moments per unit length vary along lines A–A and B–B, with the maximum occurring adjacent to the column. To simplify reinforcement placing, however, ACI Section 15.4.3 states that for square footings the reinforcement shall be distributed uniformly across the entire width of the footing. A banded arrangement is used in rectangular footings, as will be illustrated in Example 16-3.

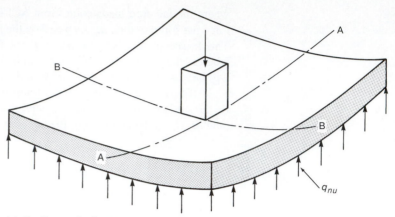

(a) Footing under load.

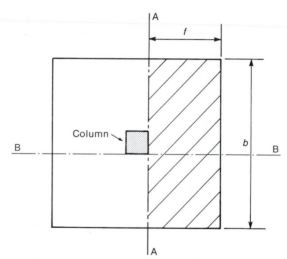

(b) Tributary area for moment at section A–A.

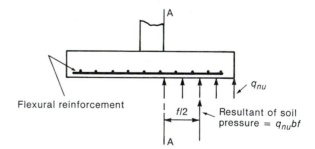

Flexural reinforcement

Resultant of soil
pressure $= q_{nu}bf$

(c) Moment About Section A–A.

Fig. 16-7
Flexural action of a spread
footing.

ACI Section 10.3.5 sets the upper limit on the flexural reinforcement in a beam equal to that corresponding to a *net tensile strain* ε_t not less than (i.e., not less tensile than) 0.004 in the *extreme-tension reinforcement*. Although this is not a beam, it is desirable that it be ductile in flexure. This can be done by limiting ε_t to the value of 0.004 allowed in ACI Section 10.3.5. When calculating ϕ, ACI Section 9.3.2 requires the value $\phi = 0.9$ be used only for "*tension-controlled sections* as defined in ACI Section 10.3.4." We shall interpret this to mean that ϕ must be calculated using the applicable equation from ACI Fig. R.9.3.2 if ε_t is less than 0.005.

If the extreme tension strain, ε_t, exceeds 0.005 in tension, $\phi = 0.9$. If ε_t is between 0.004 and 0.005, ϕ is calculated from the applicable equation from ACI Fig. R.9.3.2. This is done using ε_t from:

$$\varepsilon_t = \frac{d_t - c}{c} \times 0.003 \tag{16-10}$$

and ϕ is

$$\phi = 0.48 + 83\varepsilon_t \text{ where } 0.65 \leq \phi \leq 0.90 \tag{16-11}$$

This check can also be done by checking whether the a/d_t ration exceeds the limiting values of a/d_t from Table A-5.

ACI Section 10.5.4 states that, for footings of uniform thickness, the minimum area of flexural tensile reinforcement shall be the same as that required for shrinkage and temperature reinforcement in ACI Section 7.12. For Grade-40 steel, this requires $A_{s,min} = 0.0020bh$; for Grade-60 steel, $A_{s,min} = 0.0018bh$ is specified. This amount of steel should provide a moment capacity between 1.1 and 1.5 times the flexural cracking moment and hence should be enough to prevent sudden failures at the onset of cracking. ACI Section 10.5.4 gives the maximum spacing of the reinforcement in a footing as the lesser of three times the thickness or 18 in.

If the reinforcement required for flexure exceeds the minimum flexural reinforcement, the authors would use the maximum spacings from ACI Section 13.3.2, which calls for a maximum spacing equal to the smaller of twice the slab thickness, but not greater than 18 in.

Development of Reinforcement

The footing reinforcement is chosen by assuming that the reinforcement stress reaches f_y along the maximum-moment section at the face of the column. The reinforcement must extend far enough on each side of the points of maximum bar stress to develop this stress. In other words, the bars must extend ℓ_d from these points or be hooked at the outer ends.

Shear

A footing may fail in shear as a wide beam, as shown in Figs. 16-8a and 13-35a or as a result of punching, as shown in Fig. 16-8b and 13-35b. These are referred to as one-way shear and two-way shear and are discussed more fully in Section 13-7.

Recently, concern has been expressed about the shear strength of deep, lightly reinforced concrete members [16-7], [16-8], and [16-9]. Tests of members similar to one-way footings [16-8] suggest that, if the ratio of the length, a, of the portion of the footing that projects outward from the column or wall, to the depth of the footing, d, does not exceed 3, the crack-restraining effect of the soil pressure under the footing tends to offset the strength reduction due to size. ACI Section 11.5.5.1 requires minimum stirrups in all flexural members with V_u greater than $\frac{1}{2}\phi V_c$, except for footings, joists, or wide shallow beams. Stirrups are required in all footings for which V_u exceeds ϕV_c.

One-Way Shear

A footing failing through one-way shear is designed as a beam with (ACI Section 11.12.1.1)
$$V_u \leq \phi(V_c + V_s) \tag{6-9 and 6-14}$$

where

$$V_c = 2\sqrt{f'_c}b_w d \tag{6-8}$$

Web reinforcement is very seldom used in strip footings or spread footings, due to the difficulty in placing it, and due to the fact that it is usually cheaper and easier to deepen the

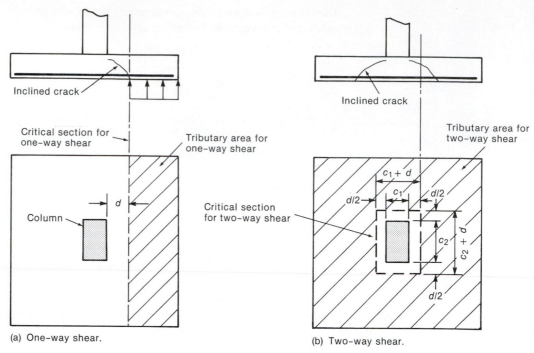

Fig. 16-8
Critical sections and tributary areas for shear in a spread footing.

footing than it is to provide stirrups. Hence, $V_s = 0$ in most cases. The inclined crack shown in Fig. 16-8a intercepts the bottom of the member about d from the face of the column. As a result, the critical section for one-way shear is located at d away from the face of the column or wall, as shown in plan view in Fig. 16-8a. For footings supporting columns with steel base plates, the critical section is d away from a line halfway between the face of the column and the edge of the base plate. The shear V_u is q_{nu} times the tributary area shown shaded in Fig. 16-8a.

Two-Way Shear

Research [16-6], [16-7] has shown that the critical section for punching shear is at the face of the column, while the critical loaded area is that lying outside the area of the portion punched through the slab. To simplify the design equations, the critical-shear perimeter for design purposes has been defined as lying $d/2$ from the face of the column, as shown by the dashed line in Fig. 16-8b (ACI Sections 11.12.1.2 and 11.12.1.3). For the column shown in Fig. 16-8b, the length, b_o, of this perimeter is

$$b_o = 2(c_1 + d) + 2(c_2 + d) \qquad (16\text{-}12)$$

where c_1 and c_2 are the lengths of the sides of the column and d is the average effective depth in the two directions. The tributary area assumed critical for design purposes is shown cross-hatched in Fig. 16-8b.

Since web reinforcement is rarely used in a footing, $V_u \leq \phi V_c$, where, from ACI Section 11.12.2.1, V_c shall be the smallest of

$$\text{(a) } V_c = \left(2 + \frac{4}{\beta_c}\right)\sqrt{f_c'}\, b_o d \qquad (13\text{-}15)$$

$$\text{(ACI Eq. 11-33)}$$

where β_c is the ratio of the long side to the short side of the column (c_2/c_1 in Fig. 16-8b) and b_o is the perimeter of the critical section,

$$\text{(b) } V_c = \left(\frac{\alpha_s d}{b_o} + 2\right)\sqrt{f'_c}\, b_o d \qquad\qquad \text{(13-16)}$$

$$\text{(ACI Eq. 11-34)}$$

where α_s is 40 for columns in the center of footing, 30 for columns at an edge of a footing, and 20 for columns at a corner of a footing, and

$$\text{(c)} \quad V_c = 4\sqrt{f'_c}\, b_o d \qquad\qquad \text{(13-17)}$$

$$\text{(ACI Eq. 11-35)}$$

Transfer of Load from Column to Footing

The column applies a concentrated load on the footing. This load is transmitted by bearing stresses in the concrete and by stresses in the dowels or column bars that cross the joint. The design of such a joint is considered in ACI Section 15.8. The area of the dowels can be less than that of the bars in the column above, provided that the area of the dowels is at least 0.005 times the column area (ACI Section 15.8.2.1) and is adequate to transmit the necessary forces. Such a joint is shown in Fig. 16-9. Generally, the column bars stop at the

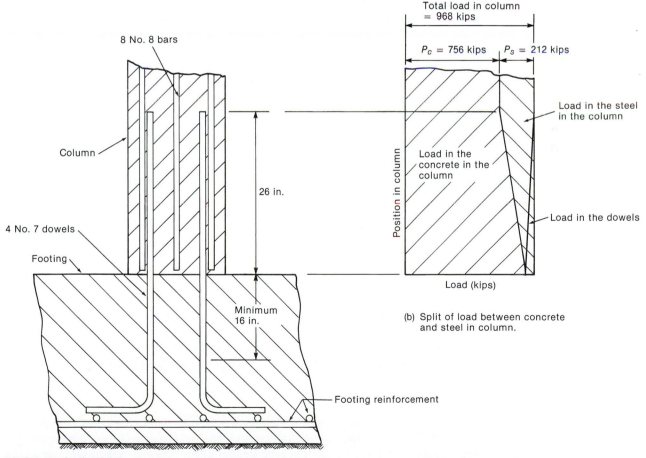

(a) Column–footing joint.

(b) Split of load between concrete and steel in column.

Fig. 16-9
Column–footing joint.

bottom of the column, and dowels are used to transfer forces across the column–footing joint. Dowels are used because it is awkward to embed the column steel in the footing, due to its unsupported height above the footing and the difficulty in locating it accurately. Figure 16-9a shows an 18 in. $\times$ 18 in. column with $f'_c = 5000$ psi and eight No. 8 bars. The column is supported on a footing made of 3000-psi concrete. There are four No. 7 Grade-60 dowels in the connection. The dowels extend into the footing a distance equal to the compression development length of a No. 7 bottom bar in 3000-psi concrete (19 in.) and into the column a distance equal to the greater of

1. the compression lap-splice length for a No. 7 bar in 5000-psi concrete (26 in.), and
2. the compression development length of a No. 8 bar in 5000-psi concrete (18 in.).

This joint could fail by reaching various limit states, including

1. crushing of the concrete at the bottom of the column, where the column bars are no longer effective,
2. crushing in the footing below the column,
3. bond failure of the dowels in the footing, and
4. failure in the column of the lap splice between the dowels and the column bars.

Bearing Strength

The total capacity of the column for pure axial load is 968 kips, of which 212 kips is carried by the steel and the rest by the concrete, as shown in Fig. 16-9b. At the joint, the area of the dowels is less than that of the column bars, and the force transmitted by the dowels is $\phi A_{sd} f_y$, where A_{sd} is the area of the dowels and ϕ is that for tied columns. As a result, the load carried by the concrete has increased. In Fig. 16-9, the dowels are hooked so that they can be supported on and tied to the mat of footing reinforcement when the footing concrete is placed. The hooks cannot be used to develop compressive force in the bars (ACI Section 12.5.5).

The maximum bearing load on the concrete is defined in ACI Section 10.17 as $\phi(0.85 f'_c A_1)$. If the load combinations from ACI Section 9.2 are used, ACI Section 9.3.2.4 gives $\phi = 0.65$ for bearing and A_1 is the area of the contact surface. When the supporting surface is wider on all sides than the loaded area, the maximum bearing load may be taken as

$$\phi(0.85 f'_c A_1)\sqrt{\frac{A_2}{A_1}} \tag{16-15}$$

but not more than $\phi(1.7 f'_c A_1)$, where A_2 is the area of the lower base of a right pyramid or cone formed by extending lines out from the sides of the bearing area at a slope of 2 horizontal to 1 vertical to the point where the first such line intersects an edge. This is illustrated in Fig. 16-10. The first intersection with an edge occurs at point B, resulting in the area A_2 shown crosshatched in Fig. 16-10b.

Two distinct cases must be considered: (1) joints that do not transmit computed moments to the footing and (2) joints that do. These will be discussed separately.

No Moment Transferred to Footing

If no moments are transmitted, or if the eccentricity falls within the kern of the column, there will be compression over the full section. The total force transferred by bearing is then calculated as $(A_g - A_{sd})$ times the smaller of the bearing stresses allowed on the column or the footing, where A_g is total area of the column and A_{sd} is the area of the bars or dowels crossing the joint. Any additional load must be transferred by dowels.

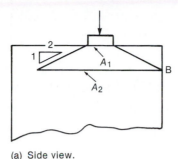

(a) Side view.

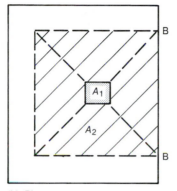

Fig. 16-10
Definition of A_1 and A_2.

(b) Plan.

Moments Are Transferred to Footing

If moments are transmitted to the footing, bearing stresses will exist over part, but not all, of the column cross section. The number of dowels required can be obtained by considering the area of the joint as an eccentrically loaded column with a maximum concrete stress equal to the smaller of the bearing stresses allowed on the column or the footing. Sufficient reinforcement must cross the interface to provide the necessary axial load and moment capacity. Generally, this requires that all the column reinforcement must cross the interface. This steel must be spliced in accordance with the requirements for column splices.

Practical Aspects

Three other aspects warrant discussion prior to the examples. The minimum cover to the reinforcement in footings cast against the soil is 3 in. (ACI Section 7.7.1). This allows for small irregularities in the surface of the excavation and for potential contamination of the bottom layer of concrete with soil. Sometimes, the bottom of the excavation for the footing is covered with a lean concrete seal coat, to prevent the bottom from becoming uneven after rainstorms and to give a level surface for placing reinforcement.

The minimum depth of the footing above the bottom reinforcement is 6 in. for footings on soil and 12 in. for footings on piles (ACI Section 15.7). ACI Section 10.6.4, covering the distribution of flexural reinforcement in beams and one-way slabs, does not apply to footings.

16-4   STRIP OR WALL FOOTINGS

A wall footing cantilevers out on both sides of the wall as shown in Figs. 16-1a and 16-11. The soil pressure causes the cantilevers to bend upward, and as a result, reinforcement is required at the bottom of the footing, as shown in Fig. 16-11. The critical sections for design

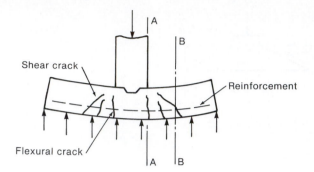

Fig. 16-11
Structural action of a strip
footing.

for flexure and anchorage are at the face of the wall (section A–A in Fig. 16-11). One-way shear is critical at a section a distance d from the face of the wall (section B–B in Fig. 16-11). The presence of the wall prevents two-way shear. Thicknesses of wall footings are chosen in 1-in. increments, widths in 2- or 3-in. increments.

EXAMPLE 16-1 Design of a Wall Footing

A 12-in.-thick concrete wall carries a service (unfactored) dead load of 10 kips per foot and a service live load of 12.5 kips per foot. The allowable soil pressure, q_a, is 5000 psf at the level of the base of the footing, which is 5 ft below the final ground surface. Design a wall footing, using $f'_c = 3000$ psi and $f_y = 60,000$ psi. The density of the soil is 120 lb/ft^3. Frequently, the strength of the concrete in the footing is lower than that in the column. Dowels are used to accommodate this change in strength. Most strip footings on soil have one mat of reinforcement.

1. **Estimate the size of the footing and the factored net pressure.** Consider a 1-ft strip of footing and wall. Allowable soil pressure is 5 ksf; allowable net soil pressure is 5 ksf − weight/ft^2 of the footing and of the soil over the footing. Since the thickness of the footing is not known at this stage, it is necessary to guess a thickness for a first trial. Generally, the thickness will be 1 to 1.5 times the wall thickness. We shall try a 12-in.-thick footing. Therefore, $q_n = 5 - (1 \times 0.15 + 4 \times 0.12) = 4.37$ ksf, and we have:

$$\text{Area required} = \frac{10 \text{ kips} + 12.5 \text{ kips}}{4.37 \text{ ksf}}$$

$$= 5.15 \text{ ft}^2 \text{ per foot of length of wall}$$

Try a footing 5 ft 2 in. = 62 in. wide.

Using the load factors in ACI Section 9.2.1:

$$\text{Factored net pressure, } q_{nu} = \frac{1.2 \times 10 + 1.6 \times 12.5}{5.167} = 6.19 \text{ ksf}$$

In the design of the concrete and reinforcement, we shall use $q_{nu} = 6.19$ ksf.

2. **Check the shear.** Shear usually governs the thickness of footings. Only one-way shear is significant in a wall footing. Check it at d away from the face of the wall (section B–B in Fig. 16-11)

$$d = 12 \text{ in.} - 3 \text{ in cover} - \tfrac{1}{2} \text{ bar diameter} \simeq 8.5 \text{ in.}$$

The tributary area for shear is shown crosshatched in Fig. 16-12a.

$$V_u = 6.19\left(\frac{16.5}{12} \times 1\right) \text{ft}^2 = 8.51 \text{ kips/ft}$$

$$\phi V_c = \phi \times 2\sqrt{f'_c}\,b_w d = 0.75 \times 2 \times \sqrt{3000} \times 12 \times 8.5/1000$$

$$= 8.38 \text{ kips/ft}$$

where, from ACI Section 9.3.2.3, $\phi = 0.75$ for shear design when the load factors from ACI Section 9.2.1 are used.

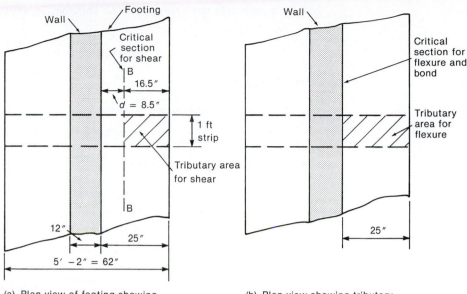

(a) Plan view of footing showing
     tributary area for shear.

(b) Plan view showing tributary
     area for moment.

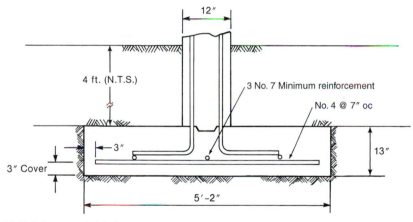

Fig. 16-12
Strip footing—Example 16-1.   (c) Reinforcement details.

Since $V_u > \phi V_c$, the footing depth is too small. If V_u is larger or considerably smaller than ϕV_c, choose a new thickness and repeat steps 1 and 2. **Try a 13-in.-thick footing 5 ft 2 in. wide.** A 13-in.-thick footing has $d = 9.50$ in. and $\phi V_c = 9.37$ kips/ft. Because ϕV_c exceeds $V_u = 8.51$ kips/ft, **a 13-in.-thick footing 5 ft 2 in. wide is adequate for shear.** Recompute d:

$$d = 13 \text{ in.} - 3 \text{ in. cover} - \frac{1}{2} \text{ bar diameter} \simeq 9.5 \text{ in.}$$

3.   Design the reinforcement. The critical section for moment is at the face of the wall (section A–A in Fig. 16-11). The tributary area for moment is shown crosshatched in Fig. 16-12b.

The required moment is

$$M_u = 6.19 \times \frac{(25/12)^2}{2} \times 1 \text{ ft-kips/ft} = 13.4 \text{ ft-kips/ft of length}$$

$$M_u = \phi M_n = \phi A_s f_y j d$$

Footings are generally very lightly reinforced. Therefore, assume that $j = 0.925$. Therefore,

$$A_s = \frac{13.4 \times 12,000}{0.9 \times 60,000(0.925 \times 9.5)} = 0.339 \text{ in.}^2/\text{ft}$$

From ACI Secs. 10.5.4 and 7.12.2

$$\text{Minimum } A_s = 0.0018bh$$
$$= 0.0018 \times 12 \times 13 = 0.281 \text{ in}^2/\text{ft}$$

Maximum spacing of bars $= 2h$, or 18 in.
We could use:

No. 5 bars at 11 in. o.c., $A_s = 0.34 \text{ in.}^2/\text{ft}$
No. 4 bars at 7 in. o.c., $A_s = 0.34 \text{ in.}^2/\text{ft}$
Try No. 4 bars at 7 in. o.c., $A_s = 0.34 \text{ in.}^2/\text{ft}$

Because the calculation of A_s was based on assumptions, recompute the moment capacity:

$$a = \frac{0.34 \times 60,000}{0.85 \times 3000 \times 12} = 0.667 \text{ in.}$$

Since $a/d = 0.667/9.5 = 0.070$ is much less than the $a/d = 0.319$ for the tension-controlled limit from Table A-4, the section is tension-controlled, and $\phi = 0.9$.

$$\phi M_n = \frac{0.9 \times 0.34 \times 60,000(9.5 - 0.667/2)}{12,000} = 14.0 \text{ ft-kips/ft}$$

Because this exceeds $M_n = 13.4$ ft-kips/ft, the moment capacity is OK.
For completeness, we could compute ε_t directly and use it to check ϕ. From similar triangles

$$\frac{c}{0.003} = \frac{d_t - c}{\varepsilon_t}$$

and

$$\varepsilon_t = \frac{(d_t - c)}{c} \times 0.003$$

where $d_t = 9.5$ in. and $c = \alpha/\beta_1 = 0.667/0.85 = 0.785$ in.
Calculations show $e_t = 0.333$.
This far exceeds ε_t limit $= 0.005$. Therefore, $f_s = f_y$ and $\phi = 0.90$.

4. **Check the development.** The clear spacing of the bars being developed exceeds $2d_b$ and the clear cover exceeds d_b. Therefore, this is case 2 development in Tables 8–1 and A–11. From Table A–11, ℓ_d for a No. 4 bottom bar in 3000-psi concrete is 21.9 in.

The distance from the point of maximum bar stress (at the face of the wall) to the end of the bar is 25 in. $-$ 3 in. cover on the ends of the bars $= 22$ in. This is more than $\ell_d = 21.9$. **Use No. 4 bars at 7 in. on centers.** This is still case 2 development, and 22 in. just satisfies ℓ_d.

5. **Select the minimum (temperature) reinforcement.** By ACI Section 7.12.2 we require that

$$A_s = 0.0018bh = 0.0018 \times 62 \times 13 \text{ in.}$$
$$= 1.45 \text{ in.}^2$$

The maximum spacing is $5 \times 12 = 60$ in. or 18 in. **Provide three No. 7 bars for shrinkage reinforcement.**

6.   Design the connection between the wall and the footing. ACI Section 15.8.2.2 requires that reinforcement equivalent to the minimum vertical wall reinforcement extend from the wall into the footing. A cross section through the wall footing designed in this example is shown in Fig. 16-12c.

The section through the strip footing shows a shear key in the top surface of the footing. This is formed by a two-by-four pushed down into the surface. The shear key is intended to resist some shear to prevent the wall from being dislocated laterally. Sometimes the shear key is omitted, and shear friction is used to resist dislocation of the wall during construction.   ∎

16-5   SPREAD FOOTINGS

Spread footings are square or rectangular pads that spread a column load over an area of soil large enough to support the column load. The soil pressure causes the footing to deflect upward, as shown in Fig. 16-7a, causing tension in two directions at the bottom. As a result, reinforcement is placed in two directions at the bottom, as shown in Fig. 16-7c. Two examples will be presented: a square axially loaded footing and a rectangular axially loaded footing.

EXAMPLE 16-2   Design of a Square Spread Footing

A square spread footing supports an 18-in.-square column supporting a service dead load of 400 kips and a service live load of 270 kips. The column is built of 5000-psi concrete and has eight No. 9 longitudinal bars with $f_y = 60,000$ psi. Design a spread footing to be constructed by using 3000-psi concrete and Grade-60 bars. The top of the footing will be covered with 6 in. of fill with a density of 120 lb/ft^3 and a 6-in. basement floor (Fig. 16-13). The basement floor loading is 100 psf. The allowable bearing pressure on the soil is 6000 psf. Use load and resistance factors from ACI Sections 9.2 and 9.3.

1.   Compute the factored loads and the resistance factors, ϕ. From ACI Section 9.2, the applicable load combinations are as follows:

$$U = 1.4(D + F) \qquad \text{(ACI Eq. 9-1)}$$

$$U = 1.2(D + F + T) + 1.6(L + H) + 0.5(L_r \text{ or } S \text{ or } R) \qquad \text{(ACI Eq. 9-2)}$$

$$U = 1.2D + 1.6(L_r \text{ or } S \text{ or } R) + (1.0L \text{ or } 0.8W) \qquad \text{(ACI Eq. 9-3)}$$

$$U = 1.2D + 1.6W + 1.0L + 0.5(L_r \text{ or } S \text{ or } R) \qquad \text{(ACI Eq. 9-4)}$$

$$U = 1.2D + 1.6W + 1.6H \qquad \text{(ACI Eq. 9-6)}$$

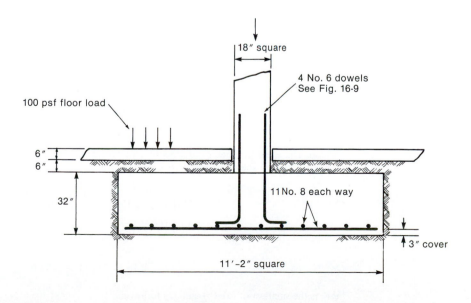

Fig. 16-13
Spread footing—
Example 16-2.

Because the statement of the problem mentioned only dead and live loads, we will assume these as the only applicable loads. In effect, we are assuming that the wind loads and roof loads are small compared to the dead loads. This reduces the set of load combinations to the following:

$$U = 1.4(D)$$

$$U = 1.2(D) + 1.6(L)$$

From these equations, the design loads are

$$U = 1.4 \times 400 = 560 \text{ kips}$$

and

$$U = 1.2 \times 400 + 1.6 \times 270 = 912 \text{ kips}$$

Strength-reduction factors ϕ are given in ACI Section 9.3. These were selected on the basis of whether flexure or shear is being considered. For shear, ACI Section 9.3.2.3 gives $\phi = 0.75$. For flexure, ϕ will be a function of the strain in the extreme-tension layer of bars, but ϕ will probably be 0.9 for footings·

2. Estimate the footing size and the factored net soil pressure. Allowable net soil pressure $q_n = 6$ ksf $-$ (weight/ft^2 of the footing and the soil and floor over the footing and the floor loading). Estimate the overall thickness of the footing at between one and two times the width of the column, say, 27 in.:

$$q_n = 6.0 - \left(\frac{27}{12} \times 0.15 + 0.5 \times 0.12 + 0.5 \times 0.15 + 0.100 \right)$$

$$= 5.43 \text{ ksf}$$

$$\text{Area required} = \frac{400 \text{ kips} + 270 \text{ kips}}{5.43 \text{ ksf}} = 123 \text{ ft}^2$$

$$\approx 11.1 \text{ ft square.}$$

Try a footing 11 ft 2 in. square by 27 in. thick:

$$\text{Factored net soil pressure} = \frac{1.2 \times 400 + 1.6 \times 270}{11.17^2}$$

$$= 7.31 \text{ ksf}$$

3. Check the thickness for two-way shear. Generally, the thickness of a spread footing is governed by two-way shear. The shear will be checked on the critical perimeter at $d/2$ from the face of the column and, if necessary, the thickness will be increased or decreased. Since there is reinforcement in both directions, the average d will be used:

$$\text{Average } d = 27 \text{ in.} - (3 \text{ in. cover}) - (1 \text{ bar diameter})$$

$$= 23 \text{ in.}$$

The critical shear perimeter (ACI Section 11.12.1.2) is shown dashed in Fig. 16-14a. The tributary area for two-way shear is shown crosshatched. We have

$$V_u = 7.31 \text{ ksf} \left[11.17^2 - \left(\frac{41}{12} \right)^2 \right] \text{ ft}^2 = 827 \text{ kips}$$

Length of critical shear perimeter:

$$b_o = 4(18 + 23) \text{ in.} = 164 \text{ in.}$$

ϕV_c is the smallest of (a), (b), and (c) following:

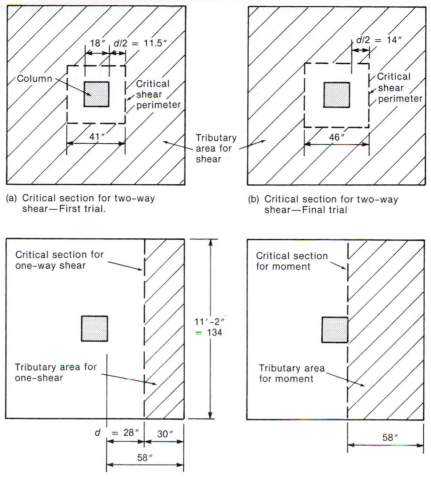

Fig. 16-14
Critical sections—Example
16-2.

(a) Critical section for two-way
shear—First trial.

(b) Critical section for two-way
shear—Final trial

(c) Critical section for one-way shear

(d) Critical section for moment

(a) $\phi V_c = \phi\left(2 + \dfrac{4}{\beta_c}\right)\sqrt{f'_c}\,b_o d$ (13-15)
(ACI Eq. 11-33)

$$\beta_c = \frac{\text{long side of column}}{\text{short side of column}} = 1.0$$

$$\phi V_c = \frac{0.75(2 + 4/1)\sqrt{3000}\times 164 \times 23}{1000} = 930 \text{ kips}$$

(b) $\phi V_c = \phi\left(\dfrac{\alpha_s d}{b_o} + 2\right)\sqrt{f'_c}\,b_o d$ (13-16)
(ACI Eq. 11-34)

where $\alpha_s = 40$ for interior columns, 30 for edge columns, and 20 for corner columns. Thus,

$$\phi V_c = 0.75\left(\frac{40 \times 23}{164} + 2\right)\frac{\sqrt{3000}\times 164 \times 23}{1000} = 1179 \text{ kips}$$

(c) $\phi V_c = \phi(4)\sqrt{f'_c}\,b_o d = 620 \text{ kips}$ (13-17)
(ACI Eq. 11-35)

Because $\phi V_c = 620$ kips is less than $V_u = 827$ kips, the footing is not thick enough. Try $h = 32$ in., $d = 28$ in., and $b_o = 184$ in. The footing is thicker, and it weighs more. Hence, a larger area may be required:

$$q_n = 6.0 - \left(\frac{32}{12} \times 0.15 + 0.5 \times 0.12 + 0.5 \times 0.15 + 0.100 \right)$$

$$= 5.37 \text{ ksf}$$

$$\text{Area required} = \frac{400 + 270}{5.37}$$

$$= 124.8 \text{ ft}^2$$

$$= 11.17 \text{ ft square}$$

Try an 11-ft-2-in.-square footing, 32 in. thick:

$$\text{Factored net soil pressure, } q_{nu} = \frac{1.2 \times 400 + 1.6 \times 270}{11.17^2}$$

$$= 7.31 \text{ ksf}$$

$$\text{Average } d = 32 - 3 - 1$$

$$= 28 \text{ in.}$$

The new critical shear perimeter and tributary area for shear are shown in Fig. 16-14b. We have

$$V_u = 7.31 \left[11.17^2 - \left(\frac{46}{12} \right)^2 \right] = 804 \text{ kips}$$

ACI Eq. (11-35) governs again:

$$\phi V_c = 0.75 \times 4 \sqrt{3000} \times (184) \times \frac{28}{1000} = 846 \text{ kips}$$

This is adequate. A check using $h = 30$ in. shows that a 30-in.-thick footing is not adequate. **Use an 11-ft-2-in.-square footing, 32 in. thick.**

    4. Check the one-way shear. Although one-way shear is seldom critical, we shall check it. The critical section for one-way shear is located at d away from the face of the column (ACI Section 11.12.1.1), as shown in Fig. 16-14c. Thus,

$$V_u = 7.31 \text{ ksf} \left(11.17 \text{ ft} \times \frac{30}{12} \text{ft} \right) = 204 \text{ kips}$$

$$\phi V_c = \phi 2 \sqrt{f_c'} b_w d = 0.75 \times 2 \sqrt{3000} \times 134 \times \frac{28}{1000}$$

$$= 308 \text{ kips}$$

Therefore, **OK in one-way shear.**

    5. Design the flexural reinforcement. The critical section for moment and anchorage of the reinforcement is shown in Fig. 16-14d. The ultimate moment is

$$M_u = 7.31 \left[11.17 \times \frac{(58/12)^2}{2} \right] = 954 \text{ ft-kips}$$

Assuming that $j = 0.9$ and $\phi = 0.90$, the area of steel required is

$$A_s = \frac{954 \times 12,000}{0.9 \times 60,000(0.9 \times 28)} = 8.41 \text{ in.}^2$$

The average value of d was used in this calculation for simplicity. The same reinforcement will be used in both directions:

$$\text{Minimum } A_s \text{ (ACI Sections 10.5.3 and 7.12.2)} = 0.0018bh$$

$$= 0.0018 \times 134 \times 32 = 7.72 \text{ in.}^2 \text{ (does not govern)}$$

Maximum spacing (ACI Section 7.6.5) $= 18$ in.

Try eleven No. 8 bars each way, $A_s = 8.69$ in.^{2}. Recompute ϕM_n as a check.

$$a = \frac{8.69 \times 60,000}{0.85 \times 3000 \times 134} = 1.53 \text{ in.}$$

because $a / d = 1.53$ in./ 28 in. $= 0.055$ is much less than a / d for the tension-controlled limit from Table A-4, the beam is tension-controlled, and $\phi = 0.90$.

$$\phi M_n = 1070 \text{ ft} - \text{kips}$$

This exceeds $M_u = 954$ ft-kip.

   6.   **Check the development.** The clear spacing of the bars being developed exceeds $2d_b$ and the clear cover exceeds d_b. Therefore, this is case 2 development in Tables 8-1 and A-11. From Table A-11, ℓ_d for a No. 8 bottom bar in 3000-psi concrete is 54.8 d_b. The development length is

$$\ell_d = 54.8 \ d_b \beta \lambda$$

where $\beta = 1.0$ for uncoated reinforcement and $\lambda = 1.0$ for normal-weight concrete. Accordingly, we have

$$\ell_d = 54.8 \times 1.00 \times 1.0 \times 1.0$$
$$= 54.8 \text{ in.}$$

The bar extension past the point of maximum moment is $(58 \text{ in.} - 3 \text{ in.}) = 55$ in. This is OK. **Use eleven No. 8 bars each way; $A_s = 8.69$ in.^{2}.**

   7.   **Design the column–footing joint.** The column–footing joint is shown in Fig. 16-13. The factored load at the base of the column is

$$1.2 \times 400 + 1.6 \times 270 = 912 \text{ kips}$$

The maximum bearing load on the bottom of the column (ACI Section 10.17.1) is $\phi 0.85 f'_c A_1$, where A_1 is the area of the contact surface between the column and the footing and f'_c is for the column. When the the contact supporting surface on the footing is wider on all sides than the loaded area, the maximum bearing load on the top of the footing may be taken as

$$0.85 \phi f'_c A_1 \sqrt{\frac{A_2}{A_1}}, \text{ but not more than } 1.7 \phi f'_c A_1 \qquad (16\text{-}15)$$

where A_2 is the area of the lower base of a right pyramid or cone as defined in Fig. 16-10. ACI Section 9.3.2.4 defines ϕ equal to 0.65 for bearing. By inspection, $\sqrt{A_2/A_1}$ for the footing exceeds 2; hence, the maximum bearing load *on the footing* is $0.85 \times 0.65 \times 3 \times 18^2 \times 2 = 1074$ kips. The allowable bearing on the base of the column is

$$\phi(0.85 f'_c A_1) = 0.65 \times 0.85 \times 5 \times 18^2$$
$$= 895 \text{ kips}$$

Thus, the maximum load that can be transferred by bearing is 895 kips, and dowels are needed to transfer the excess load. Accordingly, we have

$$\text{Area of dowels required} = \frac{912 - 895}{\phi f_y} = 0.44 \text{ in.}^2$$

where $\phi = 0.65$ has been used. This is the ϕ value from ACI Sections 9.3.2.2(b) for compression-controlled tied columns and 9.3.2.4 for bearing. The area of dowels must also satisfy ACI Section 15.8.2.1:

$$\text{Area of dowels} \geq 0.005 A_g = 1.62 \text{ in.}^2$$

Try four No. 6 dowels ($A_s = 1.76$ in^2); dowel each corner bar. The dowels must extend into the footing by the compression-development length for a No. 6 bar in 3000-psi concrete, or 16 in. The bars will be extended down to the level of the main footing steel and hooked 90°. The hooks will be tied to the main steel to hold the dowels in place. The dowels must extend into the column a distance equal to the greater of a compression splice for the dowels (23 in.) or the compression-development length of the column bars (25 in.). **Use four No. 6 dowels; dowel each corner bar. Extend dowels 25 in. into column.** (See Fig. 16-13.) ∎

Rectangular Footings

Rectangular footings may be used when there is inadequate clearance for a square footing. In such a footing, the reinforcement in the short direction is placed in the three bands shown in Fig. 16-15, with a closer bar spacing in the band under the column than in the two end bands. The band under the column has a width equal to the length of the short side of the footing, but not less than the width of the column (if that is greater) and is centered on the column. The reinforcement in the band shall be $2/(\beta + 1)$ times the total reinforcement in the short direction, where β is the ratio of the long side of the footing to the short side (ACI Section 15.4.4). The reinforcement within each band is distributed evenly, as is the reinforcement in the long direction.

EXAMPLE 16-3   Design of a Rectangular Spread Footing

Redesign the footing from Example 16-2, assuming that the maximum width of the footing is limited to 9 ft. Steps 1 and 2 of Example 16-3 would proceed in the same sequence as in Example 16-2, leading to a footing 9 ft wide by 13 ft 8 in. long by 32 in. thick. Using the load factors from ACI 318-02 Section 9.2, the factored net soil pressure is 7.31 ksf.

1.   Check the one-way shear. One-way shear may be critical in a rectangular footing and must be checked. The critical section and tributary area for one-way shear are shown in Fig. 16-16a. We have

$$V_u = 7.31\left(\frac{45}{12} \times 9\right) = 247 \text{ kips}$$

$$\phi V_c = 0.75 \times 2\sqrt{3000} \times 108 \times \frac{28}{1000} = 248 \text{ kips}$$

This is just OK in one-way shear.

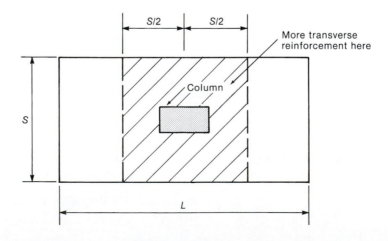

Fig. 16-15
Rectangular footing.

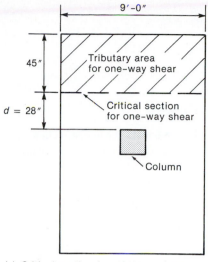

(a) Critical section for one–way shear.

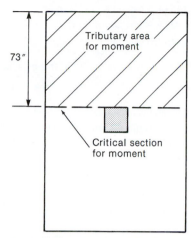

(b) Critical section for moment—
Long direction.

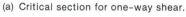

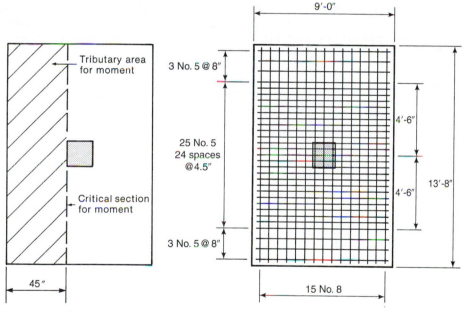

(c) Critical section for moment—
Short direction.

(d) Bar placement.

Fig. 16-16
Rectangular footing—Example 16-3.

2.   Design the reinforcement in the long direction. The critical section for moment and reinforcement anchorage is shown in Fig. 16-16b. The ultimate moment is

$$M_u = 7.31\left(9 \times \frac{(73/12)^2}{2}\right) = 1217 \text{ ft-kips}$$

Assuming that $\phi = 0.9$ and $j = 0.9$, the area of steel required is

$$A_s = \frac{1217 \times 12,000}{0.9 \times 60,000(0.9 \times 28)} = 10.73 \text{ in.}^2$$

$$A_{s(\min)} = 0.0018 \times 108 \times 32 = 6.22 \text{ in.}^2 \text{ (does not govern)}$$

We could use

$$14 \text{ No. 8 bars}, A_s = 11.1 \text{ in.}^2$$
$$11 \text{ No. 9 bars}, A_s = 11.0 \text{ in.}^2$$
$$9 \text{ No. 10 bars}, A_s = 11.4 \text{ in.}^2$$

Try 14 No. 8 bars—$\phi M_n = 1330$ ft-kips. Check development:

$$\ell_d = 54.8 \times 1.0 \times 1.0 \times 1.0 = 54.8 \text{ in.}$$

The length available is 70 in.—therefore, OK. **Use 14 No. 8 bars in the long direction.**

3. Design the reinforcement in the short direction. The critical section for moment and reinforcement anchorage is shown in Fig. 16-16c. We have

$$M_u = 7.31\left(13.67 \times \frac{(45/12)^2}{2} \right) = 702 \text{ ft-kips}$$

Assuming that $j = 0.9$, $A_s = 6.19$ in.^{2},

$$A_{s(\min)} = 0.0018 \times 164 \times 32 = 9.45 \text{ in.}^2 \text{ (This governs.)}$$

Try 12 No. 8 bars: $A_s = 9.48$ in.^{2}.

Check development. $\ell_d = 54.8$ in. and the length available is 28.5 in.—therefore, not OK. We must consider smaller bars. Try 31 No. 5 bars, $A_s = 9.61$ in.^{2}, $\ell_d = 27.4$ in.—therefore, OK. Use 31 No. 5 bars in the short direction of the footing. For the arrangement of the bars in the transverse direction (ACI Section 15.4.4.2),

$$\beta = \frac{\text{long side}}{\text{short side}} = 1.519$$

In the middle strip of width 9 ft, provide

$$\frac{2}{1.519 + 1} \times 31 \text{ bars} = 24.6 \text{ bars}$$

Provide 25 No. 5 bars in the middle strip, and provide three No. 5 bars in each end strip. The final design is shown in Fig. 16-16d. ∎

Footings Transferring Vertical Load and Moment

On rare occasions, footings must transmit both axial load and moment to the soil. The design of such a footing proceeds in the same manner as that for a square or rectangular footing, except for three things. First, a deeper slab will be necessary, since there will be shear stresses developed both by direct shear and by moment. The calculations concerning this are discussed in Chapter 13. Second, the soil pressures will not be uniform, as discussed in Section 16-2 and shown in Fig. 16-5b. Third, the design for two-way shear must consider moment and shear. (See Section 13-8.)

The uneven soil pressures will lead to a tilting settlement of the footing, which will relieve some of the moment if the moment results from compatibility at the fixed end. The tilting can be reduced by offsetting the footing so that the column load acts through the center of the footing base area. If the moment is necessary for equilibrium, it should not be relieved by rotation of the foundation.

16-6 COMBINED FOOTINGS

Combined footings are used when it is necessary to support two columns on one footing, as shown in Fig. 16-1 or 16-17. When an exterior column is so close to a property line that a spread footing cannot be used, a combined footing is often used to support the edge column and an interior column. Reference [16-10] discusses the design of combined footings.

The shape of the footing is chosen such that the centroid of the area in contact with soil coincides with the resultant of the column loads supported by the footing. Common shapes are shown in Fig. 16-17. For the rectangular footing in Fig. 16-17a, the distance from the exterior end to the resultant of the loads is half the length of the footing. If the interior column load is much larger than the exterior column load, a tapered footing may be used (Fig. 16-17b). The location of the centroid can be adjusted to agree with the resultant of the loads by dividing the area into a parallelogram or rectangle of area A_1, plus a triangle of area A_2, such that $A_1 + A_2 =$ required area and $\bar{y}_1 A_1 + \bar{y}_2 A_2 = y_R A$.

Sometimes, a combined footing will be designed as two isolated pads joined by a *strap* or stiff beam, as shown in Fig. 16-17c. Here, the exterior footing acts as a wall footing, cantilevering out on the two sides of the strap. The interior footing can be designed as a two-way footing. The strap is designed as a beam and may require shear reinforcement in it.

For design, the structural action of a combined footing is idealized as shown in Fig. 16-18a. The soil pressure is assumed to act on longitudinal beam strips, A–B–C in

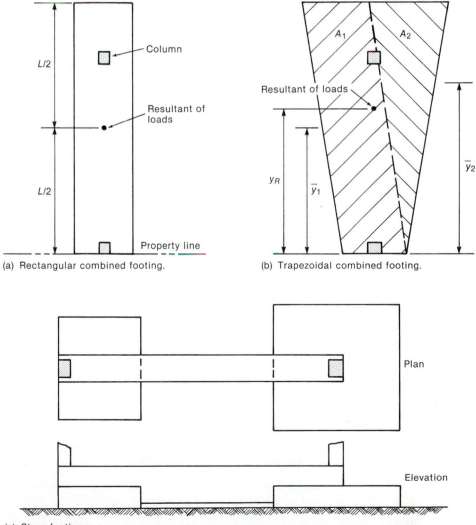

(a) Rectangular combined footing.   (b) Trapezoidal combined footing.

(c) Strap footing.

Fig. 16-17
Types of combined footings.

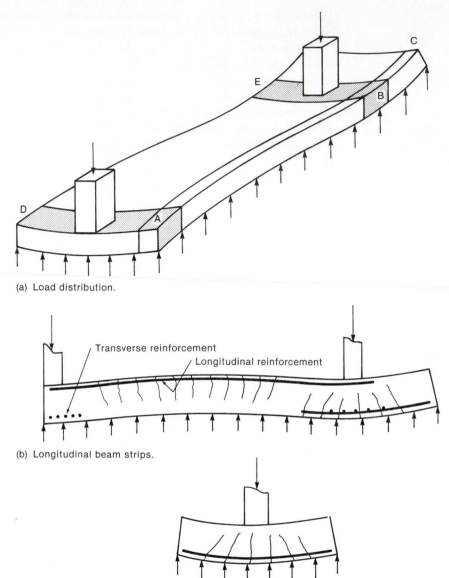

(a) Load distribution.

Transverse reinforcement

Longitudinal reinforcement

(b) Longitudinal beam strips.

(c) Transverse beam strips.

Fig. 16-18
Structural action of combined footing.

Fig. 16-18a. These transmit the load to hypothetical cross beams, *A–D* and *B–E*, which transmit the upward soil reactions to the columns. For the column placement shown, the longitudinal beam strips would deflect as shown in Fig. 16-18b, requiring the reinforcement shown. The deflected shape and reinforcement of the cross beams are shown in Fig. 16-18c. The cross beams are generally assumed to extend a distance $d/2$ on each side of the columns.

EXAMPLE 16-4 Design of a Combined Footing

A combined footing supports a 24 in. × 16 in. exterior column carrying a service dead load of 200 kips and a service live load of 150 kips, plus a 24-in.-square interior column carrying service loads of 300 kips dead load and 225 kips live load. The distance between the columns is 20 ft, center to center. The allowable soil bearing pressure is 5000 psf at a depth of 4 ft below the finished basement floor. The basement floor is 5 in. thick and supports a live load of 100 psf. The density of the fill above the footing is 120 lb/ft^3. Design the footing, assuming that $f'_c = 3000$ psi and $f_y = 60,000$ psi.

Use the load and resistance factors from ACI Appendix C, Sections C.2 and C.3.

1.   **Estimate the size and the factored net pressure.** Allowable net soil pressure $q_n = 5$ ksf $-$ (weight/ft^2 of the footing and the soil, the floor over the footing, and the basement floor loading). This can be calculated, as in the previous examples, by using the actual densities and thicknesses of the soil and concrete, or it can be approximated by using an average density for the soil and concrete. This will be taken as 140 lb/ft^3. Therefore,

$$q_n = 5.0 - (4 \times 0.14 + 0.10) = 4.34 \text{ ksf}$$

$$\text{Area required} = \frac{200 + 150 + 300 + 225}{4.34} = 202 \text{ ft}^2$$

The resultant of the column loads is located at

$$\frac{8 \text{ in.} \times 350 \text{ kips} + 248 \text{ in.} \times 525 \text{ kips}}{350 \text{ kips} + 525 \text{ kips}} = 152 \text{ in.}$$

from the exterior face of the exterior column (Fig. 16-19a). To achieve uniform soil pressures, the centroid of the footing area will be located at 152 in. from exterior edge. Thus, the footing will be 304 in. $= 25$ ft 4 in. long. The width of the footing will be 7.96 ft—say, 8 ft.
    The factored net pressure is

$$q_{nu} = \frac{1.4(200 + 300) + 1.7(150 + 225)}{25.33 \times 8} = 6.6 \text{ ksf}$$

As expected, q_{nu} at factored load level $= 6.6$ ksf exceeds $q_a = 4.34$ ksf at unfactored, service-load levels. Subsequent design will be based on q_{nu}.

2.   **Calculate the bending-moment and shear-force diagrams for the longitudinal action.** The factored loads on the footing and the corresponding bending-moment and shearing-force diagrams for the footing are shown in Fig. 16-19. These are plotted for the full 8-ft width of the footing.

3.   **Calculate the thickness required for maximum positive moment.** Because the cross section is so massive (8 ft wide by 2 to 3 ft deep), the use of more than about 0.5 percent reinforcement will lead to very large numbers of large bars. As a first trial, we shall select the depth, assuming 0.5 percent reinforcement. Since this member acts as a beam, the minimum flexural reinforcement ratio, from ACI Section 10.5.1, is $3\sqrt{f'_c}/f_y \geq 200/f_y = 0.0033$. From (4-16), try a width of 96 in. and $\rho = 0.005$:

$$\frac{M_u}{\phi k_n} = \frac{bd^2}{12,000}$$

where $k_n = f'_c \omega (1 - 0.59\omega)$, $\omega = \rho f_y / f'_c$, and M_u is in ft-kips. For $\rho = 0.005$ and $\phi k_n = 254$ (Table A-3),

$$\frac{bd^2}{12,000} = \frac{2354}{254} = 9.27$$

For $b = 96$ in., $d = 34$ in. **As a first trial, we shall assume an overall thickness of 36 in., with $d = 32.5$ in.**

4.   **Check the two-way shear at the interior column.** The critical perimeter is a square with sides $24 + 32.5 = 56.5$ in. long, giving $b_o = 4 \times 56.5 = 226$ in. The shear, V_u, is the column load minus the force due to soil pressure on the area within the critical perimeter:

$$V_u = 802.5 - 6.6 \left(\frac{56.5}{12}\right)^2 = 656 \text{ kips}$$

ϕV_c is the smallest of

(a) $\phi V_c = \phi \left(2 + \dfrac{4}{\beta_c} \right) \sqrt{f'_c} b_o d$ where $\beta_c = 1.0$ (ACI Eq. 11-33)

$$= 0.85 \left(2 + \frac{4}{1} \right) \sqrt{3000} \times 226 \times \frac{32.5}{1000} = 2050 \text{ kips}$$

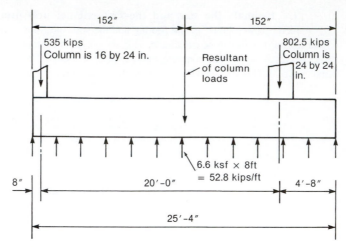

(a) Freebody diagram.

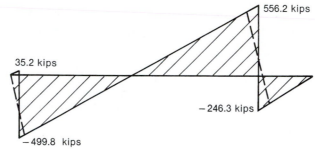

(b) Shear force diagram.

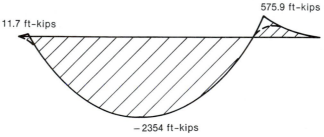

(c) Bending moment diagram.

Fig. 16-19
Bending-moment and
shearing-force diagrams—
Example 16-4.

(b) $\quad \phi V_c = \phi \left(\dfrac{\alpha_s d}{b_o} + 2 \right) \sqrt{f_c'} b_o d$ $\qquad$ (ACI Eq. 11-34)

$$= 0.85 \left(\dfrac{40 \times 32.5}{226} + 2 \right) \sqrt{3000} \times 226 \times \dfrac{32.5}{1000} = 2650 \text{ kips}$$

(c) $\quad \phi V_c = 4\sqrt{f_c'} b_o d$ $\qquad$ (ACI Eq. 11-35)

$$= 0.85 \times 4\sqrt{3000} \times 226 \times \dfrac{32.5}{1000} = 1370 \text{ kips}$$

Therefore, the depth of the footing is more than adequate for two-way shear at this column.

5. **Check the two-way shear at the exterior column.** The shear perimeter around the exterior column is three-sided, as shown in Fig. 16-20a. The distance from line A–B to the centroid of the shear perimeter is

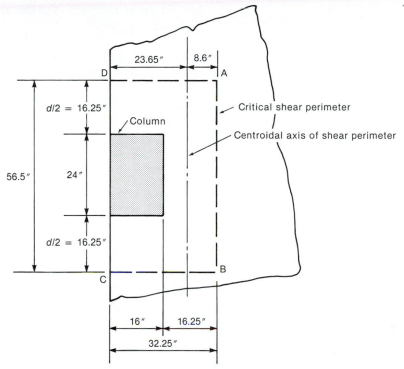

(a) Plan.

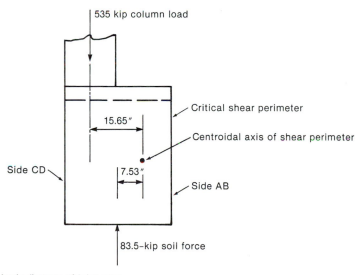

Fig. 16-20
Shear calculations at exterior
column—Example 16-4.

(b) Free-body diagram of joint area

$$\frac{2(32.5 \times 32.25) \times 32.25/2}{32.5(2 \times 32.25 + 56.5)} = 8.60 \text{ in.} \qquad (13\text{-}29)$$

The force due to soil pressure on the area within the critical perimeter is

$$6.6\left(32.25 \times \frac{56.5}{144}\right) = 83.5 \text{ kips}$$

A free-body diagram of the critical perimeter is shown in Fig. 16-20b. Summing moments about the centroid of the shear perimeter gives

$$M_u = 535 \text{ kips} \times 15.65 \text{ in.} - 83.5 \text{ kips} \times 7.53 \text{ in.}$$

$$= 7744 \text{ in.-kips}$$

This moment must be transferred to the footing by shear stresses and flexure, as explained in Chapter 13. The moment of inertia of the shear perimeter is

$$J_c = 2\left[\left(32.25 \times \frac{32.5^3}{12}\right) + \left(32.5 \times \frac{32.25^3}{12}\right)\right.$$

$$\left. + (32.25 \times 32.5)(16.125 - 8.60)^2\right] + (56.5 \times 32.5)8.60^2 \qquad (13\text{-}30)$$

$$= 620{,}700 \text{ in.}^4$$

The fraction of moment transferred by flexure is

$$\gamma_f = \frac{1}{1 + 2\sqrt{b_1/b_2}/3} \qquad (13\text{-}27)$$

$$\frac{1}{1 + 2\sqrt{32.25/56.5}/3} = 0.665$$

ACI Section 13.5.3.3 allows an adjustment in γ_f at edge columns in two-way slab structures. We shall not make this adjustment, since the structural action of this footing is quite different from that of a two-way slab.

The fraction transferred by shear is $\gamma_v = 1 - \gamma_f = 0.335$.

The shear stresses due to the direct shear and the shear due to moment transfer will add at points C and D in Fig. 16-20, giving the largest shear stresses on the critical shear perimeter:

$$v_u = \frac{V_u}{b_o d} + \frac{\gamma_v M_u c}{J_c}$$

$$= \frac{535 - 83.5}{(2 \times 32.25 + 56.5) \times 32.5} + \frac{0.335 \times 7744(32.25 - 8.60)}{620{,}700}$$

$$= 0.115 + 0.099$$

$$= 0.214 \text{ ksi}$$

ϕv_c is computed as $\phi V_c/b_o d$, where ϕV_c is the smallest value from ACI Eqs. 11-33 to 11-35:

(a) $\quad \phi v_c = 0.85\left(2 + \dfrac{4}{(24/16)}\right)\sqrt{3000} = 217 \text{ psi}$

(b) $\quad \phi v_c = 0.85\left(\dfrac{30 \times 32.5}{121} + 2\right)\sqrt{3000} = 468 \text{ psi}$

(c) $\quad \phi v_c = 0.85 \times 4\sqrt{3000} = 186 \text{ psi}$

Thus, $\phi v_c = 186$ psi or 0.186 ksi, and the thickness is not adequate for shear at the exterior column. In this case, the direct shear stress and the maximum shear stress due to moment are approximately equal. Additional calculations show that an overall depth of 40 in., with $d = 36.5$ in., is required for punching shear at the exterior column. For the final choice, the total moment transferred to the footing is 8417 in.-kips, of which 5202 in.-kips are transferred by flexure. **Use a combined footing 25 ft 4 in. by 8 ft in plan, 3 ft 4 in. thick, with effective depth 36.5 in.**

6. One-way shear in the longitudinal direction. One-way shear is critical at d from the face of the interior column, $(d + 12$ in. from the center of the column).

$$V_u = 556 \text{ kips} - \left(\frac{12 + 36.5}{12}\right)ft \times 6.6 \text{ } ksf \times 8 \text{ } ft = 343 \text{ kips}$$

And $V_u/\phi = 343/0.85 = 403$ kips.

The combined footing is 96 in. wide and $d = 36.5$ in., so

$$V_c = 2\sqrt{f'_c}b_w d = 2\sqrt{3000} \times 96 \times 36.5 = 384 \text{ kips}$$

Thus, shear reinforcement is required, but probably the minimum area requirement will govern. The minimum A_v/s is:

$$\frac{A_v}{s} \geq \frac{0.75\sqrt{f'_c}}{f_y}b_w, \text{ but not less than } \frac{50}{f_y}b_w \text{ (second term governs)}$$

$$\frac{A_v}{s} \geq \frac{50 \times 96}{60,000} = 0.080 \text{ } in.^2/in.$$

Try No. 4, eight leg stirrups at 18 in. o.c., $A_v/s = 1.60/18 = 0.089$ in.^{2}/in.
Note: we should make a check of the A_v/s provided in the transverse direction. The spacing between legs in the transverse direction is 12 in, and taking an 18 in. strip, we get:

$$\frac{A_v}{s} = \frac{0.20}{18} = 0.0167 \text{ in.}^2/in. \geq \frac{50 \times 18}{12} = 0.015 \text{ in.}^2/in. \text{ (ok)}$$

Now, checking in the longitudinal direction for $\phi V_n = \phi(V_c + V_s)$, where:

$$V_s = \frac{A_v f_{yv} d}{s} = \frac{1.60 \times 60 \times 36.5}{18} = 195 \text{ kips}$$

Then,

$$\phi(V_c + V_s) = 0.85(384 + 195) = 492 \text{ kips} \geq V_u = 343 \text{ kips (ok)}$$

Use No. 4, eight leg stirrups at a spacing of 18 in. o.c.

7. Design the flexural reinforcement in the longitudinal direction

    (a) Midspan (negative moment):

$$A_s = \frac{M_u \times 12,000}{\phi f_y jd}$$

Estimate $j = 0.95$, because ρ will be very small.

$$A_s = \frac{2354 \times 12,000}{0.9 \times 60,000 \times 0.95 \times 36.5} = 15.09 \text{ in.}^2$$

$$\text{Min. } A_s = \frac{3\sqrt{f'_c}}{f_y}bd, \text{ but not less than } \frac{200}{f_y}bd \text{ (second term governs)}$$

$$= \frac{200 \times 96 \times 36.5}{60,000}$$

$$= 11.68 \text{ in.}^2 \text{ (Does not govern.)}$$

Try 19 No. 8 bars, $A_s = 15.0$ in.^{2}.

$$\phi M_n = \frac{0.9 \times 15.0 \times 60,000(36.5 - 3.68/2)}{12,000} = 2340 \text{ ft-kips}$$

Since this is less than 1 percent less than M_u, we shall accept it. **Use 19 No. 8 top bars at midspan.**

    (b) Interior column (positive moment). The positive moment at the interior support requires that $A_s = 3.7$ in.^{2}, which is less than $A_{s,min}$. **Use 15 No. 8 bottom bars; $A_s = 11.9$ in.^{2}** at the interior column.

    8. Check the development of the top bars. ℓ_d for a No. 8 top bar in 3000-psi concrete (Table A-11) is

$$\ell_d = 71.2 \times 1.00 \times 1.0 \times 1.0 = 71.2 \text{ in.} = 5.93 \text{ ft}$$

In a normally loaded beam (loaded on the top surface and supported on the bottom surface), it is necessary to check ACI Section 12.11.3 at the positive-moment points of inflection to determine whether the rate of change of moment, and thus of bar force, exceeds the bond strength of the bottom bars. In this footing, the loading, the supports, and the shape of the moment diagram are all inverted from those found in a beam carrying gravity loads. As a result, this check is made for the top steel.

We shall extend all the bars into the column regions at both ends. At points of inflection (1.09 ft from the center of the interior column and under the exterior column), $V_u = 499$ kips, and by ACI Section 12.11.3,

$$\frac{M_n}{V_u} + \ell_a \geq \ell_d \qquad\qquad (8\text{-}20)$$
$$\text{(ACI Eq. 12-3)}$$

$$\frac{M_n}{V_u} = \frac{2340/0.9}{499} = 5.21 \text{ ft}$$

where ℓ_a must be at least $5.93 - 5.21 = 0.72$ ft to satisfy (8-20).

At the exterior support, ℓ_a is the extension beyond the center of the support. This will not exceed 5 in., which is not enough to satisfy (8-20). Therefore, the top bars will all have to be hooked at the exterior end. This is also necessary to anchor the bars transferring the unbalanced moment from the column to the footing.

At the interior column, extend the bars to the interior face (away from exterior column) of the column. This will give ℓ_a in excess of 0.72 ft, therefore, OK.

9. **Check the development of the bottom bars.** ℓ_d for a No. 8 bottom bar in 3000-psi concrete is

$$\ell_d = 54.8 \times 1.00 \times 1.0 \times 1.0 = 54.8 \text{ in.}$$

Extend the bottom bars to 3 in. from the exterior end and d past the point of inflection. Cut off the bottom bars at 4 ft 5 in. from the centerline of the interior column toward the exterior column.

10. **Design the transverse "beams."** Transverse strips under each column will be assumed to transmit the load from the longitudinal beam strips into the column, as shown in Fig. 16-18a. The width of the beam strips will be assumed to extend $d/2$ on each side of the column. The actual width is unimportant, since the moments to be transferred are independent of the width of the transverse beams. Figure 16-21 shows a section through the transverse beam under the interior column. The factored load on this column is 803 kips. This is balanced by an upward net force of 803 kips/8 ft = 100 kips/ft. The maximum moment in this transverse beam is

$$M_u = \frac{100 \times 3^2}{2} = 450 \text{ ft-kips}$$

Assuming that $j = 0.95$ and $d = 35.5$ in., the required A_s is

$$A_s = \frac{450 \times 12,000}{0.9 \times 60,000 \times 0.95 \times 35.5} = 2.97 \text{ in.}^2$$

$$A_{s(min)} = \frac{200}{60,000} \times 60.5 \times 35.5 = 7.16 \text{ in.}^2 \text{ (This governs.)}$$

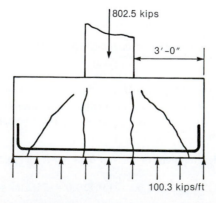

802.5 kips

3'-0"

100.3 kips/ft

Fig. 16-21
Transverse beam at interior column—Example 16-4.

Use nine No. 8 transverse bottom bars at the interior column; $A_s = 7.11$ in.^{2}. $A_{s(min)}$ also controls at the exterior column. **Use six No. 8 transverse bottom bars at the exterior column.** Because two-way shear cracks would extend roughly the entire width of the footing, we shall hook the transverse bars at both ends for adequate anchorage outside the inclined cracks.

  11. **Design the column-to-footing dowels.** This is similar to step 6 in Example 16-2 and so will not be repeated here. The complete design is detailed in Fig. 16-22. ■

16-7 MAT FOUNDATIONS

A mat foundation supports all the columns in a building, as shown in Fig. 16-1g. A mat foundation would be used when buildings are founded on soft or irregular soils in locations where pile foundations cannot be used. Design is carried out by assuming that the foundation acts as an inverted slab. The distribution of soil pressure is affected by the relative stiffness of the soil and foundation, with more pressure being developed under the columns than at points between columns. Detailed recommendations for the design of such foundations are given in [16-11].

16-8 PILE CAPS

Piles may be used when the surface soil layers are too soft to properly support the loads from the structure. The pile loads either are transmitted to a stiff bearing layer some distance below the surface or are transmitted to the soil by friction along the length of the pile. Treated timber piles have capacities of up to about 30 tons. Precast concrete and steel piles often have capacities ranging from 40 to 200 tons. The most common pile caps for high-strength piles comprise groups of 2 to 6 piles, although groups of 25 to 30 piles have been used. The center-to-center pile spacing is a function of the type and capacity of the piles, but frequently is on the order of 3 ft. It is not uncommon for a pile to be several inches to a foot away from its intended location, due to problems during pile driving. The design of pile caps is discussed in Reference [16-12].

  The structural action of a four-pile group is shown schematically in Fig. 16-23. The pile cap is a special case of a "deep beam" and can be idealized as a three-dimensional truss or strut-and-tie model, with four compression struts transferring load from the column to the tops of the piles and four tension ties equilibrating the outward components of the compression thrusts (Fig. 16-23b). The tension ties have constant force in them and must be anchored for the full horizontal tie force outside the intersection of the

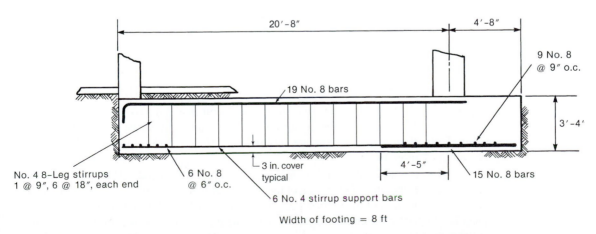

Fig. 16-22
Combined footing—Example 16-4.

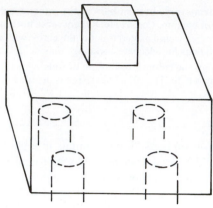

(a) Pile cap.

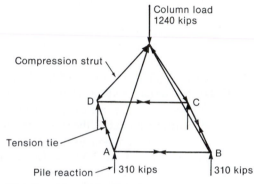

(b) Internal forces in pile cap.

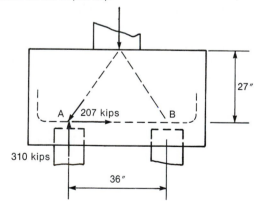

Fig. 16-23
Forces in a pile cap.

(c) Force in tie A–B.

pile and the compression strut (outside points *A* and *B* in Fig. 16-23c). Hence, the bars either must extend ℓ_d past the centerlines of the piles or must be hooked outside this point. Tests of model pile caps are reported in [16-13]. Strut-and-tie models are discussed in Chapter 18.

The modes of failure (limit states) for such a pile cap include (1) crushing under the column or over the pile, (2) bursting of the side cover where the pile transfers its load to the pile cap, (3) yielding of the tension tie, (4) anchorage failure of the tension tie, (5) two-way shear failure where the cone of material inside the piles punches downward, and (6) failure of the compression struts. Least understood of these is the two-way shear-failure mode. ACI Section 11.12 is not strictly applicable, since the failure cone differs from the one assumed in the code. Guidance in selecting allowable shear stresses is given in [16-12]. This reference is based on analyses rather than tests, however. It should be noted that [16-12] gives inadequate attention to the anchorage of the horizontal bars. ACI Section 15.5.3 applies to pile caps. The minimum thickness is sometimes governed by the development length of the dowels from the pile cap into the column.

For the pile cap shown in Fig. 16-23, the total horizontal tie force in one direction can be calculated from the force triangle shown in Fig. 16-23b. The factored downward force is 1240 kips, equilibrated by a vertical force of 310 kips in each pile. Considering the joint at A shown in Fig. 16-23c, we find that the horizontal steel force in tie A–B required to equilibrate the force system is 207 kips, corresponding to $\phi A_s f_y$, which, taking $\phi = 0.75$ for shear, requires seven No. 8 bars for tie A–B and the same number for each of the ties B–C, C–D, and A–D. The use of strut-and-tie models in design is discussed in Chapter 18.

Reference [16-12] recommends an edge distance of 15 in. measured from the center of the pile for piles up to 100-ton capacity and 21 in. for higher-capacity piles, to prevent bursting of the side cover over the piles.

PROBLEMS

Design wall footings for the following conditions.

16-1 Service dead load is 6 kips/ft, service live load is 8 kips/ft. Wall is 12 in. thick. Allowable soil pressure, q_a, is 4000 psf at 3 ft below the final grade. $f_c' = 2500$ psi, and $f_y = 60,000$ psi.

16-2 Service dead load is 18 kips/ft, service live load is 8 kips/ft. Wall is 16 in. thick. Allowable soil pressure, q_a, is 6000 psf at ground surface. $f_c' = 3000$ psi, and $f_y = 60,000$ psi.

Design square spread footings for the following conditions.

16-3 Service dead load is 350 kips, service live load is 275 kips. Soil density is 130 lb/ft^3. Allowable soil

pressure is 4500 psf at 5 ft below the basement floor. Column is 18 in. square. $f_c' = 3000$ psi, and $f_y = 60,000$ psi. Place bottom of footing at 5 ft below floor level.

16-4 Service dead load is 400 kips, service live load is 250 kips. Soil density is 120 lb/ft^3. Allowable soil pressure is 6000 psf. Column cross section is 30 in. × 12 in. $f_c' = 3000$ psi, and $f_y = 60,000$ psi. Select the elevation of the top of the footings so that there is 6 in. of soil and a 6 in. slab on grade above the footing.

17

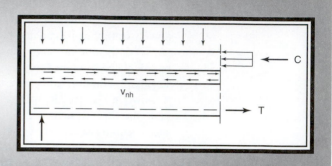

Shear Friction, Horizontal Shear Transfer, and Composite Concrete Beams

17-1 INTRODUCTION

This chapter considers the shear strength of interfaces between members or parts of members that can slip relative to one another. Among other cases, this includes the interface between a beam and a slab cast later than the beam, but expected to act in a composite manner.

17-2 SHEAR FRICTION

From time to time, shear must be transferred across an interface between two members that can slip relative to one another. The shear-carrying mechanism is known variously as *aggregate interlock, interface shear transfer*, or *shear friction*. The last of these terms is used here. The interface on which the shears act is referred to as the *shear plane* or *slip plane*.

Three methods of computing shear transfer strengths have been proposed in the literature. These include: (a) *Shear friction* models, (b) *Cohesion plus friction* models, and (c) *Horizontal shear models* as occur in composite beams.

Behavior in Shear Friction Tests

Several types of test specimens have been used to study shear transfer. The most common are the so-called *push-off specimens* similar to that shown in Fig. 17-1, which were tested in one of three ways:

1. Mattock [17-1], [17-2] and his associates reported tests of push-off specimens in which lateral expansion or contraction and crack widths were not held constant or otherwise controlled during the tests except by the reinforcement perpendicular to the slip plane. These specimens were tested either *as constructed* (uncracked) or *precracked* along the shear plane shown by the vertical dashed line in Fig. 17-1. Only the failure load and the amount of transverse reinforcement were reported.

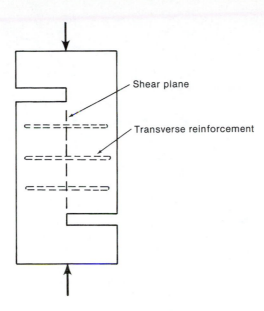

Fig. 17-1
Shear-transfer specimen.

2. Walraven [17-3], [17-4] tested push-off specimens but held the crack width constant during the tests. Loads and slip were reported.

3. A third type of test, reported by Loov and Patnaik [17-5] and others [17-6], measured the shear transferred between the web and the slab in a composite beam. This is referred to as *horizontal shear*.

An excellent review of tests and equations for shear transfer is presented by Ali and White [17-7], [17-8].

Cohesion and Friction

The results of 13 tests of uncracked specimens, and 21 tests of specimens with previously cracked shear planes are plotted in Fig. 17-2, with open circles and solid circles, respectively. The test results from [17-1] and [17-2] plotted in Fig. 17-2 suggest that

(a) The strengths of cracked and uncracked specimens can be represented by equations of the form

$$v_n = c + \sigma \tan \theta \tag{17-1a}$$

$$v_n = c + \mu\sigma \tag{17-1b}$$

where c is a *cohesion-like* term equal to the intercepts of the sloping lines on the vertical axis in Fig. 17-2, plus a *friction-like* term $\sigma \tan \theta$ that is equal to the product of σ, the compressive stress on the shear plane, and the coefficient of friction, μ. The plots in Fig. 17-2, can be idealized as straight sloping lines that intercept the vertical axis at "cohesions" of $c \approx 505$ and $c \approx 255$ psi, for the uncracked and precracked specimens, respectively, and a coefficient of friction of about 0.95. The curves terminate at a ceiling of about 1300 psi.

(b) The strengths of the uncracked specimens were higher than those of the initially precracked specimens. The uncracked and precracked specimens reached similar upper limits on shear transfer as shown in Fig. 17-2.

In Fig. 17-2, the shear stress at failure, v_u, is plotted against $\rho_v f_y$, where ρ_v is the ratio of the area of the transverse reinforcement across the shear plane to the area of the

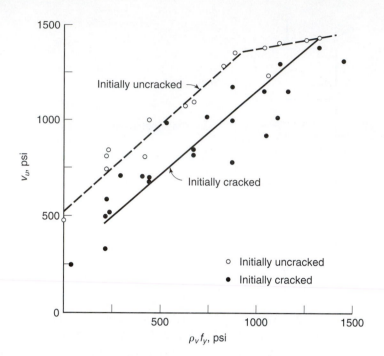

Fig. 17-2
Variation shear strength v_u with web reinforcement ratio $\rho_v f_y$. (From [17-1] and [17-2].)

shear plane. The crack widths, w, were not held constant or otherwise controlled during these test series. The shear stress at failure and the value of $\rho_v f_y$, taken as a measure of the clamping force due to the tension in the reinforcement crossing the shear plane. In Mattock's tests of initially uncracked push-off tests, the first cracks were a series of diagonal tension cracks, as shown in Fig. 17-3 . With further deformation, the compression struts between these cracks rotated at their ends (points A and B) such that point B moved downward relative to A. At the same time, the distance AC, measured across the crack, increased, stretching the transverse reinforcement. The tension in the reinforcement was equilibrated by an increase in the compression in the struts. Failure occurred when the bars yielded or the struts were crushed.

When a shear is applied to an initially cracked specimen or to a specimen formed by placing a layer of concrete on top of or against an existing layer of hardened concrete, relative slip of the layers causes a separation of the surfaces, as shown in Fig. 17-4a. If the

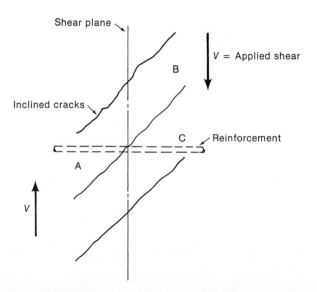

Fig. 17-3
Diagonal tension cracking along a previously uncracked shear plane. (From [17-8].)

reinforcement across the crack is developed on both sides of the crack, it is elongated by the separation of the surfaces and hence is stressed in tension. For equilibrium, a compressive stress is needed as shown in Fig. 17-4b. Shear is transmitted across the crack by (a) friction resulting from the compressive stresses [17-4], [17-5], and (b) interlock of aggregate roughness on the cracked surfaces, combined with dowel action of the reinforcement crossing the surface.

Shear-Friction Model

The original and simplest design model is the *shear-friction model* [17-1], [17-8], shown in Fig. 17-4, which ignores the cohesion-like component, c, and assumes that the shear transfer is due entirely to friction. This model is the basis of the shear-friction design procedure in ACI Section 11.7. Because the cohesion component is ignored, unusually high values of the coefficient of friction must be used to fit the test data. Design is based on the shear-friction equation, given in terms of forces as

$$V_n = A_{vf} f_y \mu \tag{17-2}$$
$$\text{(ACI Eq. 11-25)}$$

or in terms of stresses,

$$v_n = \rho_v f_y \mu \tag{17-3}$$

where ρ_v is defined above and μ is the appropriate value of the *coefficient of friction* given in ACI Section 11.7.4.3.

Shear-Friction Design Method, ACI Code Method

Figure 17-6 plots the data in Fig. 17-2 plus additional shear transfer test data from [17-6] for comparison with a series of radial lines through the origin, given by

$$v_n = \mu \sigma \tag{17-4}$$

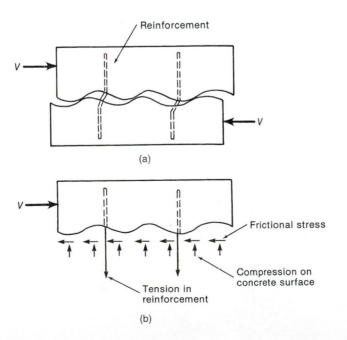

Fig. 17-4
Shear-friction model. (From [17-6].)

where σ is compressive stress on the shear plane, which is equal to $\rho_v f_y$, and is multiplied by different coefficients of friction, μ. These are expressed as functions of the roughness of the shear planes, plus a common upper limit, shown as a solid horizontal line in Fig. 17-6. Each of the three sloping lines in 17-6 corresponds to a particular surface condition and the coefficient of friction for that type of surface. The solid inclined line has $\mu = 1.4$ and applies to initially uncracked surfaces. The other two sloping lines apply to concrete placed against a hardened, roughened concrete surface with $\mu = 1.0$, and concrete placed against an as-rolled steel surface with $\mu = 0.6$. It can be seen that $\mu = 1.0$ and 1.4 approximate the trends of the strengths of the tests on initially uncracked and initially cracked specimens, respectively. An upper limit on the strengths was set at 800 psi based on the lowest strengths of the initially cracked tests.

Permanent Compression Force

A permanent compressive force, N_u, perpendicular to the slip plane, causes a normal stress on the slip plane in the concrete. If N_u is compressive, the normal stress N_u/A_{cv} adds to the compressive stress on the concrete due to the reinforcement. If tensile forces, N_u, act on the shear plane, they must be equilibrated by tensile reinforcement A_n provided in addition to the shear-friction reinforcement

$$A_n = \frac{N_u}{\phi f_y} \tag{17-5}$$

Inclined Shear-Friction Reinforcement

When the shear-friction reinforcement is inclined to the slip plane such that slip along this plane produces tensile stresses in the reinforcement, as shown in Fig. 17-5, the component of bar force perpendicular to the slip plane causes an compressive force on the plane equal to $A_{vf} f_y \sin \alpha_f$. In addition, there is a force component parallel to the crack equal to $A_{vf} f_y \cos \alpha_f$.

In stress units, the nominal shear-friction resistance at failure from Eqs. (17-4) and (17-5) is

$$v_n = \mu\sigma + \rho_v f_y \cos \alpha_f \tag{17-6}$$

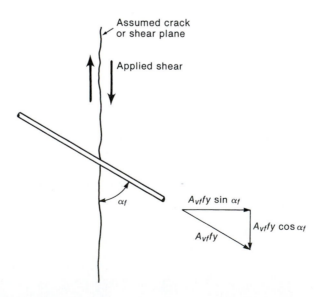

Fig. 17-5
Force components in a bar inclined to the shear plane.
(From [17-1].)

which has units of stress or, in terms of forces,

$$V_n = A_{vf} f_y (\mu \sin \alpha_f + \cos \alpha_f) \qquad (17\text{-}7)$$
$$(\text{ACI Eq. 11-26})$$

Shear-friction reinforcement that is inclined to the shear plane, such that the anticipated slip along the shear plane cause compression in the inclined bars, is not desirable. This compression tends to force the crack surfaces apart, leading to a decrease in the shear that can be transferred. Such steel is not allowed as shear-friction reinforcement.

Other Factors Affecting Shear Transfer

Lightweight Concrete

Tests [17-2] have indicated that the resistance to slip along the shear plane is smaller for lightweight-concrete specimens than for normal-weight concrete specimens. This happens because the cracks penetrate the pieces of lightweight aggregate rather than following the perimeters of the aggregate. The resulting crack faces are smoother than for normal-weight concrete [17-2]. This effect has been accounted for by multiplying the term in brackets by the correction factor for lightweight concrete, λ, given in ACI Section 11.7.4.3. Originally, λ was introduced in ACI Chapter 11 to account for the smaller tensile strength of lightweight aggregate. Here, it is used for convenience to account for the smoother crack surfaces that occur in lightweight concrete. When λ is included, (17-6) becomes

$$v_n = \lambda \mu \sigma + \rho_v f_y \cos \alpha_f \qquad (17\text{-}8)$$

where the nominal shear resistance, v_n, has units of stress. The factored shear resistance may be expressed as a force, V_n, by multiplying v_n by the area of the slip plane, A_{cv}.

Coefficients of Friction

ACI Section 11.7.4.3 gives coefficients of friction, μ, as follows:

    1.   for concrete placed monolithically, 1.4λ

    2.   for concrete placed against hardened concrete with surface intentionally roughened, as specified in ACI Section 11.7.9, 1.0λ

    3.   for concrete placed against hardened concrete not intentionally roughened, as specified in ACI Section 11.7.9, 0.6λ

    4.   for concrete anchored to as-rolled structural steel by studs or by reinforcing bars (see ACI Section 11.7.10), 0.7λ,

where $\lambda = 1.0$ for normal-weight concrete, 0.85 for sand-lightweight concrete, and 0.75 for all-lightweight concrete. The code allows a linear transition when part of the fine aggregate is lightweight and part is normal weight.

    The first term in (17-1) represents the portion of the shear transferred by shearing off surface protrusions and by dowel action. Equation 17-4, as plotted in Fig. 17-5, applies only for $\rho_v f_y$ greater than 200 psi and the radial lines in Fig. 17-6 are not plotted below this value. For Grade-60 reinforcement, this limit requires a minimum reinforcement ratio, $\rho = 0.0033$.

Design Rules in the ACI Code

ACI Section 11.7 presents design rules for cases "where it is appropriate to consider shear transfer across a given plane, such as an existing or a potential crack, an interface between

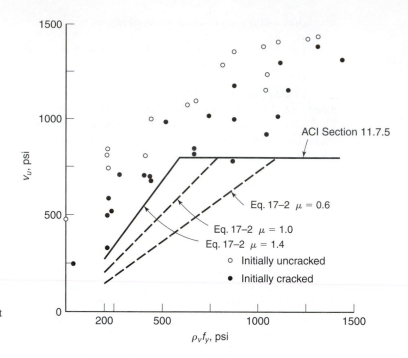

Fig. 17-6
Comparison of (17-2) with test data from [17-1], [17-2], and [17-6].

dissimilar materials, or an interface between two concretes cast at different times." Typical examples are shown in Fig. 17-7.

In design, a crack is assumed to exist along the shear plane, and reinforcement is provided across that crack. The amount of reinforcement is computed (ACI Section 11.7.4) from

$$\phi V_n \geq V_u \qquad \qquad (17\text{-}9)$$
$$(\text{ACI Eq. 11-1})$$

$$V_n = A_{vf} f_y \mu \qquad \qquad (17\text{-}2)$$
$$(\text{ACI Eq. 11-25})$$

where $\phi = 0.75$ for designs carried out by using the load factors in ACI 318-02 Section 9.2 or $\phi = 0.85$ for design by ACI 318-02 Section C.2 and where μ is the coefficient of friction, taken equal to 0.6λ to 1.4λ depending on the surfaces in contact.

In cases 2 and 3 defined previously, the surface must be clean and free of *laitance* (a weak layer on the top surface of a concrete placement due to bleed water collecting at the surface). ACI Section 11.7.10 requires that, in case 4, the steel must be clean and free of paint. Case 2 applies to concrete placed against hardened concrete that has been roughened to a "full amplitude" (wave height) of approximately $\frac{1}{4}$ in. but does not specify a "wave length" for the roughened surface. This was done to allow some freedom in satisfying this requirement. It was intended, however, that the wave length be on the same order of magnitude as the full amplitude, say $\frac{1}{4}$ to $\frac{3}{4}$ in.

ACI Section 11.7.5 sets the upper limit on V_n from (17-2) as $0.2 f'_c A_c$ or $800 A_c$ pounds, whichever is smaller. Equation (17-2) and its upper limits are plotted in Fig. 17-6 for comparison with the test data. Equation (17-2), with its upper limits, is conservative.

Upper Limit on Shear Friction

The test data plotted in Figs. 17-2 and 17-6 approach an upper bound of about 1200 to 1400 psi. ACI Section 11.7.5 specifies the smaller of $0.2 f'_c$ and 800 psi, plotted in Fig. 17-6, as an

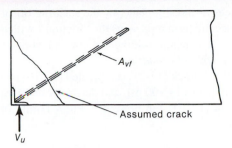

(a) Precast beam bearing.

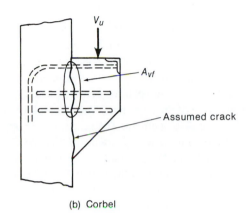

(b) Corbel

Fig. 17-7
Examples of shear friction.
(From [17-9].)

(c) Shear wall

upper limit on v_n. This much lower limit is needed in the shear-friction method to offset the overestimate of the coefficient of friction.

Cohesion-plus-Friction Model

For the usual case of transverse reinforcement perpendicular to the shear plane, Mattock [17-2] rewrote (17-1) as

$$V_n = K_1 A_{cv} + 0.8 A_{vf} f_y \tag{17-10}$$

(unnumbered equation in ACI Commentary R11.7.3)

where $K_1 = 400$ psi for normal-weight concrete, 200 psi for all-lightweight concrete, and 250 psi for sand-lightweight concrete. The first term on the right-hand side of (17-10) represents the shear transferred by "cohesion," which is caused by pieces of aggregate bearing on the surfaces of the slip plane, by the shearing off of surface protrusions, and by dowel action. The second term represents the "friction," with the coefficient of friction taken to be 0.8 for cracked concrete sliding on cracked concrete. Equation (17-10) is called the *modified shear-friction equation.*

In 2001, Mattock [17-11] reevaluated (17-10) on the basis of 199 tests, with f'_c ranging from 2450 psi to 14,400 psi, and derived a set of equations that retained the numerical constant 0.8 in (17-10), but presented a new family of values of K_1 for various types of shear planes.

Walraven Model

Walraven [17-3] idealized concrete as a series of size-graded spherical pieces of coarse aggregate embedded in a matrix of hardened concrete paste. A crack was assumed to cross the matrix between pieces of aggregate until it reached a piece of aggregate, at which time it followed the aggregate–matrix interface around the piece of aggregate. The strength of the interface between the cement paste and the aggregate was assumed to be less than the strength of the aggregate. Such a crack is shown in Fig. 17-8a, after cracking, but before shearing displacement. The radius of the aggregate particle is R, the diameter is a, and the crack width is w. It is assumed that the crack width is held constant while a shearing load is applied. After a small slip parallel to the crack, the spherical piece of aggregate comes into contact with the matrix in the dark shaded area, allowing shear to be transferred across the crack (Fig. 17-8b). Further slip mobilizes rigid–plastic stresses at the point of contact, and the crack is restrained against further slip until the crack surfaces deteriorate (see Fig. 17-8c). The maximum shearing force transferred by a single aggregate particle occurs at the stage shown in Fig. 17-8c, corresponding to the largest area of plastic stresses. By assuming randomly chosen gradations of aggregate sizes, Walraven developed expressions for the shear forces transferred for various crack widths, w, and maximum aggregate diameters, a.

Walraven [17-3] tested 88 push-off shear-transfer specimens similar to Fig. 17-1. The major difference between these tests and those by Mattock and others was that Walraven kept the crack widths, w, constant throughout each test. Vecchio and Collins [17-12] fitted (17-6) and (17-10) to Walraven's test data and obtained

$$v_{ci} = 0.18v_{ci\,max} + 1.64f_{ci} - 0.82 \frac{f_{ci}^2}{v_{ci\,max}} \tag{17-11}$$

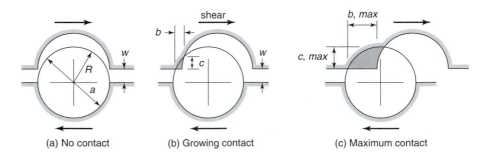

Fig. 17-8
Walraven crack model.

(a) No contact   (b) Growing contact   (c) Maximum contact

where v_{ci} is the shear stress transferred across the crack, f_{ci} is the compression stress required across the crack in psi, and $v_{ci\,max}$ is the maximum shear stress that can be transmitted across a given crack.

The maximum shear stress, $v_{ci\,max}$, that can be transferred across a crack when its width is held at w in. is given by

$$v_{ci\,max} = \frac{2.16\sqrt{f'_c}}{0.3 + \left(\dfrac{24w}{a + 0.63}\right)} \tag{17-12}$$

where a is the diameter of the coarse aggregate in the cracked concrete in inches. The crack width, w in inches, is computed from the spacing, s_θ, of the inclined crack as

$$w = \varepsilon_1 s_\theta \tag{17-13}$$

where ε_1 is the principal tensile strain, which is assumed to act perpendicular to the crack, and s_θ is the spacing of the cracks, measured perpendicular to the cracks. Equation (17-13) assumes that all the strain is concentrated in the crack. Both ε_1 and s_θ are discussed in Chapter 6, where (17-11) is considered in several modern analyses of the shear transfer mechanisms in a beam.

Walraven's tests were limited to concrete strengths, f'_c, from 2900 to 8200 psi. For high-strength concretes with cylinder strengths in excess of 10,000 psi, the cracks tend to cross the individual pieces of aggregate rather than going around them. As a result, the crack surface was smoother than for weaker concretes. For this case, the effective size of the aggregate, a, decreases, approaching zero. Angelakos, Bentz, and Collins [17-13] handled this by arbitrarily reducing the effective aggregate size, a, in (17-12) from the nominal diameter, a, to zero as the concrete strength increases from 8500 psi to 10,000 psi. For concrete strengths of 10,000 psi or higher, they took a equal to zero.

Ali and White [17-7] carried out similar analyses, assuming an undulating crack path, to illustrate the force transfer that develops when the two sides of the crack come into bearing.

Loov and Patnaik — Composite Beams

From tests of composite beams, Loov and Patnaik [17-5] derived the following equation for shear transfer across cracks and for horizontal shear in composite beams (see Fig. 17-11, discussed later):

$$v_{ci\,max} = \lambda k \sqrt{\sigma f'_c} + \rho_v f_y \cos \alpha_f \tag{17-14}$$

In (17-14), λ accounts for lightweight concrete and k is a constant equal to 0.5 for concrete placed against hardened concrete and 0.6 for concrete placed monolithically. The square-root term allows this equation to fit the test data more closely than do other models. Equation (17-14) is plotted in Fig. 17-9 for comparison with test results and with (17-10).

Because the reinforcement is assumed to yield in order to develop the necessary forces, the yield strength of the steel is limited to 60,000 psi. Each bar must be anchored on both sides of the crack to develop the bar. The steel must be placed approximately uniformly across the shear plane, so that all parts of the crack are clamped together.

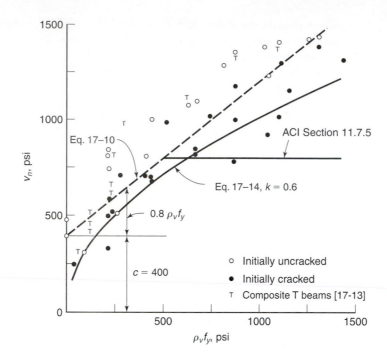

Fig. 17-9
Comparison of test results
from [17-1] and [17-5] and
design equations (17-10) and
(17-14).

Comparison of Design Rules with Test Results

In Fig. 17-9, test data for push-off tests of initially uncracked specimens with reinforcement perpendicular to the shear plane [17-1] are compared with

(a) Equation (17-10), with $K_1 = 400$ psi, and

(b) Equation (17-14), with $k = 0.5$, for initially cracked concrete

The test data are compared with the nominal strengths, with $\phi = 1.0$. The specimens had concrete strengths ranging from 3840 psi to 4510 psi. The nominal strengths from (17-10) are plotted, with $f'_c = 4000$ psi and $k = 0.5$ for concrete placed against hardened concrete or for initially cracked concrete.

Test data for push-off specimens with a precracked interface [17-1] and an average concrete strength of 4060 psi are plotted in Fig. 17-9, along with data from tests of composite beams with average concrete strengths of 5710 psi for the webs and 5160 psi for the flanges [17-5]. The nominal strengths computed from (17-14), with $k = 0.5$ for a precracked interface, and $f'_c = 4350$ psi, fit the data quite well.

EXAMPLE 17-1 Design of the Reinforcement in the Bearing Region of a Precast Beam

Figure 17-10 shows the support region of a precast concrete beam. The factored beam reactions are 62 kips vertical force and a horizontal tension force of 12 kips. The horizontal force arises from restraint of the shrinkage of the precast beam. The cross section of the beam is 12 in. wide by 18 in. deep. Use $f'_c = 3000$ psi, assume normal-density concrete, and use $f_y = 60,000$ psi.

1. Assume the cracked plane. The crack plane giving the maximum area, A_{cv}, is a vertical crack. This will tend to overestimate the cohesion component. Assume that the support region of the beam is enclosed by a 6-in. × 6-in. angle, as shown in Fig. 17-10, and assume that the crack is at 60° to the horizontal. It intercepts the end of the beam at 10.2 in. above the bottom and has a length of 12 in. We shall take $A_{cv} = 12$ in. × 12 in. = 144 in.^{2}.

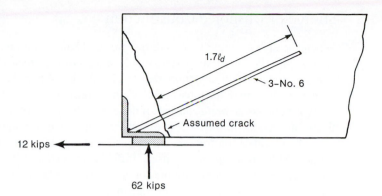

Fig. 17-10
Example 17-1.

2.   Compute the area of steel required. Resolving the forces onto the inclined plane gives a normal force of 20.6 kips compression and a shear force of 61.2 kips for a 60° plane. Each assumed crack angle will result in a different combination of normal and shear forces. However, it is quick and conservative to assume that the shear force is equal to the vertical reaction of 62 kips and that the normal force is equal to the horizontal reaction $N_u = -12$ kips (negative in tension). In addition, we shall assume that $\alpha_f = 90°$.

From (17-2),

$$V_n = A_{vf} f_y \mu$$

where $V_u = \phi V_n$. Therefore,

$$A_{vf} = \frac{V_u}{\phi f_y \mu}$$

where $V_u = 68.1$ kips, $\mu = 1.4\lambda$ (since the crack plane is in monolithically placed concrete), and $\lambda = 1.0$ (since normal-weight concrete is used). We thus have

$$A_{vf} = \frac{62}{0.75 \times 60 \times 1.4} = 0.984 \text{ in.}^2$$

The tensile force must also be transferred across the crack by reinforcement (17-5):

$$A_n = \frac{12}{0.75 \times 60} = 0.267 \text{ in.}^2$$

Therefore, the total steel across the crack must be $(0.984 + 0.267 \text{ in.}^2) = 1.25 \text{ in.}^2$

Provide three No. 6 bars across the assumed crack. These must be anchored on both sides of the crack. This is done by welding them to the bearing angle and by extending them $1.7\ell_d$, as recommended in [17-14].   ∎

17-3   COMPOSITE CONCRETE BEAMS

Frequently, precast beams or steel beams have a slab cast on top of them and are designed assuming that the slab and beam act as a monolithic unit to support loads. Such a beam-and-slab combination is referred to as a *composite beam*. This discussion will deal only with composite beams where the beam is precast concrete or other concrete cast at an earlier time than the slab.

Shored or Unshored Construction

When the slab concrete is placed, the precast beam can either be shored or unshored. If shored with the shores supporting the dead load of the precast beam, the dead load of the beam and slab is initially supported by the shores. When the strength of the concrete is high enough to resist the forces induced, the shores are removed, and the dead load is resisted by the composite beam and slab. If the precast beam is not shored when the slab is placed, the beam supports its own weight plus the weight of the slab and the slab forms. ACI Section 17.2 allows either construction process and requires that each element be strong enough to support all loads it supports by itself. If the beam is shored, ACI Section 17.3 requires that the shores be left in place until the composite section has a strength adequate to support all loads and to limit deflections and cracking.

Tests have shown that the ultimate strength of a composite beam is the same whether the member was shored or unshored during construction. For this reason, ACI Section 17.2.4 allows strength computations to be made that consider only the final composite member.

Horizontal Shear

In the beam shown in Fig. 17-11a, there are no horizontal shear stresses transferred from the slab to the beam. They act as two independent members. In Fig. 17-11b, horizontal shear stresses act on the interface, and as a result, the slab and beam act in a composite manner. The ACI Code provisions for horizontal shear are given in ACI Section 17.5. Although the mechanism of horizontal shear transfer and that of shear friction are similar, if not identical, there is a considerable difference between the two sets of provisions. The difference results from the fact that (17-10), ACI Section 17.5.2.3, and ACI Section 11.7 are all empirical attempts to fit test data. Equation (17-10) is valid for relatively short shear planes with lengths up to several feet, but is believed to give shear strengths that are too high for long shear-transfer regions if the maximum stress is localized. It is also unconservative for low values of $\rho_{vf}f_y$.

The term *horizontal shear stress* is used to describe the shear stresses acting on the interface in Fig. 17-11b. In a normal composite beam, these stresses are horizontal, as is shown in Fig. 17-11b. If the beam were vertical, however, the term "horizontal shear"

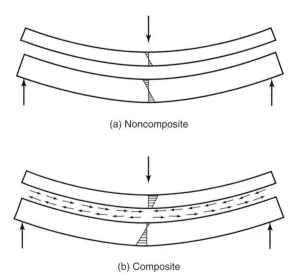

(a) Noncomposite

(b) Composite

Fig. 17-11
Horizontal shear transfer in a composite beam.

would still be used to distinguish between this shear and the orientation of shear stresses in a beam. Tests of horizontal shear in composite beams are reported in [17-3], [17-5], [17-6], [17-15], and [17-16]. The tests reported in [17-6] included members with and without shear keys along the interface. The presence of shear keys stiffened the connection at low slips but had no significant effect on its strength.

Computation of Horizontal Shear Stress

From strength of materials, the horizontal shear stresses, v_h, on the contact surface between an uncracked elastic precast beam and a slab can be computed from

$$v_{uh} = \frac{VQ}{I_c b_v} \qquad (17\text{-}15)$$

where

V = shear force acting on the section in question

Q = first moment of the area of the slab or flange about the centroidal axis of the composite section

I_c = moment of inertia of the composite section

b_v = width of the interface between the precast beam and the cast-in-place slab

Equation (17-15) applies to uncracked elastic beams and is only an approximation for cracked concrete beams.

The ACI Code gives two ways of calculating the horizontal shear stress.

ACI Section 17.5.2 defines the nominal horizontal shear force, V_{nh}, to be transferred as

$$\phi V_{nh} \geq V_u \qquad (17\text{-}16)$$
$$(\text{ACI Eq. 17-1})$$

Rearranging this gives

$$v_{nh} = \frac{V_u / \phi}{b_v d} \qquad (17\text{-}17\text{a})$$

This is based on the observation that the shear stresses on opposite pairs of sides of an element located at the top of the web are equal in magnitude, as shown in Fig. 6-3a, but are arranged to give couples in opposite directions. For an element taken from directly over the beam web at the interface between the web and flange, the shear stresses on the top and bottom sides of the element are v_{nh}, and the shear stresses on the vertical sides of the element are

$$v_{nv} = \frac{V_u / \phi}{b_v d} \qquad (17\text{-}17\text{b})$$

where V_u is the factored shear force acting on the cross section of the beam as obtained from a shear-force diagram for the beam. If the shear stresses on the left and right faces of the element form a counter-clockwise couple, those on the bottom and top of the element must form a clockwise couple. For equilibrium, the shear stresses on four sides of an element must be equal in magnitude. Thus,

$$v_{nh} = v_{nv} \qquad (17\text{-}18)$$

Calculation of Horizontal Shear Stresses from C and T Forces

Alternatively, ACI Section 17.5.3 allows horizontal shear to be computed from the change in compressive or tensile force in the slab in any segment of its length. Figure 17-12 illustrates this clause. At midspan, the force in the compression zone is C, as shown in Fig. 17-12a. All of this force acts above the interface. At the end of the beam, the force in the flange is zero. Thus, the horizontal shear force to be transferred across the interface between midspan and the support is

$$V_h = C \tag{17-19}$$

A similar derivation could be carried out by considering the force in the reinforcement,

$$V_{uh} = T_u \tag{17-20}$$

where T_u is the tensile force in the reinforcement related to M_u.

ACI Section 17.5.3.1 says that when ties are provided to resist the horizontal shear calculated using (17-19) or (17-20), the spacing of the ties should approximately reflect the distribution of shear forces along the member. This is specified because the slip of the slab relative to the web at the onset of a horizontal shear failure is too small to allow much redistribution of horizontal shear stresses. In [17-5], the measured slip at maximum horizontal shear stress was on the order of 0.02 in., and the slip at failure ranged from 0.08 to 0.3 in. We shall calculate the horizontal shear stresses by dividing the beam into a series of segments and computing the value of C_u or T_u at both ends of each segment. Then,

$$v_{uh} = \frac{T_{u1} - T_{u2}}{b_v \ell_s} \tag{17-21}$$

where T_{u1} and T_{u2} are the larger and smaller tension forces on the two ends of a segment, and ℓ_s is the length of the segment. The tension force can be taken as $T_u = M_u/jd$. If jd is assumed to be constant, T_u varies as the M_u diagram. For a uniformly loaded simple beam, the M_u diagram is a parabola, and the moments at the $\frac{1}{4}$ and $\frac{1}{8}$ points of the span are 0.75 and 0.438 times the maximum moment, respectively. These fractions of the moment diagram can be calculated from the shear-force diagram because the change in moment from one section to another is equal to the area of the shear-force diagram between the two sections.

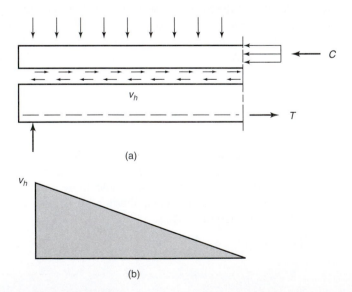

Fig. 17-12
Horizontal shear stresses in a composite beam.

The two procedures give similar results, as will be seen in Example 17-2. The limits on V_{nh} (V_{uh}/ϕ) from ACI Sections 17.5.2.1 to 17.5.2.3 are given in Table 17-1.

In composite beams, the contact surfaces must be clean and free from laitance. The words "intentionally roughened" imply that the surface has been roughened with a "full amplitude" of $\frac{1}{4}$ in., as discussed in connection with shear friction. When the factored shear force, $V_u = \phi V_{nh}$, at the section exceeds $\phi(500 b_v d)$ psi, ACI Section 17.5.2.4 requires design be based on shear friction, in accordance with ACI Section 11.7.4. This limit reflects the range of test data used to derive ACI Section 17.5.2.3.

ACI Section 17.6 requires that the ties provided for horizontal shear be not less than the minimum stirrups required for shear, given by

$$A_v = \frac{0.75\sqrt{f_c'}\, b_w s}{f_y}, \text{ and } \geq \frac{50\, b_w s}{f_y} \qquad \begin{array}{c} (17\text{-}22) \\ (\text{ACI Eq. }11\text{-}13) \end{array}$$

The tie spacing shall not exceed 4 times the least dimension of the supported element, which is usually the thickness of the slab, but not more than 24 in. The ties must be fully anchored both in the beam stem and in the slab.

Deflections

The beam cross section considered when calculating deflections depends on whether the beam was shored or unshored when the composite slab is placed. If it is shored so that the full dead load of both the precast beam and the slab is carried by the composite section, ACI Section 9.5.5.1 allows the designer to consider the loads to be carried by the full composite section when computing deflections. The modulus of elasticity should be based on the strength of the concrete in the compression zone, while the modulus of rupture should be based on the strength of the concrete in the tension zone. For nonprestressed beams constructed with shores, it is not necessary to check deflections if the overall height of the composite section satisfies ACI Table 9.5(a).

ACI Section 9.5.5.2 covers the calculation of deflections for unshored construction of nonprestressed beams. If the thickness of the precast member satisfies ACI Table 9.5(a), it is not necessary to consider deflections. If the thickness of the composite section satisfies the table, but the thickness of the precast member does not, it is not necessary to compute deflections occurring after the section becomes composite, but it is necessary to compute the instantaneous deflections and that part of the sustained load deflections occurring prior to the beginning of effective composite action. The latter can be assumed to occur when the modulus of elasticity of the slab reaches 70 to 80 percent of its 28-day value, usually about four to seven days after the slab is placed.

ACI Section 9.5.5.1 states that if deflections are computed, they should account for the curvatures induced by the differential shrinkage between the slab and the precast beam. Shrinkage of the slab relative to the beam causes the slab to shorten relative to the beam.

TABLE 17-1   Calculation of V_{nh}

ACI Section	Contact Surfaces	Ties	V_{nh}
17.5.2.1	Intentionally roughened	None	$80 b_v d$
17.5.2.2	Not roughened	Minimum ties from ACI Section 17.6	$80 b_v d$
17.5.2.3	Intentionally roughened	$A_{vh} f_y$	$\left(260 + \dfrac{0.6 A_{vh} f_y}{b_v s}\right)\lambda b_v d$ but not more than $500 b_v d$

Because the slab and beam are joined together, this relative shortening causes the beam to deflect downward, adding to the deflections due to loads. Some of the shrinkage of the concrete in the beam will have occurred before the beam is erected in the structure. All of the slab shrinkage occurs after the slab is cast. As the slab shrinks relative to the beam, tensile stresses are induced in the slab and compressive stresses in the beam. These are redistributed to some degree by creep of the concrete in the slab and beam. This effect can be modeled by using an *age-adjusted effective modulus*, E_{caa}, and an *age-adjusted transformed section* in the calculations, as discussed in Section 3-4 and in [17-17] and [17-18].

EXAMPLE 17-2 Design of a Composite Beam

Precast, simply supported beams that span 24 ft and are spaced 10 ft on centers are composite with a slab that supports an unfactored live load of 100 psf, a partition load of 20 psf, and a superimposed dead load of 10 psf. Design the beams and the composite beam and slab. Use $f'_c = 3000$ psi for the slab, 5000 psi for the precast beams, and $f_y = 60,000$ psi. Use load factors from ACI section 9.2.1.

   1. **Select the trial dimensions.** For the end span of the slab, ACI Table 9.5(a) gives the minimum thickness of a one-way slab as $h = \ell/24 = 120$ in./24 = 5 in. For a simply supported beam, the table gives $h = \ell/16 = (24 \times 12)/16 = 18$ in. Deflections of the composite beam may be a problem if the overall depth is less than 18 in. However, for unshored construction, ACI Section 9.5.5.2 requires that deflections of the precast member be considered if its overall depth is less than that given by Table 9.5(a). To avoid this, we shall try an 18-in.-deep precast beam 12 in. wide to allow the steel to be in one layer, plus a 5-in. slab. For the precast beam, $d = 18 - 2.5 = 15.5$ in.

   2. **Compute the factored loads on the precast beam.** Since the floor will be constructed in an unshored fashion, the precast beam must support its own dead load, the dead load of the slab, the weight of the forms for the slab, assumed to be 10 psf, and some construction live load, assumed to be 50 psf. Thus, we have the following data:

Dead loads:

$$\text{Beam stem} \quad w = \frac{12 \times 18}{144} \times 0.150 \text{ kcf} = 0.225 \text{ kip/ft}$$

$$\text{Slab} \quad w = 5 \text{ in.}/12 \times 10 \text{ ft} \times 0.150 \text{ kcf} = 0.625 \text{ kip/ft}$$

$$\text{Forms} \quad w = 10 \text{ ft} \times 0.010 \text{ ksf} = 0.100 \text{ kip/ft}$$

$$\text{Total} = 0.950 \text{ kip/ft}$$

$$\text{Live load:} \quad w = 10 \text{ ft} \times 0.050 \text{ ksf} = 0.500 \text{ kip/ft}$$

The ACI Code does not specifically address the load factors for this construction-load case. We shall take $U = 1.2D + 1.6L$, giving

$$w_u = 1.2 \times 0.950 + 1.6 \times 0.500 = 1.94 \text{ kips/ft}$$

   3. **Compute the size of the precast member required for flexure.**

$$M_u = \frac{1.94 \times 24^2}{8} = 140 \text{ ft-kips}$$

We shall select a steel percentage close to, but less than, the tension-controlled limit in the precast beam, so that $\phi = 0.9$, and later check whether that provides enough steel for flexure in the composite section. From Table A-3 for 5 ksi, the value of ρ corresponding to the tension-controlled limit is 0.021. We shall try $\rho = 0.02$. For this value of ρ, $\phi k_n = 927$ and $j = 0.858$, so that

$$\frac{bd^2}{12,000} = \frac{M_u}{\phi k_n} = \frac{140}{927}$$

$$= 0.151$$

From Table A-9, for $b = 12$ in., we require that $d = 12.3$ in. and $h = 12.3 + 2.5 = 14.8$ in. Therefore, the size chosen for deflection control is adequate.

For the 12-by-18-in. beam, $d = 18 - 2.5 = 15.5$

$$A_s = \frac{M_u}{\phi f_y jd} = \frac{140 \times 12,000}{0.9 \times 60,000 \times 0.858 \times 15.5}$$

$$= 2.34 \text{ in.}^2$$

Try a 12-by-18-in. precast beam with two No. 8 bars plus two No. 7 bars; $A_s = 2.78$ in.^{2}. Then

$$a = \frac{2.78 \times 60,000}{0.85 \times 5000 \times 12} = 3.27 \text{ in.}$$

$$\frac{a}{d_t} = \frac{3.27}{15.5} = 0.211$$

This is less than $a/d_t = 0.300$ for the tension-controlled limit. Hence, $\phi = 0.9$, and we have

$$\phi M_n = \frac{0.9 \times 2.78 \times 60,000(15.5 - 3.27/2)}{12,000} = 173 \text{ ft-kips}$$

Therefore, OK.

4. Check the capacity of the composite member in flexure. The factored loads on the composite member are as follows:

Dead load:

$$\text{Precast stem } w = 0.225 \text{ kip/ft}$$

$$\text{Slab } w = 0.625 \text{ kip/ft}$$

Superimposed dead load

$$w = 0.100 \text{ kip/ft}$$

$$\text{Total } w = 0.950 \text{ kip/ft} \qquad \text{Factored} = 1.2 \times 0.950 = 1.14 \text{ kips/ft}$$

Live load:

$$\text{Floor load } w = 10 \text{ ft} \times 0.100 \text{ ksf} = 1.0 \text{ kip/ft}$$

$$\text{Partitions } w = 10 \text{ ft} \times 0.020 \text{ ksf} = 0.2 \text{ kip/ft}$$

$$\text{Total } w = 1.20 \text{ kips/ft} \qquad \text{Factored} = 1.6 \times 1.20 = 1.92 \text{ kips/ft}$$

$$\text{Total factored load} = 3.06 \text{ kips/ft}$$

$$M_u = \frac{3.06 \times 24^2}{8} = 220 \text{ ft-kips}$$

Compute ϕM_n. From ACI Section 8.10.2, the effective flange width is 72 in. The overall height is $18 + 5 = 23$ in.; $d = 23 - 2.5 = 20.5$ in. Assuming rectangular beam action, where f'_c in the slab is 3000 psi, we have

$$a = \frac{2.78 \times 60,000}{0.85 \times 3000 \times 72} = 0.908 \text{ in.}$$

Since a is less than the flange thickness, rectangular beam action exists, and $a = 0.908$ in. Also, a/d_t is much less than a/d_t for the tension-controlled limit; thus, $\phi = 0.9$ for flexure, and we have

$$\phi M_n = \frac{0.9 \times 2.78 \times 60,000(20.5 - 0.908/2)}{12,000}$$

$$= 251 \text{ ft-kips}$$

Therefore, the steel chosen is adequate to resist the moments acting on the composite section. **Use a 12-by-18-in. precast section with two No. 8 plus two No. 7 longitudinal bars and a 5-in. cast-in-place slab.**

5. Check vertical shear. V_u at d from the support $= 3.06$ kips/ft $\times$ $(12 - 20.5/12)$ ft $= 31.5$ kips.

$$V_c = 2\sqrt{f_c'}b_w d \text{ (where we shall use the smaller of the two concrete strengths) (ACI Eq. 11-3)}$$

$$= 2\sqrt{3000} \times 12 \times 20.5 = 26.9 \text{ kips} \qquad \phi V_c = 20.2 \text{ kips}$$

Since $V_u > \phi V_c/2$, we need stirrups. Thus,

$$V_s = \frac{V_u}{\phi} - V_c = \frac{31.5}{0.75} - 26.9$$

$$= 15.1 \text{ kips at } d \text{ from the support}$$

Try No. 3 U stirrups, and let $A_v = 0.22$ in.^{2}. Then

$$s = \frac{A_v f_y d}{V_s} = \frac{0.22 \times 60 \times 20.5}{15.1} \qquad \text{(ACI Eq. 11-15)}$$

$$= 17.9 \text{ in. on centers}$$

$$\text{Maximum spacing} = d/2 = 10.25 \text{ in., say, 10 in.}$$

$$\text{Minimum stirrups (50 psi governs): } s = \frac{A_v f_y}{50 b_w} = \frac{0.22 \times 60,000}{50 \times 12} \qquad \text{(ACI Eq. 11-13)}$$

$$= 22.0 \text{ in. on centers}$$

We will select the stirrups after considering horizontal shear.

6. Compute the horizontal shear. Horizontal shear may be computed according to ACI Section 17.5.2 or Section 17.5.3. We shall do the calculations both ways and compare the results. We shall assume that the interface is clean, free of laitance, and intentionally roughened.

ACI Section 17.5.2: From (17-16) (ACI Eq. (17-1)), $\phi V_{nh} \geq V_u = 31.5$ kips at d from the support.

From ACI Sec. 17.5.2.2, an intentionally roughened surface without ties is adequate for

$$\phi V_{nh} = \phi 80 b_v d = 0.75 \times 80 \times 12 \times 20.5 = 14.8 \text{ kips}$$

Therefore, ties are required.

From ACI Section 17.5.2.3, if minimum ties are provided according to ACI Section 17.6 and the interface is intentionally roughened,

$$\phi V_{nh} = \phi(260 + 0.6\rho_v f_y)\lambda b_v d$$

The maximum tie spacing allowed by ACI Section 17.6.1 is 4×5 in. $= 20$ in. but not more than 24 in. The maximum spacing for vertical shear governs. Assume that the ties are No. 3 two-leg stirrups at the maximum spacing allowed for shear, 10 in. Then

$$\rho_v = \frac{A_v}{b_v s} = \frac{0.22}{12 \times 10}$$

$$= 0.00183$$

and

$$\phi V_{nh} = 0.75(260 + 0.6 \times 0.00183 \times 60,000) \times 1.0 \times 12 \times 20.5$$
$$= 60.1 \text{ kips}$$

Thus, minimum stirrups provide more than enough ties for horizontal shear. We shall use closed stirrups to better anchor them into the top slab. **Use No. 3 closed stirrups at 10 in. on centers throughout the length of the beam.**

ACI Section 17.5.3: Alternatively, horizontal shear could be considered by using ACI Section 17.5.3. The compression zone is entirely in the slab. Therefore, the force transferred across the interface is

$$V_{nh} = C = T = A_s f_y = 2.78 \text{ in.}^2 \times 60 \text{ ksi} = 167 \text{ kips}$$

From (17-21),

$$v_{uh} = \frac{T_{u1} - T_{u2}}{b_v \ell_s} \tag{17-21}$$

Consider the section of the flange between the midspan and the quarter point of the span. At midspan, $T_{u1} = M_u / jd = M_u / 0.85d = 152$ kips. The distance from the support to the quarter point is $24/4 = 6$ ft. The moment at the quarter point is

$$M_u = Rx - wx^2/2 = (3.06 \text{ kip/ft} \times 12 \text{ ft}) \times 6 \text{ ft} - 3.06 \times 6^2/2$$
$$= 220 - 55.1 = 165 \text{ ft-kips}$$
$$T_{u2} = M_u / 0.85d = 114 \text{ kips}$$

The average horizontal shear stress between the midspan and the quarter point is

$$v_{nh} = \frac{(152 - 114) \text{ kips} \times 1000}{12 \text{ in} \times (6 \times 12)} = 44.4 \text{ psi}$$

Since this is less than 80 psi, minimum ties are sufficient between the midspan and the quarter points. Now consider the portion of the beam between the quarter point and the eighth point:

M_u at the eighth point $= 96.4$ ft-kips

T_{u1} (at the quarter point) $= 114$ kips, and T_{u2} (at the eighth point) $= 66.4$ kips

The average horizontal shear stress between the quarter and the eighth point of the span is

$$v_{nh} = \frac{(114 - 66.4) \times 1000}{12 \times (3 \times 12)} = 110 \text{ psi}$$

Since this value exceeds 80 psi, stirrups are required to satisfy ACI Section 17.5.2.1. Minimum ties, according to ACI Section 17.6, are equivalent to No. 3 two-legged stirrups at 10 in. on centers, with $\rho_v = 0.00183$. For minimum ties, ACI Section 17.5.2.3 gives

$$v_{nh} = (260 + 0.6\rho_v f_y) \text{ psi} = (260 + 0.6 \times 0.00183 \times 60,000)$$
$$= 326 \text{ psi}$$

Minimum stirrups are more than enough. Consider the portion of beam between the support and the eighth point. At the eighth point, $T_{u1} = 66.4$ kips. At the support, $T_{u2} = $ zero. The average horizontal shear stress on this segment is

$$v_{nh} = \frac{(66.4 - 0) \times 1000}{12 \times (3 \times 12)} = 154 \text{ psi}$$

Thus, minimum ties are satisfactory. We shall use closed stirrups to better anchor them into the top slab. **Use No. 3 closed stirrups at 10 in. on centers throughout the length of the beam.** ∎

Discontinuity Regions and Strut-and-Tie Models

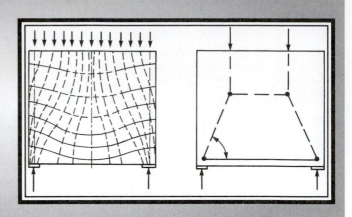

18-1 INTRODUCTION

Definition of Discontinuity Regions

Structural members may be divided into portions called *B-regions*, in which beam theory applies, including linear strains and so on, and other portions called *discontinuity regions*, or *D-regions*, adjacent to discontinuities or disturbances, where beam theory does not apply. D-regions can be *geometric discontinuities*, adjacent to holes, abrupt changes in cross section, or direction, or *statical discontinuities*, which are regions near concentrated loads and reactions. Corbels, dapped ends, and joints are affected by both statical and geometric discontinuities. Up to this point, most of this book has dealt with B-regions.

For many years, D-region design has been by "good practice," by rule of thumb or empirical. Three landmark papers by Professor Schlaich of the University of Stuttgart and his coworkers [18-1], [18-2], [18-3] have changed this. This chapter will present rules and guidance for the design of D-regions, based largely on these and other recent papers.

Saint Venant's Principle and Extent of D-regions

St. Venant's principle suggests that the localized effect of a disturbance dies out by about one member-depth from the point of the disturbance. On this basis, D-regions are assumed to extend one member-depth each way from the discontinuity. This principle is conceptual and not precise. However, it serves as a quantitative guide in selecting the dimensions of D-regions.

Figure 18-1 shows D-regions in a number of structures, some of which have B-regions (bending regions) between two D-regions. Figure 18-2 shows examples of D-regions. The D-regions in Fig. 18-2b and c extend one member-width from the discontinuity as suggested by St. Venant's principle. Occasionally, D-regions are assumed to fill the overlapping region common to two members meeting at a joint. This definition is used in the traditional definition of a joint region.

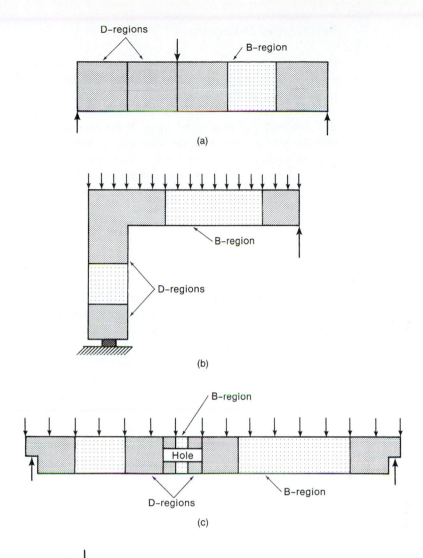

Fig. 18-1
B-regions and D-regions.

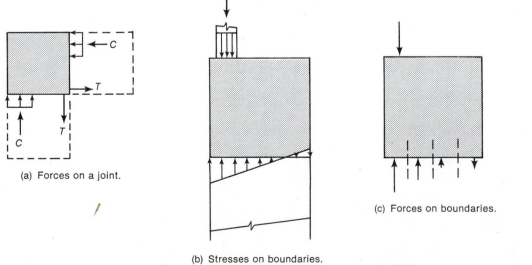

Fig. 18-2
Forces on boundaries of D-regions.

Behavior of D-Regions

Prior to any cracking, an elastic stress field exists, which can be quantified with an elastic analysis, such as a finite-element analysis. Cracking disrupts this stress field, causing a major reorientation of the internal forces. After cracking, the internal forces can be modeled via a *strut-and-tie model* consisting of concrete compression struts, steel tension ties, and joints referred to as nodal zones. If the compression struts are narrower at their ends than they are at midsection, the struts may, in turn, crack longitudinally. For unreinforced struts this may lead to failure. On the other hand, struts with transverse reinforcement to restrain the cracking can carry further load and will fail by crushing, as shown in Fig. 6-19. Failure may also occur by yielding of the tension ties, failure of the bar anchorage, or failure of the nodal zones. As always, failure initiated by yield of the steel tension ties tends to be more ductile and is desirable.

Strut-and-Tie Models

A strut-and-tie model for a deep beam is shown in Fig. 18-3. It consists of two concrete compressive *struts*, longitudinal reinforcement serving as a tension *tie*, and joints referred to as *nodes*. The concrete around a node is called a *nodal zone*. The nodal zones transfer the forces from the inclined struts to other struts, to ties and to the reactions.

ACI Section 11.1.1 allows D-regions to be designed using strut-and-tie models according to the requirements in ACI Appendix A, *Strut-and-tie models*. This Appendix is new in the 2002 ACI Code. The derivation of Appendix A is summarized in [18-4]. The examples in [18-5] were solved using the appendix as part of the internal check of Appendix A, by members of ACI Committee 318 Subcommittee E and ACI Committee 445.

A strut-and-tie model is a model of a portion of the structure that satisfies the following:

(a) it embodies a system of forces that is in equilibrium with a given set of loads, and

(b) the factored-member forces at every section in the struts, ties, and nodal zones do not exceed the corresponding factored-member strengths for the same sections.

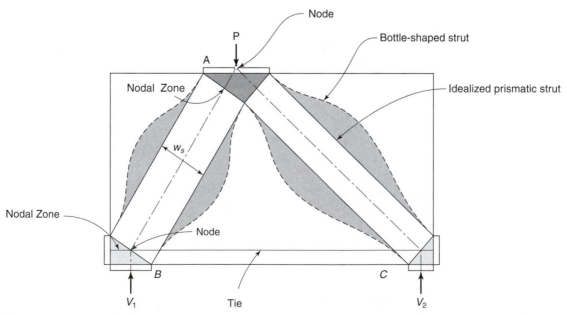

Fig. 18-3
Strut-and-tie model of a deep beam.

The *lower-bound theorem of plasticity* states that the capacity of a system of members, supports, and applied forces that satisfies both (a) and (b) is a lower bound on the strength of the structure. For the lower-bound theorem to apply, the structure must have:

(c) The structure has sufficient ductility to make the transition from elastic behavior to enough plastic behavior to redistribute the factored internal forces into a set of forces that satisfy items (a) and (b).

The combination of factored loads acting on the structure and the distribution of factored internal forces is a *lower bound on the strength of the structure*, provided that no element is loaded beyond its capacity. (The lower-bound theorem was introduced in Section 2-6.) For this reason, strut-and-tie models are chosen so that the internal forces in the struts, ties, and nodal zones are somewhere between the elastic distribution and the fully plastic set of internal forces.

18-2   DESIGN EQUATION AND METHOD OF SOLUTION

Before we embark on a design example, we will review the strengths of struts, ties, and nodal zones, and factors affecting the layout of strut-and-tie models. In most applications of strut-and-tie models the internal forces, F_u, due to the factored loads, and the struts, ties, and nodal zones are proportioned using

$$\phi F_n \geq F_u \qquad (18\text{-}1a)$$

Alternatively, the structural analysis is carried out for loads equal to F_u/ϕ, and the members are proportioned for F_n.

$$F_n \geq F_u/\phi \qquad (18\text{-}1b)$$

Equations (18-1a) and (18-1b) are called the *design equations*.

When design is based on a strut-and-tie model, either the load and resistance factors in ACI 318-02 Sections 9.2 and 9.3 are used, or the factors in Appendix C, Section C.2 and C.3 are used.

18-3   STRUTS

In a strut-and-tie model, the struts represent concrete compressive stress fields with the compression stresses acting parallel to the strut. Although they are frequently idealized as prismatic or uniformly tapering members, as shown in Fig.18-4a, struts generally vary in cross section along their length, as shown in Fig. 18-4b and c. This is because the concrete stress fields are wider at midlength of the strut than at the ends. Struts that change in width along the length of the member are sometimes idealized as *bottle-shaped* due to their shape, as shown in Fig. 18-4b, or are idealized using *local* strut-and-tie models, as shown in Fig. 18-4c. The spreading of the compression forces gives rise to transverse tensions in the strut which may cause it to crack longitudinally. A strut without transverse reinforcement may fail after this cracking occurs. If adequate transverse reinforcement is provided, the strength of the strut will be governed by crushing.

Strut Failure by Longitudinal Cracking

Figure 18-5a shows one end of a bottle-shaped strut. The width of the bearing area is a, and the thickness of the strut is t. At midlength the strut has an effective width b_{ef}. Reference [18-1] assumes that the bottle-shaped region at one end of a strut extends approximately $1.5b_{ef}$ from the end of the strut and in examples used $b_{ef} = \ell/3$ but not less than a, where

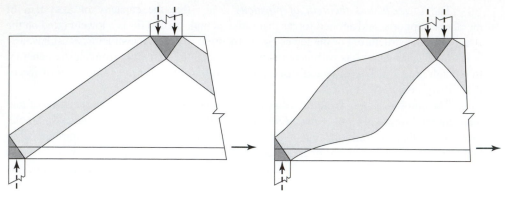

(a) Idealized prismatic strut.

(b) Bottle-shaped strut.

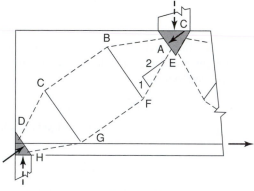

(c) Strut-and-tie model of a bottle-shaped strut.

Fig. 18-4
Bottle-shaped strut.

ℓ is the length of the strut from face to face of the nodes. For short struts, the limit that b_{ef} not be less than a often governs. We shall assume that in a strut with bottle-shaped regions at each end

$$b_{ef} = a + \ell/6 \quad \text{but not more than the available width} \tag{18-2}$$

Figures 18-4c and 18-5b show strut-and-tie models for the bottle-shaped region. It is based on the assumption made in [18-6], that the longitudinal projection of the inclined struts is equal to $b_{ef}/2$. The transverse tension force T at one end of the strut is

$$T = \frac{C}{2}\left(\frac{b_{ef}/4 - a/4}{b_{ef}/2}\right)$$

or

$$T = \frac{C}{4}\left(1 - \frac{a}{b_{ef}}\right) \tag{18-3}$$

The force T causes transverse stresses in the concrete, which may cause cracking. The transverse tensile stresses are distributed as shown by the curved line in Fig. 18-5c.

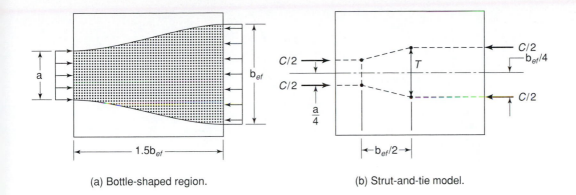

(a) Bottle-shaped region.

(b) Strut-and-tie model.

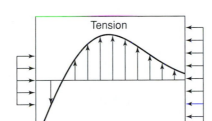

(c) Transverse tensions and compressions.

Fig. 18-5
Spread of stresses and transverse tensions in a strut.

Analyses by Adebar and Zhou [18-7] suggest that the tensile stress distributions at the two ends of a strut are completely separate when ℓ/a exceeds about 3.5 and overlap completely when ℓ/a is between 1.5 and 2. Assuming a parabolic distribution of transverse tensile stresses spread over a length of $1.6b_{ef}$ in a strut of length $2b_{ef}$ and equilibrating a tensile force of $2T$ indicates that the minimum load, C_n, at cracking is 0.51 to $0.57at\,f'_c$ for a strut with $a/b_{ef} = 1/2$. This analysis and [18-2] and [18-6] suggest that longitudinal cracking of the strut may be a problem if the bearing forces on the ends of the strut exceed:

$$C_n = 0.55at f'_c \qquad (18\text{-}4)$$

where at is the loaded area at the end of the strut.

   In tests of cylindrical specimens loaded axially through circular bearing plates with diameters less than that of the cylinders, failure occurred at 1.2 to 2 times the cracking loads [18-7].

   The maximum load on an unreinforced strut in a wall-like member such as the deep beam in Fig. 18-3, if governed by cracking of the concrete in the strut, is given by (18-3). This assumes that the compression force spreads in only one direction. If the bearing area does not extend over the full thickness of the member, there will also be transverse tensile stresses through the thickness of the strut that will require reinforcement through the thickness as shown in Fig. 18-6. This would require a reanalysis of the support region to design the transverse ties shown in Fig. 18-6a.

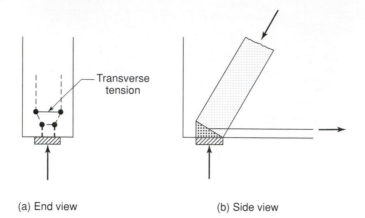

Fig. 18-6
Transverse spread of forces
through the thickness of a
strut.

(a) End view   (b) Side view

Compression Failure of Strut

The crushing strength of the concrete in a strut is referred to as the *effective strength*,

$$f_{cu} = \nu f'_c \tag{18-5}$$

where ν is an *efficiency factor* having a value between 0 and 1.0. ACI Section A.3.2 calls f_{cu} the *effective compressive strength*. Various sources give differing values of the efficiency factor [18-3], [18-8] to [18-12], and [18-15]. The major factors affecting the effective compression strength are:

    1.   **The concrete strength.** Concrete becomes more brittle and ν tends to decrease as the concrete strength increases.

    2.   **Load duration effects.** The strength of concrete beams and columns tends to be less than the cylinder strength, f'_c. Various reasons are given for this lower strength, including the observed reduction in compressive strength under sustained load, the weaker concrete near the tops of members due to vertical migration of bleed water during the placing of the concrete, and the different shapes of compression zones and test cylinders. For flexural members, ACI Section 10.2.7.1 accounts for this, in part, by taking the maximum stress in the equivalent rectangular stress block as $0.85f'_c$. For struts, load duration effects are accounted for by rewriting (18-5) as $f_{cu} = 0.85\beta_s f'_c$. Nodal zones are treated similarly except that β_s is replaced by β_n.

    3.   **Tensile strains transverse to the strut,** result from tensile forces in the reinforcement crossing the cracks [18-11] to [18-14]. In tests of uniformly strained concrete panels, such strains were found to reduce the compressive strength of the panels, as discussed in Section 3-2. The AASHTO Specification bases f_{cu} on this concept [18-14].

    4.   **Cracked struts.** Struts crossed by cracks inclined to the axis of the strut are weakened by the cracks. ACI Section A.3.1 presents the nominal compressive strength of a strut as:

$$F_{ns} = f_{cu}A_c \tag{18-6a}$$
$$\text{(ACI Eq. A-2)}$$

where subscript n = nominal, s = strut, A_c is the cross-sectional area at the end of the strut, and f_{cu} is:

$$f_{cu} = 0.85\beta_s f'_c \tag{18-7a}$$
$$\text{(ACI Eq. A-3)}$$

TABLE 18-1   ACI Code Values of β_s and β_n for Struts and Nodal Zones

Struts, $f_{cu} = 0.85\beta_s f_c'$

ACI Section A.3.2.1 For struts in which the area of the midsection cross section
is the same as the area at the nodes, such as the compression zone of a beam. . . . . $\beta_s = 1.0$

ACI Section A.3.2.2 For struts located such that the width of the midsection of the strut
is larger than the width at the nodes (bottle-shaped struts):

    (a) with reinforcement satisfying A.3.3. . . . . $\beta_s = 0.75$
    (b) without reinforcement satisfying A.3.3. . . . . $\beta_s = 0.60\lambda$

ACI Section A.3.2.3 For struts in tension members or the tension flanges
of members. . . . . $\beta_s = 0.40$

ACI Section A.3.2.4 For all other cases . . . . $\beta_s = 0.60$

Nodal zones, $f_{cu} = 0.85\beta_n f_c'$

ACI Section A.5.2.1 In nodal zones bounded on all sides by struts or bearing areas, or both. . $\beta_n = 1.0$

ACI Section A.5.2.2 In nodal zones anchoring a tie in one direction. . . . . $\beta_n = 0.80$

ACI Section A.5.2.3 In nodal zones anchoring two or more ties. . . . . $\beta_n = 0.60$

As stated in item 2 of this list, the term 0.85 in (18-7a) is equivalent to the 0.85 in ACI
Section 10.2.7.1. Values of β_s are given in Table 18-1. For nodal zones, (18-6a) and (18-7a)
become

$$F_{nn} = f_{cu}A_n \qquad (18\text{-}6b)$$
$$(\text{ACI Eq. A-7})$$

and

$$f_{cu} = 0.85\beta_n f_c' \qquad (18\text{-}7b)$$
$$(\text{ACI Eq. A-8})$$

Values of β_n are also given in Table 18-1. These were derived by ACI Committee 318E
[18-4]. Examples of the use of ACI Appendix A are given in [18-5].

Explanation of Types of Struts Described in Table 18-1

Case A.3.2.1 applies to a strut equivalent to a rectangular stress block of depth, a, and
thickness, b, as occurs in the compression zones of beams or eccentrically loaded columns.
In this case β_s is equal to 1.0. The corresponding neutral axis depth is $c = a/\beta_1$. The strut
is assumed to have a depth of a and the resultant compressive force in the rectangular stress
block, $C = f_{cu}ab$, acts at $a/2$ from the most compressed face of the beam or column as
shown in Fig. 18-7.

    A.3.2.2(a) applies to bottle-shaped struts similar to those in Fig. 18-4b which contain
reinforcement crossing the potential splitting cracks. Although such struts tend to split longitudinally, the opening of a splitting crack is restrained by the reinforcement allowing the
strut to carry additional load after the splitting cracks develop. For this case $\beta_s = 0.75$. If
there is no reinforcement to restrain the opening of the crack, the strut is assumed to fail
upon cracking, or shortly after, and a lower value of β_s is used.

    The yield strength of the reinforcement required to restrain the crack is taken equal
to the tension force that is lost when the concrete cracks. This is computed using a localized strut-and-tie model of the cracking in the strut as shown in Fig. 18-4c. As discussed
later, the slope of the load-spreading struts is taken as 2 to 1 (parallel to axis of strut, to perpendicular to axis):

$$T_n = \frac{C_n}{2}\left(\frac{b_{ef}/4 - a/4}{b_{ef}/2}\right) \qquad (18\text{-}8)$$

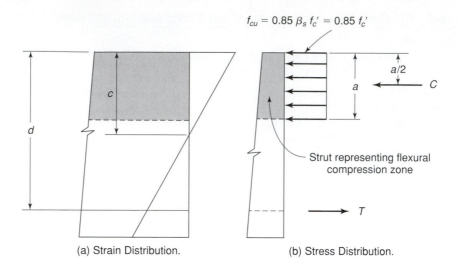

$f_{cu} = 0.85 \, \beta_s \, f_c' = 0.85 \, f_c'$

Strut representing flexural
compression zone

Fig. 18-7
Strut representing the compression stress block in a
beam

(a) Strain Distribution.   (b) Stress Distribution.

Rearranging and setting T_n equal to $A_{si}f_y$ gives the transverse tension force T_n at the ends of the bottle-shaped strut at cracking as

$$A_s f_y \geq \Sigma \left[\frac{C_n}{4} \left(1 - \frac{a}{b_{ef}} \right) \right] \qquad (18\text{-}9)$$

where C_n is the nominal compressive force in the strut and a is the width of the bearing area at the end of the strut, computed from the distance between the longitudinal struts at the end of the strut-and-tie model of the bottle-shaped strut, shown as $a/4$ in Fig. 18-5b. The width of the bottle-shaped strut, b_{ef}, is computed from the distance between the longitudinal struts and the axis of the strut at midlength of the strut-and-tie model of the strut, $b_{ef}/4$, also shown in Fig. 18-5b. The summation Σ implies the sum of the values at the two ends of the strut. If the reinforcement is at an angle θ to the axis of the strut, $A_s f_y$ should be multiplied by $\sin \theta$. This reinforcement will be referred to as *crack control reinforcement*.

In lieu of using a strut-and-tie model to compute the necessary amount of crack control reinforcement, ACI Section A.3.3.1 allows the crack control reinforcement to be determined using (18-10):

$$\Sigma \frac{A_{si}}{b s_i} \sin \gamma_i \geq 0.003 \qquad (18\text{-}10)$$

(ACI Eq. A-4)

where A_{si} refers to the crack control reinforcement adjacent to the two faces of the member at an angle γ_i to the crack, as shown in Fig. 18-8. The arrangement of the crack control reinforcement is specified in ACI Section A.3.3.2.

Equation (18-10) was written in terms of a reinforcement ratio rather than the tie force to simplify the presentation. This is acceptable for concrete strengths not exceeding 6000 psi. (See ACI Section A.3.3). For higher concrete strengths the committee felt the load-spreading should be computed. A tensile strain in bar 1, ε_{s1}, in Fig. 18-8 results in a tensile strain of $\varepsilon_{s1} \sin \gamma_1$ perpendicular to the axis of the strut. Similarly for bar 2, the strain perpendicular to the axis of the strut is $\varepsilon_{s2} \sin \gamma_2$, where $\gamma_1 + \gamma_2 = 90°$.

A.3.2.2(b) In mass concrete members such as pile caps for more than two piles, it may be difficult to place the crack control reinforcement. ACI Section A.3.2.2(b) specifies a lower value of f_{cu} in such cases. Because the struts are assumed to fail shortly after

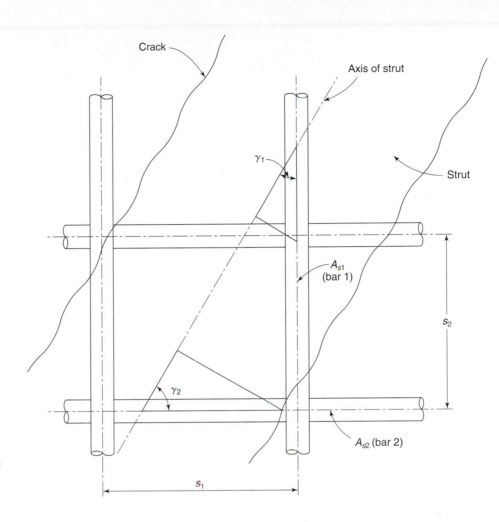

Fig. 18-8
Crack control reinforcement
crossing a strut in a cracked
web.

longitudinal cracking occurs, β_s is multiplied by the correction factor, λ, for lightweight concrete when such concrete is used. Values of λ are given in ACI Section 11.7.4.3. It is 1.0 for normal-weight concrete.

A.3.2.3 The value of β_s in ACI Section A.3.2(3) is used in proportioning struts in strut-and-tie models used to design the reinforcement for the tension flanges of beams, box girders and the like. It accounts for the fact that such cracks will tend to be wider than cracks in beam webs.

A.3.2.4 The value of β_s in ACI Section A.3.2.4 applies to all other types of struts not covered in A.3.2.1, A.3.2.2, and A.3.2.3. This includes struts in the web of a beam where more or less parallel cracks divide the web into parallel struts. It also includes struts likely to be crossed by cracks at an angle to the struts.

Design of Struts

Once the strut-and-tie model has been laid out, the strength of a strut is computed as follows:

1. If there is no transverse reinforcement in the strut, the strength should be taken as the compression causing cracking, computed as

$$C_{cr} = 0.85(0.60\lambda)f'_c \, at \tag{18-11}$$

where a and t, respectively, are the width and thickness of the nodal zone.

2. If the strut is crossed by reinforcement satisfying (18-9) or (18-10), the nominal strength of the strut should be based on the smallest cross-sectional area of the strut and on the applicable effective concrete strength from Table 18-1.

18-4 TIES

The second major component of a strut-and-tie model is the tie. A tie represents one or several layers of reinforcement in the same direction. Design is based on

$$\phi F_{nt} \geq F_{ut} \qquad (18\text{-}12)$$

where the subscript t refers to "tie," and F_{nt} is the nominal strength of the tie, taken as

$$F_{nt} = A_{st} f_y + A_{ps}(f_{se} + \Delta f_p) \qquad (18\text{-}13)$$
$$(\text{ACI Eq. A-6})$$

The second term in the parentheses on the right-hand side of (18-13) is for prestressed ties. It drops out if the ties are not prestressed.

ACI Section A.4.2 requires that the axis of the reinforcement in a tie coincide with the axis of the tie. In the layout of a strut-and-tie model, ties consist of the reinforcement plus a prism of concrete concentric with the longitudinal reinforcement making up the tie. The width of the concrete prism surrounding the tie is referred to as the *effective width* of the tie, w_t. ACI Sections A.4.2 and R.A.4.2 give limits on w_t. The lower limit is a width equal to twice the distance from the surface of the concrete to the centroid of the tie reinforcement. In a hydrostatic C–C–T nodal zone (defined in Section 18-5), the stresses on all faces of the nodal zone should be equal. As a result, the upper limit on the width of a tie is taken equal to

$$w_{t,\max} = F_{nt}/(f_{cu}b) \qquad (18\text{-}14)$$

The concrete is included in the tie to establish the widths of the faces of the nodal zones acted on by ties. The concrete in a tie does not resist any load. It aids in the transfer of loads from struts to ties or to bearing areas through bond with the reinforcement. The concrete surrounding the tie steel increases the axial stiffness of the tie by tension stiffening. Tension stiffening may be used in modeling the axial stiffness of the ties in a serviceability analysis.

Ties may fail due to lack of end anchorage. The anchorage of the ties in the nodal zones is a major part of the design of a D-region using a strut-and-tie model. Ties are shown as solid lines in strut-and-tie models.

18-5 NODES AND NODAL ZONES

The points at which the forces in struts-and-ties meet in a strut-and-tie model are referred to as *nodes*. Conceptually, they are idealized as pinned joints. The concrete in and surrounding a node is referred to as a *nodal zone*. In a planar structure, three or more forces must meet at a node for the node to be in equilibrium, as shown in Fig. 18-9. This requires that

$$\Sigma F_x = 0 \qquad \Sigma F_y = 0 \qquad \text{and } \Sigma M = 0 \qquad (18\text{-}15)$$

The $\Sigma M = 0$ condition implies that the lines of action of the forces must pass through a common point, or must be able to be resolved into forces that act through a common point. The two compressive forces shown in Fig. 18-9a meet at an angle and are not in equilibrium unless a third force is added, as shown in Fig. 18-9b or c. Nodal zones are

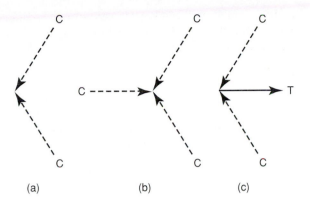

Fig. 18-9
Forces acting on nodes.

       (a)   (b)   (c)

classified as C–C–C if three compressive forces meet, as in Fig. 18-9b, and as C–C–T if one of the forces is tensile as shown in Fig. 18-9c. C–T–T joints and T–T–T joints may also occur.

Hydrostatic Nodal Zones

Two common ways of laying out nodal zones are illustrated in Figs. 18-10 and 18-11. The prismatic compression struts in Figs. 18-3 and 18-4a are assumed to be stressed in uniaxial compression. A section perpendicular to the axis of a strut is acted on only by compression stresses, while sections at any other angle have combined compression and shear stresses. One way of laying out nodal zones is to orient the sides of the nodes at right angles to the axes of the struts or ties meeting at that node, as shown in Fig. 18-10, and to have the same bearing pressure on each side of the node. When this is done for a C–C–C node, the ratio of the lengths of the sides of the node, $w_1 : w_2 : w_3$, is the same as the ratio of the forces in the three members meeting at the node, $C_1 : C_2 : C_3$, as shown in Fig. 18-10a. Nodal zones laid out in this fashion are sometimes referred to as *hydrostatic nodal zones* since the in-plane stresses in the node are the same in all directions. In such a case, the Mohr's circle for the in-plane stresses reduces to a point. If one of the forces is tensile, the width of that side of the node is calculated from a *hypothetical bearing plate* on the end of the tie, which is assumed to exert a bearing pressure on the node equal to the compressive stress in the struts at that node, as shown in Fig. 18-10b. Alternatively, the reinforcement may extend through the nodal zone to be anchored by bond, hooks, or mechanical anchorage before the reinforcement reaches point A on the right-hand side of the extended nodal zone, as shown by Fig. 18-10c. Such a nodal zone approaches being a hydrostatic C–C–C nodal zone. However, the strain incompatibility resulting from the tensile steel strain adjacent to the compressive concrete strain reduces the strength of the nodal zone. Thus, this type of joint should be designed as a C–C–T joint with $\beta_n = 0.80$.

Geometry of Hydrostatic Nodal Zones

Because the stresses are equal or close to equal on all faces of a hydrostatic nodal zone that are perpendicular to the plane of the structure, equations can be derived relating the lengths of the sides of the nodal zone to the forces in each side of the nodal zone. Figure 18-10a shows a hydrostatic C–C–C node. For a nodal zone with a 90° corner, as shown, the horizontal width of the bearing area is $w_3 = \ell_b$. The height of the vertical side of the nodal

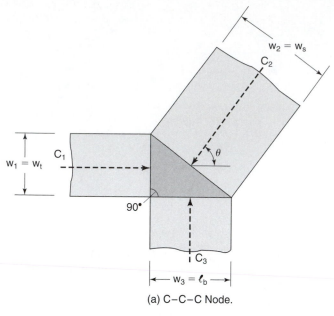

(a) C–C–C Node.

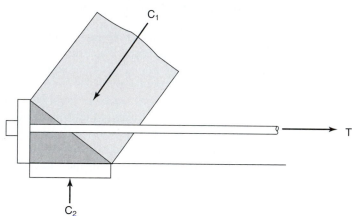

(b) C–C–T Node.

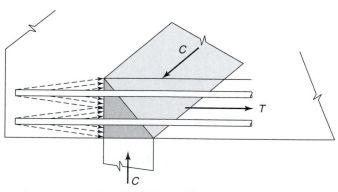

Fig. 18-10
Hydrostatic nodal zones in
planar structures.

(c) *C–C–T* Node, *T* anchored by Bond.

zone is $w_1 = w_t$. The angle between the axis of the inclined strut and the horizontal is θ. The width of the third side, the strut, w_2, $= w_s$, can be computed as

$$w_s = w_t \cos \theta + \ell_b \sin \theta \qquad (18\text{-}16)$$

This equation can also be applied to a C–C–T node, as shown in Fig. 18-10b. If the width of the strut, w_s, computed from the strut force by using (18-6a), is larger than the width given by (18-16), it is necessary to increase either w_t or ℓ_b or both, until the width from (18-16) equals or exceeds the width calculated from the strut forces.

Extended Nodal Zones

The use of hydrostatic nodes can be tedious in design, except possibly for C–C–C nodes. More recently, the design of nodal zones has been simplified by considering the nodal zone to comprise that concrete lying within extensions of the members meeting at the joint as shown in Fig. 18-11 [18-3], [18-18]. This allows different stresses to be assumed in the struts and over bearing plates, for example. Two examples are given in Fig. 18-11. Figure 18-11a shows a C–C–T node. The bars must be anchored within the nodal zone or to the left of point A, which ACI Section A.4.3.2 describes as "the point where the centroid of the reinforcement in the tie leaves the extended nodal zone." The length, ℓ_d, in which the bars of the tie must be developed is shown. The vertical face of the node is acted on by a stress equal to the tie force T divided by the area of the vertical face. The stresses on the three faces of the node can all be different, provided that

1. the resultants of the three forces coincide,
2. the stresses are within the limits given in Table 18-1, and
3. the stress is constant on any one face.

An extended nodal zone consists of the node itself, plus the concrete in extensions of the struts, bearing areas, and ties that meet at a joint. Thus in Fig. 18-11a the darker shaded region indicates the nodal zone extends into the area occupied by the struts and ties at this node. This layout of a nodal zone contains much of the concrete stressed in compression

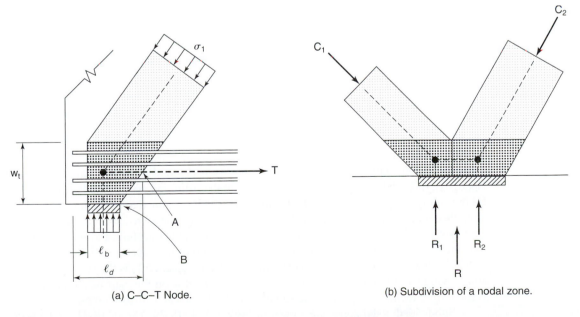

(a) C–C–T Node.

(b) Subdivision of a nodal zone.

Fig. 18-11
Extended nodal zones.

over a reaction. An alternate and sometimes easier nodal zone to use is shown in Fig. 18-10b, where the assumed nodal zone is the smallest size possible for this node because it does not include any concrete that is not common to the struts, bearing areas, and ties at the node. The advantage of the nodal zones in Fig. 18-11a and b comes from the fact that ACI Section A.4.3.2 allows the length available for bar development to anchor the tie bars to be taken out to point A in Fig. 18-11a rather than point B at the edge of the bearing plate. This extended anchorage length recognizes the beneficial effect of the compression from the re-action and the struts for improving bond between the concrete and the tie reinforcement.

Strength of Nodal Zones

Nodal zones are assumed to fail by crushing. Anchorage of the tension ties is a matter of design consideration. If a tension tie is anchored in a nodal zone there is a strain incompat-ibility between the tensile strains in the bars and the compressive strain in the concrete of the node. This tends to weaken the nodal zone. ACI Section A.5.1 limits the effective con-crete strengths, f_{cu}, for nodal zones as:

$$F_{nn} = f_{cu}A_n$$
$$\text{(18-6b)}$$
$$\text{(ACI Eq. A-7)}$$

where A_n is the area of the face of the node that the strut or tie acts on, taken perpendicular to the axis of the strut or tie, or the area of a section through the nodal zone, and f_{cu} is the effective compression strength of the concrete

$$f_{cu} = 0.85\,\beta_n f'_c$$
$$\text{(18-7b)}$$
$$\text{(ACI Eq. A-8)}$$

where the term 0.85 is equivalent to the 0.85 term in ACI Section 10.2.7.1. ACI Section A.5.1 gives the following three values of β_n for nodal zones. (See also Table 18-1.)

1. $\beta_n = 1.0$ in C–C–C nodal zones bounded by compressive struts and bearing areas.
2. $\beta_n = 0.80$ in C–C–T nodal zones anchoring a tension tie in only one direction.
3. $\beta_n = 0.65$ in C–T–T nodal zones anchoring tension ties in more than one direction.

Tests of C–C–T and C–T–T nodes reported in [18-16] and [18-17] developed $\beta_n = 0.95$ in properly detailed nodal zones.

Subdivision of Nodal Zones

Frequently, it is easier to lay out the size and location of nodal zones if they are subdivided into several parts, each of which is assumed to transfer a particular component of the load through the nodal zone. In Fig. 18-11b, the reaction R has been divided into two compo-nents R_1, which equilibrates the vertical component of C_1, and R_2, which equilibrates C_2. Generally, this subdivision simplifies the layout of the struts and nodes. Subdivision is use-ful in dealing with the dead load of a beam which can be assumed to be applied as a series of equivalent concentrated loads, each of which is transferred to a reaction by an individual strut as shown in Fig. 18-13a. In this figure, the dead load is transferred to the support by four inclined struts, which are supported by the portion of the support nodal zone labeled as V_s. The horizontal width of the part of the nodal zone labelled V_s is the sum of the hori-zontal widths of the four struts that support the dead loads on this half of the beam. The vertical truss members represent the subdivided dead loads and stirrup forces.

Subdivided nodal zones are shown in Figs. 18-24, 18-29, 18-30, 18-31, and 18-35. Equation 18-16 is used to compute the widths perpendicular to the axis of the struts of inclined

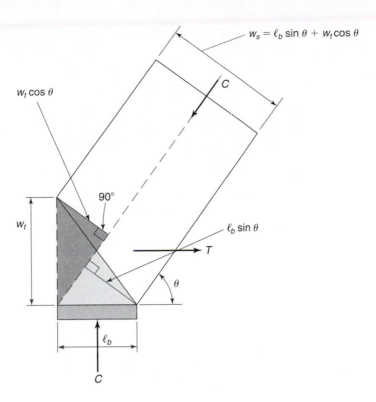

$$w_s = \ell_b \sin\theta + w_t \cos\theta$$

Fig. 18-12
Width of inclined strut at a
C–C–T nodal zone.

struts in extended nodal zones, as shown in Fig. 18-12, even though these equations derived for hydrostatic nodal zones. These equations are useful in adjusting the width of an inclined strut if the original width is found to be inadequate. The lower limit on the width of a tie in an extended nodal zone corresponds to the bars in the tie being placed in one layer with a minimum thickness of concrete equal to that of the cover concrete both above and below the bar. The width of such a tie is defined in ACI Section RA.4.2 as the diameter of the bars plus twice the cover to the surface of the bars $w_{t,min} = d_b + 2c$. The maximum width is:

$$w_{t,max} = F_{nt}/(f_{cu}b) \tag{18-17}$$

Resolution of Forces Acting on a Nodal Zone

If more than three forces act on a nodal zone in a planar strut-and-tie model, it is sometimes advantageous to resolve several strut or tie forces so that three forces remain. Figure 18-14 shows a hydrostatic nodal zone that is in equilibrium with four strut forces. The forces on sides A–E and E–C can be replaced with a single strut acting on the side AC. A–C. This is equivalent to including the small nodal zone A–C–E–A. For the nodal zone to be in equilibrium, the forces on the final nodal region A–B–C–A, must pass through a common point, as must the forces on the second nodal zone A–C–E–A.

Another example is shown in Fig. 18-11b. This may be divided into two subnodes. It is necessary to ensure that the stresses in the members entering the node, the stress over the bearing plate, and the stress on any vertical line that divides the two subnodes are within the limits in Table 18-1.

Anchorage of Ties in Nodal Zones

A challenge in design using strut-and-tie models is the anchorage of the tie forces at the nodal zones at the edges or ends of a strut-and-tie model. This problem is independent of

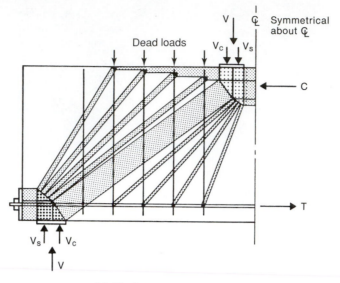

(a) Plastic truss model.

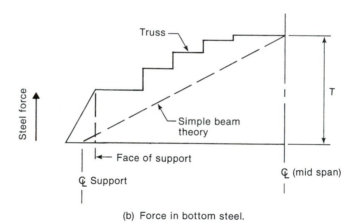

Fig. 18-13
Strut-and-tie model of a deep
beam with dead load and stirrups.

(b) Force in bottom steel.

the type of analysis used in design. It occurs equally in structures designed by elastic analyses or strut-and-tie models. In fact, one of the advantages of strut-and-tie models comes from the attention that the strut-and-tie model places on the anchorage of ties as described in ACI Section A.4.3. For nodal zones anchoring one tie, the tie must be developed by bond, by hooks, or by mechanical anchorage between the free end of the bar and the point at which the centroid of the tie reinforcement leaves the compressed portion of the nodal zone. This corresponds to point *A* in Fig. 18-11a. If the bars are anchored by hooks, the hooks should be confined within reinforcement extending into the member from the supporting column, if applicable. This location was chosen by committee consensus. The more usual location of the end of the anchorage length at point *B* was felt to be too short, taking into account the approximate size and confinement of the extended nodal zone.

European practice [18-18] sometimes uses lap splices between the tie bars and U bars lying horizontally. Typically two layers of U bars are used to anchor one layer of tie bars. Each layer of U bars is designed to anchor one-third of the total bar force, leaving one-third to be anchored by bond stresses on the tie bars.

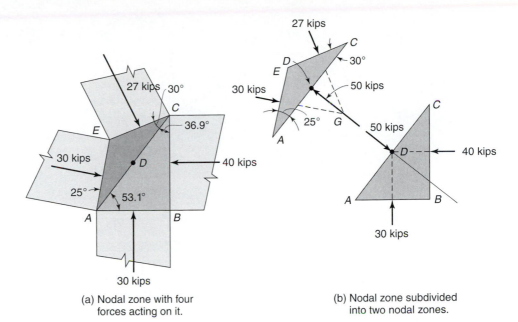

Fig. 18-14
Resolution of forces acting
on a nodal zone.

(a) Nodal zone with four
forces acting on it.

(b) Nodal zone subdivided
into two nodal zones.

Nodal Zone Anchored by a Bent Bar

Sometimes the two tension ties in a C–T–T node are both provided by a bar bent through $90°$
as shown in Fig. 18-15. The compressive force in the strut can be anchored by bearing and
shear stresses transferred from the strut to the bent bar. Figure 8-13a shows the bar forces measured in a standard $90°$ bend with an inside radius of 3 bar diameters. In the test shown in
that figure, over half of the bar stress was dissipated between the start and end of the bent portion. Such a detail must satisfy the laws of statics and limits on the bearing stresses on the concrete inside the bent bar.

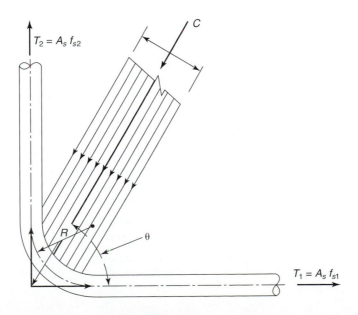

Fig. 18-15
C–T–T node anchored by a
bent bar.

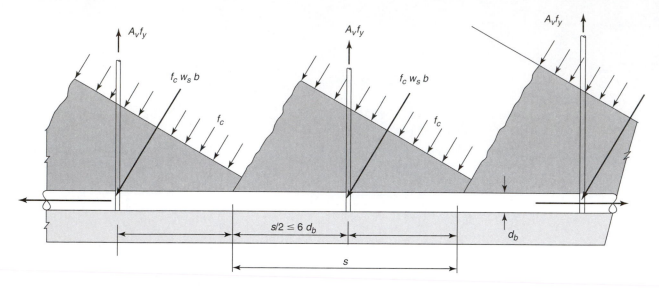

Fig. 18-16
Struts anchored by stirrups and longitudinal bars.

Strut Anchored by Reinforcement

Sometimes, diagonal struts in the web of a truss model of a flexural member are anchored by longitudinal reinforcement that, in turn, is supported by a stirrup, as shown in Fig. 18-16. Reference [18-11] recommends that the length of longitudinal bar able to support the strut be limited to 6 bar diameters each way from the center of the strut.

The Use of a Strut-and-Tie Model in Design

Example 18-1 considers the design of a wall loaded and supported by columns. The purpose of the example is to illustrate the choice and use of a strut-and-tie model, to demonstrate the choice of D-regions, and to discuss reasons for making certain assumptions.

EXAMPLE 18-1   Design of D-Regions in a Wall.

The 14 in. thick wall shown in Fig. 18-17 supports a 14 in. by 20 in. column carrying unfactored loads of 100 kips dead load and 165 kips live load, plus the weight of the wall. The wall in Fig. 18-17a supports this column and is supported on two other columns which are 14 by 14 in. The floor slabs (not shown) provide stiffness against out-of-plane buckling. Design the reinforcement. Use $f'_c = 3000$ psi and $f_y = 60,000$ psi. The primary design equation is $\phi F_n \geq F_u$ where F_n is the nominal capacity of the element, and F_u is the force on the element due to the factored loads.

1.   **Isolate the D-regions.** The loading discontinuities due to the column load on the wall dissipate in a distance of approximately one member dimension from the location of the discontinuity. Based on this, the wall will be divided into two D-regions separated by a B-region as shown in Fig. 18-17a. The wall has two statical discontinuities:

(i) the first is under the column load at the top, and
(ii) the second is over the two columns supporting the bottom of the wall.

Using St. Venant's principle, the D-regions are assumed to extend a distance equal to the width of the wall (8 ft) down from the top and the same distance up from the tops of the two columns that support

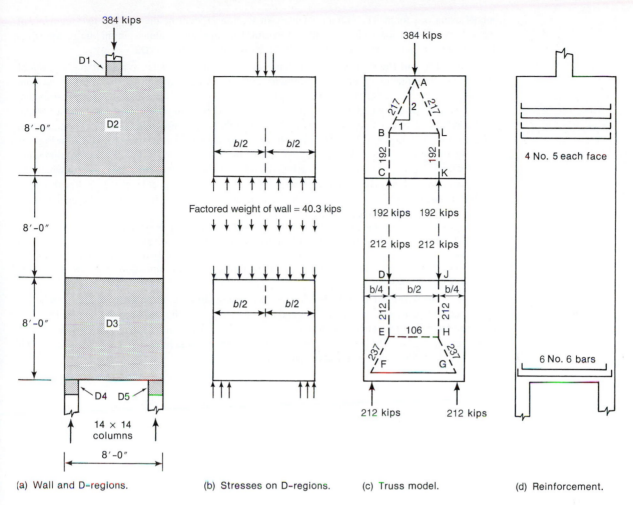

(a) Wall and D–regions.

(b) Stresses on D–regions.

(c) Truss model.

(d) Reinforcement.

Fig. 18-17
D-regions in a wall—Example 18-1.

the wall. There are three more D-regions at the ends of the columns, which have little effect on the wall and will not be considered in this example.

The self weight of the wall is $(24 \text{ ft} \times 8 \text{ ft} \times 14/12 \text{ ft} \times 0.150 \text{ kips/ft}^3) = 33.6$ kips. We shall assume that this acts as a uniformly distributed load acting on the structure at midheight of the wall as shown in Fig. 18-17b and c.

2.   **Compute the factored loads.** Using the load factors in ACI Sec. 9.2, the factored load on the upper column is the larger of:

$$U = 1.4 \times 100 \text{ kips} = 140 \text{ kips} \qquad \text{(ACI Eq. 9-1)}$$

$$U = 1.2 \times 100 \text{ kips} + 1.6 \times 165 \text{ kips} = 384 \text{ kips} \qquad \text{(ACI Eq. 9-2)}$$

By inspection, the rest of the ACI Load Combinations (ACI Eq. 9-3 to 9-7) do not govern the vertical loads on the wall. The factored weight of the wall is $1.2 \times 33.6 = 40.3$ kip.

3.   **Subdivide the boundaries of the D-regions and compute the force resultants on the boundaries of the D-region.** For $D2$, we can represent the load on the top boundary by a single force of 384 kips at the center of the column, or as two forces of 384 kips/2 $= 192$ kips acting at the quarter points of the width of the column at the interface with the wall. We shall draw the strut-and-tie

model using one force. The bottom boundary of Region $D2$ will be divided into two segments of equal lengths, $b/2$ each with its resultant force of 192 kips acting along the middle of the struts loaded by the column above. This gives uniform stress on the bottom of $D2$.

3.   **Lay out the strut-and-tie models** Two strut-and-tie models are needed, one in each of $D2$ and $D3$. The function of the upper strut-and-tie model of $D2$ is to transfer the column load from the center of the top of $D2$ to the bottom of $D2$, where the load is essentially uniformly distributed. Figure 18-21a (discussed later) shows the stress trajectories from an elastic analysis of a vertical plate loaded with in-plane loads. The dashed lines in Fig. 18-21a represent the flow of compression stresses, and the solid lines show the directions of the tensile stresses. Struts A-B and B-C in Fig. 18-17c replace the stress trajectories in the left half of the D-region in Fig. 18-21a. The compression stresses fan out from the column, approaching a uniformly distributed stress at the height where the struts pass through the quarter points of the section. In $D2$ this occurs at level B-L. Below this level struts B-C and L-K are vertical and pass through the quarter points of the width of the section. This gives uniform compression stresses over the width.

For $D3$, similarly, the load on the top of D-region $D3$ will be represented by struts at the quarter points of the top of the D-region. The stress trajectories in $D3$ are equivalent to those in Fig. 18-27a (discussed later). The strut-and-tie model in $D3$ transfers the uniformly distributed loads, including the dead load of the wall, from the top of $D3$ down to the two concentrated loads where the wall is supported by the columns.

4.   **Draw the Strut-and-Tie Models.** In drawing strut-and-tie models, compression struts will always be plotted using dashed lines and tensile members with solid lines. In Section 18-6 it is recommended that load-spreading strut-and-tie models with struts at a (2 to 1) slope relative to the axis of the applied load be used, i.e., struts at (2 units parallel to the force that is spreading) to (1 unit perpendicular to the force). These correspond to θ.

$$\theta = \arctan 1/2 = 26.6°$$

from the axis of the force. The strut-and-tie models are shown in Fig. 18-17c. The forces in the struts and ties in the wall are listed in Table 18-2.

5.   **Compute the forces and strut widths in both two strut-and-tie models.** The calculations are given in Table 18-2.

6.   **D-region D2.**

(a) **Node A and Struts A–B and A–L:** Treating Node A as a hydrostatic node, either the node at A or one of the struts A–B and A–L will control.

Table 18–2   Calculation of Forces in the Strut-and-Tie Models—Example 18-1.

D-Region (1)	Member (2)	Vertical Force Component, kips (3)	Horizontal Force Component, kips (4)	Axial Force, kips (5)	Effective Concrete Strength, f_{cu}, psi (6)	Width of Strut or Nodal Zone w_s^-, in. (7)
D2	Node A	384	0	384	1910	19.1
	A-B	192	97	217	1910	10.8
	B-C	192	0	192	2040	8.96
	A-L	192	97	217	1910	10.8
	L-K	192	0	192	2040	8.96
	B-L	0	96	96	2040	4.53
D3						
	D-E	212	0	212	2040	9.90
	E-F	212	106	237	2040	11.1
	F-G	0	106	106	2040	4.95
	G-H	212	106	237	2040	11.1
	H-J	212	0	212	2040	9.90
	E-H	0	106	106	2040	4.95

Node A: Because this node is compressed on all in-plane faces, $\beta_n = 1.0$ and the effective compression stress for node A from Table 18-1 is:

$$f_{cu} = 0.85 \times 1.0 \times 3000 \text{ psi} = 2550 \text{ psi}$$

Struts A–B and A–L: Because the stresses in the concrete beside struts A–B and A–L are low, a portion of the stress in the struts is resisted by the concrete adjacent to the idealized prismatic struts, making these bottle-shaped struts. We will provide vertical and horizontal reinforcement satisfying ACI Sec. A.3.3, thereby allowing ACI Sec. A.3.2.2(a) to apply with $\beta_s = 0.75$. This allows f_{cu} in Strut A–B or Strut A–C to be

$$f_{cu} = 0.85 \times 0.75 \times 3000 \text{ psi} = 1910 \text{ psi}.$$

Because this is less than $f_{cu} = 2550$ psi for the node, 1910 psi governs. For the factored load in the column at A, $P_u = 384$ kips

an area of $\dfrac{384 \text{ kips} \times 1000 \text{ lbs/kip}}{0.75 \times 1910 \text{ psi}} = 268 \text{ in.}^2$ is required. The column loading the wall is $14 \times 20 = 280 \text{ in.}^2$, therefore the column is large enough.

(b) Minimum dimensions for Nodes B and L: These are C-C-T nodes, so the effective compressive stress from Table 18-1 is:

$$f_{cu} = 0.85 \times 0.80 \times 3000 \text{ psi} = 2040 \text{ psi}$$

This will control the base dimension of nodes B and L because struts B–C and L–K are prismatic struts that can be designed by using $\beta_s = 1.0$. Thus, the base dimension of node B and the width of strut B–C is:

$$W_s = \frac{192{,}000 \text{ lbs}}{0.75 \times 2040 \text{ psi} \times 14 \text{ in.}} = 8.96 \text{ in.}$$

This is much less than b/2 (4 ft), so the node easily fits within the dimensions of the wall. The height of node B is of interest for tie B–L. So,

$$W_t = \frac{97{,}000 \text{ lbs}}{0.75 \times 2040 \text{ psi} \times 14 \text{ in.}} = 4.53 \text{ in.}$$

This is a very small dimension and the reinforcement for tie B–L will be spread over a larger distance. Essentially, the dimensions of nodes B and L will be much larger than the minimum values calculated here.

(c) Required Area of Reinforcement for Tie B–L

$$\text{Tie force } T_u = \frac{192 \text{ kips}}{\tan \theta} = 97.0 \text{ kips}$$

$$\text{Required } A_s = \frac{T_u}{\phi f_y} = \frac{97.0 \text{ kips}}{0.75 \times 60 \text{ ksi}} = 2.16 \text{ in.}^2$$

We will choose the steel after the minimum reinforcement has been computed. Essentially, a band of transverse steel having this area should be provided across the full width of the wall extending about 25 percent of the width of the wall above and below the position of tie B–L so that the centroid of the areas of the bars is close to tie B–L. (See Fig. 18-17d.) Both ends of each bar should be hooked.

7.   D-region D3   Nodes F and G are C-C-T nodes, similar to nodes B and L. We will use the effective compressive stress for these nodes ($f_{cu} = 2040$ psi) to determine the minimum dimensions for all the struts, ties, and nodes in D-region D3, as given in column 7 of Table 18-2. Clearly, all of these element dimensions easily fit within the dimensions of the wall and supporting columns.

(a) Required Area of Reinforcement for Tie *F–G*

$$\text{Tie force } T_u = 106 \text{ kips}$$

$$\text{Required } A_s = \frac{106 \text{ kips}}{0.75 \times 60 \text{ ksi}} = 2.36 \text{ in.}^2$$

Thus, use 6 No. 6 bars, $A_s = 2.64$ in.², placed in two layers of three bars per layer. This should put the centroid of these bars approximately at the mid-height of tie *F–G*, whose height (width) is given in Table 18-2. All of these bars must be hooked at the edges of the wall, as shown in Fig. 18-17d.

8. Minimum distributed wall reinforcement Minimum requirements for wall reinforcement are covered in detail in Chapter 19. For this problem, we will assume the requirements of ACI Section 14.3 govern. From ACI Section 14.3.3, the minimum percentage of Grade 60 horizontal reinforcement is 0.0020, with a maximum spacing not to exceed three times the wall thickness or 18in. (ACI Section 14.3.5). For a wall width of 14 in., reinforcement will be required in each face. Thus, throughout the height of the wall, except for the locations of ties *B–L* and *F–G*, provide No. 4 bars in each face at a vertical spacing of 14in. o.c. (horizontal reinforcement ratio = 0.00204).

Reinforcement for tie *F–G* was selected in step 7. From step 6, the area of reinforcement required for tie *B–L* was 2.16 in.². **Use 8 No. 5 bars ($A_s = 2.48$ in.²) at a vertical spacing of 12 in., half in each face and hooked at both ends (Fig. 18-17d).** This spacing provides a tie width of approximately 4ft, as recommended in step 6. ∎

18-6 COMMON STRUT-AND-TIE MODELS

Compression Fans

A compression fan is a series of compression struts that radiate out from a concentrated applied force to distribute that force to a series of localized tension ties, such as the stirrups. Fans are shown over the reaction and under the load in Fig. 18-18. The failure of a compression fan is shown in Fig. 6-23.

Compression Fields

A compression field is a series of parallel compression struts combined with appropriate tension ties and compression chords, as shown in Fig. 18-18. Compression fields are shown between the compression fans in Fig. 18-18 and Fig. 6-20b.

Force Whirls, U-Turns

In Fig. 18-19 the column load causes the stresses shown by the shaded areas at the bottom of the D-region. The stresses on the bottom edge, *A–I*, have been computed with the use of the formula $\sigma = (P/A) + (My/I)$. The neutral axis (axis of zero strains) is at *G*, 26.7 in. from the right edge of the wall. The widths of *E–G* and *G–I* have both been chosen as 26.7

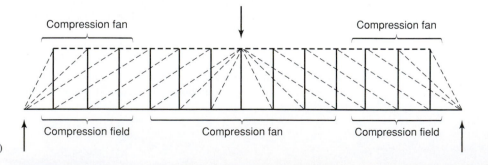

Fig. 18-18
Compression fans and compression fields.(From [18-19].)

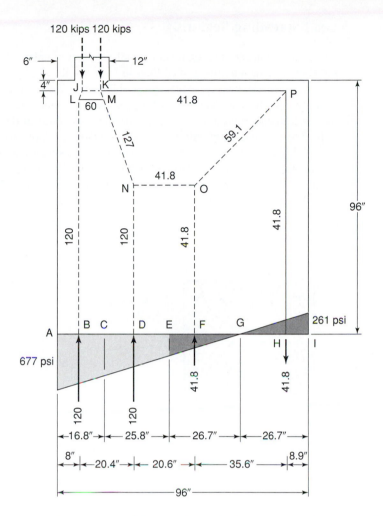

Fig. 18-19
Column supported near one
end of a wall, showing a
strut-and-tie model with a
U-turn.

in. so that the upward force at *F* equals the downward force at *H*. The widths of the other
two parts (*A–C* and *C–E*) were chosen so that the forces in them were equal. The right-
hand two reactions are each 41.8 kips. They cause the force whirl made up of compression
members *F–O* and *O–P* and tension member *H–P*. The reinforcement computed from the
strut-and-tie model is shown in Fig. 18-20. The strut-and-tie model plotted in Figs. 18-19
and 20 is solved in [18-18].

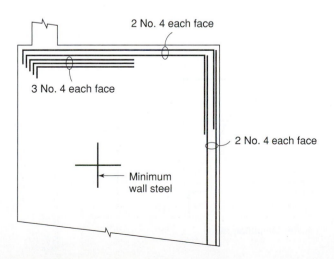

Fig. 18-20
Reinforcement in the wall in
Fig 18-19.

Load-Spreading Regions

Frequently, concentrated loads act on walls or other member. These loads spread out in the member, as shown by the dashed lines in Figs. 18-4c, 18-5b, and 18-21a. For equilibrium, transverse tension ties are needed for equilibrium of the joints at, for example, points B and F in Fig. 18-21c. The magnitude of the tensile force in the ties depends on the slope of the load-spreading struts. In Figs. 18-4c and 18-5b, the slope of these struts has been assumed to be 2 to 1, that is (2 parallel) to (1 perpendicular), relative to the axis of the force being spread. If half of the applied load C is resisted by each branch of the load-spreading strut-and-tie model, as shown in Figs. 18-4c and 5b, the tie force B–F will be $C/4$. On the other hand, if the slope were 1 to 1, the tie force would double, to $C/2$.

Elastic analyses of edge-loaded elastic plates of width b subjected to in-plane loads applied to one edge show that the load-spreading angle is primarily a function of the ratio

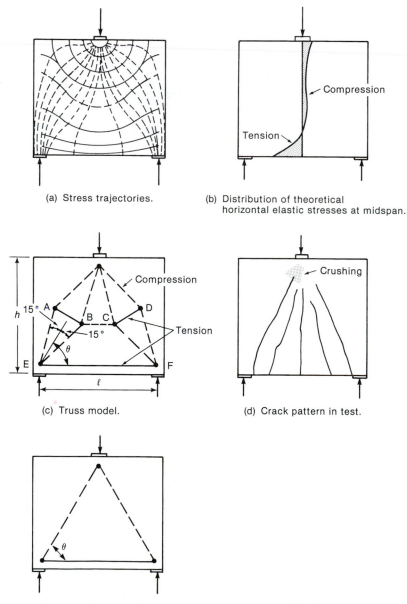

(a) Stress trajectories.

(b) Distribution of theoretical horizontal elastic stresses at midspan.

(c) Truss model.

(d) Crack pattern in test.

(e) Simplified truss.

Fig. 18-21
Single-span deep beam supporting a concentrated load. (Adapted from [18-1].)

of the width of the loading plate, a, to that of the loaded plate, b. Using analyses, [18-1] shows that the angle, θ, between the load and the inclined struts varies from 28° for a concentric load with $a/b = 0.10$, to 19° for $a/b = 0.2$, and down to about 12° for $a/b = 0.5$. As a result, the transverse tie in Fig. 18-5c would correspond to strut slopes from 1.9 to 1 for $a/b = 0.1$, up to 2.9 to 1 for $a/b = 0.2$, and to 4.7 to 1 for $a/b = 0.5$. Similar values are obtained for other cases of load spreading, such as concentrated loads acting near one edge of a plate, or multiple, concentrated loads.

A strut slope of 2:1 (longitudinal to transverse) is conservative for a wide range of cases, and we shall assume a 2-to-1 slope in all similar cases.

18-7   LAYOUT OF STRUT-AND-TIE MODELS

Factors Affecting the Choice of Strut-and-Tie Models

A general procedure for laying out strut-and-tie models was presented in Section 18-2 and illustrated in Example 18-1. Additional factors that affect the choice of strut-and-tie models include:

Equilibrium

1. The strut-and-tie model must be in equilibrium with the loads. There must be a clearly laid out load path.

Direction of Struts and Ties

2. The strut-and-tie model for a simply supported beam with unsymmetrically applied, concentrated loads consists of an arch, made of straight line segments or a hanging cable, that has the same shape as the bending-moment diagram for the loaded beam as shown in Fig. 18-22. This is also true for uniformly loaded beams, except that the moment diagram and the strut-and-tie model have parabolic sections.

3. The strut-and-tie model should represent a realistic flow of forces from the loads through the D-region to the reactions. Frequently this can be determined by observation. From an elastic stress analysis, such as a finite element analysis, it is possible to derive the stress trajectories in an uncracked D-region, shown in Fig. 18-21a for a deep beam. Principal compression stresses act parallel to the dashed lines, which are known as *compressive stress trajectories*. Principal tensile stresses act parallel to the solid lines, which are called *tensile stress trajectories*. Such a diagram shows the flow of internal forces and is a useful, but by no means essential, step in laying out a strut-and-tie model. The compressive struts should roughly follow the direction of the compressive stress trajectories, as shown by the refined and simple strut-and-tie models in Fig. 18-21c and e. Generally, the strut direction should be within ±15° of the direction of the compressive stress trajectories [18-1].

Because a tie consists of a finite arrangement of reinforcing bars which are usually placed orthogonally in the member, there is less restriction on the conformance of ties with the tensile stress trajectories. However, they should be in the general direction of the tension stress trajectories.

4. Struts cannot cross or overlap because the width of the individual struts has been calculated assuming they are stressed to the maximum.

5. Ties can cross struts.

6. It is generally assumed that the structure will have enough plastic deformation capacity to adapt to the directions of the struts and ties chosen in design if they are within ±15° of the elastic stress trajectories. The load-spreading reinforcement from ACI

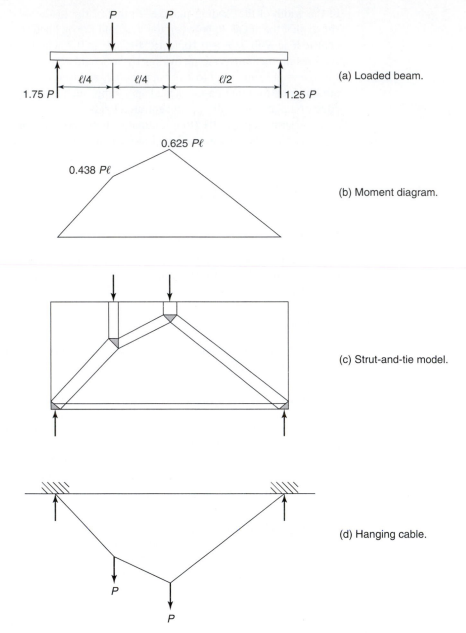

(a) Loaded beam.

(b) Moment diagram.

(c) Strut-and-tie model.

(d) Hanging cable.

Fig. 18-22
Statical equivalence of various types of structures.

Section A.3.3.3 is intended to allow the moment redistribution needed to accommodate this change in angles.

Ties

7. In addition to generally corresponding to the tensile stress trajectories, ties should be located to give a practical reinforcement layout. Wherever possible, the reinforcement should involve groups of orthogonal bars which are straight except for hooks needed to anchor the bars.

8. If photographs of test specimens are available, the crack pattern may assist one in selecting the best strut-and-tie model. Figure 18-42a (which will be discussed later) shows the crack pattern in a dapped end at the support of a precast beam. Figure 18-42b, c, and d show possible models for this region. Compression strut *B–D* in Fig. 18-42d crosses a zone of cracking in the test specimen, which makes it an unsuitable location for a compression strut.

Load-Spreading Regions

9. Elastic analyses of plates of width *b* subjected to in-plane loads applied to one edge show that the load-spreading angle is primarily a function of the ratio of the width of the loading plate, *a*, to the width of the member, *b*. A strut slope of 2-to-1 (parallel to the axis of the load-to-perpendicular to the axis) is conservative for a wide range of cases. A slope of 2-to-1 will be used in all similar cases in this book.

10. Angles, θ, between the struts and attached ties at a node should be large, in the order of 45°, and never less than the 25° specified in ACI Section A.2.5. The size of compression struts is sensitive to the angle θ between the strut and the reinforcement in a tie. To illustrate this, consider a strut carrying a force with a vertical component of $V_u = 200$ kips in a deep beam made of 3000-psi concrete. The thickness of the beam is 12 in.

If $\theta = 65°$, the axial force in the strut needed to transfer this shear is 221 kips. Assuming a bottle-shaped strut, the effective strength of the concrete in the strut is $f_{cu} = 0.85 \times 0.75 \times 3000$ psi $= 1912$ psi, and the width of the strut must be

$$w_s = \frac{V_u}{\phi f'_c \times b} = \frac{221,000}{0.75 \times 1912 \times 12}$$

$$= 12.8 \text{ in.}$$

For $\theta = 45°$, the axial force in the strut is 283 kips, and the width of the strut must be 16.4 in.

For $\theta = 25°$, the axial force in the strut is 474 kips, and the strut must be 27.5 in. wide. In some cases it will be difficult to fit a strut of this width within the space available. In the examples of deep beams which follow, θ has been limited arbitrarily to 40°, or larger, to keep the width of the strut within reasonable limits, even though this is not required by ACI.

Minimum Steel Content

11. The loads will try to follow the path involving the least forces and deformations. Because the tensile ties are more deformable than the compression struts, the model with the least and shortest ties is the best. Thus the strut-and-tie model in Fig. 18-23a is a better model than the one in Fig. 18-23b because Fig. 18-23a more closely approaches the elastic stress trajectories in Fig. 18-21a. Schlaich et al. [18-2], [18-3] propose the following criterion for guidance in selecting a good model:

$$\Sigma F_i \ell_i \epsilon_{mi} = \text{minimum}$$

where F_i, ℓ_i, and ϵ_{mi} are the force, length, and mean strain in strut or tie i, respectively. Because the strains in the concrete are small, the struts can be ignored in the summation.

Suitable Strut-and-Tie Layouts

12. The finite widths of struts and ties must be considered. The axis of a strut representing the compression zone in a deep flexural member should be located about *a*/2 from the compression face of the beam where *a* is the depth of the rectangular stress block, as shown in Fig. 18-7. Similarly, if hydrostatic nodal zones are used, the axis of a tension

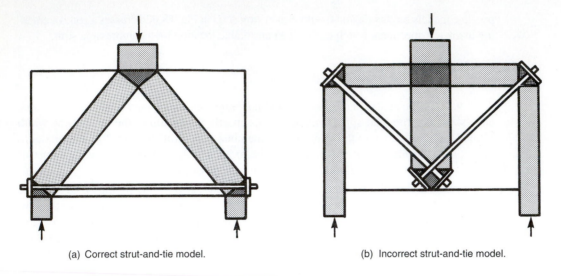

(a) Correct strut-and-tie model.   (b) Incorrect strut-and-tie model.

Fig. 18-23
Suitable and unsuitable strut-and-tie models.

tie should be about $a/2$ from the tensile face of the beam. One of the first steps in modeling a beam-like member is to locate the nodes in the strut-and-tie model. This can be done by estimating values for $a/2$.

A possible strut-and-tie model for the beam shown in Fig. 18-24 consists of two trusses, one utilizing the lower steel as its tension tie, the other using the upper steel. For an ideally plastic material, the capacity would be the sum of the shears transmitted by the two trusses, $V_1 + V_2$. Tests [18-8] have shown, however, that the upper layer of steel has little, if any, effect on the strength. In part this is due to the very low angle θ_2 in the upper truss. When this beam is loaded, the bottom tie yields first.

13. As the angle, θ, between the struts and ties decreases, it is often desirable to include web reinforcement in addition to the confinement reinforcement required by ACI Section A.3.3. The following equation is used in European design standards [18-18] to

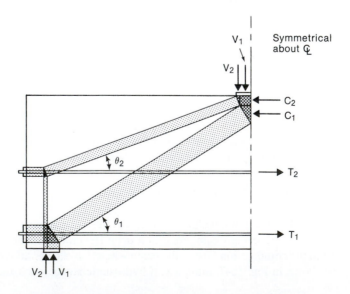

Fig. 18-24
Strut-and-tie model for beam
with horizontal web rein-
forcement at midheight.
(From [18-8].)

require stirrup reinforcement in beams that approach the lower limit on the angle between struts and ties in ACI Section A.2.5:

$$F_{nw} \approx \frac{2a/jd - 1}{3 - N_n/F_n} F_n \qquad (18\text{-}18)$$

Here F_{nw} is the yield force $\Sigma A_v f_y$ in the web reinforcement in the shear span; F_n is the concentrated load and also the reaction; N_n is the axial force acting on the beam, if any; a is the distance between the axes of the load and reaction; and jd is the lever arm between the resultant compression force and the resultant force in the longitudinal tie. For a beam with $N_u = 0$, and having a/jd equal to 2, (18-18) requires that all the shear be carried by shear reinforcement. At $a/jd = 0.5$, all the shear is resisted by the compression strut.

14. Subdividing the nodal zone, assuming each part of the nodal zone can be assigned to a particular force or reaction, makes the truss easier to lay out.

15. Sometimes a better representation of the real stress flow is obtained by adding two possible simple models, each of which is in equilibrium with a part of the applied load provided the struts do not overlap or cross.

18-8   DEEP BEAMS

Definition of Deep Beam

The term *deep beam* is defined in ACI Section 10.7.1 as a member

(a) loaded on one face and supported on the opposite face so that compression struts can develop between the loads and the supports and

(b) having either

(i) clear spans, ℓ_n, equal to or less than four times the overall member height, h, or

(ii) regions loaded with concentrated loads within $2d$ from the face of the support. The ACI Code does not quantify the magnitude of the concentrated load needed for the beam to act as a deep beam. ACI Committee 445 has suggested that a concentrated load that causes 30 percent or more of the reaction at the support of the beam in question would qualify.

ACI Section 11.8.2, *Deep Beams*, gives essentially the same requirements. Both sections require that deep beams be designed via nonlinear analyses or by strut-and-tie models. The design equations for V_s given in previous ACI Codes were deleted from the 2002 code, because they did not have a clearly defined load path and had serious discontinuities as the span-to-depth ratio was varied.

Most typically, deep beams occur as *transfer girders*, which may be single span (Fig. 18-25) or continuous (Fig. 18-26). A transfer girder supports the load from one or more columns, transferring it laterally to other columns. Deep-beam action also occurs in some walls and in pile caps. Although such members are not uncommon, satisfactory design methods are only recently being developed.

Analyses and Behavior of Deep Beams

Elastic analyses of deep beams in the uncracked state are meaningful only prior to cracking. In a deep beam, cracking will occur at one-third to one-half of the ultimate load. After

Fig. 18-25
Single-span deep beam.
(Photograph courtesy of J. G.
MacGregor.)

cracks develop, a major redistribution of stresses is necessary since there can be no tension across the cracks. The results of elastic analyses are of interest primarily because they show the distribution of stresses which cause cracking and hence give guidance as to the direction of cracking and the flow of forces after cracking. In Figs. 18-21a, 18-27a and 18-28a, the dashed lines are *compressive stress trajectories* drawn parallel to the directions of the principal compressive stresses, and the solid lines are *tensile stress trajectories* parallel to the principal tensile stresses. Cracks would be expected to occur perpendicular to the solid lines (i.e., parallel to the dashed lines).

In the case of a single-span beam supporting a concentrated load at midspan (Fig. 18-21a), the principal compressive stresses act roughly parallel to the dashed lines joining the load and the supports and the largest principal tensile stresses act parallel to the bottom of the beam. The horizontal tensile and compressive stresses on a vertical plane at midspan are shown in Fig. 18-21b. For a beam with aspect ratio about 1.0 the resultant compressive force is closer to the tension tie than a straight-line distribution of stresses would predict. Although it cannot be seen from the figures, it is important to note that the flexural stress at the bottom is constant over much of the span. The stress trajectories in Fig. 18-21a can be simplified to the pattern given in Fig. 18-21c. Again, dashed lines represent compression struts and solid lines denote tension ties. The angle θ varies approximately linearly from 68° (2.5:1 slope) for $\ell/d = 0.80$ or smaller, to 40° (0.85:1 slope) at $\ell/d = 1.8$. If such a beam were tested, the crack pattern would be as shown in Fig. 18-21d. Note that each of the three tension ties in Fig. 18-21c (*AB, CD,* and *EF*) has cracked. At failure, the shaded region in Fig. 18-21d would crush, or the anchorage zones at *E* and *F* would fail. The strut-and-tie model in Fig. 18-21c could be simplified further to

Fig. 18-26
Three-span deep beam, Brunswick building, Chicago. (Photograph courtesy of J. G. MacGregor.)

the model shown in Fig. 18-21e. This model does not explain the formation of the inclined cracks, however.

An uncracked, elastic, single-span beam supporting a uniform load has the stress trajectories shown in Fig. 18-27a. The distribution of horizontal stresses on vertical sections at midspan and the quarter point are plotted in Fig. 18-27b. The stress trajectories can be represented by the simple truss in Fig. 18-27c or the slightly more complex truss in Fig. 18-27d. In the first case, the uniform load is divided into two parts, each represented by its resultant. In the second case, four parts were used. The angle θ varies from about 68° for $\ell/d = 1.0$ or smaller to about 55° for $\ell/d = 2.0$ [18-1]. The crack pattern in such a beam is shown in Fig. 18-27e.

Figure 18-28a shows the stress trajectories for a deep beam supporting a uniform load acting on ledges at the lower face of the beam. The compression trajectories form an arch, with the loads hanging from it, as shown in Fig. 18-28b and c. The crack pattern in Fig. 18-28d clearly shows that the load is transferred upward by reinforcement until it acts on the compression arch, which then transfers the load down to the supports.

The distribution of the flexural stresses in Fig. 18-28b and the truss models in Figs. 18-21, 18-27 and 18-28 all suggest that the force in the longitudinal tension ties will be constant along the length of the deep beam. This implies that this force must be anchored at the joints over the reactions. Failure to do so is a major cause of distress in deep beams. ACI Section 12.10.6 alludes to this.

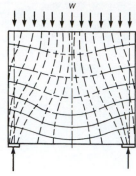

(a) Stress trajectories.

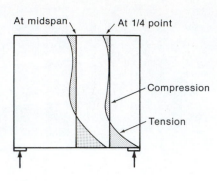

(b) Distribution of theoretical horizontal elastic stresses.

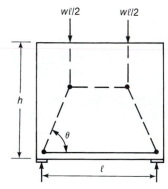

(c) Truss model.
$\theta = 68°$ if $\ell/h \le 1$
$= 54°$ if $\ell/h = 2$

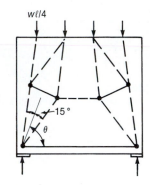

(d) Refined truss model.

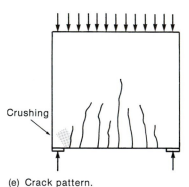

(e) Crack pattern.

Fig. 18-27
Uniformly loaded deep beam.
(Adapted from [18-1].)

Strut-and-Tie Models for Deep Beams

Figure 18-3 shows a simple strut-and-tie model for a single-span deep beam. The loads, reactions, struts, and ties in Fig. 18-3 are all laid out such that the centroids of each truss member and the lines of action of all externally applied loads coincide at each joint. This is necessary for joint equilibrium. In Fig. 18-3, the bars are shown with external end anchor plates. In a reinforced concrete beam, the anchorage would be accomplished with horizontal or vertical hooks or, in extreme cases, with an anchor plate, as shown.

The strut-and-tie model shown in Fig. 18-3 can fail in one of four ways: (1) the tie could yield, (2) one of the struts could crush when the stress in the strut exceeded f_{cu}, (3) a

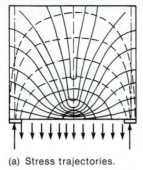
(a) Stress trajectories.

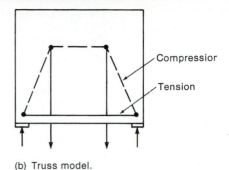
(b) Truss model.

Compression

Tension

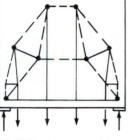
(c) Refined truss model.

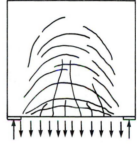
(d) Crack pattern.

Fig. 18-28
Deep beam uniformly loaded
on the bottom edge.
(Adapted from [18-1].)

node could fail by being subjected to stresses that are greater than its effective compressive strength, or (4) the enchorage of the tie could fail. Frequently, this involves a bearing failure at the loads or reactions. Since a tension failure of the steel will be more ductile than either a strut failure or a node failure, the beam should be proportioned so that the strength of the steel governs.

A second example consisting of a simple span beam with vertical stirrups subjected to a concentrated load at midspan is shown in Fig. 18-13. This is the sum of several trusses. One truss uses a direct compression strut running from the load to the support. This truss carries a shear V_c. The other truss uses the stirrups as vertical tension members and has compression fans under the load and over the reactions. The vertical force in each stirrup is computed by assuming that the stirrup has yielded. The vertical force component in each of the small compression struts must be equal to the yield strength of its stirrup for the joint to be in equilibrium. The farthest-left stirrup is not used, since one cannot draw a compression diagonal from the load point to the bottom of this stirrup without encroaching on the direct compression strut.

The compression diagonals radiating from the load point intersect the stirrups at the level of the centroid of the bottom steel, because a change in the force in the bottom steel is required to equilibrate the horizontal component of the force in the compression diagonal. The force in the bottom steel is reduced at each stirrup by the horizontal component of the compression diagonal intersecting at that point. This is illustrated in Fig. 18-13b, where the stepped line shows the resulting tensile force in the bottom steel. The tensile force computed from beam theory, M/jd, is shown by a dashed line in the same figure. Note that the tensile force from the plastic truss analogy exceeds that from beam theory. This is in accordance with test results (see Fig. 6-27) [18-8].

Design Using Strut-and-Tie Models

The design of a deep beam using a strut-and-tie model involves laying out a truss that will transmit the necessary loads. Once a satisfactory truss has been found, the joints and members of the truss are detailed to transmit the necessary forces. The overall dimensions of the beam must be such that the entire truss fits within the beam and has adequate cover.

Continuous deep beams are very stiff elements and, as such, are very sensitive to differential settlement of their supports due to foundation movements, or differential shortening of the columns supporting the beam. The first stage in the design of such a beam is to estimate the range of reactions and use this to compute shear and moment envelopes. Although some redistribution of moment and shear may occur, the amount will be limited.

ACI Section 11.8.3 limits V_n in deep beams to $10\sqrt{f'_c}b_w d$. One can obtain an initial trial section on the basis of limiting $V_u = \phi V_n$ to $V_n = \phi \,(7 \text{ to } 10)\, \sqrt{f'_c}b_w d$.

EXAMPLE 18-2 Design of a Single-Span Deep Beam

Design a deep beam to support an unfactored column load of 300 kips dead load and 340 kips live load from a 20-in.-by-20-in. column. The axes of the supporting columns are 6.5 ft and 10 ft from the axis of the loading column as shown in Fig. 18-29. The supporting columns are also 20 in. square. Use 4000-psi concrete and Grade-60 steel. Use the load and resistance factors from ACI 318-02 Chapter 9.

1. **Select a strut-and-tie model—Single-span deep beam**

  1(a) Select shape and flow of forces. The strut-and-tie model is shown in Fig. 18-29. It consists of two inclined struts, three nodal zones, and one tie. We shall assume that a portion of the column load equal to the left reaction flows down through strut $A-B$ to the reaction at A. The rest of the column load is assumed to flow down strut $B-C$ to the reaction at the right end of the beam.

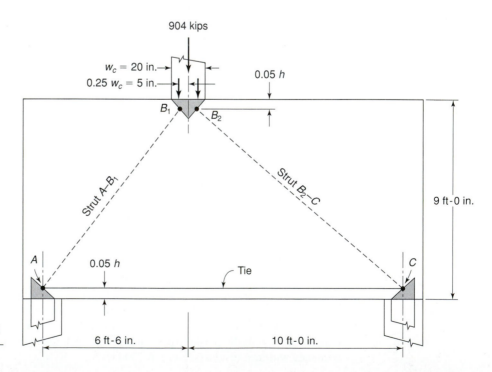

Fig. 18-29
First trial strut-and-tie model of a single-span deep beam—Example 18-2.

1(b)   Compute the factored load and the reactions—Single-span deep beam. The factored load from the column is:

$$P_u = 1.2D + 1.6L = 1.2 \times 300 + 1.6 \times 340 = 904 \text{ kips}$$

The reactions are 548 kips at A and 356 kips at C. The dead load of the beam will be added in after the size of the beam is known.

1(c)   Design format. In design we will use $\phi V_n = V_u$ where, from ACI Section 9.3.2.6, $\phi = 0.75$.

2.   Estimate the size of the beam. This will be done in two different ways.

2(a)   Limit ϕV_n to $\phi \times (0.7 \text{ to } 1.0) \times 10\sqrt{f'_c}\, b_w d$. ACI Section 11.8.3 limits the shear in a deep beam to $10\sqrt{f'_c}\, b_w d$. To allow a little leeway and to leave some capacity for the dead load, we will select the beam using maximum shear forces of $\phi 7\sqrt{f'_c}\, b_w d$ to $\phi 10\sqrt{f'_c}\, b_w d$. The maximum shear ignoring the dead load of the beam is 548 kips, so

$$b_w d = \frac{V_n}{\phi \times (7 \text{ to } 10)\sqrt{f'_c}} = \frac{548,000 \text{ lb}}{0.75(7 \text{ to } 10)\sqrt{4000}} = 1650 \text{ to } 1160 \text{ in.}^2$$

Try b_w equal to the width of the columns, $b_w = 20$ in. This gives d from 82.6 to 57.8 in. Assuming that $h \approx d/0.9$ gives h from 64.2 to 91.7 in. Try a 20-by-96-in. beam.

2(b) Select height to keep the flattest strut at an angle of $\approx 40°$ from the tie. Note that a limit of 40° is more restrictive than the limit of 25° in ACI Section A.2.5. The strut-and-tie forces begin to grow rapidly for angles less than 40°. For the 120-in. shear span, the minimum height center to center of nodes is $120 \tan 40° = 100.7$ in. Assuming that the nodes at the bottom of this strut are located at midheight of a tie with an effective tie height of $0.1h$, this puts the lower nodes at $0.05h = 5.4$ in. from the bottom of the beam. We shall locate the node at B the same distance below the top, giving the required height of the beam as $h \approx 100.7/0.9 = 112$ in.

For a first trial, we shall assume $b_w = 20$ in. and $h = 108$ in. with the lower chord located at $0.05h = 5.4$ in. above the bottom of the beam and the center of the top node at $0.05h = 5.4$ in. below the top of the beam.

The weight of the beam is

$$[(20 \text{ in.}/12 \text{ in./ft.}) \times (108/12) \text{ ft} \times (16.5 + 1.67) \text{ ft} \times 0.150 \text{ kips/ft}^3)] = 40.9$$

The factored weight is $1.2 \times 40.9 = 49.1$ kips, say. To simplify the calculations we shall add this to the column load giving a total factored load of 954 kips. The corresponding reactions are 578 kips at A and 376 kips at C.

First trial design is based on a beam with $b_w = 20$ in. and $h = 108$ in., with $\phi P_n = P_u = 954$ kips, and $\phi V_n = V_u$ equal to 578 kips in shear span A–B, and 376 kips in shear span B–C. (See Fig. 18-29.)

3.   Compute effective compression strengths, f_{cu}, for the nodal zones and struts—Single-span deep beam

3(a)   Nodal zones

Nodal zone at A. This is a C–C–T node. From ACI Section A.5.2:

$$f_{cu} = 0.85\beta_n f'_c = 0.85 \times 0.80 \times 4000 = 2720 \text{ psi}$$

Nodal zone at B. This is a C–C–C node. From ACI Section A.5.2,

$$f_{cu} = 0.85\beta_n f'_c = 0.85 \times 1.0 \times 4000 = 3400 \text{ psi}$$

Nodal zone at C. This is a C–C–T node, $f_{cu} = 2720$ psi.

3(b)   Struts

Strut A–B: This strut has room for the width of the strut to be bottle-shaped. We will provide reinforcement satisfying ACI Section A.3.3. From ACI Section A.3.2

$$f_{cu} = 0.85\beta_s f'_c = 0.85 \times 0.75 \times 4000 = 2550 \text{ psi}$$

Strut B–C: This also is a bottle-shaped strut, and $f_{cu} = 2550$ psi

The effective compression strengths of the struts govern. Use $f_{cu} = \mathbf{2550}$ **psi** throughout.

4. **Locate nodes—First trial—Single span deep beam**

Nodal Zones at A and C: In step 2(a), we assumed that the nodes at A and C are located at $0.05h = 5.4$ in. above the bottom of the beam. Use this location in the first trial strut-and-tie model.

Nodal Zone at B: Assume Node B is located at $0.05h = 5.4$ in. below the top of the beam. The node at B is subdivided into two subnodes, as shown in Fig. 18-30, one assumed to transmit a factored load of 578 kips to the left reaction and the other transmitting 376 kips to the right reaction. Each of the subnodes is assumed to receive load from a vertical strut in the column over B. Both of the subnodes are C–C–C nodes. The effective strength of the nodal zone at B is $f_{cu} = 3400$ psi. For struts A–B_1 and B_2–C, $f_{cu} = 2550$ psi. We shall use $f_{cu} = 2550$ psi for the faces of the nodal zone loaded by bottle-shaped struts, and $f_{cu} = 3400$ psi for the face of the nodal zone that is loaded by compression stresses in the column and for the vertical struts in the column over point B.

$$w_{B1} = \frac{578,000}{0.75 \times 3400 \times 20} = 11.3 \text{ in.}$$

$$w_{B2} = 7.37 \text{ in.}$$

The total width of the vertical struts in the column at B is $\Sigma w_{BS} = 18.7$ in.

This will fit in the 20-in.-square column. If it did not, ACI Section A.3.5 allows us to include the capacity of the column bars, $\phi A_s f_y$, when computing the capacity of the strut.

Because the vertical load components acting at B_1 and B_2 are different, the vertical division of node B is not located at the center of the column. The resultant force in the left-hand strut acts at point B_1, which is 3.70 in. to the left of the center of the column, and the right-hand strut acts at point B_2, located 5.67 in. to the right of the center. See Fig. 18-30.

The horizontal distance from A to B is 78 in.; that from B to C is 120 in.

5. **Compute the lengths and widths of the struts and ties and draw the first trial strut-and-tie model—Single-span deep beam**

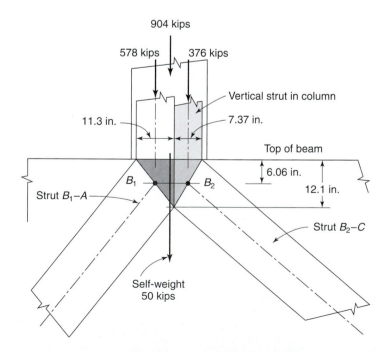

Fig. 18-30
Struts and nodal zones adjacent to the column at B—Example 18-2.

TABLE 18-3   Geometry and Forces in Struts and Tie

Strut or Tie (1)	Horiz. Proj. in. (2)	Vertical Proj. in. (3)	Angle (4)	Vert. Comp. kips (5)	Horiz. Comp. kips (6)	Axial Force kips (7)	Width of Strut in. (8)
Vert. at A	0.0	—	Vertical	578	0	578	15.1
$A-B_1$	$78.0 - 0 - 3.7$ $= 74.3$ in.	$108 - 5.4 - 5.4$ $= 97.2$ in.	52.6	578	442	728	19.0
B_2-C	$120 - 5.67 - 0.0$ $= 114.3$ in.	97.2	40.4	376	442	580	15.1
$A-C$	—	—	—	—	442 at A 442 at C	442 at A 442 at C	$w_t = 10.83$
Vert. at C	0.0	—	Vertical	376	0	376	9.83

5(a)   Geometry and forces in struts and tie.

Strut $A-B_1$: The calculations for strut $A-B_1$ in Table 18-3 are:

<u>Column (4)</u> The angle between the axis of the strut and horizontal $= \arctan \dfrac{97.2}{74.3} = 52.6°$.

<u>Column (5)</u> The vertical component is equal to the shear $= 578$ kips.

<u>Column (6)</u> The horizontal component $= \dfrac{\text{Vertical component}}{\tan 52.6°} = \dfrac{578}{1.307} = 442$ kips.

<u>Column (7)</u> The axial force in strut $A-B_1 = \dfrac{578 \text{ kips}}{\sin 52.6°} = 728$ kips.

<u>Column (8)</u>

$$w_s = \frac{728,000}{0.75 \times 2550 \times 20} = 19.0 \text{ in.}$$

Vertical strut at A

The width of the strut under the node at A is

$$w_{sA} = \frac{578,000}{0.75 \times 2550 \times 20} = 15.1 \text{ in.}$$

Check the width of strut $A-B_1$ at A, using (18-16), where angle θ is 52.6°, $\ell_b = 15.1$ in. and $w_t = 10.83$.

$$w_s = w_t \cos\theta + l_b \cos\theta = 10.83 \cos 52.6° + 15.1 \sin 52.6°$$
$$= 18.57 \text{ in.}$$

Because the 18.57 in. is less than the 19.0 in. required from column 8 of Table 18-3, we should increase h_t or ℓ_b to increase w_s to at least 19.0 in. By increasing the width, ℓ_b, of the vertical strut under node A from 15.1 in. to 15.7 in., the 19.0 width of strut $A-B_1$ can be accomodated.

Vertical strut at C

The width of the strut under the node at C is

$$w_{sC} = \frac{376,000}{0.75 \times 2550 \times 20} = 9.83 \text{ in.}$$

Strut B_2-C

The calculations for strut B_2-C are given in Table 18-3.

Tie $A-C$:

The force in tie $A-C$ is calculated in Table 18-3. The axial (horizontal) force in the tie is equal to the horizontal forces in the struts at the two ends of the tie.

The axial force in tie based on the geometry at $A = 442$ kips.

The axial force based on the geometry at $C = 442$ kips.

These should be the same. Slight differences occur due to approximations in the location of the nodes. The agreement is a partial check on the solution. The effective height of the tie A–C is

$$w_t = \frac{442,000}{0.75 \times 2720 \times 20} = 10.83 \text{ in.}$$

The upper limit on the height of the centroid of the tie reinforcement is $10.83/2 = 5.42$ in.

5(b)   Check whether the beam is a deep beam. ACI Section 11.8.1 defines a deep beam as one with shear span-to-height ratio less than or equal to 2. ACI Section 10.7.1 also requires that there be a compression strut between the load and the reaction. From step 5(a) in Table 18-3, the horizontal projection of the left shear span is 73.4 in. The height is 108 in. The shear span-to-depth ratio is $73.4/108 = 0.68$. This is less than 2 and the shear span is deep. The right shear span is also deep.

6.   Compute tie forces, reinforcement, and the effective width of the tie—Single-span deep beam

6(a)   Trial choice of reinforcement in tie A–C. From Table 18-3, the force in tie A–C is 442 kips.

$$A_s = \frac{442,000}{0.75 \times 60,000} = 9.82 \text{ in.}^2$$

We could use 10 No. 9 bars, $A_s = 10.0$ in.^{2}, development length, $\ell_d = 53.3$ in., minimum web width if bars are in two layers is $b_{w,min} = 14.5$ in.

13 No. 8 bars, $A_s = 10.27$ in.^{2}, development length, $\ell_d = 47.4$ in.; $b_{w,min}$ for two layers is 17.5 in.

18 No. 7 bars, $A_s = 10.8$ in.^{2}, development length, $\ell_d = 41.5$ in.; $b_{w,min}$ is 20.5 in.

Try 13 No. 8 bars. No. 8 bars will be easier to anchor than No. 9 bars. The bars will fit into two layers, although there is relatively little free space for inserting vibrators when consolidating the concrete. We will use one layer of 5 bars and two layers of 4 bars and we will use No. 10 bars as spacers between the layers.

6(b)   Minimum flexural reinforcement. ACI Section 10.7.3 requires that deep beams have at least the minimum flexural reinforcement required by ACI Section 10.5. This is

$$A_{s,min} = \frac{3\sqrt{f'_c}}{f_y} b_w d \qquad \text{but not less than } 200 b_w d / f_y \qquad (18\text{-}20)$$

$$= \frac{3\sqrt{4000}}{60,000} \times 20 \times 104.5 = 6.61 \text{ in.}^2 \qquad (\text{ACI Eq. 10-3})$$

but not less than $\dfrac{200 b_w d}{f_y} = 6.97$ in.^{2}.

The reinforcement provided in the tie is 13 No. 8 bars, $A_s = 10.3$ in.^{2}. Therefore, OK.

6(c)   Effective height of tie A–C, w_t. ACI Section R.A.4.2 suggests that the tie be spread over a height ranging from a maximum equal to the tie force divided by the effective compression strength for the nodes at A and C to a minimum of half of this. The maximum height of the tension tie is $w_t = \dfrac{442,000}{0.75 \times 2720 \times 20} = 10.83$ in. where is f_{cu} for the nodal zones at A and C is 2720 psi. The maximum height of the centroid of the reinforcement is $10.83/2 = 5.42$ in. above the bottom of the beam.

If the bars are placed in three layers with No. 10 bars as spacers between the layers, the centroid of the tie reinforcement is:

$$y = \frac{\begin{array}{c}5(1.5 + 0.5 + 0.5) + 4(1.5 + 0.5 + 1.0 + 1.27 + 0.5) \\ + 4(1.5 + 0.5 + 1.0 + 1.27 + 1.0 + 1.27 + 0.5)\end{array}}{13}$$

$$= 4.60 \text{ in. above the bottom.}$$

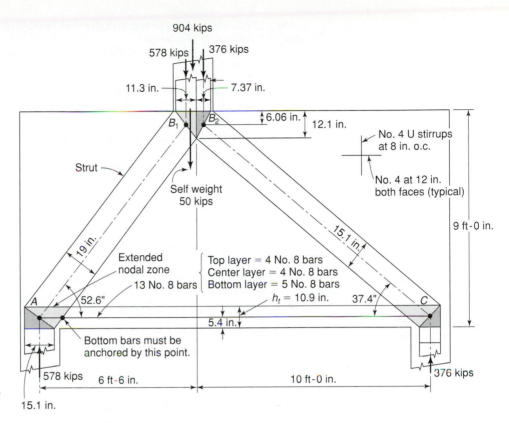

Fig. 18-31
Final design single-span deep
beam—Example 18-2.

This corresponds to a height of tie, $w_t = 9.20$ in.

For tie A–C try 13 No. 8 bars with one layer of 5 bars and two layers of 4 bars with 1.27 in. (No. 10 bar) vertical spaces between layers. This exceeds the minimum reinforcement calculated in step 6(b), OK.

6(d)   Anchor the tie reinforcement at A. The tie reinforcement must be anchored for the full yield strength at the point where the axis of the tie enters the struts at A and C. Assuming an extended nodal zone with node A directly over the center of the column, the length available to anchor the tie is (see Fig. 18-31):

Distance from the center of the 20 in. column to the outside face of the cutoff or hooked bars = $10 - (1.5 + 0.5) = 8.0$ in., The anchorage length available is:

$$= 8.0 \text{ in.} + \left(\frac{15.1}{2} \text{ in.}\right) + (5.4/\tan 52.6°) \text{ in.} = 19.7 \text{ in.}$$

Since $\ell_d = 47.4$ in, the length available to develop anchorage of straight bars is insufficient. We must use hooks or some similar method to anchor the bars.

Try a standard 90° hook with $\ell_{dh} = 19$ in. multiplied by $\beta = 1.0$ and $\lambda = 1.0$ from ACI Section 12.5.2, and multipliers from:

ACI Section 12.5.3(a) for tail cover = 2.0 in., and side cover = $1.5 + 0.5 + 1.0$ = 3.0 in. (assuming the bars in the column are No. 8 bars and the tie bars are inside the column bars) $\times$ 0.7

ACI Section 12.5.3(b) We shall ignore the effect of ties in the joint, thus, $\times$ 1.0.

$\ell_{dh} = 19.0 \times 0.7 \times 1.0 = 13.3$ in. Check whether this will fit into a 20-in.-square column.

We could use 13 No. 8 bars with 90° hooks in three layers, with 2 in. tail cover on the lower layer and (1.5 in. cover) + (0.5 in. tie) + 1.0 in. bar + (1.27 in. space between layers)= 4.27 in. cover on the tails of the second layer, leaving 15.7 in. for the hooks on the second layer, and 4.27 in. + 1.0 + 1.27 in. = 6.54 in. cover on the tails of the hooks in the third layer, leaving 13.5 in. for the third layer of hooks.

Try 13 No. 8 bars, 5 in the bottom layer, and two layers of 4 bars, with 90° hooks inside the column reinforcement.

Alternatively, we could lap splice the tie bars with the legs of horizontal hairpin bars. Assuming the five bars in the lower layer have 90° hooks the remaining bars could be anchored using two No. 8 horizontal hairpin bars lying flat per layer, with the legs of each hairpin lap spliced $1.3\ell_d = 61.6$ in., say 62 in., with two of the bars making up the tie. From ACI Section 7.2.1 the total out-to-out width of a No. 8 hairpin bar is $8d_b = 8.0$ in. Two of these would fit between the stirrups in the 20-in. width of the beam.

6(e) Check height of Node B. Summing horizontal forces on a vertical section through the entire beam and the nodal zone at B shows that the horizontal force in the struts meeting at B equals the horizontal force in tie $A-C$ which will be taken equal to $\phi A_s f_y$, thus, $\phi F_{nn} = 0.75 \times 10.3$ in.$^2 \times 60{,}000$ psi $= 463{,}500$ lb. Total depth of the nodal zone at B is $463{,}500/(0.75 \times 2550 \times 20) = 12.1$ in. The top nodes, B_1 and B_2, are located at $12.1/2 = 6.06$ in. below the top of the beam. (See Fig. 18-31).

7. Draw second trial strut-and-tie model. This drawing has been omitted here. It is similar to the drawing of the finished design in Fig. 18-31.

8. Recompute strut-and-tie model using computed locations of the nodes from the second trial—Single-span deep beam

Node B_1 is 3.70 in. to the left of the column center and 6.06 in. below the top of the beam.

Node B_2 is 5.67 in. to the right and 6.06 in. down.

8(b) Geometry and forces in struts and tie—Single-span deep beam.

The calculations in Table 18-4 are as follows:

Strut $A-B_1$:

<u>Column (4)</u> The angle between the axis of the strut and horizontal $= \arctan \dfrac{97.3}{74.3} = 52.6°$.

<u>Column (5)</u> The vertical component is equal to the shear $= 578$ kips.

<u>Column (6)</u> The horizontal component $= \dfrac{Vertical\ component}{\tan 52.6} = \dfrac{578}{1.31} = 441$ kips.

<u>Column (7)</u> The axial force in strut $A-B_1 = \dfrac{578\ \text{kips}}{\sin 52.6} = 728$ kips.

The axial (horizontal) force in the tie is equal to the horizontal forces in the struts at the two ends of the tie.

TABLE 18-4 Geometry and Forces in Struts and Tie, Second Trial—Single-Span Deep Beam

Strut or Tie	Horiz. Proj. in.	Vertical Proj. in	Angle	Vert. Comp. kips	Horiz. Comp. kips	Axial Force kips	Width of Strut in.
(1)	(2)	(3)	(4)	(5)	(6)	(7)	(8)
Vert. at A	0.0	—	Vertical	578	0	578	15.1
$A-B_1$	$78.0 - 0 - 3.7$ $= 74.3$ in.	$108 - 6.06 - 4.60$ $= 97.3$ in.	52.6	578	441	728	19.0
B_2-C	$120 - 5.67 - 0.0$ $= 114.3$ in.	97.3	40.4	376	441	580	15.1
$A-C$	—	—	—	—	441 at A 441 at C	441 at A 441 at C	$w_t = 10.87$
Vert. at C	0.0	—	Vertical	376	0	376	9.83

Column (8)

$$w_s = \frac{728,000}{0.75 \times 2550 \times 20} = 19.0 \text{ in.}$$

$$w_t = \frac{441,000}{0.75 \times 2720 \times 20} = 10.8 \text{ in.}$$

where 2720 psi is f_{cu} for the nodal zones at A and C. The upper limit on the height of the centroid of the tie reinforcement is $10.83/2 = 5.40$ in.

Nodes A and C will be assumed to be at the centers of the columns and 4.60 in. above the bottom of the beam.

9. Compute the required crack control reinforcement.

9(a)   Minimum reinforcement from ACI Sections 11.8.4 and 11.8.5. ACI Section 11.8.4 requires vertical reinforcement not less than $0.0025 b_w s$ at a maximum spacing of $d/5 = 104.5/5 = 21$ in. but not more than 12 in.

Try No. 4 bars at 8 in. on centers each face. $A_v/b_w s = \dfrac{2 \times 0.20 \text{ in.}^2}{20 \text{ in.} \times 8 \text{ in.}} = 0.0025$, exactly right.

ACI Section 11.8.5 requires horizontal reinforcement not less than $0.0015 b_w s$ at a maximum spacing of $d/5 = 104.5/5 = 21$ in. but not more than 12 in.

Try No. 4 bars at 12 in. on centers each face. $A_{vh}/b_w s = \dfrac{2 \times 0.20}{20 \times 12} = 0.0017$, OK.

9(b)   Minimum reinforcement from ACI Section A.3.3. ACI Section A.3.3 requires an orthogonal grid of bars at each face if the value of f_{cu} for bottle-shaped struts is used. The amount of steel is either computed using a local strut-and-tie model for the end of a bottle-shaped strut, or if f'_c is not greater than 6000 psi, the steel is adequate if it satisfies:

$$\frac{\Sigma A_{si}}{b s_i} \sin \gamma_i \geq 0.003 \qquad\qquad\qquad (18\text{-}10)$$
$$\text{(ACI Eq. A-4)}$$

Left end. The angle between the axis of the strut and the horizontal.

$$\gamma_i = \arctan \frac{97.3}{74.3} = 52.6°$$

Vertical steel: Try No. 4 bars at 8 in. on centers vertically on two faces.

The angle between the vertical steel and the axis of the struts is $90° - 52.6° = 37.4°$. Therefore,

$$A_s/(b_w s_i) \sin \gamma_i = 2\left(\frac{0.20}{20 \times 8}\right) \sin 37.4° = 0.00152$$

Horizontal steel: The angle between the strut and the horizontal is $52.6°$. Thus, $\gamma_i = 52.6°$. Try No. 4 bars at 12 in. on centers, horizontally on two faces, to get

$$A_s/(b_w s_i) \sin \gamma_i = 2\left(\frac{0.20}{20 \times 12}\right) \sin 52.6° = 0.00132$$

Total steel:

$$\Sigma A_s/(b_w s_i) \sin \gamma_i = 0.00152 + 0.00132 = 0.0028$$

This is too low. Try No. 4 at 7.0 in. Vertically

$$A_s/(b_w s_i) \sin \gamma_i = 2\left(\frac{0.20}{20 \times 7.0}\right) \sin 37.4° = 0.00174$$

Total steel:

$$\Sigma A_s/(b_w s_i) \sin \gamma_i = 0.00174 + 0.00132 = 0.00306$$

10(c)   Alternate solution for crack control reinforcement using a local strut-and-tie model. The region in which the strut stresses spread out across the width of the strut will be modeled using a local strut-and-tie model. Figure 18-29 shows strut $A - B_1$. The axial force in the strut is 728 kips. Half of this (364 kips) will be assumed to act at the center points of each half of the width of the end of the strut (at the quarter points of the width). The resulting strut-and-tie model is similar to Fig. 18-4c. ACI Section A.3.3 assumes the resultant forces spread at a slope of 2 longitudinally and 1 transverse. Thus, the transverse force at one end of the strut is $364/2 = 182$ kips. The sum of the transverse forces at the two ends of the strut is $182 + 182 = 364$ kips. Using No. 4 vertical bars at 8 in. on centers at an angle of 52.9° to the crack and No. 4 horizontal bars at 12 in. on centers at an angle of 37.1° to the crack, there will be $(108 \text{ in.})/(12 \text{ in./bar}) = 9$ horizontal bars per face, which can resist a force of

$$2 \times 9 \times 0.20 \times 60 \times \sin 52.9° = 172 \text{ kips}$$

perpendicular to the crack.

In the left shear span, there will be $(78/8) = 9$ bars per face, which can resist a force of

$$2 \times 9 \times 0.20 \times 60 \times \sin 37.4° = 130 \text{ kips}$$

perpendicular to the crack, for a total of $172 + 130 = 302$ kips. This must be increased to 364 kips. Reducing the spacing of the horizontal bars to 9 in. and reducing the spacing of the vertical bars to 7 in. increases the total resistance perpendicular to the axis of the strut to $210 + 159 = 369$ kips. This is now satisfactory, but it can be seen that using this approach leads to a more conservative design for this deep beam.

Summary of Design—Single-span deep beam.   **Use a 20-in.-by-108-in. beam with f'_c = 4000 psi and f_y = 60,000 psi. Provide 13 No. 8 bars in three layers, 5 bars in the bottom layer, plus two layers of 4 bars each. Use No. 10 bars as spacers between the layers. Use 90° hooks on all the No. 8 bars, and place the bars inside the vertical reinforcement from the supporting columns. Use No. 4 at 7 in. on centers vertically on each face and No. 4 bars at 12 in. on centers horizontally on each face.** ∎

18-9   CONTINUOUS DEEP BEAMS

Reactions of Continuous Deep Beams

A deep beam is a very stiff element, and as a result, the reactions are strongly affected by the assumptions made in the structural analysis and by differential settlements of the supports. At one extreme, a very flexible member on rigid supports will have reactions similar to those computed from an elastic-beam analysis. At the other extreme, the three reactions for a flexurally stiff, two-span beam supported on axially soft columns will approach three equal reactions. In the case of short deep beams, a proper analysis will include both shearing deflections and flexural deflections.

EXAMPLE 18-3   Design a Two-Span Continuous Deep Beam

A beam spans between three supporting columns with centerlines 20 ft apart and supports a column at the middle of each span, as shown in Fig. 18-32. Both upper columns support a factored load of P_u = 1500 kips. The supporting columns are 24 in. × 20 in., 24 in. × 45 in., and 24 in. × 20 in. The loading columns are 24 in. × 33 in. Use a strut-and-tie model to design the beam. Use f'_c = 4000 psi and f_y = 60 ksi, and use the load and strength-reduction factors from ACI 318-02 Sections 9.2.1, 9.3.2.2, and 9.3.2.6.

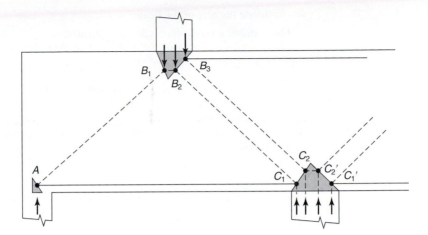

Fig. 18-32
Initial strut-and-tie model, two-span deep beam— Example 18-3.

1.   **General—Select a strut-and-tie model—Two-span deep beam**

   1(a)   Select the first trial strut-and-tie model—Two-span deep beam. In general, the best strut-and-tie models minimize the amount of reinforcement. This will occur when compression forces are transmitted by struts directly to the nearest supports. The left end of an appropriate strut-and-tie model is shown in Fig. 18-32. It consists of three triangular trusses, each consisting of two compression struts and one tie. The load supported by each triangular truss depends on the division of the applied loads between the three reactions.

   The load from the column at B is divided into three parts, each roughly a third of the column load. One part is transferred to the reaction at A by strut $A-B_1$. The change in direction of this load at A results in tension in the tie $A-C_1$ at A. The second part of the column load is transferred to the reaction at C_1 by strut B_2-C_1. This also gives a tension force in tie $A-C_1$. The tension forces at the two ends of tie $A-C_1$ should be equal. The final portion of the column load at B is transferred to the reaction at C_2 by strut B_3-C_2. For equilibrium at B_3, a tie B_3-B_3' is required.

   1(b)   Compute the reactions—Two-span deep beam. The reactions were estimated by using an elastic analysis of the reactions from an elastic propped-cantilever beam, loaded with a concentrated load, P_u, at midspan, supported on a hinged support at the exterior end and a fixed interior support. The second span will be taken as a mirror image of the first. Each span will have a reaction of $0.312 P_u$ at the exterior supports and a reaction of $0.688 P_u$ from each of the spans at the interior support, as shown in Fig. 18-33. In design, it usually is necessary to consider several distributions of loads and reactions with or without support settlements and to provide concrete reinforcement for the highest forces in each strut or tie.

   1(c)   Design equation. Design will be based on

$$\phi F_n \geq F_u$$

(18-1a)
(ACI Eq. A-1)

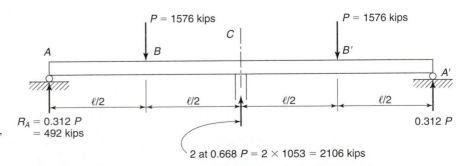

Fig. 18-33
Loads and assumed reactions, two-span deep beam— Example 18-3.

2. **Estimate the Size of the Beam.**

(2a) Based on maximum allowable shear stress—ACI Section 11.8.3 limits the nominal shear in a deep beam to $\phi V_n = \phi 10 \sqrt{f'_c}\, b_w d$. As a first trial, we shall limit V_n to the range from 0.6 to 0.9 times the maximum:

$$\text{Trial } \phi V_n = (0.6 \text{ to } 0.9) \times \phi 10 \sqrt{f'_c}\, b_w d$$

Here, from ACI Section 9.3.2.6, $\phi = 0.75$ for strut-and-tie models. Ignoring the weight of the beam, the maximum shear, V_u is 1002 kips (0.668×1500 kips). We will add the weight of the beam later.

$$d = \frac{1{,}002{,}000}{0.75 \times (0.6 \text{ to } 0.9) \times 10 \sqrt{4000} \times 24} = 147 \text{ to } 98 \text{ in.}$$

And $h \approx d/0.9 = 163$ to 109 in.

(2b) Based on limiting the strut angle to not less than 40°—Assuming that the nodes at A, C_1, and B_3 are $0.05h$ above the bottom or below the top of the beam and that nodes B_1, B_2, C_2, and C'_2 are $3 \times 0.05h = 0.15h$ from the top or bottom of the beam, the first trial vertical projection of strut $A-B_1$ is

$$h - 0.05h - 0.15h = 0.8h$$

and the horizontal projection is $a = 109$ in. (center to center of columns minus 33/3 in.), where the loading columns have a width of 33 in.

(2c) Based on flexure at the point of maximum moment—The effective depth of the beam is $d = h - h_t/2$, where h_t is the effective height of the tie, and the lever arm $jd = (d - a/2)$, where $a = w_s$ is similar to the depth of the rectangular stress block for flexure. The moment acting on a vertical section cut through the deepest part of the nodal zone at B and the rest of the beam is equilibrated by the internal compression and tension forces in the struts and tie acting on this section. The heights a and h_t can be related to f_{cu} and used to solve for the position of the nodes at A and B.

Try $b_w = 24$ in. and $h = 120$ in. The dead load of one 20-ft span is

$$W = (2 \text{ ft} \times 10.0 \text{ ft} \times 21 \text{ ft}) \times 0.150 \text{ kips/ft}^3 = 63.0 \text{ kips}$$

$$W_f = 1.2 \times 63.0 = 75.6 \text{ kips, say } 76 \text{ kips}$$

2(d) Check whether this is a deep beam—Two-span deep beam. $a = 109$ in., $h = 120$ in., and $a/h = 109/120 = 0.91$. ACI Sections 10.7.1 and 11.8.1 define a deep beam as one that is loaded on the top surface, is supported on the bottom of the beam, and has a/h not greater than 2. Thus, this beam is a deep beam. Further checking shows that all four shear spans are deep.

3. **Compute effective compression strengths, f_{cu}, for the nodel zones and struts—Two-span deep beam.**

3(a) Check the capacities of the columns—Two-span deep beam. Are the columns big enough for the loads?

Columns at A and A' The columns at A and A' are 24 in. by 20 in. in section. The factored loads in the columns at A and A' are $P_u = 492$ kips. From ACI Section 10.3.6.2, the maximum axial load capacity of a 24×20 in. column with 1 percent longitudinal reinforcement is

$$\phi P_{n(\max)} = 0.80\, \phi\, [0.85 f'_c (A_g - A_{st}) + f_y A_{st}] \tag{18-21}$$

where ϕ is the strength-reduction factor for compression-controlled tied columns ($\phi = 0.65$). So, for the columns at A and A' we obtain:

$$\phi P_{n(\max)} = 0.8 \times 0.65[0.85 \times 4000(480 - 4.8) + 60{,}000 \times 4.8]$$

$$\phi P_{n(\max)} = 990 \text{ kips}$$

Because $\phi P_{n(\max)}$ is less then P_u, we must increase the size of the supporting columns at A and A'. Using column dimensions of 22×24 in. results in $\phi P_{n(\max)} = 1090$ kips $> P_u = 1078$ kips.

Columns at B and B' Similar calculations give the factored capacity $\phi P_{n(\max)} = 1630$ *kips*, which exceeds both the 1500-kip load at the bottoms of the loading columns and the 1576-kip load in the part of the beam loaded by the column, which includes the factored dead load of the deep beam.

Again, $\phi P_n > P_u$. The columns over B and B' are adequate with 1 percent steel.

Column at C The factored load on each half of column C is $P_u = 1084$ kips, for a total load of 2168 kips. Assuming a 45-in.-by-24-in. column with 1 percent steel, the column capacity is 2230, which is more than the factored load of 2170 kips. If this were not true, it would be necessary to enlarge the column or to increase its capacity by increasing the column reinforcement.

3(b)   Compute the effective compression strengths of the nodal zones—First model—Two-span deep beam

Nodal zones A and B anchor one tension tie. They are C–C–T nodes. From A.5.2.2, $\beta_n = 0.80$ and $f_{cu} = 0.85 \times 0.80 \times 4000$ psi $= 2720$ psi.

Nodal zone C anchors two ties, one from each side of the line of symmetry. Because it can be subdivided into two C–C–T nodes, we shall base f_{cu} on the C–C–T nodal zones. From ACI Section A.5.2.2, $f_{cu} = 2720$ psi.

3(c)   Compute the effective compression strengths of the struts—First model—Two-span deep beam. For struts, (18-5) (ACI Eq. (A-3)) gives $f_{cu} = 0.85\beta_s f'_c$.

Struts A–B_1, B_2–C_1, and B_3–C_2 are all "bottle-shaped" because the struts can spread laterally into the unstressed concrete adjacent to the struts on at least one side. **Crack control reinforcement must be provided** to allow the use of f_{cu} from ACI Section A.3.2.2(a). From ACI Section A.3.2.2(a), $\beta_s = 0.75$ and $f_{cu} = 0.85 \times 0.75 \times 4000$ psi $= 2550$ psi.

4. Locate the nodes and compute the strut-and-tie forces—First model—Two-span deep beam

4(a)   Subdivide the nodal zone at B—First model—Two-span deep beam. We will subdivide the nodal zone at B into three parts, as shown in Fig.18-32:

(1) a vertical strut inside column B, with a vertical force equal to the nominal vertical reaction at A ($R_{An} = 492$ kips), acting in a strut centered over node B_1, as shown in Fig. 18-32, and

(2) and **(3)** two vertical struts carrying the remainder of the column load from the left span of the beam ($R_{Cn} = 1084$ kips), acting in the two vertical struts centered over nodes B_2 and B_3. As a first trial, assume these last two forces are equal and are each 0.5×1084 kips $= 542$ kips.

4(b)   Compute the widths and locations of the vertical struts in column B—First model—Two-span beam.

$$\phi F_n \geq F_u \qquad w_s = \frac{F_u}{\phi f_{cu} b}$$

where $f_{cu} = 0.85\beta_s f'_c$ for struts and $f_{cu} = 0.85\beta_n f'_c$ for nodes.

Struts in column over node B
For struts like the vertical struts in the column over B, ACI Section A.3.2.1 specifies $\beta_s = 1.0$ because these struts have parallel vertical stress trajectories:

$$f_{cu} = 0.85 \times 1.0 \times 4000 = 3400 \text{ psi}$$

Nodal zone at B   The node at B anchors one tension tie. From A.5.2.2, $\beta_n = 0.80$ and

$$f_{cu} = 0.85 \times 0.80 \times 4000 = 2720 \text{ psi.}$$

We shall assume that this value of f_{cu} controls the strength of the entire nodal zone at B.

The width of the vertical strut above B_1 is

$$w_s = \frac{492,000}{0.75 \times 2720 \times 24} = 10.05 \text{ in.}$$

TABLE 18-5 Summary of Locations of the Nodes—First Model—Two-Span Deep Beam

Point	Left of Center of Column A, in.	Left of Center of Column B, in.	Left of Center of Column C, in.	Below Top of Beam, in.	Above Bottom of Beam, in.
A	0	—	—	—	6.4
B_1	—	11.07	—	19.2	—
B_2	—	0.52	—	19.2	—
B_3	—	−10.56	—	6.4	—
C_1	—	—	16.61	—	6.4
C_2	—	—	5.4	—	19.2

At each of B_2 and B_3, $F_n = 542$ kips and $w_s = 11.07$ in.

The total width of the three struts is $(10.05 + 11.07 + 11.07) = 32.19$ in. These fit into the 33-in.-wide column, leaving a space of 0.81 in. We shall assume that half of this space occurs on each edge of the column.

Assuming that the sum of the widths of the three struts is centered on the axis of the column at B, the locations of the axes of the three vertical struts are as follows (see Fig. 18-32):

node B_1 is 11.07 in. to the left of the centerline of column B,

node B_2 is 0.52 in. to the left, and

node B_3 is 10.56 in. to the right.

Below the nodal zone, the total factored load at B is 1576 kips.

4(c) Width of the strut in column A—First model—Two-span deep beam. By observation, this is the same width as the column strut at B_1, 10.05 in. This fits inside the 22-in. column width at A.

4(d) Widths and locations of the vertical struts in the column under C—First model—Two-span deep beam. The total factored vertical reaction at C from one span is 1084 kips. Assume that each vertical strut in the column at C resists half of this, 542 kips. f_{cu} for the struts under C is 3400 psi, for the nodal zone is 2720 psi. The nodal zone governs, so $f_{cu} = 2720$ psi, both for the nodal zone and for the struts in the column under node C_1. The width of each of the four vertical struts at C is 11.07 in., for a total width of 44.3 in. This will fit into the 45-in. column, with a gap of 0.18 in. on each side. Node C_1 is 16.61 in. left of the column centerline. Node C_2 is 5.54 in. to the left. (See Fig. 18-32.)

4(e) Assume the vertical positions of the nodes—First model—Two-span deep beam. We will assume that the nodal zones at B and C each consist of two layers, each with a height of $0.10h$, as shown in Fig. 18-32 and Table 18-6. This is arbitrarily based on an assumed depth of flexural compression, a, equal to $0.2h$. Assume that the tie reinforcement is spread over the height of the layer assumed to be the tie; then the centroid of tie $A–C_1$ can be taken to be $0.05h = 6.4$ in. above the bottom of the beam, and that of tie $B_3–B_3'$ to be 6.4 in. below the top of the beam. The centroid of the second layer is at $0.15h = 19.2$ in. above or below the top of the beam.

TABLE 18-6 Geometry and Forces in Struts and Ties—Second Model—Two-Span Deep Beam

Strut or Tie (1)	Horiz. Proj. in. (2)	Vertical Proj. in. (3)	Angle θ (4)	Vert. Comp. kips (5)	Horiz. Comp. kips (6)	Axial Force kips (7)
Strut $A–B_1$	120 − 0 − 11.07 = 108.9	120 − 6.4 − 19.2 = 94.4	40.9°	492	568	751
Strut $B_2–C_1$	120 + 0.52 − 16.6 = 103.9	120 − 6.4 − 19.2 = 94.4	42.3°	542	596	805
Strut $B_3–C_2$	120 − 10.56 − 5.54 = 103.9	120 − 6.4 − 19.2 = 94.4	42.3°	542	596	805
Tie $A–C_1$	—	—	—	—	568 at A 596 at C_1	
Tie $B_3–B_3'$	—	—	—	—	—	596

4(f)   Summary of locations of the nodes—First model—Two-span deep beam. The calculations for the widths of the struts and ties are given in Table 18-6 as follows:

5.   Compute forces in the struts and ties due to factored loads—Second model—Two-span deep beam

5(a)   Forces in struts and ties—Second model—Two-span deep beam. These are computed in Table 18-6. We have

<u>Column (4)</u> Angle $= \arctan \dfrac{94.4}{108.9} = 40.9°$

<u>Column (5)</u> Vertical component is the factored shear in shear span A–B, or 492 kips.

<u>Column (6)</u> Horizontal component $= \dfrac{\text{Vertical component}}{\tan 40.9} = 568$ kips

<u>Column (7)</u> Axial force $= \dfrac{\text{Vertical component}}{\sin 40.9} = 751$ kips

5(b)   Balance the tie forces if necessary—Second model—Two-span deep beam. Tie A—C_1 From Table 18-6, the tie force from strut A–B_1 is 568 kips, and that from strut B_2–C_1 is 596 kips. These should be the same. If not, several strategies are possible. We will try to maintain the reactions chosen earlier. Because of this, we will not change the load in strut A–B_1. Reduce the horizontal component of the force in strut B_2–C_1 to 568 kips, so that the tie forces are the same at the two ends of tie A–C_1. This will be done by reducing the vertical force in B_2–C_1 and increasing the vertical force in B_3–C_2 to keep the reaction at C equal to 1084 kips. We thus have

Vertical force at $B_2 = \dfrac{568}{596} \times 542 = 517$ kips

and

Vertical force at $B_3 = 1084 - 517 = 567$ kips

The geometry at node B changes: For the struts in the column over B, $f_{cu} = 2720$ psi; for the vertical strut at B_1, $w_s = 10.05$ in.,

w_s at $B_2 = 10.56$ in.

w_s at $B_3 = 11.58$ in.

Total $w_s = 10.05 + 10.56 + 11.58 = 32.2$ in. This fits into a 24-by-33-in. column with 0.4 in. space on each side. From the geometry of the vertical struts at B (see Fig. 18-34a), B_1 is 11.06 in. left of the centerline of the column, B_2 is 0.76 in. left, and B_3 is 10.30 in. right.

After the tie forces have been balanced, the vertical forces at C_1 and C_2 are 517 and 567 kips, giving struts of widths 10.56 in. below C_1 and 11.58 in. below C_2, which locate C_1 at 16.86 in. left of the center of the column at C and C_2 at 5.79 in. left. After the tie forces are balanced, the axial forces in B_2–C_1 and B_3–C_2 are 768 and 842 kips respectively. These values have been used in Table 18-7.

The balanced model with updated node locations will be called the second model.

5(d)   Compute the widths of struts and ties due to factored loads—Second model—Two-span deep beam. The calculations in Table 18-7 proceed as follows:

<u>Column (2)</u> is based on Table 18-6, the axial force in B_2–C_1 is 768 kips and that in B_3–C_2 is 842 kips, both after the balancing of the tie force in A–C_1.

<u>Column (3)</u> is the value of f_{cu} for strut A–B_1 at end A.

<u>Column (4)</u> is the value of f_{cu} for the nodal zone at A. The values not crossed out govern.

<u>Column (5)</u> is the computed width of the strut, for the governing value of f_{cu} from columns (3) and (4).

There are two entries for each strut in Table 18-7, to accommodate different values of f_{cu} at the two ends of the struts.

5(e)   Effective height of ties—Second model—Two-span beam. ACI Section R.A.4.2, limits the effective height of the tie, w_t, to

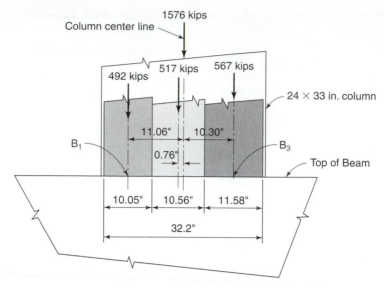

(a) Column over the beam at B.

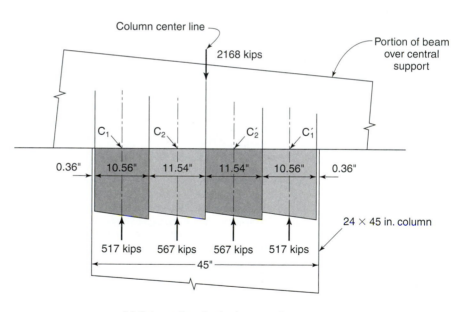

Fig. 18-34
The widths of vertical struts
in the columns at B and C—
Two-span deep beam—
Example 18-3.

(b) Column C under the beam at C.

(a)   twice the minimum height from the bottom or top of the beam to the centroid of the tie reinforcement, and

(b)   the height, w_t, computed from the tie force by assuming that the tie is stressed to the value f_{cu} for the nodal zones at the ends of the tie.

From (b), the effective height of the concrete concentric with Tie A–C_1 is

$$w_{t,\mathrm{max}} = \frac{568{,}000}{0.75 \times 2720 \times 24} = 11.60 \text{ in.}$$

where 2720 psi is f_{cu} for a C–C–T node. The height of the centroid of such a tie would be at $11.60/2$ in. $= 5.80$ in.

TABLE 18-7 Compute the Widths of Struts and Ties—Second Model—Two Span Deep Beam

Member (1)	Axial Force, kips (2)	Strut f_{cu} psi (3)	Node f_{cu} psi (4)	Width in. (5)
Vertical strut @ A	492	~~3400~~	2720	10.05
Strut $A–B_1$ @ A	751	2550	~~2720~~	16.36
@ B_1	751	2550	~~2720~~	16.36
Vertical strut @ B_1	492	~~3400~~	2720	10.05
Vertical strut @ B_2	517	~~3400~~	2720	10.56
Vertical strut @ B_3	567	~~3400~~	2720	11.58
Strut $B_2–C_1$ @ B_2	768	2550	~~2720~~	16.73
@ C_1	768	2550	~~2720~~	16.73
Strut $B_3–C_2$ @ B_3	842	2550	~~2720~~	18.34
Strut $B_3–C_2$ @ C_2	842	2550	~~2720~~	18.34
Vertical strut @ C_1	517	~~3400~~	2720	10.56
Vertical strut @ C_2	567	~~3400~~	2720	11.54
Tie $A–C_1$	568	—	2720	$w_t = 11.60$ in. $w_t/2 = 5.80$ in. $A_s = 12.62$ in.^{2}
Tie $B_3–B_3'$	623		2720	$w_t = 12.72$ in. $w_t/2 = 6.36$ in. $A_s = 13.84$ in.^{2}

5(f) Tie reinforcement—Second model—Two-span beam. Tie $A–C_1$—The tie forces at A and C_1 are each 568 kips. The area of steel required for a factored tie force of 568 kips is

$$A_s = \frac{568,000}{0.75 \times 60,000} = 12.62 \text{ in.}^2.$$

We could use 16 No. 8 bars, $A_s = 12.64$ in.^{2}.

Tie $A–C_1$— 16 No. 8 bars in four layers of 4 bars each, with 1-in. spaces between layers, with centroid at $1.5 + 0.5 + 1.0 + 1.0 + 1.0 + 0.5 = 5.5$ in. above the bottom. The effective height of the tie is 11.0 in.

Several other choices were tried before the final selection of the reinforcement to get the centroid of the steel the desired distance from the bottom of the beam to the center of the tie. The anchorage of the tie reinforcement in the nodal zone at A is shown in Fig 18-36.

6. Recompute w_t, and revise the location of the nodes—Second model—Two-span deep beam.

Nodes A and C_1 will be located 5.5 in. above the bottom of the beam, and the effective height of tie $A–C_1$ is 11.0 in. Node C_2 will be taken at 1.5 times the effective height of tie $A–C_1$ above the bottom of the beam—that is, at $1.5 \times 11.0 = 16.5$ in. above the bottom.

Tie $B3–B3'$

The revised axial force in tie $B_3–B_3'$ is $\dfrac{567}{\tan 42.3°} = 623$ kips,

$$A_s = \frac{623,000}{0.75 \times 60,000} = 13.84 \text{ in.}^2$$

$$w_t = \frac{623,000}{0.75 \times 2720 \times 24} = 12.72 \text{ in.}$$

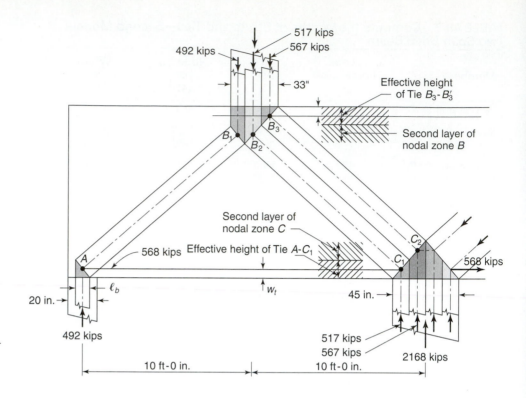

Fig. 18-35
Selection of height of ties—
two-span deep beam—
Example 18-3.

and

$$w_t/2 = 6.36 \text{ in.}$$

6(a)   Tie $B3-B3'$—Use 18 No. 8 bars in four layers; a top layer of 6 bars plus three layers of 4 bars each, with transverse No. 11 bars as spacers between layers. The centroid of the steel at B_3 is 6.11 in. below the top of the beam, and nodes B_1 and B_2 are $1.5(2 \times 6.11) = 18.33$ in. below the top. Placing 4 or 6 bars in a layer also allows vertical spaces to ease the vibrating of the concrete.

6(b)   Check agreement between the widths of the struts and the nodal zones—Third model—Two-span deep beam

Equation (18-16) relates the widths of the struts, the bearing lengths, and the heights of ties at nodal zones.

Node A–Strut $A-B_1$

Minimum $w_s = \ell_b \cos \theta_A + w_t \sin \theta_A$

Bearing length: $\ell_b = 10.05$ in.

Height of tie (from steel arrangement chosen in step 5(e)): $w_t = 11.0$ in.

Angle between strut and horizontal: $\theta_A = 41.8°$

In order to maintain the reactions chosen earlier; we will not change the load in strut $A-B_1$.

$$w_s = 10.05 \cos 41.8° + 11.0 \sin 41.8°$$
$$= 7.49 + 7.33 = 14.82 \text{ in.}$$

Since $w_s = 16.36$ in. from Table 18-7 exceeds the minimum from (18-16), namely, 14.82 in., the strut $A-B_1$ is not adequate. Increase ℓ_b to 13.0 in., giving minimum strut width = 17.0 in. This exceeds the $w_s = 16.36$ from Table 18-7 and therefore is adequate.

Node C, Struts $B2$–$C1$ and $B3$–$C2$

Bearing length: $\ell_b = 10.56 + 11.58 = 22.14$ in.

Total height of nodal zone from steel chosen: $h_t = 2 (2 \times 6.11) = 24.44$ in.

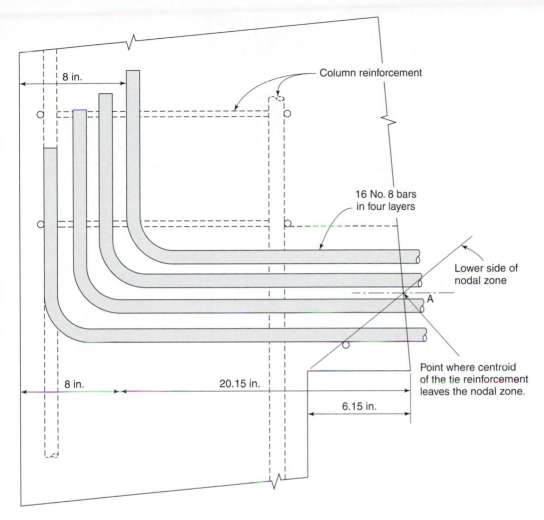

Fig. 18-36
Anchorage of reinforcement from tie *A–C* at node *A*—Two-span deep beam—Example 18-3.

Angle: 43.2°

$$w_s = 22.14 \cos 43.2° + 24.44 \sin 43.2° = 16.14 + 16.73$$
$$= 32.8 \text{ in.}$$

Because the widths of the two struts combined from Table 18-7, which is 16.73 + 18.34 = 35.07 in., exceeds the calculated width available at Node C, $w_s = 32.8$ in., the total too small and high height of the nodal zone at C must be increased. This means the width of the Tie $B_3 - B_3'$ must be increased by changing the arrangement of the No. 8 bars. We will change to five layers of bars with four bars per layer, giving us a total of 20 No. 8 bars. Using No.11 bars as too small and high spacers between the layers, this moves the centroid of the reinforcement in Tie $B_3 - B_3'$ to 7.32 in. below the top of the beam. Thus, the total width of the strut, w_t, is 14.64 in. and the total height of Node C is twice that value, 29.3 in. Recalculating the width available at Node C for the inclined strut gives $w_s = 36.2$ in, which exceeds the required value of 35.07 in.

6(c)   Draw the strut-and-tie model—Third model—Two-span deep beam. Figure 18-37 shows the left span of the third model. The widths of the struts and nodal zones are compatible.

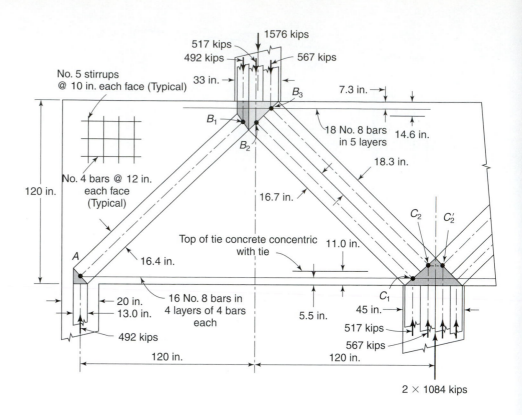

Fig. 18-37
Final design, two-span deep
beam—Example 18-3.

7. Select the reinforcement and details—Two-span deep beam

7(a) Minimum flexural reinforcement—Two-span deep beam. ACI Section 10.7.3 requires that the flexural reinforcement satisfy the minimum from ACI Section 10.5, which in turn requires that

$$A_{s,min} = \frac{3\sqrt{f_c'}}{f_y} b_w d, \text{ but not less than } \frac{200 \; psi}{f_y} b_w d$$

$$A_{s,min} = 8.69 \text{ in.}^2, \text{ but not less than } 9.16 \text{ in.}^2$$

where $d = 120$ in. -5.5 in. $= 114.5$ in. The tie reinforcement chosen previously exceeds these values, so the minimum reinforcement requirement does not govern.

7(b) Anchor the ties—Third model—Two-span deep beam

Tie $A-C_1$ at A
ACI Section A.4.3(a) requires the tie force at the face of the support to be developed by the point where the centroid of the tie reinforcement leaves the extended nodal zone. In our case, this is where the centroid of the 16 bars composing the tie leave the bottom side of the inclined strut in the extended nodal zone. (See Fig. 18-11a) This occurs at

$$\left(\frac{10.05 \text{ in.}}{2} \right) + \frac{5.50}{\tan 41.8°} = 11.2 \text{ in.}$$

to the right of the center of the column at A. The total width available to anchor the tie reinforcement measured from the exterior face of column A is $(22/2 \text{ in.} + 11.4 \text{ in.}) = 22.4$ in. The development length for a straight Grade-60 No. 8 bottom bar, inside stirrups, in 4000-psi concrete, is 47.4 in. The length of a $90°$ hook on the same bars is 19 in. The distance from the exterior face of the column to the tail of the bend on the top row of bars is $(1.5 \text{ in. cover} + 1/2 \text{ in. stirrup} + 3 \text{ bars and 3 spaces of 1 in. each}) = 8$ in. (see Fig. 18-36), leaving 14.4 in. to accommodate development length or hook length. The hook-development length can be reduced by the factor in ACI Section 12.5.3.2 to $0.7 \times 19 \text{ in.} = 13.3$ in. if the side cover to the hooks is set equal to 2.5 in. and the cover to the tail of the hook is at least 2 in. We will place the tie reinforcement inside the column reinforcement. There is just enough space to develop the bars with hooks.

The anchorage of tie $A-C_1$, at A, shown in Fig. 18-36, uses 90° hooks with the tails vertical. This would be acceptable only if the hooked bars were inside the column cages at the two exterior supports.

Tie $A-C_1$ at C_1

The tie $A-C_1$ will be anchored at C_1 by extending it continuously through the support at C.

Tie B_3-B_3' at B_3

The development length of No. 8 top bar in 4000-psi concrete is 61.7 in. Anchor tie B_3-B_3' by extending the bars through the node at B_3. Node B_3 is 10.20 in. right of the center of the column at B. Extend the bars 61.7 in. $-$ 10.20 in. = 51.5 in.—say, 52 in.—past the centers of the columns at B and B'.

8.   Compute the required minimum crack control reinforcement and load-spreading reinforcement—Two-span deep beam

8(a)   Minimum crack control reinforcement—Two-span deep beam. ACI Sections 11.8.4 and 11.8.5 require vertical reinforcement with $A_v \geq 0.0025b_w s$ and horizontal reinforcement with $A_{vh} \geq 0.0015b_w s$, respectively. Try one layer of vertical No. 5 bars in each face. The maximum horizontal spacing of the bars is

$$s_h = \frac{2 \times 0.31}{0.0025 \times 24} = 10.33 \text{ in.}$$

and for one layer of horizontal No. 4 bars on each face, the maximum vertical spacing is

$$s_v = \frac{2 \times 0.20}{0.0015 \times 24} = 11.1 \text{ in.}$$

The reinforcement required by ACI Sections 11.8.4 and 11.8.5 is No. 5 U stirrups at 10 in. o.c. plus horizontal No. 4 bars at 10 in. on each face.

8(b)   Reinforcement required for bottle-shaped struts—Two-span deep beam. For bottle-shaped struts, to satisfy

$$\Sigma \frac{A_{si}}{b_w s_i} \sin \gamma_i \geq 0.003 \tag{18-10}$$

ACI Section A.3.3 requires layers of reinforcement near both faces of the web of the deep beam.

For strut $A-B_1$, the angle between the horizontal reinforcement and the strut is 42.8°, and $\sin \gamma_i = 0.680$. For the vertical reinforcement, $\gamma_i = 47.2°$, and $\sin \gamma_i = 0.734$. Thus,

$$\Sigma \frac{A_{si}}{bs} \sin \gamma_i = \frac{2 \times 0.31}{24 \times 10} \times 0.680 + \frac{2 \times 0.20}{24 \times 10} \times 0.734 = 0.00298 \text{—OK}$$

Provide No. 5 vertical bars on each face at 10 in. on centers and No. 4 horizontal bars on each face at 12 in. on centers.

9.   Check other details—Two-span deep beam

9(a)   Check the stress on a vertical section through the nodal zone at C—Two-span beam. The node at C has a height of 29.3 in. This height is loaded with the horizontal components of the compressive forces in struts B_2-C_1 and B_3-C_2. The tie reinforcement will be ignored, because part or all of it is anchored in the node or passes through the node. The stress on the dividing line between the two halves of this node is as follows:

Horizontal component of the force in strut B_2-C_1: 560 kips
Horizontal component of the force in strut B_3-C_2: 614 kips

Stress on vertical plane through nodal zone: $\dfrac{560 + 614 \text{ kips}}{29.3 \text{ in.} \times 24 \text{ in.}} \times 1000 = 1670 \text{ psi}$

f_{cu} for nodal zone C is 2720 psi—therefore, OK.

9(b)   Check lateral buckling—Two-span deep beam. ACI Section 10.7.1 requires that lateral buckling of the deep beam be considered. The dimensions of this beam (2 ft wide by 10 ft high) will prevent lateral buckling of the beam itself. We shall assume the bracing of the building prevents relative lateral displacement of the top and bottom of the deep beam.

10. Summary of design—Two-span deep beam

- Use a 24 in.-by-10-ft beam with $f'_c = 4000$ psi and $f_y = 60,000$ psi.
- **Bottom flexural reinforcement: Provide 16 No. 8 bottom longitudinal bars in four layers of 4 bars each for full length of beam, with bar layers to be spaced by transverse No. 8 bars. Anchor the No. 8 longitudinal bars into the exterior columns with 90° hooks.**
- **Top flexural reinforcement: Provide 20 No. 8 top bars in five layers spaced by transverse No.11 bars with 4 bars per layer. Anchor these by extending them 52 in. past the centers of the columns at *B* and *B'*.**
- **Crack control reinforcement: Provide No.5 vertical bars on each face at 10 in. on centers and No.4 horizontal bars on each face at 10 in. on centers.** ■

18-10 BRACKETS AND CORBELS

A *bracket* or *corbel* is a short member that cantilevers out from a column or wall to support a load. The corbel is generally built monolithically with the column or wall, as shown in Fig. 18-38. The term "corbel" is generally restricted to cantilevers having shear span-to-depth ratios, *a/d*, less than or equal to 1.

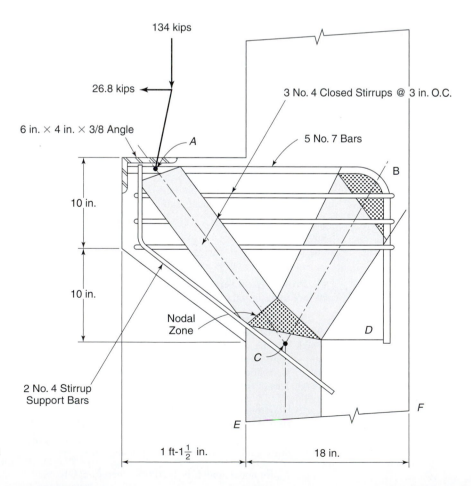

Fig. 18-38
Final strut-and-tie model of a
corbel—Example 18-4.

Structural Action

A strut-and-tie model for a corbel supported by a column is shown in Fig. 18-38. Within the corbel itself, the structural action consists of an inclined compression strut, A–C, and a tension tie, A–B. Shears induced in the columns above and below the corbel are resisted by tension in the column bars and ties and by compression forces in struts between the ties.

In tests, [18-20], [18-21] corbels display several typical modes of failure, the most common of which are yielding of the tension tie; failure of the end anchorages of the tension tie, either under the load point or in the column; failure of the compression strut by crushing or shearing; and local failures under the bearing plate. If the tie reinforcement is hooked downward, as shown in Fig.18-39a, the concrete outside the hook may split off, causing failure. The tie should be anchored by welding it to a crossbar or plate. Bending the tie bars in a horizontal loop at the outer face of the corbel is also possible, but may be difficult to do because of bends in two directions and may require extra cover. If the corbel is too shallow at the outside end, there is a danger that cracking will extend through the corbel, as shown in Fig. 18-39b. For this reason, ACI Section 11.9.2 requires the depth measured at the outside edge of the bearing area must be at least one-half the depth at the face of the column.

Design of Corbels

ACI Section 11.9.1 requires corbels having a/d between 1 and 2 to be designed using strut-and-tie models, where a is the distance from the load to the face of the column, and d is the depth of the corbel below the tie, measured at the face of the column. Corbels having a/d between 0 and 1 may be designed either using strut-and-tie models, or by the closely related traditional ACI design method, which is based in part on the strut-and-tie model and part on shear friction. This procedure was limited to a/d ratios less than or equal to 1.0 because

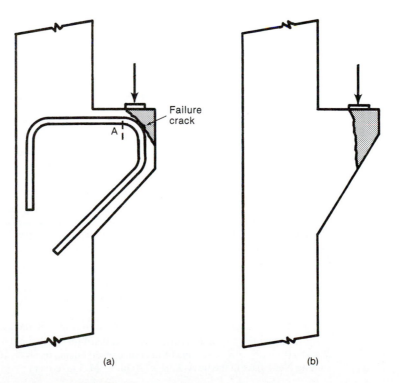

Fig. 18-39
Failure of corbels due to poor detailing.

(a)   (b)

little test data was available for longer corbels. Regardless of the design method used, the general requirements in ACI Sections 11.9.2, 11.9.3.2.1 and 11.9.3.2.2, 11.9.5, 11.9.6, and 11.9.7 must be satisfied.

Two closely related design procedures for corbels will be presented: design using strut-and-tie models, and design according to ACI Section 11.9. The strut-and-tie method is a little more versatile than the ACI method, but both give essentially the same results within the range of application of the ACI Code.

EXAMPLE 18-4   Design of a Corbel via a Strut-and-Tie Model

Design a corbel to support the reaction from a precast beam (Fig. 18-38). The end of the beam is 12 in. wide. The column is 16 in. square. The unfactored beam reaction is 60 kips dead load and 39 kips live load. The beam does not support fluid loads, soil loads, wind loads, or roof loads. The beam is partially or fully restrained against longitudinal shrinkage. Use $f'_c = 5000$ psi and $f_y = 60,000$ psi. Use the load factors and strength-reduction factors from ACI 318-02 Sections 9.2 and 9.3.2.

1. **Compute the factored load—Corbel**
The following load combinations from ACI Section 9.2.1 are applicable:

$$U = 1.4(D + F) \qquad \text{(ACI Eq. 9-1)}$$

$$U = 1.2(D + F + T) + 1.6(L + H) + 0.5(L_r \text{ or } S \text{ or } R) \qquad \text{(ACI Eq. 9-2)}$$

Note that F, T, H, L_r, S, and R all are zero.

$$U = 1.4D = 1.4 \times 60 = 84 \text{ kips}$$

$$U = 1.2D + 1.6L = 1.2 \times 60 + 1.6 \times 39 = 134 \text{ kips}$$

Thus, the factored vertical load on the corbel is 134 kips.

2. **Compute the distance from the face of the column to the beam reaction—Corbel**
Assume a 12-in.-wide bearing plate. From ACI Section 10.17.1, the maximum bearing stress is $0.85f'_c = 0.85 \times 5000 = 4250$ psi. The required width of the bearing plate is

$$\frac{134,000 \text{ lb}}{0.65 \times 4250 \text{ psi} \times 12} = 4.04 \text{ in.}$$

where $\phi = 0.65$ is the strength-reduction factor for bearing.

ACI Section 7.7.1 requires 1.5-in. cover to stirrups or main reinforcement. Try a bearing plate 5 in. by 12 in. by 1.5 in. thick, with the top surface flush with the top of the corbel. This will provide 1.5-in. cover to the principal reinforcement at the top of the corbel.

Assume that the beam extends 7.5 in. past the center of the bearing plate. The design gap between the end of the beam and the face of the column is 1 in. Erection tolerances could make this as large as 2 in. or as small as 0 in. Thus, the beam reaction could act as far as 9.5 in. from the face of the column.

3. **Establish the depth of the corbel**
ACI Section 11.9 does not give guidance in choosing the size of a corbel. ACI Section 11.9.3.2.1 limits the interface shear transfer stress to the smaller of $0.2f'_c$ or 800 psi. For 5000-psi concrete, the second limit governs. We shall use this as guidance, but use a corbel somewhat larger than the minimum. To simplify forming, the corbel will have the same width as the column, $b = 16$ in. Then

$$d = \frac{134,000 \text{ lb}}{0.75 \times 800 \text{ psi} \times 16 \text{ in.}} = 14.0 \text{ in.}$$

where the strength-reduction factor for the strut-and-tie model, $\phi = 0.75$, is used.

The smallest corbel that satisfies ACI Section 11.9.3.2.1 would have $b = 16$ in., $h = 16$ in., and $d = 16.0 - 1.5 - d_b/2$—say, 14 in. For conservatism, try $b = 16$ in. and $h = 20$ in.; then, assuming No. 8 bars in the tie A–B, $d = 20$ in. $- (1.5$ in. cover$) - (1/2$ bar diameter$) = 18$ in.

Try $b = 16$ in., $h = 20$ in., $d = 18$ in.

4.   Select design method—Corbel

ACI Section 11.9.1 allows corbels with a/d between 1.0 and 2.0 to be designed by using Appendix A. Corbels with a/d less than 1.0 can be designed by using Appendix A or by using the traditional ACI corbel design method in 11.9. Here, $a/d = 9.5/18 < 1.0$, so either design method may be used. We shall use Appendix A.

5.   Select a first trial strut-and-tie model—Corbel

ACI Section 11.9.3.4 requires corbels to be designed for a shear, V_u, equal to the factored beam reaction and for a factored tensile force, N_{uc}, equal to the actual tension force acting on the corbel, but not less than $0.2V_u$. This force represents the forces induced by restrained shrinkage of the structure supported by the corbel. **Design the corbel for $V_u = 134$ kips and $N_{uc} = 26.8$ kips.** (See Fig. 18-38.) The forces V_u and N_{uc} can be resolved into an inclined force that intercepts the centroid of the tie reinforcement at node A, which is located at $9.5 + 0.2 \times (1.5 + 0.5)$ in. $= 9.9$ in., from the face of the column.

Figure 18-38 shows the final strut-and-tie model. Assuming (a) that the column reinforcement is No. 9 bars with No. 3 ties and (b) that the nodes B and D are in the plane of the column reinforcement located at 1.5 in. cover + 0.375 in. ties + 1.270/2 = 2.5 in. from the right-hand face of the column, gives the distance from A to B as 9.9 in. + 16 in. − 2.5 in. = 23.4 in.

Locate node C. Node C is located at a distance $a / 2$ from the left side of the column, where a is the depth of the stress block resisting the force in the strut C–E. ACI Section A.5.2 gives the effective compression strength of the nodal zone as

$$f_{cu} = 0.85\beta_n f'_c \qquad\qquad (18\text{-}7b)$$
$$(\text{ACI Eq. A-8})$$

The node at C anchors three compression struts and a tension tie. From ACI Section A.5.2, a nodal zone anchoring one tie has $\beta_n = 0.80$, so that

$$f_{cu} = 0.85 \times 0.80 \times 5000 \text{ psi} = 3400 \text{ psi}$$

and

$$a = \ = \frac{P_{u,C-E} \text{ kips} \times 1000 \text{ lb/kip}}{0.75 \times 3400 \text{ psi} \times 16 \text{ in.}} = 0.0245 P_{u,C-E} \text{ in.}$$

where $P_{u,C-E}$ is in kips and the 0.75 is the ϕ factor for strut-and-tie models from ACI Section 9.3.2.6. Summing moments about D gives

$$\Sigma M_D = 0 = 134 \text{ kips} \times 23.4 \text{ in.} + 26.8 \text{ kips} \times 18 \text{ in.} - P_{u,C-E} \times (13.5 \text{ in.} - a/2)$$

For $a = 0.0245 P_{u,C-E}$, this becomes

$$0 = 296{,}000 - 1102 P_{u,C-E} + (P_{u,C-E})^2$$

$$P_{u,C-E} = \frac{1102 \pm \sqrt{1102^2 - 4 \times 1 \times 296{,}000}}{2} = 638 \text{ or } 464 \text{ kips}$$

Thus, $P_{u,C-E} = 464$ kips and $a = 0.0245 \times 464 = 11.36$ in. Node C is $11.36/2 = 5.68$ in. from the left edge of the column adjacent to C, and C–D is 13.5 in. − 5.68 in. = 7.82 in.

The nodal zone at C takes up 11.36 in. out of the 13.5 in. width between the compression face of the column and node D. For comparison, the rectangular stress block in a beam would have a depth of about $0.3d$. This indicates that the column is too small. We shall increase the width of the column to 18 in. The distance from A to B increases to 25.4 in.

6.   Recompute a for the new depth of column—Second model—Corbel

Summing moments about D gives

$$\Sigma M_D = 0 = 134 \times 25.4 + 26.8 \times 18 - P_{u,C-E} \times (15.5 - a/2)$$
$$0 = 317{,}000 - 1265 P_{u,C-E} + (P_{u,C-E})^2$$

which gives $P_{u,C-E} = 345$ kips. Thus, $P_{u,C-E}$ drops from 464 kips to 345 kips, corresponding to $a = 8.45$ in., and $a/2 = 4.23$ in. The distance C–D is 15.5 in. − 4.23 in. = 11.27 in.

7. Solve for forces in the struts and ties—Second model—Corbel

Node A

Figure 18-40a shows the forces acting at node A.

Strut A—C has a horizontal projection of $25.4 - 11.27 = 14.13$ in. and a vertical projection of 18 in. It supports a factored vertical force component of 134 kips, a horizontal force component of

$$\frac{14.13}{18} \times 134 \text{ kips} = 105.2 \text{ kips}$$

and an axial force of $\sqrt{134^2 + 105.2^2} = 170.4$ kips compression. The angle between A–C and the tie A–B is $\arctan(18/14.13) = 51.9°$.

Tie A–B: Summing horizontal forces at node A gives $P_{u,A-B} = 105.2 + 26.8 = 132$ kips tension.

Node B

Figure 18-40b shows the forces acting at node B.

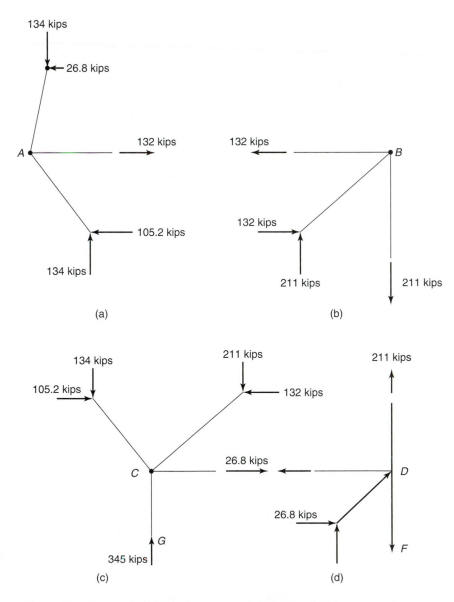

(a)

(b)

Fig. 18-40
Calculation of forces in struts
and ties—
Corbel—Example 18-4.

(c)

(d)

Strut B–C: Member B–C has a vertical projection of 18 in. and a horizontal projection of 15.5 in. $-$ $a/2$, where $a/2 = 5.68$ in., making the horizontal projection of B–C 11.27 in. Summing horizontal forces at joint B gives the horizontal component of the force in B–C as 132 kips. The vertical force component in B–C is

$$\frac{18}{11.27} \times 132 = 211 \text{ kips}$$

and the axial force is 249 kips.

The angle between B–C and tie B–D is $\arctan(11.27/18) = 32.1°$.

Tie B–D: Summing vertical forces at node B gives $P_{u,B-D} = 211$ kips tension.

Node C

Figure 18-40c shows the forces acting at node C. The sum of vertical forces is 134 kips down + 211 kips down + 345 kips up = 0. This should total zero—OK. Summing horizontal forces gives a tension of 26.8 kips in tie C–D. The angle between strut A–C and tie C–D is 51.9° and the angle between strut B–C and tie C–D is 57.9°.

Node D

Figure 18-40d shows the forces at node D. An inclined strut D–E is required in the column for equilibrium. A tension tie D–F is also required.

8.   **Compute reinforcement required for ties—Corbel**

Tie A–B:

$$A_s = \frac{132,000}{0.75 \times 60,000} = 2.93 \text{ in.}^2$$

Possible choices are (a) four No. 8 bars, $A_s = 3.16$ in.^{2}, which will fit into a width of 11.5 in. and have a basic hook-development length of 17 in., and (b) five No. 7 bars, $A_s = 3.00$ in.^{2}, which will fit into a width of 13 in. and have a basic hook-development length of 17 in. Use five No. 7 bars for tie A–B. Hook these with the vertical tails of the hooks in the plane of the right-hand layer of column steel located 1.5 + 0.5 in. from the right-hand side of the column.

Tie B–D:

$$A_s = \frac{211,000}{0.75 \times 60,000} = 4.69 \text{ in.}^2$$

The reinforcement for tie B–D will be selected by considering the statics and resistance of the column as a whole. The longitudinal column reinforcement will be sized to provide the required reinforcement for this tie. It may be necessary to enlarge the column further.

Tie C–D:

$$A_s = \frac{26,800}{0.75 \times 60,000} = 0.60 \text{ in.}^2$$

Use two No. 4 closed column ties, $A_s = 0.80$ in.^{2}.

9.   **Compute f_{cu} and the widths of the struts—Third model—Corbel**

These calculations are summarized in Table 18-8.

10.   **Draw the strut-and-tie model to scale—Third model—Corbel**

This is done to see whether the struts fit within the space available. Figure 18-38 shows that they do with very little extra space.

11.   **Provide reinforcement to confine struts—Corbel**

When f_{cu} was computed for the struts, they were all assumed to be bottle-shaped. Such struts must have transverse reinforcement satisfying ACI Section A.3.3 or A.3.3.1. These sections allow two methods of calculating the amount required. First, try the method from ACI Section A.3.3.

The splitting force at the two ends of the strut is $2(C/4)$, where the 2 comes from there being two ends and C is from Table 18-10: 170.4 kips.

TABLE 18-8 Widths of Struts and Ties—Corbel—Example 18-4

Member	Axial Force kips	f_{cu} for Strut psi	f_{cu} for Node psi	w_s in.
A–C at A	170.4	0.85 × 0.75 × 5000 = 3190	0.85 × 0.8 × 5000 = ~~3400~~	4.45
A–C at C	170.4	3190	= ~~3400~~	3.34
B–C at B	249	~~3190~~	0.85 × 0.60 × 5000 = 2550	6.10
B–C at C	249	3190	~~3400~~	4.88
C–G at C	345	~~4250~~	3190	6.76

Note: Two rows are provided for each strut, to allow different values of f_{cu} at the two ends of a strut. Also, use an effective corbel width of 12 in. under the bearing plate at Node A. At all other nodes use a width of 16 in.

Strut A–B:

$$\phi A_s f_y = 2(C/4) = 2 \times 170{,}400 \text{ lb}/4 = 85{,}200 \text{ lb.}$$

$$A_s \text{ required} = \frac{85{,}200}{0.75 \times 60{,}000} = 1.89 \text{ in.}^2$$

Alternatively, ACI Section A.3.3.1 requires that the strut be crossed by steel satisfying

$$\Sigma \frac{A_{si}}{bs_i} \sin \gamma_i \geq 0.003$$

where A_{si} is the area of transverse steel at an angle γ_i to the axis of the strut, which cannot be taken as less than 40°. The computed angle is arctan (18 in. / 14.13 in.) = 51.9° where the horizontal projection of strut A–C is 14.13 in. and the vertical projection is 18 in. The length of the strut is 22.9 in. For four No. 4 closed stirrups at s_i = 4 in., spread over the length and width of strut A–C,

$$\Sigma \frac{A_{si}}{bs_i} \sin \gamma_i = 4 \times \frac{2 \times 0.2}{16 \times 22.9} \sin 51.9° = 0.00344$$

This slightly exceeds the required 0.003. The total area of transverse reinforcement is 1.6 in.^{2}, compared with the 1.89 in.^{2} required in this tie for the forces arising out of the statics of the strut-and-tie model. This will be considered in step 12.

12. Satisfy the detailing requirements—Third model—Corbel

ACI Section 11.9 presents a number of detailing requirements for corbels.

11.9.2—The depth under the outer edge of the bearing plate shall not be less than half the height at the face of the column—OK. This requirement was introduced to prevent failures like the one shown in Fig. 18-39b.

11.9.3.4 and 11.9.3.5—Satisfied in step 3 of the corbel design—OK.

11.9.4—Requires closed stirrups or ties parallel to tie A–B with a total area, A_h, not less than $0.5(A_s - A_n)$, where A_s is the portion of the area of the main tie A–B provided to resist the tie forces resulting from the vertical load on the corbel and A_n is the portion of the area of reinforcement provided to resist the horizontal force, N_{uc}, that acts, or is assumed to act, on the corbel, in accordance with ACI Section 11.9.3.4. Thus, this section requires stirrups parallel to tie A–B with a tensile capacity of 132 kips minus N_{uc} = 26.8 kips. The stirrups should be distributed within $\frac{2}{3}d$ of tie A–B = 12 in. and have an area of

$$0.5 \times \left(\frac{132 - 26.8}{132} \times 3.00 \right) = 1.20 \text{ in.}^2$$

In step 11, the confining steel was computed as A_s = 1.60 in.^{2} Either way, the steel would satisfy 11.9.4 and A.3.3.

Use four No. 4 two-legged closed stirrups at 4 in. o.c. with $A_s = 1.60$ in.^{2}. Place the first one at 4 in below the centroid of tie A–B.

11.9.5—The ratio $A_s/bd = 3.00/(16 \times 18) = 0.0104$, calculated at the face of the support, shall not be less than $0.04(f'_c/f_y) = 0.04 \times (5/60) = 0.0033$—OK.

11.9.6—Probably the most important detailing requirement is that the tie A–B be anchored for the tension tie force at the front face of the corbel (at A). Because the tie in the truss in Fig. 18-38 is assumed to be stressed to f_y in tension over the whole distance from the loading plate to the column, it must be anchored outside the loading plate for that tension. This is done in one of several ways. It can be anchored

    (a) by welding the bars making up the tie to a transverse angle or bar, which may also serve as a bearing plate,

    (b) by welding the bars making up the tie to a transverse bar of the same diameter as the bars making up the tie, or

    (c) by bending the bar in a horizontal loop.

Although the tension tie could also be anchored by bending the bars in a vertical bend, as illustrated in Fig. 18-39a, this is discouraged, since failures have occurred when this detailing was used, as shown in that figure. We shall assume that the $(4 \times 12 \times 1.5)$-in. bearing plate welded across the ends of the five No. 7 bars, as chosen in step 8, will anchor the tie. Because welding is required, the five No. 7 bars must be specified as Grade-60W steel. Alternatively, a steel angle could be used to anchor tie A–B at A, as shown in Fig. 18-38. The plate is easier to weld, and easier to concrete under, but the angle provides resistance to damage to the outer edge of the bearing area.

11.9.7—requires that the bearing area of the load either

    (a) not project beyond the start of the bend of the top tie bar, if it is anchored at node A by a hook, or

    (b) not project past the interior face of the transverse anchor bar, if that detail is used.

Welding the $(4 \times 12 \times 1.5)$-in. bearing plate across the five No. 7 bars with the center of the plate under the center of the reaction bearing plate would anchor the tie. Alternatively, an angle welded to the ends of the bars would also anchor the tie.

13.   Check the moments in the column—Third model—Corbel

The loads on the corbel cause a moment of

$$134 \times (9.9 + 18/2) = 2530 \text{ in.-kips}$$

about the center of the column on a line joining nodes A and B. This moment should be divided between the columns above and below the joint in proportion to the stiffnesses of each of these columns. The strut-and-tie model in Fig. 18-41 gives a more complete idea of the corbel action.   ■

Design of Corbels by ACI Code Method

ACI Section 11.9 presents a design procedure for brackets and corbels. It is based in part on the strut-and-tie truss model and in part on shear friction. The design procedure is limited to a/d ratios of 1.0 or less. At the time it was included in the code there was little test data for longer brackets.

In the ACI design method, the section at the face of the support is designed to resist the shear V_u, the horizontal tensile force N_{uc}, and a moment of $[V_u a + N_{uc}(h - d)]$, where the moment has been calculated relative to the tension steel at point A in Fig. 18-38. The maximum shear strength, V_n, shall not be taken greater than

$0.2f'_c b_w d$, but not more than $800 b_w d$ lb, for normal-weight concrete.

$(0.2 - 0.07a/d)f'_c b_w d$, but not more than $(800 - 280a/d)b_w d$ lb, for all-lightweight or sand-lightweight concrete.

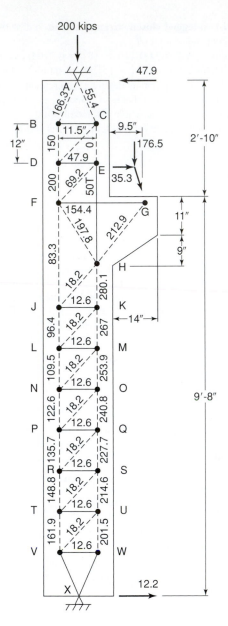

Fig. 18-41
Global strut-and-tie model of a corbel and the supporting column.

In design, the size of the corbel is selected so that $V_u \leq \phi V_n$, based on the maximum shear strength. If a high value of V_n is used, cracking at service loads may lead to serviceability problems. The designer then calculates the following:

1. the area, A_{vf}, of shear-friction steel required by (18-22):

$$V_n = A_{vf} f_y \mu \qquad (18\text{-}19)$$
$$(\text{ACI Eq. 11-25})$$

2. the area, A_f, of flexural reinforcement required to support a moment of $[V_u a + N_{uc}(h - d)]$, based on ACI Chapter 10 (Chapter 4 of this book); and

3. the area, A_n, of direct-tension reinforcement required to resist the tension force N_{uc}, where

$$\phi A_n f_y \geq N_{uc} \qquad (18\text{-}20)$$

In all these calculations, ϕ is taken equal to the value for shear, 0.75, which is also the value for strut-and-tie models.

The resulting area of tensile steel, A_s, and the placement of the reinforcement within the corbel are specified in ACI Sections 11.9.3.5 and 11.9.4. In the corbel tests reported in [18-20], [18-21], and [18-19], the best behavior was obtained in corbels that had some horizontal stirrups in addition to the tension tie shown in Fig. 18-38. Accordingly, ACI Section 11.9.3.5 requires that two reinforcement patterns be considered and the one giving the greater area, A_s, be used:

1. a tension tie having area $A_s = A_f + A_n$, plus horizontal stirrups having area $A_f/2$;

2. a tension tie having area $A_s = (2A_{vf}/3) + A_n$, plus horizontal stirrups having area $A_{vf}/3$.

The horizontal stirrups are to be placed within $\frac{2}{3}d$ below the tension tie.

Because the tension tie in the truss in Fig. 18-38 is assumed to be stressed to f_y in tension between the loading plate and the column, it must be anchored outside the loading plate for that tension. This is done in one of several ways. It can be anchored by welding to an angle or bar at right angles to the tie (Fig. 18-38), or by welding to a transverse reinforcing bar of the same diameter as the tie, or by bending the bar in a horizontal loop. Although the tension tie could also be anchored by bending the bar in a vertical hook, as shown in Fig. 18-39a, this is discouraged, since failures have occurred, as shown in that figure. ACI Section 11.9.7 requires that the outer edge of the bearing plate be inside the cross-bar or, in the case of the detail shown in Fig. 18-39a, inside the start of the bend (inside point A). The use of a welded transverse reinforcing bar requires special welding techniques (see ACI Section 3.5.2.) and weldable grade reinforcement.

EXAMPLE 18-5 Design of a Corbel—Traditional ACI Code Method

Design a corbel to transfer a precast-beam reaction to a supporting column. The factored shear to be transferred is 150 kips. The column is 16 in. square. The beam being supported is restrained against longitudinal shrinkage. Use $f'_c = 5000$ psi and $f_y = 60,000$ psi. Use ACI 318-02 Sections 9.2 and 9.3.

1. **Compute the distance, a, from the column to V_u.** Assume a 12-in.-wide bearing plate. From ACI Section 10.17.1, the allowable bearing stress is

$$\phi 0.85 f'_c = 0.65 \times 0.85 \times 5 \text{ ksi} = 2.76 \text{ ksi}$$

The required width of the bearing plate is

$$\frac{150 \text{ kips}}{2.76 \times 12 \text{ in.}} = 4.53 \text{ in.}$$

Use a 12-in. × 5-in. bearing plate. Assume that the beam overhangs the bearing plate by 6 in., that a 1-in. gap is left between the end of the beam and the face of the column, and that the distance a is 9.5 in.

2. **Compute the minimum depth, d.** Base this calculation on ACI Section 11.9.3.2.1:

$$\phi V_n \geq V_u$$
$$\text{max. } V_n = 0.2 f'_c b_w d \text{ but not more than } V_n = 800 b_w d \text{ lb}$$

For 5000-psi concrete, $0.2f'_c$ exceeds 800 psi; therefore, the second equation governs. Thus,

$$\text{minimum } d = \frac{V_u}{\phi \times 800 b_w}$$
$$= \frac{150,000 \text{ lb}}{0.75 \times 800 \times 16} = 15.6 \text{ in.}$$

Hence, the smallest corbel we could use is a corbel with $b = 16$ in., $h = 16$ in., and $d = 14$ in. For conservatism, we shall use $h = 20$ in. and $d = 20$ in. $- \left(1\frac{1}{2}\text{ in. cover} + \frac{1}{2}\text{ bar diameter}\right)$, which equals 18 in. The corbel will be the same width as the column (16 in. wide), to simplify forming.

3. **Compute the forces on the corbel.** The factored shear is 150 -kips. Since the beam is restrained against shrinkage, we shall assume the normal force to be (ACI Section 11.9.3.4)

$$N_{uc} = 0.2V_u = 30 \text{ kips}$$

The factored moment is

$$M_u = V_u a + N_{uc}(h - d)$$
$$= 150 \text{ kips} \times 9.5 \text{ in.} + 30 \text{ kips}(20 \text{ in.} - 18 \text{ in.})$$
$$= 1485 \text{ in.-kips}$$

4. **Compute the shear friction steel, A_{vf}.** From (6-14) and (18-19),

$$\phi V_n \geq V_u$$
$$A_{vf} = \frac{V_n}{\mu f_y} = \frac{V_u}{\phi \mu f_y}$$

where $\mu = 1.4\lambda$ for a shear plane through monolithic concrete and $\lambda = 1.0$ for normal-weight concrete. Therefore,

$$A_{vf} = \frac{150,000 \text{ lb}}{0.75(1.4 \times 1.0)60,000 \text{ psi}} = 2.38 \text{ in.}^2$$

5. **Compute the flexural reinforcement, A_f.** A_f is computed from (4–11b) (with A_s replaced by A_f):

$$M_u = \phi A_f f_y \left(d - \frac{a}{2}\right)$$

Here, $\phi = 0.75$ (ACI Section 11.9.3.1) and

$$a = \frac{A_f f_y}{0.85 f_c' b}$$

As a first trial, we shall assume that $(d - a/2) = 0.9d$. Thus,

$$A_f = \frac{M_u}{\phi f_y (0.9d)}$$
$$= \frac{1485 \text{ in.-kips}}{0.75 \times 60 \times 0.9 \times 18} = 2.04 \text{ in.}^2$$

Since this is based on a guess for $(d - a/2)$, we shall compute a and recompute A_f:

$$a = \frac{2.04 \times 60}{0.85 \times 5 \times 16} = 1.80 \text{ in.}$$
$$A_f = \frac{1485 \text{ in.-kips}}{0.75 \times 60 \text{ ksi } (18 - 1.80/2) \text{ in.}}$$
$$= 1.93 \text{ in.}^2$$

Therefore, use $A_f = 1.93 \text{ in.}^2$.

6. **Compute the reinforcement, A_n, for direct tension.** From ACI Section 11.9.3.4,

$$A_n = \frac{N_{uc}}{\phi f_y} = \frac{30 \text{ kips}}{0.75 \times 60 \text{ ksi}}$$

$$= 0.67 \text{ in.}^2$$

7. **Compute the area of the tension-tie reinforcement, A_s.** From ACI Section 11.9.3.5, A_s shall be the larger of

$$(A_f + A_n) = 1.93 + 0.67 = 2.60 \text{ in.}^2, or$$

$$\left(\frac{2A_{vf}}{3} + A_n\right) = 1.59 + 0.67 = 2.26 \text{ in.}^2$$

Minimum A_s (ACI Section 11.9.5):

$$A_{s(\text{min})} = \frac{0.04 f'_c}{f_y} b_w d = 0.96 \text{ in.}^2$$

Therefore, $A_s = 2.60 \text{ in.}^2$. Try three No. 9 bars, giving $A_s = 3.00 \text{ in.}^2$.

8. **Compute the area of horizontal stirrups.**

$$0.5(A_s - A_n) = 2.60 - 0.67 = 0.97 \text{ in.}^2$$

Select three No. 4 double-leg stirrups, area = 1.20 in.^{2}; ACI Section 11.9.4 requires that these be placed within $(2/3)d$, measured from the tension tie.

9. **Establish the anchorage of the tension tie into the column.** The column is 16 in. square. Try a 90° standard hook. From ACI Section 12.5.1,

$$\ell_{dh} = \frac{1200 d_b}{\sqrt{f'_c}} \times 0.7 \times 0.8 = 10.7 \text{ in.}$$

measured from the face of the column. Therefore, **use three No. 9 bars hooked into the column.** The hooks are inside the column cage.

10. **Establish the anchorage of the outer end of the bars.** The outer end of the bars must be anchored to develop $A_s f_y$. This can be done by welding the bars to a transverse plate, angle, or bar. If the horizontal force, N_{uc}, is required for equilibrium of the structure, some direct connection would be required from the beam base plate to the tension tie. This is not the case here, and a welded cross-angle will be provided, as shown in Fig. 18-38.

11. **Consider all other details.** To prevent cracks similar to those shown in Fig. 18-39b, ACI Section 11.9.2 requires that the depth at the outside edge of the bearing area be at least $0.5d$. ACI Section 11.9.4 requires that the stirrups be placed within $\frac{2}{3}d$ of A_s. ACI Section 11.9.7 requires that the anchorage of the tension tie be outside the bearing area. Finally, two No. 4 bars are provided to anchor the front ends of the stirrups. All of these aspects are satisfied in the final corbel layout which is similar to Fig. 18-38.   ■

Comparison of the Strut-and-Tie Method and the ACI Method

The strut-and-tie method required more steel in the tension tie and less confining reinforcement than the ACI method. The strut-and-tie method explicitly considered the effect of the corbel on the forces in the column. The strut-and-tie method could also be used for corbels that have a/d greater than the limit of 1.0 given in ACI Section 11.9.1. For

$a/d > 1$, the confining stirrups would be more efficient in restraining the splitting of the strut if they were vertical.

18-11 DAPPED ENDS

The ends of precast beams are sometimes supported on an end projection that is reduced in height, as shown in Fig. 18-42. Such a detail is referred to as a *dapped end*. Although several design procedures exist, the best method of design is by means of strut-and-tie models. Tests of such regions are reported in [18-23] and [18-24].

Four common strut-and-tie models for dapped-end regions are compared to the crack pattern observed in tests in Fig. 18-42. Cracking originates at the reentrant corner of the notch, point *A* in Fig. 18-42a. The strut-and-tie models in Fig. 18-42b to d all involve a vertical tie *B–C* at the end of the deeper portion of the beam and an inclined strut *A–B* over the reaction. In tests, specimens with tie *B–C* composed of closed vertical stirrups with 135° bends around longitudinal bars in the top of the beam performed better than specimens with open-topped stirrups [18-34]. The horizontal component of the compressive force in *A–B* is equilibrated by the tension tie *A–D*. The three strut-and-tie models differ in the manner in which the horizontal tie is anchored at *D*. The model in Fig. 18-42c has the advantage that the force in tie *C–E* is lower, and hence easier to anchor, than the corresponding force in tie *C–F* in Fig. 18-42b. In Fig. 18-42d, tie *A–D* is anchored by strut *B–D*, which will be crossed by cracks, as shown in Fig. 18-42a. This suggests that Fig. 18-42d is not a feasible model.

The strut-and-tie model in Fig. 18-42e has an inclined hanger tie *B–C* and a vertical strut over the reaction. Care must be taken to anchor the tie *B–C* at its upper end. It is customary to provide a horizontal tie at *A* to resist any tensile forces due to restrained shrinkage of the precast beam. In tests [18-24], dapped ends designed by using the models in Fig. 18-42b or c performed just as well as ends designed via the model in Fig. 18-42e. A compound model, designed by assuming that half the reaction was resisted by each of these two types of strut-and-tie models, also performed well in tests.

In laying out a dapped-end support, it is good practice to have the depth of the extended part of the beam be at least half of the overall height of the beam. The extended part of the beam should be deep enough that the inclined compression strut *A–B* at the support

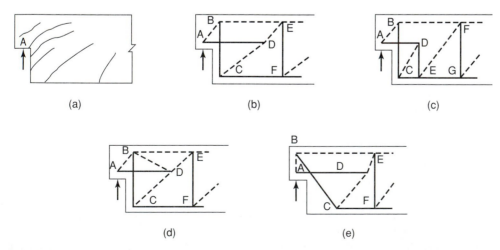

Fig. 18-42
Comparison of strut-and-tie models for dapped ends.

is no flatter than 45°. Otherwise, the forces in this strut and in the tie that meets it at the support become too large to deal with in a simple manner. Great care must be taken to anchor the bars in the vicinity of the dap.

EXAMPLE 18-6   Design of a Dapped-End Support

A precast beam supports an unfactored dead load of 2 kips/ft and an unfactored live load of 2.5 kips/ft on a 20-ft span. The beam is 30 in. deep by 15 in. wide and is made from 3000-psi concrete and Grade-60 reinforcement. The longitudinal steel is four No. 8 bars. Design the reinforcement in the support region.

1.   Isolate the D-region; compute the reactions and the forces on the boundaries of the D-region—Dapped end

1(a)   Design equation. Design will be based on

$$\phi F_n \geq F_u \qquad (18\text{-}1a)$$
$$(\text{ACI Eq. A-1})$$

where, from ACI Section 9.3.2.6, $\phi = 0.75$, F_n is the nominal resistance, and F_u is the factored-load effect.

1(b)   Compute factored loads. $U = 1.2 \times 2 + 1.6 \times 2.5 = 6.4$ kips per ft. This gives a vertical reaction of 64.0 kips and a horizontal reaction of $0.2 \times 64.0 = 12.8$ kips.

1(c)   Isolate the D-region. The D-region will be assumed to extend 30 in. from the lower corner of the full depth portion of the beam. Assuming the center of the support is 2 in. from the end of the beam (Fig. 18-43), the dead and live loads in the D-region are replaced by a single force equal to the sum of the factored dead and live loads acting on the D-region equal to $(34/12)$ ft $\times$ 6.4 kips/ft $= 18.1$ kips is applied to the beam at convenient place in the D-region. We shall apply it to the beam at node B.

Moment equilibrium at the right end of the D-region gives a bending moment of 155.5 ft-kips $= 1866$ in-kips acting on the vertical section to the right of node F. This can be subdivided into a flexural compression and a flexural tension of $1866/(0.9 \times 27) = 76.8$ kips at Nodes F and E, respectively. The horizontal reaction force of 12.8 kips causes tensions of 6.4 kips at Nodes E and F giving forces $76.8 - 6.4 = 70.4$ kips at F and $76.8 + 6.4$ kips $= 83.2$ kips at E.

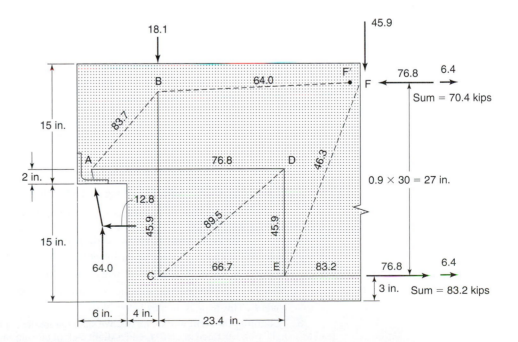

Fig. 18-43
Strut-and-tie model for a dapped end—Example 18-5.

2. **Select a strut-and-tie model—Dapped end.** A strut-and-tie model similar to that in Fig. 18-42c will be used.

3. **Compute the effective compressive strength of the nodal zones and struts—Dapped end.** The effective compressive strength of the nodal zones, f_{cu}, is given in ACI Section A.5.2 as

$$f_{cu} = 0.85\beta_n f'_c \qquad (18\text{-}7b)$$
$$(\text{ACI Eq. A-3})$$

Nodal zone A: Nodal zone A anchors compressive forces from the reaction and from strut A–B and tension from the tie A–D. From A.5.2.2, $\beta_n = 0.80$, and

$$f_{cu} = 0.85 \times 0.80 \times 3000 = 2040 \text{ psi}$$

Nodal zone B: Nodal zone B is a C–C–T node. ACI Section A.5.2 gives β_n as 0.80, and $f_{cu} = 2040$ psi.

Nodal zone F: Nodal zone F is a C–C–C node. ACI Section A.5.2 gives $\beta_n = 1.0$, and thus $f_{cu} = 2550$ psi.

Nodal zones C, D, and E: These nodes all anchor more than one tie; thus, $\beta_n = 0.60$, and $f_{cu} = 1530$ psi at all of them.

Struts A–B, C–D, and E–F: Since there is room beside the struts A–B, C–D, and E–F for the struts to expand sideways into the unstressed concrete, we shall assume that all three are bottle-shaped struts with $\beta_s = 0.75$. Thus,

$$f_{cu} = 0.85 \times 0.75 \times 3000 = 1912 \text{ psi}$$

Strut B–F: This strut is the compression zone of the beam. ACI Section A.3.2.1 gives $\beta_s = 1.0$. Thus, $f_{cu} = 0.85 \times 1.0 \times 3000 = 2550$ psi.

4. **Estimate the locations of the nodes—Dapped end**

Node A: Node A is located above the reaction. The size of the bearing plate at the support will be based on the bearing strength from 10.17.1, but must not be less than the nodal zone strength, F_{nn}, from ACI Section A.5.2. From ACI Section 10.17.1, the bearing strength is $0.85f'_c$. The required bearing area is

$$\frac{64,000 \text{ lb}}{0.65 \times 0.85 \times 3000 \text{ psi}} = 38.6 \text{ in.}^2$$

where $\phi = 0.65$ for bearing, but not less than the area based on the strength of the nodal zone given by:

$$f_{cu} = 0.85 \times 0.80 \times 3000 = 2040 \text{ psi}$$

and

$$\frac{64,000}{0.75 \times 2040} = 41.8 \text{ in.}^2$$

in which $\phi = 0.75$ for strut-and-tie models.

Use a 4 by 4 by 1/2 in. angle, 15 in. long at the bearings.

We shall assume that the reaction acts 2 in. from the end of the member. Node A is located at the intersection of the inclined reaction and the bar A–D, which we shall assume is located 2 in. above the bottom of the extended part of the beam.

Node B: This node is located at the upper ends of strut A–B and tie B–C.

Node C: This node is at the lower right corner of the 30 in. deep portion of the beam.

Node D: This node is to the right horizontally from Node A.

Node E: This node is directly below Node D.

5. **Compute the strut and tie forces.** We shall ignore the slight slope in strut B–F. This will lead to a slightly higher force in tie B–C and a slight lack of closure in the force diagram. Use can be

made of a scale drawing of the strut-and-tie model in Fig. 18-43 in computing the internal forces. The drawing should be large enough to scale lengths from.

Node B: This node is located at the upper end of strut A–B, which is assumed to be acting at a 45° angle. The vertical reaction from Node A is 64 kips, resulting in a the total force in strut A–B of 90.5 kips. Thus, strut A–B produces a vertical force of 64 kips and a horizontal force of 64 kips at Node B. The factored dead and live load force acting on Node B, from step 1(c), is a vertical downward load of 18.1 kips. For equilibrium at the joint, the force in the strut B–F must be equal to 64 kips and the force in tie B–C must be equal to 64 − 18.1 = 45.9 kips.

Node A: Strut A–B exerts a horizontal force component of 64 kips, acting to the left at Node A. A horizontal reaction of 12.8 kips also acts to the left at Node A. Summing horizontal forces at Node A results in a force in tie A–D equal to 64 + 12.8 = 76.8 kips.

Node C and D: The tie B–C applies a vertical upward force of 45.9 kips at Node C. Thus, the vertical force component of the inclined strut C–D must be 45.9 kips. Similarly, at Node D tie A–D applies a horizontal force of 76.8 kips, acting to the left. Thus, the horizontal force component in strut C–D must be 76.8 kips. From these values, the total force in strut C–D is:

$$\sqrt{(45.9)^2 + (76.8)^2} = 89.5 \text{ kips}$$

From Fig. 18-43, the vertical distance from Node C to Node D is 14.0 in. Thus, the horizontal distance between these nodes must be 14.0 × 76.8/45.9 = 23.4 in. To complete the equilibrium at Node C, the force in tie C–E must be 76.8 kips. To complete the equilibrium at Node D, the force in tie D–E must be 45.9 kips.

Node E: To satisfy horizontal equilibrium at this node, strut E–F must have a horizontal force component of 83.2 − 76.8 = 6.40 kips. For vertical equilibrium, strut E–F must have a vertical force component of 45.9 kips. Thus, the total force in strut E–F is:

$$\sqrt{(6.40)^2 + (45.9)^2} = 46.3 \text{ kips}$$

Node F: Summing horizontal and vertical forces shows that Node F is in equilibrium.

6. **Compute the strut widths, and check whether they will fit—Dap.** The first estimate of strut and tie forces is shown in Fig. 18-43. It is now necessary to compute the widths of the struts to see whether they will fit into the available space without overlapping. From step 3, the effective concrete strength for struts A–B, C–D, and E–F will be taken as f_{cu} = 1912 psi. The struts will be assumed to have a thickness equal to the thickness of the beam, 15 in. (except for A–B and B–F). In tests, the cover over the sides of the stirrups spalled off at node B. As a result, we shall assume that struts A–B and B–F are 12 in. thick.

Strut A–B   From step 3 for nodes A and B, f_{cu} = 2040 psi. The f_{cu} for strut A–B governs, and we have

$$\text{width: } \frac{90,500 \text{ lb}}{0.75 \times 1912 \text{ psi} \times 12 \text{ in.}} = 5.26 \text{ in.}$$

Strut B–F   From step 3 for nodal zone F, f_{cu} = 2550 psi and for strut B–F, f_{cu} = 2550 psi. The f_{cu} for strut B–F will be used, and we have

$$\text{width: } \frac{64,000}{0.75 \times 2550 \times 12} = 2.79 \text{ in.}$$

Strut C–D   From step 3 for nodes C and D, f_{cu} = 1530 psi and for strut C–D, f_{cu} = 1912 psi. The f_{cu} = 1530 psi governs, and

$$\text{width: } \frac{89,500}{0.75 \times 1530 \times 15} = 5.20 \text{ in.}$$

7. **Provide reinforcement to control of cracking—Dapped end.** When f_{cu} was selected for the struts, they were assumed to be bottle-shaped struts. Such struts must have transverse reinforcement satisfying ACI Section A.3.3 or Section A.3.3.1. These sections allow two methods of calculating the amount of steel that is required. We shall use the method from A.3.3.1. The amount of this steel is given by the following equation:

$$\sum \frac{A_{si}}{bs_i} \sin \gamma_i \geq 0.003 \tag{18-10}$$

Try two No. 5 U-shaped bars enclosing the strut A–B. In (18-10), we will use the total area provided by both bars in the numerator and in the denominator we will use the projected vertical length of the strut, 15 in., in place of s_i. The angle between the axis of the strut and a horizontal bar is 45°, and sin 45° is 0.707. Substituting into (18-10) gives:

$$\frac{4 \times 0.31}{15 \times 15} \times 0.707 = 0.00390$$

This is more than the required value of 0.003. **Use two No. 5 U-shaped horizontal bars in the upper portion of the dap extension.**

For strut C–D, try three No. 4 vertical U-shaped stirrups and two No. 4 horizontal stirrups. As above, use the total vertical and horizontal steel areas in the numerator of (18-10) and replace s_i in the denominator with the projected horizontal and vertical lengths of strut C–D. The angle between the axis of the strut and a horizontal bar is:

$$\text{Arctan}\left(\frac{14.0}{23.4}\right) = 30.9°$$

and sin 30.9° = 0.513. The angle between the axis of the strut and the vertical bars is 59.1° and sin 59.1° = 0.858. So,

$$\frac{4 \times 0.20}{15 \times 14} \times 0.513 + \frac{6 \times 0.20}{15 \times 23.4} \times 0.858 = 0.00488 \text{ (o.k.)}$$

Strut E–F is crossed by both the U-shaped No. 5 bars in the upper half of the beam and the U-shaped No. 4 bars in the lower half of the beam. It is clear that this is sufficient without doing additional calculations.

Use two No. 4 U-shaped horizontal bars in the lower portion of the beam near the dap and U-shaped vertical stirrups at a spacing of 8 in. within the D-region.

8. **Compute the steel required in the ties—Dap**

Tie A–D:

$$A_s = \frac{\text{tie force}}{\phi f_y} = \frac{76.8 \text{ kips}}{0.75 \times 60 \text{ ksi}}$$

$$= 1.71 \text{ in.}^2$$

Use four No. 6 bars welded to the angle. Development length $\ell_d = 32.9$ in. Theoretically, the bars should be anchored toward midspan from node D. Extend the bars 33 in. past D.

Tie B–C:

$$A_s \geq \frac{45.9 \text{ kips}}{0.75 \times 60 \text{ ksi}} = 1.02 \text{ in.}^2$$

Use four No. 4 closed stirrups. Use longitudinal No.4 bars inside the top corners of the stirrups.

Tie D–E: The force requirement is the same as for tie B–C, so four No. 4 double legged stirrups can be used. We shall arbitrarily spread them out over a longer length of the beam than was done for tie B–C. Also, their upper anchorage is less critical, so normal U-shaped stirrups with 135° too small too high hooks can be used. **Use No. 4 U-shaped stirrups at a spacing of 6 in. on centers throughout the D-region.** Note, this essentially provides more vertical steel across strut C–D.

Tie C–E: Assume that at midspan, the bottom steel is four No. 8 bars for flexure. This is enough for the 76.8 kip force in C–E. However, it is necessary to develop this force within the node at C. The length available for development of the bars within the node is 9.7 in., as shown in Fig. 18-44. The development length of a No. 8 bar is 54.8 in. The force that can be developed in the No. 8 bars is

$$\frac{9.7 \text{ in.}}{54.8 \text{ in.}} \times 4 \times 0.79 \text{ in.}^2 \times 60 \text{ ksi} = 30.6 \text{ kips}$$

This is not enough. Either provide some sort of mechanical anchor for these bars, or provide horizontal U bars to anchor the force. We shall provide horizontal U bars. The area required is

$$A_s = \frac{76.8}{0.75 \times 60} = 1.71 \text{ in.}^2$$

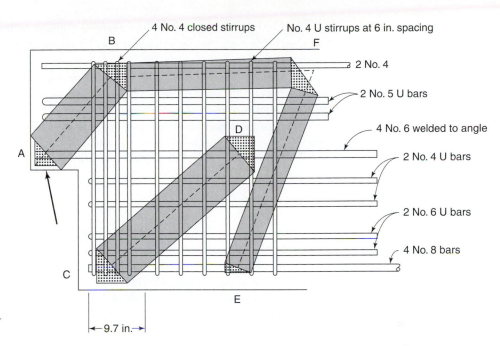

Fig. 18-44
Final strut-and-tie model for a dapped end showing the congestion due to the widths of the struts and reinforcement—Example 18-6.

Use **two No. 6 U bars** with bends adjacent to the end of the beam; total area is 1.76 in.² Place these above the No. 8 bars with clear spaces of 1 in. between the bars. Lap splice these $1.3\ell_d = 42.8$ in.— say, 3 ft 8 in.—into the beam.

The final reinforcement is shown in Fig. 18-44.

9.   Check the stresses on the sides of the nodes—Dap. It is generally not necessary to check bearing stresses on nodes, but we shall check the height of Node *C*.

Height of node *C*: The steel should be at a height approximately equal to the height of the tension tie, to anchor the tie force based on concrete stressed at $f_{cu} = (0.85 \times 0.60 f'_c) = 1530$ psi.

$$\text{Nodal area required} = \frac{76{,}800 \text{ psi}}{0.75 \times 1530} = 66.9 \text{ in.}^2$$

The width is 15 in., so a height of 4.46 in. is required. The No. 8 bars and No. 6 U bars take more than this—therefore, OK. ∎

18-12   BEAM–COLUMN JOINTS

In Chapters 4, 5, 10, and 11, beams and columns were discussed as isolated members on the assumption that they can somehow be joined together to develop continuity. The design of the joints requires a knowledge of the forces to be transferred through the joint and the likely ways in which this transfer can occur. The ACI Code touches on joint design in several places:

1.   ACI Section 7.9 requires enclosure of splices of continuing bars and of the end anchorages of bars terminating in connections of primary framing members, such as beams and columns.

2.   ACI Section 11.11.2 requires a minimum amount of lateral reinforcement (ties or stirrups) in beam–column joints if the joints are not restrained on all four sides by beams or slabs of approximately equal depth. The amount required is the same as the minimum stirrup requirement for beams (ACI Eq. (11-13)).

3. ACI Section 12.12.1 requires negative-moment reinforcement in beams to be anchored in, or through, the supporting member by embedment length, hooks, or mechanical anchorage.

4. ACI Section 12.11.2 requires that, in frames forming the primary lateral load-resisting system, a portion of the positive-moment steel should be anchored in the joint to develop the yield strength, f_y, in tension at the face of the support.

None of these sections gives specific guidelines for design. Design guidance can be obtained from [18-1], [18-23], [18-24], and [18-25]. Extensive tests of beam–column joints have been reported in [18-26] and [18-27].

Corner Joints: Opening

In considering joints at the intersection of a beam and column at a corner of a rigid frame, it is necessary to distinguish between joints that tend to be *opened* by the applied moments (Fig. 18-45) and those that tend to be *closed* by the applied moments (Fig. 18-47). Opening joints occur at the corners of frames and in L-shaped retaining walls. In bridge abutments, the joint between the wing-walls and the abutment is normally an opening joint.

The elastic distribution of stresses before cracking is illustrated in Fig. 18-45b. Large tensile stresses occur at the reentrant corner and in the middle of the joint. As a result,

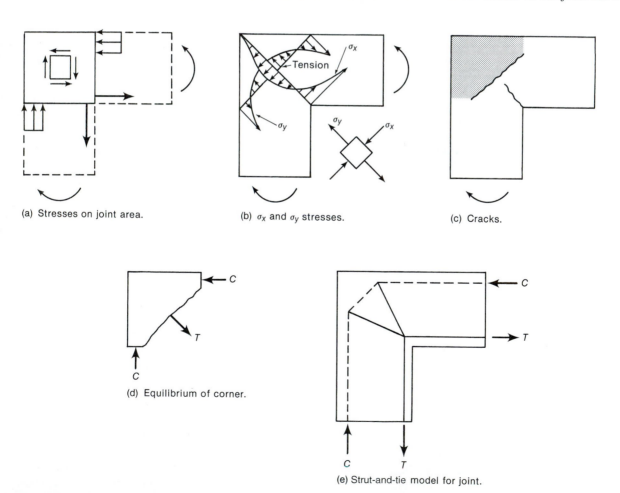

(a) Stresses on joint area. (b) σ_x and σ_y stresses. (c) Cracks.

(d) Equilibrium of corner.

(e) Strut-and-tie model for joint.

Fig. 18-45
Stresses in an opening joint.

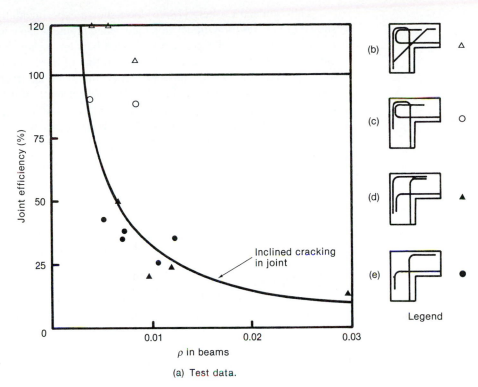

Fig. 18-46
Measured efficiency of opening joints.

(a) Test data.

cracking develops as shown in Fig. 18-45c. A free-body diagram of the portion outside the diagonal crack is shown in Fig. 18-45d. The force T is necessary for equilibrium. If reinforcement is not provided to develop this force, the joint will fail almost immediately after the development of the diagonal crack. A truss model of the joint is shown in Fig. 18-45e.

Figure 18-46a compares the measured efficiency of a series of corner joints reported in [18-26] and [18-27]. The *efficiency* is defined as the ratio of the failure moment of the joint to the moment capacity of the members entering the joint. The reinforcement was detailed as shown in Fig. 18-46, b to e. The solid curved line corresponds to the computed moment at which diagonal cracking is expected to occur in such a joint. Typical beams have reinforcement ratios of about 1 percent. At this reinforcement ratio, the common joint details shown in Fig. 18-46d and e can transmit at most 25 to 35 percent of the moment capacity of the beams.

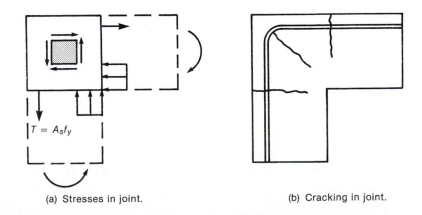

Fig. 18-47
Closing joints.

(a) Stresses in joint.   (b) Cracking in joint.

Nilsson and Losberg [18-26] have shown experimentally that a joint reinforced as shown in Fig. 18-46b will develop the needed moment capacity without excessive deformations. The joint consists of two hooked bars enclosing the corner and diagonal bars and having a total cross-sectional area half that of the beam reinforcement. The tension in the hooked bars has a component across the diagonal crack, helping to provide the T force in Fig. 18-45d. The inclined bar limits the growth of the crack at the reentrant corner, slowing the propagation of cracking in the joint. The open symbols in Fig. 18-46 show the efficiency of the joints shown in Fig. 18-46a and b, with and without the diagonal corner bar. It can be seen that the corner bar is needed to develop the full efficiency in the joint.

Corner Joints: Closing

The elastic stresses in a closing corner joint are exactly opposite to those in an opening corner joint. The forces at the ends of the beams load the joint in shear as shown in Fig. 18-47a. As a result, cracking of such a joint occurs as shown in Fig. 18-47b, with a major crack on the diagonal. Such joints generally have efficiencies between 80 and 100 percent. Problems arise from the bearing inside the bent bars in the corner, since these bars must transmit a force of $\sqrt{2}A_s f_y$ to the concrete on the diagonal of the joint. For this reason, it may be desirable to increase the radius of this bend above minimum values given in ACI Section 7.2.

Frequently, the depth of the beam will be greater than that of the column, as shown in Fig. 18-48a. In such a case, the internal lever arm in the beam is larger than that in the column, and as a result, the tension force in the column steel will be larger than that in the beam. In the case shown, T_2 is three times T_1. For simplicity, the effects of the shears in the beam and column have been omitted in drawing Fig. 18-48. Although the strut-and-tie model in Fig. 18-48a is in overall equilibrium, the bar force jumps suddenly by a factor of 3 at A. A strut-and-tie model that accounts for the change in bar force in such a region is shown in Fig. 18-48b. Stirrups are required in the joint to achieve the increase in tension force in the column reinforcement. It can be seen from this strut-and-tie model that the stirrups in the joint region must provide a tie force of

$$\Sigma T_3 = T_2 - T_1$$

The reinforcement is detailed as shown in Fig. 18-48c.

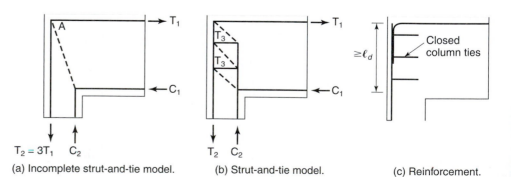

Fig. 18-48
Closing joint, beam deeper
than column.

(a) Incomplete strut-and-tie model.

(b) Strut-and-tie model.

(c) Reinforcement.

T Joints

T joints occur at exterior column–beam connections, at the base of retaining walls, and where roof beams are continuous over columns. The forces acting on such a joint can be idealized as shown in Fig. 18-49a. Two different reinforcement patterns for column-to-roof beam joints are shown in Fig. 18-49b and c, and their measured efficiencies are shown in Fig. 18-50. The most common detail is that shown in Fig. 18-49b. This detail produces unacceptably low joint efficiencies. Joints reinforced as shown in Fig. 18-49c and d had much better performance in tests [18-24]. The hooks in these two patterns act to restrain the opening of the inclined crack and to anchor the diagonal compressive strut in the joint. (See Fig. 18-49a.)

In the case of a retaining wall, the detail shown in Fig. 18-49d is satisfactory to develop the strength of the wall, provided that the toe is long enough to develop bar A–B. The diagonal bar, shown in dashed lines, can be added if desired, to control cracking at the base of the wall at C.

Beam–Column Joints in Frames

The function of a beam–column joint in a frame is to transfer the loads and moments at the ends of the beams into the columns. Again, force diagrams can be drawn for such joints. The exterior joint in Fig. 18-51 has the same flow of forces as the T joint in Fig. 18-49a and cracks in the same way. An interior joint under gravity loads transmits the tensions and

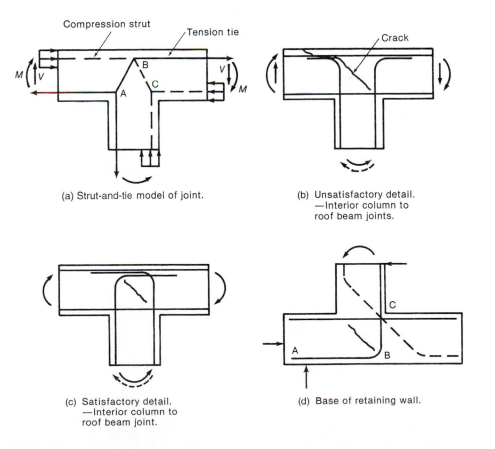

(a) Strut-and-tie model of joint.

(b) Unsatisfactory detail.
—Interior column to roof beam joints.

(c) Satisfactory detail.
—Interior column to roof beam joint.

(d) Base of retaining wall.

Fig. 18-49
T joints.

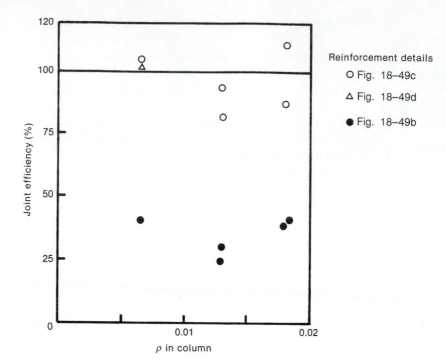

Reinforcement details

O Fig. 18–49c

△ Fig. 18–49d

● Fig. 18–49b

Fig. 18-50
Measured efficiency of T joints.

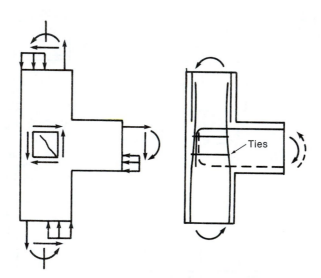

Fig. 18-51
Exterior beam column joint.

compressions at the ends of the beams and columns directly through the joint, as shown in Fig. 18-52a. An interior joint in a laterally loaded frame requires diagonal tensile and compressive forces within the joint, as shown in Figs. 18-49b and c, and Fig. 18-52b. Cracks develop perpendicular to the tension diagonal in the joint and at the faces of the joint where the beams frame into the joint. Although the reinforcing pattern shown in Fig. 18-49b is common, it should not be used.

Design of Nonseismic Joints According to ACI 352

Type of Joints

The ACI Committee 352 report [18-25] on the design of reinforced concrete beam-column joints divides joints into two groups depending on the deformations the joints are subjected to:

(a) Structures that are not apt to be subjected to large inelastic deformations and do not need to be designed according to ACI Chapter 21 are referred to as *nonseismic* structures. Such structures have *Type-1* beam-columns joints, and

(b) Structures that must be able to accommodate large inelastic deformations and as a result must satisfy ACI Chapter 21 are referred to as *seismic* structures. Such structures have *Type-2* beam-column joints.

Calculation of Shear Forces in Joints

The shaded areas of Fig. 18-54a and c each show the upper half of the joint regions in beam-column joints in reinforced concrete frames that are deflecting to the left in response to loads. Figures 18-54b and d are free body diagrams of the portions of the joints above the neutral axes of the beams entering the beam-columns joint. For the exterior joint shown in Fig. 18-54a, the horizontal shear at the midheight of the joint is given by:

$$V_{u,\,joint} = T_n - V_{col} \tag{18-21a}$$

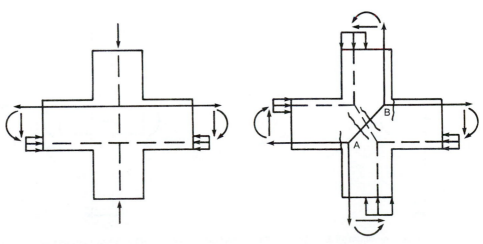

Fig. 18-52
Interior beam–column joint.

(a) Forces due to gravity loads.

(b) Forces due to lateral loads.

where the joint shear is equal to the nominal force in the top steel in the joint, minus the shear in the columns due to sway of the columns. An interior beam has beams on both sides that contribute to the shear in the joint.

$$V_{u, \text{joint}} = T_{n1} + C_{n2} - V_{col} \tag{18-21b}$$

where C_{n2} is found from section equilibrium of the beam section at the left side of the joint, and thus should be equal to the force in the tension reinforcement at the bottom of that beam. The column shears, V_{col}, can be obtained from a frame analysis; for most practical cases, they are estimated from the free-body diagrams in Fig. 18-54a and c, where points of contraflexure are assumed at the midheight of each story. The force T_n is the tension in the reinforcement in the beam at its nominal capacity. Thus,

$$T_n = \alpha A_s f_y \tag{18-22}$$

For an interior column, the free-body diagram in Fig. 18-54d includes both tension and compression forces in the steel at the top of the joint, where C_n is usually different from T_n (because the negative- and positive-moment capacities of the beams are usually different).

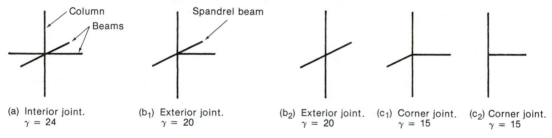

(a) Interior joint.
$\gamma = 24$

(b₁) Exterior joint.
$\gamma = 20$

(b₂) Exterior joint.
$\gamma = 20$

(c₁) Corner joint.
$\gamma = 15$

(c₂) Corner joint.
$\gamma = 15$

Fig. 18-53
Classification of joints—ACI 352. (γ Values are for Type-1 joints.)

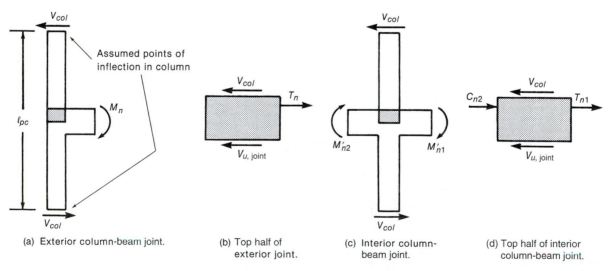

(a) Exterior column-beam joint.

(b) Top half of exterior joint.

(c) Interior column-beam joint.

(d) Top half of interior column-beam joint.

Fig. 18-54
Calculation of shear in joints.

The factor α is intended to account for the fact that the actual yield strength of a bar is larger than the specified strength, in most cases. It is taken to be at least 1.0 for Type-1 frames, where only limited ductility is required, and at least 1.25 for Type-2 frames, which require considerable ductility.

The ACI Committee 352 design procedure for Type-1 (nonseismic) joints consists of three main stages:

1. Provide confinement to the joint region by means of beams framing into the sides of the joint or by a combination of the confinement, from the column bars and from the ties in the joint region. The confinement allows the compression diagonal to form within the joint and intercepts the inclined cracks. For the joint to be properly confined, the beam steel must be inside the column steel.

2. Limit the shear in the joint.

3. Limit the bar size in the beams to a size that can be developed in the joint.

For best joint behavior, the longitudinal column reinforcement should be uniformly distributed around the perimeter of the column core. For Type-1 joints, ACI Committee 352 recommends that at least two layers of transverse reinforcement (ties) be provided between the top and the bottom levels of the longitudinal reinforcement in the deepest beam framing into the joint. The vertical center-to-center spacing of the transverse reinforcement should not exceed 12 in. in frames resisting gravity loads and should not exceed 6 in. in frames resisting nonseismic lateral loads. In nonseismic regions, the transverse reinforcement can be closed ties, formed either by U-shaped ties and cap ties or by U-shaped ties lap spliced within the joint.

The hoop reinforcement can be omitted within the depth of the shallowest beam entering an interior joint, provided that at least three-fourths of the column width is masked by the beams on each side of the column.

Shear Strength of Type-1 Joints (Nonseismic)

The shear strength on a horizontal plane at midheight of the joint is:

$$V_n = \gamma\sqrt{f_c'}b_j h_{col} \tag{18-23a}$$

where γ refers to a set of constants related to the confinement of the joint given by ACI Committee 352, h_{col} is the column dimension parallel to the shear force in the joint, and b_j is the effective width of the joint as defined in (18-24a) with reference to Fig. 18-55.

$$b_j = \frac{b_b + b_{col}}{2} \le b_b + h_{col} \tag{18-24a}$$

where b_b is the width of the beam running parallel to the applied shear force and b_{col} is the dimension of the column perpendicular to the applied shear force. When beams of different widths frame into the opposite sides of the column, b_b should be taken as the average width of the two beams.

Values for the quantity γ are given in Fig. 18-53 for various classifications of Type-1 joints. These values have been empirically derived from test results. If lightweight concrete is used in the joint, the shear capacity should be multiplied by 0.85 for *sand-lightweight* concrete and by 0.75 for *all-lightweight* concrete.

The nominal shear strength of the joint defined in (18-23a) must satisfy the normal strength requirement that $\phi V_n \ge V_u$, where $\phi = 0.75$ and V_u is computed from (18-21a or b). If this is not satisfied, either the size of the column will need to be increased or the amount of shear being transferred to the joint will need to be decreased.

Beam reinforcement terminating in a Type-1 joint should have standard 90° hooks with a development length, ℓ_{dh} given by ACI Section 12.5. The critical section for developing tension in the beam reinforcement is taken at the face of the joint in the case of Type-1 joints. If ℓ_{dh} is too large to fit into the joint (column), it is necessary to either decrease the size of the bar or increase the size of the column.

Shear Strength of Type-2 Joints (Seismic)

The shear strength on a horizontal plane at midheight of the joint is:

$$V_n = \gamma \sqrt{f_c'} A_j \tag{18-23b}$$

where γ refers to a set of constants related to the configuration and confinement of the joint given in ACI Section 21.5.3 and A_j is the effective area of the joint, similar to the product of b_j and h_{col} used in (18-23a), but it cannot exceed the area of the column. The definition of b_j is modified for seismic design, as discussed in ACI Commentary Section R21.5.3 and as given in (18-24b).

$$b_j = b_b + 2x \le b_b + h_{col} \tag{18-24b}$$

where x is the smaller of the distances measured from either side face of the beam, which runs parallel to the applied shear force, to the corresponding side face of the column.

Values for the quantity γ, which have been empirically derived from test results, are given below for various classifications of Type-2 joints.

$\gamma = 20$ for confined interior joints,

$\gamma = 15$ for exterior joints confined on three faces or on two opposite faces, and

$\gamma = 12$ for corner of other Type-2 joints.

For a joint to qualify as an interior joint, the beams on the four faces of the joint must cover at least three-quarters of the width and depth of the joint face, where the depth of the joint is taken as the depth of the deepest beam framing into the joint. Joints that have a interior configuration (beams framing into all four faces), but the beams do not satisfy these size requirements, should be evaluated using the γ value for exterior joints. For joints with an exterior configuration (beams framing into three faces or two opposite faces), the width of the beams on the two opposite joint faces must be at least three-quarters of the width of the joint face and the depth of the shallower of these two beams must be at least three-quarters of the depth of the deeper beam. Joints that have an exterior configuration, but the beams do not satisfy these size requirements, should be evaluated using the γ value for corner joints.

If lightweight concrete is used in the joint, ACI Section 21.5.3.2 states that the shear capacity from (18-24b) should be multiplied by 0.75.

The nominal shear strength of the joint defined in (18-23b) must satisfy the normal strength requirement that $\phi V_n \ge V_u$, where V_u is computed from (18-21a or b). However, (18-22) must be used to calculate the T_n and corresponding C_n values used in (18-21a and b), with α set equal to at least 1.25 to account for overstrength of the reinforcement and the high probability that those bars will go into strain hardening if plastic hinges form in the beams adjacent to the column faces. Because the joint shear loads are based on increased beam capacities, the appropriate ϕ factor for this *capacity-design* approach is 0.85. If this shear strength requirement is not satisfied, either the size of the column will need to be increased or the amount of shear being transferred to the joint will need to be decreased.

Rules for developing beam reinforcement terminating in a Type-2 joint are given in ACI Sections 21.5.1.3 and 21.5.4. For bars terminating in a standard 90° hook, the development length, ℓ_{dh} is given by ACI Equation (21-6). This equation considers the beneficial effect of anchoring the bar in the well-confined joint core, and also the detrimental effect of subjecting the bar to load reversals during earthquake loading. For simplicity, the critical section for developing the hooked bars is at the face of the joint, although several researchers state that effective anchorage starts at the face of the joint core.

For beam reinforcement extending through a beam-column joint, ACI Section 21.5.1.4 requires that the dimension of the column parallel to the beam bars must be greater than or equal to twenty times the diameter of the largest bar. This length is not sufficient to fully anchor the bars in tension, but rather is intended to delay a potential breakdown in bond between the bars and the concrete in the joint when the beam reinforcement is subjected to load reversals during earthquake loading.

The requirements for straight bars in ACI Section 21.5.1.4 and for hooked bars in ACI Section 21.5.4.1 are both increased if the joint is constructed with lightweight concrete.

EXAMPLE 18-7   Design of Joint Reinforcement

An exterior joint in a braced frame is shown diagrammatically in Fig. 18-56a. The concrete and steel strengths are 3000 psi and 60,000 psi, respectively. The story-to-story height is 12 ft 6 in.

1.   Check the distribution of the column bars, and lay out the joint ties. For Type-1 joints, no specific column-bar spacing limits are given by ACI Committee 352. The column bars should be well distributed around the perimeter of the joint. Figure 18-56b shows an acceptable arrangement of column bars and ties.

Since the frame is braced, the frame is not the primary lateral load-resisting mechanism. Hence, the spacing of the joint ties can be $s \leq 12$ in., with at least two sets of ties between the top and bottom steel in the deepest beam. The required area of these ties will be computed in step 3.

2.   Calculate the shear force on the joint. We will check this in the direction perpendicular to the edge. Since the frame does not resist lateral loads, there is no possibility of a sway mechanism due to lateral loads parallel to the edge. A free-body cut through the joint is similar to Fig. 18-54b. The column loads have been omitted to simplify Fig. 18-54. The initial beam design used 4 No.11 top bars, so the shear in the joint is

$$V_{u(\text{joint})} = T_n - V_{col} \qquad (18\text{-}26a)$$

where

$$T_n = A_s \alpha f_y \text{ and } \alpha = 1.0 \text{ for a Type-1 joint}$$

$$= 4 \times 1.0 \times 1.56 \text{ in.}^2 \times 60 \text{ ksi} = 374 \text{ kips}$$

To compute V_{col}, consider the free-body diagram in Fig. 18-54a where $\ell_{pc} = 12.5$ ft. The nominal moment capacity of the beam is

$$M_n = A_s f_y \left(d - \frac{a}{2} \right) = 4 \times 1.56 \times 60 \left(25 - \frac{7.34}{2} \right)$$

$$= 7986 \text{ in.-kips} = 665 \text{ ft-kips}$$

$$V_{col} = \frac{M_n}{12.5} = 53.2 \text{ kips}$$

Therefore,

$$V_{u(\text{joint})} = 374 - 53.2 = 321 \text{ kips}$$

3.   Check the shear strength of the joint. From Fig. 18-55, the width of the joint is

$$b_j \leq \tfrac{1}{2}(20 + 24) = 22 \text{ in.}$$
$$\leq 20 + 22 \text{ in.}$$

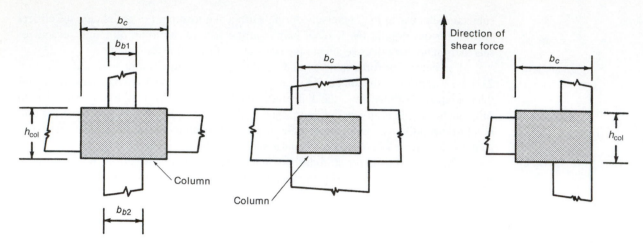

Fig. 18-55
Width of joint, b_j.

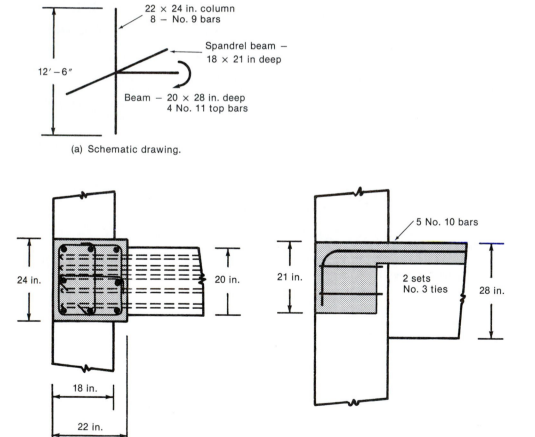

(a) Schematic drawing.

(b) Plan of joint.

(c) Elevation of joint.

Fig. 18-56
Joint design—Example 18-7.

Use $b_j = 22$ in. The thickness of the joint, h_{col}, is equal to the column dimension parallel to the shear force in the joint. Use $h_{col} = 22$ in. The equation is

$$V_n = \gamma \sqrt{f'_c} b_j h_{col}$$

Check the joint classification: This will be an exterior joint, provided that all of the beams are at least three-fourths as wide as the corresponding column face and the shallowest beam is at least three-fourths of the depth of the deepest beam. Therefore, this is an exterior joint, and $\gamma = 20$, so that

$$V_n = \frac{20\sqrt{3000} \times 22 \times 22}{1000} = 530 \text{ kips}$$

$$\phi V_n = 0.85 \times 530 = 451 \text{ kips}$$

Use $\phi = 0.85$ to be consistent with the capacity design procedure in ACI Comm. 352 report. Since this value exceeds $V_{u(\text{joint})} = 321$ kips, the joint is acceptable in shear.

For No. 3 ties, the area provided by the joint ties shown in Fig. 18-56 is

$$\text{area per set} = 3 \times 0.11 \text{ in.}^2 = 0.33 \text{ in.}^2$$

The required spacing to satisfy ACI section 11.11.2 and ACI Eq. (11-13) is

$$s = \frac{0.33 \times 60,000}{50 \times 20} = 19.8 \text{ in.}$$

but not more than 12 in. (to satisfy ACI 352). Provide two sets of ties in the joint.

4. Check the bar anchorages. From ACI Section 12.5, the basic development length of a Grade 60 hooked bar is

$$\ell_{dh} = \frac{1200 d_b}{\sqrt{f'_c}} = \frac{1200 \times 1.41 \text{ in.}}{\sqrt{3000}}$$

$$= 30.9 \text{ in.}$$

If the beam bars are inside the column bars and have 2 in. of tail cover, ACI Section 12.5.3 allows

$$\ell_{dh} = 0.7 \times 30.9 = 21.6 \text{ in.}$$

The development length available is 22 in. $- 1\frac{1}{2}$ in. (cover) $- \frac{3}{8}$ in. (tie) $= 20.125$ in., but not more than 22 in. $- 2$ in. (cover on tail) $= 20$ in.

Since ℓ_{dh} exceeds the space available, we must either increase the column size to 24 in. or reduce the bar size in the beam to No. 10 or smaller. Try five No. 10 bars:

$$\ell_{dh} = 0.7\left(\frac{1200 \times 1.27}{\sqrt{3000}}\right) = 19.5 \text{ in.}$$

Therefore, OK. Note that changing the bar area will change T_n and hence change $V_{u(\text{joint})}$ to 326 kips, which is still less than V_n. **Use the joint as detailed in Fig. 18-56b and c with two sets of ties in the joint.** ■

Joints between Wide Beams and Narrow Columns

Occasionally, a beam will be considerably wider than the column supporting it. The detailing of such a joint must provide a clear force path.

At beam–column joints, the tensile force T_n (see Figs. 18-54a and b) must be transferred to the joint core. This can be done either by grouping the top bars inside the column bars or by providing closely placed stirrups around the entire joint to create a horizontal "truss" action to transmit the tension into the joint.

If a wide deep beam is supported on a narrow column, the tensile force in the longitudinal bars outside the column will not be equilibrated by a direct compression strut, since this strut will exist only over the column. Hence, these bars will tend to shear the overhanging portions off the transverse beam. Again, the region should be confined with stirrups to provide a horizontal truss to anchor these bars.

Use of Strut-and-Tie Models in Joint Design

Although strut-and-tie models can be used to design joint regions, the models sometimes become complex when the effects of shears in the beams and columns are included. For some purposes, it is adequate to consider only the flexural forces in the steel and concrete within the joint drawing the strut-and-tie models, as was done in Figs. 18-45e, 18-48, and 18-52. If it is necessary to include the effects of beam and column shears, it is generally necessary to consider a strut-and-tie model that includes portions of the beams between the zero-shear locations and portions of the columns.

18-13 BEARING STRENGTH

Frequently, the load from a column or beam reaction acts on a small area on the top of a wall or a pedestal. As the load spreads out in the wall or pedestal, transverse tensile stresses develop, which may cause splitting and, possibly, failure of the wall or pedestal.

This section starts with a review of the internal stresses and forces developed in such a region, followed by a review of the ACI design requirements for bearing strength. The discussion of internal forces is based in large part on the excellent design manual by Schlaich [18-1].

Internal Forces near Bearing Areas

Figure 18-57a shows the stress trajectories due to a concentrated load, F, acting at the center of the top of an isolated wall of length ℓ, height h, and thickness t, where $\ell = h$ and t is small compared to ℓ and h. As before, compressive-stress trajectories are shown by dashed lines, tensile by solid lines. The horizontal stresses across a vertical line under the load are shown in Fig. 18-57b. These tend to cause a splitting crack directly under the load. For the case shown, where the width of the load is 0.1ℓ, the maximum tensile stress is about 0.4ρ, where $\rho = F/\ell t$.

For calculation purposes, this state of stress can be idealized as shown in Fig. 18-57c. Here, the uniform compressive stress on the bottom surface is replaced by two concentrated forces, $F/2$ at the quarter-points of the base points D and E. The inclined struts act at an angle θ, which, for load size $w/h = 0.1$ is 65° for $\ell/h = 1.0$ or smaller. Flatter angles will occur if ℓ/h increases, reaching about 55° for a load with $w/h = 0.10$ on a wall having $\ell/h = 2.0$. For design, a slope of 2 vertical to 1 horizontal may be assumed, as done in Example 18-1 (Fig. 18-4c).

For equilibrium, the truss needs a horizontal tension tie, shown by the solid line $B–C$ in Fig. 18-57c. This is made up of well-anchored horizontal steel placed over the zone of horizontal tensile stress (Fig. 18-57b) and having $A_s f_y$ sufficient to serve as the horizontal tie in the truss.

A similar situation occurs when a series of concentrated loads acts on a continuous wall, as shown in Fig. 18-57d. The horizontal force in the diagonal struts must be equilibrated by the tension tie $B–C$ and a compression strut, $C–B'$. For force equilibrium on a vertical plane between C and B', the force in $A–A'$ must be equal and opposite to $C–B'$.

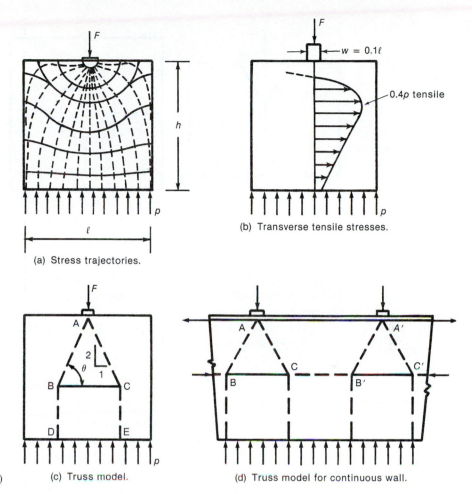

Fig. 18-57
Internal forces near bearing areas. (Adapted from [18-1].)

(a) Stress trajectories.

(b) Transverse tensile stresses.

(c) Truss model.

(d) Truss model for continuous wall.

Elastic analyses suggest that the tension in $B–C$ is roughly twice the compression in $C–B'$. Reinforcement should be placed in the direction of the tension ties.

The two cases shown in Fig. 18-57 involved concentrated loads acting on thin walls. If the concentrated loads were to act on a square pedestal, similar stresses would develop in a three-dimensional fashion. Although the stresses would not be as large, it may be necessary to reinforce for them.

ACI Code Requirements for Bearing Areas

The ACI Code treats bearing on concrete in ACI Section 10.15, for dealing with normal situations, and ACI Section 18.13.1, for dealing with prestress anchorage zones. Section 18.13.1 requires consideration of the spread of forces in the anchorage zone and requires reinforcement where this leads to large internal stresses.

ACI Section 10.15 is based on tests by Hawkins [18-27] on unreinforced concrete blocks supported on a stiff support and loaded through a stiff plate. A section through such a test is shown in Fig. 18-58a. As load is applied, the crack labeled 1 occurs in the center of the block at a point under the load. This then progresses to the surface, as crack 2. The resulting conical wedge is forced into the body, causing circumferential tension in the surrounding concrete. When this occurs, the radial crack 3 forms (Fig. 18-58b), the block breaks, and a "bearing failure" occurs. Two solutions are available: (1) Provide reinforcement

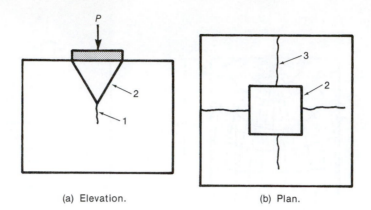

Fig. 18-58
Failure in a bearing test.

(a) Elevation.

(b) Plan.

to replace the tension lost when crack 1 formed, as is done in Example 18-1, or (2) limit the bearing stresses so that internal cracking does not occur. ACI Section 10.15 follows the latter course.

The permissible bearing stress is set at $0.85f'_c$ if the bearing area is equal to the area of the supporting member; it can be increased to

$$f_b = 0.85f'_c\sqrt{\frac{A_2}{A_1}}, \text{ but not more than } 1.7\sqrt{f'_c} \qquad (18\text{-}25)$$

if the support has a larger area than the actual bearing area. In (18-25), A_1 refers to the actual bearing area, and A_2 is the area of the base of a frustrum of a pyramid or cone with its upper area equal to the actual bearing area and having sides extending at 2 horizontal to 1 vertical until they first reach the edge of the block, as illustrated in Fig. 16-10.

The maximum bearing load, B_{max}, is computed from

$$B_{max} = \phi f_b A_1 \qquad (18\text{-}26)$$

where ϕ is 0.70, f_b is from (18-25), and A_1 is the bearing area.

The 2:1 rule used to define A_2 does not imply that the load spreads at this rate; it is merely an empirical relationship derived by Hawkins [18-27]. It should not be confused with the 2-vertical-to-1-horizontal slope assumed in the truss model in Fig. 18-57c.

18-14 T-BEAM FLANGES

In Fig. 5-5, a rudimentary strut-and-tie model was used to illustrate the spread of forces in the compression flange of a T beam. The forces in T-beam flanges will be examined more closely in strut-and-tie models of the beam web and flange. Figure 18-59a shows a strut-and-tie model of half of a simply supported beam loaded with a concentrated load at midspan (J). The horizontal components of the compression forces in the diagonal struts in the web apply loads to the top flange at B, D, and so on, as shown in Fig. 18-59b. Compression struts and transverse tension ties in the flange act to spread this compression across the width of the flange. If the web struts are inclined at 45°, except at the reaction

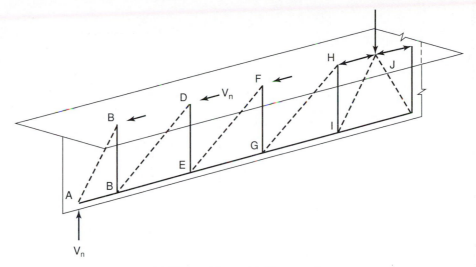

(a) Strut-and-tie model of beam web.

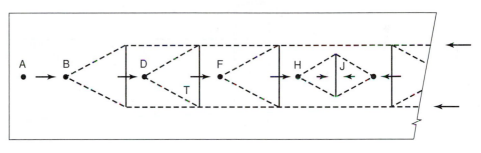

Fig. 18-59
Strut-and-tie models of a
T beam with the flange in
compression.

(b) Strut-and-tie model of compression flange.

and the concentrated load, the horizontal components of the forces acting on the flange at
D, F, and H are all equal to the shear V_n. If the flange struts are at a 2:1 slope, the transverse
tensions arising from the forces acting on the flange at D, F, and H are $T = V_n/4$ and
require transverse reinforcement with a capacity of

$$A_s f_y = V_n/4$$

For the force at F, for example, this steel would be distributed over the length of flange extending about $jd/2$ each way from the locations of the transverse tie resulting from the
force acting on the flange at F. If transverse steel were needed for cross-bending of the
flanges by loads on the overhanging flanges, it would be added to the steel needed to
spread the compressive forces. Ghali [18-29] has reviewed the load transfer in concrete
T beams.

  The strut-and-tie model in Fig. 18-59b indicates that there will not be any compression in the flange to the left of B and that there will be a concentration of horizontal compressive force in the vicinity of the load at J, because the horizontal component in strut I–J
cannot spread across the flange.

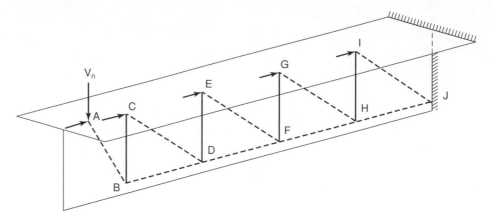

(a) Strut-and-tie model of beam web.

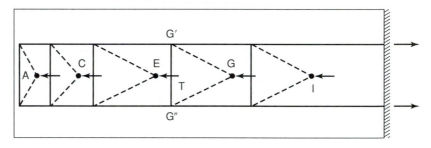

Fig. 18-60
Strut-and-tie models of a
T beam with the flange in
tension.

(b) Strut-and-tie model of tension flange.

The situation in a tension flange is more extreme, as is shown in Fig. 18-60. Here, a cantilever T beam is loaded with a single concentrated load equal to V_n. Again, when the web struts are at 45°, except at the concentrated load and the support, the horizontal components of the strut forces acting on the flange at C, E, G, and I are equal to V_n. Figure 18-60b is drawn by assuming that the flexural tensile reinforcement in the flange is spread evenly over the width of the flange and hence can be represented by two bars at the quarter points of the width of the flange. The force acting on the flange at G is spread by compression struts to engage longitudinal steel at G' and G''. If the struts are at 2:1, a transverse tension tie G'–G'' is required to resist a transverse tension of $V_n/4$. The longitudinal forces acting on the flange at A and C are too close to the free end to be spread by 2:1 struts. As a result, the transverse-tie forces, and hence the amounts of transverse reinforcement, are larger in this region than in the rest of the beam.

Figure 18-60b was drawn by assuming that the longitudinal tension steel in the flange was evenly spread over the width of the flange. Using steel close to the web to resist the longitudinal forces introduced into the flange at A and C allows the amounts of transverse steel needed at the end of the beam to be reduced.

Figures 6-22a and 8-17c illustrated the effects of shear in displacing the longitudinal steel force diagram away from the point of maximum moment in a rectangular beam. This effect is even more pronounced in the tension flange of a T beam. The forces in the flange steel at G' and G'' in Fig. 18-60b are controlled by the horizontal force applied to the flange at G. The horizontal force in the flange at G would be computed by cutting a section through the web and summing moments about H.

PROBLEMS

18-1 The deep beam shown in Fig. P18-1 supports a factored load of 1450 kips. The beam and columns are 24 in. wide. Draw a truss model neglecting the effects of stirrups and the dead load of the wall. Check the strength of the nodes and struts, and design the tension tie. Use $f'_c = 4000$ psi and $f_y = 60,000$ psi.

18-2 Repeat Problem 18-1, but include the dead load of the wall. Assume that stirrups crossing the lines AB and CD have a capacity $\phi \Sigma A_v f_y$ equal to one-third or more of the shear due to the column load.

18-3 Design a corbel to support a factored vertical load of 120 kips acting at 5 in. from the face of a column. The column and corbel are 14 in. wide. The

concrete in the column and corbel was cast monolithically. Use 5000-psi normal-weight concrete and $f_y = 60,000$ psi.

18-4 Repeat Problem 18-3, but with a factored vertical load of 100 kips and a factored horizontal load of 40 kips.

18-5 Figure P18-5 shows the dapped support region of a simple beam. The factored reaction is 100 kips, $f'_c = 5000$ psi, and $f_y = 60,000$ psi (weldable). The beam is 16 in. wide.

    (a) Isolate the D-region.
    (b) Draw a truss to support the reaction.
    (c) Detail the reinforcement.

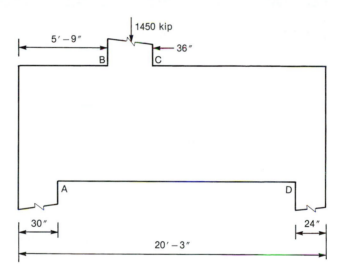

Fig. P18-1

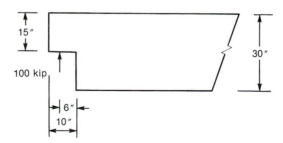

Fig. P18-5

19

(a) Hinged coupling beams.   (b) Stiff coupling beams.

Walls and Shear Walls

19-1   INTRODUCTION

Definitions—Walls and Wall Loadings

ACI Section 2.1 defines a wall as follows:

"*Wall*—Member, usually vertical, used to enclose or separate spaces."

This definition fails to consider the structural actions of walls. ACI Section 2.1 also defines the term "structural walls":

"*Structural walls*—Walls proportioned to resist combinations of shears, moments, and axial loads induced by earthquake motions. A shear wall is a structural wall."

The second definition is better, but it is restricted to walls subjected to earthquake forces.

Major factors that affect the design of structural walls include the following:

(a)   the structural function of the wall relative to the rest of the structure

- the way the wall is supported and braced by the rest of the structure
- the way the wall supports and braces the rest of the structure

(b)   the types of loads the wall resists

(c)   the location and amount of reinforcement

Two frequent characteristics of walls are their slenderness, h_w/t, where h_w is the total height of the wall and t is the wall thickness, generally higher than for columns, and the reinforcement ratios, generally about a fifth to a tenth of those in columns.

Types of Walls

Structural walls can be classified as:

(a)   *Bearing walls*—walls that are laterally supported and braced by the rest of the structure, that resist primarily in-plane vertical loads acting downward on the top of the wall. These will be referred to as *bearing walls*. The vertical load may act eccentrically on the wall, causing *weak-axis bending*. (See Fig. 19-1a.)

(b)   *Shear walls*—walls that resist lateral loads due to wind or earthquakes acting on the building are called *shear walls* or *structural walls*. These walls provide lateral bracing for the rest of the structure. (See Fig. 19-1b.) They resist gravity loads transferred to the wall by the parts of the structure tributary to the wall, plus lateral-load moments about the *strong axis* of the wall, together with the shear forces necessary to equilibrate the moments.

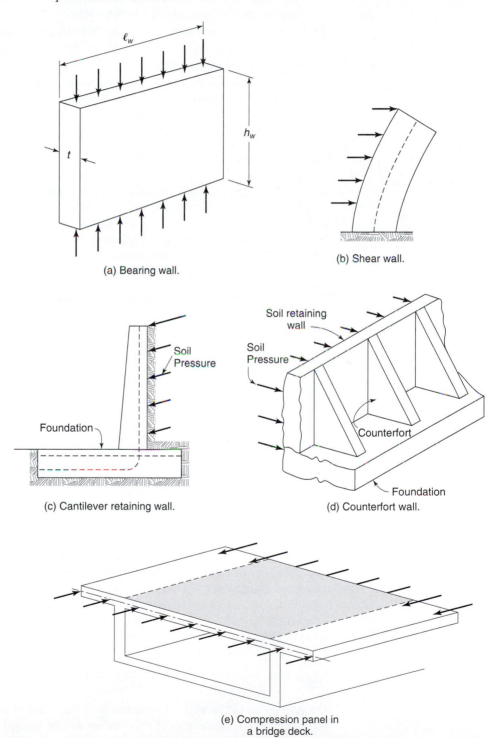

(a) Bearing wall.

(b) Shear wall.

(c) Cantilever retaining wall.

(d) Counterfort wall.

(e) Compression panel in a bridge deck.

Fig. 19-1
Types of walls.

(c) *Nonbearing walls*—walls that do not support gravity in-plane loads other than their own weight. These walls may resist shears and moments due to pressures or loads acting on one or both sides of the wall. Examples are basement walls and retaining walls used to resist lateral soil pressures. (See Fig. 19-1c and d.)

(d) *Tilt-up walls*—are very slender walls that are concreted in a horizontal position adjacent to the structure. They are then tilted into their intended vertical position and fastened together to resist lateral loads.

(e) *A wall assembly*—is a group of walls that are interconnected to act as a single cantilever to resist lateral forces. Its three-dimensional shape serves to enclose stairwells and elevator shafts.

(f) Although they are not walls as such, plates that resist in-plane compression, such as the compression flanges or the decks of box girder bridges, display some of the characteristics of walls. (See Fig. 19-1e.)

One-Way and Two-Way Walls

Walls may be supported and restrained against lateral deflections along one to four sides. *Cantilever retaining walls* are generally supported solely along the lower edge of the wall. Such walls act as vertical flexural cantilevers that resist lateral loads from the adjoining soil. *Bearing walls* are generally laterally supported and restrained against deflection along two opposite sides, usually the top and bottom supports. Cantilever retaining walls and bearing walls transfer load in one direction: to supports at the top and bottom of the wall, for example or to supports at the east and west edges. In the terminology used for one-way and two-way slabs, these are referred to as *one-way walls*. One-way walls may be designed as wide columns spanning between the top and bottom supports, using ACI Sections 10 and 11, or they may be designed using ACI Eq. (14-1) ((19-12) in this book), which approximates one-way wall behavior. Walls that transfer load in more than one direction are called *two-way walls*. Walls supported on three sides may occur in open-topped, rectangular tanks; in storage bins for bulk materials; or in counterfort retaining walls (see Fig. 19-1d). Walls supported on four sides are used to resist forces or pressures applied perpendicular to the walls.

Wall Assemblies

Shear walls may be planar walls, standing in one vertical plane, or three-dimensional assemblies of planar walls or wall segments. The latter occur as elevator shafts in buildings where four or more vertical walls enclose a stairwell or a group of elevators. A section through such a wall assembly is shown shaded in the floor plan in Fig. 10-14. Figure 19-2 is a photograph of a wall assembly that will enclose an elevator shaft and will brace a steel-frame building that will be built around it. The frame will be attached to the wall assembly by welding or otherwise fastening the frame to steel plates embedded in the wall concrete during construction of the walls.

In the design of a wall assembly, it is necessary to consider the transfer of shear forces from the wall segments serving as webs of the assemblies, to wall segments which act as flanges.

Notation

The orientation of the walls in a vertical direction leads to ambiguity in the notation for height and width. While most of the notation will be defined where it is first used, a few key symbols are defined here. Some of these are illustrated in Fig. 19-1a.

Fig. 19-2
Wall assembly in a building
under construction. (Photograph courtesy of J.K.Wight.)

f_c'' is the strength of the concrete in a compression stress block, similar to the 0.85 f_c' in ACI Section 10.2.7.1.

f_2 is the minor (compressive) principal stress

$f_{2,\max}$ is the compressive strength of the concrete

h_b is the height of a beam

h_z is the height of a location in a building

h_w is the overall height of a wall

ℓ_c is the clear height of a wall between lateral supports

ℓ_w is the horizontal length of a wall

t is the thickness of a wall

19-2   TESTS OF WALLS WITH IN-PLANE LOADS

Tests of One-Way Walls

A one-way wall corresponds to a wide beam-column laterally supported at the top and bottom of the column, bent about the weak axis of the wall cross section. The basic design equation for one-way walls, (19-12) (ACI Eq. (14-1)), derived later, was based on 54 wall tests reported by Oberlender and Everard. [19-1]. The test specimens were loaded as one-way walls with slenderness ratios h_w/t varying from 8 to 28, loaded either with concentric loads or eccentric loads. The walls were loaded at the top and bottom edges with line loads applied parallel with the long side of the cross sections of the wall. Half the walls were loaded concentrically and the other half were loaded at an eccentricity of one-sixth of the wall thickness, ($e = t/6$), measured from the midplane of the walls, i.e., at the kern of the wall section. All the specimens had two layers of reinforcement with total reinforcement ratios of 0.0033 and 0.0047. The test results showed no effect of the reinforcement ratios on the strength of the walls. Concentrically loaded specimens with $h_w/t = 28$ and eccentrically loaded specimens with h_w/t of 16 or more, exhibited buckling failures.

Tests of Models of Hollow Reinforced Concrete Bridge Piers

Taylor, Rowell, and Breen [19-2], [19-3] reported tests of 12 one-fifth-size models of hollow box piers. Figure 19-3 shows the overall dimensions of a box pier specimen. The hollow rectangular test specimens were 72 in. long vertically between concrete end blocks provided to anchor the vertical reinforcement at each end of the hollow section and to impose linear strains at the ends of the specimens. The specimens had out-to-out cross-sectional dimensions of 51 in. by 102 in. The measured out-of-plane deflections of the compression flange of the model box pier with the largest wall slenderness are shown in Fig. 19-3, at distorted and enlarged scale. The specimen shown had a wall thickness of 2.9 in. [19-3]. The slenderness ratio of the wall that failed was $102/2.9 = 35.2$.

The box pier models were loaded with concentrated loads at an eccentricity with respect to the *weak axis* of the pier models, as shown schematically in Fig. 19-3. As a result, the highest compression stresses occurred in wall A–B–C–D–A which was parallel to the axis of bending. Failures were due to buckling and crushing of this face. Near failure the specimen shown in Fig. 19-3 developed buckles alternately inward and outward as shown.

If the height of the wall, h_w is several times the width, b, the wall tries to buckle in a series of square bulges. Two such buckles can be seen in Fig. 19-3. The lowest critical loads for this wall configuration occur when h_w/b has integer values, 1, 2, 3, and so on, and the wall buckling constant, K, is equal to 4 [19-4]. Later in this chapter the critical load of a compressed wall will be related to the number of buckles.

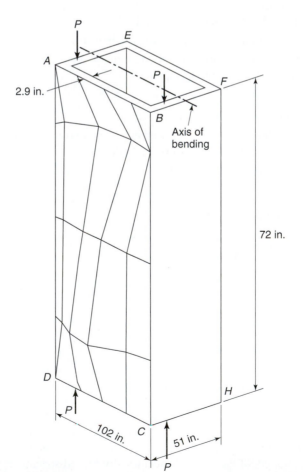

Fig. 19-3
Geometry, dimensions, and buckled shape of the compression flange of a hollow pier test specimen. (Note (a) that the concrete end blocks are not shown and (b) the deflections are plotted to an enlarged scaling)

Tests of Shear Walls

Shear walls fall into two classes, *squat* or *short shear walls* with height-to-horizontal-length aspect ratios of 2 or less [19-5], [19-6], [19-7], and *slender* or *flexural shear walls* with aspect ratios greater than 2 [19-8], [19-9], [19-10]. Reference [19-6] presents an example of the design of a squat wall using a strut-and-tie model which was compared to the test behavior. References [19-5] and [19-8] observed that the behavior of walls with aspect ratios of 2 or less involved in-plane strut-and-tie forces, while walls with aspect ratios greater than 2 displayed flexural behavior.

Wood compared the measured and computed axial and in-plane lateral-load capacities of shear-wall tests reported in the literature. Her studies included 147 squat shear walls [19-7] and 37 approximately one-third-scale flexural walls [19-10].

Short Shear Walls: Three-quarters of the 147 shear walls studied by Wood had $(M/V\ell_w)$ between 0.5 and 1.0 corresponding to shear-span-to-depth ratios a/d from 0.5 to 1.0. The test results were compared to the 1983 ACI Code Eq. (A-7) for the design of short walls for in-plane shears given by (19-1):

$$v_n = \alpha_c \sqrt{f_c'} + \rho_n f_y \tag{19-1}$$

where $\alpha_c = 3.0$ for: $h_w/\ell_w = 1.5$,

with α_c reducing linearly to

$$\alpha_c = 2.0 \text{ for } h_w/\ell_w = 2.0$$

and staying constant at $\alpha_c = 2.0$ for $h_w/\ell_w \geq 2.0$.

The symbol, ρ_n represents the reinforcement ratio for the horizontal reinforcement in the wall.

Squat or *short shear walls:* Equation (19-1) was a reasonable lower bound to the observed shear strengths of lightly reinforced walls with orthogonal shear reinforcement. Wood also observed that (19-2) gave a reasonable lower bound to the shear strength of short or squat shear walls. (See Fig. 19-9a.)

$$v_n = 6\sqrt{f_c'} \tag{19-2}$$

The maximum failure shear stress tended to increase with an increase in the vertical tie reinforcement. This increase could be approximated by a shear friction model, with the reinforcement in the boundary element establishing the limit on shear transmitted across the plane at the base of the wall.

Wood proposed an upper strength limit of:

$$v_n = 10\sqrt{f_c'} \tag{19-3}$$

Slender or flexural walls: Tests of slender or flexural shear walls are reported by Cardenas et al, [19-8], [19-9] and others. Cardenas's specimen with the lowest amount of longitudinal flexural steel, failed by a sudden and complete fracture of the vertical reinforcement when the first crack appeared. When this crack developed, the first layers of steel it crossed had inadequate tensile strength to replace the concrete tensile stresses lost when the concrete cracked.

Wood studied the strengths of 37 approximately one-third size models, reported in the literature, that were taller than twice their horizontal length. These walls resisted in-plane lateral loads by flexure [19-10]. Fifteen out of 37 specimens failed by fracture of the flexural tension reinforcement. Wood defined two behavioral indices

(a)   The first index was a *shear force index* (SFI)

$$\text{SFI} = v_{max}/v_n \tag{19-4}$$

where v_{max} was the shear stress at failure and v_n was the shear strength from Eq. (19-1)

(b)   The second index was the *flexural strain index* which was the calculated strain, e_t, in the extreme flexural reinforcement at the nominal flexural capacity. The strain was computed using ACI Sections 10.2 and 10.3 with $\varepsilon_{cu} = 0.003$ with the ACI rectangular stress block replaced by the Hognestad stress block. Twenty of the 24 beams with SFI greater than 0.75 failed in shear. Twelve of the 13 walls that had SFI less than 0.75, failed in flexure.

The vertical reinforcement fractured in 10 of 11 specimens that both (a) failed in flexure and (b) had calculated extreme-tensile steel strains greater than 2.5 percent and (c) had nominal flexural capacities greater than 0.75 times the shear force index.

Other observations were:

(i)   Reinforcement fractured in specimens that were not susceptible to shear failure (SFI $\leq$ 0.75), for which the calculated strain in the extreme tensile fiber exceeded 4 percent at the nominal strength.

(ii)   Reinforcement did not fracture in walls with calculated neutral axis heights, c/d_t, less than 15 percent.

Another index, referred to as the *flexural strength index* was used to separate the groups of tests

$$\text{FSI} = \rho_v f_y + (P/A)/f_c' \tag{19-5}$$

where ρ_v reprents the vertical reinforcement ratio in the wall. Failures involving fracture of the flexural reinforcement were observed in walls with FSI less than 15 percent. Finally, slender shear walls with a total reinforcement ratio less than 1 percent were identified as being susceptible to fracture of the reinforcement.

Four tests by Zhang and Wang [19-11] considered two levels of axial-load ratios, $N/(f_c'A_g)$. Two tests used a high axial load ratio equal to 0.35, and two had, $N/(f_c'A_g)$ equal to 0.25. The four walls all had horizontal lengths of 700 mm and vertical heights of 1500 mm for an aspect ratio of 2.14. The plan dimensions were 100 mm thick by 700 mm horizontal length for a cross-sectional aspect ratio of 7. The specimens were loaded by a single load parallel to the plane of the wall applied at the top of the specimens, plus a vertical axial load. The axial load was applied in increments up to its full value and then held constant while the horizontal load was cycled, increasing from cycle to cycle.

The tests with axial-load ratios of 0.25, and previous tests with axial-load ratios of 0.11 quoted by Zhang and Wang [9-11], failed in flexure at loads slightly above the calculated flexural strengths. The specimens tested at an axial-load ratio of 0.35 developed yielding of the flexural reinforcement at about 106 percent of the predicted flexural yield loads. Failure was due to out-of-plane buckling, which led to a premature failure at a low ductility. Wallace and Oracal [19-12] reported two tests of walls, one T-shaped, the other rectangular. Both were designed using an approximate calculation for the flexural steel. Reasonable agreement with the 1999 ACI Code requirements were reported.

Tests of Isolated Two-Way Walls

Aghayere [19-13] tested reinforced concrete walls subjected to in-plane line loads acting on the perimeter of the wall, and transverse loads that simulated a uniform pressure acting perpendicular to the surface of the wall. Walls loaded in this manner occur less frequently than those subjected to in-plane compressive loads or shearing loads acting parallel to slabs with uniform transverse loads. Transversely loaded walls are considered here because their behavior represents a general case that is relatively easy to visualize. An example of such an element in service is the bridge deck shown shaded in Fig. 19-1e. The effects of the boundary conditions and the loading history on the behavior of similar test specimens was discussed by Ghoneim [19-14].

The resistance to transverse pressures acting on a compressed wall that is laterally supported on four edges can be represented by crossing beam strips, similar to those shown in Figs. 13-9, 10, and 15-16, except that the walls are in a vertical position. In the absence of in-plane forces, this crossing-beam analogy is the same as for two-way slabs, shown in Fig. 13-9.

Compressive in-plane loads have two offsetting effects on the stiffnesses of the crossing-beam strips referred to in the preceding paragraph.

1. The flexural stiffnesses of the beam strips increase as the in-plane load in the strips increases from zero to the balanced failure load of the beam strip. At loads greater than the balanced failure load, the stiffness of the beam strips stays about the same as that at balanced failure load, or it decreases.

2. The product of the transverse deflections of the plate and the axial forces in the slab strips results in second-order moments, $P-\delta$, in the axially loaded slab strips. In a two-way slab, the increase in stiffness due to the axial forces, described in item 1, is partially offset by the effective decrease in stiffness due to second-order effects. In a plate carrying in-plane vertical loads and transverse pressure, the portion of the transverse pressure transmitted to the side supports by the horizontal beam strips increases when the vertical in-plane load is increased. As this occurs, the fraction of the load transmitted to the top and bottom supports decreases. The behavior in wall tests also depends on such things as the width-to-height ratio of the panel and the levels of in-plane forces in the horizontal and vertical crossing beams. See [19-14].

In transversely loaded *one-way* walls, laterally supported only at the top and bottom edges, all the transverse load on the wall is transferred vertically to the top and bottom edges of the wall, and there is no redistribution to the supports on the sides of the specimen.

19-3   CRITICAL LOADS FOR AXIALLY LOADED WALLS

Buckling of Compressed Columns

The critical stress for buckling of a hinged column or a one-way wall, with a rectangular cross sectional area $(b \times t)$ is

$$\sigma_{cr} = \frac{P_{cr}}{bt} = \frac{\pi^2 EI}{(k\ell)^2}\left(\frac{1}{bt}\right) \tag{19-6}$$

where b is the width of the column, and $k\ell$ is the *effective length* of the column. The flexural stiffness of a column section of width b and thickness (height of cross section) t is.

$$EI = \frac{Ebt^3}{12}$$

Buckling of Compressed Walls

The corresponding stiffness of a slab or a two-way wall, per unit of width of wall, is

$$D = \frac{Et^3}{12(1 - \mu^2)} \tag{19-7}$$

For an axially loaded and uncracked concrete column with little or no reinforcement the critical stress, σ_{cr}, is a function of the tangent modulus of elasticity, E_T, evaluated at the critical stress, σ_{cr}. Substituting E_T into (19-6), replacing ℓ^2 with b^2, replacing EI with the plate stiffness, D, and including an appropriate edge restraint factor, K, gives the critical stress for a two-way wall:

$$\sigma_{cr} = \frac{P_{cr}}{bt} = K\frac{\pi^2 E_T}{12(1 - \mu^2)}\left(\frac{t}{b}\right)^2 \tag{19-8}$$

In this equation, b is the effective width of the panel, taken equal to the smaller of the width and height of the wall between lateral supports, t is the thickness of the wall, μ is Poisson's ratio, and K is an *edge restraint factor* that varies depending on the degree of fixity of the edges of the plate. It is similar to the effective length term k^2 in (19-6). The term K is plotted in Fig. 19-4. Values of K at buckling for other types of compressed plates are given in books on structural stability [19-4]. Equations 19-6 and 19-8 have been included to show the similarity between the critical loads for columns and walls.

Properties of Concrete Affecting the Stability of Concrete Walls

Compressive Stress–Strain Curve for the Concrete

The tangent modulus of the concrete in the wall will be used to estimate the tangent modulus buckling loads. A good approximation of the stress–strain curve for the wall is the second-degree parabola given by (19-9) and shown in Fig. 19-5a, with a peak stress of $f_{2,\max} = 0.90 f'_c$. The apex of the parabola will be taken at a strain of $\varepsilon_{co} = 0.0020$ and:

$$f_2 = f_{2,\max}\left[2\left(\frac{\varepsilon_2}{\varepsilon_{co}}\right) - \left(\frac{\varepsilon_2}{\varepsilon_{co}}\right)^2\right] \tag{19-9}$$

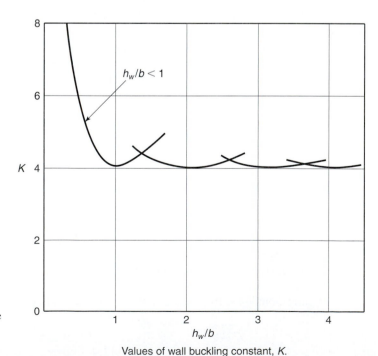

Fig. 19-4
Wall buckling factor, K, for buckling of a compressed plate analogous to a compressed wall in a hollow rectangular bridge pier. (From [19-4].)

Values of wall buckling constant, K.

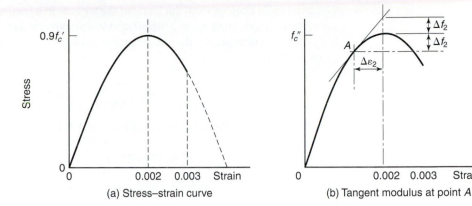

Fig. 19-5
Second-order parabolic
stress–strain curve for compressed concrete.

(a) Stress–strain curve

(b) Tangent modulus at point A

where f_2 is the compressive stress in the most compressed wall of the box pier; ε_2 is the corresponding strain, assumed constant over the thickness of the wall; and $f_{2,\,max}$ is the maximum stress in the stress–strain curve, taken equal to $0.9f'_c$. The strain at any level of stress in (19-9) is given by (19-10).

$$\varepsilon_c = \varepsilon_{co}\left(1 - \sqrt{1 - \frac{f_2}{f_{2,\,max}}}\right) \tag{19-10}$$

The tangent modulus of elasticity, E_T, is the slope of the tangent to the stress–strain curve at some particular stress, in our case, at the critical stress, σ_{cr}. It is computed using

$$E_T = d\sigma/d\varepsilon$$

evaluated at the critical stress, σ_{cr}. For example, the tangent modulus at point A in Fig. 19-5b is calculated as

$$E_T = \frac{2\Delta f_2}{\Delta \varepsilon_2} \tag{19-11}$$

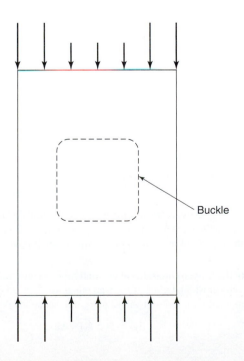

Buckle

Fig. 19-6
Stresses in a plate after
buckling.

The 2 in (19-11) comes from the fact that the tangent to a second-order parabola intersects the vertical axis through ε_{co} at 2 times the stress increment between f_2 and $f_{2,max}$, as shown in Fig. 19-5b. For example, the initial tangent modulus at $f_2 = 0$ for the stress–strain curve given by (19-11) is

$$E_T = \frac{2 \times 3600 \text{ psi}}{0.0020} = 3{,}600{,}000 \text{ psi}$$

This is very close to the value obtained from ACI Section 8.5:

$$E = 57{,}000\sqrt{f_c'} = 3{,}605{,}000 \text{ psi}$$

The amount of reinforcement in a wall is generally so small that it can be ignored when determining E_T.

Post Buckling Behavior of Concrete plates

After a uniaxially loaded wall panel hinged on four edges buckles, the strips of wall along the edges parallel to the load continue to resist a portion of the in-compressive in-plane load as shown in Fig. 19-6. At the same time, the load resisted by the center portion of the buckled plate decreases. Failure generally comes from excess bending of the buckled parts. The net effect is that the plate does not suddenly fail. Instead, strips along the edges of the wall continue to support a stress that may approach the buckling stress. This is offset by a drop in the stress resisted by the center of the plate.

EXAMPLE 19-1   Compute the Buckling Load of a Wall

A bridge pier consists of a hollow, rectangular concrete tube similar to the one shown in Fig. 19-3. Side A–B–C–D–A has the largest slenderness ratios of any of the sides of the box pier. It has $b/t = 25$. The concrete strength is 4000 psi. The applied loads cause axial load and bending about the weak axis of the box as shown in Fig. 19-3. The stress–strain curve of the concrete is assumed to be given by (19-9) and (19-10), with a peak compressive stress, $f_{2,max}$, equal to $f_c'' = 0.9f_c'$, where the 0.9 is similar to the 0.85 in ACI Section 10.2.7.1.

Compute the buckling load for a side wall having a slenderness ratio b/t equal to 25.

1.   Assume a stress–strain curve for the concrete. We will use the second-degree parabolic stress–strain curve given by (19-9) with a peak stress and strain of $f_{2,max} = 0.90f_c'$ and $\varepsilon_{co} = 0.0020$:

$$f_2 = f_{2,\,max}\left[2\left(\frac{\varepsilon_2}{\varepsilon_{co}}\right) - \left(\frac{\varepsilon_2}{\varepsilon_{co}}\right)^2 \right] \tag{19-9}$$

The strain at any level of stress in (19-10) is given by:

$$\varepsilon_c = \varepsilon_{co}\left(1 - \sqrt{1 - \frac{f_2}{f_{2\,max}}} \right) \tag{19-10}$$

Assume the wall thickness is small compared to the overall distance from the extreme-compression fiber to the neutral axis in the pier. This allows us to assume the stress is uniform across the section at the value at the middle of the compressed wall. We shall ignore the effect of reinforcement on the stiffness of the concrete.

2.   Compute the tangent modulus of elasticity for various values of $f_2/f_{2,max}$. We shall take the maximum stress at a crushing failure of the wall equal to $f_c'' = 0.90 \times 4000 \text{ psi} = 3600 \text{ psi}$. Values of ε_c will be computed for a series of values of f_2 and, hence, of $f_2/f_{2,max}$.

$$E_T = 2(\Delta f_2)/(\Delta \varepsilon_2) \tag{19-11d}$$

where Δf_2 is the increase in stress from f_2 to $f_{2,\max}$ and $\Delta \varepsilon_2$ is the increase in strain as the stress increases from f_2 to $f_{2,\max}$, and the strain increases from ε_2 to ε_{co}.

3.   **Compute the critical wall slenderness ratios for buckling at various stress ratios,** $f_2/f_{2,\max}$. Use f_2 to $f_{2,\max}$ ratios of 0.85, 0.90, 0.925, 0.95, and 0.975.

The equation for the critical stress of a wall hinged on all four sides and subjected to in-plane load in one direction is:

$$\sigma_{cr} = \frac{P_{cr}}{bt} = K \frac{\pi^2 E_T}{12(1 - \mu^2)}\left(\frac{t}{b}\right)^2 \qquad (19\text{-}8)$$

A plate with height ℓ (parallel to the load) and width $b = \ell$ will buckle in a series of square bulges, alternately on one side of the plate, and then the other. One bulge will form if $h/\ell = 1$, two bulges if $h/\ell = 2$ and so on. At integer values of b/ℓ, the buckling constant, $K = 4$.

Substituting $\mu = 0.20$; $f_c'' = 0.90 f_c' = 3600$ psi; $\sigma_{cr} = 0.85 f_c'$ and assuming that $K = 4$, the critical slenderness ratios, h/ℓ, are computed in Table 19-1 for the most highly stressed side of the box (side A–B–C–D–A) in Fig. 19-3. This is done for a series of values of $f_2/f_{2,\max}$ for $f_c'' = 3600$ psi loaded to the stress ratios $f_2/f_{2,\max}$ given in Table 19-1.

The calculated values of h/t in Table 19-1 indicate that a side wall with $h/t = 25$ would buckle at a stress of 94 percent of the computed flexural capacity.

The statement of the example specified the slenderness ratio of the critical wall of the bridge pier as $h/\ell = 25$. From the last line in Table 19-1, this slenderness corresponds to plate buckling at a stress, f_{cr}, equal to $0.942 f_{2,\max} = 3390$ psi.

In the bridge pier tests described in [19-2] and [19-3], the failure strengths of piers with wall thickness ratios between 15 and 35 were reduced by plate buckling. For $h/t = 25$, Table 19-1 predicts, a buckling stress of 94 percent of the flexural capacity. This suggests that plates, this thin or thinner, should not be used in situations where such walls are highly stressed in compression. It is difficult to apply this directly to the stability of an assemblage of plates such as those that might enclose an elevator shaft, because the floor slabs and elevator door sills brace the wall at various points along the height of the wall.   ◼

EXAMPLE 19-2   Compute the Buckling Load of a Compressed Wall Flange

Stability of Compressed Flanges

Figure 19-7a shows a portion of the projecting flange A–C–D–F–A of an H-shaped wall. When such a flange is loaded in compression, it may buckle in a series of up-and-down bulges at the free edge of the flange, as shown in Fig. 19-7b, and the line where the flange joins the web is assumed to act as a hinge. For analysis, the flange can be idealized as a rectangular plate, connected to the web by a hinge A–B–C along one long edge of length ℓ, free along the other long edge D–E–F, and hinged along the two short edges, A–F and C–D. The flange projects a distance b from the web and has a thickness t. The buckling load of such a flange can be computed from (19-8). In this case, books on elastic stability show that K is nearly constant and is equal to 0.5 when $b/h \geq 4$ [19-4].

Substituting $K = 0.5$ into (19-8), the critical slenderness ratio for an outstanding flange is found to range from 7 to 12 as the average compressive stress in the flange, f_2, increases from 0.85 to 0.975 times f_c''. The calculations are summarized in Table 19-2. The values of the critical slenderness ratios, b/t for $f_2/f_{2,\max} = 0.90$ and 0.925 have been rounded off to the nearest integer, in this

TABLE 19-1 Critical Slenderness Ratios for a Wall Subjected to Various of $f_2/f_{2,\max}$

$f_2/f_{2,\max} =$	0.85	0.90	0.925	0.942	0.95	0.975
$\varepsilon_c =$	0.00095	0.00113	0.00124	0.00133	0.00138	0.00156
E_T psi $=$	1,025,100	825,900	710,600	622,100	576,400	405,000
h/t	34	30	27	**25**	24	20

Origin: Equation 19-8, Example 19-1.

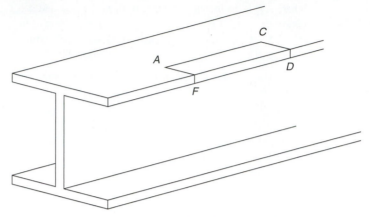

(a) Location of portion of the flange considered.

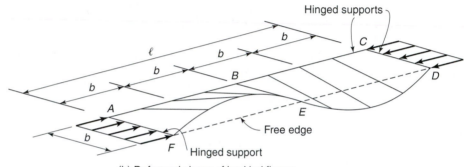

Fig. 19-7
Buckling of an outstanding
flange of a wall assemblage.

(b) Deformed shape of buckled flange.

TABLE 19-2 Critical Slenderness Ratios, b/t, for a Wall Flange
Subjected to Various Compression Stress Ratios, $f_2/f_{2,max}$

$f_2/f_{2,max}$	.85	0.90	0.925	0.95	0.975
$\varepsilon_c =$	0.000946	0.001130	0.001240	0.001380	0.001556
E_T, psi $=$	1,025,000	825,900	710,600	578,000	405,000
$b/t =$	12	10	10	8	7

Origin: Equation 19-8, Example 19-2.

case to 10. The analysis given in Table 19-2 predicts that an outstanding flange with a width-to-thick-ness ratio of $b/t = 10$ is in danger of buckling at flange stresses of 92.5 percent of the maximum stress f''_c in the stress–strain curve. This may limit the flexural capacity of flanged shear walls. These are similar to the limits on the flange widths allowed in codes for the design of steel structures. ■

19-4   BEARING WALLS

Walls used primarily to support gravity loads in buildings are referred to as *bearing walls*. Design is by ACI Section 14.5, which was derived specifically to apply to walls subjected to axial loads and moments due to the axial loads acting at an eccentricity of one-sixth of the thickness of the wall from the midplane of the wall (i.e., at the kern of the wall). The resulting moments are referred to as *weak-axis bending* moments. ACI Section 14.4 allows the design of bearing walls to be carried out:

1.   either by using the one-way column design and slenderness requirements in ACI Sections 10.11, 10.12, and 10.13, or

2.   by the so-called *empirical design method* in ACI Section 14.5.

Walls with *strong-axis* moments and significant in-plane shear forces acting parallel to the wall, referred to as *shear walls* or *structural walls*, are not covered by ACI Chapter 14, although the code does not state this. Shear walls will be discussed in Section 19-7.

Axial-Load Capacity—ACI Eq. (14-1) was based on the results of 54 wall tests reported by Oberlender and Everard [19-1]. The wall tests were described in Section 19-2. The test results showed no effect of the reinforcement ratios. Concentrically loaded test specimens with $h_w/t = 28$, and eccentrically loaded specimens with $h_w/t = 16$ or more, exhibited buckling failures.

About the same time, Kripanarayanan [19-15] discussed the strength of slender walls based on analytical studies. He observed that reinforcement amounts of $\rho = 0.75$ to 1.0 percent were needed for the reinforcement to affect the failure loads of slender walls.

Equation 19-12 (ACI Eq. (14-1)) was derived in a two-step procedure. First, the capacity of a short wall was derived. Then this was multiplied by a factor reflecting the effects of slenderness on the axial-load capacity.

The largest eccentricity at which a load can be applied to a plain concrete column without the wall cracking is at one-sixth of the wall thickness from the midthickness of the wall (at the kern point of the section). This load case can be approximated by a rectangular

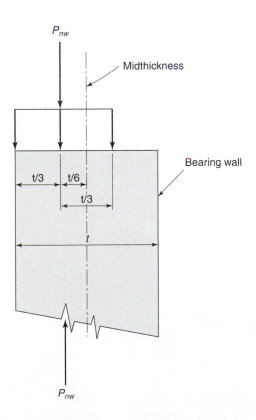

Fig. 19-8
Cross Section through the top of a bearing wall showing the flexural compression zone.

stress block extending from the compressed face of the wall for a distance of two-thirds of the thickness of the wall as shown in Fig. 19-8. The force per horizontal length of wall, ℓ_w, is:

$$P_{n,short} = 0.85f'_c \times \frac{2}{3}t \times \ell_w = 0.567f'_c \times t \times \ell_w$$

This was rounded off to $0.55f'_c t\ell$ and then multiplied by the term in the square brackets in (19-12) to account for the slenderness of the wall. The slenderness term was derived to give reasonable agreement with the slenderness effects in ACI Sections 10.11 through 10.12. The equation for the axial-load capacity of a bearing wall is

$$\phi P_{nw} = 0.55\phi f'_c A_g \left[1 - \left(\frac{k\ell_c}{32t} \right)^2 \right] \qquad (19\text{-}12)$$
$$(\text{ACI Eq. } 14\text{-}1)$$

where

ℓ_c is the clear vertical distance between lateral supports,

k is the effective length factor for a column, taken as

0.8 if the wall is braced against translation at both ends and the top or bottom end or both is restrained against rotation;

1.0 if both ends are effectively hinged; and

2.0 for walls which are not effectively braced against lateral translation at the top, and therefore must be considered to be free-standing

h is the overall thickness of the wall

ϕ is the strength-reduction factor taken equal to 0.7.

Thickness, Reinforcement and Sustained Loads—Equation 19-12 is not affected by the amount of wall reinforcement and does not allow for creep under sustained axial loads. This is in contrast to design by Chapters 10 and 11.

ACI Section 14.5.3.1 limits the minimum thicknesses of walls designed using the so-called empirical design method to 1/25 of the unsupported height or length, whichever is shorter. ACI Chapter 14 does not require wall reinforcement to be designed for the loads on the wall. Instead, ACI Sections 14.3.2 and 14.3.3 give the minimum vertical and horizontal reinforcement ratios.

These reinforcement ratios can be written in terms of the maximum spacing of the bars. Thus, for No. 5 or smaller bars with f_y not less than 60 ksi, the maximum horizontal and vertical spacings are as follows:

Vertical steel:

$$s_{h,max} = A_v/(0.0012t) \qquad (19\text{-}13a)$$

Horizontal steel:

$$s_{v,max} = A_h/(0.0020t) \qquad (19\text{-}13b)$$

If the reinforcement is in two layers, A_v is the total area of vertical bars within the spacing s_h, and similarly for the horizontal bars.

ACI Section 14.3 requires more reinforcement horizontally than vertically. This reflects the greater chance that vertical cracks in walls might form as a result of restrained horizontal shrinkage or temperature stresses, compared with a lower chance that vertical cracks will form as a result of restrained vertical stresses. Generally, if shrinkage occurs in the vertical direction, the shrinkage stresses are dissipated by vertical shortening of the wall.

EXAMPLE 19-3   Compute the Capacity of a Bearing Wall

A wall with a clear vertical height between lateral supports of 16 ft and a horizontal length of 25 ft between intersecting walls supports a uniformly distributed factored gravity load of 41 kips/ft, including the self-weight of the wall. The wall is supported on a strip footing that prevents lateral movement of the bottom of the wall. The wall supports a wooden-frame roof deck, which acts as a diaphragm to restrain lateral displacement of the top of the wall. Is an 8-in.-thick wall adequate if $f'_c = 3000$ psi? If so, select reinforcement for the wall. Use the load and resistance factors from ACI 318-02 Sections 9.2 and 9.3.

1. **Check whether the wall thickness is sufficient.** ACI Section 14.5.3.1 limits the thickness of walls designed by the empirical design method to 1/25 of the shorter of the unsupported height or the length. Thus, the minimum thickness is $(16 \times 12)/25$ in. $= 7.68$ in. An 8-in.-thick wall satisfies the minimum thickness given in ACI Section 14.5.3.1. **Use an 8-in. wall.**

2. **Compute the capacity of a 1-ft-wide strip of wall.**

$$\phi P_{nw} = 0.55\phi f'_c A_g \left[1 - \left(\frac{k\ell_c}{32t} \right)^2 \right] \tag{19-12}$$

ACI Section 14.5.2 gives $k = 0.8$ for the end restraints described in the statement of the problem. Walls will seldom have spiral reinforcement. As a result, the wall is an "other" type of member, and ACI 318-02 Section 14.5.2 specifies $\phi = 0.70$ for *bearing walls*. We thus have

$$\phi P_{nw} = 0.55 \times 0.7 \times 3000 \text{ psi} \times 8 \text{ in.} \times 12 \text{ in.} \left[1 - \left(\frac{0.8 \times 16 \times 12}{32 \times 8} \right)^2 \right]$$

$$= 111{,}000 \text{ lb} \times [0.64] \text{ per foot of wall length}$$

$$= 71.0 \text{ kips/ft}$$

This value exceeds the applied factored gravity load of 41.0 kips per ft; thus, the wall has adequate capacity.

3. **Select reinforcement.** ACI Sections 14.3.2 and 14.3.3 require minimum areas of $0.0012A_g$ and $0.0020A_g$ for vertical and horizontal reinforcement, respectively. ACI Section 14.3.4 allows the reinforcement to be placed in one layer, or "curtain," because the wall thickness is less than 10 in. ACI Section 14.3.5 gives the maximum bar spacing parallel to the wall as the smaller of 3 times the wall thickness—3 times 8 in. $= 24$ in.—and an upper limit of 18 in. The maximum spacing of the reinforcement is as follows:

Horizontal spacing of vertical reinforcement—from ACI Section 14.3.2:

$$s_{h,\max} = A_v/(0.0012t) \tag{19-13a}$$

If the required vertical steel is placed in a single layer of vertical No. 4 bars, $A_v = 0.20$ in.^{2}, and the spacing is

$$s_{h,\max} = 0.20 \text{ in.}^2/(0.0012 \times 8 \text{ in.}) = 20.8 \text{ in. on centers. Use 18 in. on centers.}$$

Vertical spacing of horizontal reinforcement—from ACI Section 14.3.3:

$$s_{v,\max} = A_h/(0.0020t) \tag{19-13b}$$

In a similar way, the spacing of the minimum horizontal reinforcement is

$$s_{v,\max} = 0.20/(0.0020 \times 8) = 12.5 \text{ in. on centers}$$

The reinforcement can be chosen with the use of Table A-7 in Appendix A.

Because the vertical reinforcement is not specifically used in the strength design, ACI Section 14.3.6 does not require the vertical bars to be tied in nonseismic regions provided that

(a) the area of vertical steel is less than 0.01 times the gross area of the wall—
$A_{s,\min} = 0.01 \times 8 \text{ in.} \times 12 \text{ in.} = 0.96 \text{ in.}^2/\text{ft.—or}$

(b) the steel is not used as compression steel.

The steel provided has $A_s = 0.2$ in.$^2/(8 \times 18)$ in.$^2 = 0.0014$ times the gross area of the wall.

It is good practice to provide a No. 5 bar vertically at each end of each curtain of wall steel.

Use an 8-in.-thick wall with $f'_c = 3000$ psi, with one curtain of bars at the center of the wall with No. 4 vertical Grade-60 bars at 18 in. o.c. and No. 4 horizontal Grade-60 bars at 12-in. o.c. Add one No. 5 vertical bar at each end of the wall. ∎

19-5 RETAINING WALLS

Reinforced concrete walls are used to resist the horizontal soil pressures on a wall when the surface of the ground is higher on one side of the wall than on the other. These are called *retaining walls*. The most common type is the *cantilever retaining wall* shown in Fig. 19-1c, which consists of a vertical cantilever wall that resists the horizontal earth pressure and a footing that resists the moments at the base of the cantilever and transfers the forces to the ground. ACI Section 14.1.2 specifies that flexural design of retaining walls should be in accordance with the flexural design provisions of ACI Chapter 10, with minimum horizontal wall reinforcement in accordance with ACI Section 14.3.3.

The lateral soil pressure acting on the wall and the bearing capacity of the soil under the wall footing are obtained by consultation with a geotechnical engineer, especially if the wall resists loads acting on the upper ground surface. It is important to provide drainage through the wall to minimize hydrostatic pressure behind the wall.

19-6 TILT-UP WALLS

Industrial buildings and warehouses are frequently constructed from tall thin concrete walls that are cast in a horizontal position on a slab on grade that will become the floor slab for the completed building. After the walls have gained adequate strength, a crane is used to tilt them up to their final vertical position at which time they are connected together. Such walls and buildings are referred to as *tilt-up walls* and *tilt-up buildings*. These constitute a form of construction common in the western states. Most aspects of the design of tilt-up buildings is covered by the report of the ACI Committee 551, *Tilt-up Concrete Construction* [19-16]. ACI Section 14.8, *Alternative Design of Slender Walls*, presents design requirements for the very slender walls used in tilt-up construction. Guidance is also found in the Canadian Concrete Design Manual [19-17].

Nathan [19-18] analyzed thin one-way walls. His design charts are widely used in the design of tilt-up buildings.

19-7 SHEAR WALLS

The term *shear wall* is used to describe a wall that resists lateral wind or earthquake loads acting parallel to the plane of the wall in addition to the gravity loads from the floors and roof adjacent to the wall. Such walls are referred to as *structural walls* in ACI Chapter 21.

Short, one- or two-story shear walls generally carry lateral loads acting as D-regions, as shown in Fig. 19-9a, and can be designed using strut-and-tie models. These walls are called *short or squat shear walls*. If the wall is more than 3 or 4 stories in height, lateral load is resisted mainly by flexural action of the vertical cantilever wall (see Fig. 19-9b), rather than by in-plane strut-and-tie forces. ACI Section 21.1 refers to this type of wall as a *flexural wall*; we shall use the term *slender shear wall*. Although shear walls may be simple planar walls, commonly several wall segments are connected together to act as a three-dimensional unit.

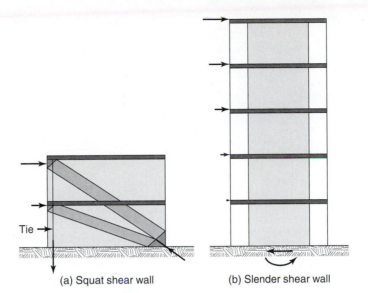

Fig. 19-9
Shear walls.

(a) Squat shear wall   (b) Slender shear wall

Such *wall assemblies* have regular or irregular C, T, L, or H-shaped cross sections with webs and flanges that may enclose spaces in buildings, such as stair wells or elevator shafts.

19-8   LATERAL-LOAD-RESISTING SYSTEMS FOR BUILDINGS

Three common systems for resisting wind or earthquake lateral loads are:

1.   *Moment-resisting frames* are made up of interconnected beams and columns. Lateral loads are resisted by bending of the beams and columns (see Fig. 19-10a and 19-10b). Such frames undergo relatively large lateral deflections. Typically the deflected shape of a moment-resisting frame is as shown in Fig. 19-10a. If all stories have beams and columns with sizes proportional to the shear in the story, the lateral deflection of each story would be similar. To simplify construction, however, the sizes of the beams and columns selected for the lower stories are commonly used throughout, or are changed only every third or fourth story. Hence, the beams and columns in a building tend to be oversized in the upper stories. Moment-resisting frames are used for buildings up to 8 to 10 stories.

In a moment-resisting frame, deflections of the columns and beams both contribute to the sway deflections of the frame. If we isolate the shaded portion of the frame in Fig. 19-10a, the deflection of A relative to B is given by (19-13):

$$\Delta_{AB} = \frac{V\ell^3}{24E_cI_c} + \frac{V\ell^2L}{24E_bI_b} \tag{19-14}$$

where L is the span of the beam, center to center of supports, and ℓ is the height of the column from midheight of the story above the beam to midheight of the story below the beam. The first term on the right-hand side of (19-14) is the deflection due to the bending of the column. The second term is due to the bending of the beam. Shear deformations are not included in (19-14) because they are very small compared to bending deformations in a typical frame member.

Assuming $L = 2\ell$ and $E_cI_c = E_bI_b$, the portion of the relative horizontal deflection due to bending of the beam is *two-thirds* of the total deflection. Frequently in tall frames, it

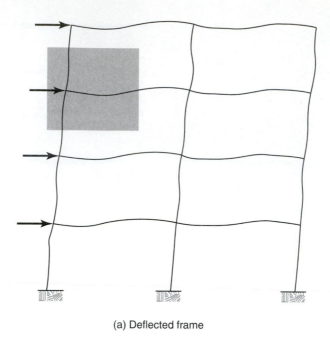

(a) Deflected frame

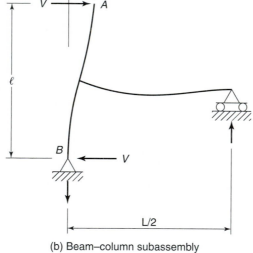

(b) Beam–column subassembly
from part (a)

Fig. 19-10
Moment-resisting frame.

may be impossible to make the beams stiff 1enough to prevent large deflections. In such cases, walls or other stiffening elements are used to control lateral deflections.

2. *Bearing-wall systems* are used for apartment buildings or hotels. A bearing-wall building has a series of parallel transverse shear walls between rooms or apartments. The walls resist lateral loads by flexural action and deflect as vertical cantilevers.

3. *Shear-wall–frame buildings,* shown schematically in Fig. 19-11, are used in buildings ranging from about 8 to about 30 stories. The lateral load is resisted in part by the wall and in part by the frame. Some of the most slender shear-wall–frame structures ever built are described by Grossman [19-19]. He presents some of the wind modeling rationale and two case histories of buildings with heights up to 10 times the least width at ground level.

4.   *Very tall concrete buildings* Structural systems for very tall concrete buildings are discussed in [19-20]. The bottom stories of such a building are shown in Fig. 18-26. The closely spaced columns in the upper stories transfer vertical loads much like a continuous closed *tube* would. At the top of the second floor the vertical loads are transferred to a continuous deep beam called a *transfer beam*. It, in turn, transfers the vertical loads to the ten large columns on the perimeter of the ground floor. In this region the more-or-less uniform compression in the tube is disrupted.

19-9   SHEAR-WALL–FRAME INTERACTION

The division of lateral load between the wall and frame in a shear-wall–frame building can be analyzed by using a frame analysis and a model similar to that in Fig. 19-11. The frame members in the model represent the sums of the stiffnesses of the columns and beams in the building in the bays parallel to the plane of the wall. Similarly, the wall in the model represents the sum of the walls in the structure. The wall and frame are connected by axially stiff link beams at every floor. The link beams in the model shown in Fig. 19-11 may or may not be hinged. In computing internal forces and moments due to the factored loads, the flexural stiffnesses, EI, from ACI Section 10.11.1 may be used. The model in

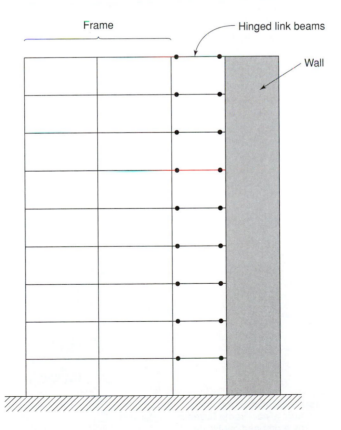

Fig. 19-11
Analytical model of a shear-wall–frame building.

Fig. 19-11 may be acceptable for buildings that are symmetrical in plan. A three-dimensional model is required for an unsymmetrical building, and one would normally be used for all buildings.

The lateral-force analysis of shear-wall–frame buildings must account for the different deformed shapes of the frame and the wall. Due to the incompatibility of the deflected shapes of the wall and the frame, the fractions of the total lateral load resisted by the wall and frame differ from story to story. Near the top of the building, the lateral deflection of the wall in a given story tends to be larger than that of the frame in the same story and the frame pushes back on the wall. This alters the forces acting on the frame in these stories. At some floors the forces change direction, as shown schematically by the range of possible moment diagrams in the wall as shown in Fig. 19-12c. As a result, the frame resists a larger fraction of the lateral loads in the upper stories than it does in the lower stories.

Bounds on the Forces in a Shear-Wall–Frame Building

The relative portions of the lateral loads resisted by the walls and frames in a shear-wall–frame building can be estimated by considering the wall and the frame as two vertical cantilevers, fixed at the bottom and connected via a single extensionally rigid link beam joining the wall and frame at the top, as shown in Fig. 19-12a [19-21]. If the frame is so stiff that it prevents horizontal deflection of the top of the wall, the reaction of the loaded frame at the top of the wall is $3/8 wh_w$, where w is the lateral load per foot of height and h_w is the height of the wall. This is equivalent to the reaction at the pinned end of a uniformly loaded beam having a constant EI that is fixed at one end and pinned at the other. As the lateral stiffness of the frame decreases relative to the lateral stiffness of the wall, the reaction at the top of the wall decreases, approaching zero for a very flexible frame combined with a stiff wall. As a result, the shear-force and bending-moment diagrams for the wall can vary between the approximate limits shown in Fig. 19-12b and c. The total of the shear in the frame and the shear in the wall in a given story must equal the shear due to the applied loads.

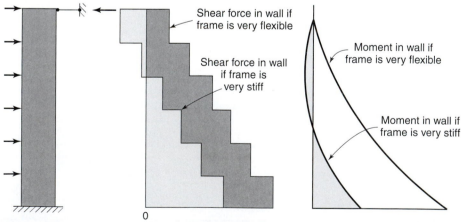

Fig. 19-12
Effect of frame stiffness on shear and moment in the shear wall.

(a) Idealization of the frame from a wall–frame building as a propped cantilever.

(b) Range of shear-force diagrams for wall.

(c) Range of moment diagrams for wall.

19-10 COUPLED SHEAR WALLS

Two or more shear walls in the same plane (or two wall assemblies) are sometimes connected at floor levels by *coupling beams*, so that the walls act as a unit when resisting lateral loads, as shown in Fig. 19-13. The coupling beams frame into the edges of the walls, as shown in this figure. The discussion will be limited to the case of two walls separated by a single vertical line of openings, which are spanned by reinforced concrete *coupling beams*. Walls with more than two lines of openings, as shown in Fig. 19-2, or walls with coupling beams arranged in an irregular fashion need special attention, especially if the widths and heights of the line of openings are irregular.

When the coupled wall deflects, the axes of both parts of the wall at A and A' deflect laterally and rotate through an angle θ, as shown in Fig. 19-14. This results in shearing deflections of the coupling beams than join the two walls. Localized cracking of the beam-to-wall joint reduces the angle the coupling beam must go through where it is attached to the wall. It is customary to assume the effect of these localized deflections can be represented by moving the connection in from the face of the wall by about $h_b/2$, where h_b is the height of the coupling beam.

We shall assume the walls are joined by coupling beams spanning from B to B'. The downward deflection of point B is

$$\Delta_B = \left(\frac{b_w}{2} - \frac{h_b}{2} \right) \tan \theta \tag{19-15}$$

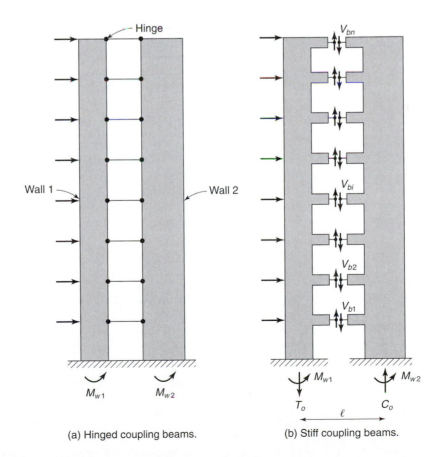

Fig. 19-13
Coupled walls.

(a) Hinged coupling beams.   (b) Stiff coupling beams.

where b_w is the width of the wall. Point B is assumed to be located at half the coupling beam height $(0.5h_b)$ in from the opening. This allows for the reduction in the stiffness of the coupling beam in the vicinity of the point where the coupling beam is attached to the wall [19-21].

The moments and axial forces in the two wall segments in Fig. 19-14 must be in equilibrium with the axial forces, shears, and moments in the entire coupled wall system. The signs of the moments and shears may change over the height of the building, as shown by Fig. 19-12b and c.

Statical System

Figure 19-13a shows two prismatic walls, wall 1 and wall 2, connected at each floor level by link beams hinged at each end. The moments at the bases of the two walls are equal to

$$M_{w1} = M_o \frac{I_{w1}}{I_{w1} + I_{w2}} \tag{19-16}$$

and

$$M_{w2} = M_o \frac{I_{w2}}{I_{w1} + I_{w2}} \tag{19-17}$$

where I_{w1} and I_{w2} are the wall moments of inertia and M_o is the moment at the base of the wall due to factored lateral loads. The total lateral load moments in the walls equal

$$M_{w1} + M_{w2} = M_o \tag{19-18}$$

Figure 19-13b shows the same two walls, except that the walls are now coupled by beams that are continuous with the walls at every floor level and have some flexural stiffness. As the walls deflect laterally, the coupling beams deflect as shown in Fig. 19-14, and shears and moments are generated in the coupling beams. A free-body diagram through the coupling beams halfway between the faces of the two walls has shear forces V_{bi} in each coupling beam, as shown in Fig. 19-13b or Fig. 19-14. There are also axial forces in the beams. For equilibrium of vertical forces, an axial tension, T_o, must be added to the axial forces in the

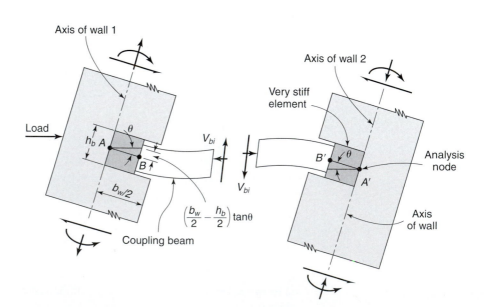

Fig. 19-14
Effect of shear wall deflections on the forces in a coupling beam.

walls at the centroid of the bottom of wall 1 and an axial compression, C_o, at the centroid of the bottom of wall 2, where

$$T_o = \sum_{i=1}^{n} V_{bi} = C_o \tag{19-19}$$

Taking the distance between the lines of action of the forces T_o and C_o as ℓ, we find that the total moment at the base of the coupled wall system is

$$M_o = M_{w1} + M_{w2} + T_o \times \ell \tag{19-20}$$

Equation 19-20 is plotted in Fig. 19-15 for a coupled wall consisting of two walls with $EI_{w1} = 2EI_{w2}$ [19-22]. The vertical axis is the slenderness of the beam, h_b/ℓ_b, where h_b and ℓ_b are the height and the adjusted span of the coupling beams, respectively. The beam slenderness is used here as a measure of the stiffness of the coupling beams. A coupling beam with h/ℓ_b equal to zero has no flexural stiffness, and as a result, the wall moments are divided in proportion to the ratio of the wall stiffnesses, as given by (19-16) and (19-17). As the flexural stiffness of the coupling beams increases, the shears in them increase. As a result, the fraction of the overturning moment resisted by the axial force couple $T_o \times \ell$ increases asymptotically. A major effect of the coupling beams is to reduce the moments M_{w1} and M_{w2} at the bases of the two walls. This makes it easier to transmit the wall reactions to the foundation. The coupling beams also act to reduce the lateral deflections. If the beams are perfectly rigid, the two walls act as one wall. Figure 19-15 illustrates trends only. The actual division of M_o into M_{w1}, M_{w2}, and $T_o \times \ell$ depends on more variables than just h_b/ℓ_b.

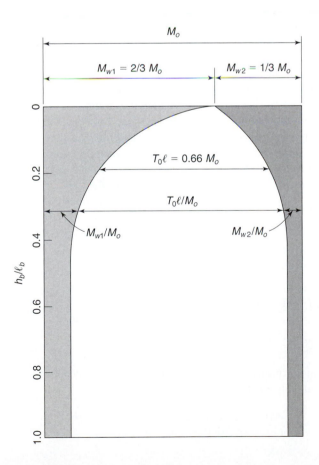

Fig. 19-15
Effect of the stiffness of the coupling beams on the moments in walls 1 and 2, $I_{w1} = 2I_{w2}$. (From [19-22].)

Coupling beams may be rectangular beams, T beams, or portions of the floor slab [19-23], [19-24]. In seismic regions, short coupling beams may have diagonally placed reinforcement. This is discussed in Chapter 20.

For seismic design, the Canadian concrete code [19-17] defines a *coupled wall* as one in which $T_o \times \ell$ is at least 66 percent of M_o. In Fig. 19-15, this occurs when h_b/ℓ_b is about 0.2. If $T_o \times \ell$ is less than 66 percent of M_o, the wall is called a *partially coupled wall*.

Analysis of Coupled Walls

Before modern structural analysis programs, the individual coupling beams shown in Figs. 19-13b and 19-14 were replaced for analysis by a series of closely spaced *laminae*, each having a unit height and stiffness I_b/h_s, where h_s is the story height. This allowed a closed-form solution of the forces in the laminae and the walls. Such analyses were limited to a few uniform wall layouts [19-22]. Computerized structural-analysis programs have made it possible to model coupled structural walls, as shown in Fig. 19-13b, and to compute the forces and moments in the coupling beams directly. As a result, laminar analyses are seldom used now.

In a frame analysis to determine factored moments for design, the member stiffnesses may be based on ACI Section 10.11.1. The coupling beams are joined to hypothetical members with high values of the moment of inertia, I, between the face of the wall and the centerline of the wall, as shown in Fig. 19-14. Short, deep coupling beams develop both flexural and shear deflections. The shear deflections can be included by replacing the I_b of the coupling beam between the walls with

$$I_{eff} = \frac{I_b}{1 + 2.8\left(\dfrac{h_b}{\ell_b}\right)^2} \qquad (19-21)$$

This equation comes from adding the moment deflections and shear deflections of the beam, where h_b and ℓ_b are the depth and the span of the coupling beam from face to face of the walls. The second term in the denominator of (19-21) accounts for the shear deflections of the beam.

Floor slabs may serve as coupling beams. Their stiffness can be based on a slab with a width perpendicular to the wall equal to the wall thickness plus half of the width of the opening, $\ell_b/2$, between the walls, added on each side of the opening [19-23], [19-24], [19-25]. In tests of shear walls coupled by slabs, the specimens failed by punching-shear failures in the slab around the ends of the walls. Under cyclic loads, the stiffness of slabs serving as coupling beams decreased rapidly.

Figure 19-16a, b, and c shows the distributions of moments in the walls, the axial-force couple, and the shears in the coupling beams for a typical coupled wall [19-17] where $I_{w1} = I_{w2}$. Typically, the maximum shear in the coupling beams occurs at about a third of the height above the base. The sawtooth shape of Fig. 19-16a results from the end moments in the coupling beams.

19-11 DESIGN OF STRUCTURAL WALLS—GENERAL

Layout of Building

Major considerations in selecting a structural system for a multistory building with structural walls are the following:

(a) The building must have enough rigidity to withstand the service loads without excessive deflections or vibrations.

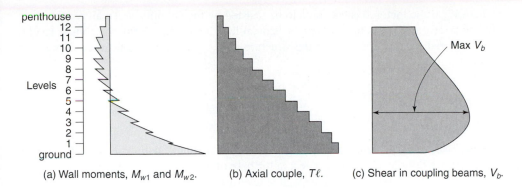

(a) Wall moments, M_{w1} and M_{w2}.   (b) Axial couple, $T\ell$.   (c) Shear in coupling beams, V_b.

Fig. 19-16
Typical distribution of wall moments, axial-force couple, and shear in the coupling beams, $I_{w1} = I_{w2}$. (From [19-17].)

    (b)   It is desirable that the wall be loaded with enough vertical load to resist any uplift of parts of the wall foundations due to lateral loads.

    (c)   The locations of frames and walls should minimize torsional deformations of the building about the vertical axis of the building. (See also Section 20-3.)

    (d)   The walls must have adequate strength in shear, and in combined flexure and axial loads.

    (e)   The wall thickness or cover on the reinforcement may be governed by the fire code.

Diaphragms

Lateral loads are transferred to the lateral-force resisting system by the floor and roof slabs serving as horizontal *diaphragms* that act as wide, flat beams in the plane of the floor or roof system. The diaphragms must have a tension flange, a compression flange, and a web capable of transmitting the lateral loads. Because the direction of the lateral load changes to and fro during wind or earthquake cycles, the compression and tension flanges of the diaphragm must be able to accommodate changes in the sign of the loading. Notches or other discontinuities in the tension and compression flanges of the diaphragm should be avoided or reinforced to transmit the forces around the discontinuities. Diaphragms are also discussed in Section 20-10.

Forces on Walls

For simplicity, a designer may assume that all the lateral load is resisted by the walls and none by the frames. This assumption is acceptable for those buildings up to about 12 stories with more or less symmetric plans. Design must consider the factored shear, V_u, the moment, M_u, and the axial force, P_u, at each floor level, i. At any level x, measured from the base of the wall in a building with n floors or roofs, the factored shear is

$$V_{ux} = \sum_{i=x}^{n} H_{ui} \qquad (19\text{-}22)$$

where V_{ux} is the factored shear force to be resisted by the columns and walls of the story below level x and H_{ui} is the factored horizontal load acting on the floor or roof at level i. The factored shear at the base of the structure is V_{uo}. Similarly, the factored moment at floor level x is

$$M_{ux} = \sum_{i=x}^{n} (H_{ui}h_{xi}) \qquad (19\text{-}23)$$

where h_{xi} is the vertical distance from floor level x to the line of action of force H_{ui}, at floor level i above floor level x. The moment at the base of the structure is M_{uo}.

The factored axial force in the wall at level x is

$$P_{ux} = \sum_{i=x}^{n} W_{ui} \qquad (19\text{-}24)$$

where W_{ui} is the factored vertical load from the floor area tributary to the wall at level i, plus the factored weight of the wall from level i to level $i + 1$.

Effects of Torsion

Figure 20-5a shows the effect of torsion due to the resultant applied load being offset from the center of resistance of the building. This should be avoided as much as possible.

Wall Foundations

The foundations at the base of the wall must be able to transfer the shear, moment, and axial force as required from the base of the wall to the supporting soil or rock. If the axial gravity forces P_u in the wall are small, the moments at the base of the wall may cause uplift under one side of the shear-wall footing, as shown in Fig. 19-17a. Both an axial load and a moment are resisted by the soil, and as a result, the soil pressure will vary across the width of the wall footing. For a vertical load acting at the kern of the footing area, or for a vertical load and moment acting on the footing, the soil pressure will range from smaller than average (as low as zero) on one side of the footing to highter than the average stress (up to twice as high) at the other side larger moments will result in computed tensions between the footing and soil. Because tensile uplift stresses are difficult to resist, they must be avoided. If the footing size becomes excessive, possible solutions are:

(a) replace the rectangular footing with an H-shaped footing, to increase the radius of gyration $\left(r = \sqrt{\dfrac{Inertia}{Area}} \right)$ of the footing;

(b) use pile or caisson foundations;

(c) use coupled shear walls to widen the footprint of the wall and divide the moment to be transferred into the three components shown in Fig. 19-13b [19-15];

(d) attach the base of the wall to horizontal outrigger beams in the basement which extend the foundation to receive the vertical loads from adjacent frame columns so that there is no uplift; or the uplift is counteracted (see Fig. 19-17b) or

(e) attach the base of the wall to the ground-floor diaphragm and basement floor to provide a horizontal force couple to react to the moment at the base of the wall.

Wyllie [19-26] and Paulay and Priestly [19-6] both discuss shear-wall foundations.

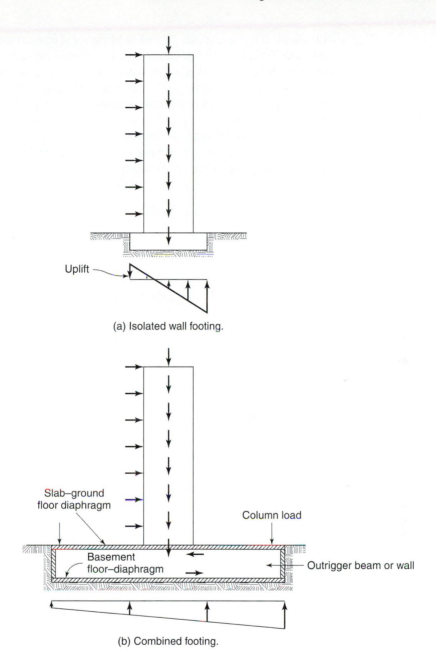

(a) Isolated wall footing.

Uplift

Slab–ground
floor diaphragm

Column load

Basement
floor–diaphragm

Outrigger beam or wall

Fig. 19-17
Wall foundations.

(b) Combined footing.

Required Size of Wall

In choosing a structural wall section for a given building, the wall must

    (a)   have enough strength to resist the factored moments, shears, and axial loads acting on it; and

    (b)   have enough stiffness to limit the lateral deflections.

There is no widely accepted way of doing this. A *rough estimate* of the minimum wall stiffness, *EI*, required to limit the lateral deflections to an acceptable value can be made by considering the walls as a vertical cantilever with a constant *EI* over the height, loaded with a constant wind load over the height, as shown in Fig. 19-17a. Wall thicknesses, and thus the structure's *EI* would generally decrease as the moments and shears decrease near the top of the structure, we shall assume it is constant over the building height and equal to the sum of the EI values of the walls in the bottom story. This cantilever will be loaded with a constant, uniform service wind load of w_s lb per foot of height. The wind load is the product of the length and height of the building and the wind pressure per foot of height, taken equal to the algebraic sum of the windward wind pressures and leeward wind suctions (see Example 19-4) evaluated at the top of the building. It is customary to limit the relative horizontal deflection of the top and bottom of a story to a fraction of the height of the story, expressed as $\Delta = h_s/(200 \text{ to } 500)$, where Δ/h_s is called the *story drift*. This limit is expressed in terms of the slope of the wall relative to the vertical in any story throughout the height of the building. The largest slope in a shear wall or shear-wall–frame building is generally at the top of the structure and is given by [19-19]

$$\text{(slope)} = \text{(relative lateral deflection, top and bottom of story)/(height of story)} \quad (19\text{-}25)$$

$$\text{(slope at top)} = \text{(rotation of base of wall)} + \text{(area of } M/EI \text{ diagram from base to top)}$$
$$(19\text{-}26)$$

If the base is fixed against rotation and assuming that $EI = EI_{base}$ is constant over the height of the building, h_w gives the slope at the top relative to the undeflected position:

$$\text{slope at top} = \frac{w_s h_w^3}{6E_c I_g} \quad (19\text{-}27)$$

Here,

> h_w is the height of the top of the wall
>
> E_c is the modulus of elasticity of concrete, psi (since all other units are in pounds and feet, the modulus of elasticity in psi is multiplied by $(144 \text{ in.}^2/\text{ft}^2)$ to change the psi units to psf)
>
> I_g is the uncracked moment of inertia of the wall or walls at the bottom of the structure (an uncracked wall is taken to be representative of service-load levels, because codes limit service-load deflections and the walls are generally not severely cracked under service loads); ACI Section 10.11.1 gives $I = 0.70I_g$ ft^4 for an uncracked wall
>
> w_s is the unfactored wind load per vertical foot of wall, evaluated at the top of the wall, in lb/ft. of height

Limits on Story Drift

The relative lateral deflection of the top and bottom of a story divided by the height of the story is called the *story drift*. ASCE 7 and the 2002 ACI Code do not limit the lateral deflection or the story drift. The National Building Code of Canada [19-17] limits the maximum *story drift* under unfactored loads to less than 1/500 of the story height. The maximum acceptable drift depends on the ability for occupants of the building to perceive the motion of the building during a major windstorm. This storm may be chosen as a rare event, such as the maximum annual windstorm or the maximum windstorm in one tenancy of the building—say, every three to eight years. Grossman discusses design for wind induced movement of slender concrete buildings [19-19].

If a wall having height h_w and constant wall stiffness EI is fixed at the base, the maximum drift will be in the top story. Setting the service-load story drift from (19-27) equal to 1/500, we can compute the minimum total I_g for the walls parallel to the direction of the wind load as

$$\Sigma(0.70I_g) = \frac{500w_s h_n^3}{6E_c} \quad (19\text{-}28)$$

Once the minimum ΣI_g has been estimated, walls can be selected to give ΣI_g equal to or greater than what is required by (19-28). Although this analysis is intentionally simplified, it gives an order-of-magnitude estimate of story drift. It does not consider torsional load-ings on the structure, and it assumes that the wind pressure is constant over the height of the building. At the stage in design where the sizes of shear walls are initially chosen, these details are not yet known. A better estimate of Δ/h can be obtained by calculating the sec-ond term on the right-hand-side of Eq. (19-26) more accurately based on a better estimate of the EI and moment.

For planar walls, I_g is the sum over all the walls and wall segments at the base in the direction being modelled.

$$\Sigma I_g = \frac{\sum t \ell_w^3}{12} \tag{19-29}$$

Minimum Wall Thickness

The minimum thicknesses given in ACI Section 14.5.3 are intended to apply only to bear-ing walls designed via the empirical design method from ACI Section 14.5. Although the ACI Code does not limit the thickness of other types of walls, Examples 19-1 and 19-2 suggest that minimum thickness limits are needed to prevent buckling of the walls or the flanges. The author would limit the thickness of walls with rectangular cross sections to the absolute minimum thickness of 1/20 of the unsupported height of the wall, ℓ_u, given in ACI Section 14.5.3. Preferably, the wall thickness should not be less than 1/15 of the un-supported height. The wall thickness must be large enough to allow the concrete to be placed without honeycombing.

Reinforcement in Shear Walls

Distributed and Concentrated Reinforcement

The reinforcement in a shear wall is generally made up of:

(a) *Distributed* horizontal and vertical reinforcement spread uniformly over the length between the boundary elements and over the height of the wall.

(b) *Concentrated* vertical reinforcement is located in boundary elements at or near the edges of the wall and is tied in much the same way that column cages are.

Minimum Wall Reinforcement

In addition to the distributed reinforcement required by ACI Section 14.3, several other ACI sections require minimum amounts of distributed horizontal and vertical steel that may apply to walls. These are given in Table 19-3.

ACI Section 14.3 requires minimum areas for distributed vertical reinforcement in walls to be equal to $0.0012A_g$ and the minimum area of distributed horizontal rein-forcement in walls to be equal to $0.0020A_g$. This steel can be placed in two layers or *curtains* of distributed vertical and horizontal reinforcement with the bars in the two curtains tied together at intervals, but not enclosed in ties. One reinforcement curtain is allowed at the midplane of walls having thicknesses of 10 in. or less (ACI Section 14.3.4). This steel cannot be tied because there is only one curtain. Thicker walls require two cur-tains of reinforcement, each consisting of not less than half of the total minimum rein-forcement required in each direction. Each of these two layers is placed not more than one-third of the wall thickness from the surface. It is desirable to have the steel in two layers close to the sides of the wall because the internal lever arm, $j_u d$, for weak-axis bending is much smaller if the reinforcement is at the center of the wall. ACI Section 14.3.5 requires that distributed vertical and horizontal bars be spaced not further apart

TABLE 19-3 Minimum Reinforcement in Walls Compared to Other Members[a]

Reason	ACI Section	Requirement	Maximum Spacing
Shrinkage and temperature	7.12.2.1	(b) Slabs where Grade-60 deformed bars or welded-wire fabric (plain or deformed) are used: 0.0018	Five times the slab thickness, no farther apart than 18 in.
Minimum flexural steel	10.5.4	The minimum area of tensile reinforcement in the direction of the span is the same as by 7.12	Three times the thickness, no farther apart than 18 in.
Minimum flexural steel	10.5.3	The requirements of 10.5.1 and 10.5.2 need not be applied if at every section the area of reinforcement provided is at least one-third greater than that required by analysis.	Same as in 10.5.1 and 10.5.2
Deep beams	10.7.3	Minimum flexural tension reinforcement shall conform to 10.5.	
Deep beams	11.8.4	The area of shear reinforcement perpendicular to the span shall not be less than $0.0025b_w s$	s shall not exceed $d/5$ or 12 in.
	11.8.5	The area of shear reinforcement parallel to the span not be less than $0.0015b_w s_2$	s_2 shall not exceed $d/5$ or 12 in.
Walls	11.10.9.2	Ratio ρ_n of horizontal shear reinforcement area to gross concrete area shall not be less than 0.0025.	Spacing of horizontal shear reinforcement s_2 shall not exceed $\ell_w/5$, $3t$, or 18 in.
	11.10.9.4	Ratio ρ_n of vertical shear reinforcement shall not be less than $\rho_n = 0.0025 + 0.5(2.5 - h_w/\ell_w) \times (\rho_h - 0.0025)$ nor 0.0025, but need not be greater than the required horizontal shear reinforcement.	Spacing of vertical shear reinforcement s_1 shall not exceed $\ell_w/3$, $3t$, or 18 in.
Slab reinforcement	13.3.1	Area of reinforcement in each direction shall be determined from moments at critical sections, but shall not be less than required by 7.12	Spacing of reinforcement at critical sections shall not exceed two times the slab thickness
Minimum reinforcement— Walls	14.3.2	Minimum ratio of vertical reinforcement area to gross concrete area shall be: (a) 0.0012 for deformed bars not larger than No. 5 with a specified yield strength not less than 60,000 psi.	14.3.5 Vertical and horizontal reinforcement shall not be spaced farther apart than three times the wall thickness, nor farther apart than 18 in.
	14.3.3	Minimum ratio of horizontal reinforcement area to gross concrete area shall be: (a) 0.0020 for deformed bars not larger than No. 5 with a specified yield strength not less than 60,000 psi.	

[a]If more than one of these sections apply, the sections requiring the largest minimum area and the smallest spacing shall govern.

than three times the wall thickness or 18 in., whichever is less. ACI Section 11.10.9.3 allows the maximum spacing of horizontal shear reinforcement to be the smallest of $\ell_w/5$, $3t$, and 18 in. ACI Section 11.10.9.5 allows the maximum spacing of vertical shear reinforcement to be the smallest of $\ell_w/3$, $3t$, and 18 in.

Although it can be argued that ACI Section 10.7.3 does not apply to shear walls, this section requires deep beams to have minimum flexural reinforcement satisfying ACI Sections 10.5.1 and 10.5.2. These sections require a significant amount of steel. Frequently this steel can be reduced, if at every section, the area of flexural reinforcement provided is at least one-third greater than the area of flexural steel required by analysis. Whether this is mandatory or not, the principle of providing extra resistance in areas of unrestrained cracking is sound.

In calculating the cracking load, it should be noted that the axial loads increase about linearly with the distance below the top of the building, while the moments increase at a power of this distance. The relative magnitudes of axial load and moment will affect the cracking load at points near the base of the wall.

Fracture of reinforcement has been observed in shear walls. (See Section 19-2.)

Ties for Vertical Reinforcement

ACI Section 14.3.6 specifies that the distributed vertical reinforcement *need not* be enclosed by lateral ties

    (a)   if the vertical reinforcement area is not greater than 0.01 times the gross concrete area or

    (b)   if the vertical reinforcement is not required as compression reinforcement.

Many designers interpret part (b) of ACI Section 14.3.6 to mean that wall steel satisfying (a) in this list that is not enclosed in transverse ties should be ignored in strength calculations if:

    (b1)   if it is stressed in compression under static loads, or

    (b2)   if it is stressed by cyclic loads that cause compression in the bars.

Distributed wall steel placed in two separate curtains can be tied through the thickness of the wall using stirrups or through-the-wall cross-ties engaging reinforcement on both faces of the wall. Although ACI 318-02 does not specify what fraction of the bars should be tied in this manner, it is customary for such ties to engage every second or third bar each way on both faces. (See [19-2].)

In many cases the distributed vertical reinforcement has enough moment capacity to resist the wind load moments. If not, concentrated reinforcement is provided in boundary elements at the ends of the walls or at the intersections of walls. These elements contain vertical steel satisfying the minimum reinforcement requirements from ACI Section 10.9.1 based on the area of the boundary elements, A_{gb} rather than on the gross area of the wall, A_g. This steel is enclosed by ties satisfying ACI Section 7.10.5. As a result, the steel is assumed to be restrained against buckling, and thus able to resist compressive bar forces. The boundary elements may be the same thickness as the rest of the wall, or they may be thicker.

Although ACI Chapter 14 does not require concentrated reinforcement at the ends of the walls it is good practice to provide some.

Transfer of Wall Load through Floor Systems

In tall buildings, the strength of the concrete in the walls may be higher than the strength of the concrete in the floor system. Walls may be of high-strength concrete to reduce the lateral deflections and the wind-induced sway vibrations of the building. On the basis of tests of column–floor joints, ACI Section 10.15 allows an increase in the effective strength of the floor concrete in locations where the floor concrete is confined on all sides by the higher-strength column concrete. This effect is smaller in edge and corner column joints than in interior joints, because there is less confinement of the joint concrete at edge or corner columns. The lack of confinement is even more pronounced in wall–floor joints, because a greater length of joint concrete is unconfined. In the absence of tests, it is recommended that the strength of wall-to-floor joints be based on the lower of the wall and floor strengths.

19-12 FLEXURAL STRENGTH OF SHEAR WALLS

Slender Shear Walls and Squat Shear Walls

The behavior and strength of shear walls separates them into two groups for design, as shown in Fig. 19-9.

1. Walls that extend $2\ell_w$ or more above the point of maximum moment in the wall resist lateral forces by flexural action and beam-like shear action. These are referred to as *slender shear walls*. (See Fig. 19-9b.) They are designed according to ACI Chapters 10 and 11.

2. Walls that extend less than $2\ell_w$ above the point of maximum moment in the wall are assumed to resist lateral forces by strut-and-tie action [19-20], [19-21], [19-22]. These are referred to as *squat walls* and are designed by using strut-and-tie models according to ACI Appendix A.

Flexure—Nominal Strength, Factored Strength, and Resistance Factors

Cross sections in a wall are designed to satisfy

$$\phi M_n \geq M_u \tag{19-30}$$

$$\phi P_n \geq P_u \tag{19-31}$$

$$\phi V_n \geq V_u \tag{19-32}$$

where M_n is the nominal resistance based on the specified material strengths, M_u is the required resistance computed from the factored loads, U, and so on. The strength-reduction factor, ϕ, comes from ACI 318-02 Section 9.3.2 for flexure and axial loads and from ACI 318-02 Section 9.3.2.3 for shear. The factored loads are from ACI 318-02 Section 9.2.1. Alternatively, the ϕ factors and load factors may come from ACI 318-02 Sections C.2 and C.3. Example 19-4 uses the load and ϕ factors from ACI-02 Sections 9.2 and 9.3.

Strength-Reduction Factors for Flexure and Axial Load—ACI 318-02 Section 9.3.2

The strength-reduction (ϕ) factors for combined flexure and axial loads for a flexural shear wall vary, depending on the maximum steel strains anticipated at ultimate load. As explained in Sections 4-3, 5-2, 5-3 and 11-4, the calculation of strength-reduction factors, ϕ, is based on the strain, ε_t, in the *extreme-tension layer* of steel at the depth, d_t, of the layer of steel located farthest from the extreme-compression fiber.

Axial load at the tension controlled limit load on a rectangular wall. The strength-reduction factor, ϕ, can be computed directly from the computed strain in the extreme-tension layer of reinforcement, if this is noted during design. Alternatively, the choice of whether a given wall cross section is a tension-controlled section, a transition section, or a compression-controlled section can be established by comparing the factored axial load on the section, P_u, with the axial load corresponding to the *tension-controlled limit load*, P_{TCL}, and comparing it with the *compression-controlled limit load*, P_{CCL}, for the given wall. ACI Section 10.2.3 defines the tension-controlled limit load as the load causing a strain distribution having a maximum strain of 0.0030 in./in. *in compression* on the most compressed face of the member, when the steel strain in the extreme layer of tension reinforcement reaches

0.0050 in./in. tension. (See ACI Section 10.3.4.) For Grade-60 steel, from similar triangles, the neutral axis will be located at $c/d_t = 0.375d_t$ from the compressed face, where d_t is the distance from the extreme-compression fiber to the centroid of the layer of bars farthest from the compression face of the member. Generally, the reinforcement will be located symmetrically about the midlength of a rectangular shear wall and the extreme-tension layer of steel will be the layer closest to the tension face of the wall or the boundary element. For a given load direction, the distributed vertical reinforcement will yield in compression at one end of the wall and the steel will yield in tension at the other. The axial forces in the steel partially cancel each other out, leaving the axial load capacity $P_{TCL} \approx C_{TCL}$ at the tension-controlled axial-load capacity of the concrete for the assumed strain diagram given by Fig. 4-18e.

For a rectangular cross section, the compression axial load in the concrete, C_{tcl}, at the tension-controlled limit is

$$C_{TCL} = 0.85f_c'\beta_1(0.375d_t)t \tag{19-33a}$$

$$= 0.319\beta_1 f_c'd_t t \tag{19-33b}$$

which can be rounded off to

$$C_{TCL} \approx 0.32\beta_1 f_c'd_t t \tag{19-33c}$$

If the factored axial load P_u on the wall with rectangular cross section is less than or equal to the axial load, C_{TCL}, from (19-33a) or (19-33b), the wall is tension-controlled, and $\phi = 0.90$.

For nonrectangular sections, the strain diagram for the tension-controlled limit is used to locate the depth of the neutral axis and the corresponding area of the compression zone is used to compute the compression and tension forces at ultimate load.

Compression-controlled limit load on a rectangular wall. The compression-controlled limit corresponds to a strain distribution with the neutral axis at $0.6d_t$ and, for a rectangular cross section, a concrete compression force of

$$C_{CCL} = 0.85\beta_1 f_c'(0.60d_t)t \tag{19-34a}$$

which becomes

$$C_{CCL} = 0.51\beta_1 f_c'd_t t \tag{19-34b}$$

This can be rounded off to

$$C_{CCL} \approx 0.50\beta_1 f_c'd_t t \tag{19-34c}$$

Compression-controlled sections with the compression zone confined by spirals have $\phi = 0.70$. Other walls have $\phi = 0.65$ where the term "other" refers to columns without spiral reinforcement. The higher ϕ-value reflects the belief that a spiral column is more ductile than a tied column.

Flexural Strength of Rectangular Walls with Uniform Curtains of Vertical Distributed Reinforcement

Code guidance on the use of vertical distributed reinforcement in walls loaded cyclically is ambiguous. In seismic regions, the loads resisted by vertical bars that are not tied, are ignored.

ACI Section 14.3.6 is more lenient. It requires that vertical bars be tied (a) if the total distributed vertical reinforcement, A_s, equals or exceeds $0.01A_g$, or (b) if the vertical reinforcement is either not used or not included in the calculations. The following strength analysis applies to walls with two curtains of reinforcement with ties through the wall to the other curtain of bars. The equations in this subsection also apply if the area of steel is less than 0.01, or if the forces in the bars satisfy ACI Section 14.3.6 or are not stressed in compression.

ACI Section 21.4.4.3 suggests that, in seismic regions, column reinforcement would be adequately tied if the center-to-center spacing of cross-ties or the legs of hoops did not exceed 14 in. Given this guidance, the author believes that the vertical reinforcement in a nonseismic wall can be taken as "tied" if at least every second bar in a curtain of reinforcement is tied through the wall to a bar in the other curtain of steel, near the opposite face.

Figure 19-18a shows a wall with a rectangular cross section. Because the steel in this wall is placed in one layer and is not tied, it may not be included in some strength calculations. It will be idealized as a steel plate (Fig. 19-18b) with a total area, A_s, equal to the total area of the distributed vertical steel. Cardenas and Magura [19-8], [19-9] present the following equations for the moment capacity of the wall in Fig. 19-18b:

Location of neutral axis. The ratio of the distance from the extreme-compression fiber to the neutral axis (axis of zero strain), c, to the horizontal length of the wall, ℓ_w, is

$$\frac{c}{\ell_w} = \frac{\omega + \alpha}{2\omega + 0.85\beta_1} \tag{19-35}$$

where β_1 is the stress-block parameter from ACI Section 10.2.7.3.

$$\omega = \frac{A_s f_y}{\ell_w t f'_c} \tag{19-36}$$

and

$$\alpha = \frac{P_u}{\ell_w t f'_c} \tag{19-37}$$

Nominal moment capacity of rectangular walls. Cardenas and Magura [19-8] developed (19-38) for planar walls reinforced as shown in Fig. 19-18.

$$M_n = 0.5A_s f_y \ell_w \left(1 + \frac{P_u}{A_s f_y}\right)\left(1 - \frac{c}{\ell_w}\right) \tag{19-38}$$

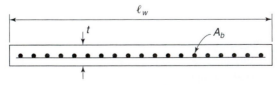

(a) Wall with one curtain of reinforcement.

Fig. 19-18
Plan of a wall with one curtain of reinforcement.

(b) Wall with reinforcement idealized as a steel plate.

Good agreement was reported [19-8], [19-9] between the strengths measured in tests and the strengths computed from (19-38).

Once the reinforcement has been selected, the value of M_n could be checked against a strain-compatibility solution, as given in Section 5-5 of this book.

Moment Resistance of Wall Assemblies, Walls with Overhanging Flanges, and Walls with Boundary Elements

Frequently, shear walls have webs and flanges that act together to form H-, C-, T-, and L-shaped wall cross sections referred to as wall *assemblies*. The effective flange widths can be taken from ACI Sections 8.10.2 and 8.10.3. In regions subject to earthquakes, ACI Section 21.7.5.2 limits the flange widths to the smaller of

(a) half the distance to an adjacent web and

(b) 25 percent of the total height of the wall above the section under consideration.

These constraints give flange widths comparable to the width-to-thickness ratios from 7 to 12 computed in Example 19-2. We shall use the limiting flange thicknesses from ACI Section 21.7.5.2 for both seismic and nonseismic walls.

Frequently, the thickness is increased at the ends of a wall to give room for concentrated vertical reinforcement that is tied like a tied column (see ACI Section 7.10.5). Increased thickness also helps to prevent buckling of the flanges [19-20]. Regions containing concentrated and tied reinforcement are known as *boundary elements*, regardless of whether or not they are thicker than the rest of the wall.

Nominal Moment Capacity of Wall Assemblies

Paulay and Priestly [19-6] present the following method of computing the required reinforcement in a wall assembly or a nonplanar wall. Figure 19-19 shows a plan of a wall consisting of a web and two flanges. The wall is loaded with a factored axial load, P_{ua}; a factored load and a moment, M_{ua}, that causes compression in flange 1; and a shear, V_u, parallel to the

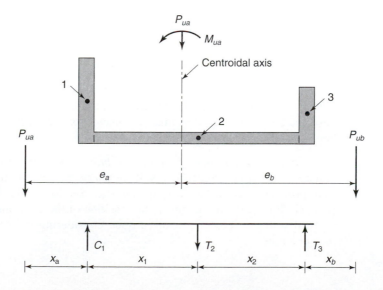

Fig. 19-19
Reinforcement Forces acting on a wall segment. (From *Seismic Design of Reinforced Concrete and Masonry Buildings*, Paulay and Priestly, Copyright John Wylie & Sons, 1992. Reprinted by permission of John Wylie & Sons, Inc.)

web. The axial load and the moment act through the centroid of the area of the wall. The moment can be replaced by an eccentric axial load located at

$$e_a = \frac{M_{ua}}{P_{ua}}$$
(19-39)

from the centroid, which is equivalent to it acting at $x_a = e_a - x_1$ from the centroid of the flange that is in compression.

Because of the shear, V_u, the vertical reinforcement in the web will be assumed to be the minimum steel required by ACI Section 11.10.9.4. The calculations will be simplified by assuming that all of the web steel yields in tension. The compression strength of this steel will be ignored. We can assume this because under cyclic loads, the neutral axis will alternately be close to the left and the right ends of the web. The steel in the web resists an axial force of

$$T_2 = A_{s2}f_y$$
(19-40)

where A_{s2} is the area of distributed steel in wall element 2, the web. Summing moments about the centroid of the compression flange gives the following equation for the tension, T_3, in the right-hand flange:

$$T_3 \approx \frac{P_{ua}x_a - T_2x_1}{x_1 + x_2}$$
(19-41)

In a similar manner, the force in the tension reinforcement, required in the left-hand flange for the axial load and the moment M_{ub}, causing compression in the right-hand flange can be computed as

$$T_1 \approx \frac{P_{ua}x_b - T_2x_2}{x_1 + x_2}$$
(19-42)

The required area of concentrated vertical reinforcement in the flanges can be computed from equations similar to (19-41) and (19-42).

After the reinforcement has been chosen, the moment resistance of the wall should be computed by using the strain-compatibility analysis given in Section 5-5. The factored nominal moment capacity of the wall should be

$$\phi M_n \geq M_u$$
(19-30)

where ϕ depends on the steel strain at ultimate load.

Biaxially loaded walls. A wall is said to be biaxially loaded if it resists axial load plus moments about two axes. One method of computing the strength of such walls is the equivalent eccentricity method presented in Section 11-7. In this method, a fraction between 0.4 and 0.8 times the weak-axis moment is added to the strong-axis moment. The wall is then designed for the axial load and the combined biaxial moment treated as a case of uniaxial bending and compression.

Inelastic resistance of wall assemblies. Strictly speaking, the elastic moment resistance of an unsymmetrical wall should be computed allowing for moments about both principal axes of the cross section (See Section 11-7). This is not widely done in practice. It is generally assumed that cracking of the walls and the proportioning of the vertical wall reinforcement can be done considering moments about one orthogonal axis at a time.

Three moment conditions are encountered as a shear wall is loaded to failure:

(a)   Initially, elastic moments occur. These moments act around the principal axes of the gross concrete cross section.

(b)   After cracking, the axes about which the moments occur are controlled in part by the placement of the reinforcement. If the steel is placed to resist bending about the elastic axes, the resisting moments will tend to be about axes corresponding to the elastic axes.

(c)   If the steel is placed in concentrations away from the elastic axes, the yield moments may require a significant shift in axes.

Shear Transfer between Wall Segments in Wall Assemblies

For the flanges to work with the rest of the cross section of a wall assembly, so-called "vertical shear stresses" must exist on the interface between the flange and web, even when the wall and the wall segments are constructed monolithically. The stresses to be transferred are calculated in the same manner as for a composite beam, by using (17-15) and (17-17). The reinforcement should satisfy ACI Section 11.7, *Shear Friction*.

19-13 SHEAR STRENGTH OF SHEAR WALLS

The design of structural walls for shear in seismic regions is covered in ACI Section 11.10, *Special Provisions for Walls*. The basis for this code section was a series of tests of one-third-size, planar, shear walls, reported in [19-5], [19-8], [19-9]. The test specimens were divided between flexural shear walls with ratios of $M_u/(V_u \ell_w)$ of 1.0, 2.0, and higher and short shear walls with $M_u/(V_u \ell_w) = 0.50$. The basic shear-design equations are similar to those for the shear design of prestressed concrete beams [19-27], [19-28]:

$$\phi V_n \geq V_u \tag{19-32}$$
$$\text{(ACI Eq. 11-1)}$$

$$V_n \geq V_c + V_s \tag{19-43}$$
$$\text{(ACI Eq. 11-2)}$$

$$V_s \geq \left(\frac{V_u}{\phi} - V_c \right) \tag{19-44}$$

V_c for Slender Shear Walls

The design procedure differentiates between *slender shear walls*, with $h_w/\ell_w \geq 2$, and *squat walls*, with $h_w/\ell_w < 2$ (See Fig. 19-9.) Sometimes, similar limits are expressed in terms of whether $M_u/(V_u \ell_w)$ is greater than 1.0 or less than 1.0, respectively. In tests, short walls displayed D-region behavior in which loads are resisted by a strut-and-tie mechanism. Slender shear walls developed flexure-shear cracks in much the same manner as prestressed concrete beams do. Equation (19-45) assumes that the shear carried by the concrete, V_c, is equal to the inclined cracking shear and further assumes that the shears carried by the concrete and steel are fully additive. Equation (19-45) assumes that the first flexural crack occurs at a distance $\ell_w/2$ above the section of maximum moment in the wall. The shear, V_c, carried by the concrete is taken equal to the following sum:

(a) the shear required for a flexural crack to form at $\ell_w/2$ from the section of maximum moment, given by the second term in the square brackets in (19-45), plus

(b) a small increment of shear $(0.6\sqrt{f_c'}td)$ that is required for the flexural crack to become an inclined crack, which is given by the first term in the square brackets in:

$$V_c = \left[0.6\sqrt{f_c'} + \frac{\ell_w\left(1.25\sqrt{f_c'} + 0.2\dfrac{N_u}{\ell_w h}\right)}{\dfrac{M_u}{V_u} - \dfrac{\ell_w}{2}} \right] td \qquad (19\text{-}45)$$

(ACI Eq. 11-30)

where N_u is negative for tension. When $M_u/V_u - \ell_w/2$ is negative, (19-47) does not apply. Cross sections closer to the base than the smaller of $\ell_w/2$ or half the wall height are designed for the V_c computed at $\ell_w/2$ or at half the total height from the base of the wall.

Wall assemblies may have walls oriented at angles other than 90° to the neutral axis. In such cases, t is taken as the least web width, and d is taken as the distance between the resultant tension and compression forces, measured perpendicular to the neutral axis.

In hollow rectangular or circular wall assemblies, reinforcing bars bent around the inside radius of the curve or around inside corners must be anchored to prevent them from being pulled out through the surface of the concrete.

V_c for Squat Shear Walls

For squat walls, (19-46) predicts that inclined cracking will occur when the principal tension stress in the web reaches $\sigma_1 = 4\sqrt{f_c'}$ at the centroidal axis [19-27]. In ACI Section 11.10, this is approximated by (19-46),

$$V_c = 3.3\sqrt{f_c'}td + \frac{N_u \cdot d}{4\ell_w} \qquad (19\text{-}46)$$

(ACI Eq. 11-29)

where V_c is the shear carried by the concrete, taken equal to the shear at inclined cracking. ACI Section 11.10.3 limits the shear in shear walls due to lateral loads to

$$V_n \leq 10\sqrt{f_c'}td, \qquad (19\text{-}47)$$

where t is the thickness of the wall and d can be taken as $0.8\ell_w$ (ACI Section 11.10.4).

Paulay and Priestly use strut-and-tie models for short walls [19-6].

Shear Reinforcement

In addition to restraining the widths of cracks, the minimum distributed horizontal reinforcement, as is required by ACI Section 11.10.9, serves as shear reinforcement for shear forces parallel to the wall. This steel must be anchored at each end of the wall to develop the yield strength of the horizontal steel. If boundary elements are provided, the horizontal wall steel must be anchored within the vertical steel in the boundary elements. The shear carried by the horizontal reinforcement is:

$$V_s = \frac{A_v f_y d}{s_v} \qquad (19\text{-}48)$$

where V_s is the shear resisted by the distributed, horizontal reinforcement, and s_v is the vertical spacing between successive horizontal bars or groups of horizontal bars. Spacing limits for horizontal and vertical bars are given in ACI Sections 11.10.9.3, 11.10.9.5, and 14.3.5.

A_v is the total area of distributed horizontal reinforcement within a vertical spacing of s_v. When the factored shear force V_u exceeds $V_c/2$, horizontal shear reinforcement shall be provided in accordance with ACI Section 11.10.9. When V_u is less than $V_c/2$, horizontal shear reinforcement shall satisfy ACI Section 14-3.

Shear Transfer across Construction Joints

The shear force transferred across construction joints can be designed using shear friction. The clamping force is the sum of the tensile reinforcement force components of all the tensile bar forces and permanent compressive forces acting on the joint. This includes all the reinforcement perpendicular to the joint, regardless of the primary use of this steel.

19-14 DESIGN OF A SHEAR WALL FOR WIND LOADS

EXAMPLE 19-4   Design a Shear Wall

The plan of a ten-story office building in Minneapolis is shown in Fig. 19-20. All the columns are 16 in. by 16 in. The ground story is 15 ft high; the other stories are 12 ft high, top of slab to top of slab. The roof is flat with a parapet extending 20 in. above the roof surface. The floors are 8-in.-thick flat slabs except for the basement floor, which is a 5-in. slab on grade. The ground-floor slab acts as a diaphragm to transfer the wind load shears to the end walls of the basement, which in turn transfer them to the foundations. The live load is 100 psf on all floors, f_c' is 3000 psi, and f_y is 60,000 psi. The allowable bearing pressure on the subsoil is 5 tons (10,000 lb) per square foot. Design the wall using load and strength reduction factors from ACI 318-02 Sections 9.2 and 9.3, wind loads from ASCE 7, and the serviceability load factors from ASCE 7 Section 2.4.1.

    1.   Select wind design method. ASCE 7-98 [19-29] has three methods for computing wind loads. They are

        Method 1—Simplified Method. This is limited to buildings with height less than 60 ft, among other things.

        Method 2—Analytical Procedure. This is applicable for most urban buildings. The building must be "rigid," which ASCE 7 defines as having a natural frequency, F, not less than 1 Hz.

        Method 3—Wind Tunnel Testing.

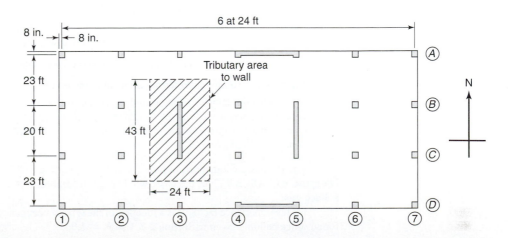

Fig. 19-20
Plan of building —
Example 19-4.

2. **Compute the natural frequency of the building.** Chapter 9 of ASCE 7 gives equations for the periods of buildings. For a moment-resisting frame with N stories:

$$T = 0.1N \qquad (19\text{-}49a)$$

where $N = 10$, which gives a fundamental period of $T = 1$ sec. Alternatively,

$$T = C_T h_w^{3/4} \qquad (19\text{-}49b)$$

where h_w is the total height, and $C_T = 0.020$ for concrete structures, other than moment-resisting frames, giving $T = 0.739$ sec, equivalent to $F = 1.35$ Hz. Thus, this building is "rigid." **Use Wind Design Method 2, Analytical Procedure.**

3. **Select coefficients and pressures needed to compute the wind load.** The *design pressure p* is:

$$p = q(GC_p) - q_i(GC_{pi}) \qquad (19\text{-}50)$$

where p is the *design pressure*, and q and q_i are the *velocity pressures* for exterior and interior surfaces, respectively. The first term on the right-hand side of (19-52) is the exterior wind pressure, the second term is the interior wind pressure. The gust factor, G, accounts for the turbulence of the wind. For simplicity, we shall use the *exterior pressure coefficients*, C_p, from Fig. 6-3 of ASCE 7-98 (See Fig. 2-12 of this book.) This figure gives pressure coefficients for gable roofs, hip roofs, or sloping shed roofs. Our building has a flat roof, which is a "shed roof" with a slope of $\theta = 0$.

For brevity, we will ignore the requirement in ASCE 7 Section 6.5.12.3 that *full and partial wind loads* be considered. These subdivide the surface of the building into parts that sustain the full wind pressure p, and other parts loaded with part of the full wind pressure. This is an attempt to simulate torsion due to wind loads.

From ASCE 7, the velocity pressure, q_z, is

$$q_z = 0.00256 K_z K_{zt} K_d V^2 I \text{ psf} \qquad (19\text{-}51)$$

where:

0.00256 is a constant needed to convert from velocity pressure q to design wind pressure p. It is a function of the density of the air and the mixed units used in the calculations.

K_z is the velocity pressure coefficient at z ft above grade. This accounts for the increase in velocity with height. It is a function of the roughness of the ground surface for a distance of about 2000 ft upwind.

The wind exposures are:

A—centers of large cities,
B—urban and suburban areas,
C—open terrain with scattered obstructions, and
D—flat unobstructed areas exposed to wind coming over water.

These exposures are illustrated in Fig. 2-11.

In this example, the building site is surrounded by moderately closely spaced medium-height buildings extending a minimum of 1500 ft upwind. **Use exposure B.**

K_{zt} accounts for the effects of hills, valleys, and escarpments on the wind velocity.

Use $K_{zt} = 1.0,$ assuming level terrain adjacent to the building.

K_d = wind directionality factor used to account for the reduced probability that the strongest wind will align itself with the weakest axis of the structure. Section 9.2.1(b) of the ACI Commentary requires a wind load factor of 1.6 if $K_d = 0.85$. If $K_d = 1.0$, the load factor on wind load is reduced to 1.3.

V is the basic wind speed in miles/hour corresponding to a 3-sec gust at 33 ft above grade in exposure C. ASCE 7 gives the basic wind speed in Minneapolis as **90 miles per hour** based on a 3-sec gust.

I is the importance factor. It is a function of the type of occupancy, for office buildings $I = 1.0$ for an office building given in Section 2-7. **Use $I = 1.0.$**

The increase in wind velocity with height for the windward wall, is accounted for by K_z:

For $z < 15$ ft,

$$K_z = 2.01\left(\frac{15}{z_g}\right)^{2/\alpha} \tag{19-52a}$$

For $15 \leq z \leq z_g$:

$$K_z = 2.01\left(\frac{z}{z_g}\right)^{2/\alpha} \tag{19-52b}$$

where, in both cases, for exposure B, $z_g = 1200$ ft and $\alpha = 7.0$. The term z_g refers to the thickness of the boundary layer of turbulent air adjacent to the surface of the ground in a windstorm, as shown by the horizontal lines in Fig. 2-11.

4. Compute the wind loads acting on the building. The overall height of the main wind-resisting structure is $h_w = 123$ ft. For this example, the length and width of the building are $B = 145$ ft 4 in. perpendicular to the wind, and $L = 67$ ft 4 in. parallel to the wind, respectively. The wind loads for the assumed wind direction are calculated in Table 19-4. In the calculations, the following subscripts are used:

> z refers to the height above the ground surface
>
> n refers to the top story
>
> h refers to a function evaluated at the top of the structure
>
> w refers to the windward wall
>
> ℓ refers to the leeward wall

Wind Forces on Windword Wall

Column 3, Table 19-4 The tributary height is the height between lines of zero shear in the cladding above and below the floor under consideration.

Column 4, Table 19-4 The velocity pressure coefficient is from (19-52a) and (19-52b).

Column 5, Table 19-4 The velocity pressure, q_z, at the roof level ($z = 123$ ft) is:

$$q_z = 0.00256 \times 1.048 \times 1.0 \times 0.85 \times 90^2 \times 1.0 = 18.47 \text{ psf} \tag{19-51}$$

At the 10 th floor level the computed value of K_z is 1.018 and $q = 17.94$ psf.

Column 6, Table 19-4 The design wind pressure is

$$p_{z,w} = q(GC_p) - q_i(GC_{pi}) \tag{19-50}$$

In this building the wall cladding spans vertically from floor to floor. As a result, the exterior wind pressures are transferred to the frame at the floor levels. Because the interior wind pressure or suction is equal on all interior surfaces in a closed building, the interior pressures on the windward and leeward walls cancel each other out. As a result, (19-50) reduces to

$$p_{z,w} = q(GC_p) \tag{19-50a}$$

Because the building is "rigid" (See step 2), ASCE 7 specifies the gust factor $G = 0.85$ for structures designed on the basis of 3-sec gusts.

We shall assume that the concentrated wind-load reactions from the cladding act on the frame at the levels of the top of the floors.

In Fig. 6-3 of ASCE 7 (see Fig. 2-12), C_p is +0.8 on the exterior face of the windward wall and varies between -0.2 and -0.5 on the exterior face of the leeward wall as a function of the length and width of the building. (The minus sign denotes suction.) In (19-50) the pressure coefficient, C_p, on the exterior face of the windward wall is +0.80. The unfactored wind pressure at the top of the windward face of the wall in the 10th floor level is

$$p_{z,w} = q(GC_p) = 17.94 \times 0.85 \times (+0.8) = +12.20 \text{ psf}$$

TABLE 19-4 Computation of Wind Loads at the Roof and Floor Levels—Example 19-4

1	2	3	4	5	6	7	8	9	10	11	12
Floor	Elevation of Floor z, ft	Tributary Height h trib, ft	Windward Pressure Coef K_z	Design Windward Velocity Pressure q_z psf	Design Windward Pressure $p_w = q_z(C_pG)$ psf	Windward Wind Force H_{hw} kips	Leeward Pressure Coef $K_{h,l}$	Design Leeward Velocity Pressure $q_{h,l}$	Design Leeward Velocity Pressure $p_f = q_h(C_gG)$ psf	Design Leeward Force $H_{h,w}$ kips	Total Wind Force H_z kips
Roof	123	7.67	1.048	18.47	12.56	14.00	1.048	18.47	−7.85	−8.7	22.7
10	111	12.0	1.018	17.94	12.20	21.28	1.048	18.47	−7.85	−13.7	34.9
9	99	12.0	0.985	17.36	11.81	20.59	1.048	18.47	−7.85	−13.7	34.2
8	87	12.0	0.950	16.73	11.38	19.85	1.048	18.47	−7.85	−13.7	33.5
7	75	12.0	0.910	16.04	10.91	19.02	1.048	18.47	−7.85	−13.7	32.7
6	63	12.0	0.866	15.26	10.38	18.10	1.048	18.47	−7.85	−13.7	31.8
5	51	12.0	0.815	14.37	9.77	17.04	1.048	18.47	−7.85	−13.7	30.7
4	39	12.0	0.755	13.31	9.05	15.78	1.048	18.47	−7.85	−13.7	29.4
3	27	12.0	0.680	11.98	8.14	14.21	1.048	18.47	−7.85	−13.7	27.9
2	15	13.5	0.575	10.13	6.88	13.51	1.048	18.48	−7.85	−15.4	28.9
Ground	0										

Column 7, Table 19-4 The total exterior North–South wind force acting on the frame $H_{z,w}$ at the height z on the windward wall is

$$H_{z,w} = p_{z,w} \times \text{Column 3} \times L \tag{19-53}$$

For the 10th story,

$$H_{z,w} = \frac{12.20 \text{ psf} \times 12.0 \text{ ft} \times 145.3 \text{ ft}}{1000} = 21.28 \text{ kips}$$

Wind Forces on Leeward Wall

Column 8, Table 19-4 The design wind pressure on the leeward wall, $p_{h,\ell}$ is calculated using (19-52), with K_z evaluated at $z = h_n = 123$ ft. $K_z = K_{h,\ell}$ is constant at 1.048 over the height of the wall.
    Column 9, Table 19-4 The velocity pressure on the leeward wall is

$$q_{h,\ell} = 0.00256 \times 1.048 \times 1.0 \times 0.85 \times 90^2 \times 1.0 = 18.47 \text{ psf}$$

This is constant over the height of the building.
    Column 10, Table 19-4 The design wind pressure on the leeward wall is

$$p_{h\ell} = q_{h,\ell}(GC_p) = 18.47 \text{ psf} \times (0.85 \times (-0.50))$$

$$= -7.85 \text{ psf, constant over the entire height.}$$

The negative sign indicates that this is a suction.
    Column 11, Table 19-4 Leeward wind force:

$$H_{w\ell} = p_{h\ell} \times (\text{tributary area})$$

$$= p_{h\ell}(\text{tributary height from column 3}) \times \text{length}$$

At the roof, the leeward wind force is

$$H_{w\ell} = (-7.85) \times 7.67 \text{ ft} \times 145.3 \text{ ft}) = -8.75 \text{ kips}$$

At the 10th floor, $H_{h,\ell} = 13.70$ kips

At the 2nd floor, $H_{h,\ell} = (6.44 \times 13.5 \times 145.3)$ kips $= 15.40$ kips

Column 12 Table 19-4 The total wind load H_z at a given height is the algebraic sum of column 7 plus column 11. In the 10th story, $H_z = 14.0$ kips force (to the right in Fig. 2-12) and -8.7 kips suction force (also to the right) $= 14.0 - (-8.7) = 22.7$ kips.

5. Select the Wall Thickness. The floor plan of the building in Fig. 19-20 shows two walls in each direction. ACI Section 14.5.3.1 limits the slenderness of walls designed using the empirical design method to 1/25 of the unsupported height or width. If this was a bearing wall, the minimum thickness for story 1 would be

$$h = (15 - 0.667) \text{ ft}/25 = 6.9 \text{ in. This is too thin.}$$

Based on the horizontal length of the wall, 1/25 times the horizontal length of the wall is $256/25 = 10.2$ in. Three possible thicknesses are

(a) the width of the columns (16 in.) or

(b) 10.2 in., or

(c) an even 12 in.

Try 12 in.

6. Estimate the size of walls required. From (19-28),

$$\Sigma(0.70I_g) = 500w_s h_n{}^3/6E_c \tag{19-28}$$

where units are pounds and feet.

$$E_c = 57{,}000\sqrt{f'_c} = 3{,}122{,}000 \text{ lb/in.}^2 \times 144 \text{ in.}^2/\text{ft}^2 = 449 \times 10^6 \text{ psf}$$

w_s = wind load per 1 ft of height across the entire width at the top of the building.

$$= (p_{n,w} - p_{n,\ell}) \text{ psf} \times (L \text{ ft} \times 1 \text{ ft})$$

$$= 12.56 - (-7.85) \text{ psf} \times (145.3 \text{ ft} \times 1 \text{ ft}) = 2966 \text{ lb/ft of height}$$

For wind blowing against the long side of the building,

$$\Sigma(0.70I_g) = \frac{500 w_s h_n^3}{6E_c} = \frac{500 \times 2966 \text{ lb/ft} \times 123^3 \text{ ft}^3}{6 \times 449 \times 10^6 \text{ psf}} = 1024 \text{ ft}^4$$

thus $\Sigma I_g = 1024/0.70 = 1463 \text{ ft}^4$. Each of the two similar planar N–S walls must provide at least half of this, or 732 ft^4. Previously it was determined that $b = 12$ in. $= 1$ ft.

$$\ell_w{}^3 = (732 \text{ ft}^4 \times 12)/1 \text{ ft} = 8780 \text{ ft}^3$$

$$\ell_w \geq 20.6 \text{ ft}$$

Try two walls, each 12 in. thick by 23.33 ft (280 in.) in length, located on column lines 3 and 5. These are symmetrical about the axis through which the wind shears act, in Fig. 19-20.

7.   Compute shears and overturning moments over the height of the structure. The story shears and moments are calculated in Table 19-5.

Columns 1 and 2, Table 19-5 These columns differentiate between the floor levels and the stories. Thus, the 5th story extends from the 5th floor to the 6th floor.

TABLE 19-5 Computation of Total Unfactored Story Shears and Moments at Floor Levels—Example 19-4[a]

1	2	3	4	5	6
Floor	Story	Unfactored Wind Load kips	Story Height ft	Story Shear kips	Moment at Floor Level ft-kips
Roof		22.76			
	10		12	22.76	
10		34.98			273.1
	9		12	57.74	
9		34.29			966.0
	8		12	92	
8		33.55			2,070
	7		12	126	
7		32.72			3,577
	6		12	158	
6		31.80			5,477
	5		12	190	
5		30.74			7,758
	4		12	221	
4		29.48			10,408
	3		12	250	
3		27.91			13,412
	2		12	278	
2		28.92			16,750
	1		15	307	
Ground					21,400

[a]These loads and moments are for the entire building. They are divided equally between the two North–South walls.

Column 3, Table 19-5 The unfactored lateral load at each floor level comes from column 12 of Table 19-4.

Column 5, Table 19-5 The 10th story extends from the top of the 10th floor to the top of the roof slab. The shear in the 10th story is the horizontal force acting at the roof level from column 12, Table 19-4. The shear in the 9th story is the shear in the 10th story plus the horizontal force acting on the 10th floor.

$$V_9 = 22.76 \text{ kips} + 34.98 \text{ kips} = 57.74 \text{ kips}$$

Column 6, Table 19-5 The overturning moment at a given floor level is the area of the shear force diagram from the top of the building down to the floor level in question. At the 10th floor,

$$M_{10} = 0 + (22.76 \text{ kips} \times 12 \text{ ft}) = 273 \text{ ft-kips}$$

Enlarged boundary elements can be provided at the ends of the wall if a constant thickness wall is inadequate. Two more walls will be provided in the E–W direction. These will be 24 ft + 1.33 in = 25.33 ft long. They have a very small I_g (about 2 ft^4) effective in resisting the N–S wind and will be ignored when computing wind resistance for wind perpendicular to the long side of the building. Due to the symmetry of the building, half of the moments and shears at each level in Table 19-5 are assigned to each wall.

8. Compute dead and live loads on a wall. One of the North–South walls supports the gravity loads from the roof and floor loads on the tributary area supported by the wall, shown shaded in Fig. 19-20. The tributary area for computing the dead loads of the floors and the roof loads is

$$A_t = (22 \text{ ft} + 2 \times 22 \text{ ft}/2) \times 24 \text{ ft} = 1056 \text{ ft}^2$$

The tributary area for slab live loads is 1056 ft^2 minus the area of the wall. Assuming the wall is 1 ft thick throughout its length as selected in step 5:

$$A_t = 1056 - (1 \text{ ft} \times 23.33 \text{ ft}) = 1031 \text{ ft}^2.$$

Column 2, Table 9-6

The weights of roofing and insulation (9 psf), ceiling (5 psf), and mechanical ducts supported by the roof (5 psf) = 9 + 5 + 5 = 19 psf, on the roof.

The weights of ceiling, mechanical ducts, and flooring (10 psf) total 18 psf, say 20 psf on a typical floor.

The unfactored dead load of the slab = 1056 ft^2 × 2/3 ft × 0.15 kcf = 105.6 kips.

Superimposed dead load (roof) = 19 psf × 1056 ft^2 = 20.1 kips

Superimposed dead load (floor) = 20 × 1031 = 20.6 kips

Dead load of shear wall per story = $(1 \times 23.33) \times (12.0 - 0.667) \times 0.15 = 39.7$ kips in stories 2 through 10, and 50.2 kips in the first story.

Total dead load (typical floor) = 105.6 + 20.6 + 39.7 = 165.9 kips

The calculations of the dead and live loads are presented story by story in Table 19-6. The vertical load in a story is the gravitational force acting on the bottom of the walls and columns in the story because the wind moments and weight of wall are a maximum at that point. The forces in the wall are computed as:

The dead load in story 9 is the dead load in story 10, plus the weight of the walls, ceilings, floors, and mechanical loads between floors 9 and 10, which totals:

$$D_9 = 165.4 \text{ kips} + (\text{weight of floor 10}) + (\text{superimposed dead load of floor 10})$$

$$+ (\text{weight of wall from the 9th floor to the 10th floor})$$

$$= 165.4 + (105.6 + 20.6 + 39.7) = 331.3 \text{ kips}$$

TABLE 19-6 Compute Unfactored Dead, Live, and Snow Loads on One Wall—Example 19-4

1	2	3	4	5	6	7	8	9	10	11
			Non-Reducable		Cumulative Live Loads, Reducable					Total Live Loads
Story	Dead One Story kips	Dead Cumul kips	Cumulative Loads Snow kips	Cumulative Loads Partition kips	Unreduced Live Use and Occ kips	Tributary Area ft²	LLRF	Reduced LL Use and Occ One Story kips	Reduced LL Cumulative kips	Total Unfactored Live Load kips
Roof	0	0	37						0.0	
10	165.4	165.4	37	20.6	103.1	1031	0.717	73.9	73.9	94.5
9	165.9	331.3	37	41.2	103.1	2062	0.580	59.8	133.7	175
8	165.9	497.2	37	61.9	103.1	3093	0.519	53.5	187.2	249
7	165.9	663.1	37	82.5	103.1	4124	0.483	49.8	237.0	320
6	165.9	829.0	37	103.1	103.1	5155	0.459	47.3	284.3	387
5	165.9	995	37	123.7	103.1	6186	0.441	45.5	329.8	454
4	165.9	1161	37	144.3	103.1	7217	0.426	43.9	373.7	518
3	165.9	1327	37	165.0	103.1	8248	0.415	42.8	416.5	582
2	165.9	1493	37	185.6	103.1	9279	0.406	41.9	458.4	644
1	176.4	1669	37							

The dead load at the first floor is the sum of the dead loads of the stories above, plus the weight of the wall between floors 1 and 2.

Column 4 Table 19-6 ASCE 7 gives the snow load as

$$p_f = 0.70 \times C_e \times C_t \times I \times p_g = 0.70 \times 1.0 \times 1.0 \times 1.0 \times 50 \text{ psf} = 35 \text{ psf} \quad (19\text{-}53)$$

Total snow load supported by one wall $= S = 0.035 \text{ ksf} \times 1056 \text{ ft}^2 = 37.0 \text{ kips}$

Snow is considered separately from live load because:

(a)   it results from a different physical process than live load and hence is not likely to reach its maximum load simultaneously with maximum live load

(b)   the live load reduction factor does not apply to the snow load

(c)   the ASCE 7 load combinations include S as a separate load with a different load factor than L

The 10th-story wall supports the largest of the snow load, S; a use and occupancy load for the roof, L_r, which ASCE 7 sets between 12 and 20 psf; or a rain load, R. The snow load is larger than L_r and thus governs. The snow load on the roof must be supported by the walls in every story from the 10th to the foundation. The snow load in column 4 is included in the summations of the wall loads for each floor. The calculations of the dead and live loads are presented story by story in Table 19-6. The controlling vertical load in a story is the gravity load acting on the bottom of the walls and columns in the story because the wind moments are a maximum at that point. The forces in the frame are computed as:

Column 5, Table 19-6 ASCE 7 classifies partition loads (20 psf) as live loads, but they are not reduced by live-load reduction factors. The 9th-story wall supports a use and occupancy live load of $0.100 \text{ ksf} \times 1031 \text{ ft}^2 = 103.1 \text{ kips}$ from the 10th floor. The values for each story are the total live loads not including partition loads on stories above and including the story in question.

Column 6, Table 19-6 The values shown for each story are the live loads due to use and occupancy for all stories above the story except the roof.

Column 7, Table 19-6 This is the tributary area supporting live load, not counting the area of the roof which supports the snow load which is not reduced.

Column 7, Table 19-6 The live load reduction factor is:

$$LLRF = 0.25 + \frac{15}{\sqrt{K_{LL}A_t}} \quad (19\text{-}54)$$

where A_t is the tributary area in ft^2 supporting use and occupancy live loads. K_{LL} is the correction factor to convert tributary area to influence area. Values of K_{LL} are given in Table 4-2 of ASCE 7. This factors equals 4 for interior and edge columns, 2 for edge beams and interior beams, and 1 for two-way slabs. Equation (19-54) was originally derived using influence areas, $A_i = K_{LL} \times A_t$, and was converted to tributary loads in the 1998 edition of ASCE 7 [19-29].

For walls supporting snow on the roof, $LLRF = 1.0$ (no reduction) because the reduction in the snow load with area is assumed to be negligible.

Column 7, Table 19-5 The tributary area for the 9th-story wall is 1031 ft^2 and $K_{LL} = 1$.

$$LLRF = 0.25 + \frac{15}{\sqrt{1 \times 1031}} = 0.717$$

but LLRF is not less than 0.5 for a member supporting one floor, and not less than 0.4 for members supporting more than one floor.

The tributary area of the wall at the bottom of the 7th floor is 1031 ft^2 from each of the 10th, 9th, and 8th floors, for a total of $A_T = 3093 \text{ ft}^2$.

Column 8, Table 19-6 The LLRF for the walls supporting the 7th story, $A_T = 3093 \text{ ft}^2$. LLRF $= 0.519$ but not less than 0.40, so LLRF $= 0.519$.

Column 9, Table 19-6 This is the product of columns 6 and 8.

Column 11, Table 19-6 Snow loads and partition loads are not reduced by the *LLRF*. Column 11 is the sum of columns 5 and 10.

9. Compute the factored loads acting on one wall—forces at bottom of wall in first story—Steps 9 through 14 consider the critical section at the bottom of the walls. In step 15 we will consider the foundations for one wall. However, the force path between the first floor and the foundation has not been selected. Using the load factors from ACI 318-02 Section 9.2, the load combinations that must be considered are the following.

Load Combination 1

$$U = 1.4(D + F) \qquad \text{(ACI Eq. 9-1)}$$

Load Combination 2

$$U = 1.2(D + F) + 1.6(L + H) + 0.5(L_r \text{ or } S \text{ or } R) \qquad \text{(ACI Eq. 9-2)}$$

Load Combination 3

$$U = 1.2(D + F) + 1.6(L_r \text{ or } S \text{ or } R) + (1.0L \text{ or } 0.8W) \qquad \text{(ACI Eq. 9-3)}$$

Load Combination 4

$$U = 1.2D + 1.6W + 1.0L + 0.5(L_r \text{ or } S \text{ or } R) \qquad \text{(ACI Eq. 9-4)}$$

Load Combination 5

$$U = 0.9D + 1.6W + 1.6H \qquad \text{(ACI Eq. 9-6)}$$

The term $(L_r \text{ or } S \text{ or } R)$ means that only the largest of L_r, S, or R is used. In our case, this is S. ACI 318-02 Section 9.2.1(a) allows the load factor for the companion load L in load combinations (ACI Eqs. (9-3), (9-4)) to be reduced from 1.0 to 0.5, except for garages, assembly areas, and all areas where L exceeds 100 psf (ACI 319-02 Section 9.2.1(a)). The wind load, W, is based on wind loads multiplied by K_d to account for directionality. This type of wind load has a load factor of 1.6 rather than 1.3 (ACI 318-02 Section 9.3.1(b)).

The maximum unfactored axial loads, moments, and shears on the bottom of one shear wall at the ground-floor level are given in Table 19-7.

9(a) Factored load effects, axial load—Base of wall at ground floor. The load combinations reduce to

LC 1

$$U = 1.4D = 1.4 \times 1670 = 2340 \text{ kips}$$

LC 2

$$U = 1.2D + 1.6L + 0.5S$$
$$= 1.2 \times 1670 + (1.6 \times 644) + (0.5 \times 37) = \textbf{3050 kips}$$

LC 3a

$$U = 1.2D + 1.6S + 0.5L = 1.2 \times 1670 + (1.6 \times 37) + (0.5 \times 644)$$
$$= 2390 \text{ kips}$$

LC 3b

$$U = 1.2D + 1.6S + 0.8W = 1.2 \times 1670 + 1.6 \times 37 + 0 = 2060 \text{ kips}$$

TABLE 19-7 Maximum Unfactored Loads at the Bottom of the Wall in the First Story

Axial load, kips	Moment, ft-kips	Shear, kips
$D = 1670$ kips	—	—
$L = 644$ kips	—	—
$S = 37$ kips		—
$W =$	$21{,}400/2 = 10{,}700$	$307/2 = 153.5$

LC 4

$$U = 1.2D + 1.6W + 0.5L + 0.5S$$
$$= 1.2 \times 1670 + (1.6 \times 0) + (0.5 \times 644) + (0.5 \times 37) = 2340 \text{ kips}$$

LC 5

$$U = 0.9D + 1.6W = 0.9 \times 1670 + 1.6 \times 0 = 1500 \text{ kips}$$

9(b) Factored load effects—Moments about the strong axis of the wall. The strong-axis moments result from wind-loads, W. The largest wind load moments are at the bottom of the walls at the ground floor level. These are for one wall.

LC 3b

$$U = 1.2D + 1.6S + 0.8W \qquad M_{uw} = 0.8 \times 10,700 \text{ ft-kips} = 8560 \text{ ft-kips}$$

LC 4

$$M_{uw} = 1.6 \times 10,700 \qquad M_{uw} = 17,120 \text{ ft-kips}$$

LC 5

$$M_{uw} = 1.6 \times 10,700 \qquad M_{uw} = 17,120 \text{ ft-kips}$$

9(c) Factored load effects—Shear parallel to the length of the wall. These result from wind load. They for one wall.

LC 3b

$$V_{uw} = 0.8 \times 153.5 = 123 \text{ kips}$$

LC 4

$$V_{uw} = 1.6 \times 153.5 = 246 \text{ kips}$$

LC 5

$$V_{uw} = 246 \text{ kips}$$

Arranging these by load combination gives Table 19-8.

10.   Check the stiffness of the shear wall. Up to this point in the design of the shear wall, all the calculations have been based on the assumption that the wall section chosen in step 6, based on Fig. 19-21, will have enough stiffness to limit the lateral deflections and vibrations.

This must be checked once the foundation structure has been selected. We shall omit this step for brevity.

TABLE 19-8 Maximum Factored Load Effects on One Wall at the Ground Floor Level

	Axial load, kips	Moment, ft-kips	Shear, kips
LC 1	2340	—	—
LC 2	**3050**	—	—
LC 3a	2390	—	—
LC 3b	2060	8,560	123
LC 4	2340	**17,120**	**246**
LC 5	1500	**17,120**	**246**

Load combination 2 gives the largest axial load.
Load combination 4 gives the largest moment and shear with the largest axial load in combination with moment and shear.
Load combination 5 gives the least axial load together with the largest moment and shear.
Each factored load combination is used in its entirety. It is not correct to use, say, the factored axial loads from LC 2 with the factored moment and shear from LC 4.

11. Minimum distributed reinforcement. Because a significant portion of the applied loads and moments could be resisted by minimum reinforcement, we shall select the minimum reinforcement before designing the reinforcement for flexure. See Table 19-3.

Required by ACI Section 14.3

ACI Section 14.3.4 requires that the reinforcement be in two curtains (layers) if the wall is more than 10 in. thick. Thus two curtains are needed. In a shear wall the minimum reinforcement requirements for shear in Chapter 11 will normally govern.

Distributed Wall reinforcement required by ACI Section 11.10.9

ACI Section 11.10.9.2 requires $\rho_h = 0.0025$.

ACI Section 11.10.9.3 The maximum spacing of horizontal shear reinforcement shall not exceed (a) $\ell_w/5 = 280/5 = 56$ in. but (b) not more than three times the wall thickness $= 36$ in. or 18 in.

ACI Section 11.10.9.4 A variable spacing is specified with horizontal reinforcement becoming less effective for low walls. When h_w/ℓ_w is less than 0.5, the amount of vertical reinforcement is set equal to the horizontal reinforcement. ACI Section 11.10.9.4 requires

$$\rho_n \text{ (ratio of vertical steel to } A_g) = 0.0025 + 0.5\left(2.5 - \frac{h_w}{\ell_w}\right)(\rho_h - 0.0025) \quad (19\text{-}55)$$

$$\rho_n = 0.0025 + 0.5\left(2.5 - \frac{123 \times 12}{280}\right)(\rho_h - 0.0025) \quad \text{(ACI Eq. 11-32)}$$

To solve (19-55) we need a value of ρ_h which we shall take equal to the 0.0025 minimum required in ACI Section 11.10.9.2.

$$\rho_n = 0.0025 + 0.5(-2.77)(0.0025 - 0.0025) = 0.0025, \text{ but not less than } 0.0025.$$

From ACI Section 11.10.9.5 the maximum spacing of vertical shear reinforcement shall not exceed $\ell_w/3 = 280/3 = 93$ in. but not more than three times the wall thickness $= 36$ in. or 18 in. From ACI Section 11.10.9 the vertical reinforcement is as worked out in the next paragraph.

Using 18 in. spaces, $(280/18)$ in. $= 15.6$ spaces are required. Round this off to fifteen 18-in. spaces (16 pairs of bars) that will take 270 in. centered on the middle of the wall, leaving 5 in. at each end of the wall. Add 1 No. 5 bar each end of each curtain at 2 in. from the end of the wall.

Use 16 No. 5 vertical distributed bars, at 18 in. on centers, each face, plus 1 extra No. 5 bar at each end of each face. Total steel is 36 No. 5 bars, total area, $A_{st} = 11.2$ in.^{2}

The vertical spacing, s_v, of the distributed horizontal reinforcement is

$$s_v = \Sigma A_b/(0.0025 \times b_w) = 2 \times 0.20/(0.0025 \times 12) = 13.3 \text{ in. but not more than three}$$

times the wall thickness or 18 in.

Use No. 4 distributed horizontal reinforcement, at 12 in. o.c., each face

12. Compute the axial load resistance of the chosen wall. The design up to this point has considered strong-axis bending. If weak-axis bending exists, the effect of weak-axis bending on the strong-axis strength should also be considered. Ideally, this should involve a computation based on biaxial bending of the wall.

For this wall the axial-load capacity for weak-axis bending will be estimated from ACI Eq. (14-1):

$$\phi P_{nw} = 0.55\phi f'_c A_g\left[1 - \left(\frac{k\ell_c}{32t}\right)^2\right] \quad (19\text{-}12)$$
$$\text{(ACI Eq. 14-1)}$$

From ACI Section 14.5.2, $\phi = 0.70$ for bearing walls.

$$\phi P_{nw} = 0.55 \times 0.70 \times 3000 \text{ psi} \times 280 \text{ in.} \times 12 \text{ in.}\left[1 - \left(\frac{0.8 \times 138 \text{ in.}}{32 \times 12 \text{ in.}}\right)^2\right]$$

$$= 3880 \times [1 - 0.083] = 3560 \text{ kips}$$

This exceeds the maximum factored axial load of 3050 kps from LC 2, Table 19-8: OK.

13.   Design for shear—Use ACI Section 11.10 to design for shear. Shear design is based on the axial load, moment, and shear from the load combination requiring the most shear reinforcement. We shall assume that the controlling load combination is LC 5 because of the lower axial load. The factored shear, V_u, in the bottom story is 246 kips per wall. From load combination 5, the corresponding factored axial load and moment are $P_u = 1500$ kips and $M_u = 17,120$ ft-kips, respectively.

Design will be based on (19-45), (19-46), and (19-48).

$$\phi V_n \geq V_u \tag{19-32}$$
$$\text{(ACI Eq. 11-1)}$$

$$V_n = V_c + V_s \tag{19-43}$$
$$\text{(ACI Eq. 11-2)}$$

$$V_s \geq \frac{V_u}{\phi} - V_c \tag{19-44}$$

The shear acts horizontally and parallel to the axis of the wall.

The effective depth, d, from ACI Section 11.10, is $d = 0.80\ell_w = 0.8 \times 23.33 \times 12 = 224$ in. From ACI Section 9.3.2.3, $\phi = 0.75$ for shear, and ACI Section 11.10.3 limits the maximum allowable shear to

$$\phi V_{n,\max} = 10\phi \sqrt{f'_c}\, d \times t \tag{19-47}$$

where t is the thickness of the wall.

$$= 10 \times 0.75 \times \sqrt{3000} \times 224 \text{ in.} \times 12 \text{ in.} = 1104 \text{ kips}$$

The maximum allowable ϕV_n is 1104 kips. The maximum factored shear is $V_u = 246$ kips per wall—therefore, OK.

For walls subject to compressive axial forces P_u, ACI Section 11.10.5 allows V_c to be taken as $V_c = 2\sqrt{f'_c}\, d \times t = 294$ kips for walls subject to a compressive axial force N_u, unless a more detailed calculation is made in accordance with ACI Section 11.10.6, which, in turn, defines V_c as the smaller of the values given by (19-48) and (19-47) where:

$$V_c = 3.3\sqrt{f'_c}\, td + \frac{P_u d}{4\ell_w t} \tag{19-46}$$
$$\text{(ACI Eq. 11-29)}$$

or

$$V_c = \left[0.6\sqrt{f'_c} + \frac{\ell_w\left(1.25\sqrt{f'_c}\right) + 0.2\dfrac{P_u}{\ell_w t}}{\dfrac{M_u}{V_u} - \dfrac{\ell_w}{2}} \right] td \tag{19-45}$$
$$\text{(ACI Eq. 11-30)}$$

where $d = 224$ in., $t = 12$ in., $\ell_w = 280$ in. and $P_u = 1,500,000$ lbs.
From (19-46),

$$V_c = 3.3\sqrt{3000} \times 12 \times 224 + \frac{1,500,000 \times 224}{4 \times 280 \times 12}$$

$$V_c = 486 \text{ kips} + 25.0 \text{ kips} = 511 \text{ kips}$$

From (19-45),

$$V_c = \left[0.6\sqrt{3000} + \frac{280\left(1.25\sqrt{3000} + 0.2\dfrac{1,500,000}{280 \times 12}\right)}{\dfrac{17,120 \times 12,000}{246 \times 1000} - \dfrac{280}{2}} \right] 12 \times 224$$

$$= [32.9 + 63.5] \times 12 \times 224 = 259,000 \text{ lbs} = 259 \text{ kips}$$

So, $V_c = 259$ kips.

$V_c = 259$ kips and $V_u = 246$ kips. Shear reinforcement is required because $V_u > \phi V_c/2$. Hence,

$$V_s = \frac{V_u}{\phi} - V_c = 246/0.75 - 259 = 69 \text{ kips} \qquad (19\text{-}44)$$

V_s is computed for the horizontal reinforcement because that steel is parallel to the shear force:

$$V_s = \frac{A_v f_y d}{s_2} \qquad (19\text{-}48)$$

In step 11, we chose No. 4 bars at 12 in. o.c., each face. With this $V_s = 336$ kips. This is more than enough.

The No. 4 horizontal bars at 12 in. on centers, in both faces, from step 11 are adequate as shear reinforcement. They must be anchored at the two ends of the wall.

14.   Design the wall for flexure

14 (a) Is the wall a slender shear wall or a squat shear wall? A slender shear wall has h_w/ℓ_w greater than or equal to 2. In this beam, $h_w/\ell_w = (123 \times 12)/288 = 5.13$. **Therefore, this is a slender shear wall.**

14 (b) Do the vertical bars need to be tied? ACI Section 14.3.6 states that vertical reinforcement does not need to be tied

(i)   if it has an area less than or equal to 0.01 times the gross area of the concrete, or

(ii) if the vertical reinforcement is not required as compression steel.

This applies for this wall if we are in a nonseismic region. In a seismic region, the vertical reinforcement in boundary elements must be tied in accordance with ACI Section 21.7.6. Therefore, assume the distributed vertical reinforcement can be included in the moment calculation.

In step 11, a total area of 11.2 in.^{2} of distributed vertical reinforcement was chosen, giving

$$\rho_v = 11.2 \text{ in.}^2/[12 \text{ in.} \times (23.33 \times 12) \text{ in.}] = 0.00333$$

Load combination 5 will require the most reinforcement, because LC 5 has the largest M_u combined with the smallest axial load, P_u. Because the factored axial load, P_u, is probably less than that for the C_{CCL} limit ignoring the reinforcement, this section will correspond to a point on the interaction diagram below the balanced load. In this part of the interaction diagram, the amount of reinforcement required to resist a given moment decreases as the axial load increases.

From Table 19-8 the largest and smallest factored design forces at ultimate are axial load $P_u = 1500$ kips and moment $M_u = 17,120$ ft-kips.

14 (c) Check whether the wall is compression-controlled or tension-controlled.

Nominal tension-controlled limit:

$$C_{CTL} \approx 0.32\beta_1 f'_c d_t t \qquad (19\text{-}33)$$

$$= 0.32 \times 0.85 \times 3000 \times (280 - 3) \times 12 = 2710 \text{ kips}$$

Compression-controlled limit:

$$C_{CCL} \approx 0.50\beta_1 f'_c d_t t \qquad (19\text{-}34)$$

$$= 0.50 \times 0.85 \times 3000 \times (280 - 3) \times 12 = 4.24 \times 10^6 \text{ lb}$$

$$= 4240 \text{ kips}$$

Since $P_u = 1500$ kips is less than $C_{CCL} = 4240$ kips and $C_{CTL} = 2120$ kips, the section is tension-controlled and $\phi = 0.90$.

The strong-axis moment resistance of the wall with the two curtains of 18 No. 5 vertical bars, selected in step 11, with total $A_s = 11.2$ in.^{2}, can be computed using 19-37 with:

$$\omega = \frac{A_s f_y}{f'_c \ell_w t} = \frac{11.2 \text{ in.}^2 \times 60,000 \text{ psi}}{3000 \text{ psi} \times 280 \text{ in.} \times 12 \text{ in.}} = 0.0667 \qquad (19\text{-}36)$$

$$\alpha = \frac{P_u}{f_c'\ell_w t} = \frac{1,500,000 \text{ lb}}{3000 \text{ psi} \times 280 \text{ in.} \times 12 \text{ in.}} = 0.149 \qquad (19\text{-}37)$$

$$\frac{c}{\ell_w} = \frac{\omega + \alpha}{2\omega + \alpha_1\beta_1} = \frac{0.0667 + 0.149}{2 \times 0.0667 + 0.85 \times 0.85} = 0.252 \qquad (19\text{-}35)$$

For 3000-psi concrete, from ACI Sections 10.2.7.1 and 10.2.7.3, $\alpha_1 = 0.85$ and $\beta_1 = 0.85$, and the neutral-axis depth is

$$c = 0.252 \times 280 = 70.6 \text{ in.}$$

$$M_n = 0.5A_s f_y \ell_w \left(1 + \frac{P_u}{A_s f_y}\right)\left(1 - \frac{c}{\ell_w}\right) \qquad (19\text{-}38)$$

$$= 0.5 \times 11.2 \text{ in.}^2 \times 60,000 \text{ psi} \times 280 \text{ in.}\left(1 + \frac{1,500,000 \text{ lb}}{11.2 \text{ in} \times 60,000 \text{ psi}}\right)\left(1 - \frac{70.6 \text{ in.}}{280 \text{ in.}}\right)$$

$$= 227 \times 10^6 \text{ in.-lb} = 18,900 \text{ ft-kips}$$

But the factored $M_u = 17,120$ ft-kips per wall and the required $M_n = M_u/\phi = 19,000$ ft-kips where ϕ is taken as 0.90. Because ϕM_n is essentially equal to M_u, we will use the vertical reinforcement selected in step 11.

Thus, Provide 36 No. 5 bars for flexural reinforcement.

The cross section of the walls is shown in Fig. 19-21. The horizontal bars are hooked 90° at each end to anchor the horizontal reinforcement. In this example, much of the reinforcement was minimum in both directions and this reinforcement is used over the height of the wall.

15.   Load transfer at wall foundation. The size of the wall footing is chosen using working-stress load combinations as done in Chapter 16. Once the size has been chosen, the footing is designed using strength design.

The 1998 edition of ASCE 7 included the following set of load combinations for the serviceability limit states

SLS load combination 1	D	(SLS–1)
SLS load combination 2	$D + L + F + H + T + (L_r \text{ or } S \text{ or } R)$	(SLS–2)
SLS load combination 3	$D + (W \text{ or } 0.7E) + (L_r \text{ or } S \text{ or } R)$	(SLS–3)
SLS load combination 4	$0.6D + W + H$	(SLS–4)

The SLS factored loads are computed below and summarized in Table 19-9.

SLS–LC 1   Axial load = $1.0 \times 1670 = 1670$ kips
There are no moments or shears due to SLS–LC 1 loads

SLS–LC 2   Axial load = $1.0(D + L + F + H + T + (L_r \text{ or } S \text{ or } R))$
= $1.0(1670 + 644 + 37) = 2350$ kips
There are no moments or shears due to SLS–LC 2 loads

SLS–LC 3   Axial load = $D + (W \text{ or } 0.7E) + (L_r \text{ or } S \text{ or } R)$ (SLS–3)
= $1.0 \times 1670 + 0 + 1.0 \times 37 = 1710$ kips

Moment = $1.0 \times 0 + 1.0 \times 10,700 + 1.0 \times 0 = 10,700$ ft-kips

Shear = $1.0 \times 0 + 1.0 \times 154 = 154$ kips

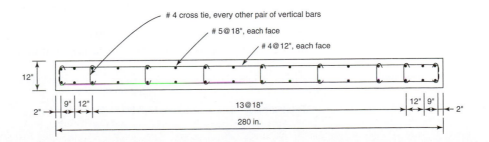

Fig. 19-21
Cross Section of Wall—
Example 19-4.

TABLE 19-9 Maximum Factored SLS Load Effects on One Wall at the Ground-Floor Level

	Axial load, kips	Moment, ft-kips	Shear, kips
SLS–LC 1	1670		
SLS–LC 2	2350		
SLS–LC 3	1710	10,700	154
SLS–LC 4	1002	10,700	154

Notes

1. Load combination 2 gives the largest axial load.

2. Load combination 3 gives the largest combination of moment and shear with the largest axial load in a combination.

3. Load combination 4 gives the largest combination of moment and shear with the smallest axial load in a combination.

SLS–LC 4   Axial load $= 0.6 \times 1670 + 1.0 \times 0 = 1002$ kips
Moment $= 0.6 \times 0 + 1.0 \times 10,700 = 10,700$ ft-kips
Shear $= 0.6 \times 0 + 1.0 \times 154 = 154$ ft-kips

Compute footing size for axial wall load only.

In SLS–2 the unfactored dead load at the ground floor is 2350 kips. We shall assume that the ground under the footing must also support the weight of the ground-floor tributary to the wall plus live load on this floor plus the weight of the footing itself. We will start with a trial footing size for axial load. The distributed vertical wall reinforcement is all No. 5 bars. in the ends. All the bars will extend vertically downward into the footing, where they must be have at least one development length embedment in the footing.

This is classed as a bottom bar for anchorage purposes. From Table A-11 the tension development length of a Grade-60, No. 5 bottom bar in 3000-psi concrete is 43.8 bar diameters or 27.4 in. The corresponding straight compression development length is 14 in. The embedment length of a 90° hook is 14 in., of which 4 bar diameters = 2.5 in. is the hook itself. The straight portion of a hooked anchorage between the surface of the concrete and the point where the hook begins is 11.5 in. ACI Section 12.5.5 does not allow hooks in compression. We shall hook all the vertical steel from the boundary elements with 90° hooks with a full compression development length before the start of the hook. The total depth needed to do this is 2.5 in. plus 14 in. of hook, for a total embedment of 16.5 in., say, 18 in. The hooks will sit on top of the flexural steel in the footing. Allowing 3 in. clear cover to the transverse footing steel plus 2 in. for the two layers of footing bars gives a total footing thickness of 5 in. + 18 in. = 23 in. as the thickness of the footing. We shall round this up to 24 in. The trial footing weighs 0.30 ksf. The basement floor load is 0.1 ksf. The net soil pressure is $(10 \text{ ksf} - 0.30 - 0.10) = 9.7$ ksf.

The area required for SLS–LC 2 is $2350/9.7 = 242$ ft^2. The length of the wall is 23.33 ft. Possible footing sizes are

23.33 by 10.5 ft; Area $= 245$ ft^2

1-ft wall plus 4-ft projection on each side $= 9$ ft by 28 ft $= 252$ ft^2

Overhang 3 ft 6 in. on all four sides $= 8$ ft by 33 ft; Area $= 264$ ft^2—Try this one.

The I for strong axis bending of the contact area is $I = 8$ ft $\times 33$ ft$^3/12 = 23,960$ ft^4.

Compute the maximum and minimum stresses for the SLS load combinations.

SLS–LC 2: Axial load $= 2350/264 = 8.90$ ksf

SLS–LC 3: Axial stress $= 1710$ kips$/264$ ft^2 kips; Soil stress $= 6.48$ ksf

$$\text{Stress due to moments} = \frac{10,700 \text{ ft-kips} \times 33/2 \text{ ft}}{23,960 \text{ ft}^4} = 7.37 \text{ ksf}$$

Total stress $= 6.48 + 7.37 = 13.8$ ksf—This is too large.

Try a 10-ft-by-35-ft footing—Area = 350 ft^2, P/A = 1710/350 = 4.89 ksf

$I = 10 \times 35^3/12 = 35{,}730$ ft^4

Moment stress = 10,700 × 17.5/35,730 = 5.24 ksf

Total stress = 4.89 + 5.24 = 10.1 ksf—Still too big.

Try a 12-by-36-ft footing:

$$I = 46{,}650 \qquad A = 432 \text{ ft}^2 \qquad \text{Stress} = 3.96 + 4.13 = 8.09 \text{ ksf}$$

Use a 12-ft-by-36-ft wall footing.

Further calculations are required to complete the footing design. ∎

PROBLEMS

19-1 Design the East–West walls in the building for wind loads. All dimensions and loadings are the same, except the building is located in St. Louis where the 1/30-year 3-sec gust wind velocity is 90 miles per hour.

20

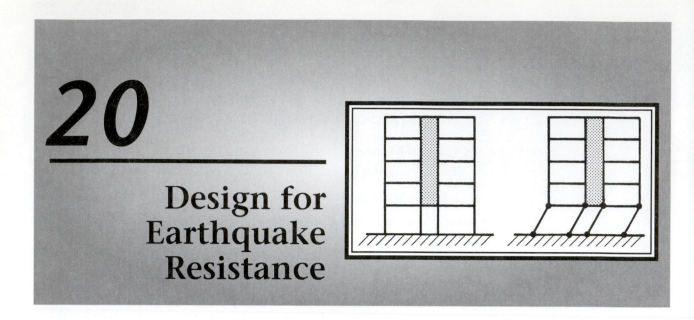

Design for Earthquake Resistance

20-1   INTRODUCTION—EARTHQUAKES AND SEISMIC RESPONSE SPECTRA

An earthquake causes the ground under a structure to move rapidly back and forth and thereby impart accelerations, a, to the base of the structure. If the structure were completely rigid, forces of magnitude $f = ma$ would be generated in it, where m is the mass of the structure. Because real structures are not rigid, the actual forces generated will differ from this value by amounts depending on the period of the building and the dominant periods of the earthquake. The determination of the force, f, is made more complicated because any given earthquake contains a wide and unpredictable range of frequencies and intensity base acceleration.

Seismic Design Requirements

The design of structures for seismic effects is in a state of rapid development. Major improvements and major changes in the philosophy of code requirements appear every year or so. For this reason, portions of design methods in effect when this edition was revised will be out of date by the time this book is published. Because of this, the presentation on seismic effects will be qualitative and somewhat incomplete. The design of members for equivalent lateral seismic forces has not changed as rapidly. This chapter includes examples of the design of members and joints that use the load combinations and strength-reduction factors from ACI Sections 9.2 and 9.3.

Plate Tectonics

Plate tectonics theory visualizes the earth as consisting of a viscous molten magma core with a number of lower-density rock plates floating on it. The exposed surfaces of the plates form the continents and the bottoms of the oceans. As time goes by, the plates move relative to each other, breaking apart in some areas and jamming together in others. Where the plates are moving apart, this movement causes cracks (or rifts) to form, generally in the ocean beds. In some cases, molten magma flows out of these rifts. In regions where the plates are either moving towards each other or are sliding adjacent to each other, compression and shear stresses are generated in the plates. The resulting strain energy builds up in the tectonic plates near the faults. The strain energy continues to build up until the stresses and strain energy at a locked fault surpass the limiting resistance to crushing or slipping along the fault. Once started, the energy is released rapidly, causing intense vibrations in the vicinity of the fault. Three main types of vibrations travel through the rock layers primary (compression) waves, secondary (shear) waves, and surface waves, each at a different speed. As a result, the effects of these vibrations differ from location to location.

Definitions

Size of earthquake—The size of an earthquake is quantified in terms of a Richter magnitude.

Magnitude—The *magnitude* of an earthquake is an estimate of the total energy release during the earthquake event. It is given by the *Richter magnitude, M*. An increase in magnitude by one digit, from 6 to 7, for example, involves an increase in energy released of $10^{1.5}$ times.

Intensity—The intensity of an earthquake is a largely subjective measure of the shaking at a given site, presented in terms of visual clues. This measure is most meaningful in estimating seismic forces on structures. A commonly used scale is called the Modified Mercalli, MM scale. Unfortunately many people find the verbal descriptions too vague.

Location of earthquake—The location of the place that the earthquake is initiated is called the *hypocenter*. The location of the hypocenter is defined by the latitude and longitude and depth below the surface. The *epicenter* is the point on the surface of the earth directly over the hypocenter. Earthquakes may involve regions of slip along surface faults ranging from several feet to several miles.

20-2   SEISMIC RESPONSE SPECTRA

The effect of the size and type of vibration waves released during a given earthquake can be organized so as to be more useful in design in terms of a *response spectrum* for a given earthquake or family of earthquakes. Figure 20-1a shows a family of inverted, damped pendulums, each of which has a different period of vibration, T. To derive a point on a response spectrum, one of these hypothetical pendulum structures is analytically subjected to the vibrations recorded during a particular earthquake. The largest acceleration of this pendulum structure during the entire record of a particular earthquake can be plotted as shown in Fig. 20-1b. Repeating this for each of the other pendulum structures shown in Fig. 20-1a and plotting and connecting the peak values for each of the pendulum structures produces an *acceleration response spectrum*.

Generally, the vertical axis of the spectrum is normalized by expressing the computed accelerations in terms of the acceleration due to gravity. If, for example, the ordinate of a point on the response spectrum is 2 for a given period T, it means that the peak acceleration of the pendulum structure for that value of T and for that earthquake was twice that due to gravity. The random wave content of an earthquake causes the derived acceleration response spectrum to plot as a jagged line, as shown in Fig. 20-2c. The spectra in Fig. 20-1b has been smoothed.

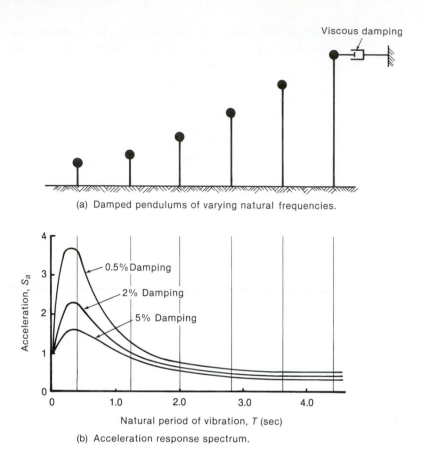

(a) Damped pendulums of varying natural frequencies.

(b) Acceleration response spectrum.

Fig. 20-1
Earthquake response
spectrum.

Velocity and Displacement Spectra

Following the procedure used to obtain an acceleration spectrum, but plotting the peak *velocity* relative to the ground during the entire earthquake against the periods of the family of pendulum structures, gives a *velocity response spectrum*. A plot of the maximum *displacements* of the structure relative to the ground during the entire earthquake is called a *displacement response spectrum*. These three spectra for a particular earthquake are shown in Fig. 20-2. Each of these curves represents idealized structural response to the measured earthquake motion. Individual structures are most affected by different parts of the spectra. Structures with periods of 0.2 to 0.5 seconds are almost rigid and are most affected by ground accelerations. Structures with medium periods (from 0.5 sec to 2.5 sec) are affected most by velocities. Structures with long periods, T, greater than 2.5 sec, such as tall buildings or long span bridges, are most affected by displacements.

Factors Affecting Peak Response Spectra

Period of Building

Lateral seismic forces are closely related to the fundamental period of vibration of the building. At periods less than about 0.5 sec, the maximum effect on the structure results from the magnification of the acceleration, as shown by the highest spikes in Fig. 20-2c. Comparing the three response spectra in Fig. 20-2 shows that, for periods less than about 0.5 sec, the highest spikes in the three diagrams were in the acceleration spectra. At medium periods (from about 0.5 sec to about 2.5 sec), the largest peak structural effects appear

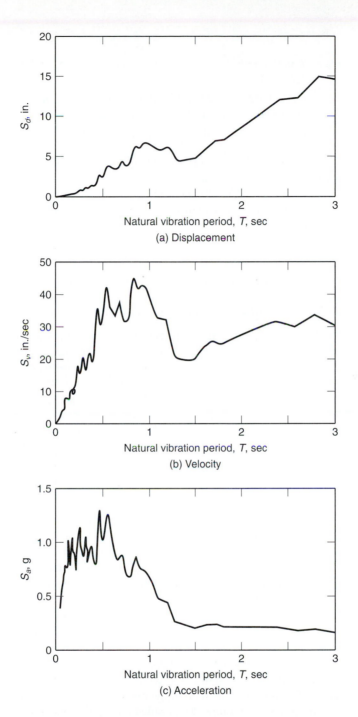

Fig. 20-2
Displacement, velocity, and
acceleration spectra for a
given earthquake [20-1].

in Figure 20-2b, the velocity response spectrum. Finally, at long periods (above about 2.5 sec), the dominant structural effects appear in the displacement spectra.

The period of the first mode of vibration, referred to as the *fundamental or natural period*, can be estimated from dynamic calculations, from empirical equations given in ASCE 7 [20-2], or from Rayleigh's method [20-1]. For concrete structures, the period, T, in seconds can be estimated from the formula

$$T = C_T h_n^{3/4}$$

(20-1a)

where

> T is the period in seconds, h_n is the height of the building above exterior grade in feet, and
>
> $C_T = 0.030$ for concrete buildings with moment-resisting frames providing 100 percent of the required lateral force resistance, or
>
> $C_T = 0.0020$ for all other concrete structures.

The exponent 3/4 in (20-1a) was changed to a variable exponent, x, in the 2002 version of ASCE 7, where x is given in a table.

Alternatively, the fundamental period of buildings not exceeding 12 stories and consisting entirely of concrete moment-resisting frames with a story height of at least 10 ft can be estimated from

$$T = 0.1N \qquad (20\text{-}1b)$$

where N is the number of stories above the exterior grade.

Effect of Damping on Response Spectrum

Each of the curves in Fig. 20-1b corresponds to a particular degree of damping. Damping, a measure of the dissipation of energy in the structure, is due to cracking, friction on the cracks, and so on. As the damping increases, the ordinates of the response spectrum decrease. Typically, a reinforced concrete building will have 1 to 2 percent of critical damping prior to the building being exposed to an earthquake. As cracking and structural and nonstructural damage develop during the earthquake, the damping increases to about 5 percent. By definition, *critical damping* acts to damp out structural vibrations.

Effect of Ductility on Seismic Forces

As an undamped elastic pendulum is deflected to the right, energy is stored in it in the form of strain energy. The stored energy is equal to the shaded area under the load-deflection diagram shown in Fig. 20-3a. When the pendulum is suddenly released, this energy reenters the system as velocity energy and helps drive the pendulum to the left. This pendulum will oscillate back and forth along the load-deflection diagram shown.

If the pendulum were to develop a plastic hinge at its base, the load-deflection diagram for the same lateral deflection would be as shown in Fig. 20-3b. When this pendulum is suddenly released, only the energy indicated by the triangle *a–b–c* reenters the system as velocity energy, the rest being dissipated by friction, heat, crack development, and so on.

Studies of hypothetical elastic and elastic–plastic buildings subjected to a number of different earthquake records suggest that the maximum lateral deflections of elastic and elastic–plastic structures are roughly the same. Figure 20-4 compares the load-deflection diagrams for an elastic structure and an elastic–plastic structure subjected to the same lateral deflection, Δ_u. The ratio of the maximum deflection, Δ_u, to the deflection at yielding for the inelastic structure, Δ_y, is called the *displacement ductility ratio*, μ:

$$\mu = \frac{\Delta_u}{\Delta_y} \qquad (20\text{-}2)$$

From Fig. 20-4, it can be seen that, for a ductility ratio of 4, the lateral load acting on the elastic–plastic structure would be $\frac{1}{\mu} = \frac{1}{4}$ of that on the elastic structure and the energy recovered in each cycle would be $\frac{1}{16}$ as great. Thus, if a structure is ductile, it can be designed for lower seismic forces.

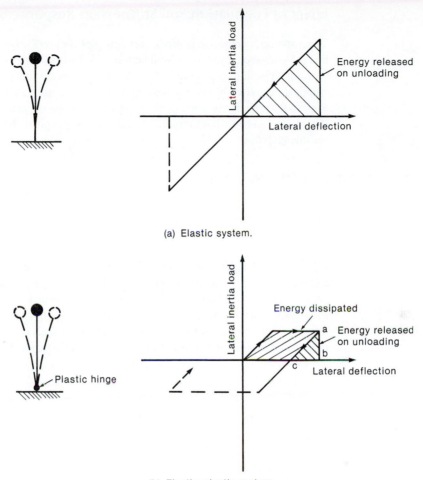

(a) Elastic system.

(b) Elastic–plastic system.

Fig. 20-3
Energy in vibrating pendu-
lums. (From [20-3], copyright
John Wiley & Sons, Inc.)

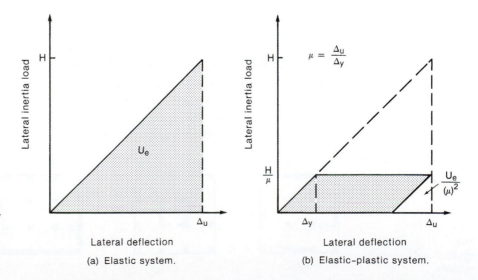

Fig. 20-4
Effect of ductility ratio, μ, on
lateral force and strain energy
in structures deflected to the
same Δ_u. (From [20-3],
copyright John Wiley & Sons,
Inc.)

(a) Elastic system.

(b) Elastic–plastic system.

Effect of Foundation Soil Stiffness on Response Spectrum

The response spectrum at bedrock normally plots below the response spectrum of a structure that is founded on layers of subsoil between the bedrock and the building footings. For structures on subsoil, the increase in the ordinates of the response spectrum is a function of the amplication caused by the various soil layers. Some soils, such as the lake bed sediments in Mexico City, tend to accentuate the seismic vibrations imparted to the structure. Since 1995, seismic design codes have required that designers recognize the effects of foundation soils on seismic response.

20-3   SEISMIC DESIGN PHILOSOPHY

Effect of Building Configuration

One of the most important steps in the design of a building for seismic effects is the choice of the building configuration—that is, the distribution of masses and stiffnesses in the building and the choice of load paths by which lateral loads will eventually reach the ground. In recent years, seismic design codes [20-2], [20-4] have classified buildings as *regular* or *irregular*. Irregularities include many aspects of structural design that are conducive to seismic damage. Irregular buildings require a more detailed structural analysis, design provisions to reduce the impact of each irregularity, and more detailing requirements than do regular buildings. Irregularities are classified as *plan irregularities* or *vertical irregularities*, as summarized here.

Plan Irregularities

1a and 1b. Torsional Irregularities.   Ideally, a building subjected to earthquakes should be symmetrical—or, at least, the distance between the center of mass (the point through which the seismic forces act on a given floor) and the center of resistance should be minimized. If there is an eccentricity, as illustrated in Fig. 20-5a, the building will

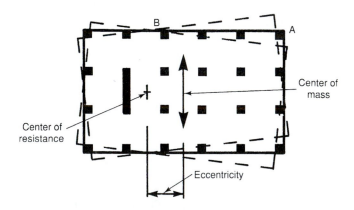

(a) Eccentricity of earthquake forces

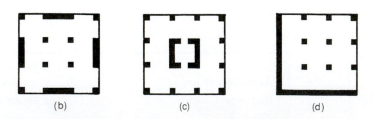

Fig. 20-5
Eccentricities and torsional deformations.

(b)   (c)   (d) undergo torsional deflections, as shown. The column at A in Fig. 20-5a will then experience larger shears than the column at B. The location of the center of resistance is affected by the presence of both structural and "nonstructural" elements.

The computed relative deflection of the top and bottom of a story is referred to as the *story drift*, δ_{max}. A category 1a torsional irregularity exists when the maximum story drift, at one end of a story, is more than 1.2 times the average story drift in the same story. This definition applies only to buildings with rigid or semirigid diaphragms.

A category 1b torsional irregularity exists when the ratio of maximum to average elastic computed drifts exceeds 1.4.

Irregular buildings should have significant torsional resistances and stiffnesses. Because the individual walls in Fig. 20-5b are farther from the center of resistance than those in Fig. 20-5c, they provide more torsional resistance. The core of the building in Fig. 20-5c is almost a closed tube, which tends to be stiffer in torsion than the four walls in the building shown in Fig. 20-5b. The plan in Fig. 20-5d is particularly unsuitable. It has a large eccentricity and very little torsional resistance. The relationship between the arrangement of gravity loads on a building and the first torsional buckling mode is discussed briefly in Section 12-9.

2. Reentrant Corner Irregularity.   If the plan has reentrant corners and the floor system projects beyond the reentrant corner by more than 15 percent of the plan dimension of the building in the same direction, the building is said to have a *reentrant corner irregularity*. For the building on the left in Fig. 20-6, one solution is to separate the two wings by a joint that is wide enough so that the wings can vibrate separately without banging together. If this is not practical, the region joining the two parts must be strengthed to resist the tendency to pull apart.

3. Diaphragm Discontinuity Irregularity.   Figure 20-7 shows a plan of a floor diaphragm transmitting seismic forces to shear walls at each end of a building. The diaphragm acts as a wide flat beam that develops tension and compression on its edges. Abrupt discontinuities or changes in the diaphragms, such as a notch in a flange, may lead to damage.

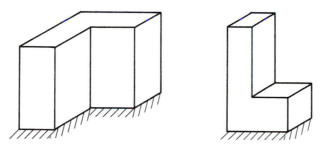

Fig. 20-6
Geometric irregularities.

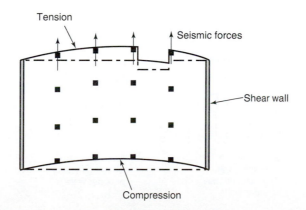

Fig. 20-7
Diaphragm discontinuities.

If there are abrupt changes in the stiffness of the diaphragms, including a cutout or open areas comprising more than 50 percent of the diaphragm or cross-sectional area, or 50 percent from one story to the next, the building is said to have a *diaphragm discontinuity irregularity*. The loss of cross-sectional area for the discontinuity shown in Fig. 20-7 is much smaller than 50 percent and would not qualify as a diaphragm irregularity.

Vertical Irregularities

Vertical irregularities are abrupt changes in the geometry, strength, or stiffness of a structure from floor to floor.

1a and 1b. Stiffness Irregularity—Soft Story.   A category 1a soft story is one in which the lateral stiffness is between 70 and 80 percent of that of the story above or below it. This becomes an *extreme soft-story irregularity* (1b) if the lateral stiffnesses are from 60 to 70 percent of those of the adjacent stories. The soft story created by terminating or greatly reducing the stiffness of the shear walls of the ground story, shown in Fig. 20-8, concentrates lateral deformations at this level.

2. Weight (Mass) Irregularity.   A mass irregularity exists where the effective mass of any story exceeds 150 percent of the effective mass of an adjacent story.

3. Vertical Geometric Discontinuity.   This type of irregularity occurs when the horizontal dimension of the lateral-force-resisting system in any story is more than 130 percent of that in an adjacent story.

4. In-Plane Discontinuity in Vertical, Lateral-Force-Resisting Elements.   An in-plane discontinuity is considered to exist where an in-plane offset of the lateral-force-resisting elements is greater than the length of those elements, as shown in Fig. 20-9c, or where the stiffness of the resisting element in the story below is smaller than that for the story in question.

5. Discontinuity in Lateral Strength: Weak Story.   A weak story exists if the lateral resistance of a story is less than 80 percent that of the story above. The lateral resistance of a story is the total strength of all lateral-force-resisting elements in the story.

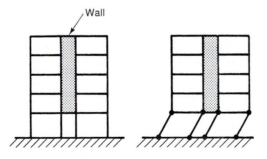

Fig. 20-8
Soft story due to discontinued shear walls.

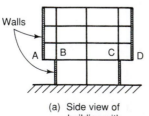

(a) Side view of building with discontinuous shear walls

(b) End view showing rocking of upper walls

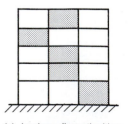

(c) In-plane discontinuities

Fig. 20-9
Discontinuous shear walls.

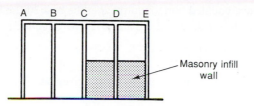

Fig. 20-10
Differences in column
stiffnesses.

Variations in the column stiffnesses attract forces to the stiffer columns. Because of their different free lengths, the lateral stiffness of column D in Fig. 20-10 would be four times that of column B for the same cross section. Initially, column D would be called upon to resist four times the shear force in column B. Frequently, such a column will fail in shear above the wall. Sometimes the change in column stiffness is caused by the restriction of free movement caused by nonstructural elements, such as the masonry walls shown shaded in the figure.

20-4   SEISMIC FORCES ON STRUCTURES

The ACI Code does not specify earthquake ground motions for a given site or give details about how structures should be analyzed for seismic actions. These details are provided in the general building code for the area. Currently, the general building codes allow several levels of seismic analysis. The two most common methods for determining seismic forces are variations of the *equivalent lateral force procedure* and the *modal analysis* procedure. The former idealizes the inertia forces on the structure with a series of equivalent horizontal static loads; the latter superimposes the computed lateral forces from several modes of vibration of the structure.

Vertical and Horizontal Components of E

American earthquake design codes [20-2], [20-4] published since 1997 have divided the earthquake load, E, into horizontal and vertical force components, E_h and E_v, as follows:

$$E = E_h + E_v \qquad (20\text{-}3)$$

$$E = \rho Q_E + 0.2 S_{DS} D \qquad (20\text{-}4)$$

In (20-4), ρ is a *reliability/redundancy factor* that varies from 1.0 to 1.5 with the extent of structural redundancy present in the building, and Q_E is the horizontal load effect caused by E. In (20-3), E_v is the vertical component of E, taken equal to $S_{DS}D$, the design spectral acceleration, S_{DS}, at a short period such as 0.2 sec, multiplied by the dead load, D. Assume that S_{DS} is about 0.30, then $E_v \simeq 0.2 \times 0.3D$. This is equivalent to having a vertical earthquake load of approximately 6 percent of the dead load.

Equivalent Lateral Force Method for Computing Earthquake Forces

Typically, the equivalent lateral force method is permitted for regular buildings up to about 20 stories. Sometimes it can be used for irregular buildings if special attention is given to the types of irregularities.

Geotechnical tests of the subsoil at the site help the designer estimate the degree to which soil–structure interaction will modify the seismic effects on the structure.

Seismic Base Shear, V_b

The seismic base shear is calculated as

$$V_b = C_s W \qquad (20\text{-}5)$$

where V_b is the computed horizontal shear at the base of the building, C_s is the seismic response coefficient for the building, and W is the effective seismic weight of the building.

Seismic Response Coefficient, C_s

Typically, the seismic response coefficient is given by

$$C_s = \frac{S_{DS}}{R/I} \qquad (20\text{-}6)$$

where S_{DS} quantifies the response spectrum as a function of the period, T, the damping, and the foundation stiffness; R is a *response-modification factor* that accounts for the reduction in seismic loads caused by inelastic action and energy dissipation; and I is the earthquake importance factor for the building and its occupancy.

A typical plot of C_s as a function of the period, T, is given in Fig. 20-11. Sometimes, this curve is drawn with three branches, adding a steep rising section at low periods. The actual shape used depends on the factors incorporated in C_s.

Effective Seismic Weight of the Building, W

The weight, W, of the building used to compute V_b is intended to represent the gravity loads likely to be present when an earthquake occurs. In ASCE 7, W is calculated as follows:

W = 100 percent of the unfactored dead load,
+ 100 percent of the partition load based on the weight of the partitions (or a minimum weight of 10 psf),
+ at least 25 percent of the unfactored live load,
+ not less than 20 percent of the unfactored snow load on a roof,
+ 100 percent of the unfactored load from the full contents of any tanks, and
+ the weight of permanent equipment

Force-Modification Factor, R

The force-modification factor, R, reflects

(a) the ability of the structure to dissipate energy through inelastic action, as shown in Fig. 20-4b, and

(b) the redundancy of the structure.

It is assumed that the level of ductility governs the reduction in seismic forces for the various families of lateral-force-resisting systems. Typical values are given in Table 20-1.

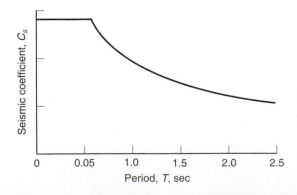

Fig. 20-11
Variation of seismic response coefficient, C_s, with period, T.

TABLE 20-1 Response-Modification Coefficient, R, for Seismic Resistance

Basic Structural System	Seismic Force Resisting System	Response Modification Coefficient R
Bearing wall system	Reinforced concrete shear walls	4.5
Building frame system	Reinforced concrete shear walls	5.5
Moment-resisting frame system	Special moment frames (SMF)	8
	Intermediate moment frames (IMF)	5
	Ordinary moment frames (OMF)	3
Dual system with a SMF capable of resisting at least 25 percent of prescribed seismic forces	Reinforced concrete shear walls	7.5
Dual system with a IMF capable of resisting at least 25 percent of prescribed seismic forces	Reinforced concrete shear walls	6
Inverted pendulum structures	Special moment frames	2.5

Source: Abridged from [20-2].

Distribution of Lateral Forces over the Height of a Building

The base shear, V_b, from (20-5) is distributed as a series of lateral forces at each floor level and at the roof. The distribution of the lateral forces is assumed to be similar to the deflected shape for the first mode of vibration, which corresponds to an inverted triangular distribution of lateral forces at the floor levels. To account for the higher modes of vibration the force F_x is broken into two parts:

(i) a single concentrated load, F_t, acting at the top on the order of 0 to 15 percent of the base shear, V_b, and

(ii) the rest of the lateral load, $(V_b - F_t)$, distributed at each floor level including the top as an inverted triangle.

Thus, the lateral force at a given floor level, x, is

$$F_t = \text{extra load at the top}$$

$$F_x = (V_b - F_t) C_{vx} \qquad (20\text{-}7)$$

where

$$C_{vx} = \frac{w_x h_x}{\sum\limits_{i=1}^{n} w_i h_i} \qquad (20\text{-}8)$$

in which F_x, w_x, and h_x are the lateral force, weight, and height, respectively, at level x above grade; $i = 1$ refers to the first level of the building above grade; and $i = n$ refers to the top level (roof). Typical lateral load distributions are shown in Fig. 20-12.

Some codes present (20-7) without the force F_t. Instead, the terms in the numerator and denominator of (20-8) are taken to a power.

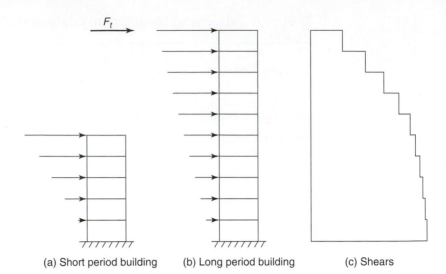

Fig. 20-12
Distribution of equivalent lateral forces and shears.

(a) Short period building   (b) Long period building   (c) Shears

Story Shear

The shear, V_x, in any story x is the sum of the lateral forces, F_i, acting *above* that story:

$$V_x = \sum_{i=x}^{n} F_i \qquad (20\text{-}9)$$

Story Torsion

The story shear, V_x, in a story is assumed to act horizontally through the center of mass of the story, resulting in a torsional moment, M_{tx}, equal to the product of the seismic forces times the horizontal eccentricity, e_x or e_y, between the center of mass and the center of resistance, measured perpendicular to the line of action of the seismic force. (See Fig. 20-5a.) To account for accidental torsion, a moment, M_{tax}, equal to the story shear times a distance e_a is added. Some codes take e_{ax} equal to $\pm 0.05 L_b$, where L_b is the horizontal length of the building perpendicular to the assumed direction of the applied forces. In that case,

$$M_{tx} = V_x \times e_y + V_x \times e_{ay} \qquad (20\text{-}10)$$

where M_{tx} is the torsional moment due to earthquake forces in the x direction, e_y is the distance in the y direction between the center of mass and the center of resistance, and e_{ay} is 5 percent of the length in the x direction. The second term represents an accidental increase in the torsional effects. Such an increase may occur if, for example, a corner column, such as A in Fig. 20-5a, cracks and loses some of its stiffness before the other columns crack. When this occurs, the center of resistance moves toward the stiffer columns (to the left in Fig. 20-5a), thereby increasing the torsional effects. Each element in the building is then designed for the most severe effects of the accidental torsions due to forces in the x direction and the y direction.

Overturning-Moment Reduction Factor

The distribution of lateral forces from (20-7) gives an envelope of the lateral story shears for a building. At any given time during an earthquake, some of these forces may act to the left and others to the right, although (20-7) assumes that they are all in the same direction. When the overturning moments in the structure are computed from these forces, the moments are too high, because some of the forces should have negative signs. To account for this, some codes multiply the overturning moments in each story by an *overturning-moment reduction factor.*

The total lateral forces are used in a linearly elastic structural analysis of the frame. For regular structures, independent two-dimensional models may be used. For concrete buildings, cracked-section properties are assumed in the analysis. For irregular structures, three-dimensional analyses must be used. Where the diaphragms are flexible relative to the lateral-force-resisting members, that flexibility must be represented in the analysis.

In summary, two concepts are important here. First, the force developed in the structure does not have a fixed value, but instead results from the stiffness of the structure and its response to a ground vibration. Second, if a structure is detailed so that it can respond in a ductile fashion to the ground motion, the earthquake forces are reduced from the elastic values.

20-5   DUCTILITY OF REINFORCED CONCRETE

Factors affecting the ductility of reinforced concrete beams under monotonically applied loadings have been discussed in Sections 4-2, 4-3, 5-3, and 11-2. (See Figs. 4-16, 5-15, 5-16, and 11-5.) The ductility of a beam increases as the ratio ρ/ρ_b goes down and as ρ'/ρ goes up, where ρ_b is the reinforcement ratio for balanced failure and ρ' is the ratio of compression reinforcement.

When a reinforced concrete member is subjected to load, flexural and shear cracks develop, as shown in Fig. 20-13a. When the load is reversed, these cracks close and new cracks form [20-5]. After several cycles of loading, the member will resemble Fig. 20-13b. The left end of the beam is divided into a series of blocks of concrete held together by the reinforcing cage. If the beam cracks through its depth, as shown in Fig. 20-13b, shear is transferred across the crack at low rotations by dowel action of the longitudinal reinforcement and grinding friction along the crack. If the concrete cover crushes, the longitudinal bars will buckle unless restrained by closely spaced stirrups or hoops. The hoops also provide confinement of the core concrete, increasing the beam's ductility.

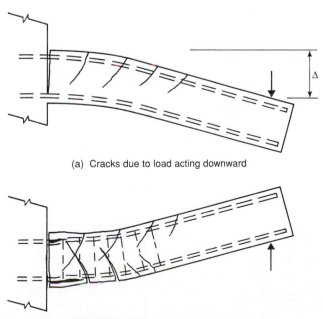

(a) Cracks due to load acting downward

Fig. 20-13
Beam subjected to cyclic loads.

(b) Cracks due to load acting upwards

It should be noted that the ductility ratio μ is defined in (20-2) in terms of the deflection Δ at the end of the beam shown in Fig. 20-13 or at a top of a building. Because most of the deformation is concentrated in the cracked and hinging regions, the curvature ductility, ϕ_u/ϕ_y, in the hinging regions of the beam is significantly larger than the required deflection ductility, μ.

In Section 3-3, it was shown that concrete subjected to triaxial compressive stresses increases in both strength and ductility (Fig. 3-15). In a spiral column, the lateral expansion of the concrete inside the spiral stresses the spiral in tension, and this in turn causes a confining pressure on the core concrete leading to an increase in the strength and ductility of the core (Fig. 11-5). ACI Chapter 21 requires that beams, columns, and the ends of shear walls have *hoops* in regions where the flexural reinforcement is expected to yield. Hoops are closely spaced closed ties or continuously wound ties or spirals, the ends of which have 135° hooks with six-bar-diameter (but not less than 3-in.) extensions into the confined core. The hoops must enclose the longitudinal reinforcement and give lateral support to those bars in the manner required for column ties in ACI Section 7.10.5.3. Although hoops can be circular, they most often are rectangular, as shown in Fig. 20-14, because most beams and columns have rectangular cross sections. The core concrete shown shaded in Fig. 20-14a is confined by the hoop. As a result, it tends to be more ductile and a little stronger than the unconfined concrete. In addition to confining the core concrete, the hoops restrain the buckling of the longitudinal bars and act as shear reinforcement.

Special moment frame members designed using ACI Chapter 21 can achieve deflection ductilities in excess of 5 and well detailed flexural walls can achieve about 4, compared to 1 or 2 for conventional concrete frames.

The response-modification coefficients, R, given in Table 20-1 are a measure of the deflection ductilities various types of structures can attain.

20-6 GENERAL ACI CODE PROVISIONS FOR SEISMIC RESISTANCE

Applicability

Seismic design provisions presented in ACI Chapter 21 were extended in the 2002 code to apply to cast-in-place and precast structures.

ACI Sections 21.2.1.2, 21.2.1.3, and 21.2.1.4 give the design requirements for regions of low, intermediate or moderate, and high seismic risk, respectively. These requirements are summarized in Table 20-2.

ACI Chapter 21 refers to a moment-resisting frame designed by using ACI Chapters 1 to 18 as an ordinary moment frame (OMF), and to a moment-resisting frame designed using ACI Chapters 1 to 18 plus ACI Section 21.12, which requires special detailing, as an intermediate moment frame (IMF). A moment-resisting frame designed using ACI Chapters 1 to 18 plus ACI Sections 21.5 to 21.5 is called a special moment frame (SMF).

(a)

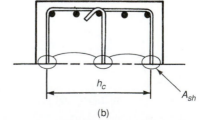

(b)

Fig. 20-14
Confinement by hoops.

TABLE 20-2   ACI 318-02 Sections Applicable to Various Seismic Risk Levels

Seismic Risk Level	Assigned Seismic Performance or Design Category	Moment Resisting Frames	Beams, Columns, and Joints	Structural Walls Coupling Beams and Diaphragms	Foundations
Low, 21.2.1.2					
21.2.1.2	Low	ACI Chapters 1 to 18 and 22 as modified in Chapter 21.	ACI Chapters 1 to 18 and 22 as modified in Chapter 21.	ACI Chapters 1 through 18 and 22 as modified in Chapter 21.	None ACI Chapters 1 through 18
21.2.1.2	Low, but design for seismic loads from Chapter 21 is required.	Must satisfy Chapter 21 for special systems as applicable.	Must satisfy Chapter 21 for special systems as applicable.	Must satisfy Chapter 21 for special systems as applicable.	
Moderate or Intermediate, 21.2.1.3 and 21.12					
		Intermediate moment frames 21.12.	Intermediate moment frames 21.12.		
High, 21.2.1.4					
Special moment frames, diaphragms, walls and trusses.			Chapters 1 through 18 plus 21.2, 21.3, 21.4, and 21.5. Frame members not part of lateral-force-resisting system 21.11.	Cast-in-place walls 21.2, 21.7 plus for precast walls, 21.8.	21.2.2.3 and must comply with Chapter 21.10

Materials

The compressive strength of the concrete shall not be less than 3000 psi (ACI Section 21.2.4.1). Because some high-strength lightweight concretes display brittle crushing failures (see Fig. 3-25), the strength of lightweight concrete shall not exceed 4000 psi unless good behavior of the particular high-strength lightweight aggregate concrete is documented.

Reinforcement resisting earthquake-induced stresses in frame members and the boundary elements of walls shall comply with ASTM A 706, *Specification for Low-Alloy Steel Deformed Bars for Concrete Reinforcement*. Specially graded A 615 steel may also be used.

Load Factors, Load Combinations, and Strength-Reduction Factors

Design will be based on

$$\phi R_n \geq R_u \tag{20-11}$$

where R_u is the sum of the factored load effects for an applicable load combination, U. R_n is the nominal resistance of the member and ϕ is is the applicable strength-reduction factor from ACI 318-02 Sections 9.2 and 9.3, or from ACI 318-02 Sections C.2 and C.3.

The examples in this chapter are based on the load and resistance factors from ACI 318-02, Sections 9.2 and 9.3.

Load and Resistance Factors— ACI 318-02 Sections 9.2 and 9.3

ACI 318-02 Section 9.2.1 presents seven load combinations, including two which include earthquake loads, E:

Load combination 9–5 $U = 1.2D + 1.0E + 1.0L + 0.2S$ (ACI Eq. 9-5)

and

Load combination 9–7 $U = 0.9D + 1.0E + 1.6H$ (ACI Eq. 9-7)

where D is an unfactored dead load effect, L is an unfactored live load effect, S is an unfactored snow load effect, E is an unfactored earthquake load effect, as calculated in Section 20-4, and H is an unfactored load effect due to the weight and pressure of soil, including water in soil or other materials. (An *effect* is the result of a force acting on a structure.)

Load combination 9–7 is used when the dead load stabilizes a structure subjected to overturning loads or stress reversals.

ACI Section 9.3.4 defines special reduction factors, ϕ, for three types of *shear-sensitive* members encountered in seismic design. In part (a) for structural members with a nominal shear strength less than the shear corresponding to the development of the nominal flexural strength of the member, the factor ϕ shall be taken as 0.60. The calculation of the shear corresponding to the development of the flexural strength of a member is discussed in Section 20-7 for beams and 20-8 for columns. In part (b) for diaphragms, the ϕ factor for shear strength shall not exceed the lowest ϕ factor in shear used for the vertical components of the primary lateral-force-resisting system. In part (c), the ϕ factor for shear in joints and diagonally reinforced coupling beams is set at 0.85. The shear forces in these members are determined by a *capacity design* procedure, as discussed in the next section, and the ϕ factor was selected to be consistent with prior editions of the ACI Code.

Capacity Design

The reinforcing details required to ensure adequate ductility of hinging regions in a laterally loaded structure tend to be tedious to design and expensive to place. This complexity is reduced if the structure is designed so that only a few cross sections form hinges under the seismic loads, while the rest of the structure has enough reinforcement to remain elastic. Consider, for example, a special structural wall with the shear and moment diagrams shown in Fig. 20-15. This wall is loaded by vertical loads totaling P_{uw} and horizontal loads totaling E. In a nonseismic design, the size of the wall and the wall reinforcement would be chosen to have the desired stiffness plus a strength equal or greater than the effect of the factored loads.

Because this wall acts as a vertical cantilever, the first plastic hinge is expected to occur at the base of the wall. In *capacity design* this section is designed to hinge at a lateral load level that will allow the hinge to be detailed for ductile behavior under flexure and compressive axial loads. At the same time, however, the sections away from the hinging region are designed to remain elastic throughout the loading history, thereby avoiding the need for seismic detailing at those sections.

Because a shear failure of the hinge region would not be ductile, the reinforcement providing the shear resistance at the plastic hinge at the bottom of the wall is chosen so that it will remain elastic at the shear expected at hinging of the wall, as shown by the moment and shear envelopes in Fig. 20-15. Thus, the hinging capacities of all sections not chosen to be hinges exceed the moments from the assumed hinging mechanism in the structure. This process is called *capacity design* [20-6].

The structure is not merely designed to resist the applied load effects; instead it is proportioned so that the hinging capacities of all nonyielding elements in the structure exceed the loads corresponding to yielding of the critical elements that the designer has designed to yield first.

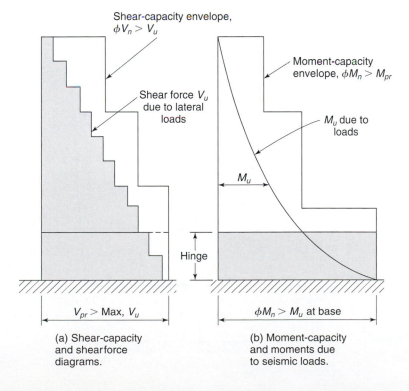

Fig. 20-15
Capacity design of a shear wall.

(a) Shear-capacity and shear force diagrams.

(b) Moment-capacity and moments due to seismic loads.

Strong Column–Weak Beam Design

If plastic hinges form in columns, the axial force can cause a rapid degradation of the ability of the hinge to absorb energy while undergoing cyclic motions. As a result, the design of ductile moment-resisting frames attempts to force the structure to respond in what is referred to as strong column–weak beam action in which the plastic hinges induced by the seismic forces form at the ends of the beams, as shown in Fig. 20-16. The hinging regions are detailed to maintain their shear capacity while the plastic hinges to undergo flexural yielding in both positive and negative bending.

Service-load-level and strength-level earthquakes. Prior to the 1999 ACI Code, earthquake design was carried out using *service-load-level* earthquake forces in 1995 ACI Eqs. (9-2) and (9-3) with a load factor of 1.4 on the earthquake load, E. In those codes the design earthquake had a 10 percent probability of exceedance in 50 years. More recent codes, such as ASCE 7-98 and the 2002 ACI Code, are based on *strength-level* earthquakes, which correspond to earthquakes with a 2 percent probability of exceedance in 50 years. The strength-level earthquakes are larger than the service-level earthquakes, and as a result, the load factor on E in 1999 ACI Code Eqs. (9-2) and (9-3) was reduced to $1.0E$.

20-7 FLEXURAL MEMBERS IN SPECIAL MOMENT FRAMES

Geometric Limits on Beam Cross Sections

ACI Section 21.3 defines a flexural member as a member proportioned to resist primarily flexure and having either no axial load or a factored axial compressive force less than $(A_g f'_c/10)$. Geometric limitations are placed on the span-to-depth ratio ($\ell_n \geq 4h$) to avoid deep beam action, except that this limit does not apply to coupling beams in shear walls. The widths of flexural members in special moment frames shall not be less than (a) 0.3 times the depth of the beam, or (b) 10 in., or (c) not more than the width of the

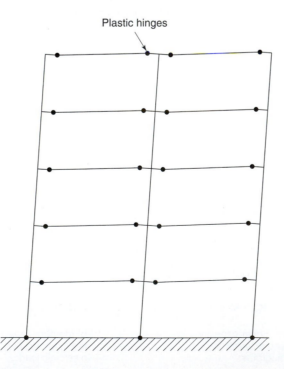

Plastic hinges

Fig. 20-16
Strong-column–weak-beam
behavior.

supporting member plus $\frac{3}{4}d$ on either side. All of these limitations are intended to provide members that will be large enough to contain adequately anchoral hoops.

Classification of Resisting Moments

Two levels of resisting moments are used in seismic design:

$M_n = $ *nominal moment resistance*, calculated by using $\phi = 1.0$, the specified yield strength, f_y, and the specified concrete strength, f'_c. The most economical solution is to set $M_n = M_u/\phi$. The nominal moment resistance is used in ACI Section 21.4.2.2 to ensure that the columns are stronger than the beams meeting at a joint.

$M_{pr} = $ *probable moment resistance*, calculated by using $\phi = 1.0$, but replacing f_y with $1.25 f_y$ because the average yield strength tends to be greater than f_y and because beam longitudinal reinforcement will likely go into strain hardening in plastic hinging zones. The probable moment resistance is used in ACI Section 21.3.4.1 to ensure that the shear strengths of beams or columns meeting at a joint exceed the shears that equilibrate flexural hinging at the ends of the beams at a joint. It is also used in ACI Section 21.4.5.1 to compute shears in columns.

Sign Convention for Moments

The moments in this chapter, including those in Example 20-1, are given in a *beam sign convention*, in which a moment is positive when it causes compression in the top fiber of the beam and negative when it causes compression in the bottom fiber. During an earthquake, a building will sway right and left. Sway to the right, as shown in Fig. 20-17, gives positive sway moments at the left end of beam *A–B* corresponding to tension in the bottom bars, and negative sway moments at the right end, with tension in the top bars. The opposite occurs when the frame sways to the left.

Computation of Moment Capacity of Sections

In calculating the moment capacities of beams subjected to earthquake forces, it is widespread practice to ignore both the compression flange of the beam and any reinforcement in

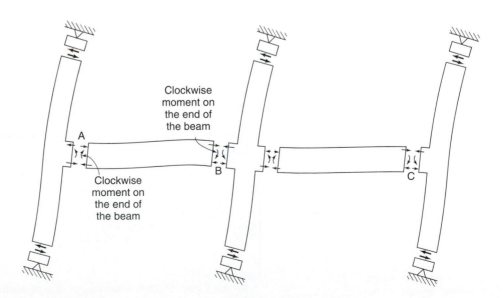

Fig. 20-17
Moments in a frame that is swaying to the right.

the flange or steel in the compression side of the beam. Ignoring the flange and the slab reinforcement results in the calculated strength of the beam being less than it would be if these effects were included in the calculations [20-7]. In rectangular beams the error introduced by this is relatively small, as is shown in Figs. 5-13 and 5-15. The overstrength in flexure, however, does use up some of the cushion of shear strength provided to make shear failures less likely than flexural failures. Also, ACI Section 21.4.2.2 now requires that the steel in the beam flanges be considered during the computation of the required column strengths. We will allow for the additional moment capacity due to the reinforcement in flanges with bars in tension.

Longitudinal (Horizontal) Reinforcement

Seismic loads cause the moment diagram shown by the solid line in Fig. 20-18b when the frame is swaying to the right and an opposite diagram shown by the dashed line when the frame is swaying to the left. To this must be added the dead- and live-load moments shown in Fig. 20-18c, giving the moment envelope in Fig. 20-18d. The maximum moments in the span normally occur at the face of a column. In addition to providing adequate moment resistance,

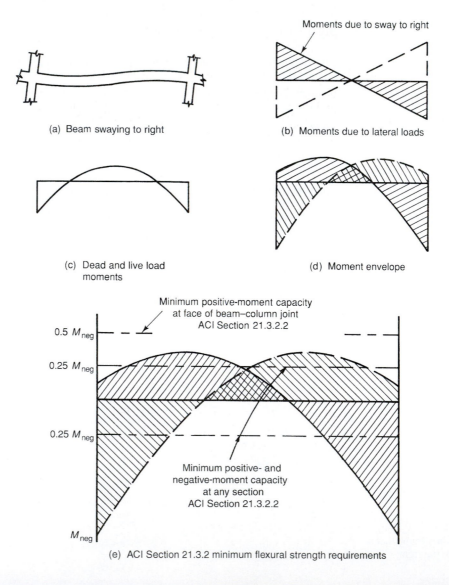

(a) Beam swaying to right

(b) Moments due to lateral loads

Moments due to sway to right

(c) Dead and live load moments

(d) Moment envelope

Minimum positive-moment capacity at face of beam–column joint ACI Section 21.3.2.2

$0.5\ M_{neg}$

$0.25\ M_{neg}$

$0.25\ M_{neg}$

Minimum positive- and negative-moment capacity at any section ACI Section 21.3.2.2

M_{neg}

(e) ACI Section 21.3.2 minimum flexural strength requirements

Fig. 20-18
Moment diagram due to gravity loads and seismic loads.

flexural reinforcement must satisfy the following detailing requirements from ACI Section 21.3.2 to provide adequate ductility:

1.   At least two bars must be provided continuously top and bottom.

2.   The areas of each of the top and bottom reinforcement at every section shall not be less than given by (ACI Eq. 10-3), nor less than $200b_wd/f_y$. The reinforcement ratio, $\rho = A_s/bd$, shall not exceed 0.025 for either the top or bottom reinforcement. Normally, ρ would not exceed about 0.015.

3.   The positive-moment strength of the beam section at the face of the beam–column joint shall not be less than half the negative-moment strength. (See Fig. 20-18e). This provides $\rho' \approx 0.5\rho$, which allows the beam to develop large curvatures at hinging regions and greatly improves the ductility of the ends of the beams.

4.   At every section, the positive and negative moment capacity shall not be less than one-fourth the maximum moment capacity provided at the face of either joint. This is also plotted in Fig. 20-18e.

The upper limit on ρ of 0.025 in item 2 is greater than the ρ that would normally be used in a nonseismic beam with Grade-60 steel and most concrete strengths. It is set this high because there will always be confinement reinforcement and compression steel with ρ' equal to at least 0.25ρ.

Development and Splicing of Flexural Reinforcement

The development lengths and splice lengths specified in ACI Sections 12.2 and 12.15 apply to frames resisting seismic forces except as altered in ACI Section 21.5.4, which deals with development of bars in beam–column joints. ACI Section 21.3.2.3 *prohibits* lap splices

1.   within joints;

2.   within 2h from the faces of joints;

3.   in locations where flexural yielding can occur due to lateral deformations of the frame.

Lap splices must be enclosed by hoops or spirals at the smaller of 4 in. or $d/4$. Mechanical splices can be used as limited by ACI Section 21.3.2.4. Welding is not encouraged; tack welding of bars for assembly purposes embrittles the bars locally, and thus is not permitted.

Transverse Reinforcement

Transverse reinforcement is required

1.   to confine the concrete,

2.   to prevent buckling of the compression bars in the hinging areas (ACI Section 21.3.3),

3.   to provide adequate shear strength (ACI Section 21.3.4), and as mentioned in the preceding section, and

4.   to confine lap splices.

Confinement Reinforcement

Hoops for confinement and to control buckling of the longitudinal reinforcement are required

1.   over a length equal to 2h from the face of supports and

2.   within 2h on each side of other locations where hinging can result due to lateral deformations of the frame.

The spacing of the hoops is specified in ACI Section 21.3.3.2. In the rest of the beam, either stirrups or hoops are required at a maximum spacing of $d/2$.

ACI Section 21.1 defines a *seismic hook* as a hook on a stirrup, hoop, or cross-tie having a bend not less than 135° with a six-diameter (but not less than 3-in.) extension that engages the longitudinal reinforcement and projects into the confined concrete in the interior of the stirrup or hoop.

A *cross-tie* is defined as a continuous reinforcing bar having a seismic hook at one end and a hook not less than 90° with at least a six-diameter extension at the other end, as shown in Fig. 20-19a. Both hooks shall engage peripheral longitudinal bars. The 90° hooks of two successive cross-ties engaging the same longitudinal bars are alternated end for end, except as allowed in ACI Section 21.3.3.6.

A *hoop* is a closed tie, as shown in Fig. 20-19c, or a continuously wound tie. A closed tie can be made up of several reinforcing bars, each having seismic hooks at one or both ends. This allows the use of a number of bars or interlocking sheets of welded-wire fabric to make up a cage of hoops and longitudinal bars for a beam or column. In flexural members, ACI Section 21.3.3.6 allows hoops to be made up of a cross-tie as shown in Fig. 20-19a plus a stirrup with seismic hooks at each end, as shown in Fig. 20-19b. If the longitudinal bars secured by the cross-ties are confined by a slab on only one side of the beam, as shown in Fig. 20-19d, the 90° hooks on the cross-ties are placed on that side.

Shear Reinforcement

When the frame is displaced laterally through the inelastic deformations required to develop the ductility of the structure, the reinforcement at the ends of the beam will yield unless the moment capacity is several times the moment due to seismic loads. The yielding of the reinforcement sets an upper limit on the moments that can be developed at the ends of the beam. The design shear forces, V_e, are based on the shears due to factored dead and live loads (Fig. 20-20c) plus the shears due to hinging at the two ends of the beam for the frame swaying to the right or to the left, as shown in Fig. 20-20a, where M_{pr} is the probable moment capacity of the members, based on the dimensions and reinforcement at the joint and assuming a tensile strength of $1.25f_y$ and $\phi = 1.0$. For a rectangular beam without axial loads, ACI Section 21.4.5.1 requires that beams be designed for the sum of

$$V_{sway} = \frac{M_{pr1} + M_{pr2}}{\ell_n} \tag{20-12}$$

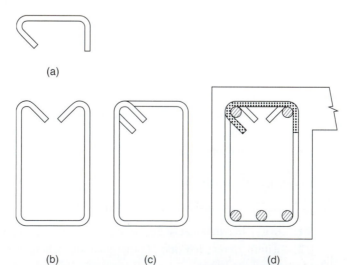

(a)

(b)   (c)   (d)

Fig. 20-19
Hoops and cross-ties.

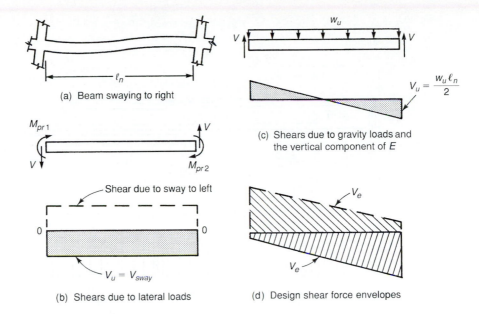

Fig. 20-20
Shear force diagrams due to gravity loads and seismic loads.

(a) Beam swaying to right

(b) Shears due to lateral loads

(c) Shears due to gravity loads and the vertical component of E

(d) Design shear force envelopes

and

$$V_g = \frac{w_u \ell_n}{2}$$

(20-13)

Design shear

$$V_e = V_g \pm V_{sway}$$

(20-14)

where

$$M_{pr} = 1.25 f_y A_s \left(d - \frac{1.25 A_s f_y}{2 \times 0.85 f'_c b} \right)$$

(20-15)

The beam is then designed for the resulting shear force envelope with $V_u = V_e$ in the normal way, except that if

(a) the shear, V_{sway}, due to the moments M_{pr1} and M_{pr2} is half or more of the total shear, V_e, *and*

(b) the factored axial compressive force (if any) including earthquake effects is less than $(A_g f'_c / 20)$,

then V_c is taken equal to zero. The damage to the hinging area due to repeated load reversals greatly reduces the ability of the cross section to resist shear, requiring more transverse reinforcement [20-3]. Hoops and stirrups provided to satisfy ACI Section 21.3.3 can also serve as shear reinforcement.

EXAMPLE 20-1   Design of Flexural Member in a Special Moment-Resisting Frame

The beam shown in Fig. 20-21a and b is a typical floor beam in the special moment-resisting frame of an office building. The beam supports a uniform unfactored dead load of 4.0 kips/ft and a uniform unfactored live load of 2 kips/ft. The concrete and reinforcement strengths are 4000 psi and 60,000 psi. Design the reinforcement.

1. **Select the level of seismic design.** Techniques for calculating earthquake loads are changing rapidly, so the seismic-force calculations will not be completed in detail. We shall arbitrarily choose $S_{DS} = 0.30$ g.

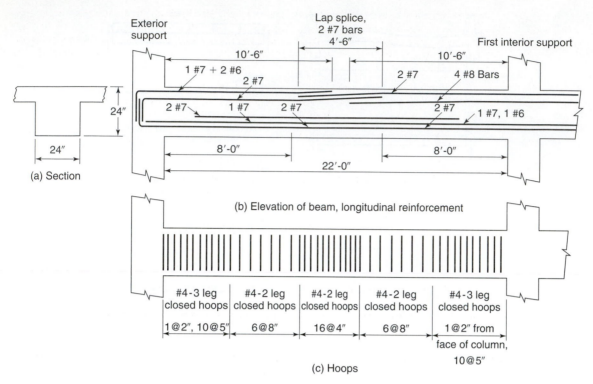

Fig. 20-21
Beam—Example 20-1.

2. Compute Factored Load Combinations from ACI 318-02 Section 9.2.1. ACI Section 9.2.1 presents seven load combinations for the structural design of buildings. Two of these apply specifically to earthquake loads. Often, the first selection of the beam reinforcement is made by using the nonseismic load combinations from ACI Section 9.2.1, because the beam must be able to resist everyday loads while it waits for an earthquake. Following this, the seismic detailing requirements from ACI Chapter 21 will be used to choose steel at other locations where hinges will form under seismic loads. By substituting the live-load factor 0.5 from ACI Section 9.2.1(a) in ACI Eq. (9-5) and replacing E in ACI Eqs. (9-5) and (9-7) with (20-3), and by similar substitutions in the other load combination equations, the factored loads can be computed as follows:

> 1. LC 9–1 $U = 1.4 \times 4.0 = 5.60$ kips/ft (ACI Eq. 9-1)
> 2. LC 9–2 $U = 1.2 \times 4.0 + 1.6 \times 2.0 = 8.0$ kips/ft (ACI Eq. 9-2)
> 3. LC 9–3 $U = 1.2 \times 4.0 + 1.0 \times 2.0 = 6.8$ kips/ft (ACI Eq. 9-3)

Seismic Load Combinations

> 5. LC 9–5 $U = 1.2D + 1.0E + 0.5L + 0.2S$ (ACI Eq. 9-5)
> 7. LC 9–7 $U = 0.9D + 1.0E + 1.6H$ (ACI Eq. 9-7)

Setting S and H equal to zero and substituting (20-3) into the two combinations with E loads gives

> 5. LC 9–5 $U = 1.2D + 1.0(\rho\,Q_E) + 1.0(0.2S_{DS}D) + 0.5L$ (ACI Eq. 9-5a)
> 7. LC 9–7 $U = 0.9D + 1.0(\rho\,Q_E) + 0.2S_{DS}D$ (ACI Eq. 9-7a)

where $\rho\,Q_E$ is the horizontal component of E and $0.2S_{DS}D$ is the vertical component. The horizontal component will be included in the lateral forces when the lateral-force-resisting system is analyzed. The vertical component will be added to the other vertical loads on a floor beam. We then have

> 5. LC 9–5 $U = 1.2 \times 4.0 + 1.0(0.2 \times 0.3 \times 4.0) + 0.5 \times 2.0 = 6.04$ kips/ft (ACI Eq. 9-5b)
> 7. LC 9–7 $U = 0.9D + 1.0E_h + 0.2S_{Ds}D$ (ACI Eq. 9-7b)

ACI load combinations 9-6 and 9-7 apply in cases where dead load stabilizes a structure that is later-ally loaded by wind or seismic loads that are apt to cause overturning. This is not normally a problem in beam design, and LC 9-6 and 9-7 will normally not be applied. Cases where these two load com-binations may apply include that of a beam with an overhang, where the reaction at the noncantilever end that anchors the moment from the overhang may change sign from upward on the end of the beam to downward (resisting uplift).

The largest factored uniform loads on the beam being designed are

1. LC 9–2: $w_u = 8.0$ kips/ft
2. LC 9–5: $w_u = 6.04$ kips/ft

The largest vertical load is $w_u = 8$ kips/ft (from LC 9–2).

3. Does the beam satisfy the definition of a flexural member? ACI Section 21.3.1 requires flexural members to have

(i) a factored compression force less than $0.1 A_g f_c'$. There is very little, if any, axial load—OK.
(ii) a clear span not less than four times the effective depth, $\ell_n/d = 22 \times 12/21.5 = 12.3$ —OK.
(iii) a width
  (a) not less than 10 in.—OK;
  (b) not more than width of column plus $\frac{3}{4}h$ on each side of column—OK.

Thus, the beam satisfies the requirements of a beam. If it did not, it would be necessary to change the dimensions of the beam.

4. Calculate the design moments. At all sections, the beam must have $\phi M_n \geq M_u$, where M_u is the moment due to the factored gravity loads plus the vertical component of the seismic loads. We shall use ACI Section 8.3.3 to compute the gravity-load moments at critical sections of an exterior span for maximum negative and positive moments. Normally, this would be part of the frame analysis.

Exterior negative gravity-load moment for LC 9–2: $\dfrac{w\ell_n^2}{16} = \dfrac{8.0 \times 22^2}{16}$

$$M_u = -242 \text{ ft-kips}$$

Midspan positive gravity-load moment LC 9–2: $M_u = 277$ ft-kips
Interior-support negative gravity-load moment LC 9–2: $M_u = -387$ ft-kips

The reinforcement chosen for each cross section, the ultimate moment, and the nominal and probable moment capacities are listed in Table 20-3. The various sections are referred to in the cal-culations by the case number from the first column of that table.

TABLE 20-3 Reinforcement and Moment Capacities, End Span

Case	Location	Sway Direction	M_u ft-kips	Reinforcement	A_s in.^{2}	ϕM_n ft-kips	M_{pr} ft-kips
1.	Interior End Negative M	Right	−387	4 No. 8 plus 2 No. 7	4.36	390	531 cw
2.	Exterior End, Negative M	Left	−242	3 No. 7 plus 2 No. 6	2.68	247	339 cw
3.	Exterior End Positive	Right	+124	3 No. 7	1.80	169	233 ccw
4.	Interior End Positive	Left	+195	3 No. 7 plus 1 No. 6	2.24	208	286 ccw
5.	Midspan, Positive M	Either	−277	4 No. 7 plus 2 No. 6	3.28	300	410

Note: "cw" and "ccw" stand for "clockwise" and "counterclockwise," respectively.

5. Calculate the steel required for flexure. Case 1, Table 20-3. Interior support, negative moment, sway to right. Assuming rectangular beam action, the required reinforcement for $M_u = -387$ ft-kips is computed. Assume one layer of steel, $d = 24 - 2.5 = 21.5$ in. We shall ignore compression steel (if any).

$$A_s = \frac{M_u}{\phi f_y j d}$$

Assume that $j = 0.90$ and $\phi = 0.9$. Then

$$A_s = \frac{387 \times 12,000}{0.9 \times 60,000 \times 0.9 \times 21.5}$$
$$= 4.44 \text{ in.}^2$$

For a first estimate, try six No. 8 bars, $A_s = 4.74$ in.^{2}. These will fit into one layer. ACI Section 21.5.1.4 requires the bar sizes be limited so that the column width or depth parallel to the bars is at least $20d_b$. The columns are 24 in. square. This sets the maximum bar diameter as $24/20 = 1.2$ in. Thus, No. 9 bars are the largest that can be used, and the No. 8 bars are OK.

However, the moment capacity of a beam with six No. 8 bars is 12 percent higher than is needed. We thus have

$$a = \frac{A_s f_y}{0.85 f'_c b} = \frac{4.74 \times 60,000}{0.85 \times 4000 \times 24}$$
$$= 3.49 \text{ in.}$$

$$\phi M_n = \phi A_s f_y(d - a/2)$$
$$= \frac{0.9 \times 4.74 \times 60,000(21.5 - 3.49/2)}{12,000}$$
$$= 421 \text{ ft-kips}$$

Try four No. 8 bars plus two No. 7 bars, $A_s = 4.36$ in.^{2}. These bars give a $\phi M_n = 390$ ft-kips, which just satisfies $M_u = -387$ ft-kips.

ACI Section 21.3.2.1

Check whether $A_s \geq A_{s,\min}$.

$$A_{s,\min} = \frac{3\sqrt{f'_c}}{f_y} b_w d, \text{ but not less than } \frac{200 b_w d}{f_y} \qquad \text{(ACI Eq. 10-3)}$$

$$= 1.63 \text{ in.}^2, \text{ but} \geq 1.72 \text{ in.}^2 - A_{s,\min} = 1.72 \text{ in.}^2$$

$$A_s = 4.36 \text{ in.}^2 > A_{s,\min} = 1.72 \text{ in.}^2 - \text{OK.}$$

$$\text{Check whether } \rho = \frac{4.74}{24 \times 21.5} = 0.0092 \leq 0.025 - \text{OK.}$$

ACI Section 10.3.3: Check whether section is tension-controlled.

$$\frac{a}{d_t} = \frac{3.07}{21.5} = 0.143 \qquad \frac{a_{tc\ell}}{d_t} = 0.375\beta_1 = 0.391 \qquad (4\text{-}23)$$

Since $a/d_t = 0.143$ is less than $a_{TCL}/d_t = 0.319$, the section is tension-controlled and $\phi = 0.9$. This check could also be made directly, by using similar triangles to compute the extreme tensile strains.

Case 1—Interior support, negative moment. Use four No. 8 plus two No. 7 top bars at the interior support. $A_s = 4.36$ in.^{2}, $\phi M_n = -390$ ft-kips.

Case 2—Exterior support, negative moment. From step 4, $M_u = -242$ ft-kips. This moment is required to support the factored gravity loads at times other than during an earthquake. **Try three No. 7 bars plus two No. 6 bars,** $A_s = 2.68$ in.^{2}, $\phi M_n = 247$ ft-kips—OK.

At these two sections, cases 1 and 2, the gravity loads control the steel selection. For the rest of the span, longitudinal reinforcement is provided to satisfy detailing rules in ACI Section 21.3.2.

Case 3—Exterior support, positive moment. ACI Section 21.3.2.2 requires the positive-moment capacity at the face of the support to be at least 0.5 times the ϕM_n for the negative-moment steel at the face of the support: 0.5×247 ft-kips $= 124$ ft-kips. Design for 124 ft-kips. **Try three No. 7 bars,** $A_s = 1.80$ in.^{2}, $a = 1.32$ in., and $\phi M_n = 169$ ft-kips. From step 4, the minimum A_s was 1.72 in.^{2}. We shall limit flexural steel to $A_s \geq 1.72$ in.^{2} As a result, it is not possible to use fewer than three No. 7 bars.

Case 4—Interior support, positive moment. ACI Section 21.3.2.2 requires that the positive-moment capacity at the face of the joint not be less than 0.5 times the negative-moment capacity ϕM_n provided by the negative-moment reinforcement at the face of the same joint. Thus, the minimum positive-moment capacity required by ACI Section 21.3.2.2 is $0.5 \times 390 = 195$ ft-kips. Design the interior support for a positive moment of $\phi M_n = 195$ ft-kips. **Use three No. 7 bars plus one No. 6, with the No. 6 bar cut off as shown in Fig. 20-21a.** Before the cutoff, $A_s = 2.24$ in.^{2}, and $\phi M_n = 208$ ft-kips.

Case 5—Midspan, positive moment: From step 4, the maximum positive moment at midspan is 277 ft-kips. **Try four No. 7 bars plus two No. 6 bars,** $A_s = 3.28$ in.^{2}, $\phi M_n = 300$ ft-kips; A_s satisfies the minimums. The extreme-tensile strain, ε_t, exceeds 0.004, so this is a beam. It also exceeds 0.005, so the beam is tension-controlled and ϕ can be taken as 0.9.

Minimum positive- and negative-moment capacities: ACI Section 21.3.2.2 requires that the minimum positive- and negative-moment capacities at any section along the beam not be less than 0.25 times the maximum negative-moment capacity provided at either joint: 0.25×390 ft-kips $= 97.5$ ft-kips. Two No. 7 bars are adequate as minimum steel.

The steel chosen for flexure and to suit the detailing requirements is shown in Fig. 20-21a.

5.   Compute the probable moment capacities, M_{pr}. The seismic shears in the beam are computed by assuming that plastic hinges form at each end of the beam with the reinforcement stressed to $1.25 f_y$ and $\phi = 1.0$. (See (20-13).)

Moments for frame swaying to the right (See Fig. 20-17a.)

Case 1—Probable interior-end negative moment, four No. 8 plus two No. 7 bars, $A_s = 4.36$ in.^{2} top steel. The probable depth of the stress block, a_{pr}, and the moment capacity, M_{pr}, for the top steel, with $1.25 f_y$ and $\phi = 1.0$, are

$$a_{pr} = \frac{(1.25 \times 60,000 \times 4.36)}{0.85 \times 4000 \times 24} = 4.01 \text{ in.}$$

$$M_{pr} = \frac{(1.25 \times 60,000) \times 4.36(21.5 - a_{pr}/2)}{12,000} = 531 \text{ ft-kips}$$

(clockwise on the interior end of the beam).

Case 2—Probable exterior-end negative-moment capacity for top steel: three No. 7 plus two No. 6 bars, $A_s = 2.68$ in.^{2}. Then

$$M_{pr} = \frac{(1.25 \times 60,000) \times 2.68\left(21.5 - \dfrac{1.25 \times 60,000 \times 2.68}{1.7 \times 4000 \times 24}\right)}{12,000} = 339 \text{ ft-kips}$$

(clockwise on the exterior end of the beam).

Moments for frame swaying to the left (See Fig. 20-22c.)

Case 4—Probable interior positive-moment capacity, three No. 7 bars plus one No. 6 bar, $A_s = 2.24$ in.^{2}; probable moment capacity is $M_{pr} = 286$ ft-kips clockwise on interior end of beam.

6.   Compute the shear-force envelope and design the stirrups. Figure 20-22(a) shows the moments and uniform load for load combination 9–5 acting on the beam with the frame swaying to the right. The reactions consist of two parts:

Hinges develop at both ends of the beam, with hinging moments of $M_{pr} = 339$ ft-kips at the exterior end and 531 ft-kips at the interior-end negative-moment region.

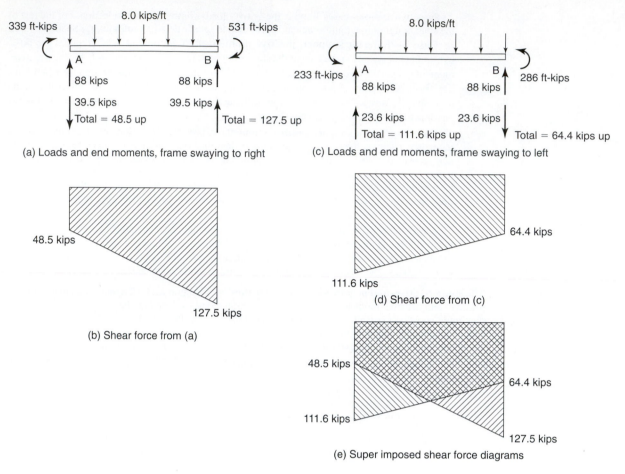

(a) Loads and end moments, frame swaying to right

(c) Loads and end moments, frame swaying to left

(b) Shear force from (a)

(d) Shear force from (c)

(e) Super imposed shear force diagrams

Fig. 20-22
Computation of shear forces—Example 20-1.

Reactions due to gravity loads: $w_u \ell_n/2 = 8.0 \times 22/2 = 88$ kips upward at each end.
Reactions due to M_{pr} at each end, *frame swaying to right*: $(M_{pr1} + M_{pr2})/\ell_n$

$$= (339 + 531)/22$$
$$= 39.5 \text{ kips down at left end}$$

Total reaction at left end   = 88 kips up + 39.5 kips down
$$= 48.5 \text{ kips up at left end}$$
Total reaction at right end = 127.5 kips up at right end.
The shear-force diagram is plotted in Fig. 20-22b.

Reactions due to M_{pr} at each end, *frame swaying to left*: $(233 + 286)/22$
$$= 23.6 \text{ kips up at left end.}$$

Total reaction at left end = 88 kips up + 23.6 kips up
$$= 111.6 \text{ kips up at left end}$$
$$= 64.4 \text{ kips up at right end}$$

The shear-force diagrams are plotted in Figs. 20-22b and 22d. These diagrams are superimposed in Fig. 20-22e to show the maximum shear at every section.

Stirrups for shear: ACI Section 21.3.4.2 states that V_c shall be taken equal to zero if

(a)   the shear V_{sway} due to plastic hinging at the two ends of the beam exceeds half or more of the maximum shear, V_u; and

(b) the factored axial compression force, including earthquake effects, is less than $A_g f'_c/20$.
Otherwise, V_c has the regular value $V_c = 2\sqrt{f'_c}\, b_w d$ —or the effects of axial load on V_c can be included by using ACI Eq. 11-4.

The end reactions for gravity loads are 88 kips upward at both ends of the beam, independently of the sway direction. For sway to the right, shown in Fig. 20-22a and b, the shear due to sway moments plus the gravity reactions, total 48.5 kips at A and 127.5 kips at B. $V_{sway} = 39.5$ kips exceeds half of the total end shear at end A. For sway to the left (see Fig. 20-22c and 22d), $V_{sway} = 23.6$ kips, and the total end reactions are 64.4 kips and 112 kips. In summary, V_e exceeds half of the end shear only at end A for the frame swaying to the right. In accordance with ACI Section 21.3.4.2, V_c will be taken equal to zero within the length of the hinging region at end A for the frame swaying right. At all other sections, V_c takes its usual value for beams (or ACI Eq. (11-4) for axially loaded members).

Exterior end: Maximum shear $V_u = 112$ kips

$$V_s = V_u/\phi - V_c = 112/0.75 - 0 = 149 \text{ kips}$$

ACI Section 11.5.6.8 sets the maximum $V_s = 8\sqrt{f'_c} b_w d = 261$ kips—OK. Then

$$\frac{A_v}{s} = \frac{149 \text{ kips}}{f_y d} = 0.116$$

Try No. 3, four-leg stirrups, $A_v = 0.44$ in.^{2}, and $s = 0.44/0.116 = 3.8$ in., or try No. 4, three-leg stirrups, $A_v = 0.60$ in.^{2}, and $s = 0.60/0.116 = 5.17$ in. The hoops will be selected after the confinement stirrups are calculated.

Hoops for confinement: ACI Section 21.3.3.1 requires hoops over a distance of $2h = 48$ in. measured from the face of the column. Every corner and alternate longitudinal bars in the beam must be at the corner of a stirrup in accordance with ACI Section 7.10.5.3. The stirrups and hoops have the shapes shown in Fig. 20-23. ACI Section 21.3.3.2 requires the first hoop to be 2 in. from the face of the column and the subsequent spacing of hoops of to be

(a) $d/4 = 5.38$ in.
(b) 8 times the maximum longitudinal bar diameter $= 8 \times 1.0$ in. $= 8$ in.
(c) 24 times the diameter of the hoop bars $= 24 \times 0.5 = 12$ in., or
(d) 12 in.

The maximum shear, V_u, at the ends of the beam is 127.5 kips at interior end. ACI Section 21.3.3.1 defines the length of the plastic hinge as $2h = 48$ in. from the face of the support. Try No. 4 three-leg hoops, one at 2 in. from the face of the column at each end, plus ten at 5 in. o.c., total 52 in. each end. $V_c = 0$ within the hinge. The hinge ends at 48 in. from the face of the support. At this section, we shall revert to $V_c = 2\sqrt{f'_c} b_w d = 65.3$ kips.

The maximum shear, V_u, at the end of the hinging region, 48 in. from the end, is 127.5 kips $-$ 4 ft $\times$ 8 kips/ft $=$ 127.5 kips $-$ 32 kips $=$ 95.5 kips. Also,

$$V_c = 2\sqrt{f'_c} b_w d = 65.3 \text{ kips} \qquad \text{(ACI Eq. 11-3)}$$
$$V_s = (95.5 \text{ kips}/0.75) - 65.3 = 62.0 \text{ kips}$$

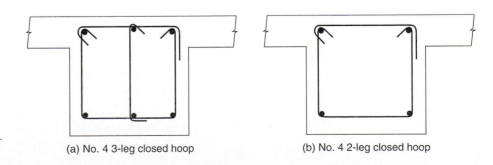

Fig. 20-23
Closed hoops and stirrups—
Example 20-1.

(a) No. 4 3-leg closed hoop   (b) No. 4 2-leg closed hoop

From ACI Eq. (11-13),

$$\frac{A_y}{s} = \frac{V_s}{f_y d} = \frac{62.0}{60 \times 21.5} = 0.0481 \qquad \text{(ACI Eq. 11-13)}$$

For No. 4 two-leg stirrups, $0.40/0.0481 = 8.32$ in. Maximum spacing from ACI Section 21.3.3.4 is $d/2 = 10.75$ in.

Use No. 4 three-leg hoops, one at 2 in. from the face of the column at each end, plus ten at 5 in. o.c., plus ten No. 4 two-leg closed stirrups at 8 in.

ACI Section 21.3.2.1 requires that at least two bars be made continuous, top and bottom. At the top we will accomplish this by lap splicing two No. 7 bars. The moment in the midspan region will either be positive or a very small negative value. Thus, the stress in the top bars will either be compression or a very small tension value. For these conditions, ACI Section 12.15.2 allows a Class A lap splice of length $1.0 \, \ell_d$, where $\ell_d = 61.7 \, d_b = 53.9$ in. ACI Section 21.3.2.3 requires that the lap splice be enclosed in hoops spaced at a maximum spacing of the smaller of $d/4 = 5.38$ in. or 4 in. Thus, we will provide No. 4 closed hoops at a 4 in. spacing along the 53.9 in. (say 54 in.) length of the lap splice for the top No. 7 bars. Sufficient continuous bottom steel is already in place to satisfy ACI Section 21.3.2.1. Any possible splicing of bottom steel shoul not be done at midspan becuase this is the location of maximum positive bending, i.e. maximum tension stress in the bottom bars.

Summary — From each column face use No. 4 three-leg hoops, one at 2 in. from the face of the column, plus ten at 5 in. (covers plastic hinging zones), plus six No. 4 two-leg closed stirrups at 8 in. Then, in the midspan region use sixteen No. 4 two-leg stirrups at a 4 in. spacing. (Fig. 20-21)

Summary—Lap splice two No. 7 top bars at midspan for 70 in., say 6 ft, and two No. 7 bottom bars for 54 in. at midspan.

Use No. 4 three-leg hoops, one at 2 in. from the face of the column at each end, plus ten at 5 in. o.c., plus seven No. 4 two-leg closed stirrups at 8 in., plus twelve No. 4 hoops at 4 in. o.c. at each end enclosing the lap splices located at midspan.

7. **Compute the cutoff points for the flexural reinforcement.** The bar cutoffs are calculated by assuming that the ends of the beam are hinging at a moment of $\pm M_{pr}$, due to the frame swaying to the right, and that the moment at the cutoff point is ϕM_n for the bars remaining after the cutoff point.

Figure 20-24 shows one-half of an interior span of a bending-moment diagram between midspan at A and the right-hand support at H. If the beam is loaded only with concentrated loads at midspan and the supports, the moment diagram is the sloping straight line A–H. If the beam supports a factored uniform load of $w_u = 8.0$ kips/ft from LC 9–2, the moment diagram is curved, as shown by the dashed line in Fig. 20-24. Line B–C–D represents a group of bars with a moment capacity corresponding to the height 0–B.

If $w_u = 0$, the bars must extend from B to C, with the flexural cutoff point at C. If there is a uniform load of w_u—in this case, 8.0 kips/ft—the cutoff point is at D. This shows that, in a positive-moment region, a uniform load increases the length to the cutoff by the distance C–D.

The opposite is true for negative-moment bars: line G–E represents the distance from the support to the flexural cutoff point for $w_u = 0$ kips/ft. Line E–F represents the distance from the support to the moment diagram for $w_u = 8.0$ kips/ft. The distance from G to F is shorter than the distance from G to E. Thus, in beams, the presence of a uniform load w_u increases the length to the flexural cutoff point for positive-moment steel by the distance C–D. For negative-moment reinforcement, the distance G–F is shorter than G–E. This shows that the negative-moment bars must extend farther from the support if there is no uniform load w_u than they would have to if the full w_u were present.

Frame swaying to the right—top (negative-moment) bars at interior support. There are four No. 8 bars plus two No. 7 top bars at the interior (right) support. (See Fig. 20-21.) These are in tension when the frame sways to the right as shown in Fig. 20-17. ACI Section 21.3.2.1 requires two continuous bars top and bottom. We will use two of the No. 7 top bars as continuous bars; two of No. 8 bars will be cut off when they are no longer needed for negative moment at the right end. The probable moment capacity at the left end support is $M_{pr} = -339$ ft-kips. At the right end support, M_{pr} is -531 ft-kips. The negative-moment capacity after the bars are cut off is $\phi M_n = -256$ ft-kips. Figure 20-25 is a free-body diagram of the interior end of the span. The uniform load has been taken as 8.0 kips/ft., and x is the distance in feet from the right end of the beam to the flexural cutoff.

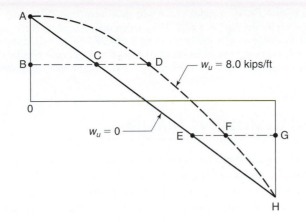

Fig. 20-24
Effect of uniform load on
cutoff points.

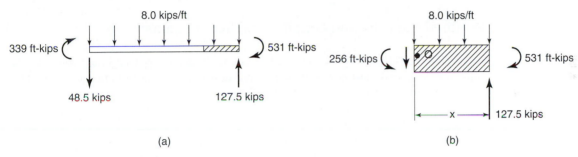

(a)   (b)

Fig. 20-25
Calculation of cutoff points—frame swaying to the right—Example 20-1.

Summing moments about O gives

$$4x^2 - 127.5x + 275 = 0$$

$$x = \frac{-b \pm \sqrt{b^2 - 4ac}}{2a} = 2.33 \text{ ft}$$

The flexural cutoff point for the two No. 8 bars is located 2.33 ft = 28 in. from the face of the support.

ACI Section 12.10.3 requires that bars extend the longest of $d = 21.5$ in., and 12 bar diameters (=12 in.) past the flexural cutoff point.

ACI Section 12.10.4 requires reinforcement to extend ℓ_d from the face of the column, where, for No. 8 top bars,

$$\ell_d = \left(\frac{f_y \alpha \beta \lambda}{20\sqrt{f_c'}}\right) d_b = \left(\frac{60{,}000 \times 1.3 \times 1.0 \times 1.0}{20\sqrt{4000}}\right) \times 1.0 \qquad (8\text{-}12)$$

$$= 61.7 \text{ in.} \text{---say, 5 ft 2 in.}$$

Extend the bars the longer of $28 + 21.5 = 49.5$ in., or 28 in. + 12 in. = 40 in., and $\ell_d = 5$ ft, 2 in. past the face of the support at B.

ACI Sec. 12.10.5 requires extra stirrups at cutoff points, unless the shear, V_u, at the cutoff points is less than or equal to two-thirds of the shear capacity, $0.667\phi(V_n)$, where $V_n = (V_c + V_s)$.

Before deciding to use this cutoff point, we will need to check the shear strength requirement at cutoff points in a tension zone. ACI Section 12.10.5 requires extra stirrups at such cutoff points, unless the shear, V_u, at the cutoff point is less than or equal to two-thirds of the nominal shear capacity, V_n, where $V_n = V_c + V_s$. A quick check will show that $(0.667)\phi V_n$ does not exceed V_u at the calculated cutoff point for the top No. 8 bars. Thus, rather than extend the three-leg No. 4 stirrups further from the face of the columns, it is easier to extend the No. 8 bars past the point of inflection before terminating them, following the requirements of ACI Section 12.12.3. A similar decision was made when determining the cutoff points for the negative- and positive-moment reinforcement at other points in the span.

20-8 COLUMNS IN SPECIAL MOMENT FRAMES

ACI Section 21.4 applies to columns in frames resisting earthquake forces and supporting a factored axial force exceeding $(A_g f'_c/10)$. Columns in frames in regions of high seismic risk must satisfy two geometric requirements: the smallest dimension through the centroid of the column must be at least 12 in., and the ratio of the shortest to the longest cross-sectional dimension shall not be less than 0.4. These limits ensure a minimum robustness and produce a cross section that can be confined using practical hoop layouts, which might be difficult with highly rectangular columns.

Required Capacity and Longitudinal Reinforcement

It is highly desirable that plastic hinges form in the beams rather than in the columns. Because the dead load must always be transferred down through the columns, damage to the columns should be minimized. ACI Section 21.4.2 strongly encourages the use of a strong-column–weak-beam design. In the event that this is not possible, the columns in question are disregarded in the structural analysis (i.e., assumed to have failed) if they add to the stiffness and strength of the building. (If inclusion of such columns in the analysis has a negative effect on the stiffness or strength, they should be included in the analysis.)

Strong-column—weak-beam behavior is made more likely by requiring (Fig. 20-26) that

$$\Sigma M_c \geq 1.2\Sigma M_g \qquad (20\text{-}16)$$

$$(\text{ACI Eq. 21-1})$$

where M_c is the *nominal* flexural capacity of the columns corresponding to the factored seismic load combination leading to the lowest axial load and hence the lowest flexural strength, and M_g is the nominal flexural capacity of the girders at that joint.

Columns that do not satisfy 20-16 must have transverse reinforcement satisfying ACI Section 21.4.4 over their entire length.

Longitudinal reinforcement is designed for the axial loads and moments in the same way as in a nonseismic column. It may range from $\rho = 0.01$ to 0.06. Generally, it is difficult to place and splice much more than 2 to 3 percent reinforcement in a column.

Because the cover concrete will probably spall in hinging regions, that may form near the ends of the column, longitudinal bars that are to be lap spliced must be spliced in the

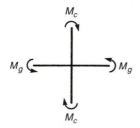

Fig. 20-26
Moments at a beam–column
joint—general.

center half of the column height. Such splices must be designed as tension splices, because the alternating moments due to the cyclic loads alternately stress the bars on each side of the column in tension and compression. Furthermore, there is frequently a possibility of uplift forces. The considerable length required for tension lap splices may require small-diameter bars or mechanical splices.

Transverse Reinforcement

Confinement Reinforcement

Transverse reinforcement in the form of spirals or hoops must be provided over a height of ℓ_o from each end of the column to confine the concrete and restrain the longitudinal bars from buckling. The height ℓ_o is the greater of (ACI Section 21.4.4.4)

  (a)   the depth of the column, h at the face of the joint,

  (b)   one-sixth of the height of the column, and

  (c)   18 in.

Within the length ℓ_o, ACI Section 21.4.4.2 requires that the spacing of the transverse reinforcement shall not exceed

  (a)   one-quarter of the minimum dimension, b or h, of the column cross section;

  (b)   six times the diameter of the longitudinal bar diameter; and

  (c)   the distance

$$s_x = 4 + \left(\frac{14 - h_x}{3} \right)$$

(20-17)
(ACI Eq. 21-5)

where h_x = maximum horizontal spacing between hoop or crosstie legs on all faces of the column, but not less than 4 in. nor more than 6 in. Transverse reinforcement also serves as shear reinforcement and has to conform to minimum stirrup spacings.

If spirals are used, they are designed as outlined in Section 11-5 by using ACI Eq. (10-5). An additional lower limit on the ratio of spiral reinforcement is given by ACI Eq. (21-2). This will govern if A_g/A_c is less than 1.27, which, for $1\frac{1}{2}$ in. cover, will occur for columns larger than 24 in. in diameter.

Because the pressure on the sides of the hoops causes the sides to deflect outward, hoops are less efficient than spirals at confining the core concrete (Fig. 20-14a). The equation for the required area of hoops, ACI Eq. (21-3), was based on the equation for spirals, ACI Eq. (10-5), but the constant was selected to give hoops with about one-third more cross-sectional area than would be required for spirals; that is,

$$A_{sh} = 0.3 \frac{s h_c f'_c}{f_{yh}} \left(\frac{A_g}{A_{ch}} - 1 \right)$$

(20-18)
(ACI Eq. 21-3)

but not less than

$$A_{sh} = 0.09 \frac{s h_c f'_c}{f_{yh}}$$

(20-19)
(ACI Eq. 21-4)

where

  A_{ch} = cross-sectional area of the core of the column measured out-to-out of the hoops

  A_g = gross area of the section

  A_{sh} = total cross-sectional area of all the legs of the hoops and cross-ties within a spacing s and perpendicular to the dimension h_c (see Fig. 20-14b)

h_c = cross-sectional dimension of the column core, measured center to center of outer legs of the hoops

s = spacing of the hoops measured parallel to the axis of the column

A_{sh} is checked separately for each direction.

Figure 20-14 shows typical hoop arrangement for a column. The maximum distance between hoop or cross-tie legs in the plane of the cross section is 14 in. The hoops must also satisfy ACI Section 7.10.5.3, which requires that every corner bar and alternate side bars be at the corner of a tie. Equation (20-19) gives a lower limit on the amount of confining reinforcement for columns larger than about 24 in. square.

Columns supporting discontinued shear walls are extremely susceptible to seismic damage. ACI Section 21.4.4.5 requires hoops or spirals over the full height of such members. These hoops must extend into the wall from the face of the column and the footing or other member under the column [20-6].

Shear Reinforcement

The transverse reinforcement must also be designed for shear. The design shear force V_{sway} is computed by assuming inelastic action in either the columns or the beams and is given by:

(a) the shear corresponding to plastic hinges at each end of the column given by

$$V_{sway} = \frac{M_{prc\ top} + M_{prc\ btm}}{\ell_u}$$

(20-20)

where $M_{prc\ top}$ and $M_{prc\ btm}$ are the probable moment capacities at the top and bottom of the column and ℓ_u is the clear height of the column. These are obtained from an interaction diagram for the probable strength, $P_{pr} - M_{pr}$ of the column, for the range of factored loads on the member for the load combination under consideration.

(b) but it need not be more than

$$V_{sway} = \frac{\Sigma M_{prb\ top}DF_{top} + \Sigma M_{prb\ btm}DF_{btm}}{\ell_u}$$

(20-21)

where $\Sigma M_{prb\ top}$ and $M_{prb\ btm}$ are the sum of the probable moment capacities of the beams framing into the joints at the top and bottom of the column for the frame swaying to the left or right, and DF_{top} and DF_{btm} are the moment-distribution factors at the top and bottom of the column being designed. This reflects the strong-column–weak-beam philosophy and (20-16), which makes the beams weaker than the columns.

(c) but not less than the factored shear from a frame analysis.

Transverse reinforcement is designed for shear according to ACI Section 11.1.1, and V_c may be increased to allow for the effect of axial loads (see Section 6-8), except that, within the length ℓ_o, defined in the discussion of confinement reinforcement, V_c shall be taken equal to zero when the earthquake-induced shear force makes up half or more of the maximum shear force in the lengths ℓ_o and *if* the factored compression force is less than $A_g f'_c/20$ (ACI Section 21.4.5.2). Columns with such low axial loads essentially behave like a beam. Thus, the concrete contribution to shear, V_c, is set equal to zero in potential plastic hinging zones at the ends of a column, just as was done for plastic hinging zones at the end of beams. Thus, we can use $V_c = 2\sqrt{f'_c}\,b_w d$ for axial loads greater than $0.05f'_c A_g$, and $V_c = 0$ for axial loads less than this.

It should be noted that, although the axial load increases V_c, it also increases the rate of shear degradation [20-8]. For this reason, it may be prudent to ignore V_c when a major portion of the shear results from earthquake loads.

EXAMPLE 20-2   Design of a Column

The column supporting the interior end of the beam designed in Example 20-1 is 24 in. square and is constructed of 4000-psi concrete and 60,000-psi steel. The floor-to-floor height is 12 ft, with 24-in.-deep beams in each floor, giving a clear column height of 10 ft. The column size and the floor-to-floor heights are the same in the stories above and below the column being designed. The unfactored moments, shears, and axial loads from an elastic analysis for earthquake loads are given in Table 20-4. Design the reinforcement in the column. Use load factors from ACI Sections 9.2.1 and 9.2.1(a), except that wind loads (W), fluid loads (F), soil loads (H), roof loads (L_r or S or R), and loads arising from restrained deformations (T) will not be considered.

1.   **Select the design procedure, method of analysis, and the load combinations to be used.** In some building codes the design method depends on the size of the design earthquake and it is necessary to choose the method to be used to analyze the seismic effects.

2.   **Compute the factored loads and moments.** An examination of the relative size of the loads suggested that the governing load combinations are 9–1, 9–2, and the two seismic load combinations, 9–5 and 9–7:

> Load combination 9–1 $U = 1.4D$
> Load combination 9–2 $U = 1.2D + 1.6L$
> Load combination 9–5 $U = 1.2D + 1.0E + 0.5L$
> Load combination 9–7 $U = 0.9D + 1.0E$

The calculations are summarized in Table 20-5.

3.   **Does the column satisfy the definition of a column?** ACI Section 21.4.1 lists four requirements for a member to be designed as a column under ACI Section 21.4:

(a)   Column resists earthquake-induced forces—OK.
(b)   Factored axial force exceeds $A_g f'_c/10 = 24 \times 24 \times 4/10 = 230$ kips—OK.
(c)   Shortest cross-sectional dimension is not less than 12 in.—OK.
(d)   Ratio of cross-sectional dimensions is not less than 0.4—OK.

Therefore, design the column according to ACI Section 21.4. If these were not satisfied, it would be necessary to modify the column dimensions.

4.   **Initial selection of column steel.** As a first trial, we shall select a 24×24 in. column with 12 No. 8 bars, $A_{st} = 9.48$ in.^{2}:

$$\rho_g = \frac{9.48}{24 \times 24} = 0.0165$$

ACI Section 21.4.3.1 limits ρ_g to not less than 0.01 or more than 0.06—OK. No. 8 bars were chosen to avoid excessive splice lengths. Interaction diagrams for $\phi P_n - \phi M_n$ and for $P_{pr} - M_{pr}$ are given in Fig. 20-28.

TABLE 20-4 Unfactored Axial Forces, Moments, and Shears in Column

	Dead Load	Live Load	Earthquake
Axial load, kips			
Column in story above	510	140	±5
Column being designed	560	154	±5
Column in story below	610	168	±6
Moments, ft-kips[a]			
Top of column	−4	−1	±195
Bottom of column	−4	−1	±210
Shears, kips	0	0	40

[a] Counterclockwise moment on the end of a member is positive.

TABLE 20-5 Factored Forces and Moments on Columns

	Axial Load (kips)	Top Moment[a] (ft-kips)	Bottom Moment (ft-kips)	Shear (kips)
Column in story above				
Load combination 9–1	714			
Load combination 9–2	836			
Load combination 9–5	687			
Load combination 9–7	464			
Column being designed				
Load combination 9–1	784	−6	−6	
Load combination 9–2	918	−6	−6	
Load combination 9–5				
Sway to right	749 + 5	195 − 5	210 − 5	40
Sway to left	749 − 5	−195 − 5	−210 − 5	40
Load combination 9–7				
Sway to right	504 + 5	195 − 5	210 − 5	40
Sway to left	504 − 5	−195 − 5	−210 − 5	40
Column in story below				
Load combination 9–1	854			
Load combination 9–2	1001			
Load combination 9–5	816 ± 6			
Load combination 9–7	549 ± 6			

[a]Counterclockwise moment on the end of the column is positive.

5. **Check whether the column strengths satisfy $\Sigma M_c \geq 1.2 \, \Sigma M_g$.** ACI Section 21.4.2.2 requires that the flexural strengths, ϕM_n, of the columns satisfy

$$\Sigma M_c \geq 1.2 \, \Sigma M_g \qquad\qquad (20\text{-}16)$$
$$(\text{ACI Eq. 21-1})$$

where ΣM_e is the sum of the ϕM_n strengths for the two columns meeting at a floor joint corresponding to the factored axial loads in the columns and ΣM_g is the sum of the ϕM_n strengths of the beams meeting at the joint.

For the frame swaying to the right, the moments ϕM_n at the ends of the beams meeting at the top of the column, from step 4 of Example 20-1, are as shown in Fig. 20-27a, and

$$1.2 \, \Sigma M_g = 1.2 \times (208 \text{ ft-kips} + 390 \text{ ft-kips})$$
$$= 718 \text{ ft-kips}$$

For load combination 9-2, the axial load in the column in the story above the column being designed is 836 kips. From the interaction diagram for $\phi P_n - \phi M_n$ in Fig. 20-28, the moment capacity corresponding to $\phi P_n = 836$ kips is $\phi M_n = 520$ ft-kips. The axial load in the column being designed

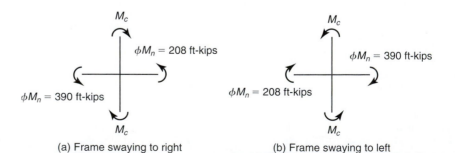

Fig. 20-27
Moments at a beam–column
joint—Example 20-2.

(a) Frame swaying to right   (b) Frame swaying to left

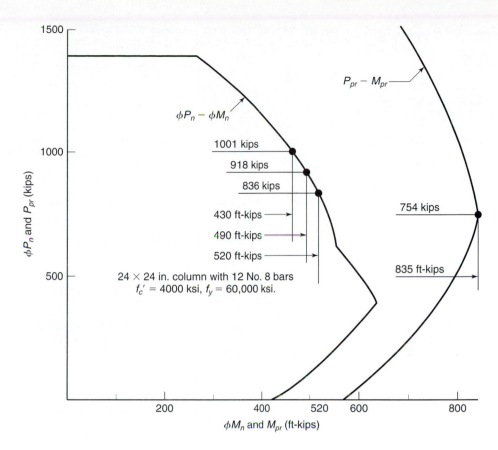

Fig. 20-28
Interaction diagrams for
$\phi P_n - \phi M_n$ and for $P_{pr} - M_{pr}$.

for the same load combination is 918 kips, corresponding to a moment capacity $\phi M_n = 490$ ft-kips. Thus, at the top of the column,

$$\Sigma M_c = 520 \text{ ft-kips} + 490 \text{ ft-kips} = 1010 \text{ ft-kips}$$

This column cross section satisfies the requirement that $\Sigma M_c \geq 1.2\Sigma M_g$ at the top of the column.

Assuming that the beams at the bottom of the column are the same as those at the top, $1.2\,\Sigma M_g = 718$ ft-kips. The moment capacity of the column being designed is 490 ft-kips. The axial load in the column in the story below the column being designed is 1001 kips, corresponding to a moment capacity $\phi M_n = 430$ ft-kips. Here, $\Sigma M_c = 430$ ft-kips $+ 490$ ft-kips $= 920$ ft-kips, which is still greater than $1.2\Sigma M_g$—OK.

6. Design the confinement reinforcement. ACI Section 21.4.4.1(b) requires that the total cross-sectional area of hoop reinforcement not be less than the larger of

$$A_{sh} = 0.3\left(\frac{sh_c f'_c}{f_{yh}}\right)\left(\frac{A_g}{A_{ch}} - 1\right) \qquad \text{(20-18)}$$
$$\text{(ACI Eq. 21-3)}$$

and

$$A_{sh} = \frac{0.09sh_c f'_c}{f_{yh}} \qquad \text{(20-19)}$$
$$\text{(ACI Eq. 21-4)}$$

where h_c is the cross-sectional dimension of the core, measured from center to center of hoops (see Fig. 20-14) $h_c = 24$ in. $- 2 \times (1.5 + 0.5/2)$ in. $= 20.5$ in., and A_{ch} is the cross-sectional area of the core of the column, measured out-to-out of the transverse reinforcement: $(24 - 2 \times 1.5)^2$ in.^{2} = 441 in.^{2}. Rearranging (20-18) and (20-19) and solving gives

$$\frac{A_{sh}}{s} = 0.3\left(\frac{20.5 \times 4000}{60,000}\right)\left(\frac{24 \times 24}{441} - 1\right)$$

$$= 0.126 \text{ in.}^2/\text{in.}$$

and

$$\frac{A_{sh}}{s} = \frac{0.09 \times 20.5 \times 4000}{60,000}$$

$$= 0.123 \text{ in.}^2/\text{in.}$$

ACI Section 21.4.4.2 sets the maximum spacing as the smaller of

    (a)   0.25 times the minimum cross-sectional dimension, 0.25×24 in. = 6 in.

    (b)   six times the longitudinal bar diameter, 6×1 in. = 6 in.

    (c)   from ACI Eq. (20-17)

$$s_x \leq 4 + \frac{14 - h_x}{3} \tag{20-17}$$

From Fig. 20-29, h_x is approximately two-thirds of the core dimension, i.e., 0.667×20.5 in. = 13.7 in. (which is less than the maximum permissible spacing of 14 in.). With this value for h_x, s_x = 4.11 in., which is between the limits of 4 in. and 6 in. Thus use s = 4 in.

For s = 4 in., the required A_{sh} = 0.126×4 = 0.504 in.^{2}. Use No. 4 hoops with three legs in each direction as shown in Fig. 20-29, giving A_{sh} = 0.60 in.^{2} in each direction. This layout satisfies ACI Section 21.4.4.3.

ACI Section 21.4.4.4 requires hoop reinforcement over a length ℓ_o adjacent to each end of the column, where ℓ_o is the largest of

    (a)   the depth of the member at the joint face: 24 in.,

    (b)   one-sixth of the clear height of the column, or 120 in./6 = 20 in., and

    (c)   18 in.

Thus, ℓ_o = 24 in. Throughout the rest of the height of the column, ACI Section 21.4.4.6 requires hoops at 6 in.

    7. **Design the shear reinforcement.** The design shear force V_e shall be

    (a)   the shear corresponding to plastic hinges at each end of the column, given by

$$V_{sway} = \frac{M_{\text{prc top}} + M_{\text{prc btm}}}{\ell_u} \tag{20-20}$$

    (b)   but need not be more than

$$V_{sway} = \frac{\Sigma M_{\text{prb top}}DF_{\text{top}} + \Sigma M_{\text{prb btm}}DF_{\text{btm}}}{\ell_u} \tag{20-21}$$

    (c)   but not less than the factored shear from a frame analysis.

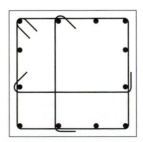

Fig. 20-29
Hoops—Example 20-2.

For load combination 9-5, the corresponding factored axial loads in the column being designed are 744 and 754 kips for sway to the left and right, respectively. From the interaction diagram for P_{pr}–M_{pr} in Fig. 20-28, the maximum value of M_{pr} for the column is 835 ft-kips. Substituting into (20-20) gives

$$V_{sway} = \frac{835 \text{ ft-kips} + 835 \text{ ft-kips}}{10 \text{ ft}} = 167 \text{ kips}$$

From step 5 of Example 20-1, the probable moment capacities of the beams framing into the joints at the top and bottom of the interior column are 531 ft-kips and 286 ft-kips. Because the columns in the stories above and below and the column being designed all have the same stiffness, DF_{top} and DF_{btm} are both 0.5. Substituting into (20-24) gives

$$V_{sway} = \frac{(531 + 286) \text{ ft-kips} \times 0.5 + (531 + 286) \text{ ft-kips} \times 0.5}{10 \text{ ft}} = 81.7 \text{ kips}$$

The shear V_e shall not be less than the factored shear from the analysis, 40 kips. Therefore, the design shear $V_e = 81.7$ kips.

As stated earlier, it may be prudent to set V_c equal to zero if the earthquake-induced shear represents half or more of the total design shear, regardless of axial loads. For the column being designed, this is true; hence, V_c is set equal to zero. We thus have

$$V_s = \frac{V_u}{\phi} - V_c$$

$$V_u = V_e = 81.7 \text{ kips} \qquad V_c = 0$$

Therefore,

$$V_s = \frac{81.7}{0.75} = 109 \text{ kips}$$

$$\frac{A_v}{s} = \frac{V_s}{f_y d} = \frac{109}{60 \times 21.5}$$

$$= 0.0844 \text{ in.}^2/\text{in.}$$

For $s = 4$ in., $A_v = 0.34$ in.^{2}. The hoops for confinement have $A_v = 0.60$ in.^{2}—OK.

Outside of the lengths ℓ_o, V_c is given by ACI Eq. (11–4):

$$V_c = 2\left(1 + \frac{N_u}{2000A_g}\right)\sqrt{f'_c}b_w d \qquad\qquad\qquad \text{(6-17a)} \\ \text{(ACI Eq. 11-4)}$$

where N_u is conservatively taken as the lowest value from the load combinations. Thus, $N_u = 464$ kips.

$$V_c = 2\left(1 + \frac{464 \text{ kips} \times 1000}{2000 \times 576 \text{ in.}^2}\right)\sqrt{4000} \times 24 \text{ in.} \times 21.5 \text{ in.}/1000$$

$$= 91.6 \text{ kips}$$

Since V_c exceeds V_u outside the length ℓ_o, stirrups are not needed for shear and instead will be provided for confinement.

Provide No. 4 hoops as shown in Fig. 20-29 at 2 in. from end of column, and five at 4 in. on centers at each end; provide similar No. 4 hoops at 6 in. on centers over the rest of the height.

8.   **Design lap splices for the column bars.** ACI Section 21.4.3.2 requires that splices be in the midheight of the column, be designed as tension splices, and be enclosed within transverse reinforcement conforming to ACI Sections 21.4.4.2 and 21.4.4.3. Thus, over the length of the lap splice the spacing between the layers of transverse reinforcement must be reduced to 4 in. o.c.

ACI Section 12.17.2.2 requires a Class B tension lap splice if all the bars are spliced at the same location. For a vertical No. 8 bar,

$$\ell_d = 47.4 \text{ in.}$$

A Class B splice has a length of $1.3\ell_d = 1.3 \times 47.4$ in. $= 61.7$ in. ACI Section 12.17.2.4 allows this length to be multiplied by 0.83 if the ties throughout the splice length have an effective area of not less than $0.0015hs$, which, for the $s = 4$ in. is 0.14 in.^{2}. The hoops have an area of 0.60 in.^{2}; therefore, they are adequate to allow this reduction. The lap length becomes 0.83×61.7 in. $= 51.2$ in. —say, 4 ft 4 in.

Lap splice all vertical bars with a 4 ft 4 in. lap splice at midheight of the column. ∎

20-9 JOINTS OF SPECIAL MOMENT FRAMES

The flow of forces within beam–column joints and the design of such joints has been discussed in Section 18-7, and an example of the design of an exterior nonseismic joint is given in Example 18-7. Code provisions for joints in special moment-resisting frames (SMFs) are given in ACI Section 21.5. These differ from the design recommendations for nonseismic joints in a number of areas.

ACI Section 21.5.1.1 requires that joint forces be calculated by taking the stress in the flexural reinforcement in the beams as $1.25f_y$. This is analogous to using the probable strength in the calculations of shear in columns and beams in special-moment frames.

ACI Section 21.5.1.4 limits the diameter of the longitudinal beam reinforcement that passes through a joint to $\frac{1}{20}$ of the width of the joint parallel to the beam bars. When hinges form in the beams, the beam reinforcement is stressed to the actual yield strength of the bar on one side of the joint and is stressed in compression on the other side. This results in very large bond stresses in the joint, possibly leading to slipping of the bar in the joint. The minimum bonded length of such a bar in a joint is thus $20d_b$, which is considerably less than is required by the development-length equations in ACI Chapter 12. The minimum bonded length was selected from test results of joints tested under cyclic loads to limit, but not entirely eliminate slip of the beam bar in the joint.

ACI Section 21.5.2.1 requires hoop reinforcement around the column reinforcement in all joints in special moment-resisting frames. In joints confined on all four sides by beams satisfying ACI Section 21.5.2.2, the amount of hoop reinforcement is reduced, and its spacing is less restrictive within the depth of the shallowest beam entering the joint.

ACI Section 21.5.3.1 gives upper limits on the shear strength of joints. As indicated in Section 18-7, these are lower than the joint shear strengths recommended in nonseismic joints. This reflects the possible damage to joints resulting from cyclic loads.

ACI Section 21.5.4 gives special development lengths for hooks and straight bars in joints. These are shorter than the development lengths given in ACI Chapter 12 because the effects of the joint confinement by hoops have already been included.

EXAMPLE 20-3 Design an Interior Beam–Column Joint

Design the interior beam–column joint connecting the beams and columns from Examples 20-1 and 20-2. Beams, which are 24 in. by 24 in. in section, frame into the 24-in.-by-24-in. column on all four sides.

1. Define the size of the joint. The joint has width, depth, and vertical height of 24 in. The area of a horizontal section through the joint, A_j (see definition in ACI Section 21.0), is $A_j = 24 \times 24 = 576$ in.^{2}.

ACI Section 21.5.1.4 requires the length of the joint measured parallel to the flexural steel causing the joint shear to be at least 20 times the diameter of those bars, $(20 \times 1$ in.$)$—OK.

2. Determine the transverse reinforcement for confinement. ACI Section 21.5.2.1 requires confinement steel within the joint. Because the joint has beams on all four sides, ACI Section 21.5.2.2 sets the amount of confinement steel as half of the confinement steel required in the ends of the columns, given by (20-18) and (20-19) (ACI Eqs. (21-3) and (21-4)). In the column, (20-18)

(ACI Eq. (21-3)) governed (see step 6 of Example 20-2) and required that $A_{sh}/s = 0.126$ in.^{2}/in. Within the height of the joint, we require that

$$\frac{A_{sh}}{s} = 0.5 \times 0.126 \text{ in.}^2/\text{in.} = 0.063 \text{ in.}^2/\text{in.}$$

The vertical spacing of the hoops from ACI Section 21.5.2.2 is permitted to be 6 in.

The clear distance between the top and bottom beam steel is 18 in. Provide three sets of hoops, the first at 3 in. below the top steel. The required A_{sh} is $6 \times 0.063 = 0.378$ in.^{2}. Use No. 3 three-legged hoops in the arrangement shown in Fig. 20-29.

3.   **Compute the shear on the joint, and check the shear strength.** Figure 20-30 is a series of free-body diagrams of the joint for the frame swaying to the right. The beams entering the joint have probable moment capacities of -531 ft-kips and $+339$ ft-kips. At the joint, the stiffnesses of the columns above and below the joint are the same, giving distribution factors of $DF = 0.5$ for each column. Thus, the moment in the column above is

$$M_e = 0.5(531 + 339) = 435 \text{ ft-kips}$$

The shear in the column above is

$$V_{sway} = \frac{435 + 435}{10 \text{ ft}} = 87.0 \text{ kips} \qquad \text{(See step 7 of Example 20-2.)}$$

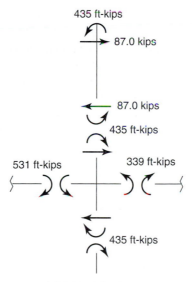

435 ft-kips

87.0 kips

87.0 kips

435 ft-kips

531 ft-kips   339 ft-kips

435 ft-kips

(a) Joint and columns

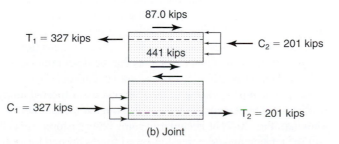

87.0 kips

$T_1 = 327$ kips $C_2 = 201$ kips

441 kips

$C_1 = 327$ kips $T_2 = 201$ kips

(b) Joint

Fig. 20-30
Free-body diagrams of
joint—Example 20-3.

The area of the top steel at the interior support is four No. 8 bars plus two No. 7 bars, so $A_s = 4.36$ in.^{2} The force in the steel in the beam on the left of the joint is

$$T_1 = 1.25 A_s f_y = 1.25 \times 4.36 \text{ in.}^2 \times 60 \text{ ksi}$$
$$= 327 \text{ kips}$$

The compression force in the beam to the left is $C_1 = T_1 = 327$ kips. Similarly, T_2 and C_2 in the beam to the right of the joint are $1.25 \times 2.68 \times 60 = 201$ kips.

Summing horizontal forces gives the shear in the joint as

$$V_j = V_{sway} - T_1 - C_2 \qquad\qquad (20\text{-}22)$$
$$= 87.0 \text{ kips to the right} - 327 \text{ kips to the left} - 201 \text{ kips to the left}$$
$$= 441 \text{ kips to the left}$$

From ACI Section 21.5.3.1, the nominal shear strength of a joint confined on all four sides is

$$V_n = 20\sqrt{f_c'}A_j = 20 \times \sqrt{4000} \text{ psi} \times 576 \text{ in.}^2$$
$$= 729 \text{ kips}$$
$$\phi V_n = 0.85 \times 729 = 619 \text{ kips}$$

Therefore, the joint has adequate shear strength.
Provide three No. 3 three-legged hoops at 6 in. on centers in the joint. ■

20-10   STRUCTURAL DIAPHRAGMS

Floor slabs, roof slabs, and cast-in-place toppings on precast concrete floors may all be used as diaphragms to transfer horizontal forces acting in the building to the lateral-force-resisting system and, eventually, to the foundations, as shown in Fig. 20-7. In effect, they act as deep flexural members lying in horizontal planes. Cast-in-place toppings serving as diaphragms may or may not be composite with the floor or roof members. ACI Section 21.9.4 allows the use of 2-in.-thick composite topping slabs or $2\frac{1}{2}$-in. noncomposite toppings. ACI Section 21.9.3 requires that noncomposite toppings be designed for the forces transferred.

Flexural Strength

Diaphragms function as deep beams lying on their sides with the tensile and compressive chords needed for flexure located near the edges of the diaphragm. ACI Section 21.9.8.1 requires flexural reinforcement for the chord. Chord forces are computed as the sum of (a) the applicable portion of the factored axial load on the diaphragm, if any, plus (b) the force obtained by dividing the factored moment at the section by the distance between the chords or boundary elements of the diaphragm. Transversely reinforced boundary elements are required if the elastically computed flexural compression stresses, neglecting cracking, exceed $0.2f_c'$ at any section. (See ACI Section 21.9.5.3.) The transverse reinforcement in the boundary elements can be discontinued when these stresses drop below $0.15f_c'$. These two limiting stresses are easy-to-compute index values that reflect the need to protect against buckling of the reinforcement located along the edges of the diaphragm with properly tied reinforcement in boundary elements. If a diaphragm is notched (as shown in Fig. 20-7) or irregular in some other fashion, special details are required to transmit the tension and compression chord forces around the notch.

Shear Strength

ACI Section 21.9.7 gives expressions for the nominal shear resistance of walls and diaphragms. The nominal shear strength of monolithic structural diaphragms shall not exceed

$$V_n = A_{cv}(2\sqrt{f'_c} + \rho_n f_y) \tag{20-23}$$
(ACI Eq. 21-10)

This is equivalent to $V_n = (V_c + V_s)$, where V_c and V_s have been divided by the shear area, A_{cv}, defined as the area bounded by the web thickness and the length of the section in the direction of the shear force being considered.

For cast-in-place composite-topping-slab diaphragms and cast-in-place noncomposite topping slabs on precast floors or roofs, V_c is taken equal to zero, to reflect the likelihood that the topping will be cracked by shrinkage or other effects during construction and service. The required slab steel in this case is computed as

$$V_n = A_{cv}\rho_n f_y \tag{20-24}$$
(ACI Eq. 21-11)

This was derived by using shear friction.

Effect of Diaphragm Stiffness on Lateral-Load Distribution

Figure 20-31 illustrates the effect of diaphragm stiffness on the distribution of lateral loads to the lateral-load-resisting elements. The building shown in the figure has three walls of equal lateral stiffness. If the diaphragm is essentially rigid in plane and there is no torsion, all three walls will displace by the same amount, and each wall will resist one-third of the total lateral load, as shown in Fig. 20-31a. On the other hand, if the diaphragm is flexible relative to the walls, the two end walls will each resist a quarter of the lateral shear and the center wall will resist half of it, as shown in Fig. 20-31b.

The following derivations are presented to give an idea of the factors affecting the relative stiffnesses of the walls and diaphragms. The lateral stiffness, K_ℓ, of a cantilever of height ℓ that is fixed at the base is

$$K_\ell = \frac{V}{\Delta} \tag{20-25}$$

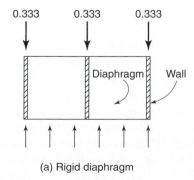

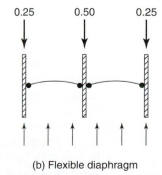

Fig. 20-31
Plan view of a building showing the effect of diaphragm stiffness on distribution of lateral loads to walls in a building.

(a) Rigid diaphragm

(b) Flexible diaphragm

where Δ is the lateral deflection on the top of the cantilever due to the load V at the top, equal to

$$\Delta = \frac{V\ell^3}{3EI} + \frac{1.2V\ell}{AG} \tag{20-26}$$

The first term represents the flexural deflections, the second the shear deflections. Substituting (20-26) into (20-25) and taking $G = E/2$ gives

$$K_\ell = \frac{3EI}{\ell^3} + \frac{AE}{2.4\ell} \tag{20-27}$$

A similar expression can be derived for the lateral stiffness of a piece of diaphragm between two walls.

Benjamin [20-9] has shown that if the stiffness, $K_{\ell d}$, of the diaphragm exceeds about two times that of the walls, $K_{\ell w}$, the diaphragm will act as a rigid diaphragm in transmitting loads to the walls. If $K_{\ell d}/K_{\ell w} = 0$, the diaphragm is fully flexible.

20-11 STRUCTURAL WALLS

Structural walls or *shear walls* are frequently used to resist a major fraction of the design seismic shears. These are designed according to ACI Section 21.7. The factored shears, moments, and axial forces to be considered in the design of the wall are obtained from the frame analysis. Chapter 19 deals with the layout and design of nonseismic structural walls.

Design of Shear Walls

Chapter 19 identified two types of shear walls: flexural shear walls, and squat (or short) shear walls. Flexural shear walls are designed using the theory of flexure for members subjected to axial loads and bending. The design of squat or short walls is based, in part, on strut-and-tie models.

1. The first step in the design of a shear wall is to select the size and shape of the wall, using stiffness, building geometry, and the design flexure and shear. In this stage, the geometry of the wall is chosen. The wall will either be a flexural shear wall or a short shear wall. Ideally, a flexural shear wall should be proportioned for flexure by using a strain-compatibility solution, as given in Section 5-5 or Section 11-4. Currently, flexural design follows the computations in Section 19-12.

2. The foundations for the walls may be required to transfer very large overturning moments to the soil or rock under the building. A new section 21.10 presents requirements for the design of foundations resisting earthquake forces. ACI Section 21.10.2 requires that special attention be given to the anchorage of the wall reinforcement into the foundations.

3. Once the concrete section is established, it is necessary to investigate the need for boundary elements. Boundary elements are regions at the ends of the cross section of the wall that are reinforced as columns, with the reinforcement enclosed by hoop reinforcement, as shown in Fig. 20-32. Boundary elements strengthen and confine the edges of the walls to resist stress reversals and prevent reinforcement buckling near the edges. They generally are thicker than the walls although ACI Section 21.1 allows them to have the same thickness as the wall. ACI Sections 21.7.6.1 and 21.7.6.2 give two methods of determining the need for boundary elements.

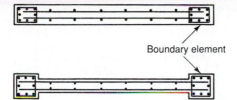

Boundary element

Fig. 20-32
Plan views of structural walls
with boundary elements.

(a)   ACI Section 21.7.6.2 applies to walls that are effectively continuous from
the base of the structure to the top of the wall and are designed to have a single critical section for axial loads and bending at the base of the wall. Boundary elements are
required if

$$c > \frac{\ell_w}{600(\delta_u/h_w)}$$

(20-28)
(ACI Eq. 21-8)

where

c = depth from the neutral axis to the extreme compression fiber

ℓ_w = horizontal length of entire wall or of a segment of wall considered in the
direction of the shear force

h_w = height of the entire wall, or the segment of wall considered

δ_u = design displacement, defined as the total lateral displacement deflection of the
top of the building for the design-basis earthquake

The quantity δ_u/h_w shall not be taken less than 0.007.

Where boundary elements are required by (20-28), the special boundary reinforcement shall extend upward from the critical section a distance not less than the
larger of ℓ_w or $M_u/4V_u$. This is an example of capacity design of the wall. The hinging section is confined, but the rest of the wall does not need special details.

(b)   In ACI Section 21.7.6.2(a), the uncracked wall is analyzed by using

$$\sigma = \frac{P}{A} \pm \frac{My}{I}$$

(20-29)

where A and I are based on the gross concrete section. If the computed maximum
compressive stress in the extreme fiber exceeds $\sigma = 0.2f'_c$ at any point, ACI Section
21.7.6.3 requires boundary elements over that portion of the height where the ex-
treme-fiber stress exceeds $\sigma = 0.15f'_c$. This calculation is simply a convenient way to
establish a limit value and is not intended to be representative of the true wall behavior or the stresses in the wall.

If boundary elements are required by either ACI Section 21.7.6.2 or ACI Section
21.7.6.3, they must satisfy ACI Section 21.7.6.4.

Design for flexure should follow ACI Sections 10.2 and 10.3 except that the nonlinear strains mentioned in ACI Section 10.2.2 shall not apply. In addition, the axial-load capacity computed by ACI Section 10.3.6 does not apply. A simpler method of design is
presented in ACI Section 21.6.6.3. It is assumed that when the wall is displaced laterally,
only the boundary element transmits compression to the next lower level as shown in
Fig. 20-33. The boundary element at A is designed for a compressive force of $(W_u + M_u/z)$
as shown in Fig. 20-33 and a tensile force of (M_u/z) corresponding to sway in the opposite

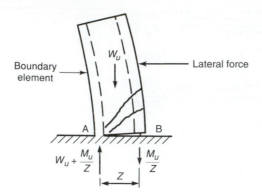

Fig. 20-33
Forces in boundary elements.

direction. It is assumed to act as an axially loaded short column or tension member and design is based on ACI Eq. (10-2) with $\phi = 0.9$, if the boundary element has hoop reinforcement. An example of the use of (20-28) and (20-29) is given in [20-10].

Design for Shear

The design of structural walls for shear is given in ACI Section 21.7.4. The basic design equation is essentially the same as the $\phi(V_c + V_s)$ procedure used in beam design. The basic shear (v_c) stress carried by the concrete is $2\sqrt{f'_c}$ psi, except for short stubby walls, for which a higher stress is allowed. The horizontal reinforcement in the wall must be anchored in the boundary elements, as specified in ACI Section 21.7.6.4.

Strength-Reduction Factor

The strength-reduction factor for flexure or flexure and axial loads is 0.65 for compression controlled walls with a transition to $\phi = 0.90$ for tension-controlled sections. Since walls generally have axial load ratios $N_u/A_g F_c$ between 0.1 and 0.3, they are generally tension-controlled. The ϕ factor for shear is 0.75 for shear unless the nominal shear strength is less than the shear corresponding to development of the nominal flexural strength of the wall. In such a case, ϕ is taken as 0.6 (ACI Section 9.3.4(a)).

Coupled Walls

Frequently, two shear walls are *coupled* by beams or slabs spanning across a doorway or similar opening. Depending on the stiffness of the coupling beams, the walls act as two independent cantilevers as the coupling-beam stiffness approaches zero or as one solid cantilever if the coupling-beam stiffness is high. The coupling beams transmit shear from one wall to the other and undergo large shearing deformations, as shown in Fig. 20-34a. As a result, these beams may degrade rapidly in shear during an earthquake. Paulay [20-3] has demonstrated experimentally that the reinforcement pattern shown in Fig. 20-34c transmits cyclic shear load much better than conventional top and bottom steel and stirrups. This diagonal steel acts as a truss member developing forces T_u and C_u, which transmit a moment and a shear, where

$$T_u = C_u = \phi A_s f_y \tag{20-30}$$

$$V_u = 2T_u \sin \alpha = 2\phi A_s f_y \sin \alpha \tag{20-31}$$

$$M_u = (\phi A_s f_y \cos \alpha)(h - 2d') \tag{20-32}$$

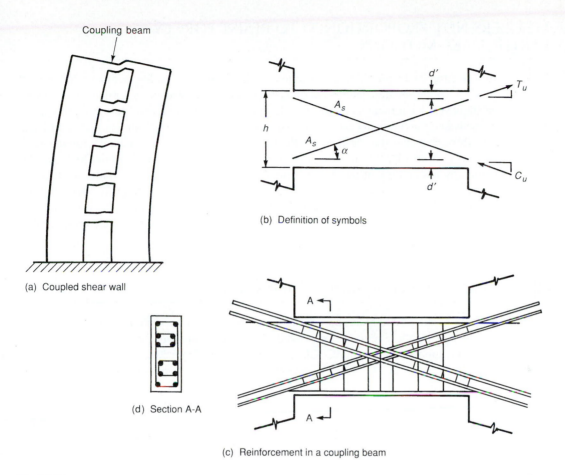

(a) Coupled shear wall

(b) Definition of symbols

(d) Section A-A

(c) Reinforcement in a coupling beam

Fig. 20-34
Coupled shear walls and coupling beams.

ACI Section 21.7.7.2 allows coupling beams like Fig. 20-34c for aspect ratios $l_n/d < 4$. ACI Section 21.7.7.3 requires diagonally reinforced coupling beams if $V_u > 4\sqrt{f_c'}\, b_w d$ and $l_n/d < 2$.

The diagonal bars must be tied to form intersecting column cages. Nominal top and bottom steel and stirrups are provided to prevent spalling of large pieces of concrete during cyclic deformations.

Coupled walls may need boundary elements at the outside ends only or at both ends of both walls, depending on the stiffness of the coupling beams. This would be determined by an analysis according to ACI Section 21.7.6.1 and (20-29), using the moments and axial forces assigned to each part of the coupled wall.

Wall Foundations

A major stage in the structural design is the design of a foundation for the wall. Frequently, overturning moments from wind or seismic loads require special structures in the basement of the building necessary to transfer these forces to the ground.

20-12 FRAME MEMBERS NOT PROPORTIONED TO RESIST FORCES INDUCED BY EARTHQUAKE MOTIONS

ACI Section 21.11 provides less stringent design requirements for members that are not part of the designated lateral-load-resisting system in structures subjected to severe earthquakes. Such members must be able to resist the factored axial forces due to gravity loads and the moments and shears induced in them when the frame is deflected laterally through twice the elastically calculated lateral deflections under factored lateral loads. Traditionally, the so-called *building frame system* defined in Table 20-1 was designed by assuming that the frame members supported only gravity loads, while shear walls resisted all of the lateral loads. In the 1994 Northridge earthquake, columns in a number of buildings of this type failed when subjected to the lateral displacements imposed by that earthquake. In the 2002 ACI Code, Section 21.11 was made considerably more stringent.

20-13 FRAMES IN REGIONS OF INTERMEDIATE SEISMIC RISK

Portions of the United States have been designated as regions of intermediate or moderate seismic risk. In these regions, the relaxed seismic design provisions given in ACI Section 21.12 are applied along with ACI Chapters 1 to 18. See Table 20-2.

In such regions, the shear V_u in beams, columns, and two-way slabs is calculated as the larger of

(a) The shear resulting from reaching the nominal moment capacity $(\phi = 1.0, f_s = 1.0f_y)$ at each end of the member in question, as shown in Fig. 20-20b, except that M_n is substituted for M_{pr}, plus, in the case of beams and slabs, the shear resulting from a gravity load.

(b) Special details are given in ACI Section 21.12.6 for the design of two-way slabs without beams. Such structures are not permitted as part of the lateral-load-resisting frame in regions of high seismic risk, but can be used in regions of moderate seismic risk. It should be noted, however, that the lateral stiffness of unbraced two-way slab–column frames is very low and the lateral deflections under even moderate earthquakes may be very large. It should also be noted that, under some circumstances, there may be a moment reversal at the ends of slab spans that will require special detailing.

20-14 SPECIAL PRECAST STRUCTURES

ACI Section 21.6 is a major addition to Chapter 21 of the 2002 ACI Code. It presents a set of requirements for precast frame members and precast shear walls designed for severe earthquakes. These take the form of minimum shear strengths and minimum requirements for the strength of connections. Precast beams are required to have clear spans not less than four times the effective depth. Either *ductile connections* (ACI Section 21.6.1) or *strong connections* (ACI Section 21.6.2) are required. These are required to have specified minimum design strengths. Ductile connections are expected to experience flexural yielding in connection regions. Localized capacity-design principles are used to ensure that strong connections develop hinging outside of the connections.

Special structural precast walls shall satisfy all the requirements of ACI Section 21.7 for cast-in-place concrete walls, plus Section 21.13 concerning intermediate precast structural walls.

20-15 FOUNDATIONS

ACI Section 21.10 deals with foundations for seismic structures, including footings, mat foundations, pile caps, piles, piers, and caissons. The major emphasis is on the pull-out strength of reinforcement extending from the structure into the foundations. Minimum confinement reinforcement and shear reinforcement is required in piles, piers, and caissons.

APPENDIX *A*

Design Aids

TABLE A-1   Areas, Weights, and Dimensions of Reinforcing Bars

Bar Size Designation No.[b]	Grades[c]	Weight (lb/ft)	Nominal Dimensions[a]	
			Diameter (in.)	Cross-Sectional Area (in.^{2})
3	40, 60	0.376	0.375	0.11
4	40, 60	0.668	0.500	0.20
5	40, 60	1.043	0.625	0.31
6	40, 60, 75	1.502	0.750	0.44
7	60, 75	2.044	0.875	0.60
8	60, 75	2.67	1.000	0.79
9	60, 75	3.40	1.128	1.00
10	60, 75	4.30	1.270	1.27
11	60, 75	5.31	1.410	1.56
14	60, 75	7.65	1.693	2.25
18	60, 75	13.60	2.257	4.00

[a]The nominal dimensions of a deformed bar are equivalent to those of a plain round bar having the same weight per foot as the deformed bar.

[b]Bar numbers are based on the number of eighths of an inch included in the nominal diameter.

[c]Grade is nominal yield strength in ksi.

TABLE A-1M Areas, Weights, and Dimensions of Reinforcing Bars—SI Units

Bar Size Designation No.[b]	Grades[c]	Nominal Mass (kg/m)	Nominal Dimensions[a]	
			Diameter (mm)	Cross-Sectional Area (mm²)
10	300, 420	0.560	9.5	71
13	300, 420	0.994	12.7	129
16	300, 420	1.552	15.9	199
19	300, 420, 520	2.235	19.1	284
22	420, 520	3.042	22.2	387
25	420, 520	3.973	25.4	510
29	420, 520	5.060	28.7	645
32	420, 520	6.404	32.3	819
36	420, 520	7.907	35.8	1006
43	420, 520	11.38	43	1452
57	420, 520	20.24	57.3	2581

[a]The nominal dimensions of a deformed bar are equivalent to those of a plain round bar having the same mass per meter as the deformed bar.

[b]Bar-designation numbers are the nominal diameter, rounded off to the nearest mm. The sequence of the bar-designation numbers is the sequence of US Customary diameters rounded off to the nearest mm.

[c]Grade is nominal yield strength in MPa.

TABLE A-2   Welded-Wire Reinforcement

(a) Wires

Wire Size Number[a]		Nominal Diameter (in.)	Area (in.^{2} per ft of width for center-to-center spacing, in.)			
Smooth	Deformed		4	6	10	12
W31	D31	0.628	0.93	0.62	0.372	0.31
W11	D11	0.374	0.33	0.22	0.132	0.11
W10	D10	0.356	0.30	0.20	0.12	0.10
W9	D9	0.338	0.27	0.18	0.108	0.09
W8	D8	0.319	0.24	0.16	0.096	0.08
W7	D7	0.298	0.21	0.14	0.084	0.07
W6	D6	0.276	0.18	0.12	0.072	0.06
W5.5		0.264	0.165	0.11	0.066	0.055
W5	D5	0.252	0.15	0.10	0.06	0.05
W4	D4	0.225	0.12	0.08	0.048	0.04
W3.5		0.211	0.105	0.07	0.042	0.035
W2.9		0.192	0.087	0.058	0.035	0.029
W2.5		0.178	0.075	0.05	0.03	0.025
W2.1		0.162	0.063	0.042	0.025	0.021
W1.4		0.135	0.042	0.028	0.017	0.014

[a]Wire size number is 100 times the wire area in in.^{2}.

(b) Common-Stock Welded-Wire Reinforcement

Style Designation[a]	Steel Area (in.^{2}/ft)		Approximate Weight (lb/100 ft^2)
	Longitudinal	Transverse	
6 × 6—W2.9 × W2.9	0.058	0.058	42
4 × 4—W2.1 × W2.1	0.062	0.062	44
6 × 6—W4 × W4	0.080	0.080	58
4 × 4—W2.9 × W2.9	0.087	0.087	62
6 × 6—W5.5 × W5.5	0.110	0.110	80
4 × 4—W4 × W4	0.120	0.120	85
4 × 4—W5.5 × W5.5	0.165	0.165	119

[a]The numbers in the style designation refer to (longitudinal wire spacing × transverse wire spacing)–(longitudinal wire size × transverse wire size).

TABLE A-3 Values of ϕk_n and j [a]

$$\omega = \frac{\rho f_y}{f_c'} \qquad \phi k_n = \phi[f_c'\,\omega(1-0.59\omega)] \qquad \frac{M_n}{\phi k_n} = \frac{bd^2}{12{,}000} \qquad j = 1 - 0.59\omega \qquad A_s = \frac{M_u}{\phi f_y j d} \qquad \phi = 0.90$$

| | $f_y = 40{,}000$ psi | | | | $f_y = 60{,}000$ psi | | | | | | | | | |
| | $f_c' = 3000$ psi | | 3750 psi | | $f_c' = 3000$ psi | | 3750 psi | | 4000 psi | | 5000 psi | | 6000 psi | |
ρ	ϕk_n	j	ϕk_n	j	ϕk_n	j	ϕk_n	j	ϕk_n	j	ϕk_n	j	ϕk_n	j
0.0033					171	0.961	173	0.969	173	0.971	174	0.977	175	0.981
0.004					206	0.953	208	0.962	208	0.965	210	0.972	211	0.976
0.005	173	0.961	174	0.969	254	0.941	257	0.953	258	0.956	260	0.965	262	0.971
0.006	206	0.953	208	0.962	301	0.929	306	0.943	307	0.947	310	0.958	313	0.965
0.007	238	0.945	241	0.956	347	0.917	353	0.934	355	0.938	359	0.950	362	0.959
0.008	270	0.937	274	0.950	391	0.906	399	0.924	401	0.929	408	0.943	412	0.953
0.009	301	0.929	306	0.943	434	0.894	445	0.915	447	0.920	455	0.936	460	0.947
0.010	332	0.921	337	0.937	476	0.882	489	0.906	492	0.912	502	0.929	508	0.941
0.011	362	0.913	369	0.931	517	0.870	532	0.896	536	0.903	548	0.922	555	0.935
0.012	391	0.906	399	0.924	556	0.858	575	0.887	579	0.894	593	0.915	602	0.929
0.013	420	0.898	430	0.918	594	0.847	616	0.877	621	0.885	637	0.908	648	0.923
0.014	448	0.890	460	0.912			656	0.868	662	0.876	681	0.901	694	0.917
0.015	476	0.882	489	0.906			695	0.858	702	0.867	724	0.894	738	0.912
0.016	504	0.874	518	0.899			734	0.849	742	0.858	766	0.887	782	0.906
0.017	530	0.866	547	0.893					780	0.850	808	0.880	826	0.900
0.018	556	0.858	575	0.887					817	0.841	848	0.873	869	0.894
0.019	582	0.851	602	0.880							888	0.865	911	0.888
0.020	607	0.843	629	0.874							927	0.858	953	0.882
0.021			656	0.868							965	0.851	993	0.876
0.022			682	0.862									1030	0.870
0.023			708	0.855									1070	0.864
0.024			734	0.849										
0.025			758	0.843										

[a] Upper line in each column is below the entry for $\rho = 0.35\rho_b$; lower line is below the entry for $\rho = 0.5\rho_b$; bottom entry is the tension-controlled limit. Larger values of ρ can be used, but they require that ϕ be evaluated.

TABLE A-3M   Values of ϕk_n and j—SI Units[a]

$$\omega = \frac{\rho f_y}{f'_c} \qquad \phi k = \phi[f'_c\,\omega(1-0.59\omega)] \qquad \frac{M_u}{\phi k_n} = \frac{bd^2}{10^6} \qquad j = 1 - 0.59\omega \qquad \phi = 0.90 \qquad A_s = \frac{M_u}{\phi f_y\, j d}$$

	$f_y = 300$ MPa				$f_y = 420$ MPa									
	$f'_c = 20$ MPa		$f'_c = 25$ MPa		$f'_c = 20$ MPa		$f'_c = 25$ MPa		$f'_c = 30$ MPa		$f'_c = 35$ MPa		$f'_c = 40$ MPa	
ρ	ϕk_n	j	ϕk_n	j	ϕk_n	j	ϕk_n	j	ϕk_n	j	ϕk_n	j	ϕk_n	j
0.0033					1.20	0.959	1.21	0.967	1.21	0.973	1.22	0.977	1.22	0.980
0.004					1.44	0.950	1.45	0.960	1.46	0.967	1.47	0.972	1.47	0.975
0.005	1.29	0.956	1.30	0.965	1.77	0.938	1.80	0.950	1.81	0.959	1.82	0.965	1.83	0.969
0.006	1.53	0.947	1.55	0.958	2.10	0.926	2.13	0.941	2.16	0.950	2.17	0.958	2.18	0.963
0.007	1.77	0.938	1.80	0.950	2.42	0.913	2.46	0.931	2.49	0.942	2.51	0.950	2.53	0.957
0.008	2.01	0.929	2.04	0.943	2.72	0.901	2.78	0.921	2.82	0.934	2.85	0.943	2.87	0.950
0.009	2.24	0.920	2.28	0.936	3.02	0.888	3.10	0.911	3.15	0.926	3.19	0.936	3.21	0.944
0.010	2.46	0.911	2.51	0.929	3.31	0.876	3.41	0.901	3.47	0.917	3.51	0.929	3.55	0.938
0.011	2.68	0.903	2.74	0.922	3.59	0.864	3.70	0.891	3.78	0.909	3.83	0.922	3.87	0.932
0.012	2.90	0.894	2.96	0.915	3.86	0.851	4.00	0.881	4.09	0.901	4.15	0.915	4.20	0.926
0.013	3.11	0.885	3.19	0.908	4.12	0.839	4.28	0.871	4.39	0.893	4.46	0.908	4.52	0.919
0.014	3.31	0.876	3.41	0.901			4.56	0.861	4.68	0.884	4.77	0.901	4.83	0.913
0.015	3.51	0.867	3.62	0.894			4.83	0.851	4.97	0.876	5.07	0.894	5.14	0.907
0.016	3.71	0.858	3.83	0.887			5.09	0.841	5.25	0.868	5.36	0.887	5.45	0.901
0.017	3.90	0.850	4.04	0.880					5.52	0.860	5.65	0.880	5.75	0.894
0.018	4.09	0.841	4.24	0.873					5.79	0.851	5.94	0.873	6.05	0.888
0.019			4.44	0.865					6.05	0.843	6.22	0.865	6.34	0.882
0.020			4.64	0.858					6.31	0.835	6.49	0.858	6.62	0.876
0.021			4.83	0.851							6.76	0.851	6.91	0.870
0.022			5.01	0.844							7.02	0.844	7.18	0.864
0.023													7.46	0.858
0.024													7.72	0.851

[a]Upper line in each column is below the entry for $\rho = 0.35\rho_b$; lower line is below the entry for $\rho = 0.5\rho_b$; bottom entry is the tension-controlled limit. Larger values of ρ can be used, but they require that ϕ be evaluated.

TABLE A-4 Ratio of Depth of Rectangular Stress Block for Balanced Failure (a_b), Compression-Controlled Limit (a_{CCL}), and Tension-Controlled Limit (a_{TCL}) to Effective Depth (d) or Depth to Extreme-Tension Steel Layer (d_t)[a]

f_y (psi)		f'_c (psi)			
		Less than or equal to 4000	5000	6000	8000
40,000	a_b/d, a_{CCL}/d_t	0.582	0.548	0.514	0.445
	$0.75a_b/d$	0.437	0.411	0.385	0.334
	a_{TCL}/d_t	0.319	0.300	0.281	0.244
	$0.50a_b/d$	0.291	0.274	0.257	0.223
	$0.35a_b/d$	0.204	0.192	0.180	0.156
60,000	a_b/d, a_{CCL}/d_t	0.503	0.474	0.444	0.385
	$0.75a_b/d$	0.377	0.355	0.333	0.288
	a_{TCL}/d_t	0.319	0.300	0.281	0.244
	$0.50a_b/d$	0.252	0.237	0.222	0.192
	$0.35a_b/d$	0.176	0.166	0.155	0.135
	β_1	0.85	0.80	0.75	0.65

[a] a_b/d from (4-20); desirable range for beams a/d from 0.35 to 0.50 a_b/d; a_{CCL}/d_t and a_{TCL}/d_t from (4-21) and (4-23). ACI 318-02 Sections 9.3.2.1 and 10.3.4 require $\phi < 0.9$ if $a/d_t > a_{TCL}/d_t$.

TABLE A-4M Ratio of Depth of Rectangular Stress Block for Balanced Failure (a_b), Compression-Controlled Limit (a_{CCL}), and Tension Controlled Limit (a_{TCL}) to Effective Depth (d) or Depth to Extreme-Tension Steel Layer (d_t)—SI Units[a]

f_y (MPa)		f'_c (MPa)			
		Less than or equal to 30	35	40	50
300	a_b/d, a_{CCL}/d_t	0.567	0.540	0.513	0.460
	$0.75a_b/d$	0.425	0.405	0.385	0.345
	a_{TCL}/d_t	0.319	0.304	0.289	0.259
	$0.50a_b/d$	0.283	0.270	0.256	0.230
	$0.35a_b/d$	0.198	0.189	0.180	0.161
420	a_b/d, a_{CCL}/d_t	0.500	0.477	0.453	0.406
	$0.75a_b/d$	0.375	0.357	0.339	0.304
	a_{TCL}/d_t	0.313	0.298	0.283	0.254
	$0.50a_b/d$	0.250	0.238	0.227	0.203
	$0.35a_b/d$	0.175	0.167	0.159	0.142
	β_1	0.85	0.80	0.77	0.69

[a] a_b/d from (4-20M); desirable range of a/d for beams from 0.35 to 0.50 a_b/d; a_{CCL}/d_t and a_{TCL}/d_t from (4-21M) and (4-23). ACI 318-02 Sections 9.3.2.1 and 10.3.4 require $\phi < 0.9$ if $a/d_t > a_{TCL}/d_t$.

TABLE A-5   Steel Ratios at Balanced Condition (ρ_b), Compression-Controlled Limit (ρ_{CCL}), and Tension-Controlled Limit (ρ_{TCL}) for Rectangular Beams with Tension Reinforcement Only[a]

f_y (psi)		f'_c (psi)					
		3000	3750	4000	5000	6000	8000
40,000	ρ_b, ρ_{CCL}	0.0371	0.0464	0.0495	0.0582	0.0655	0.0703
	$0.75\rho_b$	0.0278	0.0348	0.0371	0.0437	0.0491	0.0527
	ρ_{TCL}	0.0203	0.0254	0.0271	0.0319	0.0359	0.0414
	$0.50\rho_b$	0.0186	0.0232	0.0247	0.0291	0.0328	0.0352
	$0.35\rho_b$	0.0130	0.0162	0.0173	0.0204	0.0229	0.0246
60,000	ρ_b, ρ_{CCL}	0.0214	0.0267	0.0285	0.0335	0.0377	0.0405
	$0.75\rho_b$	0.0161	0.0200	0.0214	0.0251	0.0283	0.0307
	ρ_{TCL}	0.0135	0.0169	0.0181	0.0213	0.0239	0.0276
	$0.50\rho_b$	0.0107	0.0134	0.0143	0.0168	0.0189	0.0202
	$0.35\rho_b$	0.0075	0.0094	0.0100	0.0117	0.0132	0.0142
	β_1	0.85	0.85	0.85	0.80	0.75	0.65

[a]$\rho = A_s/bd$, ρ_b from (4-25); desirable steel ratio for beams, $\rho = 0.35$ to $0.50\rho_b$.

TABLE A-5M   Steel Ratios at Balanced Condition (ρ_b), Compression-Controlled Limit (ρ_{CCL}), and Tension-Controlled Limit (ρ_{TCL}) for Rectangular Beams with Tension Reinforcement Only—SI Units

f_y (MPa)		f'_c (MPa)				
		20	25	30	35	40
300	ρ_b, ρ_{CCL}	0.0321	0.0401	0.0482	0.0535	0.0582
	$0.75\rho_b$	0.0241	0.0301	0.0361	0.0401	0.0436
	ρ_{TCL}	0.0181	0.0226	0.0271	0.0301	0.0327
	$0.5\rho_b$	0.0160	0.0200	0.0241	0.0267	0.0291
	$0.35\rho_b$	0.0112	0.0140	0.0169	0.0187	0.0204
420	ρ_b, ρ_{CCL}	0.0202	0.0252	0.0304	0.0338	0.0367
	$0.75\rho_b$	0.0152	0.0189	0.0228	0.0254	0.0275
	ρ_{TCL}	0.0202	0.0253	0.0304	0.0337	0.0367
	$0.5\rho_b$	0.0101	0.0126	0.0152	0.0169	0.0275
	$0.35\rho_b$	0.0071	0.0088	0.0106	0.0118	0.0128
	β_1	0.85	0.85	0.85	0.81	0.77

[a]$\rho = A_s/bd$, ρ_b from (4-25M); desirable ρ for beams, $\rho = 0.35$ to $0.50\rho_b$; ACI 318-02 Section 4.3.2.1 requires $\phi < 0.9$ for $\rho > \rho_{TCL}$.

TABLE A-6 Minimum Beam Web Widths, b_w, for Various Bar Combinations, Interior Exposure; Minimum Bar Spacing (in.)[a,b,c]

No. of Bars	A Bar No.	0	5	B Bar No.	1	2	3	4	5	C Bar No.	1	2	3	4	5
1		5.5	13.0		7.0	8.5	9.5	11.0	12.5						
2		7.0	14.5		8.5	9.5	11.0	12.5	14.0						
3	4	8.5	16.0	3	10.0	11.0	12.5	14.0	15.5						
4		10.0	17.5		11.5	12.5	14.0	15.5	17.0						
5		11.5	19.0		13.0	14.0	15.5	17.0	18.5						
1		5.5	13.5		7.0	8.5	10.0	11.5	13.0		7.0	8.5	9.5	11.0	12.5
2		7.0	15.0		8.5	10.0	11.5	13.0	14.5		8.5	10.0	11.0	12.5	14.0
3	5	8.5	17.0	4	10.0	11.5	13.0	14.5	16.0	3	10.0	11.5	13.0	14.0	15.5
4		10.5	18.5		12.0	13.5	15.0	16.5	18.0		11.5	13.0	14.5	16.0	17.0
5		12.0	20.0		13.5	15.0	16.5	18.0	19.5		13.5	14.5	16.5	17.5	19.0
1		5.5	14.0		7.0	9.0	10.5	12.0	13.5		7.0	8.5	10.0	11.5	13.0
2		7.0	16.0		9.0	10.5	12.0	13.5	15.5		8.5	10.0	11.5	13.0	14.5
3	6	9.0	17.5	5	10.5	12.0	14.0	15.5	17.0	4	10.5	12.0	13.5	15.0	16.5
4		10.5	19.5		12.5	14.0	15.5	17.0	19.0		12.0	13.5	15.0	16.5	18.0
5		12.5	21.0		14.0	15.5	17.5	19.0	20.5		14.0	15.5	17.0	18.5	20.0
1		5.5	15.0		7.5	9.0	11.0	12.5	14.5		7.0	9.0	10.5	12.0	13.5
2		7.5	16.5		9.0	11.0	12.5	14.5	16.0		9.0	10.5	12.0	14.0	15.5
3	7	9.0	18.5	6	11.0	12.5	14.5	16.0	18.0	5	11.0	12.5	14.0	15.5	17.5
4		11.0	20.5		13.0	14.5	16.5	18.0	20.0		12.5	14.5	16.0	17.5	19.0
5		13.0	22.5		14.5	16.5	18.0	20.0	21.5		14.5	16.0	18.0	19.5	21.0
1		5.5	15.5		7.5	9.5	11.0	13.0	15.0		7.5	9.0	11.0	12.5	14.5
2		7.5	17.5		9.5	11.0	13.0	15.0	17.0		9.0	11.0	12.5	14.5	16.0
3	8	9.5	19.5	7	11.5	13.0	15.0	17.0	19.0	6	11.0	13.0	14.5	16.5	18.0
4		11.5	21.5		13.5	15.0	17.0	19.0	21.0		13.0	15.0	16.5	18.5	20.0
5		13.5	23.5		15.5	17.0	19.0	21.0	23.0		15.0	17.0	18.5	20.5	22.0
1		5.5	17.0		7.5	9.5	11.5	13.5	15.5		7.5	9.5	11.5	13.0	15.0
2		8.0	19.0		10.0	12.0	14.0	16.0	18.0		9.5	11.5	13.5	15.5	17.0
3	9	10.0	21.5	8	12.0	14.0	16.0	18.0	20.0	7	12.0	14.0	15.5	17.5	19.5
4		12.5	23.5		14.5	16.5	18.5	20.5	22.5		14.0	16.0	18.0	20.0	21.5
5		14.5	26.0		16.5	18.5	20.5	22.5	24.5		16.5	18.5	20.0	22.0	24.0
1		5.5	18.0		8.0	10.0	12.5	14.5	17.0		8.0	10.0	12.0	14.0	16.0
2		8.0	20.5		10.5	12.5	15.0	17.0	19.5		10.0	12.0	14.0	16.0	18.0
3	10	10.5	23.5	9	13.0	15.0	17.5	19.5	22.0	8	12.5	14.5	16.5	18.5	20.5
4		13.0	26.0		15.5	17.5	20.0	22.0	24.5		15.0	17.0	19.0	21.0	23.0
5		15.5	28.5		18.0	20.0	22.5	24.5	27.0		17.5	19.5	21.5	23.5	25.5
1		5.5	19.5		8.0	10.5	13.0	15.5	18.0		8.0	10.5	12.5	15.0	17.0
2		8.5	22.5		11.0	13.5	16.0	18.5	21.0		10.5	13.0	15.0	17.5	19.5
3	11	11.0	25.0	10	13.5	16.0	19.0	21.5	24.0	9	13.5	15.5	18.0	20.0	22.5
4		14.0	28.0		16.5	19.0	21.5	24.0	26.5		16.0	18.5	20.5	23.0	25.0
5		17.0	31.0		19.5	22.0	24.5	27.0	29.5		19.0	21.5	23.5	26.0	28.0

[a]Clear cover, $1\frac{1}{2}$ in.; No. 3 double-leg stirrup; $\frac{3}{4}$ in.-maximum-size aggregate.

[b]This table consists of three basic parts: Part A lists the web widths required for 1 to 5 bars of the size given in the left margin of part A. *Example:* 3 No. 5, min b_w = 8.5 in. Part A also lists the minimum b_w for 1 to 5 bars of the size given in the left margin plus 5 bars of the same size. *Example:* 3 No. 5 plus 5 No. 5, min b_w = 17.0 in. Part B lists the minimum b_w of 1 to 5 bars of the size given in the left margin of part A plus 1 to 5 bars of the size listed in the left margin of part B. *Example:* 3 No. 5 plus 2 No. 4, min b_w = 11.5 in. Part C is similar to part B.

[c]In many cases it is desirable to provide on extra 2 in. web width, b_w, to allow vibrators to penetrate the layers of steel.

Source: Based on a table from [4-11], used with the permission of the American Concrete Institute.

TABLE A-6M   Minimum Beam Web Widths, b_w, for Various Bar Combinations, Interior Exposure; Minimum Bar Spacing (mm)[a,b]

No. of Bars	A Bar No.	0	5	B Bar No.	1	2	3	4	5	C Bar No.	1	2	3	4	5
1		140	330		180	220	245	280	320						
2		180	370		220	245	280	320	360						
3	13	220	410	10	260	280	320	360	400						
4		260	450		295	320	360	400	435						
5		295	490		330	360	400	430	470						
1		140	350		180	220	260	295	330		180	220	250	280	320
2		180	390		220	260	295	330	370		220	260	280	320	360
3	16	220	440	13	260	295	330	370	410	10	260	295	330	360	400
4		270	470		310	350	385	420	460		295	330	370	410	435
5		305	510		350	385	420	460	500		350	370	420	450	485
1		140	360		180	230	270	310	350		180	220	260	300	330
2		180	410		230	270	310	345	400		220	260	295	330	370
3	19	230	450	16	270	310	360	400	435	13	270	310	350	390	410
4		270	500		320	360	400	435	485		310	350	385	420	460
5		320	540		360	400	450	485	520		360	400	440	470	510
1		140	390		200	230	280	320	370		180	230	270	310	350
2		190	420		230	280	320	370	410		230	270	310	360	400
3	22	230	470	19	280	320	370	410	460	16	280	320	360	400	450
4		280	530		330	370	420	460	510		320	370	410	450	485
5		330	580		370	420	460	510	550		370	410	460	500	540
1		140	400		200	250	280	330	390		200	230	280	320	370
2		190	450		250	280	330	385	435		230	280	320	370	410
3	25	240	500	22	295	330	385	435	485	19	280	330	370	420	460
4		295	550		350	385	435	485	540		330	390	410	470	510
5		395	600		410	435	485	540	590		390	435	470	520	560
1		140	440		200	250	300	350	400		200	250	300	340	390
2		210	490		260	310	360	410	460		250	300	350	400	440
3	29	260	550	25	310	360	410	460	510	22	310	360	400	450	500
4		320	600		370	410	480	530	580		360	410	460	510	590
5		370	670		420	480	530	580	630		410	480	510	560	630
1		140	460		210	260	320	370	440		210	260	310	360	410
2		210	530		270	320	390	440	500		260	310	360	410	460
3	32	270	600	29	340	390	450	500	560	25	320	370	410	480	530
4		340	670		400	450	510	560	630		390	440	490	540	590
5		400	730		460	510	580	630	690		450	500	550	600	660
1		140	500		210	270	340	400	460		210	270	320	390	440
2		220	580		280	350	410	480	540		270	340	390	450	500
3	36	280	640	32	350	410	490	550	620	29	350	400	460	510	580
4		360	720		430	490	550	630	680		410	480	530	590	640
5		440	790		500	560	630	690	760		490	550	600	670	720

[a]Clear cover, 40 mm No. 3 double-leg stirrup; 19 mm maximum-s aggregate.

[b]This table consists of three basic parts: Part A lists the web widths required for 1 to 5 bars of the size given in the left margin of part A. *Example:* 3 No. 16, min b_w = 220 mm. Part A also lists the minimum b_w for 1 to 5 bars of the size given in the left margin plus 5 bars of the same size. *Example:* 3 No. 16 plus 5 No. 16, min b_w = 440 mm. Part B lists the minimum b_w of 1 to 5 bars of the size given in the left margin of part A plus 1 to 5 bars of the size listed in the left margin of part B. *Example:* 3 No. 16 plus 2 No. 13, min b_w = 295 mm. Part C is similar to part B.

[c]In many cases it is desirable to provide an extra 50 mm web width, b_w, to allow vibrators to penetrate the layers of steel.

Source: Based on a table from [4-11], used with the permission of the American Concrete Institute.

TABLE A-7 Values of $bd^2/12,000$ for Use in Choosing Beam Sizes

$$\frac{\phi M_n}{\phi k_n} = \frac{bd^2}{12,000} \quad \text{or} \quad \frac{M_u}{\phi k_n} = \frac{bd^2}{12,000} \quad \text{where } \phi k_n \text{ is from Table A-3}$$

d	b (in.)																	
	6	7	8	9	10	11	12	14	16	18	20	22	24	26	28	30	36	48
5	0.013	0.015	0.017	0.019	0.021	0.023	0.025	0.027	0.033	0.037	0.042	0.046	0.050	0.054	0.058	0.062	0.075	0.100
6	0.018	0.021	0.024	0.027	0.030	0.033	0.036	0.042	0.048	0.054	0.060	0.066	0.072	0.078	0.084	0.090	0.108	0.144
7	0.025	0.029	0.033	0.037	0.041	0.045	0.049	0.057	0.065	0.073	0.082	0.090	0.098	0.106	0.114	0.123	0.147	0.196
8	0.032	0.037	0.043	0.048	0.053	0.059	0.064	0.075	0.085	0.096	0.107	0.117	0.128	0.139	0.149	0.160	0.192	0.256
9	0.041	0.047	0.054	0.061	0.068	0.074	0.081	0.095	0.108	0.122	0.135	0.149	0.162	0.176	0.189	0.203	0.243	0.324
10	0.050	0.058	0.067	0.075	0.083	0.092	0.100	0.117	0.133	0.150	0.167	0.183	0.200	0.217	0.233	0.250	0.300	0.400
11	0.061	0.071	0.081	0.091	0.101	0.111	0.121	0.141	0.161	0.181	0.202	0.222	0.242	0.262	0.282	0.303	0.363	0.484
12	0.072	0.084	0.096	0.108	0.120	0.132	0.144	0.168	0.192	0.216	0.240	0.264	0.288	0.312	0.336	0.360	0.432	0.576
13	0.085	0.099	0.113	0.127	0.141	0.155	0.169	0.197	0.225	0.253	0.282	0.310	0.338	0.366	0.394	0.423	0.507	0.676
14	0.098	0.114	0.131	0.147	0.163	0.180	0.196	0.229	0.261	0.294	0.327	0.359	0.392	0.425	0.457	0.490	0.588	0.784
15	0.113	0.131	0.150	0.169	0.188	0.206	0.225	0.263	0.300	0.338	0.375	0.413	0.450	0.488	0.525	0.563	0.675	0.900
16	0.128	0.149	0.171	0.192	0.213	0.235	0.256	0.299	0.341	0.384	0.427	0.469	0.512	0.555	0.597	0.640	0.768	1.02
18	0.162	0.189	0.216	0.243	0.270	0.297	0.324	0.378	0.432	0.486	0.540	0.594	0.648	0.702	0.756	0.810	0.972	1.30
20	0.200	0.233	0.267	0.300	0.333	0.367	0.400	0.467	0.533	0.600	0.667	0.733	0.800	0.867	0.933	1.00	1.20	1.60
22		0.282	0.323	0.363	0.403	0.444	0.484	0.565	0.645	0.726	0.807	0.887	0.968	1.05	1.13	1.21	1.45	1.94
24		0.336	0.384	0.432	0.480	0.528	0.576	0.672	0.768	0.864	0.960	1.06	1.15	1.25	1.34	1.44	1.73	2.30
26			0.451	0.507	0.563	0.620	0.676	0.789	0.901	1.01	1.13	1.24	1.35	1.46	1.58	1.69	2.03	2.70
28			0.523	0.588	0.653	0.719	0.784	0.915	1.04	1.18	1.31	1.44	1.57	1.70	1.83	1.96	2.35	3.14
30				0.675	0.750	0.825	0.900	1.05	1.20	1.35	1.50	1.65	1.80	1.95	2.10	2.25	2.70	3.60
32				0.768	0.853	0.939	1.02	1.19	1.37	1.54	1.71	1.88	2.05	2.22	2.39	2.56	3.07	4.10
34					0.963	1.06	1.16	1.35	1.54	1.73	1.93	2.12	2.31	2.50	2.70	2.89	3.47	4.62
36					1.08	1.19	1.30	1.51	1.73	1.94	2.16	2.38	2.59	2.81	3.02	3.24	3.89	5.18
38						1.32	1.44	1.68	1.93	2.17	2.41	2.65	2.89	3.13	3.37	3.61	4.33	5.78
40						1.47	1.60	1.87	2.13	2.40	2.67	2.93	3.20	3.47	3.73	4.00	4.80	6.40
45							2.03	2.36	2.70	3.04	3.38	3.71	4.05	4.39	4.73	5.06	6.08	8.10
50							2.50	2.92	3.33	3.75	4.17	4.58	5.00	5.42	5.83	6.25	7.50	10.0
55							3.03	3.53	4.03	4.54	5.04	5.55	6.05	6.55	7.06	7.56	9.07	12.1
60						3.30	3.60	4.20	4.80	5.40	6.00	6.60	7.20	7.80	8.40	9.00	10.8	14.4

Source: Based on a table in [4-12], used with the permission of the American Concrete Institute.

TABLE A-7M   Values of $bd^2/10^6$ for Use in Choosing Beam Sizes—SI Units

$$\frac{\phi M_n}{\phi k_n} = \frac{bd^2}{10^6} \quad \text{or} \quad \frac{M_u}{\phi k_n} = \frac{bd^2}{10^6} \quad \text{where } \phi k_n \text{ is from Table A-3M}$$

	b (mm)																
d (mm)	150	175	200	225	250	275	300	350	400	450	500	600	700	800	900	1000	1500
125	2.34	2.73	3.13	3.52	3.91	4.30	4.69	5.47	6.25	7.03	7.81	9.38	10.94	12.50	14.06	15.63	23.44
150	3.38	3.94	4.50	5.06	5.63	6.19	6.75	7.88	9.00	10.13	11.25	13.50	15.75	18.00	20.25	22.50	33.75
175	4.59	5.36	6.13	6.89	7.66	8.42	9.19	10.72	12.25	13.78	15.31	18.38	21.44	24.50	27.56	30.63	45.94
200	6.00	7.00	8.00	9.00	10.00	11.00	12.00	14.00	16.00	18.00	20.00	24.00	28.00	32.00	36.00	40.00	60.00
225	7.59	8.86	10.13	11.39	12.66	13.92	15.19	17.72	20.25	22.78	25.31	30.38	35.44	40.50	45.56	50.63	75.94
250	9.38	10.94	12.50	14.06	15.63	17.19	18.75	21.88	25.00	28.13	31.25	37.50	43.75	50.00	56.25	62.50	93.75
275	11.34	13.23	15.13	17.02	18.91	20.80	22.69	26.47	30.25	34.03	37.81	45.38	52.94	60.50	68.06	75.63	113
300	13.50	15.75	18.00	20.25	22.50	24.75	27.00	31.50	36.00	40.50	45.00	54.00	63.00	72.00	81.00	90.00	135
325	15.84	18.48	21.13	23.77	26.41	29.05	31.69	36.97	42.25	47.53	52.81	63.38	73.94	84.50	95.06	106	158
350	18.38	21.44	24.50	27.56	30.63	33.69	36.75	42.88	49.00	55.13	61.25	73.50	85.75	98.00	110	123	184
375	21.09	24.61	28.13	31.64	35.16	38.67	42.19	49.22	56.25	63.28	70.31	84.38	98.44	113	127	141	211
400	24.00	28.00	32.00	36.00	40.00	44.00	48.00	56.00	64.00	72.00	80.00	96.00	112	128	144	160	240
450	30.38	35.44	40.50	45.56	50.63	55.69	60.75	70.88	81.00	91.13	101	122	142	162	182	203	304
500	37.50	43.75	50.00	56.25	62.50	68.75	75.00	87.50	100	113	125	150	175	200	225	250	375
550	45.38	52.94	60.50	68.06	75.63	83.19	90.75	106	121	136	151	182	212	242	272	303	454
600	54.00	63.00	72.00	81.00	90.00	99.00	108	126	144	162	180	216	252	288	324	360	540
650	63.38	73.94	84.50	95.06	106	116	127	148	169	190	211	254	296	338	380	423	634
700	73.50	85.75	98.00	110	123	135	147	172	196	221	245	294	343	392	441	490	735
800	96.00	112	128	144	160	176	192	224	256	288	320	384	448	512	576	640	960
900	122	142	162	182	203	223	243	284	324	365	405	486	567	648	729	810	1215
1000			200	225	250	275	300	350	400	450	500	600	700	800	900	1000	1500
1200					360	396	432	504	576	648	720	864	1008	1152	1296	1440	2160
1400					490	539	588	686	784	882	980	1176	1372	1568	1764	1960	2940
1500					563	619	675	788	900	1013	1125	1350	1575	1800	2025	2250	3375
2000							1200	1400	1600	1800	2000	2400	2800	3200	3600	4000	6000

TABLE A-8   Cross-Sectional Areas, A_s, for Various Combinations of Bars (in.^{2})[a]

No. of Bars	A Bar No.	0	5	B Bar No.	1	2	3	4	5	C Bar No.	1	2	3	4	5
1	4	0.20	1.20	3	0.31	0.42	0.53	0.64	0.75						
2		0.40	1.40		0.51	0.62	0.73	0.84	0.95						
3		0.60	1.60		0.71	0.82	0.93	1.04	1.15						
4		0.80	1.80		0.91	1.02	1.13	1.24	1.35						
5		1.00	2.00		1.11	1.22	1.33	1.44	1.55						
1	5	0.31	1.86	4	0.51	0.71	0.91	1.11	1.31	3	0.42	0.53	0.64	0.75	0.86
2		0.62	2.17		0.82	1.02	1.22	1.42	1.62		0.73	0.84	0.95	1.06	1.17
3		0.93	2.48		1.13	1.33	1.53	1.73	1.93		1.04	1.15	1.26	1.37	1.48
4		1.24	2.79		1.44	1.64	1.84	2.04	2.24		1.35	1.46	1.57	1.68	1.79
5		1.55	3.10		1.75	1.95	2.15	2.35	2.55		1.66	1.77	1.88	1.99	2.10
1	6	0.44	2.64	5	0.75	1.06	1.37	1.68	1.99	4	0.64	0.84	1.04	1.24	1.44
2		0.88	3.08		1.19	1.50	1.81	2.12	2.43		1.08	1.28	1.48	1.68	1.88
3		1.32	3.52		1.63	1.94	2.25	2.56	2.87		1.52	1.72	1.92	2.12	2.32
4		1.76	3.96		2.07	2.38	2.69	3.00	3.31		1.96	2.16	2.36	2.56	2.76
5		2.20	4.40		2.51	2.82	3.13	3.44	3.75		2.40	2.60	2.80	3.00	3.20
1	7	0.60	3.60	6	1.04	1.48	1.92	2.36	2.80	5	0.91	1.22	1.53	1.84	2.15
2		1.20	4.20		1.64	2.08	2.52	2.96	3.40		1.51	1.82	2.13	2.44	2.75
3		1.80	4.80		2.24	2.68	3.12	3.56	4.00		2.11	2.42	2.73	3.04	3.35
4		2.40	5.40		2.84	3.28	3.72	4.16	4.60		2.71	3.02	3.33	3.64	3.95
5		3.00	6.00		3.44	3.88	4.32	4.76	5.20		3.31	3.62	3.93	4.24	4.55
1	8	0.79	4.74	7	1.39	1.99	2.59	3.19	3.79	6	1.23	1.67	2.11	2.55	2.99
2		1.58	5.53		2.18	2.78	3.38	3.98	4.58		2.02	2.46	2.90	3.34	3.78
3		2.37	6.32		2.97	3.57	4.17	4.77	5.37		2.81	3.25	3.69	4.13	4.57
4		3.16	7.11		3.76	4.36	4.96	5.56	6.16		3.60	4.04	4.48	4.92	5.36
5		3.95	7.90		4.55	5.15	5.75	6.35	6.95		4.39	4.83	5.27	5.71	6.15
1	9	1.00	6.00	8	1.79	2.58	3.37	4.16	4.95	7	1.60	2.20	2.80	3.40	4.00
2		2.00	7.00		2.79	3.58	4.37	5.16	5.95		2.60	3.20	3.80	4.40	5.00
3		3.00	8.00		3.79	4.58	5.37	6.16	6.95		3.60	4.20	4.80	5.40	6.00
4		4.00	9.00		4.79	5.58	6.37	7.16	7.95		4.60	5.20	5.80	6.40	7.00
5		5.00	10.00		5.79	6.58	7.37	8.16	8.95		5.60	6.20	6.80	7.40	8.00
1	10	1.27	7.62	9	2.27	3.27	4.27	5.27	6.27	8	2.06	2.85	3.64	4.43	5.22
2		2.54	8.89		3.54	4.54	5.54	6.54	7.54		3.33	4.12	4.91	5.70	6.49
3		3.81	10.16		4.81	5.81	6.81	7.81	8.81		4.60	5.39	6.18	6.97	7.76
4		5.08	11.43		6.08	7.08	8.08	9.08	10.08		5.87	6.66	7.45	8.24	9.03
5		6.35	12.70		7.35	8.35	9.35	10.35	11.35		7.14	7.93	8.72	9.51	10.30
1	11	1.56	9.36	10	2.83	4.10	5.37	6.64	7.91	9	2.56	3.56	4.56	5.56	6.56
2		3.12	10.92		4.39	5.66	6.93	8.20	9.47		4.12	5.12	6.12	7.12	8.12
3		4.68	12.48		5.95	7.22	8.49	9.76	11.03		5.68	6.68	7.68	8.68	9.68
4		6.24	14.04		7.51	8.78	10.05	11.32	12.59		7.24	8.24	9.24	10.24	11.24
5		7.80	15.60		9.07	10.34	11.61	12.88	14.15		8.80	9.80	10.80	11.80	12.80

[a]For directions on how to use this table, see Table A-6, footnote b.

Source: Based on a table from [4-11]; used with the permission of the American Concrete Institute.

TABLE A-9   Areas of Bars in a Section 1 ft Wide (in.^{2}/ft)

Bar Spacing (in.)	Bar No.				
	3	4	5	6	7
4	0.33	0.60	0.93	1.32	1.80
$4\frac{1}{2}$	0.29	0.53	0.83	1.17	1.60
5	0.26	0.48	0.74	1.06	1.44
$5\frac{1}{2}$	0.24	0.44	0.68	0.96	1.31
6	0.22	0.40	0.62	0.88	1.20
$6\frac{1}{2}$	0.20	0.37	0.57	0.81	1.11
7	0.19	0.34	0.53	0.75	1.03
$7\frac{1}{2}$	0.18	0.32	0.50	0.70	0.96
8	0.17	0.30	0.47	0.66	0.90
$8\frac{1}{2}$	0.16	0.28	0.44	0.62	0.85
9	0.15	0.27	0.41	0.59	0.80
$9\frac{1}{2}$	0.14	0.25	0.39	0.56	0.76
10	0.13	0.24	0.37	0.53	0.72
$10\frac{1}{2}$	0.13	0.23	0.35	0.50	0.69
11	0.12	0.22	0.34	0.48	0.65
$11\frac{1}{2}$	0.11	0.21	0.32	0.46	0.63
12	0.11	0.20	0.31	0.44	0.60
13	0.10	0.18	0.29	0.41	0.55
14	0.09	0.17	0.27	0.38	0.51
15	0.09	0.16	0.25	0.35	0.48
16	0.08	0.15	0.23	0.33	0.45
17	0.08	0.14	0.22	0.31	0.42
18	0.07	0.13	0.21	0.29	0.40

TABLE A-9M   Areas of Bars in a Section 1 m Wide (mm²/m)

Bar Spacing (mm)	Bar No.			
	10	15	20	25
100	1000	2000	3000	5000
110	909	1818	2727	4545
120	833	1667	2500	4167
130	769	1538	2308	3846
140	714	1429	2143	3571
150	667	1333	2000	3333
160	625	1250	1875	3125
180	556	1111	1667	2778
200	500	1000	1500	2500
220	455	909	1364	2273
240	417	833	1250	2083
250	400	800	1200	2000
260	385	769	1154	1923
280	357	714	1071	1786
300	333	667	1000	1667
350	286	571	857	1429
400	250	500	750	1250
450	222	444	667	1111
500	200	400	600	1000

TABLE A-10   Limiting Values of d'/a for Checking if Compression Steel Yields[a]

f_y (psi)	f_c' (psi)			
	≤ 4000	5000	6000	8000
40,000	0.636	0.675	0.720	0.831
50,000	0.500	0.532	0.567	0.654
60,000	0.365	0.388	0.414	0.477

[a]If d'/a exceeds the appropriate value given in this table, the compression steel will not yield before failure; calculated from (5-7).

TABLE A-10M   Limiting Values of d'/a for Checking if Compression Steel Yields—SI Units[a]

f_y (MPa)	f_c' (MPa)			
	≤ 30	35	40	50
300	0.588	0.617	0.649	0.725
420	0.353	0.370	0.390	0.435

[a]If d'/a exceeds the values given in this table, the compression steel will not yield before failure; calculated from (5-7M).

TABLE A-11 Basic Tension Development-Length Ratio, ℓ_{db}/d_b (in./in.)

$$\ell_d = \frac{\ell_{db}}{d_b} \times \beta\lambda \times d_b \text{ but not less than 12 in.[a]}$$

Bar No.	$f'_c = 3000$ psi Bottom Bar	Top Bar	$f'_c = 3750$ psi Bottom Bar	Top Bar	$f'_c = 4000$ psi Bottom Bar	Top Bar	$f'_c = 5000$ psi Bottom Bar	Top Bar	$f'_c = 6000$ psi Bottom Bar	Top Bar
Case 1: Clear spacing of bars being developed or spliced not less than d_b, clear cover not less than d_b, and stirrups or ties not less than the ACI Code minimum, throughout ℓ_d										
or										
Case 2: Clear spacing of bars being developed or spliced not less than $2d_b$ and clear cover not less than d_b.										
$f_y = 60{,}000$ psi, uncoated bars, normal-weight concrete										
3 to 6	43.8	57.0	39.2	50.9	37.9	49.3	33.9	44.1	31.0	40.3
7 to 18	54.8	71.2	49.0	63.7	47.4	61.7	42.4	55.2	38.7	50.3
$f_y = 40{,}000$ psi, uncoated bars, normal-weight concrete										
3 to 6	29.2	38.0	26.1	34.0	25.3	32.9	22.6	29.4	20.7	26.9
Other Cases										
$f_y = 60{,}000$ psi, uncoated bars, normal-weight concrete										
3 to 6	65.7	85.5	58.8	76.4	56.9	74.0	50.9	66.2	46.5	60.5
7 to 18	82.2	106.8	73.5	95.6	71.1	92.6	63.6	82.8	58.1	75.5
$f_y = 40{,}000$ psi, uncoated bars, normal-weight concrete										
3 to 6	43.8	57.0	39.2	51.0	38.0	49.4	33.9	44.1	31.1	40.4

[a] β, coating factor; λ, lightweight-concrete factor.

TABLE A-11M Basic Tension Development-Length Ratio, ℓ_{db}/d_b (mm/mm)

$$\ell_d = \frac{\ell_{db}}{d_b} \times \beta\lambda \times d_b, \text{ but not less than 300 mm}^a$$

Bar No.	$f_c' = 20$ MPa		$f_c' = 25$ MPa		$f_c' = 30$ MPa		$f_c' = 35$ MPa		$f_c' = 40$ MPa	
	Bottom Bar	Top Bar	Bottom Bar	Top Bar	Bottom Bar	Top Bar	Bottom Bar	Top Bar	Bottom Bar	Top Bar

Case 1: Clear spacing of bars being developed or spliced not less than d_b, clear cover not less than d_b, and stirrups or ties not less than the ACI Code minimum, throughout ℓ_d

or

Case 2: Clear spacing of bars being developed or spliced not less than $2d_b$ and clear cover not less than d_b.

$f_y = 420$ MPa, uncoated bars, normal-weight concrete

Bar No.	Bottom	Top	Bottom	Top	Bottom	Top	Bottom	Top	Bottom	Top
10 to 20	43.8	56.9	39.2	50.9	35.8	46.5	33.2	43.0	31.0	40.3
25 to 45	54.8	71.2	49.0	63.6	44.7	58.1	41.4	53.8	38.7	50.3

$f_y = 300$ MPa, uncoated bars, normal-weight concrete

Bar No.	Bottom	Top	Bottom	Top	Bottom	Top	Bottom	Top	Bottom	Top
10 to 20	32.2	41.9	28.8	37.4	26.3	34.2	24.3	31.6	222.8	29.6

Other Cases

$f_y = 420$ MPa, uncoated bars, normal-weight concrete

Bar No.	Bottom	Top	Bottom	Top	Bottom	Top	Bottom	Top	Bottom	Top
10 to 20	45.1	58.6	40.3	74.9	52.6	68.4	48.7	63.3	45.5	59.2
25 to 45	56.3	104.6	72.0	93.6	65.7	85.4	60.9	79.1	56.9	74.0

$f_y = 300$ MPa, uncoated bars, normal-weight concrete

Bar No.	Bottom	Top	Bottom	Top	Bottom	Top	Bottom	Top	Bottom	Top
10 to 20	48.3	62.8	43.2	56.2	39.4	51.3	36.5	47.5	34.2	44.4

$^a\beta$, coating factor; λ, lightweight-concrete factor.

TABLE A-12 Basic Compression Development Length, ℓ_{dbc} (in.)^a

$$\ell_{dc} = \ell_{dbc} \times \text{(Factors in ACI Section 12.3.3)}$$
$$f_c' \text{ (psi)}$$

Bar No.	3000	4000	5000 psi and up
	$f_y = 60,000$ psi		
3	8	8	8
4	11	9	9
5	14	12	11
6	16	14	14
7	19	17	16
8	22	19	18
9	25	21	20
10	28	24	23
11	34	27	25
14	37	32	30
18	49	43	41
	$f_y = 40,000$ psi		
3	8	8	8
4	8	8	8
5	9	8	8
6	11	9	9

^{a}Lengths may be reduced if excess reinforcement is anchored or if the splice is enclosed in a spiral. See ACI Section 12.3.3. Reduced length shall not be less than 8 in.

TABLE A-12M Basic Compression Development Length, ℓ_{dbc} (mm)[a]

$$\ell_{dc} = \ell_{dbc} \times \quad \text{(Factors in ACI Section 12.3.3)}$$

Bar No.	f_c' (MPa)				
	20	25	30	35	40
	$f_y = 420$ (MPa)				
10	235	210	192	177	168
13	305	273	249	231	216
16	376	336	307	284	266
19	446	399	364	337	315
22	517	462	422	390	365
25	587	525	479	444	415
29	681	609	556	515	481
32	751	672	613	568	531
36	845	756	690	639	598
43	1010	903	824	763	714
57	1338	1197	1093	1012	946
	$f_y = 300$ (MPa)				
10	200	200	200	200	200
15	252	225	205	200	200
20	335	300	274	254	240

[a]Lengths may be reduced if excess reinforcement is anchored or if the splice is enclosed in a spiral. See ACI Section 12.3.3. Reduced length shall not be less than 200 mm.

TABLE A-13 Basic Development Lengths for Hooked Bars, ℓ_{hb} (in.)

$$\ell_{dh} = \ell_{hb} \times \quad \text{(Factors in 12.5.3)[a]}$$

Normal-weight concrete, $f_y = 60,000$ psi

Standard 90° or 180° Hooks

Bar No.	f_c' (psi)			
	3000	4000	5000	6000
3	8.2	7.1	6.4	5.8
4	11	9.5	8.5	7.8
5	13.7	11.9	10.6	9.7
6	16.4	14.2	12.7	11.6
7	19.2	16.6	14.9	13.6
8	22	19	17	15.5
9	25	21	19	17.5
10	28	24	22	20
11	31	27	24	22
14	37	32	29	26
18	49	43	38	35

[a]ℓ_{dh} is defined in Fig. 8-12a. The development length of a hook, ℓ_{dh}, is the product of ℓ_{hb} from this table and factors relating to bar yield strength, cover, presence of stirrups, and type of concrete, given in ACI Section 12.5.3. The resulting length, ℓ_{dh}, shall not be less than the larger of 8 bar diameters or 6 in.

TABLE A-13M   Basic Development Lengths for Hooked Bars, ℓ_{hb} (mm)

$\ell_{dh} = \ell_{hb} \times$ (Factors in 12.5.3)[a]
Normal-weight concrete, $f_y = 400$ MPa
Standard 90° or 180° hooks

Bar No.	f_c' (MPa)				
	20	25	30	35	40
10	224	200	183	169	158
13	291	260	237	220	206
16	358	320	292	270	253
19	425	380	347	321	300
22	492	440	402	372	348
25	559	500	456	423	395
29	648	580	529	490	459
32	716	640	584	541	506
36	805	720	657	609	569
43	962	860	785	727	680
57	1275	1140	1041	963	901

[a] ℓ_{dh} is defined in Fig. 8-12a. The development length of a hook, ℓ_{dh}, is the product of ℓ_{hb} from this table and factors relating to bar yield strength, cover, presence of stirrups, and type of concrete, given in ACI Section 12.5.3. The resulting length, ℓ_{dh}, shall not be less than the larger of 8 bar diameters or 150 mm.

TABLE A-14   Minimum Thicknesses of Nonprestressed Beams or One-Way Slabs Unless Deflections Are Computed

Exposure	Member	Minimum Thickness, h				Source
		Simply Supported	One End Continuous	Both Ends Continuous	Cantilever	
Not supporting or attached to partitions or other construction likely to be damaged by large deflections	Solid one-way slabs	$\ell/20$	$\ell/24$	$\ell/28$	$\ell/10$	ACI Table 9.5(a)
	Beams or ribbed one-way slabs	$\ell/16$	$\ell/18.5$	$\ell/21$	$\ell/8$	ACI Table 9.5(a)
Supporting or attached to partitions or other construction likely to be damaged by large deflections	All members: $\omega \leq 0.12^a$ and $\dfrac{\text{sustained load}}{\text{total load}} < 0.5$	$\ell/10$	$\ell/13$	$\ell/16$	$\ell/4$	[9-20]
	All members: $\dfrac{\text{sustained load}}{\text{total load}} > 0.5$	$\ell/6$	$\ell/8$	$\ell/10$	$\ell/3$	[9-20]

[a] $\omega = \rho f_y / f_c'$

TABLE A-15   Maximum Allowable Spiral Pitch, s (in.), for Circular Spiral Columns, Grade-60 Spirals[a]

Column Diameter (in.)	Core Diameter (in.)	f'_c (psi)						
		4000 Spiral Size		5000 Spiral Size		6000 Spiral Size		
		No. 3	No. 4	No. 3	No. 4	No. 3	No. 4	No. 5
12	9	2	$3\frac{1}{2}$	$1\frac{1}{2}$*	$2\frac{3}{4}$	—	$2\frac{1}{4}$	$3\frac{1}{2}$
14	11	2	$3\frac{1}{2}$	$1\frac{1}{2}$*	3	—	$2\frac{1}{4}$	$3\frac{1}{2}$
16	13	2	$3\frac{1}{2}$	$1\frac{1}{2}$*	3	—	$2\frac{1}{2}$	$3\frac{1}{2}$
18	15	2	$3\frac{1}{2}$	$1\frac{1}{2}$*	3	—	$2\frac{1}{2}$	$3\frac{1}{2}$
20	17	2	$3\frac{1}{2}$	$1\frac{3}{4}$	3	—	$2\frac{1}{2}$	$3\frac{1}{2}$
22	19	2	$3\frac{1}{2}$	$1\frac{3}{4}$	3	—	$2\frac{1}{2}$	$3\frac{1}{2}$
24	21	2	$3\frac{1}{2}$	$1\frac{3}{4}$	3	—	$2\frac{1}{2}$	$3\frac{1}{2}$
26	23	$2\frac{1}{4}$	$3\frac{1}{2}$	$1\frac{3}{4}$	$3\frac{1}{4}$	$1\frac{1}{2}$*	$2\frac{3}{4}$	$3\frac{1}{2}$
28	25	$2\frac{1}{4}$	$3\frac{1}{2}$	$1\frac{3}{4}$	$3\frac{1}{4}$	$1\frac{1}{2}$*	$2\frac{3}{4}$	$3\frac{1}{2}$
30	27	$2\frac{1}{4}$	$3\frac{1}{2}$	$1\frac{3}{4}$	$3\frac{1}{4}$	$1\frac{1}{2}$*	$2\frac{3}{4}$	$3\frac{1}{2}$
32	29	$2\frac{1}{4}$	$3\frac{1}{2}$	$1\frac{3}{4}$	$3\frac{1}{4}$	$1\frac{1}{2}$*	$2\frac{3}{4}$	$3\frac{1}{2}$
34	31	$2\frac{1}{4}$	$3\frac{1}{2}$	$1\frac{3}{4}$	$3\frac{1}{4}$	$1\frac{1}{2}$*	$2\frac{3}{4}$	$3\frac{1}{2}$
36	33	$2\frac{1}{4}$	$3\frac{1}{2}$	$1\frac{3}{4}$	$3\frac{1}{4}$	$1\frac{1}{2}$*	$2\frac{3}{4}$	$3\frac{1}{2}$
38	35	$2\frac{1}{4}$	$3\frac{1}{2}$	$1\frac{3}{4}$	$3\frac{1}{4}$	$1\frac{1}{2}$*	$2\frac{3}{4}$	$3\frac{1}{2}$
40	37	$2\frac{1}{4}$	$3\frac{1}{2}$	$1\frac{3}{4}$	$3\frac{1}{4}$	$1\frac{1}{2}$*	$2\frac{3}{4}$	$3\frac{1}{2}$

[a]The pitch is measured center to center of consecutive turns. Cover $1\frac{1}{2}$ in. to spiral. The tabulated values can be used with 1-in.-maximum-size aggregate, except that values marked with an asterisk require $\frac{3}{4}$-in.-maximum aggregate.

Source: From [11-4], the *Design Handbook*, reprinted with permission of the American Concrete Institute.

TABLE A-16   Maximum Number of Bars That Can Be Placed in Square Columns with the Same Number of Bars in Each Face, Based on Normal (Radial) Lap Splices; Minimum Bar Spacing[a]

b (in.)	A_g (in.^{2})		Bar No. 5	6	7	8	9	10	11
10	100	n_{max}	8	4	4	4	4		
		A_{st}	2.48	1.76	2.40	3.16	4.00		
		ρ_t	0.025	0.018	0.024	0.032	0.040		
12	144	n_{max}	12	8	8	8	4	4	4
		A_{st}	3.72	3.52	4.80	6.32	4.00	5.08	6.24
		ρ_t	0.026	0.024	0.033	0.044	0.028	0.035	0.043
14	196	n_{max}	16	12	12	12	8	8	4
		A_{st}	4.96	5.28	7.20	9.48	8.00	10.16	6.24
		ρ_t	0.025	0.027	0.037	0.048	0.041	0.052	0.032
16	256	n_{max}	—	16	16	12	12	8	8
		A_{st}		7.04	9.60	9.48	12.00	10.16	12.48
		ρ_t		0.028	0.038	0.037	0.047	0.040	0.049
18	324	n_{max}	—	20	20	16	16	12	12
		A_{st}		8.80	12.00	12.64	16.00	15.24	18.72
		ρ_t		0.027	0.037	0.039	0.049	0.047	0.058
20	400	n_{max}	—	—	20	20	16	16	12
		A_{st}			12.0	15.80	16.00	20.32	18.72
		ρ_t			0.030	0.039	0.040	0.051	0.047
22	484	n_{max}	—	—	24	24	20	16	16
		A_{st}			14.40	18.96	20.00	20.32	24.96
		ρ_t			0.030	0.039	0.041	0.042	0.052
24	576	n_{max}	—	—	28	28	24	20	16
		A_{st}			16.80	22.12	24.00	25.40	24.96
		ρ_t			0.029	0.038	0.042	0.044	0.043
26	676	n_{max}	—	—	32	28	24	20	20
		A_{st}			19.20	22.12	24.00	25.40	31.20
		ρ_t			0.028	0.033	0.036	0.038	0.046
28	784	n_{max}	—	—	36	32	28	24	20
		A_{st}			21.60	25.28	28.00	30.48	31.20
		ρ_t			0.028	0.032	0.036	0.039	0.040
30	900	n_{max}	—	—	—	36	32	28	24
		A_{st}				28.44	32.00	35.56	37.44
		ρ_t				0.032	0.036	0.039	0.042
32	1024	n_{max}	—	—	—	40	32	28	28
		A_{st}				31.60	32.00	35.56	43.68
		ρ_t				0.031	0.031	0.035	0.043

[a]Based on 1-in.-maximum-size aggregate.

Source: From [11-4], the *Design Handbook*, reprinted with permission of the American Concrete Institute.

TABLE A-17   Maximum Number of Bars That Can Be Placed in Circular Columns, Based on Normal (Radial) Lap Splices; Minimum Bar Spacing[a]

Diameter (in.)	A_g (in.²)		Bar Size						
			5	6	7	8	9	10	11
12	113	n_{max}	8	7	6	6	—	—	—
		A_{st}	2.48	3.08	3.60	4.74			
		ρ_t	0.022	0.027	0.032	0.042			
14	154	n_{max}	11	10	9	8	7		
		A_{st}	3.41	4.40	5.40	6.32	7.00		
		ρ_t	0.022	0.029	0.035	0.041	0.046		
16	201	n_{max}	14	13	12	11	9	7	6
		A_{st}	4.34	5.72	7.20	8.69	9.00	8.89	9.36
		ρ_t	0.022	0.029	0.036	0.043	0.045	0.044	0.047
18	254	n_{max}	—	16	14	13	11	9	8
		A_{st}		7.04	8.40	10.27	11.00	11.43	12.48
		ρ_t		0.028	0.033	0.040	0.043	0.045	0.049
20	314	n_{max}	—	—	17	16	13	11	10
		A_{st}			10.20	12.64	13.00	13.97	15.60
		ρ_t			0.033	0.040	0.041	0.044	0.050
22	380	n_{max}	—	—	20	18	16	13	12
		A_{st}			12.00	14.22	16.00	16.51	18.72
		ρ_t			0.032	0.037	0.042	0.043	0.049
24	452	n_{max}	—	—	22	21	18	15	13
		A_{st}			13.20	16.59	18.00	19.05	20.28
		ρ_t			0.029	0.037	0.040	0.042	0.045
26	531	n_{max}	—	—	25	23	20	17	15
		A_{st}			15.00	18.17	20.00	21.59	23.40
		ρ_t			0.028	0.034	0.038	0.041	0.044
28	616	n_{max}	—	—	28	26	22	19	17
		A_{st}			16.80	20.54	22.00	24.13	26.52
		ρ_t			0.027	0.033	0.036	0.039	0.043
30	707	n_{max}	—	—	—	28	25	21	19
		A_{st}				22.12	25.00	26.67	29.64
		ρ_t				0.031	0.035	0.038	0.042
32	804	n_{max}	—	—	—	31	27	23	21
		A_{st}				24.29	27.00	29.21	32.76
		ρ_t				0.031	0.034	0.036	0.041

[a]Based on No. 4 spirals or ties, 1-in.-maximum-size aggregate, and $1\frac{1}{2}$-in. clear cover to spirals.

Source: This table is an abridged version of a table in [11-4] and is printed with the permission of the American Concrete Institute.

TABLE A-18 Number of Bars Required to Provide a Given Area of Steel[a]

Area (in.²)	Bar No. 5 (0.31)	Bar No. 6 (0.44)	Bar No. 7 (0.60)	Bar No. 8 (0.79)	Bar No. 9 (1.00)
1.24	4				
1.76		4			
1.86	6				
2.17	7				
2.40			4		
2.48	8				
2.64		6			
3.08		7			
3.16				4	
3.41	11				
3.52		8			
3.60			6		
3.72	12				
3.96		9			
4.00	13				4
4.20			7		
4.40		10			
4.65	15				
4.74				6	
4.80			8		
5.28		12			
5.40			9		
5.53				7	
5.72		13			
6.00			10		6
6.32				8	
7.00					7
7.04		16			
7.20			12		
7.48		17			
7.80			13		
7.90				10	
8.00					8
8.40			14		
8.69				11	
8.80		20			
9.00			15		9
9.48				12	
9.60			16		
10.00					10
10.80			18		

Area (in.²)	Bar No. 8 (0.79)	Bar No. 9 (1.00)	Bar No. 10 (1.27)	Bar No. 11 (1.56)	Bar No. 14 (2.25)
10.16			8		
10.27	13				
11.00		11			
11.06	14				
11.43			9		
11.85	15				
12.00		12			
12.48				8	
12.64	16				
12.70			10		
13.00		13			
13.43	17				
13.50					6
13.97			11		
14.00		14			
14.22	18				
15.00	19	15			
15.24			12		
15.60				10	
15.75					7
15.80	20				
16.00		16			
16.51			13		
16.59	21				
17.00		17			
17.16				11	
17.78			14		
18.00		18			8
18.72				12	
19.00		19	15		
20.00		20			
20.25				13	9
20.32			16		
21.00		21			
21.59			17		
21.84				14	
22.00		22			
22.50					10
22.86			18		
24.00		24			
24.96				16	

[a]Bold figures denote combinations that will give an equal number of bars in each side of a square column.

TABLE A-19   Lap-Splice Lengths for Grade-60 Bars in Columns (in.)

f'_c (psi)	Bar No.						
	5	6	7	8	9	10	11
Compression lap splices							
Lap splice length = (length from table) × (factors in note a)							
<3000	26	31	35	40	46	51	56
≥3000	19	23	26	30	34	38	42
Tension lap splices							
Lap splice length = (length from table) × $\beta\lambda^b$							
Class A tension lap splice: half or fewer of the bars spliced at any location and $0 \leq f_s \leq 0.5f_y$ in tension (ACI Section 12.17.2.2)							
3000	27.4	32.9	48.0	54.8	61.8	69.6	77.3
4000	23.7	28.4	41.5	47.4	53.5	60.2	66.8
5000	21.2	25.4	37.1	42.4	47.8	53.8	59.8
6000	19.4	23.3	33.9	38.7	43.7	49.1	54.6
Class B tension lap splices: more than half of the bars spliced at any section and/or f_s greater than $0.5f_y$ in tension (ACI Section 12.17.2.2)							
3000	35.6	42.7	62.3	71.2	80.4	90.5	100.4
4000	30.8	37.0	53.9	61.6	69.5	78.3	86.9
5000	27.5	33.0	48.2	55.1	61.2	70.0	77.7
6000	25.2	30.2	44.0	50.3	56.7	63.9	70.9

[a]Compression lap splices may be multiplied by 0.83 or 0.75 if enclosed by ties or spirals satisfying ACI Sections 12.17.2.4 or 12.17.2.5.

[b]β = coating factor, λ = lightweight-concrete factor.

TABLE A-20 Moment-Distribution Factors for Slabs without Drop Panels[a]

FEM (uniform load w) $= Mw\ell_2\ell_1^2$ $\qquad$ K (stiffness) $= kE\ell_2t^3/12\ell_1$

Carryover factor $= COF$

c_1/ℓ_1		c_2/ℓ_2					
		0.00	0.05	0.10	0.15	0.20	0.25
0.00	M	0.083	0.083	0.083	0.083	0.083	0.083
	k	4.000	4.000	4.000	4.000	4.000	4.000
	COF	0.500	0.500	0.500	0.500	0.500	0.500
0.05	M	0.083	0.084	0.084	0.084	0.085	0.085
	k	4.000	4.047	4.093	4.138	4.181	4.222
	COF	0.500	0.503	0.507	0.510	0.513	0.516
0.10	M	0.083	0.084	0.085	0.085	0.086	0.087
	k	4.000	4.091	4.182	4.272	4.362	4.449
	COF	0.500	0.506	0.513	0.519	0.524	0.530
0.15	M	0.083	0.084	0.085	0.086	0.087	0.088
	k	4.000	4.132	4.267	4.403	4.541	4.680
	COF	0.500	0.509	0.517	0.526	0.534	0.543
0.20	M	0.083	0.085	0.086	0.087	0.088	0.089
	k	4.000	4.170	4.346	4.529	4.717	4.910
	COF	0.500	0.511	0.522	0.532	0.543	0.554
0.25	M	0.083	0.085	0.086	0.087	0.089	0.090
	k	4.000	4.204	4.420	4.648	4.887	5.138
	COF	0.500	0.512	0.525	0.538	0.550	0.563
$x = (1 - c_2/\ell_2^3)$		1.000	0.856	0.729	0.613	0.512	0.421

[a]c_1 and c_2 are the widths of the column measured parallel to ℓ_1 and ℓ_2.
Source: [14-8].

TABLE A-21   Moment-Distribution Factors for Slabs with Drop
Panels; $h_1 = 1.25h^a$

FEM (uniform load w) = $Mw\ell_2\ell_1^2$ K (stiffness) = $kE\ell_2 t^3/12\ell_1$

Carryover factor = COF

c_1/ℓ_1		c_2/ℓ_2						
		0.00	0.05	0.10	0.15	0.20	0.25	0.30
0.00	M	0.088	0.088	0.088	0.088	0.088	0.088	0.088
	k	4.795	4.795	4.795	4.795	4.795	4.795	4.795
	COF	0.542	0.542	0.542	0.542	0.542	0.542	0.542
0.05	M	0.088	0.088	0.089	0.089	0.089	0.089	0.090
	k	4.795	4.846	4.896	4.944	4.990	5.035	5.077
	COF	0.542	0.545	0.548	0.551	0.553	0.556	0.558
0.10	M	0.088	0.088	0.089	0.090	0.090	0.091	0.091
	k	4.795	4.894	4.992	5.039	5.184	5.278	5.368
	COF	0.542	0.548	0.553	0.559	0.564	0.569	0.573
0.15	M	0.088	0.089	0.090	0.090	0.091	0.092	0.092
	k	4.795	4.938	5.082	5.228	5.374	5.520	5.665
	COF	0.542	0.550	0.558	0.565	0.573	0.580	0.587
0.20	M	0.088	0.089	0.090	0.091	0.092	0.093	0.094
	k	4.795	4.978	5.167	5.361	5.558	5.760	5.962
	COF	0.542	0.552	0.562	0.571	0.581	0.590	0.600
0.25	M	0.088	0.089	0.090	0.091	0.092	0.094	0.095
	k	4.795	5.015	5.245	5.485	5.735	5.994	6.261
	COF	0.542	0.553	0.565	0.576	0.587	0.598	0.609
0.30	M	0.088	0.089	0.090	0.092	0.093	0.094	0.095
	k	4.795	5.048	5.317	5.601	5.902	6.219	6.550
	COF	0.542	0.554	0.567	0.580	0.593	0.605	0.618

[a]h, Slab thickness; h_1, total thickness in drop panel.

Source: [14-8].

TABLE A-22   Moment-Distribution Factors for Slabs with
Drop Panels; $h_1 = 1.5h^a$

FEM (uniform load w) $= Mw\ell_2\ell_1^2$ K (stiffness) $= kE\ell_2 t^3/12\ell_1$

Carryover factor $= COF$

c_1/ℓ_1		c_2/ℓ_2					
		0.00	0.05	0.10	0.15	0.20	0.25
0.00	M	0.093	0.093	0.093	0.093	0.093	0.093
	k	5.837	5.837	5.837	5.837	5.837	5.837
	COF	0.589	0.589	0.589	0.589	0.589	0.589
0.05	M	0.093	0.093	0.093	0.093	0.094	0.094
	k	5.837	5.890	5.942	5.993	6.041	6.087
	COF	0.589	0.591	0.594	0.596	0.598	0.600
0.10	M	0.093	0.093	0.094	0.094	0.094	0.095
	k	5.837	5.940	6.024	6.142	6.240	6.335
	COF	0.589	0.593	0.598	0.602	0.607	0.611
0.15	M	0.093	0.093	0.094	0.095	0.095	0.096
	k	5.837	5.986	6.135	6.284	6.432	6.579
	COF	0.589	0.595	0.602	0.608	0.614	0.620
0.20	M	0.093	0.093	0.094	0.095	0.096	0.096
	k	5.837	6.027	6.221	6.418	6.616	6.816
	COF	0.589	0.597	0.605	0.613	0.621	0.628
0.25	M	0.093	0.094	0.094	0.095	0.096	0.097
	k	5.837	6.065	6.300	6.543	6.790	7.043
	COF	0.589	0.598	0.608	0.617	0.626	0.635

[a]h, Slab thickness; h_1, total thickness in drop panel.
Source: [14-8].

TABLE A-23   Stiffness and Carryover Factors for Columns

$$K_c = k \frac{EL_c}{\ell_c}$$

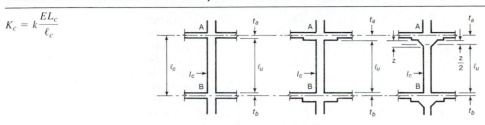

t_a/t_b		ℓ_c/ℓ_u								
		1.05	1.10	1.15	1.20	1.25	1.30	1.35	1.40	1.45
0.00	k_{AB}	4.20	4.40	4.60	4.80	5.00	5.20	5.40	5.60	5.80
	C_{AB}	0.57	0.65	0.73	0.80	0.87	0.95	1.03	1.10	1.17
0.2	k_{AB}	4.31	4.62	4.95	5.30	5.65	6.02	6.40	6.79	7.20
	C_{AB}	0.56	0.62	0.68	0.74	0.80	0.85	0.91	0.96	1.01
0.4	k_{AB}	4.38	4.79	5.22	5.67	6.15	6.65	7.18	7.74	8.32
	C_{AB}	0.55	0.60	0.65	0.70	0.74	0.79	0.83	0.87	0.91
0.6	k_{AB}	4.44	4.91	5.42	5.96	6.54	7.15	7.81	8.50	9.23
	C_{AB}	0.55	0.59	0.63	0.67	0.70	0.74	0.77	0.80	0.83
0.8	k_{AB}	4.49	5.01	5.58	6.19	6.85	7.56	8.31	9.12	9.98
	C_{AB}	0.54	0.58	0.61	0.64	0.67	0.70	0.72	0.75	0.77
1.0	k_{AB}	4.52	5.09	5.71	6.38	7.11	7.89	8.73	9.63	10.60
	C_{AB}	0.54	0.57	0.60	0.62	0.65	0.67	0.69	0.71	0.73
1.2	k_{AB}	4.55	5.16	5.82	6.54	7.32	8.17	9.08	10.07	11.12
	C_{AB}	0.53	0.56	0.59	0.61	0.63	0.65	0.66	0.68	0.69
1.4	k_{AB}	4.58	5.21	5.91	6.68	7.51	8.41	9.38	10.43	11.57
	C_{AB}	0.53	0.55	0.58	0.60	0.61	0.63	0.64	0.65	0.66
1.6	k_{AB}	4.60	5.26	5.99	6.79	7.66	8.61	9.64	10.75	11.95
	C_{AB}	0.53	0.55	0.57	0.59	0.60	0.61	0.62	0.63	0.64
1.8	k_{AB}	4.62	5.30	6.06	6.89	7.80	8.79	9.87	11.03	12.29
	C_{AB}	0.52	0.55	0.56	0.58	0.59	0.60	0.61	0.61	0.62
2.0	k_{AB}	4.63	5.34	6.12	6.98	7.92	8.94	10.06	11.27	12.59
	C_{AB}	0.52	0.54	0.56	0.57	0.58	0.59	0.59	0.60	0.60
2.2	k_{AB}	4.65	5.37	6.17	7.05	8.02	9.08	10.24	11.49	12.85
	C_{AB}	0.52	0.54	0.55	0.56	0.57	0.58	0.58	0.59	0.59
2.4	k_{AB}	4.66	5.40	6.22	7.12	8.11	9.20	10.39	11.68	13.08
	C_{AB}	0.52	0.53	0.55	0.56	0.56	0.57	0.57	0.58	0.58
2.6	k_{AB}	4.67	5.42	6.26	7.18	8.20	9.31	10.53	11.86	13.29
	C_{AB}	0.52	0.53	0.54	0.55	0.56	0.56	0.56	0.57	0.57
2.8	k_{AB}	4.68	5.44	6.29	7.23	8.27	9.41	10.66	12.01	13.48
	C_{AB}	0.52	0.53	0.54	0.55	0.55	0.55	0.56	0.56	0.56
3.0	k_{AB}	4.69	5.46	6.33	7.28	8.34	9.50	10.77	12.15	13.65
	C_{AB}	0.52	0.53	0.54	0.54	0.55	0.55	0.55	0.55	0.55
3.5	k_{AB}	4.71	5.50	6.40	7.39	8.48	9.69	11.01	12.46	14.02
	C_{AB}	0.51	0.52	0.53	0.53	0.54	0.54	0.54	0.53	0.53
4.0	k_{AB}	4.72	5.54	6.45	7.47	8.60	9.84	11.21	12.70	14.32
	C_{AB}	0.51	0.52	0.52	0.53	0.53	0.52	0.52	0.52	0.52
4.5	k_{AB}	4.73	5.56	6.50	7.54	8.69	9.97	11.37	12.89	14.57
	C_{AB}	0.51	0.52	0.52	0.52	0.52	0.52	0.51	0.51	0.51
5.0	k_{AB}	4.75	5.59	6.54	7.60	8.78	10.07	11.50	13.07	14.77
	C_{AB}	0.51	0.51	0.52	0.52	0.51	0.51	0.51	0.50	0.49
6.0	k_{AB}	4.76	5.63	6.60	7.69	8.90	10.24	11.72	13.33	15.10
	C_{AB}	0.51	0.51	0.51	0.51	0.50	0.50	0.49	0.49	0.48
7.0	k_{AB}	4.78	5.66	6.65	7.76	9.00	10.37	11.88	13.54	15.34
	C_{AB}	0.51	0.51	0.51	0.50	0.50	0.49	0.48	0.48	0.47
8.0	k_{AB}	4.78	5.68	6.69	7.82	9.07	10.47	12.01	13.70	15.54
	C_{AB}	0.51	0.51	0.50	0.50	0.49	0.49	0.48	0.47	0.46
9.0	k_{AB}	4.80	5.71	6.74	7.89	9.18	10.61	12.19	13.93	15.83
	C_{AB}	0.50	0.50	0.50	0.49	0.48	0.48	0.47	0.46	0.45

Source: [14-7], courtesy of the Portland Cement Association.

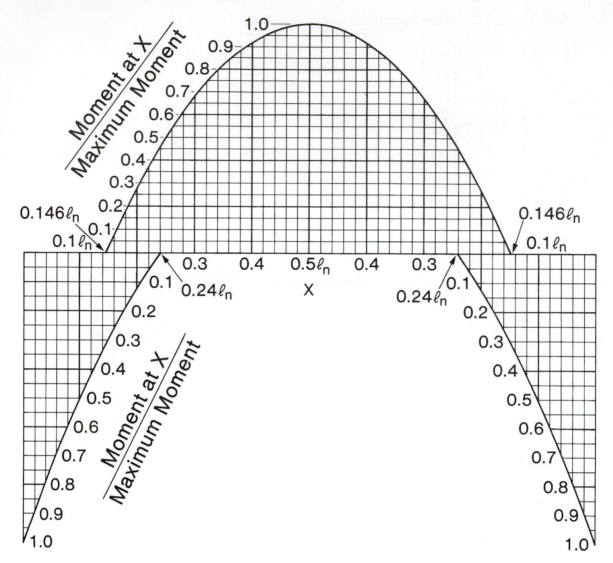

Fig. A-1
Bending-moment envelope for typical interior span (moment coefficients: $-1/11$, $+1/16$, $-1/11$).

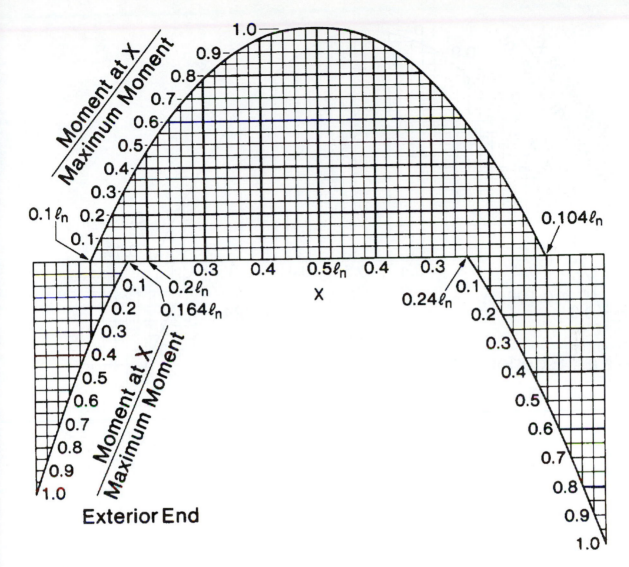

Fig. A-2
Bending-moment envelope for exterior span with exterior support built integrally with a column (moment coefficients: $-1/16$, $+1/14$, $-1/10$).

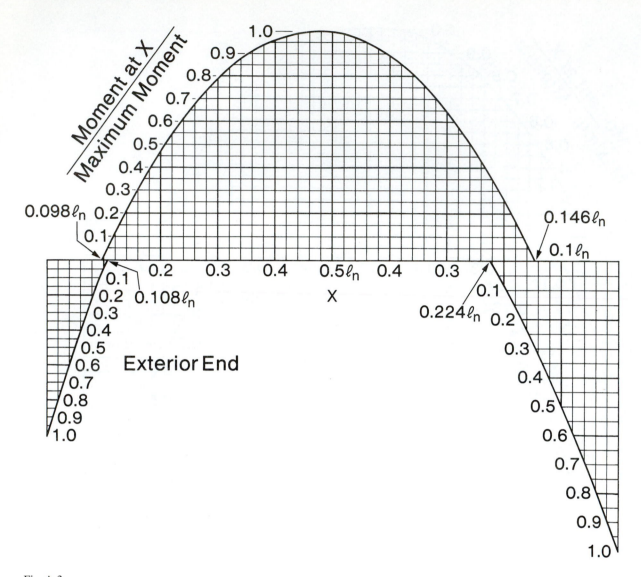

Fig. A-3
Bending-moment envelope for exterior span with exterior support built integrally with a spandrel beam or girder (moment coefficients: $-1/24$, $+1/14$, $-1/10$).

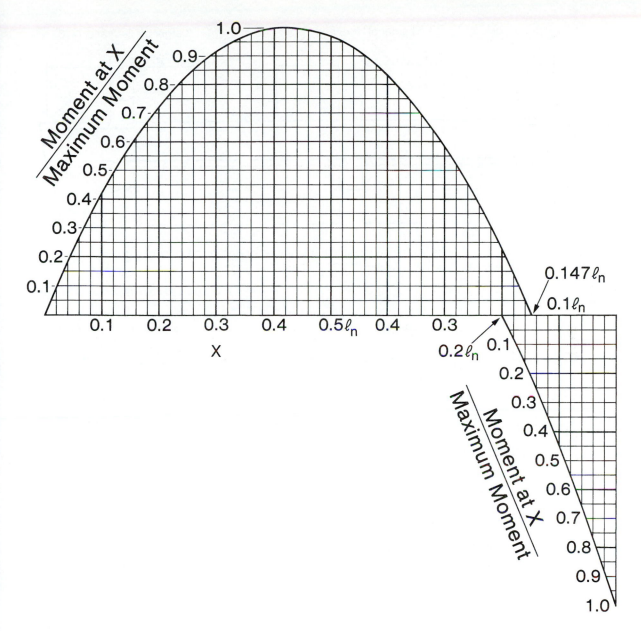

Fig. A-4
Bending-moment envelope for exterior span with discontinuous end unrestrained (moment coefficients: 0, $+1/11$, $-1/10$).

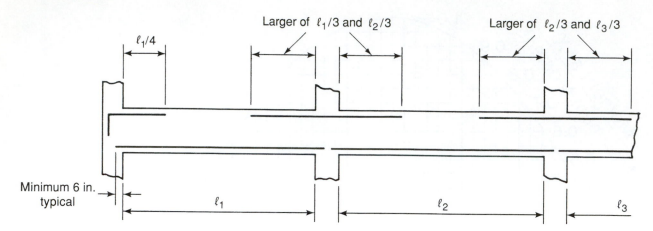

(a) Beam with U stirrups with 135° hooks enclosing longitudinal top bars, or one-piece closed stirrups.
      If closed stirrups are not provided, see ACI Section 7.13.

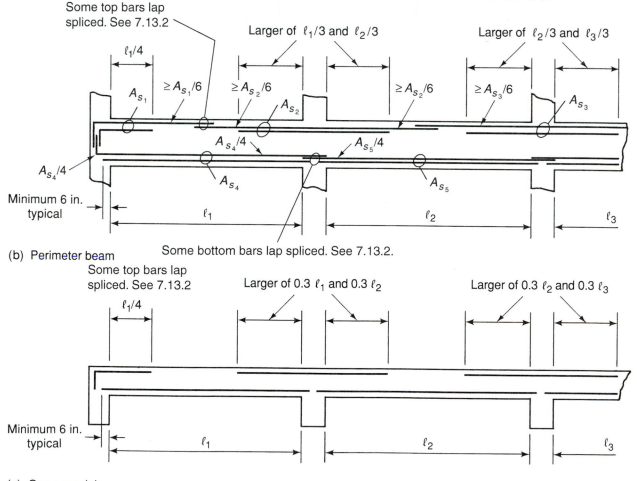

(b) Perimeter beam

(c) One-way slab

Fig. A-5
Standard bar details.

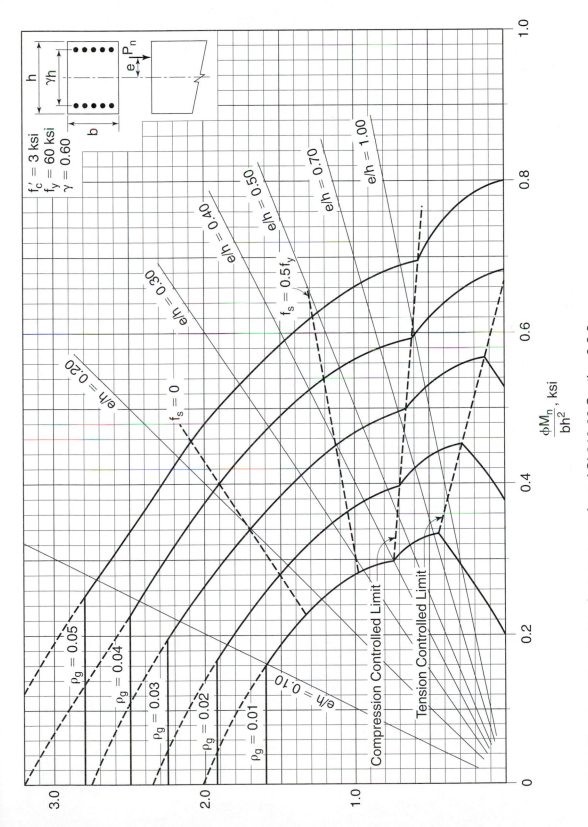

Note: The strength-reduction factors, φ, are from ACI 318-02 Section 9.3.2.
They must be used with load combinations from Section 9.2.

Fig. A-6
Nondimensional interaction diagram for tied columns with bars in two faces; $\gamma = 0.60$.

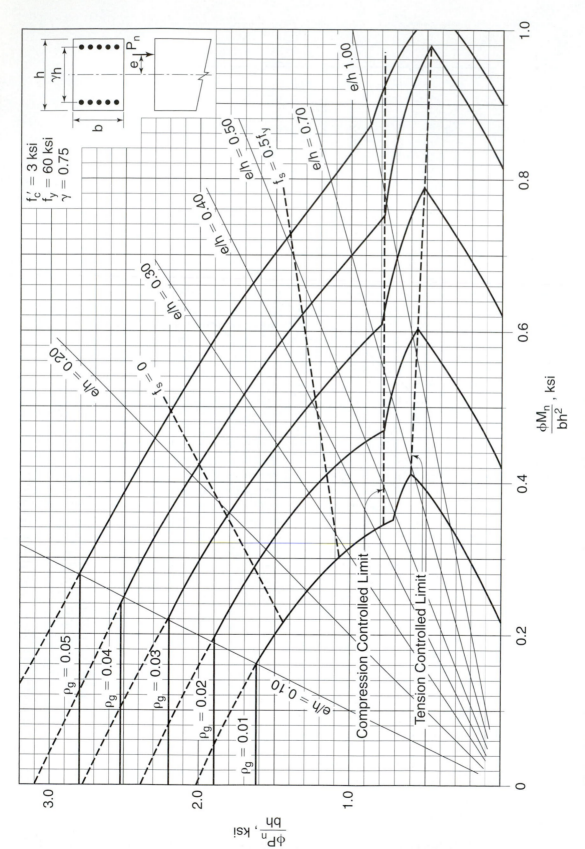

f'_c = 3 ksi
f_y = 60 ksi
γ = 0.75

Note: The strength-reduction factors, ϕ, are from ACI 318-02 Section 9.3.2.
They must be used with load combinations from Section 9.2.

Fig. A-7
Nondimensional interaction diagram for tied columns with bars in two faces; γ = 0.75.

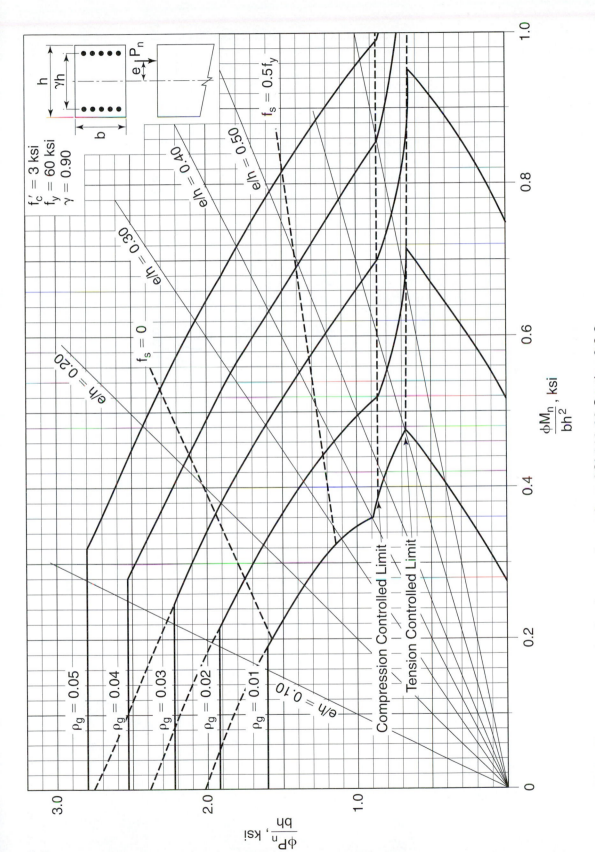

Fig. A-8
Interaction diagram for tied columns with bars in two faces; $\gamma = 0.90$.

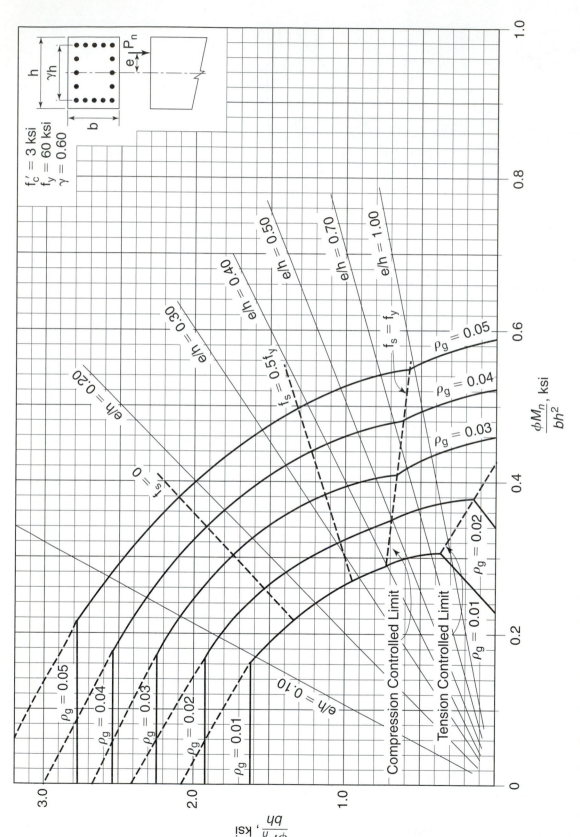

Fig. A-9
Nondimensional interaction diagram for tied columns with bars in four faces; $\gamma = 0.60$.

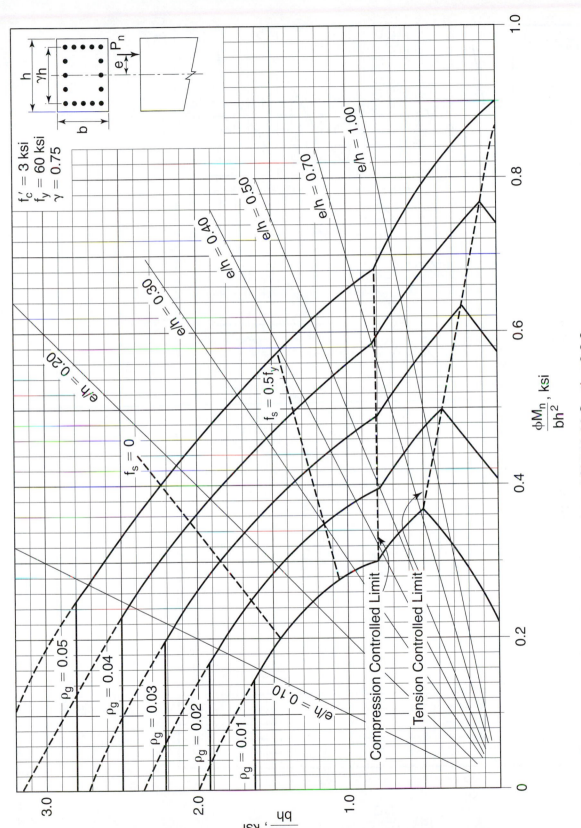

$$\frac{\phi M_n}{bh^2}, \text{ksi}$$

Note: The strength-reduction factors, ϕ, are from ACI 318-02 Section 9.3.2.
They must be used with load combinations from Section 9.2.

Fig. A-10
Interaction diagram for tied columns with bars in four faces; $\gamma = 0.75$.

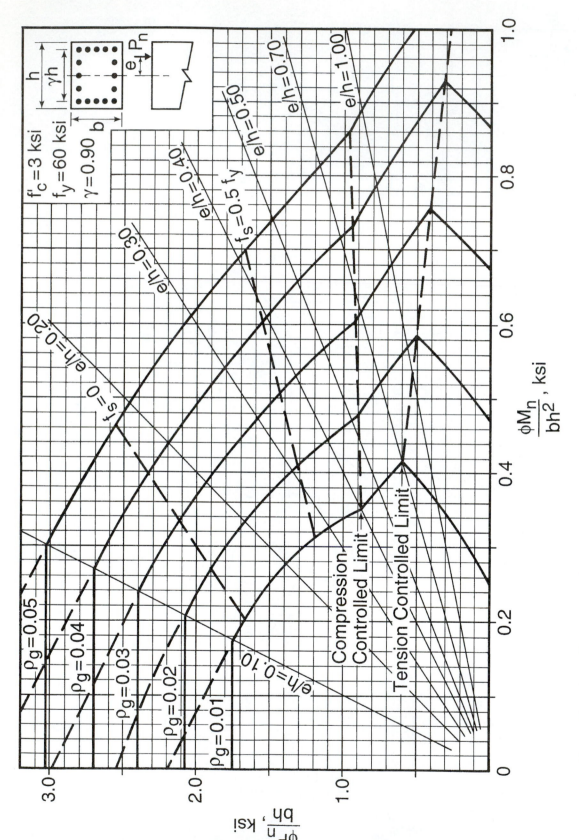

$\dfrac{\phi P_n}{bh}$, ksi

$\dfrac{\phi M_n}{bh^2}$, ksi

Note: The strength-reduction factors, ϕ, are from ACI 318-02 Section 9.3.2.
They must be used with load combinations from Section 9.2.

Fig. A-11
Interaction diagram for tied columns with bars in four faces; $\gamma = 0.90$.

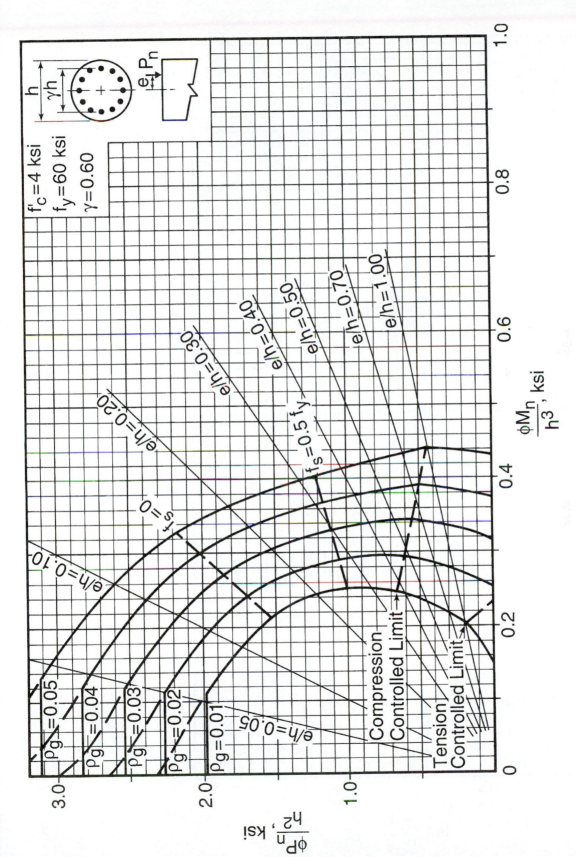

Note: The strength-reduction factors, ϕ, are from ACI 318-02 Section C.3.2
They must be used with the load combinations from Section C.2.

Fig. A-12
Interaction diagrams for *spiral* columns; $\gamma = 0.60$.

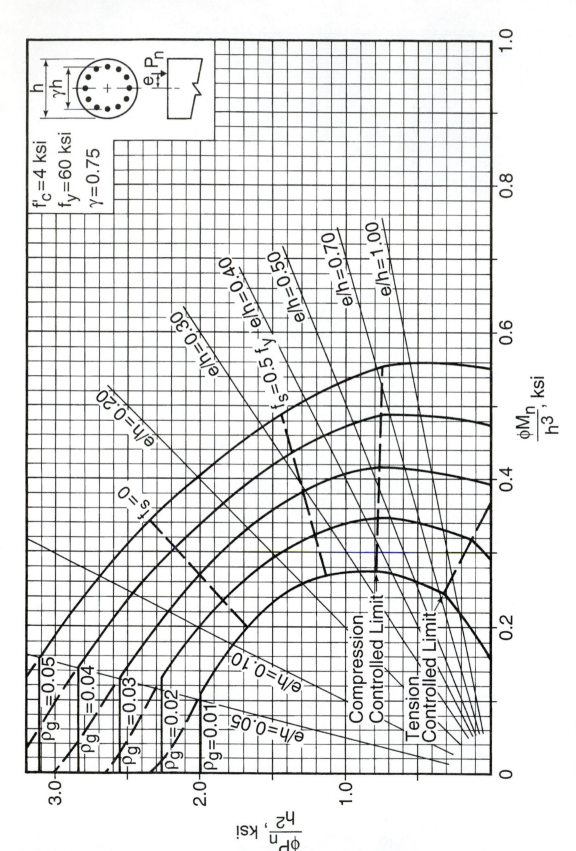

Note: The strength-reduction factors, φ, are from ACI 318-02 Section C.3.2
They must be used with the load combinations from Section C.2.

Fig. A-13
Interaction diagrams for *spiral* columns; $\gamma = 0.75$.

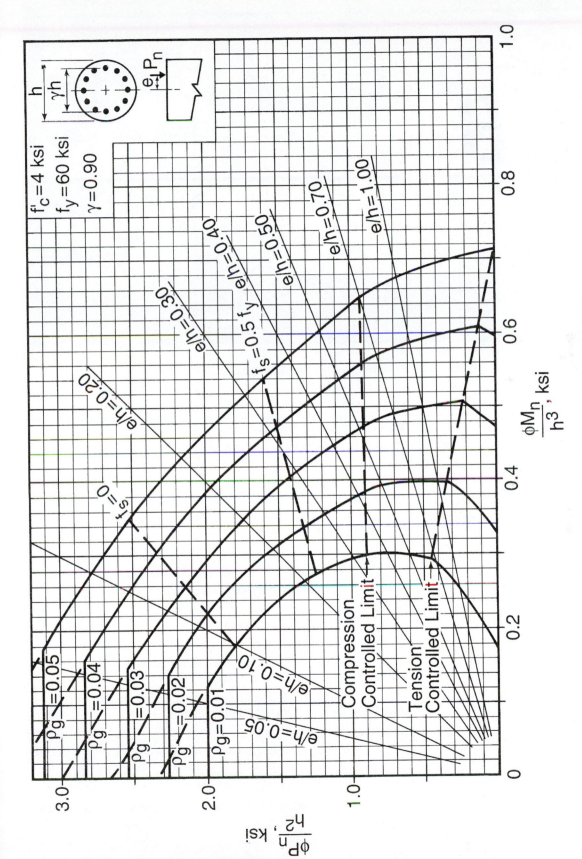

Note: The strength-reduction factors, φ, are from ACI 318-02 Section C.3.2
They must be used with the load combinations from Section C.2.

Fig. A-14
Interaction diagrams for *spiral* columns; $\gamma = 0.90$.

APPENDIX B

Notation

Note: US customary units are used to define common units of terms. Alternate definitions with SI units are also possible.

a = depth of equivalent rectangular stress block, in.

a = shear span, distance between concentrated load and face of support, in.

a_b = depth of rectangular stress block corresponding to balanced strain conditions, in.

a_{CCL} = depth of rectangular stress block at the compression-controlled limit, in.

a_{TCL} = depth of rectangular stress block at the tension-controlled limit, in.

A_b = area of an individual bar, in.^{2}

A_c = area of core of spirally reinforced compression member measured to outside diameter of spiral, in.^{2}

A_c = area of concrete section resisting shear transfer, in.^{2}

A_c = the effective area at one end of a strut in a strut-and-tie model, taken perpendicular to the axis of the strut, in.^{2}

A_{cp} = area enclosed by outside perimeter of concrete cross section, in.^{2}

A_{cv} = gross area of concrete section bounded by web thickness and length of section in the direction of shear force considered, in.^{2}

A_g = gross area of section, in.^{2} (For a hollow section, A_g is the area of concrete only and does not include the area of the void(s).)

A_I = influence area (see Section 2-7)

A_ℓ = total area of longitudinal reinforcement to resist torsion, in.^{2}

A_n = area of a face of a nodal zone or a section through a nodal zone, in.^{2}

A_o = gross area enclosed by shear flow path, in.^{2}

A_{oh} = area enclosed by centerline of the outermost closed transverse torsional reinforcement, in.^{2}

A_{ps} = area of prestressed reinforcement in a tie, in.^{2}

A_s = area of nonprestressed tension reinforcement, in.^{2}

A'_s = area of compression reinforcement, in.^{2}

A'_s = area of compression reinforcement in a strut, in.^{2}

A_{sf} = area of tension reinforcement balancing the compression force in the overhanging flanges of a T beam, in.^{2}

A_{si} = area of surface reinforcement in the ith layer crossing a strut, in.^{2}

A_{sk} = area of skin reinforcement per unit height in one side face, in.^{2}/ft.

A_{st} = area of nonprestressed reinforcement in a tie, in.^{2}

A_{sw} = area of tension reinforcement balancing the compression force in the web of a T beam, in.^{2}

A_{s1} = area of tension reinforcement balancing the compression force in the compression reinforcement, in.^{2}

A_{s2} = area of tension reinforcement balancing the compression force in the concrete in a beam with compression reinforcement, in.^{2}

A_t = area of one leg of a closed stirrup resisting torsion within a distance s, in.^{2}

A_{tr} = total cross-sectional area of all transverse reinforcement that is within the spacing s and crosses the potential plane of splitting through the reinforcement being developed, in.^{2}

A_v = area of shear reinforcement within a distance s, in.^{2}

A_{vf} = area of shear-friction reinforcement, in.^{2}

A_1 = loaded area in bearing, in.^{2}

A_2 = area of the lower base of the largest frustrum of a pyramid, cone, or tapered wedge contained wholly within the support, having for its upper base the loaded area and having side slopes of 1 vertical to 2 horizontal (see Fig. 16-10), in.^{2}

b = width of compression face of member, effective compressive flange width of a T beam, in.

b_o = perimeter of critical section for two-way shear in slabs and footings, in.

b_w = web width, or diameter of circular section, in.

b_1 = length of critical shear perimeter for two-way shear, measured parallel to the span ℓ_1, in.

b_2 = length of critical shear perimeter for two-way shear perpendicular to b_1, in.

c = spacing or cover dimension, in.

c = distance from extreme-compression fiber to neutral axis, in.

c_b = distance from extreme-compression fiber to neutral axis corresponding to balanced strain conditions, in.

c_c = clear cover from the nearest surface in tension to the surface of the flexural tension reinforcement, in.

c_{CCL} = distance from extreme-compression fiber to the neutral axis when the strain in the extreme-tension steel is ϵ_y in tension, in.

c_{TCL} = distance from extreme-compression fiber to the neutral axis when the strain in the extreme-tension steel is 0.005 in tension, in.

c_1 = size of rectangular or equivalent rectangular column, capital, or bracket, measured parallel to the span ℓ_1, in.

c_2 = size of rectangular or equivalent rectangular column, capital, or bracket, measured parallel to ℓ_2, in.

C = compressive force in cross section; in subscripts, c = concrete, s = steel

C = cross-sectional constant to define torsional properties, in.^{4}

C_m = factor relating the actual moment diagram of a slender column to an equivalent uniform moment diagram (Chapter 12)

C_m = moment coefficient (Chapter 10)

C_v = shear coefficient

d = effective depth = distance from extreme-compression fiber to centroid of tension reinforcement, in.

d' = distance from extreme-compression fiber to centroid of compression reinforcement, in.

d_b = nominal diameter of bar, wire, or prestressing strand, in.

d_t = distance from extreme-compression fiber to extreme-tension steel, in.

D = dead loads, or related internal moments and forces

D = diagonal compression force in the web of a beam or in a compression strut in a D-region, lb

e = eccentricity of axial load on a column = M/P

E_c = modulus of elasticity of concrete, psi

EI = flexural stiffness of compression member, lb$-$in.^{2}

E_s = modulus of elasticity of reinforcement, psi

f'_c = specified compressive strength of concrete, psi

$\sqrt{f'_c}$ = square root of specified compressive strength of concrete, psi

f_{cd} = diagonal compressive stress in the web of a beam, psi

f'_{cr} = required average compressive strength for f'_c to meet statistical acceptance criteria, psi

f_{ct} = splitting tensile strength of concrete, psi

f_{cu} = effective compressive strength of the concrete in a strut or a nodal zone, psi.

f_r = modulus of rupture of concrete, psi

f_s = calculated stress in reinforcement at service loads, ksi

f'_s = stress in compression reinforcement, psi

f_{si} = stress in the ith layer of surface reinforcement, psi

f_y = specified yield strength of nonprestressed reinforcement, psi

F = loads due to weight and pressures of fluids with well-defined densities and controllable heights

F_n = nominal strength of a strut, tie, or nodal zone, lb

F_{nn} = nominal strength of a face of a nodal zone, lb

F_{ns} = nominal strength of a strut, lb

F_{nt} = nominal strength of a tie, lb

F_u = factored force acting on a strut, tie, bearing area, or nodal zone in a strut-and-tie model, lb

h = height of member, in.

h = overall thickness of member, in.

h_t = effective height of tie, in.

$h_{t,max}$ = maximum effective height of concrete concentric with a tie, in.

h_w = total height of wall from base to top, in.

H = loads due to weight and pressure of soil, water in soil, or other materials, or related internal moments and forces

I = moment of inertia of section; in.^{4} subscripts, b = beam, c = column, s = slab

I_{cr} = moment of inertia of cracked section transformed to concrete, in.^{4}

I_e = effective moment of inertia for computation of deflection, in.^{4}

I_g = moment of inertia of gross concrete section, neglecting reinforcement, in.^{4}

I_{gt} = moment of inertia of gross transformed uncracked section, in.^{4}

I_{se} = moment of inertia of the reinforcement in a column about centroidal axis of gross member cross section, in.^{4}

jd = distance between the resultants of the internal compressive and tensile forces on a cross section, in.

J_c = property of assumed critical section for two-way shear, analogous to polar moment of inertia, in.^{4}

k = effective length factor for compression members

k_n = factor used in calculating flexural capacity of a section

K = flexural stiffness: moment per unit rotation; in used with subscripts, b = beam, c = column, ec = equivalent column, s = slab

K_ℓ = lateral stiffness of a bracing element in a building: force/displacement

K_t = torsional stiffness of torsional member; moment per unit rotation

K_{tr} = transverse reinforcement index—see Chapter 8

ℓ = span length of beam or one-way slab, generally center to center of supports; clear projection of cantilever, in.

ℓ_a = additional embedment length at support or at point of inflection, in.

ℓ_a = length in which anchorage of a tie must take place, in.

ℓ_b = length of bearing area parallel to the plane of the strut-and-tie model, in.

ℓ_c = length of compression member in a frame, measured from center to center of the joints in the frame

ℓ_d = development length, in.

ℓ_{db} = basic development length in tension before modification, in.

ℓ_{dc} = development length in compression, in.

ℓ_{dh} = development length of standard hook in tension, measured from critical section to outside end of hook, in.

ℓ_{hb} = basic development length of standard hook in tension, in.

ℓ_n = clear span measured face to face of supports:

 = clear span for positive moment, negative moment at exterior support, or shear

 = average of adjacent clear spans for negative moment

ℓ_o = minimum length, measured from joint face along axis of structural member, over which transverse reinforcement must be provided, in.

ℓ_u = unsupported length of compression member

ℓ_1 = length of span of two-way slab in direction that moments are being determined, measured center to center of supports

ℓ_2 = length of span of two-way slab transverse to ℓ_1, measured center to center of supports

L = live loads, or related internal moments and forces

L_r = specified live load for roof, or related internal forces or moments

m = moment per unit width; in subscripts, xy = twisting moment; ft-kips/ft

m_r = resisting moment per unit width, ft-kips/ft

M_a = maximum moment in member at the stage for which deflections are being computed—see Section 9-4

M_c = factored moment to be used for design of a slender compression member, in-lb.

M_c = moment at the face of the joint, corresponding to the nominal flexural strength of the column framing into that joint, calculated for the factored axial force, consistent with the direction of the lateral forces considered, resulting in the lowest flexural strength, in-lb.

M_{cr} = cracking moment

M_g = moment at the face of the joint, corresponding to the nominal flexural strength of the girder including slab where it is in tension, framing into that joint, in-lb.

M_n = nominal moment strength, in-lb.

M_o = total factored static moment, in-lb.

M_{pr} = probable flexural moment strength—see Equation (20-15)

M_s = moment in column due to loads causing appreciable sidesway

M_u = moment due to factored loads

M_1 = smaller factored end moment on a compression member; positive if the member is bent in single curvature, negative if bent in double curvature, in-lb.

M_{1ns} = factored end moment on a compression member at the end at which M_1 acts, due to loads that cause no appreciable sidesway, in-lb.

M_{1s} = factored end moment on a compression member at the end at which M_1 acts, due to loads that cause appreciable sidesway, calculated by using a first-order elastic frame analysis, in-lb.

M_2 = larger factored end moment on a compression member, always positive, in-lb.

M_{2ns} = factored end moment on a compression member at the end at which M_2 acts, due to loads that cause no appreciable sidesway, in-lb.

M_{2s} = factored end moment on a compression member at the end at which M_2 acts, due to loads that cause appreciable sidesway, calculated using a first-order elastic-frame analysis, in-lb.

n = modular ratio = E_s/E_c

n = number of bars in a layer being spliced or developed at a critical section

N_u = factored axial load normal to cross section occurring simultaneously with V_u. The signs of this force and similar forces change from code section to section; N_u is positive in compression in ACI Sections 11.3.1.2, 11.3.2.3, and 11.10.7, but the similarly defined N_{uc} is positive in tension in ACI Section 11.9.3

N_v = axial tension force due to shear, lb

p_{cp} = outside perimeter of the concrete cross section, in.

p_h = perimeter of the centerline of the outermost closed transverse torsional reinforcement, in.

P_b = nominal axial load strength at balanced strain conditions

P_c = critical load

P_E = buckling load of an elastic, hinged-end column

P_n = nominal axial load strength at given eccentricity

P_o = nominal axial load strength at zero eccentricity

P_u = axial force due to factored loads

Q = stability index for a story—see Chapter 12

r = radius of gyration of cross section of a compression member, in.

R = rain load, or related internal moments and forces

s = standard deviation

s = spacing of shear or torsion reinforcement measured along the longitudinal axis of the structural member, in.

s_i = spacing of reinforcement in the ith layer adjacent to the surface of a strut-and-tie model, in.

s_x = longitudinal spacing of transverse reinforcement within the length ℓ_o, in.

S = snow load, or related internal forces and moments

s_{sk} = spacing of skin reinforcement, in.

t = thickness of wall or strut-and-tie model, in.

T = cumulative effect of temperature, creep, shrinkage, differential settlement, and shrinkage-compensating cement

T_n = nominal torsional moment strength, in-lb.

T_s = nominal torsional moment strength provided by torsion reinforcement, in-lb.

T_u = factored torsional moment at section, in-lb.

U = strength required to resist factored loads or related internal moments and forces

v = shear stress

v_c = nominal shear stress carried by concrete, psi.

v_n = nominal shear stress, psi.

V_c = nominal shear force carried by concrete, lb.

V_e = design shear force in beams or columns due to dead, live, and seismic loads, lb.

V_n = nominal shear strength, lb.

V_{nh} = nominal horizontal shear strength, lb.

V_s = nominal shear strength provided by shear reinforcement, lb.

V_{sway} = the shear in the span of a beam or column, resulting from a plastic hinging mechanism in the columns and beams of the frame due to seismically in-duced sidesway deflections of the frame, lb.

V_u = factored shear force at section, lb.

w_c = weight of concrete, lb/ft^3.

w_D = factored dead load per unit area of slab or per unit length of beam, lb/ft or lb/ft^2.

w_L = factored live load per unit area of slab or per unit of length of beam

w_s = effective width of a strut, in.

w_T = Effective width of a tie, in.

w_u = total factored load per unit length of beam or per unit area of slab

x = shorter overall dimension of rectangular part of cross section, in.

y = longer overall dimension of rectangular part of cross section, in.

y_t = distance from centroidal axis of cross section, neglecting reinforcement, to extreme fiber in tension, in.

z = quantity limiting distribution of flexural reinforcement

α = angle between inclined stirrups and longitudinal axis of member

α = ratio of flexural stiffness of beam section to flexural stiffness of a width of slab bounded laterally by centerlines of adjacent panels (if any) on each side of the beam = $E_{cb}I_b/(E_{cs}I_s)$

α = reinforcement location factor for development-length calculation

α_m = average value of α for all beams on edges of a panel

α_s = coefficient used to compute V_c in slabs

α_1 = α in direction of ℓ_1

α_2 = α in direction of ℓ_2

β = coating factor for development-length calculation

β = ratio of clear spans in long to short direction of two-way slabs

β = ratio of long side to short side of footing

β_a = ratio of dead load per unit area to live load per unit area (in each case, without load factors)

β_b = ratio of area of reinforcement cut off to total area of tension reinforcement at section

β_c = ratio of long side to short side of column or concentrated load

β_d = (a) for nonsway frames, β_d is the ratio of the maximum factored axial dead load to the total factored axial load

(b) for sway frames, except as required in (c), β_d is the ratio of the maximum factored sustained shear in a story to the total factored shear in that story

(c) for stability checks of sway frames carried out in accordance with ACI Section 10.13.6, β_d is the ratio of the maximum factored axial dead load to the total factored axial load

β_n = factor to account for the effect of the anchorage of ties on the effective compressive strength of a nodal zone

β_s = factor to account for the effect of cracking and confining reinforcement on the effective compressive strength of the concrete in a strut

β_t = ratio of torsional stiffness of edge beam section to flexural stiffness of a width of slab equal to span length of beam, center to center of supports = $E_{cb}C/(2E_{cs}I_s)$

β_1 = ratio of depth of rectangular stress block, a, to depth to neutral axis, c

γ = ratio of the distance between the outer layers of reinforcement in a column to the overall depth of the column (see Fig. 11-30)

γ = reinforcement size factor for development-length calculations

γ_f = fraction of unbalanced moment transferred by flexure at slab–column connections

γ_i = angle between the axis of a strut and the bars in the ith layer of reinforcement crossing that strut

γ_v = fraction of unbalanced moment transferred by eccentricity of shear at slab–column connections = $1 - \gamma_f$

δ_{ns} = moment-magnification factor for frames braced against sidesway, to reflect effects of member curvature between ends of compression member

δ_s = moment-magnification factor for frames not braced against sidesway, to reflect lateral drift resulting from lateral and gravity loads

ϵ = strain

ϵ_c = strain in concrete

ϵ_{cu} = compressive strain at crushing of concrete

ϵ_s = strain in steel

ϵ_t = net tensile strain in extreme-tension steel at nominal strength

λ = multiplier for additional long-term deflection

λ = correction factor related to unit weight of concrete

μ = coefficient of friction

ν = ratio of effective concrete strength in web of beam or compression strut to f'_c

ξ = time-dependent factor for sustained load deflections

ρ = ratio of nonprestressed tension reinforcement = A_s/bd

ρ' = ratio of nonprestressed compression reinforcement = A'_s/bd

ρ_b = reinforcement ratio corresponding to balanced strain conditions

ρ_g = ratio of total reinforcement area to cross-sectional area of column

ρ_s = ratio of volume of spiral reinforcement to total volume of core (out-to-out spirals) of a spirally reinforced compression member

ρ_w = $A_s/b_w d$

ϕ = strength-reduction factor

θ = angle of compression struts in web of beam

ω = mechanical-reinforcement ratio = $\rho f_y/f'_c$

References

Chapter 1

1-1 Hedley E. H. Roy, "Toronto City Hall and Civic Square," *ACI Journal, Proceedings*, Vol. 62, No. 12, December 1965, pp. 1481–1502.

1-2 CRSI Subcommittee on Placing Reinforcing Bars, *Reinforcing Bar Detailing*, Concrete Reinforcing Steel Institute, Chicago, 1971, 280 pp.

1-3 Robert Mark, *Light, Wind, and Structure: The Mystery of the Master Builders*, MIT Press, Boston, 1990, pp. 52–67.

1-4 Michael P. Collins, "In Search of Elegance: The Evolution of the Art of Structural Engineering in the Western World," *Concrete International*, July 2001, Vol. 23, No. 7, July 2001, pp. 57–72.

1-5 Committee on Concrete and Reinforced Concrete, "Standard Building Regulations for the Use of Reinforced Concrete," *Proceedings, National Association of Cement Users*, Vol. 6, 1910, pp. 349–361.

1-6 Special Committee on Concrete and Reinforced Concrete, "Progress Report of Special Committee on Concrete and Reinforced Concrete," *Proceedings of the American Society of Civil Engineers*, 1913, pp. 117–135.

1-7 Special Committee on Concrete and Reinforced Concrete, "Final Report of Special Committee on Concrete and Reinforced Concrete," *Proceedings of the American Society of Civil Engineers*, 1916, pp. 1657–1708.

1-8 Frank Kerekes and Harold B. Reid, Jr., "Fifty Years of Development in Building Code Requirements for Reinforced Concrete," *ACI Journal, Proceedings*, Vol. 50, No. 8, February 1954, pp. 441–472.

1-9 *Uniform Building Code*, International Conference of Building Officials, Whittier, Calif., various editions.

1-10 *Standard Building Code*, Southern Building Code Congress, Birmingham, Ala., various editions.

1-11 *Basic Building Code*, Building Officials and Code Administrators International, Chicago, various editions.

1-12 *International Building Code*, International Code Council, Falls Church, VA, 2000.

1-13 ACI Committee 318, *Building Code Requirements for Structural Concrete* (ACI 318–02) and *Commentary* (ACI 318R–02), American Concrete Institute, Formington Hills MI, 2002, 444 pp.

1-14 *Standard Specifications for Highway Bridges*, American Association of State Highway and Transportation Officials, Washington, D.C., various editions.

1-15 American Association of State Highway and Transportation Officials, *AASHTO, LRFD Bridge Design Specifications*, 2nd Edition, Washington, 1998, 1216 pp.

1-16   *CEB-FIP Model Code 1990*, Thomas Telford Services Ltd., London, for Comité Euro-International du Béton, Laussane, 1993, 437 pp.

1-17   *Eurocode No. 2, Design of Concrete Structures, Part 1: Part 1: General Rules and Rules for Buildings*, Commission of the European Community, Brussels, 1991, 255 pp.

1-18   *Minimum Design Loads for Buildings and Other Structures (ASCE 7–98)*, American Society of Civil Engineers, Reston, VA, 2000, 337 pp. 337.

Chapter 2

2-1   Donald Taylor, "Progressive Collapse," *Canadian Journal of Civil Engineering*, Vol. 2, No. 4, December 1975, pp. 517–529.

2-2   *Minimum Design Loads for Buildings and Other Structures*, ASCE 7-02, American Society of Civil Engineers, New York, 2002, 214 pp.

2-3   C. Allan Cornell, "A Probability Based Structural Code," *ACI Journal, Proceedings*, Vol. 66, No. 12, December 1969, pp. 974–985.

2-4   Alan H. Mattock, Ladislav B. Kriz, and Eivind Hognestad, "Rectangular Concrete Stress Distribution in Ultimate Strength Design," *ACI Journal, Proceedings*, Vol. 57, No. 8, February 1961, pp. 875–928.

2-5   Jong-Cherng Pier and C. Allan Cornell, "Spatial and Temporal Variability of Live Loads," *Proceedings ASCE, Journal of the Structural Division*, Vol. 99, No. ST5, May 1973, pp. 903–922.

2-6   James G. MacGregor, "Safety and Limit States Design for Reinforced Concrete," *Canadian Journal of Civil Engineering*, Vol. 3, No. 4, December 1976, pp. 484–513.

2-7   Bruce Ellingwood, Theodore V. Galambos, James G. MacGregor, and C. Allan Cornell, *Development of a Probability Based Load Criterion for American National Standard A58*, NBS Special Publication 577, National Bureau of Standards, U.S. Department of Commerce, Washington, D.C., June 1980, 222 pp.

2-8   James G. MacGregor, "Load and Resistance Factors for Concrete Design," *ACI Journal, Proceedings*, Vol. 80, No. 4, July–August 1983, pp. 279–287.

2-9   *International Building Code*, International Code Council, Falls Church, various editions.

2-10 Nowak, A.S., and Szerszen, M.M., "Calibration of Design Code for Buildings (ACI 318): Part1—Statistical Models for Resistance," *ACI Structural Journal*, Vol. 100, May–June 2003, pp 377–382, "Calibration of Design Code for Buildings (ACI 318): Part 2, --Reliability Analysis and Resistance Factors," May–June 2003, pp 383–389.

2-11 Robert F. Mast, "Unified Design Provisions for Reinforced and Prestressed Prestressed Concrete Flexural and Compression Members," *ACI Structural Journal*, Vol. 89, No. 2, Mar.–Apr. 1992, pp. 185–199.

2-12 Naaman, A.E., "Unified Design Recommendations for Reinforced, Prestressed, and Partially Prestressed Concrete Bending and Compression Members," ACI Structural Journal, Vol. 89, No. 2, March–April 1992, pp. 200–210.

2-13 Donald A. Sawyer, "Ponding of Rainwater on Flexible Roof Systems," *Proceedings ACSE, Journal of the Structural Division*, Vol. 93, No. ST1, February 1967, pp. 127–147.

2-14 Allan G. Davenport, "Gust Loading Factors," *Proceedings ASCE, Journal of the Structural Division*, Vol. 93, No. ST6, June 1967, pp. 12–34.

2-15 Amin Ghali, A., Walter H. Dilger, and Adam Neville, "Time Dependent Forces Induced by Settlement of Supports in Continuous Reinforced Concrete Beams," *ACI Journal*, Vol. 66, No., 1969, pp. 907–915.

2-16 *Material and Cost Estimating Guide for Concrete Floor and Roof Systems*, Publication PA 136.01B, Portland Cement Association, Skokie, Ill., 1974, 14 pp.

2-17 Percival E. Pereiru, *McGraw-Hill's Dodge Construction Systems Costs 1980*, McGraw-Hill Information Systems Company, New York, 1980, 281 pp.

2-18 ACI Committee 340, *Design Handbook in Accordance with the Strength Design Method of ACI 318-89*, Vol. 1, *Beams, One-Way Slabs, Brackets, Footings, and Pile Caps*, ACI Publication SP-17(91), American Concrete Institute, Detroit, 1991.

2-19 ACI Committee 340, *Design Handbook in Accordance with the Strength Design Method of ACI 318-89*, Vol. 2, *Columns*, ACI Publication SP-17A(90), American Concrete Institute, Detroit, 1990.

2-20 ACI Committee 340, *Design Handbook in Accordance with the Strength Design Method of ACI 318-89*, Vol. 3, *Two-Way Slabs*, ACI Publication SP-17(91)(S), American Concrete Institute, Detroit, 1991.

2-21 *CRSI Handbook, 1992*, Concrete Steel Reinforcing Steel Institute, Schaumberg, Ill., 1992, 710 pp.

2-22 ACI Committee 315, *ACI Detailing Manual—1994*, ACI Publication SP-66 (94), American Concrete Institute, Detroit, 1994, 244 pp.

2-23 ACI Committee 301, *Specifications for Structural Concrete for Buildings*, ACI 301-95, American Concrete Institute, Detroit, 1995, 36 pp.

2-24 Mary K. Hurd and ACI Committee 347, *Formwork for Concrete*, 6th ed., ACI Publication SP-4, American Concrete Institute, Detroit, 1995, 500 pp.

2-25 *Manual of Concrete Practice*, Vol. 1 to 5, American Concrete Institute, Detroit, published annually.

2-26 ACI Committee 117, "Standard Specifications for Tolerances for Concrete Construction and Materials (ACI 117-90) and Commentary (ACI 117R-90)," *ACI Manual of Concrete Practice*, American Concrete Institute, Detroit, 1991 and later editions.

Chapter 3

3-1 Thomas T. C. Hsu, F. O. Slate, G. M. Sturman, and George Winter, "Micro-cracking of Plain Concrete and the Shape of the Stress-Strain Curve," *ACI Journal, Proceedings*, Vol. 60, No. 2, February 1963, pp. 209–224.

3-2 K. Newman and J. B. Newman, "Failure Theories and Design Criteria for Plain Concrete," Part 2 in M. Te 'eni (ed.), *Solid Mechanics and Engineering Design*, Wiley-Interscience, New York, 1972, pp. 83/1–83/33.

3-3 F. E. Richart, A. Brandtzaeg, and R. L. Brown, *A Study of the Failure of Concrete under Combined Compressive Stresses*, Bulletin 185, University of Illinois Engineering Experiment Station, Urbana, Ill., November 1928, 104 pp.

3-4 Hubert Rüsch, "Research toward a General Flexural Theory for Structural Concrete," *ACI Journal, Proceedings*, Vol. 57, No. 1, July 1960, pp. 1–28.

3-5 Llewellyn E. Clark, Kurt H. Gerstle, and Leonard G. Tulin, "Effect of Strain Gradient on Stress–Strain Curve of Mortar and Concrete," *ACI Journal, Proceedings*, Vol. 64, No. 9, September 1967, pp. 580–586.

3-6 Comité Euro-International du Béton, *CEB-FIP Model Code 1990*, Thomas Telford Services Ltd. London, 1993, 437 pp.

3-7 Aïtcin, P.-C., Miao, B., Cook, W.D., and Mitchell, D. "Effects of Size and Curing on Cylinder Compressive Strength of Normal and High-Strength Concretes," *ACI Materials Journal*, Vol. 91, No. 4, July–August 1994, pp. 349–354.

3-8 ACI Committee 363, "State-of-the-Art Report on High-Strength Concrete," *ACI Manual of Concrete Practice*, Vol. 1, American Concrete Institute, Detroit, 1993 and later editions, pp. 363R-1 to 363R-55.

3-9 ACI Committee 214, *Recommended Practice for Evaluation of Strength Test Results of Concrete*, ACI 214–77, American Concrete Institute, Detroit, 1977, 14 pp.

3-10 Nowak, A.S., and Szerszen, M.M., "Calibration of Design Code for Buildings (ACI 318): Part1— Statistical Models for Resistance," *ACI Structural Journal*, Vol. 100, May-June 2003, pp 377–382, "Calibration of Design Code for Buildings (ACI 318): Part 2,—Reliability Analysis and Resistance Factors," May–June 2003, pp 383–389.

3-11 Andrzej S. Nowak and Maria Szerszen, "Reliability-Based Calibration for Structural Concrete, Phase 1," Report UMCEE 01–04, University of Michigan, 2001, 73 pp.

3-12 Michael L. Leming, "Probabilities of Low Strength Events in Concrete," *ACI Structural Journal*, Vol. 96, No. 3, May–June 1999, pp. 369–376.

3-13 H. F. Gonnerman and W. Lerch, *Changes in Characteristics of Portland Cement as Exhibited by Laboratory Tests over the Period 1904 to 1950*, ASTM Special Publication 127, American Society for Testing and Materials, Philadelphia, 1951.

3-14 ACI Committee 211, "Standard Practice for Selecting Proportions for Normal, Heavyweight, and Mass Concrete" (ACI 211.1-91), *ACI Manual of Concrete Practice*, American Concrete Institute, Detroit, 1992 and later editions, pp. 211.1–1 to 211.1–38.

3-15 ACI Committee 226, "Use of Fly Ash in Concrete" (ACI 226.3R-87), *ACI Manual of Concrete Practice*, American Concrete Institute, Detroit, 1988 and later editions, pp. 226.3R–1 to 226.3R–38.

3-16 ACI Committee 226, "Ground Granulated Blast Furnace Slag as a Cementitious Constituent in Concrete" (ACI 226.1R-87), ACI *Manual of Concrete Practice*, American Concrete Institute, Detroit, 1988 and later editions, pp. 226.1R–1 to 226.1R–16.

3-17 Walter H. Price, "Factors Influencing Concrete Strength," *ACI Journal, Proceedings*, Vol. 47, No. 6, December 1951, pp. 417–432.

3-18 Paul Klieger, "Effect of Mixing and Curing Temperature on Concrete Strength," *ACI Journal, Proceedings*, Vol. 54, No. 12, June 1958, pp. 1063–1081.

3-19 Adam M. Neville, "Water, Cinderella Ingredient of Concrete," *Concrete International*, Vol. 22, No. 9, pp. 66–71.

3-20 Adam M. Neville, "Seawater in the Mixture," *Concrete International*, Vol. 23, No. 1, January 2001, pp. 48–51.

3-21 ACI Committee 209, "Prediction of Creep, Shrinkage and Temperature Effects in Concrete Structures," *Designing for Creep and Shrinkage in Concrete Structures*, ACI Publication SP-76, American Concrete Institute, Detroit, 1982, pp. 193–300.

3-22 V. M. Malhotra, "Maturity Concept and the Estimation of Concrete Strength: A Review," *Indian Concrete Journal*, Vol. 48, No. 4, April 1974, pp. 122–126 and 138; No. 5, May 1974, pp. 155–159 and 170.

3-23 H. S. Lew and T. W. Reichard, "Prediction of Strength of Concrete from Maturity," *Accelerated Strength Testing*, ACI Publication SP-56, American Concrete Institute, Detroit, 1978, pp. 229–248.

3-24 Adam M. Neville, "Core Tests: Easy to Perform, not Easy to Interpret", *Concrete International*, Vol. 23, No. 11, November, 2001.

3-25 Bartlett, F.M. and MacGregor, J.G., "Effect of Moisture Content on Concrete core Strengths," *ACI Materials Journal*, Vol. 91, No. 3, May–June 1994, pp. 227–236.

3-26 F. Michael Bartlett and James G. MacGregor, "Equivalent Specified Concrete Strength from Core Test Data," *Concrete International*, Vol. 17, No. 3, March 1995, pp. 52–58.

3-27 F. Michael Bartlett and James G. MacGregor, "Statistical Analysis of the Compressive Strength of Concrete in Structures," *ACI Materials Journal*, Vol. 93, in press.

3-28 Jerome M. Raphael, "Tensile Strength of Concrete," *ACI Journal, Proceedings*, Vol. 81, No. 2, March–April 1984, pp. 158–165.

3-29 D. J. McNeely and Stanley D. Lash, "Tensile Strength of Concrete," *Journal of the American Concrete Institute, Proceedings*, Vol. 60, No. 6, June 1963, pp. 751–761.

3-30 *Proposed Complements to the CEB-FIP International Recommendations—1970*, Bulletin d'Information 74, Comité Européen du Béton, Paris, March 1972 revision, 77 pp.

3-31 H. S. Lew and T. W. Reichard, "Mechanical Properties of Concrete at Early Ages," *ACI Journal, Proceedings*, Vol. 75, No. 10, October 1978, pp. 533–542.

3-32 H. Kupfer, Hubert K. Hilsdorf, and Hubert Rüsch, "Behavior of Concrete under Biaxial Stress," *ACI Journal, Proceedings*, Vol. 66, No. 8, August 1969, pp. 656–666.

3-33 Frank J. Vecchio and Michael P. Collins, *The Response of Reinforced Concrete to In-Plane Shear and Normal Stresses*, Publication 82-03, Department of Civil Engineering, University of Toronto, Toronto, March 1982, 332 pp.

3-34 Frank J. Vecchio and Michael P. Collins, "The Modified Compression Field Theory for Reinforced Concrete Elements Subjected To Shear," *ACI Journal, Proceedings*, Vol. 83, No. 2, March–April 1986, pp. 219–231.

3-35 J. A. Hansen, "Strength of Structural Lightweight Concrete under Combined Stress," *Journal of the Research and Development Laboratories, Portland Cement Association*, Vol. 5, No. 1, January 1963, pp. 39–46.

3-36 Eivind Hognestad, Norman W. Hanson, and Douglas McHenry, "Concrete Stress Distribution in Ultimate Strength Design," *ACI Journal, Proceedings*, Vol. 52, No. 4, December 1955, pp. 475–479.

3-37 Paul H. Kaar, Norman W. Hanson, and H. T. Capell, "Stress-Strain Characteristics of High-Strength Concrete," *Douglas McHenry International Symposium on Concrete and Concrete Structures*, ACI Publication SP-55, American Concrete Institute, Detroit, 1978, pp. 161–186.

3-38 Adrian Pauw, "Static Modulus of Elasticity as Affected by Density," *ACI Journal, Proceedings*, Vol. 57, No. 6, December 1960, pp. 679–683.

3-39 Eivind Hognestad, *A Study of Combined Bending and Axial Load in Reinforced Concrete Members*, Bulletin 399, University of Illinois Engineering Experiment Station, Urbana, Ill., November 1951, 128 pp.

3-40 Claudio E. Todeschini, Albert C. Bianchini, and Clyde E. Kesler, "Behavior of Concrete Columns Reinforced with High Strength Steels," *ACI Journal, Proceedings*, Vol. 61, No. 6, June 1964, pp. 701–716.

3-41 Popovics, S., "A Review of Stress-Strain Relationships for Concrete, *ACI Journal, Proceedings*, Vol. 67, No. 3, March 1970, pp. 243–248.

3-42 Thorenfeldt, E., Tomaszewicz, A. and Jensen, J. J., "Mechanical Properties of High Strength Concrete and Application to Design," *Proceedings of the Symposium: Utilization of High-Strength Concrete*," Stavanger, Norway, June 1987, Tapir, Trondheim, pp. 149–159.

3-43 Collins, M.P. and Mitchell, D., *Prestressed Concrete Structures*, Prentice Hall, Englewood Cliffs, 1991, 766 pp.

3-44 S. H. Ahmad and Surendra P. Shah, "Stress–Strain Curves of Concrete Confined by Spiral Reinforcement," *ACI Journal, Proceedings*, Vol. 79, No.6. November–December 1982, pp. 484–490.

3-45 B. P. Sinha, Kurt H. Gerstle, and Leonard G. Tulin, "Stress–Strain Relations for Concrete under Cyclic Loading," *ACI Journal, Proceedings*, Vol. 61, No. 2, February 1964, pp. 195–212.

3-46 Surendra P. Shah and V. S. Gopalaratnam, "Softening Responses of Plain Concrete in Direct Tension," *ACI Journal, Proceedings*, Vol. 82, No. 3, May–June 1985, pp. 310–323.

3-47 Zedenek P. Bazant, "Prediction of Concrete Creep Effects Using Age-Adjusted Effective Modulus Method," *ACI Journal, Proceedings*, Vol. 69, No. 4, April 1972, pp. 212–217.

3-48 Walter H. Dilger, "Creep Analysis of Prestressed Concrete Structures Using Creep-Transformed Section Properties," *PCI Journal*, Vol. 27, No. 1, January–February 1982, pp. 99–118.

3-49 Amin Ghali and Rene Favre, *Concrete Structures: Stresses and Deformations*, Chapman & Hall, New York, 1986, 348 pp.

3-50 *Structural Effects of Time-Dependent Behaviour of Concrete*, Bulletin d'Information, 215, Comité Euro-International du Béton, Laussane, March 1993, pp. 265–291.

3-51 ACI Committee 216, "Guide for Determining the Fire Endurance of Concrete Elements," *Concrete International: Design and Construction*, Vol. 3, No. 2, February 1981, pp. 13–47.

3-52 Said Iravani, "Mechanical Properties of High-Performance Concrete," *ACI Materials Journal*, Vol. 93, No. 5, September–October 1996, pp. 416–426.

3-53 Said Iravani and James G. MacGregor, "Sustained Load Strength and Short-Term Strain Behavior of High-Strength Concrete," *ACI Materials Journal*, Vol. 95, No. 5, September–October 1998, pp. 636–647.

3-54 Boris Bresler, "Lightweight Aggregate Reinforced Concrete Columns," *Lightweight Concrete.* ACI Publication SP-29, American Concrete Institute, Detroit, 1971, pp. 81–130.

3-55 ACI Committee 201, "Guide to Durable Concrete," (ACI 201.2R-92), *ACI Manual of Concrete Practice*, American Concrete Institute, Detroit, 1993 and later editions, pp. 201R-1 to 201.2R-41.

3-56 ACI Committee 222, "Corrosion of Metals in Concrete" (ACI 222R–85), *ACI Journal, Proceedings*, Vol. 82, No. 1, January–February 1985, pp. 3–32.

3-57 Adam M. Neville, *Properties of Concrete*, 3rd Edition, Pitman, 1981, 779 pp.

3-58 PCI Committee on Durability, "Alkali–Aggregate Reactivity—A Summary," *PCI Journal*, Vol. 39, No. 6, November–December, 1994, pp. 26–35.

3-59 ACI Committee 515, "A Guide to the Use of Waterproofing, Dampproofing, Protective, and Decorative Barrier Systems for Concrete" (ACI 515.R-79), *ACI Manual of Concrete Practice*, American Concrete Institute, Detroit, 1993 and later editions, pp. 515.1R-1 to 515.1R-44.

3-60 Monfore, G.E. and Lentz, A.E., Physical Preperties of Concrete at Very Low Temperatures," *Journal of the PCA Research and Development Laboratories*, Vol. 4, No. 2, May 1962, pp. 33–39.

3-61 ACI Committee 506, "Guide to Shotcrete (ACI 506R-90)," *ACI Manual of Concrete Practice*, 1991 and later editions, American Concrete Institute, Farmington Hills, pp. 506R-1 to 506R-41.

3-62 Neville, Adam M., "A 'New' Look at High-Alumina Cement," *Concrete International*, August 1998, pp. 51–55.

3-63 Sher Al Mirza and James G. MacGregor, "Variability of Mechanical Properties of Reinforcing Bars," *Proceedings ASCE, Journal of the Structural Division*, Vol. 105, No. ST5, May 1979, pp. 921–937.

3-64 T. Helgason and John M. Hanson, "Investigation of Design Factors Affecting Fatigue Strength of Reinforcing Bars—Statistical Analysis," *Abeles Symposium on Fatigue of Concrete*, ACI Publication SP-41, American Concrete Institute, Detroit, 1974, pp. 107–137.

3-65 ACI Committee 215, "Considerations for Design of Concrete Structures Subjected to Fatigue Loading," (ACI 215R-74, revised 1992), *ACI Manual of Concrete Practice*, American Concrete Institute, Detroit, 1993 and later editions, Detroit, pp. 215R-1 to 215R-24.

3-66 Mirza, S.A. and MacGregor, J.G., "Strength and Ductility of Concrete Slabs Reinforced with Welded Wire Fabric," *ACI Journal, Proceedings*, Vol. 78, No. 5, September–October 1981, pp. 374–380.

3-67 Griezic, A., Cook, W.D., and Mitchell, D., "Tests to Determine Performance of Deformed Welded Wire Fabric Stirrups," *ACI Structural Journal*, Vol. 91, No. 2, March–April 1994, pp. 213–219.

3-68 Guimaraes, G.N. and Kreger, M.E., "Evaluation of Joint-Shear Provisions for Interior Beam-Column Connections Using High-Strength Materials,' *ACI Structural Journal*, Vol. 89, No. 1, January–February 1992, pp. 89–98.

3-69 ACI Committee 440, "Guide for the Design and Construction of Concrete Reinforced with FRP Bars," American Concrete Institute, in press.

Chapter 4

4-1 Minimum Design Loads for Buildings and Other Structures, ASCE 7–98, American Society of Civil Engineers, Reston, VA, 2000, 337 pp.

4-2 Eivind Hognestad, *A Study of Combined Bending and Axial Load in Reinforced Concrete Members*, Bulletin 399, University of Illinois Engineering Experiment Station, Urbana, Ill., November 1951, 128 pp.

4-3 Hubert Rüsch, "Research toward a General Flexural Theory for Structural Concrete," *ACI Journal, Proceedings*, Vol. 57, No. 1, July 1960, pp. 1–28; Discussion, Vol. 57, No. 9, March 1961, pp. 1147–1164.

4-4 Alan H. Mattock, Ladislav B. Kriz, and Eivind Hognestad, "Rectangular Concrete Stress Distribution in Ultimate Strength Design," *ACI Journal, Proceedings*, Vol. 57, No. 8, February 1961, pp. 875–926.

4-5 Paul H. Kaar, Norman W. Hanson, and H. T. Capell, "Stress–Strain Characteristics of High Strength Concrete," *Douglas McHenry International Symposium on Concrete Structures*, ACI Publication SP-55, American Concrete Institute, Detroit, 1978, pp. 161–185.

4-6 *CEB-FIP Model Code 1990*, Thomas Telford Services Ltd., London, for Comité Euro-International du Béton, Laussane, 1993, 437 pp.

4-7 J. Stephen Ford, D. C. Chang, and John E. Breen, "Design Indications from Tests of Unbraced Multipanel Concrete Frames," *Concrete International; Design and Construction*, Vol. 3, No. 3, March 1981, pp. 37–47.

4-8 John E. Breen, "Computer Use in Studies of Frames with Long Columns," *Flexural Mechanics of Reinforced Concrete, ACI Special Publication, SP—12*, American Concrete Institute, Detroit, 1965, pp 535–556.

4-9 Dudley Charles Kent and Robert Park, "Flexural Members with Confined Concrete," *Proceedings ASCE, Journal of the Structural Division*, Vol. 97, No. ST7, July 1971, pp. 1969–1990; Closure to Discussion, Vol. 98, No. ST12, December 1972, pp. 2805–2810.

4-10 Robert F. Mast, "Unified Design Provisions for Reinforced and Prestressed Concrete Flexural and Compression Members," *ACI Structural Journal, Proceedings*, Vol. 89, No. 2, March–April 1992, pp. 185–199.

4-11 Antoine E. Naaman, "Limits of Reinforcement in the 2002 ACI Code: Transition, Flaws, and Solution," *ACI Structural Journal*, Vol. 101, No. 2, March–April 2004, pp 209–218.

4-12 David Darwin, et al., "ACI318-02 Discussions and Closure," *Concrete International*, Vol. 24, No. 1, January 2002, pp 93–103.

4-13 P. W. Birkeland and L. J. Westhoff, "Dimensional Tolerance—Concrete," State-of-Art Report 5, Technical Committee 9, *Proceedings of International Conference on Planning and Design of Tall Buildings*, Vol. Ib, American Society of Civil Engineers, New York, 1972, pp. 845–849.

4-14 Charles S. Whitney, "Plastic Theory in Reinforced Concrete Design," *Transactions, ASCE*, Vol 107, 1942, pp 251–326.

4-15 Sher-Ali Mirza and James G. MacGregor, "Variations in Dimensions of Reinforced Concrete Members," *Proceedings ACSE, Journal of the Structural Division*, Vol. 105, No. ST4, April 1979, pp. 751–766.

4-16 ACI Committee 309, Guide to Concrete Consolidation in Congested Areas," *ACI Structional Journal*, Vol. 89, No.5, September–October 1992, pp 577–586.

4-17 ACI Committee 340, *Design Handbook*, Vol. 1, *Beams, One-Way Slabs, Brackets, Footings, and Pile Caps, in Accordance with the Strength Design Method of ACI 318–91*, 4th ed., ACI Publication SP-17(91), American Concrete Institute, Detroit, 1991, 374 pp.

4-18 *CRSI Handbook 1992*, Concrete Reinforcing Steel Institute, Schaumberg, Ill., 1992, 710 pp.

Chapter 5

5-1 CRSI Subcommittee on Placing Reinforcing Bars, *Placing Reinforcing Bars*, Concrete Reinforcing Steel Institute, Chicago, 1971, 201 pp.

5-2 Stephen Timoshenko and J. N. Goodier, *Theory of Elasticity*, 2nd ed., McGraw-Hill, New York, 1951, pp. 171–177.

5-3 Gottfried Brendel, "Strength of the Compression Slab of T-Beams Subject to Simple Bending," *ACI Journal, Proceedings*, Vol. 61, No. 1, January 1964, pp. 57–76.

5-4 A. Ghani Razapur and Amin Ghali, "Design of Transverse Reinforcement in Flanges of T-Beams," *ACI Journal, Proceedings*, Vol. 83, No. 4, July-August 1986, pp. 680–689.

5-5 G. W. Washa and P. G. Fluck, "Effect of Compressive Reinforcement on the Plastic Flow of Reinforced Concrete Beams," *ACI Journal, Proceedings*, Vol. 49, No. 4, October 1952, pp. 89–108.

5-6 Mircea Z. Cohn and S. K. Ghosh, "Flexural Ductility of Reinforced Concrete Sections," *Publications*, International Association of Bridge and Structural Engineers, Zurich, Vol. 32-II, 1972, pp. 53–83.

5-7 Hubert Rüech, "Research toward a General Flexural Theory for Structural Concrete," *ACI Journal, Proceedings*, Vol 57, No. 1, July 1960, pp 1–28.

5-8 ACI Committee 440, "Guide for the Design and Construction of Concrete Reinforced with FRP bars," *Manual of Concrete Practice*, American Concrete Institute, Farmington Hills, MI, various editions since 2001.

Chapter 6

6-1 Boyd G. Anderson, "Rigid Frame Failures," *ACI Journal, Proceedings*, Vol. 53, No. 7, January 1957, pp. 625–636.

6-2 ACI Committee 445, "Recent Approaches to Shear Design of Structural Concrete," *Journal of Structural Engineering*, ASCE, Vol. 124, No. 12, December 1998.

6-3 Howard P. J. Taylor, "Investigation of Forces Carried across Cracks in Reinforced Concrete Beams in Shear by Interlock of Aggregate," TRA 42.447, Cement and Concrete Association, London, 1970, 22 pp.

6-4 Robert Park and Thomas Paulay, *Reinforced Concrete Structures*, A Wiley-Interscience Publication, Wiley, New York, 1975, 769 pp.

6-5 ACI-ASCE Committee 426, "The Shear Strength of Reinforced Concrete Members—Chapters 1 to 4," *Proceedings ASCE, Journal of the Structural Division*, Vol. 99, No. ST6, June 1973, pp. 1091–1187.

6-6 Jörg Schlaich, Kurt Schaefer, and Mattias Jennewein, "Towards a Consistent Design of Reinforced Concrete Structures," *Journal of the Prestressed Concrete Institute*, Vol. 32, No. 3, May–June 1987.

6-7 ACI-ASCE Committee 426, *Suggested Revisions to Shear Provisions for Building Codes*, American Concrete Institute, Detroit, 1978, 88 pp; abstract published in *ACI Journal, Proceedings*, Vol. 75, No. 9, September 1977, pp. 458–469; Discussion, Vol. 75, No. 10, October 1978, pp. 563–569.

6-8 Michael P. Collins and Dan Kuchma, "How Safe are our Large, Lightly Reinforced Concrete Beams, Slabs, and Footings?" *ACI Structural Journal, Proceedings*, Vol. 96, No. 4, July–August 1999, pp. 482–490.

6-9 Z.P. Bazant and J.K. Kim, "Size Effect in Shear Failure of Longitudinally Reinforced Beams," *ACI Journal, Proceedings*, Vol. 81, No. 5, April 1984, pp. 456–468.

6-10 David M. Rogowsky and James G. MacGregror, "Design of Reinforced Concrete Deep Beams," *Concrete International: Design and Construction*, Vol. 8. No. 8, August 1986, pp. 49–58.

6-11 Peter Marti, "Basic Tools of Beam Design," *ACI Journal, Proceedings*, Vol. 82, No. 1, January–February 1985, pp. 46–56.

6-12   Peter Marti, "Truss Models in Detailing," *Concrete International Design and Construction*, Vol. 7, No. 12, December 1985, pp. 66–73.

6-13   Michael P. Collins and Denis Mitchell, *Prestressed Concrete Structures*, Prentice Hall, Englewood Cliffs, N. J., 1991, 765 pp.

6-14   Michael P. Collins and Denis Mitchell, "Design Proposals for Shear and Torsion," *Journal of the Prestressed Concrete Institute*, Vol. 25, No. 5, September–October 1980, 70 pp.

6-15   *Bruchwiderstand und Bemessung von Stahlbeton-und Spannbetontragwerken (Ultimate Limit States and Design of Reinforced Concrete and Prestressed Concrete Structures)*, Schweizerischer Ingenieur-und Architekten-Verein (SIA), Zurich, 1976.

6-16   *CEB-FIP Model Code 1990*, First Draft, Bulletin d'Information 195 and 196, Comité Euro-International du Béton, Lausanne, March 1990, 348 pp.

6-17   ACI-ASCE Committee 326, Shear and Diagonal Tension," *ACI Journal, Proceedings*, Vol. 59, Nos. 1–3, January–March 1962, pp. 1–30, 277–344, and 352–396.

6-18   Theodore C. Zsutty, "Shear Strength Prediction for Separate Categories of Simple Beam Tests," *ACI Journal, Proceedings*, Vol. 68, No. 2 February 1971, pp. 138–143.

6-19   M.J. Faradji, and Roger Diaz de Cossio, "Diagonal Tension in Concrete Members of Circular Section," (in Spanish) Institut de Ingeniria, Mexico (translation by Portland Cement Association, Foreign Literature Study No. 466).

6-20   J.U. Khalifa, and Michael P. Collins, "Circular Members Subjected to Shear," Publications No. 81–08, Department of Civil Engineering, University of Toronto, December 1981.

6-21   Fritz Leonhardt and Rene Walther, *The Stuttgart Shear Tests, 1961*, Translation 111, Cement and Concrete Association, London, 1964, 110 pp.

6-22   Young-Soo Yoon, William D. Cook, and Denis Mitchell, "Minimum Shear Reinforcement in Normal, Medium and High-Strength Concrete Beams," *ACI Structural Journal*, Vol. 93, No. 5, September–October 1996, pp. 576–584.

6-23   A. G. Mphonde and Gregory C. Frantz, "Shear Tests for High- and Low-Strength Concrete Beams without Stirrups," *ACI Journal, Proceedings*, Vol. 81, No. 4, July–August 1984, pp. 350–357.

6-24   A. H. Elzanaty, Arthur H. Nilson, and Floyd O. Slate, "Shear Capacity of Reinforced Concrete Beams Using High Strength Concrete," *ACI Journal, Proceedings*, Vol. 83, No. 2, March–April 1986, pp. 290–296.

6-25   Technical Committee on Reinforced Concrete Design, *Design of Concrete Structures, A23.3-M84*, Canadian Standards Association, Rexdale, 1984.

6-26   Technical Committee on Reinforced Concrete Design, *Design of Concrete Structures, A23.3–94*, Canadian Standards Association, Rexdale, Ontario, 1994.

6-27   *LRFD Bridge Specifications and Commentary*, 2nd Edition, American Association of State Highway and Transportation Officials, Washington, 1998, 1216 pp.

6-28   Technical Committee on Reinforced Concrete Design, *Design of Concrete Structures, A23.3-04*, Canadian Standards Association, Rexdale, Ontario, 2004, (in press).

6-29   Frank J. Vecchio and Michael P. Collins, "Modified Compression Field Theory for Reinforced Concrete Elements Subjected to Shear," *ACI Structural Journal*, Vol. 83, No. 2, March-April 1986, pp. 219–231.

6-30   Michael P. Collins, Denis Mitchell, Perry Adebar, and Frank J. Vecchio, "A General Shear Design Method," *ACI Structural Journal*, Vol. 93, No. 1, January–February 1996, pp. 36–45.

6-31   Frank J. Vecchio and Michael P. Collins, "The Response of Reinforced Concrete to In-Plane Shear and Normal Stresses," *Publication* No. 82–03, Department of Civil Engineering, University of Toronto, March 1982, 332 pp.

6-32   Joost C. Walraven, "Fundamental Analysis of Aggregate Interlock," *Journal of the Structural Division*, American Society of Civil Engineers, Vol. 107, No. ST11, November 1981, pp. 2245–2270.

6-33   Robert E. Loov, "Review of A23.3-94, 'Simplified Method' of Shear Design, and Comparison with Test Results Using Shear Friction," *Canadian Journal of Civil Engineering*, Vol. 25, No. 3, June 1998, pp 437–450.

6-34   H. Wagner, "Metal Beams with Very Thin Webs," *Zeitscrift für Flugteknik und Motorluftschiffahr*, Vol. 20, No. 8 to 12, 1929.

6-35   Dino Angelakos, Evan C. Bentz, and Michael P. Collins, "The Effect of Concrete Strength and Minimum Stirrups on the Shear Strength of Large Members," *ACI Structural Journal*, Vol. 98, No. 3, May–June 2001, pp 290–296.

6-36   Robert E. Loov and Anil K. Patnaik, "Horizontal Shear Strength of Composite Concrete Beams with a Rough Interface," *PCI Journal*, January–February 1994, Vol. 39, No. 1, pp 369–390.

6-37   Alan H. Mattock and J.F. Shen, "Joints Between Reinforced Concrete Members of Similar Depth," *ACI Structural Journal*, Proceedings Vol. 89, No. 3, May–June 1992, pp 290–295.

6-38   Pawan R. Gupta and Michael P. Collins, "Evaluation of Shear Design Procedures for Reinforced Concrete Members under Axial Compression," *ACI Structural Journal*, Vol. 98, No. 4, July–August 2001, pp 537–547.

Chapter 7

7-1 Stephen P. Timoshenko and J. N. Goodier, *Theory of Elasticity*, McGraw-Hill, New York, 1951, pp. 275–288.

7-2 E. P. Popov, *Mechanics of Materials, SI Version*, 2nd ed., Prentice Hall, Englewood Cliffs, N. J., 1978, 590 pp.

7-3 Christian Menn, *Prestressed Concrete Bridges*, Birkhäuer Verlag, Basel, 1990, 535 pp.

7-4 Thomas T. C. Hsu, "Torsion of Structural Concrete—Behavior of Reinforced Concrete Rectangular Members," *Torsion of Structural Concrete*, ACI Publication SP-18, American Concrete Institute, Detroit, 1968, pp. 261–306.

7-5 Ugor Ersoy and Phil M. Ferguson, "Concrete Beams Subjected to Combined Torsion and Shear—Experimental Trends," *Torsion of Structural Concrete*, ACI Publication SP-18, American Concrete Institute, Detroit, 1968, pp. 441–460.

7-6 N. N. Lessig, *Determination of the Load Carrying Capacity of Reinforced Concrete Elements with Rectangular Cross-Section Subjected to Flexure with Torsion*, Work 5, Institute Betona i Zhelezobetona, Moscow, 1959, pp. 4–28; also available as Foreign Literature Study 371, PCA Research and Development Labs, Skokie, Ill.

7-7 Paul Lampert and Bruno Thürlimann, "Ultimate Strength and Design of Reinforced Concrete Beams in Torsion and Bending," *Publications*, International Association for Bridge and Structural Engineering, Zurich, Vol. 31-I, 1971, pp. 107–131.

7-8 Paul Lampert and Michael P Collins, "Torsion, Bending, and Confusion—An Attempt to Establish the Facts," *ACI Journal, Proceedings* , Vol. 69, No. 8 August 1972, pp. 500–504.

7-9 *CEB-FIP Model Code 1990*, First Draft, Bulletin d'Information 195 and 196, Comité Euro-International du Béton, Lausanne, March 1990, 348 pp.

7-10 James G. MacGregor and Mashour G. Ghoneim, "Design for Torsion," *ACI Structural Journal*, Vol. 92, No. 2, March–April 1995, pp. 211–218.

7-11 Thomas T. C. Hsu, "Torsion of Structural Concrete—Plain Concrete Rectangular Sections," *Torsion of Structural Concrete*, ACI Publication SP-18, American Concrete Institute, Detroit, 1968, pp. 203–238.

7-12 Michael P. Collins and Denis Mitchell, "Design Proposals for Shear and Torsion," *PCI Journal*, Vol. 25, No. 5, September–October 1980, 70 pp.

7-13 Denis Mitchell and Michael P. Collins, "Detailing for Torsion," *ACI Journal, Proceedings*, Vol. 73, No. 9, September 1976, pp. 506–511.

7-14 Michael P. Collins and Paul Lampert, "Redistribution of Moments at Cracking—The Key to Simpler Torsion Design?" *Analysis of Structural Systems for Torsion*, ACI Publication SP-35, American Concrete Institute, Detroit, 1973, pp. 343–383.

7-15 Denis Mitchell and Michael P. Collins, "Detailing for Torsion," *ACI Journal, Proceedings*, Vol. 73, No. 9, September 1976, pp. 506–511.

7-16 ACI Committee 315, *ACI Detailing Manual*, ACI Publication SP-66(88), American Concrete Institute, Detroit, 1988, 218 pp.

Chapter 8

8-1 ACI Committee 408, "Bond Stress—The State of the Art," *ACI Journal, Proceedings*, Vol. 63, No. 11, November 1966, pp. 1161–1190; Discussion, pp. 1569–1570.

8-2 C. O. Orangun, J. O. Jirsa, and J. E. Breen, "A Reevaluation of Test Data on Development Length and Splices," *ACI Journal, Proceedings*, Vol. 74, No. 3, March 1977, pp. 114–122; Discussion, pp. 470–475.

8-3 J. O. Jirsa, L. A. Lutz, and P. Gergely, "Rationale for Suggested Development, Splice and Standard Hook Provisions for Deformed Bars in Tension," *Concrete International: Design and Construction*, Vol. 1, No. 7, July 1979, pp. 47–61.

8-4 *Bond Action and Bond Behaviour of Reinforcement, State-of-the-Art Report*, Bulletin d'Information 151, Comité Euro-International du Béton, Paris, April 1982, 153 pp.

8-5 Paul R. Jeanty, Denis Mitchell, and Saeed M. Mirza, "Investigation of Top Bar Effects in Beams," *ACI Structural Journal*, Vol. 85, No. 3, May–June 1988, pp. 251–257.

8-6 Robert A. Treece and James O. Jirsa, "Bond Strength of Epoxy-Coated Reinforcing Bars," *ACI Materials Journal*, Vol. 86, No. 2, March–April 1989, pp. 167–174.

8-7 J. L. G. Marques and J. O. Jirsa, "A Study of Hooked Bar Anchorages in Beam-Column Joints," *ACI Journal, Proceedings*, Vol. 72, No. 5, May 1975, pp. 198–209.

8-8 G. Rehm, "Kriterien zur Beurteilung von Bewehrungsstäben mit hochwertigem Verbund (Criteria for the Evaluation of High Bond Reinforcing Bars)" *Stahlbetonbau-Berichte aus Forschung und Praxis-Hubert Rüsch gewidmet*, Berlin, 1969, pp. 79–85

8-9 Joint Committee on Standard Specifications for Concrete and Reinforced Concrete, *Recommended Practice and Standard Specifications for Concrete and Reinforced Concrete*, American Concrete Institute, Detroit, June 1940, 140 pp.

8-10 M. Baron, "Shear Strength of Reinforced Concrete Beams at Point of Bar Cutoff," *ACI Journal, Proceedings*, Vol. 63, No. 1, January 1966, pp. 127–134.

8-11 Y. Goto and K. Otsuka, "Experimental Studies on Cracks Formed in Concrete around Deformed Tension Bars," *Technological Reports of Tohoku University*, Vol. 44, June 1979, pp. 49–83.

8-12 J. F. Pfister and A. H. Mattock, "High Strength Bars as Concrete Reinforcement," Part 5, "Lapped Splices in Concentrically Loaded Columns," *Journal of the Research and Development Laboratories*, Portland Cement Association, Vol. 5, No. 2, May 1963, pp. 27–40.

8-13 F. Leonhardt and K. Teichen, "Druck-Stoesse von Bewehrungsstaeben (Compression Splices of Reinforcing Bars)," *Deutscher Ausschuss fuer Stahlbeton*, Bulletin 222, Wilhelm Ernst, Berlin, 1972, pp. 1–53.

8-14 ACI Committee 439, "Mechanical Connections of Reinforcing Bars," *Concrete International: Design and Construction*, Vol. 5, No. 1, January 1983, pp. 24–35.

Chapter 9

9-1 David, Darwin, et al., "ACI 318–02 Discussion and Closure," *Concrete International*, Vol. 24, No. 1, January 2002, pp. 93–103.

9-2 *Minimum Design loads for buildings and Other Structures*, ASCE Standard, ASCE 7–98, American Society of Civil Engineers, Renton, VI, 2000, 337 pp.

9-3 *Design of concrete structures*, Norwegian Standard, N.S. 3473, Norwegian Council for Building Standardization, Oslo, 1990.

9-4 Zedenek P. Bazant, "Prediction of Concrete Creep Effects Using Age-Adjusted Effective Modulus Method," *ACI Journal, Proceedings*, Vol. 69, No. 4, April 1972, pp. 212–217.

9-5 Walter H. Dilger, "Creep Analysis of Prestressed Concrete Structures Using Creep-Transformed Section Properties," *PCI Journal*, Vol. 27, No. 1, January–February 1982, pp. 99–118.

9-6 Amin Ghali and Rene Favre, *Concrete Structures: Stresses and Deformations*, Chapman & Hall, New York, 1986, 348 pp.

9-7 Fritz Leonhardt, "Crack Control in Concrete Structures," *IABSE Surveys*, IABSE Periodica, 3/1977, International Association for Bridge and Structural Engineering, Zurich, 1977, 26 pp.

9-8 ACI Committee 224, "Control of Cracking in Concrete Structures" (ACI 224R-90), *ACI Manual of Concrete Practice,* American Concrete Institute, Detroit, 1991 and later editions, pp. 224R-1 to 224R-43.

9-9 ACI Committee 224, "Causes, Evaluation, and Repair of Cracks in Concrete Structures" (ACI 224.1R-90), *ACI Manual of Concrete Practice,* American Concrete Institute, Detroit, 1991 and later editions, pp. 224.1R-1 to 224.1R-20.

9-10 Comité Euro-International du Béton, *CEB Manual—Cracking and Deformations,* Ecole Polytechnique Fédérale de Lausanne, 1985, 232 pp.

9-11 Peter Gergely and Leroy A. Lutz, "Maximum Crack Width in Reinforced Concrete Flexural Members," *Causes, Mechanism and Control of Cracking in Concrete,* ACI Publication SP-20, American Concrete Institute, Detroit, 1973, pp. 87–117.

9-12 Andrew W. Beeby, "Cracking, Cover, and Corrosion of Reinforcement," *Concrete International: Design and Construction,* Vol. 5, No. 2, February 1983, pp. 35–41.

9-13 David Darwin, David G. Manning, Eivind Hognestad, Andrew W. Beeby, Paul F. Rice, Abdoul Q. Ghowrwal, "Debate: crack width, cover, and corrosion", *Concrete International*, Vol. 7, No. 5, May 1985, pp. 20–35.

9-14 ACI Committee 222, "Corrosion of Metals in Concrete" (ACI 222R-89), *ACI Manual of Concrete Practice,* American Concrete Institute, Detroit, 1990 and later editions, pp. 222R-1 to 222R-30.

9-15 ACI Committee 350, "Environmental Engineering Structures, ACI 350R-89", *Manual of Concrete Practice*, American Concrete Institute, Farmington Hills, MI, 1989, 24 pp.

9-16 R. J. Frosch, "Another Look at Cracking and Crack Control in Reinforced Concrete," *ACI Structural Journal*, Vol. 96, No. 3, May-June 1999, pp. 437–442.

9-17 *CEB-FIP Model Code 1990*, Thomas Telford Service Ltd., London, for Comité Euro-International du Béton, Lausanne, 1993, 437 pp.

9-18 Gregory C. Frantz and John E. Breen, "Design Proposal for Side Face Crack Control Reinforcement for Large Reinforced Concrete Beams," *Concrete International: Design and Construction,* Vol. 2, No. 10, October 1980, pp. 29–34.

9-19 R. I. Gilbert, "Shrinkage Cracking in Fully Restrained Concrete Members," *ACI Structural Journal*, Vol. 89, No. 2, Mar.-Apr. 1992, pp. 141–149.

9-20 Andrew W. Beeby, Discussion of R. I. Gilbert, "Shrinkage Cracking in Fully Restrained Reinforced Concrete Members", *ACI Structural Journal, Proceedings*, Vol. 90, No. 1. January-February 1993, pp. 123–126.

9-21 Amin Ghali, "Deflection of Reinforced Concrete Members: A Critical Review," *ACI Structural Journal,* Vol. 90, No. 4, July–August 1993, pp. 364–373.

9-22 ACI Committee 435, Deflections of Reinforced Concrete Flexural Members," *ACI Journal, Proceedings*, Vol. 63 No. 6 June 1966, pp. 637–674.

9-23 Dan E. Branson, "Compression Steel Effect on Long-Time Deflections," *ACI Journal, Proceedings,* Vol. 68, No. 8, August 1971, pp. 555–559.

9-24 H. Mayer and Hubert Rüsch, "Damage to Buildings Resulting from Deflection of Reinforced Concrete Members," *Deutscher Ausschuss für Stahlbeton,* Bulletin 193, Wilhelm Ernst, Berlin, 1967, 90 pp.; English translation: Technical Translation 1412, National Research Council of Canada, Ottawa, 1970, 115 pp.

9-25 D. A. Sawyer, "Ponding of Rainwater on Flexible Roof Systems," *Proceedings ASCE, Journal of the Structural Division,* Vol. 93, No. ST1, February 1967, pp. 127–147.

9-26 Amin Ghali, Walter Dilger and Adam M. Neville, "Time-Dependent Forces Induced by Settlement of Supports in Continuous Reinforced Concrete Beams", *Journal of the American Concrete Institute*, Vol. 66, No. 11, November 1969, pp. 907–915.

9-27 ACI Committee 435, "Allowable Deflections," *ACI Journal, Proceedings,* Vol. 65, No. 6, June 1968, pp. 433–444.

9-28 Robert Ramsey, Sher-Ali Mirza, and James G. MacGregor, "Variability of Deflections of Reinforced Concrete Beams," *ACI Journal, Proceedings,* Vol. 76, No. 8, August 1979, pp. 897–918.

9-29 Jacob S. Grossman, "Simplified Computations for Effective Moment of Inertia $[\&I\_\{e\}\&]$ and Minimum Thickness to Avoid Deflection Computations," *ACI Journal, Proceedings,* Vol. 78, No. 6, November–December 1981, pp. 423–440.

9-30 B. S. Choi, B. H. Oh, and Andrew Scanlon, "Probabilistic Assessment of ACI 318 Minimum Thickness Requirements for One-Way Members," *ACI Structural Journal*, May-June 2002, pp. 344–351.

9-31 "Motion Perception and Tolerance," Chapter PC-13, *Planning and Environmental Criteria for Tall Buildings,* Monograph on Planning and Design of Tall Buildings, Vol. PC, American Society of Civil Engineers, New York, 1981, pp. 805–862.

9-32 Fazlur R. Khan and Mark Fintel, "Effects of Column Exposure in Tall Structures," *ACI Journal, Proceedings,* Vol. 62, No. 12, December 1965, pp. 1533–1556; Vol. 63, No. 8, August 1966, pp. 843–864; Vol. 65, No. 2, February 1968, pp. 99–110.

9-33 Mark Fintel and S. K. Ghosh, *Column Shortening in Tall Structures—Prediction and Compensation,* EB108.01D, Portland Cement Association, Skokie, Ill., 1986, 35 pp.

9-34 "Commentary A, Serviceability Criteria for Deflections and Vibrations," *Supplement to the National Building Code of Canada 1990,* NRCC 30629, National Research Council of Canada, Ottawa, 1990, pp. 134–140.

9-35 David E. Allen, J. Hans Rainer, and G. Pernica, "Vibration Criteria for Long-Span Concrete Floors," *Vibrations of Concrete Structures,* ACI Publication SP-60, American Concrete Institute, Detroit, 1979, pp. 67–78.

9-36 ACI Committee 215, "Considerations for the Design of Concrete Structures Subjected to Fatigue Loading" (ACI 215R-74, revised 1992), *ACI Manual of Concrete Practice,* American Concrete Institute, Detroit, 1993 and later editions, pp. 215R-1 to 215R-24.

9-37 ACI Committee 343, "Analysis and Design of Reinforced Concrete Bridge Structures" (ACI 343R-88), *ACI Manual of Concrete Practice,* American Concrete Institute, Detroit, 1989 and later editions, pp. 343R-1 to 343R-162.

9-38 *Bond Action and Bond Behavior of Reinforcement,* Bulletin d'Information 151, Comité Euro-International du Béton, Paris, April 1982, 153 pp.

Chapter 10

10-1 ACI Committee 216, "Guide for Determining the Fire Endurance of Concrete Elements," *Concrete International: Design and Construction*, Vol. 3, No. 2, February 1981, pp. 13–47.

10-2 R. Ian Gilbert, "Shrinkage Cracking in Fully-Restrained Concrete Members," *ACI Structural Journal*, Vol. 89, No. 2, March–April 1992, pp. 141–150.

10-3 *Building Movements and Joints*, Engineering Bulletin EB 086.10B, Portland Cement Association, Skokie, Ill., 1982, 64 pp.

10-4 *Minimum Design Loads for Buildings and Other Structures*, ASCE 7-98, American Society of Civil Engineers, New York, 1998, 214 pp.

10-5

10-6 Bianchini, A.C., Woods, Robert E., and Kesler, C.E., "Effect of Floor Concrete Strength on Column Strength," ACI Journal, Proceedings, Vol. 56, No. 11, May 1960, pp. 1149–1169.

10-7 Carlos E. Ospina and Scott D.B. Alexander, "Transmission of Interior Concrete Column Loads through Floors," *ASCE Journal of Structural Engineering*, Vol. 124, No. 6, 1998.

10-7 ACI-ASCE Committee 428, "Progress Report on Code Clauses for Limit Design," *ACI Journal, Proceedings*, Vol. 65, No. 9, September 1968, pp. 713–720.

10-8 Richard W. Furlong, "Design of Concrete Frames by Assigned Limit Moments," *ACI Journal, Proceedings*, Vol. 67, No. 4, April 1970, pp. 341–353.

10-9 Alan H. Mattock, "Redistribution of Design Bending Moments in Reinforced Concrete Continuous Beams," *Proceedings, Institution of Civil Engineers, London*, Vol. 13, 1959, pp. 35–46.

Chapter 11

11-1 Eivind Hognestad, *A Study of Combined Bending and Axial Load in Reinforced Concrete Members*, Bulletin 399, University of Illinois Engineering Experiment Station, Urbana, Ill., June 1951, 128 pp.

11-2 ACI Committee 105, "Reinforced Concrete Column Investigation," *ACI Journal, Proceedings*, Vol. 26, April 1930, pp. 601–612; Vol. 27, February 1931, pp. 675–676; Vol. 28, November 1931, pp. 157–578; Vol. 29, September 1932, pp. 53–56; Vol. 30, September–October 1933, pp. 78–90; November–December 1933, pp. 153–156.

11-3 ACI-ASCE Committee 327, "Report on Ultimate Strength Design," *Proceedings ASCE*, Vol. 81, October 1955, Paper 809. See also *ACI Journal, Proceedings*, Vol. 2, No. 7, January 1956, pp. 505–524.

11-4 Robert F. Mast, "Unified Design Provision for Reinforced and Prestressed Concrete Members," *ACI Structural Journal*, Vol. 89, No. 2, March–April, pp. 185–199.

11-5 Ravi K. Devalapura and Maher K. Tadros; Charles W. Dolan; John V. Loscheider, Discussion of Paper by Robert Mast, *ACI Structural Journal*, Vol. 89, No. 5, September–October, 1992, pp 591–601.

11-6 ACI Committee 340, *Design Handbook Volume 2—Columns* (ACI 340.2R-91), ACI Publication SP-17A(90), American Concrete Institute, Detroit, 250 pp.

11-7 Albert C. Bianchini, R. E. Woods, and Clyde E. Kesler, "Effect of Floor Concrete Strength on Column Strength," *ACI Journal, Proceedings*, Vol. 56, No. 11, May 1960, pp. 1149–1169.

11-8 Carlos E. Ospina and Scott D.B. Alexander, "Transmission of Interior Concrete Column Loads through Floors," *ASCE Journal of Structural Engineering*, Vol. 124, No. 6, 1998.

11-9 J.F. Pfister, "Influence of Ties on the Behavior of Reinforced Concrete Columns," *ACI Journal, Proceedings*, Vol. 61, No. 5, May 1964, pp. 521–537.

11-10 Fred M. Hudson, "Reinforced Concrete Columns: Effect of Lateral Tie Spacing on Ultimate Strength, *ACI Special Publication SP 13*, American Concrete Institute, pp 235–244, 1966.

11-11 Boris Bresler and P.H. Gilbert, "Tie Requirements for Reinforced Concrete Columns," *ACI Journal*, Vol. 58, No. 5, November 1961, pp. 555–570.

11-12 ACI Committee 363, "State-of-the-Art Report on High-Strength Concrete," *ACI Manual of Concrete Practice*, American Concrete Institute, Detroit, 1993 and later editions, pp. 363R-1 to 363R-55.

11-13 P.H. Ziehl, J.E. Cloyd, and Michael E. Kreger, "Investigation of Minimum Longitudinal Reinforcements for Concrete Columns Using Present-Day Construction Materials," *ACI Structural Journal*, Vol. 101, No. 2, March-April 2004, pp 165–175.

11-14 Robert F. Warner, B.V. Rangan, and A.S. Hall, *Reinforced Concrete*, Pitman Australia, 1976, 475 pp.

11-15 Troels Brondiem-Nielsen, "Ultimate Limit States of Cracked Arbitrary Concrete Sections under Axial Load and Biaxial Loading," *Concrete International: Design and Construction*, Vol. 4, No. 11, November 1982, pp. 51–55.

11-16 Richard W. Furlong, C.-T.T. Hsu, and Sher A. Mirza, "Analysis and Design of Concrete Columns for Biaxial Bending—Overview," *ACI Structural Journal*, Vol. 101, No. 3, May–June 2004, pp 413–423.

11-17 James G. MacGregor, "Simple Design Procedures for Concrete Columns," Introductory Report, Symposium on Design and Safety of Reinforced Concrete Compression Members, *Reports of the Working Commissions*, Vol. 15, International Association of Bridge and Structural Engineering, Zurich, April 1973, pp. 23–49.

11-18 Boris Bresler, "Design Criteria for Reinforced Concrete Columns under Axial Load and Biaxial Bending," *ACI Journal, Proceedings*, Vol. 57, No. 5, November 1960, pp. 481–490; Discussion, pp. 1621–1638.

11-19 Albert J. Gouwens, "Biaxial Bending Simplified," *Reinforced Concrete Columns*, ACI Publication SP-50, American Concrete Institute, Detroit, 1975, pp. 233–261.

Chapter 12

12-1 James G. MacGregor, John E. Breen, and Edward O. Pfrang, "Design of Slender Columns," *ACI Journal, Proceedings*, Vol. 67, No. 1, January 1970, pp. 6–28.

12-2 Bengt Broms and Ivan M. Viest, "Long Reinforced Concrete Columns—A Symposium," *Transactions ASCE*, Vol. 126, Part 2, 1961, pp. 308–400.

12-3 *Commentary on Specification for the Design, Fabrication and Erection of Structural Steel for Buildings*, American Institute of Steel Construction, New York, 1970.

12-4 Walter J. Austin, "Strength and Design of Metal Beam-Columns," *Proceedings ASCE, Journal of the Structural Division*, Vol. 87, No. ST4, April 1961, pp. 1–34.

12-5 James G. MacGregor, Urs H. Oelhafen, and Sven E. Hage, "A Reexamination of the *El* Value for Slender Columns," *Reinforced Concrete Columns*, ACI Publication SP-50, American Concrete Institute, Detroit, 1975, pp. 1–40.

12-6 A. A. Gvozdev and E. A. Chistiakov, "Effect of Creep on Load Capacity of Slender Compressed Elements," State-of-Art Report 2, Technical Committee 23 ASCE-IABSE International Conference on Tall Buildings, *Proceedings*, Vol. III-23, August 1972, pp. 537–554.

12-7 Roger Green and John E. Breen, "Eccentrically Loaded Columns under Sustained Load,"*ACI Journal, Proceedings*, Vol. 66, No. 11, November 1969, pp. 866–874.

12-8 Adam M. Neville, Walter H. Dilger, and J. J. Brooks, *Creep of Plain and Structural Concrete*, Construction Press, New York, 1983, pp. 347–349.

12-9 Robert F. Warner, "Physical-Mathematical Models and Theoretical Considerations," *Introductory Report, Symposium on Design and Safety of Reinforced Concrete Compression Members*, International Association on Bridge and Structural Engineering, Zurich, 1974, pp. 1–21.

12-10 Sher Ali Mirza, P. M. Lee, and D. L. Morgan, "ACI Stability Resistance Factor for RC Columns," *ASCE Structural Engineering*, Vol. 113, No. 9, September 1987, pp. 1963–1976.

12-11 Richard W. Furlong and Phil M. Ferguson, "Tests of Frames with Columns in Single Curvature," *Symposium on Reinforced Concrete Columns*, ACI Publication SP-13, American Concrete Institute, Detroit, 1966, pp. 55–73.

12-12 William B. Cranston, *Analysis and Design of Concrete Columns*, Research Report 20, Paper 41.020, Cement and Concrete Association, London, 1972, 54 pp.

12-13 Robert F. Manuel and James G. MacGregor, "Analysis of Restrained Reinforced Concrete Columns under Sustained Load, *ACI Journal, Proceedings*, Vol. 64, No. 1, January 1967, pp. 12–23.

12-14 Shu-Ming Albert Lai, James G. MacGregor, and Jostein Hellesland, "Geometric Non-linearities in Non-sway Frames, *Proceedings ASCE, Journal of the Structural Division*, Vol. 109, No. ST12, December 1983, pp. 2770–2785.

12-15 Thomas C. Kavenagh, "Effective Length of Framed Columns," *Transactions ASCE*, Vol. 127, Part 2, 1962, pp. 81–101.

12-16 William McGuire, *Steel Structures*, Prentice Hall, Englewood Cliffs, N.J., 1968, 1110 pp.

12-17 *PCI Design Handbook*, 3rd ed., Prestressed Concrete Institute, Chicago, 1985, 521 pp.

12-18 Emilio Rosenblueth, "Slenderness Effects in Buildings," *Journal of the Structural Division*, American Society of Civil Engineers, Vol. 91, No. ST1, Proceedings Paper 4235, pp 229–252.

12-19 L.K. Stevans, "Elastic Stability of Practical Multi-Storey Frames," *Proceedings*, Institution of Civil Engineers, London, Vol. 36, pp 99–117.

12-20 James G. MacGregor and Sven E. Hage, "Stability Analysis and Design of Concrete Frames," *Proceedings ASCE, Journal of the Structural Division*, Vol. 103, No. ST10, October 1977, pp. 1953–1970.

12-21 J. Steven Ford, D. C. Chang, and John E. Breen, "Design Implications from Tests of Unbraced Multipanel Concrete Frames," *Concrete International: Design and Construction*, Vol. 3, No. 3, March 1981, pp. 37–47.

12-22 ACI Committee 318, *Commentary on Building Code Requirements for Reinforced Concrete (ACI 318R-89)*, American Concrete Institute, Farmington Hills, MI, 1989, 369 pp.

12-23 Shu-Ming Albert Lai and James G. MacGregor, "Geometric Non-Linearities in Multi-story Frames," *Proceedings ASCE, Journal of the Structural Division*, Vol. 109, No. ST11, November 1983, pp. 2528–2545.

12-24 James G. MacGregor, "Design of Slender Columns—Revisited," *ACI Structural Journal*, Vol. 90, No. 3, May–June 1993, pp. 302–309.

12-25 Minimum Design Loads for Buildings and Other Structures, ASCE 7-98, American Society of Civil Engineers, Reston, VA, 2000, 214 pp.

12-26 W. G. Godden, *Numerical Analysis of Beam and Column Structures*, Prentice Hall, Englewood Cliffs, N. J., 1965, 320 pp.

12-27 Randal W. Poston, John E. Breen, and Jose M. Roesset, "Analysis of Nonprismatic or Hollow Slender Concrete Bridge Piers," *Journal of the American Concrete Institute*, Vol. 82, No. 5, September–October 1985, pp. 731–739.

Chapter 13

13-1 Mete A. Sozen and Chester P. Siess, "Investigation of Multiple Panel Reinforced Concrete Floor Slabs: Design Methods—Their Evolution and Comparison," *ACI Journal, Proceedings*, Vol. 60, No. 8, August 1963, pp. 999–1027.

13-2 John R. Nichols, "Statical Limitations upon the Steel Requirements in Reinforced Concrete Flat Slab Floors," *Transactions ASCE*, Vol. 77, 1914, pp. 1670–1681.

13-3 Discussions of Ref. 13–2, *Transactions ASCE*, Vol. 77, 1914, pp. 1682–1736.

13-4 H. M. Westergaard and N. A. Slater, "Moments and Stresses in Slabs," *ACI Proceedings*, Vol. 17, 1921, pp. 415–538.

13-5 ACI Committee 318, *Building Code Requirements for Structural Concrete* (ACI 318-95) and *Commentary* (ACI 318R-95), American Concrete Institute, Detroit, 1995, 369 pp.

13-6 *Notes on ACI 318–XX, Building Code Requirements for Reinforced Concrete, with Design Applications*, Portland Cement Association, Skokie, Ill. various editions, 860 pp.

13-7   David P. Thompson and Andrew Scanlon, "Minimum Thickness Requirements for Control of Two-Way Slab Deflections," *ACI Structural Journal*, Vol. 85, No. 1, January–February 1988, pp. 12–22.

13-8   James O. Jirsa, Mete A. Sozen, and Chester P. Siess, "Pattern Loadings on Reinforced Concrete Floor Slabs," *Proceedings ASCE, Journal of the Structural Division*, Vol. 95, No. ST6, June 1969, pp. 1117–1137.

13-9   R. E. Woodring and Chester P. Siess, *An Analytical Study of the Moments in Continuous Slabs Subjected to Concentrated Loads*, Structural Research Series 264, Department of Civil Engineering, University of Illinois, Urbana, May 1963, 151 pp.

13-10 Scott D. B. Alexander and Sidney H. Simmonds, "Ultimate Strength of Column–Slab Connections," *ACI Structural Journal, Proceedings*, Vol. 84, No. 3, May–June 1987, pp. 255–261.

13-11 ACI-ASCE Committee 426, "The Shear Strength of Reinforced Concrete Members," Chapter 5, "Shear Strength of Slabs," *Proceedings ASCE, Journal of the Structural Division*, Vol. 100, No. ST8, August 1974, pp. 1543–1591.

13-12 Paul E. Regan and Michael W. Braestrup, *Punching Shear in Reinforced Concrete*, Bulletin d'Information 168, Comité Euro-International du Béton, Lausanne, January 1985, 232 pp.

13-13 Johannes Moe, *Shearing Strength of Reinforced Concrete Slabs and Footings under Concentrated Loads*, Development Department Bulletin D47, Portland Cement Association, Skokie, Ill., April 1961.

13-14 ACI-ASCE Committee 326, "Shear and Diagonal Tension, Slabs," *ACI Journal, Proceedings*, Vol. 59, No. 3, March 1962, pp. 353–396.

13-15 ACI-ASCE Committee 426, "Suggested Revisions to Shear Provisions for Building Codes," *ACI Journal, Proceedings*, Vol. 75, No. 9, September 1978, pp. 458–469.

13-16 Neil M. Hawkins, H. B. Fallsen, and R. C. Hinojosa, "Influence of Column Rectangularity on the Behavior of Flat Plate Structures," *SP-30 Cracking, Deflection and Ultimate Load of Concrete Slab Systems*, American Concrete Institute, Detroit, 1971, pp. 127–146.

13-17 M. Daniel Vanderbilt, "Shear Strength of Continuous Plates, *Proceedings ASCE, Journal of the Structural Division*, Vol. 98, No. ST5, May 1972, pp. 961–973.

13-18 Sami, Megally and Amin Ghali, "Cautionary Note on Shear Capitals," *Concrete International*, Vol 24, No. 3, March 2002, pp 75–82.

13-19 W. Gene Corley and Neil M. Hawkins, "Shearhead Reinforcement for Slabs," *ACI Journal, Proceedings*, Vol. 65, No. 10, October 1968, pp. 811–824.

13-20 Walter H. Dilger and Amin Ghali, "Shear Reinforcement for Concrete Slabs," *Proceedings ASCE, Journal of the Structural Division*, Vol. 107, No. ST12, December 1981, pp. 2403–2420.

13-21 Adel E. Elgabry and Amin Ghali, "Design of Stud-Shear Reinforcement for Slabs," *ACI Structural Journal*, Vol. 87, No. 3, May–June 1990, pp. 350–361.

13-22 ACI Committee 421, "Shear Reinforcement for Slabs" (ACI 421.1R), *ACI Structural Journal*, Vol. 89, No. 5, September–October 1992, pp. 587–589.

13-23 Neil M. Hawkins, "Shear Strength of Slabs with Moments Transferred to Columns," *Shear in Reinforced Concrete*, Vol. 2, ACI Publication SP-42, American Concrete Institute, Detroit, 1974, pp. 817–846.

13-24 Sidney H. Simmonds and Scott D. B. Alexander, "Truss Model for Edge Column–Slab Connections," *ACI Structural Journal*, Vol. 84, No. 4, July–August 1987, pp. 296–303.

13-25 Walter H. Dilger and Alaa Sherif, Discussion of "Proposed Revisions to Building Code Requirements for Reinforced Concrete," *Concrete International*, Vol. 17, No. 7, July 1995, pp. 70–73.

13-26 Joseph DiStasio and M. P. Van Buren, "Transfer of Bending Moment between Flat Plate Floor and Column," *ACI Journal, Proceedings*, Vol. 57, No. 3, September 1960, pp. 299–314.

13-27 Norman W. Hanson and John M. Hanson, "Shear and Moment Transfer between Concrete Slabs and Columns," *Journal of the Research and Development Laboratories, Portland Cement Association*, Vol. 10, No. 1, January 1968, pp. 2–16.

13-28 Amin Ghali and Adel Elgabry, Discussion of "Proposed Revisions to Building Code Requirements for Structural Concrete," *Concrete International*, Vol. 17, No. 7, July 1995, pp. 75–82.

13-29 ACI Committee 352, "Recommendations for Design of Slab-Column Connections in Monolithic Reinforced Concrete Structures," *ACI Structural Journal*, Vol. 85, No. 6, November–December 1988, pp. 675–696.

13-30 Jack P. Moehle, "Strength of Slab–Column Edge Connections," *ACI Structural Journal*, Vol. 85, No. 1, January–February 1988, pp. 89–98; Discussions and Closure, Vol. 85, No. 6, November–December 1988, pp. 703–709.

13-31 Denis Mitchell and William D. Cook, "Preventing Progressive Collapse of Slab Structures," *Proceedings ASCE, Journal of the Structural Division*, Vol. 110, No. ST7, July 1984, pp. 1513–1532.

13-32 R. K. Agerwal and Noel J. Gardner, "Form and Shore Requirements for Multistory Flat Plate Type Buildings, *ACI Journal, Proceedings*, Vol. 71, No. 11, November 1974, pp. 559–569.

13-33 John A. Sbarounis, "Multistory Flat Plate Buildings," *Concrete International: Design and Construction*, Vol. 81, No. 2, February 1984, pp. 70–77; No. 4, April 1984, pp. 62–70; No. 8, August 1984, pp. 31–35.

Chapter 14

14-1 Dean Peabody, Jr., "Continuous Frame Analysis of Flat Slabs," *Journal, Boston Society of Civil Engineers*, January 1948.

14-2 Chester P. Siess and Nathan M. Newmark, "Rational Analysis and Design of Two-Way Concrete Slabs," *ACI Journal, Proceedings*, Vol. 45, 1949, pp. 273–315.

14-3 Chester P. Siess and Nathan M. Newmark, *Moments in Two-Way Concrete Floor Slabs*, Bulletin 385, University of Illinois Engineering Experiment Station, Urbana, Ill., 1950, 124 pp.

14-4 W. Gene Corley and James O. Jirsa, "Equivalent Frame Analysis for Slab Design," *ACI Journal, Proceedings*, Vol. 67, No. 11, November 1970, pp. 875–884.

14-5 M. Daniel Vanderbilt, "Equivalent Frame Analysis of Unbraced Concrete Frames," *Significant Developments in Engineering Practice and Research—A Tribute to Chester P. Siess*, ACI Publication SP-72, American Concrete Institute, Detroit, 1981, pp. 219–246.

14-6 C. K. Wang, *Statically Indeterminate Structures*, McGraw-Hill, New York, 1953.

14-7 *Notes on ACI-318-95, Building Code Requirements for Reinforced Concrete with Design Applications*, Portland Cement Association, Skokie, Ill., 1996, 860 pp.

14-8 Sidney H. Simmonds and Janko Misic, "Design Factors for the Equivalent Frame Method," *ACI Journal, Proceedings*, Vol. 68, No. 11, November 1971, pp. 825–831.

14-9 Sidney H. Simmonds, *Effects of Supports on Slab Behavior*, meeting preprint 1697, presented at ASCE National Structural Engineering Meeting, Cleveland, Ohio, April 24–28, 1972.

14-10 ACI Committee 318, *Building Code Requirements for Reinforced Concrete*, (ACI 318–89), and *Commentary* (ACI 318R–89), American Concrete Institute, Detroit,1989, 353 pp.

Chapter 15

15-1 Arne Hillerborg, *Strip Method of Design*, E. & F., N. Spon, London, 1975, 258 pp.

15-2 Randal H. Wood and G. S. T. Armer, "The Theory of the Strip Method for Design of Slabs," *Proceedings, Institution of Civil Engineers, London*, Vol. 41, October 1968, pp. 285–313.

15-3 Kurt W. Johansen, *Yield-Line Theory*, Cement and Concrete Association, London, 1962, 181 pp.

15-4 Leonard L. Jones and Randal H. Wood, *Yield Line Analysis of Slabs*, Elsevier, New York, 1967.

15-5 Roger P. Johnson, *Structural Concrete*, McGraw-Hill, London, 1967, 271 pp.

15-6 Robert Park and William L. Gamble, *Reinforced Concrete Slabs*, Wiley-Interscience, New York, 1980, 618 pp.

15-7 Arne Hillerborg, "Equilibrium Theory for Concrete Slabs" (in Swedish), *Betong*, Vol. 41, No. 4, 1956, pp. 171–182.

15-8 Arne Hillerborg, "The Advanced Strip Method—A Simple Design Tool," *Magazine of Concrete Research*, Vol. 34, No. 121, December 1982, pp. 175–181.

Chapter 16

16-1 Canadian Commission on Building and Fire Codes, "Commentary L–Foundations, "*User's Guide–NBC 1995 Structural Commentaries (Part 4)*, National Research Council Canada, Ottawa, pp. 103–129.

16-2 Ontario, Ministry of Transportation, *Ontario Highway Bridge Design Code–Commentary*, 3rd Edition, Downsview, Ontario, 1991, 344 pp.

16-3 *ASCE Standard A-7, Design Loads for Buildings and Other Structures,* American Society of Civil Engineers, Washington, 2002.

16-4 Arthur N. Talbot, *Reinforced Concrete Wall Footings and Column Footings*, Bulletin 67, University of Illinois Engineering Experiment Station, Urbana, Ill., 1913, 96 pp.

16-5 Frank E. Richart, "Reinforced Concrete Wall and Column Footings," *ACI Journal, Proceedings*, Vol. 45, October 1948, pp. 97–128; November 1948, pp. 327–360.

16-6 ACI-ASCE Committee 326, "Shear and Diagonal Tension," Part 3, "Slabs and Footings," *ACI Journal, Proceedings*, Vol. 59, No. 3 March 1962, pp. 353–396.

16-7 Michael P. Collins and Dan Kuchma, "How Safe are our Large, Lightly Reinforced Concrete Beams, Slabs, and Footings?" *ACI Structural Journal, Proceedings*, Vol. 96, No. 4, July-August 1999, pp. 482–490.

16-8 CRSI Handbook, 1992, Concrete Steel Reinforcing Steel Institute, Schaumberg, Ill., 1992, 710 pp.

16-9 ACI Committee 336, "Suggested Analysis and Design Procedures for Combined Footings and Mats, ACI 336.2R-88," *ACI Manual of Concrete Practice*, 1989 and later editions, American Concrete Institute, Farmington Hills, MI, pp. 336.2R-1 to 336.2R-21 plus discussion.

16-10 F. Kramrisch and P. Rogers, "Simplified Design of Combined Footings," *Proceedings ASCE, Journal of the Soil Mechanics Division*, Vol. 87, No. SM5, October 1961, pp. 19–44.

16-11 Perry Adebar, Daniel Kuchma, and Michael P. Collins, "Strut-and-Tie Models for the Design of Pile Caps: An Experimental Study" *ACI Structural Journal*, Vol. 87, No. 1, January-February 1990, pp. 81–92.

Chapter 17

17-1 J. A. Hofbeck, I. A. Ibrahim, and Alan H. Mattock, "Shear Transfer in Reinforced Concrete," *ACI Journal, Proceedings*, Vol. 66, No. 2, February 1969, pp. 119–128.

17-2 Alan H. Mattock and Neil M. Hawkins, "Shear Transfer in Reinforced Concrete–Recent Research," *Journal of the Prestressed Concrete Institute*, Vol. 17, No. 2, March-April 1972, pp. 55–75.

17-3 J.C. Walraven, "Fundamental Analysis of Aggregate Interlock," *Journal of the Structural Division, Proceedings of the American Society of Civil Engineers*, Vol. 107, No. ST11, November 1981, pp. 2245–2271.

17-4 H.W. Reinhardt and J.C. Walraven, "Cracks in Concrete Subject to Shear," *Journal of the Structural Division, Proceedings of the American Society of Civil Engineers*, Vol. 108, No. ST1, January, 1982, pp. 225–244.

17-5 Robert E. Loov and A. K. Patnaik, "Horizontal Shear Strength of Composite Concrete Beams with a Rough Interface," *PCI Journal*, Vol. 39, No. 1, January–February 1994, pp. 48–69.

17-6 Norman W. Hanson, "Precast-Prestressed Concrete Bridges," Part 2, "Horizontal Shear Connections," *Journal of the Research and Development Laboratories, Portland Cement Association*, Vol. 2, No. 2, May 1960, pp. 38–58.

17-7 M.A. Ali and R.N. White, "Enhanced Contact Model for Shear Friction of Normal and High-Strength Concrete," *ACI Structural Journal*, May-June 1999, Vol. 96, No. 3, pp. 348–360

17-8 ACI-ASCE Committee 426, "The Shear Strength of Reinforced Concrete Members Slabs, *Proceedings ASCE, Journal of the Structural Division*, Vol. 99, No. ST6, June 1973, pp. 1148–1157.

17-9 P. W. Birkeland and H. W. Birkeland, "Connections in Precast Concrete Construction, *ACI Journal, Proceedings*, Vol. 63, No. 3, March 1966, pp. 345–368.

17-10 Mast, R.F., "Auxiliary Reinforcement in Precast Connections," *Proceedings, Journal of the Structural Division* ASCE, Vol. 94, ST6, June 1968, pp. 1485–1504.

17-11 A.H. Mattock, "Shear Friction and High-Strength Concrete," *ACI Structural Journal*, Vol. 98, No. 1, January–February 2001, pp. 50–59.

17-12 F.J. Vecchio and M.P. Collins, "The Modified Compression Field Theory," *ACI Journal, Proceedings*, Vol. 83, No. 2, March–April 1986, pp. 219–231.

17-13 Dino Angelakos, Evan C. Bentz, and Michael P. Collins, "The Effect of Concrete Strength and Minimum Stirrups on the Shear Strength of Large Members," *ACI Structural Journal*, Vol. 98, No. 3, May–June 2001, pp 290–296.

17-14 *PCI Design Handbook–Precast and Prestressed Concrete*, 3rd ed., Prestressed Concrete Institute, Chicago, 1985, 521 pp.

17-15 Paul H. Kaar, L. B. Kriz, and Eivind Hognestad, "Precast-Prestressed Bridges: (1) Pilot Tests of Continuous Girders," *Journal of the Research and Development Laboratories, Portland Cement Association*, Vol. 2, No. 2, May 1960, pp. 21–37.

17-16 J. C. Saemann and George W. Washa, "Horizontal Shear Connections Between Precast Beams and Cast-in-Place Slabs," *ACI Journal, Proceedings*, Vol. 61, No. 11, November 1964, pp. 1383–1409.

17-17 Amin Ghali and Rene Favre, *Concrete Structures: Stresses and Deformations*, Chapman & Hall, New York, 1986, 348 pp.

17-18 Walter H. Dilger, "Creep Analysis of Prestressed Concrete Structures Using Creep-Transformed Section Properties," *PCI Journal*, Vol. 27, No. 1, January–February 1982, pp. 99–118.

Chapter 18

18-1 Jörg Schlaich and Dieter Weischede, *Detailing of Concrete Structures* (in German), Bulletin d'Information 150, Comité Euro-International du Béton, Paris, March 1982, 163 pp.

18-2 Jörg Schlaich, Kurt Schäfer, and Mattias Jennewein, "Toward a Consistent Design of Structural Concrete," *Journal of the Prestressed Concrete Institute*, Vol. 32, No. 3, May–June 1987, pp. 74–150.

18-3 Jörg Schlaich and Kurt Schäfer, "Design and Detailing of Structural Concrete Using Strut-and-Tie Models," *The Structural Engineer*, Vol. 69, No. 6, March 1991, 13 pp.

18-4 James G. MacGregor, "Derivation of Strut-and-Tie Models for the 2002 ACI Code" ACI Publication, SP–208, *Examples for the Design of Structural Concrete with Strut-and-Tie Models*, American Concrete Institute, Farmington Hills, MI, 2002, pp. 7–40.

18-5 Karl-Heinz Reineck, Editor, *Examples for the Design of Structural Concrete with Strut-and-Tie Models*, ACI Publication, SP–208, American Concrete Institute, Farmington Hills, MI, 2002, pp. 7–40.

18-6 David M. Rogowsky and Peter Marti, "Detailing for Post-Tensioning," *VSL Report Series*, No. 3, VSL International Ltd., Bern, 1991, 49 pp.

18-7 Perry Adebar and Zongyu Zhou, "Bearing Strength of Compressive Struts Confined by Plain Concrete," *ACI Structural Journal*, Vol. 90, No. 5, September–October 1993, pp. 534–541.

18-8 David M. Rogowsky and James G. MacGregor, "Design of Deep Reinforced Concrete Continuous Beams," *Concrete International: Design and Construction*, Vol. 8, No. 8, August 1986, pp. 49–58.

18-9 *CEB-FIP Model Code 1990*, Thomas Telford Services, Ltd., London, for Comité Euro-International du Béton, Lausanne, 1993, 437 pp.

18-10 M. P. Nielsen, M. N. Braestrup, B. C. Jensen, and F. Bach, *Concrete Plasticity, Beam Shear—Shear in Joints—Punching Shear*, Special Publication of the Danish Society for Structural Science and Engineering, Technical University of Denmark, Lyngby/Copenhagen, 1978, 129 pp.

18-11 CSA Technical Committee on Reinforced Concrete Design, *A23.3-94 Design of Concrete Structures*, Canadian Standards Association, Rexdale, Ontario, December 1994, 199 pp.

18-12 Michael P. Collins and Denis Mitchell, "Design Proposals for Shear and Torsion," *Journal of the Prestressed Concrete Institute*, Vol. 25, No. 5, September-October 1980, 70 pp.

18-13 Frank Vecchio, and Michael P Collins, *The Response of Reinforced Concrete to In-Plane Shear and Normal Stresses*, Publication 82–03, Department of Civil Engineering, University of Toronto, March 1982, 332 pp.

18-14 AASHTO, *LRFD Bridge Specifications and Commentary*, 2nd Edition, American Association of state Highway and Transportation Officials, Washington, 1998, 1216 pp.

18-15 Julio Ramirez and John E. Breen, "Evaluation of a Modified Truss-Model Approach for Beams in Shear," *ACI Structural Journal*, Vol. 88, No. 5, September–October, 1991, pp. 562–571.

18-16 Konrad Bergmeister, John E. Breen, and James O. Jirsa, "Dimensioning of the Nodes and Development of Reinforcement," *Structural Concrete, IABSE Colloquium, Stuttgart 1991, Report*, International Association for Bridge and Structural Engineering, Zurich, 1991, pp. 551–556.

18-17 James O. Jirsa, Konrad Bergmeister, Robert Anderson, John E. Breen, David Barton, and Hakim Bouadi, "Experimental Studies of Nodes in Strut-and-Tie Models," *Structural Concrete IABSE Colloquium Stuttgart, 1991, Report*, International Association for Bridge and Structural Engineering, Zurich, 1991, pp. 525–532.

18-18 FIP Recommendations, *Practical Design of Structural Concrete*, FIP Commission 3, Practical Design, September 1996, Publ. SETO, London, September 1999. (distributed by *fib* Lausanne.)

18-19 James G. MacGregor, *Reinforced Concrete: Mechanics and Design*, 3rd ed. Prentice-Hall, Inc., Upper Saddle River, New Jersey, 1997, 939 pp.

18-20 Ladislav Kriz and Charles H. Raths, "Connections in Precast Concrete Structures—Strength of Corbels," *Journal of the Prestressed Concrete Institute*, Vol. 10, No. 1, February 1965, pp. 16–47.

18-21 Alan H. Mattock, K. C. Chen, and K. Soongswany, "The Behavior of Reinforced Concrete Corbels," *Journal of the Prestressed Concrete Institute*, Vol. 21, No. 2, March-April 1976, pp. 52–77.

18-22 *PCI Design Handbook—Precast and Prestressed Concrete*, 3rd ed., Prestressed Concrete Institute, Chicago, 1985, 521 pp.

18-23 Alan H. Mattock and T. Theryo, *Strength of Members with Dapped Ends*, Research Project 6, Prestressed Concrete Institute, 1980, 25 pp.

18-24 William D. Cook and Denis Mitchell, "Studies of Disturbed Regions near Discontinuities in Reinforced Concrete Members," *ACI Structural Journal*, Vol. 85, No. 2, March–April 1988, pp. 206–216.

18-25 ACI-ASCE Committee 352, "Recommendations for Design of Beam-Column Joints in Monolithic Reinforced Concrete Structures," *ACI Journal, Proceedings*, Vol. 82, No. 3, May–June 1985, pp. 266–283.

18-26 Ingvar H. E. Nilsson and Anders Losberg, "Reinforced Concrete Corners and Joints Subjected to Bending Moment," *Proceedings ASCE, Journal of the Structural Division*, Vol. 102, No. ST6, June 1976, pp. 1229–1254.

18-27 P. S. Balint and Harold P. J. Taylor, *Reinforcement Detailing of Frame Corner Joints with Particular Reference to Opening Corners*, Technical Report 42.462, Cement and Concrete Association, London, February, 1972, 16 pp.

18-28 Neil M. Hawkins, "The Bearing Strength of Concrete Loaded through Rigid Plates," *Magazine of Concrete Research*, Vol. 20, No. 62, March 1968, pp. 31–40.

Chapter 19

19-1 Oberlender, G.D., and Everard, N.J., "Investigation of Reinforced Concrete Walls," *ACI Journal, Proceedings* Vol. 74, No. 6, June 1977, pp 256-263.

19-2 Andrew W. Taylor, Randall B. Rowell, and John E. Breen, "Behavior of Thin-Walled Concrete Box Piers," *ACI Structural Journal*, Vol. 92, No. 3, May–June 1995, pp 319–333.

19-3 Andrew W. Taylor and John E. Breen, Design Recommendations for Thin-Walled Box Piers and Pylons," *Concrete International*, American Concret Institute, Vol. 16, No. 12, December 1994, pp 36–41.

19-4 Alexander Chajes, *Principles of Structural Stability Theory*, Prentice Hall, Englewood Cliffs, N.J., 1974, 336 pp.

19-5 Felix Barda, Hanson, John M., and Corley, W. Gene., "Shear Strength of Low-RiseWalls with Boundary Elements," *Reinforced Concrete Structures in Seismic Zones*, SP-53, American Concrete Institute, Detroit, 1977, pp 149–202.

19-6 Thomas Paulay and M.J. Nigel Priestley, *Seismic Design of Reinforced Concrete and Masonry Buildings*, Wiley Interscience, New York, 1992, 744 pp.

19-7 Sharon Wood, "Shear Strength of Low-Rise Reinforced Concrete Walls, *ACI Structural Journal*, Vol. 87, No. 1, January–February 1990, pp 99–107.

19-8 Alex E. Cardenas, and Donald D. Magura, "Strength of High-Rise Shear Walls—Rectangular Cross Section," *Response of Multistory Concrete Structures to Lateral Forces*, SP-36, American Concrete Institute, Detroit, 1973, pp 119–150.

19-9 Cardenas, A.E.; Hanson, J.M.; Corley, W.G., and Hognestad, E., "Design Provisions for Shear Walls," *ACI Journal, Proceedings*, Vol. 70, No. 3, March 1973, pp 221–230

19-10 Sharon Wood, "Minimum Tensile Reinforcement Requirements in Walls," *ACI Structural Journal*, Vol. 86, No. 4, September–October 1989, pp 582–591.

19-11 Y. Zhang and Z. Wang, Seismic Behavior of Reinforced Concrete Shear Walls Subjected to High Axial Loading," *ACI Structural Journal*, Vol. 97, No. 5, September–October 2000, pp 739–750.

19-12 John W. Wallace and Kutay Orakcal. "ACI 318-99 Provisions for Seismic Design of Structural Walls," *ACI Structural Journal*, Vol. 99, No. 4, July–August 2002, pp 499–508.

19-13 Aghayere, A.O. and MacGregor, James G. MacGregor, "Tests of Reinforced Concrete Plates under Combined Inplane and Lateral Loads," *ACI Structural Journal*, Vol 87, No. 6, November–December, 1990.

19-14 Mashour G. Ghoneim and James G. MacGregor, "Behavior of Reinforced Concrete Plates under Combined inplane and Lateral Loads," *ACI Structural Journal*, Vol. 91, No.2, March–April 1994, pp 188–198.

19-15 Kripanarayanan, K.M., "Interesting Aspects of the Empirical Wall Design Equation," *ACI Journal, Proceedings*, Vol. 74, No. 5, May 1997, pp 204–207.

19-16 ACI Committee 551, "Tilt-Up Concrete Structures, (ACI 551R-92)," *ACI Manual of Concrete Practice*, American Concrete Institute, Farmington Hills, MI, Vol. 5, 1993 and later editions, pp 551R-1 to 551R-46.

19-17 *Concrete Design Manual*, Canadian Portland Cement Association, Ottawa, 1995, 350 pp.

19-18 Noel D. Nathan, "Slenderness of Prestressed Concrete Columns," *PCI Journal*, Vol. 28, No. 2, March–April 1983, pp 50–77.

19-19 Jacob S. Grossman, "Slender Concrete Structures—The New Edge," *ACI Structural Journal*, Vol.87, No. 1, January–February 1990, pp 39–52.

19-20 ACI Committee 442, "Response of Concrete Buildings to Lateral Forces, (ACI 442R-88)," *ACI Manual of Concrete Practice*, American Concrete Institute, Farmington Hills, MI, Vol. 3, 1989 and later editions, pp 442R–1 to 442R–36.

19-21 Iain A. MacLeod, *Shear Wall-Frame Interaction*, Special Publication SP011.01D, Portland Cement Association, Skokie, Illinois, 1971, 62 pp.

19-22 A.R. Santhakumar, and Thomas Paulay (Supervisor), *Ductility of Coupled Shearwalls*, Ph.D. Thesis, University of Canterbury, Christchurch, New Zealand, October 1974.

19-23 Alexander Coull and J.R. Choudhury, "Analysis of Coupled Shear Walls," ACI *Journal, Proceedings*, Vol. 64, No. 9, September 1967, American Concrete Institute, Detroit, pp 587–593.

19-24 Joseph Schwaighofer and Michael P. Collins, "Experimental Study of the Behavior of Reinforced Concrete Coupling Slabs," *ACI Journal, Proceedings*, Vol. 74, No. 3, March 1977, pp 123–127.

19-25 Thomas Paulay and R.G. Taylor, "Slab Coupling of Earthquake-Resisting Shearwalls," *ACI Journal, Proceedings*, Vol. 78, No. 2, March–April 1981, pp 130–140.

19-26 Loring A. Wyllie, Jr., "Chapter 7, Structural Walls and Diaphragms-How they Function," *Building Structural Design Handbook*, Wiley–Interscience, New York, 1987, pp 188–215.

Chapter 20

20-1 "NEHRP Recommended Provisions for Seismic Regulations for New Buildings and Other Structures," Part 1: Provisions and Part @: Commentary, Building seismic Safety Council, Washington, D.C., Various Editions.

20-2 Minimum Design Loads for Buildings and Other Structures (ASCE 7-XX), American Society of Civil Engineers, New York, Various Editions.

20-3 Robert Park and Thomas Paulay, *Reinforced Concrete Structures*, Wiley-Interscience, New York, 1975, 768 pp.

20-4 Tom Paulay and M.J. Nigel Priestley, *Seismic Design of Reinforced Concrete and Masonry Buildings*, John Wiley & Sons, Inc. 1992, 744 pp.

20-5 Anil K. Chopra, *Dynamics of Structures–A Primer*, Earthquake Engineering Research Institute, Berkeley, Calif., 1981, 126 pp.

20-6 Egor P. Popov, Vitelmo V. Bertero and H. Krawinkler, *Cyclic Behavior of Three R/C Flexural Members with High Shear*, EERC Report 72–5, Earthquake Engineering Research Center, University of California, Berkeley, October 1972.

20-7   Cathy W. French and Jack Moehle, "Effect of Floor Slab on Behavior of Slab-Beam-Column Connections," ACI Sp 123, *Design of Beam-Column Joints for Seismic Resistance*, American Concrete Institute, Farmington Hills, MI, 1991, pp. 225–258.

20-8   James K. Wight and Mete A. Sozen, "Shear Strength Decay of RC Columns under Shear Reversals," *Proceedings ASCE, Journal of the Structural Division*, Vol. 101, No. ST5, May 1975, pp. 1053–1065.

20-9   Jack R. Benjamin, *Statically Indeterminate Structures*, McGraw-Hill, New York, 1959, 350 pp.

Index